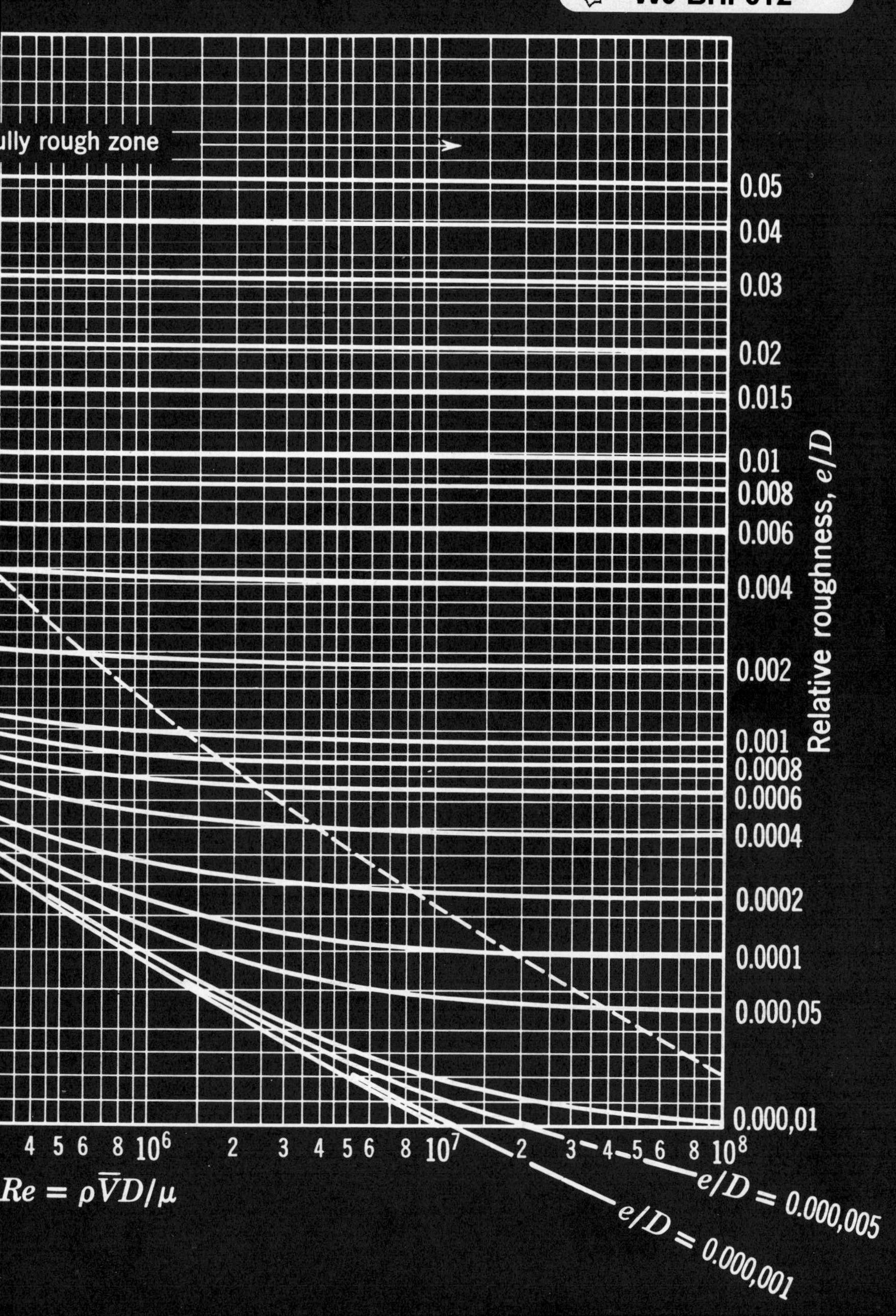

ully rough zone
0.05
0.04
0.03
0.02
0.015
0.01
0.008
0.006
0.004
0.002
0.001
0.0008
0.0006
0.0004
0.0002
0.0001
0.000,05
0.000,01
Relative roughness, e/D
4 5 6 8 10^6 2 3 4 5 6 8 10^7 2 3 4 5 6 8 10^8
$Re = \rho\overline{V}D/\mu$
$e/D = 0.000,005$
$e/D = 0.000,001$

INTRODUCTION TO FLUID MECHANICS

Third Edition

ROBERT W. FOX
ALAN T. McDONALD

School of Mechanical Engineering
Purdue University

JOHN WILEY & SONS

New York · Chichester · Brisbane · Toronto · Singapore

On the cover: *The Illuminated Vortex*

The flow field within the cylinder of an internal combustion engine can be measured using Laser Doppler Anemometer (LDA) instrumentation. The LDA requires no probe in the flow. This non-invasive measurement technique allows simultaneous measurement of mean velocity and turbulence velocity components. The cover illustration suggests the strong vortex nature of the flow within the cylinder, and the location of the core of the vortex away from the cylinder center.

(Reprinted with the permission of General Motors Corporation; photograph by Michel Tcherevkoff.)

Library of Congress Cataloging in Publication Data:

Fox, Robert W., 1934-
Introduction to fluid mechanics.

Includes bibliographies and indexes.
1. Fluid mechanics. I. McDonald, Alan T. II. Title.
TA357.F69 1985 532 84-15282
ISBN 0-471-88598-3

Printed in the United States of America

10 9 8 7 6 5 4 3 2 1

PREFACE

This text was written for an introductory course in fluid mechanics. In the Third Edition, the international system of units (SI) is used in approximately 70 percent of the example problems and problem exercises after each chapter. (English engineering units have been retained in the remaining problems to provide experience with this traditional system and to highlight methods of conversion among unit systems.)

Our approach to the subject in the Third Edition is unchanged. The physical concepts of fluid mechanics and methods of analysis beginning from basic principles are emphasized throughout. The primary objective of this book is to help students develop an orderly approach to problem solving. Thus we start from basic equations, state assumptions clearly, and relate results to expected physical behavior. The approach is illustrated by the 108 example problems in the text. Solutions to these examples have been prepared to demonstrate good solution techniques and to explain troublesome points. The example problems are set apart from the text in format, so they are particularly easy to follow.

Complete explanations in the chapter discussions, together with numerous detailed examples, have in our experience made this book understandable for students. This allows the instructor freedom to depart from conventional "lecture" teaching methods. Classroom time can be used to bring in outside material, expand on special topic areas (such as blood flow, non-Newtonian flow, or measurement methods), solve example problems, or explain any difficult points of the assigned homework problems. Each class period thus can be used in the manner most appropriate to satisfy student needs.

The material in this book has been selected carefully. There is a detailed presentation of a broad range of topics suitable for a one- or two-semester course in fluid mechanics at the junior or senior level. Desirable prerequisites are introductory courses in rigid-body dynamics and mathematics through integral calculus. Some background in heat power or thermodynamics is desirable for the study of one-dimensional compressible flow.

The presentation is organized into seven broad topic areas:

- Introductory concepts, scope of fluid mechanics, and fluid statics (Chapters 1, 2, and 3)
- Development and application of control volume forms of basic equations (Chapter 4)
- Development and application of differential forms of basic equations (Chapters 5 and 6)
- Dimensional analysis and correlation of experimental data (Chapter 7)
- Applications for incompressible flow (internal flows in Chapter 8 and external flows in Chapter 9)
- Analysis and applications of flow in open channels (Chapter 10)
- Analysis and applications in one-dimensional compressible flow (Chapters 11 and 12).

"Summary Objectives" are included at the end of each chapter. They specifically indicate what students should be able to do after they have studied each chapter.

More than 500 new problems have been added for homework assignment and student exercises. The Third Edition now contains 1161 problems, so that at least six semesters can be covered without repeating problem assignments. (All problem statements follow the 70/30-percent ratio of SI to traditional units.)

The Solutions Manual that accompanies the text contains complete and detailed solutions for each of the 1161 homework problems; the format used in the example problems has been followed in preparing the Manual. Solutions may be photocopied for classroom or library use, eliminating the labor of problem solving for the instructor using the text.

Many fine instructional films and film loops are available to further clarify and demonstrate basic principles in fluid mechanics. We refer to these films in the text where their use is appropriate; a complete list of suppliers and titles is included in Appendix C.

When our students have finished this course, we expect them to be able to apply the basic equations to a variety of problems, including new problems that they have not encountered previously. We emphasize physical understanding throughout to make students aware of the variety of phenomena that occur in real fluid flow situations. By minimizing the number of "magic formulas" and emphasizing the fundamental approach, we believe students will develop confidence in their ability to apply the material and will be able to reason out solutions to challenging problems.

The book also is well suited for independent study by students or practicing engineers. Its readability and clear examples help to build student confidence. The summary objectives at the end of each chapter may be used for review and to assess the achievement of educational goals.

We recognize that no single approach can satisfy all needs. We are grateful to the many students and faculty whose comments have helped us improve the Third Edition, and we welcome criticisms and suggestions from interested readers or users of this book.

Robert W. Fox
Alan T. McDonald

CONTENTS

Chapter 1 INTRODUCTION 1

1-1 Note to Students 1
1-2 Definition of a Fluid 3
1-3 Scope of Fluid Mechanics 4
1-4 Basic Equations 4
1-5 Methods of Analysis 5
1-5.1 System and Control Volume 5
1-5.2 Differential versus Integral Approach 7
1-5.3 Methods of Description 8
1-6 Dimensions and Units 10
1-6.1 Systems of Dimensions 10
1-6.2 Systems of Units 11
1-6.3 Preferred Systems of Units 12
1-7 Summary Objectives 12
Problems 13

Chapter 2 FUNDAMENTAL CONCEPTS 18

2-1 Fluid as a Continuum 18
2-2 Velocity Field 20
2-2.1 One-, Two-, and Three-Dimensional Flows 21
2-2.2 Timelines, Pathlines, Streaklines, and Streamlines 22
2-3 Stress Field 25
2-4 Viscosity 28
2-4.1 Newtonian Fluid 29
2-4.2 Non-Newtonian Fluids 31
2-5 Description and Classification of Fluid Motions 33
2-5.1 Viscous and Inviscid Flows 33
2-5.2 Laminar and Turbulent Flows 38
2-5.3 Compressible and Incompressible Flows 39
2-5.4 Internal and External Flows 40
2-6 Summary Objectives 41
Problems 41

Chapter 3 FLUID STATICS 49

3-1 The Basic Equation of Fluid Statics 49
3-1.1 Pressure Variation in a Static Fluid 53
3-2 Absolute and Gage Pressures 58
3-3 The Standard Atmosphere 58
**3-4 Hydraulic Systems 61
3-5 Hydrostatic Force on Submerged Surfaces 61
3-5.1 Hydrostatic Force on a Plane Submerged Surface 61
3-5.2 Hydrostatic Force on Curved Submerged Surfaces 69
**3-6 Buoyancy and Stability 73
**3-7 Fluids in Rigid-Body Motion 74
3-8 Summary Objectives 80
Problems 81

Chapter 4 BASIC EQUATIONS IN INTEGRAL FORM FOR A CONTROL VOLUME 95

4-1 Basic Laws for a System 95
4-1.1 Conservation of Mass 95
4-1.2 Newton's Second Law 96
**4-1.3 Moment of Momentum 96
4-1.4 The First Law of Thermodynamics 96
4-1.5 The Second Law of Thermodynamics 97
4-2 Relation of System Derivatives to the Control Volume Formulation 97
4-2.1 Derivation 98
4-2.2 Physical Interpretation 103
4-3 Conservation of Mass 104
4-3.1 Special Cases 105
4-4 Momentum Equation for Inertial Control Volume 111
**4-4.1 Differential Control Volume Analysis 123
4-4.2 Control Volume Moving with Constant Velocity 128
4-5 Momentum Equation for Control Volume with Rectilinear Acceleration 130
**4-6 Momentum Equation for Control Volume with Arbitrary Acceleration 139
**4-7 Moment of Momentum 143
4-7.1 Equation for Fixed Control Volume 143
4-7.2 Application to Turbomachinery 144
4-7.3 Equation for Rotating Control Volume 152
4-8 The First Law of Thermodynamics 158
4-8.1 Rate of Work Done on a Control Volume 159
4-8.2 Control Volume Equation 160
4-9 The Second Law of Thermodynamics 165
4-10 Summary Objectives 166
Problems 167

Chapter 5 INTRODUCTION TO DIFFERENTIAL ANALYSIS OF FLUID MOTION 201

5-1 Conservation of Mass 201
5-1.1 Rectangular Coordinate System 201
5-1.2 Cylindrical Coordinate System 207

**5-2 Stream Function for Two-Dimensional Incompressible Flow 210
5-3 Motion of a Fluid Element (Kinematics) 213
5-3.1 Acceleration of a Fluid Particle in a Velocity Field 214
5-3.2 Fluid Rotation 219
5-3.3 Fluid Deformation 220
5-4 Momentum Equation 228
5-4.1 Forces Acting on a Fluid Particle 228
5-4.2 Differential Momentum Equation 230
5-4.3 Newtonian Fluid: Navier–Stokes Equations 230
5-5 Summary Objectives 231
Problems 232

Chapter 6 INCOMPRESSIBLE INVISCID FLOW 241

6-1 Momentum Equation for Frictionless Flow: Euler's Equations 241
6-2 Euler's Equations in Streamline Coordinates 242
6-3 Bernoulli Equation—Integration of Euler's Equation Along a Streamline for Steady Flow 246
6-3.1 Derivation Using Streamline Coordinates 246
**6-3.2 Derivation Using Rectangular Coordinates 247
6-3.3 Applications 249
6-3.4 Cautions on Use of the Bernoulli Equation 255
6-4 Static, Stagnation, and Dynamic Pressures 256
6-5 Relation Between the First Law of Thermodynamics and the Bernoulli Equation 259
**6-6 Unsteady Bernoulli Equation—Integration of Euler's Equation Along a Streamline 266
**6-7 Irrotational Flow 269
6-7.1 Bernoulli Equation Applied to Irrotational Flow 269
6-7.2 Velocity Potential 271
6-7.3 Stream Function and Velocity Potential for Two-Dimensional, Irrotational, Incompressible Flow; Laplace's Equation 272
6-7.4 Elementary Plane Flows 274
6-7.5 Superposition of Elementary Plane Flows 278
6-8 Summary Objectives 284
Problems 284

Chapter 7 DIMENSIONAL ANALYSIS AND SIMILITUDE 295

7-1 Nature of Dimensional Analysis 295
7-2 Buckingham Pi Theorem 296
7-3 Determining the Π Groups 297
7-4 Dimensionless Groups of Significance in Fluid Mechanics 303
7-5 Flow Similarity and Model Studies 305
7-5.1 Incomplete Similarity 308
7-5.2 Scaling Laws for Turbomachinery 315
7-5.3 Comments on Model Testing 318
7-6 Nondimensionalizing the Basic Differential Equations 319

7-7 Summary Objectives 322
References 322
Problems 323

Chapter 8 INTERNAL INCOMPRESSIBLE VISCOUS FLOW 331

8-1 Introduction 331
PART A. FULLY DEVELOPED LAMINAR FLOW 333
8-2 Fully Developed Laminar Flow Between Infinite Parallel Plates 333
8-2.1 Both Plates Stationary 333
8-2.2 Upper Plate Moving with Constant Speed, *U* 339
8-3 Fully Developed Laminar Flow in a Pipe 344
PART B. FLOW IN PIPES AND DUCTS 351
8-4 Shear Stress Distribution in Fully Developed Pipe Flow 351
8-5 Turbulent Velocity Profiles in Fully Developed Pipe Flow 353
8-6 Energy Considerations in Pipe Flow 356
8-6.1 Kinetic Energy Coefficient 358
8-6.2 Head Loss 358
8-7 Calculation of Head Loss 359
8-7.1 Major Losses: Friction Factor 359
8-7.2 Minor Losses 365
8-8 Solution of Pipe Flow Problems 371
8-8.1 Single-Path Systems 371
**8-8.2 Multiple-Path Systems 384
**8-8.3 Noncircular Ducts 389
PART C. FLOW MEASUREMENT 390
8-9 Direct Methods 390
8-10 Restriction Flow Meters for Internal Flows 391
8-10.1 The Orifice Plate 394
8-10.2 The Flow Nozzle 395
8-10.3 The Venturi 397
8-10.4 The Laminar Flow Element 398
8-11 Linear Flow Meters 402
8-12 Traversing Methods 403
8-13 Summary Objectives 404
References 405
Problems 406

Chapter 9 EXTERNAL INCOMPRESSIBLE VISCOUS FLOW 425

PART A. BOUNDARY LAYERS 426
9-1 The Boundary-Layer Concept 426
9-2 Boundary-Layer Thicknesses 428
**9-3 Laminar Flat-Plate Boundary Layer: Exact Solution 431
9-4 Momentum Integral Equation 436
9-5 Use of the Momentum Integral Equation for Zero Pressure Gradient Flow 442
9-5.1 Laminar Flow 443
9-5.2 Turbulent Flow 447
9-6 Pressure Gradients in Boundary-Layer Flow 450

PART B. FLUID FLOW ABOUT IMMERSED BODIES 454
9-7 Drag 454
9-7.1 Flow over a Flat Plate Parallel to the Flow: Friction Drag 455
9-7.2 Flow over a Flat Plate Normal to the Flow: Pressure Drag 459
9-7.3 Flow over a Sphere and Cylinder: Friction and Pressure Drag 461
9-7.4 Streamlining 466
9-8 Lift 469
9-9 Summary Objectives 483
References 484
Problems 485

Chapter 10 FLOW IN OPEN CHANNELS 501

10-1 Characteristics of Open Channels 502
10-2 Propagation of Surface Waves 504
10-2.1 Wave Speed 504
10-2.2 The Froude Number 508
10-3 Energy Equation for Open-Channel Flow 508
10-3.1 Specific Energy 510
10-4 Frictionless Flow: Effect of Area Change 513
10-4.1 Flow over a Bump 513
10-4.2 Flow through a Sluice Gate 516
10-5 Flow at Normal Depth: Uniform Flow 520
10-5.1 Basic Equations 520
10-5.2 The Manning Correlation for Velocity 522
10-5.3 Optimum Channel Cross Section 527
10-5.4 Critical Normal Flow 530
10-6 Flow with Gradually Varying Depth 532
10-6.1 Classification of Surface Profiles 533
10-6.2 Calculation of Surface Profiles 537
10-7 The Hydraulic Jump 540
10-7.1 Basic Equations 541
10-7.2 Depth Increase across a Hydraulic Jump 542
10-7.3 Head Loss across a Hydraulic Jump 543
10-8 Measurements in Open-Channel Flow 546
10-8.1 Sharp-Crested Weirs 546
10-8.2 Broad-Crested Weirs 550
10-8.3 Sluice Gates 551
10-8.4 Critical Flumes 551
10-9 Summary Objectives 552
References 553
Problems 554

Chapter 11 INTRODUCTION TO COMPRESSIBLE FLOW 561

11-1 Review of Thermodynamics 561
11-2 Propagation of Sound Waves 568
11-2.1 Speed of Sound 568
11-2.2 Types of Flow—The Mach Cone 573

11-3 Reference State: Local Isentropic Stagnation Properties 574
11-3.1 Local Isentropic Stagnation Properties for the Flow of an Ideal Gas 575
11-3.2 Critical Conditions 583
11-4 Summary Objectives 584
Problems 584

Chapter 12 STEADY ONE-DIMENSIONAL COMPRESSIBLE FLOW 592

12-1 Basic Equations for Isentropic Flow 592
12-2 Effect of Area Variation on Properties in Isentropic Flow 596
12-3 Isentropic Flow of an Ideal Gas 599
12-3.1 Basic Equations 599
12-3.2 Reference Conditions for Isentropic Flow of an Ideal Gas 600
**12-3.3 Tables for Computation of Isentropic Flow of an Ideal Gas 603
12-3.4 Isentropic Flow in a Converging Nozzle 605
12-3.5 Isentropic Flow in a Converging-Diverging Nozzle 611
12-4 Adiabatic Flow in a Constant-Area Duct with Friction 617
12-4.1 Basic Equations 618
12-4.2 The Fanno Line 620
**12-4.3 Tables for Computation of Fanno Line Flow of an Ideal Gas 625
12-5 Frictionless Flow in a Constant-Area Duct with Heat Transfer 634
12-5.1 Basic Equations 634
12-5.2 The Rayleigh Line 637
**12-5.3 Tables for Computation of Rayleigh Line Flow of an Ideal Gas 643
12-6 Normal Shocks 647
12-6.1 Basic Equations 648
**12-6.2 Tables for Computation of Normal Shocks in an Ideal Gas 656
12-6.3 Flow in a Converging-Diverging Nozzle 661
12-7 Summary Objectives 662
Problems 663

Appendix A FLUID PROPERTY DATA 681

Appendix B EQUATIONS OF MOTION IN CYLINDRICAL COORDINATES 692

Appendix C FILMS AND FILM LOOPS FOR FLUID MECHANICS 694

Appendix D TABLES FOR COMPUTATION OF COMPRESSIBLE FLOW 700

Appendix E ANALYSIS OF EXPERIMENTAL UNCERTAINTY 716

Appendix F SI UNITS, PREFIXES, AND CONVERSION FACTORS 724

Answers to Selected Even-Numbered Problems 726
Index 735

Chapter 1

INTRODUCTION

The goal of this textbook is to provide a clear, concise introduction to the subject of fluid mechanics. In beginning the study of any subject, a number of questions may come to mind. Students in the first course in fluid mechanics might ask:

What is fluid mechanics all about?

Why do I have to study it?

Why should I want to study it?

How does it relate to subject areas with which I am already familiar?

In this chapter we shall try to present at least a qualitative answer to these and similar questions. This should serve to establish a base and a perspective for our study of fluid mechanics. Before proceeding with the definition of a fluid, we digress for a moment with a few pointed comments to students.

1-1 NOTE TO STUDENTS

In writing this book we have kept you, the student, uppermost in our minds; the book is written for you. It is our strong feeling that classroom time should not be devoted to a regurgitation of textbook material by the instructor. Instead, the time should be used to amplify the textbook material through discussion of related material and the application of basic principles to the solution of problems. The necessary conditions for accomplishing this goal are: (1) a clear, concise presentation of the fundamentals that you, the student, can read and understand, and (2) your willingness to read the text material before going to class. We have assumed responsibility for meeting the first condition. You must assume responsibility for satisfying the second condition. There probably will be times when we fall short of satisfying these objectives. If so, we would appreciate hearing of these shortcomings either directly or through your instructor.

It goes without saying that an introductory text is not all-inclusive. Your instructor undoubtedly will expand on the material presented, suggest alternative approaches to a topic, and introduce additional new material. We encourage you to refer to the many other available fluid mechanics textbooks; where another text presents a particularly good discussion of a given topic, we shall refer to it directly. We assume that you have had an introductory course in thermodynamics, and prior courses in statics and dynamics, and differential and integral calculus. No attempt will be made to restate this subject material; however, the pertinent aspects of this previous study will be reviewed briefly when appropriate.

It is our strong belief that one learns best by *doing*. This is true whether the subject under study is fluid mechanics, thermodynamics, or golf. The fundamentals in any of these cases are few, and mastery of them comes through practice. *Thus it is extremely important, in fact essential, that you solve problems.* The numerous problems included at the end of each chapter provide the opportunity to gain facility in applying fundamentals to the solution of problems. You should avoid the temptation to adopt a "plug and chug" approach to solving problems. Most of the problems are such that this approach simply will not work. To solve problems we strongly recommend that you proceed using the following logical steps:

1. State briefly and concisely (in your own words) the information given.
2. State the information to be found.
3. Draw a schematic of the system or control volume to be used in the analysis. Be sure to label the boundaries of the system or control volume and label appropriate coordinate directions.
4. Give the appropriate mathematical formulation of the *basic* laws that you consider necessary to solve the problem.
5. List the simplifying assumptions that you feel are appropriate in the problem.
6. Carry the analysis to completion algebraically before substituting numerical values.
7. Substitute numerical values (using a consistent set of units) to obtain a numerical answer. The significant figures in the answer should be consistent with the given data.
8. Check the answer and review the assumptions made in the solution to make sure they are reasonable.
9. Label the answer.

In your initial work this problem format may seem unnecessary. However, such an orderly approach to the solution of problems will reduce errors, save time, and permit a clearer understanding of the limitations of a particular solution. This format is used in all example problems presented in this text; answers to example problems are given to three significant figures.

Most engineering calculations involve measured values or physical property data. Every measured value has associated with it an experimental uncertainty. The uncertainty in a measurement can be reduced with care and application of more precise measurement techniques. The cost and time needed to obtain data rise sharply as measurement precision is increased. Therefore, few engineering data are sufficiently precise to justify the use of more than three significant figures.

The principles of specifying the experimental uncertainty of a measurement and of estimating the uncertainty of a calculated result are reviewed in Appendix E. These should be understood thoroughly by anyone who performs laboratory work. We suggest you take time to review Appendix E before performing laboratory work or solving the homework problems at the end of this chapter.

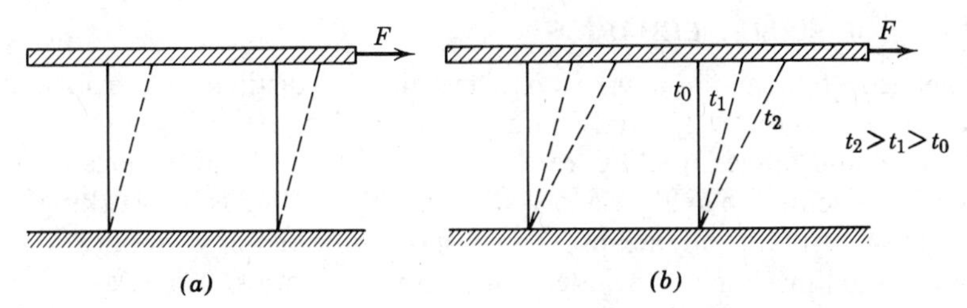

Fig. 1.1 Behavior of (a) solid and (b) fluid, under the action of a constant shear force.

1-2 DEFINITION OF A FLUID

Fluid mechanics deals with the behavior of fluids at rest and in motion. It is logical to begin with a definition of a *fluid*: a fluid is a substance that deforms continuously under the application of a shear (tangential) stress no matter how small the shear stress may be.

Thus fluids comprise the liquid and gas (or vapor) phases of the physical forms in which matter exists. The distinction between a fluid and the solid state of matter is clear if you compare fluid and solid behavior. A solid deforms when a shear stress is applied, but it does not deform continuously.

In Fig. 1.1 the behavior of a solid (Fig. 1.1*a*) and a fluid (Fig. 1.1*b*) under the action of a constant shear force are contrasted. In Fig. 1.1*a* the shear force is applied to the solid through the upper of two plates to which the solid has been bonded. When the shear force is applied to the plate, the block is deformed as shown. From our previous work in mechanics, we know that, provided the elastic limit of the solid material is not exceeded, the deformation is proportional to the applied shear stress, $\tau = F/A$, where A is the area of the surface in contact with the plate.

To repeat the experiment with a fluid between the plates, use a dye marker to outline a fluid element as shown by the solid lines (Fig. 1.1*b*). When the force, F, is applied to the upper plate, the fluid element continues to deform as long as the force is applied. The shape of the fluid element, at successive instants of time, $t_2 > t_1 > t_0$, is shown (Fig. 1.1*b*) by the dashed lines, which represent the positions of the dye markers at successive times. The fluid in direct contact with the solid boundary has the same velocity as the boundary itself; there is no slip at the boundary. This is an experimental fact based on numerous observations of fluid behavior.[1]

Because the fluid motion continues under the application of a shear stress, we may alternatively define a fluid as a substance that cannot sustain a shear stress when at rest.

[1] The no-slip condition is demonstrated in the film loops S-FM003, *Shear Deformation of Viscous Fluids*, and S-FM006, *Boundary-Layer Formation*. These loops were produced by Educational Services, Inc., Watertown, Mass., and the National Committee for Fluid Mechanics Films. The films are distributed by Encyclopaedia Britannica Educational Corporation. (A complete list of fluid mechanics film titles and sources is given in Appendix C.)

1-3 SCOPE OF FLUID MECHANICS

Having defined a fluid and noted the characteristics that distinguish it from a solid, we might ask the question: "Why study fluid mechanics?"

Knowledge and understanding of the basic principles and concepts of fluid mechanics are essential to analyze any system in which a fluid is the working medium. The design of virtually all means of transportation requires application of the principles of fluid mechanics. Included are aircraft for both subsonic and supersonic flight, ground effect machines, hovercraft (now in service for channel crossings between France and England), vertical takeoff and landing aircraft requiring minimum runway length, surface ships, submarines, and automobiles. In recent years automobile manufacturers have given more consideration to aerodynamic design. This has been true for some time for the designers of both racing cars and boats. The design of propulsion systems for space flight as well as for toy rockets is based on the principles of fluid mechanics. The collapse of the Tacoma Narrows Bridge some years ago is evidence of the possible consequences of neglecting the basic principles of fluid mechanics.[2] It is commonplace today to perform model studies to determine the aerodynamic forces on and flow fields around buildings and structures. These include studies of skyscrapers, baseball stadiums, smokestacks, and shopping plazas.

The design of all types of fluid machinery including pumps, fans, blowers, compressors, and turbines clearly requires knowledge of the basic principles of fluid mechanics. Lubrication is an area of considerable importance in fluid mechanics. Heating and ventilating systems for private homes, large office buildings, and underground tunnels, and the design of pipeline systems are further examples of technical problem areas requiring knowledge of fluid mechanics. The circulatory system of the body is essentially a fluid system. It is not surprising that the design of artificial hearts, heart-lung machines, breathing aids, and other such devices must rely on the basic principles of fluid mechanics.

Even some of our recreational endeavors are directly related to fluid mechanics. The slicing and hooking of golf balls can be explained by the principles of fluid mechanics (although they can be corrected only by a golf pro!).

The list of applications of the principles of fluid mechanics could be extended considerably. Our main point here is that fluid mechanics is not a subject studied for purely academic interest; rather, it is a subject with widespread importance both in our everyday experiences and in modern technology.

Clearly, we cannot hope to consider in detail even a small percentage of these and other specific problems of fluid mechanics. Instead, the purpose of this text is to present the basic laws and associated physical concepts that provide the basis or starting point in the analysis of any problem in fluid mechanics.

1-4 BASIC EQUATIONS

Analysis of any problem in fluid mechanics necessarily begins, either directly or indirectly, with statements of the basic laws governing the fluid motion. The basic laws,

[2] For dramatic evidence of aerodynamic forces in action, see the Ohio State University film, *Collapse of the Tacoma Narrows Bridge*.

which are applicable to any fluid, are:

1. Conservation of mass.
2. Newton's second law of motion.
3. Moment of momentum.
4. The first law of thermodynamics.
5. The second law of thermodynamics.

Clearly, not all basic laws always are required to solve any one problem. In some problems, it is necessary to bring into the analysis additional relations, in the form of equations of state or constitutive equations that describe the behavior of physical properties of fluids under given conditions.

You probably recall studying properties of gases in thermodynamics. The *ideal gas* equation of state

$$p = \rho RT \tag{1.1}$$

is a model that relates density to pressure and temperature for most gases for calculations of engineering accuracy. In Eq. 1.1, R is the gas constant. Values of R are given in Appendix A for several common gases; p and T in Eq. 1.1 are the absolute pressure and absolute temperature, respectively. Example Problem 1.1 illustrates use of the ideal gas equation of state.

It is obvious that the basic laws with which we shall deal are the same as those used in mechanics and thermodynamics. Our task will be to formulate these laws in suitable forms to solve fluid flow problems and to apply them to a wide variety of problems.

We must emphasize that there are, as we shall see, many apparently simple problems in fluid mechanics that cannot be solved analytically. In such cases we must resort to experiments and experimental observations.

1-5 METHODS OF ANALYSIS

As we have indicated, the basic laws that are used to analyze problems in fluid mechanics are the same ones used in thermodynamics and basic mechanics. The first step in solving a problem is to define the system that you are attempting to analyze. In basic mechanics, extensive use was made of the free-body diagram. In thermodynamics closed or open systems were considered. In this text we use the terms *system* and *control volume*. The importance of defining the system or control volume before applying the basic equations in the analysis of a problem cannot be overemphasized. At this point we review the definitions of systems and control volumes.

1-5.1 System and Control Volume

A system is defined as a fixed, identifiable quantity of mass; the system boundaries separate the system from the surroundings. The boundaries of the system may be fixed or movable; however, there is no mass transfer across the system boundaries.

In the familiar piston-cylinder assembly from thermodynamics, Fig. 1.2, the gas in the cylinder is the system. If a high-temperature source is brought in contact with the

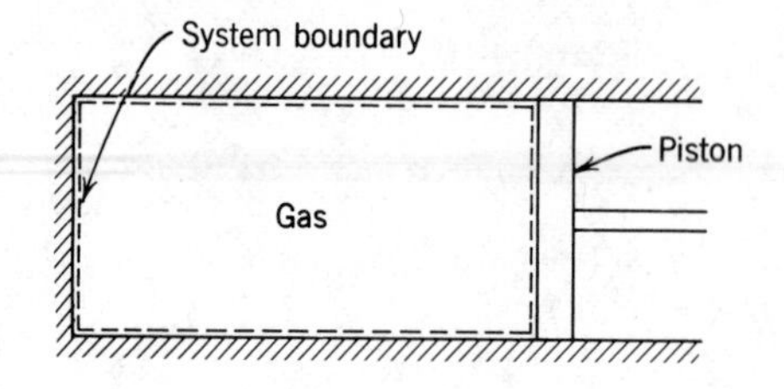

Fig. 1.2 Piston-cylinder assembly.

left end of the cylinder, the piston will move to the right; the boundary of the system thus moves. Heat and work may cross the boundaries of the system, but the quantity of matter within the system boundaries remains fixed. There is no mass transfer across the system boundaries.

Example 1.1

A piston-cylinder device contains 0.95 kg of oxygen initially at a temperature of 27 C and a pressure of 150 kPa. Heat is added to the gas and it expands at constant pressure to a temperature of 627 C. Determine the amount of heat added during the process.

EXAMPLE PROBLEM 1.1

GIVEN: Piston-cylinder containing O_2, $m = 0.95$ kg.

$T_1 = 27$ C $T_2 = 627$ C

$p = \text{constant} = 150$ kPa (abs)

FIND: $Q_{1 \to 2}$.

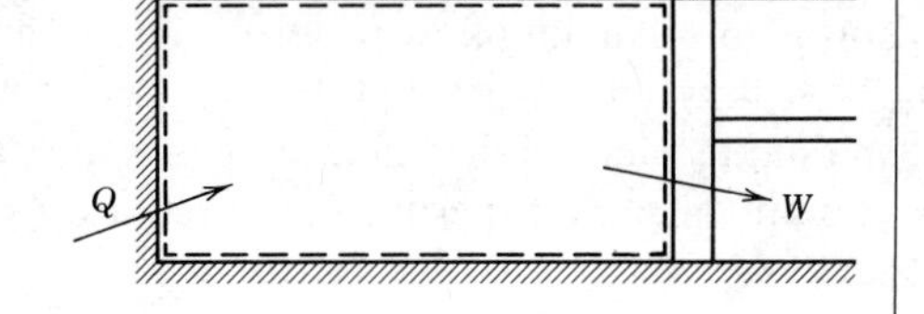

SOLUTION:

We are dealing with a system, $m = 0.95$ kg.

Basic equation: First law for the system, $Q_{12} + W_{12} = E_2 - E_1$

Assumptions: $E = U$, since the system is stationary
Ideal gas with constant specific heats

Under the above assumptions,

$$E_2 - E_1 = U_2 - U_1 = m(u_2 - u_1) = mc_v(T_2 - T_1)$$

The work done during the process is moving boundary work

$$W_{12} = -\int_{V_1}^{V_2} p\, dV = p(V_1 - V_2)$$

For an ideal gas, $pV = mRT$. Hence $W_{12} = mR(T_1 - T_2)$. Then from the first law equation,

$$Q_{12} = E_2 - E_1 - W_{12} = mc_v(T_2 - T_1) + mR(T_2 - T_1)$$

$$Q_{12} = m(T_2 - T_1)(c_v + R)$$

$$Q_{12} = mc_p(T_2 - T_1) \qquad \{R = c_p - c_v\}$$

From the Appendix, Table A.6, for O_2, $c_p = 909.4$ J/kg · K. Solving for Q_{12}, we obtain

$$Q_{12} = 0.95 \text{ kg} \times 909.4 \frac{\text{J}}{\text{kg} \cdot \text{K}} \times 600 \text{ K}$$

$$Q_{12} = 518 \text{ kJ}$$

Q_{12}

The purpose of this problem was to review the use of:
(i) the first law of thermodynamics for a system, and
(ii) the equation of state for an ideal gas.

In mechanics courses you made extensive use of the free-body diagram (system approach). This was logical because you were dealing with an easily identifiable rigid body. However in fluid mechanics, we normally are concerned with the flow of fluids through devices such as compressors, turbines, pipelines, nozzles, and so on. In these cases it is difficult to focus attention on a fixed identifiable quantity of mass. It is much more convenient, for analysis, to focus attention on a volume in space through which the fluid flows. Therefore, we use the control volume approach.

A control volume is an arbitrary volume in space through which fluid flows. The geometric boundary of the control volume is called the control surface. The control surface may be real or imaginary; it may be at rest or in motion. Figure 1.3 shows a possible control surface for analysis of flow through a pipe. Here the inside surface of the pipe, a real physical boundary, comprises part of the control surface. However, the vertical portions of the control surface are imaginary. There is no corresponding physical surface; these imaginary boundaries are selected arbitrarily for accounting purposes. Since the location of the control surface has a direct effect on the accounting procedure in applying the basic laws, it is extremely important that the control surface be clearly defined before beginning any analysis.

1-5.2 Differential versus Integral Approach

The basic laws that we apply in our study of fluid mechanics can be formulated in terms of infinitesimal or finite systems and control volumes. As you might suspect, the equations will look different in each case. Both approaches are important in the study of fluid mechanics and both will be developed in the course of our work.

In the first case the resulting equations are differential equations. Solution of the differential equations of motion provides a means of determining the detailed (point by point) behavior of the flow.

Frequently, in the problems under study, the information sought does not require a detailed knowledge of the flow. We often are interested in the gross behavior of a

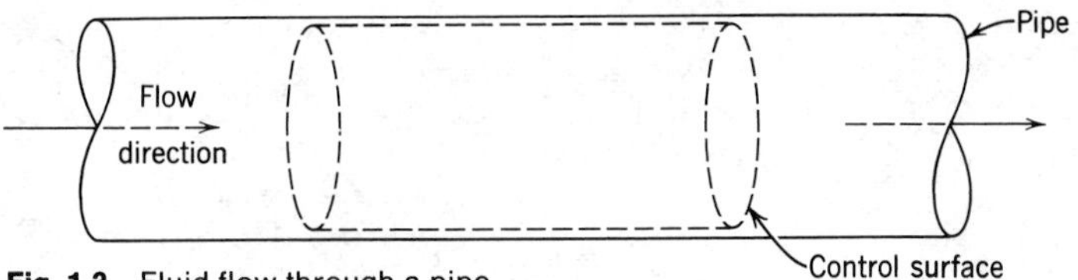

Fig. 1.3 Fluid flow through a pipe.

device; in such cases it is more appropriate to use the integral formulation of the basic laws. The integral formulation, using finite systems or control volumes, usually is easier to treat analytically. Since mechanics and thermodynamics deal with the formulation of the basic laws in terms of finite systems, these formulations are the basis for deriving the control volume equations in Chapter 4.

1-5.3 Methods of Description

Mechanics deals almost exclusively with systems; you have made extensive use of the basic equations applied to a fixed, identifiable quantity of mass. In attempting to analyze thermodynamic devices, you often found it necessary to use a control volume (open system) analysis. Clearly, the type of analysis depends on the problem. Where it is easy to keep track of identifiable elements of mass (e.g. in particle mechanics), we utilize a method of description that follows the particle. This sometimes is referred to as the *Lagrangian* method of description.

Consider, for example, the application of Newton's second law to a particle of fixed mass, m. Mathematically, we can write Newton's second law for a system of mass, m, as

$$\sum \vec{F} = m\vec{a} = m\frac{d\vec{V}}{dt} = m\frac{d^2\vec{r}}{dt^2} \tag{1.2}$$

In Eq. 1.2, $\sum \vec{F}$ is the sum of all external forces acting on the system, $\vec{a}$ is the acceleration of the center of mass of the system, $\vec{V}$ is the velocity of the center of mass of the system, and $\vec{r}$ is the position vector of the center of mass of the system relative to a fixed coordinate system.

Example 1.2

The air resistance on a 200 g ball in free flight is given by $f = 2 \times 10^{-4}\,v^2$, where f is in newtons and v is in meters per second. If the ball is dropped from rest 500 m above the ground, determine the speed at which it hits the ground. What percentage of the terminal speed is the result?

EXAMPLE PROBLEM 1.2

GIVEN: Ball, $m = 0.2$ kg, released from rest at $y_0 = 500$ m
Air resistance, $f = kv^2$, where $k = 2 \times 10^{-4}\ \text{N}\cdot\text{sec}^2/\text{m}^2$

Units: f(N), v(m/sec)

FIND: (a) The speed at which the ball hits the ground.
(b) Ratio of speed to terminal speed.

SOLUTION:

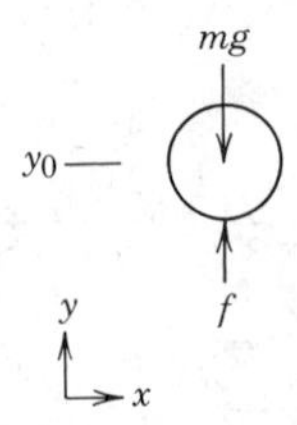

Basic equation: $\sum \vec{F} = m\vec{a}$

The motion of the ball is governed by the equation

$$\sum F_y = ma_y = m\frac{dv}{dt}$$

Since $v = v(y)$, $\sum F_y = m \dfrac{dv}{dy}\dfrac{dy}{dt} = mv \dfrac{dv}{dy}$

$$\sum F_y = f - mg = kv^2 - mg = mv \frac{dv}{dy}$$

Separating variables and integrating,

$$\int_{y_0}^{y} dy = \int_0^v \frac{mv\,dv}{kv^2 - mg}$$

$$y - y_0 = \left[\frac{m}{2k}\ln(kv^2 - mg)\right]_0^v = \frac{m}{2k}\ln\frac{kv^2 - mg}{-mg}$$

Taking antilogarithms, we obtain

$$kv^2 - mg = -mg\, e^{\left[\frac{2k}{m}(y - y_0)\right]}$$

Solving for v gives

$$v = \left[\frac{1}{k}mg\left(1 - e^{\left\{\frac{2k}{m}(y - y_0)\right\}}\right)\right]^{1/2}$$

Substituting numerical values with $y = 0$ yields

$$v = \left[0.2\text{ kg} \times 9.81\frac{\text{m}}{\text{sec}^2} \times \frac{\text{m}^2}{2 \times 10^{-4}\text{ N}\cdot\text{sec}^2} \times \frac{\text{N}\cdot\text{sec}^2}{\text{kg}\cdot\text{m}}\left(1 - e^{\left\{\frac{2 \times 2 \times 10^{-4}}{0.2}(-500)\right\}}\right)\right]^{1/2}$$

$v = 78.7$ m/sec

At terminal speed, $a_y = 0$ and $\sum F_y = 0 = kv_t^2 - mg$

Then $v_t = \left[\dfrac{mg}{k}\right]^{1/2} = \left[0.2\text{ kg} \times 9.81\dfrac{\text{m}}{\text{sec}^2} \times \dfrac{\text{m}^2}{2 \times 10^{-4}\text{ N}\cdot\text{sec}^2} \times \dfrac{\text{N}\cdot\text{sec}^2}{\text{kg}\cdot\text{m}}\right]^{1/2}$

$$v_t = 99.0\text{ m/sec} \quad\text{and}\quad \frac{v}{v_t} = \frac{78.7}{99.0} = 79.5\%$$

$\dfrac{v}{v_t}$

{This problem is included as a reminder of the method of description used in particle mechanics.}

We may consider a fluid to be composed of a very large number of particles whose motion must be described; keeping track of the motion of each fluid particle separately would become a horrendous bookkeeping problem. Consequently, a particle description becomes unmanageable. Often we find it convenient to use a different type of description. Particularly with control volume analyses, it is convenient to use the field, or *Eulerian*, method of description, which focuses attention on the properties of a flow at a given point in space as a function of time. In the Eulerian method of description, the properties of a flow field are described as functions of space coordinates and time. We shall see in Chapter 2 that this method of description is a logical outgrowth of the assumption that fluids may be treated as continuous media.

1-6 DIMENSIONS AND UNITS

Engineering problems are solved to answer specific questions. It goes without saying that the answer must include units. (It makes a difference whether a pipe diameter required is 1 meter or 1 foot!) Consequently, it is appropriate to present a brief review of dimensions and units. We say "review" because the topic is familiar from your earlier work in mechanics.

We refer to physical quantities such as length, time, mass, and temperature as *dimensions.* In terms of a particular system of dimensions all measurable quantities can be subdivided into two groups—primary quantities and secondary quantities. We refer to a small group of dimensions from which all others can be formed as primary quantities. Primary quantities are those for which we set up arbitrary scales of measure; secondary quantities are those quantities whose dimensions are expressible in terms of the dimensions of the primary quantities.

Units are the arbitrary names (and magnitudes) assigned to the primary dimensions adopted as standards for measurement. For example, the primary dimension of length may be measured in units of meters, feet, yards, or miles. These units of length are related to each other through unit conversion factors (1 mile = 5280 feet = 1609 meters).

1-6.1 Systems of Dimensions

Any valid equation that relates physical quantities must be dimensionally homogeneous; each term in the equation must have the same dimensions. We recognize that Newton's second law ($\vec{F} \propto m\vec{a}$) relates the four dimensions, F, M, L, and t. Thus force and mass cannot both be selected as primary dimensions without introducing a constant of proportionality that has dimensions (and units).

Length and time are primary dimensions in all dimensional systems in common use. In some systems, mass is taken as a primary dimension. In others, force is selected as a primary dimension; a third system chooses both force and mass as primary dimensions. Thus we have three basic systems of dimensions, corresponding to the different ways of specifying the primary dimensions.

a. Mass $[M]$, length $[L]$, time $[t]$, temperature $[T]$.
b. Force $[F]$, length $[L]$, time $[t]$, temperature $[T]$.
c. Force $[F]$, mass $[M]$, length $[L]$, time $[t]$, temperature $[T]$.

In system *a*, force $[F]$ is a secondary dimension and the constant of proportionality in Newton's second law is dimensionless. In system *b*, mass $[M]$ is a secondary dimension, and again the constant of proportionality in Newton's second law is dimensionless. In system *c*, both force $[F]$ and mass $[M]$ have been selected as primary dimensions. In this case the constant of proportionality, g_c, in Newton's second law (written $\vec{F} = m\vec{a}/g_c$) is not dimensionless. The dimensions of g_c must in fact be $[ML/Ft^2]$ for the equation to be dimensionally homogeneous. The numerical value of the constant of proportionality depends on the units of measure chosen for each of the primary quantities.

1-6.2 Systems of Units

There is more than one way of selecting the unit of measure for each primary dimension. We shall present only the most common engineering system of units for each of the basic systems of dimensions.

a. MLtT

SI, which is the official abbreviation in all languages for the Système International d'Unités[3], is an extension and refinement of the traditional metric system. More than 30 countries have declared it to be the only legally accepted system. The United States most likely will adopt it within the next decade.

In the SI system of units, the unit of mass is the kilogram (kg), the unit of length is the meter (m), the unit of time is the second (sec), and the unit of temperature is the kelvin (K). Force is a secondary dimension, and its unit, the newton (N), is defined from Newton's second law as

$$1 \text{ N} \equiv 1 \text{ kg} \cdot \text{m/sec}^2$$

In the Absolute Metric system of units, the unit of mass is the gram, the unit of length is the centimeter, the unit of time is the second, and the unit of temperature is the Kelvin. Since force is a secondary dimension, the unit of force, the dyne, is defined in terms of Newton's second law as

$$1 \text{ dyne} \equiv 1 \text{ g} \cdot \text{cm/sec}^2$$

b. FLtT

In the British Gravitational system of units, the unit of force is the pound (lbf), the unit of length is the foot, the unit of time is the second, and the unit of temperature is the Rankine (R). Since mass is a secondary dimension, the unit of mass, the slug, is defined in terms of Newton's second law as

$$1 \text{ slug} \equiv 1 \text{ lbf} \cdot \text{sec}^2/\text{ft}$$

c. FMLtT

In the English Engineering system of units, the unit of force is the pound force (lbf), the unit of mass is the pound mass (lbm), the unit of length is the foot, the unit of time is the second, and the unit of temperature is the Rankine. Since both force and mass are chosen as primary dimensions, Newton's second law is written as

$$\vec{F} = \frac{m\vec{a}}{g_c}$$

A force of one pound (1 lbf) is the force that gives a pound mass (1 lbm) an acceleration equal to the standard acceleration of gravity on Earth, 32.17 ft/sec^2. From

[3] American Society for Testing and Materials, *ASTM Standard for Metric Practice*, E380-82. Philadelphia: ASTM, 1982.

Newton's second law we see that (to three significant figures)

$$1 \text{ lbf} \equiv \frac{1 \text{ lbm} \times 32.2 \text{ ft/sec}^2}{g_c}$$

or

$$g_c \equiv 32.2 \text{ ft} \cdot \text{lbm/lbf} \cdot \text{sec}^2$$

The constant of proportionality, g_c, has both dimensions and units. The dimensions arose because we selected both force and mass as primary dimensions; the units (and the numerical value) are a consequence of our choices for the standards of measurement.

Since a force of 1 lbf accelerates 1 lbm at 32.2 ft/sec^2, it would accelerate 32.2 lbm at 1 ft/sec^2. A slug also is accelerated at 1 ft/sec^2 by a force of 1 lbf. Therefore,

$$1 \text{ slug} \equiv 32.2 \text{ lbm}$$

1-6.3 Preferred Systems of Units

In this text we shall use both the SI and the British Gravitational systems of units. In either case, the constant of proportionality in Newton's second law is dimensionless and has a value of unity. Consequently, Newton's second law is written as $\vec{F} = m\vec{a}$. In these systems, it follows that the gravitational force (the "weight"[4]) on an object of mass, m, is given by $W = mg$.

SI units and prefixes are summarized in Appendix F.

1-7 SUMMARY OBJECTIVES

After completing study of Chapter 1, you should be able to do the following:

1. Give operational definitions of:
   - fluid
   - no-slip condition
   - system
   - control volume
   - Lagrangian method of description
   - Eulerian method of description
   - dimensions
   - units
   - dimensional homogeneity
   - weight
2. Give examples in which fluid mechanics is important to an understanding of phenomena from everyday experience and modern technology.
3. List the five basic laws governing the motion of fluids.
4. State the three basic systems of dimensions.
5. Give typical units of physical quantities in the SI, British Gravitational, and English Engineering systems of units.
6. Solve the problems at the end of the chapter that relate to the material you have studied.

[4] Note that in the English Engineering system, the weight of an object is given by $W = mg/g_c$.

PROBLEMS

1.1 A number of common substances are

Tar	Sand
"Silly Putty"	Jello
Modeling clay	Toothpaste
Wax	Shaving cream

Some of these materials exhibit characteristics of both solid and fluid behavior under different conditions. Explain and give examples.

1.2 Give a word statement of each of the five basic conservation laws stated in Section 1–4, as they apply to a system.

1.3 A tank of compressed oxygen for flame cutting is to contain 10 kg of oxygen at a pressure of 14 MPa (the temperature is 35 C). How large must be the tank volume?

1.4 A tank for scuba diving is designed to contain 50 standard cubic feet (SCF) of air when filled to a pressure of 3000 pounds per square inch (gage) at an ambient temperature of 80 F. Calculate the interior volume of the tank. A standard cubic foot of gas occupies one cubic foot at standard temperature and pressure ($T = 15$ C and $p = 101.3$ kPa absolute).

1.5 Moist air is a mixture of water vapor and air. At 100 percent relative humidity, the water vapor is at its saturation pressure. The atmospheric pressure given by the barometer equals the sum of the partial pressures of (dry) air and of the water vapor. Thus

$$p_{atm} = p_{air} + p_{vapor}$$

The saturation pressure of water at 15 C is 1.70 kPa. Use Dalton's law of partial pressures to calculate the density of moist air with 100 percent relative humidity at standard temperature and pressure (STP conditions are $T = 15$ C and $p = 101.3$ kPa absolute). Compare with the density of dry air at the same conditions.

1.6 Air at a pressure of 40 psia and a temperature of 70 F is moving with a speed of 100 ft/sec. Calculate:

(a) The kinetic energy per unit mass of the air
(b) The kinetic energy per unit volume of the air

1.7 Air at an absolute pressure of 300 kPa and a temperature of 20 C is moving at a speed of 30 m/sec. Calculate:

(a) The kinetic energy per unit mass of the air
(b) The kinetic energy per unit volume of the air

1.8 A compressed air tank in a service station holds 0.2 m^3 of compressed air at 800 kPa (gage). Determine the amount of energy required to compress this much air isothermally from atmospheric pressure, assuming a frictionless process. (Note that the release of this much energy would be catastrophic if the tank were ruptured.)

1.9 Air trapped in a bicycle tire pump is compressed suddenly to $\frac{1}{5}$ of its original volume. Both heat transfer and friction may be neglected as a first approximation. Determine the final temperature of the air in the pump if the initial temperature is 20 C.

1.10 The catapult on an aircraft carrier has a piston diameter of 2 ft and a stroke of 100 ft. Initially, the cylinder contains steam at 250 psia, 600 F in a volume of 100 ft^3. Assume that heat transfer and friction are negligible as the piston moves through its stroke. Determine:

(a) The temperature and pressure of the steam at the end of the stroke
(b) The work done by the steam during the expansion process

1.11 A projectile is fired with velocity, $\vec{V}_0$, and elevation angle, θ, above the horizon. Air resistance may be neglected. Express the range of the projectile in terms of V_0 and θ. Determine the angle that gives the maximum range.

1.12 Parametric equations for the motion of a particle are

$$x = A\cos\omega t \qquad A > B$$
$$y = B\sin\omega t$$

Determine the velocity and acceleration of the particle as functions of time. Indicate the particle trajectory in the xy plane on a sketch, and locate the point(s) of maximum velocity and acceleration.

1.13 The English perfected the longbow as a weapon after the Medieval period. In the hands of a skilled archer, the longbow was reputed to be accurate at ranges to 100 meters or more. If the maximum altitude of an arrow is less than 10 m while traveling to a target 100 m away from the archer, and neglecting air resistance, estimate the speed and angle at which the arrow must leave the bow.

1.14 Very small particles moving in fluids experience a drag force proportional to speed. Consider a particle of net weight, W, dropped in a fluid. The particle experiences a drag force, kV, where V is the particle speed. Determine the time required for the particle to accelerate from rest to 95 percent of its terminal speed, V_t, in terms of k, W, and g.

1.15 A sky diver with a mass of 75 kg is in free fall at an altitude of 2 km. The aerodynamic drag force acting on the sky diver is known to be $F_D = kV^2$, where $k = 0.228\ \text{N}\cdot\text{sec}^2/\text{m}^2$. Determine the maximum speed of free fall for the sky diver.

1.16 A sky diver with mass, $m = 80$ kg, drops from a slow moving aircraft and falls straight down. The aerodynamic drag force acting on the diver is $F_D = kV^2$, where $k = 0.27\ \text{N}\cdot\text{sec}^2/\text{m}^2$, and V is the speed relative to the air. Evaluate the terminal speed of the sky diver. Estimate the vertical distance required for the sky diver to reach 95 percent of terminal speed. Compare with the distance required to reach the same speed if air resistance were neglected.

1.17 A droplet of SAE 30 oil with $D = 0.7$ mm is placed in water at 20 C. Since the droplet is small, its resistance to motion may be characterized approximately by $F_D = kV$, where $k = 6.60 \times 10^{-6}\ \text{N}\cdot\text{sec/m}$. Because of buoyancy, the net weight of the droplet is $F_N = 0.0505\ \mu\text{N}$. Calculate the terminal speed of the droplet.

1.18 A small particle moving in water experiences drag force, $F_D = kV$, where the dimensions of k are force per unit speed. A particle of mass, m, is set in motion horizontally with initial speed, V_0. Show that the horizontal distance traveled before the particle stops is finite and equal to $s = mV_0/k$.

1.19 A spear of mass, $m = 0.3$ kg, is propelled horizontally from a spear gun by a scuba diver. The initial speed of the spear is $V_0 = 30$ m/sec. The force that resists its motion through the water is given by $F_D = kV^2$, where $k = 0.033\ \text{N}\cdot\text{sec}^2/\text{m}^2$. The spear is effective against sharks when its speed is above $V = 10$ m/sec. Estimate the effective range of the spear.

1.20 A ball is thrown vertically upward with initial speed, V_0. Air resistance on the ball is proportional to the square of its speed, $F_D = kV^2$. Analyze the motion to obtain expressions for the velocity of the ball and its height as functions of time. Express the time to reach maximum height for the ball in terms of V_0, m, k, and g. Compare with the value for no air resistance.

1.21 The aerodynamic drag force on a tractor-trailer rig moving in still air is given by $F_D = kV^2$, where $k = 0.135\ \text{lbf}\cdot\text{sec}^2/\text{ft}^2$. Calculate the force required to overcome aerodynamic drag at a speed of 55 mph. Evaluate the power saving if the aerodynamic drag were reduced 6 percent by installing a fairing on the cab roof.

1.22 A swimmer in fresh water can move at a maximum steady speed of $V_s = 1.5$ m/sec in still water. To do so, the swimmer must produce enough power to overcome water resistance. The drag force acting on the swimmer is estimated to be $F_D = kV^2$, where $k = 30\ \text{N} \cdot \text{sec}^2/\text{m}^2$, and V is the speed of the swimmer *relative* to the water. Evaluate the power produced by the swimmer in still water. If the swimmer can maintain the same power output in a river current that moves at 3 km/hr, estimate the maximum speeds the swimmer can reach: (a) swimming upstream, and (b) swimming downstream.

1.23 For each quantity listed, indicate dimensions using the MLtT system of dimensions, and give typical SI and English units:

(a) Power
(b) Pressure
(c) Modulus of elasticity
(d) Angular velocity
(e) Energy
(f) Momentum
(g) Shear stress
(h) Specific heat
(i) Thermal expansion coefficient

1.24 For each quantity listed, indicate dimensions using the FLtT system of dimensions, and give typical SI and English units:

(a) Power
(b) Pressure
(c) Modulus of elasticity
(d) Angular velocity
(e) Energy
(f) Moment of a force
(g) Momentum
(h) Shear stress
(i) Strain

1.25 The stylus of a high-quality stereo system is adjusted to balance a mass of 0.75 g on a small equal arm beam balance. What force (in lbf) does the stylus exert on a record? What is the force expressed in newtons?

1.26 The unit of pressure in the SI system is the pascal (Pa). How many pounds force per square inch (psi) correspond to 1 Pa?

1.27 A U.S. gallon, by definition, contains a volume of 231 cubic inches. Determine: (a) the number of gallons in a cubic foot, and (b) the number of liters in a gallon.

1.28 Derive the following conversion factors:

(a) Convert a volume flow rate in cubic meters per second to cubic feet per second.
(b) Convert a volume flow rate in cubic feet per second to gallons per minute.
(c) Convert a volume flow rate of water in gallons per hour to grams per minute.
(d) Convert a volume flow rate of air in standard cubic feet per minute (SCFM) to pounds per hour. A standard cubic foot of gas occupies one cubic foot at standard temperature and pressure ($T = 15$ C and $p = 101.3$ kPa absolute).

1.29 At an institution known for its basketball prowess, a new set of physical units has been suggested. The basic unit of length is to be the "three-pointer," which is 21 ft; the basic unit of time is the "shot clock," which is 30 sec; the basic unit of force is the "basketball," which is 21 oz. Determine the conversion factors among these units and their SI equivalents. How are the units of mass in the two systems related?

1.30 At an institution known for its baseball prowess, it has been suggested that force, velocity, and length be considered as basic dimensions. The basic unit of force is the "baseball," which is 5.1 oz; the basic unit of velocity is the "fastball," which is 90 mph; the basic unit of length is the "homerun," which is 385 ft. Determine the conversion factors among these units and their SI equivalents. What is the unit of mass in the new system? What is the conversion factor between this unit of mass and the SI unit?

1.31 A can of pet food has the following internal dimensions: 102 mm height and 73 mm diameter (each ± 1 mm at odds of 20 to 1). The label lists the mass of the contents as 397 g. Evaluate the magnitude and estimated uncertainty of the density of the pet food if the mass value is accurate to ± 1 g at the same odds.

1.32 The mass of the standard American golf ball is 1.62 ± 0.01 oz and its mean diameter is 1.68 ± 0.01 in. Determine the density and specific gravity of the American golf ball. Estimate the uncertainties in the calculated values.

1.33 The mass of the standard British golf ball is 1.62 ± 0.01 oz and its mean diameter is 1.62 ± 0.01 in. Determine the density and specific gravity of the British golf ball. Estimate the uncertainties in the calculated values.

1.34 A container weighs 2.9 lbf when empty. When filled with water at 90 F, the mass of the container and its contents is 1.95 slug. Find the weight of water in the container, and its volume in cubic feet, using data from Appendix A.

1.35 Determine the specific gravity of mercury at 4 C.

1.36 Truck weight laws in Michigan allow a gross combination weight of 130,000 lbf. A tractor-trailer tank truck weighs 36,000 lbf empty. Calculate the number of gallons of gasoline that it can carry legally. Use data for gasoline from Appendix A.

1.37 A super-tanker carries a cargo of 400,000 long tons (1 long ton = 2240 lbm) of crude oil. Assume that the specific gravity of the oil is SG = 0.72. Calculate the number of barrels of oil in the tanker's cargo (the petroleum industry defines 1 bbl as 42 U.S. gallons).

1.38 Evaluate the change in specific weight of mercury, in lbf/ft^3, as its temperature changes from 70 to 90 F. Use the data in Appendix A.

1.39 The density of mercury is given as 26.3 slug/ft^3. Calculate the specific gravity and the specific volume in m^3/kg of the mercury. Calculate the specific weight in lbf/ft^3 on Earth and on the moon. Acceleration of gravity on the moon is 5.47 ft/sec^2.

1.40 The variation in density of ice with temperature is approximately linear (Gross, M. Grant, *Oceanography*, 2nd ed., Englewood Cliffs, N.J.: Prentice-Hall, 1977) as given by

$$\frac{\partial \rho}{\partial T} = -0.0001 \frac{\text{g}}{\text{cm}^3 \cdot \text{C}}$$

The specific gravity of ice at 0 C (referred to water at 4 C) is SG = 0.917. Evaluate the density of ice at 0 C and its specific gravity at $T = -10$ C.

1.41 Typical measured data for automatic transmission fluid are (*SAE Handbook*, 1976):

Temperature (C)	16	99	149
Specific gravity	0.880	0.821	0.789

The same reference gives the volume coefficient of thermal expansion for the fluid as $\alpha = 0.00072\ \text{m}^3/\text{m}^3 \cdot \text{C}$ at $T = 16$ C. Starting with the value at 16 C and assuming α to be constant, calculate specific gravity values at 99 and 149 C. Compare with the measured values.

1.42 Two vectors, $\vec{a}$ and $\vec{b}$, are given by the expressions:

$$\vec{a} = xy\hat{i} + y^2\hat{j} + 2\hat{k} \qquad \vec{b} = x^2\hat{i} - xy\hat{j} + z\hat{k}$$

and a scalar, ϕ, is given by $\phi = \dfrac{x^2}{2} - \dfrac{y^2}{2}$

Evaluate each of the following:

(a) $\vec{a} \cdot \vec{b}$ (b) $\vec{a} \times \vec{b}$

(c) $\partial \vec{a}/\partial x$ (d) $\nabla \phi$

(e) $\nabla \cdot \vec{a}$ (f) $\nabla \times \vec{b}$

1.43 Two vectors, $\vec{c}$ and $\vec{d}$, are given by the expressions:

$$\vec{c} = xyz\hat{\imath} + 2\hat{\jmath} + y^2\hat{k} \qquad \vec{d} = x^2\hat{\imath} + y^2\hat{\jmath} + x\hat{k}$$

and a scalar, ψ, is given by $\psi = xy$

Evaluate each of the following:

(a) $\vec{c} \cdot \vec{d}$ (b) $\vec{c} \times \vec{d}$

(c) $\partial \vec{c}/\partial x$ (d) $\nabla \psi$

(e) $\nabla \times \vec{c}$ (f) $\nabla \cdot \vec{d}$

1.44 Two vectors, $\vec{r}$ and $\vec{s}$, are given by the expressions:

$$\vec{r} = x^2y\hat{\imath} + z^2\hat{k} \qquad \vec{s} = xz\hat{\imath} + xyz\hat{\jmath} + x^2y\hat{k}$$

and a scalar, ζ, is given by $\zeta = \frac{1}{2}(x^2 - y^2)$

Evaluate each of the following:

(a) $\vec{r} \cdot \vec{s}$ (b) $\vec{r} \times \vec{s}$

(c) $\partial \vec{s}/\partial z$ (d) $\nabla \times \vec{r}$

(e) $\nabla \cdot \vec{s}$ (f) $\nabla^2 \zeta$

Chapter 2

FUNDAMENTAL CONCEPTS

In Chapter 1 we indicated that our study of fluid mechanics will build on earlier studies in mechanics and thermodynamics. To develop a unified approach, we review some familiar topics and introduce some new concepts and definitions. The purpose of this chapter is to develop these fundamental concepts.

2-1 FLUID AS A CONTINUUM

In our definition of a fluid, no mention was made of the molecular structure of fluids. All fluids are composed of molecules in constant motion. However, in most engineering applications we are interested in the average or macroscopic effects of many molecules. It is these macroscopic effects that we can perceive and measure. We thus treat a fluid as an infinitely divisible substance, a *continuum*, and do not concern ourselves with the behavior of individual molecules.

The concept of a continuum is the basis of classical fluid mechanics. The continuum assumption is valid in treating the behavior of fluids under normal conditions. However, it breaks down whenever the mean free path of the molecules (approximately 10^{-7} mm for gas molecules that show ideal behavior at STP)[1] becomes the same order of magnitude as the smallest significant characteristic dimension of the problem. In problems such as rarefied gas flow (e.g. as encountered in flights into the upper reaches of the atmosphere), we must abandon the concept of a continuum in favor of the microscopic and statistical points of view.

As a consequence of the continuum assumption, each fluid property is assumed to have a definite value at each point in space. Thus fluid properties such as density, temperature, velocity, and so on, are considered to be continuous functions of position and time.

To illustrate the concept of a property at a point, consider the manner in which we determine the density at a point. A region of fluid is shown in Fig. 2.1. We are interested

[1] STP (Standard Temperature and Pressure) for air are 15 C (59 F) and 101.3 kPa absolute (14.696 psia), respectively.

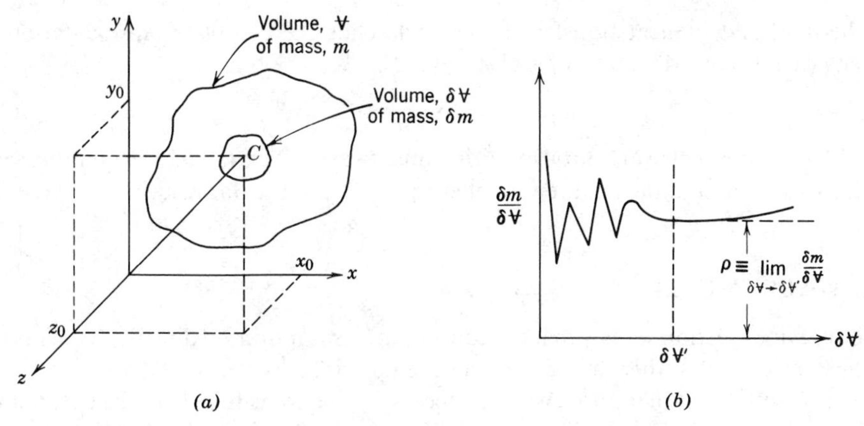

Fig. 2.1 Definition of density at a point.

in determining the density at the point C, whose coordinates are x_0, y_0, and z_0. The density is defined as mass per unit volume. Thus the mean density within the volume, $\forall$, would be given by $\rho = m/\forall$. In general, this will not be equal to the value of the density at point C. To determine the density at point C, we must select a small volume, $\delta\forall$, surrounding point C and determine the ratio, $\delta m/\delta\forall$. The question is how small can we make the volume, $\delta\forall$? Let us answer this question by plotting the ratio, $\delta m/\delta\forall$, and allowing the volume to shrink continuously in size. Assuming that the volume, $\delta\forall$, is initially relatively large (but still small compared to the volume, $\forall$) a typical plot of $\delta m/\delta\forall$ might appear as in Fig. 2.1*b*. The average density tends to approach an asymptotic value as the volume is shrunk to enclose only homogeneous fluid in the immediate neighborhood of point C. When $\delta\forall$ becomes so small that it contains only a small number of molecules, it becomes impossible to fix a definite value for $\delta m/\delta\forall$; the value will vary erratically as molecules cross into and out of the volume. Thus there is a lower limiting value of $\delta\forall$, designated $\delta\forall'$ in Fig. 2.1*b*, allowable for use in defining fluid density at a point.[2] The density at a point is then defined as

$$\rho \equiv \lim_{\delta\forall \to \delta\forall'} \frac{\delta m}{\delta\forall} \tag{2.1}$$

Since point C was arbitrary, the density at any point in the fluid could be determined in a like manner. If density determinations were made simultaneously at an infinite number of points in the fluid, we would obtain an expression for the density distribution as a function of the space coordinates, $\rho = \rho(x, y, z)$, at the given instant in time. Clearly, the density at a point may vary with time as a result of work done on or

[2] The size of $\delta\forall'$ is extremely small. For example 1 m^3 of air at STP contains approximately 2.5×10^{25} molecules. Thus the number of molecules in a volume of 10^{-12} m^3 (about the size of a grain of sand) would be 2.5×10^{13}. This number is certainly large enough to insure that the average mass within $\delta\forall'$ will be constant.

by the fluid and/or heat transfer to the fluid. Thus the complete representation of density (the *field* representation) is given by

$$\rho = \rho(x, y, z, t) \tag{2.2}$$

Since density is a scalar quantity, requiring only the specification of a magnitude for a complete description, the field represented by Eq. 2.2 is a scalar field.

2-2 VELOCITY FIELD

In the previous section we saw that the continuum assumption led directly to the notion of the density field. Other fluid properties are described by fields.

To deal with fluids in motion, we shall necessarily be concerned with the description of a velocity field. Refer again to Fig. 2.1*a*. Define the fluid velocity at point C as the instantaneous velocity of the center of gravity of the volume, $\delta \mathcal{V}'$, instantaneously surrounding point C. Define a *fluid particle* as the small mass of fluid of fixed identity of volume, $\delta \mathcal{V}'$. Thus we define the velocity at point C as the instantaneous velocity of the fluid particle which, at a given instant, is passing through point C. The velocity at any point in the flow field is defined similarly. At a given instant the velocity field, $\vec{V}$, is a function of the space coordinates x, y, z. The velocity at any point in the flow field might vary from one instant to another. Thus the complete representation of velocity (the velocity field) is given by

$$\vec{V} = \vec{V}(x, y, z, t) \tag{2.3}$$

The velocity vector, $\vec{V}$, can be written in terms of its three scalar components. Denoting the components in the x, y, and z directions by u, v, and w, then

$$\vec{V} = u\hat{i} + v\hat{j} + w\hat{k} \tag{2.4}$$

In general, each of the components u, v, and w will be a function of x, y, z, and t.

If properties at each point in a flow field do not change with time, the flow is termed *steady*. Stated mathematically, the definition of steady flow is

$$\frac{\partial \eta}{\partial t} = 0$$

where η represents any fluid property. For steady flow,

$$\frac{\partial \rho}{\partial t} = 0, \qquad \text{or} \qquad \rho = \rho(x, y, z)$$

and

$$\frac{\partial \vec{V}}{\partial t} = 0 \qquad \text{or} \qquad \vec{V} = \vec{V}(x, y, z)$$

Thus, in steady flow, any property may vary from point to point in the field, but all properties remain constant with time at each point.

2-2.1 One-, Two-, and Three-Dimensional Flows

A flow is classified as one-, two-, or three-dimensional depending on the number of space coordinates required to specify the velocity field.[3] Equation 2.3 indicates that the velocity field may be a function of three space coordinates and time. Such a flow field is termed *three-dimensional* (it is also *unsteady*) because the velocity at any point in the flow field depends on the three coordinates required to locate the point in space.

Not all flow fields are three-dimensional. Consider, for example, the steady flow through a long straight pipe of constant cross section. Far from the entrance to the pipe the velocity distribution may be described by

$$u = u_{\max}\left[1 - \left(\frac{r}{R}\right)^2\right] \tag{2.5}$$

This profile is shown in Fig. 2.2, where cylindrical coordinates r, θ, and x are used to locate any point in the flow field. The velocity field is a function of r only; it is independent of the coordinates x and θ. Thus this is a one-dimensional flow.

An example of a two-dimensional flow is illustrated in Fig. 2.3; the velocity distribution is depicted for a flow between diverging straight walls that are imagined to be infinite in extent (in the z direction). Since the channel is considered to be infinite in the z direction, the velocity field will be identical in all planes perpendicular to the z axis. Consequently, the velocity field is a function only of the space coordinates x and y; the flow field is classified as two-dimensional.

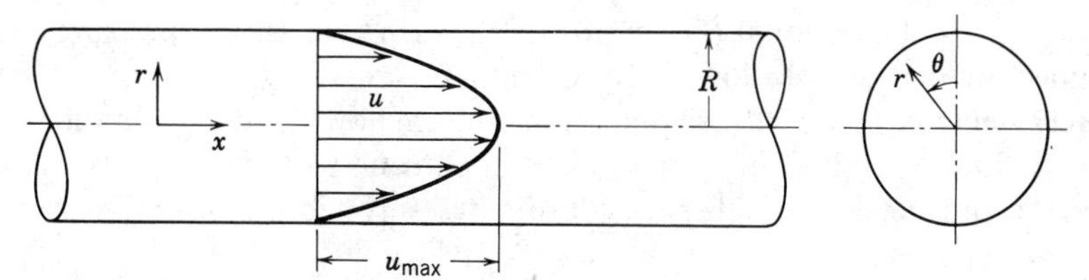

Fig. 2.2 Example of one-dimensional flow.

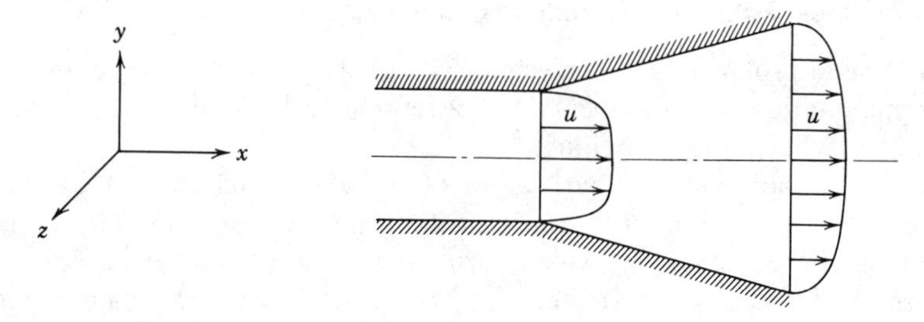

Fig. 2.3 Example of two-dimensional flow.

[3] Some authors choose to classify a flow as one-, two-, or three-dimensional on the basis of the number of space coordinates required to specify all fluid properties. In this text, classification of flow fields will be based on the number of space coordinates required to specify the velocity field only.

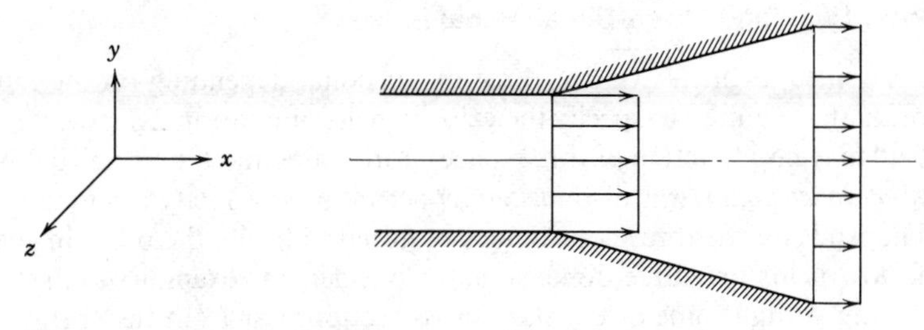

Fig. 2.4 Example of uniform flow at a section.

As you might suspect, the complexity of analysis increases considerably with the number of dimensions of the flow field. The simplest to analyze is one-dimensional flow. For many problems encountered in engineering, a one-dimensional analysis is adequate to provide approximate solutions of engineering accuracy.

Since all fluids satisfying the continuum assumption must have a zero relative velocity at a solid surface (to satisfy the no-slip condition), most flows are inherently two- or three-dimensional. For purposes of analysis it often is convenient to introduce the notion of *uniform flow* at a given cross section. In a flow that is uniform at a given cross section, the velocity is constant across any section normal to the flow. Under this assumption,[4] the two-dimensional flow of Fig. 2.3 is modeled as the flow shown in Fig. 2.4. In the flow of Fig. 2.4, the velocity field is a function of x alone, and thus the flow model is one-dimensional. (Other properties, such as density or pressure, also may be assumed uniform at a section, if appropriate.)

The term *uniform flow field* (as opposed to uniform flow at a cross section) is used to describe a flow in which the magnitude and direction of the velocity vector are constant, i.e. independent of all space coordinates, throughout the entire flow field.

2-2.2 Timelines, Pathlines, Streaklines, and Streamlines

In the analysis of problems in fluid mechanics, frequently it is advantageous to obtain a visual representation of a flow field. Such a representation is provided by timelines, pathlines, streaklines, and streamlines.[5]

If a number of adjacent fluid particles in a flow field are marked at a given instant, they form a line in the fluid at that instant; this line is called a *timeline*. Subsequent observations of the line may provide information about the flow field. For example, in discussing the behavior of a fluid under the action of a constant shear force

[4] Convenience alone does not justify this assumption; often results of acceptable accuracy are obtained. Sweeping assumptions such as uniform flow at a cross section should always be reviewed carefully to be sure they provide a reasonable analytical model of the real flow.

[5] Timelines, pathlines, streaklines, and streamlines are demonstrated in the film loops: S-FM047, *Pathlines, Streaklines, Streamlines, and Timelines in Steady Flow*, and S-FM048, *Pathlines, Streaklines, and Streamlines in Unsteady Flow*. These two loops are taken from the film *Flow Visualization*, S. J. Kline, principal.

(Section 1-2) timelines were introduced to demonstrate the deformation of a fluid at successive instants.

A *pathline* is the path or trajectory traced out by a moving fluid particle. To make a pathline visible, we might identify a fluid particle at a given instant, e.g. by the use of dye, and then take a long exposure photograph of its subsequent motion. The line traced out by the particle is a pathline.

On the other hand, we might choose to focus our attention on a fixed location in space and identify, again by the use of dye, all fluid particles passing through this point. After a short period of time we would have a number of identifiable fluid particles in the flow, all of which had, at some time, passed through one fixed location in space. The line joining these fluid particles is defined as a *streakline.*

Streamlines are lines drawn in the flow field so that at a given instant they are tangent to the direction of flow at every point in the flow field. Since the streamlines are tangent to the velocity vector at every point in the flow field, there can be no flow across a streamline.

In steady flow, the velocity at each point in the flow field remains constant with time and, consequently, the streamlines do not vary from one instant to the next. This implies that a particle located on a given streamline will remain on the same streamline. Furthermore, consecutive particles passing through a fixed point in space will be on the same streamline and, subsequently, will remain on this streamline. Thus in a steady flow, pathlines, streaklines, and streamlines are identical lines in the flow field.

The shape of the streamlines may vary from instant to instant if the flow is unsteady. In the case of unsteady flow, pathlines, streaklines, and streamlines do not coincide.

Example 2.1

A velocity field is given by $\vec{V} = ax\hat{i} - ay\hat{j}$; the units of velocity are m/sec; x and y are given in meters; $a = 0.1\ \text{sec}^{-1}$.

(a) Determine the equation for the streamline passing through the point $(x_0, y_0, 0) = (2, 8, 0)$.

(b) Determine the velocity of a particle at the point $(2, 8, 0)$.

(c) If the particle passing through the point $(x_0, y_0, 0)$ is marked at time $t_0 = 0$, determine the location of the particle at time $t = 20$ sec.

(d) What is the velocity of the particle at $t = 20$ sec?

(e) Show that the equation of the particle path (the pathline) is the same as the equation of the streamline.

EXAMPLE PROBLEM 2.1

GIVEN: Velocity field, $\vec{V} = ax\hat{i} - ay\hat{j}$; x and y in meters, $a = 0.1\ \text{sec}^{-1}$

FIND:
(a) Equation of streamline through point $(2, 8, 0)$.
(b) Velocity of particle at point $(2, 8, 0)$.
(c) Position at $t = 20$ sec of particle located at $(2, 8, 0)$ at $t = 0$.
(d) Velocity of particle at position found in (c).
(e) Equation of pathline of particle located at $(2, 8, 0)$ at $t = 0$.

SOLUTION:

(a) Streamlines are lines drawn in the flow field such that, at a given instant, they are tangent to the direction of flow at every point.

Consequently,

$$\left.\frac{dy}{dx}\right)_{\text{streamline}} = \frac{v}{u} = \frac{-ay}{ax} = \frac{-y}{x}$$

Separating variables and integrating, we obtain

$$\int \frac{dy}{y} = -\int \frac{dx}{x} \qquad \text{or} \qquad \ln y = -\ln x + c_1$$

This can be written as $xy = c$.

For the streamline passing through the point $(x_0, y_0, 0) = (2, 8, 0)$ the constant, c, has a value of 16 and the equation of the streamline through the point $(2, 8, 0)$ is

$$xy = x_0 y_0 = 16 \text{ m}^2$$

(b) The velocity field is $\vec{V} = ax\hat{i} - ay\hat{j}$. At the point $(2, 8, 0)$

$$\vec{V} = a(x\hat{i} - y\hat{j}) = 0.1 \text{ sec}^{-1}(2\hat{i} - 8\hat{j})\text{m} = 0.2\hat{i} - 0.8\hat{j} \text{ m/sec}$$

(c) A particle moving in the flow field will have velocity given by

$$\vec{V} = ax\hat{i} - ay\hat{j}$$

Thus

$$u_p = \frac{dx}{dt} = ax \qquad \text{and} \qquad v_p = \frac{dy}{dt} = -ay$$

Separating variables and integrating (in each equation) gives

$$\int_{x_0}^{x} \frac{dx}{x} = \int_0^t a\,dt \qquad \text{and} \qquad \int_{y_0}^{y} \frac{dy}{y} = \int_0^t -a\,dt$$

Then

$$\ln \frac{x}{x_0} = at \qquad \text{and} \qquad \ln \frac{y}{y_0} = -at$$

or

$$x = x_0 e^{at} \qquad \text{and} \qquad y = y_0 e^{-at}$$

At $t = 20$ sec

$$x = 2 \text{ m } e^{(0.1)20} = 14.8 \text{ m} \qquad \text{and} \qquad y = 8 \text{ m } e^{-(0.1)(20)} = 1.08 \text{ m}$$

At $t = 20$ sec, particle is at $(14.8, 1.08, 0)$m

(d) At the point $(14.8, 1.08, 0)$m

$$\vec{V} = a(x\hat{i} - y\hat{j}) = 0.1 \text{ sec}^{-1}(14.8\hat{i} - 1.08\hat{j})\text{m} = 1.48\hat{i} - 0.108\hat{j} \text{ m/sec}$$

(e) To determine the equation of the pathline, we use the parametric equations

$$x = x_0 e^{at} \qquad \text{and} \qquad y = y_0 e^{-at}$$

and eliminate t. Solving for e^{at} from both equations

$$e^{at} = \frac{y_0}{y} = \frac{x}{x_0} \qquad \therefore\ xy = x_0 y_0 = 16\ \text{m}^2$$

Note: (i) the equation of the streamline through $(x_0, y_0, 0)$ and the equation of the pathline traced out by the particle passing through $(x_0, y_0, 0)$ are the same for this steady flow.

(ii) in following a particle (Lagrangian method of description), both the coordinates of the particle (x, y) and the components of the particle velocity ($u_p = dx/dt$ and $v_p = dy/dt$) are functions of time.

2-3 STRESS FIELD

Surface and body forces are encountered in the study of continuum fluid mechanics. *Surface forces* include all forces acting on the boundaries of a medium through direct contact. Forces developed without physical contact, and distributed over the volume of the fluid, are termed *body forces*. Gravitational and electromagnetic forces are examples of body forces arising in a fluid.

The gravitational body force acting on an element of volume, $d\mathcal{V}$, is given by $\rho\vec{g}\ d\mathcal{V}$, where ρ is the density (mass per unit volume) and $\vec{g}$ is the local gravitational acceleration. Thus the gravitational body force per unit volume is $\rho\vec{g}$ and the gravitational body force per unit mass is $\vec{g}$.

Stresses in a medium result from forces acting on some portion of the medium. The concept of stress provides a convenient means to describe the manner in which forces acting on the boundaries of the medium are transmitted through the medium. Since force and area are both vector quantities, we might anticipate that the stress field will not be a vector field. We shall show that, in general, nine quantities are required to specify the state of stress in a fluid. (Stress is a tensor quantity of second order.)

In a flowing fluid, consider a portion, $d\vec{A}$, of the surface passing through the point C. The orientation of $d\vec{A}$ is given by the unit vector, $\hat{n}$, as shown in Fig. 2.5. The direction of $\hat{n}$ is normal to the surface.

The force, $\delta\vec{F}$, acting on $\delta\vec{A}$ may be resolved into two components, one normal and the other tangential to the area. A normal stress σ_n and a shear stress τ_n are then defined

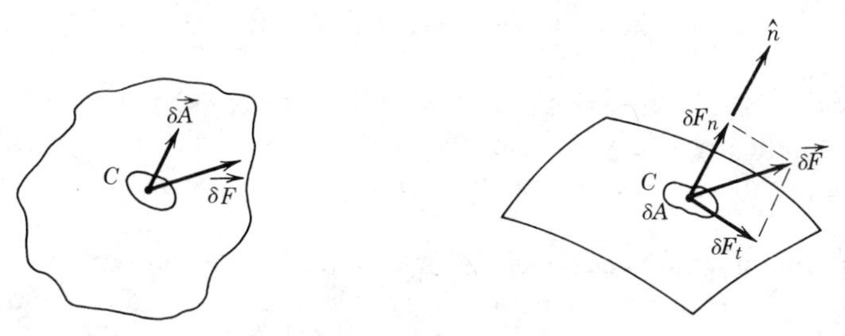

Fig. 2.5 The concept of stress in a continuum.

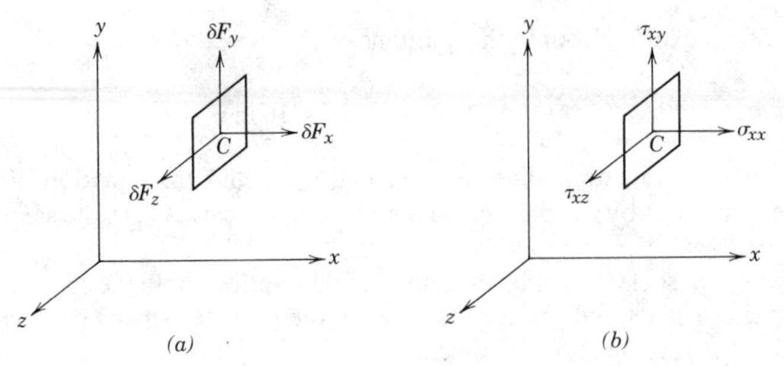

Fig. 2.6 (a) Force components, and (b) stress components, on the element of area, δA_x.

as

$$\sigma_n = \lim_{\delta A_n \to 0} \frac{\delta F_n}{\delta A_n} \tag{2.6}$$

and

$$\tau_n = \lim_{\delta A_n \to 0} \frac{\delta F_t}{\delta A_n} \tag{2.7}$$

The subscript, n, on the stress is included as a reminder that the stresses are associated with a particular surface $\delta \vec{A}$ through C, namely, the one having an outer normal in the $\hat{n}$ direction through C. For any other surface through C the values of the stresses could be different.

For purposes of analysis we usually reference the area to some coordinate system. In rectangular coordinates we might consider the stresses acting on planes whose outward drawn normals are in the x, y, or z directions. In Fig. 2.6 we consider the stress on the element, δA_x, whose outward drawn normal is in the x direction. The force, $\delta \vec{F}$, has been resolved into components along each of the coordinate directions. Dividing the magnitude of each force component by the area and taking the limit as δA_x approaches zero, we define the three stress components shown in Fig. 2.6*b*:

$$\sigma_{xx} = \lim_{\delta A_x \to 0} \frac{\delta F_x}{\delta A_x} \tag{2.8}$$

$$\tau_{xy} = \lim_{\delta A_x \to 0} \frac{\delta F_y}{\delta A_x} \qquad \tau_{xz} = \lim_{\delta A_x \to 0} \frac{\delta F_z}{\delta A_x}$$

We have used a double subscript notation to label the stresses. The first subscript (in this case, x) indicates the plane on which the stress acts (in this case, a surface perpendicular to the x axis). The second subscript indicates the direction in which the stress acts.

Consideration of an area element, δA_y, would lead to the definitions of the stresses, σ_{yy}, τ_{yx}, and τ_{yz}; use of area element, δA_z, would similarly lead to the definitions of σ_{zz}, τ_{zx}, and τ_{zy}.

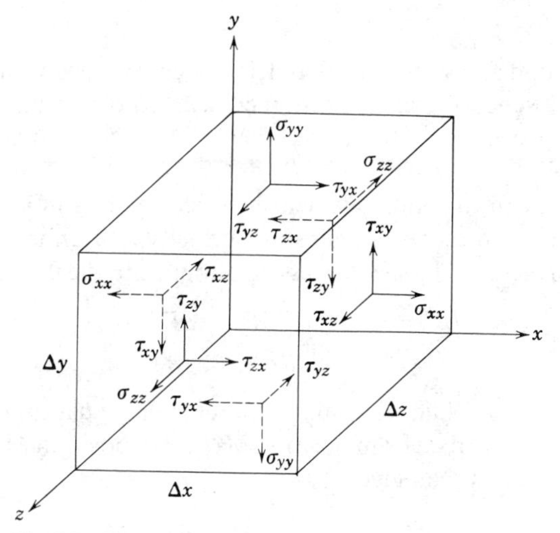

Fig. 2.7 Notation for stress.

An infinite number of planes can be passed through point C, resulting in an infinite number of stresses associated with that point. Fortunately, the state of stress at a point can be described completely by specifying the stresses acting on three mutually perpendicular planes through the point. The stress at a point is specified by the nine components

$$\begin{bmatrix} \sigma_{xx} & \tau_{xy} & \tau_{xz} \\ \tau_{yx} & \sigma_{yy} & \tau_{yz} \\ \tau_{zx} & \tau_{zy} & \sigma_{zz} \end{bmatrix}$$

where σ has been used to denote a normal stress and shear stresses are denoted by the symbol, τ. The notation for designating stress is shown in Fig. 2.7.

Referring to the infinitesimal element shown in Fig. 2.7, we see that there are six planes (two x planes, two y planes, and two z planes) on which stresses may act. In order to designate the plane of interest, we could use terms like front and back, top and bottom, or left and right. However, it is more logical to name the planes in terms of the coordinate axes. The planes are named and denoted as positive or negative according to the direction of the outward drawn normal to the plane. Thus the top plane, for example, is a positive y plane and the back plane is a negative z plane.

It also is necessary to adopt a sign convention for the stress. A stress component is considered positive when the direction of the stress component and the plane on which it acts are both positive or both negative. Thus $\tau_{yx} = 5\ \text{lbf/in.}^2$ represents a shear stress on a positive y plane in the positive x direction or a shear stress on a negative y plane in the negative x direction. In Fig. 2.7 all stresses have been drawn as positive stresses. Stress components are negative when the direction of the stress component and the plane on which it acts are of opposite sign.

2-4 VISCOSITY

We have defined a fluid as a substance that deforms continuously under the action of a shear stress. In the absence of a shear stress, there will be no deformation. Fluids may be broadly classified according to the relation between the applied shear stress and the rate of deformation.

Consider the behavior of a fluid element between the two infinite plates shown in Fig. 2.8. The upper plate moves at constant velocity, δu, under the influence of a constant applied force, δF_x. The shear stress, τ_{yx}, applied to the fluid element is given by

$$\tau_{yx} = \lim_{\delta A_y \to 0} \frac{\delta F_x}{\delta A_y} = \frac{dF_x}{dA_y}$$

where δA_y is the area of the fluid element in contact with the plate. During time interval δt, the fluid element is deformed from position $MNOP$ to position $M'NOP'$. The rate of deformation of the fluid is then given by

$$\textit{deformation rate} = \lim_{\delta t \to 0} \frac{\delta \alpha}{\delta t} = \frac{d\alpha}{dt}$$

To calculate the shear stress, τ_{yx}, it is desirable to express $d\alpha/dt$ in terms of readily measurable quantities. This can be done easily. The distance, δl, between the points M and M' is given by

$$\delta l = \delta u \, \delta t$$

or alternatively, for small angles,

$$\delta l = \delta y \, \delta \alpha$$

Equating these two expressions for δl gives

$$\frac{\delta \alpha}{\delta t} = \frac{\delta u}{\delta y}$$

Taking the limit of both sides of the equality, we obtain

$$\frac{d\alpha}{dt} = \frac{du}{dy}$$

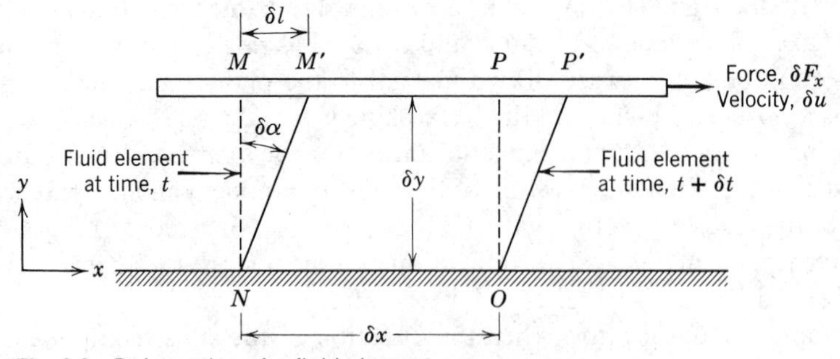

Fig. 2.8 Deformation of a fluid element.

Thus, the fluid element of Fig. 2.8, when subjected to shear stress, τ_{yx}, experiences a rate of deformation (or *shear rate*) given by du/dy. Fluids in which shear stress is directly proportional to rate of deformation are termed *Newtonian fluids*. The term *non-Newtonian* is used to classify all fluids in which shear stress is not directly proportional to shear rate.

2-4.1 Newtonian Fluid

Most common fluids such as water, air, and gasoline are Newtonian under normal conditions. If the fluid of Fig. 2.8 is Newtonian, then

$$\tau_{yx} \propto \frac{du}{dy} \tag{2.9}$$

The shear stress acts on a plane normal to the y axis. If one considers the deformation of two different Newtonian fluids, say glycerin and water, one recognizes that they will deform at different rates under the action of the same applied shear stress. Glycerin exhibits a much larger resistance to deformation than water. Thus we say it is much more viscous. The constant of proportionality in Eq. 2.9 is the *absolute* (or *dynamic*) *viscosity*, μ. Thus in terms of the coordinates of Fig. 2.8, Newton's law of viscosity is given for one-dimensional flow by

$$\tau_{yx} = \mu \frac{du}{dy} \tag{2.10}$$

Note that since the dimensions of τ are $[F/L^2]$ and the dimensions of du/dy are $[1/t]$, then μ has dimensions $[Ft/L^2]$. Since the dimensions of force, F, mass, M, length, L, and time, t, are related by Newton's second law of motion, the dimensions of μ can also be expressed as $[M/Lt]$. In the British Gravitational system, the units of viscosity are $\text{lbf} \cdot \text{sec/ft}^2$ or $\text{slug/ft} \cdot \text{sec}$. In the Absolute Metric system, the basic unit of viscosity is called a poise ($\text{poise} \equiv \text{g/cm} \cdot \text{sec}$); in the SI system the units of viscosity are $\text{kg/m} \cdot \text{sec}$ or $\text{Pa} \cdot \text{sec}$ ($= \text{N} \cdot \text{sec/m}^2$). The calculation of viscous shear stress is illustrated in Example Problem 2.2.

In fluid mechanics the ratio of absolute viscosity, μ, to density, ρ, often arises. This ratio is given the name *kinematic viscosity* and is represented by the symbol, ν. Since density has dimensions $[M/L^3]$, the dimensions of ν are $[L^2/t]$. In the Absolute Metric system of units, the unit for ν is a stoke ($\text{stoke} \equiv \text{cm}^2/\text{sec}$).

Viscosity data for a number of common Newtonian fluids are given in Appendix A. Note that for gases viscosity increases with temperature while for liquids, viscosity decreases with increasing temperature.

In gases the resistance to deformation is primarily due to the transfer of molecular momentum. Molecules from regions of high bulk velocity collide with molecules moving with lower bulk velocity, and vice versa. These collisions transport momentum from one region of fluid to another. Since the random molecular motions increase with increasing temperature, viscosity also increases with temperature.

For liquids, where molecules are much more closely packed, resistance to deformation is primarily controlled by cohesive forces among molecules. These cohesive forces decrease with increasing temperature and hence the viscosity of liquids decreases with temperature.

Example 2.2

An infinite plate is moved over a second plate on a layer of liquid as shown. For small gap width, d, we assume a linear velocity distribution in the liquid. The liquid viscosity is 0.65 centipoise and its specific gravity is 0.88. Calculate:

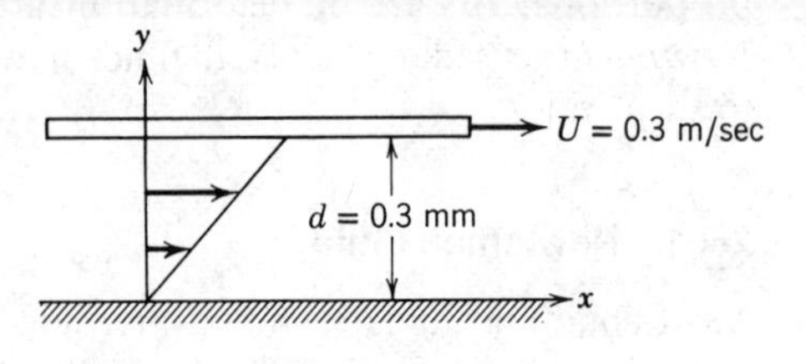

(a) The absolute viscosity of the liquid, in $\text{lbf} \cdot \text{sec/ft}^2$
(b) The kinematic viscosity of the liquid, in m^2/sec
(c) The shear stress on the upper plate, in lbf/ft^2
(d) The shear stress on the lower plate, in Pa
(e) Indicate the direction of each shear stress calculated in parts (c) and (d)

EXAMPLE PROBLEM 2.2

GIVEN: Linear velocity profile in the liquid between infinite parallel plates as shown.

$\mu = 0.65$ cp (1 poise = 1 g/cm · sec)

SG = 0.88

y
U = 0.3 m/sec
d = 0.3 mm
x

FIND:
(a) μ in units of $\text{lbf} \cdot \text{sec/ft}^2$.
(b) ν in units of m^2/sec.
(c) τ on upper plate in units of lbf/ft^2.
(d) τ on lower plate in units of Pa.
(e) Direction of stress in parts (c) and (d).

SOLUTION:

Basic equation: $\tau_{yx} = \mu \dfrac{du}{dy}$ Definition: $\nu = \dfrac{\mu}{\rho}$

(a) $$\mu = \frac{0.65 \text{ cp}}{} \times \frac{\text{poise}}{100 \text{ cp}} \times \frac{\text{g}}{\text{cm} \cdot \text{sec} \cdot \text{poise}} \times \frac{\text{lbm}}{453.6 \text{ g}} \times \frac{\text{slug}}{32.2 \text{ lbm}} \times \frac{30.48 \text{ cm}}{\text{ft}} \times \frac{\text{lbf} \cdot \text{sec}^2}{\text{slug} \cdot \text{ft}}$$

$$\mu = 1.36 \times 10^{-5} \text{ lbf} \cdot \text{sec/ft}^2 \qquad \mu$$

(b) $$\nu = \frac{\mu}{\rho} = \frac{\mu}{\text{SG}\rho_{\text{H}_2\text{O}}}$$

$$= 1.36 \times 10^{-5} \frac{\text{lbf} \cdot \text{sec}}{\text{ft}^2} \times \frac{\text{ft}^3}{(0.88)1.94 \text{ slug}} \times \frac{\text{slug} \cdot \text{ft}}{\text{lbf} \cdot \text{sec}^2} \times \frac{(0.3048)^2 \text{ m}^2}{\text{ft}^2}$$

$$\nu = 7.40 \times 10^{-7} \text{ m}^2/\text{sec} \qquad \nu$$

(c) $\tau_{\text{upper}} = \tau_{yx,\,\text{upper}} = \mu \left.\dfrac{du}{dy}\right)_{y=d}$. Since u varies linearly with y,

$$\frac{du}{dy} = \frac{\Delta u}{\Delta y} = \frac{U-0}{d-0} = \frac{U}{d} = \frac{0.3\ \text{m}}{\text{sec}} \times \frac{1}{0.3\ \text{mm}} \times \frac{1000\ \text{mm}}{\text{m}} = 1000\ \text{sec}^{-1}$$

$$\tau_{\text{upper}} = \mu \frac{U}{d} = \frac{1.36 \times 10^{-5}\ \text{lbf}\cdot\text{sec}}{\text{ft}^2} \times \frac{1000}{\text{sec}} = 0.0136\ \text{lbf/ft}^2$$

τ_{upper}

(d) $$\tau_{\text{lower}} = \mu \frac{U}{d} = \frac{0.0136\ \text{lbf}}{\text{ft}^2} \times \frac{4.448\ \text{N}}{\text{lbf}} \times \frac{\text{ft}^2}{(0.3048)^2\ \text{m}^2} \times \frac{\text{Pa}\cdot\text{m}^2}{\text{N}} = 0.651\ \text{Pa}$$

τ_{lower}

(e) Direction of shear stress on upper and lower plates.

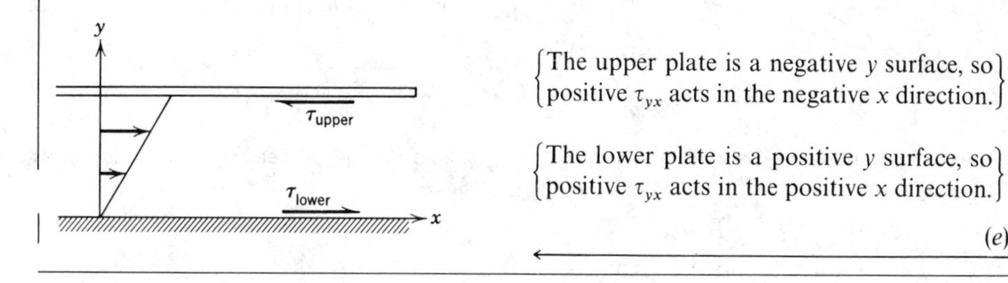

{The upper plate is a negative y surface, so positive τ_{yx} acts in the negative x direction.}

{The lower plate is a positive y surface, so positive τ_{yx} acts in the positive x direction.}

(e)

2-4.2 Non-Newtonian Fluids

Many common fluids exhibit non-Newtonian behavior. Two familiar examples are toothpaste and Lucite[6] paint. The latter is very "thick" when in the can, but becomes "thin" when sheared by brushing. Toothpaste behaves as a "fluid" when squeezed from the tube. However, it does not run out by itself when the cap is removed. There is a threshold or yield stress below which toothpaste behaves as a solid. Strictly speaking, our definition of a fluid is valid only for materials that have zero yield stress. The term non-Newtonian is used to classify all fluids in which shear stress is not directly proportional to deformation rate. Such fluids commonly are classified as having time-independent, time-dependent, or viscoelastic behavior. Four examples of time-independent behavior are shown in the rheological diagram of Fig. 2.9.

Numerous empirical equations have been proposed to model the observed relations between τ_{yx} and du/dy for time-independent fluids. They may be adequately represented for many engineering applications by the power law model, which for one-dimensional flow becomes

$$\tau_{yx} = k\left(\frac{du}{dy}\right)^n \tag{2.11}$$

where the exponent, n, is called the flow behavior index and k, the consistency index. This equation reduces to Newton's law of viscosity for $n = 1$ with $k = \mu$.

[6] Trademark, E. I. du Pont de Nemours & Company.

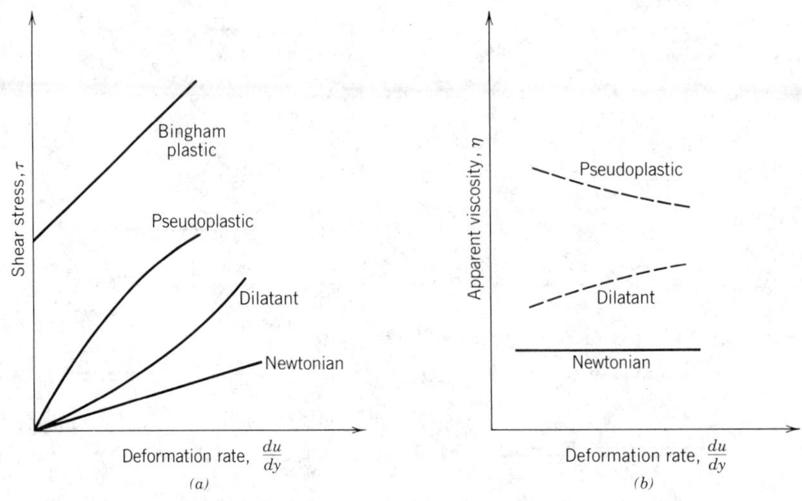

Fig. 2.9 (a) Shear stress, τ, and (b) apparent viscosity, η, as a function of deformation rate for one-dimensional flow of various non-Newtonian fluids.

If Eq. 2.11 is rewritten in the form,

$$\tau_{yx} = k\left|\frac{du}{dy}\right|^{n-1}\frac{du}{dy} = \eta\frac{du}{dy} \tag{2.12}$$

then $\eta = k|du/dy|^{n-1}$ is referred to as the *apparent viscosity*.

Fluids in which the apparent viscosity decreases with increasing deformation rate ($n < 1$) are called *pseudoplastic* (or shear thinning) fluids. Most non-Newtonian fluids fall into this group; examples include polymer solutions, colloidal suspensions, and paper pulp in water. If the apparent viscosity increases with increasing deformation rate ($n > 1$) the fluid is termed *dilatant* (or shear thickening). Suspensions of starch and of sand are examples of dilatant fluids.

A "fluid" that behaves as a solid until a minimum yield stress, τ_y, is exceeded and subsequently exhibits a linear relation between stress and rate of deformation is referred to as an ideal or *Bingham plastic*. The shear stress model is then

$$\tau_{yx} = \tau_y + \mu_p\frac{du}{dy} \tag{2.13}$$

Clay suspensions, drilling muds, and toothpaste are examples of substances exhibiting this behavior.

Most non-Newtonian fluids have apparent viscosities that are relatively high compared to the viscosity of water.

The study of non-Newtonian fluids is further complicated by the fact that the apparent viscosity may be time-dependent.[7] Fluids that show a decrease in η with time

[7] Examples of time-dependent fluids are illustrated in the film, *Rheological Behavior of Fluids*, H. Markovitz, principal.

under a constant applied shear stress are called *thixotropic*; many paints are thixotropic. Fluids that show an increase in η with time are termed *rheopectic*. In addition, some fluids after deformation partially return to their original shape when the applied stress is released; such fluids are called *viscoelastic*.

2-5 DESCRIPTION AND CLASSIFICATION OF FLUID MOTIONS

In Chapter 1 we listed a wide variety of typical problems encountered in fluid mechanics and outlined our method of approach to the subject. Before proceeding with our detailed study, we shall attempt a broad classification of fluid mechanics on the basis of observable physical characteristics of flow fields. Since there is much overlap in the types of flow fields encountered, there is no universally accepted classification scheme. One possible classification is shown in Fig. 2.10.

2-5.1 Viscous and Inviscid Flows

The main subdivision indicated is between inviscid and viscous flows. In an inviscid flow the fluid viscosity, μ, is assumed to be zero. Fluids with zero viscosity do not exist; however, there are many problems where an assumption that $\mu = 0$ will simplify the analysis and, at the same time, lead to meaningful results. (While simplification of the analysis is always desirable, the results must be reasonably accurate if the solution is to be of value.)

All fluids possess viscosity and, consequently, viscous flows are of paramount importance in the study of continuum fluid mechanics. We shall study viscous flows in some detail later; here we consider a few examples of viscous flow phenomena.

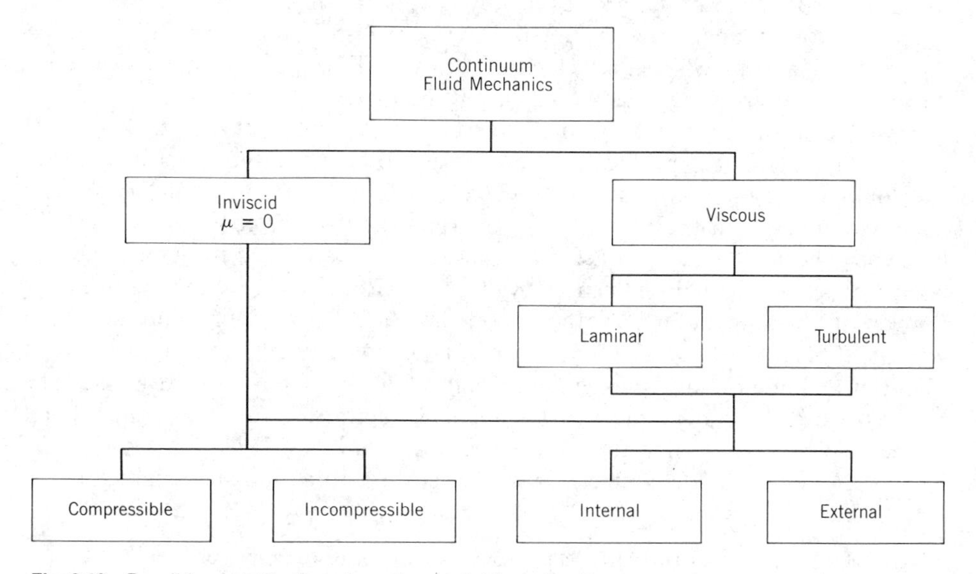

Fig. 2.10 Possible classification of continuum fluid mechanics.

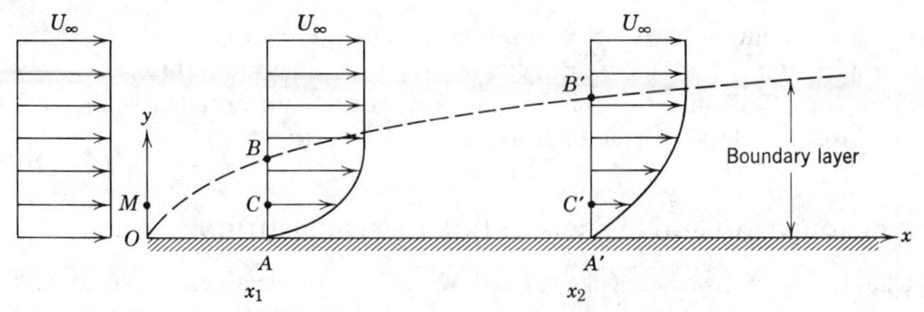

Fig. 2.11 Incompressible laminar viscous flow over a semi-infinite flat plate.

In our discussion following the definition of a fluid (Section 1-2), we noted that in any viscous flow, the fluid in direct contact with a solid boundary has the same velocity as the boundary itself; there is no slip at the boundary. For the one-dimensional viscous flow of Fig. 2.8, the shear stress[8] was given by Eq. 2.10,

$$\tau_{yx} = \mu \frac{du}{dy} \tag{2.10}$$

Since the fluid velocity at a stationary solid surface in a moving fluid is zero, but the bulk fluid is moving, velocity gradients and hence shear stresses must be present in the flow. These stresses in turn affect the motion.

As a practical case, consider the fluid motion around a thin wing or ship hull. Such a flow might be represented crudely by the flow over a flat plate, as shown in Fig. 2.11. The flow approaching the plate is of uniform velocity, U_∞. We are interested in providing a qualitative picture of the velocity distribution at various locations along the plate. Two such locations are denoted by x_1 and x_2. Consider first location, x_1. In order to arrive at a qualitative picture of the velocity distribution, we start by labeling the y coordinates at which the velocity is known. (For clarity, distances in the y direction have been exaggerated greatly in Fig. 2.11.)

From the no-slip condition, we know the velocity at point A must be zero; we have one point on the velocity profile. Can we locate any other points on the profile? Let us stop for a minute and ask ourselves, "What is the effect of the plate on the flow?" The plate is stationary and, therefore, exerts a retarding force on the flow; it slows the fluid in the neighborhood of the surface. At a y location sufficiently far from the plate, say point B, the flow will not be influenced by the presence of the plate. If the pressure does not vary in the x direction (as is the case for flow over a semi-infinite flat plate) the velocity at point B will be U_∞. It seems reasonable to expect the velocity to increase smoothly and monotonically from the value $u = 0$ at $y = 0$ to $u = U_\infty$ at $y = y_B$. The profile has been so drawn; thus at some point, C, intermediate between points A and B, the velocity has a value that lies between zero and U_∞. For $0 \leq y \leq y_B$, then $0 \leq u \leq U_\infty$. From these characteristics of the velocity profile and our definition of the

[8] In general, $\tau_{yx} = \mu\left(\frac{\partial u}{\partial y} + \frac{\partial v}{\partial x}\right)$ for flows that are not one-dimensional, see Chapter 5.

shear stress,[9] we see that within the region $0 \leq y \leq y_B$, shear stresses are present; for $y > y_B$, the velocity gradient is zero and hence no shear stresses are present.

What about the velocity profile at location, x_2? Is it exactly the same as the profile at x_1? A look at Fig. 2.11 suggests that it is not. At least it has not been drawn that way! While it is qualitatively the same, why is it not exactly the same? We might guess that the plate would influence a greater region of the flow field as we move farther down the plate. Looking again at the profile at location x_1, we see that the slower moving fluid adjacent to the plate exerts a retarding force on the faster moving fluid above it. We can see this by considering the shear stress on the y plane through the point C. Since we are interested in the stress exerted on the faster moving fluid above the plane, we are looking for the direction of the shear stress on a negative y plane through the point C. Since $\partial u/\partial y > 0$, τ_{yx} on the plane through point C has a positive numerical value; consequently the shear stress must be in the negative x direction.

To establish the qualitative picture of the velocity profile at x_2, we recognize that the no-slip condition requires the velocity at the wall to be zero; this fixes the velocity at A' as zero. Since, at location, x_1, the slower moving fluid exerts a retarding force on the fluid above it, we would expect the distance out to the point where the velocity is U_∞ to be increased at location, x_2; i.e. $y_{B'} > y_B$. Furthermore, it is reasonable to expect that $u_{C'} < u_C$.

From our qualitative picture of the flow field, we see that we can divide the flow into two general regions. In the region adjacent to the boundary, shear stresses are present; this region is called the boundary layer.[10] Outside the boundary layer the velocity gradient is zero and hence the shear stresses are zero. In this region we may use inviscid flow theory to analyze the flow.

Before leaving our discussion of the viscous flow over a semi-infinite flat plate, we should stop and reflect on two points. In our qualitative description of the flow field, we were only concerned about the behavior of the x component of velocity, the component, u. What about the y component of velocity, the component, v? Is it zero throughout the flow field? We also might ask if the edge of the boundary layer is a streamline.

To answer these questions, consider the streamlines of the flow. Rather than consider all possible streamlines, let us consider the streamline through the point M. Recalling that a streamline is defined as a line drawn tangent to the velocity vector at every point in the flow, our first inclination might be to depict the streamline through M as a straight line parallel to the x axis. However, this would violate the requirement that there can be no flow across a streamline. Because there can be no flow across a streamline, the mass flow between adjacent streamlines (or between a streamline and a solid boundary) must be a constant. For the incompressible viscous flow of Fig. 2.11, we recognize that the streamline through the point M cannot be a straight line parallel to the x axis.

[9] For the two-dimensional boundary-layer flow of Fig. 2.11, the shear stress is given closely by $\tau_{yx} = \mu \dfrac{\partial u}{\partial y}$.

[10] The formation of a boundary layer is demonstrated in the film loop, S-FM006, *Boundary-Layer Formation.*

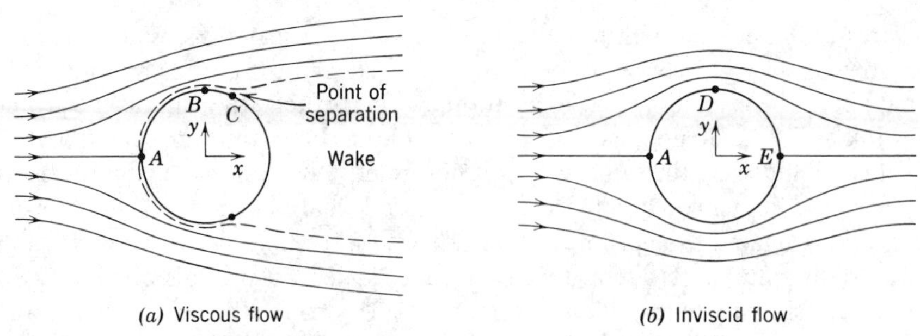

Fig. 2.12 Qualitative picture of incompressible flow over a cylinder.

The spacing between the streamline through the point M and the x axis must increase continuously as we move along the plate. Therefore, although small, the y component of velocity is not zero. The streamline through M crosses the dashed line we have used to denote the edge of the boundary layer. Consequently, we conclude that the edge of the boundary layer is not a streamline, and that there is flow into the boundary layer as we move down the plate. Indeed, if the boundary layer is to grow, there must be flow across the edge of the boundary layer.

For a given freestream velocity, U_∞, the size of the boundary layer will depend on the properties of the fluid. Since the shear stress is directly proportional to the viscosity, we expect the size of the boundary layer to depend on the viscosity of the fluid. In Chapter 9, we shall develop expressions for determining the rate of boundary-layer growth.

We have used incompressible flow over a semi-infinite flat plate to establish a qualitative picture of the viscous flow over a solid boundary. In that example, we had to consider only the effect of shear forces; the pressure was constant throughout the flow field. Now let us consider a steady flow field (the incompressible flow over a cylinder) where both pressure forces and viscous forces are important. For steady flow, pathlines, streaklines, and streamlines all are identical. If we were to use some means of flow visualization, we would find the flow field to be of the general character shown in Fig. 2.12*a*.[11]

We see that the streamlines are symmetric about the x axis. The fluid along the central streamline impinges on the cylinder at point A, divides, and flows around the cylinder. Point A on the cylinder is called a *stagnation point*. As in flow over a flat plate, a boundary layer develops in the neighborhood of the solid surface. The velocity distribution outside the boundary layer can be determined from the spacing of the streamlines. Since there can be no flow across a streamline, we would expect the flow velocity to increase in regions where the spacing between streamlines decreases. Conversely, an increase in streamline spacing implies a decrease in flow velocity.

Consider for a moment the incompressible flow field around a cylinder calculated assuming an inviscid flow, as shown in Fig. 2.12*b*; this flow is symmetric about both the

[11] The details of the flow will depend on the various flow properties. For all but very low-speed flows, the qualitative picture will be as shown.

x and y axes. The velocity around the cylinder increases to a maximum at point D and then decreases as we move further around the cylinder. For inviscid flow, an increase in velocity is accompanied by a decrease in pressure; conversely, a decrease in velocity is accompanied by an increase in pressure. Thus in the case of an incompressible inviscid flow, the pressure along the surface of the cylinder decreases as we move from point A to point D and then increases again from point D to point E. Since the flow is symmetric with respect to both the x and y axes, we would also expect the pressure distribution to be symmetric with respect to these axes. This is indeed the case for inviscid flow.

Since no shear stresses are present in an inviscid flow, the pressure forces are the only forces we need consider in determining the net force on the cylinder. The symmetry of the pressure distribution leads to the conclusion that for an inviscid flow, there is no net force on the cylinder in either the x or y directions. The net force in the x direction is termed the *drag*. Thus for an inviscid flow over a cylinder, we are led to the conclusion that the drag is zero; this conclusion is contrary to experience, for we know that all bodies experience some drag when placed in a real flow. In treating the inviscid flow over a body we have, by the definition of inviscid flow, neglected the presence of the boundary layer. Let us go back and look again at the real flow situation.

In the real flow, Fig. 2.12*a*, experiments show the boundary layer to be thin between points A and C. Since the boundary layer is thin, it is reasonable to assume that the pressure field is qualitatively the same as in the inviscid flow case. Since the pressure decreases continuously between points A and B, a fluid element inside the boundary layer experiences a net pressure force in the direction of flow. In the region between A and B, this net pressure force is sufficient to overcome the resisting shear force and motion of the element in the flow direction is maintained.

Now consider an element of fluid inside the boundary layer on the back of the cylinder beyond point B. Since the pressure increases in the direction of flow, the fluid element experiences a net pressure force opposite to its direction of motion. Finally the momentum of the fluid in the boundary layer is insufficient to carry the element further into the region of increasing pressure. The fluid layers adjacent to the solid surface will be brought to rest and the flow will *separate* from the surface.[12] The point at which this occurs is called the point of separation. Boundary-layer separation results in the formation of a relatively low pressure region behind a body; this region, which is deficient in momentum, is called the wake. Thus, for separated flow over a body, there is a net unbalance of pressure forces in the direction of flow; this results in a pressure drag on the body. The greater the size of the wake behind a body, the greater is the pressure drag.

It is logical to ask how one might reduce the size of the wake and thus reduce the pressure drag. Since a large wake results from boundary-layer separation, which in turn is related to the presence of an adverse pressure gradient (increase of pressure in the direction of flow), reducing the adverse pressure gradient should delay the onset of separation and, hence, reduce the drag.

[12] The flow over a variety of models, illustrating flow separation, is demonstrated in the NCFMF film loops: S-FM012, *Flow Separation and Vortex Shedding*; S-FM004, *Separated Flows—Part I*; and S-FM005, *Separated Flows—Part II*.

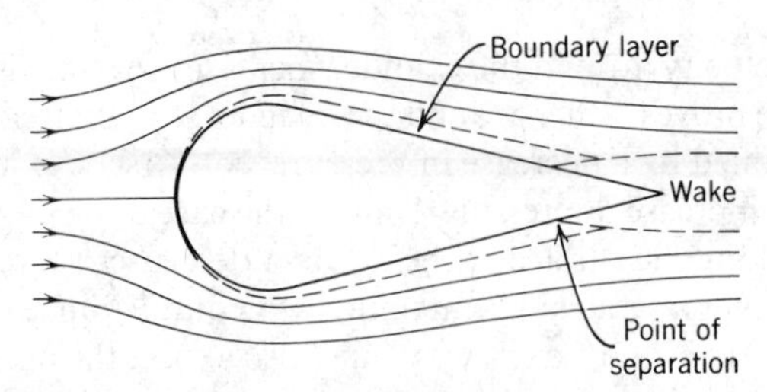

Fig. 2.13 Flow over a streamlined object.

Streamlining a body reduces the adverse pressure gradient by spreading a given pressure rise over a larger distance. For example, if a gradually tapered rear section were added to the cylinder of Fig. 2.12, the flow field would appear qualitatively as shown in Fig. 2.13. Streamlining the body delays the onset of separation; although the surface area of the body and, hence, the total shear force acting on the body is increased, the drag is reduced significantly.[13]

Flow separation also may occur in internal flows (flows through ducts) as a result of rapid or abrupt changes in duct geometry.[14]

2-5.2 Laminar and Turbulent Flows

Viscous flow regimes are classified as laminar or turbulent on the basis of internal flow structure. In the laminar regime, flow structure is characterized by smooth motion in laminae or layers. Flow structure in the turbulent regime is characterized by random, three-dimensional motions of fluid particles superimposed on the mean motion.

In laminar flow there is no macroscopic mixing of adjacent fluid layers. A thin filament of dye injected into a laminar flow appears as a single line; there is no dispersion of dye throughout the flow, except the slow dispersion due to molecular motion. On the other hand, a dye filament injected into a turbulent flow disperses quickly throughout the flow field; the line of dye breaks up into myriad entangled threads of dye. This behavior of turbulent flow is due to small velocity fluctuations superimposed on the mean motion; the macroscopic mixing of fluid particles from adjacent layers of fluid results in rapid dispersion of the dye. The straight filament of smoke rising from a cigarette in still surroundings gives a clear picture of laminar flow. As the smoke continues to rise, it breaks up into random, haphazard motions; this is an example of turbulent flow.[15]

One can obtain a more quantitative picture of the difference between laminar and turbulent flow by examining the output from a sensitive velocity-measuring device immersed in the flow. If one measures the x component of velocity at a fixed location in a pipe for both laminar and turbulent steady flow, the traces of velocity versus time

[13] The effect of streamlining a body is demonstrated in the film loop, S-FM004, *Separated Flows—Part I.*

[14] Examples of separation in internal flows are shown in the film loop, S-FM015, *Incompressible Flow through Area Contractions and Expansions.*

[15] Several examples illustrating the nature of laminar and turbulent flows are shown in the film loop S-FM008, *The Occurrence of Turbulence.*

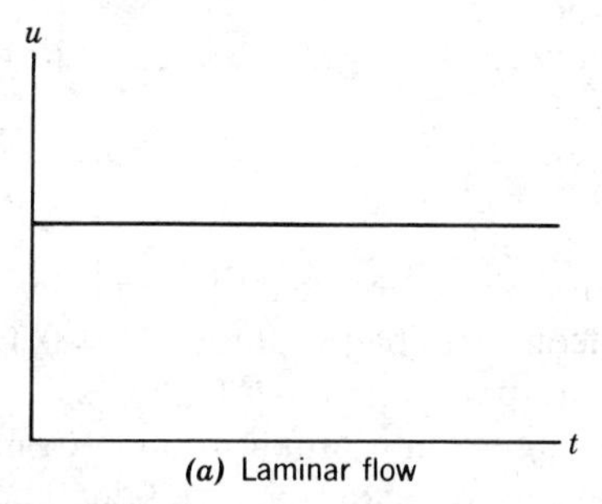

(a) Laminar flow

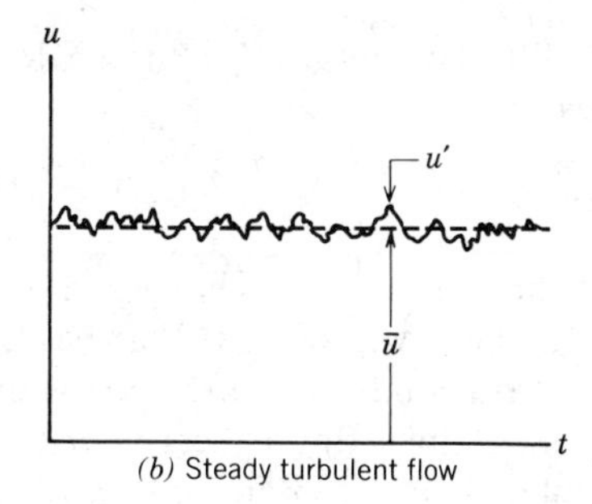

(b) Steady turbulent flow

Fig. 2.14 Variation of axial velocity with time.

appear as shown in Fig. 2.14. For steady laminar flow, the velocity at a point remains constant with time. In turbulent flow the velocity trace indicates random fluctuations of the instantaneous velocity, u, about the time mean velocity, $\bar{u}$. We can consider the instantaneous velocity, u, as the sum of the time mean velocity, $\bar{u}$, and the fluctuating component, u',

$$u = \bar{u} + u'$$

Because the flow is steady, the mean velocity, $\bar{u}$, does not vary with time.

Although many turbulent flows of interest are steady in the mean ($\bar{u}$ is not a function of time), the presence of the random, high-frequency velocity fluctuations makes the analysis of turbulent flows extremely difficult. In a one-dimensional laminar flow, the shear stress is related to the velocity gradient by the simple relation

$$\tau_{yx} = \mu \frac{du}{dy} \tag{2.10}$$

For a turbulent flow in which the mean velocity field is one-dimensional, no such simple relation is valid. Random, three-dimensional velocity fluctuations (u', v', and w') transport momentum across the mean flow streamlines, increasing the effective shear stress. Consequently, in turbulent flow there is no universal relationship between the stress field and the mean-velocity field. Thus in turbulent flows we must rely heavily on semi-empirical theories and on experimental data.

2-5.3 Compressible and Incompressible Flows

Flows in which variations in density are negligible are termed *incompressible*; when density variations within a flow are not negligible, the flow is called *compressible*. If one considers the two states of matter, liquid and gas, included within the definition of a fluid, one is tempted to make the general statement that all liquid flows are incompressible flows and all gas flows are compressible flows. For many practical cases the first portion of the statement is correct; most liquid flows are essentially incompressible. However, water hammer and cavitation are examples of the importance of compressibility effects in liquid flows. Gas flows also may be considered incompressible provided the flow speeds are small relative to the speed of sound; the ratio of the flow speed, V, to the local speed of sound, c, in the gas is defined as the Mach number,

$$M \equiv \frac{V}{c}$$

For values of $M < 0.3$, changes in density are less than 2 percent of the mean value. Thus gas flows with $M < 0.3$ can be treated as incompressible; a value of $M = 0.3$ in air at standard conditions corresponds to a speed of approximately 100 m/sec.

Compressible flows occur frequently in engineering applications. Common examples include compressed air systems used to power shop tools and dental drills, transmission of gases in pipelines at high pressure, and pneumatic or fluidic control and sensing systems. Compressibility effects are very important in the design of modern high-speed aircraft and missiles, power plants, fans, and compressors.

2-5.4 Internal and External Flows

Flows completely bounded by solid surfaces are called internal or duct flows. Internal flows may be laminar or turbulent, compressible or incompressible.

In the case of incompressible flow through a pipe, the nature of the flow (laminar or turbulent) is determined by the value of a dimensionless parameter, the Reynolds number, $Re = \rho \bar{V} D/\mu$, where ρ is the density of the fluid, $\bar{V}$ the average flow velocity, D the pipe diameter, and μ the viscosity of the fluid. Pipe flow is laminar when $Re \leq 2300$; it may be turbulent for larger values. (The Reynolds number and other important dimensionless parameters encountered in fluid mechanics will be discussed in Chapter 7.) Chapter 8 will be devoted to a study of internal incompressible flow.

In the case of internal compressible flows, proper duct design is necessary to attain supersonic flow. The variation of fluid properties within a variable-area flow passage is not the same for supersonic flow ($M > 1$) as it is for subsonic flow ($M < 1$). Likewise the boundary conditions on the flow at the exit of an internal flow (e.g. the discharge from a nozzle) are different in the two cases. For subsonic flow discharge, the pressure in the exit plane of the nozzle is ambient pressure. For sonic flow, the nozzle exit pressure may be greater than ambient. For a supersonic jet, the pressure in the exit plane of the nozzle may be greater than, equal to, or less than ambient pressure. One-dimensional, steady compressible flow will be treated in Chapters 11 and 12.

External flows occur over bodies immersed in an unbounded fluid. The flow over a semi-infinite flat plate (Fig. 2.11) and the flow over a cylinder (Fig. 2.12*a*) are examples of external flows.

Boundary-layer flows also may be laminar or turbulent; the definitions of laminar and turbulent flows given earlier also apply to boundary-layer flows; the details of a flow field may be significantly different depending on whether the boundary layer is laminar or turbulent. In Chapter 9, boundary-layer flows and flow over immersed bodies will be discussed in detail.

Flows of liquids in which the duct does not flow full—where there is a free surface subject to a constant pressure—are termed *open-channel* flows. Common examples of open-channel flow include flow in rivers, irrigation ditches, and aqueducts. Open-channel flow will be treated in Chapter 10.

2-6 SUMMARY OBJECTIVES

After completing study of Chapter 2, you should be able to do the following:

1. Give operational definitions of:
   - continuum
   - property at a point
   - scalar field
   - vector field
   - steady flow
   - uniform flow at a section
   - timeline
   - pathline
   - streakline
   - streamline
   - body force
   - surface force
   - shear stress
   - normal stress
   - Newtonian fluid
   - non-Newtonian fluid
   - viscosity
   - kinematic viscosity
   - apparent viscosity
   - dilatant fluid
   - pseudoplastic fluid
   - Bingham plastic
   - viscous flow
   - inviscid flow
   - boundary layer
   - stagnation point
   - drag
   - point of separation
   - wake
   - laminar flow
   - turbulent flow
   - compressible flow
   - incompressible flow
   - Mach number
   - internal flow
   - external flow
   - open-channel flow
2. Give examples of one-, two-, and three-dimensional flows.
3. State the convention for designating the nine components of the stress field.
4. Write Newton's law of viscosity and determine the shear stress and shear force that correspond to a given one-dimensional velocity profile.
5. Solve the problems at the end of the chapter that relate to the material you have studied.

PROBLEMS

2.1 For the velocity fields given below, determine:
(a) Whether the flow field is one-, two-, or three-dimensional, and why
(b) Whether the flow is steady or unsteady, and why
(The quantities a and b are constants.)

(1) $\vec{V} = [ae^{-bx}]\hat{i}$
(2) $\vec{V} = ax^2\hat{i} + bx\hat{j}$
(3) $\vec{V} = [ax^2e^{-bt}]\hat{i}$
(4) $\vec{V} = ax\hat{i} - by\hat{j}$
(5) $\vec{V} = (ax + t)\hat{i} - by^2\hat{j}$
(6) $\vec{V} = ax^2\hat{i} + bxz\hat{j}$
(7) $\vec{V} = a(x^2 + y^2)^{1/2}(1/z^3)\hat{k}$
(8) $\vec{V} = axy\hat{i} - byzt\hat{j}$

2.2 For the velocity fields given below, determine:
(a) Whether the flow field is one-, two-, or three-dimensional, and why
(b) Whether the flow is steady or unsteady, and why
(The quantities a, b, and c are constants.)

(1) $\vec{V} = [ae^{-by}]\hat{i}$
(2) $\vec{V} = by^2\hat{j} + cy\hat{k}$
(3) $\vec{V} = [ay^2e^{-bt}]\hat{j}$
(4) $\vec{V} = by\hat{i} - ax\hat{j}$
(5) $\vec{V} = ax\hat{i} + (t - by)\hat{j}$
(6) $\vec{V} = ax^2\hat{i} + bxy\hat{k}$
(7) $\vec{V} = ax^2\hat{i} + by\hat{j} + cxz\hat{k}$
(8) $\vec{V} = ax\hat{i} - by\hat{j} + (t - cz)\hat{k}$

2.3 For the velocity fields given below, determine:
(a) Whether the flow field is one-, two-, or three-dimensional, and why
(b) Whether the flow is steady or unsteady, and why
(The quantities a, b, and c are constants.)
(1) $\vec{V} = [ae^{-bz}]\hat{k}$ (2) $\vec{V} = az^2\hat{i} + bz\hat{k}$
(3) $\vec{V} = ax\hat{i} + bx^2e^{-ct}\hat{j}$ (4) $\vec{V} = ax\hat{i} + by\hat{j} + cx^2\hat{k}$
(5) $\vec{V} = ax\hat{i} + by^2\hat{j} + cyt\hat{k}$ (6) $\vec{V} = ax\hat{i} + bx^2\hat{j} - cx^2\hat{k}$
(7) $\vec{V} = ay^2\hat{i} + bx\hat{j} + cxy\hat{k}$ (8) $\vec{V} = ax\hat{i} + by^2\hat{j} + czt\hat{k}$

2.4 The air density field in the vicinity of a powerplant exhaust is approximated by

$$\rho = \rho_0 + \frac{\Delta\rho}{2}\left[\frac{r_0}{r_0 + (x^2 + y^2)^{1/2}} + e^{-kz}\right]$$

Is the field one-, two-, or three-dimensional? Is it steady or unsteady?

2.5 The density field in the exhaust pipe of a Diesel engine is approximated by

$$\rho = a[1 + be^{-cx}\cos(\omega t)]$$

Is the field one-, two-, or three-dimensional? Is it steady or unsteady?

2.6 Parametric equations for the position of a particle in a flow field are given as

$$x_p = c_1e^{at} \qquad \text{and} \qquad y_p = c_2e^{-bt}$$

Find the equation of the pathline for a particle location at $(x, y) = (1, 2)$ at $t = 0$. Compare with a streamline through the same point, found in Problem 2.10.

2.7 Consider the flow field $\vec{V} = ax(1 + bt)\hat{i} + cy\hat{j}$, where $a = c = 1 \text{ sec}^{-1}$, and $b = 0.2 \text{ sec}^{-1}$. Plot the path of the particle that passes through the point $(x, y) = (1, 1)$ at the instant, $t = 0$, during the interval from $t = 0$ to $t = 3$ sec. Compare with the streamline through the same point at the instant, $t = 0$.

2.8 A velocity field is given by $\vec{V} = ax\hat{i} + ay\hat{j} + bxyt\hat{k}$, where $a = 2 \text{ sec}^{-1}$, and $b = 1 \text{ m}^{-1}\cdot\text{sec}^{-2}$. Determine the number of dimensions of the flow field. Is it steady? Find the slope of the streamline through the point $(x, y, z) = (1, 2, 0)$ at $t = 0$.

2.9 A velocity field is given by $\vec{V} = ay\hat{i} + bx\hat{j} + c\hat{k}$, where $a = 2 \text{ sec}^{-1}$, $b = 1 \text{ sec}^{-1}$, and $c = 2$ m/sec. Determine the number of dimensions of the flow field. Is it steady? Determine the velocity components u, v, w, at the point $(1, 2, 0)$. Determine the slope in the xy plane of the streamline through the point $(1, 2, 0)$.

2.10 The velocity field $\vec{V} = ax\hat{i} - by\hat{j}$, where $a = b = 1 \text{ sec}^{-1}$, can be interpreted to represent flow in a corner. Find an equation for the flow streamlines. Plot several streamlines in the first quadrant, including the one that passes through the point $(x, y) = (0, 0)$.

2.11 The velocity distribution in a certain region is given by $\vec{V} = 2x\hat{i} - ay\hat{j} + (3t - bz)\hat{k}$. Determine the number of dimensions of the flow field. Is it steady? Find the equation of the streamline through the point $(x, y, z) = (1, 1, 3)$ at $t = 0$ and $t = 1$.

2.12 A tornado can be represented in polar coordinates by the velocity field

$$\vec{V} = -\frac{a}{r}\hat{i}_r + \frac{b}{r}\hat{i}_\theta$$

where $\hat{i}_r$ and $\hat{i}_\theta$ are unit vectors in the r and θ directions, respectively. Recall that an element of distance along the θ direction is $dl = r\,d\theta$. Show that streamlines form logarithmic spirals,

$$r = ce^{-\frac{a}{b}\theta}$$

2.13 A streamline is a line drawn in the flow field so that at each instant it is tangent to the velocity vector at each point. Consider a two-dimensional, steady flow in the xy plane with velocity field given by $\vec{V} = u\hat{i} + v\hat{j}$. Define an element of distance along a streamline as $d\vec{s} = dx\hat{i} + dy\hat{j}$. Use the vector cross product, $\vec{V} \times d\vec{s} = 0$, to show that $dx/u = dy/v$ for the streamline.

2.14 Streaklines are traced out by neutrally buoyant marker fluid injected into a flow field from a fixed point in space. A particle of the marker fluid that is at point (x, y) at time, t, must have passed through the injection point (x_0, y_0) at some earlier instant, $t = \tau$. The time history of a marker particle may be found by solving the pathline equations for the initial conditions that $x = x_0$, $y = y_0$ when $t = \tau$. The present locations of particles on the streakline are obtained by setting τ equal to values in the range $0 \leq \tau \leq t$. Consider the flow field $\vec{V} = ax(1 + bt)\hat{i} + cy\hat{j}$, where $a = c = 1\ \text{sec}^{-1}$, and $b = 0.2\ \text{sec}^{-1}$. Evaluate the streakline that passes through the initial point, $(x_0, y_0) = (1, 1)$, during the interval from $t = 0$ to $t = 3$ sec. Compare with the streamline through the same point at the instant, $t = 0$.

2.15 The two-dimensional steady flow described by the velocity field, $\vec{V} = ax\hat{i} - ay\hat{j}$, was analyzed in Example Problem 2.1, with x and y given in meters, and $a = 0.1\ \text{sec}^{-1}$. Consider this flow field again. Assume that a timeline is marked in the flow at $t = 0$, connecting points along the line, $y =$ constant, from $(x, y) = (1, 8)$ to $(3, 8)$. Calculate the position of the timeline at $t = 10$ sec. What general conclusion can you draw about the motion of any timeline that initially is horizontal in this flow?

2.16 The density distribution in the fluid column shown is given by $\rho = \rho_0(1 + ky)$, where $\rho_0 = 0.00238\ \text{slug/ft}^3$ and $k = 0.10\ \text{ft}^{-1}$. Compute the body force acting on the volume.

2.17 A body force distribution is given as $\vec{B} = ax\hat{i} + b\hat{j} + cz\hat{k}$ per unit mass of the material acted on. The density of the material is given as $\rho = lx^2 + ry + nz$. All coordinates are measured in meters. Determine the resultant body force in the region shown when: $a = 0$, $b = 0.1$ N/kg, $c = 0.5$ N/kg · m; $l = 2.0\ \text{kg/m}^5$, $r = 0$, and $n = 1.0\ \text{kg/m}^4$.

2.18 For the region shown in Problem 2.17, determine the resultant body force when: $a = 5.0$ N/kg · m, $b = 0$, $c = 10.0$ N/kg · m; $l = 0.1\ \text{kg/m}^5$, $r = 0.5\ \text{kg/m}^4$, and $n = 0$.

2.19 For the region shown in Problem 2.17, determine the resultant body force when: $a = 1.0$ N/kg · m, $b = 2.0$ N/kg, $c = 0$; $l = 0$, $r = 1.0\ \text{kg/m}^4$, and $n = 2.0\ \text{kg/m}^4$.

2.20 Use double index notation to label the six shear stresses shown.

2.21 On the element shown, indicate all possible stresses represented by

$$\tau_{zx} = -10 \text{ lbf/ft}^2 \qquad \text{and} \qquad \sigma_{yy} = 15 \text{ lbf/ft}^2$$

2.22 As shown in the film *Flow Visualization*, small hydrogen bubbles frequently are used as markers to visualize streaklines and timelines in liquid flow. The diameter of bubbles produced by a hydrogen bubble generating wire is about the same as the wire diameter. Surface tension produces a stress along the interface between a liquid and a gas. Surface tension is given the symbol, σ, and expressed as force per unit length of interface. Assume that the surface tension for a hydrogen-water interface is the same as that for an air-water interface. Evaluate the pressure within a hydrogen bubble as a function of bubble diameter for diameters in the range from 0.01 mm to 1 mm. Evaluate the mass of the bubbles as a function of diameter.

2.23 The following stress levels are known to exist at a point in a continuous medium:

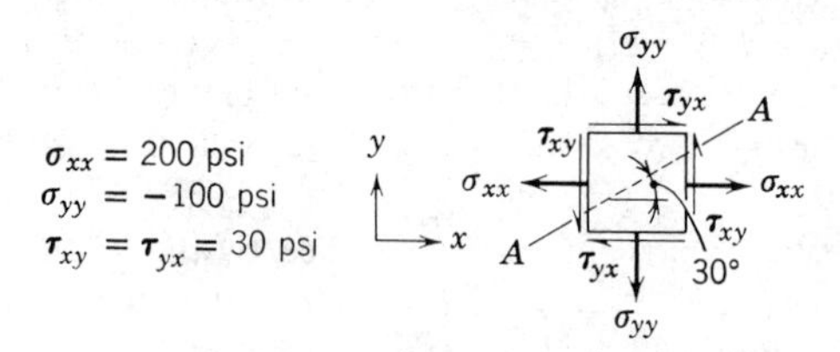

Find the magnitudes of the normal and shear stresses acting on plane AA.

2.24 Obtain the conversion factor for converting the viscosity, μ, from units of newtons, seconds, and meters to units of pounds force, seconds, and feet. Check your answer using Appendix A.

2.25 The temperature dependence of the viscosity of water is represented well by the empirical equation

$$\mu = Ae^{B/T}$$

where T is absolute temperature. The following data were measured in a laboratory viscometer: at 10 C, $\mu = 0.0013$ kg/m · sec, and at 20 C, $\mu = 0.0010$ kg/m · sec. Use the method suggested in Appendix A to evaluate constants A and B in the correlating equation. Check against data from Fig. A.2.

2.26 The variation with temperature of the viscosity of air is correlated well by the empirical Sutherland equation

$$\mu = \frac{bT^{1/2}}{1 + S/T}$$

Best-fit values of b and S are given in Appendix A for use with SI units. Use these values to develop an equation for calculating air viscosity in British Gravitational units as a function of absolute temperature in degrees Rankine. Check your result using data from Appendix A.

2.27 The variation with temperature of the viscosity of air is represented well by the empirical Sutherland correlation

$$\mu = \frac{bT^{1/2}}{1 + S/T}$$

Best-fit values of b and S are given in Appendix A. Develop an equation in SI units for kinematic viscosity versus temperature for air at atmospheric pressure. Assume ideal gas behavior. Check using data from Appendix A.

2.28 Data for the temperature variation of the viscosity of water are correlated well by the equation

$$\mu = 2.414 \times 10^{-5} \exp\left(\frac{570.6}{T - 140}\right)$$

where μ is dynamic viscosity in $\text{N} \cdot \text{sec/m}^2$ and T is absolute temperature in kelvins (Reference 9 of Appendix A). Calculate the viscosity of water at 0, 20, and 40 C. Compare with values obtained from the plots in Appendix A.

2.29 Viscosity data for gear lubricants, when plotted as $\log_{10} \mu$ versus temperature, are fitted well by straight lines for temperatures below -20 F. Consider the following data for SAE 85W gear lube (*SAE Handbook*, 1976):

Temperature (F)	−20	−30	−45
Viscosity (Pa · sec)	10	29.5	175

Use the extreme points to develop an equation for a straight line curve fit to these data. Evaluate the error in fit at the -30 F datum point. Express your answer in percent.

2.30 An analysis developed in Chapter 8 for the shear stress and velocity profile for laminar flow in a circular tube may be extended. The results for a power-law non-Newtonian fluid may be written in the form

$$\frac{u}{u_{\text{mean}}} = \left(\frac{3n + 1}{n + 1}\right)\left[1 - \left(\frac{r}{R}\right)^{\frac{n+1}{n}}\right]$$

Plot velocity profiles, u/u_{mean} versus r/R, for a Newtonian fluid with $n = 1$, a dilatant fluid with $n = 3$, and a pseudoplastic fluid with $n = \frac{1}{3}$. Compare features of the curves.

2.31 Consider again the velocity profile for laminar flow of a power-law non-Newtonian fluid in a long circular tube given in Problem 2.30. Investigate the limiting cases of infinite dilatancy and infinite pseudoplasticity, represented by $n = \infty$ and $n = 0$, respectively. Plot u/u_{mean} versus r/R for these cases and compare to the Newtonian case for which $n = 1$.

2.32 Viscometric data for heavy cream show it to behave as a pseudoplastic fluid that can be modeled well by a power-law relationship between shear stress and rate of shearing strain at low shear rates. Assume the following data:

Shear stress, τ_{yx} (dyne/cm^2)	0.1	1.0
Shear rate, $\frac{du}{dy}$ (sec^{-1})	0.023	0.75

Use these data to fit a power-law model. Evaluate the flow behavior and consistency indexes using SI units. Estimate the shear stress that would be expected at a shear rate of 0.1 sec^{-1}. Compare the apparent viscosity of cream at this shear rate to that of water.

2.33 Dilatant behavior sometimes is found when dilute suspensions are tested at high shear rates. The following data were measured in a test of a suspension containing 12 percent solids by volume:

Shear stress, τ (N/m^2)	6.5	4.8	2.7	1.7
Shear rate, $\frac{du}{dy}$ (sec^{-1})	600	470	300	200

Evaluate the consistency index and flow behavior index for this suspension by fitting the data to a power-law model.

2.34 Pulverized coal mixed with water to form a slurry often is transported over long distances in pipelines. When the weight percent of solids exceeds approximately 45 percent, the slurry behaves as a Bingham plastic. Data from a coal slurry with 50 percent solids by weight showed a yield stress value of 3.1 N/m^2 and a plastic viscosity of 0.003 $N \cdot sec/m^2$. Use the Bingham plastic model to evaluate the shear stress that would be expected in this suspension at a shear rate of 500 sec^{-1}. Compare to the shear stress for pure water at the same shear rate.

2.35 The velocity distribution for laminar flow between parallel plates is given by

$$\frac{u}{u_{max}} = 1 - \left(\frac{2y}{h}\right)^2$$

where h is the distance separating the plates, and the origin is placed midway between the plates. Consider a flow of water at 15 C, with $u_{max} = 0.30$ m/sec, and $h = 0.50$ mm. Calculate the shear stress on the upper plate and give its direction.

2.36 Consider the incompressible laminar viscous flow over a semi-infinite flat plate shown in Fig. 2.11. Sketch the variation in shear stress, τ_{yx}, as a function of y at the locations x_1 and x_2. Also sketch the variation in shear stress along the plate surface ($y = 0$) as a function of x.

2.37 The velocity distribution for laminar flow between parallel plates is given by

$$\frac{u}{u_{max}} = 1 - \left(\frac{2y}{h}\right)^2$$

where h is the distance separating the plates and the origin is placed midway between the plates. Consider flow of water at 15 C with maximum speed of 0.05 m/sec and $h = 5$ mm. Calculate the force on a 0.3 m^2 section of the lower plate and give its direction.

2.38 A concentric cylinder viscometer may be formed by rotating the inner member of a pair of closely fitting cylinders. The gap must be made small so that a linear velocity profile will exist. Consider a viscometer with an inner cylinder of 3 in. diameter and 6 in. height, and a clearance gap of 0.001 in. width, filled with castor oil at 90 F. Determine the torque required to turn the inner cylinder at 250 rpm.

2.39 A concentric cylinder viscometer may be formed by rotating the inner member of a pair of closely fitting cylinders. For small clearances, a linear velocity profile may be assumed in the liquid filling the clearance gap. A viscometer has an inner cylinder of 75 mm diameter and 150 mm height, with a clearance gap width of 0.02 mm. A torque of 0.021 $N \cdot m$ is required to turn the inner cylinder at 100 rpm. Determine the viscosity of the liquid in the clearance gap of the viscometer.

2.40 A shaft with outside diameter of 18 mm turns at 20 revolutions per second inside a stationary journal bearing 60 mm long. A thin film of oil 0.2 mm thick fills the concentric annulus between the shaft and journal. The torque needed to turn the shaft is 0.0036 $N \cdot m$. Estimate the viscosity of the oil that fills the gap.

2.41 Capillary-tube viscometers are used widely for industrial purposes, particularly for petroleum products and lubricants. The American Society for Testing and Materials (ASTM) has standardized dimensions and test procedures for Saybolt Universal and Saybolt Furol (contraction for "fuel and motor oils") capillary viscometers. These instruments are simple to use: the liquid to be tested is placed in a cylinder that has a short small-bore tube and stopper at its lower end. The cylinder is surrounded by a constant-temperature bath. After the sample temperature stabilizes, the stopper is removed. The time is measured for 60 cubic centimeters of liquid to flow out. The Saybolt reading is the flow time in seconds. These "viscosity" readings may be converted to engineering units using the empirical equations (Streeter, V. L., *Handbook of Fluid Dynamics*, New York: McGraw-Hill, 1961, Section 1.11):

Viscometer	Time Range (sec)	Kinematic Viscosity ν (St)
Saybolt Universal:	$32 \leq t \leq 100$	$0.00226t - \frac{1.95}{t}$
	$100 < t < 1000$	$0.00220t - \frac{1.35}{t}$
Saybolt Furol:	$25 \leq t \leq 40$	$0.0224t - \frac{1.84}{t}$
	$40 < t$	$0.0216t - \frac{0.60}{t}$

Calculate the ranges of kinematic viscosity (in SI units) for which each instrument may be used.

2.42 The cone and plate viscometer shown is an instrument used frequently to characterize non-Newtonian fluids. It consists of a flat plate and a rotating cone with a very obtuse angle (typically θ is less than 0.5 degrees). The apex of the cone just touches the plate surface and the liquid to be tested fills the narrow gap formed by the cone and plate. Derive an expression for the shear rate in the liquid that fills the gap in terms of the geometry of the system. Evaluate the torque on the driven cone in terms of the shear stress and geometry of the system.

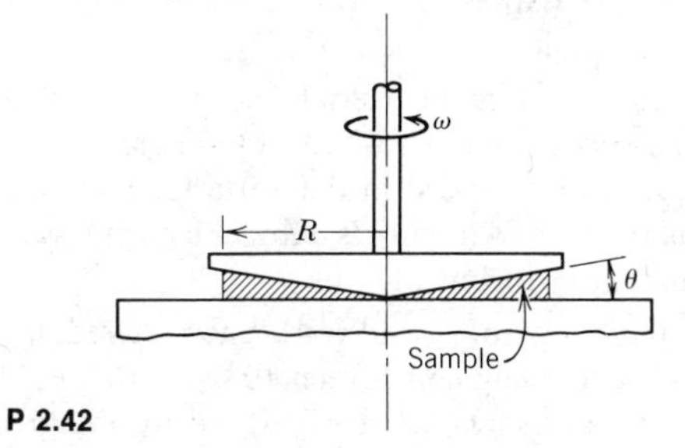

P 2.42

2.43 A block has a mass of 2 kg and is 0.2 m square. It slides down a smooth incline on a thin film of oil. The slope is 30° from the horizontal. The oil is SAE 30 at 20 C, the film is 0.02 mm thick, and the velocity profile may be assumed linear. Calculate the terminal speed of the block.

2.44 A block weighing 10 lbf and having dimensions 10 in. on each edge is pulled up an inclined surface on which there is a film of SAE 10 oil at 100 F. If the speed of the block is 5 ft/sec and the oil film is 0.001 in. thick, find the force required to pull the block. Assume the velocity distribution in the oil film to be linear. The surface is inclined at an angle of 15° from the horizontal.

2.45 Recording tape is to be coated on both sides with lubricant by drawing it through a narrow gap. The tape is 0.015 in. thick and 1.00 in. wide. It is centered in the gap with a clearance of 0.012 in. on each side. The lubricant, of viscosity $\mu = 0.021$ slug/ft · sec, completely fills the space between the tape and gap, for a length of 0.75 in. along the tape. If the tape can withstand a maximum tensile force of 7.5 lbf, determine the maximum speed with which it can be pulled through the gap.

2.46 Magnet wire is to be coated with varnish for insulation by drawing it through a circular die of 0.9 mm diameter. The wire diameter is 0.8 mm and it is centered in the die. The varnish ($\mu = 20$ centipoise) completely fills the space between the wire and the die for a length of 20 mm. The wire is drawn through the die at a speed of 50 m/sec. Determine the force required to pull the wire.

2.47 The air hockey game at Rocky's Rec Room has pucks of mass 30 g, with a diameter of 100 mm. The air film under a puck is 0.1 mm thick. Calculate the time required after impact for a puck to lose 10 percent of its initial speed.

2.48 The air hockey table of Problem 2.47 has been installed poorly; its surface slopes at 2° from the horizontal. A player fires a puck directly uphill at an initial speed of 10 m/sec. Determine its speed at the instant it passes through the goal, 2 m away.

2.49 Air at standard conditions flows through a pipe of 1 in. diameter. The average velocity is 1 ft/sec. Is the flow laminar or turbulent?

2.50 Water at 15 C flows in a tube with an inside diameter of 50 mm. Determine the maximum value of average velocity for which the flow would be laminar.

2.51 The Reynolds number, which is an important parameter in viscous flow phenomena, is defined by the equation

$$Re = \frac{\rho V L}{\mu}$$

where ρ and μ are the fluid density and viscosity, and V and L are characteristic velocity and length, respectively. Express each variable in terms of its basic dimensions. Show that the Reynolds number is dimensionless.

2.52 The fluid mechanics of the human arterial system is of vital importance to our health and safety. The specific gravity and viscosity of blood are about 1.06 and 3.3 centipoise (cp), respectively. The mean flow speed in the aorta (30 mm i.d.) of a large human is about 0.15 m/sec. Calculate the flow Reynolds number using these properties. Would you expect this flow to be laminar or turbulent?

2.53 Estimates place the range of shear rates encountered in the lubricating oil during winter cranking of a cold engine between 10^4 and 10^5 sec^{-1} (*SAE Handbook*, 1976). A concentric-cylinder viscometer is to be designed to obtain shear rates in the necessary range. Constraints on the viscometer include a maximum speed for the outer cylinder of 12,000 rpm and a maximum Reynolds number for the lubricant in the clearance gap of 1200, based on the linear speed of the outer cylinder and the gap width. Using viscosity data for SAE 10W-30 motor oil (from Appendix A), calculate the gap width required. Assess the feasibility of the design.

Chapter 3

FLUID STATICS

By definition, a fluid must deform continuously when a shear stress of any magnitude is applied. The absence of relative motion (and thus, angular deformation) implies the absence of shear stresses. Therefore, fluids either at rest or in "rigid-body" motion are able to sustain only normal stresses. Analysis of hydrostatic cases is thus appreciably simpler than for fluids undergoing angular deformation (see Section 5–3.3).

Mere simplicity does not justify our study of a subject. Normal forces transmitted by fluids are important in many practical situations. Using the principles of hydrostatics, we can compute forces on submerged objects, develop instruments for measuring pressures, and deduce properties of the atmosphere and oceans. The principles of hydrostatics also may be used to determine forces developed by hydraulic systems in applications such as industrial presses or automobile brakes.

In a static fluid, or in a fluid undergoing rigid-body motion, a fluid particle retains its identity for all time. Since there is no relative motion within the fluid, a fluid element does not deform. We may apply Newton's second law of motion to evaluate the reaction of the particle to the applied forces.

3-1 THE BASIC EQUATION OF FLUID STATICS

Our primary objective is to obtain an equation that will enable us to determine the pressure field within the fluid. To do this, we choose a differential element of mass, dm, with sides dx, dy, and dz as shown in Fig. 3.1. The fluid element is stationary relative to the stationary rectangular coordinate system shown. (Fluids in rigid-body motion will be treated in Section 3–7.)

From our previous discussion, recall that two general types of forces may be applied to a fluid: body forces and surface forces. The only body force that must be considered in most engineering problems is due to gravity. In some situations body forces due to electric or magnetic fields might be present; they will not be considered in this text.

For a differential fluid element, the body force, $d\vec{F}_B$, is

$$d\vec{F}_B = \vec{g}\, dm = \vec{g}\, \rho\, d\forall$$

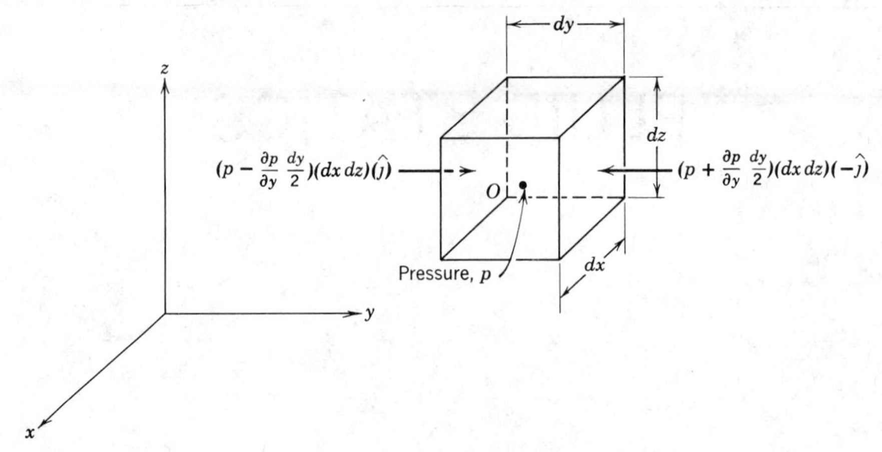

Fig. 3.1 Differential fluid element and pressure forces in the y direction.

where $\vec{g}$ is the local gravity vector, ρ is the density, and $d\forall$ is the volume of the element. In Cartesian coordinates $d\forall = dx\,dy\,dz$, so

$$d\vec{F}_B = \rho\vec{g}\,dx\,dy\,dz$$

In a static fluid no shear stresses can be present. Thus the only surface force is the pressure force. Pressure is a field quantity, $p = p(x, y, z)$; the pressure varies with position within the fluid. The net pressure force that results from this variation can be evaluated by summing the forces that act on the six faces of the fluid element.

Let the pressure at the center, O, of the element be p. To determine the pressure at each of the six faces of the element, we use a Taylor series expansion of the pressure about the point O. The pressure at the left face of the differential element is

$$p_L = p + \frac{\partial p}{\partial y}(y_L - y) = p + \frac{\partial p}{\partial y}\left(-\frac{dy}{2}\right) = p - \frac{\partial p}{\partial y}\frac{dy}{2}$$

(Terms of higher order are omitted because they will vanish in the subsequent limiting process.) The pressure on the right face of the differential element is

$$p_R = p + \frac{\partial p}{\partial y}(y_R - y) = p + \frac{\partial p}{\partial y}\frac{dy}{2}.$$

The pressure *forces* acting on the two y surfaces of the differential element are shown in Fig. 3.1. Each pressure force is a product of three terms. The first is the magnitude of the pressure. The magnitude is multiplied by the area of the face to give the pressure force, and a unit vector is introduced to indicate direction. Note also in Fig. 3.1 that the pressure force on each face acts *against* the face. A positive pressure corresponds to a *compressive* stress.

Pressure forces on the other faces of the element are obtained in the same way. Combining all such forces gives the net surface force acting on the element. Thus

$$d\vec{F}_S = \left(p - \frac{\partial p}{\partial x}\frac{dx}{2}\right)(dy\,dz)(\hat{i}) + \left(p + \frac{\partial p}{\partial x}\frac{dx}{2}\right)(dy\,dz)(-\hat{i})$$
$$+ \left(p - \frac{\partial p}{\partial y}\frac{dy}{2}\right)(dx\,dz)(\hat{j}) + \left(p + \frac{\partial p}{\partial y}\frac{dy}{2}\right)(dx\,dz)(-\hat{j})$$
$$+ \left(p - \frac{\partial p}{\partial z}\frac{dz}{2}\right)(dx\,dy)(\hat{k}) + \left(p + \frac{\partial p}{\partial z}\frac{dz}{2}\right)(dx\,dy)(-\hat{k})$$

Collecting and canceling terms, we obtain

$$d\vec{F}_S = \left(-\frac{\partial p}{\partial x}\hat{i} - \frac{\partial p}{\partial y}\hat{j} - \frac{\partial p}{\partial z}\hat{k}\right) dx\,dy\,dz$$

or,

$$d\vec{F}_S = -\left(\frac{\partial p}{\partial x}\hat{i} + \frac{\partial p}{\partial y}\hat{j} + \frac{\partial p}{\partial z}\hat{k}\right) dx\,dy\,dz \tag{3.1a}$$

The term in parentheses is called the gradient of the pressure or simply the pressure gradient and may be written grad p or ∇p. In rectangular coordinates

$$\text{grad}\, p \equiv \nabla p \equiv \left(\hat{i}\frac{\partial p}{\partial x} + \hat{j}\frac{\partial p}{\partial y} + \hat{k}\frac{\partial p}{\partial z}\right) \equiv \left(\hat{i}\frac{\partial}{\partial x} + \hat{j}\frac{\partial}{\partial y} + \hat{k}\frac{\partial}{\partial z}\right)p$$

The gradient can be viewed as a vector operator; taking the gradient of a scalar field gives rise to a vector field. Using the gradient designation, Eq. 3.1a can be written as

$$d\vec{F}_S = -\text{grad}\, p(dx\,dy\,dz) = -\nabla p\, dx\,dy\,dz \tag{3.1b}$$

From Eq. 3.1b,

$$\text{grad}\, p = \nabla p = -\frac{d\vec{F}_S}{dx\,dy\,dz}$$

Physically the gradient of pressure is the negative of the surface force per unit volume due to pressure. We note that the level of pressure is not important in evaluating the net pressure force. Instead, what matters is the rate at which pressure changes occur with distance, the *pressure gradient*. We shall find this term very useful throughout our study of fluid mechanics.

Since no other kinds of force may be present in a static fluid, we can combine the formulations for surface and body forces that we have developed to obtain the total force acting on a fluid element. Thus

$$d\vec{F} = d\vec{F}_S + d\vec{F}_B = (-\text{grad}\, p + \rho\vec{g})\, dx\,dy\,dz$$

or on a unit volume basis

$$\frac{d\vec{F}}{d\mathcal{V}} = \frac{d\vec{F}}{dx\,dy\,dz} = -\text{grad}\, p + \rho\vec{g} \tag{3.2}$$

For a fluid particle, Newton's second law gives $d\vec{F} = \vec{a}\, dm = \vec{a}\rho\, d\mathcal{V}$. For a static fluid, $\vec{a} = 0$. Thus

$$\frac{d\vec{F}}{d\mathcal{V}} = \rho\vec{a} = 0$$

Substituting for $d\vec{F}/d\mathcal{V}$ from Eq. 3.2, we obtain

$$-\operatorname{grad} p + \rho\vec{g} = 0 \tag{3.3}$$

Let us review briefly our derivation of this equation. The physical significance of each term is

$$-\operatorname{grad} p \quad + \quad \rho\vec{g} \quad = 0$$

$$\left\{\begin{array}{l}\text{pressure force}\\ \text{per unit volume}\\ \text{at a point}\end{array}\right\} + \left\{\begin{array}{l}\text{body force per}\\ \text{unit volume}\\ \text{at a point}\end{array}\right\} = 0$$

This is a vector equation, which means that it consists of three component equations that must be satisfied individually. The components are

$$\left.\begin{array}{ll} -\dfrac{\partial p}{\partial x} + \rho g_x = 0 & x \text{ direction} \\[2ex] -\dfrac{\partial p}{\partial y} + \rho g_y = 0 & y \text{ direction} \\[2ex] -\dfrac{\partial p}{\partial z} + \rho g_z = 0 & z \text{ direction} \end{array}\right\} \tag{3.4}$$

Equations 3.4 describe the pressure variation in each of the three coordinate directions in a static fluid. To simplify further, it is logical to choose a coordinate system such that the gravity vector is aligned with one of the axes. If the coordinate system is chosen such that the z axis is directed vertically, then $g_x = 0$, $g_y = 0$, and $g_z = -g$. Under these conditions, the component equations become

$$\frac{\partial p}{\partial x} = 0, \qquad \frac{\partial p}{\partial y} = 0, \qquad \frac{\partial p}{\partial z} = -\rho g \tag{3.5}$$

Equations 3.5 indicate that under the assumptions made, the pressure is independent of coordinates x and y; it depends on z alone. Thus since p is a function of a single variable, a total derivative may be used instead of a partial derivative. With these simplifications, Eqs. 3.5 finally reduce to

---

$$\frac{dp}{dz} = -\rho g \equiv -\gamma \tag{3.6}$$

Restrictions: (1) Static fluid
(2) Gravity is the only body force
(3) The z axis is vertical

This equation is the basic pressure-height relation of fluid statics. It is subject to the restrictions noted. Therefore it must be applied only where these restrictions are reasonable for the physical situation. To determine the pressure distribution in a static fluid, Eq. 3.6 may be integrated and appropriate boundary conditions applied.

3-1.1 Pressure Variation in a Static Fluid

Although ρg may be defined as the *specific weight*, γ, it has been written as ρg in Eq. 3.6 to emphasize that *both* ρ and g must be considered variables. In order to integrate Eq. 3.6 to find the pressure distribution, assumptions must be made about variations in both ρ and g.

For most practical engineering situations, the variation in g will be negligible. Only for a situation such as computing very precisely the pressure change over a large elevation difference would the variation in g need to be included. For our purposes we shall assume g to be constant with elevation at any given location.

In many practical engineering problems the variation in ρ will be appreciable, and accurate results will require that it be accounted for. Several types of variation are easy to treat analytically. The simplest is the idealization of an incompressible fluid.

a. Incompressible Fluid

For an incompressible fluid, $\rho = \rho_0 =$ constant. Then for constant gravity,

$$\frac{dp}{dz} = -\rho_0 g = \text{constant}$$

To determine the pressure variation, we must integrate this equation and apply appropriate boundary conditions. If the pressure at the reference level, z_0, is designated as p_0, then the pressure, p, at location z is found by integration

$$\int_{p_0}^{p} dp = -\int_{z_0}^{z} \rho_0 g \, dz$$

or

$$p - p_0 = -\rho_0 g(z - z_0) = \rho_0 g(z_0 - z)$$

For liquids, it is often convenient to take the origin of the coordinate system at the free surface (reference level) and to measure distances as positive downward from the free surface, as shown in Fig. 3.2. With h measured positive downward, then

$$z_0 - z = h$$

and

$$p = p_0 + \rho_0 g h \tag{3.7}$$

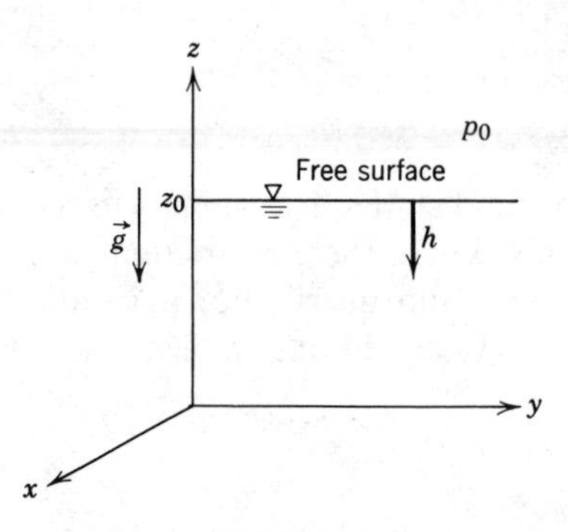

Fig. 3.2 Coordinates for determination of pressure variation in a static liquid.

This form of the basic pressure-height relation often is used to solve manometer problems. Students sometimes have trouble analyzing multiple tube manometer situations. The following rules of thumb are useful:

1. Any two points at the same elevation in a continuous length of the same liquid are at the same pressure.
2. Pressure increases as one goes *down* a liquid column (remember the pressure change on diving into a swimming pool).

Example 3.1

Water flows through pipes *A* and *B*. Oil, with specific gravity 0.8, is in the upper portion of the inverted U. Mercury (specific gravity 13.6) is in the bottom of the manometer bends. Determine the pressure difference, $p_A - p_B$, in units of lbf/in.^{2}

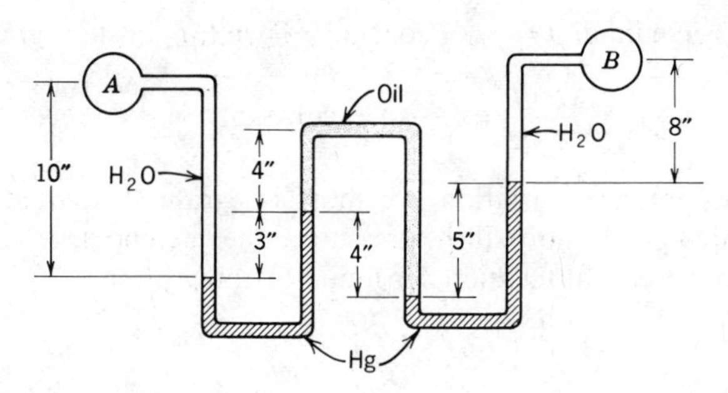

EXAMPLE PROBLEM 3.1

GIVEN: Multiple tube manometer as shown. Specific gravity of oil is 0.8; specific gravity of mercury is 13.6.

FIND: The pressure difference, $p_A - p_B$, in lbf/in.^{2}

SOLUTION:

Basic equations: $\dfrac{dp}{dz} = -\dfrac{dp}{dh} = -\rho g = -\gamma \qquad SG = \dfrac{\rho}{\rho_{H_2O}} = \dfrac{\gamma}{\gamma_{H_2O}}$

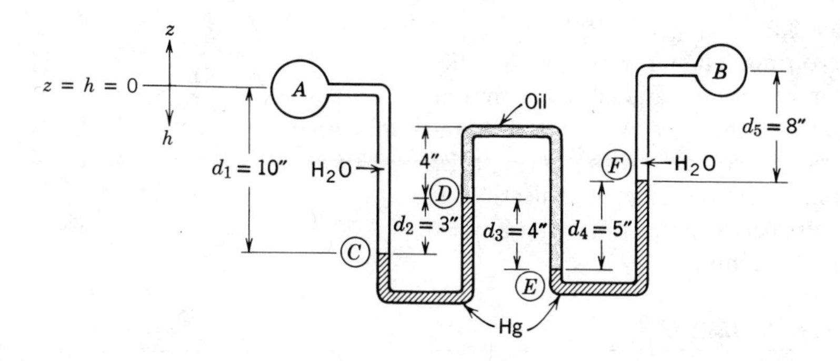

Then

$$dp = \gamma \, dh \qquad \text{and} \qquad \int_{p_1}^{p_2} dp = \int_{h_1}^{h_2} \gamma \, dh$$

For γ = constant

$$p_2 - p_1 = \gamma(h_2 - h_1)$$

Beginning at point A and applying the equation between successive points around the manometer gives

$$p_C - p_A = +\gamma_{H_2O} d_1$$

$$p_D - p_C = -\gamma_{Hg} d_2$$

$$p_E - p_D = +\gamma_{oil} d_3$$

$$p_F - p_E = -\gamma_{Hg} d_4$$

$$p_B - p_F = -\gamma_{H_2O} d_5$$

$$\begin{aligned} p_A - p_B &= (p_A - p_C) + (p_C - p_D) + (p_D - p_E) + (p_E - p_F) + (p_F - p_B) \\ &= -\gamma_{H_2O} d_1 + \gamma_{Hg} d_2 - \gamma_{oil} d_3 + \gamma_{Hg} d_4 + \gamma_{H_2O} d_5 \end{aligned}$$

Substituting $\gamma = SG\gamma_{H_2O}$ yields

$$\begin{aligned} p_A - p_B &= -\gamma_{H_2O} d_1 + 13.6\gamma_{H_2O} d_2 - 0.8\gamma_{H_2O} d_3 + 13.6\gamma_{H_2O} d_4 + \gamma_{H_2O} d_5 \\ &= \gamma_{H_2O}(-d_1 + 13.6 d_2 - 0.8 d_3 + 13.6 d_4 + d_5) \\ &= \gamma_{H_2O}(-10 + 40.8 - 3.2 + 68 + 8) \text{ in.} \\ &= \gamma_{H_2O} \times 103.6 \text{ in.} \\ &= \frac{62.4 \text{ lbf}}{\text{ft}^3} \times 103.6 \text{ in.} \times \frac{\text{ft}}{12 \text{ in.}} \times \frac{\text{ft}^2}{144 \text{ in.}^2} \end{aligned}$$

$$p_A - p_B = 3.74 \text{ lbf/in.}^2$$

$p_A - p_B$

Manometers are simple and inexpensive devices used frequently for pressure measurements. Because the liquid level change is small at low pressure differential, a U-tube manometer may be difficult to read accurately. The level change can be increased by changing the manometer design or by using two liquids of slightly different density. Analysis of a typical reservoir manometer design is illustrated in Example Problem 3.2.

Example 3.2

A reservoir manometer is built with a tube diameter of 10 mm and a reservoir diameter of 30 mm. The manometer liquid is Meriam red oil with SG = 0.827. Determine the manometer deflection in millimeters per millimeter of water applied pressure differential.

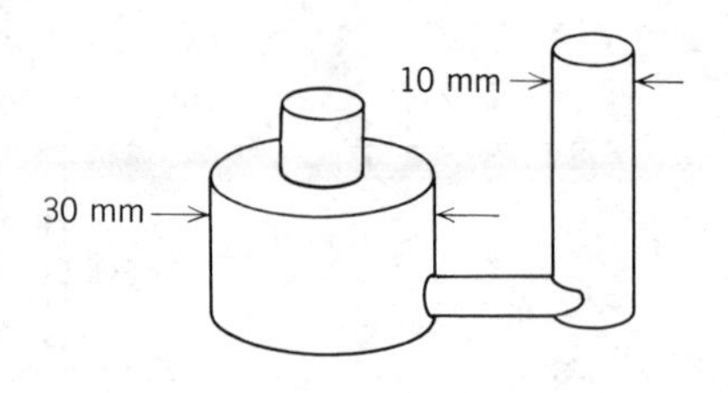

EXAMPLE PROBLEM 3.2

GIVEN: Reservoir manometer as shown.
$d = 10$ mm
$D = 30$ mm

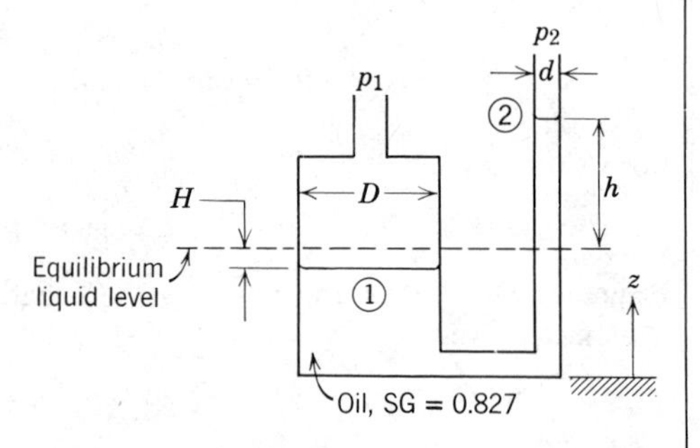

FIND: Liquid deflection, h, in millimeters per millimeter of water applied pressure differential.

SOLUTION:

Basic equations:

$$\frac{dp}{dz} = -\rho g, \qquad SG = \frac{\rho}{\rho_{H_2O}}$$

Then

$$dp = -\rho g\, dz \qquad \text{and} \qquad \int_{p_1}^{p_2} dp = -\int_{z_1}^{z_2} \rho g\, dz$$

For $\rho =$ constant

$$p_2 - p_1 = -\rho g(z_2 - z_1)$$

or

$$p_1 - p_2 = \rho g(z_2 - z_1) = \rho_{\text{oil}} g(h + H)$$

To eliminate H, note that the *volume* of manometer liquid must remain constant. Thus the volume displaced from the reservoir must be the same as that which rises into the tube,

$$\frac{\pi}{4} D^2 H = \frac{\pi}{4} d^2 h \qquad \text{or} \qquad H = \left(\frac{d}{D}\right)^2 h$$

Substituting gives

$$p_1 - p_2 = \rho_{\text{oil}} g h \left[1 + \left(\frac{d}{D}\right)^2\right]$$

This equation can be simplified by expressing the applied pressure differential as an equivalent water column of height Δh_e

$$p_1 - p_2 = \rho_{H_2O} g \, \Delta h_e$$

and noting that $\rho_{oil} = SG_{oil}\rho_{H_2O}$. Then

$$\rho_{H_2O} g \, \Delta h_e = SG_{oil}\rho_{H_2O} g h \left[1 + \left(\frac{d}{D} \right)^2 \right]$$

or

$$\frac{h}{\Delta h_e} = \frac{1}{SG_{oil}[1 + (d/D)^2]}$$

Evaluating, we obtain

$$\frac{h}{\Delta h_e} = \frac{1}{0.827[1 + (10/30)^2]} = 1.09 \qquad \frac{h}{\Delta h_e}$$

{This problem illustrates the effects of manometer design and choice of gage liquid on sensitivity.}

b. Compressible Fluid

We have seen that pressure variation in any static fluid is described by the basic pressure-height relation

$$\frac{dp}{dz} = -\rho g \tag{3.6}$$

Pressure variation in a compressible fluid also can be evaluated by integrating Eq. 3.6. Before this can be done, density must be expressed as a function of one of the other variables in the equation. Property information or an equation of state may be used to obtain the required relation for density.

For many liquids, density is only a weak function of temperature. Pressure and density of liquids are related by the *bulk compressibility modulus*, or modulus of elasticity,

$$E_v \equiv \frac{dp}{(d\rho/\rho)} \tag{3.8}$$

If the bulk modulus is assumed constant, then density is only a function of pressure (the fluid is *barotropic*) and Eq. 3.8 provides the additional density relation needed to integrate the basic pressure-height relation. Bulk modulus data for some common liquids are given in Appendix A.

The density of gases generally depends on pressure and temperature. The ideal gas equation of state

$$p = \rho R T \tag{1.1}$$

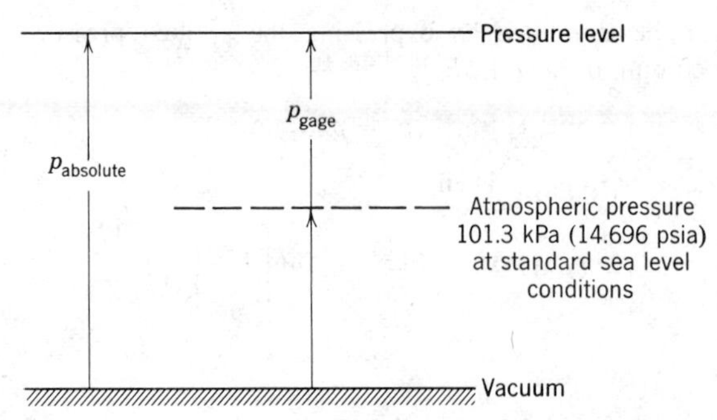

Fig. 3.3 Absolute and gage pressures, showing reference levels.

where R is the gas constant (see Appendix A) and T the absolute temperature, accurately models the behavior of most gases under engineering conditions. However, the use of Eq. 1.1 introduces the gas temperature as an additional variable. Therefore, an additional assumption must be made about temperature variation before Eq. 3.6 can be integrated.

3-2 ABSOLUTE AND GAGE PRESSURES

Pressure values must be stated with respect to a reference level. If the reference level is a vacuum, pressures are termed *absolute*, as shown in Fig. 3.3.

Most pressure gages actually read a pressure *difference*—the difference between the measured pressure level and the ambient level (usually atmospheric pressure). Pressure levels measured with respect to atmospheric pressure are termed *gage* pressures.

Absolute pressures must be used in all calculations with the ideal gas or other equations of state. Thus

$$p_{\text{absolute}} = p_{\text{gage}} + p_{\text{atmosphere}}$$

Atmospheric pressure may be obtained from a *barometer*, in which the height of a mercury column is measured. The measured height may be converted to engineering units using Eq. 3.7 and the data for specific gravity of mercury given in Appendix A. For precise work, temperature and altitude corrections must be applied to the measured level.

3-3 THE STANDARD ATMOSPHERE

Several International Congresses for Aeronautics have been held so that aviation experts around the world might better be able to communicate. The outcome of one such Congress was an internationally accepted definition of the Standard Atmosphere. The sea level conditions of the U.S. Standard Atmosphere are summarized in Table 3.1.

Table 3.1 Sea Level Conditions of the U.S. Standard Atmosphere

Property	Symbol	SI	English
Temperature	T	288 K	59 F
Pressure	p	101.3 kPa (abs)	14.696 psia
Density	ρ	1.225 kg/m^3	0.002377 slug/ft^3
Specific weight	γ	—	0.07651 lbf/ft^3
Viscosity	μ	1.781×10^{-5} kg/m · sec (Pa · sec)	3.719×10^{-7} lbf · sec/ft^2

The temperature profile of the U.S. Standard Atmosphere is shown in Fig. 3.4. Additional property values are tabulated as functions of elevation in Appendix A.

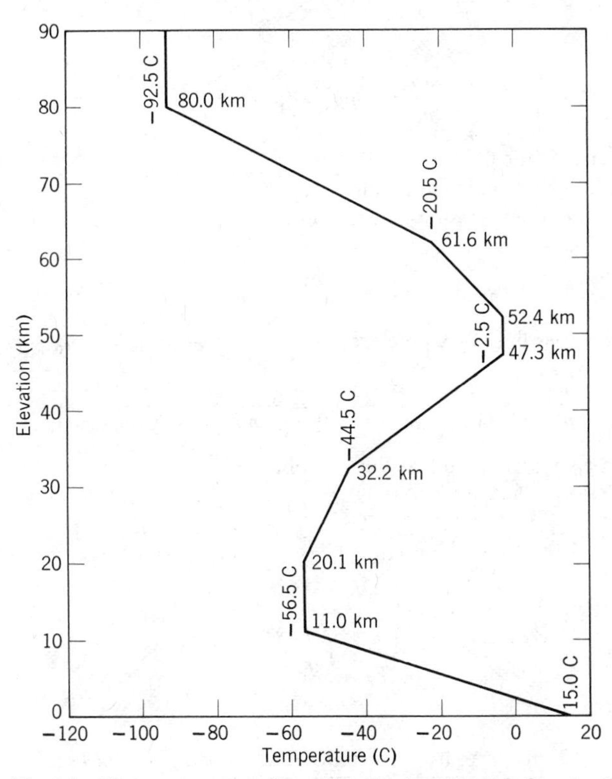

Fig. 3.4 Temperature variation with altitude in the U.S. Standard Atmosphere.

Example 3.3

The maximum power output capability of an internal combustion engine decreases with altitude because the air density and hence the mass flow rate of air decrease. A truck leaves Denver (elevation 5280 ft) on a day when the local temperature and barometric pressure are 80 F and 24.8 in. of mercury, respectively. It travels through Vail Pass (elevation 10,600 ft). The temperature decreases at the rate of 3 F/1000 ft of elevation change. Determine the local barometric pressure at Vail Pass.

EXAMPLE PROBLEM 3.3

GIVEN: Truck travels from Denver to Vail Pass.

Denver: $z = 5{,}280$ ft, $p = 24.8$ in. Hg, $T = 80$ F

Vail Pass: $z = 10{,}600$ ft, $\dfrac{dT}{dz} = -0.003\,\dfrac{\text{F}}{\text{ft}}$

FIND: Atmospheric pressure at Vail Pass.

SOLUTION:

Basic equations:
$$\frac{dp}{dz} = -\rho g \qquad p = \rho R T$$

Assumptions: (1) Static fluid
(2) Air behaves as an ideal gas

Substituting into the basic pressure-height relation yields
$$\frac{dp}{dz} = -\frac{p}{RT}g \qquad \text{or} \qquad \frac{dp}{p} = -\frac{g\,dz}{RT}$$

But temperature varies linearly with elevation, $dT/dz = -m$, so $T = T_0 - m(z - z_0)$. Thus
$$\frac{dp}{p} = -\frac{g\,dz}{R[T_0 - m(z - z_0)]} = -\frac{g\,md(z - z_0)}{mR[T_0 - m(z - z_0)]}$$

Integrating from p_0 in Denver to p at Vail, we obtain
$$\ln\left(\frac{p}{p_0}\right) = \frac{g}{mR}\ln\left[\frac{T_0 - m(z - z_0)}{T_0}\right] = \frac{g}{mR}\ln\left(\frac{T}{T_0}\right)$$

or
$$\frac{p}{p_0} = \left(\frac{T}{T_0}\right)^{g/mR}$$

Evaluating gives
$$\frac{g}{mR} = \frac{32.2\ \text{ft}}{\text{sec}^2} \times \frac{\text{ft}}{0.003\ \text{F}} \times \frac{\text{lbm}\cdot R}{53.3\ \text{ft}\cdot\text{lbf}} \times \frac{\text{slug}}{32.2\ \text{lbm}} \times \frac{\text{lbf}\cdot\text{sec}^2}{\text{slug}\cdot\text{ft}} = 6.25$$

and
$$\frac{T}{T_0} = \left[1 - \frac{0.003\ \text{F}}{\text{ft}} \times (10{,}600 - 5{,}280)\ \text{ft} \times \frac{1}{(460 + 80)\,R}\right] = 0.970$$

{Note that T_0 must be expressed as an absolute temperature because it came from the ideal gas equation of state.}

Thus

$$\frac{p}{p_0} = \left(\frac{T}{T_0}\right)^{g/mR} = (0.970)^{6.25} = 0.827$$

and

$$p = 0.827 p_0 = (0.827)24.8 \text{ in. Hg} = 20.5 \text{ in. Hg} \qquad p$$

{This problem is included to illustrate use of the ideal gas equation of state with the basic pressure-height relation to evaluate the pressure distribution in the atmosphere.}

**3-4 HYDRAULIC SYSTEMS

Hydraulic systems are characterized by very high pressures. As a consequence of these high system pressures, hydrostatic pressure variations often may be neglected. Automobile hydraulic brakes develop pressures up to 10 MPa (1500 psi); aircraft and machinery hydraulic actuation systems frequently are designed for pressures up to 30 MPa (4500 psi), and jacks use pressures to 70 MPa (10,000 psi). Special-purpose laboratory test equipment is commercially available for use at pressures to 1000 MPa (150,000 psi)!

Although liquids generally are considered incompressible at ordinary pressures, density changes may be appreciable at high pressures. Compressibility moduli of hydraulic fluids also may vary sharply at high pressures. In problems involving unsteady flow, both compressibility of the fluid and elasticity of the boundary structure must be considered. Analysis of problems such as noise and vibration in hydraulic systems, actuators, and shock absorbers quickly becomes complex and is beyond the scope of this book.

3-5 HYDROSTATIC FORCE ON SUBMERGED SURFACES

Now that we have determined the manner in which the pressure varies in a static fluid, we can examine the force on a surface submerged in a liquid.

In order to determine completely the force acting on a submerged surface, we must specify:

1. The magnitude of the force.
2. The direction of the force.
3. The line of action of the resultant force.

We shall consider both plane and curved submerged surfaces.

3-5.1 Hydrostatic Force on a Plane Submerged Surface

A plane submerged surface, on whose upper face we wish to determine the resultant hydrostatic force, is shown in Fig. 3.5. The coordinates have been chosen so that the surface lies in the xy plane.

** This section may be omitted without loss of continuity in the text material.

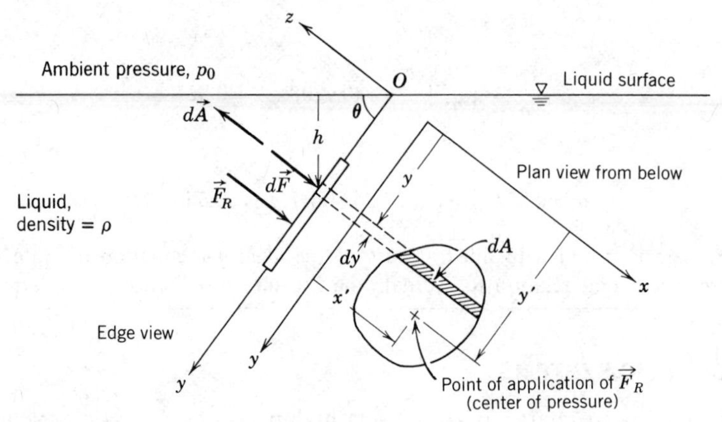

Fig. 3.5 Plane submerged surface.

Since there can be no shear stresses in a static fluid, the hydrostatic force on any element of the surface must act normal to the surface. The pressure force acting on an element of the upper surface, $d\vec{A}$, is given by

$$d\vec{F} = -p\,d\vec{A} \tag{3.9}$$

The positive direction of the vector $d\vec{A}$ is the outward drawn normal to the area; the negative sign in Eq. 3.9 indicates that the force, $d\vec{F}$, acts *against* the surface in a direction opposite to that of $d\vec{A}$. The *resultant* force acting on the surface is found by summing the contribution of the infinitesimal forces over the entire area. Thus

$$\vec{F}_R = \int_A -p\,d\vec{A} \tag{3.10}$$

In order to evaluate the integral in Eq. 3.10, both the pressure, p, and the element of area, $d\vec{A}$, must be expressed in terms of the same variables. The basic pressure-height relation for a static fluid can be written as

$$\frac{dp}{dh} = \rho g$$

where h is measured positive downward from the liquid free surface. Then, if the pressure at the free surface ($h = 0$) is p_0, we may integrate the pressure-height relation to obtain an expression for the pressure, p, at any depth, h. Thus

$$p = p_0 + \int_0^h \rho g\,dh = p_0 + \rho g h$$

This expression for p then can be substituted into Eq. 3.10. The geometry of the surface is expressed in terms of x and y; since the depth h is expressible in terms of y ($h = y \sin\theta$), the equation can be integrated to determine the resultant force.

The point of application of the resultant force must be such that the moment of the resultant force about any axis is equal to the moment of the distributed force about the same axis. If the position vector, from an arbitary origin of coordinates to the point of application of the resultant force, is designated as $\vec{r}\,'$, then

$$\vec{r}\,' \times \vec{F}_R = \int \vec{r} \times d\vec{F} = -\int_A \vec{r} \times p\, d\vec{A} \tag{3.11}$$

Referring to Fig. 3.5, we see that $\vec{r}\,' = \hat{i}x' + \hat{j}y'$, $\vec{r} = \hat{i}x + \hat{j}y$, and $d\vec{A} = dA\hat{k}$. Since the resultant force, $\vec{F}_R$, acts against the surface (in a direction opposite to that of $d\vec{A}$), then $\vec{F}_R = -F_R\hat{k}$. Substituting into Eq. 3.11 gives

$$(\hat{i}x' + \hat{j}y') \times -F_R\hat{k} = \int (\hat{i}x + \hat{j}y) \times d\vec{F} = -\int_A (\hat{i}x + \hat{j}y) \times p\, dA\hat{k}$$

Evaluating the cross product, we obtain

$$\hat{j}x'F_R - \hat{i}y'F_R = \int_A (\hat{j}xp - \hat{i}yp)\, dA$$

This is a vector equation, so the components must be equal. Thus

$$y'F_R = \int_A yp\, dA \qquad \text{and} \qquad x'F_R = \int_A xp\, dA \tag{3.12}$$

where x' and y' are the coordinates of the point of application of the resultant force. Note that Eqs. 3.10 and 3.11 can be used to determine the magnitude of the resultant force and its point of application on any plane submerged surface. They do not require that the density be a constant or that the free surface of the liquid be at atmospheric pressure.

Equations 3.10 and 3.12 are mathematical statements of basic principles familiar to you from previous courses in physics and statics:

1. The resultant force is the sum of the infinitesimal forces (Eq. 3.10).
2. The moment of the resultant force about any axis is equal to the moment of the distributed force about the same axis (Eq. 3.12).

In evaluating the hydrostatic force acting on a plane submerged surface, we have used vector notation to emphasize that forces and moments are vector quantities. Since all elements of the force are parallel, the use of vectors is not essential. Summarizing:

1. The magnitude of $\vec{F}_R$ is given by

$$F_R = |\vec{F}_R| = \int p\, dA$$

2. The direction of $\vec{F}_R$ is normal to the surface.

3. For a surface in the xy plane, the line of action of $\vec{F}_R$ passes through the point x', y' (the center of pressure), where

$$y'F_R = \int_A yp\,dA \qquad \text{and} \qquad x'F_R = \int_A xp\,dA$$

Example 3.4

Consider a plane submerged surface with free surface at atmospheric pressure. Using the notation of Fig. 3.5, (a) show that the hydrostatic force is equal to the pressure at the centroid times the area of the surface, and (b) derive expressions for the coordinates of the center of pressure in terms of the geometric parameters of the surface.

EXAMPLE PROBLEM 3.4

GIVEN: Plane submerged surface as shown, with centroid of area at x_c, y_c. Free surface at ambient pressure (zero gage pressure).

FIND: (a) Show that $F_R = p_c A$.
(b) Determine expressions for coordinates of center of pressure.

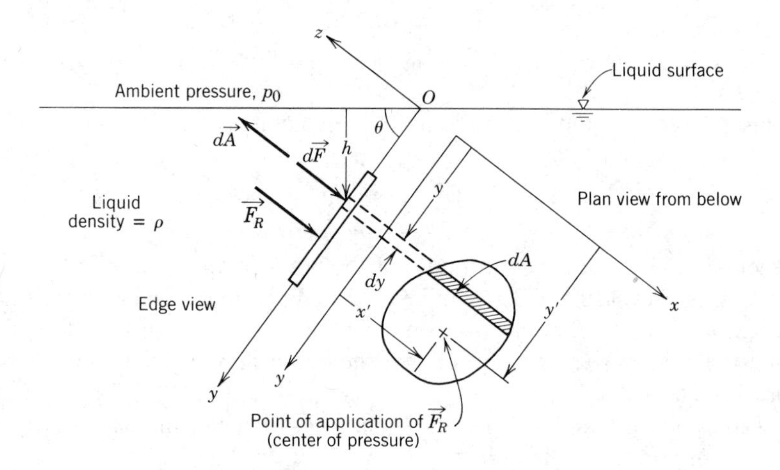

SOLUTION:

Basic equations:

$$F_R = \int p\,dA \qquad \frac{dp}{dh} = \rho g$$

For an incompressible fluid, integrating the pressure-height relation from the free surface ($h = 0, p = p_0$) gives

$$p = p_0 + \rho g h$$

Since we are interested in the hydrostatic force of the water on the gate, then we drop p_0. The force F_R is then

$$F_R = \int_A p\,dA = \int_A \rho g h\,dA = \int_A \rho g y \sin\theta\,dA = \rho g \sin\theta \int_A y\,dA$$

The $\int_A y\,dA$ is the first moment of the surface area about the x axis,

$$\int_A y\,dA = y_c A$$

where y_c is the y coordinate of the centroid of the area. Hence

$$F_R = \rho g \sin\theta y_c A = \rho g h_c A = p_c A$$

{Note: This result is valid for any value of the pressure p_0 at the free surface of the liquid.}

To find expressions for the coordinates of the center of pressure we recognize that the moment of the resultant force about any axis must be equal to the moment of the distributed force about the same axis. Taking moments about the x axis gives

$$y' F_R = \int_A yp\,dA$$

Substituting $F_R = \rho g \sin\theta y_c A$, $p = \rho g h$, and $h = y \sin\theta$, we obtain

$$y' \rho g \sin\theta y_c A = \int_A y\rho g h\,dA = \int_A y^2 \rho g \sin\theta\,dA = \rho g \sin\theta \int_A y^2\,dA$$

Recognizing that $\int_A y^2\,dA = I_{xx}$, the second moment of the area about the x axis, we find that

$$y' = I_{xx}/A y_c$$

From the parallel axis theorem, $I_{xx} = I_{\hat{x}\hat{x}} + A y_c^2$, where $I_{\hat{x}\hat{x}}$ is the second moment of the area about the centroidal $\hat{x}$ axis,

$$y' = y_c + \frac{I_{\hat{x}\hat{x}}}{A y_c}$$

Taking moments about the y axis gives $x' F_R = \int_A xp\,dA$.

Substituting for F_R, p and h as above results in

$$x' \rho g \sin\theta y_c A = \int_A x\rho g h\,dA = \int_A xy\rho g \sin\theta\,dA = \rho g \sin\theta \int_A xy\,dA$$

Recognizing that $\int_A xy\,dA = I_{xy}$, the area product of inertia, we obtain

$$x' = I_{xy}/A y_c$$

From the parallel axis theorem, $I_{xy} = I_{\hat{x}\hat{y}} + A x_c y_c$, where $I_{\hat{x}\hat{y}}$ is the area product of inertia with respect to the centroidal $\hat{x}\hat{y}$ axis. Then

$$x' = x_c + \frac{I_{\hat{x}\hat{y}}}{A y_c}$$

Note: The equations derived for x' and y' are valid only when the pressure at the free surface is atmospheric.

{This problem is included to illustrate the derivation of computing equations that would be convenient to use if a number of such problems were to be solved.}

Example 3.5
The inclined surface shown, hinged along A, is 5 m wide. Determine the resultant force, $\vec{F}_R$, of the water on the inclined surface.

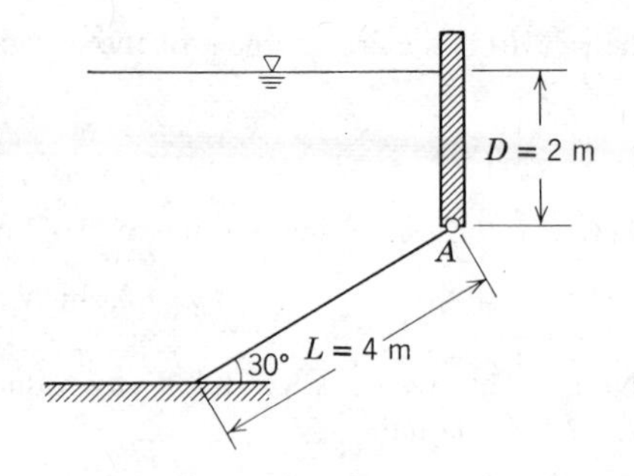

EXAMPLE PROBLEM 3.5

GIVEN: Rectangular gate, hinged along A, $w = 5$ m.

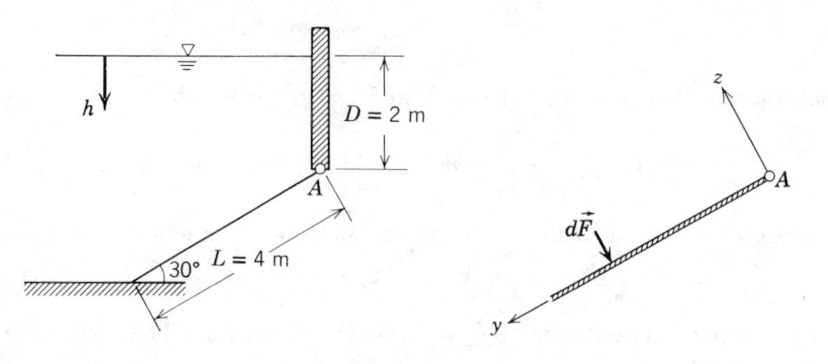

FIND: Resultant force, $\vec{F}_R$, of the water on the gate.

SOLUTION:
In order to completely determine $\vec{F}_R$, we must specify: (a) the magnitude, (b) the direction, and (c) the line of action, of the resultant force.

Basic equations: $$\vec{F}_R = -\int p\,d\vec{A} \qquad \frac{dp}{dh} = \rho g$$

Consider the gate, hinged at A, lying in the xy plane, with coordinates as shown.

$$\vec{F}_R = -\int_A p\,d\vec{A} = -\int_A pw\,dy\,\hat{k} \qquad (d\vec{A} = w\,dy\,\hat{k})$$

We now need p as a function of y to perform the integration. From the basic pressure-height relation,

$$\frac{dp}{dh} = \rho g \qquad \text{so} \qquad dp = \rho g\,dh \qquad \text{and} \qquad \int_{p_a}^{p} dp = \int_0^h \rho g\,dh$$

By assuming $\rho = \text{constant}$,

$$p = p_a + \rho g h \qquad \{\text{This gives } p = p(h). \text{ We need } p = p(y).\}$$

From the diagram

$$h = D + y \sin 30^\circ \qquad \text{where } D = 2 \text{ m}$$

Since we are interested in the force of the water on the gate, then we drop p_a and obtain

$$p = \rho g(D + y \sin 30°)$$

Thus

$$\vec{F}_R = -\int_A p\, d\vec{A} = -\int_0^L \rho g(D + y\sin 30°) w\, dy\, \hat{k}$$

$$= -\rho g w \left[Dy + \frac{y^2}{2} \sin 30° \right]_0^L \hat{k} = -\rho g w \left[DL + \frac{L^2}{2} \sin 30° \right] \hat{k}$$

$$= \frac{-999 \text{ kg}}{\text{m}^3} \times \frac{9.81 \text{ m}}{\text{sec}^2} \times 5 \text{ m} \left[2 \text{ m} \times 4 \text{ m} + \frac{16 \text{ m}^2}{2} \times \frac{1}{2} \right] \frac{\text{N} \cdot \text{sec}^2}{\text{kg} \cdot \text{m}} \hat{k}$$

$$\vec{F}_R = -588\, \hat{k} \text{ kN}$$

{Force acts in negative z direction} $\vec{F}_R$

To find the line of action of the resultant force, $\vec{F}_R$, we recognize that the line of action of the resultant force must be such that the moment of the resultant force about an axis through point A equals the moment of the distributed force about the same axis. Considering moments about the x axis, we obtain

$$F_R y' = \int_A y p\, dA$$

Then

$$y' = \frac{1}{F_R} \int_A y p\, dA = \frac{1}{F_R} \int_0^L y p w\, dy = \frac{\rho g w}{F_R} \int_0^L y(D + y \sin 30°)\, dy$$

$$= \frac{\rho g w}{F_R} \left[\frac{D}{2} y^2 + \frac{y^3}{3} \sin 30° \right]_0^L = \frac{\rho g w}{F_R} \left[\frac{DL^2}{2} + \frac{L^3}{3} \sin 30° \right]$$

$$= \frac{999 \text{ kg}}{\text{m}^3} \times \frac{9.81 \text{ m}}{\text{sec}^2} \times \frac{5 \text{ m}}{5.88 \times 10^5 \text{ N}} \left[\frac{2 \text{ m} \times 16 \text{ m}^2}{2} + \frac{64 \text{ m}^3}{3} \times \frac{1}{2} \right] \frac{\text{N} \cdot \text{sec}^2}{\text{kg} \cdot \text{m}}$$

$$y' = 2.22 \text{ m}$$

Also, from consideration of moments about the y axis,

$$x' = \frac{1}{F_R} \int_A x p\, dA$$

In calculating the moment of the distributed force (right side), recall from your earlier courses in statics, that the centroid of the area element must be used for "x". Since the area element is of constant width, $x = w/2$, then

$$x' = \frac{1}{F_R} \int_A \frac{w}{2} p\, dA = \frac{w}{2F_R} \int_A p\, dA = \frac{w}{2} = 2.5 \text{ m}$$

$$\vec{r}\,' = 2.5\hat{i} + 2.22\hat{j} \text{ m}$$

{Line of action of $\vec{F}_R$ is along negative z axis through $\vec{r}\,'$} $\vec{r}\,'$

{This problem illustrates the procedure used to determine the resultant force, $\vec{F}_R$, equivalent to a distributed force on a plane submerged surface.}

Example 3.6

The door shown in the side of the tank is hinged along its bottom. A pressure of 100 psfg is applied to the liquid free surface. Find the force, F, required to hold the door shut.

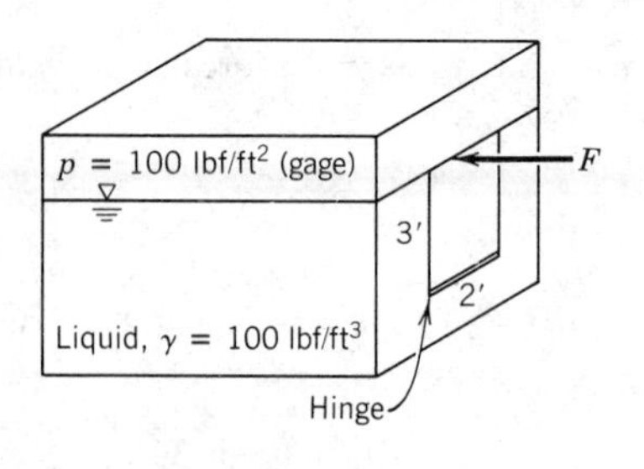

EXAMPLE PROBLEM 3.6

GIVEN: Door as shown in the figure; x axis is along the hinge.

FIND: Force required to keep door shut.

SOLUTION:

Basic equations: $F_R = \int p\, dA \qquad \frac{dp}{dh} = \rho g \qquad \sum \vec{M} = 0$

Summing moments about the hinge axis yields

$$\sum M_x = 0 = LF - \int z\, dF = 0$$

$$\therefore\ F = \frac{1}{L}\int z\, dF = \frac{1}{L}\int zp\, dA = \frac{1}{L}\int_0^L zpb\, dz$$

To solve for F, we need to know p as a function of z:

$$\frac{dp}{dh} = \rho g = \gamma \qquad \text{and} \qquad dp = \gamma\, dh$$

Then

$$p - p_0 = \int_{p_0}^{p} dp = \int_0^h \gamma\, dh$$

$$p = p_0 + \gamma h$$

Since atmospheric pressure acts on the outside of the door, the pressure p_0 in the above expression should be gage pressure. With $p = p_0 + \gamma h$ and $h = L - z$,

$$F = \frac{1}{L}\int_0^L z[p_0 + \gamma(L - z)]b\, dz = \frac{b}{L}\int_0^L p_0 z\, dz + \frac{\gamma b}{L}\int_0^L (Lz - z^2)\, dz$$

$$= \left.\frac{p_0 b z^2}{2L}\right]_0^L + \frac{\gamma b}{L}\left[\frac{Lz^2}{2} - \frac{z^3}{3}\right]_0^L$$

$$= \frac{p_0 bL}{2} + \gamma bL^2\left[\frac{1}{2} - \frac{1}{3}\right] = \frac{p_0 bL}{2} + \frac{\gamma bL^2}{6}$$

$$= 100\,\frac{\text{lbf}}{\text{ft}^2} \times 2\text{ ft} \times 3\text{ ft} \times \frac{1}{2} + 100\,\frac{\text{lbf}}{\text{ft}^3} \times 2\text{ ft} \times 9\text{ ft}^2 \times \frac{1}{6}$$

$F = 600$ lbf $\qquad$ F

This problem illustrates:
(i) inclusion of a nonzero gage pressure at the free surface of the liquid.
(ii) direct use of the distributed moment without evaluating the resultant force and line of application separately.

3-5.2 Hydrostatic Force on Curved Submerged Surfaces

Determining the hydrostatic force on a curved submerged surface is slightly more involved than calculating the force on a plane surface. The hydrostatic force on an infinitesimal element of a curved surface, $d\vec{A}$, acts normal to the surface. However, the differential pressure force on each element of the surface acts in a different direction because of the surface curvature. Accounting for this change in direction makes the problem a little more involved.

What do we normally do when we wish to sum a series of force vectors acting in different directions? The usual procedure is to sum the components of the vectors relative to a convenient coordinate system.

Consider the curved surface shown in Fig. 3.6. The pressure force acting on the element of area, $d\vec{A}$, is given by

$$d\vec{F} = -p\, d\vec{A} \tag{3.9}$$

The resultant force is again given by

$$\vec{F}_R = -\int_A p\, d\vec{A} \tag{3.10}$$

We can write

$$\vec{F}_R = \hat{i}F_{R_x} + \hat{j}F_{R_y} + \hat{k}F_{R_z} \tag{3.13}$$

where F_{R_x}, F_{R_y}, and F_{R_z} are the components of $\vec{F}_R$ in the positive x, y, and z directions, respectively.

To evaluate the component of the force in a given direction, we take the dot product of the force with the unit vector in the given direction. For example, taking the dot

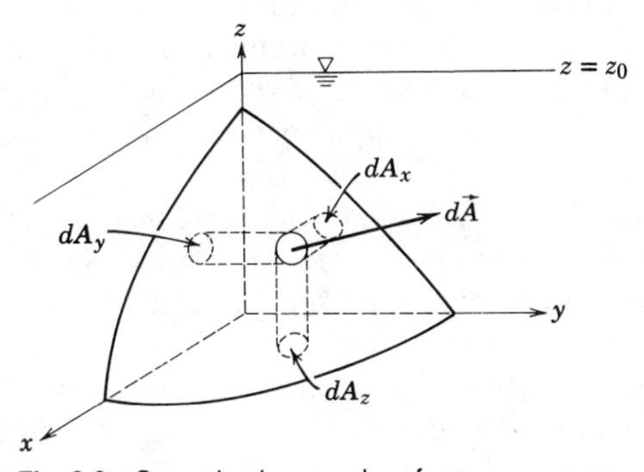

Fig. 3.6 Curved submerged surface.

product of each side of Eq. 3.10 with the unit vector $\hat{\imath}$ gives

$$F_{R_x} = \int dF_x = \vec{F}_R \cdot \hat{\imath} = \int d\vec{F} \cdot \hat{\imath} = -\int_A p\,d\vec{A} \cdot \hat{\imath} = -\int_{A_x} p\,dA_x$$

Since, in any problem, the direction of the force component can be determined by inspection, the use of vectors is not necessary. In general, the magnitude of the component of the resultant force in the l direction is given by

$$F_{R_l} = \int_{A_l} p\,dA_l \tag{3.14}$$

where dA_l is the projection of the area element dA on a plane perpendicular to the l direction. The line of action of each component of the resultant force is found by recognizing that the moment of the resultant force component about a given axis must be equal to the moment of the corresponding distributed force component about the same axis.

In considering the vertical component, F_{R_z}, of the resultant force, we note that the pressure exerted by the liquid is given by

$$p = \int_{z_s}^{z_0} \rho g\,dz$$

where z_s is the vertical coordinate of the surface, and z_0 is the vertical coordinate of the free surface. Then (refer to Fig. 3.6)

$$dF_z = -p\,dA_z = -\left(\int_{z_s}^{z_0} \rho g\,dz\right) dA_z$$

The integral represents the weight of a differential cylinder of liquid above the element of surface area, dA_z; the cylinder extends from the curved surface to the free surface. The vertical component of the resultant force is obtained by integrating over the entire surface,

$$F_z = -\int_{A_z} \int_{z_s}^{z_0} \rho g\,dz\,dA_z$$

The magnitude of the vertical component of the resultant force is equal to the total weight of the liquid directly above the surface. The minus sign indicates that a curved surface with a positive dA_z projection is subjected to a force in the negative z direction. It can be shown that the line of action of the vertical force component passes through the center of gravity of the volume of liquid between the surface and the free surface of the liquid.

We have shown that the resultant hydrostatic force on a curved submerged surface is specified in terms of its components. To determine the components and their corresponding lines of action, we proceed for each component just as we did for plane submerged surfaces. Because we are dealing with a curved surface, the lines of action of the components of the resultant force will not necessarily coincide; the complete resultant may not be expressed as a single force. In most problems, it is the components parallel and perpendicular to the liquid free surface that are of interest.

Example 3.7

The gate shown has a constant width, w, of 5 m. The equation of the surface is $x = y^2/a$, where $a = 4$ m. The depth of water to the right of the gate is 4 m. Find the horizontal and vertical components of the resultant force due to the water and the line of action of each.

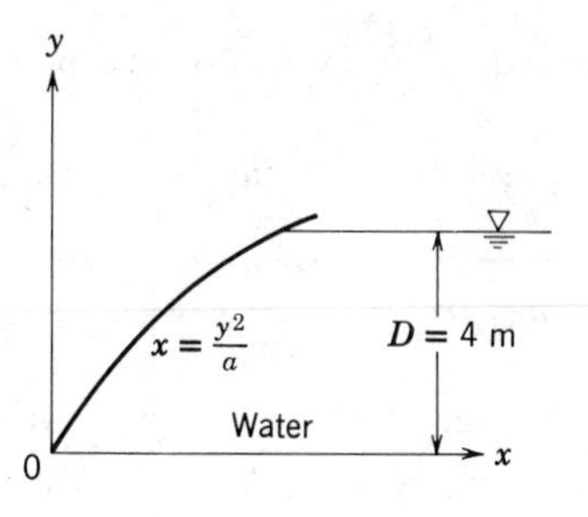

EXAMPLE PROBLEM 3.7

GIVEN: Gate of constant width, $w = 5$ m.
Equation of surface in xy plane is $x = y^2/a$, where $a = 4$ m.
Water stands at depth, D, of 4 m to the right of the gate.

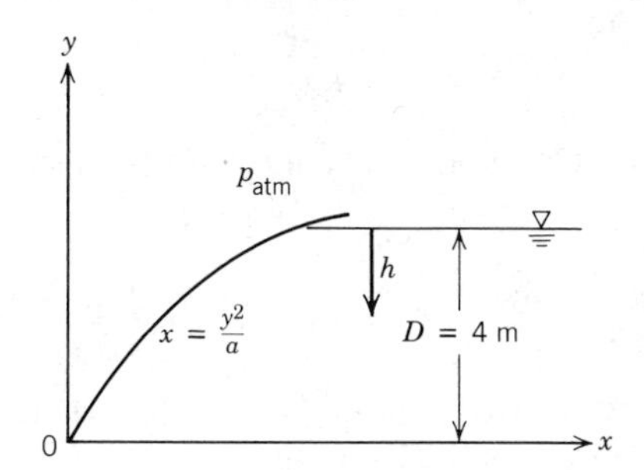

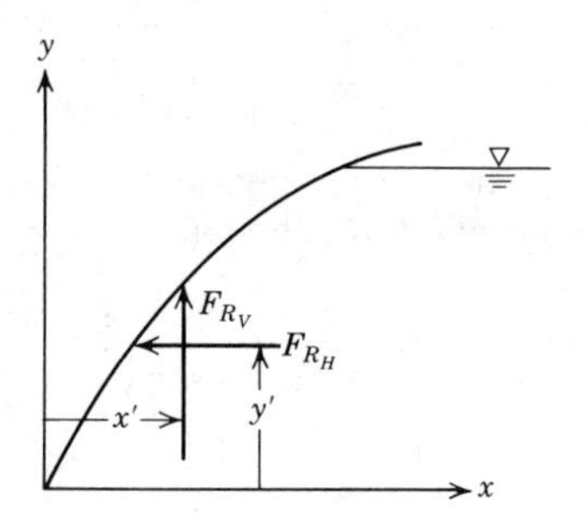

FIND: F_{R_H}, F_{R_V} and line of action of each.

SOLUTION:

Basic equations: $\vec{F}_R = -\int p\,d\vec{A} \qquad \frac{dp}{dh} = \rho g$

$$F_{R_H} = \int_0^D pw\,dy \qquad F_{R_V} = \int_0^{D^2/a} pw\,dx$$

In order to integrate, we need expressions for $p(y)$ and $p(x)$ along the surface of the gate.

$$\frac{dp}{dh} = \rho g, \quad dp = \rho g\,dh \qquad \text{and} \qquad \int_{p_a}^{p} dp = \int_0^h \rho g\,dh$$

If we assume ρ = constant, then

$$p = p_a + \rho g h$$

Since atmospheric pressure acts on both the top of the gate and the free surface of the liquid, there is no net contribution of the atmospheric pressure force. Thus, in determining the force due to the liquid, we take $p = \rho g h$.

We now need an expression for $h = h(y)$ and $h = h(x)$ along the surface of the gate. Along the surface of the gate, $h = D - y$. Since the equation of the gate surface is $x = y^2/a$, then along the gate $y = \sqrt{a}x^{1/2}$ and thus h can also be written as $h = D - \sqrt{a}x^{1/2}$. Substituting the appropriate equations for h into the expressions for F_{R_H} and F_{R_V} gives

$$F_{R_H} = \int_0^D pw\,dy = \int_0^D \rho g h w\,dy = \rho g w \int_0^D h\,dy = \rho g w \int_0^D (D - y)\,dy$$

$$= \rho g w \left[Dy - \frac{y^2}{2} \right]_0^D = \rho g w \left[D^2 - \frac{D^2}{2} \right] = \frac{\rho g w D^2}{2}$$

$$F_{R_H} = \frac{999\text{ kg}}{\text{m}^3} \times \frac{9.81\text{ m}}{\text{sec}^2} \times 5\text{ m} \times \frac{(4)^2\text{ m}^2}{2} \times \frac{\text{N}\cdot\text{sec}^2}{\text{kg}\cdot\text{m}} = 392\text{ kN}$$

F_{R_H}

$$F_{R_V} = \int_0^{D^2/a} pw\,dx = \int_0^{D^2/a} \rho g h w\,dx = \rho g w \int_0^{D^2/a} h\,dx = \rho g w \int_0^{D^2/a} (D - \sqrt{a}x^{1/2})\,dx$$

$$= \rho g w \left[Dx - \frac{2}{3}\sqrt{a}x^{3/2} \right]_0^{D^2/a} = \rho g w \left[\frac{D^3}{a} - \frac{2}{3}\sqrt{a}\frac{D^3}{a^{3/2}} \right] = \frac{\rho g w D^3}{3a}$$

$$F_{R_V} = \frac{999\text{ kg}}{\text{m}^3} \times \frac{9.81\text{ m}}{\text{sec}^2} \times 5\text{ m} \times \frac{(4)^3\text{ m}^3}{3} \times \frac{1}{4\text{ m}} \times \frac{\text{N}\cdot\text{sec}^2}{\text{kg}\cdot\text{m}} = 261\text{ kN}$$

F_{R_V}

To find the line of action of F_{R_H}, the moment of F_{R_H} about O must be equal to the sum of moments of dF_H about O.

$$y'F_{R_H} = \int_{A_x} yp\,dA_x \qquad \text{and} \qquad y' = \frac{1}{F_{R_H}} \int_{A_x} yp\,dA_x$$

$$y' = \frac{1}{F_{R_H}} \int_0^D ypw\,dy = \frac{1}{F_{R_H}} \int_0^D y\rho g h w\,dy = \frac{w\rho g}{F_{R_H}} \int_0^D y(D - y)\,dy$$

$$= \frac{w\rho g}{F_{R_H}} \left[\frac{D}{2} y^2 - \frac{y^3}{3} \right]_0^D$$

$$y' = \frac{\rho g w D^3}{6F_{R_H}} = \frac{\rho g w D^3}{6} \left[\frac{2}{\rho g w D^2} \right] = \frac{D}{3} = \frac{4\text{ m}}{3} = 1.33\text{ m}$$

y'

To find the line of action of F_{R_V}, the moment of F_{R_V} about O must be equal to the sum of moments of dF_V about O.

$$x'F_{R_V} = \int_{A_y} xp\,dA_y \qquad \text{and} \qquad x' = \frac{1}{F_{R_V}} \int_{A_y} xp\,dA_y$$

$$x' = \frac{1}{F_{R_V}} \int_0^{D^2/a} xpw\,dx = \frac{1}{F_{R_V}} \int_0^{D^2/a} x\rho g h w\,dx = \frac{w\rho g}{F_{R_V}} \int_0^{D^2/a} x(D - \sqrt{a}x^{1/2})\,dx$$

$$= \frac{w\rho g}{F_{R_V}} \left[\frac{D}{2} x^2 - \frac{2}{5}\sqrt{a}x^{5/2} \right]_0^{D^2/a} = \frac{\rho g w}{F_{R_V}} \left[\frac{D^5}{2a^2} - \frac{2}{5}\sqrt{a}\frac{D^5}{a^{5/2}} \right] = \frac{\rho g w D^5}{10F_{R_V}a^2}$$

$$= \frac{\rho g w D^5}{10a^2} \left[\frac{3a}{\rho g w D^3} \right]$$

$$x' = \frac{3D^2}{10a} = \frac{3}{10} \times (4)^2\text{ m}^2 \times \frac{1}{4\text{ m}} = 1.2\text{ m}$$

x'

{This problem illustrates the calculation of resultant force components on a curved submerged surface.}

**3-6 BUOYANCY AND STABILITY

If an object is immersed in or floating on the surface of a liquid, the force acting on it due to liquid pressure is termed *buoyancy*. Consider the object shown in Fig. 3.7, immersed in static liquid.

The vertical force on the body due to hydrostatic pressure may be found most easily by considering cylindrical volume elements similar to the one shown in Fig. 3.7. For a static fluid

$$\frac{dp}{dh} = \rho g$$

Integrating for constant ρ gives

$$p = p_0 + \rho g h$$

The net vertical force on the element is

$$dF_z = (p_0 + \rho g h_2)\, dA - (p_0 + \rho g h_1)\, dA = \rho g (h_2 - h_1)\, dA$$

But $(h_2 - h_1)\, dA = d\forall$, the volume of the element. Thus

$$F_z = \int dF_z = \int_{\forall} \rho g \, d\forall = \rho g \forall \tag{3.15}$$

where $\forall$ is the volume of the object. Thus the net vertical pressure force, or buoyancy of the object, equals the force of gravity on the liquid displaced by the object. This relation reportedly was used by Archimedes in 220 B.C. to determine the gold content in the crown of King Hiero II. Consequently, it is often called "Archimedes' Principle." In more current technical applications, Eq. 3.15 is used to design displacement vessels, flotation gear, and bathyscaphes.

The line of action of the buoyancy force may be found using the methods of Section 3-5.2. Since floating bodies are in equilibrium under body and buoyancy forces, the location of the line of action of the buoyancy force determines stability, as shown in Fig. 3.8.

The body force due to gravity on an object acts through its center of gravity, CG. In Fig. 3.8*a*, the buoyant force is offset and produces a couple that tends to right the craft.

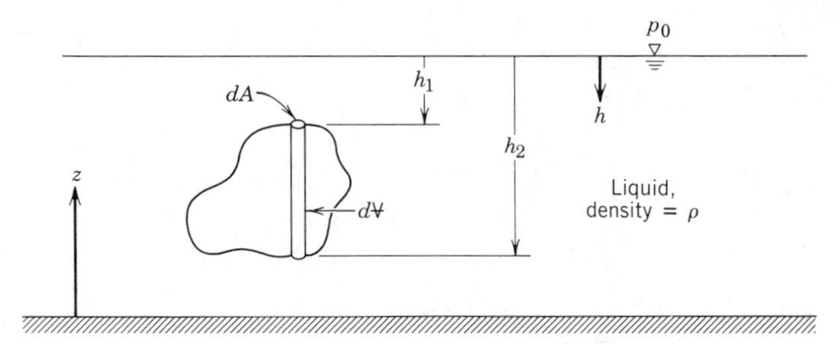

Fig. 3.7 Immersed body in static liquid.

** This section may be omitted without loss of continuity in the text material.

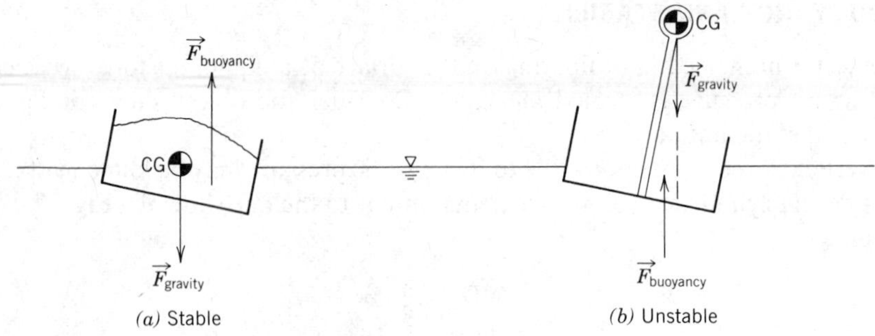

Fig. 3.8 Stability of floating bodies.

In Fig. 3.8*b*, the couple tends to capsize the craft. In sailing, wind loads bring additional forces onto a boat that must be considered in analyzing stability.

**3-7 FLUIDS IN RIGID-BODY MOTION

A fluid in rigid-body motion moves without deformation as though it were a solid body. Since there is no deformation, there can be no shear stress. Consequently, the only surface stress on each element of fluid is that due to pressure.

A fluid particle retains its identity in rigid-body motion because the fluid does not deform. As in the case of the static fluid, we may apply Newton's second law of motion to determine the pressure field that results from a specified rigid-body motion.

In Section 3-1 we derived an expression for the total force due to pressure and gravity acting on a fluid particle of volume, dV. We obtained

$$d\vec{F} = (-\text{grad}\, p + \rho\vec{g})\, dV$$

or

$$\frac{d\vec{F}}{dV} = -\text{grad}\, p + \rho\vec{g} \tag{3.2}$$

Newton's second law was written

$$d\vec{F} = \vec{a}\, dm = \vec{a}\rho\, dV$$

or

$$\frac{d\vec{F}}{dV} = \rho\vec{a}$$

Substituting from Eq. 3.2, we obtain

$$-\text{grad}\, p + \rho\vec{g} = \rho\vec{a} \tag{3.16}$$

** This section may be omitted without loss of continuity in the text material.

The physical significance of each term in this equation is

$$
\begin{array}{ccccccc}
-\text{grad}\, p & + & \rho\vec{g} & = & \rho\vec{a} & & \\
\left\{\begin{array}{l}\text{pressure force}\\ \text{per unit volume}\\ \text{at a point}\end{array}\right\} & + & \left\{\begin{array}{l}\text{body force}\\ \text{per unit volume}\\ \text{at a point}\end{array}\right\} & = & \left\{\begin{array}{l}\text{mass per}\\ \text{unit}\\ \text{volume}\end{array}\right\} & \times & \left\{\begin{array}{l}\text{acceleration}\\ \text{of fluid}\\ \text{particle}\end{array}\right\}
\end{array}
$$

This vector equation consists of three component equations that must be satisfied individually. In rectangular coordinates the component equations are

$$
\left.\begin{array}{ll}
-\dfrac{\partial p}{\partial x} + \rho g_x = \rho a_x & x \text{ direction} \\[2ex]
-\dfrac{\partial p}{\partial y} + \rho g_y = \rho a_y & y \text{ direction} \\[2ex]
-\dfrac{\partial p}{\partial z} + \rho g_z = \rho a_z & z \text{ direction}
\end{array}\right\} \tag{3.17}
$$

Component equations for other coordinate systems can be written using the appropriate expression for grad p.

Example 3.8

As a result of a promotion, you are transferred from your present location. You must transport a fish tank in the back of your station wagon. The tank is 12 in. × 24 in. × 12 in. How much water should you leave in the tank to be reasonably sure that it will not spill over during the trip?

EXAMPLE PROBLEM 3.8

GIVEN: Fish tank 12 in. × 24 in. × 12 in. partially filled with water to be transported in an automobile.

FIND: Allowable depth of water for reasonable assurance that it will not spill during the trip.

SOLUTION:

The first step in the solution is to formulate the problem, by translating the general problem into a more specific one.

We recognize that there will be motion of the water surface as a result of the car's traveling over bumps in the road, going around corners, etc. However, we shall assume that the main effect on the water surface is due to linear accelerations (and decelerations) of the car; we shall neglect sloshing.

Thus we have reduced the problem to one of determining the effect of a linear acceleration on the free surface. We have not yet decided on the orientation of the tank relative to the direction of motion. Choosing the x coordinate in the direction of motion, should we align the tank with the long side parallel, or perpendicular, to the direction of motion?

If there will be no relative motion in the water, we must assume we are dealing with a constant acceleration, a_x. What is the shape of the free surface under these conditions?

Let us restate the problem to answer the original questions without making any restrictive assumptions at the outset.

GIVEN: Tank partially filled with water (to a depth d in.) subject to constant linear acceleration, a_x. Tank height is 12 in.; length parallel to direction of motion is b in. Width perpendicular to direction of motion is c in.

FIND: (a) Shape of free surface under constant a_x.
(b) Allowable water height, d, to avoid spilling as a function of a_x and tank orientation.
(c) Optimum tank orientation and allowable depth.

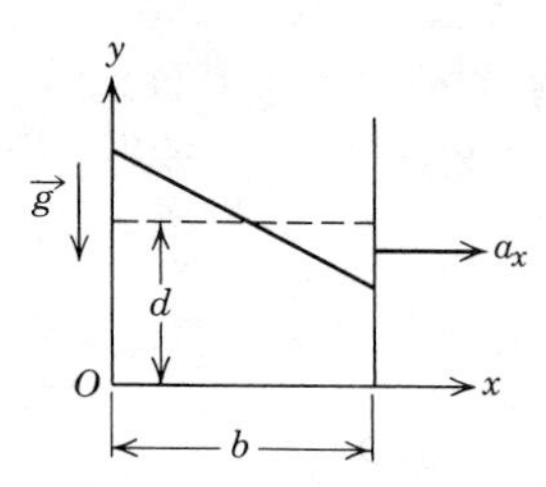

SOLUTION:

Basic equation:

$$-\nabla p + \rho \vec{g} = \rho \vec{a}$$

$$-\left(\hat{i}\frac{\partial p}{\partial x} + \hat{j}\frac{\partial p}{\partial y} + \hat{k}\frac{\partial p}{\partial z}\right) + \rho(\hat{i}g_x + \hat{j}g_y + \hat{k}g_z) = \rho(\hat{i}a_x + \hat{j}a_y + \hat{k}a_z)$$

Since p is not a function of z, $\partial p/\partial z = 0$. Also, $g_x = 0$, $g_y = -g$, $g_z = 0$ and $a_y = a_z = 0$.

$$\therefore -\hat{i}\frac{\partial p}{\partial x} - \hat{j}\frac{\partial p}{\partial y} - \hat{j}\rho g = \hat{i}\rho a_x$$

The component equations are:

$$\frac{\partial p}{\partial x} = -\rho a_x$$

$$\frac{\partial p}{\partial y} = -\rho g$$

{Recall that a partial derivative means that all other independent variables are held constant in the differentiation.

The problem now is to find an expression for $p = p(x, y)$. This would enable us to find the equation of the free surface. But perhaps we do not have to do that.

Since the pressure, $p = p(x, y)$, the difference in pressure between two points (x, y) and $(x + dx, y + dy)$ is

$$dp = \frac{\partial p}{\partial x}\,dx + \frac{\partial p}{\partial y}\,dy$$

Since the free surface is a line of constant pressure, then along the free surface, $p =$ constant, so $dp = 0$ and

$$0 = \frac{\partial p}{\partial x}\,dx + \frac{\partial p}{\partial y}\,dy = -\rho a_x\,dx - \rho g\,dy$$

Therefore,

$$\left.\frac{dy}{dx}\right)_{\text{free surface}} = -\frac{a_x}{g}$$

{The free surface is a straight line.}

In the diagram below,

d = original depth

e = height above the original depth

b = tank length parallel to direction of motion

$$e = \frac{b}{2}\tan\theta = \frac{b}{2}\left(-\frac{dy}{dx}\right)_{\text{free surface}} = \frac{b}{2}\frac{a_x}{g} \qquad \left\{\text{Only valid for } d \geq \frac{b}{2}\right\}$$

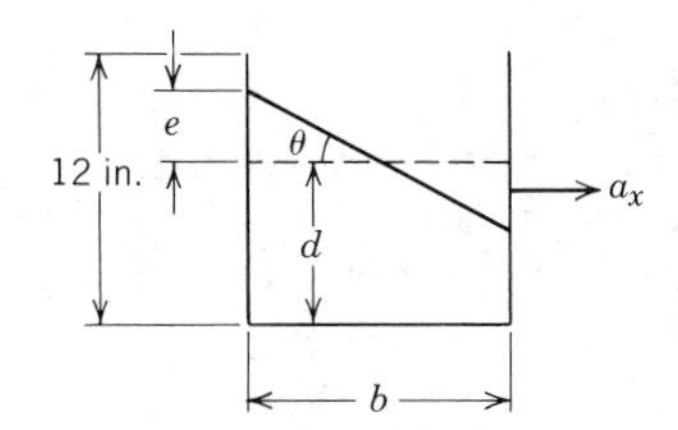

Since we want e to be smallest for a given a_x, the tank should be aligned with b as small as possible. We should align the tank with the long side perpendicular to the direction of motion, that is, choose $b = 12$ in.

With $b = 12$ in.

$$e = 6\frac{a_x}{g}\text{ in.}$$

The maximum allowable value of $e = 12 - d$ in. Thus

$$12 - d = 6\frac{a_x}{g} \qquad \text{and} \qquad d_{\max} = 12 - 6\frac{a_x}{g}$$

If the maximum a_x is assumed to be $\frac{2}{3}g$, then allowable $d = 8$ in.

To allow a margin of safety, perhaps we should select $d = 6$ in.

Recall that a steady acceleration was assumed in this problem. The car would have to be driven *very* carefully.

This problem has been included to demonstrate:
(i) not all problems are clearly defined, nor do they have unique answers, and
(ii) the application of the equation, $-\nabla p + \rho\vec{g} = \rho\vec{a}$.

Example 3.9

A cylindrical container, partially filled with liquid, is rotated at a constant angular velocity, ω, about its axis as shown in the diagram. After a short time there is no relative motion; the liquid rotates with the cylinder as if the system were a rigid body. Determine the shape of the free surface.

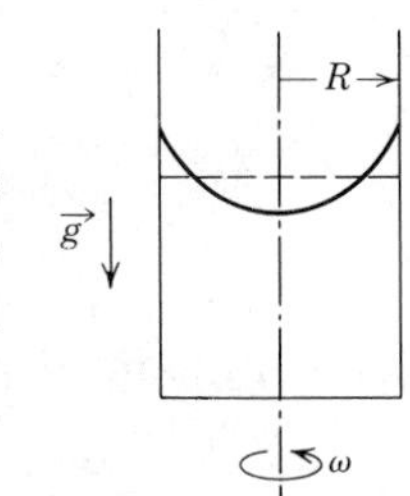

EXAMPLE PROBLEM 3.9

GIVEN: A cylinder of liquid in solid-body rotation with angular velocity, ω, about its axis.

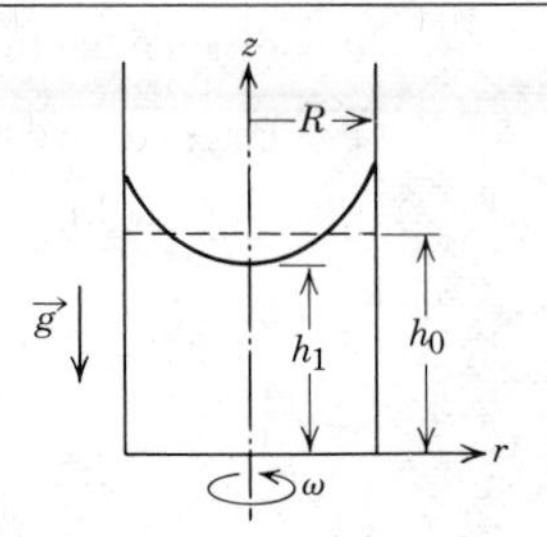

FIND: The shape of the free surface.

SOLUTION:
It is convenient to use a cylindrical coordinate system, r, θ, z. Since there is circumferential symmetry in this problem, the pressure will not be a function of θ. Then $p = p(r, z)$. The free surface is a surface of constant pressure; the problem is to find the equation of the surface.

Since $p = p(r, z)$, the differential change, dp, in pressure between two points with coordinates (r, θ, z) and $(r + dr, \theta, z + dz)$ is given by

$$dp = \left.\frac{\partial p}{\partial r}\right)_z dr + \left.\frac{\partial p}{\partial z}\right)_r dz$$

Consequently, we see that we need to obtain expressions for $\partial p/\partial z)_r$ and $\partial p/\partial r)_z$. This we can do by writing Newton's second law in the z and r directions, respectively, for an infinitesimal fluid element.

From Eq. 3.17 we have, for the z direction

$$-\left.\frac{\partial p}{\partial z}\right)_r + \rho g_z = \rho a_z$$

Since $g_z = -g$ and $a_z = 0$, then $\partial p/\partial z)_r = -\rho g$.

To obtain an expression for $\partial p/\partial r)_z$, we apply Newton's second law in the r direction to a suitable differential element.

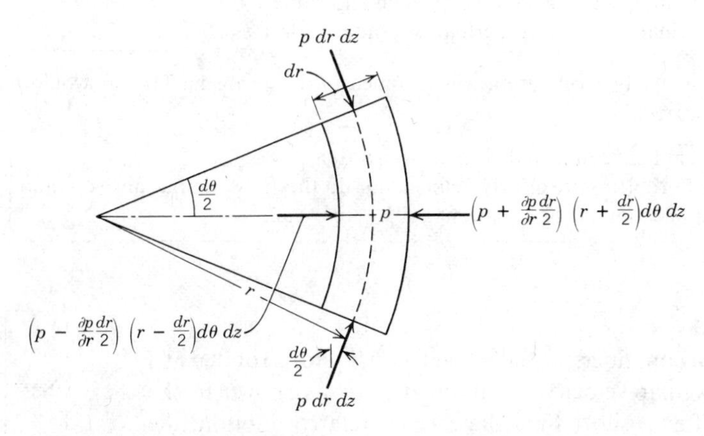

The pressure at the center of the element is p. Using a Taylor series expansion, we express forces acting in the $r\theta$ plane on the element as shown in the diagram. Writing Newton's second law in the r direction, we have

$$\sum dF_r = a_r\, dm = a_r \rho\, d\forall = -\omega^2 r \rho\, d\forall = -\omega^2 r \rho r\, d\theta\, dr\, dz$$

From the figure

$$\sum dF_r = \left(p - \frac{\partial p}{\partial r}\frac{dr}{2}\right)\left(r - \frac{dr}{2}\right) d\theta\, dz - \left(p + \frac{\partial p}{\partial r}\frac{dr}{2}\right)\left(r + \frac{dr}{2}\right) d\theta\, dz + 2p\, dr\, dz \sin\frac{d\theta}{2}$$

Expanding and canceling like terms, recognizing $\sin d\theta/2 = d\theta/2$ (small angles) gives

$$\sum dF_r = d\theta\, dz\left\{\cancel{pr} - \cancel{p\frac{dr}{2}} - r\frac{\partial p}{\partial r}\frac{dr}{2} + \cancel{\frac{\partial p}{\partial r}\left(\frac{dr}{2}\right)^2} - \cancel{pr} - \cancel{p\frac{dr}{2}} - r\frac{\partial p}{\partial r}\frac{dr}{2} - \cancel{\frac{\partial p}{\partial r}\left(\frac{dr}{2}\right)^2} + \cancel{p\, dr}\right\}$$

$$\sum dF_r = d\theta\, dz\left\{-r\frac{\partial p}{\partial r}\, dr\right\}$$

Then

$$-r\frac{\partial p}{\partial r}\, dr\, d\theta\, dz = -\omega^2 r\rho r\, d\theta\, dr\, dz$$

Dividing both sides by $-r\, dr\, d\theta\, dz$ results in

$$\frac{\partial p}{\partial r} = \rho\omega^2 r$$

Since

$$dp = \left.\frac{\partial p}{\partial r}\right)_z dr + \left.\frac{\partial p}{\partial z}\right)_r dz$$

Then

$$dp = \rho\omega^2 r\, dr - \rho g\, dz$$

To obtain the pressure difference between a reference point (r_1, z_1), where the pressure is p_1, and the arbitrary point (r, z), where the pressure is p, we must integrate

$$\int_{p_1}^{p} dp = \int_{r_1}^{r} \rho\omega^2 r\, dr - \int_{z_1}^{z} \rho g\, dz$$

$$p - p_1 = \frac{\rho\omega^2}{2}(r^2 - r_1^2) - \rho g(z - z_1)$$

Taking the reference point on the cylinder axis at the free surface gives

$$p_1 = p_{\text{atm}} \qquad r_1 = 0 \qquad z_1 = h_1$$

Then

$$p - p_{\text{atm}} = \frac{\rho\omega^2 r^2}{2} - \rho g(z - h_1)$$

Since the free surface is a surface of constant pressure ($p = p_{\text{atm}}$), the equation of the free surface is given by

$$0 = \frac{\rho\omega^2 r^2}{2} - \rho g(z - h_1)$$

or

$$z = h_1 + \frac{(\omega r)^2}{2g}$$

The equation of the free surface is a parabola with vertex on the axis at $z = h_1$.

We can solve for the height h_1 under conditions of rotation in terms of the original surface height, h_0, in the absence of rotation. To do this, we use the fact that the volume of fluid must remain constant. With no rotation

$$\forall = \pi R^2 h_0$$

With rotation

$$\forall = \int_0^R \int_0^z 2\pi r\, dz\, dr = \int_0^R 2\pi z r\, dr = \int_0^R 2\pi\left(h_1 + \frac{\omega^2 r^2}{2g}\right) r\, dr$$

$$\forall = 2\pi\left[h_1 \frac{r^2}{2} + \frac{\omega^2 r^4}{8g}\right]_0^R = \pi\left[h_1 R^2 + \frac{\omega^2 R^4}{4g}\right]$$

Then

$$\pi R^2 h_0 = \pi\left[h_1 R^2 + \frac{\omega^2 R^4}{4g}\right]$$

and

$$h_1 = h_0 - \frac{(\omega R)^2}{4g}$$

Finally,

$$z = h_0 - \frac{(\omega R)^2}{4g} + \frac{(\omega r)^2}{2g}$$

$$z = h_0 - \frac{(\omega R)^2}{2g}\left[\frac{1}{2} - \left(\frac{r}{R}\right)^2\right] \qquad z(r)$$

{This problem illustrates the application of Newton's second law to a differential element and the physical behavior of a liquid with a free surface undergoing solid-body rotation.}

3-8 SUMMARY OBJECTIVES

After completing study of Chapter 3, you should be able to do the following:

1. Write the basic equation of fluid statics in vector form and indicate the physical significance of each term.
2. Write the basic pressure-height relation for a static fluid and integrate it to determine the pressure variation for any given fluid property variations.
3. Define temperature and pressure conditions for the standard atmosphere.
4. State the relation between absolute and gage pressures.
5. For a plane submerged surface:
   (a) Determine the resultant force due to the fluid acting on the surface and its line of action.
   (b) Determine the external force(s) required to maintain the surface in equilibrium.

6. For a submerged surface with curvature in one plane:
   (a) Determine the components of the resultant force due to the fluid acting on the surface and their lines of action.
   (b) Determine the external force(s) required to maintain the surface in equilibrium.

** 7. Determine the buoyancy force on a body immersed in or floating on the surface of a liquid; determine the stability of a floating object.

** 8. Apply the basic hydrostatic equation to determine the pressure field and/or free surface shape for any body of fluid in rigid-body motion.

9. Solve the problems at the end of the chapter that relate to the material you have studied.

PROBLEMS

3.1 A hydraulic press for stamping sheet metal automotive parts can exert a force of 36 MN(meganewtons). The actuation stroke is 0.2 m. The press is hydraulically actuated by a system whose design pressure is 30 MPa. Neglecting friction, determine the minimum piston area needed to produce the clamping force. How much hydraulic oil must be supplied per press cycle?

3.2 A pneumatic hoist is to be designed for a service station. Shop air is available at a gage pressure of 600 kPa. The hoist must be capable of lifting automobiles up to 3000 kg. Friction in the piston-cylinder mechanism and seals causes a force of 980 N opposing the piston motion. Determine the piston diameter necessary to provide the lift force. What pressure should be maintained in the lift cylinder to lower smoothly a VW Rabbit with a mass of 895 kg?

3.3 Pipe for the Alaskan pipeline has an internal diameter of 1.22 m. Wall thicknesses of 11 and 14 mm are used. The pipe was tested hydrostatically to a pressure of 10 MPa; the maximum pressure in service is expected to be 7.79 MPa. Calculate the maximum tensile stress in the pipe wall that is to be expected in service. Will the direction of the maximum stress in the pipe wall be axial or circumferential?

3.4 Compressed nitrogen is shipped in a cylindrical tank of diameter, $D = 0.25$ m, and length $L = 1.3$ m. The gas in the tank is at an absolute pressure of 20 MPa and 20 C. Calculate the mass of gas in the tank. If the maximum allowable stress in the tank wall is 210 MPa, determine the theoretical minimum thickness of the cylinder wall.

3.5 A CO_2 cartridge for an air rifle is 60 mm long and has 16.5 mm inside diameter. The wall thickness is 0.5 mm. The label states it contains 12 g of CO_2. Estimate the maximum pressure inside a fully charged cartridge. Assuming a biaxial stress state in the cylinder wall, compute the maximum axial, circumferential, and shear stresses in the cylinder wall.

3.6 On integrating the pressure-height equation for a static incompressible fluid, it was assumed that the gravitational acceleration, g, was constant. The Law of Gravitational Attraction is

$$g = g_0\left(\frac{R}{R + h}\right)^2$$

where R is the radius of the Earth and h is altitude above the surface. Find the percent variation in g for the following two cases (take $R = 4000$ miles):
(a) $h = 6$ mile altitude
(b) $h = -4$ mile altitude

** These objectives apply to sections that may be omitted without loss of continuity in the text material.

3.7 A mercury barometer is read at the same location on different days. Each time, the pressure reading is 29.5 in. of mercury, but the ambient temperatures are 70 and 95 F on the two days. Determine the actual atmospheric pressures on the two days, in lbf/ft^2, and the difference between them, in psi. (Use data from Appendix A.)

3.8 The vapor pressure of mercury is $p_v = 2.5 \times 10^{-5}$ psia at 70 F. Calculate the error in barometer height due to neglecting the vapor pressure of mercury. Would this be detectable for engineering calculations?

3.9 A closed container contains water at a 5 m depth. The absolute pressure above the water surface is 0.3 atm. Calculate the absolute pressure on the inside of the bottom surface of the container.

3.10 Determine the gage pressure in psig at point *a*, if liquid *A* has a specific gravity of 0.75 and liquid *B* has a specific gravity of 1.20. The liquid surrounding point *a* is water, and the tank on the left is open to the atmosphere.

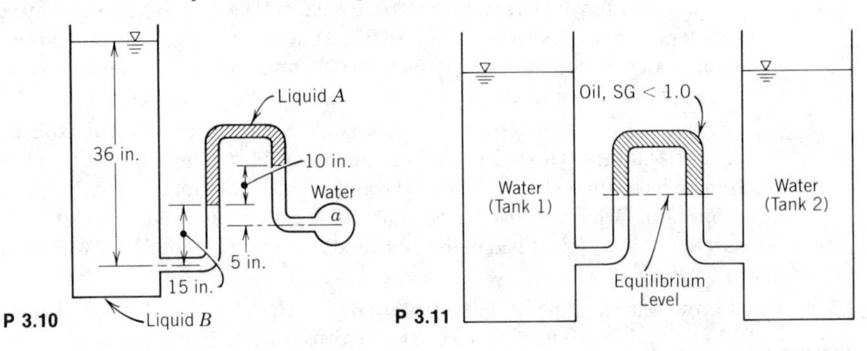

P 3.10

P 3.11

3.11 The NIH Corporation's engineering department is evaluating a sophisticated \$80,000 laser system to measure the difference in water level between two large water storage tanks. It is important that small differences be measured accurately. You suggest that the job can be done with a \$200 manometer arrangement. An oil less dense than water can be used to give a 10:1 amplification of meniscus movement; a small difference in level between the tanks will cause 10 times as much deflection in the oil levels in the manometer. Determine the specific gravity of the oil required for 10:1 amplification.

3.12 A rectangular tank, open to the atmosphere, is filled with water to a depth of 2.5 m as shown. A U-tube manometer is connected to the tank at a location 0.7 m above the tank bottom. If the zero level of the manometer fluid, Meriam blue (specific gravity 1.75), is 0.2 m below the connection, determine the deflection l after the manometer is connected and all the air has been removed from the connecting leg.

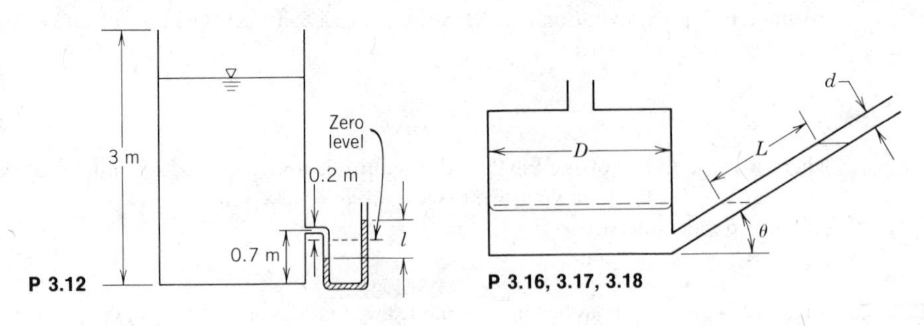

P 3.12

P 3.16, 3.17, 3.18

3.13 If the tank of Problem 3.12 is sealed tightly and water drains slowly from the bottom of the tank, determine the deflection, l, after the system has attained equilibrium.

3.14 The manometer fluid of Problem 3.12 is replaced with mercury (same zero level). The tank is sealed and the air pressure is increased to a gage pressure of 0.5 atm. Determine the deflection, l.

3.15 A reservoir manometer is calibrated for use with a fluid of specific gravity 0.827. The reservoir diameter is $\frac{5}{8}$ in. and the (vertical) tube diameter is $\frac{3}{16}$ in. Calculate the required distance between marks on the vertical scale for 1 in. of water pressure difference.

3.16 The inclined manometer shown has reservoir diameter, D, of 90 mm and measuring tube diameter, d, of 6 mm; the manometer fluid is Meriam red oil. The length of the measuring tube is 0.6 m; $\theta = 30°$. Determine the maximum pressure, in Pa, that can be measured with the manometer.

3.17 The inclined manometer shown has reservoir diameter, D, of 3 in., and measuring tube diameter, d, of 0.25 in., and is filled with gage oil (SG = 0.897). Compute the angle, θ, that will give a 5 in. oil deflection along the inclined tube for an applied pressure of 1 in. of water (gage).

3.18 The inclined manometer shown has reservoir diameter, D, of 96 mm and measuring tube diameter, d, of 8 mm. Determine the angle, θ, required to provide a 5:1 increase in liquid deflection, L, compared to the total deflection in a regular U-tube manometer.

3.19 Surface tension causes the meniscus to rise in a water-filled manometer. This phenomenon is termed *capillary rise* and it becomes significant in tubes of small diameter. Develop an expression for the capillary rise of water in a tube. Show that the result is

$$\Delta h = \frac{2\sigma \cos\theta}{\gamma R}$$

where R is the tube radius and θ is the contact angle. Evaluate and plot the results for water in tubes from 1 to 10 mm inside diameter.

3.20 Solve Problem 3.19 for the capillary depression of mercury in a tube. Evaluate and plot for tubes between 1 and 10 mm inside diameter.

3.21 A model has been proposed that accounts for specific weight variations in the Earth's atmosphere according to the equation:

$$\gamma = \gamma_0 - kz$$

where $k = 1.93 \times 10^{-6}$ lbf/ft^4, z = altitude above Earth's surface, and γ_0 = specific weight of air at sea level (0.0765 lbf/ft^3). Assuming a sea level pressure of 14.7 psia, compute the pressure at an altitude of 20,000 ft. Compare with the Standard Atmosphere.

3.22 If the variation in specific weight of atmospheric air between sea level and an altitude of 3600 ft were given by $\gamma = \gamma_0 - k\sqrt{z}$, where γ_0 is the specific weight of air at sea level, z is the altitude above sea level, and $k = 0.00020$ lbf/ft$^3 \cdot$ ft$^{1/2}$, determine the pressure in psia at an altitude of 3600 ft when sea level conditions are 14.7 psia, 59 F.

3.23 As a result of changes in temperature, salinity, and pressure, the density of seawater increases with increasing depth in accordance with the equation

$$\rho = \rho_s + bh$$

where ρ_s is the density at the surface, h is depth below the surface, and b is a positive constant. Develop an algebraic equation for the pressure as a function of depth.

3.24 An inverted cylindrical container is lowered slowly beneath the surface of a pool of water. Air trapped in the container is compressed isothermally as the hydrostatic pressure increases. Develop an expression for the water height inside the container, y, in terms of the container height, H, and depth of submersion, h.

3.25 Water is usually assumed to be an incompressible fluid when evaluating static pressure variations. Actually, it is about 100 times more compressible than steel. The bulk modulus is defined as $E_v = dp/(d\rho/\rho)$ and it may be assumed constant (see Appendix A for values). Compute the percent change in density for water raised to a gage pressure of 100 atm if the original density at atmospheric pressure is 999 kg/m^3.

3.26 Water is usually assumed to be incompressible when evaluating static pressure variations. Actually, its compressibility can be important in the design of submersible vehicles. Assume that the bulk modulus of water is constant. Compute the pressure and water density at a depth of 4 miles in seawater. The density at the seawater surface is 64 lbm/ft^3.

3.27 Oceanographic research vessels have descended to depths approaching 10 km below sea level. At these extreme depths, the compressibility of seawater can be significant. One may model the behavior of seawater by assuming that its bulk compressibility modulus remains constant. Using this assumption, evaluate the deviations in density and pressure compared to values computed using the incompressible assumption at a depth of 10 km in seawater. Express your answers in percent.

3.28 High-speed jets of water are used for industrial processes to cut concrete and composite materials, e.g. for aircraft components. The maximum pressures used are in the vicinity of 50,000 psi. Would you expect the assumption of constant density to be reasonable for engineering calculations?

3.29 Lubricating oil is used as the working fluid in a high-pressure hydraulic system. Estimate the percentage change in the density of the oil as its pressure is raised from ambient conditions to 300 atm (gage). Is constant density a reasonable model for the oil?

3.30 Power plants designed for Ocean Thermal Energy Conversion (OTEC) draw cool seawater at 5 C from depths as great as 1000 m below sea level. The hydrostatic pressure at this depth is about 10 MPa. Estimate the increase in density at this depth caused by pressure, compared to the sea surface value (assume that the bulk modulus for seawater is constant). Estimate the density change due to temperature compared to the sea surface where the temperature is 25 C.

3.31 Evaluate dp/dz for standard air on a windless day at sea level and at 5 km altitude. (The coordinate z is measured positive upward.)

3.32 Determine the change of elevation necessary to effect a 15 percent reduction in density for an isothermal atmosphere at 20 C.

3.33 At ground level in Denver, Colorado, the atmospheric pressure and temperature are 83.2 kPa and 25 C. Calculate the pressure on Pike's Peak at an elevation of 2690 m above the city assuming (a) an incompressible, and (b) an adiabatic atmosphere.

3.34 If air is assumed to be an ideal gas, knowledge of the temperature variation with altitude allows the determination of pressure at any elevation when conditions are known at a reference elevation, z_0.

(a) For $T = T_0(1 + mz)$, derive the equation for the variation of pressure as a function of altitude if the pressure at the reference elevation is p_0.

(b) Using the results of part (a), show that the variation of pressure for the isothermal case ($m \rightarrow 0$) is given by

$$\frac{p}{p_0} = e^{-(g/RT_0)(z - z_0)}$$

3.35 Because the pressure falls, water boils at a lower temperature with increasing altitude. Consequently, cake mixes and boiled eggs, among other foods, must be cooked different lengths of time. Determine the boiling temperature of water at 1000 and 2000 m elevation on a standard day, and compare with the sea level value.

3.36 Automobiles suffer a power loss with altitude as a result of the decrease in air density. If the volumetric efficiency of an engine remains constant, and the carburetion is adjusted to maintain the same air-fuel ratio, determine the percentage loss in power for an engine at 3000 m elevation, compared to sea level on a standard day.

3.37 Pressure variations that result from altitude changes can cause ear "popping" and discomfort to airplane passengers or those driving in the mountains. Each individual is affected differently, but one ear "pop" per 75 m of elevation change might be a reasonable average figure. Determine the pressure change, expressed in millimeters of water, that corresponds to this elevation difference on a standard day at an altitude of 2000 m.

3.38 A device known as a deadweight tester can be used as a standard for calibration of mechanical pressure gages (the useful range is about 30 kPa to 35 MPa). Known pressures are generated by loading weights on a vertical piston-cylinder arrangement. The weighted piston is rotated to minimize frictional effects. The maximum convenient load is 100 kg. Determine an appropriate piston size to cover the pressure range given.

3.39 Many recreation facilities are currently being built using inflatable "bubble" structures. A tennis bubble to enclose four courts is shaped roughly like a circular semicylinder with a diameter of 30 m and a length of 60 m. The blowers used to inflate the structure can maintain the air pressure inside the bubble at 10 mm of water above ambient pressure. The fabric "skin" of the bubble is of uniform thickness. Determine the maximum material density, in mass per unit area, that can be used to fabricate a pressure-supported bubble.

3.40 One of the major tire companies has put into operation a tire-curing press 6.0 m in diameter. A hemispherical dome is used to contain the mold and tire while steam is introduced under pressure. The dome mass is 130 metric tons. Determine the steam pressure (gage) at which the mass of the dome is exactly balanced. What is the corresponding saturation temperature?

3.41 A mechanical pressure gage attached to the closed reservoir tank of an air compressor indicates a pressure of 827 kPa on a day when the barometer height is 750 mm of mercury. Calculate the absolute pressure in the tank. What pressure would the gage indicate if the barometer reading changed to 775 mm of mercury?

3.42 A door 1 m wide and 1.5 m high is located in a plane vertical wall of a water tank. The door is hinged along its upper edge, which is 1 m below the water surface. Atmospheric pressure acts on the outer surface of the door and at the water surface. Determine the total resultant force due to all fluids acting on the door.

3.43 If, in Problem 3.42, the water surface gage pressure is raised to 0.3 atm, determine the total resultant force from all fluids acting on the door.

3.44 A 1 ft cube is submerged as shown. Calculate the actual force of the water on the bottom surface, and the net vertical force on the cube.

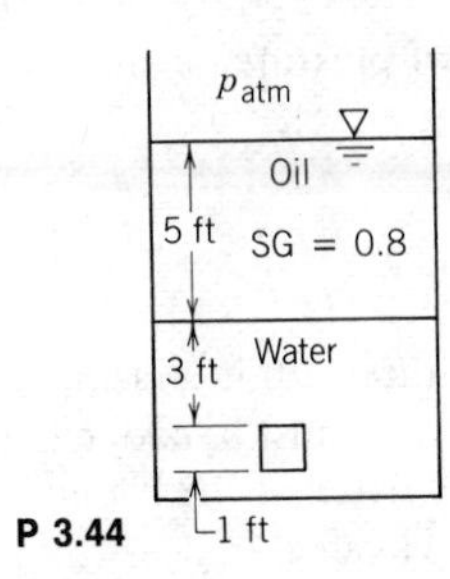

P 3.44

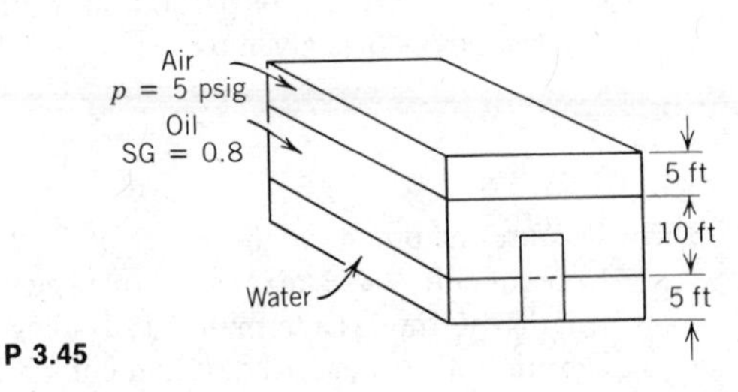

P 3.45

3.45 The door shown is 5 ft wide and 10 ft high. Find the resultant force from all fluids acting on the door.

3.46 Most large dams contain flood gates that can be raised to release stored water. The gate shown slides against a plate on each side. The gate mass is 5000 kg.
(a) Find the normal force on the gate due to the water.
(b) If μ_s (coefficient of static friction) = 0.4 between the gate and the supports, determine the magnitude of the force, R, required to start the gate in motion.

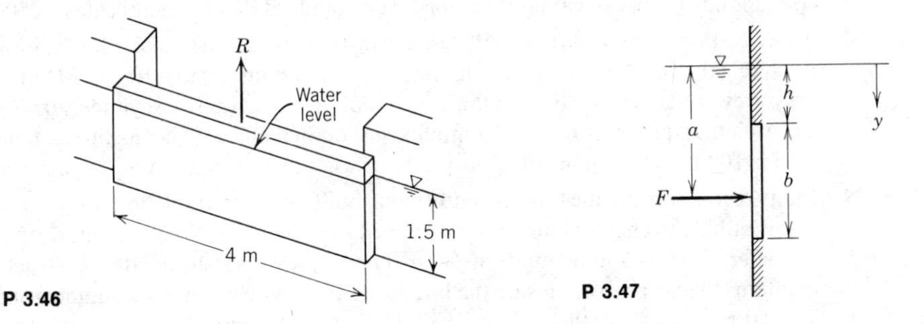

P 3.46

P 3.47

3.47 A vertical plane surface is submerged in liquid as shown. The width of the surface is w. The density of the incompressible liquid is ρ. Find:
(a) A general expression for the resultant force, F
(b) A general expression for the vertical distance, a

3.48 An open tank with a rectangular vertical side 2 ft wide and 6 ft high is filled with a liquid of variable specific weight, γ, with $\gamma = 50 + 2y$ (lbf/ft^3), where y is measured vertically downward from the free surface. Find the magnitude of the force on the side of the tank.

3.49 A door 1 m wide and 1.5 m high is located in a plane vertical wall of a water tank. The door is hinged along its upper edge, which is 1 m below the water surface. Atmospheric pressure acts on the outer surface of the door. If the pressure at the water surface is atmospheric, what force must be applied at the lower edge of the door in order to keep the door from opening?

3.50 If, in Problem 3.49, the gage pressure at the water surface is 0.5 atm, what force must be applied at the lower edge of the door in order to keep the door from opening?

3.51 An aquarium at Marineland has a window located as shown. The resultant force from seawater ($\gamma = 64$ lbf/ft^3) on the window is 1280 lbf. Determine the line of application of the resultant force, in feet below the top of the window.

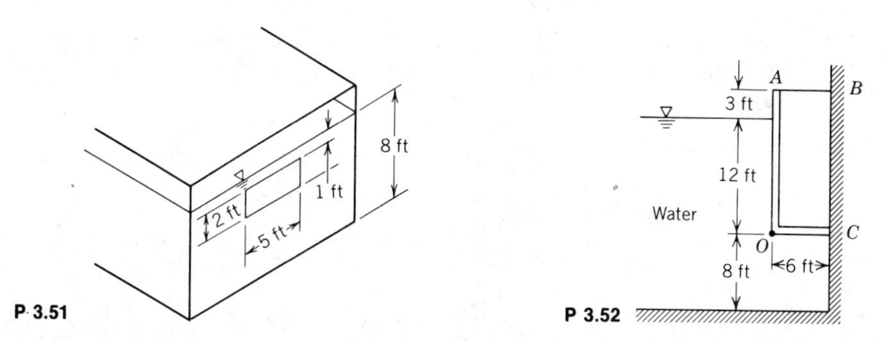

P 3.51

P 3.52

3.52 The gate AOC shown is 6 ft wide and is hinged at point O. Neglecting the weight of the gate, determine the force in the bar AB.

3.53 As water rises on the left side of the rectangular gate, the gate will open automatically. At what depth above the hinge will this occur? Neglect the mass of the gate.

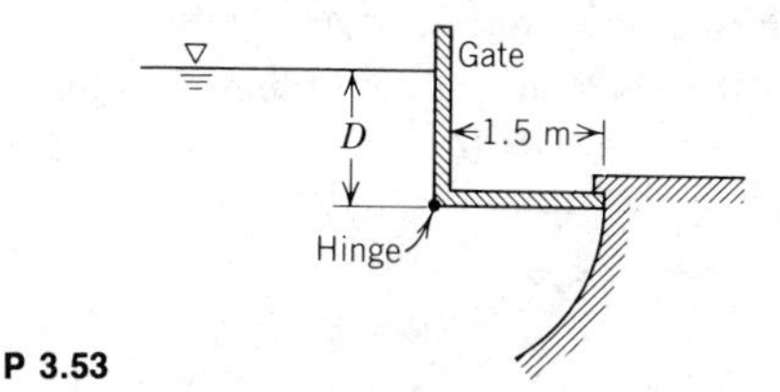

P 3.53

3.54 A tank with a center partition has a small "door" 0.5 m wide by 1 m high at the bottom. This door is hinged along the top edge. The left side has 0.6 m of water while the right side contains 1 m of nitric acid (SG 1.5). What force (magnitude and direction) is required at the lower edge of the door to hold it closed?

3.55 The circular access port in the side of a water standpipe has a diameter of 0.6 m and is held in place by eight bolts evenly spaced around the circumference. If the standpipe has a diameter of 7 m and the center of the port is located 12 m below the free surface of the water, determine (a) the total force on the port and (b) the appropriate bolt diameter.

3.56 The gate shown is hinged at H. The gate is 2 m wide normal to the plane of the diagram. Calculate the force required at A to hold the gate closed.

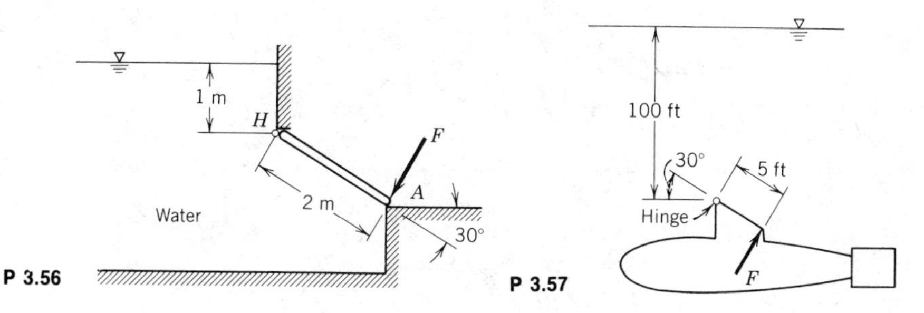

P 3.56

P 3.57

3.57 A submarine is 100 ft below the surface as shown. Find the net force, F, required to open the circular hatch when applied as shown. The pressure inside the submarine is equal to atmospheric pressure.

3.58 The gate shown is 3 m wide and for the purposes of analysis can be considered massless. For what depth of water will this rectangular gate be in equilibrium at an angle of 60° as shown?

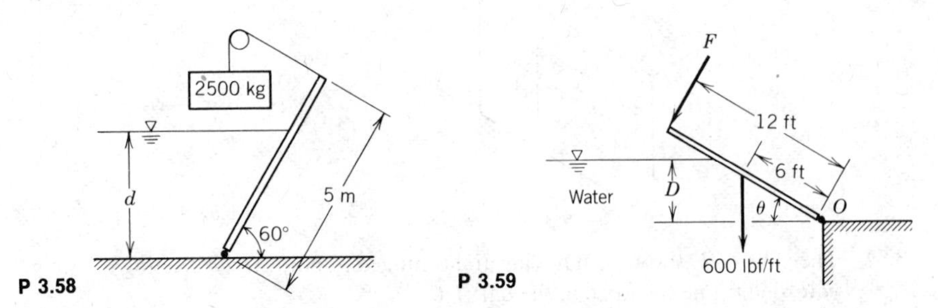

P 3.58 **P 3.59**

3.59 A plane gate is held in equilibrium by the uniformly distributed force per unit width, F, as shown. The gate weighs 600 lbf/ft of gate width and its center of gravity is 6 ft from the hinge at O. Find the unknown force per unit width, F, when $D = 5$ ft and $\theta = 30°$.

3.60 A gate of mass 2000 kg is mounted on a frictionless hinge along the lower edge. The length of the reservoir and gate (perpendicular to the plane of view) is 8 m. For the equilibrium conditions shown, compute the width, b, of the gate.

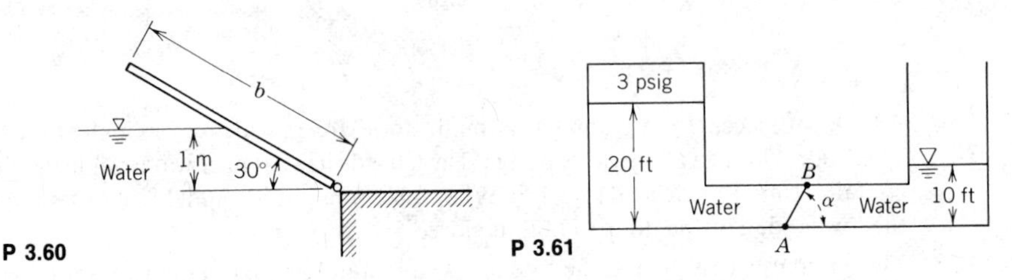

P 3.60 **P 3.61**

3.61 Gate AB is 3 ft wide and 2 ft long. It is inclined at $\alpha = 60°$ when closed. Find the moment about hinge A exerted by the water.

3.62 The rectangular gate AB, as shown, is 2 m wide. Find the force per unit width exerted against the stop at A. Assume that the gate mass is negligible.

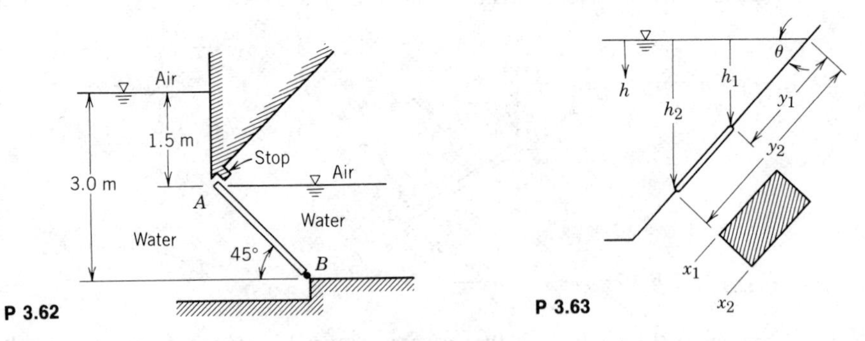

P 3.62 **P 3.63**

3.63 Due to temperature gradients in a liquid in a large tank, the density is not constant. The density variation is given by

$$\rho = \rho_0(1 + h^{1/5})$$

Find:

(a) The resultant force on the area indicated due only to the liquid

(b) The position y' at which the resultant force acts

3.64 The parabolic gate is 2 m wide. Determine the magnitude and line of action of the horizontal force on the gate due to the water; $c = 0.25\ \text{m}^{-1}$.

3.65 For the conditions of Problem 3.64, determine the magnitude and line of action of the vertical force on the gate due to the water.

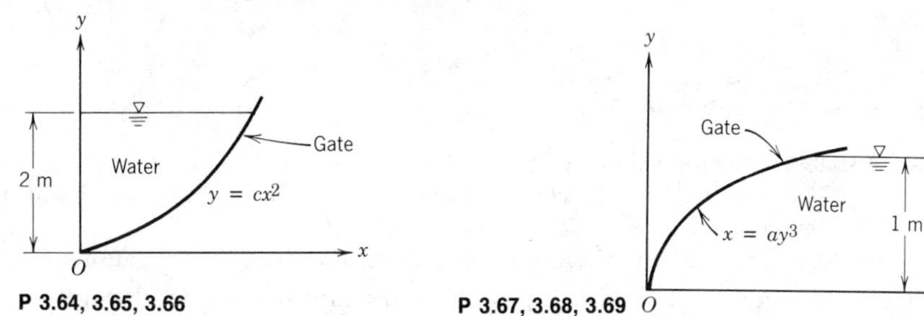

P 3.64, 3.65, 3.66 **P 3.67, 3.68, 3.69**

3.66 The depth of water to the right of the gate of Problem 3.64 is increased from zero to L m. Determine the depth, L, required to reduce the moment about O to 50 percent of the value for $L = 0$.

3.67 The gate shown is 1.5 m wide. Determine the magnitude and moment of the vertical component of the force about O. The liquid is water; $a = 1.0\ \text{m}^{-2}$.

3.68 For the conditions of Problem 3.67, determine the magnitude and line of action of the horizontal component of the force.

3.69 If water stands at a depth of 0.5 m to the left of the gate of Problem 3.67, determine the total moment about O.

3.70 Determine the magnitude and line of action of the vertical force on the curved section AB. The liquid is water and the section AB is 1 ft wide. Atmospheric pressure acts at the free surface; $k = 1.0\ \text{ft}^{-1}$.

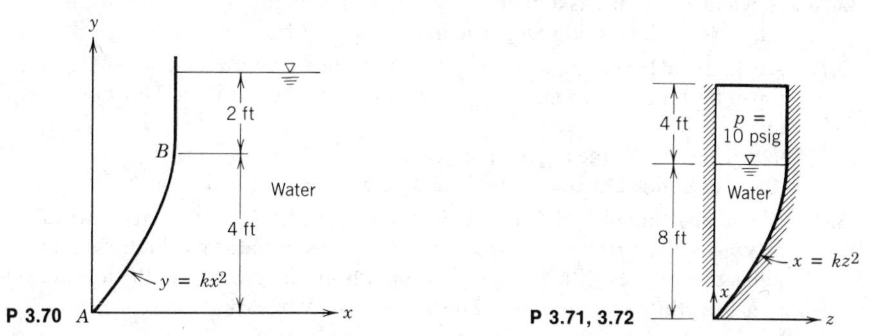

P 3.70 **P 3.71, 3.72**

3.71 The tank shown is 2 ft wide (perpendicular to the xz plane). It is filled with water to a depth of 8 ft. The air between the top of the tank and the water is pressurized to 10 psig. Determine the magnitude and the line of action of the vertical force on the curved portion of the tank; $k = 0.5\ \text{ft}^{-1}$.

3.72 If the water depth in the tank of Problem 3.71 is reduced to 4 ft and the air pressure is maintained at 10 psig, determine the magnitude and line of action of the vertical force on the curved portion of the tank.

3.73 An open tank is filled with water to the depth indicated. Atmospheric pressure acts on all outer surfaces of the tank. Determine the magnitude and line of action of the vertical component of the force of the water on the curved part of the tank bottom.

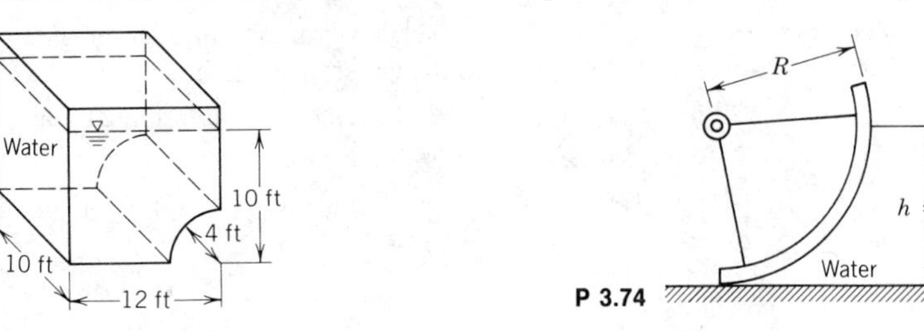

P 3.73 P 3.74

3.74 A spillway gate formed in the shape of a circular arc is w m wide. Find:

(a) The magnitude and direction of the vertical component of the force due to all fluids acting on the gate

(b) A point on the line of action of the total resultant force on the gate due to all fluids

3.75 A gate, which is in the form of a quarter-cylinder, hinged at A and sealed at B, is 2 m wide normal to the paper. The bottom of the gate is 3 m below the water surface. Determine:

(a) The magnitude of the horizontal force

(b) The line of action of the horizontal force

(c) The magnitude of the vertical force

(d) The line of action of the vertical force

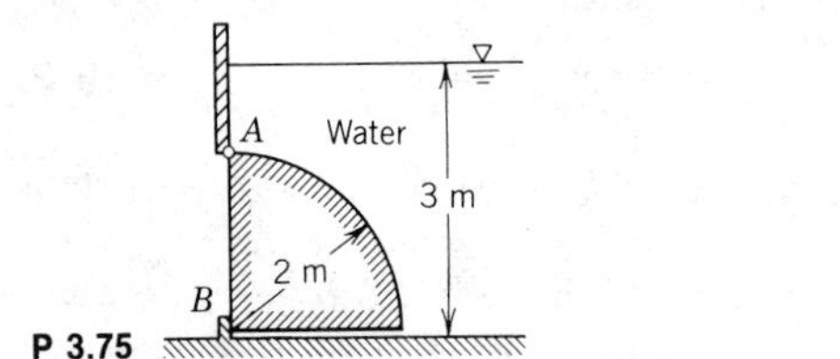

P 3.75

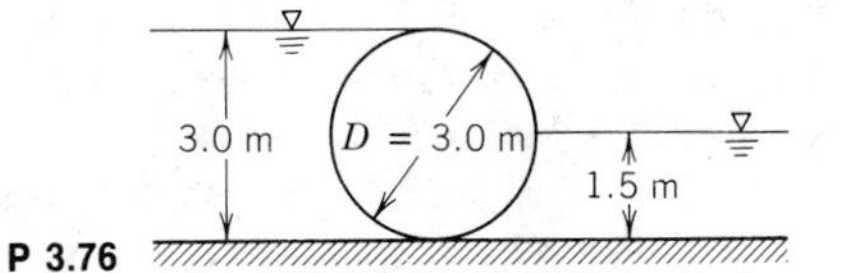

P 3.76

3.76 A cylindrical weir has a diameter of 3 m and a length of 6 m. Find the magnitude and direction of the resultant force acting on the weir from the water.

3.77 A cylindrical log of diameter, D, rests against the top of a dam. The water is level with the top of the log and the center of the log is level with the top of the dam. Obtain expressions for:

(a) The mass of the log per unit length

(b) The contact force per unit length between the log and dam

3.78 The tennis "bubble" of Problem 3.39 is subjected to a wind that blows at a speed of 50 km/hr in a direction perpendicular to the axis of the semicylindrical shape. By using polar coordinates with angle, θ, measured from the ground on the upwind side of the structure, the resulting pressure distribution may be expressed as

$$\frac{p - p_\infty}{\frac{1}{2}\rho V_w^2} = 1 - 4\sin^2\theta$$

where p is the pressure at the surface, p_∞ the atmospheric pressure, and V_w the wind speed. Determine the net vertical force exerted on the structure.

3.79 A hydrometer is a specific gravity indicator, the value being indicated by the level at which the free surface intersects the stem when floating in a liquid. The 1.0 mark is the level when in distilled water. For the unit shown the immersed volume in distilled water is 15 cm^3. The stem is 6 mm in diameter. Find the distance, h, from the 1.0 mark to the surface when the hydrometer is placed in a nitric acid solution of specific gravity 1.5.

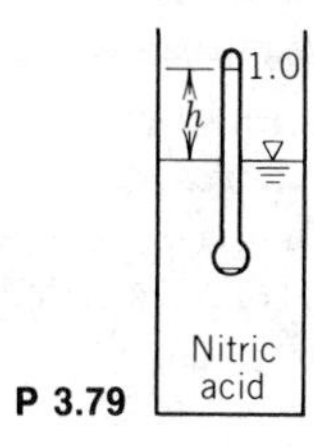

P 3.79

3.80 Find the specific weight of the sphere shown if its volume is 1 ft^3. State all assumptions. Is the weight necessary to float the sphere?

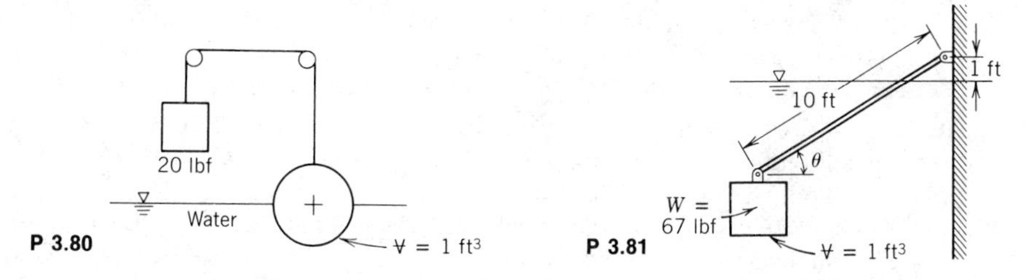

P 3.80 P 3.81

3.81 One cubic foot of material weighing 67 lbf is allowed to sink in the water as shown. A circular wooden rod 10 ft long and 3 in.^{2} in cross section is attached to the weight and also to the wall. If the rod weighs 3 lbf, what will be the angle, θ, for equilibrium?

3.82 The fat-to-muscle ratio of a person may be determined from a specific gravity measurement. The measurement is made by immersing the body in a tank of water and measuring the net weight. Develop an expression for the specific gravity of a person in terms of their weight in air, net weight in water, and SG $= f(T)$ for water.

3.83 Quantify the statement, "Only the tip of an iceberg shows (in seawater)." For ice at 0 C, SG $= 0.92$.

3.84 A manufacturer's catalog of aluminum air tanks for diving claims that the tanks are neutrally buoyant when empty. A tank holds 50 standard cubic feet of air when filled to 3000 psig, is 6.9 in. outside diameter, has a wall thickness of 0.467 in., and is 19 in. long. The specific gravity of aluminum is 2.7. Evaluate the claim.

3.85 A manufacturer's catalog lists a buoyancy compensator, BC (similar to a life vest) for scuba diving. The BC claims a lift up to 40 lbf, obtained from an inflation cartridge that contains 25 g of carbon dioxide. Evaluate the manufacturer's claim if the mass of BC is negligible. To what depth in seawater can the BC produce the lift claimed?

3.86 A modern supertanker has a tank capacity of a half million metric tons of Arabian crude oil with SG $= 0.86$. The ship is essentially rectangular with a length of 400 m and a beam (width) of 65 m. The mass of the ship is approximately 230,000 metric tons. When the ship is unloaded, it is necessary to take on seawater ballast to maintain sufficient draft for stability and to keep the propeller submerged. A minimum draft of 20 m is required. Determine the maximum draft of the fully loaded tanker in seawater. Also determine what fraction of the tanks must be filled with seawater ballast when traveling unloaded.

3.87 The World Almanac lists dimensions of many of the world's larger ships. An oil tanker with a carrying capacity of 400,000 deadweight (long) tons (1 long ton = 2240 lbm) is listed as 1200 ft in length and 206 ft in beam (breadth). Compute the change in draft (depth) that occurs between empty and fully loaded conditions in (a) seawater, and (b) fresh water. Assume that the ship's profile at the water line is rectangular.

3.88 A soda straw is made of plastic with specific gravity, SG = 1.1. The straw is 5 mm inside diameter and the thickness of the plastic is 0.4 mm. Its length is 250 mm. Experiments have shown that when placed in a glass of soft drink (SG = 1.055), the straw remains submerged. Estimate the external force required to support a straw submerged vertically in soft drink to a depth of 100 mm. Assume surface tension, σ, for the soft drink is similar to that of water.

3.89 Scientific balloons operating at pressure equilibrium with the surroundings have been used to lift instrument packages to extremely high altitudes. One such balloon, constructed of polyester with a skin thickness of 0.013 mm (0.5 mil), lifted a payload of 230 kg to an altitude of approximately 49 km. At that altitude, atmospheric conditions are 0.95 mbar and -20 C. The helium gas in the balloon was at a temperature of approximately -10 C. The specific gravity of the skin material is 1.28. Determine the diameter and mass of the balloon. Assume that the balloon is spherical.

3.90 A pressurized helium balloon is to be designed to lift a payload to an altitude of 40 km, where the atmospheric pressure and temperature are 3.0 mbar and -25 C, respectively. The balloon skin is polyester (Mylar) with specific gravity of 1.28 and thickness of 0.015 mm (0.6 mil). To maintain a spherical shape, it has been decided to pressurize the helium in the balloon to a gage pressure of 0.45 mbar. Determine the maximum balloon diameter if the allowable tensile stress in the balloon skin is limited to 62 MN/m^2. What payload can be carried?

3.91 The Goodyear "blimp" used for advertising is a nonrigid airship with a total gas volume of 150,000 ft^3. Estimate the maximum lifting force available if the airship is filled with helium. Assume standard conditions.

3.92 Hot-air ballooning is becoming a popular sport. According to a recent article, "hot-air volumes must be large because air heated to 150 F over ambient lifts only 0.018 lbf/ft^3 compared to 0.066 and 0.071 for helium and hydrogen, respectively." Check these statements for sea-level conditions. Calculate the effect of increasing the hot-air maximum temperature to 250 F above ambient.

3.93 Consider a long object with square cross section (a m on a side) floating horizontally on the surface of a liquid. Assume that the object initially is one-fourth submerged. Considering small angular displacements only, develop an expression for the torque that tends to return the object to a horizontal position. Evaluate the range of positions of the symmetric center of gravity (CG) for which the object will remain stable.

3.94 A cylindrical timber with $D = 0.3$ m and $L = 4$ m is weighted on its lower end so that it floats vertically with 3 m submerged in seawater. When displaced vertically from its equilibrium position, the timber oscillates or "heaves" in a vertical direction upon release. Estimate the frequency of oscillation in this heave mode. Neglect viscous effects and water motion.

3.95 A cubical box, 1 m on a side, half-filled with oil (SG = 0.80), is given a constant horizontal acceleration of 0.2 g. Determine the slope of the free surface and the pressure along the bottom of the box.

3.96 A rectangular container of water undergoes constant acceleration down an incline as shown. Determine the slope of the free surface using the coordinate system shown.

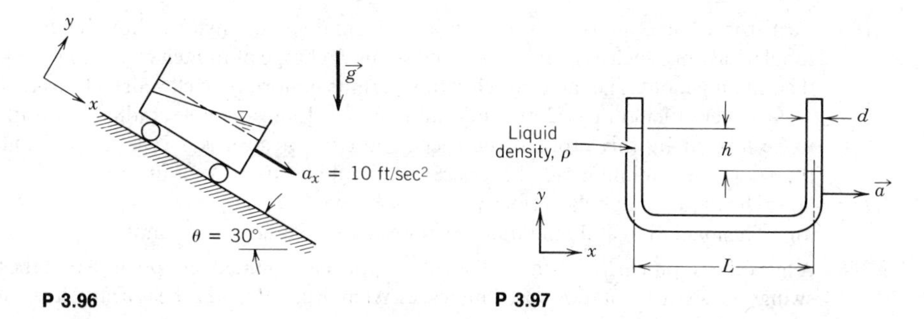

P 3.96

P 3.97

3.97 A crude accelerometer can be made from a liquid-filled U-tube as shown. Derive an expression for the acceleration, $\vec{a}$, in terms of liquid level variation, h, tube geometry, and fluid properties.

3.98 A rectangular container of base dimensions 0.4 m × 0.2 m and height 0.4 m is filled with water to a depth of 0.2 m; the mass of the empty container is 10 kg. The container is placed on a plane inclined at 30° to the horizontal. If the coefficient of sliding friction between the container and the plane is 0.3, determine the angle of the water surface relative to the horizontal.

3.99 If the container of Problem 3.98 slides without friction, determine the angle of the water surface relative to the horizontal. What is the slope of the free surface for the same acceleration up the plane?

3.100 A rectangular container of base dimensions 0.4 m × 0.2 m and height 0.5 m is filled with water to a depth of 0.2 m; the mass of the empty container is 10 kg. The container is placed on a horizontal surface and is subjected to a constant horizontal force of 150 N. If the coefficient of sliding friction between the container and the surface is 0.25 and the tank is aligned with the short dimension along the direction of motion, determine:
(a) The force of the water on each end of the tank
(b) The force of the water on the bottom of the tank

3.101 A pail, 1 ft in diameter and 1 ft deep, weighs 3 lbf, and contains 8 in. of water. The pail is swung in a vertical circle of 3 ft radius at a speed of 15 ft/sec. The water may be assumed to move as a solid body. At the instant when the pail is at the top of its trajectory, compute the tension in the string and the pressure on the bottom of the pail from the water.

3.102 A sealed chamber, which contains manometer oil (SG = 0.8), rotates about its axis with angular velocity, ω. Derive an expression for the radial pressure gradient in the oil, $\partial p/\partial r$, in terms of radius, r, and angular velocity, ω.

3.103 A cylindrical container, similar to that analyzed in Example Problem 3.9, is rotated at constant angular velocity about its axis. The cylinder is 1 ft in diameter, and initially contains water that is 4 in. deep. Determine the maximum rate at which the container can be rotated before the liquid free surface just touches the bottom of the tank. Is your answer dependent on the density of the liquid? Explain.

3.104 An automobile traveling at 90 km/hr rounds a long sweeping curve of radius 250 m. The air conditioner is on and the windows are rolled up so that the air within the car may be considered to move essentially as a solid body. A child in the back seat holds the string of a balloon filled with helium. On a straight road the string was vertical, but in the curve it is not. Determine the magnitude and direction of the string angle as measured from the vertical.

3.105 Cast iron or steel molds are used in a horizontal-spindle machine for the casting of tubular castings such as liners, tubes, and so on. A charge of molten metal is poured into the spinning mold. The radial acceleration permits uniformly thick wall sections to form. A steel liner of length $L = 2$ m, outer radius $r_0 = 0.15$ m, and inner radius $r_i = 0.10$ m is to be formed by this process; the specific gravity of steel is 7.8. To insure uniform thickness, the minimum radial acceleration should be 10 g. Determine:

(a) The required angular velocity

(b) The maximum and minimum pressures on the surface of the mold

3.106 A test tube is spun in a centrifuge. The tube support is mounted on a pivot so that the tube swings outward as rotation speed increases. At high speeds, the tube is nearly horizontal. Find:

(a) An expression for the radial component of acceleration of a liquid element located at radius, r

(b) The radial pressure gradient, $\partial p/\partial r$

(c) The maximum pressure on the bottom of the test tube if it contains water. The free surface and bottom radii are 50 and 130 mm, respectively.

3.107 Gas centrifuges are used in one process to produce enriched uranium for nuclear fuel rods. The maximum peripheral speed of a gas centrifuge is limited by stress considerations to about 300 m/sec. Assume a gas centrifuge containing uranium hexafluoride gas with molecular mass, $M = 352$, and ideal gas behavior. Develop an expression for the ratio of maximum pressure to pressure at the centrifuge axis. Evaluate the pressure ratio for a gas temperature of 325 C.

3.108 A centrifugal micromanometer can be used to create small and accurate differential pressures in air for precise measurement work. The device consists of a pair of parallel disks that rotate to develop a radial pressure difference. There is no flow between the disks. Obtain an expression for pressure difference in terms of rotation speed, radius, and air density. Evaluate the speed of rotation required to develop a differential pressure of 8 μm of water using a device with 50 mm radius.

Chapter 4

BASIC EQUATIONS IN INTEGRAL FORM FOR A CONTROL VOLUME

We shall begin our study of fluids in motion by developing the basic equations in integral form for application to control volumes. Why the control volume formulation rather than the system formulation? There are two basic reasons. First, since fluid media are capable of continuous distortion and deformation, often it is extremely difficult to identify and follow the same mass of fluid at all times (as must be done to apply the system formulation). Second, we often are interested, not in the motion of a given mass of fluid, but rather in the effect of the fluid motion on some device or structure. Thus it is more convenient to apply the basic laws to a fixed volume in space, using a control volume analysis.

The basic laws for a system should be familiar from mechanics and thermodynamics. Our approach to developing the mathematical formulation of these laws for a control volume will be to develop a general formulation that will allow us to convert from a system analysis to a control volume analysis.

4-1 BASIC LAWS FOR A SYSTEM

The basic laws for a system are summarized briefly; for reasons that will become apparent in the next section, each of the basic equations for a system is written as a rate equation.

4-1.1 Conservation of Mass

Since a system is, by definition, an arbitrary collection of matter of fixed identity, a system is composed of the same quantity of matter at all times. The conservation of mass states that the mass, M, of the system is constant. On a rate basis, we have

$$\left.\frac{dM}{dt}\right)_{\text{system}} = 0 \tag{4.1a}$$

where

$$M_{\text{system}} = \int_{\text{mass (system)}} dm = \int_{\forall\text{(system)}} \rho \, d\forall \tag{4.1b}$$

4-1.2 Newton's Second Law

Newton's second law states that for a system moving relative to an inertial reference frame, the sum of all external forces acting on the system is equal to the time rate of change of linear momentum of the system,

$$\vec{F} = \left.\frac{d\vec{P}}{dt}\right)_{\text{system}} \tag{4.2a}$$

where the linear momentum, $\vec{P}$, of the system is given by

$$\vec{P}_{\text{system}} = \int_{\text{mass (system)}} \vec{V} \, dm = \int_{\forall\text{(system)}} \vec{V}\rho \, d\forall \tag{4.2b}$$

**4-1.3 Moment of Momentum

The moment of momentum equation for a system states that the rate of change of angular momentum is equal to the sum of all torques acting on the system,

$$\vec{T} = \left.\frac{d\vec{H}}{dt}\right)_{\text{system}} \tag{4.3a}$$

where the angular momentum of the system is given by

$$\vec{H}_{\text{system}} = \int_{\text{mass (system)}} \vec{r} \times \vec{V} \, dm = \int_{\forall\text{(system)}} \vec{r} \times \vec{V}\rho \, d\forall \tag{4.3b}$$

Torque can be produced by surface and body forces, and also by shafts that cross the system boundary,

$$\vec{T} = \vec{r} \times \vec{F}_s + \int_{\text{mass (system)}} \vec{r} \times \vec{g} \, dm + \vec{T}_{\text{shaft}} \tag{4.3c}$$

4-1.4 The First Law of Thermodynamics

The first law of thermodynamics is a statement of conservation of energy for a system,

$$\delta Q + \delta W = dE$$

In rate form the equation can be written as

$$\dot{Q} + \dot{W} = \left.\frac{dE}{dt}\right)_{\text{system}} \tag{4.4a}$$

** This section may be omitted without loss of continuity in the text material.

where the total energy of the system is given by

$$E_{\text{system}} = \int_{\text{mass (system)}} e \, dm = \int_{\forall\,(\text{system})} e\rho \, d\forall \tag{4.4b}$$

and

$$e = u + \frac{V^2}{2} + gz \tag{4.4c}$$

In Eq. 4.4a the rate of heat transfer, $\dot{Q}$, is positive when heat is added to the system from the surroundings; the rate of work, $\dot{W}$, is positive when work is done on the system by its surroundings.

4-1.5 The Second Law of Thermodynamics

If an amount of heat, δQ, is transferred to a system at temperature, T, the second law of thermodynamics states that the change in entropy, dS, of the system is given by

$$dS \geq \frac{\delta Q}{T}$$

On a rate basis we can write

$$\left.\frac{dS}{dt}\right)_{\text{system}} \geq \frac{1}{T}\dot{Q} \tag{4.5a}$$

where the total entropy of the system is given by

$$S_{\text{system}} = \int_{\text{mass (system)}} s \, dm = \int_{\forall\,(\text{system})} s\rho \, d\forall \tag{4.5b}$$

4-2 RELATION OF SYSTEM DERIVATIVES TO THE CONTROL VOLUME FORMULATION

In the previous section we summarized the basic equations for a system. We found that when written on a rate basis, each equation involved the time derivative of an extensive property of the system (the total mass, momentum, moment of momentum, energy, or entropy of the system). In developing the control volume formulation of each basic law from the system formulation, we shall use the symbol, N, to designate any arbitrary extensive property of the system. The corresponding intensive property (extensive property per unit mass) will be designated by η. Thus

$$N_{\text{system}} = \int_{\text{mass (system)}} \eta \, dm = \int_{\forall\,(\text{system})} \eta\rho \, d\forall \tag{4.6}$$

Comparing Eq. 4.6 with Eqs. 4.1b, 4.2b, 4.3b, 4.4b, and 4.5b, we see that if:

$$N = M, \qquad \text{then } \eta = 1$$

$$N = \vec{P}, \qquad \text{then } \eta = \vec{V}$$

$$N = \vec{H}, \qquad \text{then } \eta = \vec{r} \times \vec{V}$$

$$N = E, \qquad \text{then } \eta = e$$

$$N = S, \qquad \text{then } \eta = s$$

The major task in going from the system to the control volume formulation of the basic laws is to express the rate of change of the arbitrary extensive property, N, for a system, in terms of time variations of this property associated with a control volume. Since mass crosses the boundaries of a control volume, time variations of the property, N, associated with the control volume involve the mass flux and the properties convected with it. A convenient way to account for mass flux is to use a limiting process involving a system and a control volume that coincide at a certain instant. Flux quantities in regions of overlap and regions surrounding the control volume are then formulated approximately, and the limiting process is applied to obtain exact results. The final equation relates the rate of change of the arbitrary extensive property, N, for a system to the time variations of this property associated with a control volume.

4-2.1 Derivation

The system and control volume to be used in the analysis are shown in Fig. 4.1. The flow field $\vec{V}(x, y, z, t)$, is arbitrary relative to coordinates x, y, and z. The control volume is fixed in space; by definition, the system always must consist of the same fluid particles, and consequently it must move with the flow field. In Fig. 4.1 the boundaries of the system are shown at two different instants, t_0 and $t_0 + \Delta t$. At t_0, the boundaries of the system and the control volume coincide; at $t_0 + \Delta t$, the system occupies regions II and III. The system has been chosen so that the mass within region I enters the control volume during the interval, Δt, and the mass in region III leaves the control volume during the same interval.

Recall that our objective is to relate the rate of change of any arbitrary extensive property, N, of the system to the time variations of this property associated with the

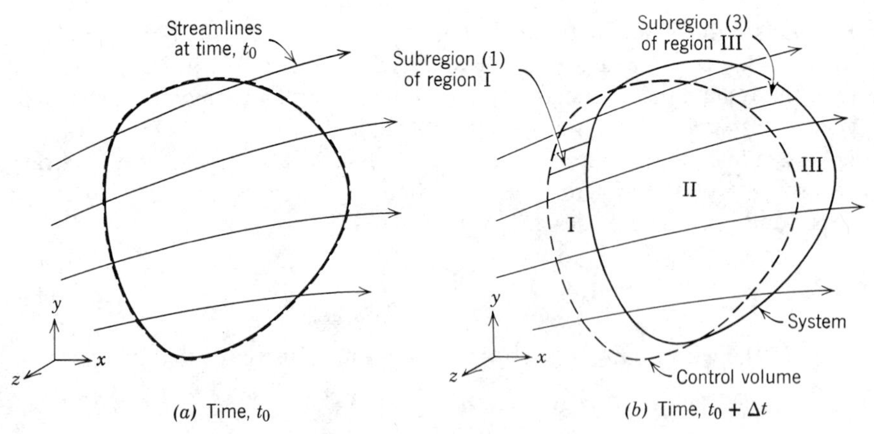

Fig. 4.1 System and control volume configuration.

control volume. From the definition of a derivative, the rate of change of N_{system} is given by

$$\left.\frac{dN}{dt}\right)_{\text{system}} \equiv \lim_{\Delta t \to 0} \frac{N_s)_{t_0+\Delta t} - N_s)_{t_0}}{\Delta t} \tag{4.7}$$

For convenience, the subscript, s, has been used to denote the system in the definition of a derivative in Eq. 4.7.

At $t_0 + \Delta t$, the system occupies regions II and III; at t_0, the system and the control volume coincide. Since

$$N_{\text{system}} = \int_{\text{mass (system)}} \eta \, dm = \int_{\forall\,(\text{system})} \eta\rho \, d\forall \tag{4.6}$$

we can write

$$N_s)_{t_0+\Delta t} = (N_{\text{II}} + N_{\text{III}})_{t_0+\Delta t} = (N_{\text{CV}} - N_{\text{I}} + N_{\text{III}})_{t_0+\Delta t}$$

$$= \left[\int_{\text{CV}} \eta\rho \, d\forall\right]_{t_0+\Delta t} - \left[\int_{\text{I}} \eta\rho \, d\forall\right]_{t_0+\Delta t} + \left[\int_{\text{III}} \eta\rho \, d\forall\right]_{t_0+\Delta t}$$

and

$$N_s)_{t_0} = (N_{\text{CV}})_{t_0} = \left[\int_{\text{CV}} \eta\rho \, d\forall\right]_{t_0}$$

Substituting these expressions into the definition of the system derivative, Eq. 4.7, we obtain

$$\left.\frac{dN}{dt}\right]_s = \lim_{\Delta t \to 0} \frac{\left[\int_{\text{CV}} \eta\rho \, d\forall\right]_{t_0+\Delta t} + \left[\int_{\text{III}} \eta\rho \, d\forall\right]_{t_0+\Delta t} - \left[\int_{\text{I}} \eta\rho \, d\forall\right]_{t_0+\Delta t} - \left[\int_{\text{CV}} \eta\rho \, d\forall\right]_{t_0}}{\Delta t} \tag{4.8}$$

Since the limit of a sum is equal to the sum of the limits, we can write

$$\left.\frac{dN}{dt}\right]_s = \underbrace{\lim_{\Delta t \to 0} \frac{\left[\int_{\text{CV}} \eta\rho \, d\forall\right]_{t_0+\Delta t} - \left[\int_{\text{CV}} \eta\rho \, d\forall\right]_{t_0}}{\Delta t}}_{①}$$

$$+ \underbrace{\lim_{\Delta t \to 0} \frac{\left[\int_{\text{III}} \eta\rho \, d\forall\right]_{t_0+\Delta t}}{\Delta t}}_{②} - \underbrace{\lim_{\Delta t \to 0} \frac{\left[\int_{\text{I}} \eta\rho \, d\forall\right]_{t_0+\Delta t}}{\Delta t}}_{③} \tag{4.9}$$

Our task now is to evaluate each of the three terms in Eq. 4.9.

Term ① in Eq. 4.9 simplifies to

$$\lim_{\Delta t \to 0} \frac{\left[\int_{CV} \eta\rho\, dV\!\!\!\!-\right]_{t_0+\Delta t} - \left[\int_{CV} \eta\rho\, dV\!\!\!\!-\right]_{t_0}}{\Delta t} = \lim_{\Delta t \to 0} \frac{N_{CV})_{t_0+\Delta t} - N_{CV})_{t_0}}{\Delta t}$$

$$= \frac{\partial N_{CV}}{\partial t} = \frac{\partial}{\partial t}\int_{CV} \eta\rho\, dV\!\!\!\!-$$

Term ② in Eq. 4.9 simplifies to

$$\lim_{\Delta t \to 0} \frac{\left[\int_{III} \eta\rho\, dV\!\!\!\!-\right]_{t_0+\Delta t}}{\Delta t} = \lim_{\Delta t \to 0} \frac{N_{III})_{t_0+\Delta t}}{\Delta t}$$

To evaluate $N_{III})_{t_0+\Delta t}$, let us look at an enlarged view of a typical subregion of region III as shown in Fig. 4.2. Vector $d\vec{A}$ has magnitude equal to the element of area, dA, of the control surface; the direction of $d\vec{A}$ is that of the normal drawn outward from the element of control surface area. The angle α is the angle between $d\vec{A}$ and the velocity vector, $\vec{V}$. Since the mass in region III is that which flows *out* of the control volume during the interval, Δt, the angle α will always be less than $\pi/2$ over the entire area of the control surface bounding region III.

For subregion (3) we can write

$$dN_3)_{t_0+\Delta t} = (\eta\rho\, dV\!\!\!\!-)_{t_0+\Delta t} = [\eta\rho(\Delta l \cos\alpha\, dA)]_{t_0+\Delta t}$$

since $dV\!\!\!\!- = \Delta l \cos\alpha\, dA$. Then for the entire region III,

$$N_{III})_{t_0+\Delta t} = \left[\int_{CS_{III}} \eta\rho\, \Delta l \cos\alpha\, dA\right]_{t_0+\Delta t}$$

where CS_{III} is the surface common to region III and the control volume. In this expression, Δl is the distance traveled by a particle on the system surface during the interval, Δt, along a streamline that existed at t_0.

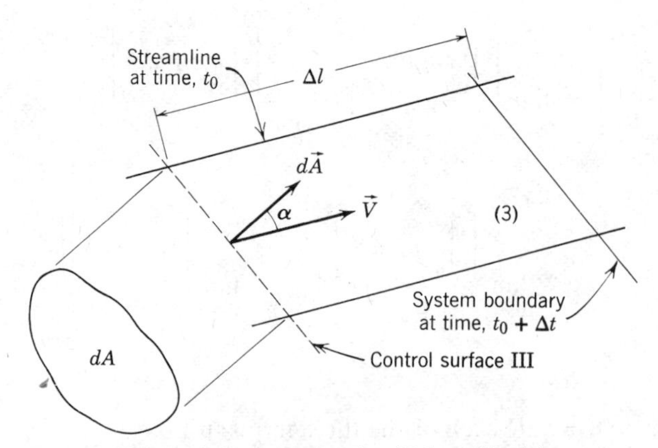

Fig. 4.2 Enlarged view of subregion (3) from Fig. 4.1.

Now that we have an expression for $N_{\text{III}})_{t_0+\Delta t}$, we can evaluate term ② in Eq. 4.9:

$$\lim_{\Delta t \to 0} \frac{\left[\int_{\text{III}} \eta \rho \, d\mathcal{V}\right]_{t_0+\Delta t}}{\Delta t} = \lim_{\Delta t \to 0} \frac{N_{\text{III}})_{t_0+\Delta t}}{\Delta t}$$

$$= \lim_{\Delta t \to 0} \frac{\int_{\text{CS}_{\text{III}}} \eta \rho \, \Delta l \cos \alpha \, dA}{\Delta t} = \lim_{\Delta t \to 0} \int_{\text{CS}_{\text{III}}} \eta \rho \frac{\Delta l}{\Delta t} \cos \alpha \, dA$$

$$= \int_{\text{CS}_{\text{III}}} \eta \rho |\vec{V}| \cos \alpha |d\vec{A}|$$

The last equality follows from the fact that

$$\lim_{\Delta t \to 0} \frac{\Delta l}{\Delta t} = |\vec{V}| \qquad \text{and} \qquad dA = |d\vec{A}|$$

Term ③ in Eq. 4.9 simplifies to

$$-\lim_{\Delta t \to 0} \frac{\left[\int_{\text{I}} \eta \rho \, d\mathcal{V}\right]_{t_0+\Delta t}}{\Delta t} = -\lim_{\Delta t \to 0} \frac{N_{\text{I}})_{t_0+\Delta t}}{\Delta t}$$

To evaluate $N_{\text{I}})_{t_0+\Delta t}$, look at an enlarged view of a typical subregion of region I as shown in Fig. 4.3.

The vector, $d\vec{A}$, has magnitude equal to the area element, dA, of the control surface; the direction of $d\vec{A}$ is that of the outward drawn normal from the element of control surface area. Angle α is the angle between $d\vec{A}$ and the velocity vector, $\vec{V}$. Since the mass in region I flows *into* the control volume during the time interval, Δt, angle α will always be greater than $\pi/2$ over the entire area of the control surface bounding region I.

For subregion (1) we can write

$$dN_1)_{t_0+\Delta t} = (\eta \rho \, d\mathcal{V})_{t_0+\Delta t} = [\eta \rho \, \Delta l(-\cos \alpha) \, dA]_{t_0+\Delta t}$$

since $d\mathcal{V} = \Delta l(-\cos \alpha) \, dA$. Why the minus sign? Recall that volume is a scalar quantity that must have a positive numerical value. Since $\alpha > \pi/2$, then $\cos \alpha$ will be negative. Hence the need for the minus sign.

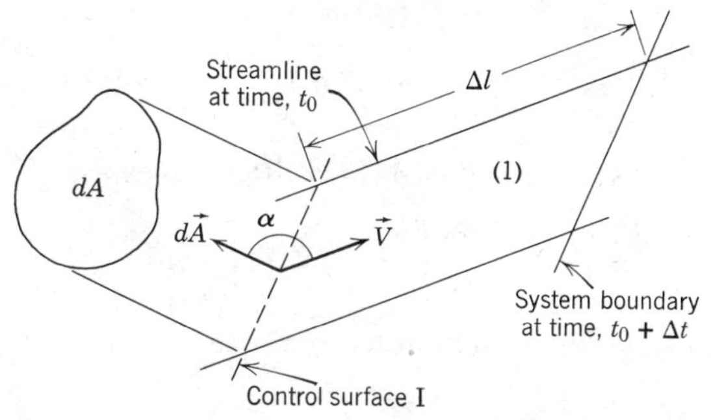

Fig. 4.3 Enlarged view of subregion (1) from Fig. 4.1.

Then, for the entire region I,

$$N_{\mathrm{I}})_{t_0+\Delta t} = \left[\int_{\mathrm{CS_I}} -\eta\rho\,\Delta l\cos\alpha\,dA\right]_{t_0+\Delta t}$$

where $\mathrm{CS_I}$ is the surface common to region I and the control volume. In this expression, Δl is the distance traveled by a particle on the system surface during interval, Δt, along a streamline that existed at t_0.

Now that we have an expression for $N_{\mathrm{I}})_{t_0+\Delta t}$, we can evaluate term ③ in Eq. 4.9:

$$-\lim_{\Delta t\to 0}\frac{\left[\int_{\mathrm{I}}\eta\rho\,d\mathcal{V}\right]_{t_0+\Delta t}}{\Delta t} = -\lim_{\Delta t\to 0}\frac{N_{\mathrm{I}})_{t_0+\Delta t}}{\Delta t}$$

$$= -\lim_{\Delta t\to 0}\frac{\int_{\mathrm{CS_I}} -\eta\rho\,\Delta l\cos\alpha\,dA}{\Delta t} = \lim_{\Delta t\to 0}\int_{\mathrm{CS_I}}\eta\rho\frac{\Delta l}{\Delta t}\cos\alpha\,dA$$

$$= \int_{\mathrm{CS_I}}\eta\rho|\vec{V}|\cos\alpha|d\vec{A}|$$

The last equality follows from the fact that

$$\lim_{\Delta t\to 0}\frac{\Delta l}{\Delta t} = |\vec{V}| \qquad \text{and} \qquad dA = |d\vec{A}|$$

Now that we have obtained expressions for each of the three terms on the right side, Eq. 4.9 can be written

$$\left.\frac{dN}{dt}\right)_{\text{system}} = \frac{\partial}{\partial t}\int_{\mathrm{CV}}\eta\rho\,d\mathcal{V} + \int_{\mathrm{CS_I}}\eta\rho|\vec{V}|\cos\alpha|d\vec{A}| + \int_{\mathrm{CS_{III}}}\eta\rho|\vec{V}|\cos\alpha|d\vec{A}|$$

Referring to Fig. 4.1, we see that the entire control surface, CS, consists of three surfaces,

$$\mathrm{CS} = \mathrm{CS_I} + \mathrm{CS_{III}} + \mathrm{CS}_p$$

where CS_p is characterized by no flow across the surface, where either $\alpha = \pi/2$ or $\vec{V} = 0$.

Consequently, we can write

$$\left.\frac{dN}{dt}\right)_{\text{system}} = \frac{\partial}{\partial t}\int_{\mathrm{CV}}\eta\rho\,d\mathcal{V} + \int_{\mathrm{CS}}\eta\rho|\vec{V}|\cos\alpha|d\vec{A}| \tag{4.10}$$

Recognizing that $|\vec{V}|\cos\alpha|d\vec{A}| = \vec{V}\cdot d\vec{A}$, Eq. 4.10 becomes

$$\left.\frac{dN}{dt}\right)_{\text{system}} = \frac{\partial}{\partial t}\int_{\mathrm{CV}}\eta\rho\,d\mathcal{V} + \int_{\mathrm{CS}}\eta\rho\vec{V}\cdot d\vec{A} \tag{4.11}$$

Equation 4.11 is the relation we set out to obtain between $dN/dt)_{\text{system}}$ and a control volume formulation.

4-2.2 Physical Interpretation

We have taken several pages to derive Eq. 4.11. Recall that our objective was to obtain a general relation between the rate of change of any arbitrary extensive property, N, of a system and time variations of this property associated with the control volume. The main reason for deriving it was to reduce the algebra required to obtain the control volume formulations of the basic equations. Because the working form of each basic equation for application to control volumes is developed from Eq. 4.11, we consider the equation itself to be "basic" and rewrite it to emphasize its importance:

$$\left.\frac{dN}{dt}\right)_{\text{system}} = \frac{\partial}{\partial t}\int_{\text{CV}} \eta\rho\, d\forall + \int_{\text{CS}} \eta\rho\vec{V}\cdot d\vec{A} \tag{4.11}$$

It is important to recall that in deriving Eq. 4.11, the limiting process (taking the limit as $\Delta t \to 0$) ensured that the relation is valid at the instant when the system and the control volume coincide. In using Eq. 4.11 to go from the system formulations of the basic laws to the control volume formulations, we recognize that Eq. 4.11 relates the rate of change of any extensive property, N, of a system to time variations of this property associated with a control volume at the instant when the system and the control volume coincide; this is true since, as $\Delta t \to 0$, the system and the control volume occupy the same volume and have the same boundaries.

Before using Eq. 4.11 to develop control volume formulations of the basic laws, let us make sure we understand each of the terms and symbols in the equation:

$\left.\frac{dN}{dt}\right)_{\text{system}}$ is the total rate of change of any arbitrary extensive property of the system.

$\frac{\partial}{\partial t}\int_{\text{CV}} \eta\rho\, d\forall$ is the time rate of change of the arbitrary extensive property, N, within the control volume.

: η is the intensive property corresponding to N; $\eta = N$ per unit mass.

: $\rho\, d\forall$ is an element of mass contained in the control volume.

: $\int_{\text{CV}} \eta\rho\, d\forall$ is the total amount of the extensive property, N, contained within the control volume.

$\int_{\text{CS}} \eta\rho\vec{V}\cdot d\vec{A}$ is the net rate of efflux of the extensive property, N, through the control surface.

: $\rho\vec{V}\cdot d\vec{A}$ is the rate of mass efflux through area element $d\vec{A}$ per unit time (we recognize that the dot product is a scalar product; the sign of $\rho\vec{V}\cdot d\vec{A}$ depends on the direction of the velocity vector, $\vec{V}$, relative to the area vector, $d\vec{A}$).

: $\eta\rho\vec{V}\cdot d\vec{A}$ is the rate of efflux of the extensive property, N, through the area, $d\vec{A}$.

Two additional points about Eq. 4.11 should be made. It should be clear from the derivation that velocity $\vec{V}$ is measured relative to the control volume. In developing

Eq. 4.11, we considered a control volume fixed relative to the reference coordinates, x, y, and z. The velocity field was specified relative to the same reference coordinates. Since, in our development, the system moved in the specified velocity field, then the time rate of change of the arbitrary extensive property, N, within the control volume must be evaluated by an observer fixed in the control volume.

We shall further emphasize these points in deriving the control volume formulation of each of the basic laws. In each case we begin with the familiar system formulation and use Eq. 4.11 to relate system derivatives to time variations associated with a control volume at the instant when the system and the control volume coincide.

4-3 CONSERVATION OF MASS

The first physical principle to which we apply the relation between system and control volume formulations is conservation of mass. It is intuitive that mass neither can be created nor destroyed; if the flow rate of mass into a control volume exceeds the rate of flow out, mass will accumulate within the CV.

Recall that conservation of mass states simply that the mass of a system is constant,

$$\left.\frac{dM}{dt}\right)_{\text{system}} = 0 \tag{4.1a}$$

where

$$M_{\text{system}} = \int_{\text{mass (system)}} dm = \int_{\maltese\,(\text{system})} \rho \, d\maltese \tag{4.1b}$$

The system and control volume formulations are related by Eq. 4.11

$$\left.\frac{dN}{dt}\right)_{\text{system}} = \frac{\partial}{\partial t}\int_{\text{CV}} \eta\rho \, d\maltese + \int_{\text{CS}} \eta\rho\vec{V}\cdot d\vec{A} \tag{4.11}$$

where

$$N_{\text{system}} = \int_{\text{mass (system)}} \eta \, dm = \int_{\maltese\,(\text{system})} \eta\rho \, d\maltese \tag{4.6}$$

To derive the control volume formulation of the conservation of mass, we set

$$N = M \qquad \text{and} \qquad \eta = 1$$

With this substitution, we obtain

$$\left.\frac{dM}{dt}\right)_{\text{system}} = \frac{\partial}{\partial t}\int_{\text{CV}} \rho \, d\maltese + \int_{\text{CS}} \rho\vec{V}\cdot d\vec{A} \tag{4.12}$$

Comparing Eqs. 4.1a and 4.12, we arrive at the control volume formulation of the conservation of mass:

$$0 = \frac{\partial}{\partial t}\int_{\text{CV}} \rho \, d\maltese + \int_{\text{CS}} \rho\vec{V}\cdot d\vec{A} \tag{4.13}$$

In Eq. 4.13 the first term represents the rate of change of mass within the control volume; the second term represents the net rate of mass efflux through the control surface. Conservation of mass requires that the sum of the rate of change of mass within the control volume and the net rate of mass outflow through the control surface be zero.

We emphasize that the velocity, $\vec{V}$, in Eq. 4.13 is measured relative to the control surface. Furthermore, the dot product, $\rho\vec{V} \cdot d\vec{A}$, is a scalar product. The sign depends on the direction of the velocity vector, $\vec{V}$, relative to the area vector, $d\vec{A}$. Referring back to the derivation of Eq. 4.11, we see that the dot product, $\rho\vec{V} \cdot d\vec{A}$, is positive where flow is out through the control surface, negative where flow is in through the control surface, and zero where flow is tangent to the control surface.

4-3.1 Special Cases

In special cases it is possible to simplify Eq. 4.13. Consider first the case of incompressible flow, in which the density remains constant. When ρ is a constant, it may be taken out from under the integral signs in Eq. 4.13. For this case we write Eq. 4.13 as

$$0 = \frac{\partial}{\partial t}\rho \int_{\mathrm{CV}} d\forall + \rho \int_{\mathrm{CS}} \vec{V} \cdot d\vec{A} \tag{4.14a}$$

The integral of $d\forall$ over the control volume is simply the volume of the control volume. Thus we write Eq. 4.14a as

$$0 = \frac{\partial}{\partial t}[\rho\forall] + \rho \int_{\mathrm{CS}} \vec{V} \cdot d\vec{A} \tag{4.14b}$$

For incompressible flow, ρ is a constant throughout the control volume. Since the size of the control volume is fixed, the first term in Eq. 4.14b is zero. The conservation of mass for incompressible flow becomes

$$0 = \int_{\mathrm{CS}} \vec{V} \cdot d\vec{A} \tag{4.15}$$

The dimensions of the integrand in Eq. 4.15 are L^3/t. The integral of $\vec{V} \cdot d\vec{A}$ over a section of the control surface is commonly called the *volume flow rate* or *volume rate of flow*.

Note that we have not assumed the flow to be steady in reducing Eq. 4.13 to the form 4.15. We have only imposed the restriction of incompressible flow. Thus Eq. 4.15 is a statement of the conservation of mass for an incompressible flow that may be steady or unsteady.

Consider now the general case of steady flow that is not incompressible. Since the flow is steady, this means that at most $\rho = \rho(x, y, z)$. By definition none of the fluid properties varies with time in a steady flow. Consequently, the first term of Eq. 4.13 must be zero and, hence, for steady flow the statement of conservation of mass reduces to

$$0 = \int_{\mathrm{CS}} \rho\vec{V} \cdot d\vec{A} \tag{4.16}$$

As we noted in our previous discussion of velocity fields in Section 2-2, the idealization of *uniform flow at a section* frequently provides an adequate flow model. Uniform flow at a section implies the velocity is constant across the entire area at a section. When the density also is constant at a section, the flux integral in Eq. 4.13 may be replaced by a product. Thus, when uniform flow at section n is assumed,

$$\int_{A_n} \rho \vec{V} \cdot d\vec{A} = \rho_n \vec{V}_n \cdot \vec{A}_n$$

or using scalar magnitudes

$$\int_{A_n} \rho \vec{V} \cdot d\vec{A} = \pm |\rho_n V_n A_n|$$

Again note that when $\rho \vec{V} \cdot d\vec{A}$ is negative, mass flows in through the control surface. Mass flows out through the control surface in regions where $\rho \vec{V} \cdot d\vec{A}$ is positive. This fact provides a quick check of the signs on the various flux terms in an analysis.

Example 4.1

Consider the steady flow of water ($\rho = 1.94$ slug/ft^3) through the device shown in the diagram. The areas are: $A_1 = 0.2$ ft^2, $A_2 = 0.5$ ft^2, and $A_3 = A_4 = 0.4$ ft^2. Mass flow out through section ③ is given as 3.88 slug/sec. The volume flow rate in through section ④ is given as 1 ft^3/sec, and $\vec{V}_1 = 10\hat{i}$ ft/sec. If properties are assumed uniform across all inlet and outlet flow sections, determine the flow velocity at section ②.

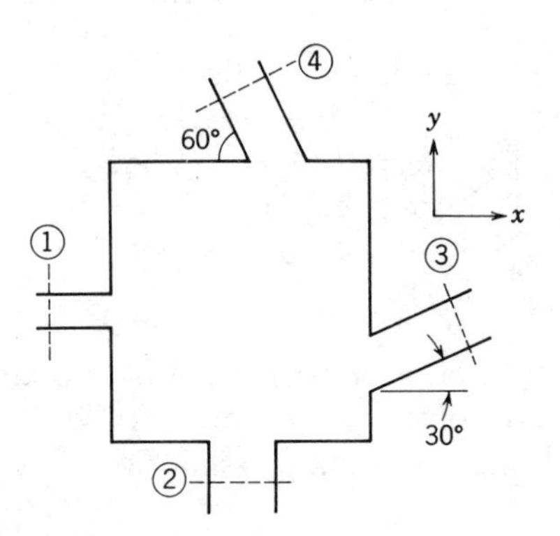

EXAMPLE PROBLEM 4.1

GIVEN: Steady flow of water through the device. Properties uniform at all ports.

$A_1 = 0.2$ ft^2 $\quad A_2 = 0.5$ ft^2

$A_3 = A_4 = 0.4$ ft^2 $\quad \rho = 1.94$ slug/ft^3

$\dot{m}_3 = 3.88$ slug/sec (outflow)

$\vec{V}_1 = 10\hat{i}$ ft/sec

Volume flow rate in at ④ = 1.0 ft^3/sec

FIND: Velocity at section ②.

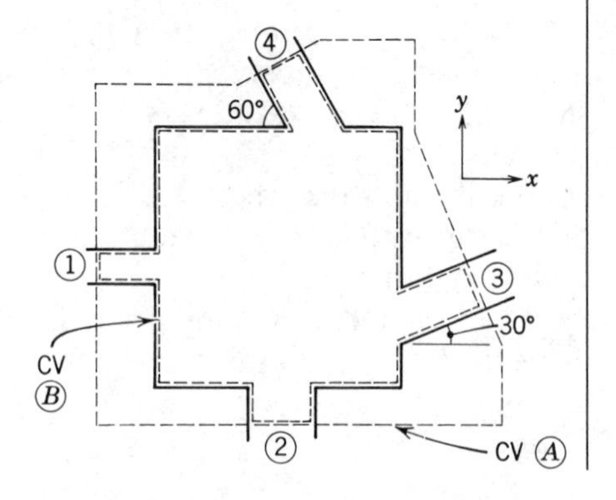

SOLUTION:
Choose a control volume. Two possibilities are shown by dashed lines.

Basic equation:
$$0 = \frac{\partial}{\partial t}\int_{CV} \rho \, d\forall + \int_{CS} \rho \vec{V} \cdot d\vec{A}$$

For steady flow, the first term is zero by definition, so

$$0 = \int_{CS} \rho \vec{V} \cdot d\vec{A}$$

In looking at either control volume, we see that there are four sections where mass flows across the control surface. Thus we write

$$\int_{CS} \rho \vec{V} \cdot d\vec{A} = \int_{A_1} \rho \vec{V} \cdot d\vec{A} + \int_{A_2} \rho \vec{V} \cdot d\vec{A} + \int_{A_3} \rho \vec{V} \cdot d\vec{A} + \int_{A_4} \rho \vec{V} \cdot d\vec{A} = 0 \qquad (1)$$

Let us look at these integrals one at a time, recognizing that properties are uniform over each area and ρ = constant.

$$\int_{A_1} \rho \vec{V} \cdot \vec{A} = -\int_{A_1} |\rho V \, dA| = -|\rho V_1 A_1|$$

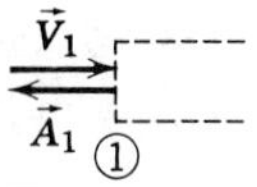

$\vec{V}_1$, $\vec{A}_1$, ①

{sign of $\vec{V} \cdot d\vec{A}$ is negative at surface ①.}

{With the absolute value signs indicated, we have accounted for the directions of $\vec{V}$ and $d\vec{A}$ in taking the dot product.}

Since we do not know the direction of $\vec{V}_2$, we shall leave section ② for the moment.

$$\int_{A_3} \rho \vec{V} \cdot d\vec{A} = \int_{A_3} |\rho V \, dA| = |\rho V_3 A_3| = \dot{m}_3$$

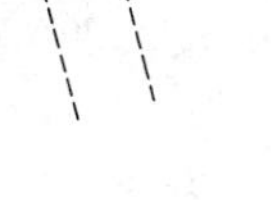

$\vec{V}_3$, $\vec{A}_3$, ③

{sign of $\vec{V} \cdot d\vec{A}$ is positive at surface ③, since flow is out.}

$$\int_{A_4} \rho \vec{V} \cdot d\vec{A} = -\int_{A_4} |\rho V dA| = -|\rho V_4 A_4|$$
$$= -\rho|V_4 A_4| = -\rho|Q_4|$$

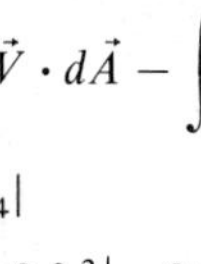

$\vec{V}_4$, $\vec{A}_4$, ④

{sign of $\vec{V} \cdot d\vec{A}$ is negative at surface ④.}

where Q is the volume flow rate.

From Eq. 1 above,

$$\int_{A_2} \rho \vec{V} \cdot d\vec{A} = -\int_{A_1} \rho \vec{V} \cdot d\vec{A} - \int_{A_3} \rho \vec{V} \cdot d\vec{A} - \int_{A_4} \rho \vec{V} \cdot d\vec{A}$$
$$= +|\rho V_1 A_1| - \dot{m}_3 + \rho|Q_4|$$
$$= \left|\frac{1.94 \text{ slug}}{\text{ft}^3} \times \frac{10 \text{ ft}}{\text{sec}} \times 0.2 \text{ ft}^2\right| - \frac{3.88 \text{ slug}}{\text{sec}} + \frac{1.94 \text{ slug}}{\text{ft}^3}\left|\frac{1.0 \text{ ft}^3}{\text{sec}}\right|$$

$$\int_{A_2} \rho \vec{V} \cdot d\vec{A} = 1.94 \text{ slug/sec}$$

Since this is positive, $\vec{V} \cdot d\vec{A}$ at section ② is positive. Flow is out as shown in the sketch:

$$\int_{A_2} \rho \vec{V} \cdot d\vec{A} = \int_{A_2} |\rho V \, dA| = |\rho V_2 A_2| = 1.94 \text{ slug/sec}$$

$$|V_2| = \frac{1.94 \text{ slug/sec}}{\rho A_2} = \frac{1.94 \text{ slug}}{\text{sec}} \times \frac{\text{ft}^3}{1.94 \text{ slug}} \times \frac{1}{0.5 \text{ ft}^2} = 2 \text{ ft/sec}$$

Since V_2 is in the negative y direction, then

$$\vec{V}_2 = -2\hat{j} \text{ ft/sec}$$

$\left\{ \text{This problem illustrates the procedure recommended for evaluating } \int_{\text{CS}} \rho \vec{V} \cdot d\vec{A}. \right\}$

Example 4.2

A tank of 0.05 m^3 volume contains air at 800 kPa (absolute) and 15 C. At $t = 0$, air escapes from the tank through a valve with a flow area of 65 mm^2. The air passing through the valve has a speed of 311 m/sec and a density of 6.13 kg/m^3. Properties in the rest of the tank may be assumed to be uniform at each instant. Determine the instantaneous rate of change of density in the tank at $t = 0$.

EXAMPLE PROBLEM 4.2

GIVEN: Tank of volume $\mathcal{V} = 0.05 \text{ m}^3$ contains air at $p = 800$ kPa (absolute), $T = 15$ C. At $t = 0$, air escapes through a valve. Air leaves with speed $V = 311$ m/sec, and density $\rho = 6.13 \text{ kg/m}^3$ through area $A = 65 \text{ mm}^2$.

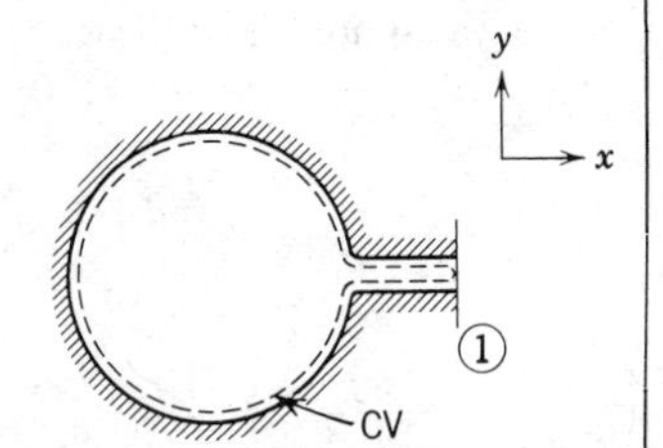

FIND: Rate of change of air density in the tank at $t = 0$.

SOLUTION:

Choose a control volume as shown by the dashed line.

Basic equation: $$0 = \frac{\partial}{\partial t} \int_{\text{CV}} \rho \, d\mathcal{V} + \int_{\text{CS}} \rho \vec{V} \cdot d\vec{A}$$

Since properties are assumed uniform in the tank at any instant, we can take ρ out from within the integral of the first term,

$$\frac{\partial}{\partial t} \left[\rho_{\text{CV}} \int_{\text{CV}} d\mathcal{V} \right] + \int_{\text{CS}} \rho \vec{V} \cdot d\vec{A} = 0$$

Now, $\int_{\text{CV}} d\mathcal{V} = \mathcal{V}$ and hence

$$\frac{\partial}{\partial t} (\rho \mathcal{V})_{\text{CV}} + \int_{\text{CS}} \rho \vec{V} \cdot d\vec{A} = 0$$

The only place where mass crosses the boundary of the control volume is at surface ①. Hence

$$\int_{\mathrm{CS}} \rho \vec{V} \cdot d\vec{A} = \int_{A_1} \rho \vec{V} \cdot d\vec{A} \qquad \text{and} \qquad \frac{\partial}{\partial t}(\rho \mathbf{V}) + \int_{A_1} \rho \vec{V} \cdot d\vec{A} = 0$$

At surface ① the sign of $\rho \vec{V} \cdot d\vec{A}$ is positive, so

$$\frac{\partial}{\partial t}(\rho \mathbf{V}) + \int_{A_1} |\rho V \, dA| = 0$$

If it is assumed that properties are uniform over surface ①, then

$$\frac{\partial}{\partial t}(\rho \mathbf{V}) + |\rho_1 V_1 A_1| = 0 \qquad \text{or} \qquad \frac{\partial}{\partial t}(\rho \mathbf{V}) = -|\rho_1 V_1 A_1|$$

Since the volume, $\mathbf{V}$, of the tank is not a function of time

$$\mathbf{V} \frac{\partial \rho}{\partial t} = -|\rho_1 V_1 A_1| \qquad \text{and} \qquad \frac{\partial \rho}{\partial t} = -\frac{|\rho_1 V_1 A_1|}{\mathbf{V}}$$

At $t = 0$,

$$\frac{\partial \rho}{\partial t} = \frac{-6.13 \text{ kg}}{\text{m}^3} \times \frac{311 \text{ m}}{\text{sec}} \times 65 \text{ mm}^2 \times \frac{1}{0.05 \text{ m}^3} \times \frac{\text{m}^2}{10^6 \text{ mm}^2}$$

$$\frac{\partial \rho}{\partial t} = -2.48 \text{ kg/m}^3\text{/sec} \qquad \{\text{the density is decreasing}\}$$

{This problem illustrates the application of the control volume formulation of conservation of mass to an unsteady flow.}

Example 4.3

The fluid in direct contact with a stationary solid boundary has zero velocity; there is no slip at the boundary. Thus the flow over a flat plate adheres to the plate surface and forms a boundary layer, as depicted below. The flow ahead of the plate is uniform with velocity, $\vec{V} = U\hat{i}$; $U = 30$ m/sec. The velocity distribution within the boundary layer $(0 \leq y \leq \delta)$ at cd is approximated as $u/U = 2(y/\delta) - (y/\delta)^2$.

The boundary-layer thickness at this location is 5 mm. The fluid is air with density, $\rho = 1.24 \text{ kg/m}^3$. Assuming the plate width to be 0.6 m, calculate the mass flow rate across the surface bc of control volume $abcd$.

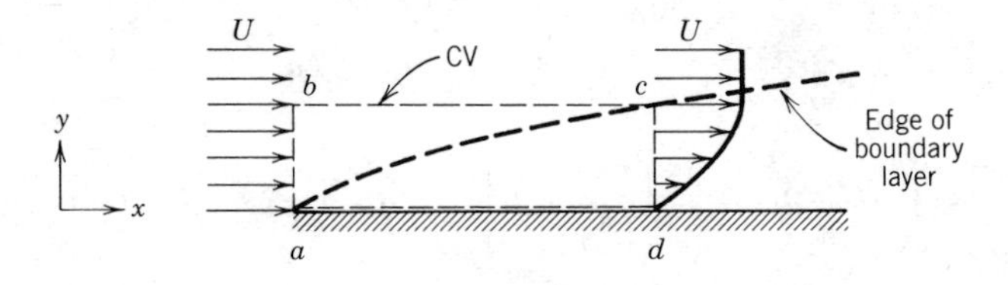

EXAMPLE PROBLEM 4.3

GIVEN: Steady, incompressible flow over a flat plate, $\rho = 1.24\ \text{kg/m}^3$.
Width of plate, $w = 0.6$ m.
Velocity ahead of plate is uniform: $\vec{V} = U\hat{i}$, $U = 30$ m/sec

At $x = x_d$:

$\delta = 5$ mm

$$\frac{u}{U} = 2\left(\frac{y}{\delta}\right) - \left(\frac{y}{\delta}\right)^2$$

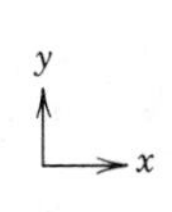

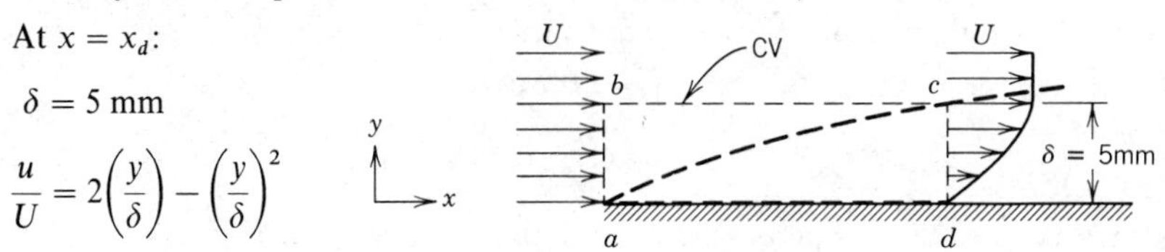

FIND: The mass flow rate across the surface bc.

SOLUTION:
The control volume is selected as shown by the dashed lines.

Basic equation:
$$0 = \frac{\partial}{\partial t}\int_{\text{CV}} \rho\, d\forall + \int_{\text{CS}} \rho \vec{V} \cdot d\vec{A}$$

For steady flow,

$$\frac{\partial}{\partial t}\int_{\text{CV}} \rho\, d\forall = 0 \quad \text{and hence} \quad \int_{\text{CS}} \rho \vec{V} \cdot d\vec{A} = 0$$

If it is assumed that there is no flow in the z direction, then

$$0 = \int_{\text{CS}} \rho \vec{V} \cdot d\vec{A}$$

$$= \int_{A_{ab}} \rho \vec{V} \cdot d\vec{A} + \int_{A_{bc}} \rho \vec{V} \cdot d\vec{A} + \int_{A_{cd}} \rho \vec{V} \cdot d\vec{A} + \int_{A_{da}} \rho \vec{V} \cdot d\vec{A}$$

$= 0$ (no flow across da)

$$\therefore \dot{m}_{bc} = \int_{A_{bc}} \rho \vec{V} \cdot d\vec{A} = -\int_{A_{ab}} \rho \vec{V} \cdot d\vec{A} - \int_{A_{cd}} \rho \vec{V} \cdot d\vec{A}$$

{We need to evaluate the integrals on the right side of the equation.}

$$\int_{A_{ab}} \rho \vec{V} \cdot d\vec{A} = -\int_{A_{ab}} |\rho u\, dA| = -\int_{y_a}^{y_b} |\rho u w\, dy|$$

{$\vec{V} \cdot d\vec{A}$ is negative, $dA = w\, dy$}

$$= -\int_0^{\delta} |\rho u w\, dy| = -\left|\int_0^{\delta} \rho U w\, dy\right| \qquad \{u = U \text{ over area } ab\}$$

$$= -|[\rho U w y]_0^{\delta}| = -\rho U w \delta$$

$$\int_{A_{cd}} \rho\vec{V} \cdot d\vec{A} = \int_{A_{cd}} |\rho u\, dA| = \int_{y_d}^{y_c} |\rho uw\, dy|$$

(c) dA, V (d)

$$\left\{\begin{array}{l}\vec{V} \cdot d\vec{A} \text{ is positive} \\ dA = w\, dy\end{array}\right\}$$

$$= \int_0^\delta |\rho uw\, dy| = \int_0^\delta \left|\rho wU\left[2\left(\frac{y}{\delta}\right) - \left(\frac{y}{\delta}\right)^2\right] dy\right|$$

$$= \left|\rho wU\left[\frac{y^2}{\delta} - \frac{y^3}{3\delta^2}\right]_0^\delta\right| = \left|\rho wU\delta\left[1 - \frac{1}{3}\right]\right|$$

$$= \frac{2\rho Uw\delta}{3}$$

$$\therefore \dot{m}_{bc} = \int_{A_{bc}} \rho\vec{V} \cdot d\vec{A} = -\int_{A_{ab}} \rho\vec{V} \cdot d\vec{A} - \int_{A_{cd}} \rho\vec{V} \cdot d\vec{A} = \rho Uw\delta - \frac{2\,\rho Uw\delta}{3}$$

$$= \frac{\rho Uw\delta}{3} = \frac{1}{3} \times 1.24\,\frac{\text{kg}}{\text{m}^3} \times 30\,\frac{\text{m}}{\text{sec}} \times 0.6\text{ m} \times 5\text{ mm} \times \frac{\text{m}}{1000\text{ mm}}$$

$$\int_{A_{bc}} \rho\vec{V} \cdot d\vec{A} = 0.0372\text{ kg/sec}$$

$\left\{\begin{array}{l}\text{Positive sign indicates flow out} \\ \text{across surface } bc.\end{array}\right\}$ $\dot{m}_{bc}$

{This problem illustrates the application of the control volume formulation of the conservation of mass to the case of nonuniform flow at a section.}

4-4 MOMENTUM EQUATION FOR INERTIAL CONTROL VOLUME

We wish to develop a mathematical formulation of Newton's second law suitable for application to a control volume. In this section our derivation will be restricted to an inertial control volume that is not accelerating relative to a stationary frame of reference (inertial coordinate system).

In deriving the control volume formulation, the procedure is analogous to the procedure followed in deriving the mathematical formulation for the conservation of mass applied to a control volume. We begin with the mathematical formulation for a system and then use Eq. 4.11 to go from the system to the control volume formulation.

Recall that Newton's second law for a system moving relative to an inertial coordinate system was given by Eq. 4.2a as

$$\vec{F} = \left.\frac{d\vec{P}}{dt}\right)_{\text{system}} \tag{4.2a}$$

where the linear momentum, $\vec{P}$, of the system is given by

$$\vec{P}_{\text{system}} = \int_{\text{mass (system)}} \vec{V}\, dm = \int_{\forall\text{ (system)}} \vec{V}\rho\, d\forall \tag{4.2b}$$

and the resultant force, $\vec{F}$, includes all surface and body forces acting on the system,

$$\vec{F} = \vec{F}_S + \vec{F}_B$$

The system and control volume formulations are related by Eq. 4.11

$$\left.\frac{dN}{dt}\right)_{\text{system}} = \frac{\partial}{\partial t}\int_{\text{CV}} \eta\rho\, d\maltese + \int_{\text{CS}} \eta\rho\vec{V}\cdot d\vec{A} \tag{4.11}$$

where

$$N_{\text{system}} = \int_{\text{mass (system)}} \eta\, dm = \int_{\maltese\text{ (system)}} \eta\rho\, d\maltese \tag{4.6}$$

To derive the control volume formulation of Newton's second law, we set

$$N = \vec{P} \qquad \text{and} \qquad \eta = \vec{V}$$

From Eq. 4.11, with this substitution, we obtain

$$\left.\frac{d\vec{P}}{dt}\right)_{\text{system}} = \frac{\partial}{\partial t}\int_{\text{CV}} \vec{V}\rho\, d\maltese + \int_{\text{CS}} \vec{V}\rho\vec{V}\cdot d\vec{A} \tag{4.17}$$

From Eq. 4.2a

$$\left.\frac{d\vec{P}}{dt}\right)_{\text{system}} = \left.\vec{F}\right)_{\text{on system}} \tag{4.2a}$$

Since, in deriving Eq. 4.11, the system and the control volume coincided at t_0,

$$\vec{F}\,\big]_{\text{on system}} = \vec{F}\,\big]_{\text{on control volume}}$$

Equations 4.2a and 4.17 yield the control volume formulation of Newton's second law for a nonaccelerating control volume

$$\vec{F} = \vec{F}_S + \vec{F}_B = \frac{\partial}{\partial t}\int_{\text{CV}} \vec{V}\rho\, d\maltese + \int_{\text{CS}} \vec{V}\rho\vec{V}\cdot d\vec{A} \tag{4.18}$$

This equation states that the sum of all forces (surface and body forces) acting on a nonaccelerating control volume is equal to the sum of the rate of change of momentum inside the control volume and the net rate of efflux of momentum through the control surface.

The derivation of the momentum equation for a control volume was straightforward. Application of this basic equation to the solution of problems will not be difficult if you exercise care in using the equation.

In using any basic equation for a control volume analysis, the first step should be to draw the boundaries of the control volume and label appropriate coordinate directions. In Eq. 4.18, the force, $\vec{F}$, represents all forces acting on the control volume. It includes both surface forces and body forces. If we denote the body force per unit mass as $\vec{B}$, then

$$\vec{F}_B = \int \vec{B}\, dm = \int_{\mathrm{CV}} \vec{B}\rho\, d\mathcal{V}$$

When the force of gravity is the only body force acting, then the body force per unit mass is $\vec{g}$. The surface force due to pressure is given by

$$\vec{F}_S = \int_A -p\, d\vec{A}$$

The nature of the forces acting on the control volume undoubtedly will influence the choice of the control volume boundaries.

All velocities, $\vec{V}$, in Eq. 4.18 are measured relative to the control volume. The momentum flux, $\vec{V}\rho\vec{V}\cdot d\vec{A}$, through an element of the control surface area, $d\vec{A}$, is a vector. The sign of the scalar product, $\rho\vec{V}\cdot d\vec{A}$, depends on the direction of the velocity vector, $\vec{V}$, relative to the area vector, $d\vec{A}$. The sign of the vector velocity, $\vec{V}$, depends on the coordinate system chosen.

The momentum equation is a vector equation. As with all vector equations, it may be written in scalar component equations. Relative to an xyz coordinate system, the scalar components of Eq. 4.18 are

$$F_x = F_{S_x} + F_{B_x} = \frac{\partial}{\partial t}\int_{\mathrm{CV}} u\rho\, d\mathcal{V} + \int_{\mathrm{CS}} u\rho\vec{V}\cdot d\vec{A} \tag{4.19a}$$

$$F_y = F_{S_y} + F_{B_y} = \frac{\partial}{\partial t}\int_{\mathrm{CV}} v\rho\, d\mathcal{V} + \int_{\mathrm{CS}} v\rho\vec{V}\cdot d\vec{A} \tag{4.19b}$$

$$F_z = F_{S_z} + F_{B_z} = \frac{\partial}{\partial t}\int_{\mathrm{CV}} w\rho\, d\mathcal{V} + \int_{\mathrm{CS}} w\rho\vec{V}\cdot d\vec{A} \tag{4.19c}$$

To use the scalar equations, it again is necessary to select a coordinate system at the outset. The positive directions of the velocity components, u, v, and w, and the force components, F_x, F_y, F_z, are then established relative to the selected coordinate system. As we have previously pointed out, the sign of the scalar product, $\rho\vec{V}\cdot d\vec{A}$, depends on the direction of the velocity vector, $\vec{V}$, relative to the area vector, $d\vec{A}$. Thus the flux term in either Eq. 4.18 or Eqs. 4.19 is a product of two quantities, both of which have algebraic signs. We suggest that you proceed in two steps to determine the momentum flux through any portion of a control surface:

1. The first step is to determine the sign of $\rho\vec{V}\cdot d\vec{A}$,

$$\rho\vec{V}\cdot d\vec{A} = \rho|V\, dA|\cos\alpha = \pm|\rho V\, dA\cos\alpha|$$

2. The second step is to determine the sign for each velocity component, u, v, and w. The sign, which depends on the choice of coordinate system, should be accounted for when substituting numerical values into the terms $u\rho\vec{V}\cdot d\vec{A} = u\{\pm|\rho V\, dA\cos\alpha|\}$, and so on.

Example 4.4

Water from a stationary nozzle strikes a flat plate as shown. The velocity of the water leaving the nozzle is 15 m/sec; the nozzle area is 0.01 m^2. Assuming the water is directed normal to the plate, and flows along the plate, determine the horizontal force on the support.

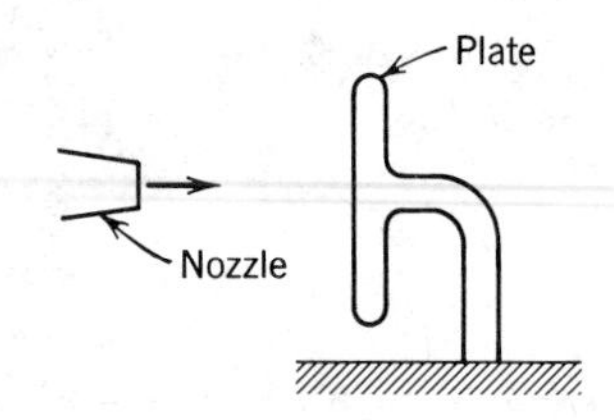

EXAMPLE PROBLEM 4.4

GIVEN: Water from a stationary nozzle is directed normal to the plate; subsequent flow is parallel to plate.

Jet velocity, $\vec{V} = 15\hat{i}$ m/sec

Nozzle area, $A_n = 0.01$ m^2

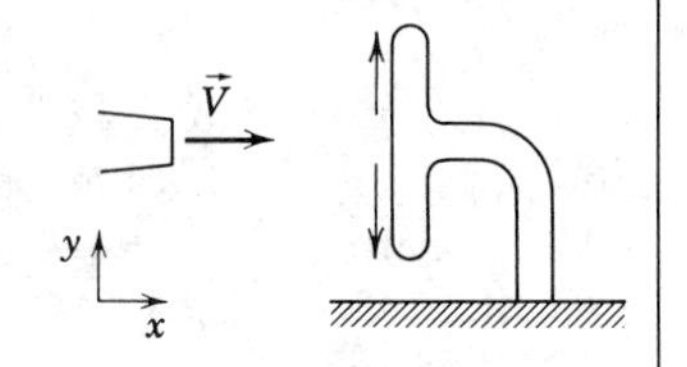

FIND: Horizontal force on the support.

SOLUTION:

We chose a coordinate system in defining the problem above. We must now choose a suitable control volume. A number of possible choices are shown by the dashed lines below.

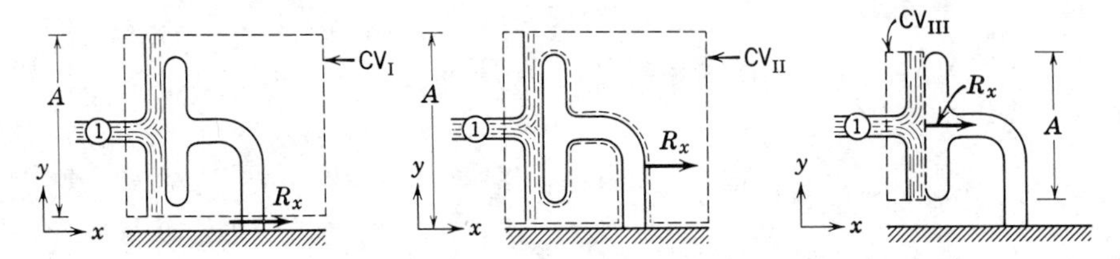

In all three cases the water from the nozzle crosses the control surface through area A_1 (assumed equal to the nozzle area) and the water is assumed to leave the control volume tangent to the plate surface in the $+y$ or $-y$ direction. Before trying to decide which is the "best" control volume to use, let us write the basic equations.

$$\vec{F} = \vec{F}_S + \vec{F}_B = \frac{\partial}{\partial t}\int_{CV} \vec{V}\rho\, d\forall + \int_{CS} \vec{V}\rho\vec{V}\cdot d\vec{A} \quad \text{and} \quad \frac{\partial}{\partial t}\int_{CV} \rho\, d\forall + \int_{CS} \rho\vec{V}\cdot d\vec{A} = 0$$

Regardless of our choice of control volume, the flow is steady and the basic equations become

$$\vec{F} = \vec{F}_S + \vec{F}_B = \int_{CS} \vec{V}\rho\vec{V}\cdot d\vec{A} \quad \text{and} \quad \int_{CS} \rho\vec{V}\cdot d\vec{A} = 0$$

Evaluating the momentum flux term will lead to the same result for all of the control volumes. We should choose the control volume that allows the most straightforward evaluation of the forces.

Remember in applying the momentum equation that the force, $\vec{F}$, represents all forces acting *on* the control volume.

Let us solve the problem using each of the three control volumes.

CV_I

The control volume has been selected so that the area of the left surface is equal to the area of the right surface. Denote this area by A.

The control volume cuts through the support. We denote the force of the support on the control volume as R_x and assume it to be positive. (The force of the control volume on the support is equal and opposite to R_x.)

Since we are looking for the horizontal force, we write the x component of the steady flow momentum equation

$$F_{S_x} + F_{B_x} = \int_{CS} u\rho \vec{V} \cdot d\vec{A}$$

There are no body forces in the x direction. $F_{B_x} = 0$ and

$$F_{S_x} = \int_{CS} u\rho \vec{V} \cdot d\vec{A}$$

To evaluate F_{S_x}, we must include all surface forces acting on the control volume

$F_{S_x} =$	$p_a A$	$-$	$p_a A$	$+$	R_x
	force due to atmospheric pressure acts to right (positive direction) on left surface		force due to atmospheric pressure acts to left (negative direction) on right surface		force of support on control volume (assumed positive)

Consequently, $F_{S_x} = R_x$, and

$$R_x = \int_{CS} u\rho \vec{V} \cdot d\vec{A} = \int_{A_1} u\rho \vec{V} \cdot d\vec{A} \qquad \left\{\begin{array}{l}\text{for mass crossing top and bottom}\\ \text{surfaces, } u = 0.\end{array}\right\}$$

$$= \int_{A_1} u\{-|\rho V_1\, dA|\} \qquad \left\{\begin{array}{l}\text{at (1) } \rho \vec{V} \cdot d\vec{A} = -|\rho V_1\, dA|\text{, since direction of } \vec{V}_1 \text{ and}\\ d\vec{A}_1 \text{ are } 180^\circ \text{ apart.}\end{array}\right\}$$

$$= -u_1|\rho V_1 A_1| \qquad \{\text{properties uniform over } A_1\}$$

$$= -\frac{15\text{ m}}{\text{sec}}\left|\frac{999\text{ kg}}{\text{m}^3} \times \frac{15\text{ m}}{\text{sec}} \times 0.01\text{ m}^2\right|\frac{\text{N}\cdot\text{sec}^2}{\text{kg}\cdot\text{m}} \qquad \{u_1 = 15\text{ m/sec}\}$$

$$R_x = -2.25\text{ kN} \qquad \{R_x \text{ acts opposite to positive direction assumed.}\}$$

The force on the support, $K_x = -R_x = 2.25$ kN {force on support acts to the right}

CV_{II}

The control volume has been selected so the area of the left surface is equal to the area of the right surface. Denote this area by A.

The control volume does not cut through the support. However, the control volume is in contact with the support over several portions of the area of the control surface. There is a force exerted by the support on the control surface. We denote the x component of this force as R_x.

Then for this control volume the x component of the momentum equation leads directly (in the same step by step manner) to the same solution as for CV_I above.

CV_{III}

The control volume has been selected so the areas of the left surface and of the right surface are equal to the area of the plate. Denote this area by A.

As in the case of CV_{II} the x component of the force of the support on the CV is denoted by R_x.

Then the x component of the momentum equation,

$$F_{S_x} = \int_{CS} u\rho\vec{V}\cdot d\vec{A}$$

yields

$$F_{S_x} = p_a A + R_x = \int_{A_1} u\rho\vec{V}\cdot d\vec{A} = \int_{A_1} u\{-|\rho V_1\, dA|\} = -2.25 \text{ kN}$$

Then

$$R_x = -p_a A - 2.25 \text{ kN} \qquad \text{and} \qquad K_x = -R_x = p_a A + 2.25 \text{ kN}$$

To determine the net force on the plate, we need a free-body diagram of the plate.

y K_x $p_a A$ x

$$F_{\text{net}} = K_x - p_a A \qquad \{\text{since atmospheric pressure acts on the back of the plate}\}$$

$$F_{\text{net}} = p_a A + 2.25 \text{ kN} - p_a A = 2.25 \text{ kN}$$

{This problem illustrates the application of the momentum equation to an inertial control volume, with emphasis on choosing a suitable control volume.}

Example 4.5

A metal container 2 ft high with an inside cross-sectional area of 1 ft^2 weighs 5 lbf when empty. The container is placed on a scale and water flows in through an opening in the top and out through the two equal area openings in the sides as shown in the diagram. Under steady flow conditions the height of the water in the tank is 1.9 ft. Determine the reading on the scale.

$$A_1 = 0.1 \text{ ft}^2$$

$$\vec{V}_1 = -20\hat{j} \text{ ft/sec}$$

$$A_2 = A_3 = 0.1 \text{ ft}^2$$

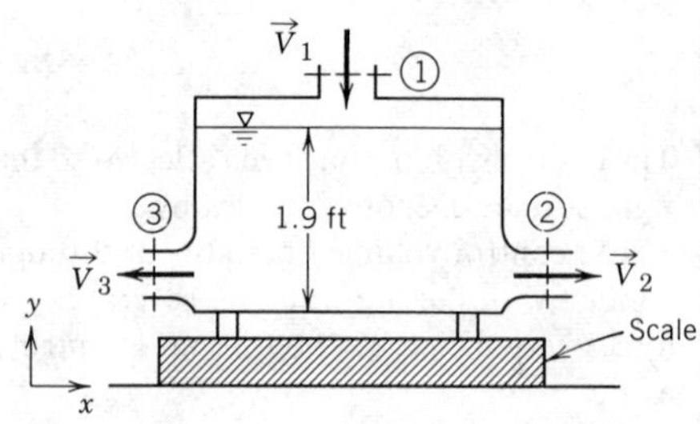

EXAMPLE PROBLEM 4.5

GIVEN: Metal container of height 2 ft and cross-sectional area, $A = 1\ \text{ft}^2$, weighs 5 lbf when empty. Container rests on scale. Under steady flow conditions water depth is 1.9 ft. Water enters vertically at section ① and leaves horizontally through sections ② and ③.

$$A_1 = 0.1\ \text{ft}^2$$

$$\vec{V}_1 = -20\hat{j}\ \text{ft/sec}$$

$$A_2 = A_3 = 0.1\ \text{ft}^2$$

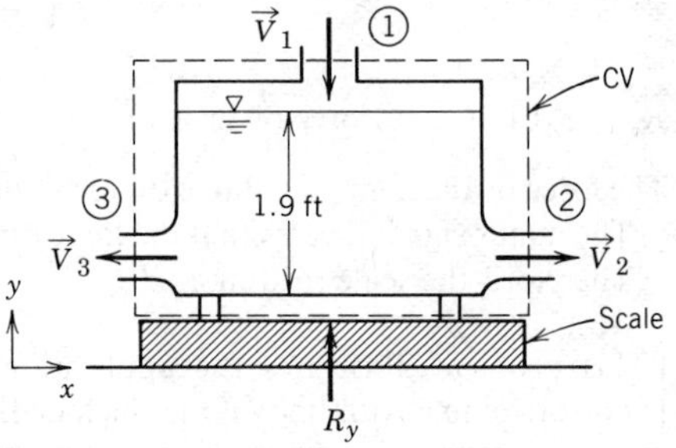

FIND: The reading on the scale.

SOLUTION:

Choose a control volume as shown; R_y is the force of the scale on the control volume and is assumed positive.

Basic equations:

$$\vec{F}_S + \vec{F}_B = \frac{\partial}{\partial t}\int_{CV} \vec{V}\rho\, d\forall + \int_{CS} \vec{V}\rho\vec{V}\cdot d\vec{A} \qquad \left(\frac{\partial}{\partial t}\int_{CV} \vec{V}\rho\, d\forall = 0\ \text{(steady flow)}\right)$$

$$0 = \frac{\partial}{\partial t}\int_{CV} \rho\, d\forall + \int_{CS} \rho\vec{V}\cdot d\vec{A} \qquad \left(\frac{\partial}{\partial t}\int_{CV} \rho\, d\forall = 0\ \text{(steady flow)}\right)$$

We write the y component of the momentum equation

$$F_{S_y} + F_{B_y} = \int_{CS} v\rho\vec{V}\cdot d\vec{A} \tag{1}$$

$F_{S_y} = R_y$ {There is no net force due to atmospheric pressure.}
$F_{B_y} = -W_{\text{tank}} - W_{H_2O}$ {Both body forces act in negative y direction.}

$$W_{H_2O} = \rho g \forall = \gamma A h$$

$$\int_{CS} v\rho\vec{V}\cdot d\vec{A} = \int_{A_1} v\rho\vec{V}\cdot d\vec{A} = \int_{A_1} v\{-|\rho_1 V_1\, dA|\} \qquad \left\{\begin{array}{l}\vec{V}\cdot d\vec{A}\ \text{is negative at ①.}\\ v = 0\ \text{at sections ② and ③.}\end{array}\right\}$$

$$= -v_1|\rho_1 V_1 A_1| \qquad \{\text{We are assuming uniform properties at ①.}\}$$

Substituting into Eq. 1 gives

$$R_y - W_{\text{tank}} - \gamma A h = -v_1|\rho_1 V_1 A_1|$$

or

$$R_y = W_{\text{tank}} + \gamma A h - v_1 |\rho_1 V_1 A_1|$$

Substituting numbers with $v_1 = -20$ ft/sec gives

$$R_y = 5 \text{ lbf} + \frac{62.4 \text{ lbf}}{\text{ft}^3} \times 1 \text{ ft}^2 \times 1.9 \text{ ft} - \left(-20 \frac{\text{ft}}{\text{sec}}\right) \left| \frac{1.94 \text{ slug}}{\text{ft}^3} \left(-20 \frac{\text{ft}}{\text{sec}}\right) 0.1 \text{ ft}^2 \right| \frac{\text{lbf} \cdot \text{sec}^2}{\text{slug} \cdot \text{ft}}$$

$R_y = 201$ lbf {Force of scale on CV is upward.}

The force of the control volume on the scale is $K_y = -R_y = -201$ lbf.

The minus sign indicates that the force on the scale is downward in our coordinate system. Therefore, the scale reading is 201 lbf.

{This problem illustrates the application of the momentum equation to an inertial control volume with body forces included.}

Example 4.6

Water in an open channel flows under a sluice gate as shown in the sketch. The flow is incompressible and uniform at sections ① and ②. Hydrostatic pressure distributions may be assumed at sections ① and ② because the flow streamlines are essentially straight there. Determine the magnitude and direction of the force exerted on the gate by the flow.

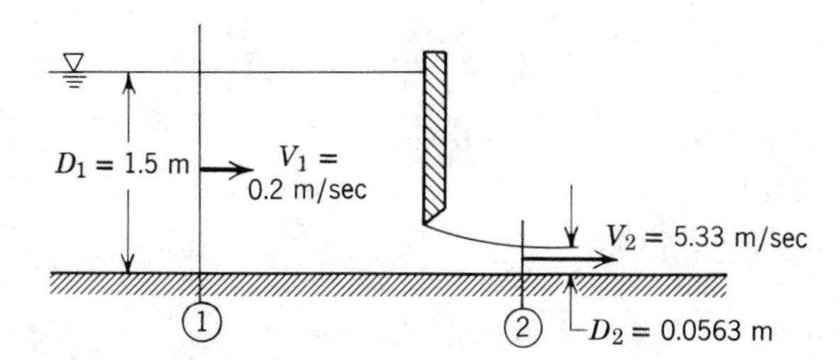

EXAMPLE PROBLEM 4.6

GIVEN: Flow under sluice gate. Width $= w$.

FIND: Force exerted (per unit width) on the gate.

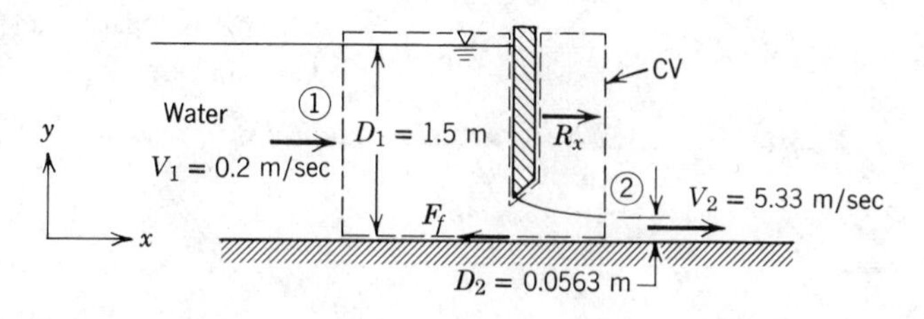

SOLUTION:
Choose the CV and coordinate system shown for analysis. Apply the x component of the momentum equation.

Basic equation:

$$F_{S_x} + \cancel{F_{B_x}}^{=0(2)} = \cancel{\frac{\partial}{\partial t}\int_{CV} u\rho\, d\forall}^{=0(3)} + \int_{CS} u\rho \vec{V} \cdot d\vec{A}$$

Assumptions:
(1) F_f negligible (neglect friction on channel bottom)
(2) $F_{B_x} = 0$
(3) Steady flow
(4) Incompressible flow
(5) Uniform flow at each section
(6) Hydrostatic pressure distributions at ① and ②

Then

$$F_{S_x} = u_1\{-|\rho V_1 w D_1|\} + u_2\{|\rho V_2 w D_2|\}$$

The surface forces acting on the CV are due to pressure and the unknown force, R_x. From assumption (6),

$$\frac{dp}{dy} = -\rho g; \qquad p = p_0 + \rho g(y_0 - y) = p_{atm} + \rho g(D - y)$$

Evaluating F_{S_x} gives

$$F_{S_x} = \int_0^{D_1} p_1\, dA_1 - \int_0^{D_2} p_2\, dA_2 - p_{atm}(D_1 - D_2)w + R_x$$

$$= \int_0^{D_1} [p_{atm} + \rho g(D_1 - y)]w\, dy - \int_0^{D_2} [p_{atm} + \rho g(D_2 - y)]w\, dy - p_{atm}(D_1 - D_2)w + R_x$$

$$F_{S_x} = \cancel{p_{atm}D_1 w} + \frac{\rho g D_1^2}{2}w - \cancel{p_{atm}D_2 w} - \frac{\rho g D_2^2}{2}w - \cancel{p_{atm}D_1 w} + \cancel{p_{atm}D_2 w} + R_x$$

or

$$F_{S_x} = R_x + \frac{\rho g w}{2}(D_1^2 - D_2^2)$$

Substituting into the momentum equation, with $u_1 = V_1$ and $u_2 = V_2$, gives

$$R_x + \frac{\rho g w}{2}(D_1^2 - D_2^2) = -V_1|\rho V_1 w D_1| + V_2|\rho V_2 w D_2|$$

or

$$R_x = \rho w(V_2^2 D_2 - V_1^2 D_1) - \frac{\rho g w}{2}(D_1^2 - D_2^2)$$

and

$$\frac{R_x}{w} = \rho(V_2^2 D_2 - V_1^2 D_1) - \frac{\rho g}{2}(D_1^2 - D_2^2)$$

$$= \frac{999 \text{ kg}}{\text{m}^3}\left[(5.33)^2(0.0563) - (0.2)^2(1.5)\right]\frac{\text{m}^2}{\text{sec}^2}\text{ m} \times \frac{\text{N}\cdot\text{sec}^2}{\text{kg}\cdot\text{m}}$$

$$-\frac{1}{2} \times \frac{999 \text{ kg}}{\text{m}^3} \times \frac{9.81 \text{ m}}{\text{sec}^2}\left[(1.5)^2 - (0.0563)^2\right]\text{m}^2 \times \frac{\text{N}\cdot\text{sec}^2}{\text{kg}\cdot\text{m}}$$

$$\frac{R_x}{w} = -9.47 \text{ kN/m}$$

R_x is the unknown external force acting *on* the control volume. It is applied to the CV by the gate. Therefore, the force from all fluids *on* the gate is K_x, where $K_x = -R_x$. Thus

$$\frac{K_x}{w} = -\frac{R_x}{w} = 9.47 \text{ kN/m} \qquad \{\text{applied to the right.}\} \qquad \frac{K_x}{w}$$

{This problem illustrates application of the momentum equation to a control volume in which the pressure is not uniform over the entire control surface.}

Example 4.7

Water flows steadily through the 90° reducing elbow shown in the diagram. At the inlet to the elbow the absolute pressure is 221 kPa and the cross-sectional area is 0.01 m^2. At the outlet the cross-sectional area is 0.0025 m^2 and the velocity is 16 m/sec. The pressure at the outlet is atmospheric. Determine the force required to hold the elbow in place.

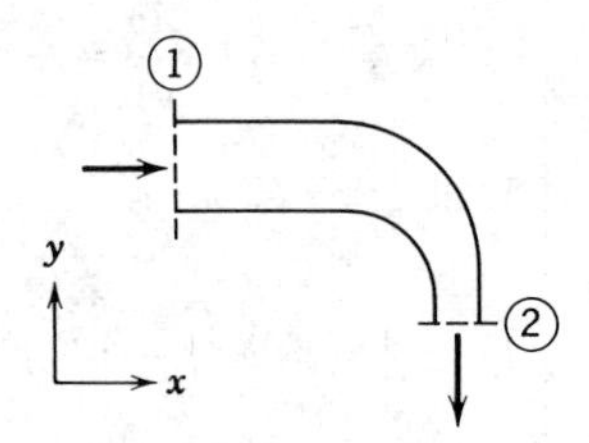

EXAMPLE PROBLEM 4.7

GIVEN: Steady flow of water through 90° reducing elbow.

$p_1 = 221$ kPa (abs) $\qquad A_1 = 0.01$ m^2

$\vec{V}_2 = -16\hat{j}$ m/sec $\qquad A_2 = 0.0025$ m^2

FIND: Force required to hold elbow in place.

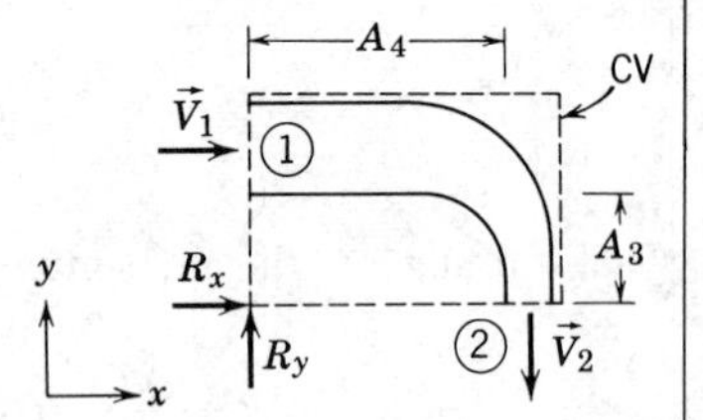

SOLUTION:

Choose control volume as shown by the dashed line.

R_x, R_y are components of force required to hold the elbow in place. (They are assumed positive and are forces acting on CV.)

A_3 is the area of vertical sides of CV excluding A_1; $A_{\text{vertical sides}} = A_1 + A_3$.

A_4 is the area of horizontal sides of CV excluding A_2; $A_{\text{horizontal sides}} = A_2 + A_4$.

Basic equations:

$$\vec{F} = \vec{F}_S + \vec{F}_B = \frac{\partial}{\partial t}\int_{CV} \vec{V}\rho\, d\forall + \int_{CS} \vec{V}\rho\vec{V}\cdot d\vec{A} \quad \left(\frac{\partial}{\partial t}\int_{CV} = 0 \text{ (steady flow)}\right)$$

$$0 = \frac{\partial}{\partial t}\int_{CV} \rho\, d\forall + \int_{CS} \rho\vec{V}\cdot d\vec{A} \quad \left(\frac{\partial}{\partial t}\int_{CV} = 0 \text{ (steady flow)}\right)$$

Assumptions:
(1) Uniform flow at each section
(2) Atmospheric pressure, $p_a = 101$ kPa
(3) Incompressible flow

Writing the x component of the momentum equation results in

$$F_{S_x} = \int_{CS} u\rho\vec{V}\cdot d\vec{A} = \int_{A_1} u\rho\vec{V}\cdot d\vec{A} \qquad \{F_{B_x} = 0 \text{ and } u_2 = 0\}$$

$$p_1A_1 + p_aA_3 - p_a(A_1 + A_3) + R_x = \int_{A_1} u\rho\vec{V}\cdot d\vec{A} \qquad \left\{\begin{array}{l}\text{Pressure over right side is}\\ p_a.\text{ Pressure over left side}\\ \text{is } p_1 \text{ on } A_1 \text{ and } p_a \text{ on } A_3.\end{array}\right\}$$

$$(p_1 - p_a)A_1 + R_x = \int_{A_1} u\{-|\rho V_1\, dA|\} \qquad \{\vec{V}\cdot d\vec{A} \text{ is negative at } A_1\}$$

$$R_x = -p_{1_g}A_1 - u_1|\rho V_1 A_1| \qquad \{p_1 - p_a = p_{1_{\text{gage}}}\}$$

To find V_1, use the continuity equation:

$$\int_{CS} \rho\vec{V}\cdot d\vec{A} = 0 = \int_{A_1} \rho\vec{V}\cdot d\vec{A} + \int_{A_2} \rho\vec{V}\cdot d\vec{A}$$

$$\therefore 0 = -\int_{A_1} |\rho V\, dA| + \int_{A_2} |\rho V\, dA| = -|\rho V_1 A_1| + |\rho V_2 A_2|$$

and

$$|V_1| = |V_2|\frac{A_2}{A_1} = \frac{16\text{ m}}{\text{sec}} \times \frac{0.0025}{0.01} = 4\text{ m/sec} \qquad \therefore \vec{V}_1 = 4\hat{i}\text{ m/sec}$$

$$R_x = -p_{1_g}A_1 - u_1|\rho V_1 A_1|$$

$$= -\frac{1.20\times 10^5\text{ N}}{\text{m}^2} \times 0.01\text{ m}^2 - \frac{4\text{ m}}{\text{sec}}\left|\frac{999\text{ kg}}{\text{m}^3} \times \frac{4\text{ m}}{\text{sec}} \times 0.01\text{ m}^2 \times \frac{\text{N}\cdot\text{sec}^2}{\text{kg}\cdot\text{m}}\right|$$

$R_x = -1.36$ kN $\quad$ {R_x to hold elbow acts to left.} $\qquad R_x$

Writing the y component of the momentum equation gives

$$F_{S_y} + F_{B_y} = \int_{CS} v\rho\vec{V}\cdot d\vec{A} = \int_{A_2} v\rho\vec{V}\cdot d\vec{A} \qquad \{v_1 = 0\}$$

$$p_aA_4 + p_aA_2 - p_aA_4 - p_aA_2 + F_{B_y} + R_y = \int_{A_2} v\{|\rho \vec{V} \cdot d\vec{A}|\}$$ $\left\{\begin{array}{l}\text{Pressure is } p_a \text{ over} \\ \text{top and bottom of CV.} \\ \vec{V} \cdot d\vec{A} \text{ is positive at } ②.\end{array}\right\}$

$$F_{B_y} + R_y = v_2|\rho V_2 A_2|$$

$$R_y = -F_{B_y} + v_2|\rho V_2 A_2|$$ $\left\{\begin{array}{l}\text{Since we do not know the volume of} \\ \text{the elbow, we cannot evaluate } F_{B_y}.\end{array}\right\}$

Substituting numbers, recognizing $\vec{V}_2 = -16\hat{j}$ m/sec, so $v_2 = -16$ m/sec

$$R_y = -F_{B_y} + \left(-16\,\frac{\text{m}}{\text{sec}}\right)\left|999\,\frac{\text{kg}}{\text{m}^3}\left(-16\,\frac{\text{m}}{\text{sec}}\right)0.0025\,\text{m}^2\right| \times \frac{\text{N}\cdot\text{sec}^2}{\text{kg}\cdot\text{m}}$$

$$R_y = -F_{B_y} - 639\ \text{N}$$

Neglecting F_{B_y} gives

$$R_y = -639\ \text{N} \quad \{R_y \text{ to hold elbow acts down.}\} \qquad R_y$$

{This problem illustrates the application of the momentum equation to an inertial control volume in which the pressure is not atmospheric across the entire control surface.

Since the pressure forces over the entire control surface must be included in the analysis, use of gage pressure on all surfaces gives correct (and often more direct) results.}

Example 4.8

A horizontal conveyor belt moving at 3 ft/sec receives sand from a hopper. The sand falls vertically from the hopper to the belt at a speed of 5 ft/sec and a flow rate of 500 lbm/sec (the density of sand is approximately 2700 lbm/cubic yard). The conveyor belt is initially empty but begins to fill with sand. If friction in the drive system and rollers is negligible, find the tension required to pull the belt while the conveyor is filling.

EXAMPLE PROBLEM 4.8

GIVEN: Conveyor and hopper shown in sketch.

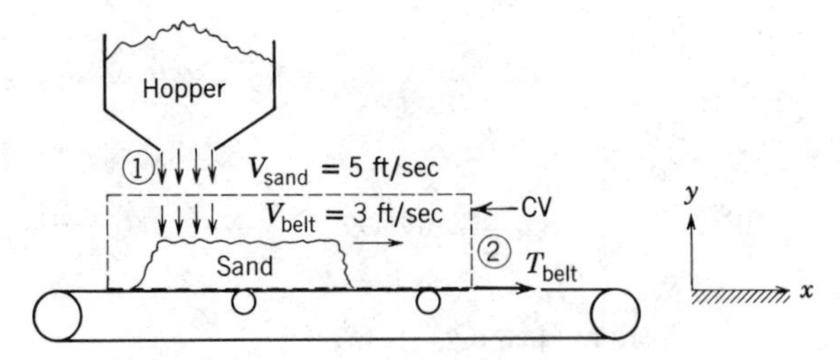

FIND: T_{belt} at the instant shown.

SOLUTION:

Use the control volume and coordinates shown. Apply the x component of the momentum equation.

Basic equations:

$$F_{S_x} + \cancel{F_{B_x}}^{=0(2)} = \frac{\partial}{\partial t}\int_{CV} u\rho\, d\forall + \int_{CS} u\rho \vec{V}\cdot d\vec{A}; \qquad 0 = \frac{\partial}{\partial t}\int_{CV} \rho\, d\forall + \int_{CS} \rho\vec{V}\cdot d\vec{A}$$

Assumptions: (1) $F_{S_x} = T_{\text{belt}} = T$
(2) $F_{B_x} = 0$
(3) Uniform flow at section ①
(4) All sand on belt moves with $V_{\text{belt}} = V_b$

Then

$$T = \frac{\partial}{\partial t}\int_{CV} u\rho\, d\forall + u_1\{-|\rho V_1 A_1|\} + u_2\{|\rho V_2 A_2|\}$$

Since $u_1 = 0$, and there is no flow at section ②, then

$$T = \frac{\partial}{\partial t}[V_b M_s] = M_s \cancel{\frac{\partial V_b}{\partial t}}^{0} + V_b \frac{\partial M_s}{\partial t}$$

where M_s is the mass of sand on the belt. From the continuity equation,

$$\frac{\partial}{\partial t}\int_{CV} \rho\, d\forall = \frac{\partial}{\partial t} M_s = -\int_{CS} \rho\vec{V}\cdot d\vec{A} = \dot{m}_s = 500 \text{ lbm/sec}$$

Then

$$T = V_b \dot{m}_s = \frac{3 \text{ ft}}{\text{sec}} \times \frac{500 \text{ lbm}}{\text{sec}} \times \frac{\text{slug}}{32.2 \text{ lbm}} \times \frac{\text{lbf}\cdot\text{sec}^2}{\text{slug}\cdot\text{ft}}$$

$$T = 46.6 \text{ lbf}$$

{This problem illustrates an application of the momentum equation to a problem in which the rate of change of momentum within the control volume is not equal to zero.}

**4-4.1 Differential Control Volume Analysis

We have considered a number of examples in which conservation of mass and the momentum equation have been applied to finite control volumes. The control volume chosen for analysis need not be finite in size.

Application of the basic equations to a differential control volume leads to a differential equation describing the relationship among properties in the flow field. In some cases the differential equation can be solved to give detailed information about property variations in the flow field.

To illustrate the use of differential control volumes, let us apply the continuity and momentum equations to a steady incompressible flow without friction, as shown in

** This section may be omitted without loss of continuity in the text material.

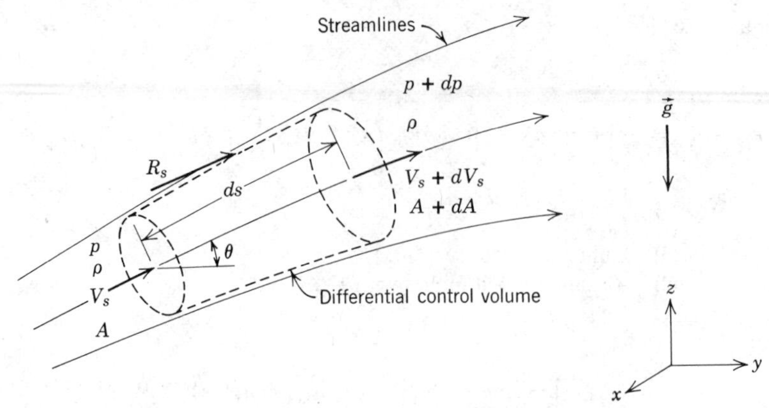

Fig. 4.4 Differential control volume for momentum analysis of flow along a streamline.

Fig. 4.4. For the case of steady, incompressible, frictionless flow along a streamline, integration of the differential equation leads to a useful relationship among velocity, pressure, and elevation in a flow field. The control volume chosen is fixed in space and bounded by flow streamlines, and is thus an element of a stream tube. The length of the control volume is ds.

Because the control volume is bounded by streamlines, the only flow across the bounding surfaces occurs at the end sections. These are located at coordinates s and $s + ds$, measured along the central streamline.

Properties at the inlet section are assigned arbitrary symbolic values. Properties at the outlet section are assumed to increase by a differential amount. Thus at $s + ds$, the flow speed is assumed to be $V_s + dV_s$, and so forth. The differential changes, dp, dV_s, and dA, all are assumed to be positive in formulating the problem. (As in a free-body analysis in statics or dynamics, the actual algebraic sign of each differential change will be determined from the results of the analysis.)

Now let us apply the continuity equation and the s component of the momentum equation to the control volume of Fig. 4.4.

a. Continuity Equation

Basic equation:

$$0 = \overbrace{\frac{\partial}{\partial t} \int_{\mathrm{CV}} \rho \, d\forall}^{=0(1)} + \int_{\mathrm{CS}} \rho \vec{V} \cdot d\vec{A} \tag{4.13}$$

Assumptions: (1) Steady flow
(2) No flow across bounding streamlines
(3) Incompressible flow, ρ = constant

Then

$$0 = \{-|\rho V_s A|\} + \{|\rho(V_s + dV_s)(A + dA)|\}$$

or

$$\rho V_s A = \rho(V_s + dV_s)(A + dA) \tag{4.20}$$

b. *s* Component of the Momentum Equation

Basic equation:

$$F_{S_s} + F_{B_s} = \overset{=0(1)}{\cancel{\frac{\partial}{\partial t}}} \int_{CV} u_s \rho \, d\forall + \int_{CS} u_s \rho \vec{V} \cdot d\vec{A} \tag{4.21}$$

Assumption: (4) No friction: $R_s = 0$ and F_{S_s} is due to pressure forces only.

The pressure force will have three terms:

$$F_{S_s} = pA - (p + dp)(A + dA) + \left(p + \frac{dp}{2}\right)(dA) \tag{4.22a}$$

The first and second terms in Eq. 4.22a are the pressure forces on the end faces of the control surface. The third term is the pressure force acting in the s direction on the bounding stream surface of the control volume. Its magnitude is the product of the average pressure acting on the stream surface, $p + \frac{1}{2}dp$, times the area component of the stream surface in the s direction, dA. Equation 4.22a simplifies to

$$F_{S_s} = -A\,dp - \tfrac{1}{2}dp\,dA \tag{4.22b}$$

The body force component in the s direction is

$$F_{B_s} = \rho g_s \, d\forall = \rho(-g\sin\theta)\left(A + \frac{dA}{2}\right) ds$$

But $\sin\theta \, ds = dz$, so that

$$F_{B_s} = -\rho g\left(A + \frac{dA}{2}\right) dz \tag{4.22c}$$

The momentum flux will be

$$\int_{CS} u_s \rho \vec{V} \cdot d\vec{A} = V_s\{-|\rho V_s A|\} + (V_s + dV_s)\{|\rho(V_s + dV_s)(A + dA)|\}$$

since there is no mass flux across the stream surfaces. The terms in braces are equal from continuity, so

$$\int_{CS} u_s \rho \vec{V} \cdot d\vec{A} = V_s(-\rho V_s A) + (V_s + dV_s)(\rho V_s A) = \rho V_s A \, dV_s \tag{4.23}$$

Substitution of Eqs. 4.22b, 4.22c, and 4.23 into the momentum equation gives

$$-A\,dp - \tfrac{1}{2}dp\,dA - \rho g A\,dz - \tfrac{1}{2}\rho g\,dA\,dz = \rho V_s A\,dV_s$$

Dividing by ρA and noting that products of differentials are negligible compared to the remaining terms, we obtain the result

$$-\frac{dp}{\rho} - g\,dz = V_s\,dV_s = d\left(\frac{V_s^2}{2}\right)$$

or

$$\frac{dp}{\rho} + d\left(\frac{V_s^2}{2}\right) + g\,dz = 0 \tag{4.24}$$

For incompressible flow, this equation may be integrated to obtain

$$\frac{p}{\rho} + \frac{V_s^2}{2} + gz = \text{constant}$$

or dropping the subscript, s,

$$\frac{p}{\rho} + \frac{V^2}{2} + gz = \text{constant} \tag{4.25}$$

This equation is subject to the restrictions:

1. Steady flow.
2. No friction.
3. Flow along a streamline.
4. Incompressible flow.

By applying the momentum equation to an infinitesimal streamtube control volume, for steady incompressible flow without friction, we have derived a relation among the pressure, velocity, and elevation. This relationship is very powerful and useful. For example, it could have been used to evaluate the pressure at the inlet of the reducing elbow analyzed in Example Problem 4.7 or to determine the velocity of water leaving the sluice gate of Example Problem 4.6. In both of these flow situations the restrictions required to derive Eq. 4.25 are reasonable idealizations of the actual flow behavior. The restrictions must be emphasized heavily because they do not always form a realistic model for flow behavior; consequently, they must be justified carefully each time Eq. 4.25 is applied.

Equation 4.25 is a form of the Bernoulli equation. It will be derived again in detail in Chapter 6 because it is such a useful tool for flow analysis and because an alternative derivation will give added insight into the need for care in applying the equation.

Example 4.9

Water flows steadily through a horizontal nozzle, discharging to the atmosphere. At the nozzle inlet the diameter is D_1; at the nozzle outlet the diameter is D_2. Derive an expression for the minimum gage pressure required at the nozzle inlet to produce a given volume flow rate, Q. Evaluate the inlet gage pressure if $D_1 = 3.0$ in., $D_2 = 1.0$ in., and the desired flow rate is 0.7 ft^3/sec.

EXAMPLE PROBLEM 4.9

GIVEN: Steady flow of water through a horizontal nozzle, discharging to the atmosphere.

$D_1 = 3.0$ in. $\quad D_2 = 1.0$ in. $\quad p_2 = p_{\text{atm}}$

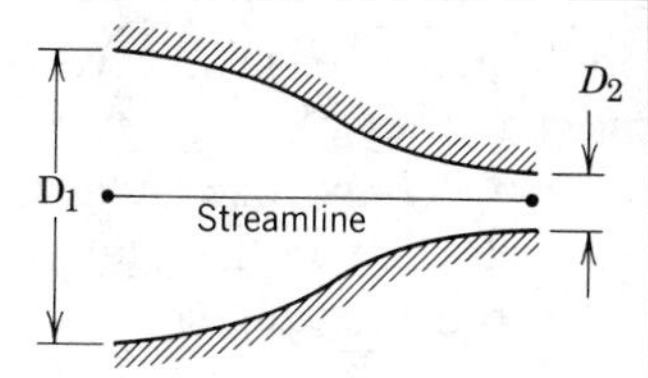

FIND: p_{1_g} as a function of volume flow rate, Q

SOLUTION:
Basic equations:

$$\frac{p_1}{\rho} + \frac{V_1^2}{2} + gz_1 = \frac{p_2}{\rho} + \frac{V_2^2}{2} + gz_2$$

$$= 0(1)$$

$$0 = \frac{\partial}{\partial t}\int_{\text{CV}} \rho \, d\forall + \int_{\text{CS}} \rho \vec{V} \cdot d\vec{A}$$

Assumptions: (1) Steady flow
(2) Incompressible flow
(3) Frictionless flow
(4) Flow along a streamline
(5) $z_1 = z_2$
(6) Uniform flow at sections ① and ②

Apply the Bernoulli equation along a streamline between points ① and ② to evaluate p_1. Then

$$p_{1g} = p_1 - p_{\text{atm}} = p_1 - p_2 = \frac{\rho}{2}(V_2^2 - V_1^2) = \frac{\rho}{2} V_1^2 \left[\left(\frac{V_2}{V_1}\right)^2 - 1\right]$$

Apply the continuity equation

$$0 = \{-|\rho V_1 A_1|\} + \{|\rho V_2 A_2|\} \qquad \text{or} \qquad V_1 A_1 = V_2 A_2 = Q$$

so that

$$\frac{V_2}{V_1} = \frac{A_1}{A_2} \qquad \text{and} \qquad V_1 = \frac{Q}{A_1}$$

Then

$$p_{1_g} = \frac{\rho Q^2}{2A_1^2}\left[\left(\frac{A_1}{A_2}\right)^2 - 1\right]$$

Since $A = \pi D^2/4$, then

$$p_{1_g} = \frac{8\rho Q^2}{\pi^2 D_1^4}\left[\left(\frac{D_1}{D_2}\right)^4 - 1\right]$$

With $D_1 = 3.0$ in., $D_2 = 1.0$ in., and $\rho = 1.94$ slug/ft^3,

$$p_{1_g} = \frac{8}{\pi^2} \times \frac{1.94 \text{ slug}}{\text{ft}^3} \times \frac{1}{(3)^4 \text{ in.}^4} \times Q^2\left[(3.0)^4 - 1\right] \frac{\text{lbf}\cdot\text{sec}^2}{\text{slug}\cdot\text{ft}} \times \frac{144 \text{ in.}^2}{\text{ft}^4}$$

$$p_{1_g} = 224\, Q^2 \frac{\text{lbf}\cdot\text{sec}^2}{\text{in.}^2\cdot\text{ft}^6}$$

p_{1_g}

With $Q = 0.7$ ft^3/sec, then $\quad p_{1_g} = 110$ lbf/in.^{2}

{This problem illustrates the application of the Bernoulli equation to a flow where the restrictions of steady, incompressible, frictionless flow along a streamline are a reasonable flow model.}

4-4.2 Control Volume Moving with Constant Velocity

In the preceding problems, which illustrate applications of the momentum equation to inertial control volumes, we have considered only stationary control volumes. A control volume (fixed relative to reference frame xyz) moving with a constant velocity, $\vec{V}_{rf}$, relative to a fixed (inertial) reference frame XYZ, is also inertial, since it has no acceleration with respect to XYZ.

Equation 4.11, which expresses system derivatives in terms of control volume variables, is valid for any constant-velocity motion of the coordinate system xyz (fixed to the control volume), provided that:

1. All velocities are measured *relative* to the control volume.
2. All time derivatives are measured *relative* to the control volume.

To emphasize this point, we rewrite Eq. 4.11 as

$$\left.\frac{dN}{dt}\right)_{\text{system}} = \frac{\partial}{\partial t}\int_{\text{CV}} \eta\rho\, d\forall + \int_{\text{CS}} \eta\rho \vec{V}_{xyz} \cdot d\vec{A} \tag{4.26}$$

Since all time derivatives must be measured relative to the control volume, in using this equation to obtain the momentum equation for an inertial control volume from the system formulation, we must set

$$N = \vec{P}_{xyz} \qquad \text{and} \qquad \eta = \vec{V}_{xyz}$$

to obtain the equation

$$\vec{F} = \vec{F}_S + \vec{F}_B = \frac{\partial}{\partial t}\int_{\text{CV}} \vec{V}_{xyz}\rho\, d\forall + \int_{\text{CS}} \vec{V}_{xyz}\rho \vec{V}_{xyz} \cdot d\vec{A} \tag{4.27}$$

Equation 4.27 is the formulation of Newton's second law applied to any inertial control volume (stationary or moving with a constant velocity). It is identical to Eq. 4.18 except that we have included the subscript xyz to emphasize that quantities must be measured relative to the control volume. (It is helpful to imagine that the velocities are those that would be seen by an observer moving at constant speed with the control volume.) The momentum equation is applied to an inertial control volume moving with constant velocity in Example Problem 4.10.

Example 4.10

The sketch shows a vane with a turning angle of 60°. The vane moves at constant speed, $U = 10$ m/sec, and receives a jet of water that leaves a stationary nozzle with speed, $V = 30$ m/sec. The nozzle has an exit area of 0.003 m^2. Determine the force that must be applied to maintain the vane speed constant.

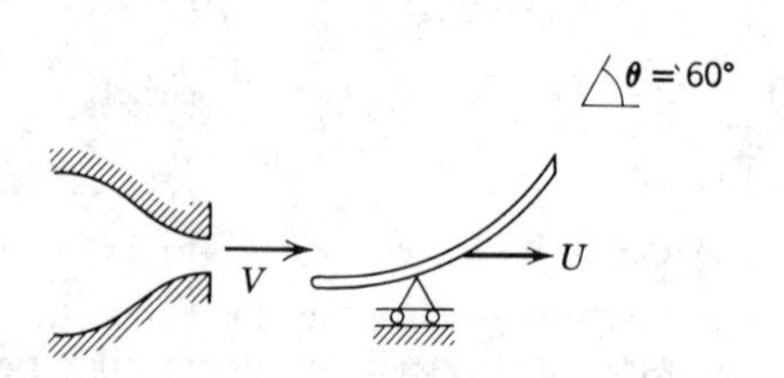

EXAMPLE PROBLEM 4.10

GIVEN: Vane having a turning angle, $\theta = 60°$, moves with constant velocity, $\vec{U} = 10\hat{i}$ m/sec. Water from a constant-area nozzle, $A = 0.003$ m^2, with velocity, $\vec{V} = 30\hat{i}$ m/sec, flows over the vane as shown.

FIND: The force that must be applied to maintain the vane speed constant.

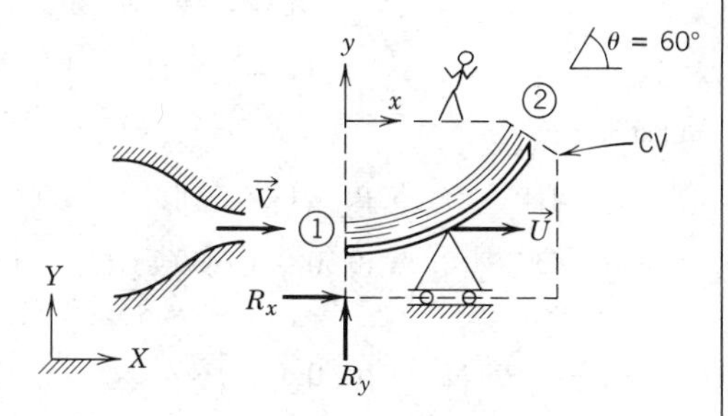

SOLUTION:
Select a control volume moving with the vane at constant velocity, $\vec{U}$, as shown by the dashed line. R_x and R_y are the components of the force required to maintain the velocity of the control volume at $10\hat{i}$ m/sec.
The control volume is inertial, since it is not accelerating (U = constant). Remember that all velocities must be measured relative to the control volume in applying the basic equations.

Basic equations: $$\vec{F}_S + \vec{F}_B = \frac{\partial}{\partial t}\int_{CV} \vec{V}_{xyz}\rho\, d\forall + \int_{CS} \vec{V}_{xyz}\rho \vec{V}_{xyz} \cdot d\vec{A}$$

$$0 = \frac{\partial}{\partial t}\int_{CV} \rho\, d\forall + \int_{CS} \rho \vec{V}_{xyz} \cdot d\vec{A}$$

Assumptions:
(1) Flow is steady relative to the vane
(2) Magnitude of relative velocity along the vane is constant: $|\vec{V}_1| = |\vec{V}_2| = V - U$
(3) Properties are uniform at sections ① and ②
(4) $F_{B_x} = F_{B_y} = 0$
(5) Incompressible flow

The x component of the momentum equation is

$$\underbrace{F_{S_x} + F_{B_x}}_{F_{B_x}=0(4)} = \underbrace{\frac{\partial}{\partial t}\int_{CV} u_{xyz}\rho\, d\forall}_{=0(1)} + \int_{CS} u_{xyz}\rho \vec{V}_{xyz} \cdot d\vec{A}$$

There is no net pressure force, since p_{atm} acts on all sides of the CV. Thus

$$R_x = \int_{A_1} u\{-|\rho V\, dA|\} + \int_{A_2} u\{|\rho V\, dA|\} = -u_1|\rho V_1 A_1| + u_2|\rho V_2 A_2|$$

(All velocities are measured relative to xyz.) From the continuity equation

$$0 = \int_{A_1}\{-|\rho V\, dA|\} + \int_{A_2}|\rho V\, dA| = -|\rho V_1 A_1| + |\rho V_2 A_2|$$

or

$$|\rho V_1 A_1| = |\rho V_2 A_2|$$

Therefore,

$$R_x = (u_2 - u_1)|\rho V_1 A_1|$$

All velocities must be measured relative to the CV, so we note that

$$u_1 = V - U \qquad u_2 = (V - U)\cos\theta$$

$$V_1 = V - U \qquad V_2 = V - U$$

Substituting yields

$$R_x = [(V-U)\cos\theta - (V-U)]|\rho(V-U)A_1| = (V-U)(\cos\theta - 1)|\rho(V-U)A_1|$$

$$= \frac{(30-10)\,\text{m}}{\text{sec}}\,(0.50-1)\Bigg|\frac{999\,\text{kg}}{\text{m}^3}\,\frac{(30-10)\,\text{m}}{\text{sec}} \times 0.003\,\text{m}^2\Bigg|\frac{\text{N}\cdot\text{sec}^2}{\text{kg}\cdot\text{m}}$$

$$R_x = -599\,\text{N} \qquad \{\text{to the left}\}$$

Writing the y component of the momentum equation, we obtain

$$F_{S_y} + \overset{=0(4)}{\cancel{F_{B_y}}} = \overset{=0(1)}{\cancel{\frac{\partial}{\partial t}\int_{\text{CV}}}} v_{xyz}\rho\, d\mathcal{V} + \int_{\text{CS}} v_{xyz}\rho \vec{V}_{xyz}\cdot d\vec{A}$$

$$R_y = \int_{\text{CS}} v\rho\vec{V}\cdot d\vec{A} = \int_{A_2} v\{\rho\vec{V}\cdot d\vec{A}\} \qquad \{v_1 = 0\} \qquad \left(\text{All velocities are measured relative to } xyz.\right)$$

$$= \int_{A_2} v|\rho V\, dA| = v_2|\rho V_2 A_2| = v_2|\rho V_1 A_1| \qquad \{\text{Recall } |\rho V_2 A_2| = |\rho V_1 A_1|\}$$

$$= (V-U)\sin\theta|\rho(V-U)A_1|$$

$$= \frac{(30-10)\,\text{m}}{\text{sec}}\,(0.866)\Bigg|\frac{999\,\text{kg}}{\text{m}^3}\,\frac{(30-10)\,\text{m}}{\text{sec}} \times 0.003\,\text{m}^2\Bigg|\frac{\text{N}\cdot\text{sec}^2}{\text{kg}\cdot\text{m}}$$

$$R_y = 1.04\,\text{kN} \qquad \{\text{upward}\}$$

Thus the force required to maintain the constant vane speed is

$$\vec{R} = -0.599\,\hat{i} + 1.04\,\hat{j}\,\text{kN}$$

$\vec{R}$

{This problem illustrates that to apply the momentum equation to an inertial control volume all velocities must be measured relative to the control volume.}

4-5 MOMENTUM EQUATION FOR CONTROL VOLUME WITH RECTILINEAR ACCELERATION

For an inertial control volume (having no acceleration relative to a stationary frame of reference), the appropriate formulation of Newton's second law is given by Eq. 4.27.

$$\vec{F} = \vec{F}_S + \vec{F}_B = \frac{\partial}{\partial t}\int_{\text{CV}} \vec{V}_{xyz}\rho\, d\mathcal{V} + \int_{\text{CS}} \vec{V}_{xyz}\rho\vec{V}_{xyz}\cdot d\vec{A} \tag{4.27}$$

Not all control volumes are inertial; a rocket must accelerate if it is to get off the ground. Since we are interested in analyzing control volumes that may accelerate relative to inertial coordinates, it is logical to ask whether Eq. 4.27 can be used for an accelerating control volume. To answer this question, let us briefly review the two major elements used in developing Eq. 4.27.

First, in relating the system derivatives to the control volume formulation (Eq. 4.26 or 4.11), the control volume was fixed relative to xyz; the flow field, $\vec{V}(x, y, z, t)$, was specified relative to the coordinates x, y, and z. No restriction was placed on the motion of the xyz reference frame. Consequently Eq. 4.26 is valid at any instant for any arbitrary motion of the coordinates x, y, and z provided that all time derivatives and velocities in the equation are measured relative to the control volume.

Second, the system equation

$$\vec{F} = \frac{d\vec{P}}{dt}\Bigg)_{\text{system}} \tag{4.2a}$$

where the linear momentum, $\vec{P}$, of the system is given by

$$\vec{P}_{\text{system}} = \int_{\text{mass (system)}} \vec{V}\,dm = \int_{\forall\text{ (system)}} \vec{V}\rho\,d\forall \tag{4.2b}$$

is valid only for velocities measured relative to an inertial reference frame. Thus, if we denote the inertial reference frame by XYZ, then Newton's second law states that

$$\vec{F} = \frac{d\vec{P}_{XYZ}}{dt}\Bigg)_{\text{system}} \tag{4.28}$$

Since the time derivatives of $\vec{P}_{XYZ}$ and $\vec{P}_{xyz}$ are not equal for a system accelerating relative to an inertial reference frame, Eq. 4.27 is not valid for an accelerating control volume.

To develop the momentum equation for a linearly accelerating control volume, it is necessary to relate $\vec{P}_{XYZ}$ of the system to $\vec{P}_{xyz}$ of the system. Then the system derivative $d\vec{P}_{xyz}/dt$ can be related to control volume variables through Eq. 4.26. Let us write Newton's second law for a system, remembering that the acceleration must be measured relative to an inertial reference frame that we have designated XYZ. We write

$$\vec{F} = \frac{d\vec{P}_{XYZ}}{dt}\Bigg)_{\text{system}} = \frac{d}{dt}\int_{\text{mass (system)}} \vec{V}_{XYZ}\,dm = \int_{\text{mass (system)}} \frac{d\vec{V}_{XYZ}}{dt}\,dm$$

$$\vec{F} = \int_{\text{mass (system)}} \vec{a}_{XYZ}\,dm \tag{4.29}$$

The only problem now is to obtain a suitable expression for $\vec{a}_{XYZ}$, for the special case in which coordinate frame xyz undergoes pure translation, without rotation, relative to inertial frame XYZ.

Since the motion of xyz is pure translation, without rotation, relative to inertial reference frame XYZ, then

$$\vec{a}_{XYZ} = \vec{a}_{xyz} + \vec{a}_{rf} \tag{4.30}$$

where

$\vec{a}_{XYZ}$ is the rectilinear acceleration of the system relative to inertial reference frame XYZ,
$\vec{a}_{xyz}$ is the rectilinear acceleration of the system relative to noninertial reference frame xyz, and
$\vec{a}_{rf}$ is the rectilinear acceleration of noninertial reference frame xyz relative to inertial frame XYZ.

For this case we write the system equation as

$$\vec{F} = \int_{\text{mass (system)}} \vec{a}_{XYZ}\, dm = \int_{\text{mass (system)}} (\vec{a}_{xyz} + \vec{a}_{rf})\, dm$$

Alternately,

$$\vec{F} - \int_{\text{mass (system)}} \vec{a}_{rf}\, dm = \int_{\text{mass (system)}} \vec{a}_{xyz}\, dm$$

Since

$$\vec{a}_{xyz} = \frac{d\vec{V}_{xyz}}{dt}$$

then

$$\vec{F} - \int_{\text{mass (system)}} \vec{a}_{rf}\, dm = \int_{\text{mass (system)}} \frac{d\vec{V}_{xyz}}{dt}\, dm$$

$$= \frac{d}{dt}\int_{\text{mass (system)}} \vec{V}_{xyz}\, dm = \left.\frac{d\vec{P}_{xyz}}{dt}\right)_{\text{system}}$$

Since $dm = \rho\, d\forall$, the system equation becomes

$$\vec{F} - \int_{\forall\,\text{(system)}} \vec{a}_{rf}\rho\, d\forall = \left.\frac{d\vec{P}_{xyz}}{dt}\right)_{\text{system}} \tag{4.31a}$$

where the linear momentum, $\vec{P}_{xyz}$, of the system is given by

$$\left.\vec{P}_{xyz}\right)_{\text{system}} = \int_{\text{mass (system)}} \vec{V}_{xyz}\, dm = \int_{\forall\,\text{(system)}} \vec{V}_{xyz}\rho\, d\forall \tag{4.31b}$$

and the force, $\vec{F}$, includes all surface and body forces acting on the system.

The system formulation is related to the formulation for a moving control volume by Eq. 4.26

$$\left.\frac{dN}{dt}\right)_{\text{system}} = \frac{\partial}{\partial t}\int_{\text{CV}} \eta\rho\, d\forall + \int_{\text{CS}} \eta\rho\vec{V}_{xyz}\cdot d\vec{A} \tag{4.26}$$

where

$$N_{\text{system}} = \int_{\text{mass (system)}} \eta\, dm = \int_{\forall\,\text{(system)}} \eta\rho\, d\forall \tag{4.6}$$

To derive the control volume formulation of Newton's second law, we set

$$N = \vec{P}_{xyz} \qquad \text{and} \qquad \eta = \vec{V}_{xyz}$$

From Eq. 4.26, with this substitution, we obtain

$$\left.\frac{d\vec{P}_{xyz}}{dt}\right)_{\text{system}} = \frac{\partial}{\partial t}\int_{\text{CV}} \vec{V}_{xyz}\rho \, dV\!\!\!\!- + \int_{\text{CS}} \vec{V}_{xyz}\rho \vec{V}_{xyz} \cdot d\vec{A} \tag{4.32}$$

From the system equation,

$$\left.\frac{d\vec{P}_{xyz}}{dt}\right)_{\text{system}} = \vec{F}_{\text{on system}} - \int_{V\!\!\!- \text{(system)}} \vec{a}_{rf}\rho \, dV\!\!\!\!- \tag{4.31a}$$

Since the system and the control volume coincided at t_0, then,

$$\vec{F}_{\text{on system}} - \int_{V\!\!\!- \text{(system)}} \vec{a}_{rf}\rho \, dV\!\!\!\!- = \vec{F}_{\text{on CV}} - \int_{\text{CV}} \vec{a}_{rf}\rho \, dV\!\!\!\!-$$

In light of this, Eqs. 4.31a and 4.32 may be combined to yield the formulation of Newton's second law for a control volume accelerating, without rotation, relative to an inertial reference frame:

$$\vec{F} - \int_{\text{CV}} \vec{a}_{rf}\rho \, dV\!\!\!\!- = \frac{\partial}{\partial t}\int_{\text{CV}} \vec{V}_{xyz}\rho \, dV\!\!\!\!- + \int_{\text{CS}} \vec{V}_{xyz}\rho \vec{V}_{xyz} \cdot d\vec{A} \tag{4.33}$$

Since $\vec{F} = \vec{F}_S + \vec{F}_B$, Eq. 4.33 becomes

$$\vec{F}_S + \vec{F}_B - \int_{\text{CV}} \vec{a}_{rf}\rho \, dV\!\!\!\!- = \frac{\partial}{\partial t}\int_{\text{CV}} \vec{V}_{xyz}\rho \, dV\!\!\!\!- + \int_{\text{CS}} \vec{V}_{xyz}\rho \vec{V}_{xyz} \cdot d\vec{A} \tag{4.34}$$

Comparing the momentum equation for a control volume with rectilinear acceleration, Eq. 4.34, to that for a nonaccelerating control volume, Eq. 4.27, we see that the only difference is the presence of one additional term in Eq. 4.34. When the control volume is not accelerating relative to an inertial reference frame, XYZ, $\vec{a}_{rf} = 0$, and Eq. 4.34 reduces to Eq. 4.27.

The precautions concerning the use of Eq. 4.27 also apply to the use of Eq. 4.34. Before attempting to apply either equation, one should draw the boundaries of the control volume and label appropriate coordinate directions. For an accelerating control volume, one must label a coordinate system (xyz) on the control volume and an inertial reference frame (XYZ).

In Eq. 4.34, $\vec{F}_S$ represents all surface forces acting on the control volume. Both the remaining terms on the left side of the equation may be functions of time, since the mass within the control volume may vary with time. Furthermore, the acceleration, $\vec{a}_{rf}$, of the reference frame xyz relative to an inertial frame in general will be a function of time.

All velocities in Eq. 4.34 are measured relative to the control volume. The momentum flux, $\vec{V}_{xyz}\rho\vec{V}_{xyz} \cdot d\vec{A}$, through an element of the control surface area, $d\vec{A}$, is a vector. The sign of the scalar product, $\rho \vec{V}_{xyz} \cdot d\vec{A}$, depends on the direction of the velocity vector, $\vec{V}_{xyz}$, relative to the area vector, $d\vec{A}$. The sign of the vector velocity, $\vec{V}_{xyz}$, depends on the coordinate system.

The momentum equation is a vector equation. As with all vector equations, it may be written in scalar component equations. The scalar components of Eq. 4.34 are

$$F_{S_x} + F_{B_x} - \int_{CV} a_{rf_x} \rho \, d\forall = \frac{\partial}{\partial t} \int_{CV} u_{xyz} \rho \, d\forall + \int_{CS} u_{xyz} \rho \vec{V}_{xyz} \cdot d\vec{A} \tag{4.35a}$$

$$F_{S_y} + F_{B_y} - \int_{CV} a_{rf_y} \rho \, d\forall = \frac{\partial}{\partial t} \int_{CV} v_{xyz} \rho \, d\forall + \int_{CS} v_{xyz} \rho \vec{V}_{xyz} \cdot d\vec{A} \tag{4.35b}$$

$$F_{S_z} + F_{B_z} - \int_{CV} a_{rf_z} \rho \, d\forall = \frac{\partial}{\partial t} \int_{CV} w_{xyz} \rho \, d\forall + \int_{CS} w_{xyz} \rho \vec{V}_{xyz} \cdot d\vec{A} \tag{4.35c}$$

Example 4.11

A vane with turning angle $\theta = 60°$ is attached to a cart. The cart and vane of mass 75 kg roll on a level track. Friction and air resistance may be neglected. The vane receives a jet of water, which leaves a stationary nozzle horizontally at 35 m/sec. The nozzle exit area is 0.003 m^2. Determine the velocity of the cart as a function of time and plot the results.

EXAMPLE PROBLEM 4.11

GIVEN: Vane and cart as sketched, with $M = 75$ kg.

FIND: (a) $U(t)$.
(b) Plot results.

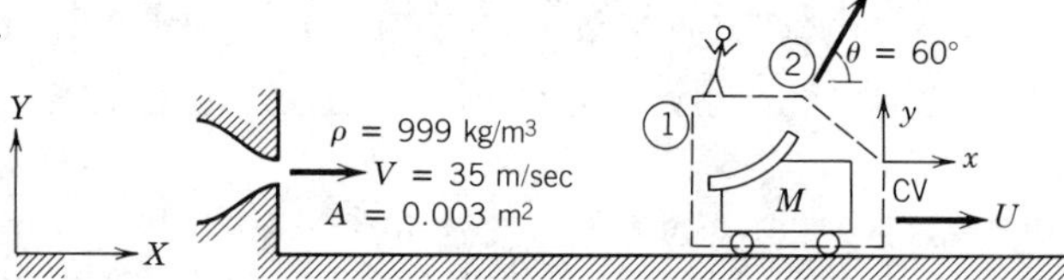

SOLUTION:

Choose the control volume and coordinate systems shown for the analysis. Note that XY is a fixed frame, while frame xy moves with the cart. Apply the x component of the momentum equation.

Basic equation:

$$\overset{=0(1)}{\cancel{F_{S_x}}} + \overset{=0(2)}{\cancel{F_{B_x}}} - \int_{CV} a_{rf_x} \rho \, d\forall = \overset{\simeq 0(3)}{\cancel{\frac{\partial}{\partial t}}} \int_{CV} u_{xyz} \rho \, d\forall + \int_{CS} u_{xyz} \rho \vec{V}_{xyz} \cdot d\vec{A}$$

Assumptions: (1) $F_{S_x} = 0$, since no resistance is present
(2) $F_{B_x} = 0$
(3) Neglect the rate of change of u_{xyz} for water in contact with the vane;

$$\frac{\partial}{\partial t} \int_{CV} u_{xyz} \rho \, d\forall \simeq 0$$

(4) Uniform flow at sections ① and ②
(5) Water stream is not slowed by friction on the vane, so $|\vec{V}_{xyz_1}| = |\vec{V}_{xyz_2}|$
(6) $A_2 = A_1 = A$

Then

$$-\int_{CV} a_{rf_x}\rho\, d\mathbf{V} = u_{xyz_1}\{-|\rho V_{xyz_1} A_1|\} + u_{xyz_2}\{|\rho V_{xyz_2} A_2|\}$$

where all velocities are measured relative to the xyz frame. Dropping subscripts rf and xyz, we obtain

$$-\int_{CV} a_x\rho\, d\mathbf{V} = u_1\{-|\rho V_1 A_1|\} + u_2\{|\rho V_2 A_2|\} \tag{1}$$

Evaluating these terms separately gives

$$-\int_{CV} a_x\rho\, d\mathbf{V} = -a_x M_{CV} = -a_x M = -\frac{dU}{dt} M$$

$$u_1\{-|\rho V_1 A_1|\} = (V - U)\{-|\rho(V - U)A|\} = -\rho(V - U)^2 A$$

$$u_2\{|\rho V_2 A_2|\} = (V - U)\cos\theta\{|\rho(V - U)A|\} = \rho(V - U)^2 A\cos\theta$$

Absolute value signs have been dropped from the flux terms, since $V \geq U$. Substitution in Eq. 1 gives

$$-M\frac{dU}{dt} = -\rho(V - U)^2 A + \rho(V - U)^2 A\cos\theta$$

or

$$-M\frac{dU}{dt} = (\cos\theta - 1)\rho(V - U)^2 A$$

Separating variables, we obtain

$$\frac{dU}{(V - U)^2} = \frac{(1 - \cos\theta)\rho A}{M}\, dt = b\, dt$$

Note that since V = constant, $dU = -d(V - U)$. Integrating between limits $U = 0$ at $t = 0$, and $U = U$ at $t = t$,

$$\int_0^U \frac{dU}{(V - U)^2} = \int_0^U \frac{-d(V - U)}{(V - U)^2} = \frac{1}{(V - U)}\Bigg]_0^U = \int_0^t b\, dt = bt$$

or

$$\frac{1}{(V - U)} - \frac{1}{V} = \frac{U}{V(V - U)} = bt$$

Solving for U, we obtain

$$\frac{U}{V} = \frac{Vbt}{1 + Vbt}$$

Evaluating the term, Vb, gives

$$Vb = V\frac{(1-\cos\theta)\rho A}{M}$$

$$Vb = \frac{35\ \text{m}}{\text{sec}} \times \frac{(1-0.5)}{75\ \text{kg}} \times \frac{999\ \text{kg}}{\text{m}^3} \times 0.003\ \text{m}^2 = 0.699\ \text{sec}^{-1}$$

Thus

$$\frac{U}{V} = \frac{0.699t}{1+0.699t} \qquad (t \text{ in sec}) \qquad U(t)$$

Plot:

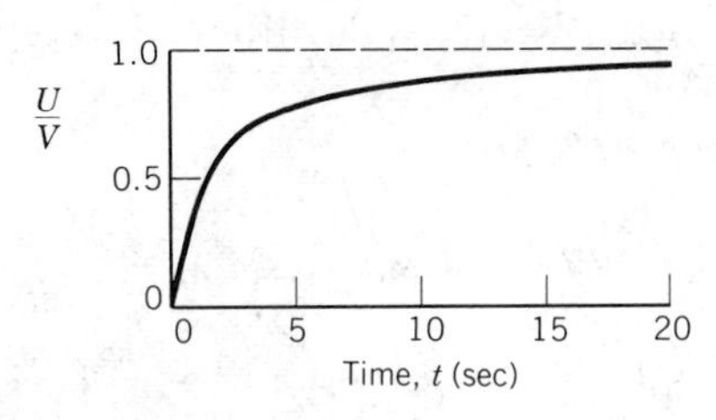

{As the vane speed, U, nears the jet speed, V, the mass flow rate crossing the control surface decreases toward zero. The plot shows the corresponding decrease in vane acceleration.}

Example 4.12

A small rocket, with an initial mass of 400 kg, is to be launched vertically. Upon ignition the rocket consumes fuel at the rate of 5 kg/sec and ejects gas at atmospheric pressure with a speed of 3500 m/sec relative to the rocket. Determine the initial acceleration of the rocket, and the rocket speed after 10 sec if air resistance is neglected.

EXAMPLE PROBLEM 4.12

GIVEN: Small rocket accelerates vertically from rest.
Air resistance may be neglected.
Rate of fuel consumption, $\dot{m}_{\text{out}} = 5$ kg/sec.
Exhaust velocity, $V_e = 3500$ m/sec, relative to rocket, leaving at atmospheric pressure.

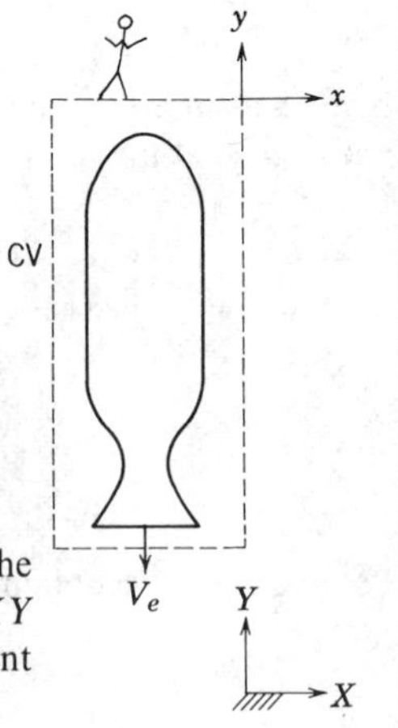

FIND: (a) Initial acceleration of the rocket.
(b) Rocket velocity after 10 sec.

SOLUTION:

Choose a control volume as shown by dashed lines. Because the control volume is accelerating, define inertial coordinate system XY and coordinate system xy attached to the CV. Apply the y component of the momentum equation.

Basic equation:

$$F_{S_y} + F_{B_y} - \int_{CV} a_{rf_y}\rho \, d\forall = \frac{\partial}{\partial t}\int_{CV} v_{xyz}\rho \, d\forall + \int_{CS} v_{xyz}\rho \vec{V}_{xyz} \cdot d\vec{A}$$

Assumptions: (1) Atmospheric pressure acts on all surfaces of the CV. Since air resistance is neglected, then $F_{S_y} = 0$
(2) Only body force is due to gravity
(3) Flow leaving the rocket is uniform and V_e is constant

Under these assumptions the momentum equation reduces to

$$\underset{Ⓐ}{F_{B_y}} - \underset{Ⓑ}{\int_{CV} a_{rf_y}\rho \, d\forall} = \underset{Ⓒ}{\frac{\partial}{\partial t}\int_{CV} v_{xyz}\rho \, d\forall} + \underset{Ⓓ}{\int_{CS} v_{xyz}\rho \vec{V}_{xyz} \cdot d\vec{A}} \tag{1}$$

Let us look at the equation term by term:

Ⓐ $$F_{B_y} = -\int_{CV} g\rho \, d\forall = -g\int_{CV} \rho \, d\forall = -gM_{CV}, \qquad \text{since } g \text{ is constant}$$

The mass of the CV will be a function of time because mass is leaving the CV at rate, $\dot{m}_e$. To determine M_{CV} as a function of time, we use the conservation of mass equation

$$\frac{\partial}{\partial t}\int_{CV} \rho \, d\forall + \int_{CS} \rho \vec{V} \cdot d\vec{A} = 0$$

Then

$$\frac{\partial}{\partial t}\int_{CV} \rho \, d\forall = -\int_{CS} \rho \vec{V} \cdot d\vec{A} = -\int_{A_e} \rho \vec{V} \cdot d\vec{A} = -\int_{A_e} \{|\rho V \, dA|\} = -|\dot{m}_e|$$

The minus sign indicates that the mass of the CV is decreasing with time. Since the mass of the CV is only a function of time, we can write

$$\frac{dM_{CV}}{dt} = -|\dot{m}_e|$$

To find the mass of the CV at any time, t, we integrate

$$\int_{M_0}^{M} dM_{CV} = -\int_0^t |\dot{m}_e| dt \qquad \text{where at } t = 0,\ M_{CV} = M_0, \text{ and at } t = t,\ M_{CV} = M$$

Then $M - M_0 = -|\dot{m}_e|t$ or $M = M_0 - \dot{m}_e t$.
{Since $\dot{m}_e$ is positive, we have dropped the absolute value sign.}
Substituting the expression for M into term Ⓐ, we obtain

$$F_{B_y} = -\int_{CV} g\rho \, d\forall = -gM_{CV} = -g(M_0 - \dot{m}_e t)$$

Ⓑ $$-\int_{CV} a_{rf_y}\rho \, d\forall$$

The acceleration, a_{rf_y}, of the CV is that seen by an observer in the XY coordinate system.

Thus a_{rf_y} is not a function of the coordinates xyz, and

$$-\int_{CV} a_{rf_y}\rho\, d\mathcal{V} = -a_{rf_y}\int_{CV} \rho\, d\mathcal{V} = -a_{rf_y}M_{CV} = -a_{rf_y}(M_0 - \dot{m}_e t)$$

Ⓒ $$\frac{\partial}{\partial t}\int_{CV} v_{xyz}\rho\, d\mathcal{V}$$

is the time rate of change of the y momentum of the fluid in the control volume measured relative to the control volume.

While the y momentum of the fluid inside the CV, measured relative to the CV, is a large number, it does not change appreciably with time. To see this, we must recognize that:

(1) The unburned fuel and the rocket structure have zero momentum relative to the rocket.
(2) The velocity of the gas at the nozzle exit remains constant with time as does the velocity at various points in the nozzle.

Consequently, it is reasonable to assume that

$$\frac{\partial}{\partial t}\int_{CV} v_{xyz}\rho\, d\mathcal{V} \approx 0$$

Ⓓ $$\int_{CS} v_{xyz}\rho\vec{V}_{xyz}\cdot d\vec{A} = \int_{A_e} v_{xyz}|\rho V_{xyz}\, dA| = v_{xyz}|\dot{m}_e|$$

Since $\vec{V}_e = -V_e\hat{j}$, then

$$v_{xyz}|\dot{m}_e| = -V_e|\dot{m}_e| = -V_e\dot{m}_e$$

Substituting terms Ⓐ through Ⓓ into Eq. 1, we obtain

$$-g(M_0 - \dot{m}_e t) - a_{rf_y}(M_0 - \dot{m}_e t) = -V_e\dot{m}_e$$

or

$$a_{rf_y} = \frac{V_e\dot{m}_e}{M_0 - \dot{m}_e t} - g \tag{2}$$

At time, $t = 0$

$$a_{rf_y}\Big)_{t=0} = \frac{V_e\dot{m}_e}{M_0} - g = \frac{3500\text{ m}}{\text{sec}} \times \frac{5\text{ kg}}{\text{sec}} \times \frac{1}{400\text{ kg}} - \frac{9.81\text{ m}}{\text{sec}^2}$$

$$a_{rf_y}\Big)_{t=0} = 33.9\text{ m/sec}^2$$

$a_{rf_y})_{t=0}$

The acceleration of the CV is by definition

$$a_{rf_y} = \frac{dV_{CV}}{dt}$$

Substituting from Eq. 2,

$$\frac{dV_{CV}}{dt} = \frac{V_e\dot{m}_e}{M_0 - \dot{m}_e t} - g$$

Separating variables and integrating gives

$$V_{CV} = \int_0^{V_{CV}} dV_{CV} = \int_0^t \frac{V_e \dot{m}_e dt}{M_0 - \dot{m}_e t} - \int_0^t g\,dt = -V_e \ln\left[\frac{M_0 - \dot{m}_e t}{M_0}\right] - gt$$

At $t = 10$ sec

$$V_{CV} = -\frac{3500\text{ m}}{\text{sec}} \times \ln\left[\frac{350\text{ kg}}{400\text{ kg}}\right] - \frac{9.81\text{ m}}{\text{sec}^2} \times 10\text{ sec}$$

$$V_{CV} = 369\text{ m/sec} \qquad V_{CV})_{t=10\text{ sec}}$$

{This problem illustrates the application of the momentum equation to a linearly accelerating control volume.}

**4-6 MOMENTUM EQUATION FOR CONTROL VOLUME WITH ARBITRARY ACCELERATION

In Section 4-5 we formulated the momentum equation for a control volume with rectilinear acceleration. The purpose of this section is to extend the formulation to include rotation and angular acceleration of the control volume in addition to translation and rectilinear acceleration.

First, we develop an expression for Newton's second law in an arbitrary, noninertial coordinate system. Then we use Eq. 4.26 to complete the formulation for a control volume. Newton's second law for a system moving relative to an inertial coordinate system is given by

$$\vec{F} = \frac{d\vec{P}_{XYZ}}{dt}\Big)_{\text{system}}$$

Since

$$\vec{P}_{XYZ}\Big)_{\text{system}} = \int_{M\,(\text{system})} \vec{V}_{XYZ}\,dm,$$

and M (system) is constant, then

$$\vec{F} = \frac{d}{dt}\int_{M\,(\text{system})} \vec{V}_{XYZ}\,dm = \int_{M\,(\text{system})} \frac{d\vec{V}_{XYZ}}{dt}\,dm$$

or

$$\vec{F} = \int_{M\,(\text{system})} \vec{a}_{XYZ}\,dm \tag{4.36}$$

The basic problem is to relate $\vec{a}_{XYZ}$ to the acceleration, $\vec{a}_{xyz}$, measured relative to a noninertial coordinate system and other variables. For this purpose, consider the noninertial reference frame, xyz, shown in Fig. 4.5.

** This section may be omitted without loss of continuity in the text material.

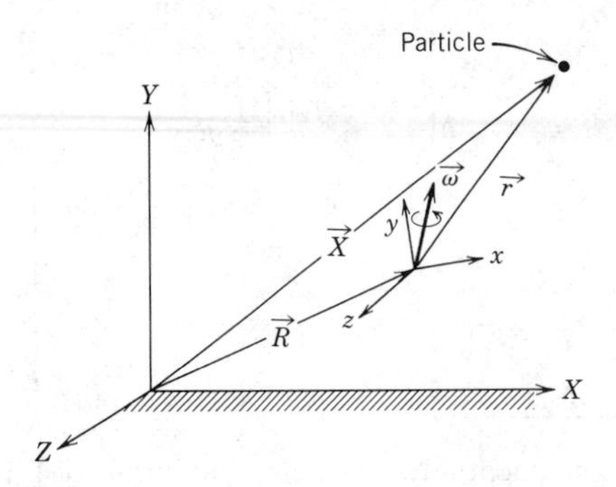

Fig. 4.5 Location of a particle in inertial (XYZ) and noninertial (xyz) reference frames.

The noninertial frame, xyz, is located by position vector $\vec{R}$ relative to the fixed frame. The noninertial frame rotates with angular velocity $\vec{\omega}$.[1] The position of a particle is located relative to the moving frame by position vector $\vec{r} = \hat{i}x + \hat{j}y + \hat{k}z$. Relative to the inertial reference frame, XYZ, the position of the particle is denoted by position vector $\vec{X}$. From the geometry of the figure, $\vec{X} = \vec{R} + \vec{r}$.

The velocity of the particle relative to an observer in the XYZ system is

$$\vec{V}_{XYZ} = \frac{d\vec{X}}{dt} = \frac{d\vec{R}}{dt} + \frac{d\vec{r}}{dt} = \vec{V}_{rf} + \frac{d\vec{r}}{dt} \tag{4.37}$$

We must be careful in evaluating $d\vec{r}/dt$ because both the magnitude $|\vec{r}|$ and the orientation of the unit vectors, $\hat{i}$, $\hat{j}$, and $\hat{k}$, are functions of time. Thus

$$\frac{d\vec{r}}{dt} = \frac{d}{dt}(x\hat{i} + y\hat{j} + z\hat{k}) = \hat{i}\frac{dx}{dt} + x\frac{d\hat{i}}{dt} + \hat{j}\frac{dy}{dt} + y\frac{d\hat{j}}{dt} + \hat{k}\frac{dz}{dt} + z\frac{d\hat{k}}{dt} \tag{4.38a}$$

The terms dx/dt, dy/dt, and dz/dt are the velocity components of the particle relative to xyz. Thus

$$\vec{V}_{xyz} = \hat{i}\frac{dx}{dt} + \hat{j}\frac{dy}{dt} + \hat{k}\frac{dz}{dt} \tag{4.38b}$$

You may recall from dynamics that for a rotating coordinate system,

$$\vec{\omega} \times \vec{r} = x\frac{d\hat{i}}{dt} + y\frac{d\hat{j}}{dt} + z\frac{d\hat{k}}{dt} \tag{4.38c}$$

Combining Eqs. 4.38a, 4.38b, and 4.38c, we obtain

$$\frac{d\vec{r}}{dt} = \vec{V}_{xyz} + \vec{\omega} \times \vec{r} \tag{4.38d}$$

Substituting into Eq. 4.37 gives

$$\vec{V}_{XYZ} = \vec{V}_{rf} + \vec{V}_{xyz} + \vec{\omega} \times \vec{r} \tag{4.39}$$

[1] Note that any arbitrary motion can be decomposed into a translation plus a rotation.

The acceleration of the particle relative to an observer in the XYZ system is

$$\vec{a}_{XYZ} = \frac{d\vec{V}_{XYZ}}{dt} = \frac{d\vec{V}_{rf}}{dt} + \frac{d\vec{V}_{xyz}}{dt} + \frac{d}{dt}(\vec{\omega} \times \vec{r})$$

or

$$\vec{a}_{XYZ} = \vec{a}_{rf} + \frac{d\vec{V}_{xyz}}{dt} + \frac{d}{dt}(\vec{\omega} \times \vec{r}) \tag{4.40}$$

Both $\vec{V}_{xyz}$ and $\vec{r}$ are measured relative to xyz, so the same caution observed in developing Eq. 4.38d applies. Thus

$$\frac{d\vec{V}_{xyz}}{dt} = \vec{a}_{xyz} + \vec{\omega} \times \vec{V}_{xyz} \tag{4.41a}$$

and

$$\frac{d}{dt}(\vec{\omega} \times \vec{r}) = \frac{d\vec{\omega}}{dt} \times \vec{r} + \vec{\omega} \times \frac{d\vec{r}}{dt}$$
$$= \dot{\vec{\omega}} \times \vec{r} + \vec{\omega} \times (\vec{V}_{xyz} + \vec{\omega} \times \vec{r})$$

or

$$\frac{d}{dt}(\vec{\omega} \times \vec{r}) = \dot{\vec{\omega}} \times \vec{r} + \vec{\omega} \times \vec{V}_{xyz} + \vec{\omega} \times (\vec{\omega} \times \vec{r}) \tag{4.41b}$$

Substituting Eqs. 4.41a and 4.41b into Eq. 4.40, we obtain

$$\vec{a}_{XYZ} = \vec{a}_{rf} + \vec{a}_{xyz} + 2\vec{\omega} \times \vec{V}_{xyz} + \vec{\omega} \times (\vec{\omega} \times \vec{r}) + \dot{\vec{\omega}} \times \vec{r} \tag{4.42}$$

The physical meaning of each term in Eq. 4.42 is

$\vec{a}_{XYZ}$: Absolute rectilinear acceleration of a particle relative to fixed reference frame XYZ

$\vec{a}_{rf}$: Absolute rectilinear acceleration of moving reference frame xyz relative to fixed frame XYZ

$\vec{a}_{xyz}$: Rectilinear acceleration of a particle *relative* to moving reference frame xyz (this acceleration would be that seen by an observer on moving frame xyz)

$2\vec{\omega} \times \vec{V}_{xyz}$: Coriolis acceleration due to motion of the particle *within* moving frame xyz

$\vec{\omega} \times (\vec{\omega} \times \vec{r})$: Centripetal acceleration due to rotation of moving frame xyz

$\dot{\vec{\omega}} \times \vec{r}$: Tangential acceleration due to angular acceleration of moving reference frame xyz

Substituting $\vec{a}_{XYZ}$, as given by Eq. 4.42, into Eq. 4.36, we obtain

$$\vec{F}_{\text{system}} = \int_{M\,(\text{system})} [\vec{a}_{rf} + \vec{a}_{xyz} + 2\vec{\omega} \times \vec{V}_{xyz} + \vec{\omega} \times (\vec{\omega} \times \vec{r}) + \dot{\vec{\omega}} \times \vec{r}]\, dm$$

or

$$\int_{M\,(\text{system})} \vec{a}_{xyz}\,dm = \vec{F} - \int_{M\,(\text{system})} [\vec{a}_{rf} + 2\vec{\omega} \times \vec{V}_{xyz} + \vec{\omega} \times (\vec{\omega} \times \vec{r}) + \dot{\vec{\omega}} \times \vec{r}]\,dm \tag{4.43a}$$

But

$$\int_{M\,(\text{system})} \vec{a}_{xyz}\,dm = \int_{M\,(\text{system})} \frac{d\vec{V}_{xyz}}{dt}\,dm = \frac{d}{dt}\int_{M\,(\text{system})} \vec{V}_{xyz}\,dm = \left.\frac{d\vec{P}_{xyz}}{dt}\right)_{\text{system}} \tag{4.43b}$$

where all time derivatives are those seen by an observer fixed in noninertial frame xyz. Combining Eqs. 4.43a and 4.43b, we obtain

$$\left.\frac{d\vec{P}_{xyz}}{dt}\right)_{\text{system}} = \vec{F} - \int_{M\,(\text{system})} [\vec{a}_{rf} + 2\vec{\omega} \times \vec{V}_{xyz} + \vec{\omega} \times (\vec{\omega} \times \vec{r}) + \dot{\vec{\omega}} \times \vec{r}]\,dm$$

or

$$\left.\frac{d\vec{P}_{xyz}}{dt}\right)_{\text{system}} = \vec{F}_S + \vec{F}_B - \int_{\forall\,(\text{system})} [\vec{a}_{rf} + 2\vec{\omega} \times \vec{V}_{xyz} + \vec{\omega} \times (\vec{\omega} \times \vec{r}) + \dot{\vec{\omega}} \times \vec{r}]\,\rho\,d\forall \tag{4.44}$$

Equation 4.44 is a statement of Newton's second law for a system. It is an expression for the rate of change of momentum, $\vec{P}_{xyz}$, measured relative to xyz, as seen by an observer in xyz. The system and control volume formulations are related by Eq. 4.26,

$$\left.\frac{dN}{dt}\right)_{\text{system}} = \frac{\partial}{\partial t}\int_{\text{CV}} \eta\rho\,d\forall + \int_{\text{CS}} \eta\rho\,\vec{V}_{xyz} \cdot d\vec{A} \tag{4.26}$$

To obtain the control volume formulation, we set $N = \vec{P}_{xyz}$ and $\eta = \vec{V}_{xyz}$. Then Eqs. 4.26 and 4.44 may be combined to give

$$\vec{F}_S + \vec{F}_B - \int_{\text{CV}} [\vec{a}_{rf} + 2\vec{\omega} \times \vec{V}_{xyz} + \vec{\omega} \times (\vec{\omega} \times \vec{r}) + \dot{\vec{\omega}} \times \vec{r}]\,\rho\,d\forall$$
$$= \frac{\partial}{\partial t}\int_{\text{CV}} \vec{V}_{xyz}\rho\,d\forall + \int_{\text{CS}} \vec{V}_{xyz}\rho\vec{V}_{xyz} \cdot d\vec{A} \tag{4.45}$$

Equation 4.45 is the most general control volume formulation of Newton's second law. Comparing the momentum equation for a control volume moving with arbitrary acceleration, Eq. 4.45, to that for a control volume moving with rectilinear acceleration, Eq. 4.34, we see that the only difference is the presence of three additional terms on the left side of Eq. 4.45. These terms result from the angular motion of noninertial reference frame xyz. Note that Eq. 4.45 reduces to Eq. 4.34, when the angular terms are zero, and to Eq. 4.27 for an inertial control volume.

The precautions concerning the use of Eqs. 4.27 and 4.34 also apply to the use of Eq. 4.45. Before attempting to apply this equation, one must draw the boundaries of the control volume and label appropriate coordinate directions. For a control volume moving with arbitrary acceleration, one must label a coordinate system (xyz) on the control volume and an inertial reference frame (XYZ).

**4-7 MOMENT OF MOMENTUM

Next we develop the moment of momentum equation for control volumes. We begin with the mathematical statement for a system and use Eq. 4.11 or 4.26 to obtain equations for fixed and rotating control volumes.

4-7.1 Equation for Fixed Control Volume

The moment of momentum equation for a system is

$$\vec{T} = \left.\frac{d\vec{H}}{dt}\right)_{\text{system}} \tag{4.3a}$$

where $\vec{T}$ = total torque exerted on the system by its surroundings, and
$\vec{H}$ = angular momentum of the system,

$$\vec{H} = \int_{M\,(\text{system})} \vec{r} \times \vec{V}\, dm = \int_{\forall\,(\text{system})} \vec{r} \times \vec{V} \rho\, d\forall \tag{4.3b}$$

All quantities in the system equation must be formulated with respect to an inertial reference frame. Reference frames at rest or moving with constant velocity are inertial, and Eq. 4.3b can be used directly to develop the control volume form of the moment of momentum equation. (Rotating reference frames are noninertial and will be treated in Section 4-7.3.)

The position vector, $\vec{r}$, locates each mass or volume element of the system with respect to the coordinate system. The torque, $\vec{T}$, applied to a system may be written

$$\vec{T} = \vec{r} \times \vec{F}_s + \int_{M\,(\text{system})} \vec{r} \times \vec{g}\, dm + \vec{T}_{\text{shaft}} \tag{4.3c}$$

The relation between the system and fixed control volume formulations is

$$\left.\frac{dN}{dt}\right)_{\text{system}} = \frac{\partial}{\partial t} \int_{\text{CV}} \eta \rho\, d\forall + \int_{\text{CS}} \eta \rho \vec{V} \cdot d\vec{A} \tag{4.11}$$

where

$$N_{\text{system}} = \int_{M\,(\text{system})} \eta\, dm$$

** This section may be omitted without loss of continuity in the text material.

If we set $N = \vec{H}$ and $\eta = \vec{r} \times \vec{V}$, then

$$\left.\frac{d\vec{H}}{dt}\right)_{\text{system}} = \frac{\partial}{\partial t}\int_{\text{CV}} \vec{r} \times \vec{V}\rho \, d\forall + \int_{\text{CS}} \vec{r} \times \vec{V}\rho\vec{V} \cdot d\vec{A} \tag{4.46}$$

Combining Eqs. 4.3a, 4.3c, and 4.46, we obtain

$$\vec{r} \times \vec{F}_s + \int_{M\,(\text{system})} \vec{r} \times \vec{g}\, dm + \vec{T}_{\text{shaft}} = \frac{\partial}{\partial t}\int_{\text{CV}} \vec{r} \times \vec{V}\rho \, d\forall + \int_{\text{CS}} \vec{r} \times \vec{V}\rho\vec{V} \cdot d\vec{A}$$

Since the system and control volume coincided at time t_0,

$$\vec{T}_{\text{system}} = \vec{T}_{\text{CV}}$$

and

$$\vec{r} \times \vec{F}_s + \int_{\text{CV}} \vec{r} \times \vec{g}\rho \, d\forall + \vec{T}_{\text{shaft}} = \frac{\partial}{\partial t}\int_{\text{CV}} \vec{r} \times \vec{V}\rho \, d\forall + \int_{\text{CS}} \vec{r} \times \vec{V}\rho\vec{V} \cdot d\vec{A} \tag{4.47}$$

Equation 4.47 is a general vector equation for moment of momentum for an inertial control volume. The left side of the equation is an expression for all the torques that act on the control volume. Terms on the right express the rate of change of angular momentum within the control volume and the rate of efflux of angular momentum from the control volume. All velocities in Eq. 4.47 are measured relative to the fixed control volume.

For analysis of rotating machinery, Eq. 4.47 is often used in scalar form by considering only the component directed along the axis of rotation. This application is illustrated in the next section.

4-7.2 Application to Turbomachinery

Fluid handling devices that direct the flow with blades or vanes attached to a rotating member are termed *turbomachines.* In contrast to positive displacement machinery, the fluid in a turbomachine is never confined completely. All work interactions between a fluid and a turbomachine rotor result from dynamic effects on the fluid stream.

Turbines extract energy from a fluid stream. The assembly of blades attached to the turbine shaft is called the *rotor*, *wheel*, or *runner*. The two most general classifications of turbines are *impulse* and *reaction* turbines. Impulse turbines are driven by one or more high-speed free jets. Each jet is accelerated in a nozzle external to the turbine wheel. If friction and gravity are neglected, neither the fluid pressure nor its speed relative to the runner change as it passes over the turbine vanes. Thus for an impulse turbine the fluid expansion from high to low pressure takes place in nozzles external to the blades, and the runner does not flow full of fluid.

In reaction turbines, part of the fluid expansion takes place externally and part within the moving blades. The external acceleration takes place and the flow is turned to enter the runner in the proper direction as it passes through nozzles or stationary blades called *guide vanes*. Because additional fluid acceleration relative to the rotor occurs within the moving blades, both the relative velocity and pressure of the stream change across the runner. The combination of a stationary blade row and a moving blade row is called a *stage*. Reaction turbines may be designed to flow full of fluid; as a consequence, reaction turbines can produce more power for a given size than impulse turbines.

Turbines range from simple windmills to complex steam and gas turbines with many stages of carefully designed blading. All of these devices can be analyzed in idealized form by applying Eq. 4.47.

Prime movers, which add energy to a fluid stream, are called *pumps* when the flow is liquid (or slurry), and *fans*, *blowers*, or *compressors* for gas and vapor handling units, depending on pressure rise. (Flow through fans is essentially incompressible; blowers have a small pressure rise; compressors are units designed for a larger pressure rise.) Pumps and turboblowers also can be analyzed in idealized form by applying Eq. 4.47.

Flow through a turbomachine may be nearly *axial*, nearly *radial*, or a combination, *mixed flow*. The basic design for a given application is usually chosen on the basis of efficiency. In general, axial machines handle the largest flows at high efficiency, followed by mixed flow and radial flow units. For pumps and blowers, radial flow units usually are designed for high pressure rise, followed by mixed flow and axial units for lower pressure rise applications.

Dimensionless parameters such as *specific speed*, *power coefficient*, *flow coefficient*, and *pressure ratio* frequently are used to characterize the performance of turbomachines. These parameters will be discussed in Chapter 7.

For turbomachinery analysis it is convenient to choose a fixed control volume enclosing the rotor for analysis of torque reactions. Torques due to surface forces may be ignored as a first approximation. The body force contribution may be neglected by symmetry. Then, for steady flow, Eq. 4.47 becomes

$$\vec{T}_{\text{shaft}} = \int_{\text{CS}} \vec{r} \times \vec{V} \rho \vec{V} \cdot d\vec{A} \tag{4.48a}$$

Let us write this equation in scalar form and illustrate its application to axial and radial flow machines.

The most practical coordinate system is one chosen with the z axis aligned with the axis of rotation of the machine. The term on the right side of Eq. 4.48a is the product of $\vec{r} \times \vec{V}$ with the mass flow rate at each section. For uniform flow into the rotor at section ① and out of the rotor at section ②, Eq. 4.48a becomes

$$T_{\text{shaft}}\hat{k} = (r_2 V_{t_2} - r_1 V_{t_1})\dot{m}\hat{k} \tag{4.48b}$$

or in scalar form,

$$T_{\text{shaft}} = (r_2 V_{t_2} - r_1 V_{t_1})\dot{m} \tag{4.48c}$$

The velocities that appear in Eq. 4.48c are the tangential components of the absolute velocity of the fluid crossing the control surface. The velocities are chosen positive when in the same direction as the blade speed, U. This sign convention gives $T_{\text{shaft}} > 0$

for pumps, fans, blowers, and compressors and $T_{shaft} < 0$ for turbines, corresponding to work input or output for these machines.

Equation 4.48c is the basic relationship between torque and moment of momentum for all forms of turbomachines, including turbines and prime movers. It often is called the *Euler turbine equation.*

The rate of work done on a turbomachine rotor is given by the dot product of the rotor angular velocity, $\vec{\omega}$, and the applied torque, $\vec{T}_{shaft}$. Using Eq. 4.48b, we obtain

$$\dot{W}_{in} = \vec{\omega} \cdot \vec{T}_{shaft} = \omega\hat{k} \cdot T_{shaft}\hat{k}$$

or

$$\dot{W}_{in} = \omega T_{shaft} = \omega(r_2 V_{t_2} - r_1 V_{t_1})\dot{m} \qquad (4.49a)$$

According to Eq. 4.49a, the moment of momentum of the fluid is increased by addition of shaft work. For a turbine, $\dot{W}_{in} < 0$ and the moment of momentum of the fluid must decrease.

Equation 4.49a may be written in two other useful forms. Introducing $U = r\omega$, the linear speed of the rotor at radius r, then

$$\dot{W}_{in} = (U_2 V_{t_2} - U_1 V_{t_1})\dot{m} \qquad (4.49b)$$

Dividing Eq. 4.49b by $\dot{m}g$, we obtain a quantity with dimensions of length, often termed the *head* added to the flow.

$$\Delta h = \frac{\dot{W}_{in}}{\dot{m}g} = \frac{1}{g}(U_2 V_{t_2} - U_1 V_{t_1}) \qquad (4.49c)$$

Equations 4.48 and 4.49 are simplified forms of the moment of momentum equation for a control volume. They all are written for a fixed control volume under the assumptions of steady flow and uniform flow at each section. The equations show that only the difference in the product rV_t or UV_t between the outlet and inlet sections is important in determining the torque applied to the rotor or the energy transfer to the fluid. No restriction has been made on geometry; the fluid may enter and leave at different radii.

The equations that we have derived also suggest the importance of clearly defining the velocity components of the fluid and rotor at inlet and outlet sections. For this purpose it is useful to develop *velocity polygons* for the inlet and outlet flows. Figure 4.6 shows two such velocity polygons and introduces the notation for blade and flow angles.

Each point on the blade of a turbomachine follows a circular path. Velocity polygons are plotted by unwrapping the blade path onto a plane surface. In the idealized situation at the design point, the flow relative to the rotor is assumed to enter and leave tangent to the blade profile at each section. (This idealized inlet condition is sometimes called *shockless* entry flow.) Blade angles, β, are measured relative to the circumferential direction, as shown in Fig. 4.6*a*. The blade angle, β_1, fixes the direction of the relative inlet velocity at design conditions.

The runner speed at inlet is $U_1 = \omega R_1$, and therefore is specified by the impeller geometry and operating speed. The absolute fluid velocity is the vector sum of the

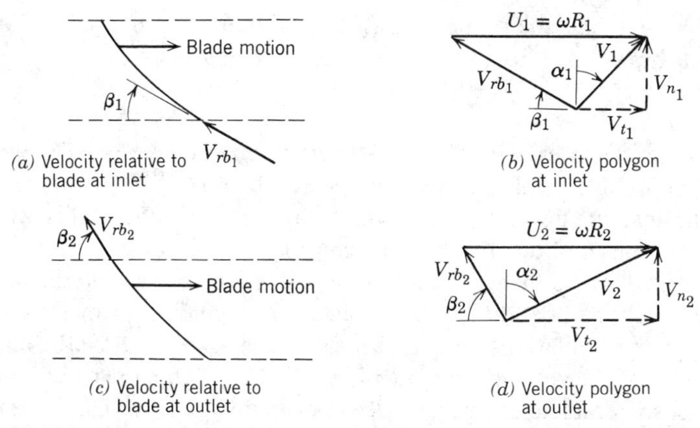

Fig. 4.6 Geometry and notation used to develop velocity polygons for a turbomachine.

impeller velocity and the flow velocity relative to the blade. The absolute velocity may be determined graphically as shown in Fig. 4.6*b*. The angle of the absolute fluid velocity, α_1, is measured from the normal as shown. The tangential component of the absolute velocity, V_{t_1}, and the component normal to the flow area, V_{n_1}, also are shown in Fig. 4.6*b*. Note from the geometry of Fig. 4.6*b* that at each section the normal component of the *absolute* velocity, V_n, and the normal component of the velocity relative to the blade, V_{rb_n}, are equal.

Velocity polygons at the outlet section are constructed in a similar fashion. The runner speed at the outlet is $U_2 = \omega R_2$, which again is known from the geometry and operating conditions of the turbomachine. The relative flow again is assumed to leave the impeller tangent to the blades, as shown in Fig. 4.6*c*. This idealizing assumption fixes its direction.

For an impulse turbine, there is no acceleration of the flow relative to the blade; the outlet and inlet velocity magnitudes are equal. For a reaction turbine or pump, the velocity relative to the blade in general changes in magnitude. The continuity equation must be applied, together with the impeller geometry, to determine the normal component of the velocity. This value, together with the outlet blade angle, is sufficient to establish the velocity relative to the blade at the impeller outlet. The velocity polygon is completed by adding the velocity relative to the blade and the wheel velocity vectorially as shown in Fig. 4.6*d*.

The inlet and outlet velocity polygons provide all the information required to calculate the torque or power absorbed or delivered by the impeller using Eqs. 4.48 or 4.49. The resulting values represent the performance of a turbomachine under idealized conditions at the design operating point, since we have assumed that all flows are uniform and that they enter and leave the rotor tangent to the blades. The idealized results represent the upper limits to performance of a turbomachine.

Performance of an actual machine is determined using the same basic approach but accounting for variations in flow properties across the blade span at inlet and outlet sections, and for deviations between blade and flow directions. Such detailed calculations are beyond the scope of this book. The idealized results that we have developed are applied next to axial and radial flow machines.

Example 4.13

An axial flow fan operates at 1200 rpm. The blade tip diameter is 1.1 m and the hub diameter 0.8 m. The blade inlet and exit angles are 30° and 60°, respectively. Inlet guide vanes give the absolute flow entering the first stage an angle of 30°. The fluid is air at standard conditions, and the flow may be considered as incompressible. There is no change in the axial component of velocity across the rotor. Assume the relative flow enters and leaves the rotor at the geometric blade angles, and use properties at the mean blade diameter for calculations. For these idealized conditions, draw the inlet velocity polygon, determine the volume flow rate of the fan, and sketch the rotor blade shapes. Using the data so obtained, draw the outlet velocity polygon, and calculate the torque and power required to drive the fan.

EXAMPLE PROBLEM 4.13

GIVEN: Flow through rotor of axial flow fan.

Tip diameter: 1.1 m
Hub diameter: 0.8 m
Operating speed: 1200 rpm
Absolute inlet angle: 30°
Blade inlet angle: 30°
Blade outlet angle: 60°

Fluid is air at standard conditions.
Use properties at mean diameter of blades.

FIND:
(a) Inlet velocity polygon.
(b) Volume flow rate.
(c) Rotor blade shape.
(d) Outlet velocity polygon.
(e) Rotor torque.
(f) Power required.

SOLUTION:
Apply the moment of momentum equation to a fixed control volume.

Computing equations:

$$\vec{T}_{\text{shaft}} = \int_{\text{CS}} \vec{r} \times \vec{V} \rho \vec{V} \cdot d\vec{A} \qquad (4.48a)$$

$$0 = \underbrace{\frac{\partial}{\partial t} \int_{\text{CV}} \rho \, d\forall}_{=0(2)} + \int_{\text{CS}} \rho \vec{V} \cdot d\vec{A}$$

Assumptions:
(1) Neglect torques due to body or surface forces
(2) Steady flow
(3) Uniform flow at inlet and outlet sections
(4) Incompressible flow
(5) No change in axial flow area
(6) Use mean radius of rotor blades, R_m

The inlet velocity polygon is

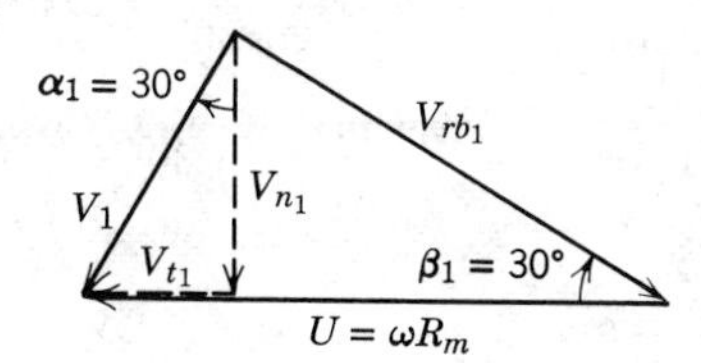

The blade shapes are

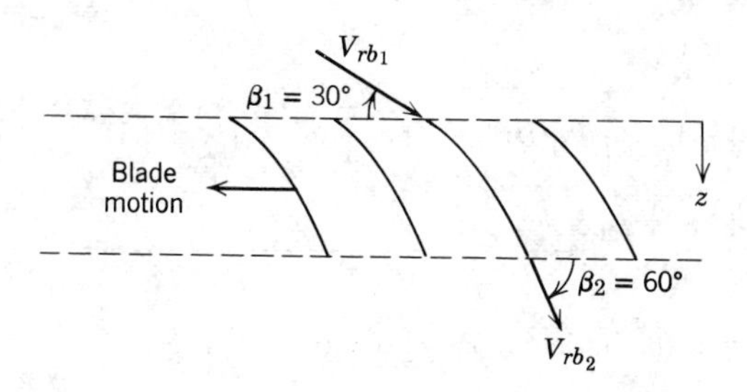

From continuity

$$0 = \{-|\rho V_{n_1} A_1|\} + \{|\rho V_{n_2} A_2|\}$$

or

$$Q = V_{n_1} A_1 = V_{n_2} A_2$$

Since $A_1 = A_2$, then $V_{n_1} = V_{n_2}$, and the outlet velocity polygon is as shown in the following figure:

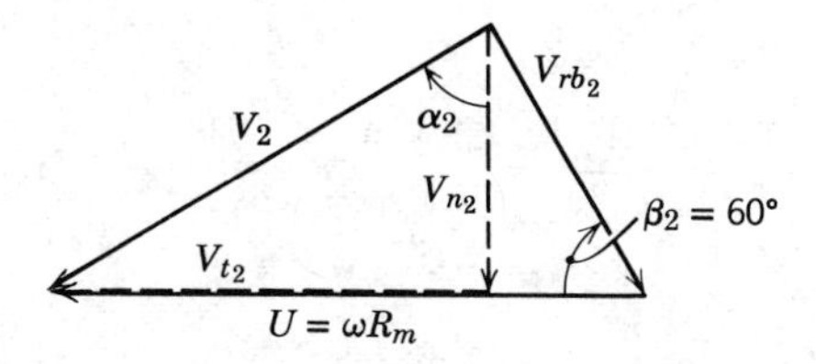

At the mean blade radius

$$U = R_m \omega = \frac{D_m}{2} \omega$$

$$U = \frac{\frac{1}{2}(1.1 + 0.8)\text{ m}}{2} \times \frac{1200\text{ rev}}{\text{min}} \times \frac{2\pi\text{ rad}}{\text{rev}} \times \frac{\text{min}}{60\text{ sec}} = 59.7\text{ m/sec}$$

From the geometry of the inlet velocity polygon,

$$U = V_{n_1}(\tan \alpha_1 + \cot \beta_1)$$

so that

$$V_{n_1} = \frac{U}{\tan \alpha_1 + \cot \beta_1} = \frac{59.7\text{ m}}{\text{sec}} \times \frac{1}{\tan 30° + \cot 30°} = 25.9\text{ m/sec}$$

Consequently,

$$V_1 = \frac{V_{n_1}}{\cos\alpha_1} = \frac{25.9\ \text{m}}{\text{sec}} \times \frac{1}{\cos 30^\circ} = 29.9\ \text{m/sec}$$

$$V_{t_1} = V_1 \sin\alpha_1 = \frac{29.9\ \text{m}}{\text{sec}} \times \sin 30^\circ = 15.0\ \text{m/sec}$$

and

$$V_{rb_1} = \frac{V_{n_1}}{\sin\beta_1} = \frac{25.9\ \text{m}}{\text{sec}} \times \frac{1}{\sin 30^\circ} = 51.8\ \text{m/sec}$$

The volume flow rate is

$$Q = V_{n_1}A_1 = \frac{\pi}{4} V_{n_1}(D_t^2 - D_h^2) = \frac{\pi}{4} \times \frac{25.9\ \text{m}}{\text{sec}} [(1.1)^2 - (0.8)^2]\ \text{m}^2$$

$$Q = 11.6\ \text{m}^3/\text{sec}$$

Q

From the geometry of the outlet velocity polygon,

$$\tan\alpha_2 = \frac{V_{t_2}}{V_{n_2}} = \frac{U - V_{n_2}\cot\beta_2}{V_{n_2}} = \frac{U - V_{n_1}\cot\beta_2}{V_{n_1}}$$

or

$$\alpha_2 = \tan^{-1}\left[\frac{\dfrac{59.7\ \text{m}}{\text{sec}} - \dfrac{25.9\ \text{m}}{\text{sec}} \times \cot 60^\circ}{25.9\ \dfrac{\text{m}}{\text{sec}}}\right] = 59.9^\circ$$

and

$$V_2 = \frac{V_{n_2}}{\cos\alpha_2} = \frac{V_{n_1}}{\cos\alpha_2} = \frac{25.9\ \text{m}}{\text{sec}} \times \frac{1}{\cos 59.9^\circ} = 51.6\ \text{m/sec}$$

Finally,

$$V_{t_2} = V_2 \sin\alpha_2 = \frac{51.6\ \text{m}}{\text{sec}} \times \sin 59.9^\circ = 44.6\ \text{m/sec}$$

The moment of momentum equation becomes

$$\vec{T} = T_z\hat{k} = \int_{\text{CS}} \vec{R}_m \times \vec{V}\rho\vec{V}\cdot d\vec{A} = \hat{k}\int_{\text{CS}} R_m V_t \rho \vec{V}\cdot d\vec{A}$$

so that for uniform flow

$$T_z = R_m V_{t_1}\{-|\rho V_{n_1} A_1|\} + R_m V_{t_2}\{|\rho V_{n_2} A_2|\} = \rho Q R_m (V_{t_2} - V_{t_1})$$

$$= \frac{1.23\ \text{kg}}{\text{m}^3} \times \frac{11.6\ \text{m}^3}{\text{sec}} \times \frac{0.95\ \text{m}}{2} \times \frac{(44.6 - 15.0)\ \text{m}}{\text{sec}} \times \frac{\text{N}\cdot\text{sec}^2}{\text{kg}\cdot\text{m}}$$

$$T_z = 201\ \text{N}\cdot\text{m}$$

T_z

Thus the torque *on* the CV is in the same sense as $\vec{\omega}$. The power required is

$$\dot{W}_{in} = \vec{\omega} \cdot \vec{T} = \omega_z T_z = \frac{1200 \text{ rev}}{\text{min}} \times \frac{2\pi \text{ rad}}{\text{rev}} \times \frac{\text{min}}{60 \text{ sec}} \times 201 \text{ N} \cdot \text{m} \times \frac{\text{W} \cdot \text{sec}}{\text{N} \cdot \text{m}}$$

$$\dot{W}_{in} = 25.3 \text{ kW}$$

$\dot{W}_{in}$

{This problem illustrates construction of velocity polygons and application of the moment of momentum equation for a fixed control volume to an axial flow machine under idealized conditions.}

Example 4.14

Water at 150 gal/min enters a mixed flow pump impeller axially through a 2 in. diameter inlet. The inlet velocity is axial and uniform. The outlet diameter of the impeller is 4 in. Flow leaves the impeller at a velocity of 10 ft/sec relative to the radial blades. The impeller speed is 3450 rpm. Determine the impeller exit width, *b*, the torque input to the impeller, and the horsepower supplied.

EXAMPLE PROBLEM 4.14

GIVEN: Flow as shown in the following figure.

FIND: (a) b_2.
(b) T_{shaft}.
(c) $\dot{W}_{in}$.

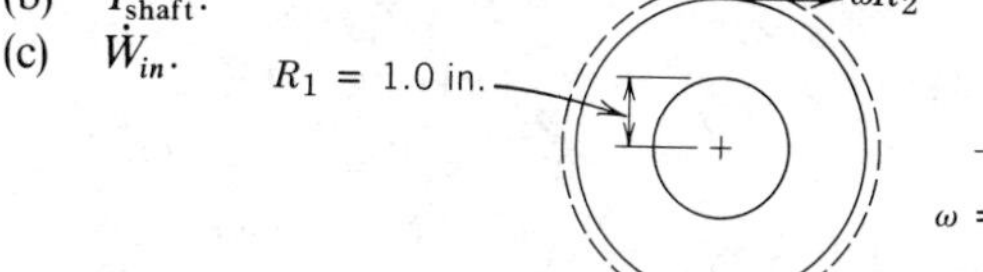

SOLUTION:

Apply the moment of momentum equation to a fixed control volume.

Computing equations:

$$\vec{T}_{\text{shaft}} = \int_{\text{CS}} \vec{r} \times \vec{V} \rho \vec{V} \cdot d\vec{A} \tag{4.48a}$$

$$= 0(2)$$

$$0 = \frac{\partial}{\partial t} \int_{\text{CV}} \rho \, d\forall + \int_{\text{CS}} \rho \vec{V} \cdot d\vec{A}$$

Assumptions: (1) Neglect torques due to body or surface forces
(2) Steady flow
(3) Uniform flow at inlet and outlet sections
(4) Incompressible flow

Then, from continuity,

$$0 = \{-|\rho V_1 \pi R_1^2|\} + \{|\rho V_{rb_2} 2\pi R_2 b_2|\}$$

or

$$\rho Q = \rho V_{rb_2} 2\pi R_2 b_2$$

so that

$$b_2 = \frac{Q}{2\pi R_2 V_{rb_2}} = \frac{1}{2\pi} \times \frac{150 \text{ gal}}{\text{min}} \times \frac{1}{2 \text{ in.}} \times \frac{\text{sec}}{10 \text{ ft}} \times \frac{\text{ft}^3}{7.48 \text{ gal}} \times \frac{\text{min}}{60 \text{ sec}} \times \frac{12 \text{ in.}}{\text{ft}}$$

$b_2 = 0.0319$ ft or 0.383 in. $\longleftarrow$ b_2

The axial inlet flow has no z component of moment of momentum. From the moment of momentum equation with uniform exit flow,

$$\hat{i}_z T_{\text{shaft}} = \vec{r}_2 \times \vec{V}_2\{|\rho Q|\}$$

At section ②,

$$\vec{r}_2 = R_2 \hat{i}_r$$

$$\vec{V}_2 = V_{rb_2} \hat{i}_r + \omega R_2 \hat{i}_\theta$$

so

$$\vec{r}_2 \times \vec{V}_2 = R_2(\omega R_2)\hat{i}_z = \omega R_2^2 \hat{i}_z$$

Thus,

$$T_{\text{shaft}} = \omega R_2^2 \rho Q = \frac{3450 \text{ rev}}{\text{min}} \times (2)^2 \text{ in.}^2 \times \frac{1.94 \text{ slug}}{\text{ft}^3} \times \frac{150 \text{ gal}}{\text{min}}$$

$$\times \frac{2\pi \text{ rad}}{\text{rev}} \times \frac{\text{min}^2}{3600 \text{ sec}^2} \times \frac{\text{ft}^3}{7.48 \text{ gal}} \times \frac{\text{ft}^2}{144 \text{ in.}^2} \times \frac{\text{lbf} \cdot \text{sec}^2}{\text{slug} \cdot \text{ft}}$$

$T_{\text{shaft}} = 6.51$ ft · lbf $\longleftarrow$ T_{shaft}

and

$$\dot{W}_{in} = \omega T_{\text{shaft}} = \frac{3450 \text{ rev}}{\text{min}} \times 6.51 \text{ ft} \cdot \text{lbf} \times \frac{2\pi \text{ rad}}{\text{rev}} \times \frac{\text{min}}{60 \text{ sec}} \times \frac{\text{hp} \cdot \text{sec}}{550 \text{ ft} \cdot \text{lbf}}$$

$\dot{W}_{in} = 4.28$ hp $\longleftarrow$ $\dot{W}_{in}$

4-7.3 Equation for Rotating Control Volume

In problems such as those involving impulse turbines or rotating spray systems, it is convenient to express all fluid velocities relative to the rotating component. The most convenient control volume is a noninertial one that rotates with the component. In this section we develop such a formulation.

Inertial and noninertial reference frames were related in Section 4-6. Figure 4.5, p. 140, showed the notation used. For a system,

$$\vec{T}_{\text{system}} = \left.\frac{d\vec{H}}{dt}\right)_{\text{system}} \tag{4.3a}$$

The angular momentum of a system in general motion must be specified relative to an inertial reference frame. Using the notation of Fig. 4.5,

$$\vec{H}_{\text{system}} = \int_{M\,(\text{system})} (\vec{R} + \vec{r}) \times \vec{V}_{XYZ}\, dm = \int_{\forall\,(\text{system})} (\vec{R} + \vec{r}) \times \vec{V}_{XYZ}\rho\, d\forall$$

With $\vec{R} = 0$ and the xyz frame restricted to rotation about XYZ, the equation becomes

$$\vec{H}_{\text{system}} = \int_{M\,(\text{system})} \vec{r} \times \vec{V}_{XYZ}\, dm = \int_{\forall\,(\text{system})} \vec{r} \times \vec{V}_{XYZ}\rho\, d\forall$$

so that

$$\vec{T}_{\text{system}} = \frac{d}{dt}\int_{M\,(\text{system})} \vec{r} \times \vec{V}_{XYZ}\, dm$$

Since the mass of a system is constant,

$$\vec{T}_{\text{system}} = \int_{M\,(\text{system})} \frac{d}{dt}(\vec{r} \times \vec{V}_{XYZ})\, dm$$

or

$$\vec{T}_{\text{system}} = \int_{M\,(\text{system})} \left(\frac{d\vec{r}}{dt} \times \vec{V}_{XYZ} + \vec{r} \times \frac{d\vec{V}_{XYZ}}{dt}\right) dm \tag{4.50}$$

From the analysis of Section 4-6,

$$\vec{V}_{XYZ} = \vec{V}_{rf} + \frac{d\vec{r}}{dt} \tag{4.37}$$

With xyz restricted to pure rotation, $\vec{V}_{rf} = 0$. The first term under the integral on the right side of Eq. 4.50 is then

$$\frac{d\vec{r}}{dt} \times \frac{d\vec{r}}{dt} = 0$$

Thus Eq. 4.50 reduces to

$$\vec{T}_{\text{system}} = \int_{M\,(\text{system})} \vec{r} \times \frac{d\vec{V}_{XYZ}}{dt}\, dm = \int_{M\,(\text{system})} \vec{r} \times \vec{a}_{XYZ}\, dm \tag{4.51}$$

From Eq. 4.42 with $\vec{a}_{rf} = 0$ (since xyz does not translate),

$$\vec{a}_{XYZ} = \vec{a}_{xyz} + 2\vec{\omega} \times \vec{V}_{xyz} + \vec{\omega} \times (\vec{\omega} \times \vec{r}) + \dot{\vec{\omega}} \times \vec{r}$$

Substituting into Eq. 4.51, we obtain

$$\vec{T}_{\text{system}} = \int_{M\,(\text{system})} \vec{r} \times [\vec{a}_{xyz} + 2\vec{\omega} \times \vec{V}_{xyz} + \vec{\omega} \times (\vec{\omega} \times \vec{r}) + \dot{\vec{\omega}} \times \vec{r}]\, dm$$

or

$$\vec{T}_{\text{system}} - \int_{M\,(\text{system})} \vec{r} \times [2\vec{\omega} \times \vec{V}_{xyz} + \vec{\omega} \times (\vec{\omega} \times \vec{r}) + \dot{\vec{\omega}} \times \vec{r}]\, dm$$

$$= \int_{M\,(\text{system})} \vec{r} \times \vec{a}_{xyz}\, dm = \int_{M\,(\text{system})} \vec{r} \times \frac{d\vec{V}_{xyz}}{dt}\, dm \qquad (4.52)$$

Using the time rate of change as observed from the system, we can write the last term as

$$\int_{M\,(\text{system})} \vec{r} \times \frac{d\vec{V}_{xyz}}{dt}\, dm = \frac{d}{dt}\int_{M\,(\text{system})} \vec{r} \times \vec{V}_{xyz}\, dm = \left.\frac{d\vec{H}_{xyz}}{dt}\right)_{\text{system}} \qquad (4.53)$$

The torque on the system is given by

$$\vec{T}_{\text{system}} = \vec{r} \times \vec{F}_S + \int_{M\,(\text{system})} \vec{r} \times \vec{g}\, dm + \vec{T}_{\text{shaft}} \qquad (4.3c)$$

The relation between the system and control volume formulations is

$$\left.\frac{dN}{dt}\right)_{\text{system}} = \frac{\partial}{\partial t}\int_{\text{CV}} \eta\rho\, d\forall + \int_{\text{CS}} \eta\rho\vec{V}_{xyz} \cdot d\vec{A} \qquad (4.26)$$

where

$$N_{\text{system}} = \int_{M\,(\text{system})} \eta\, dm$$

Setting N equal to $\vec{H}_{xyz})_{\text{system}}$ and $\eta = \vec{r} \times \vec{V}_{xyz}$ yields

$$\left.\frac{d\vec{H}_{xyz}}{dt}\right)_{\text{system}} = \frac{\partial}{\partial t}\int_{\text{CV}} \vec{r} \times \vec{V}_{xyz}\rho\, d\forall + \int_{\text{CS}} \vec{r} \times \vec{V}_{xyz}\rho\vec{V}_{xyz} \cdot d\vec{A} \qquad (4.54)$$

Combining Eqs. 4.52, 4.53, and 4.54, we obtain

$$\vec{r} \times \vec{F}_S + \int_{M\,(\text{system})} \vec{r} \times \vec{g}\, dm + \vec{T}_{\text{shaft}}$$

$$- \int_{M\,(\text{system})} \vec{r} \times [2\vec{\omega} \times \vec{V}_{xyz} + \vec{\omega} \times (\vec{\omega} \times \vec{r}) + \dot{\vec{\omega}} \times \vec{r}]\, dm$$

$$= \frac{\partial}{\partial t}\int_{\text{CV}} \vec{r} \times \vec{V}_{xyz}\rho\, d\forall + \int_{\text{CS}} \vec{r} \times \vec{V}_{xyz}\rho\, \vec{V}_{xyz} \cdot d\vec{A}$$

Since the system and control volume coincided at t_0,

$$\vec{r} \times \vec{F}_S + \int_{CV} \vec{r} \times \vec{g}\rho \, d\forall + \vec{T}_{\text{shaft}} - \int_{CV} \vec{r} \times [2\vec{\omega} \times \vec{V}_{xyz} + \vec{\omega} \times (\vec{\omega} \times \vec{r}) + \dot{\vec{\omega}} \times \vec{r}]\rho \, d\forall = \frac{\partial}{\partial t} \int_{CV} \vec{r} \times \vec{V}_{xyz}\rho \, d\forall + \int_{CS} \vec{r} \times \vec{V}_{xyz}\rho \vec{V}_{xyz} \cdot d\vec{A} \quad (4.55)$$

Equation 4.55 is the formulation of moment of momentum for a (noninertial) control volume rotating about an axis fixed in space. All fluid velocities and rates of change in Eq. 4.55 are evaluated relative to the control volume. Application of the equation to a rotating sprinkler is illustrated in Example Problem 4.15.

Example 4.15

A small lawn sprinkler is shown in the sketch below. At an inlet gage pressure of 20 kPa, the total volume flow rate of water through the sprinkler is 7.5 liters per minute and it rotates at 30 rpm. The diameter of each jet is 4 mm. Calculate the jet speed relative to each sprinkler nozzle. Evaluate the friction torque at the sprinkler pivot.

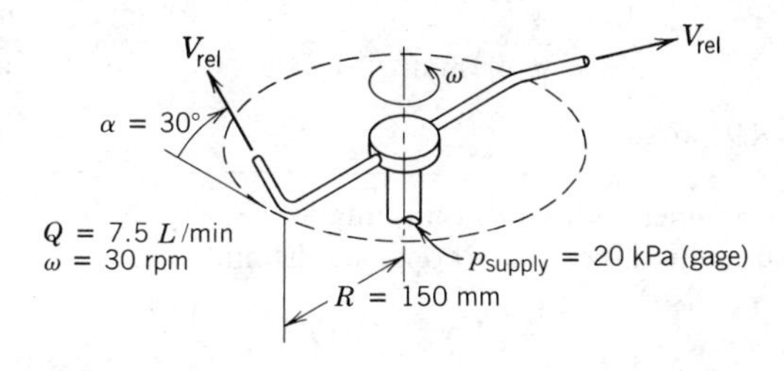

EXAMPLE PROBLEM 4.15

GIVEN: Small lawn sprinkler as shown.

FIND: (a) Jet speed relative to each nozzle.
(b) Friction torque at pivot.

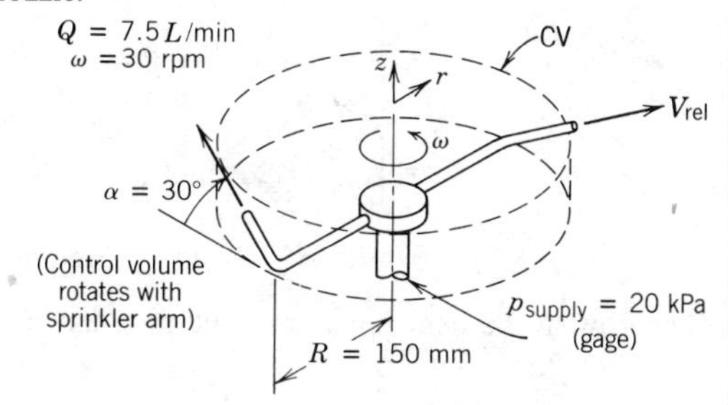

SOLUTION:

Apply continuity and moment of momentum equations using rotating control volume enclosing sprinkler arms.

Basic equations:

$$0 = \overset{=0(1)}{\frac{\partial}{\partial t}} \int_{CV} \rho \, d\forall + \int_{CS} \rho \vec{V}_{xyz} \cdot d\vec{A}$$

$$\vec{T}_{CV} - \int_{CV} \vec{r} \times [2\vec{\omega} \times \vec{V}_{xyz} + \vec{\omega} \times (\vec{\omega} \times \vec{r}) + \overset{=0(3)}{\dot{\vec{\omega}}} \times \vec{r}] \rho \, d\forall$$

$$= \overset{=0(1)}{\frac{\partial}{\partial t}} \int_{CV} \vec{r} \times \vec{V}_{xyz} \rho \, d\forall + \int_{CS} \vec{r} \times \vec{V}_{xyz} \rho \vec{V}_{xyz} \cdot d\vec{A}$$

where $\vec{T}_{CV}$ represents all external torques on the control volume.

Assumptions: (1) Steady flow
(2) Uniform flow at each section
(3) $\vec{\omega}$ = constant

From continuity

$$V_{rel} = \frac{Q}{2A_{jet}} = \frac{Q}{2} \frac{4}{\pi D_{jet}^2}$$

$$= \frac{1}{2} \times 7.5 \frac{L}{min} \times \frac{4}{\pi} \frac{1}{(4)^2 \, mm^2} \times \frac{m^3}{1000 \, L} \times \frac{10^6 \, mm^2}{m^2} \times \frac{min}{60 \, sec}$$

$$V_{rel} = 4.97 \text{ m/sec}$$

V_{rel}

Consider terms in the moment of momentum equation separately. The only external torque acting on the CV is friction in the pivot. It opposes the motion, so

$$\vec{T}_{CV} = -T_f \hat{i}_z \tag{1}$$

The integral on the left is evaluated for flow *within* the CV. Let the velocity and area within the sprinkler tubes be V_{CV} and A_{CV}, respectively. Then for *one side* the first term is

$$\int_{CV} \vec{r} \times [2\vec{\omega} \times \vec{V}_{xyz}] \rho \, d\forall = \int_0^R r\hat{i}_r \times [2\omega \hat{i}_z \times V_{CV} \hat{i}_r] \rho A_{CV} \, dr$$

$$= \int_0^R r\hat{i}_r \times 2\omega V_{CV} \hat{i}_\theta \rho A_{CV} \, dr$$

$$= \int_0^R 2\omega V_{CV} \rho A_{CV} r \, dr \hat{i}_z$$

$$= \omega R^2 \rho V_{CV} A_{CV} \hat{i}_z \qquad \text{(one side)}$$

(The flow in the bent portion of the tube has no r component of velocity, so it does not contribute to the integral.)

From continuity, $Q = 2V_{CV}A_{CV}$, so for *both sides* the integral becomes

$$\int_{CV} \vec{r} \times [2\vec{\omega} \times \vec{V}_{xyz}] \rho \, d\forall = \omega R^2 \rho Q \hat{i}_z \tag{2}$$

The second term in the integral is evaluated as

$$\int_{\mathrm{CV}} \vec{r} \times [\vec{\omega} \times (\vec{\omega} \times \vec{r})]\rho \, d\mathbf{V} = \int_{\mathrm{CV}} r\hat{i}_r \times [\omega\hat{i}_z \times (\omega\hat{i}_z \times r\hat{i}_r)]\rho \, d\mathbf{V}$$

$$= \int_{\mathrm{CV}} r\hat{i}_r \times [\omega\hat{i}_z \times \omega r\hat{i}_\theta]\rho \, d\mathbf{V}$$

$$= \int_{\mathrm{CV}} r\hat{i}_r \times \omega^2 r(-\hat{i}_r)\rho \, d\mathbf{V} = 0$$

so it contributes no torque.

The integral on the right is evaluated for flow crossing the control surface. For the right side

$$\int_{\mathrm{CS}} \vec{r} \times \vec{V}_{xyz}\rho\vec{V}_{xyz} \cdot d\vec{A}$$

$$= R\hat{i}_r \times V_{\mathrm{rel}}[\cos\alpha(-\hat{i}_\theta) + \sin\alpha\hat{i}_z]\{+\rho V_{\mathrm{rel}} A_{\mathrm{jet}}\}$$

$$= RV_{\mathrm{rel}}[\cos\alpha(-\hat{i}_z) + \sin\alpha(-\hat{i}_\theta)]\rho\frac{Q}{2}$$

In the left sprinkler arm the θ component has the same magnitude but opposite sign, so it cancels. For the complete CV,

$$\int_{\mathrm{CS}} \vec{r} \times \vec{V}_{xyz}\rho\vec{V}_{xyz} \cdot d\vec{A} = -RV_{\mathrm{rel}}\cos\alpha\,\rho Q\hat{i}_z \tag{3}$$

Combining terms (1), (2), and (3), we obtain

$$-T_f\hat{i}_z - \omega R^2\rho Q\hat{i}_z = -RV_{\mathrm{rel}}\cos\alpha\,\rho Q\hat{i}_z$$

or

$$T_f = R(V_{\mathrm{rel}}\cos\alpha - \omega R)\rho Q$$

From the data given

$$\omega R = \frac{30\ \mathrm{rev}}{\mathrm{min}} \times 150\ \mathrm{mm} \times \frac{2\pi\ \mathrm{rad}}{\mathrm{rev}} \times \frac{\mathrm{min}}{60\ \mathrm{sec}} \times \frac{\mathrm{m}}{1000\ \mathrm{mm}} = 0.471\ \frac{\mathrm{m}}{\mathrm{sec}}$$

Substituting gives

$$T_f = 150\ \mathrm{mm}\left(4.97\ \frac{\mathrm{m}}{\mathrm{sec}} \times \cos 30^\circ - 0.471\ \frac{\mathrm{m}}{\mathrm{sec}}\right) 999\ \frac{\mathrm{kg}}{\mathrm{m}^3} \times 7.5\ \frac{L}{\mathrm{min}}$$

$$\times \frac{\mathrm{m}^3}{1000\,L} \times \frac{\mathrm{min}}{60\ \mathrm{sec}} \times \frac{\mathrm{N}\cdot\mathrm{sec}^2}{\mathrm{kg}\cdot\mathrm{m}} \times \frac{\mathrm{m}}{1000\ \mathrm{mm}}$$

$$T_f = 0.0718\ \mathrm{N}\cdot\mathrm{m}$$

T_f

This problem has been included to illustrate use of the moment of momentum equation for a (noninertial) rotating control volume. It is left to you as an end-of-chapter exercise to show that the same results could be obtained by solving the problem using a fixed control volume and absolute fluid velocities.

4-8 THE FIRST LAW OF THERMODYNAMICS

The first law of thermodynamics is a statement of the conservation of energy. Recall that the system formulation of the first law was

$$\dot{Q} + \dot{W} = \left.\frac{dE}{dt}\right)_{\text{system}} \tag{4.4a}$$

where the total energy of the system is given by

$$E_{\text{system}} = \int_{M\,(\text{system})} e\, dm = \int_{\forall\,(\text{system})} e\rho\, d\forall \tag{4.4b}$$

and

$$e = u + \frac{V^2}{2} + gz$$

In Eq. 4.4a the rate of heat transfer, $\dot{Q}$, is positive when heat is added to the system from the surroundings; the rate of work, $\dot{W}$, is positive when work is done on the system by its surroundings.

The system and control volume formulations are related by

$$\left.\frac{dN}{dt}\right)_{\text{system}} = \frac{\partial}{\partial t}\int_{\text{CV}} \eta\rho\, d\forall + \int_{\text{CS}} \eta\rho\vec{V}\cdot d\vec{A} \tag{4.11}$$

where

$$N_{\text{system}} = \int_{M\,(\text{system})} \eta\, dm = \int_{\forall\,(\text{system})} \eta\rho\, d\forall \tag{4.6}$$

To derive the control volume formulation of the first law of thermodynamics, we set

$$N = E \qquad \text{and} \qquad \eta = e$$

From Eq. 4.11, with this substitution, we obtain

$$\left.\frac{dE}{dt}\right)_{\text{system}} = \frac{\partial}{\partial t}\int_{\text{CV}} e\rho\, d\forall + \int_{\text{CS}} e\rho\vec{V}\cdot d\vec{A} \tag{4.56}$$

In deriving Eq. 4.11, the system and the control volume coincided at t_0, so

$$[\dot{Q} + \dot{W}]_{\text{system}} = [\dot{Q} + \dot{W}]_{\text{control volume}}$$

In light of this, Eqs. 4.4a and 4.56 yield the control volume formulation of the first law of thermodynamics,

$$\dot{Q} + \dot{W} = \frac{\partial}{\partial t}\int_{\text{CV}} e\rho\, d\forall + \int_{\text{CS}} e\rho\vec{V}\cdot d\vec{A} \tag{4.57}$$

where

$$e = u + \frac{V^2}{2} + gz$$

Note that for steady flow the first term on the right side of Eq. 4.57 is zero. Is Eq. 4.57 the form of the first law used in thermodynamics? Even for steady flow, Eq. 4.57 is not quite the same form used in applying the first law to control volume problems. To obtain a formulation suitable and convenient for problem solutions, let us take a closer look at the work term, $\dot{W}$.

4-8.1 Rate of Work Done on a Control Volume

The term, $\dot{W}$, in Eq. 4.57 has a positive numerical value when work is done on the control volume by the surroundings. The rate of work done on the control volume is conveniently subdivided into four classifications.

$$\dot{W} = \dot{W}_s + \dot{W}_{\text{normal}} + \dot{W}_{\text{shear}} + \dot{W}_{\text{other}}$$

Let us consider these separately:

1. Shaft Work

We shall designate shaft work by W_s and the rate of work transferred in through the control surface by shaft work by $\dot{W}_s$.

2. Work Done on the Control Surface by Normal Stresses

Recall that work requires a force to be moved through a distance. Thus, in moving a force, $\vec{F}$, through an infinitesimal distance, $d\vec{s}$, the work done is given by

$$\delta W = \vec{F} \cdot d\vec{s}$$

To obtain the rate at which work is done by the force, divide by the time increment, Δt, and take the limit as $\Delta t \to 0$. Thus the rate of work done by the force, $\vec{F}$, is given by

$$\dot{W} = \lim_{\Delta t \to 0} \frac{\delta W}{\Delta t} = \lim_{\Delta t \to 0} \frac{\vec{F} \cdot d\vec{s}}{\Delta t} \qquad \text{or} \qquad \dot{W} = \vec{F} \cdot \vec{V}$$

The rate of work done on an element of area, $d\vec{A}$, of the control surface by normal stresses is given by

$$d\vec{F} \cdot \vec{V} = \sigma_{nn}\, d\vec{A} \cdot \vec{V}$$

The total rate of work done on the entire control surface by normal stresses is given by

$$\dot{W}_{\text{normal}} = \int_{\text{CS}} \sigma_{nn}\, d\vec{A} \cdot \vec{V} = \int_{\text{CS}} \sigma_{nn}\, \vec{V} \cdot d\vec{A}$$

3. Work Done on the Control Surface by Shear Stresses

Just as work is done by the normal stresses at the boundaries of the control volume, so may work be done by the shear stresses.

The shear force acting on an element of area of the control surface is given by

$$d\vec{F} = \vec{\tau}\, dA$$

where the shear stress vector, $\vec{\tau}$, is the shear stress acting in the plane of dA.

The rate of work done on the entire control surface by shear stresses is given by

$$\dot{W}_{\text{shear}} = \int_{\text{CS}} \vec{\tau}\, dA \cdot \vec{V} = \int_{\text{CS}} \vec{\tau} \cdot \vec{V}\, dA$$

This integral is better expressed as three terms

$$\dot{W}_{\text{shear}} = \int_{\text{CS}} \vec{\tau} \cdot \vec{V}\, dA$$

$$= \int_{A\,(\text{shafts})} \vec{\tau} \cdot \vec{V}\, dA + \int_{A\,(\text{solid surface})} \vec{\tau} \cdot \vec{V}\, dA + \int_{A\,(\text{ports})} \vec{\tau} \cdot \vec{V}\, dA$$

We have already accounted for the first term, since we included $\dot{W}_{\text{shaft}}$ previously. At solid surfaces, $\vec{V} = 0$, so the second term is zero (for a fixed control volume). Thus

$$\dot{W}_{\text{shear}} = \int_{A\,(\text{ports})} \vec{\tau} \cdot \vec{V}\, dA$$

This last term can be made zero by proper choice of control surfaces. If we choose a control surface that cuts across each port perpendicular to the flow, then $d\vec{A}$ is parallel to $\vec{V}$. Since $\vec{\tau}$ is in the plane of dA, then $\vec{\tau}$ is perpendicular to $\vec{V}$. Thus, for a control surface perpendicular to $\vec{V}$,

$$\vec{\tau} \cdot \vec{V} = 0 \qquad \text{and} \qquad \dot{W}_{\text{shear}} = 0$$

4. Other Work

Electrical energy could be added to the control volume. Also, electromagnetic energy, e.g. in radar or laser beams, could be absorbed. In most problems, such contributions will be absent, but we should note them in our general formulation.

With all of the terms in $\dot{W}$ evaluated, we obtain

$$\dot{W} = \dot{W}_s + \int_{\text{CS}} \sigma_{nn} \vec{V} \cdot d\vec{A} + \dot{W}_{\text{shear}} + \dot{W}_{\text{other}} \tag{4.58}$$

4-8.2 Control Volume Equation

Substituting the expression for $\dot{W}$ from Eq. 4.58 into Eq. 4.57 gives

$$\dot{Q} + \dot{W}_s + \int_{\text{CS}} \sigma_{nn} \vec{V} \cdot d\vec{A} + \dot{W}_{\text{shear}} + \dot{W}_{\text{other}} = \frac{\partial}{\partial t}\int_{\text{CV}} e\rho\, d\forall + \int_{\text{CS}} e\rho \vec{V} \cdot d\vec{A}$$

Rearranging this equation, we obtain

$$\dot{Q} + \dot{W}_s + \dot{W}_{\text{shear}} + \dot{W}_{\text{other}} = \frac{\partial}{\partial t}\int_{\text{CV}} e\rho\, d\forall + \int_{\text{CS}} e\rho \vec{V} \cdot d\vec{A} - \int_{\text{CS}} \sigma_{nn} \vec{V} \cdot d\vec{A}$$

Since $\rho = 1/v$, then

$$\int_{\mathrm{CS}} \sigma_{nn} \vec{V} \cdot d\vec{A} = \int_{\mathrm{CS}} \sigma_{nn} v \rho \vec{V} \cdot d\vec{A}$$

Hence

$$\dot{Q} + \dot{W}_s + \dot{W}_{\text{shear}} + \dot{W}_{\text{other}} = \frac{\partial}{\partial t} \int_{\mathrm{CV}} e\rho \, d\forall + \int_{\mathrm{CS}} (e - \sigma_{nn} v) \rho \vec{V} \cdot d\vec{A}$$

Viscous effects can make the normal stress, σ_{nn}, different from the negative of the thermodynamic pressure, $-p$. However, for most flows of common engineering interest, $\sigma_{nn} \simeq -p$. Then

$$\dot{Q} + \dot{W}_s + \dot{W}_{\text{shear}} + \dot{W}_{\text{other}} = \frac{\partial}{\partial t} \int_{\mathrm{CV}} e\rho \, d\forall + \int_{\mathrm{CS}} (e + pv) \rho \vec{V} \cdot d\vec{A}$$

Finally, substituting $e = u + V^2/2 + gz$ into the last term, we obtain the familiar form of the first law formulation for a control volume,

$$\dot{Q} + \dot{W}_s + \dot{W}_{\text{shear}} + \dot{W}_{\text{other}} = \frac{\partial}{\partial t} \int_{\mathrm{CV}} e\rho \, d\forall + \int_{\mathrm{CS}} \left(u + pv + \frac{V^2}{2} + gz \right) \rho \vec{V} \cdot d\vec{A} \quad (4.59)$$

Each work term in Eq. 4.59 represents the rate of work done on the control volume.

Example 4.16

Air at 14.7 psia, 70 F, enters a machine with negligible velocity and is discharged at 50 psia, 100 F through a pipe of area 1 ft^2. The flow rate is 20 lbm/sec. The power input to the machine is 600 hp. Determine the rate of heat transfer.

EXAMPLE PROBLEM 4.16

GIVEN: Air enters a compressor at ① and leaves at ② with conditions as shown. The air flow rate is 20 lbm/sec and the power input to the machine is 600 hp.

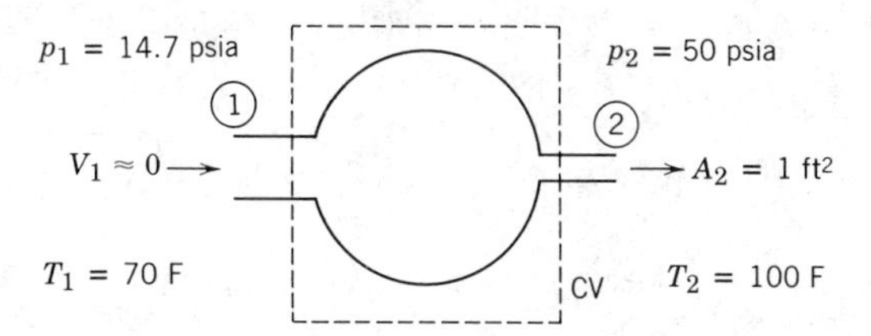

FIND: The rate of heat transfer.

SOLUTION:
Basic equations:

$$\dot{Q} + \dot{W}_s + \overset{=0(4)}{\dot{W}_{\text{shear}}} = \overset{=0(1)}{\frac{\partial}{\partial t}\int_{\text{CV}} e\rho \, d\text{V}} + \int_{\text{CS}} \left(u + pv + \frac{V^2}{2} + gz\right)\rho \vec{V} \cdot d\vec{A}$$

$$0 = \overset{=0(1)}{\frac{\partial}{\partial t}\int_{\text{CV}} \rho \, d\text{V}} + \int_{\text{CS}} \rho \vec{V} \cdot d\vec{A}$$

Assumptions:
(1) Steady flow
(2) Properties uniform over inlet and outlet sections
(3) Treat air as an ideal gas, $p = \rho RT$
(4) Area of CV at ① and ② perpendicular to velocity, thus $\dot{W}_{\text{shear}} = 0$
(5) $z_1 = z_2$
(6) Inlet kinetic energy is negligible

Under the assumptions listed, the first law becomes

$$\dot{Q} + \dot{W}_s = \int_{\text{CS}} \left(u + pv + \frac{V^2}{2} + gz\right)\rho \vec{V} \cdot d\vec{A}$$

$$\dot{Q} + \dot{W}_s = \int_{\text{CS}} \left(h + \frac{V^2}{2} + gz\right)\rho \vec{V} \cdot d\vec{A} \qquad \{h \equiv u + pv\}$$

or

$$\dot{Q} = -\dot{W}_s + \int_{\text{CS}} \left(h + \frac{V^2}{2} + gz\right)\rho \vec{V} \cdot d\vec{A}$$

For uniform properties (assumption 2) we can write

$$\dot{Q} = -\dot{W}_s + \left(h_1 + \overset{\approx 0(6)}{\frac{V_1^2}{2}} + gz_1\right)\{-|\rho_1 V_1 A_1|\} + \left(h_2 + \frac{V_2^2}{2} + gz_2\right)\{|\rho_2 V_2 A_2|\}$$

For steady flow, from conservation of mass,

$$\int_{\text{CS}} \rho \vec{V} \cdot d\vec{A} = 0$$

Therefore, $-|\rho_1 V_1 A_1| + |\rho_2 V_2 A_2| = 0$, or $|\rho_1 V_1 A_1| = |\rho_2 V_2 A_2| = \dot{m}$. Hence we can write

$$\dot{Q} = -\dot{W}_s + \dot{m}\left[(h_2 - h_1) + \frac{V_2^2}{2} + \overset{=0(5)}{g(z_2 - z_1)}\right]$$

Assume that air behaves as an ideal gas with constant c_p. Then $h_2 - h_1 = c_p(T_2 - T_1)$, and

$$\dot{Q} = -\dot{W}_s + \dot{m}\left[c_p(T_2 - T_1) + \frac{V_2^2}{2}\right]$$

From continuity $|V_2| = \dot{m}/\rho_2 A_2$. Since $p_2 = \rho_2 R T_2$, then

$$|V_2| = \frac{\dot{m}}{A_2}\frac{RT_2}{p_2} = \frac{20\text{ lbm}}{\text{sec}} \times \frac{1}{1\text{ ft}^2} \times \frac{53.3\text{ ft}\cdot\text{lbf}}{\text{lbm}\cdot\text{R}} \times 560\text{ R} \times \frac{\text{in.}^2}{50\text{ lbf}} \times \frac{\text{ft}^2}{144\text{ in.}^2}$$

$$|V_2| = 82.9\text{ ft/sec}$$

$$\dot{Q} = -\dot{W}_s + \dot{m}c_p(T_2 - T_1) + \dot{m}\frac{V_2^2}{2}$$

$$= -600\text{ hp} \times \frac{550\text{ ft}\cdot\text{lbf}}{\text{hp}\cdot\text{sec}} \times \frac{\text{Btu}}{778\text{ ft}\cdot\text{lbf}} + \frac{20\text{ lbm}}{\text{sec}} \times \frac{0.24\text{ Btu}}{\text{lbm}\cdot\text{R}} \times 30\text{ R}$$

$$+ \frac{20\text{ lbm}}{\text{sec}} \times \frac{(82.9)^2}{2}\frac{\text{ft}^2}{\text{sec}^2} \times \frac{\text{slug}}{32.2\text{ lbm}} \times \frac{\text{Btu}}{778\text{ ft}\cdot\text{lbf}} \times \frac{\text{lbf}\cdot\text{sec}^2}{\text{slug}\cdot\text{ft}}$$

$$\dot{Q} = -277\text{ Btu/sec} \qquad \{\text{heat rejection}\}$$

$\dot{Q}$

{In addition to demonstrating a straightforward application of the first law, this problem illustrates the need for keeping units straight.}

Example 4.17

A tank of volume 0.1 m^3 is connected to a high pressure air line; both line and tank are initially at a uniform temperature of 20 C. The initial tank gage pressure is 100 kPa. The absolute line pressure is 2.0 MPa; the line is large enough so that its temperature and pressure may be assumed to remain constant. The tank temperature is monitored by a fast-response thermocouple. At the instant after the valve is opened, the tank temperature rises at the rate of 0.05 C/sec. Determine the instantaneous flow rate of air into the tank if heat transfer is neglected.

EXAMPLE PROBLEM 4.17

GIVEN: Air supply pipe and tank as shown. At $t = 0^+$, $\partial T/\partial t = 0.05$ C/sec.

FIND: $\dot{m}$ at $t = 0^+$.

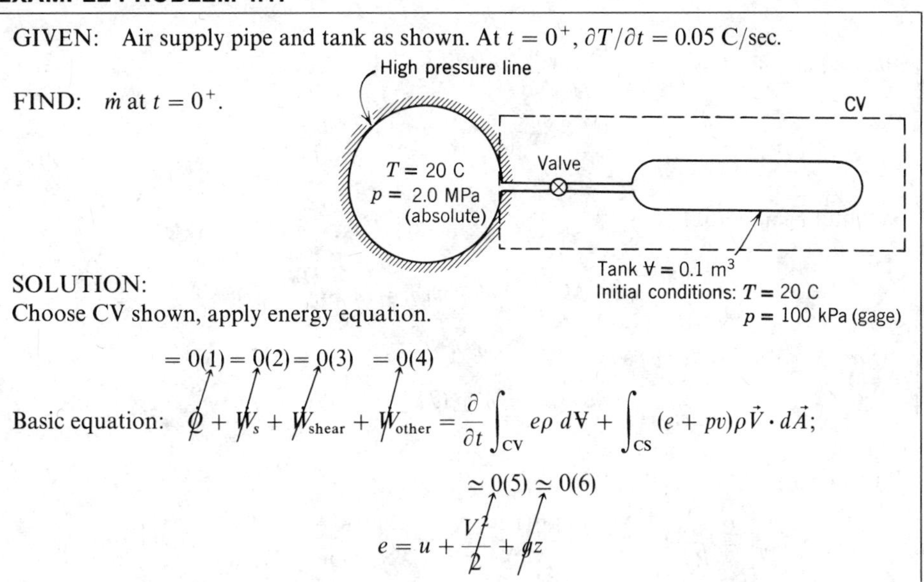

SOLUTION:

Choose CV shown, apply energy equation.

$$\text{Basic equation:} \quad \overset{=0(1)}{\dot{Q}} + \overset{=0(2)}{\dot{W}_s} + \overset{=0(3)}{\dot{W}_{\text{shear}}} + \overset{=0(4)}{\dot{W}_{\text{other}}} = \frac{\partial}{\partial t}\int_{\text{CV}} e\rho\, d\forall + \int_{\text{CS}} (e + pv)\rho \vec{V}\cdot d\vec{A};$$

$$e = u + \overset{\simeq 0(5)}{\frac{V^2}{2}} + \overset{\simeq 0(6)}{gz}$$

Assumptions: (1) $\dot{Q} = 0$ (given)
(2) $\dot{W}_s = 0$
(3) $\dot{W}_{\text{shear}} = 0$
(4) $\dot{W}_{\text{other}} = 0$
(5) Velocities in line and tank are small
(6) Neglect potential energy
(7) Uniform flow at tank inlet
(8) Properties uniform in tank
(9) Ideal gas, $p = \rho RT$, $du = c_v\, dT$

Then

$$0 = \frac{\partial}{\partial t}\int_{\text{CV}} u_{\text{tank}}\rho\, d\forall + (u_{\text{line}} + pv)\{-|\rho VA|\}$$

But initially, T is uniform, so $u_{\text{tank}} = u_{\text{line}} = u$, and

$$0 = \frac{\partial}{\partial t}\int_{\text{CV}} u\rho\, d\forall + (u + pv)\{-|\rho VA|\}$$

Since tank properties are uniform, $\partial/\partial t$ may be replaced by d/dt, and

$$0 = \frac{d}{dt}[uM] - (u + pv)\dot{m}$$

or

$$0 = u\frac{dM}{dt} + M\frac{du}{dt} - u\dot{m} - pv\dot{m} \qquad (1)$$

The term, dM/dt, may be evaluated from continuity:

Basic equation: $$0 = \frac{\partial}{\partial t}\int_{\text{CV}} \rho\, d\forall + \int_{\text{CS}} \rho \vec{V}\cdot d\vec{A}$$

$$0 = \frac{dM}{dt} + \{-|\rho VA|\} \qquad \text{or} \qquad \frac{dM}{dt} = \dot{m}$$

Substituting into Eq. 1 gives

$$0 = \cancel{u\dot{m}} + M\frac{du}{dt} - \cancel{u\dot{m}} - pv\dot{m} = Mc_v\frac{dT}{dt} - pv\dot{m}$$

or

$$\dot{m} = \frac{Mc_v(dT/dt)}{pv} = \frac{\rho\forall c_v(dT/dt)}{pv} = \frac{\rho\forall c_v(dT/dt)}{RT} \qquad (2)$$

But at $t = 0$,

$$\rho = \rho_{\text{tank}} = \frac{p_{\text{tank}}}{RT} = \frac{(1.00 + 1.01)10^5\ \text{N}}{\text{m}^2} \times \frac{\text{kg}\cdot\text{K}}{287\ \text{N}\cdot\text{m}} \times \frac{1}{293\ \text{K}} = 2.39\ \text{kg/m}^3$$

Substituting into Eq. 2, we obtain

$$\dot{m} = \frac{2.39\ \text{kg}}{\text{m}^3} \times 0.1\ \text{m}^3 \times \frac{717\ \text{N}\cdot\text{m}}{\text{kg}\cdot\text{K}} \times \frac{0.05\ \text{K}}{\text{sec}} \times \frac{\text{kg}\cdot\text{K}}{287\ \text{N}\cdot\text{m}} \times \frac{1}{293\ \text{K}} \times \frac{1000\ \text{g}}{\text{kg}}$$

$$\dot{m} = 0.102\ \text{g/sec}$$

$\dot{m}$

{This problem illustrates the application of the energy equation to an unsteady flow situation.}

4-9 THE SECOND LAW OF THERMODYNAMICS

Recall that the system formulation of the second law is

$$\left.\frac{dS}{dt}\right)_{\text{system}} \geq \frac{1}{T}\dot{Q} \tag{4.5a}$$

where the total entropy of the system is given by

$$S_{\text{system}} = \int_{\text{mass (system)}} s\, dm = \int_{\forall\,\text{(system)}} s\rho\, d\forall \tag{4.5b}$$

The relation between system and control volume formulations is

$$\left.\frac{dN}{dt}\right)_{\text{system}} = \frac{\partial}{\partial t}\int_{\text{CV}} \eta\rho\, d\forall + \int_{\text{CS}} \eta\rho\vec{V}\cdot d\vec{A} \tag{4.11}$$

where

$$N_{\text{system}} = \int_{\text{mass (system)}} \eta\, dm = \int_{\forall\,\text{(system)}} \eta\rho\, d\forall \tag{4.6}$$

To derive the control volume formulation of the second law of thermodynamics, we set

$$N = S \quad \text{and} \quad \eta = s$$

From Eq. 4.11, with this substitution, we obtain

$$\left.\frac{dS}{dt}\right)_{\text{system}} = \frac{\partial}{\partial t}\int_{\text{CV}} s\rho\, d\forall + \int_{\text{CS}} s\rho\vec{V}\cdot d\vec{A} \tag{4.60}$$

From Eq. 4.5a

$$\left.\frac{dS}{dt}\right)_{\text{system}} \geq \frac{1}{T}\dot{Q}$$

The system and the control volume coincided at t_0; thus

$$\left.\frac{1}{T}\dot{Q}\right)_{\text{system}} = \left.\frac{1}{T}\dot{Q}\right)_{\text{CV}} = \int_{\text{CS}} \frac{1}{T}\left(\frac{\dot{Q}}{A}\right) dA$$

In light of this, Eqs. 4.5a and 4.60 yield the control volume formulation of the second law of thermodynamics

$$\frac{\partial}{\partial t}\int_{\mathrm{CV}} s\rho \, d\forall + \int_{\mathrm{CS}} s\rho\vec{V}\cdot d\vec{A} \geq \int_{\mathrm{CS}} \frac{1}{T}\left(\frac{\dot{Q}}{A}\right) dA \tag{4.61}$$

In Eq. 4.61, the term $(\dot{Q}/A)$ represents the heat flux per unit area into the control volume through the area element, dA. To evaluate the term

$$\int_{\mathrm{CS}} \frac{1}{T}\left(\frac{\dot{Q}}{A}\right) dA$$

both the local heat flux, $(\dot{Q}/A)$, and local temperature, T, must be known for each area element of the control surface.

4-10 SUMMARY OBJECTIVES

After completing study of Chapter 4, you should be able to do the following:

1. Write each of the five basic laws (conservation of mass, Newton's second law, moment of momentum**, the first law of thermodynamics, and the second law of thermodynamics) for a system as a rate equation.
2. If the extensive property in the rate equations of Summary Objective 1 is designated N, define the corresponding intensive property, designated η, in each of the basic equations.
3. Write the equation that relates the rate of change of any arbitrary extensive property, N, of a system, to the time variations of the property associated with a control volume. Give the physical significance of each quantity in the equation.
4. Write the control volume formulation of the conservation of mass and state the physical meaning of each term in the equation. Apply the equation to the solution of flow problems.
5. Write the control volume formulation of the momentum equation for an inertial control volume and state the physical meaning of each term in the equation. Apply the equation to the solution of flow problems.

**6. State the relationship among fluid properties (the Bernoulli equation) that results from applying the momentum equation to a differential control volume. List the restrictions on the use of the Bernoulli equation.

7. Write the control volume formulation of the momentum equation for a control volume with rectilinear acceleration and state the physical meaning of each term in the equation. Apply the equation to the solution of flow problems.

**8. Write the control volume formulation of the momentum equation for a control volume with arbitrary acceleration and state the physical meaning of each term in the equation. Apply the equation to the solution of flow problems.

**9. Write the control volume formulation of the moment of momentum equation for (a) a fixed, and (b) a rotating control volume and state the physical meaning of each term in the equation. Apply the equation to the solution of flow problems.

** These objectives apply to sections that may be omitted without loss of continuity in the text material.

10. Write the control volume formulation of the first law of thermodynamics and state the physical meaning of each term in the equation. Apply the equation to the solution of flow problems.
11. Write the control volume formulation of the second law of thermodynamics and state the physical meaning of each term in the equation. Apply the equation to the solution of flow problems.
12. Solve the problems at the end of the chapter that relate to the material you have studied.

PROBLEMS

4.1 A body of mass, M, is at rest on a horizontal plane. At $t = 0$, the body is acted upon by a constant force, F_1, of 10 lbf. The force is applied for a period of 5 sec. Immediately upon removal of this force, the body is acted upon by a constant force, F_2, of 3 lbf in the opposite direction. There is no friction. Determine the length of time of action by F_2 required to bring the body to rest.

4.2 A police investigation of tire marks showed that a car traveling along a straight level street had skidded to a stop for a total distance of 50 m after the brakes were applied. The coefficient of friction between the tires and the pavement is estimated to be $\mu = 0.6$. What was the probable speed of the car when the brakes were applied?

4.3 The mass of an aluminum beverage can is 20 g. Its diameter and height are 65 and 120 mm, respectively. When full, the can contains 354 milliliters of soft drink with SG = 1.05. Evaluate the height of the center of gravity of the can as a function of liquid level. At what level would the can be least likely to tip over when subjected to a steady lateral acceleration? Calculate the minimum coefficient of static friction for which the full can will tip rather than slide on a horizontal surface.

4.4 A fully loaded Boeing 727 jet transport aircraft weighs 225,000 lbf. The pilot brings the 3 engines to full takeoff thrust of 18,000 lbf each before releasing the brakes. Neglecting aerodynamic and rolling resistance, estimate the minimum runway length and time needed to reach a takeoff speed of 140 mph. Assume engine thrust remains constant during ground roll.

4.5 A pistol bullet fired horizontally in air experiences an aerodynamic drag force proportional to the square of its speed, $F_D = kV^2$. Ballistic data measured at the muzzle and at a range of 50 m show that $V_0 = 260$ m/sec at the muzzle and $V = 240$ m/sec at 50 m. Estimate k if the bullet mass is 15.6 g.

4.6 A person with 75 kg total mass rides an elevator in a tall building. Calculate the force exerted by the person on the elevator floor during steady accelerations of (a) 3 m/sec^2 upward, and (b) 2 m/sec^2 downward.

4.7 A small steel ball of radius, r, placed atop a much larger sphere of radius, R, begins to roll under the influence of gravity. Rolling and air resistance are negligible. As the speed of the ball increases, it leaves the surface of the sphere and becomes a projectile. Determine the location at which the ball loses contact with the sphere.

4.8 Most modern racing cars depend heavily on aerodynamic design features to generate downforce, which increases the loading on the tires and therefore the cornering speeds. A race car running at the Indianapolis Motor Speedway has a mass of 800 kg. Its aerodynamic devices produce a downforce of 8 kN at racing speeds. Each turn at Indy has an effective radius of 250 m and is banked at 9.2°. For these conditions, estimate the maximum radial acceleration and speed that an Indy car can achieve in the turns. Assume the coefficient of friction for racing tires is 1.2.

4.9 A flywheel consisting of a steel disk 20 mm thick and 0.6 m in diameter rotates at 20,000 rpm. The flywheel is mounted in a subway car with its axis of rotation placed lengthwise. The car, traveling at a speed of 15 m/sec, rounds a curve of radius 80 m. Determine the reaction torque exerted by the flywheel assembly on the subway car. The specific gravity of steel is 7.8.

4.10 A fluid mechanics laboratory experiment consists of a cylindrical tank containing water and mounted on a turntable, as in Example 3.9. The cylinder diameter is 0.2 m; the initial water depth, h_0, is 40 mm. Measurements show that when the turntable is switched off from a speed of 78 rpm, all motion of the liquid ceases 180 sec later. Determine the average torque applied to the water during this interval.

4.11 Experimental measurements have shown that the mass and mean radius of the Earth are $m_e = 5.976 \times 10^{24}$ kg and $r_e = 6.371 \times 10^6$ m. Newton's law of gravitation, which governs the mutual attraction between bodies, is

$$F = \frac{Km_1 m_2}{r^2}$$

where F is the mutual attraction force between the two particles of masses, m_1 and m_2, and r is the distance between their centers. Consider the Earth a point mass for calculations involving gravitational attraction. Use Newton's second law to show that $g = Km_e/r_e^2$. Consider a satellite in a geosynchronous orbit to maintain position above a point on the Earth's surface at the equator. Estimate the orbit height above the Earth and the satellite speed.

4.12 A fluid particle moves through the impeller of a centrifugal pump with radial speed (relative to the impeller) given by $V_r = V_0 r_0 / r$, while the impeller turns at constant angular speed, ω. Determine the total acceleration of the particle just prior to leaving the impeller at radius, R.

4.13 Calculate the amount of work required to accelerate a 1500 kg automobile from rest to a speed of 25 m/sec on a level highway if air resistance and mechanical losses are neglected.

4.14 A 1500 kg automobile has a drivetrain that supplies 50 kW to the drive wheels. Determine the minimum time and distance required to accelerate the vehicle from rest to 25 m/sec on level road neglecting air and rolling resistance.

4.15 The distance between the airport in West Lafayette, Indiana, and O'Hare airport serving Chicago is 104 statute miles. The compass heading for a direct route is 335°. A pilot wishes to make the trip on a day when the wind is from the southwest at 25 knots (nautical miles per hour). The true air speed of his plane is 135 nautical miles per hour. What compass heading should he fly? How long will the trip take? Compare your results with corresponding values for a calm day.

4.16 The average rate of heat loss from a person to the surroundings when not actively working is about 600 Btu/hr. Suppose that in an auditorium containing 6000 people the ventilation system fails.

(a) How much does the internal energy of the air in the auditorium increase during the first 15 min after the ventilation system fails?

(b) Considering the auditorium and all the people as a system, and assuming no heat transfer to the surroundings, how much does the internal energy of the system change? How do you account for the fact that the temperature of the air increases?

4.17 Air is expanded isothermally in a nonflow process from initial conditions of 170 C, 276 kPa (abs) and 0.05 m^3 to a final volume of 0.1 m^3. Determine the heat transferred.

4.18 Air at 20 C and an absolute pressure of 1 atm is compressed adiabatically, without friction, to an absolute pressure of 3 atm. Determine the internal energy change.

4.19 A cylinder fitted with a piston contains 5 lbm of wet steam at a pressure of 200 psia and a quality of 80 percent. The piston is restrained by a spring arranged so that for zero volume in the cylinder the spring is fully extended. The spring force is proportional to the spring displacement. The weight of the piston may be neglected so that the force on the spring is exactly balanced by the pressure forces on the steam. Heat is transferred to the steam until its volume is 150 percent of the initial volume.

(a) What is the final pressure?
(b) What is the quality (if saturated) or temperature (if superheated) in the final state?
(c) Draw a $p\forall$ diagram and determine the work delivered to the spring.
(d) Determine the heat transferred.

4.20 The cooling system of an automobile contains 20 liters of coolant. The auto is started on a cold winter morning when the temperature is -10 C. During warmup, the engine rejects heat to the coolant at the rate of 5 kW. Estimate the minimum time needed for the coolant to reach 15 C, if its heat capacity is that of water.

4.21 An aluminum can of soft drink is to be cooled in a refrigerator where the temperature is $T_r = 5$ C. The rate of heat transfer from the can is

$$\dot{Q} = -k(T - T_r)$$

where $k = 0.25$ W/C. Calculate the energy that must be removed if the mass of the can and its contents is equivalent to 390 g of water and its initial temperature is 25 C. Estimate the time required to cool the can to 7 C.

4.22 In order to cool a six-pack as quickly as possible, it is placed in a freezer for a period of 1 hr. If the room temperature is 25 C and the cooled beverage is at a final temperature of 5 C, determine the change in the specific entropy of the beverage.

4.23 The velocity field in the region shown is given by $\vec{V} = az\hat{j} + b\hat{k}$, where $a = 10\ \text{sec}^{-1}$ and $b = 5$ m/sec. For a depth, w, perpendicular to the diagram, an element of area ① may be represented by $w\,dz(-\hat{j})$ and an element of area ② by $w\,dy(-\hat{k})$. (Note that both are drawn *outward* from the control volume, hence the minus sign.)

(a) Find an expression for $\vec{V} \cdot d\vec{A}_1$.
(b) Evaluate $\int_{A_1} \vec{V} \cdot d\vec{A}_1$.
(c) Find an expression for $\vec{V} \cdot d\vec{A}_2$.
(d) Find an expression for $\vec{V}(\vec{V} \cdot d\vec{A}_2)$.
(e) Evaluate $\int_{A_2} \vec{V}(\vec{V} \cdot d\vec{A}_2)$.

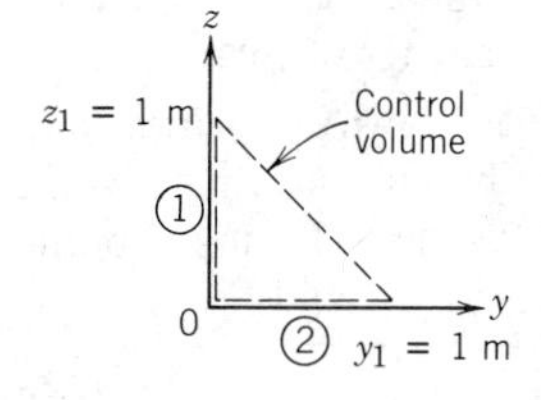

P 4.23

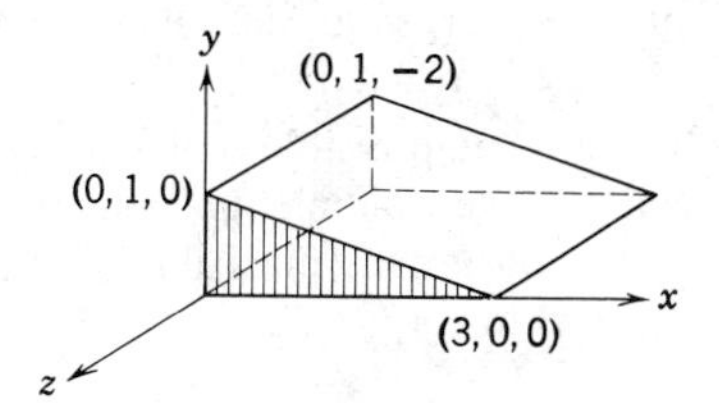

P 4.24

4.24 A flow field is given by $\vec{V} = ay\hat{i} + b\hat{k}$, where $a = 2\ \text{sec}^{-1}$ and $b = 1$ ft/sec. The volume flow rate, Q, through a surface is given by

$$Q = \int_A \vec{V} \cdot d\vec{A}$$

Evaluate the volume flow rate through the shaded surface. All dimensions are in feet.

4.25 The shaded area shown is in a flow where the velocity field is given by $\vec{V} = ax\hat{i} - by\hat{j}$, where $a = b = 1\ \text{sec}^{-1}$. Evaluate the following integrals over the shaded area.
(a) $\int \vec{V} \cdot d\vec{A}$ (b) $\int \vec{V}(\vec{V} \cdot d\vec{A})$

P 4.25

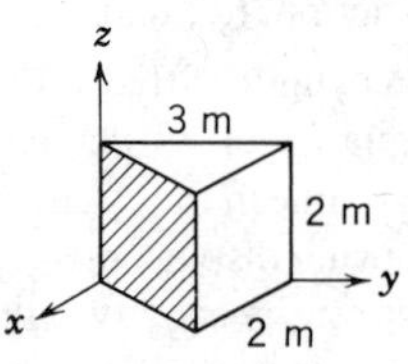

P 4.26

4.26 The area shown shaded is in a flow where the velocity field is given by $\vec{V} = -ax\hat{i} + by\hat{j} + c\hat{k}$, where $a = b = 1\ \text{sec}^{-1}$ and $c = 1$ m/sec. Write a vector expression for an element of the shaded area. Evaluate the following integrals over the shaded area.
(a) $\int \vec{V} \cdot d\vec{A}$ (b) $\int \vec{V}(\vec{V} \cdot d\vec{A})$

4.27 The velocity distribution for laminar flow in a long circular tube is given by the one-dimensional expression

$$\vec{V} = u\hat{i} = u_{\text{max}}\left[1 - \left(\frac{r}{R}\right)^2\right]\hat{i}$$

For this profile, use the area vector $d\vec{A} = dA\hat{i} = 2\pi r\,dr\,\hat{i}$ to evaluate (a) $\int \vec{V} \cdot d\vec{A}$, and (b) $\int \vec{V}(\vec{V} \cdot d\vec{A})$ for the tube cross section.

4.28 Fluid with a density of 1050 kg/m^3 is flowing steadily through the rectangular box shown. Given $A_1 = 0.05\ \text{m}^2$, $A_2 = 0.01\ \text{m}^2$, $A_3 = 0.06\ \text{m}^2$, $\vec{V}_1 = 4\hat{i}$ m/sec, and $\vec{V}_2 = -8\hat{j}$ m/sec, determine the velocity, $\vec{V}_3$.

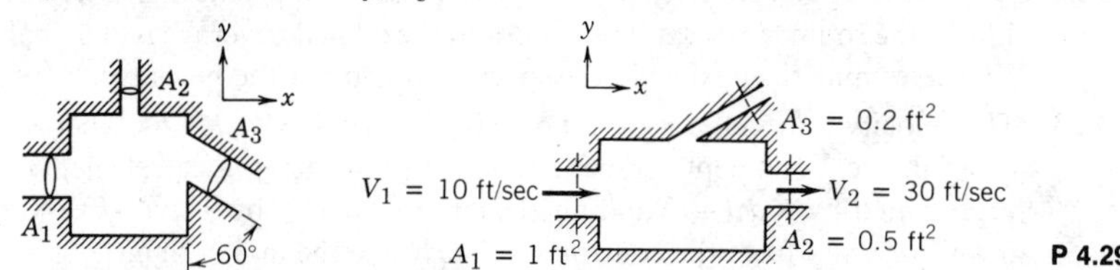

P 4.28 P 4.29

4.29 Consider steady, incompressible flow through the device shown. Determine the volume flow rate through port 3.

4.30 Air at standard atmospheric conditions enters a compressor at a rate of 20 m^3/min. The air is discharged at 800 kPa (abs) and 60 C. If the velocity in the discharge line must be limited to 20 m/sec, calculate the required diameter of the line.

4.31 In the incompressible flow through the device shown, velocities may be considered uniform over the inlet and outlet sections. If the fluid flowing is water, determine an expression for the mass flow rate at section③. The following conditions are known: $A_1 = 0.1\ \text{m}^2$, $A_2 = 0.2\ \text{m}^2$, $A_3 = 0.15\ \text{m}^2$, $V_1 = 5$ m/sec, and $V_2 = 10 + 5\cos(4\pi t)$ m/sec.

P 4.31 P 4.32

4.32 Water flowing out of a circular pipe is assumed to have a linearly varying velocity distribution as shown. What is the average velocity of the outward flow in terms of V_{max}?

4.33 Water flows steadily through a pipe of length L and radius $R = 3$ in. Calculate the value of the uniform inlet velocity, U, if the velocity distribution across the outlet is given by

$$u = 10\left[1 - \frac{r^2}{R^2}\right] \text{ft/sec}$$

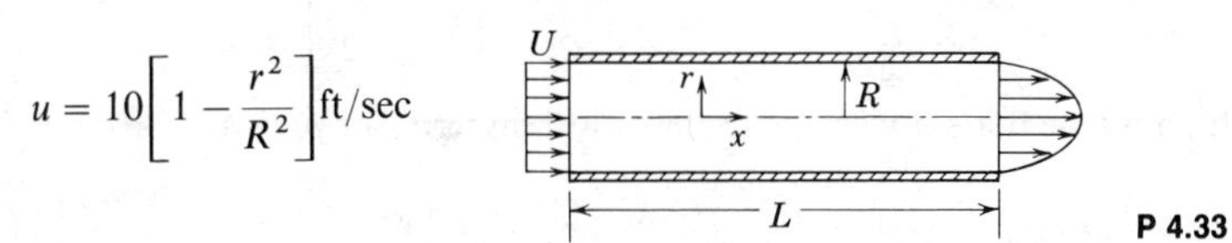

P 4.33

4.34 Water enters a wide, flat channel of height, $2h$, with a uniform velocity of 5 m/sec. At the outlet of the channel the velocity distribution is given by

$$\frac{u}{u_{\text{max}}} = 1 - \left(\frac{y}{h}\right)^2$$

The coordinate y is measured from the centerline of the channel. Determine the exit centerline velocity, u_{max}.

4.35 A two-dimensional reducing bend has a linear velocity profile at section ①. The flow is uniform at sections ② and ③. The fluid is incompressible and the flow is steady. Find the magnitude and direction of the uniform velocity at section ③.

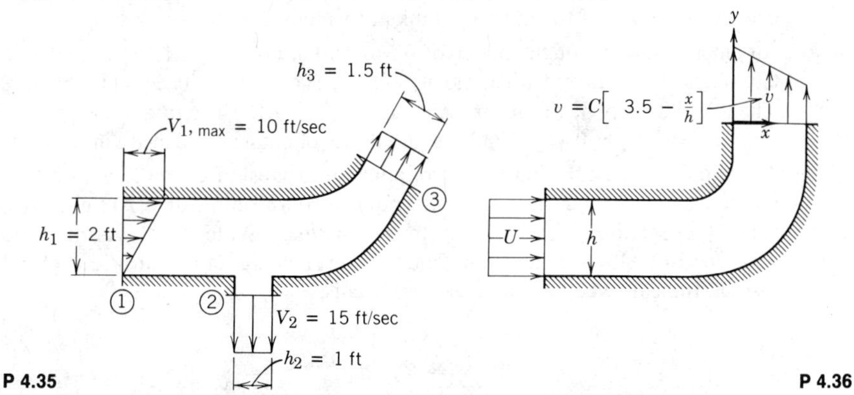

P 4.35 P 4.36

4.36 Water enters a two-dimensional channel of constant width, h, with uniform velocity, U. The channel makes a 90° bend that distorts the flow to produce the velocity profile shown at the exit. Evaluate the constant, C.

4.37 Water flows steadily past a porous flat plate. Constant suction is applied along the porous section. The velocity profile at section cd is

$$\frac{u}{U_\infty} = 3\left[\frac{y}{\delta}\right] - 2\left[\frac{y}{\delta}\right]^{1.5}$$

Evaluate the mass flow rate across section bc.

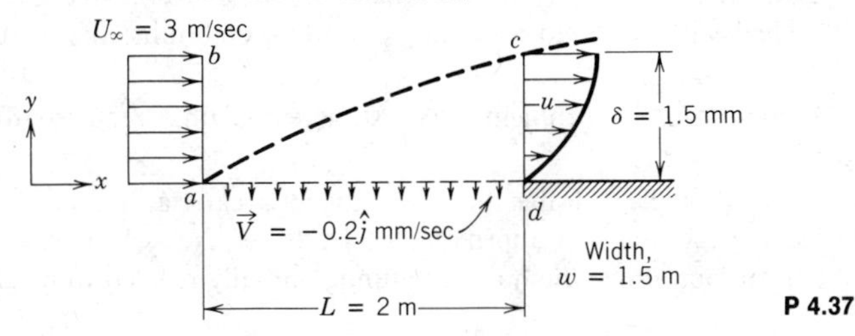

P 4.37

4.38 Oil flows steadily in a thin layer down an inclined plane. The velocity profile is

$$u = \frac{\rho g \sin\theta}{2\mu}\left[hy - \frac{y^2}{2}\right]$$

Express the mass flow rate per unit width in terms of ρ, μ, g, θ, and h.

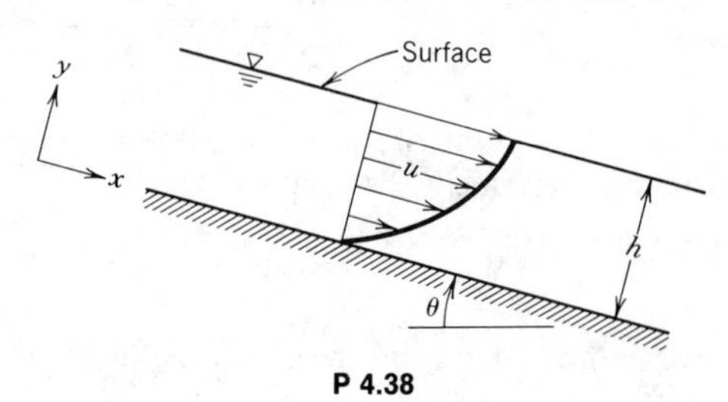

P 4.38

4.39 A tank of 0.5 m^3 volume contains compressed air. A valve is opened and air escapes with a velocity of 300 m/sec through an opening of 130 mm^2 area. Air temperature passing through the opening is -15 C and the absolute pressure is 350 kPa. Find the rate of change of density of the air in the tank at this moment.

4.40 Air enters a tank through an area of 0.2 ft^2 with a velocity of 15 ft/sec and a density of 0.03 slug/ft^3. Air leaves with a velocity of 5 ft/sec and a density equal to that in the tank. The initial density of the air in the tank is 0.02 slug/ft^3. The total tank volume is 20 ft^3 and the exit area is 0.4 ft^2. Find the initial rate of change of density in the tank.

4.41 A section of pipe carrying water contains an expansion chamber with a free surface whose area is 2 m^2. The inlet and outlet pipes are both 1 m^2 in area. At a given instant, the velocity at section ① is 3 m/sec into the chamber. Water flows out at section ② at 4 m^3/sec. Both flows are uniform. Find the rate of change of free surface level at the given instant. Indicate whether the level rises or falls.

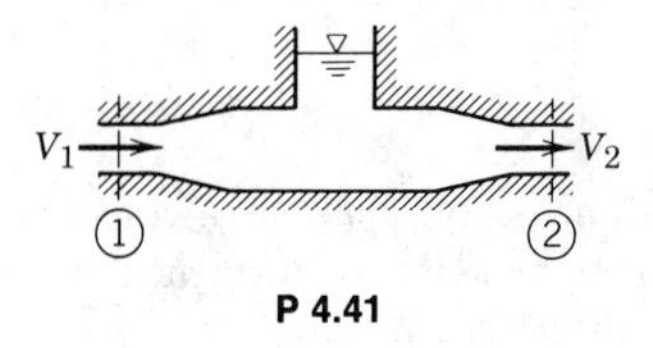

P 4.41

4.42 A cylindrical tank, 0.3 m in diameter, drains through a hole in its bottom. At the instant when the water depth is 0.6 m, the flow rate from the tank is observed to be 4 kg/sec. Determine the rate of change of water level at this instant.

4.43 A cylindrical tank of diameter, $D = 50$ mm, drains through an opening, $d = 5$ mm, in the bottom of the tank. The speed of the liquid leaving the tank may be approximated by $V = \sqrt{2gy}$, where y is the height from the tank bottom to the free surface. If the tank is initially filled with water to a depth $y_0 = 0.4$ m, determine the water depth at time, $t = 12$ sec.

4.44 For the conditions of Problem 4.43, estimate the time required to drain the tank completely.

4.45 A funnel of half angle θ, drains through a hole of area A, at the vertex. The speed of the liquid leaving the funnel is approximated by $V = \sqrt{2gy}$, where y is the height of the liquid free surface above the hole. The funnel initially is filled to height y_0. Obtain an expression for the time, t, required to drain the funnel. Express the result in terms of the initial volume, $\forall_0$, of liquid in the funnel and the initial volume flow rate, $Q_0 = A\sqrt{2gy_0} = AV_0$.

4.46 A tank of fixed volume contains brine with initial density, ρ_i, greater than that of water. Pure water enters the tank steadily and mixes thoroughly with the brine in the tank. The liquid level in the tank remains constant. Derive expressions for (a) the rate of change of density of the liquid mixture in the tank, and (b) the time required for the density to reach the value, ρ_f, where $\rho_i > \rho_f > \rho_{H_2O}$.

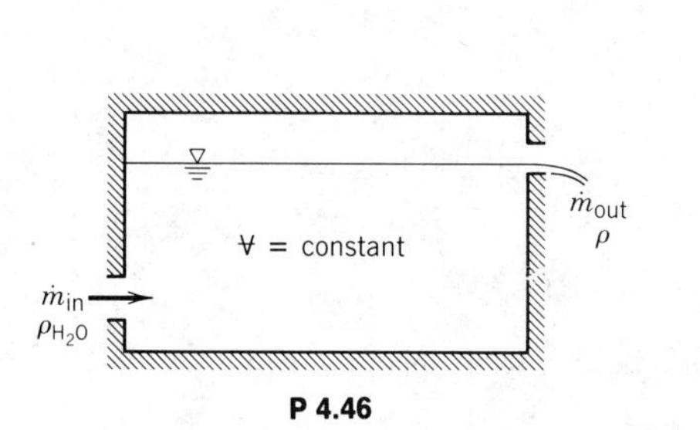

P 4.46

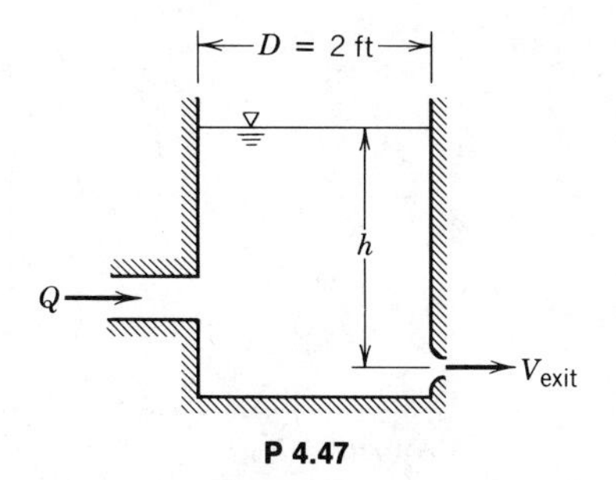

P 4.47

4.47 Water enters the tank through a pipe at rate, $Q = 0.35$ ft^3/sec. Water also leaves the tank through a smoothly rounded nozzle of 2 in. diameter. The exit velocity through this hole, which depends on the height, h, of the water level above the hole, is $V_{exit} = \sqrt{2gh}$. At time $t = 0$, $h = 9$ ft. Derive an equation that could be solved for $h(t)$ when $h > 4$ ft. For $4 < h < 9$ ft, is dh/dt greater than, less than, or equal to zero?

4.48 A continuous supply of seawater is mixed thoroughly with the brine solution already in the boiler of a seawater desalination unit. The seawater is supplied at the rate of 900 kg/min. Steam boiled from the brine solution is removed and the volume of the solution remains constant at 30 m^3. The brine density changes as seawater evaporates leaving salt behind. Derive an algebraic equation for the rate of change of density in the brine solution. If the initial density of the brine is the same as that of the entering seawater (2.5 percent salt by weight), how long must the boiler operate to double the density of the brine solution?

4.49 Motion of a hydraulic cylinder is cushioned at the end of its stroke by a piston that enters a hole as shown. The cavity and cylinder are filled with hydraulic fluid of uniform density, ρ. Obtain an expression for the velocity, V_1, at which hydraulic fluid escapes from the cylindrical hole as a function of piston displacement, x.

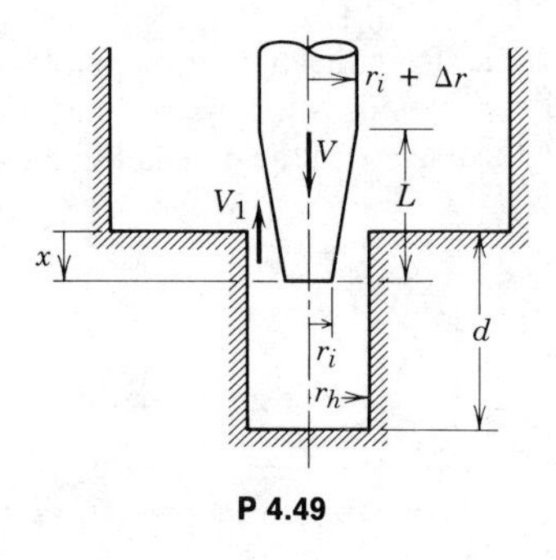

P 4.49

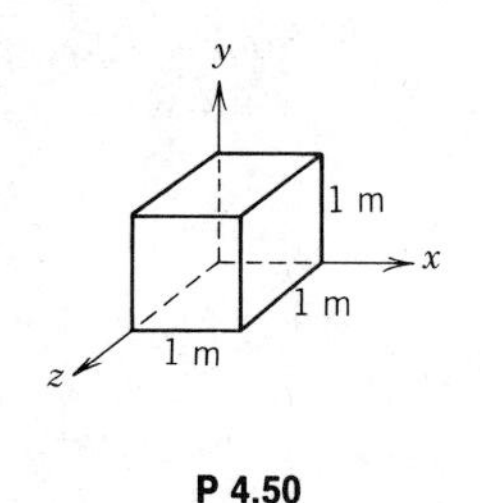

P 4.50

4.50 Consider the flow of an incompressible fluid with vector velocity field $\vec{V} = (ax + bt)\hat{i} - cy\hat{j}$, where $a = 1\ \text{sec}^{-1}$, $b = 2\ \text{m/sec}^2$ and $c = 1\ \text{sec}^{-1}$. For the control volume shown, evaluate the rate of change of momentum within the control volume.

4.51 Consider the steady flow of water between two parallel plates a distance h feet apart. At section ① the velocity is uniform across the width; the velocity distribution is assumed to be linear at section ②. The flow is identical in all planes parallel to the plane of the diagram. Calculate the ratio of the x-direction momentum flux at section ② to that at section ① for the assumed velocity distributions.

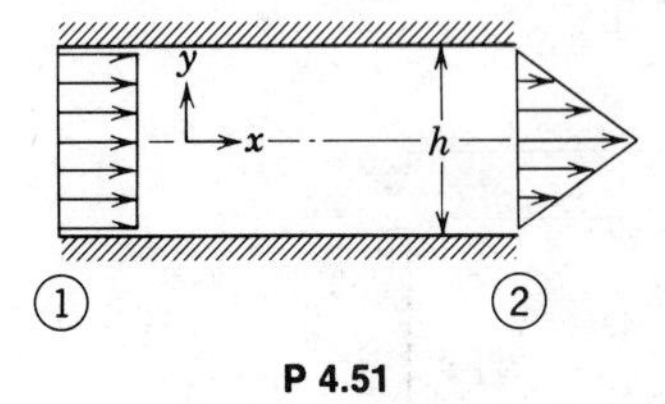

P 4.51

4.52 Evaluate the net rate of efflux of momentum through the control surface of Problem 4.28.

4.53 For the conditions of Problem 4.33, evaluate the ratio of the x-direction momentum flux at the pipe outlet to that at the inlet.

4.54 For the conditions of Problem 4.34, evaluate the ratio of the x-direction momentum flux at the channel outlet to that at the inlet.

4.55 Oil is being loaded into a barge through a 150 mm diameter pipe. The oil, $\rho = 900\ \text{kg/m}^3$, leaves the pipe with a uniform speed of 5 m/sec. Determine the force of the rope on the barge.

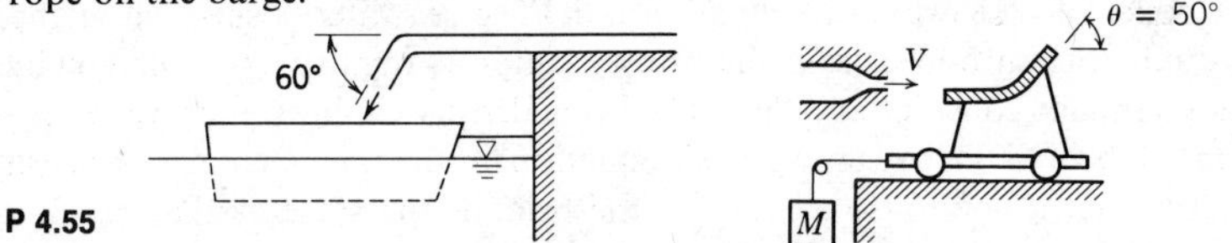

P 4.55

P 4.56, 4.57

4.56 A jet of water issuing from a stationary nozzle with a speed of 15 m/sec (jet area = 0.05 m^2) strikes a turning vane mounted on a cart as shown. The vane turns the jet through an angle $\theta = 50°$. Determine the value of M required to hold the cart stationary.

4.57 If the vane angle, θ, of Problem 4.56 is adjustable, plot the mass, M, needed to hold the cart stationary as a function of θ for $0 \leq \theta \leq 180°$.

4.58 A large tank is affixed to a cart as shown. Water issues from the tank through a 600 mm^2 nozzle at a speed of 10 m/sec. The water level in the tank is maintained constant by adding water through a vertical pipe. Determine the tension in the wire holding the cart stationary.

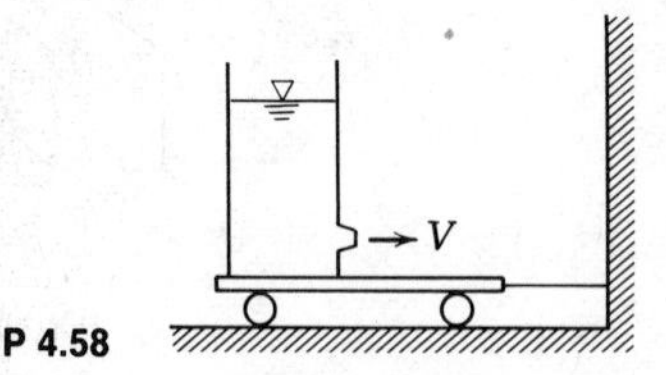

P 4.58

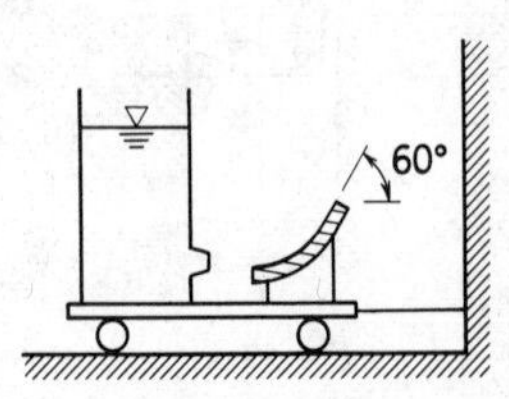

P 4.59, 4.60

4.59 A turning vane which deflects the water through an angle of 60°, is attached to the cart under the conditions of Problem 4.58. Determine the tension in the wire holding the cart stationary, and the force of the vane on the cart.

4.60 If the turning vane of Problem 4.59 deflects the water through an angle of 90°, determine the tension in the wire holding the cart stationary.

4.61 A vertical plate has a sharp-edged orifice at its center. A liquid jet of speed, V, strikes the plate concentrically. Obtain an expression for the external force needed to hold the plate in place if the jet leaving the orifice also has speed, V. Evaluate the force for $V = 5$ m/sec, $D = 100$ mm, and $d = 25$ mm.

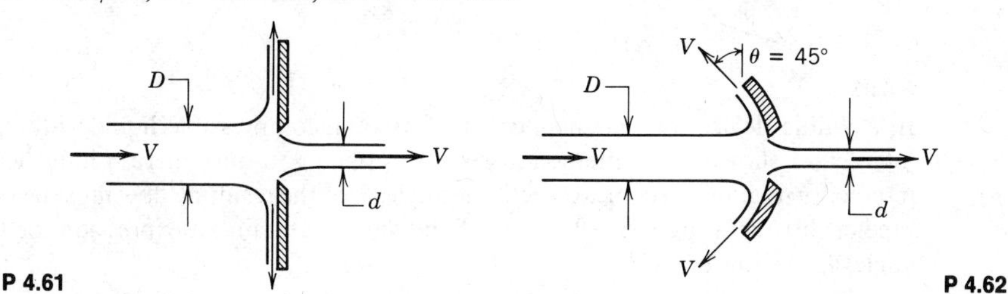

P 4.61 **P 4.62**

4.62 A shallow circular dish has a sharp-edged orifice at its center. A liquid jet of speed, V, strikes the dish concentrically. Obtain an expression for the external force needed to hold the dish in place if the jet issuing from the orifice also has speed, V. Evaluate the force for $V = 5$ m/sec, $D = 100$ mm, and $d = 20$ mm.

4.63 A sharp-edged splitter plate inserted part way into a flat stream of flowing water produces the flow pattern shown. Analyze the situation to evaluate θ as a function of α, where $0 \leq \alpha < 0.5$. Evaluate the force on the splitter plate due to the flowing water. (Neglect any friction force between the water stream and the splitter plate.)

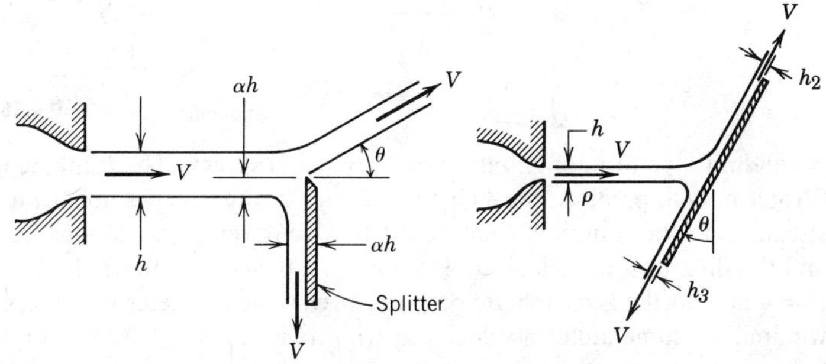

P 4.63 **P 4.64**

4.64 When a plane liquid jet strikes an inclined flat plate, it splits into two streams of equal speed but unequal thickness. For frictionless flow there can be no tangential force on the plate surface. Use this assumption to develop an expression for h_2/h as a function of the plate angle, θ. Plot your results and comment on the limiting cases, $\theta = 0$ and $\theta = 90°$.

4.65 The designer of a fluidic device wishes to deflect an air jet with mass flow rate, $\dot{m}_1$, and cross-sectional area, A_1, by 30°. The designer proposes doing this with a smaller air jet with cross-sectional area, A_2, oriented perpendicular to the original jet. What mass flow rate, $\dot{m}_2$, is needed in the smaller jet to accomplish this task? Assume incompressible flow of standard air and neglect gravity.

4.66 A circular cylinder inserted across a stream of flowing water deflects the stream through an angle θ as shown. (This is termed the "Coanda effect.") For $a = 0.5$ in., $b = 0.1$ in.,

$V = 10$ ft/sec, and $\theta = 20°$, determine the horizontal component of the force on the cylinder due to the flowing water.

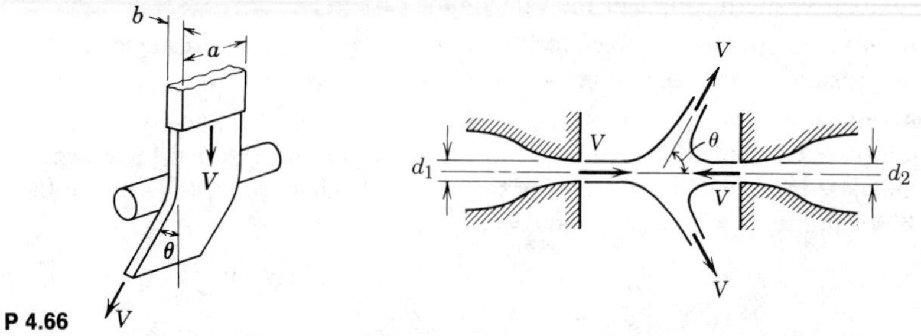

P 4.66 **P 4.67**

4.67 In a fluidic device, two circular coaxial jets of incompressible liquid with speed, V, interact as shown. The interaction region is open to atmosphere. Liquid leaves the interaction region in a conical sheet. The angle, θ, of the resulting flow must be known to predict the operating characteristics of the device. Obtain an expression for the cone angle, θ, in terms of d_2/d_1.

4.68 A propeller operating in still air is shown. A theory for uniform flow without losses developed and published by Rankine in 1885 predicts that the air speed through the propeller disk is half that of the slipstream behind the propeller. Pressure on the streamlines bounding the slipstream is close to atmospheric. Develop an expression for the thrust developed by a propeller for the conditions shown. Evaluate for standard air with $V = 30$ m/sec and $D = 3$ m.

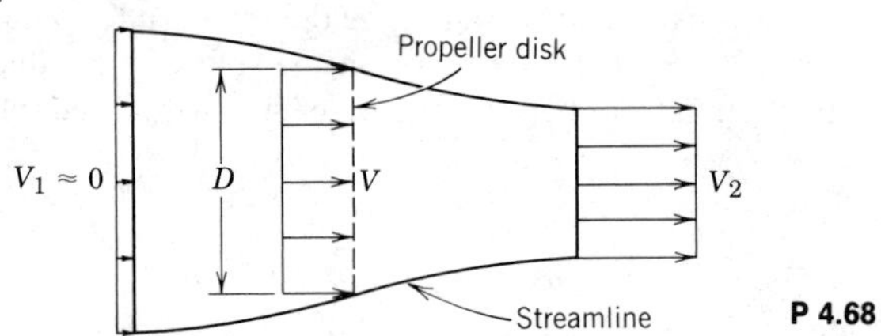

P 4.68

4.69 A windmill operating in a uniform air stream is shown. The Rankine propeller theory (Problem 4.68) predicts that half the velocity change occurs upstream and half downstream from the windmill. Analyze the flow to develop expressions for V_2, V_3, and D_3, and the thrust on the windmill. Evaluate the thrust on the windmill if $V_1 = 10$ m/sec, $D = 4$ m, and the atmospheric pressure streamline diameter is 3 m upstream from the windmill. Assume uniform velocity distributions and standard air conditions.

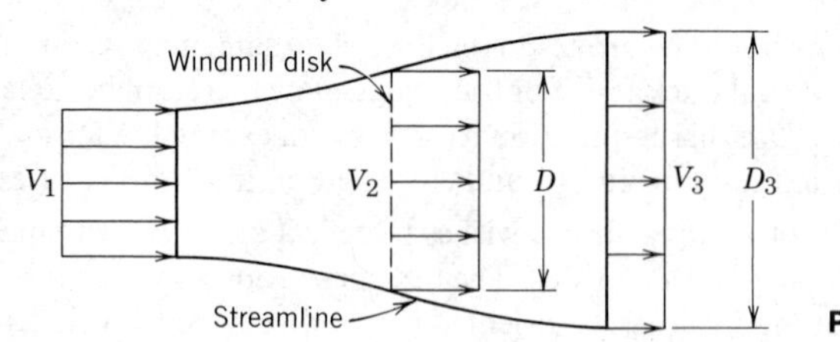

P 4.69

4.70 Dry ready-mix concrete is packaged in 90 lb sacks; the density of the dry mix is 100 lbm/ft^3. The sacks are supported by pallets as they move through the automatic bagging machine. If the filling time is 4 sec and the vertical velocity of the mix is 10 ft/sec, determine the maximum vertical force on the pallet.

4.71 A farmer purchases 675 kg of bulk grain from the local co-op. The grain is loaded into his pickup truck from a hopper with an outlet diameter of 0.3 m. The loading operator determines the payload by observing the indicated gross mass of the truck as a function of time. The grain flow from the hopper ($\dot{m} = 40$ kg/sec) is terminated when the indicated scale reading reaches the desired gross mass. If the grain density is 600 kg/m^3, determine the true payload.

4.72 A large "weigh tank" is to be used in the calibration of a flow meter. Measurements of weight as a function of time are to be made. Water enters the tank vertically from the flow-metering system at a speed of 20 ft/sec through a 1.5 in. diameter pipe. If the weight of the empty tank is 50 lbf, determine the scale reading at $t = 10$ sec. What is the true weight of the water and tank at $t = 10$ sec?

4.73 Water flows steadily through a fire hose and nozzle. The hose is 75 mm inside diameter, and the nozzle tip is 25 mm i.d.; water gage pressure in the hose is 510 kPa, and the stream leaving the nozzle is uniform. The exit speed and pressure are 32 m/sec, and atmospheric, respectively. Find the force transmitted by the coupling between the nozzle and hose. Indicate whether the coupling is in tension or compression.

4.74 Water flows through a well-made nozzle at the end of a 50 mm diameter pipe. The nozzle exit diameter is 25 mm. The water flow rate is 53.0 m^3/hr, and the pressure immediately upstream from the nozzle is 522 kPa (abs). Find the force transmitted by the coupling between the nozzle and pipe. Indicate whether the coupling is in tension or compression.

4.75 A curved nozzle assembly that discharges to the atmosphere is shown. The nozzle weighs 10 lbf and its internal volume is 150 in.^{3} The fluid is water. Determine the reaction force exerted by the nozzle on the coupling to the inlet pipe.

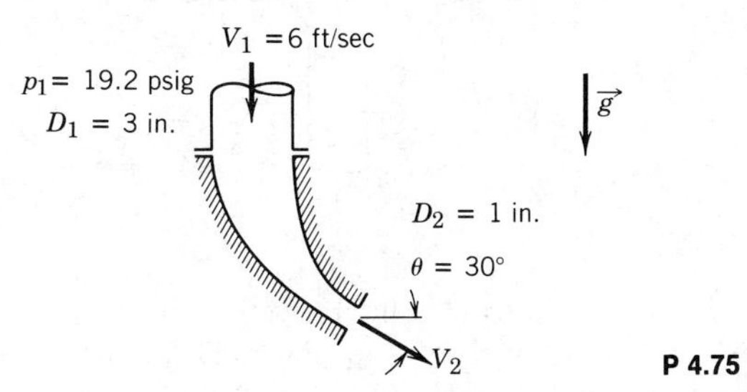

P 4.75

4.76 A reducer in a piping system is shown. The internal volume of the reducer is 0.2 m^3, and its mass is 25 kg. Evaluate the total force that must be provided by the surrounding pipes to support the reducer.

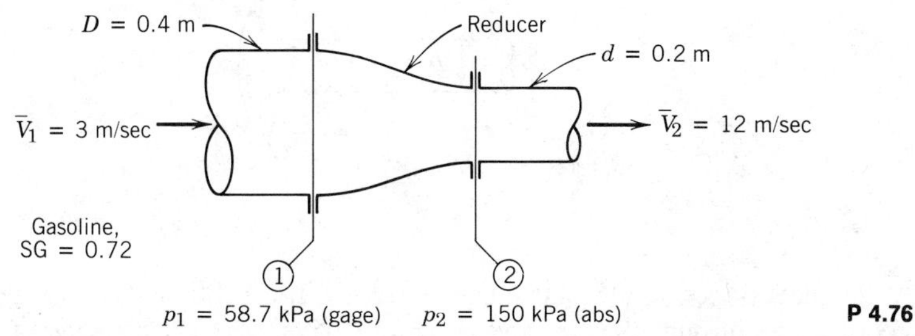

P 4.76

4.77 Water is flowing steadily through the 180° elbow shown. At the inlet to the elbow the gage pressure is 96 kPa. The water discharges to atmospheric pressure. Properties are assumed to be uniform over the inlet and outlet areas; $A_1 = 2600\ \text{mm}^2$, $A_2 = 650\ \text{mm}^2$, and $V_1 = 3.05$ m/sec. Find the horizontal component of the force required to hold the elbow in place.

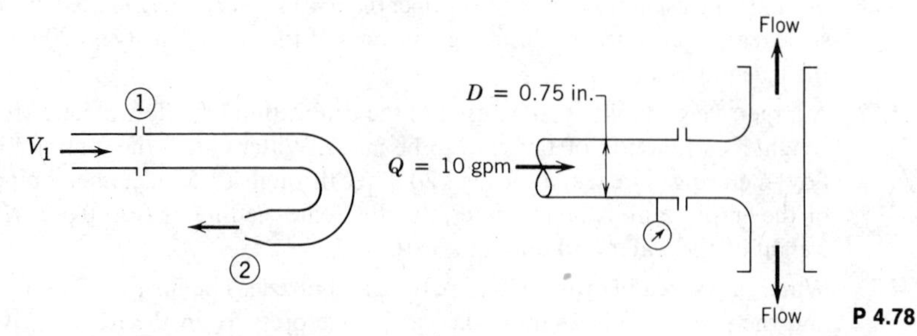

P 4.77 P 4.78

4.78 Water flows steadily at the rate of 10 gpm through a horizontal tee and discharges to atmosphere. The pressure just upstream from the tee is 15.0 psia. Determine the force exerted on the line by the tee.

4.79 The following data are given for flow of water in a horizontal tee: volume flow rates Q_1 and Q_2 are 0.09 and 0.06 m^3/sec and gage pressures p_1, p_2, and p_3 are 150, 105, and 120 kPa, respectively. Calculate the external force needed to hold the tee in place.

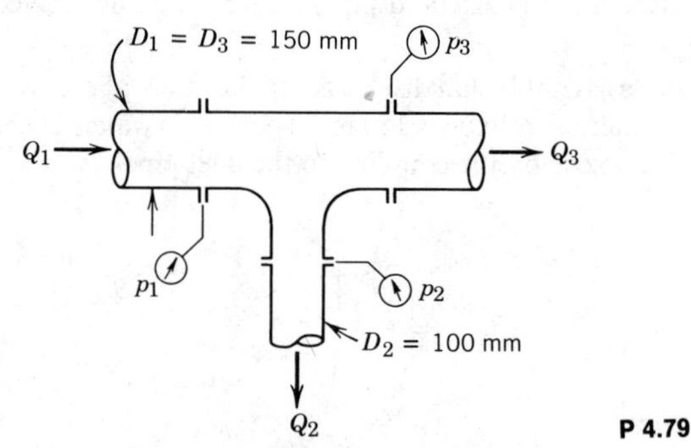

P 4.79

4.80 A conical spray head is shown. The fluid is water and the exit stream is uniform. Evaluate (a) the thickness of the spray sheet at 400 mm radius, and (b) the axial force exerted by the spray head on the supply pipe.

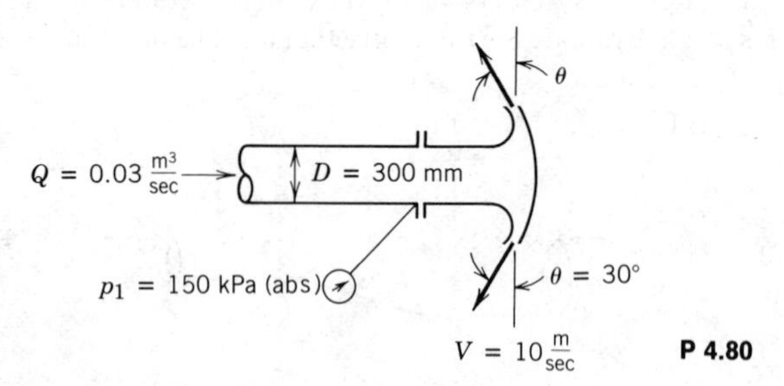

P 4.80

4.81 A flat plate orifice of 50 mm diameter is located at the end of a 100 mm diameter pipe. Water flows through the pipe and orifice at 0.05 m^3/sec. The diameter of the water jet downstream from the orifice is 35 mm. Calculate the external force required to hold the orifice in place. Neglect friction on the pipe wall.

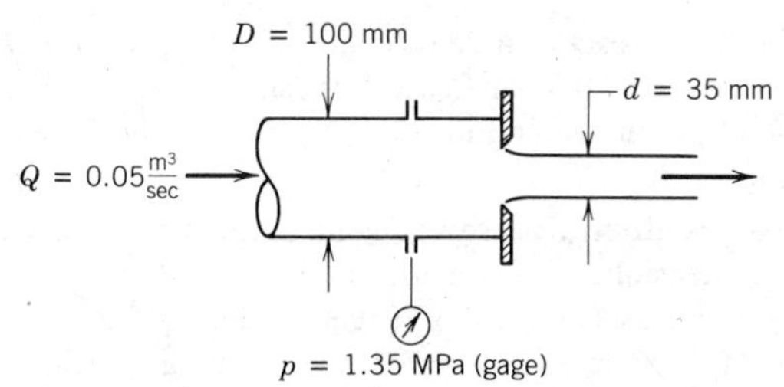

P 4.81

4.82 Consider flow through the sudden expansion shown. If the flow is incompressible and friction is neglected, show that the pressure rise, $\Delta p = p_2 - p_1$, is given by

$$\frac{\Delta p}{\frac{1}{2}\rho \bar{V}_1^2} = 2\left(\frac{d}{D}\right)^2\left[1 - \left(\frac{d}{D}\right)^2\right]$$

Hint: Assume the pressure is uniform and equal to p_1 on the vertical surface of the expansion.

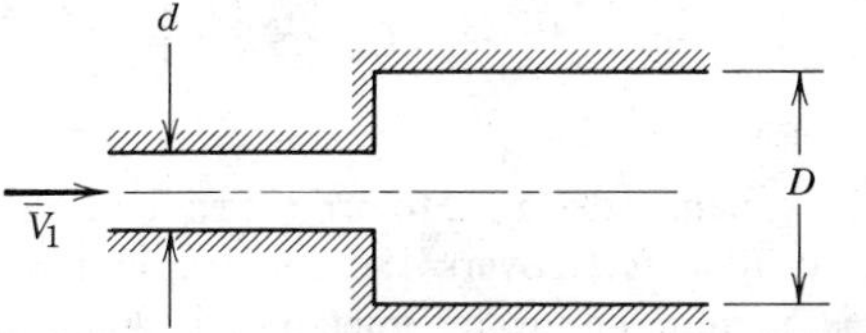

P 4.82

4.83 The General Electric CF6-50A jet engine has produced a thrust force of 52,000 lbf on the test stand. A typical installation is shown, together with some test data. Fuel enters the top of the engine vertically at a rate equal to 2 percent of the mass flow rate of the inlet air. For the given conditions, compute the air flow rate through the engine.

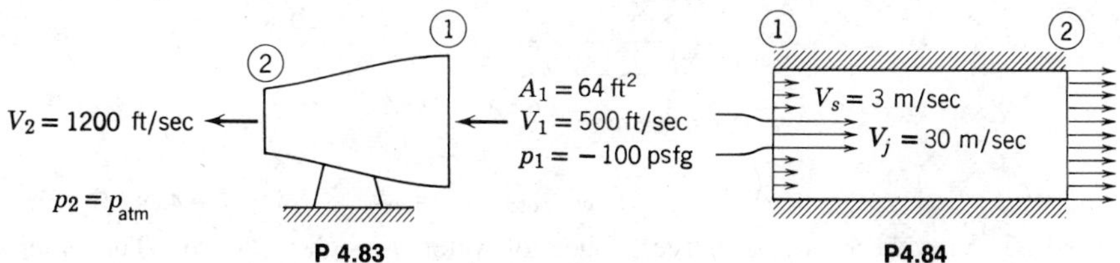

P 4.83 P4.84

4.84 A water jet pump has jet area, 0.01 m^2, and jet speed, 30 m/sec. The jet is within a secondary stream of water having a speed of 3 m/sec. The total area of the duct (the sum of the jet and secondary stream areas) is 0.075 m^2. The water is thoroughly mixed and leaves the jet pump in a uniform stream. The pressures of the jet and secondary stream are the same at the pump inlet. Determine the speed at the pump exit, and the pressure rise, $p_2 - p_1$.

4.85 Consider the steady adiabatic flow of air through a long straight pipe with a cross-sectional area of 0.5 ft^2. At the inlet, the air is at 30 psia, 140 F, and has a velocity of 500 ft/sec. At the exit, the air is at 11.3 psia and has a velocity of 985 ft/sec. Calculate the axial force of the air on the pipe. (Be sure to make the direction clear.)

4.86 A gas flows steadily through a heated porous pipe of constant 0.2 m^2 cross-sectional area. At the pipe inlet, the absolute pressure is 340 kPa, the density is 5.1 kg/m^3, and the mean velocity is 152 m/sec. The fluid passing through the porous wall leaves in a direction normal to the pipe axis, and the total flow rate through the porous wall is 29.2 kg/sec. At the pipe outlet, the absolute pressure is 280 kPa and the density is 2.6 kg/m^3. Determine the force in the axial direction of the fluid on the pipe.

4.87 A monotube boiler consists of a 20 ft length of tubing with 0.375 in. inside diameter. Water enters at the rate of 0.3 lbm/sec at 500 psia. Steam leaves at 400 psig with density of 0.024 slug/ft^3. Find the magnitude and direction of the force exerted by the flowing fluid on the tube.

4.88 Water is discharged from a narrow slot in a 150 mm diameter pipe. The resulting horizontal two-dimensional jet is 1 m long and 15 mm thick but of nonuniform velocity. The pressure at the inlet section is 30 kPa (gage). Calculate (a) the volume flow rate at the inlet section, and (b) the forces required at the coupling to hold the spray pipe in place. Neglect the mass of the pipe and the water it contains.

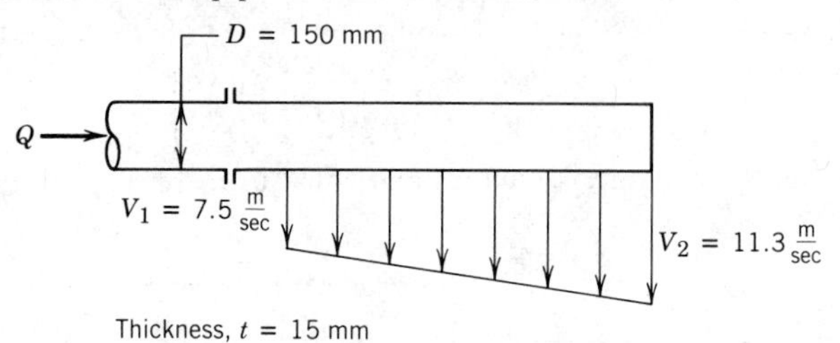

P 4.88

4.89 A nozzle for a spray system is designed to produce a flat radial sheet of water. The sheet leaves the nozzle at 10 m/sec, covers 180° of arc, and is 15 mm thick. The nozzle discharge radius is 50 mm. The water inlet pipe is 20 mm in diameter and the inlet pressure is 150 kPa (abs). Calculate the speed of the water in the inlet pipe. Evaluate the axial force exerted by the spray nozzle on the coupling.

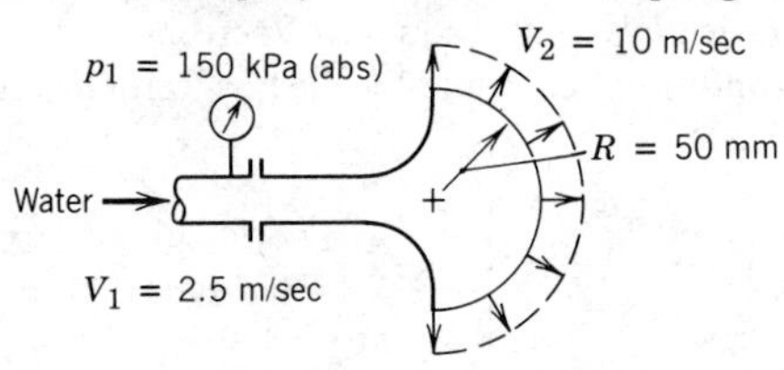

P 4.89

4.90 The nozzle shown discharges a sheet of water through a 180° arc. The water speed is 15 m/sec and the jet thickness is 30 mm at a radial distance of 0.3 m from the centerline of the supply pipe. Find (a) the volume flow rate of water in the jet sheet, and (b) the y component of force required to hold the nozzle in place.

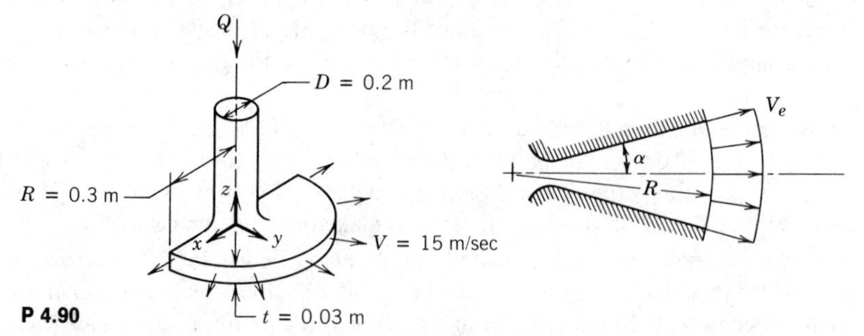

P 4.90 **P 4.91**

4.91 Gases leaving the propulsion nozzle of a rocket are modeled as flowing radially outward from a point upstream from the nozzle throat. Assume the speed of the exit flow, V_e, is constant in magnitude. Develop an expression for the axial thrust, T_a, developed by flow leaving the nozzle exit plane. Compare your result to the one-dimensional approximation, $T = \dot{m}V_e$. Evaluate the percent error for $\alpha = 15°$.

4.92 Consider the incompressible flow of fluid in a boundary layer as depicted in Example 4.3. Show that the drag force of the fluid on the surface is given by

$$D = \int_0^\delta \rho u(U - u)w\,dy$$

Evaluate the drag force for the conditions of Example 4.3.

4.93 Air at standard conditions flows along a flat plate. The undisturbed freestream speed is $U_0 = 30$ m/sec. At $L = 0.3$ m downstream from the leading edge of the plate, the boundary-layer thickness is $\delta = 1.5$ mm. The velocity profile at this location is approximated as $u/U_0 = y/\delta$. Calculate the horizontal component of force per unit width required to hold the plate stationary.

4.94 Air at standard conditions flows along a flat plate. The undisturbed freestream speed is $U_0 = 10$ m/sec. At $L = 145$ mm downstream from the leading edge of the plate, the boundary-layer thickness is $\delta = 2.3$ mm. The velocity profile at this location is

$$\frac{u}{U_0} = \frac{3}{2}\frac{y}{\delta} - \frac{1}{2}\left[\frac{y}{\delta}\right]^3$$

Calculate the horizontal component of force per unit width required to hold the plate stationary.

4.95 Consider again the flow over a flat plate with suction depicted in Problem 4.37. For the given conditions, evaluate the force per unit width required to hold the plate stationary.

4.96 Experimental measurements are made in a low-speed air jet to determine the drag force on a circular cylinder. Velocity measurements at two sections where the pressure is uniform and equal give the results shown. Evaluate the drag force on the cylinder, per unit width.

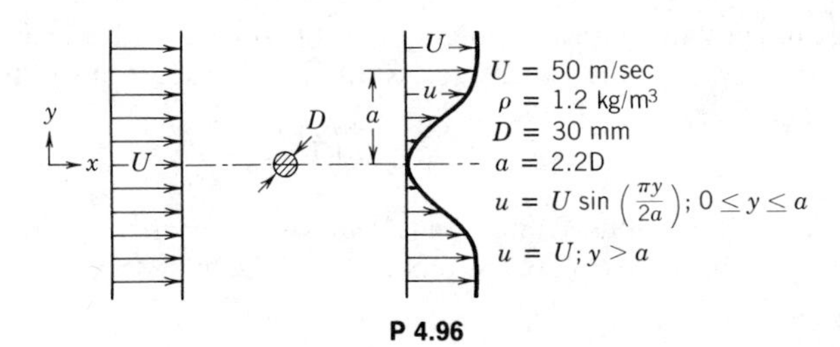

P 4.96

4.97 A small round object is tested in a 1 m diameter wind tunnel. The pressure is uniform across sections ① and ②. The upstream pressure is 20 mm H_2O (gage), the downstream pressure is 10 mm H_2O (gage), and the mean air speed is 30 m/sec. The velocity profile at section ② is linear; it varies from zero at the tunnel centerline to a maximum at the tunnel wall. Calculate (a) the mass flow rate in the wind tunnel, (b) the maximum velocity at section ②, and (c) the drag of the object and its supporting vane. Neglect viscous resistance at the tunnel wall.

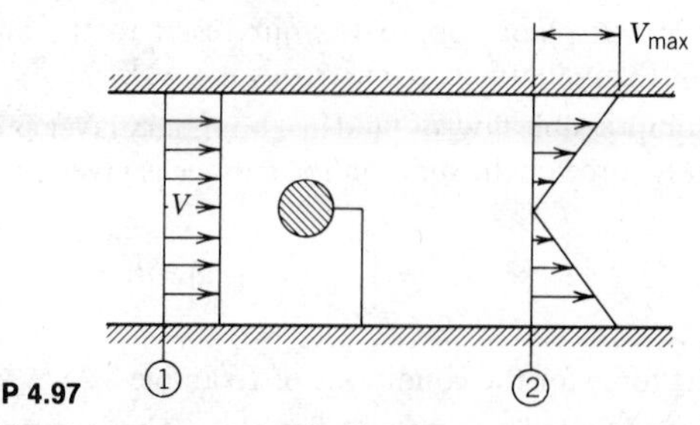

P 4.97

4.98 An incompressible fluid flows steadily in the entrance region of a two-dimensional channel of height, $2h$. The uniform velocity at the channel entrance is $U_1 = 20$ ft/sec. The velocity distribution at a section downstream is

$$\frac{u}{u_{\max}} = 1 - \left[\frac{y}{h}\right]^2$$

Evaluate the maximum velocity at the downstream section. Calculate the pressure drop that would exist in the channel if viscous friction at the walls could be neglected.

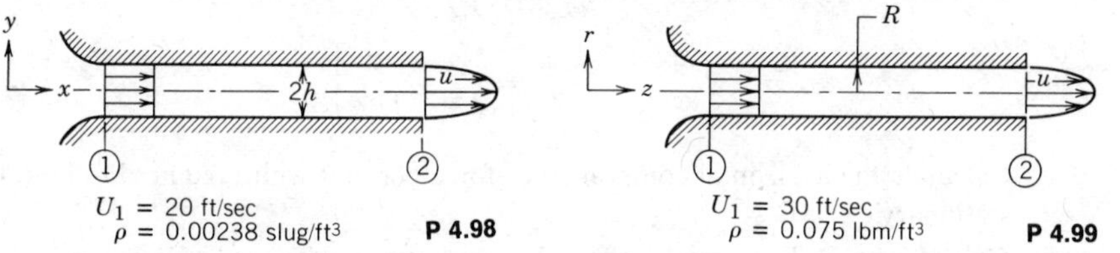

P 4.98 P 4.99

4.99 An incompressible fluid flows steadily in the entrance region of a circular tube of radius, R. The uniform velocity at the tube entrance is $U_1 = 30$ ft/sec. The velocity distribution at a section downstream is

$$\frac{u}{u_{\max}} = 1 - \left[\frac{r}{R}\right]^2$$

Evaluate the maximum velocity at the downstream section. Calculate the pressure drop that would exist in the tube if viscous friction at the walls could be neglected.

4.100 A fluid of constant density, ρ, enters a pipe of radius R with a uniform velocity, V. At a downstream section the velocity varies with radius r according to the equation

$$u = 2V\left(1 - \frac{r^2}{R^2}\right)$$

The pressure at sections ① (inlet) and ② (downstream) are p_1 and p_2, respectively. Show that the frictional force, F, of the pipe walls on the fluid between sections ① and ② is

$$F = \pi R^2[-(p_1 - p_2) + \tfrac{1}{3}\rho V^2]$$

in a direction opposing the flow.

4.101 Air flows steadily through a plane-wall diffuser and duct system. The channel width normal to the plane of the diagram is b. At section ①, the velocity profile is given by

$$\frac{u}{u_{\max}} = e^{-2\frac{|y|}{h}}$$

The velocity is uniform at section ②. The pressure is uniform across sections ① and ②. Assume incompressible flow and neglect friction at the channel walls. Calculate (a) the value of $u_{\max}$, and (b) the pressure difference, $p_2 - p_1$.

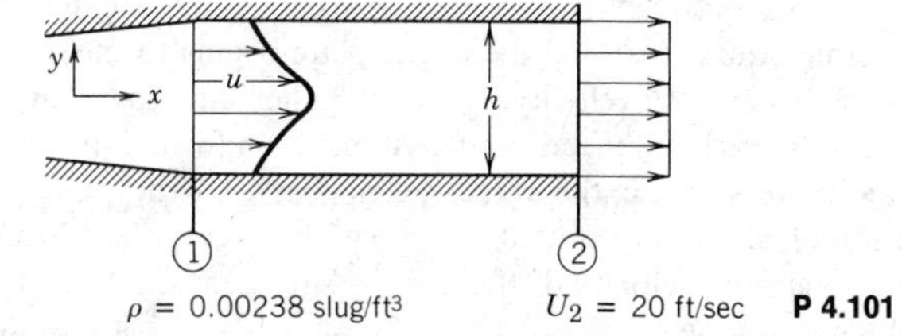

ρ = 0.00238 slug/ft³ U_2 = 20 ft/sec **P 4.101**

***4.102** Incompressible fluid of negligible viscosity is pumped at total volume flow rate, Q, through a porous surface into the small gap between closely spaced parallel plates as shown. The fluid has only horizontal motion in the gap. Assume uniform flow across any vertical section. Obtain an expression for the variation in horizontal velocity as a function of x. *Hint:* Apply conservation of mass to a control volume with outer surface located at position, x.

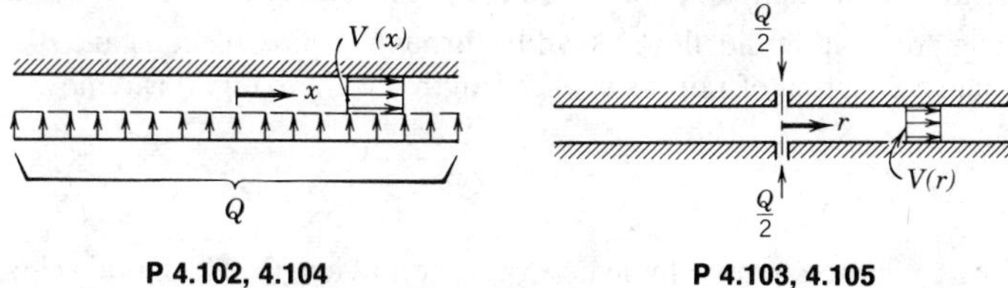

P 4.102, 4.104 **P 4.103, 4.105**

***4.103** Incompressible liquid of negligible viscosity is pumped at total volume flow rate, Q, through two small holes into the narrow gap between closely spaced parallel plates as shown. The liquid flowing away from the holes has only radial motion. Assume uniform flow across any vertical section. Obtain an expression for the variation of the radial velocity as a function of distance from the holes. *Hint:* Apply conservation of mass to a control volume with outer surface located at radius, r.

***4.104** Incompressible fluid of negligible viscosity is pumped at total volume flow rate, Q, through a porous surface into the small gap between closely spaced parallel plates as shown. The fluid has only horizontal motion in the gap. Assume uniform flow across any vertical section. Obtain an expression for the pressure variation as a function of x. *Hint:* Apply conservation of mass and the momentum equation to a differential control volume of thickness, dx, located at position, x.

***4.105** Incompressible liquid of negligible viscosity is pumped at total volume flow rate, Q, through two small holes into the narrow gap between closely spaced parallel plates as shown. The liquid flowing away from the holes has only radial motion. Assume uniform flow across any vertical section. Obtain an expression for the pressure variation as a function of distance from the holes. *Hint:* Apply conservation of mass and the momentum equation to a differential control volume of thickness, dr, located at radius, r.

***4.106** The small gap between two long narrow parallel plates initially is filled with incompressible liquid. At $t = 0$ the upper plate begins to move downward toward the lower plate with constant velocity, V_0, causing the liquid to be squeezed from the narrow gap. Neglecting viscous effects, and assuming uniform flow in the horizontal direction, develop an expression for the velocity field between the parallel plates. *Hint:* Apply conservation of mass to a control volume with outer surface located at position, x. Note that even though the velocity of the upper plate is constant, the flow is unsteady.

* These problems require material from sections that may be omitted without loss of continuity in the text material.

***4.107** The narrow gap between two closely spaced circular plates initially is filled with incompressible liquid. At $t = 0$ the upper plate begins to move downward toward the lower plate with constant velocity, V_0, causing the liquid to be squeezed from the narrow gap. Neglecting viscous effects and assuming uniform flow in the radial direction, develop an expression for the velocity field between the parallel plates. *Hint:* Apply conservation of mass to a control volume with outer surface located at radius, r. Note that even though the velocity of the upper plate is constant, the flow is unsteady.

***4.108** Incompressible liquid flows steadily through a pipe of constant diameter. The pipe contains a section of porous wall of length, L. There liquid is removed at the constant rate, q, expressed as volume flow rate per unit length. The liquid velocity in the pipe at the entrance to the porous section is V_0. The liquid removed in the porous section has no axial component of velocity. Evaluate the velocity and pressure distributions for flow through the porous section. Neglect viscous effects. *Hint:* Apply conservation of mass and the momentum equation to a differential control volume of length, dx.

***4.109** Incompressible liquid flows steadily through a pipe of constant diameter. The pipe contains a section of porous wall of length, L. There liquid is removed at the rate

$$q(x) = q_{\max} \frac{x}{L}$$

with $q_{\max}$ expressed as volume flow rate per unit length. The liquid velocity in the pipe at the entrance to the porous section is V_0. The liquid removed in the porous section has no axial component of velocity. Evaluate the velocity and pressure distributions for flow through the porous section. Neglect viscous effects. *Hint:* Apply conservation of mass and the momentum equation to a differential control volume of length, dx.

***4.110** Incompressible liquid flows steadily through a pipe of constant diameter. The pipe contains a section of porous wall of length, L, where liquid is supplied at constant rate, q, expressed as volume flow rate per unit length. The liquid velocity in the pipe at the entrance to the porous section is V_0. The liquid supplied in the porous section has no axial component of velocity. Evaluate the velocity and pressure distributions for flow through the porous section. Neglect viscous effects. *Hint:* Apply conservation of mass and the momentum equation to a differential control volume of length, dx.

***4.111** Incompressible liquid flows steadily through a pipe of constant diameter. The pipe contains a section of porous wall of length, L, where liquid is supplied at the rate

$$q(x) = q_{\max} \frac{x}{L}$$

with $q_{\max}$ expressed as volume flow rate per unit length. The liquid velocity in the pipe at the entrance to the porous section is V_0. The liquid supplied in the porous section has no axial component of velocity. Evaluate the velocity and pressure distributions for flow through the porous section. Neglect viscous effects. *Hint:* Apply conservation of mass and the momentum equation to a differential control volume of length, dx.

***4.112** Consider unsteady, one-dimensional flow of an incompressible liquid in a horizontal open channel. The free surface elevation is not constant, but the velocity distribution at any section is uniform. Depth, h, and mean velocity, V, are functions of both x and t. Assume the channel width, b, is large. Derive appropriate forms of the conservation of

* These problems require material from sections that may be omitted without loss of continuity in the text material.

mass and linear momentum equations for this flow. *Hint:* Use a differential control volume of length, dx, in the flow direction. Be sure to include the effects of hydrostatic pressure forces and friction on the channel bed.

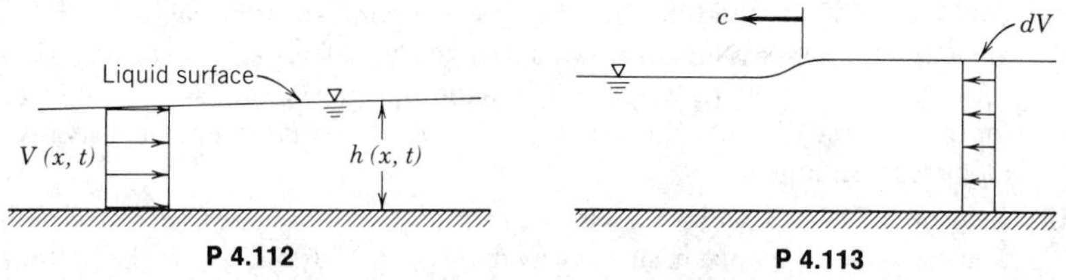

P 4.112 **P 4.113**

***4.113** Propagation of small waves on a liquid free surface may be analyzed using a differential control volume. Consider a small solitary wave moving with speed, c, from right to left. Assume a small change in water surface elevation across the wave. To make the flow appear steady, choose a differential control volume that encloses the wave and moves with it. Apply the conservation of mass and momentum equations to derive an expression for wave speed. Be sure to include in your analysis the hydrostatic pressure forces on the control surface. Neglect friction on the channel bed.

***4.114** A stream of incompressible liquid moving at low speed leaves a nozzle pointed directly downward. Assume the velocity at any cross section is uniform and neglect viscous effects. The velocity and area of the jet at the nozzle exit are V_0 and A_0, respectively. Apply conservation of mass and the momentum equation to a differential control volume of length, dz, in the flow direction. Derive expressions for the variations of jet velocity and area as functions of z. Evaluate the distance at which the jet area is half its original value. (Take the origin of coordinates at the nozzle exit.)

***4.115** A stream of incompressible liquid moving at low speed leaves a nozzle pointed directly upward. Assume the velocity at any cross section is uniform and neglect viscous effects. The velocity and area of the jet at the nozzle exit are V_0 and A_0, respectively. Apply conservation of mass and the momentum equation to a differential control volume of length, dz, in the flow direction. Derive expressions for the variations of jet velocity and area as functions of z. Evaluate the vertical distance required to reduce the jet speed to zero. (Take the origin of coordinates at the nozzle exit.)

***4.116** Liquid falls vertically into a short horizontal rectangular open channel of width, b. The total volume flow rate, Q, is distributed uniformly over area bL. Neglecting viscous effects:

(a) Obtain an expression for h_1 in terms of h_2, Q, and b. *Hint:* Choose a control volume with outer boundary located at $x = L$.

(b) Sketch the surface profile, $h(x)$. *Hint:* Use a differential control volume of width, dx.

P 4.116 **P 4.117**

* These problems require material from sections that may be omitted without loss of continuity in the text material.

***4.117** As part of an industrial process, a pipe is arranged with a small slot along its length and liquid is allowed to drain out in a continuous sheet. The pipe is tapered so the pressure and vertical flow velocity are constant along its length. The horizontal velocity component of the fluid leaving the pipe is a constant fraction, k, of the local average velocity in the pipe. Neglect friction and gravity effects as a first approximation. Show that k must be 1.0 and the velocity within the pipe must be constant. Obtain an expression for the required diameter variation, $D(x)$. *Hint:* Note that mechanical energy is conserved along a streamline.

***4.118** In ancient Egypt, circular vessels filled with water sometimes were used as crude clocks. The vessels were shaped in such a way that, as water drained from the bottom, the surface level dropped at a constant rate, s. Assume that water drains from a hole of area, A. Find an expression for the radius of the vessel, r, as a function of the water level, h. Determine the volume of water needed so that the clock will operate for n hours.

***4.119** Water flows at the rate of 0.019 m^3/sec into a tall cylindrical tank that is 1 m inside diameter. The tank bottom has a well-rounded orifice 55 mm in diameter. Water flows vertically from the orifice due to gravity. When the water in the tank reaches a certain level, the rate of outflow equals the rate of inflow. Neglecting friction, calculate the equilibrium water level.

***4.120** The tank of Problem 4.119 initially is empty. At $t = 0$ the water supply is turned on at a constant rate of 0.019 m^3/sec. Develop a differential equation for the rate of change of water level. Use numerical integration to estimate the time required for water in the tank to reach 95 percent of its equilibrium level.

***4.121** Consider a cylindrical tank of inside area, A_t, and mass, M, placed on a scale. Assume the tank is filled to level, h, with water, which drains from a well-rounded orifice of area, A, in the tank bottom. (The jet does not hit the scale platform.) Develop an expression for the percentage reduction in scale reading due to the momentum efflux through the orifice.

***4.122** Consider a cylindrical tank of inside area, A_t, and mass, M, placed on a smooth plane surface. Assume the tank is filled to level, h, with water. A small rounded orifice of area, A, is placed in the side near the tank bottom so that a jet discharges horizontally. Develop an expression for the coefficient of friction that would allow the tank to slide as a result of momentum flux from the jet. Evaluate for $A/A_t = 0.1$. Comment on your result.

***4.123** Two large tanks containing water have small smoothly contoured orifices of equal area. A jet of liquid issues from the left tank. Assume the flow is uniform and unaffected by friction. The jet impinges on a flat plate covering the opening of the right tank. Obtain an expression for the height, h, required to balance the hydrostatic force on the plate from water in the right tank.

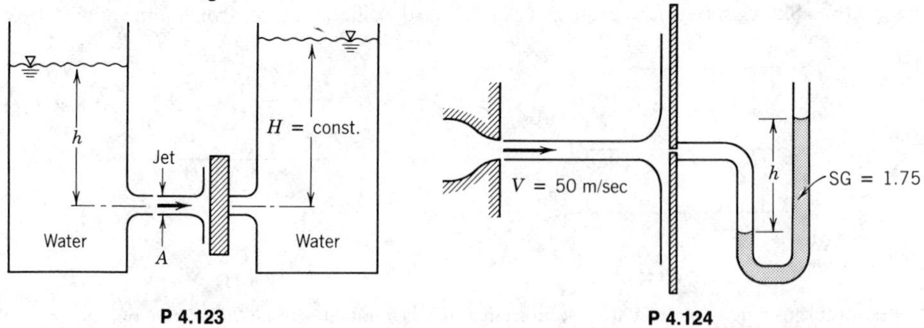

P 4.123 P 4.124

* These problems require material from sections that may be omitted without loss of continuity in the text material.

***4.124** A horizontal axisymmetric jet of air with 10 mm diameter strikes a stationary vertical disk of 200 mm diameter. The jet speed is 50 m/sec at the nozzle exit. A manometer is connected to the center of the disk. Calculate (a) the deflection, h, if the manometer liquid has SG $= 1.75$, and (b) the force exerted by the jet on the disk.

***4.125** Consider again the statement and diagram of Problem 4.68. Apply the Bernoulli equation twice to obtain expressions for the pressure just in front of and just behind the propeller disk. (Note that the thrust on the propeller may be written as the pressure difference across the disk times the disk area.) Use the momentum equation for a control volume to obtain an expression for the thrust exerted by the propeller. Equate the two expressions for thrust to show that half the velocity increase occurs ahead of and half behind the propeller.

***4.126** A uniform jet of water leaves a 15 mm diameter nozzle and flows directly downward. The jet speed at the nozzle exit plane is 1.5 m/sec. The jet impinges on a horizontal disk and flows radially outward in a flat sheet. Obtain a general expression for the velocity the liquid stream would reach at the level of the disk. Develop an expression for the force required to hold the disk stationary, neglecting the mass of the disk and water sheet. Evaluate for $h = 1.5$ m.

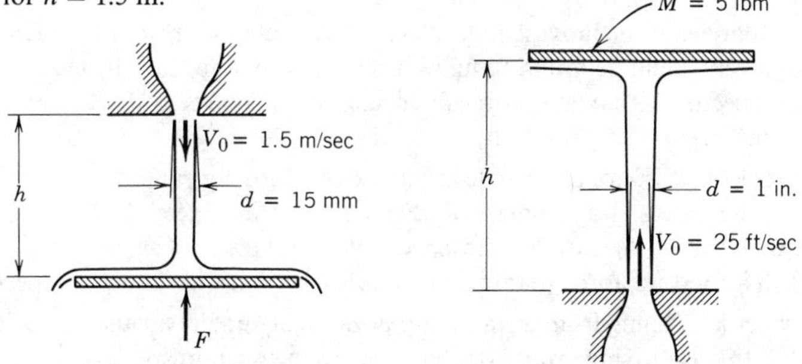

P 4.126 **P 4.127**

***4.127** A 5 lbm disk is constrained horizontally but is free to move vertically. The disk is struck from below by a vertical jet of water. The speed and diameter of the water jet are 25 ft/sec and 1 in. at the nozzle exit. Obtain a general expression for the speed of the water jet as a function of height, h. Find the height to which the disk will rise and remain stationary.

4.128 A jet of water is directed against a vane which could be a blade in a turbine or in any other piece of hydraulic machinery. The water leaves the stationary 50 mm diameter nozzle with a speed of 20 m/sec and enters the vane tangent to the surface at A. The inside surface of the vane at B makes an angle, $\theta = 150°$ with the x direction. Compute the force that must be applied to maintain the vane speed constant at $U = 5$ m/sec.

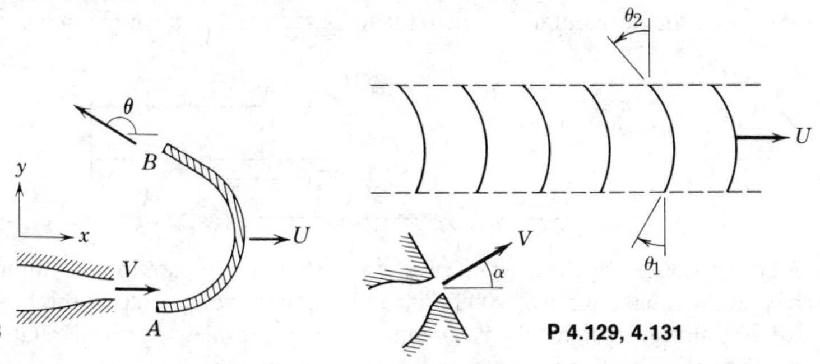

P 4.128 **P 4.129, 4.131**

* These problems require material from sections that may be omitted without loss of continuity in the text material.

4.129 Consider a series of turning vanes struck by a continuous jet of water that leaves a 50 mm diameter nozzle at constant speed, $V = 86.6$ m/sec. The vanes move with constant speed, $U = 50$ m/sec. Note that all the mass flow leaving the jet crosses the vanes. The curvature of the vanes is described by angles, $\theta_1 = 30°$, and $\theta_2 = 45°$, as shown. (The total turning angle of the vanes is the sum of these two angles.) Evaluate the nozzle angle, α, required to assure that the jet enters tangent to the leading edge of each vane. Calculate the force that must be applied to maintain the vane speed constant.

4.130 Consider a single vane with turning angle, θ, moving horizontally at constant speed, U, under the influence of an impinging jet as shown. The absolute speed of the jet is V. Obtain general expressions for the resultant force and power that the vane could produce. Show that the power is maximized when $U = V/3$.

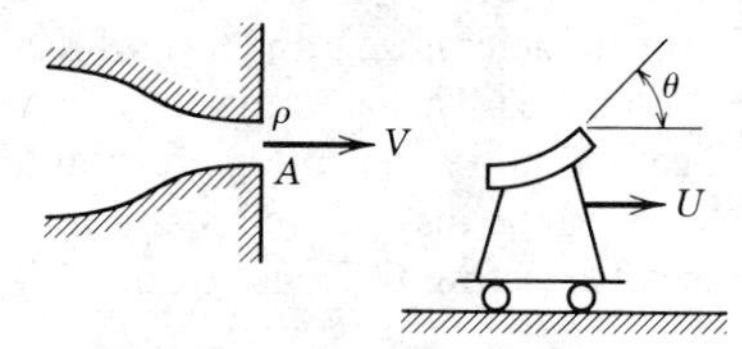

P 4.130, 4.132, 4.133, 4.134

4.131 Consider again the moving multiple-vane system described in Problem 4.129. Assuming that a way could be found to make α nearly zero (and thus, θ_1 nearly 90°), evaluate the vane speed, U, that would result in maximum power output from the moving vane system.

4.132 Water from a stationary nozzle impinges on a moving vane with a turning angle of 120°. The vane moves away from the nozzle with constant speed, $U = 30$ ft/sec, and receives a jet that leaves the nozzle with speed, $V = 100$ ft/sec. The nozzle has an exit area of 0.04 ft^2. Find the force that must be applied to maintain the vane speed constant.

4.133 A water jet, issuing from a stationary nozzle, encounters a vane curved through an angle of 90° that is moving away from the nozzle at a constant speed of 15 m/sec. The jet has a cross-sectional area of 600 mm^2 and a speed of 30 m/sec. Determine the force that must be applied to maintain the vane speed constant.

4.134 A jet of oil (SG = 0.8) strikes a curved blade that turns the fluid through an angle of 180°. The jet area is 1200 mm^2 and its speed relative to the stationary nozzle is 20 m/sec. The blade moves toward the nozzle at 10 m/sec. Determine the force that must be applied to maintain the blade speed constant.

4.135 A snow plow mounted on a truck clears a path through heavy wet snow. The plow blade is 12 ft wide. The snow is 8 in. deep and its density is 10 lbm/ft^3. The truck travels at 20 mph. Snow is discharged from the plow at an angle of 45° from the travel direction and 45° above the horizontal. Evaluate the force required to push the plow.

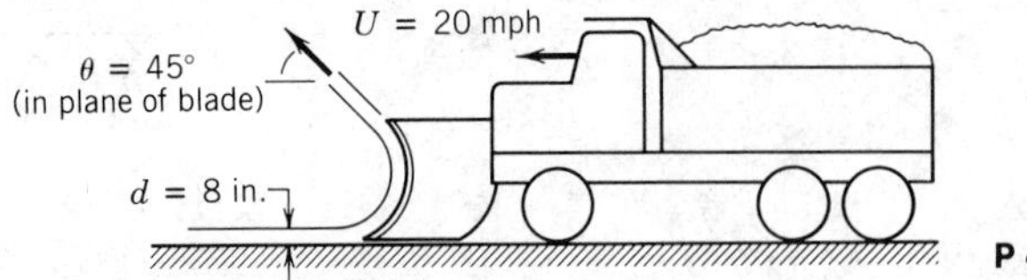

P 4.135

4.136 A boat propelled by thrust from a water jet is to achieve a speed of 30 mph. Towing tests have shown that a force of 200 lbf is required to move the boat at this speed. The water for the pump-jet system may be assumed to enter the inlet at a speed (relative to the boat) equal to the boat speed. Inlet and outlet pressures may be assumed equal to that of undisturbed water at the same level. The pump flow rate is 900 gal/min. Determine the exit jet speed relative to the boat required to sustain the desired boat speed.

4.137 Water in a 100 mm diameter jet with speed of 30 m/sec to the right is deflected by a cone that moves to the left at 15 m/sec. Determine (a) the thickness of the jet sheet at a radius of 200 mm, and (b) the external horizontal force needed to move the cone.

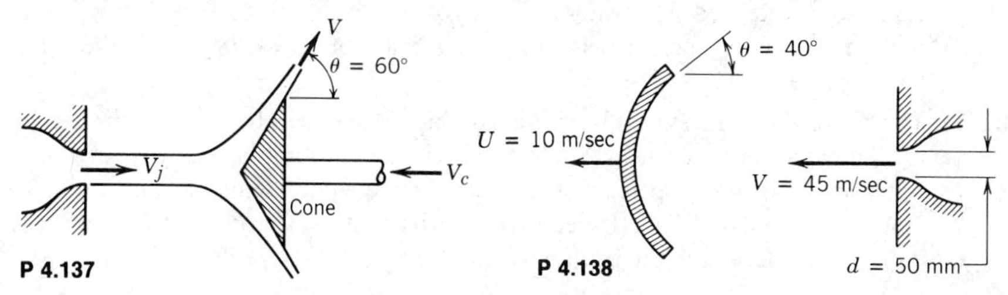

4.138 The circular dish whose cross section is shown has an outside diameter of 0.15 m. A water jet strikes the dish concentrically and then flows outward along the surface of the dish. The jet speed is 45 m/sec and the dish moves to the left at 10 m/sec. What horizontal force on the dish is required to accomplish this situation?

4.139 The circular dish whose cross section is shown has an outside diameter of 0.20 m. A water jet with speed of 30 m/sec strikes the dish concentrically. The dish moves to the left at 10 m/sec. The jet diameter is 20 mm. The dish has a hole at its center that allows a stream of water 10 mm in diameter to pass through without resistance. The remainder of the jet is deflected and flows along the dish. Calculate the force required to maintain the dish motion.

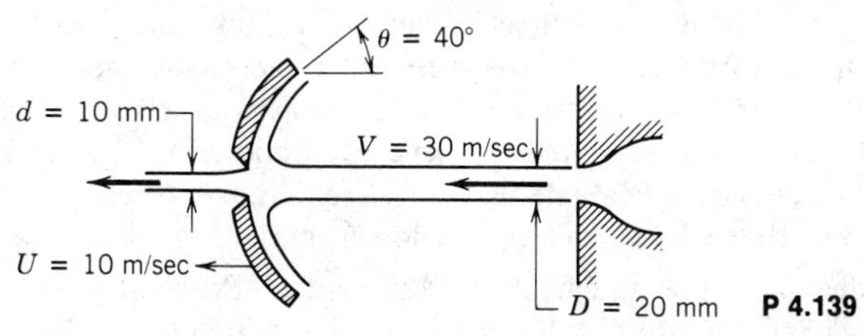

4.140 Beginning in the 1900s, water troughs called "track pans" were built between the rails to allow steam locomotives to scoop water into their tenders without stopping. Each tender equipped with a scoop could pick up 2.5 gallons of water per foot of scooping distance while traveling at 50 mph. Estimate the added drawbar force required to pull a tender while scooping water.

4.141 A jet of lye solution (SG = 1.10) 50 mm in diameter has an absolute speed of 15 m/sec. It strikes a single flat plate that is moving away from the nozzle with an absolute speed of 5 m/sec. The plate makes an angle of 60° with the horizontal. Calculate the force on the plate from the jet. Assume no friction force along the plate surface.

***4.142** Consider again the statement and diagram for Problem 4.125. Solve the problem for the case in which the propeller advances with speed, V_1, into stationary fluid.

***4.143** The propeller on an air boat used in the Florida Everglades moves air at the rate of 40 kg/sec. When at rest, the speed of the slipstream behind the propeller is 40 m/sec at a

* These problems require material from sections that may be omitted without loss of continuity in the text material.

location where the pressure is atmospheric. Calculate (a) the propeller diameter, (b) the thrust produced at rest, and (c) the thrust produced when the air boat is moving ahead at 10 m/sec, if the mass flow rate through the propeller remains constant.

4.144 The acceleration of the vane/cart assembly of Problem 4.132 is to be controlled as it accelerates from rest by changing the vane angle, θ. A constant acceleration, $a = 1.5\ \text{m/sec}^2$, is desired. The water jet leaves the nozzle of area, $A = 0.025\ \text{m}^2$, with a speed of 15 m/sec. The vane/cart assembly has a mass of 55 kg; neglect friction. Determine θ at $t = 5$ sec.

4.145 The wheeled cart shown rolls with negligible resistance. The cart is to be accelerated to the right at a *constant rate* of 2 m/sec². This is to be accomplished by "programming" the water jet area, $A(t)$, that reaches the cart. The jet speed remains constant at 10 m/sec. Obtain an expression for $A(t)$ required to produce the motion. Sketch the area variation for $t \leq 4$ sec. Evaluate the jet area at $t = 2$ sec.

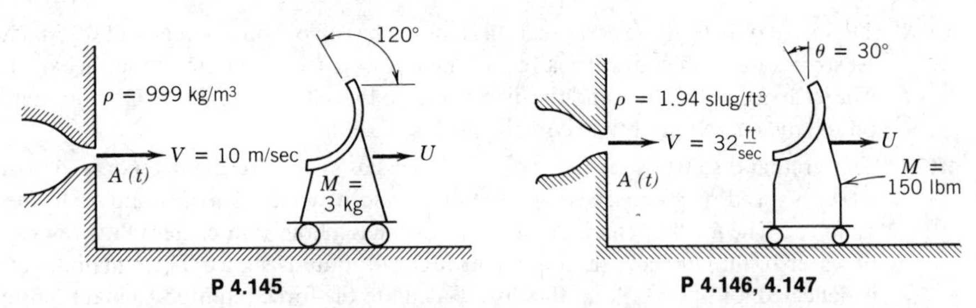

P 4.145

P 4.146, 4.147

4.146 A small vaned cart of mass, $M = 150$ lbm, is to be accelerated by a water jet. The jet speed is constant at $V = 32$ ft/sec, but its area is controlled to vary the mass flow rate. The cart is to start from rest at $t = 0$, then undergo *constant acceleration* at 4 ft/sec² until it reaches $U = 16$ ft/sec. Then it is to travel at constant speed. Neglect resistance to motion. Obtain a general expression for the jet cross-sectional area, $A(t)$, that must reach the cart during acceleration. Evaluate the jet area required at the instant the cart starts to move. Determine the time at which the jet flow from the nozzle must be cut off.

4.147 Consider again the statement and diagram of Problem 4.146. Determine the cross-sectional area variation *at the nozzle*, $A_n(t)$, required to accelerate the cart. Specify the time at which flow from the nozzle must be cut off.

4.148 A rocket sled weighing 10,000 lbf, traveling 600 mph, is to be braked by lowering a scoop into a water trough. The scoop is 6 in. wide. Determine the time required (after lowering the scoop to a depth of 3 in. into the water) to bring the sled to a speed of 20 mph.

4.149 A rocket sled is to be slowed from an initial speed of 300 m/sec by lowering a scoop into a water trough. The scoop is 0.3 m wide; it deflects the water through 150°. The trough is 800 m in length. The mass of the sled is 8000 kg. At the initial speed it experiences an aerodynamic drag force of 90 kN. The aerodynamic force is proportional to the square of the sled speed. It is desired to slow the sled to 100 m/sec. Determine the depth to which the scoop must be lowered into the water.

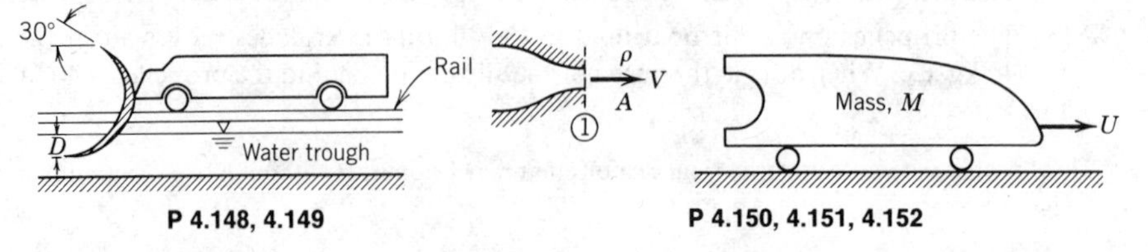

P 4.148, 4.149

P 4.150, 4.151, 4.152

4.150 Consider the vehicle shown. Starting from rest, it is propelled by a hydraulic catapult (liquid jet). The jet strikes the curved surface and makes a 180° turn, leaving horizontally. Air and rolling resistance may be neglected. Using the notation shown:
(a) Obtain an equation for the acceleration of the vehicle at any time, t.
(b) Determine the time required for the vehicle to reach $U = V/2$.

4.151 The mass of the cart in Problem 4.150 is 100 kg. The jet of water leaves the nozzle (area of 0.001 m^2) with a speed of 30 m/sec. Determine the speed of the cart 5 sec after the jet is directed against the cart.

4.152 Consider the jet and cart of Problem 4.151 again, but include an aerodynamic drag force proportional to the square of cart speed, $F_D = kU^2$, with $k = 2.0\ \text{N} \cdot \text{sec}^2/\text{m}^2$. Derive an expression for the cart acceleration as a function of cart speed and other given parameters. Evaluate the acceleration of the cart at $U = 10$ m/sec. What fraction is this speed of the terminal speed of the cart?

4.153 A jet, $\rho = 990\ \text{kg/m}^3$, issues from a stationary nozzle; the jet is deflected by a vane through an angle of 135°. The vane is attached to the cart as shown. Assume that the cart moves without friction down the 10° plane. The cart mass is 100 kg. The jet leaves the nozzle at atmospheric pressure with speed $V = 65$ m/sec relative to the nozzle. If the area of the jet is $A = 500\ \text{mm}^2$, determine the speed of the cart 5 sec after the jet is directed against the vane.

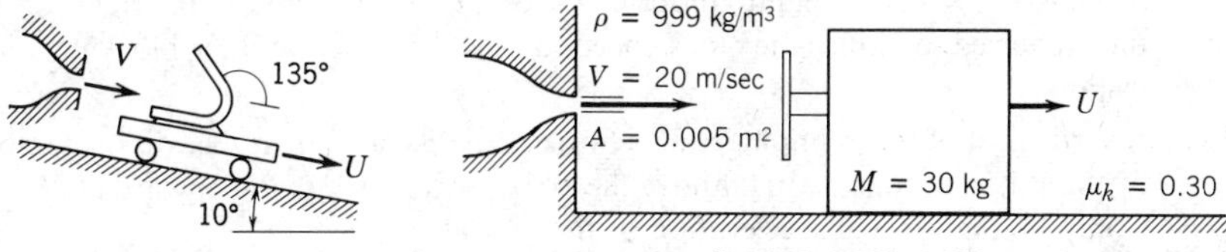

P 4.153 **P 4.154, 4.155, 4.156**

4.154 A vane/slider assembly moves under the influence of a liquid jet as shown. The coefficient of kinetic friction for motion of the slider along the surface is $\mu_k = 0.30$. Calculate (a) the acceleration of the slider at the instant when $U = 10$ m/sec, and (b) the terminal speed of the slider.

4.155 Consider again the statement and diagram of Problem 4.154. Obtain general expressions for the acceleration and speed of the slider as functions of time. Evaluate the terminal speed for the conditions given in Problem 4.154.

4.156 Solve Problem 4.154 if the vane and slider ride on a film of oil instead of sliding in contact with the surface. Assume motion resistance is proportional to speed, $F_R = kU$, with $k = 7.5\ \text{N} \cdot \text{sec/m}$.

4.157 The vane/cart assembly of mass, $M = 30$ kg, shown in Problem 4.132 is driven by a water jet. The water leaves the stationary nozzle of area, $A = 0.02\ \text{m}^2$ with a speed of 20 m/sec. The coefficient of kinetic friction between the assembly and the surface is 0.10. Plot the terminal speed of the assembly as a function of vane turning angle, θ, for $0 \leq \theta \leq \pi/2$. At what angle, θ, does the assembly begin to move if the coefficient of static friction is 0.15?

4.158 The vane angle, θ, of the vane/cart assembly in Problem 4.157 is set at 60°. Determine the cart speed at time, $t = 1$ sec.

4.159 A rectangular block of mass, M, with vertical faces, rolls without resistance along a smooth horizontal plane as shown. The block travels initially at speed, U_0. At $t = 0$ the block is struck by a liquid jet and its speed begins to slow. Obtain an algebraic expression for the acceleration of the block for $t > 0$. Solve the equation to determine the time at which $U = 0$.

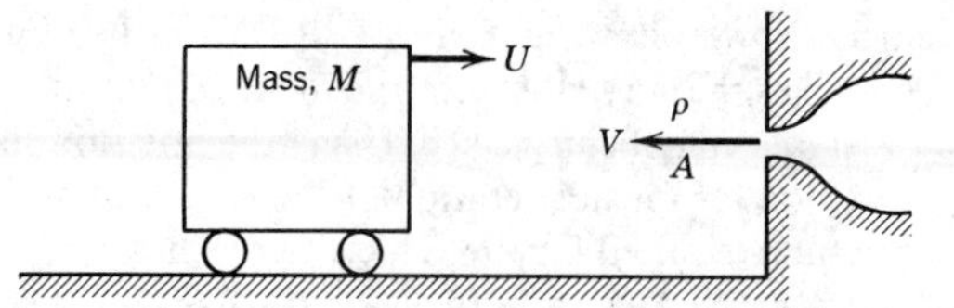

P 4.159

4.160 A rectangular block of mass, M, with vertical faces, rolls on a horizontal surface between two opposing jets as shown. At $t = 0$ the block is set into motion at speed, U_0. Subsequently, it moves without friction parallel to the jet axes with speed, $U(t)$. Neglect the mass of any liquid adhering to the block compared to the mass, M. Obtain general expressions for the acceleration of the block, $a(t)$, and the block speed, $U(t)$.

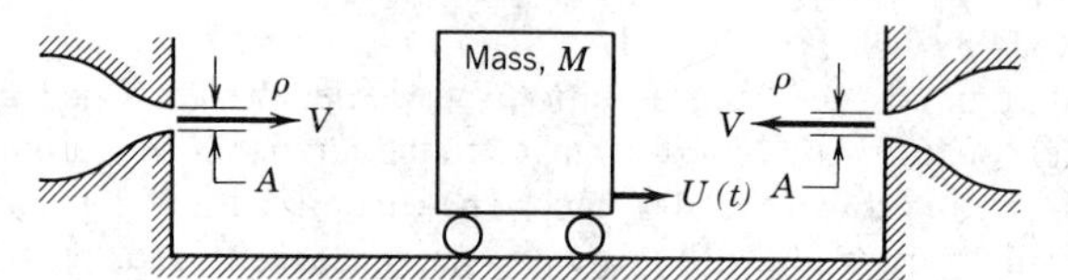

P 4.160, 4.161

4.161 Consider the statement and diagram of Problem 4.160. Assume that at $t = 0$, when the block is at $x = 0$, it is set into motion at speed, $U_0 = 10$ m/sec, to the right. Calculate the time required to reduce the block speed to $U = 0.5$ m/sec, and the block position at that instant.

***4.162** A vertical jet of water impinges on a horizontal disk as shown. The disk assembly weighs 65 lbf. When the disk is 10 ft above the nozzle exit, it is moving upward at $U = 15$ ft/sec. Compute the vertical acceleration of the disk at this instant.

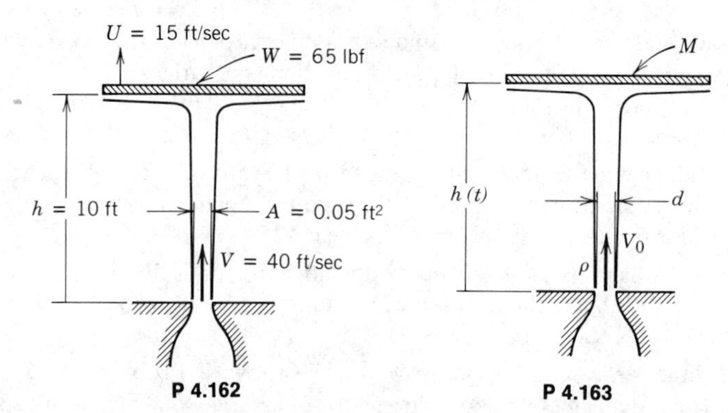

P 4.162

P 4.163

***4.163** A disk of mass, M, is constrained horizontally but is free to move vertically. A jet of water strikes the disk from below. The jet leaves the nozzle at initial speed, V_0. Obtain a differential equation for the disk height, $h(t)$, above the jet exit plane if the disk is released from large height, H. Assume that, when the disk reaches equilibrium, its height above the jet exit plane is h_0. Sketch $h(t)$ for the disk released at $t = 0$ from $H > h_0$. Explain why the sketch is as you show it.

4.164 A vehicle is moving at a speed of 50 ft/sec along level ground under the action of a constant force, $F = 100$ lbf. At time $t = 0$, mass begins leaving the vehicle through a hole

* These problems require material from sections that may be omitted without loss of continuity in the text material.

in the bottom. Assuming the mass leaves the vehicle vertically at a rate of 10 lbm/sec and the vehicle continues to move under the action of the constant force, F, determine the speed of the vehicle after 20 sec. The initial mass of the vehicle is 2000 lbm.

4.165 A hopper car is moving to the left with a speed of 15 ft/sec and is accelerating in the same direction at 0.5 ft/sec^2. A stream of grain falls into the car from a vertical stationary pipe. The velocity of the grain is 24 ft/sec in a downward direction relative to the pipe. The mass flow rate of grain leaving the pipe is 50 lbm/sec. Determine the force required to move the car as indicated, at the instant when the total mass of car and contents is 15,000 lbm. Neglect friction between the car and the track.

4.166 A container slides along a horizontal surface, where the resistance is characterized by the coefficient of kinetic friction, μ_k. The initial mass and speed of the container are M_0 and U_0. Mass leaves vertically at constant rate, $\dot{m}$. Evaluate the mass, acceleration, velocity, and position of the container as functions of time.

4.167 A container rolls along a horizontal surface with negligible friction. A restraining cord applies a constant retarding force, F_R, to the container. The initial mass and speed of the container are M_0 and U_0. Mass leaves vertically at constant rate, $\dot{m}$. Evaluate the mass, acceleration, velocity, and position of the container as functions of time.

4.168 A manned space capsule travels in level flight above the Earth's atmosphere at initial speed, $U_0 = 8.05$ km/sec. The capsule is to be slowed by a retro-rocket to $U = 5.00$ km/sec in preparation for a reentry maneuver. The initial mass of the capsule is $M_0 = 1600$ kg. The rocket consumes fuel at $\dot{m} = 8.0$ kg/sec, and exhaust gases leave at $V_e = 2940$ m/sec relative to the capsule. Evaluate the duration of the retro-rocket firing needed to accomplish this.

4.169 A rocket sled with initial mass of 3 metric tons, including 1 ton of fuel, rests on a level section of ground. At $t = 0$, the solid fuel of the rocket is ignited and the rocket burns fuel at the rate of 75 kg/sec. The exit speed of the exhaust gas relative to the rocket is 1000 m/sec. Neglecting friction and air resistance, calculate the acceleration of the sled at $t = 10$ sec.

4.170 A rocket sled with initial mass of 2000 lbm is to be accelerated on a level track. The rocket motor burns fuel at the constant rate, $\dot{m} = 30$ lbm/sec. The rocket exhaust flow is uniform and axial. Gases leave the nozzle at 9000 ft/sec relative to the nozzle. Determine the minimum mass of rocket fuel needed to propel the sled to a speed of 600 mph before burnout occurs. As a first approximation, neglect resistance forces.

4.171 A rocket sled has an initial mass of 4 metric tons including 1 ton of fuel. The motion resistance in the track on which the sled rides and that of the air total kU, where k is 75 N · sec/m, and U is the speed of the sled in m/sec. The exit speed of the exhaust gas relative to the rocket is 1500 m/sec, and it burns fuel at the rate of 90 kg/sec. Compute the speed of the sled after 10 sec.

4.172 A rocket sled of initial mass, M_0, is fired at $t = 0$, along a horizontal track. Fuel is burned at the rate, $\dot{m}$; the speed of the exhaust gas relative to the rocket is V. If the total motion resistance is given by kU^2, where U is the sled speed, determine the speed, U, of the rocket, as a function of time.

4.173 A rocket cart that is initially at rest and weighs 1610 lbf is to be fired and is to have a constant acceleration of 20 ft/sec^2. To accomplish this, the exhaust gases will be deflected through an angle θ that varies as a function of time. This will account for a frictional force that is proportional to the speed squared, $F = 0.002U^2$, where F is in lbf and U is in ft/sec. The rocket exhausts gases at the rate of 1 slug/sec at constant speed, $V_e = 1000$ ft/sec, relative to the vehicle.

(a) Find an expression for $\cos\theta$ as a function of time, t.
(b) Determine t at which $\cos\theta$ is a minimum.
(c) Find θ_{max}.

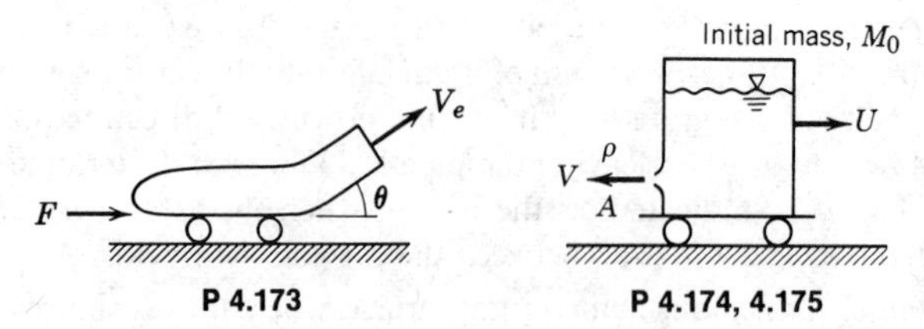

P 4.173 P 4.174, 4.175

4.174 A cart is propelled by a liquid jet issuing horizontally from a tank as shown. The track is horizontal; resistance to motion may be neglected. The tank is pressurized so that the jet speed may be considered constant. Obtain a general expression for the speed of the cart as it accelerates from rest.

4.175 A small cart is to be accelerated horizontally by a liquid jet. The acceleration rate is to be *constant* at 0.5 Gs for at least 5 sec. The jet leaves the propulsion nozzle in a uniform stream at a speed of 200 ft/sec relative to the cart. The only resistance to motion is frictional; the kinetic friction coefficient is $\mu_k = 0.15$. The initial mass of the cart and its contents is 3000 lbm. Obtain an algebraic expression for the mass flow rate of liquid needed to produce the desired acceleration. Obtain an algebraic expression for the vehicle mass as a function of time. Determine the flow rate at $t = 5$ sec.

4.176 A large two-stage liquid rocket with mass of 30,000 kg is to be launched from a sea level launch pad. The main engine burns liquid hydrogen and liquid oxygen in a stoichiometric mixture at 2450 kg/sec. The thrust nozzle has an exit diameter of 2.6 m. The exhaust gases exit the nozzle at 2270 m/sec and an exit plane pressure of 66 kPa absolute. Calculate the acceleration of the rocket at lift-off. Express your answer in Gs.

4.177 Neglecting air resistance, what speed would a vertically directed rocket attain in 10 sec if it starts from rest, has initial mass of 200 kg, burns 10 kg/sec, and ejects gas at 1200 m/sec relative to the rocket?

4.178 A "home-made" solid propellant rocket has an initial mass of 20 lbm; 15 lbm of this is fuel. The rocket is directed vertically upward from rest, burns fuel at a constant rate of 0.5 lbm/sec, and ejects exhaust gas at a speed of 3500 ft/sec relative to the rocket. Assume that the pressure at the exit is atmospheric and that air resistance may be neglected. Calculate the rocket speed after 20 sec, and the distance traveled by the rocket in 20 sec.

4.179 A small, solid fuel rocket motor is fired on a test stand. The combustion chamber is circular, with a diameter of 100 mm. Fuel, of density 1660 kg/m^3, burns uniformly at the rate of 12.7 mm/sec. Measurements show that the exhaust gases leave the rocket at ambient pressure, at a speed of 2750 m/sec. The absolute pressure and temperature in the combustion chamber are 7.0 MPa and 3610 K, respectively. The combustion products may be treated as an ideal gas with a molecular mass of 25.8. Evaluate the rate of change of mass and of linear momentum within the rocket motor. Express the rate of change of linear momentum within the motor as a percentage of the motor thrust.

4.180 A model solid propellant rocket has a mass of 69.6 g, of which 12.5 g is fuel. The rocket produces 1.3 lbf of thrust for a duration of 1.7 sec. For these conditions, calculate (a) the total impulse of the rocket motor, (b) the specific impulse of the rocket motor, and (c) the maximum speed and height attainable in the absence of air resistance. (Specific impulse is defined as the ratio of thrust to weight flow rate of propellant, expressed in seconds.)

4.181 A plastic toy rocket is shown. The rocket is propelled by a jet of water forced out the nozzle by compressed air. As a first approximation, assume that the water speed in the rocket chamber is given by $V = V_0 - kt$. The chamber and exit areas are A_c and A_e, respectively. The area ratio is approximately $A_e/A_c = 0.10$. Assume the initial mass of the rocket is M_0 and neglect the mass of the air. Find (a) the velocity of the water at the nozzle exit, (b) the mass of the rocket, M, and (c) the acceleration of the rocket, as functions of time. *Hint:* Use different control volumes for each calculation.

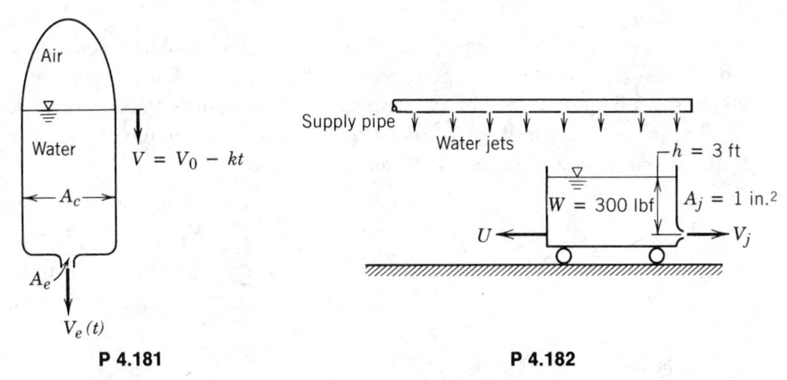

P 4.181 P 4.182

***4.182** The cart shown is supplied with water from nozzles above it. The flow rates are adjusted to maintain the water level in the cart at height, $h = 3$ ft. Water from each jet falls straight downward. The weight of the tank and water is 300 lbf. Compute the maximum speed, U, that will be attained by the cart if it rolls without friction on a level surface.

4.183 The moving tank shown is to be slowed by lowering a scoop to pick up water from a trough. The initial mass and speed of the cart and its contents are M_0 and U_0, respectively. Neglect external forces due to pressure or friction and assume that the track is horizontal. Apply the continuity and momentum equations to show that at any instant, $U = U_0M_0/M$. Obtain a general expression for U/U_0 as a function of time.

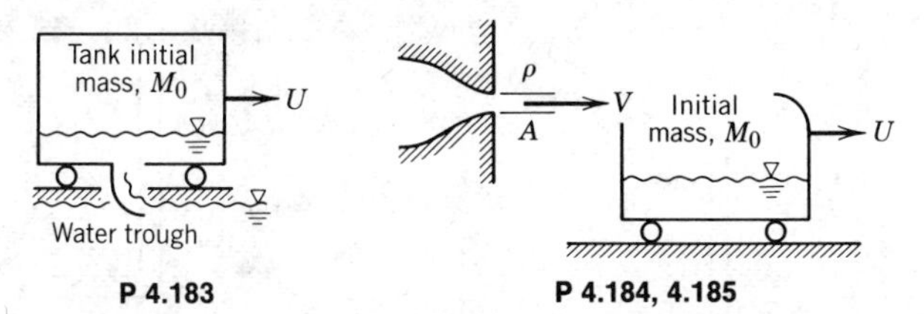

P 4.183 P 4.184, 4.185

4.184 The tank shown rolls along a level track. Water received from a jet is retained in the tank. The cart is to accelerate from rest toward the right with *constant* acceleration, a. Neglect wind and rolling resistance. Find an algebraic expression for the force (as a function of time) required to maintain the cart acceleration at the constant value, a.

4.185 The tank shown rolls with negligible resistance along a horizontal track. It is to be accelerated from rest by a liquid jet that strikes the vane and is deflected into the tank. The initial mass of the tank is M_0. Use the continuity and momentum equations to show that at any instant the mass of the vehicle and liquid contents is $M = M_0V/(V - U)$. Obtain a general expression for U/V as a function of time.

* These problems require material from sections that may be omitted without loss of continuity in the text material.

4.186 A combat airplane with initial mass of 30,000 lbm carries a 20 mm Gatling cannon that fires directly forward. The cannon has a firing rate of 2000 rounds of ammunition per minute. Each shell weighs 1.5 lb and each empty cartridge case weighs 0.2 lb. Spent cartridge cases are ejected directly downward relative to the aircraft. The muzzle velocity of a 20 mm cannon round is 1800 ft/sec. The weight and motion of burned powder gases may be neglected as a first approximation. The pilot is flying level at 1200 mph when he sights a target and begins to fire the cannon. Calculate the acceleration of the aircraft during firing.

4.187 A small "pod" propelled by a jet of air rotates on a strut about a vertical axis at a radius of 1 m from a fixed center. The pod mass is 2 kg; the jet velocity, density, and area are 200 m/sec, 1.5 kg/m^3, and 75 mm^2, respectively. At one instant, the pod moves at a speed of 50 m/sec, and the air drag force is 10 N. Determine the angular acceleration of the pod at this instant. Neglect friction and the mass of the strut.

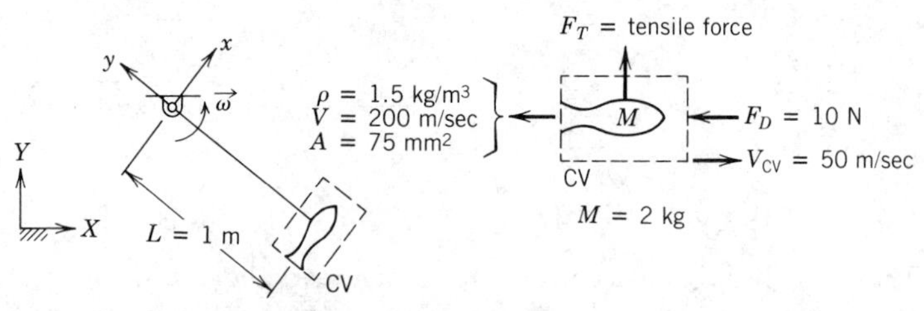

P 4.187, 4.188

4.188 Solve Problem 4.187 for the pod angular speed as a function of time starting from rest. Assume the drag force on the pod is given by $F_D = kV^2$. Evaluate k and find the time required for the pod to reach $V = 30$ m/sec.

4.189 A large irrigation sprinkler unit, mounted on a cart, discharges water with a speed of 40 m/sec at an angle of 30° to the horizontal. The 50 mm diameter nozzle is 3 m above the ground. Calculate the magnitude of the moment that tends to overturn the cart.

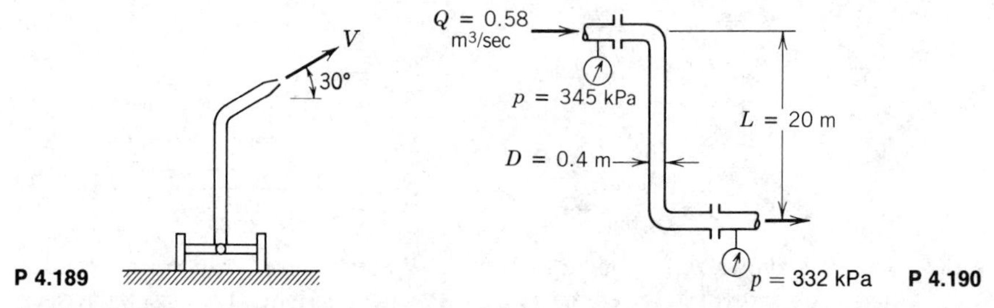

P 4.189 **P 4.190**

4.190 Crude oil (SG = 0.95) from a tanker dock flows through a pipe of 0.4 m diameter in the configuration shown. The flow rate is 0.58 m^3/sec, and the gage pressures are shown in the diagram. Determine the force and torque that are exerted by the pipe assembly on its supports.

4.191 Liquid in a thin sheet of width, w, and thickness, h, flows from a slot and strikes a stationary inclined flat plate, as shown. Experiments show that the resultant force of the liquid jet on the plate does not act through point O, where the jet centerline intersects the plate. Determine the magnitude and line of application of the resultant force as functions of θ. Evaluate the equilibrium angle of the plate if a force is applied at point O. Neglect any viscous effects.

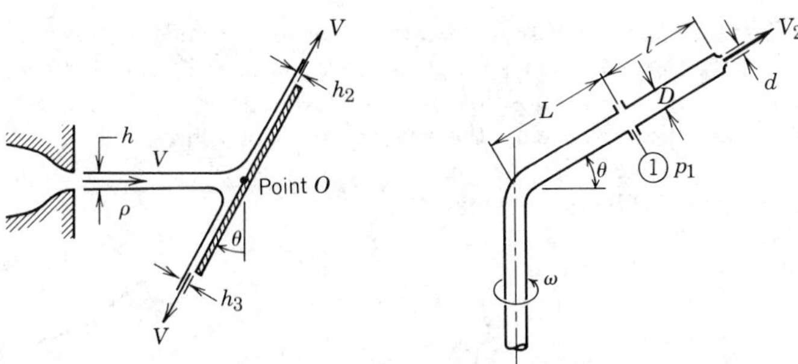

P 4.191 **P 4.192**

4.192 The large irrigation sprinkler is fabricated such that a "nozzle section" of length, l, can be mounted to the boom arm of length, L; both the boom and nozzle section have inside diameter, D; the opening at the exit of the nozzle section is of diameter, d. The boom, inclined at angle, θ, to the horizontal, rotates with angular velocity, ω, about the vertical axis. Determine the force of the nozzle section on the boom given that the weight of the nozzle section is $W = 50$ lbf and the following operating conditions: $D = 4$ in., $d = 1$ in., $\theta = 30°$, $L = 10$ ft, $l = 3$ ft, $\omega = 3$ rpm, $V_2 = 120$ ft/sec, and $p_1 = 100$ psig.

4.193 A pipe branches symmetrically into two legs of length, L, and the whole system rotates with angular speed, ω, around its axis. Each branch is inclined at angle, α, to the axis of rotation. Liquid enters the pipe steadily with zero angular momentum at volume flow rate, Q. The pipe diameter, D, is much smaller than L. Obtain an expression for the external torque required to turn the pipe. What additional torque would be required to impart an angular acceleration of value, $\dot{\omega}$?

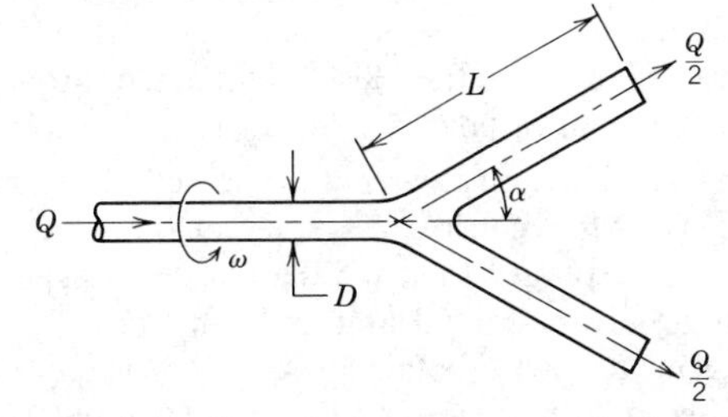

P 4.193

4.194 Water flows at the rate of 0.15 m³/sec through a nozzle assembly that rotates steadily at 30 rpm. The arm and nozzle masses are negligible compared to the water inside. Determine the torque required to drive the device and the reaction torques at the flange.

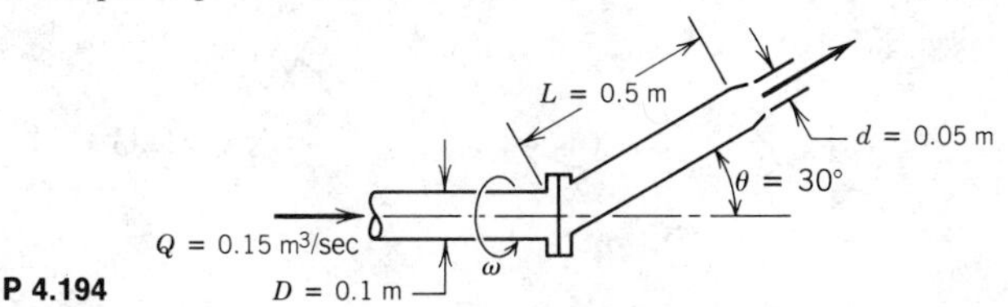

P 4.194

4.195 The hydraulic turbines at Grand Coulee Dam on the Columbia River are of the Francis, or radial inflow, type. [Francis turbines generally are preferred for locations with medium available head (80 to 600 ft) and in large installations can produce 90 to 95 percent of the theoretically possible power output.] The simplified geometry of the installation is shown schematically. Design conditions for this installation call for a turbine speed of 120 rpm at a head of 330 ft. The absolute inlet speed of the water is given by $V_1 = 0.95\sqrt{2gH}$, where H is the design head. Assume that flow is steady and uniform at each section. Assume also that the flow enters and leaves the runner tangent to the blade tips, and that the absolute velocity at the turbine outlet is purely radial. Using the conditions shown, develop the velocity polygons for the flow entering and leaving the runner. Determine the runner blade angles, β_1 and β_2, and the torque and power produced by the turbine.

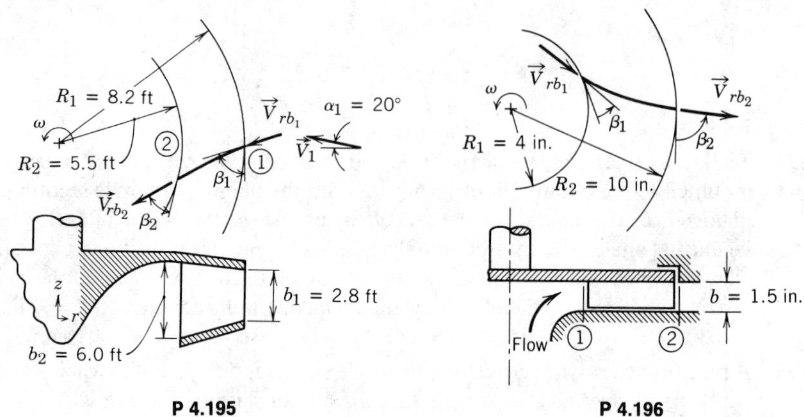

P 4.195 **P 4.196**

4.196 A centrifugal pump operates at 1750 rpm. The impeller geometry is shown. Water enters without swirl at volume flow rate, Q. The inlet and outlet radii and impeller passage height are shown on the diagram. The flow may be assumed to be steady and uniform at each section, and to enter and leave the impeller tangent to the blades. Construct the velocity polygons for the entering and leaving flows. Determine the design flow rate, in gallons per minute, for $\beta_1 = 30°$. Evaluate the torque and power required to drive the pump, as functions of flow rate, for $\beta_2 = 75$, 90, and 105°.

4.197 A Pelton wheel is a form of water turbine well adapted to situations of high head and low flow rate. The wheel consists of a series of vanes mounted on a rotor, as shown. One or more jets are arranged to strike the buckets tangentially. In practice it is possible to deflect the jet stream through angle, θ, of up to 165°. Consider the Pelton wheel and single jet arrangement shown. Obtain an expression for the torque exerted by the water stream on the wheel and the corresponding power output. (Let the bucket speed be $U = \omega R$.) Determine the value of U/V required to maximize the power produced by the wheel.

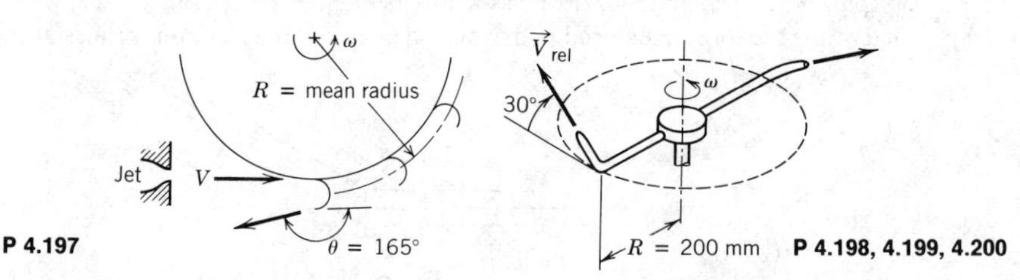

P 4.197 **P 4.198, 4.199, 4.200**

4.198 A small lawn sprinkler is shown. The sprinkler operates at a gage pressure of 140 kPa. The total flow rate of water through the sprinkler is 4 liters per minute. Each jet discharges at 17 m/sec relative to the sprinkler arm in a direction inclined 30° above the horizontal. The sprinkler rotates about a vertical axis. Friction in the bearing causes a torque of 0.18 N·m opposing rotation. Evaluate the torque required to hold the sprinkler stationary.

4.199 In Problem 4.198 calculate the initial acceleration of the sprinkler from rest if no external torque is applied, and its moment of inertia is $0.1\ \text{kg} \cdot \text{m}^2$ when filled with water.

4.200 A small lawn sprinkler is shown. The sprinkler operates at an inlet gage pressure of 140 kPa. The total flow rate of water through the sprinkler is 4.0 liters/min. Each jet discharges at a speed of 17 m/sec relative to the sprinkler arm, in a direction inclined 30° above the horizontal. The sprinkler rotates about a vertical axis. Friction in the bearing causes a torque of $0.18\ \text{N} \cdot \text{m}$ opposing rotation. Determine the steady speed of rotation of the sprinkler and the approximate area covered by the spray.

4.201 Air enters a compressor at standard conditions with a speed of 75 m/sec and leaves at an absolute pressure and temperature of 200 kPa and 345 K, with a speed of 125 m/sec. The flow rate is 1 kg/sec. The cooling water circulating around the compressor casing removes 18 kJ/kg of air. Determine the power required by the compressor.

4.202 Air enters a compressor at 14 psia, 80 F with negligible velocity and is discharged at 70 psia, 500 F with a velocity of 500 ft/sec. If the power input is 3200 hp and the flow rate is 20 lbm/sec, determine the rate of heat transfer.

4.203 Air is drawn from the atmosphere into a turbomachine. At the exit, conditions are gage pressure of 500 kPa and 130 C. The exit speed is 100 m/sec, and the mass flow rate is 0.8 kg/sec. Flow is steady and there is no heat transfer. Compute the shaft work interaction with the surroundings.

4.204 A turbine is supplied with $0.6\ \text{m}^3/\text{sec}$ of water from a 0.3 m diameter pipe; the discharge pipe has a 0.4 m diameter. Determine the pressure drop across the turbine if it delivers 60 kW.

4.205 A pump draws water from a reservoir through a 150 mm diameter suction pipe and delivers it to a 75 mm diameter discharge pipe. The end of the suction pipe is 2 m below the free surface of the reservoir. The pressure gage on the discharge pipe (2 m above the reservoir surface) reads 170 kPa. The average speed in the discharge pipe is 3 m/sec. If the pump has an efficiency of 75 percent, determine the power required to drive the pump.

4.206 A pump system in a dishwasher circulates water at 55 gal/min. The total head, $p/\rho g + V^2/2g$, leaving the pump is 120 in. of water. The head at the pump inlet may be neglected. If the power supplied to the pump is 0.4 hp, determine its efficiency.

4.207 The propulsive efficiency, η, of a jet-propelled device is defined as the ratio of the useful work output to the mechanical energy input to the fluid. Determine the propulsive efficiency for the jet boat of Problem 4.136.

***4.208** The propulsive efficiency, η, of a propeller is defined as the ratio of the useful work produced to the mechanical energy input to the fluid. Determine the propulsive efficiency of the moving air boat of Problem 4.143. What would be the efficiency of the boat if it is not moving?

4.209 The efficiency, η, of a windmill is defined as the ratio of the useful work output to the kinetic energy flux contained in a streamtube of undisturbed wind the same diameter as the windmill blades. Evaluate the efficiency of the windmill of Problem 4.69.

4.210 A jet-propelled aircraft traveling at 200 m/sec takes in 40 kg/sec of air and discharges it at 500 m/sec relative to the aircraft. Determine the propulsive efficiency (defined as the ratio of the useful work output to the mechanical energy input to the fluid) of the aircraft.

* These problems require material from sections that may be omitted without loss of continuity in the text material.

4.211 A fire truck draws water from an open constant level reservoir through a hose of 0.2 ft^2 cross section. The water is discharged from the pump through a 0.2 ft^2 hose to a surge tank pressurized to 120 psig with air. The pump adds 970 ft · lbf/slug of energy to the fluid; the water level in the surge tank is 3 ft above the pump centerline. From the surge tank the water flows steadily through a hose of 0.04 ft^2 area to a nozzle of 0.01 ft^2 area. The nozzle is held at the same level as the water free surface in the surge tank. Neglect all frictional losses and sketch the system. Find:

(a) Velocity of the water leaving the nozzle
(b) Elevation of the pump relative to the reservoir
(c) Power required to drive the pump if it has an efficiency of 80 percent
(d) Maximum height above the reservoir that the stream of water can reach

4.212 All major harbors are equipped with "fire boats" for combating ship fires. A 75 mm diameter hose is attached to the discharge of a 10 kW pump on such a boat. The nozzle attached to the end of the hose has a diameter of 25 mm. If the nozzle discharge is held 3 m above the surface of the water, determine the volume flow rate through the nozzle, the maximum height to which the water will rise, and the force on the boat if the water jet is directed horizontally over the stern.

***4.213** The total mass of the helicopter-type craft shown is 1500 kg. The pressure of the air is atmospheric at the outlet. Assume the flow is steady and one-dimensional. Treat the air as being incompressible at standard conditions and calculate for a hovering position the speed of air leaving the craft, and the minimum power that must be delivered to the air by the propeller.

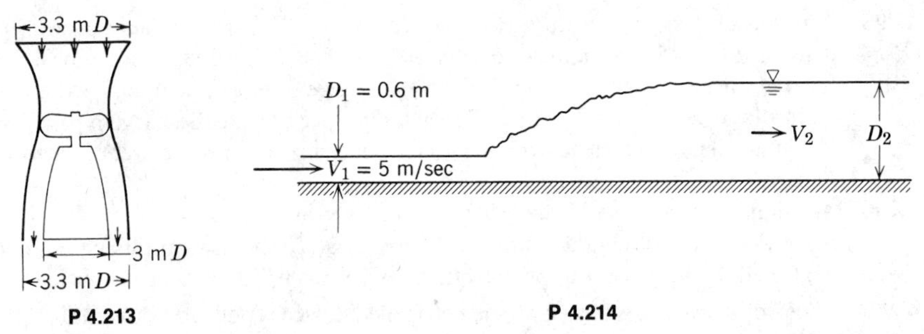

P 4.213 P 4.214

4.214 Liquid flowing at high speed in a wide, horizontal open channel under some conditions can undergo a hydraulic jump, as shown. For a suitably chosen control volume, the flows entering and leaving the jump may be considered uniform with hydrostatic pressure distribution (see Example Problem 4.6). Consider a channel of width, w, with water flow at $D_1 = 0.6$ m and $V_1 = 5$ m/sec. Show that in general, $D_2 = D_1[\sqrt{1 + 8V_1^2/gD_1} - 1]/2$. Evaluate the change in mechanical energy through the hydraulic jump. If heat transfer to the surroundings is negligible, determine the change in water temperature through the jump.

* These problems require material from sections that may be omitted without loss of continuity in the text material.

Chapter 5

INTRODUCTION TO DIFFERENTIAL ANALYSIS OF FLUID MOTION

In Chapter 4, we developed the basic equations in integral form for a control volume. The integral equations are particularly useful when we are interested in the gross behavior of a flow field and its effect on various devices. However, the integral approach does not provide detailed point by point knowledge of the flow field.

To obtain this detailed knowledge, we must apply the equations of fluid motion in differential form. In this chapter we shall develop differential equations for the conservation of mass and Newton's second law of motion. Since we are interested in formulating differential equations, our analysis will be in terms of infinitesimal systems and control volumes.

5-1 CONSERVATION OF MASS

In Chapter 2, we found that the continuum assumption—the assumption that a fluid could be treated as a continuous distribution of matter—led directly to a field representation of fluid properties. The property fields are defined by continuous functions of the space coordinates and time. The density and velocity fields are related through conservation of mass. We shall derive the differential equation for conservation of mass in rectangular and in cylindrical coordinates. In both cases the derivation is carried out by applying conservation of mass to a differential control volume.

5-1.1 Rectangular Coordinate System

In rectangular coordinates, the control volume chosen is an infinitesimal cube with sides of length dx, dy, dz as shown in Fig. 5.1. The density at the center, O, of the control volume is ρ and the velocity there is $\vec{V} = \hat{i}u + \hat{j}v + \hat{k}w$.

To evaluate the properties at each of the six faces of the control surface, we use a Taylor series expansion about point O. For example, at the right face,

$$\rho\Big)_{x+dx/2} = \rho + \left(\frac{\partial \rho}{\partial x}\right)\frac{dx}{2} + \left(\frac{\partial^2 \rho}{\partial x^2}\right)\frac{1}{2!}\left(\frac{dx}{2}\right)^2 + \cdots$$

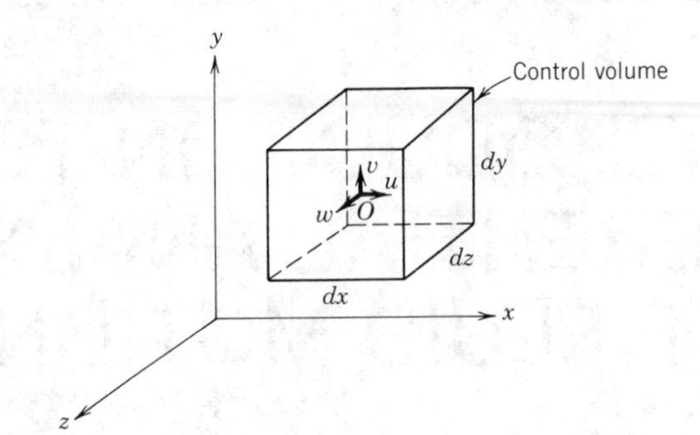

Fig. 5.1 Differential control volume in rectangular coordinates.

Neglecting higher order terms, we can write

$$\left.\rho\right)_{x+dx/2} = \rho + \left(\frac{\partial \rho}{\partial x}\right)\frac{dx}{2}$$

and

$$\left.u\right)_{x+dx/2} = u + \left(\frac{\partial u}{\partial x}\right)\frac{dx}{2}$$

The corresponding terms at the left face are

$$\left.\rho\right)_{x-dx/2} = \rho + \left(\frac{\partial \rho}{\partial x}\right)\left(-\frac{dx}{2}\right) = \rho - \left(\frac{\partial \rho}{\partial x}\right)\frac{dx}{2}$$

$$\left.u\right)_{x-dx/2} = u + \left(\frac{\partial u}{\partial x}\right)\left(-\frac{dx}{2}\right) = u - \left(\frac{\partial u}{\partial x}\right)\frac{dx}{2}$$

A word statement of conservation of mass is

$$\begin{bmatrix}\text{Net rate of mass efflux}\\ \text{through the control surface}\end{bmatrix} + \begin{bmatrix}\text{Rate of change of mass}\\ \text{inside the control volume}\end{bmatrix} = 0$$

To evaluate the first term in this equation, we must consider the mass flux through each of the six surfaces of the control surface; we must evaluate $\int_{CS} \rho \vec{V} \cdot d\vec{A}$. The details of this evaluation are shown in Table 5.1.

We see that the net rate of mass efflux through the control surface is given by

$$\left[\frac{\partial \rho u}{\partial x} + \frac{\partial \rho v}{\partial y} + \frac{\partial \rho w}{\partial z}\right] dx\,dy\,dz$$

The mass inside the control volume at any instant is the product of the mass per unit volume, ρ, and the volume, $dx\,dy\,dz$. Thus the rate of change of mass inside the control volume is given by

$$\frac{\partial \rho}{\partial t}\,dx\,dy\,dz$$

Table 5.1 Mass Flux through the Control Surface of a Rectangular Differential Control Volume

Surface	$\int \rho \vec{V} \cdot d\vec{A}$
Left $(-x)$	$= -\left[\rho - \left(\frac{\partial \rho}{\partial x}\right)\frac{dx}{2}\right]\left[u - \left(\frac{\partial u}{\partial x}\right)\frac{dx}{2}\right] dy\, dz = -\rho u\, dy\, dz + \frac{1}{2}\left[u\left(\frac{\partial \rho}{\partial x}\right) + \rho\left(\frac{\partial u}{\partial x}\right)\right] dx\, dy\, dz$
Right $(+x)$	$= \left[\rho + \left(\frac{\partial \rho}{\partial x}\right)\frac{dx}{2}\right]\left[u + \left(\frac{\partial u}{\partial x}\right)\frac{dx}{2}\right] dy\, dz = \rho u\, dy\, dz + \frac{1}{2}\left[u\left(\frac{\partial \rho}{\partial x}\right) + \rho\left(\frac{\partial u}{\partial x}\right)\right] dx\, dy\, dz$
Bottom $(-y)$	$= -\left[\rho - \left(\frac{\partial \rho}{\partial y}\right)\frac{dy}{2}\right]\left[v - \left(\frac{\partial v}{\partial y}\right)\frac{dy}{2}\right] dx\, dz = -\rho v\, dx\, dz + \frac{1}{2}\left[v\left(\frac{\partial \rho}{\partial y}\right) + \rho\left(\frac{\partial v}{\partial y}\right)\right] dx\, dy\, dz$
Top $(+y)$	$= \left[\rho + \left(\frac{\partial \rho}{\partial y}\right)\frac{dy}{2}\right]\left[v + \left(\frac{\partial v}{\partial y}\right)\frac{dy}{2}\right] dx\, dz = \rho v\, dx\, dz + \frac{1}{2}\left[v\left(\frac{\partial \rho}{\partial y}\right) + \rho\left(\frac{\partial v}{\partial y}\right)\right] dx\, dy\, dz$
Back $(-z)$	$= -\left[\rho - \left(\frac{\partial \rho}{\partial z}\right)\frac{dz}{2}\right]\left[w - \left(\frac{\partial w}{\partial z}\right)\frac{dz}{2}\right] dx\, dy = -\rho w\, dx\, dy + \frac{1}{2}\left[w\left(\frac{\partial \rho}{\partial z}\right) + \rho\left(\frac{\partial w}{\partial z}\right)\right] dx\, dy\, dz$
Front $(+z)$	$= \left[\rho + \left(\frac{\partial \rho}{\partial z}\right)\frac{dz}{2}\right]\left[w + \left(\frac{\partial w}{\partial z}\right)\frac{dz}{2}\right] dx\, dy = \rho w\, dx\, dy + \frac{1}{2}\left[w\left(\frac{\partial \rho}{\partial z}\right) + \rho\left(\frac{\partial w}{\partial z}\right)\right] dx\, dy\, dz$

Then,

$$\int_{\text{CS}} \rho \vec{V} \cdot d\vec{A} = \left[\left\{u\left(\frac{\partial \rho}{\partial x}\right) + \rho\left(\frac{\partial u}{\partial x}\right)\right\} + \left\{v\left(\frac{\partial \rho}{\partial y}\right) + \rho\left(\frac{\partial v}{\partial y}\right)\right\} + \left\{w\left(\frac{\partial \rho}{\partial z}\right) + \rho\left(\frac{\partial w}{\partial z}\right)\right\}\right] dx\, dy\, dz$$

or

$$\int_{\text{CS}} \rho \vec{V} \cdot d\vec{A} = \left[\frac{\partial \rho u}{\partial x} + \frac{\partial \rho v}{\partial y} + \frac{\partial \rho w}{\partial z}\right] dx\, dy\, dz$$

In rectangular coordinates the differential equation for the conservation of mass is then

$$\frac{\partial \rho u}{\partial x} + \frac{\partial \rho v}{\partial y} + \frac{\partial \rho w}{\partial z} + \frac{\partial \rho}{\partial t} = 0 \tag{5.1a}$$

Since the vector operator, ∇, in rectangular coordinates, is given by

$$\nabla = \hat{i}\frac{\partial}{\partial x} + \hat{j}\frac{\partial}{\partial y} + \hat{k}\frac{\partial}{\partial z}$$

then

$$\frac{\partial \rho u}{\partial x} + \frac{\partial \rho v}{\partial y} + \frac{\partial \rho w}{\partial z} = \nabla \cdot \rho \vec{V}$$

and the conservation of mass may be written as

$$\nabla \cdot \rho \vec{V} + \frac{\partial \rho}{\partial t} = 0 \tag{5.1b}$$

Two flow cases for which the differential continuity equation may be simplified are worthy of note. For incompressible flow, $\rho =$ constant; density is neither a function of space coordinates nor time. For incompressible flow, the continuity equation simplifies to

$$\frac{\partial u}{\partial x} + \frac{\partial v}{\partial y} + \frac{\partial w}{\partial z} = 0$$

or

$$\nabla \cdot \vec{V} = 0$$

Thus the velocity field, $\vec{V}(x, y, z, t)$, for incompressible flow must satisfy $\nabla \cdot \vec{V} = 0$.

For steady flow, all fluid properties are, by definition, independent of time. Thus at most $\rho = \rho(x, y, z)$, and for steady flow, the continuity equation can be written as

$$\frac{\partial \rho u}{\partial x} + \frac{\partial \rho v}{\partial y} + \frac{\partial \rho w}{\partial z} = 0$$

or

$$\nabla \cdot \rho \vec{V} = 0$$

Example 5.1

For a two-dimensional flow in the xy plane, the x component of velocity is given by $u = Ax$. Determine a possible y component for steady, incompressible flow. How many possible y components are there?

EXAMPLE PROBLEM 5.1

GIVEN: Two-dimensional flow in the xy plane for which $u = Ax$.

FIND: (a) Possible y component for steady, incompressible flow.
(b) How many y components are possible?

SOLUTION:

Basic equation: $\nabla \cdot \rho \vec{V} + \frac{\partial \rho}{\partial t} = 0$

For steady, incompressible flow, $\frac{\partial \rho}{\partial t} = 0$, and $\rho =$ constant; thus $\nabla \cdot \vec{V} = 0$. In rectangular coordinates

$$\frac{\partial u}{\partial x} + \frac{\partial v}{\partial y} + \frac{\partial w}{\partial z} = 0$$

For two-dimensional flow in the xy plane, $\vec{V} = \vec{V}(x, y)$. Then partial derivatives with respect to z are zero, and

$$\frac{\partial u}{\partial x} + \frac{\partial v}{\partial y} = 0$$

Then

$$\frac{\partial v}{\partial y} = -\frac{\partial u}{\partial x} = -A$$

which gives an expression for the rate of change of v holding x constant. This equation can be integrated to obtain an expression for v. The result is

$$v = \int \frac{\partial v}{\partial y} dy + f(x) = -Ay + f(x)$$

{The function of x appears because we had the partial derivative of v with respect to y.}

Any function $f(x)$ is allowable, since $\frac{\partial}{\partial y} f(x) = 0$. Thus any number of expressions for v could satisfy the differential continuity equation under the given conditions. The simplest expression for v would be obtained by setting $f(x) = 0$. Then

$$v = -Ay$$

and

$$\vec{V} = Ax\hat{i} - Ay\hat{j}$$

{This problem illustrates use of the differential continuity equation for steady, incompressible flow to evaluate a possible velocity component and introduces the integration of a partial derivative.}

Example 5.2

A gas-filled pneumatic strut in an automobile suspension system behaves like a piston-cylinder apparatus. At one instant when the piston is $L = 0.15$ m away from the closed end of the cylinder, the gas density is uniform at $\rho = 18\ \text{kg/m}^3$. The piston suddenly begins to move away from the closed end at $V = 12$ m/sec. The gas motion is one-dimensional and proportional to distance from the closed end; it varies linearly from zero at the end to $u = V$ at the piston. Evaluate the rate of change of gas density at this instant. Obtain an expression for the average density as a function of time.

EXAMPLE PROBLEM 5.2

GIVEN: Piston-cylinder as shown.

FIND: (a) Rate of change of density.
(b) $\rho(t)$.

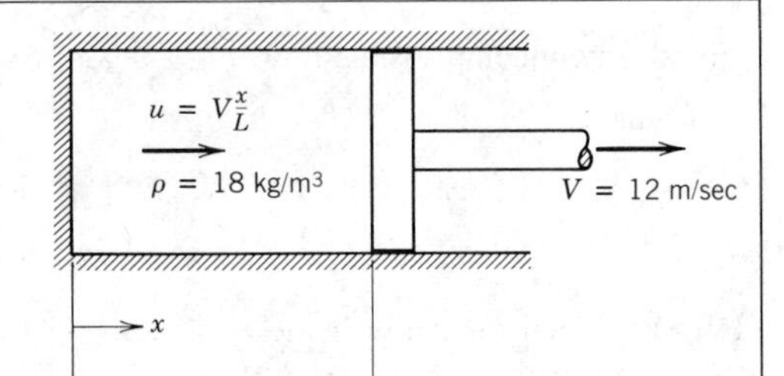

SOLUTION:

Basic equation: $\nabla \cdot \rho\vec{V} + \dfrac{\partial \rho}{\partial t} = 0$

In rectangular coordinates, $\dfrac{\partial \rho u}{\partial x} + \dfrac{\partial \rho v}{\partial y} + \dfrac{\partial \rho w}{\partial z} + \dfrac{\partial \rho}{\partial t} = 0$

Since $u = u(x)$, then partial derivatives with respect to y and z are zero, and

$$\frac{\partial \rho u}{\partial x} + \frac{\partial \rho}{\partial t} = 0$$

Then

$$\frac{\partial \rho}{\partial t} = -\frac{\partial \rho u}{\partial x} = -\rho\frac{\partial u}{\partial x} - u\frac{\partial \rho}{\partial x}$$

Since ρ is assumed uniform in the volume, then $\dfrac{\partial \rho}{\partial x} = 0$, and $\dfrac{d\rho}{dt} = -\rho\dfrac{\partial u}{\partial x}$.

Since $u = V\dfrac{x}{L}$, then $\dfrac{\partial u}{\partial x} = \dfrac{V}{L}$, and $\dfrac{d\rho}{dt} = -\rho\dfrac{V}{L}$. However, note $L = L_0 + Vt$.

Separate variables and integrate,

$$\int_{\rho_0}^{\rho} \frac{d\rho}{\rho} = -\int_0^t \frac{V}{L}\,dt = -\int_0^t \frac{V\,dt}{L_0 + Vt}$$

$$\ln\frac{\rho}{\rho_0} = \ln\frac{L_0}{L_0 + Vt} \quad \text{and} \quad \rho(t) = \rho_0\left[\frac{1}{1 + Vt/L_0}\right] \qquad \rho(t)$$

At $t = 0$

$$\frac{\partial \rho}{\partial t} = -\rho_0\frac{V}{L} = -\frac{18\ \text{kg}}{\text{m}^3} \times \frac{12\ \text{m}}{\text{sec}} \times \frac{1}{0.15\ \text{m}} = -1440\ \text{kg/m}^3 \cdot \text{sec} \qquad \frac{\partial \rho}{\partial t}$$

{This problem illustrates use of the differential continuity equation to evaluate a density variation.}

5-1.2 Cylindrical Coordinate System

In cylindrical coordinates, a suitable differential control volume is shown in Fig. 5.2. The density at the center, O, of the control volume is ρ and the velocity there is $\vec{V} = \hat{i}_r V_r + \hat{i}_\theta V_\theta + \hat{i}_z V_z$, where $\hat{i}_r$, $\hat{i}_\theta$, and $\hat{i}_z$ are unit vectors in the r, θ, and z directions, respectively, and V_r, V_θ, and V_z are the velocity components in the r, θ, and z directions, respectively. To evaluate $\int_{\text{CS}} \rho \vec{V} \cdot d\vec{A}$, we must consider the mass flux through each of the six faces of the control surface. The properties at each of the six faces of the control surface are obtained from a Taylor series expansion about point O. The details of the mass flux evaluation are shown in Table 5.2.

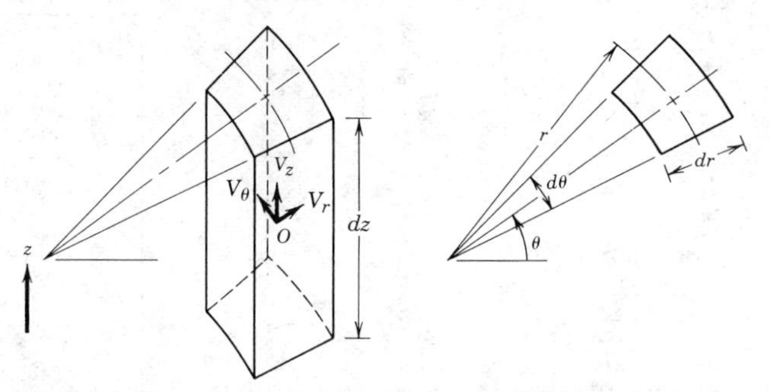

Fig. 5.2 Differential control volume in cylindrical coordinates.

We see that the net rate of mass efflux through the control surface is given by

$$\left[\rho V_r + r\frac{\partial \rho V_r}{\partial r} + \frac{\partial \rho V_\theta}{\partial \theta} + r\frac{\partial \rho V_z}{\partial z}\right] dr\, d\theta\, dz$$

The mass inside the control volume at any instant is the product of the mass per unit volume, ρ, and the volume, $r\, d\theta\, dr\, dz$. Thus the rate of change of mass inside the control volume is given by

$$\frac{\partial \rho}{\partial t} r\, d\theta\, dr\, dz$$

In cylindrical coordinates the differential equation for the conservation of mass is then

$$\rho V_r + r\frac{\partial \rho V_r}{\partial r} + \frac{\partial \rho V_\theta}{\partial \theta} + r\frac{\partial \rho V_z}{\partial z} + r\frac{\partial \rho}{\partial t} = 0$$

Dividing by r gives

$$\frac{\rho V_r}{r} + \frac{\partial \rho V_r}{\partial r} + \frac{1}{r}\frac{\partial \rho V_\theta}{\partial \theta} + \frac{\partial \rho V_z}{\partial z} + \frac{\partial \rho}{\partial t} = 0$$

or

$$\frac{1}{r}\frac{\partial r \rho V_r}{\partial r} + \frac{1}{r}\frac{\partial \rho V_\theta}{\partial \theta} + \frac{\partial \rho V_z}{\partial z} + \frac{\partial \rho}{\partial t} = 0 \tag{5.2}$$

Table 5.2 Mass Flux through the Control Surface of a Cylindrical Differential Control Volume

Surface	$\int \rho \vec{V} \cdot d\vec{A}$
Inside $(-r)$	$= -\left[\rho - \left(\frac{\partial \rho}{\partial r}\right)\frac{dr}{2}\right]\left[V_r - \left(\frac{\partial V_r}{\partial r}\right)\frac{dr}{2}\right]\left(r - \frac{dr}{2}\right) d\theta\, dz = -\rho V_r r\, d\theta\, dz + \rho V_r \frac{dr}{2}\, d\theta\, dz + \rho\left(\frac{\partial V_r}{\partial r}\right) r\frac{dr}{2}\, d\theta\, dz + V_r\left(\frac{\partial \rho}{\partial r}\right) r\frac{dr}{2}\, d\theta\, dz$
Outside $(+r)$	$= \left[\rho + \left(\frac{\partial \rho}{\partial r}\right)\frac{dr}{2}\right]\left[V_r + \left(\frac{\partial V_r}{\partial r}\right)\frac{dr}{2}\right]\left(r + \frac{dr}{2}\right) d\theta\, dz = \rho V_r r\, d\theta\, dz + \rho V_r \frac{dr}{2}\, d\theta\, dz + \rho\left(\frac{\partial V_r}{\partial r}\right) r\frac{dr}{2}\, d\theta\, dz + V_r\left(\frac{\partial \rho}{\partial r}\right) r\frac{dr}{2}\, d\theta\, dz$
Front $(-\theta)$	$= -\left[\rho - \left(\frac{\partial \rho}{\partial \theta}\right)\frac{d\theta}{2}\right]\left[V_\theta - \left(\frac{\partial V_\theta}{\partial \theta}\right)\frac{d\theta}{2}\right] dr\, dz = -\rho V_\theta\, dr\, dz + \rho\left(\frac{\partial V_\theta}{\partial \theta}\right)\frac{d\theta}{2}\, dr\, dz + V_\theta\left(\frac{\partial \rho}{\partial \theta}\right)\frac{d\theta}{2}\, dr\, dz$
Back $(+\theta)$	$= \left[\rho + \left(\frac{\partial \rho}{\partial \theta}\right)\frac{d\theta}{2}\right]\left[V_\theta + \left(\frac{\partial V_\theta}{\partial \theta}\right)\frac{d\theta}{2}\right] dr\, dz = \rho V_\theta\, dr\, dz + \rho\left(\frac{\partial V_\theta}{\partial \theta}\right)\frac{d\theta}{2}\, dr\, dz + V_\theta\left(\frac{\partial \rho}{\partial \theta}\right)\frac{d\theta}{2}\, dr\, dz$
Bottom $(-z)$	$= -\left[\rho - \left(\frac{\partial \rho}{\partial z}\right)\frac{dz}{2}\right]\left[V_z - \left(\frac{\partial V_z}{\partial z}\right)\frac{dz}{2}\right] r\, d\theta\, dr = -\rho V_z r\, d\theta\, dr + \rho\left(\frac{\partial V_z}{\partial z}\right)\frac{dz}{2}\, r\, d\theta\, dr + V_z\left(\frac{\partial \rho}{\partial z}\right)\frac{dz}{2}\, r\, d\theta\, dr$
Top $(+z)$	$= \left[\rho + \left(\frac{\partial \rho}{\partial z}\right)\frac{dz}{2}\right]\left[V_z + \left(\frac{\partial V_z}{\partial z}\right)\frac{dz}{2}\right] r\, d\theta\, dr = \rho V_z r\, d\theta\, dr + \rho\left(\frac{\partial V_z}{\partial z}\right)\frac{dz}{2}\, r\, d\theta\, dr + V_z\left(\frac{\partial \rho}{\partial z}\right)\frac{dz}{2}\, r\, d\theta\, dr$

Then,

$$\int_{\mathrm{CS}} \rho \vec{V} \cdot d\vec{A} = \left[\rho V_r + r\left\{\rho\left(\frac{\partial V_r}{\partial r}\right) + V_r\left(\frac{\partial \rho}{\partial r}\right)\right\} + \left\{\rho\left(\frac{\partial V_\theta}{\partial \theta}\right) + V_\theta\left(\frac{\partial \rho}{\partial \theta}\right)\right\} + r\left\{\rho\left(\frac{\partial V_z}{\partial z}\right) + V_z\left(\frac{\partial \rho}{\partial z}\right)\right\}\right] dr\, d\theta\, dz$$

or

$$\int_{\mathrm{CS}} \rho \vec{V} \cdot d\vec{A} = \left[\rho V_r + r\frac{\partial \rho V_r}{\partial r} + \frac{\partial \rho V_\theta}{\partial \theta} + r\frac{\partial \rho V_z}{\partial z}\right] dr\, d\theta\, dz$$

In cylindrical coordinates the vector operator, ∇, is given by

$$\nabla = \hat{i}_r \frac{\partial}{\partial r} + \hat{i}_\theta \frac{1}{r}\frac{\partial}{\partial \theta} + \hat{i}_z \frac{\partial}{\partial z}$$

Thus in vector notation the conservation of mass may be written[1]

$$\nabla \cdot \rho \vec{V} + \frac{\partial \rho}{\partial t} = 0$$

For incompressible flow, ρ = constant, and Eq. 5.2 reduces to

$$\frac{1}{r}\frac{\partial r V_r}{\partial r} + \frac{1}{r}\frac{\partial V_\theta}{\partial \theta} + \frac{\partial V_z}{\partial z} = 0$$

For steady flow, Eq. 5.2 reduces to

$$\frac{1}{r}\frac{\partial r \rho V_r}{\partial r} + \frac{1}{r}\frac{\partial \rho V_\theta}{\partial \theta} + \frac{\partial \rho V_z}{\partial z} = 0$$

Example 5.3

Consider a one-dimensional radial flow in the $r\theta$ plane, characterized by $V_r = f(r)$ and $V_\theta = 0$. Determine the conditions on $f(r)$ required for incompressible flow.

EXAMPLE PROBLEM 5.3

GIVEN: One-dimensional radial flow in the $r\theta$ plane.

$$V_r = f(r) \qquad \text{and} \qquad V_\theta = 0$$

FIND: Requirements on $f(r)$ for incompressible flow.

SOLUTION:

Basic equation: $\nabla \cdot \rho \vec{V} + \dfrac{\partial \rho}{\partial t} = 0$

For incompressible flow, ρ = constant, so $\dfrac{\partial \rho}{\partial t} = 0$. In cylindrical coordinates

$$\frac{1}{r}\frac{\partial}{\partial r}(r V_r) + \frac{1}{r}\frac{\partial}{\partial \theta} V_\theta + \frac{\partial V_z}{\partial z} = 0$$

For the given velocity field, $\vec{V} = \vec{V}(r)$. $V_\theta = 0$ and partial derivatives with respect to z are zero, so

$$\frac{1}{r}\frac{\partial}{\partial r}(r V_r) = 0$$

Integrating with respect to r gives

$$r V_r = \text{constant}$$

Thus the continuity equation shows that $V_r = \dfrac{C}{r}$.

[1] To evaluate the operation in cylindrical coordinates we must remember that

$$\frac{\partial \hat{i}_r}{\partial \theta} = \hat{i}_\theta \qquad \text{and} \qquad \frac{\partial \hat{i}_\theta}{\partial \theta} = -\hat{i}_r$$

**5-2 STREAM FUNCTION FOR TWO-DIMENSIONAL INCOMPRESSIBLE FLOW

It is convenient to have a means of describing mathematically any particular pattern of flow. An adequate description should portray the notion of the shape of the streamlines (including the boundaries) and the scale of the velocity at representative points in the flow. A mathematical device that serves this purpose is the stream function, ψ. The stream function is formulated as a relation between the streamlines and the statement of conservation of mass.

For a two-dimensional incompressible flow in the xy plane, conservation of mass, Eq. 5.1a, can be written

$$\frac{\partial u}{\partial x} + \frac{\partial v}{\partial y} = 0 \tag{5.3}$$

If a continuous function, $\psi(x, y, t)$, called the stream function, is defined such that

$$u \equiv \frac{\partial \psi}{\partial y} \qquad \text{and} \qquad v \equiv -\frac{\partial \psi}{\partial x} \tag{5.4}$$

then the continuity equation, Eq. 5.3, is satisfied exactly, since

$$\frac{\partial u}{\partial x} + \frac{\partial v}{\partial y} = \frac{\partial^2 \psi}{\partial x\, \partial y} - \frac{\partial^2 \psi}{\partial y\, \partial x} = 0$$

Recall that streamlines are lines drawn in the flow field, such that, at a given instant of time, they are tangent to the direction of flow at every point in the flow field. Thus, if $d\vec{r}$ is an element of length along a streamline, the equation of the streamline is given by

$$\begin{aligned} \vec{V} \times d\vec{r} = 0 &= (\hat{\imath} u + \hat{\jmath} v) \times (\hat{\imath}\, dx + \hat{\jmath}\, dy) \\ &= \hat{k}(u\, dy - v\, dx) \end{aligned}$$

The equation of a streamline in a two-dimensional flow is

$$u\, dy - v\, dx = 0$$

Substituting for the velocity components u and v in terms of the stream function, ψ, from Eq. 5.4, then along a streamline

$$\frac{\partial \psi}{\partial x}\, dx + \frac{\partial \psi}{\partial y}\, dy = 0 \tag{5.5}$$

Since $\psi = \psi(x, y, t)$, then at an instant, t_0, $\psi = \psi(x, y, t_0)$; at this instant, a change in ψ may be evaluated as though $\psi = \psi(x, y)$. Thus, at any instant,

$$d\psi = \frac{\partial \psi}{\partial x}\, dx + \frac{\partial \psi}{\partial y}\, dy \tag{5.6}$$

Comparing Eqs. 5.5 and 5.6, we see that along an instantaneous streamline, $d\psi = 0$; ψ is a constant along a streamline. Since the differential of ψ is exact, the integral of $d\psi$

** This section may be omitted without loss of continuity in the text material.

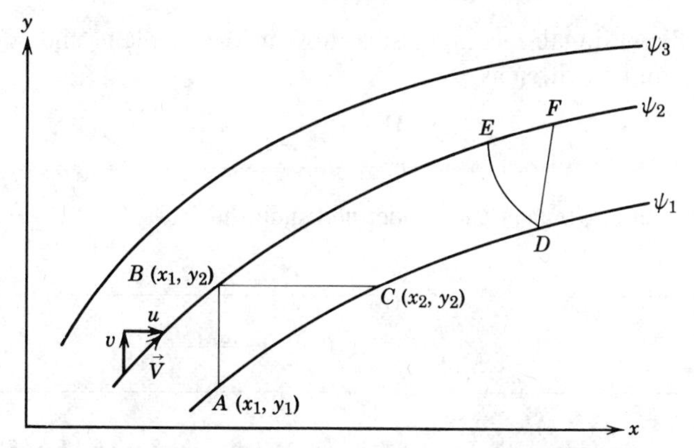

Fig. 5.3 Instantaneous streamlines in a two-dimensional flow.

between any two points in a flow field, $\psi_2 - \psi_1$, depends only on the end points of integration.

From the definition of a streamline, we recognize that there can be no flow across a streamline. Thus, if the streamlines in a two-dimensional, incompressible flow field, at a given instant are as shown in Fig. 5.3, the rate of flow between streamlines ψ_1 and ψ_2 across the lines *AB*, *BC*, *DE*, and *DF* must be equal.

The volume flow rate, Q, between streamlines ψ_1 and ψ_2 can be evaluated by considering the flow across *AB* or across *BC*. For a unit depth, the flow rate across *AB* is

$$Q = \int_{y_1}^{y_2} u\,dy = \int_{y_1}^{y_2} \frac{\partial \psi}{\partial y}\,dy$$

Along *AB*, $x =$ constant, and $d\psi = \partial\psi/\partial y\,dy$. Therefore,

$$Q = \int_{y_1}^{y_2} \frac{\partial \psi}{\partial y}\,dy = \int_{\psi_1}^{\psi_2} d\psi = \psi_2 - \psi_1$$

For a unit depth, the flow rate across *BC* is

$$Q = \int_{x_1}^{x_2} v\,dx = -\int_{x_1}^{x_2} \frac{\partial \psi}{\partial x}\,dx$$

Along *BC*, $y =$ constant, and $d\psi = \partial\psi/\partial x\,dx$. Therefore,

$$Q = -\int_{x_1}^{x_2} \frac{\partial \psi}{\partial x}\,dx = -\int_{\psi_2}^{\psi_1} d\psi = \psi_2 - \psi_1$$

Thus the volume rate of flow (per unit depth) between any two streamlines can be written as the difference between the constant values of ψ defining the two streamlines.[2]

[2] For two-dimensional steady compressible flow in the xy plane, the stream function, ψ, is defined such that

$$\rho u \equiv \frac{\partial \psi}{\partial y} \quad \text{and} \quad \rho v \equiv -\frac{\partial \psi}{\partial x}$$

The difference between the constant values of ψ defining two streamlines is the mass rate of flow (per unit depth) between the two streamlines.

For a two-dimensional, incompressible flow in the $r\theta$ plane, the conservation of mass, Eq. 5.2, can be written as

$$\frac{\partial r V_r}{\partial r} + \frac{\partial V_\theta}{\partial \theta} = 0 \tag{5.7}$$

The stream function, $\psi(r, \theta, t)$, then is defined such that

$$V_r \equiv \frac{1}{r}\frac{\partial \psi}{\partial \theta} \quad \text{and} \quad V_\theta \equiv -\frac{\partial \psi}{\partial r} \tag{5.8}$$

With ψ defined according to Eq. 5.8, the continuity equation, Eq. 5.7, is satisfied exactly.

Example 5.4

Given the velocity field for the steady, incompressible flow of Example 5.1, $\vec{V} = Ax\hat{i} - Ay\hat{j}$, determine the stream function that will yield this velocity field. Plot and interpret the streamline pattern in the first quadrant of the xy plane.

EXAMPLE PROBLEM 5.4

GIVEN: Velocity field, $\vec{V} = Ax\hat{i} - Ay\hat{j}$.

FIND: (a) Stream function, ψ.
(b) Plot in first quadrant and interpret.

SOLUTION:
The flow is incompressible, so the stream function satisfies Eq. 5.4.

From Eq. 5.4, $u = \dfrac{\partial \psi}{\partial y}$ and $v = -\dfrac{\partial \psi}{\partial x}$. From the given velocity field,

$$u = Ax = \frac{\partial \psi}{\partial y}$$

Integrating with respect to y gives

$$\psi = \int \frac{\partial \psi}{\partial y}\, dy + f(x) = Axy + f(x) \tag{1}$$

where $f(x)$ is arbitrary. The function $f(x)$ may be evaluated using the equation for v. Thus, from Eq. 1,

$$v = -\frac{\partial \psi}{\partial x} = -Ay - \frac{df}{dx} \tag{2}$$

From the given velocity field,

$$v = -Ay$$

Comparing this with Eq. 2 shows that $\frac{df}{dx} = 0$, or $f(x) = \text{constant}$. Therefore, Eq. 1 becomes

$$\psi = Axy + c$$

Lines $\psi = \text{constant}$ represent streamlines in the flow field. The constant may be chosen as any convenient value for plotting purposes. A few streamlines are plotted in the following sketch:

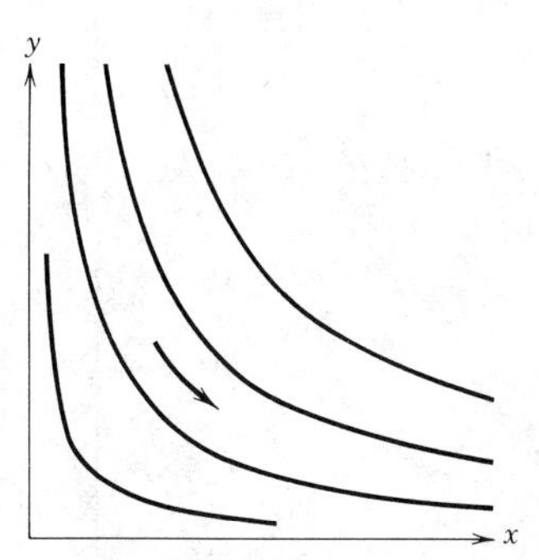

Flow velocity components are to the right and down, since $u > 0$ and $v < 0$. Regions of high-speed flow occur where the streamlines are close together. Lower speed flow occurs near the origin where streamline spacing is wider. Flow qualitatively looks like flow in a "corner," formed by a pair of walls.

5-3 MOTION OF A FLUID ELEMENT (KINEMATICS)

Before formulating the effects of forces on fluid motion (dynamics), let us consider first the motion (kinematics) of a fluid element in a flow field. For convenience, we follow an infinitesimal element of fixed identity (mass), as shown in Fig. 5.4.

As the infinitesimal element of mass, dm, moves in a flow field, several things may happen to it. Perhaps the most obvious of these is that the element translates; it undergoes a linear displacement from a location x, y, z to a different location x_1, y_1, z_1. The element may also rotate; the orientation of the element as shown in Fig. 5.4, where the sides of the element are parallel to the coordinate axes x, y, z, may change as a result of pure rotation about any one (or all three) of the coordinate axes. In addition the

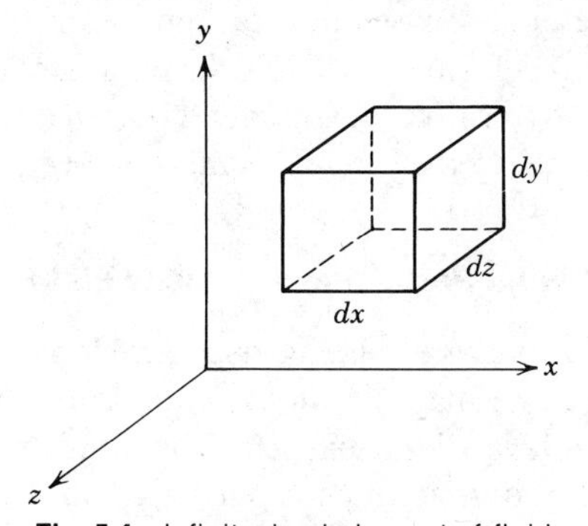

Fig. 5.4 Infinitesimal element of fluid.

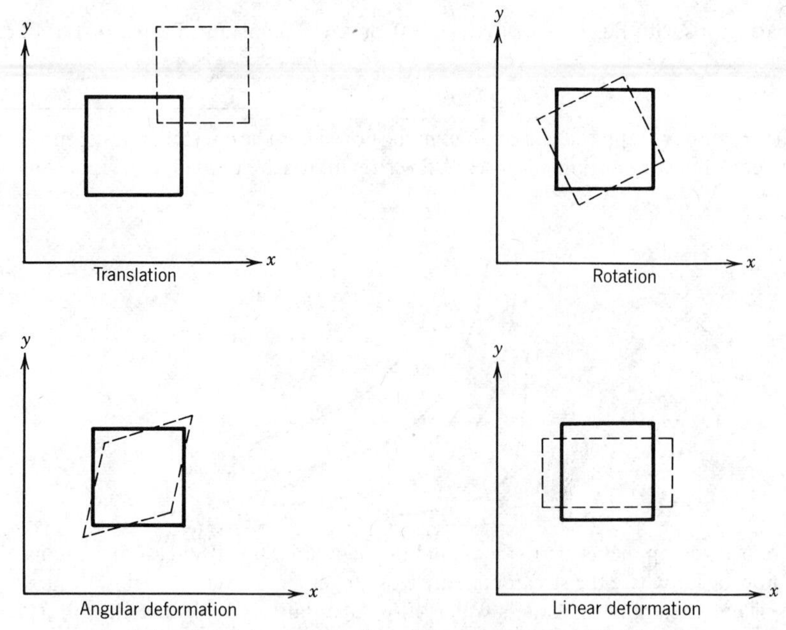

Fig. 5.5 Pictorial representation of the components of fluid motion.

element may deform. The deformation may be subdivided into two parts—linear and angular deformation. Linear deformation involves a change in shape without a change in orientation of the element: a deformation in which planes of the element that were originally perpendicular (e.g. the top and side of the element) remain perpendicular. Angular deformation involves a distortion of the element in which planes that were originally perpendicular are no longer perpendicular. In general, a fluid element may undergo a combination of translation, rotation, and linear and angular deformation during the course of its motion.

These four components of fluid motion are illustrated in Fig. 5.5 for motion in the xy plane. For a general three-dimensional flow, similar motions of the particle would be depicted in the yz and xz planes. For pure translation or rotation, the fluid element retains its shape; there is no deformation. Thus shear stresses do not arise as a result of pure translation or rotation (recall from Chapter 2 that in a Newtonian fluid the shear stress is directly proportional to the rate of angular deformation).

5-3.1 Acceleration of a Fluid Particle in a Velocity Field

Let us remember first that we are dealing with an element of fixed mass, dm. As discussed in Section 1-5.3, one may obtain the equation of motion for a particle by applying Newton's second law to that particle. The disadvantage of this approach is that a separate equation is required for each particle. Thus the bookkeeping for many particles becomes a problem.

A more general description of acceleration can be obtained by considering a particle moving in a velocity field. The basic hypothesis of continuum fluid mechanics has led us to a field description of fluid flow in which the properties of a flow field are defined by continuous functions of the space coordinates and time. In particular, the velocity field is given by $\vec{V} = \vec{V}(x, y, z, t)$. The field description is very powerful, since information for the entire flow is given by one equation.

The problem, then, is to retain the field description for fluid properties and obtain an expression for the acceleration of a fluid particle as it translates in a flow field. Stated simply, the problem is:

Given the velocity field, $\vec{V} = \vec{V}(x, y, z, t)$, find the acceleration of a fluid particle, $\vec{a}_p$.

Consider a particle moving in a velocity field. At time, t, the particle is at the position x, y, z and has a velocity corresponding to the velocity at that point in space at t,

$$\vec{V}_p]_t = \vec{V}(x, y, z, t)$$

At $t + dt$, the particle has moved to a new position, with coordinates $x + dx$, $y + dy$, $z + dz$, and has a velocity given by

$$\vec{V}_p]_{t+dt} = \vec{V}(x + dx, y + dy, z + dz, t + dt)$$

This is shown pictorially in Fig. 5.6.

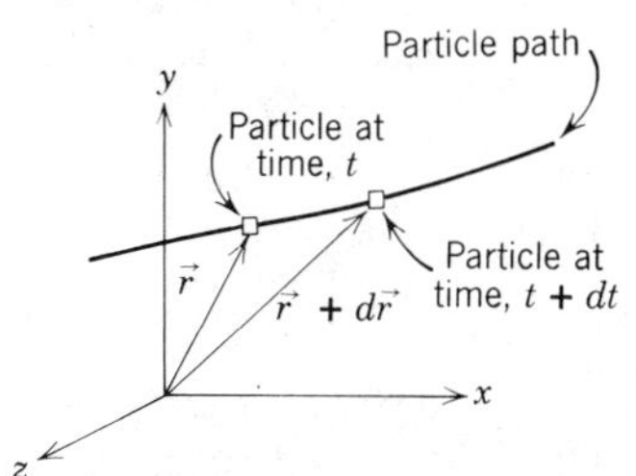

Fig. 5.6 Motion of a particle in a flow field.

The particle velocity at t (position $\vec{r}$) is given by $\vec{V}_p = \vec{V}(x, y, z, t)$. Then $d\vec{V}_p$, the change in the velocity of the particle, in moving from location $\vec{r}$ to $\vec{r} + d\vec{r}$, is given by

$$d\vec{V}_p = \frac{\partial \vec{V}}{\partial x} dx_p + \frac{\partial \vec{V}}{\partial y} dy_p + \frac{\partial \vec{V}}{\partial z} dz_p + \frac{\partial \vec{V}}{\partial t} dt$$

The total acceleration of the particle is given by

$$\vec{a}_p = \frac{d\vec{V}_p}{dt} = \frac{\partial \vec{V}}{\partial x}\frac{dx_p}{dt} + \frac{\partial \vec{V}}{\partial y}\frac{dy_p}{dt} + \frac{\partial \vec{V}}{\partial z}\frac{dz_p}{dt} + \frac{\partial \vec{V}}{\partial t}$$

Since

$$\frac{dx_p}{dt} = u, \qquad \frac{dy_p}{dt} = v, \qquad \text{and} \qquad \frac{dz_p}{dt} = w$$

then

$$\vec{a}_p = \frac{d\vec{V}_p}{dt} = u\frac{\partial \vec{V}}{\partial x} + v\frac{\partial \vec{V}}{\partial y} + w\frac{\partial \vec{V}}{\partial z} + \frac{\partial \vec{V}}{\partial t}$$

To remind us that calculation of the acceleration of a fluid particle in a velocity field requires a special derivative, it is given the symbol $D\vec{V}/Dt$. Thus

$$\frac{D\vec{V}}{Dt} \equiv \vec{a}_p = u\frac{\partial \vec{V}}{\partial x} + v\frac{\partial \vec{V}}{\partial y} + w\frac{\partial \vec{V}}{\partial z} + \frac{\partial \vec{V}}{\partial t} \tag{5.9}$$

The derivative, $D\vec{V}/Dt$, defined by Eq. 5.9, is commonly called the *substantial derivative* to remind us that it is computed for a particle of "substance." It often is called the material or particle derivative.

From Eq. 5.9 we recognize that a fluid particle moving in a flow field may undergo an acceleration for either of two reasons. It may be accelerated because it is convected into a region of higher (or lower) velocity. For example, in the steady flow through a nozzle, in which, by definition, the velocity field is not a function of time, a fluid particle will accelerate as it moves through the nozzle. The particle is convected into a region of higher velocity. If a flow field is unsteady, a fluid particle will undergo an acceleration, a "local" acceleration, because the velocity field is a function of time.

The physical significance of the terms in Eq. 5.9 is

$$\vec{a}_p = \underbrace{\frac{D\vec{V}}{Dt}}_{\substack{\text{total}\\ \text{acceleration}\\ \text{of a particle}}} = \underbrace{u\frac{\partial \vec{V}}{\partial x} + v\frac{\partial \vec{V}}{\partial y} + w\frac{\partial \vec{V}}{\partial z}}_{\substack{\text{convective}\\ \text{acceleration}}} + \underbrace{\frac{\partial \vec{V}}{\partial t}}_{\substack{\text{local}\\ \text{acceleration}}}$$

For a two-dimensional flow, say $\vec{V} = \vec{V}(x, y, t)$, Eq. 5.9 reduces to

$$\frac{D\vec{V}}{Dt} = u\frac{\partial \vec{V}}{\partial x} + v\frac{\partial \vec{V}}{\partial y} + \frac{\partial \vec{V}}{\partial t}$$

For a one-dimensional flow, say $\vec{V} = \vec{V}(x, t)$, Eq. 5.9 becomes

$$\frac{D\vec{V}}{Dt} = u\frac{\partial \vec{V}}{\partial x} + \frac{\partial \vec{V}}{\partial t}$$

Finally, for a steady flow in three dimensions, Eq. 5.9 becomes

$$\frac{D\vec{V}}{Dt} = u\frac{\partial \vec{V}}{\partial x} + v\frac{\partial \vec{V}}{\partial y} + w\frac{\partial \vec{V}}{\partial z}$$

which is not necessarily zero. Thus a fluid particle can undergo a convective acceleration due to its motion, even in a steady velocity field.

Equation 5.9 is a vector equation. As with all vector equations, it may be written in scalar component equations. Relative to an xyz coordinate system, the scalar components of Eq. 5.9 are written:

$$a_{x_p} = \frac{Du}{Dt} = u\frac{\partial u}{\partial x} + v\frac{\partial u}{\partial y} + w\frac{\partial u}{\partial z} + \frac{\partial u}{\partial t} \tag{5.10a}$$

$$a_{y_p} = \frac{Dv}{Dt} = u\frac{\partial v}{\partial x} + v\frac{\partial v}{\partial y} + w\frac{\partial v}{\partial z} + \frac{\partial v}{\partial t} \tag{5.10b}$$

$$a_{z_p} = \frac{Dw}{Dt} = u\frac{\partial w}{\partial x} + v\frac{\partial w}{\partial y} + w\frac{\partial w}{\partial z} + \frac{\partial w}{\partial t} \tag{5.10c}$$

We have obtained an expression for the acceleration of a particle anywhere in the flow field; this is the Eulerian method of description. To determine the acceleration of a particle at a particular point in the flow field, one substitutes the coordinates of the point into the field expression for acceleration. In the Lagrangian method of description, the motion (position, velocity, and acceleration) of the particle is described as a function of time. The Eulerian and Lagrangian methods of description are illustrated in Example Problem 5.5.

Example 5.5

Consider one-dimensional, steady, incompressible flow through the plane converging channel shown. The velocity field is given by $\vec{V} = V_1[1 + (x/L)]\hat{i}$. Find the x component of acceleration for a particle moving in the flow field. If we use the method of description of particle mechanics, the position of the particle, located at $x = 0$ at time $t = 0$, will be a function of time, $x_p = f(t)$. Obtain the expression for $f(t)$ and then, by taking the second derivative of the function with respect to time, obtain an expression for the x component of the particle acceleration.

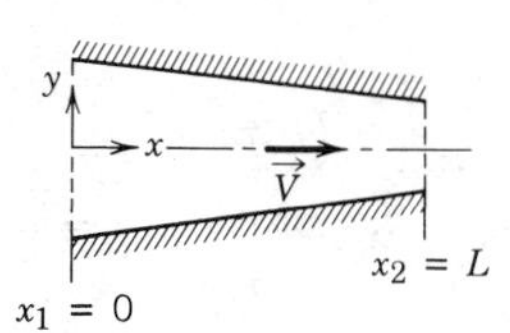

EXAMPLE PROBLEM 5.5

GIVEN: Steady, one-dimensional, incompressible flow through the converging channel shown.

$$\vec{V} = V_1\left(1 + \frac{x}{L}\right)\hat{i}$$

FIND: (a) The x component of the acceleration of a particle moving in the flow field.
(b) For the particle located at $x = 0$ at $t = 0$, obtain an expression for its
(1) position, x_p, as a function of time.
(2) x component of acceleration, a_{x_p}, as a function of time.

SOLUTION:

The acceleration of a particle moving in a velocity field is given by

$$\frac{D\vec{V}}{Dt} = u\frac{\partial \vec{V}}{\partial x} + v\frac{\partial \vec{V}}{\partial y} + w\frac{\partial \vec{V}}{\partial z} + \frac{\partial \vec{V}}{\partial t}$$

The x component of the acceleration is given by

$$\frac{Du}{Dt} = u\frac{\partial u}{\partial x} + v\frac{\partial u}{\partial y} + w\frac{\partial u}{\partial z} + \frac{\partial u}{\partial t}$$

For the flow field given,

$$v = w = 0 \qquad u = V_1\left(1 + \frac{x}{L}\right)$$

Therefore, $\frac{Du}{Dt} = u\frac{\partial u}{\partial x} = V_1\left(1 + \frac{x}{L}\right)\frac{V_1}{L} = \frac{V_1^2}{L}\left(1 + \frac{x}{L}\right)$ $\qquad \frac{Du}{Dt}$

{To determine the acceleration of a particle at any point in the flow field, we merely substitute the present location of the particle into the above result.}

In the second part of this problem we are interested in following a particular particle, namely, the one located at $x = 0$ at $t = 0$, as it flows through the channel.

The x coordinate that locates this particle will be a function of time, $x_p = f(t)$. Furthermore, $u_p = df/dt$ will be a function of time. The particle will have the velocity corresponding to its location in the velocity field. At $t = 0$, the particle is at $x = 0$, and its velocity $u_p = V_1$. At some later time, t, the particle will reach the exit, $x = L$; at that time it will have a velocity $u_p = 2V_1$. To find the expression for $x_p = f(t)$, we write

$$u_p = \frac{dx_p}{dt} = \frac{df}{dt} = V_1\left(1 + \frac{x}{L}\right) = V_1\left(1 + \frac{f}{L}\right)$$

Separating variables gives

$$\frac{df}{(1 + f/L)} = V_1\,dt$$

Since at $t = 0$, the particle in question was located at $x = 0$, and at t, this particle is located at $x_p = f$, then

$$\int_0^f \frac{df}{(1 + f/L)} = \int_0^t V_1\,dt$$

$$L\ln\left(1 + \frac{f}{L}\right) = V_1 t$$

$$\ln\left(1 + \frac{f}{L}\right) = \frac{V_1 t}{L}$$

$$1 + \frac{f}{L} = e^{V_1 t/L}$$

and

$$f = L[e^{V_1 t/L} - 1]$$

Then the position of the particle, located at $x = 0$ at $t = 0$, as a function of time is given by

$$x_p = f(t) = L[e^{V_1 t/L} - 1] \qquad x_p$$

The x component of acceleration of this particle is given by

$$a_{x_p} = \frac{d^2 x_p}{dt^2} = \frac{d^2 f}{dt^2} = \frac{V_1^2}{L}e^{V_1 t/L} \qquad a_{x_p}$$

We now have two different ways of expressing the acceleration of the particle that was located at $x = 0$ at $t = 0$. Note that although the flow field is steady, when we follow a particular particle, its position and acceleration (and velocity for that matter) are functions of time.

We check to see that both expressions for the acceleration give identical results.

$$a_{x_p} = \frac{V_1^2}{L} e^{V_1 t/L} \qquad\qquad a_{x_p} = \frac{Du}{Dt} = \frac{V_1^2}{L}\left(1 + \frac{x}{L}\right)$$

(a) At $t = 0$, $x_p = 0$ — At $t = 0$, the particle is at $x = 0$

$$a_{x_p} = \frac{V_1^2}{L} e^0 = \frac{V_1^2}{L} \quad \text{(a)} \qquad\qquad \frac{Du}{Dt} = \frac{V_1^2}{L}(1 + 0) = \frac{V_1^2}{L} \quad \text{(a)}$$

Check.

(b) When $x_p = \frac{L}{2}$, $t = t_1$, — At $x = 0.5L$

$$x_p = \frac{L}{2} = L[e^{V_1 t_1/L} - 1] \qquad\qquad \frac{Du}{Dt} = \frac{V_1^2}{L}(1 + 0.5)$$

Therefore, $e^{V_1 t_1/L} = 1.5$, and

$$\frac{Du}{Dt} = \frac{1.5V_1^2}{L} \quad \text{(b)}$$

$$a_{x_p} = \frac{V_1^2}{L} e^{V_1 t_1/L}$$

Check.

$$a_{x_p} = \frac{V_1^2}{L}(1.5) = \frac{1.5V_1^2}{L} \quad \text{(b)}$$

(c) When $x_p = L$, $t = t_2$, — At $x = L$

$$x_p = L = L[e^{V_1 t_2/L} - 1] \qquad\qquad \frac{Du}{Dt} = \frac{V_1^2}{L}(1 + 1)$$

Therefore, $e^{V_1 t_2/L} = 2$, and

$$\frac{Du}{Dt} = \frac{2V_1^2}{L} \quad \text{(c)}$$

$$a_{x_p} = \frac{V_1^2}{L} e^{V_1 t_2/L}$$

Check.

$$a_{x_p} = \frac{V_1^2}{L}(2) = \frac{2V_1^2}{L} \quad \text{(c)}$$

{This problem illustrates the Eulerian and Lagrangian methods of describing the motion of a particle.}

5-3.2 Fluid Rotation

The rotation, $\vec{\omega}$, of a fluid particle is defined as the average angular velocity of any two mutually perpendicular line elements of the particle. Rotation is a vector quantity. A particle moving in a general three-dimensional flow field may rotate about all three coordinate axes. Thus, in general,

$$\vec{\omega} = \hat{i}\omega_x + \hat{j}\omega_y + \hat{k}\omega_z$$

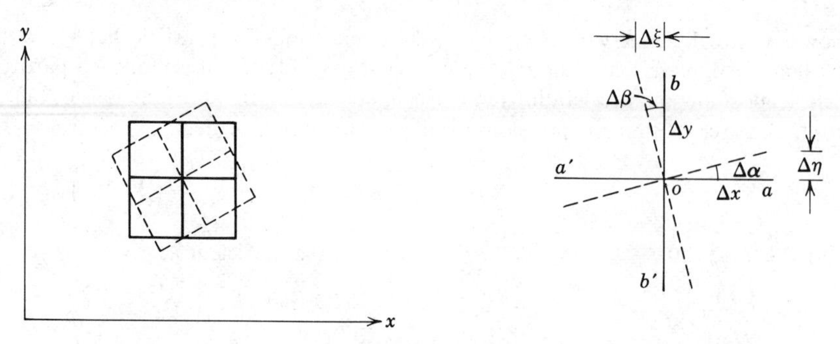

Fig. 5.7 Rotation of a fluid element in a two-dimensional flow field.

where ω_x is the rotation about the x axis, ω_y is the rotation about the y axis, and ω_z is the rotation about the z axis. The positive sense of rotation is given by the right-hand rule.

To obtain a mathematical expression for fluid rotation, consider motion of a fluid element in the xy plane. The components of the velocity at every point in the flow field are given by $u(x, y)$ and $v(x, y)$. The rotation of a fluid element in such a flow field is illustrated in Fig. 5.7. The two mutually perpendicular lines, oa and ob, will rotate to the positions shown during the interval, Δt, only if the velocities at points a and b are different from the velocity at o.

Consider first the rotation of line oa, of length, Δx. Rotation of this line is due to variations of the y component of velocity. If the y component of velocity at the point o is taken as v_o, then the y component of velocity at point a can be written, using a Taylor series expansion, as

$$v = v_o + \frac{\partial v}{\partial x} \Delta x$$

The angular velocity of line oa is given by

$$\omega_{oa} = \lim_{\Delta t \to 0} \frac{\Delta \alpha}{\Delta t} = \lim_{\Delta t \to 0} \frac{\Delta \eta / \Delta x}{\Delta t}$$

Since

$$\Delta \eta = \frac{\partial v}{\partial x} \Delta x \, \Delta t$$

$$\omega_{oa} = \lim_{\Delta t \to 0} \frac{(\partial v / \partial x) \Delta x \, \Delta t / \Delta x}{\Delta t} = \frac{\partial v}{\partial x}$$

Rotation of line ob, of length Δy, results from variations in the x component of velocity. If the x component of velocity at point o is taken as u_o, then the x component of velocity at point b can be written, using a Taylor series expansion, as

$$u = u_o + \frac{\partial u}{\partial y} \Delta y$$

The angular velocity of line *ob* is given by

$$\omega_{ob} = \lim_{\Delta t \to 0} \frac{\Delta \beta}{\Delta t} = \lim_{\Delta t \to 0} \frac{\Delta \xi / \Delta y}{\Delta t}$$

Since

$$\Delta \xi = -\frac{\partial u}{\partial y} \Delta y \, \Delta t$$

$$\omega_{ob} = \lim_{\Delta t \to 0} \frac{-(\partial u/\partial y)\Delta y \, \Delta t/\Delta y}{\Delta t} = -\frac{\partial u}{\partial y}$$

(The negative sign is introduced to give a positive value of ω_{ob}. According to our sign convention, counterclockwise rotation is positive.)

The rotation of the fluid element about the z axis is the average angular velocity of the two mutually perpendicular line elements, *oa* and *ob*, in the xy plane.

$$\omega_z = \frac{1}{2}\left(\frac{\partial v}{\partial x} - \frac{\partial u}{\partial y}\right)$$

By considering the rotation of two mutually perpendicular lines in the yz and xz planes, one can show that

$$\omega_x = \frac{1}{2}\left(\frac{\partial w}{\partial y} - \frac{\partial v}{\partial z}\right)$$

and

$$\omega_y = \frac{1}{2}\left(\frac{\partial u}{\partial z} - \frac{\partial w}{\partial x}\right)$$

Then

$$\vec{\omega} = \hat{i}\omega_x + \hat{j}\omega_y + \hat{k}\omega_z = \frac{1}{2}\left[\hat{i}\left(\frac{\partial w}{\partial y} - \frac{\partial v}{\partial z}\right) + \hat{j}\left(\frac{\partial u}{\partial z} - \frac{\partial w}{\partial x}\right) + \hat{k}\left(\frac{\partial v}{\partial x} - \frac{\partial u}{\partial y}\right)\right] \tag{5.11}$$

We recognize the term in the square brackets as

$$\text{curl } \vec{V} = \nabla \times \vec{V}$$

Then, in vector notation, we can write

$$\vec{\omega} = \tfrac{1}{2}\nabla \times \vec{V} \tag{5.12}$$

Under what conditions might we expect to have an irrotational flow? A fluid particle moving, without rotation, in a flow field cannot develop a rotation under the action of a body force or normal surface (pressure) forces. Development of rotation in a fluid particle, initially without rotation, requires the action of a shear stress on the surface of the particle. Since shear stress is proportional to the rate of angular deformation, then a particle that is initially without rotation will not develop a rotation without a simultaneous angular deformation. The shear stress is related to the rate of angular deformation through the viscosity. The presence of viscous forces means the flow is rotational.[3]

The condition of irrotationality may be a valid assumption for those regions of a flow in which viscous forces are negligible.[4] (For example, such a region exists outside the boundary layer in the flow over a solid surface.) The factor of $\frac{1}{2}$ can be eliminated in Eq. 5.12 by defining a quantity called the *vorticity*, $\vec{\zeta}$, to be twice the rotation.

$$\vec{\zeta} \equiv 2\vec{\omega} = \nabla \times \vec{V} \tag{5.13}$$

In cylindrical coordinates,

$$\vec{V} = \hat{i}_r V_r + \hat{i}_\theta V_\theta + \hat{i}_z V_z$$

and

$$\nabla = \hat{i}_r \frac{\partial}{\partial r} + \hat{i}_\theta \frac{1}{r}\frac{\partial}{\partial \theta} + \hat{i}_z \frac{\partial}{\partial z}$$

The vorticity, in cylindrical coordinates, is then[5]

$$\nabla \times \vec{V} = \hat{i}_r\left(\frac{1}{r}\frac{\partial V_z}{\partial \theta} - \frac{\partial V_\theta}{\partial z}\right) + \hat{i}_\theta\left(\frac{\partial V_r}{\partial z} - \frac{\partial V_z}{\partial r}\right) + \hat{i}_z\left(\frac{1}{r}\frac{\partial r V_\theta}{\partial r} - \frac{1}{r}\frac{\partial V_r}{\partial \theta}\right) \tag{5.14}$$

The vorticity is a measure of the rotation of a fluid element as it moves in the flow field. The *circulation*, Γ, is defined as the line integral of the tangential velocity component about a closed curve fixed in the flow,

$$\Gamma = \oint_C \vec{V} \cdot d\vec{s} \tag{5.15}$$

where $d\vec{s}$ is an elemental vector, of length ds, tangent to the curve; a positive sense corresponds to a counterclockwise path of integration around the curve. A relationship between circulation and vorticity can be obtained by considering the fluid element of Fig. 5.7. The element has been redrawn in Fig. 5.8; the velocity variations shown are consistent with those used in determining the fluid rotation.

For the closed curve *oacb*

$$d\Gamma = u\Delta x + \left(v + \frac{\partial v}{\partial x}\Delta x\right)\Delta y - \left(u + \frac{\partial u}{\partial y}\Delta y\right)\Delta x - v\,\Delta y$$

$$d\Gamma = \left(\frac{\partial v}{\partial x} - \frac{\partial u}{\partial y}\right)\Delta x\,\Delta y$$

$$d\Gamma = 2\omega_z\,\Delta x\,\Delta y$$

[3] A rigorous proof using the complete equations of motion for a fluid particle is given in W. H. Li and S. H. Lam, *Principles of Fluid Mechanics* (Reading, Mass.: Addison-Wesley, 1964), pp. 142–145.

[4] Examples of rotational and irrotational motion are shown in the film loops: S-FM014A, *Visualization of Vorticity with Vorticity Meter—Part I*; S-FM014B, *Visualization of Vorticity with Vorticity Meter—Part II*.

[5] In carrying out the curl operation, recall that $\hat{i}_r$ and $\hat{i}_\theta$ are functions of θ (see footnote[1] on p. 209).

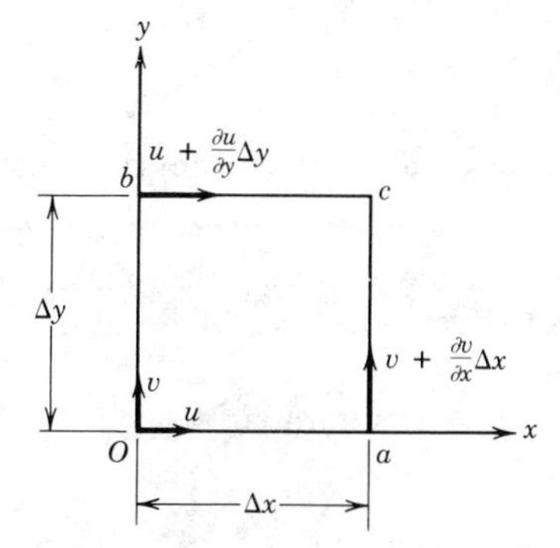

Fig. 5.8 Velocity components on the boundaries of a fluid element.

Then,

$$\Gamma = \oint_C \vec{V} \cdot d\vec{s} = \int_A 2\omega_z \, dA = \int_A (\nabla \times \vec{V})_z \, dA \tag{5.16}$$

Equation 5.16 is a statement of Stokes theorem in two dimensions. Thus the circulation around a closed contour is the sum of the vorticity enclosed within it.

Example 5.6

Consider flow fields with purely tangential motion (circular streamlines): $V_r = 0$ and $V_\theta = f(r)$. Evaluate the rotation, vorticity, and circulation for solid-body rotation, a *forced vortex*. Show that it is possible to choose $f(r)$ so that flow is irrotational, to produce a *free vortex*.

EXAMPLE PROBLEM 5.6

GIVEN: Flow field with tangential motion, $V_r = 0$ and $V_\theta = f(r)$

FIND: (a) Rotation, vorticity, and circulation for solid-body motion (a *forced vortex*).
(b) Evaluate $f(r)$ for irrotational motion (a *free vortex*).

SOLUTION:

Basic equation: $\vec{\zeta} = 2\vec{\omega} = \nabla \times \vec{V}$ (5.13)

For motion in the $r\theta$ plane, the only components of rotation and vorticity are in the z direction,

$$\zeta_z = 2\omega_z = \frac{1}{r}\frac{\partial r V_\theta}{\partial r} - \frac{1}{r}\frac{\partial V_r}{\partial \theta}$$

Because $V_r = 0$ everywhere in this field, this reduces to $\zeta_z = 2\omega_z = \frac{1}{r}\frac{\partial r V_\theta}{\partial r}$

(a) For solid-body rotation, $V_\theta = \omega r$

Then $\omega_z = \frac{1}{2}\frac{1}{r}\frac{\partial r V_\theta}{\partial r} = \frac{1}{2}\frac{1}{r}\frac{\partial}{\partial r}\omega r^2 = \frac{1}{2r}(2\omega r) = \omega$ and $\zeta_z = 2\omega$

The circulation is $\Gamma = \oint_C \vec{V} \cdot d\vec{s} = \int_A 2\omega_z \, dA$ (5.16)

Since $\omega_z = \omega =$ constant, the circulation about any closed contour is given by $\Gamma = 2\omega A$, where A is the area enclosed by the contour.

Thus for solid-body motion (a forced vortex), the rotation and vorticity are constants; the circulation depends on the area enclosed by a contour.

(b) For irrotational flow, $\frac{1}{r}\frac{\partial}{\partial r} r V_\theta = 0$. Integrating, we find

$$rV_\theta = \text{constant} \quad \text{or} \quad V_\theta = f(r) = \frac{C}{r}$$

For this flow, the origin is a singular point where $V_\theta \to \infty$. The circulation for any contour enclosing the origin is

$$\Gamma = \oint_C \vec{V} \cdot d\vec{s}$$

$$\Gamma = \int_0^{2\pi} \frac{C}{r} r \, d\theta = 2\pi C$$

The circulation around any contour not enclosing the singular point at the origin is zero.

5-3.3 Fluid Deformation

Angular deformation of a fluid element involves changes in the angle between two mutually perpendicular lines in the fluid. Referring to Fig. 5.9, we see that the rate of angular deformation of the fluid element is the rate of decrease of the angle between lines *oa* and *ob*. The rate of angular deformation is given by

$$-\frac{d\gamma}{dt} = \frac{d\alpha}{dt} + \frac{d\beta}{dt}$$

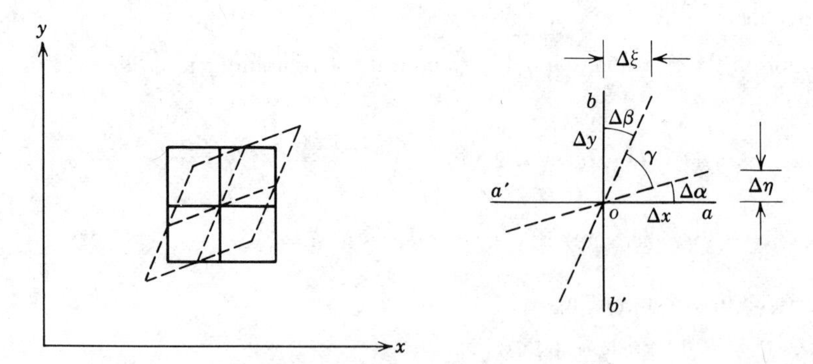

Fig. 5.9 Angular deformation of a fluid element in a two-dimensional flow field.

Now,

$$\frac{d\alpha}{dt} = \lim_{\Delta t \to 0} \frac{\Delta\alpha}{\Delta t} = \lim_{\Delta t \to 0} \frac{\Delta\eta/\Delta x}{\Delta t} = \lim_{\Delta t \to 0} \frac{(\partial v/\partial x)\Delta x\ \Delta t/\Delta x}{\Delta t} = \frac{\partial v}{\partial x}$$

and

$$\frac{d\beta}{dt} = \lim_{\Delta t \to 0} \frac{\Delta\beta}{\Delta t} = \lim_{\Delta t \to 0} \frac{\Delta\xi/\Delta y}{\Delta t} = \lim_{\Delta t \to 0} \frac{(\partial u/\partial y)\Delta y\ \Delta t/\Delta y}{\Delta t} = \frac{\partial u}{\partial y}$$

Consequently, the rate of angular deformation in the xy plane is

$$\frac{d\alpha}{dt} + \frac{d\beta}{dt} = -\frac{d\gamma}{dt} = \frac{\partial v}{\partial x} + \frac{\partial u}{\partial y} \tag{5.17}$$

The shear stress is related to the rate of angular deformation through the fluid viscosity. In a viscous flow (where velocity gradients are present) it is highly unlikely that $\partial v/\partial x$ will be equal and opposite to $\partial u/\partial y$ throughout the flow field (e.g. consider the boundary-layer flow of Fig. 2.11 and the flow over a cylinder, shown in Fig. 2.12). The presence of viscous forces means the flow is rotational.

Calculation of angular deformation is illustrated for a simple flow field in Example Problem 5.7.

Example 5.7

A viscometric flow in the narrow gap between large parallel plates is shown. The velocity field in the narrow gap is given by $\vec{V} = U(y/h)\hat{\imath}$, where $U = 4$ mm/sec and $h = 4$ mm. At $t = 0$ two lines, ac and bd, are marked in the fluid as shown. Evaluate the positions of the marked points at $t = 1.5$ sec and sketch for comparison. Calculate the rate of angular deformation and the rate of rotation of a fluid particle in this velocity field. Comment on your results.

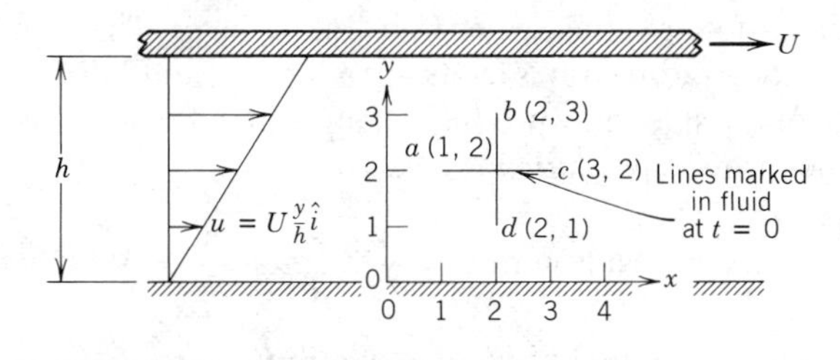

EXAMPLE PROBLEM 5.7

GIVEN: Velocity field, $V = U\frac{y}{h}\hat{\imath}$; $U = 4$ mm/sec, and $h = 4$ mm. Fluid particles marked at $t = 0$ to form cross as shown.

FIND:
(a) Positions of points a', b', c', and d' at $t = 1.5$ sec; plot.
(b) Rate of angular deformation.
(c) Rate of rotation of a fluid particle.
(d) Comment on the significance of these results.

SOLUTION:
For the given flow field, $v = 0$, so there is no vertical motion. The velocity of each point stays constant, so $\Delta x = u\,\Delta t$ for each point. At point b, $u = 3$ mm/sec, so

$$\Delta x_b = \frac{3\text{ mm}}{\text{sec}} \times 1.5\text{ sec} = 4.5\text{ mm}$$

Points a and c each move 3 mm, and point d moves 1.5 mm. The plot at $t = 1.5$ sec is

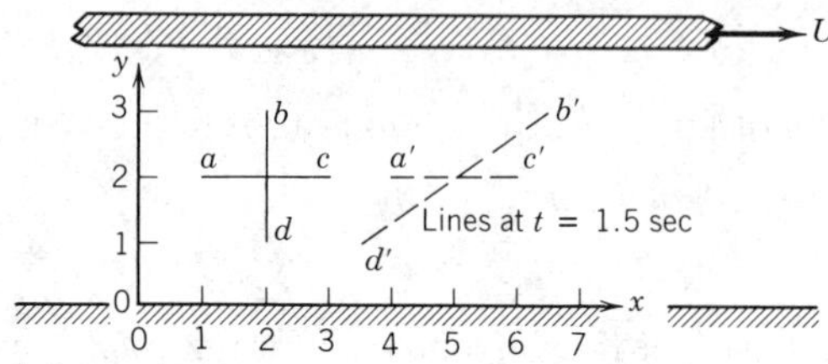

The rate of angular deformation is

$$-\dot{\gamma} = \frac{\partial u}{\partial y} + \frac{\partial v}{\partial x} = U\frac{1}{h} + 0 = \frac{U}{h} = \frac{4\text{ mm}}{\text{sec}} \times \frac{1}{4\text{ mm}} = 1\text{ sec}^{-1}$$

The rate of rotation is

$$\omega_z = \frac{1}{2}\left(\frac{\partial v}{\partial x} - \frac{\partial u}{\partial y}\right) = \frac{1}{2}\left(0 - \frac{U}{h}\right) = -\frac{1}{2} \times \frac{4\text{ mm}}{\text{sec}} \times \frac{1}{4\text{ mm}} = -0.5\text{ sec}^{-1}$$

This flow is viscous, so we expect to have both angular deformation and rotation; shape and orientation of a fluid particle both change.

During linear deformation, the shape of the fluid element, described by the angles at its vertices, remains unchanged, since all right angles continue to be right angles (see Fig. 5.5). The element will change length in the x direction only if $\partial u/\partial x$ is other than zero. Similarly, a change in the y dimension requires a nonzero value of $\partial v/\partial y$ and a change in the z dimension requires a nonzero value of $\partial w/\partial z$. These quantities represent the components of longitudinal rates of strain in the x, y, and z directions, respectively. Changes in length of the sides may produce changes in volume of the element. The rate of local instantaneous *volume dilation* is given by

$$\text{Volume dilation rate} = \frac{\partial u}{\partial x} + \frac{\partial v}{\partial y} + \frac{\partial w}{\partial z} = \nabla \cdot \vec{V} \tag{5.18}$$

For incompressible flow, the rate of volume dilation is zero.

Example 5.8

The velocity field, $\vec{V} = Ax\hat{i} - Ay\hat{j}$, represents flow in a "corner," as shown in Example Problem 5.4. Consider the case where $A = 0.3\text{ sec}^{-1}$ and the coordinates are measured in meters. A square is marked in the fluid as shown at $t = 0$. Evaluate the new positions of the four corner points when point a has moved to $x = \frac{3}{2}$ m after τ seconds. Evaluate the rates of linear deformation in the x and y directions. Compare area $a'b'c'd'$ at $t = \tau$ with area $abcd$ at $t = 0$. Comment on this result.

EXAMPLE PROBLEM 5.8

GIVEN: $\vec{V} = Ax\hat{i} - Ay\hat{j}$; $A = 0.3\ \text{sec}^{-1}$, x and y in meters.

FIND: (a) Position of square at $t = \tau$ when a is at a' at $x = \frac{3}{2}$.
(b) Rates of linear deformation.
(c) Compare area $a'b'c'd'$ with $abcd$.
(d) Comment on the results.

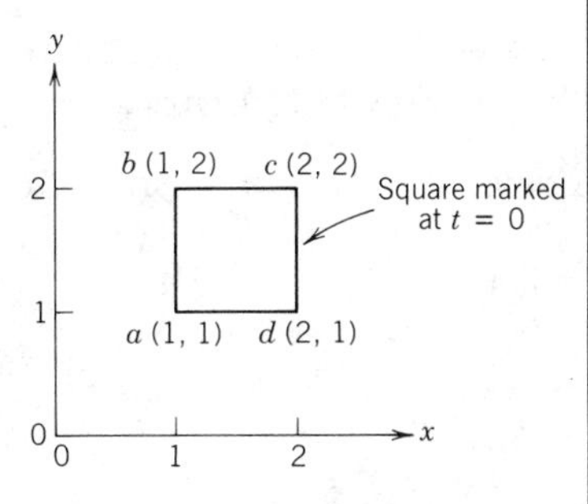

SOLUTION:
First we must find τ, so we must follow a fluid particle using Lagrangian description. Thus

$$u = \frac{dx_p}{dt} = Ax_p; \qquad \frac{dx}{x} = A\,dt; \qquad \int_{x_0}^{x} \frac{dx}{x} = \int_0^{\tau} A\,dt; \qquad \ln\frac{x}{x_0} = A\tau$$

$$\tau = \frac{\ln x/x_0}{A} = \frac{\ln(\frac{3}{2})}{0.3\ \text{sec}^{-1}} = 1.35\ \text{sec}$$

In the y direction

$$v = \frac{dy_p}{dt} = -Ay_p; \qquad \frac{dy}{y} = -A\,dt; \qquad \frac{y}{y_0} = e^{-A\tau}$$

The point coordinates at τ are:

Point	$t = 0$	$t = \tau$
a	(1, 1)	$(\frac{3}{2}, \frac{2}{3})$
b	(1, 2)	$(\frac{3}{2}, \frac{4}{3})$
c	(2, 2)	$(3, \frac{4}{3})$
d	(2, 1)	$(3, \frac{2}{3})$

The plot is:

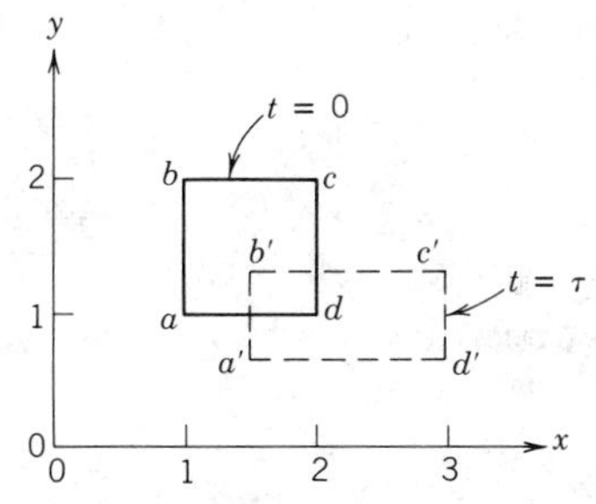

The rates of linear deformation are:

$$\frac{\partial u}{\partial x} = \frac{\partial}{\partial x} Ax = A = 0.3\ \text{sec}^{-1} \qquad \text{in the } x \text{ direction}$$

$$\frac{\partial v}{\partial y} = \frac{\partial}{\partial y}(-Ay) = -A = -0.3\ \text{sec}^{-1} \qquad \text{in the } y \text{ direction}$$

The rate of volume dilation is

$$\nabla \cdot \vec{V} = \frac{\partial u}{\partial x} + \frac{\partial v}{\partial y} = A - A = 0$$

Area $abcd = 1\ \text{m}^2$ and area $a'b'c'd' = (3 - \frac{3}{2})(\frac{4}{3} - \frac{2}{3}) = 1\ \text{m}^2$

{Note that parallel planes remain parallel; there is linear deformation but no angular deformation. The rates of linear deformation are equal and opposite, so the area of the marked region is conserved.}

We have shown in this section that the velocity field contains all information needed to determine translation, rotation, deformation, and acceleration of a particle in a flow.

5-4 MOMENTUM EQUATION

A dynamic equation describing fluid motion may be obtained by applying Newton's second law to a particle. To derive the differential form of the momentum equation, we shall apply Newton's second law to an infinitesimal fluid particle of mass, *dm*.

Recall that Newton's second law for a finite system is given by

$$\vec{F} = \left.\frac{d\vec{P}}{dt}\right)_{\text{system}} \tag{4.2a}$$

where the linear momentum, $\vec{P}$, of the system is given by

$$\vec{P}_{\text{system}} = \int_{\text{mass (system)}} \vec{V}\, dm \tag{4.2b}$$

Then, for an infinitesimal system of mass, *dm*, Newton's second law can be written

$$d\vec{F} = dm \left.\frac{d\vec{V}}{dt}\right)_{\text{system}} \tag{5.19}$$

Having obtained an expression for the acceleration of a fluid element of mass, *dm*, moving in a velocity field (Eq. 5.9), we can now write Newton's second law as the vector equation

$$d\vec{F} = dm\frac{D\vec{V}}{Dt} = dm\left[u\frac{\partial \vec{V}}{\partial x} + v\frac{\partial \vec{V}}{\partial y} + w\frac{\partial \vec{V}}{\partial z} + \frac{\partial \vec{V}}{\partial t}\right] \tag{5.20}$$

We need now to obtain a suitable formulation for the force, $d\vec{F}$, or its components dF_x, dF_y, dF_z, acting on the element.

5-4.1 Forces Acting on a Fluid Particle

Recall that the forces acting on a fluid element may be classified as body forces and surface forces; surface forces include both normal forces and tangential (shear) forces.

We shall consider the x component of the force acting on a differential element of mass, *dm*, and volume, $d\forall = dx\,dy\,dz$. Only those stresses that act in the x direction will give rise to surface forces in the x direction. If the stresses at the center of the differential element are taken to be σ_{xx}, τ_{yx}, and τ_{zx}, then the stresses acting in the x direction on each face of the element (obtained by a Taylor series expansion about the center of the element) are as shown in Fig. 5.10.

To obtain the net surface force in the x direction, dF_{S_x}, we must sum the forces in the x direction. Thus

$$dF_{s_x} = \left(\sigma_{xx} + \frac{\partial \sigma_{xx}}{\partial x}\frac{dx}{2}\right) dy\, dz - \left(\sigma_{xx} - \frac{\partial \sigma_{xx}}{\partial x}\frac{dx}{2}\right) dy\, dz$$
$$+ \left(\tau_{yx} + \frac{\partial \tau_{yx}}{\partial y}\frac{dy}{2}\right) dx\, dz - \left(\tau_{yx} - \frac{\partial \tau_{yx}}{\partial y}\frac{dy}{2}\right) dx\, dz$$
$$+ \left(\tau_{zx} + \frac{\partial \tau_{zx}}{\partial z}\frac{dz}{2}\right) dx\, dy - \left(\tau_{zx} - \frac{\partial \tau_{zx}}{\partial z}\frac{dz}{2}\right) dx\, dy$$

On simplifying, we obtain

$$dF_{s_x} = \left(\frac{\partial \sigma_{xx}}{\partial x} + \frac{\partial \tau_{yx}}{\partial y} + \frac{\partial \tau_{zx}}{\partial z}\right) dx\, dy\, dz$$

When the force of gravity is the only body force acting, then the body force per unit mass is $\vec{g}$. Then the net force in the x direction, dF_x, is given by

$$dF_x = dF_{s_x} + dF_{B_x} = \left(\rho g_x + \frac{\partial \sigma_{xx}}{\partial x} + \frac{\partial \tau_{yx}}{\partial y} + \frac{\partial \tau_{zx}}{\partial z}\right) dx\, dy\, dz \qquad (5.21a)$$

One can derive similar expressions for the force components in the y and z directions:

$$dF_y = dF_{s_y} + dF_{B_y} = \left(\rho g_y + \frac{\partial \tau_{xy}}{\partial x} + \frac{\partial \sigma_{yy}}{\partial y} + \frac{\partial \tau_{zy}}{\partial z}\right) dx\, dy\, dz \qquad (5.21b)$$

$$dF_z = dF_{s_z} + dF_{B_z} = \left(\rho g_z + \frac{\partial \tau_{xz}}{\partial x} + \frac{\partial \tau_{yz}}{\partial y} + \frac{\partial \sigma_{zz}}{\partial z}\right) dx\, dy\, dz \qquad (5.21c)$$

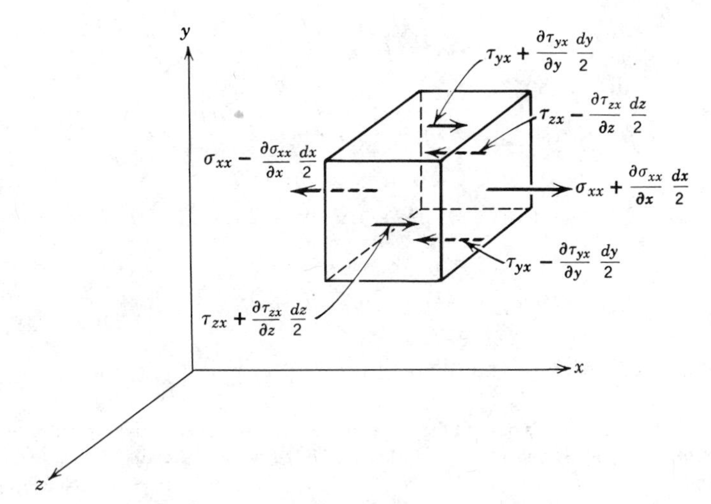

Fig. 5.10 Stresses in the x direction on an element of fluid.

5-4.2 Differential Momentum Equation

We have now formulated expressions for the components, dF_x, dF_y, and dF_z, of the force, $d\vec{F}$, acting on the element of mass, dm. If we substitute these expressions (Eqs. 5.21) for the force components into the x, y, and z components of Eq. 5.20, we obtain the differential equations of motion.

$$\rho g_x + \frac{\partial \sigma_{xx}}{\partial x} + \frac{\partial \tau_{yx}}{\partial y} + \frac{\partial \tau_{zx}}{\partial z} = \rho\left(\frac{\partial u}{\partial t} + u\frac{\partial u}{\partial x} + v\frac{\partial u}{\partial y} + w\frac{\partial u}{\partial z}\right) \tag{5.22a}$$

$$\rho g_y + \frac{\partial \tau_{xy}}{\partial x} + \frac{\partial \sigma_{yy}}{\partial y} + \frac{\partial \tau_{zy}}{\partial z} = \rho\left(\frac{\partial v}{\partial t} + u\frac{\partial v}{\partial x} + v\frac{\partial v}{\partial y} + w\frac{\partial v}{\partial z}\right) \tag{5.22b}$$

$$\rho g_z + \frac{\partial \tau_{xz}}{\partial x} + \frac{\partial \tau_{yz}}{\partial y} + \frac{\partial \sigma_{zz}}{\partial z} = \rho\left(\frac{\partial w}{\partial t} + u\frac{\partial w}{\partial x} + v\frac{\partial w}{\partial y} + w\frac{\partial w}{\partial z}\right) \tag{5.22c}$$

Equations 5.22 are the differential equations of motion for any fluid satisfying the continuum assumption. Before the equations can be used to solve problems, suitable expressions for the stresses must be obtained in terms of the velocity field.

5-4.3 Newtonian Fluid: Navier–Stokes Equations

For a Newtonian fluid the viscous stress is proportional to the rate of shearing strain (angular deformation rate). The stresses may be expressed in terms of velocity gradients and fluid properties in rectangular coordinates as follows:[7]

$$\tau_{xy} = \tau_{yx} = \mu\left(\frac{\partial v}{\partial x} + \frac{\partial u}{\partial y}\right) \tag{5.23a}$$

$$\tau_{yz} = \tau_{zy} = \mu\left(\frac{\partial w}{\partial y} + \frac{\partial v}{\partial z}\right) \tag{5.23b}$$

$$\tau_{zx} = \tau_{xz} = \mu\left(\frac{\partial u}{\partial z} + \frac{\partial w}{\partial x}\right) \tag{5.23c}$$

$$\sigma_{xx} = -p - \frac{2}{3}\mu\nabla\cdot\vec{V} + 2\mu\frac{\partial u}{\partial x} \tag{5.23d}$$

$$\sigma_{yy} = -p - \frac{2}{3}\mu\nabla\cdot\vec{V} + 2\mu\frac{\partial v}{\partial y} \tag{5.23e}$$

$$\sigma_{zz} = -p - \frac{2}{3}\mu\nabla\cdot\vec{V} + 2\mu\frac{\partial w}{\partial z} \tag{5.23f}$$

where p is the local thermodynamic pressure.

[7] The derivation of these results is beyond the scope of this book. Detailed derivations may be found in the following books: J. W. Daily and D. R. F. Harleman, *Fluid Dynamics* (Reading, Mass.: Addison-Wesley, 1966); H. Rouse, *Advanced Mechanics of Fluids* (New York: John Wiley, 1959); H. Schlichting, *Boundary-Layer Theory*, 7th ed. (New York: McGraw-Hill, 1979).

If these expressions are introduced into the differential equations of motion (Eqs. 5.22), we obtain

$$\rho \frac{Du}{Dt} = \rho g_x - \frac{\partial p}{\partial x} + \frac{\partial}{\partial x}\left[\mu\left(2\frac{\partial u}{\partial x} - \frac{2}{3}\nabla \cdot \vec{V}\right)\right] + \frac{\partial}{\partial y}\left[\mu\left(\frac{\partial u}{\partial y} + \frac{\partial v}{\partial x}\right)\right] + \frac{\partial}{\partial z}\left[\mu\left(\frac{\partial w}{\partial x} + \frac{\partial u}{\partial z}\right)\right] \tag{5.24a}$$

$$\rho \frac{Dv}{Dt} = \rho g_y - \frac{\partial p}{\partial y} + \frac{\partial}{\partial x}\left[\mu\left(\frac{\partial u}{\partial y} + \frac{\partial v}{\partial x}\right)\right] + \frac{\partial}{\partial y}\left[\mu\left(2\frac{\partial v}{\partial y} - \frac{2}{3}\nabla \cdot \vec{V}\right)\right] + \frac{\partial}{\partial z}\left[\mu\left(\frac{\partial v}{\partial z} + \frac{\partial w}{\partial y}\right)\right] \tag{5.24b}$$

$$\rho \frac{Dw}{Dt} = \rho g_z - \frac{\partial p}{\partial z} + \frac{\partial}{\partial x}\left[\mu\left(\frac{\partial w}{\partial x} + \frac{\partial u}{\partial z}\right)\right] + \frac{\partial}{\partial y}\left[\mu\left(\frac{\partial v}{\partial z} + \frac{\partial w}{\partial y}\right)\right] + \frac{\partial}{\partial z}\left[\mu\left(2\frac{\partial w}{\partial z} - \frac{2}{3}\nabla \cdot \vec{V}\right)\right] \tag{5.24c}$$

These equations of motion are called the Navier–Stokes equations. The equations are greatly simplified when applied to incompressible flows in which variations in fluid viscosity can be neglected. Under these conditions the equations reduce to

$$\rho\left(\frac{\partial u}{\partial t} + u\frac{\partial u}{\partial x} + v\frac{\partial u}{\partial y} + w\frac{\partial u}{\partial z}\right) = \rho g_x - \frac{\partial p}{\partial x} + \mu\left(\frac{\partial^2 u}{\partial x^2} + \frac{\partial^2 u}{\partial y^2} + \frac{\partial^2 u}{\partial z^2}\right) \tag{5.25a}$$

$$\rho\left(\frac{\partial v}{\partial t} + u\frac{\partial v}{\partial x} + v\frac{\partial v}{\partial y} + w\frac{\partial v}{\partial z}\right) = \rho g_y - \frac{\partial p}{\partial y} + \mu\left(\frac{\partial^2 v}{\partial x^2} + \frac{\partial^2 v}{\partial y^2} + \frac{\partial^2 v}{\partial z^2}\right) \tag{5.25b}$$

$$\rho\left(\frac{\partial w}{\partial t} + u\frac{\partial w}{\partial x} + v\frac{\partial w}{\partial y} + w\frac{\partial w}{\partial z}\right) = \rho g_z - \frac{\partial p}{\partial z} + \mu\left(\frac{\partial^2 w}{\partial x^2} + \frac{\partial^2 w}{\partial y^2} + \frac{\partial^2 w}{\partial z^2}\right) \tag{5.25c}$$

The Navier–Stokes equations in cylindrical coordinates for constant density and viscosity are given in Appendix B.

For the case of frictionless flow ($\mu = 0$) the equations of motion (Eq. 5.24 or Eq. 5.25) reduce to Euler's equation,

$$\rho \frac{D\vec{V}}{Dt} = \rho\vec{g} - \nabla p$$

We shall consider the case of frictionless flow in Chapter 6.

5-5 SUMMARY OBJECTIVES

After completing study of Chapter 5, you should be able to do the following:

1. Write the differential form of the conservation of mass in (a) vector form, (b) rectangular coordinates, and (c) cylindrical coordinates.

2. Given a velocity field, determine if the field represents a possible incompressible flow.
3. Given one component of velocity in a two-dimensional flow field, evaluate another component for steady, incompressible flow.
4. For a two-dimensional incompressible flow field define the stream function, ψ; given the velocity field, determine the stream function; given the stream function, determine the velocity field.
5. For a fluid particle moving in a given velocity field, determine the total, convective, and local accelerations.
6. For a fluid particle moving in a flow field, illustrate: translation, rotation, linear deformation, and angular deformation.
7. Define: fluid rotation, vorticity, and circulation.
8. Write the differential form of the momentum equation for viscous flow and state the physical meaning of each term in the equation.
9. Solve the problems at the end of the chapter that relate to the material you have studied.

PROBLEMS

5.1 A velocity field is given by the expression

$$\vec{V} = Ax\hat{i} - Ay\hat{j}$$

where x and y are measured in meters, and A equals 10 m/sec/m. Locate all points in the xy plane where $|\vec{V}|$ is equal to 20 m/sec.

5.2 Consider the flow described by the velocity field

$$\vec{V} = x(1 + At)\hat{i} + y\hat{j}$$

with $A = 0.5\ \text{sec}^{-1}$. For the point, $(1, 1, 0)$, calculate and plot (a) the streamline through the point at $t = 0$, and (b) the pathline traced out by the particle that passes through the point at this instant.

5.3 Consider the flow described by the velocity field

$$\vec{V} = x(1 + At)\hat{i} + y\hat{j}$$

with $A = 0.5\ \text{sec}^{-1}$. For the point, $(1, 1, 0)$, calculate and plot (a) the streamline through the point at $t = 0$, and (b) the streakline formed by particles that passed through the point at earlier times.

5.4 A three-dimensional flow is described by the velocity field

$$\vec{V} = 2Ax\hat{i} + 2Ay\hat{j} - 4Az\hat{k}$$

Obtain the equations for the streamline that passes through the point $(x, y, z) = (1, 0, 1)$.

5.5 A velocity field is given by the expression

$$\vec{V} = U\cos\theta\left[1 - \left(\frac{a}{r}\right)^2\right]\hat{i}_r - U\sin\theta\left[1 + \left(\frac{a}{r}\right)^2\right]\hat{i}_\theta$$

Find all points in the $r\theta$ plane where: (a) $V_r = 0$, (b) $V_\theta = 0$, and (c) $V_r = V_\theta = 0$.

** These objectives apply to sections that may be omitted without loss of continuity in the text material.

5.6 Which of the following sets of equations represent possible two-dimensional incompressible flow cases?

(a) $u = x + y; v = x - y$ (b) $u = x + 2y; v = x^2 - y^2$
(c) $u = 4x + y; v = x - y^2$ (d) $u = xt + 2y; v = x^2 - yt^2$
(e) $u = xt^2; v = xyt + y^2$

5.7. Which of the following sets of equations represent possible two-dimensional incompressible flow cases?

(a) $u = 2x^2 + y^2; v = x^3 - x(y^2 - 2y)$ (b) $u = 2xy - x^2 + y; v = 2xy - y^2 + x^2$
(c) $u = xt + 2y; v = xt^2 - yt$ (d) $u = (x + 2y)xt; v = (2x - y)yt$

5.8 Which of the following equations represent possible three-dimensional incompressible flow cases?

(a) $u = x + y + z^2; v = x - y + z; w = 2xy + y^2 + 4$
(b) $u = xyzt; v = -xyzt^2; w = (z^2/2)(xt^2 - yt)$
(c) $u = y^2 + 2xz; v = -2yz + x^2yz; w = \frac{1}{2}x^2z^2 + x^3y^4$

5.9 The three components of velocity in a velocity field are given by

$$u = Ax + By + Cz$$

$$v = Dx + Ey + Fz$$

$$w = Gx + Hy + Jz$$

Determine the relationship among the coefficients A through J that is necessary if this is to be a possible incompressible flow field.

5.10 For a flow in the xy plane, the y component of velocity is given by

$$v = y^2 - 2x + 2y$$

Determine a possible x component for steady, incompressible flow. Is it also valid for unsteady, incompressible flow? Why? How many possible x components are there?

5.11 A crude approximation for the x component of velocity in a laminar boundary layer is a linear variation from $u = 0$ at the surface ($y = 0$) to the freestream velocity, U, at the boundary-layer edge ($y = \delta$). The equation for the profile is

$$u = cU \frac{y}{x^{1/2}}$$

where c is a constant. Show that the simplest expression for the y component of velocity is

$$v = \frac{uy}{4x}$$

Evaluate the maximum value of the ratio, v/u, at a location where $x = 0.5$ m and $\delta = 5$ mm.

5.12 A useful approximation for the x component of velocity in a laminar boundary layer is a sinusoidal variation from $u = 0$ at the surface ($y = 0$) to the freestream velocity, U, at the edge of the boundary layer ($y = \delta$)

$$u = U \sin\left(\frac{\pi y}{2\delta}\right)$$

where $\delta = cx^{1/2}$ (c = constant). Obtain an expression for the maximum value of the ratio, v/u. Evaluate at a location where $x = 0.5$ m and $\delta = 5$ mm.

5.13 Which of the following sets of equations represent possible incompressible flow cases?
(a) $V_r = U\cos\theta$; $V_\theta = -U\sin\theta$
(b) $V_r = -q/2\pi r$; $V_\theta = K/2\pi r$
(c) $V_r = U\cos\theta[1 - (a/r)^2]$; $V_\theta = -U\sin\theta[1 + (a/r)^2]$

5.14 For an incompressible flow in the $r\theta$ plane, the r component of velocity is given as

$$V_r = -\frac{\Lambda\cos\theta}{r^2}$$

Determine a possible θ component of velocity. How many possible θ components are there?

5.15 A velocity field in cylindrical coordinates is given as

$$\vec{V} = \frac{A}{r}\hat{i}_r + \frac{A}{r}\hat{i}_\theta$$

where $A = 0.25\ \text{m}^2/\text{sec}$. Does this represent a possible incompressible flow case? Obtain the equation for a streamline passing through the point $r_0 = 1$ m, $\theta = 0$. Compare with the pathline through the same point.

5.16 A velocity field in cylindrical coordinates is given as

$$\vec{V} = \left(\frac{q}{2\pi r} + U\cos\theta\right)\hat{i}_r - U\sin\theta\hat{i}_\theta$$

where $q = 200\ \text{m}^2/\text{sec}$ and $U = 10$ m/sec. Show that this is a possible incompressible flow case. Locate the stagnation points.

5.17 A viscous liquid is sheared between two parallel disks of radius, R, one of which rotates while the other is fixed. The velocity field is purely tangential, and the velocity varies linearly with z from $V_\theta = 0$ at $z = 0$ (the fixed disk) to the velocity of the rotating disk at its surface ($z = h$). Write an expression for the velocity field between the disks.

***5.18** The stream function for a certain incompressible flow is given as $\psi = Axy$. Plot several streamlines, including $\psi = 0$. Obtain an expression for the velocity field.

5.19 A uniform flow field is inclined at angle, α, above the x axis. Evaluate the velocity components, u and v. Determine the stream function for this flow field.

5.20 The velocity field for the viscometric flow of Example Problem 5.7 is $\vec{V} = U(y/h)\hat{i}$. Find the stream function for this flow. Locate the streamline that divides the total flow rate into two equal parts.

***5.21** Derive the stream function that represents the sinusoidal approximation used to model the x component of velocity for the boundary layer of Problem 5.12. Locate streamlines at one-quarter and one-half the total volume flow rate in the boundary layer.

***5.22** Consider a flow with velocity components: $u = 0$, $v = -y^3 - 4z$, and $w = 3y^2z$.
(a) Is this one-, two-, or three-dimensional flow?
(b) Demonstrate whether this is an incompressible or compressible flow.
(c) If possible, derive a stream function for this flow.

* These problems require material from sections that may be omitted without loss of continuity in the text material.

***5.23** Determine the family of ψ functions that will yield the velocity field

$$\vec{V} = (x^2 - y^2)\hat{i} - 2xy\hat{j}$$

***5.24** An incompressible frictionless flow field is specified by the stream function

$$\psi = -6Ax - 8Ay$$

where $A = 1$ m/sec and x and y are coordinates in meters.
(a) Sketch the streamlines $\psi = 0$ and $\psi = 8$.
(b) Indicate the direction of the resultant velocity vector at the point $(0, 0)$ on the sketch of Part (a).
(c) Determine the magnitude of the flow rate between the streamlines passing through the points $(2, 2)$ and $(4, 1)$.

***5.25** In a parallel one-dimensional flow in the positive x direction, the velocity varies linearly from zero at $y = 0$ to 100 ft/sec at $y = 4$ ft. Determine an expression for the stream function, ψ. Also determine the y coordinate above which the volume flow rate is half the total between $y = 0$ and $y = 4$ ft.

5.26 Flow in a "sector" of included angle, $\alpha = \pi/n$, is represented by the stream function

$$\psi = Ur^n \sin(n\theta)$$

Show that for $n = 2$, this stream function reduces to the form found in Example Problem 5.4 for flow in a square corner ($\alpha = \pi/2$).

***5.27** The stream function for a certain incompressible flow field is given by the expression

$$\psi = -Ur\sin\theta + \frac{q}{2\pi}\theta$$

Find the point(s) where $|\vec{V}| = 0$, and show that $\psi = 0$ there. Obtain an expression for the velocity field.

5.28 Incompressible flow around a circular cylinder of radius, a, is represented by the stream function

$$\psi = -Ur\sin\theta + \frac{Ua^2\sin\theta}{r}$$

where U represents the freestream velocity. Show that $V_r = 0$ along the circle, $r = a$. Locate the points along $r = a$ where $|\vec{V}| = U$.

***5.29** A velocity field in cylindrical coordinates is given as

$$\vec{V} = \frac{A}{r}\hat{i}_r + \frac{A}{r}\hat{i}_\theta$$

Does this represent a possible incompressible flow case? If so, evaluate the stream function for the flow. If not, evaluate the rate of change of density in the flow field.

5.30 A rigid-body motion was modeled in Example Problem 5.6 by the velocity field

$$\vec{V} = r\omega\hat{i}_\theta$$

* These problems require material from sections that may be omitted without loss of continuity in the text material.

Find the stream function for this flow. Evaluate the volume flow rate per unit depth between $r_1 = 0.05$ m and $r_2 = 0.07$ m if $\omega = 0.5$ rad/sec. Sketch the velocity profile along a line of constant θ. Check the flow rate calculated from the stream function by integrating the velocity profile along this line.

***5.31** Example Problem 5.6 showed that the velocity field for a free vortex in the $r\theta$ plane is

$$\vec{V} = \frac{C}{r}\hat{i}_\theta$$

Find the stream function for this flow. Evaluate the volume flow rate per unit depth between $r_1 = 0.05$ m and $r_2 = 0.07$ m, if $C = 0.5\ \text{m}^2/\text{sec}$. Sketch the velocity profile along a line of constant θ. Check the flow rate calculated from the stream function by integrating the velocity profile along this line.

5.32 Consider the flow field given by $\vec{V} = xy^2\hat{i} - \frac{1}{3}y^3\hat{j} + xy\hat{k}$. Determine: (a) the number of dimensions of the flow, (b) if it is a possible incompressible flow, and (c) the acceleration of a particle at the point $(x, y, z) = (1, 2, 3)$.

5.33 Consider the flow field given by $\vec{V} = ax^2y\hat{i} - by\hat{j} + cz^2\hat{k}$, where $a = 1/\text{m}^2 \cdot \text{sec}$, $b = 3/\text{sec}$ and $c = 2/\text{m} \cdot \text{sec}$. Determine: (a) the number of dimensions of the flow, (b) if it is a possible incompressible flow, and (c) the acceleration of a particle at the point $(x, y, z) = (3, 1, 2)$.

5.34 A steady, two-dimensional velocity field is given by $\vec{V} = Ax\hat{i} - Ay\hat{j}$, where $A = 1\ \text{sec}^{-1}$. Show that the steamlines for this flow are rectangular hyperbolas, $xy = C$. Plot streamlines that correspond to $C = 0, 1$, and $4\ \text{m}^2$. Obtain a general expression for the acceleration of a fluid particle in this velocity field. Calculate the acceleration of fluid particles at the points $(x, y) = (\frac{1}{2}, 2), (1, 1)$, and $(2, \frac{1}{2})$, where x and y are measured in meters. Show the acceleration vectors on the streamline plot.

5.35 Consider again the steady, two-dimensional velocity field of Problem 5.34,

$$\vec{V} = Ax\hat{i} - Ay\hat{j}$$

where $A = 1\ \text{sec}^{-1}$. Obtain expressions for the particle coordinates, $x_p = f_1(t)$ and $y_p = f_2(t)$ as functions of time and the initial particle position, (x_0, y_0) at $t = 0$. Determine the time required for a particle to travel from initial position, $(x_0, y_0) = (\frac{1}{2}, 2)$ to positions $(x, y) = (1, 1)$ and $(2, \frac{1}{2})$. Compare the particle accelerations determined by differentiating $f_1(t)$ and $f_2(t)$ with those obtained in Problem 5.34.

5.36 Air flows into the narrow gap between closely spaced parallel plates through a porous surface as shown. Use a control volume with outer surface located at position, x, to show that the uniform velocity in the x direction is $u = v_0x/h$. Find an expression for the velocity component in the y direction. Evaluate the x component of acceleration for a fluid particle in the gap.

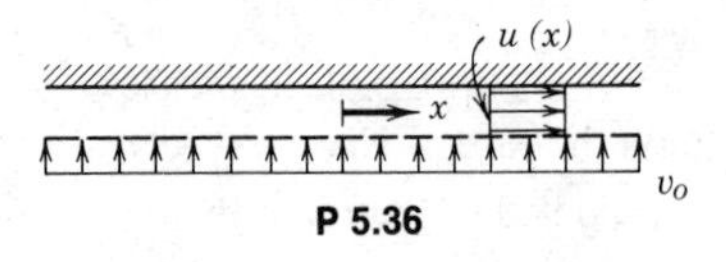

P 5.36

* These problems require material from sections that may be omitted without loss of continuity in the text material.

5.37 The differential form of the equation for conservation of mass can be used to evaluate the relative rate of change of density of a fluid particle as it moves through a flow. Show that

$$\frac{1}{\rho}\frac{D\rho}{Dt} = -\nabla \cdot \vec{V}$$

Explain the physical significance of $\nabla \cdot \vec{V}$.

5.38 In cylindrical coordinates, the velocity field for a two-dimensional flow is given by

$$\vec{V} = \vec{V}(r, \theta, t)$$

Show that the radial and tangential components of the acceleration of a particle are given by

$$a_r = V_r \frac{\partial V_r}{\partial r} + \frac{V_\theta}{r}\frac{\partial V_r}{\partial \theta} - \frac{V_\theta^2}{r} + \frac{\partial V_r}{\partial t}$$

$$a_\theta = V_r \frac{\partial V_\theta}{\partial r} + \frac{V_\theta}{r}\frac{\partial V_\theta}{\partial \theta} + \frac{V_\theta V_r}{r} + \frac{\partial V_\theta}{\partial t}$$

5.39 A cylindrical tank of radius, $R = 4$ in., is filled with water to a depth of 6 in. The tank is rotated about its vertical axis. During start up, $0 \leq t \leq \tau$, the rate of rotation is given by $\omega = \omega_0 t/\tau$, where $\tau = 2$ sec, and the steady state rotational speed is $\omega_0 = 78$ rpm. The no-slip condition requires that fluid particles at the tank wall have zero velocity relative to the wall. Using the results of Problem 5.38, determine, for a particle at the wall:
(a) The acceleration at time, $t = 1$ sec
(b) The steady state acceleration

5.40 A velocity field is given by

$$\vec{V} = -\frac{Ay}{\sqrt{x^2 + y^2}}\hat{i} + \frac{Ax}{\sqrt{x^2 + y^2}}\hat{j}$$

(a) Show that this is a possible two-dimensional, incompressible flow.
(b) For the steady state conditions of Problem 5.39, determine the value of A.
(c) Find a_x and a_y for the particle located on the tank wall at $\theta = 60°$.

5.41 The velocity field for steady inviscid flow from left to right over a circular cylinder of radius a is given by

$$\vec{V} = U\cos\theta\left[1 - \left(\frac{a}{r}\right)^2\right]\hat{i}_r - U\sin\theta\left[1 + \left(\frac{a}{r}\right)^2\right]\hat{i}_\theta$$

Obtain expressions for the acceleration of a fluid particle moving along the stagnation streamline ($\theta = \pi$) and for the acceleration along the cylinder surface ($r = a$). Determine the locations at which these accelerations reach maximum and minimum values.

5.42 Solve Problem 4.103 to show that the radial velocity in the narrow gap is $V_r = Q/2\pi rh$. Derive an expression for the acceleration of a fluid particle in the gap.

5.43 Consider the low-speed flow of air between parallel disks as shown. Assume that the flow is incompressible and inviscid, and that the velocity is purely radial and uniform at any section. The flow speed is 15 m/sec at $R = 75$ mm. Simplify the continuity equation to a form applicable to this flow field. Show that a general expression for the velocity field is $\vec{V} = V(R/r)\hat{i}_r$ for $r_i \leq r \leq R$. Calculate the acceleration of a fluid particle at the locations $r = r_i$ and $r = R$.

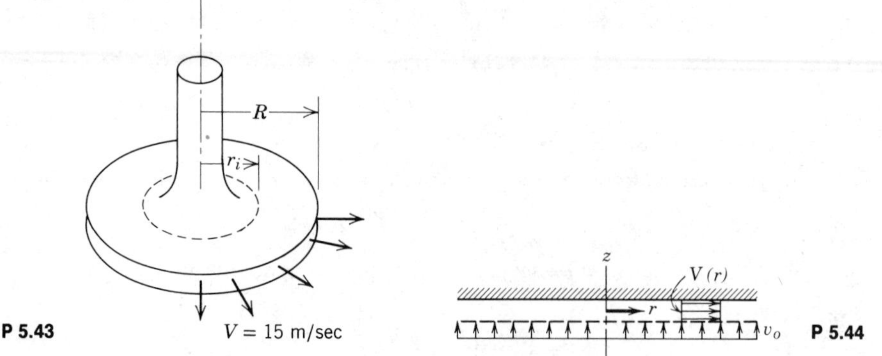

P 5.43 $V = 15$ m/sec

P 5.44

5.44 Air flows into the narrow gap between closely spaced parallel disks through a porous surface as shown. Use a control volume with outer surface located at position, r, to show that the uniform velocity in the r direction is $V = v_o r/2h$. Find an expression for the velocity component in the z direction ($v_0 \ll V$). Evaluate the r component of acceleration for a fluid particle in the gap.

5.45 An "air hockey" puck may be modeled as a circular disk supported by a layer of air that issues from multiple tiny holes in the game table. Assume that a puck floats a distance, $h = 1$ mm, above the table from which air flows at a volume flow rate per unit area of table, $q = 0.08\ \text{m}^3/\text{sec}/\text{m}^2$. Obtain an expression for radial flow speed under the puck if the flow is considered to be uniform and incompressible. If the puck diameter is 75 mm, determine the magnitude and location of the maximum radial acceleration experienced by a fluid particle under the puck.

5.46 Consider the incompressible flow of a fluid through a nozzle as shown. The area of the nozzle is given by $A = A_0(1 - bx)$ and the inlet velocity varies according to $V_0 = U(1 + at)$, where $A_0 = 1\ \text{ft}^2$, $L = 4$ ft, $b = 0.1\ \text{ft}^{-1}$, $a = 2\ \text{sec}^{-1}$, and $U = 10$ ft/sec. The flow may be assumed one-dimensional. Find the acceleration of a fluid particle at $x = L/2$ for $t = 0$ and 0.5 sec.

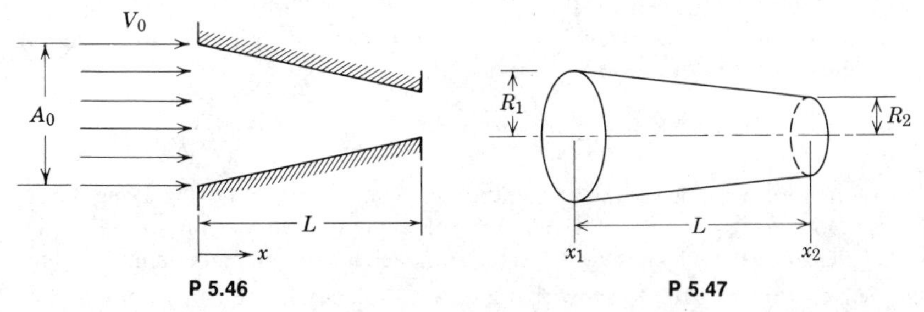

P 5.46 P 5.47

5.47 Consider the one-dimensional, incompressible flow through the circular channel shown. The velocity at section ① is given by $U = U_0 + U_1 \sin \omega t$, where $U_0 = 20$ m/sec, $U_1 = 2$ m/sec, and $\omega = 0.3$ rad/sec. The channel dimensions are $L = 1$ m, $R_1 = 0.2$ m, and $R_2 = 0.1$ m. Determine the particle acceleration at the channel exit. Plot the results as a function of time over a complete cycle.

5.48 The circular channel of Problem 5.47 is replaced by a plane channel of uniform width with the same linear dimensions. For the conditions given, determine the particle acceleration at the channel exit and plot the results as a function of time over a complete cycle.

5.49 The temperature, T, in a long tunnel is known to vary approximately as

$$T = T_0 - \alpha e^{-x/L} \sin \frac{2\pi t}{\tau}$$

where T_0, α, L, and τ are constants, and x is measured from the entrance. A particle moves into the tunnel with a constant speed, U. Find the rate of change of temperature experienced by the particle.

5.50 A flow is represented by the velocity field, $\vec{V} = 10x\hat{i} - 10y\hat{j} + 30\hat{k}$. Determine if the field is (a) a possible incompressible flow, and (b) irrotational.

5.51 A flow is represented by the velocity field, $\vec{V} = (4x^2 + 3y)\hat{i} + (3x - 2y)\hat{j}$. Determine if the field is (a) a possible incompressible flow, and (b) irrotational.

***5.52** Consider the flow field represented by the stream function, $\psi = 10xy + 17$. Is this a possible two-dimensional, incompressible flow? Is the flow irrotational?

***5.53** A velocity field for motion parallel to the x axis with constant shear is shown. The shear rate is $du/dy = A$, where $A = 0.1 \text{ sec}^{-1}$. Obtain an expression for the velocity field, $\vec{V}$. Calculate the rate of rotation. Evaluate the stream function for this flow field.

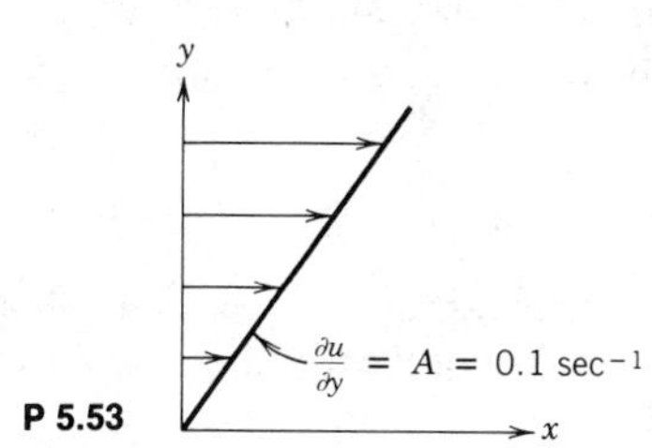

P 5.53

***5.54** A flow field is represented by the stream function, $\psi = x^2 - y^2$. Find the corresponding velocity field. Show that this flow field is irrotational.

5.55 Consider again the viscometric flow of Example Problem 5.7. Evaluate the average rate of rotation of a pair of perpendicular line segments oriented at $\pm 45°$ from the x axis. Show that the rates of rotation for these segments are the same as in the example.

***5.56** The velocity field near the core of a tornado can be approximated as

$$\vec{V} = -\frac{q}{2\pi r}\hat{i}_r + \frac{K}{2\pi r}\hat{i}_\theta$$

Is this an irrotational flow field? Obtain the stream function for this flow.

5.57 Coaxial cylinders with radii, R_1 and R_2, are rotated at angular speeds, ω_1 and ω_2. The space between the cylinders is filled with liquid that is to have irrotational motion. If the velocity at each cylinder is equal to the surface speed, find the required variation in velocity. Express ω_1/ω_2 in terms of the cylinder radii.

5.58 Consider again the sinusoidal velocity profile used to model the x component of velocity for a boundary layer in Problem 5.12. Neglect the vertical component of velocity. Evaluate the circulation around the contour bounded by $x = 0.4$ m, $x = 0.6$ m, $y = 0$, and $y = 6$ mm. What would be the result of this evaluation if it were performed 0.2 m further downstream? Assume $U = 0.5$ m/sec.

* These problems require material from sections that may be omitted without loss of continuity in the text material.

5.59 Consider the velocity field for flow in a rectangular "corner," $\vec{V} = Ax\hat{i} - Ay\hat{j}$, with $A = 0.3\ \text{sec}^{-1}$, as in Example Problem 5.8. Evaluate the circulation about the unit square from Example Problem 5.8 using Eq. 5.15.

5.60 Problem 4.34 gave the velocity profile for fully developed laminar flow between parallel plates in water as

$$\frac{u}{u_{max}} = 1 - \left(\frac{y}{h}\right)^2$$

Obtain an expression for the shear force per unit volume in the x direction for this flow. Calculate its maximum value if $u_{max} = 3$ m/sec and $h = 40$ mm.

5.61 The x component of velocity in a laminar boundary layer in water is approximated as

$$u = U \sin\left(\frac{\pi}{2}\frac{y}{\delta}\right)$$

where $U = 3$ m/sec and $\delta = 2$ mm. The y component of velocity is much smaller than u. Obtain an expression for the net shear force per unit volume in the x direction on a fluid element. Calculate its maximum value for this flow.

5.62 Problem 4.27 gave the velocity profile for fully developed laminar flow in a circular tube

$$\frac{u}{u_{max}} = 1 - \left(\frac{r}{R}\right)^2$$

Obtain an expression for the shear force per unit volume in the z direction for this flow. Evaluate its maximum value for the conditions of Problem 4.33.

Chapter 6

INCOMPRESSIBLE INVISCID FLOW

All real fluids possess viscosity. However, fluids often behave as though they were inviscid. Therefore, it is useful to investigate the dynamics of an *ideal fluid* that is incompressible and has zero viscosity. The analysis of ideal fluid motions is simpler than that of viscous flows because no shear stresses are present in inviscid flow. Normal stresses are the only stresses that must be considered in the analysis. For a nonviscous fluid in motion the normal stress at a point is the same in all directions (it is a scalar quantity). The normal stress in an inviscid flow is the negative of the thermodynamic pressure, $\sigma_{nn} = -p$. (This result is consistent with Eqs. 5.23 for $\mu = 0$.)

6-1 MOMENTUM EQUATION FOR FRICTIONLESS FLOW: EULER'S EQUATIONS

The equations of motion for frictionless flow, called Euler's equations, are obtained from the general equations of motion (Eqs. 5.22). Since, in a frictionless flow, there can be no shear stresses present and the normal stress is the negative of the thermodynamic pressure, then the equations of motion for a frictionless flow are

$$\rho g_x - \frac{\partial p}{\partial x} = \rho\left(\frac{\partial u}{\partial t} + u\frac{\partial u}{\partial x} + v\frac{\partial u}{\partial y} + w\frac{\partial u}{\partial z}\right) \tag{6.1a}$$

$$\rho g_y - \frac{\partial p}{\partial y} = \rho\left(\frac{\partial v}{\partial t} + u\frac{\partial v}{\partial x} + v\frac{\partial v}{\partial y} + w\frac{\partial v}{\partial z}\right) \tag{6.1b}$$

$$\rho g_z - \frac{\partial p}{\partial z} = \rho\left(\frac{\partial w}{\partial t} + u\frac{\partial w}{\partial x} + v\frac{\partial w}{\partial y} + w\frac{\partial w}{\partial z}\right) \tag{6.1c}$$

We can also write the above equations as a single vector equation

$$\rho\vec{g} - \nabla p = \rho\left(\frac{\partial \vec{V}}{\partial t} + u\frac{\partial \vec{V}}{\partial x} + v\frac{\partial \vec{V}}{\partial y} + w\frac{\partial \vec{V}}{\partial z}\right)$$

or

$$\rho \vec{g} - \nabla p = \rho \frac{D\vec{V}}{Dt} \tag{6.2}$$

If the z coordinate is directed vertically, then, since $\nabla z = \hat{k}$,

$$\rho \vec{g} = -\rho g \hat{k} = -\rho g \nabla z$$

and Euler's equation can be written as

$$-\frac{1}{\rho} \nabla p - g \nabla z = \frac{D\vec{V}}{Dt} \tag{6.3}$$

The momentum equation for frictionless flow also can be written in cylindrical coordinates. The equations in component form, with gravity the only body force, are

$$g_r - \frac{1}{\rho} \frac{\partial p}{\partial r} = a_r = \frac{\partial V_r}{\partial t} + V_r \frac{\partial V_r}{\partial r} + \frac{V_\theta}{r} \frac{\partial V_r}{\partial \theta} + V_z \frac{\partial V_r}{\partial z} - \frac{V_\theta^2}{r} \tag{6.4a}$$

$$g_\theta - \frac{1}{\rho r} \frac{\partial p}{\partial \theta} = a_\theta = \frac{\partial V_\theta}{\partial t} + V_r \frac{\partial V_\theta}{\partial r} + \frac{V_\theta}{r} \frac{\partial V_\theta}{\partial \theta} + V_z \frac{\partial V_\theta}{\partial z} + \frac{V_r V_\theta}{r} \tag{6.4b}$$

$$g_z - \frac{1}{\rho} \frac{\partial p}{\partial z} = a_z = \frac{\partial V_z}{\partial t} + V_r \frac{\partial V_z}{\partial r} + \frac{V_\theta}{r} \frac{\partial V_z}{\partial \theta} + V_z \frac{\partial V_z}{\partial z} \tag{6.4c}$$

If the z axis is directed vertically upward, then $g_r = g_\theta = 0$ and $g_z = -g$.

In Chapter 3 we found that if a fluid is accelerated so that there is no relative motion between adjacent layers of the fluid, the fluid moves without deformation and no shear stresses occur. We were able to determine the pressure variation within the fluid by applying the equations of motion to an appropriate free body. We considered two specific cases. For the case of rectilinear acceleration, we derived the differential equation of motion, Eq. 3.16; in Example Problem 3.8 we applied the equation to a tank of water moving as a solid body. In Example Problem 3.9 we considered the case of a fluid undergoing steady rotation about a vertical axis; we derived the equation of motion for a differential element of fluid undergoing rotation. It is left to you as an exercise to show that the use of Euler's equations to solve Example Problems 3.8 and 3.9 leads to results identical to those obtained previously.

6-2 EULER'S EQUATIONS IN STREAMLINE COORDINATES

In Chapter 2 we pointed out that the notion of a streamline, a line drawn tangent to the velocity vector at every point in the flow field, provides a convenient graphical representation. In steady flow a fluid particle will move along a streamline because, for steady flow, pathlines and streamlines coincide. Thus, in describing the motion of a fluid particle in a steady flow, the distance along a streamline is a logical coordinate to

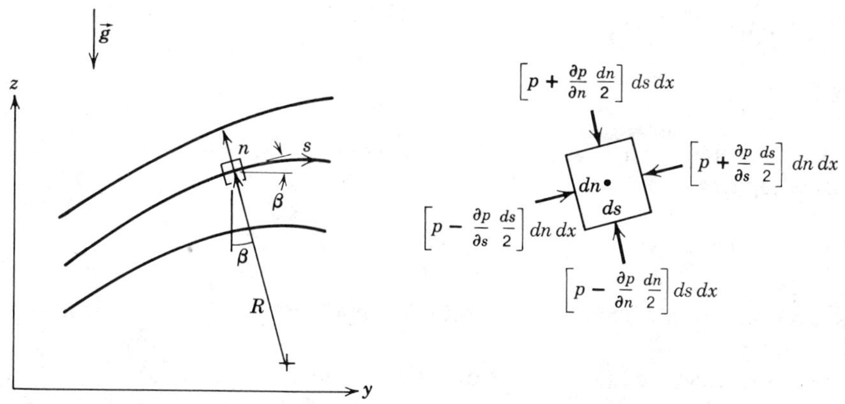

Fig. 6.1 Fluid particle moving along a streamline.

use in writing the equations of motion. "Streamline coordinates" also may be used to describe unsteady flow. Streamlines in unsteady flow give a graphical representation of the instantaneous velocity field.

For simplicity, consider the flow in the yz plane shown in Fig. 6.1. The equations of motion are to be written in terms of the coordinate, s, distance along a streamline, and the coordinate, n, distance normal to the streamline. Since the velocity vector must be tangent to the streamline, then the velocity field is given by $\vec{V} = \vec{V}(s, t)$. The pressure at the center of the fluid element is p. If we apply Newton's second law in the streamwise (the s) direction to the fluid element of volume, $ds\, dn\, dx$, then neglecting the viscous forces we obtain

$$\left(p - \frac{\partial p}{\partial s}\frac{ds}{2}\right) dn\, dx - \left(p + \frac{\partial p}{\partial s}\frac{ds}{2}\right) dn\, dx - \rho g \sin\beta\, dn\, dx\, ds = \rho\, ds\, dn\, dx\, a_s$$

where β is the angle between the tangent to the streamline and the horizontal, and a_s is the acceleration of the fluid particle along the streamline. Simplifying the equation, we obtain

$$-\frac{\partial p}{\partial s} - \rho g \sin\beta = \rho a_s$$

Since $\sin\beta = \partial z/\partial s$, we can write

$$-\frac{1}{\rho}\frac{\partial p}{\partial s} - g\frac{\partial z}{\partial s} = a_s$$

Along any streamline $V_s = V_s(s, t)$, and the total acceleration of a fluid particle in the streamwise direction is given by

$$a_s = \frac{DV_s}{Dt} = \frac{\partial V_s}{\partial t} + V_s\frac{\partial V_s}{\partial s}$$

The velocity is tangent to the streamline, so the subscript, s, on V_s is redundant and can be dropped. Euler's equation in the streamwise direction with the z axis directed vertically is then

$$-\frac{1}{\rho}\frac{\partial p}{\partial s} - g\frac{\partial z}{\partial s} = \frac{\partial V}{\partial t} + V\frac{\partial V}{\partial s} \tag{6.5a}$$

For steady flow, and neglecting body forces, Euler's equation in the streamwise direction reduces to

$$\frac{1}{\rho}\frac{\partial p}{\partial s} = -V\frac{\partial V}{\partial s} \tag{6.5b}$$

which indicates that a decrease in velocity is accompanied by an increase in pressure and conversely.[1]

To obtain Euler's equation in a direction normal to the streamlines, we apply Newton's second law in the n direction to the fluid element. Again, neglecting viscous forces, we obtain

$$\left(p - \frac{\partial p}{\partial n}\frac{dn}{2}\right) ds\,dx - \left(p + \frac{\partial p}{\partial n}\frac{dn}{2}\right) ds\,dx - \rho g \cos\beta\, dn\,dx\,ds = \rho a_n\, dn\,dx\,ds$$

where β is the angle between the n direction and the vertical, and a_n is the acceleration of the fluid particle in the n direction. Simplifying the equation, we obtain

$$-\frac{\partial p}{\partial n} - \rho g \cos\beta = \rho a_n$$

Since $\cos\beta = \partial z/\partial n$, we write

$$-\frac{1}{\rho}\frac{\partial p}{\partial n} - g\frac{\partial z}{\partial n} = a_n$$

The normal acceleration of the fluid element is toward the center of curvature of the streamline, in the minus n direction; thus in the coordinate system of Fig. 6.1, the familiar centripetal acceleration is written

$$a_n = \frac{-V^2}{R}$$

for steady flow,[2] where R is the radius of curvature of the streamline. Then, Euler's equation normal to the streamline is written for steady flow as

$$\frac{1}{\rho}\frac{\partial p}{\partial n} + g\frac{\partial z}{\partial n} = \frac{V^2}{R} \tag{6.6a}$$

[1] The relationship between variations in pressure and velocity in the streamwise direction for steady incompressible inviscid flow is illustrated in the NCFMF film loop, S-FM038, *Streamwise Pressure Gradient in Inviscid Flow.*

[2] If the flow were not steady, the streamline pattern could change with time. In that case,

$$a_n = -\frac{V_s^2}{R} + \frac{\partial V_n}{\partial t}$$

For steady flow in a horizontal plane, Euler's equation normal to a streamline becomes

$$\frac{1}{\rho}\frac{\partial p}{\partial n} = \frac{V^2}{R} \tag{6.6b}$$

which indicates that pressure increases in a direction outward from the center of curvature of the streamlines.[3] In regions where the streamlines are straight, the radius of curvature, R, of the streamlines is infinite and there is no pressure variation normal to the streamlines.

Example 6.1

The flow rate of air at standard conditions in a flat duct is to be determined by installing pressure taps across a bend. The duct is 0.3 m deep and 0.1 m wide. The inner radius of the bend is 0.25 m. The velocity profile is assumed uniform. If the measured pressure difference between the taps is 40 mm of water, compute the approximate flow rate.

EXAMPLE PROBLEM 6.1

GIVEN: Flow through duct bend as shown.

$p_2 - p_1 = \rho_{H_2O} g \, \Delta h$

where $\Delta h = 40 \text{ mm } H_2O$.
Flow uniform. Air at STP.

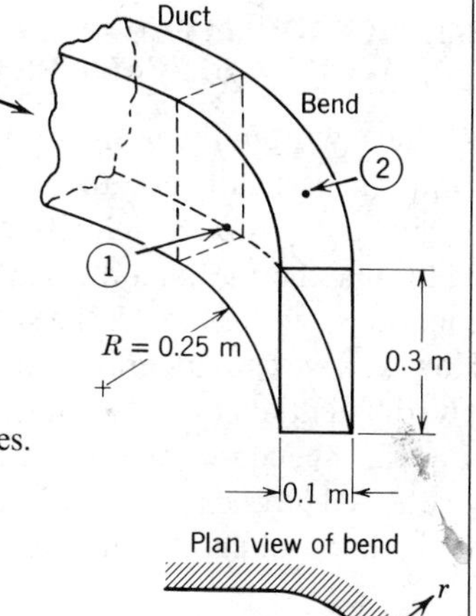

FIND: Q.

SOLUTION:

Apply Euler's n component equation across flow streamlines.

Basic equation: $$\frac{\partial p}{\partial r} = \frac{\rho V^2}{r}$$

Assumptions: (1) Frictionless flow
(2) Incompressible flow
(3) Uniform flow at measurement section

For this flow, $p = p(r)$, so

$$\frac{\partial p}{\partial r} = \frac{dp}{dr} = \frac{\rho V^2}{r} \quad \text{or} \quad dp = \rho V^2 \frac{dr}{r}$$

Integrating gives

$$p_2 - p_1 = \rho V^2 \ln r \Big]_{r_1}^{r_2} = \rho V^2 \ln \frac{r_2}{r_1}$$

[3] The effect of streamline curvature on the pressure gradient normal to a streamline is illustrated in the NCFMF film loop, S-FM037, *Streamline Curvature and Normal Pressure Gradient.*

Then

$$V = \left[\frac{p_2 - p_1}{\rho \ln(r_2/r_1)}\right]^{1/2}$$

But $\Delta p = p_2 - p_1 = \rho_{H_2O} g\, \Delta h$, so

$$V = \left[\frac{\rho_{H_2O} g\, \Delta h}{\rho \ln(r_2/r_1)}\right]^{1/2}$$

$$= \left[\frac{999\ \text{kg}}{\text{m}^3} \times \frac{9.81\ \text{m}}{\text{sec}^2} \times 0.04\ \text{m} \times \frac{\text{m}^3}{1.23\ \text{kg}} \times \frac{1}{\ln(0.35\ \text{m}/0.25\ \text{m})}\right]^{1/2}$$

$$V = 30.8\ \text{m/sec}$$

For uniform flow

$$Q = VA = \frac{30.8\ \text{m}}{\text{sec}} \times 0.1\ \text{m} \times 0.3\ \text{m} = 0.924\ \text{m}^3/\text{sec}$$ Q

{In actual applications, the velocity profile in a channel bend will not be uniform. The velocity profile in a bend tends toward a free vortex (irrotational) profile.}

6-3 BERNOULLI EQUATION—INTEGRATION OF EULER'S EQUATION ALONG A STREAMLINE FOR STEADY FLOW

We have written the momentum equations and the continuity equation in differential form. Theoretically, for an incompressible inviscid flow, these equations can be solved to give the complete velocity and pressure fields. (If the density is not constant, an additional thermodynamic relation for the density is required.) Although, in theory, the equations can be solved, the solution for a particular flow field may be very involved. However, we can integrate Euler's equation readily for steady flow along a streamline. The differential control volume analysis of Section 4-4.1 led to a differential equation (Eq. 4.24), which when integrated, led to a form of the Bernoulli equation. In order to give added physical insight about restrictions on the results, two additional derivations of the Bernoulli equation are presented.

6-3.1 Derivation Using Streamline Coordinates

Euler's equation for steady flow along a streamline is given by

$$-\frac{1}{\rho}\frac{\partial p}{\partial s} - g\frac{\partial z}{\partial s} = V\frac{\partial V}{\partial s} \tag{6.7}$$

If a fluid particle moves a distance, ds, along a streamline, then

$$\frac{\partial p}{\partial s}\, ds = dp \qquad \text{(the change in pressure along } s\text{)}$$

$$\frac{\partial z}{\partial s}\, ds = dz \qquad \text{(the change in elevation along } s\text{)}$$

$$\frac{\partial V}{\partial s}\, ds = dV \qquad \text{(the change in velocity along } s\text{)}$$

Then, after multiplying Eq. 6.7 by ds, we can write

$$-\frac{dp}{\rho} - g\,dz = V\,dV \qquad \text{(along } s\text{)}$$

or

$$\frac{dp}{\rho} + g\,dz + V\,dV = 0 \qquad \text{(along } s\text{)}$$

Integration of this equation gives

$$\int \frac{dp}{\rho} + gz + \frac{V^2}{2} = \text{constant} \qquad \text{(along } s\text{)} \tag{6.8}$$

Before Eq. 6.8 can be used, we must specify the relation between the pressure, p, and the density, ρ. For the special case of incompressible flow, $\rho = \text{constant}$, and Eq. 6.8 becomes the Bernoulli equation,

---

$$\frac{p}{\rho} + gz + \frac{V^2}{2} = \text{constant} \qquad \text{(along a streamline)} \tag{6.9}$$

Restrictions: (1) Steady flow
(2) Incompressible flow
(3) Frictionless flow
(4) Flow along a streamline

The Bernoulli equation is a powerful and useful equation because it relates pressure changes to velocity and elevation changes along a streamline. However, it gives correct results only when applied to a flow situation where all four of the restrictions are reasonable. Therefore, keep the restrictions firmly in mind when you consider using the Bernoulli equation. (In general, the Bernoulli constant in Eq. 6.9 has different values along different streamlines.[4])

**6-3.2 Derivation Using Rectangular Coordinates

The vector form of Euler's equation, Eq. 6.3, also can be integrated along a streamline. Since the velocity field, $\vec{V}$, is specified in terms of the rectangular coordinates, x, y, z, it is convenient to use vector notation. We shall restrict the derivation to steady flow; thus, the end result of our effort should be Eq. 6.8.

For steady flow, Euler's equation in rectangular coordinates becomes

$$-\frac{1}{\rho}\nabla p - g\,\nabla z = \frac{D\vec{V}}{Dt} = u\frac{\partial \vec{V}}{\partial x} + v\frac{\partial \vec{V}}{\partial y} + w\frac{\partial \vec{V}}{\partial z}$$

[4] For the case of irrotational flow, the constant has a single value throughout the entire flow field (Section 6-7.1).

** This section may be omitted without loss of continuity in the text material.

Using vector notation, we obtain

$$u \frac{\partial \vec{V}}{\partial x} + v \frac{\partial \vec{V}}{\partial y} + w \frac{\partial \vec{V}}{\partial z} = (\vec{V} \cdot \nabla) \vec{V} \tag{6.10}$$

(We suggest that you check this equality by expanding the right side of Eq. 6.10 using the familiar dot product operation.) Then, Euler's equation can be expressed as

$$-\frac{1}{\rho} \nabla p - g \nabla z = (\vec{V} \cdot \nabla) \vec{V} \tag{6.11}$$

For steady flow the velocity field is given by $\vec{V} = \vec{V}(x, y, z)$. The streamlines are lines drawn in the flow field tangent to the velocity vector at every point. Recall again that for steady flow, streamlines, pathlines, and streaklines coincide. The motion of a particle along a streamline is governed by Eq. 6.11. In the time increment, dt, the particle moves a distance, $d\vec{s}$, along the streamline.

If we take the dot product of the terms in Eq. 6.11 with the distance, $d\vec{s}$, along the streamline, we obtain a scalar equation relating the pressure, p, speed, V, and elevation, z, along the streamline. Taking the dot product of $d\vec{s}$ with Eq. 6.11 gives

$$-\frac{1}{\rho} \nabla p \cdot d\vec{s} - g \nabla z \cdot d\vec{s} = (\vec{V} \cdot \nabla) \vec{V} \cdot d\vec{s} \tag{6.12}$$

where

$$d\vec{s} = dx\hat{i} + dy\hat{j} + dz\hat{k} \qquad \text{(along } s\text{)}$$

Now we shall evaluate each of the three terms in Eq. 6.12.

$$-\frac{1}{\rho} \nabla p \cdot d\vec{s} = -\frac{1}{\rho} \left[\hat{i} \frac{\partial p}{\partial x} + \hat{j} \frac{\partial p}{\partial y} + \hat{k} \frac{\partial p}{\partial z} \right] \cdot [dx\hat{i} + dy\hat{j} + dz\hat{k}]$$

$$= -\frac{1}{\rho} \left[\frac{\partial p}{\partial x} dx + \frac{\partial p}{\partial y} dy + \frac{\partial p}{\partial z} dz \right] \qquad \text{(along } s\text{)}$$

$$-\frac{1}{\rho} \nabla p \cdot d\vec{s} = -\frac{1}{\rho} dp \qquad \text{(along } s\text{)}$$

and

$$-g \nabla z \cdot d\vec{s} = -g\hat{k} \cdot [dx\hat{i} + dy\hat{j} + dz\hat{k}]$$

$$-g \nabla z \cdot d\vec{s} = -g\, dz \qquad \text{(along } s\text{)}$$

Using a vector identity,[5] we can write the third term as

$$(\vec{V} \cdot \nabla) \vec{V} \cdot d\vec{s} = [\tfrac{1}{2} \nabla (\vec{V} \cdot \vec{V}) - \vec{V} \times (\nabla \times \vec{V})] \cdot d\vec{s}$$
$$= \{\tfrac{1}{2} \nabla (\vec{V} \cdot \vec{V})\} \cdot d\vec{s} - \{\vec{V} \times (\nabla \times \vec{V})\} \cdot d\vec{s}$$

[5] The vector identity

$$(\vec{V} \cdot \nabla) \vec{V} = \tfrac{1}{2} \nabla (\vec{V} \cdot \vec{V}) - \vec{V} \times (\nabla \times \vec{V})$$

may be verified by expanding each side into components.

The last term on the right side of this equation is zero, since $\vec{V}$ is parallel to $d\vec{s}$. Consequently,

$$(\vec{V} \cdot \nabla)\vec{V} \cdot d\vec{s} = \tfrac{1}{2}\nabla(\vec{V} \cdot \vec{V}) \cdot d\vec{s} \qquad \text{(along } s\text{)}$$

$$= \frac{1}{2}\left[\hat{i}\frac{\partial V^2}{\partial x} + \hat{j}\frac{\partial V^2}{\partial y} + \hat{k}\frac{\partial V^2}{\partial z}\right] \cdot [dx\hat{i} + dy\hat{j} + dz\hat{k}]$$

$$= \frac{1}{2}\left[\frac{\partial V^2}{\partial x}dx + \frac{\partial V^2}{\partial y}dy + \frac{\partial V^2}{\partial z}dz\right]$$

$$(\vec{V} \cdot \nabla)\vec{V} \cdot d\vec{s} = \tfrac{1}{2}\, d(V^2) \qquad \text{(along } s\text{)}$$

Substituting these three terms into Eq. 6.12 yields

$$\frac{dp}{\rho} + g\,dz + \frac{1}{2}d(V^2) = 0 \qquad \text{(along } s\text{)}$$

Integrating this equation, we obtain

$$\int \frac{dp}{\rho} + gz + \frac{V^2}{2} = \text{constant} \qquad \text{(along } s\text{)}$$

If the density is constant, we obtain the Bernoulli equation

$$\frac{p}{\rho} + gz + \frac{V^2}{2} = \text{constant} \qquad \text{(along a streamline)}$$

As expected, we see that the last two equations are identical to Eqs. 6.8 and 6.9 derived previously using streamline coordinates. The Bernoulli equation derived using rectangular coordinates is still limited by the restrictions:

1. Steady flow
2. Incompressible flow
3. Frictionless flow
4. Flow along a streamline

6-3.3 Applications

The Bernoulli equation can be applied between any two points on a streamline provided that the other three restrictions are satisfied. The result is

$$\frac{p_1}{\rho} + \frac{V_1^2}{2} + gz_1 = \frac{p_2}{\rho} + \frac{V_2^2}{2} + gz_2 \qquad (6.13)$$

where subscripts 1 and 2 represent any two points on a streamline. Applications of Eqs. 6.9 and 6.13 to typical flow situations are illustrated in Example Problems 6.2 through 6.4.

In some problems, the flow appears unsteady from one reference frame, but steady from another, which translates in the flow. Since the Bernoulli equation was derived by integrating Newton's second law for a fluid particle, it can be applied in any inertial reference frame (see the discussion of translating frames in Section 4-4.2). The procedure is illustrated in Example Problem 6.5.

Example 6.2

Air flows steadily and at low speed through a horizontal nozzle, discharging to the atmosphere. At the nozzle inlet, the area is 0.1 m^2. At the nozzle exit, the area is 0.02 m^2. The flow is essentially incompressible, and frictional effects are negligible. Determine the gage pressure required at the nozzle inlet to produce an outlet speed of 50 m/sec.

EXAMPLE PROBLEM 6.2

GIVEN: Flow through a nozzle, as shown. The air flow is steady, incompressible, and frictionless.

FIND: $p_1 - p_{atm}$

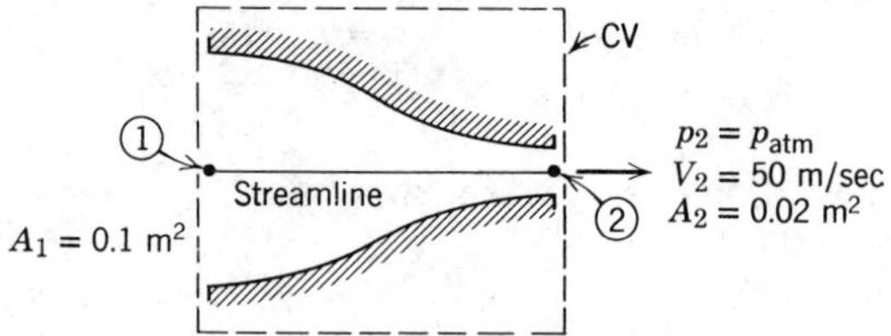

SOLUTION:

Basic equations:

$$\frac{p_1}{\rho} + \frac{V_1^2}{2} + gz_1 = \frac{p_2}{\rho} + \frac{V_2^2}{2} + gz_2$$

$$= 0(1)$$

$$0 = \frac{\partial}{\partial t}\int_{CV} \rho\, d\forall + \int_{CS} \rho \vec{V} \cdot d\vec{A}$$

Assumptions:
(1) Steady flow
(2) Incompressible flow
(3) Frictionless flow
(4) Flow along a streamline
(5) $z_1 = z_2$
(6) Uniform flow at sections ① and ②

At the exit, $M_2 = V_2/c = 50\ \text{m/sec}/340\ \text{m/sec} = 0.147$. This is less than 0.3, so the flow may be treated as incompressible.

Apply the Bernoulli equation along a streamline between points ① and ② to evaluate p_1. Then

$$p_1 - p_{atm} = p_1 - p_2 = \frac{\rho}{2}(V_2^2 - V_1^2)$$

Apply the continuity equation to determine V_1,

$$0 = \{-|\rho V_1 A_1|\} + \{|\rho V_2 A_2|\} \qquad \text{or} \qquad V_1 A_1 = V_2 A_2$$

so that

$$V_1 = V_2 \frac{A_2}{A_1} = \frac{50\ \text{m}}{\text{sec}} \times \frac{0.02\ \text{m}^2}{0.1\ \text{m}^2} = 10\ \text{m/sec}$$

For air at standard conditions, $\rho = 1.23\ \text{kg/m}^3$.

$$p_1 - p_{\text{atm}} = \frac{\rho}{2}(V_2^2 - V_1^2)$$

$$= \frac{1}{2} \times \frac{1.23\ \text{kg}}{\text{m}^3}\left[\frac{(50)^2\ \text{m}^2}{\text{sec}^2} - \frac{(10)^2\ \text{m}^2}{\text{sec}^2}\right]\frac{\text{N} \cdot \text{sec}^2}{\text{kg} \cdot \text{m}}$$

$$p_1 - p_{\text{atm}} = 1.48\ \text{kPa} \qquad \longleftarrow \qquad p_1 - p_{\text{atm}}$$

{This problem illustrates a typical application of the Bernoulli equation. Note that if the flow streamlines are straight at the nozzle inlet and exit, the pressure will be uniform at those sections.}

Example 6.3

A U tube acts as a water siphon. The bend in the tube is 1 m above the water surface; the tube outlet is 7 m below the water surface. If the flow is frictionless as a first approximation, and the fluid issues from the bottom of the siphon as a free jet at atmospheric pressure, determine (after listing the necessary assumptions) the velocity of the free jet and the absolute pressure of the fluid in the flow at the bend.

EXAMPLE PROBLEM 6.3

GIVEN: Water flowing through a siphon as shown.

FIND: (a) Velocity of water leaving as a free jet.
(b) Pressure at point Ⓐ in the flow.

SOLUTION:

Basic equation: $$\frac{p}{\rho} + \frac{V^2}{2} + gz = \text{constant}$$

Assumptions:
(1) Neglect friction
(2) Steady flow
(3) Incompressible flow
(4) Flow along a streamline
(5) Reservoir is large compared to pipe

Apply the Bernoulli equation between points ① and ②.

$$\frac{p_1}{\rho} + \frac{V_1^2}{2} + gz_1 = \frac{p_2}{\rho} + \frac{V_2^2}{2} + gz_2$$

Since $\text{area}_{\text{reservoir}} \gg \text{area}_{\text{pipe}}$, $V_1 \approx 0$. Also $p_1 = p_2 = p_{\text{atm}}$, so

$$gz_1 = \frac{V_2^2}{2} + gz_2 \qquad \text{and} \qquad V_2^2 = 2g(z_1 - z_2)$$

$$V_2 = \sqrt{2g(z_1 - z_2)} = \sqrt{2 \times \frac{9.81\ \text{m}}{\text{sec}^2} \times 7\ \text{m}} = 11.7\ \text{m/sec} \qquad \longleftarrow \qquad V_2$$

To determine the pressure at location Ⓐ, we write the Bernoulli equation between ① and Ⓐ.

$$\frac{p_1}{\rho} + \frac{V_1^2}{2} + gz_1 = \frac{p_A}{\rho} + \frac{V_A^2}{2} + gz_A$$

Again $V_1 \approx 0$ and from conservation of mass $V_A = V_2$. Hence

$$\frac{p_A}{\rho} = \frac{p_1}{\rho} + gz_1 - \frac{V_2^2}{2} - gz_A = \frac{p_1}{\rho} + g(z_1 - z_A) - \frac{V_2^2}{2}$$

$$p_A = p_1 + \rho g(z_1 - z_A) - \rho\frac{V_2^2}{2}$$

$$= \frac{1.01 \times 10^5\ \text{N}}{\text{m}^2} + \frac{999\ \text{kg}}{\text{m}^3} \times \frac{9.81\ \text{m}}{\text{sec}^2} \times (-1\ \text{m}) \times \frac{\text{N} \cdot \text{sec}^2}{\text{kg} \cdot \text{m}}$$

$$-\frac{1}{2} \times \frac{999\ \text{kg}}{\text{m}^3} \times \frac{(11.7)^2\ \text{m}^2}{\text{sec}^2} \times \frac{\text{N} \cdot \text{sec}^2}{\text{kg} \cdot \text{m}}$$

$$p_A = 22.8\ \text{kPa (abs)}$$ p_A

{This problem illustrates a straightforward application of the Bernoulli equation with elevation changes included.}

Example 6.4

Water flows under a sluice gate on a horizontal bed at the inlet to a flume. Above the gate, the water level is 1.5 ft and the velocity is negligible. At the vena contracta below the gate, the flow streamlines are straight and the depth is 2 in. Hydrostatic pressure distributions and uniform flow may be assumed at each section; friction is negligible. Determine the flow velocity downstream from the gate, and the discharge in cubic feet per second per foot of width.

EXAMPLE PROBLEM 6.4

GIVEN: Flow of water under a sluice gate. Flow is frictionless, uniform at each section, and the pressure distribution is hydrostatic at sections ① and ②.

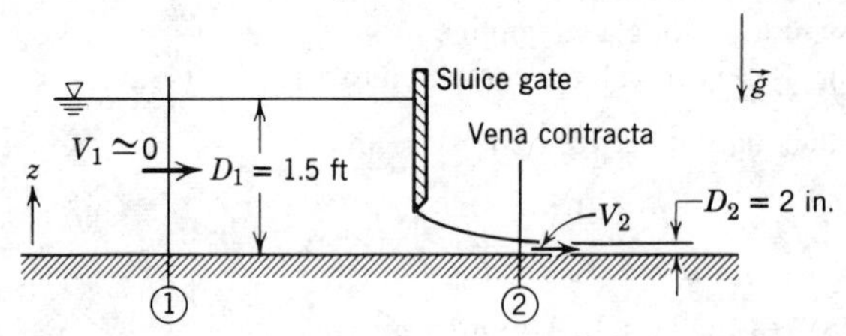

FIND: (a) V_2.
(b) Q in ft^3/sec/ft of width.

SOLUTION:

The flow satisfies all conditions necessary to apply the Bernoulli equation. The question is, what streamline do we use?

Basic equation: $$\frac{p_1}{\rho} + \frac{V_1^2}{2} + gz_1 = \frac{p_2}{\rho} + \frac{V_2^2}{2} + gz_2$$

Assumptions: (1) Steady flow
(2) Incompressible flow
(3) Frictionless flow
(4) Flow along a streamline
(5) Uniform flow at each section
(6) Hydrostatic pressure distribution

From assumption 6,

$$\frac{dp}{dz} = -\gamma \quad \text{so that,} \quad p = p_{atm} + \gamma(D - z) \quad \text{or} \quad \frac{p}{\rho} = \frac{p_{atm}}{\rho} + g(D - z)$$

Substituting this relation into the Bernoulli equation gives

$$\frac{p_{atm}}{\rho} + g(D_1 - z_1) + \frac{V_1^2}{2} + gz_1 = \frac{p_{atm}}{\rho} + g(D_2 - z_2) + \frac{V_2^2}{2} + gz_2$$

or

$$\frac{V_1^2}{2} + gD_1 = \frac{V_2^2}{2} + gD_2$$

This result implies that $V^2/2 + gD =$ constant, and the constant has the same value along *any* streamline for this flow. Solving for V_2 yields

$$V_2 = \sqrt{2g(D_1 - D_2) + V_1^2}$$

But $V_1^2 \approx 0$, so

$$V_2 = \sqrt{2g(D_1 - D_2)} = \sqrt{2 \times \frac{32.2 \text{ ft}}{\text{sec}^2}\left(1.5 \text{ ft} - 2 \text{ in.} \times \frac{\text{ft}}{12 \text{ in.}}\right)}$$

$$V_2 = 9.23 \text{ ft/sec}$$ V_2

For uniform flow, $Q = VA = VDw$, or

$$\frac{Q}{w} = VD = V_2 D_2 = \frac{9.23 \text{ ft}}{\text{sec}} \times 2 \text{ in.} \times \frac{\text{ft}}{12 \text{ in.}} = 1.58 \text{ ft}^2/\text{sec}$$

$$\frac{Q}{w} = 1.58 \text{ ft}^3/\text{sec per foot of width}$$ $\dfrac{Q}{w}$

Example 6.5

A Piper Cub flies at 150 km/hr in standard air at an altitude of 1000 m. At a certain point close to the wing, the air speed *relative* to the wing is 60 m/sec. Compute the pressure at this point.

EXAMPLE PROBLEM 6.5

GIVEN: Aircraft in flight at 150 km/hr at 1000 m altitude in standard air.

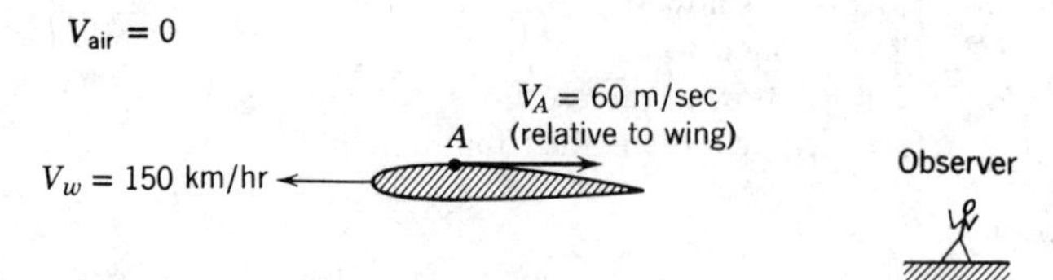

FIND: Pressure, p_A, at point A.

SOLUTION:
Flow is unsteady when observed from a fixed frame, that is, by an observer on the ground. However, an observer *on* the wing sees the following steady flow:

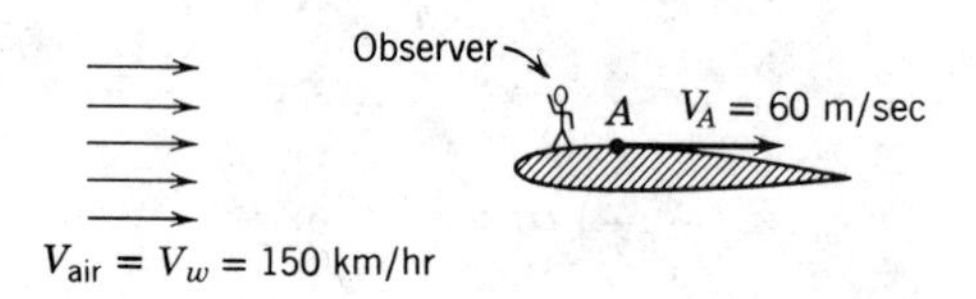

At point A, $M_A = V_A/c = 60$ m/sec/336 m/sec $= 0.178$. This is less than 0.3, so the flow may be treated as incompressible. Thus the Bernoulli equation can be applied along a streamline in the moving observer's inertial reference frame.

Basic equation:
$$\frac{p_{air}}{\rho} + \frac{V_{air}^2}{2} + gz_{air} = \frac{p_A}{\rho} + \frac{V_A^2}{2} + gz_A$$

Assumptions: (1) Steady flow
(2) Incompressible flow ($V < 100$ m/sec)
(3) Frictionless flow
(4) Flow along a streamline
(5) Neglect Δz

Values for pressure and density may be found from Table A.3. Thus, at 1000 m,

$$\frac{p}{p_0} = 0.8870 \qquad \text{and} \qquad \frac{\rho}{\rho_0} = 0.9075$$

Consequently,

$$p = 0.8870 p_0 = 0.8870 \times \frac{1.01 \times 10^5 \text{ N}}{\text{m}^2} = 8.96 \times 10^4 \text{ N/m}^2$$

and

$$\rho = 0.9075 \rho_0 = 0.9075 \times \frac{1.23 \text{ kg}}{\text{m}^3} = 1.12 \text{ kg/m}^3$$

Solving for p_A, we obtain

$$p_A = p_{\text{air}} + \frac{\rho}{2}(V_{\text{air}}^2 - V_A^2)$$

$$= 8.96 \times 10^4 \, \frac{\text{N}}{\text{m}^2}$$

$$+ \frac{1}{2} \times 1.12 \, \frac{\text{kg}}{\text{m}^3} \left[\left(150 \, \frac{\text{km}}{\text{hr}} \times 1000 \, \frac{\text{m}}{\text{km}} \times \frac{\text{hr}}{3600 \, \text{sec}} \right)^2 - (60)^2 \, \frac{\text{m}^2}{\text{sec}^2} \right] \frac{\text{N} \cdot \text{sec}^2}{\text{kg} \cdot \text{m}}$$

$$p_A = 88.6 \text{ kPa (abs)}$$

p_A

6-3.4 Cautions on Use of the Bernoulli Equation

We have seen, in Example Problems 6.2 through 6.5, several situations where the Bernoulli equation may be applied because the restrictions on its use led to a reasonable flow model. However, in some situations you might be tempted to apply the Bernoulli equation where the restrictions are not satisfied. Some subtle cases that violate the restrictions are discussed briefly in this section.

Flow through the nozzle of Example Problem 6.2 was modeled well by the Bernoulli equation. Because the pressure gradient in a nozzle is favorable, there is no separation and the boundary layers on the walls remain thin. Friction has a negligible effect on the flow velocity profile, so one-dimensional flow is a good model. The velocity at any section can be calculated from the corresponding flow area.

A diverging passage or sudden expansion in general cannot be modeled using the Bernoulli equation. Adverse pressure gradients cause rapid growth of boundary layers, severely distorted velocity profiles, and possible flow separation.[6] One-dimensional flow is a poor model for such flows. Because of area blockage due to boundary-layer growth, pressure rise in actual diffusers always is less than that predicted for inviscid one-dimensional flow.

The Bernoulli equation was a reasonable model for the siphon of Example Problem 6.3 because the entrance was well rounded, the bends were gentle, and the overall length was short. Flow separation that can occur at inlets with sharp corners and in abrupt bends[7] causes the flow to depart from that predicted by a one-dimensional model and the Bernoulli equation. Frictional effects would not be negligible if the tube were long.

Example Problem 6.4 presented an open-channel flow situation analogous to a nozzle, for which the Bernoulli equation is a good flow model. The hydraulic jump[8] is an example of an open-channel flow with adverse pressure gradient. Flow through a

[6] See the NCFMF film loops, S-FM015, *Incompressible Flow through Area Contractions and Expansions*, and S-FM049, *Flow Regimes in Subsonic Diffusers* (taken from the film *Flow Visualization*, S. J. Kline, principal).

[7] See the NCFMF film loops, S-FM016, *Flow from a Reservoir to a Duct*, and S-FM067, *Flow through Right Angle Bends*, for examples of this flow behavior.

[8] See the NCFMF film loop, S-FM142, *The Hydraulic Jump*, for examples of this behavior.

hydraulic jump is mixed violently, making it impossible to identify streamlines. Thus the Bernoulli equation cannot be used to model flow through a hydraulic jump.

The Bernoulli equation cannot be applied *through* a machine such as a propeller, pump, or windmill. The equation was derived by integrating along a streamtube (Section 4-4.1) or a streamline (Section 6-3) in the absence of moving surfaces such as blades or vanes. It is impossible to have locally steady flow or to identify streamlines during flow through a machine. As suggested in Problem 4.125, the Bernoulli equation may be applied before and following a machine if the restrictions on its use are satisfied. However, it may not be applied through a machine or across an impeller.[9]

Finally, compressibility must be considered for flow of gases. Density changes caused by dynamic compression due to motion may be neglected for engineering purposes if the local Mach number remains below about $M \approx 0.3$, as noted in Example Problems 6.2 and 6.5. Temperature changes can cause significant changes in density of a gas, even for low-speed flow. Thus the Bernoulli equation could not be applied to air flow through a heating element (e.g. of a hand-held hair dryer) where significant changes in temperature occur.

6-4 STATIC, STAGNATION, AND DYNAMIC PRESSURES

The pressure, p, which we have used in deriving the Bernoulli equation, Eq. 6.9, is the thermodynamic pressure, which is commonly called the static pressure. The static pressure is that pressure which would be measured by an instrument moving with the flow. However, such a measurement is rather difficult to make in a practical situation! How do we measure static pressure experimentally?

In Section 6-2 we showed that there was no pressure variation normal to flow streamlines when those streamlines were straight. This fact makes it possible to measure the static pressure in a flowing fluid using a wall pressure "tap," placed in a region where the flow streamlines are straight, as shown in Fig. 6.2*a*. The pressure tap is a small hole, drilled carefully in the wall, with its axis perpendicular to the surface. If the

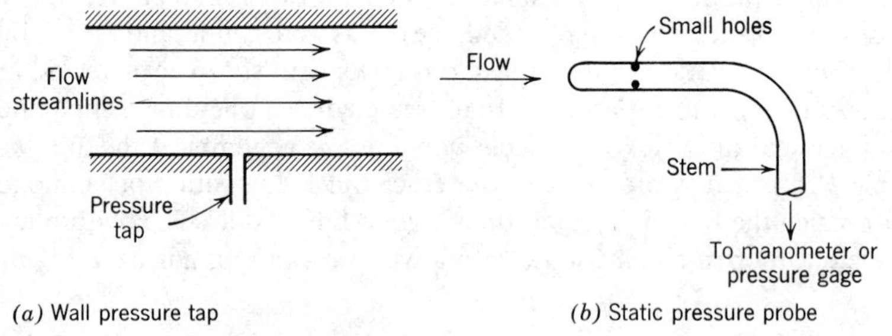

Fig. 6.2 Measurement of static pressure.

[9] See the NCFMF film loop, S-FM097, *Flow through Fans and Propellers*, for the corresponding flow patterns.

hole is perpendicular to the duct wall and free from burrs, accurate measurements of static pressure can be made by connecting the tap to a suitable measuring instrument.

In a fluid stream far from a wall, or where streamlines are curved, accurate static pressure measurements can be made by careful use of a static pressure probe, shown in Fig. 6.2*b*. Such probes must be designed so that the measuring holes are placed correctly with respect to the probe tip and stem to avoid erroneous results. In use, the measuring section must be aligned with the local flow direction.

Static pressure probes, such as that shown in Fig. 6.2*b*, and in a variety of other forms, are available commercially in sizes as small as 1.5 mm ($\frac{1}{16}$ in.) in diameter.

The stagnation pressure is that value obtained when a flowing fluid is decelerated to zero velocity by a frictionless process. In incompressible flow, the Bernoulli equation can be used to relate changes in velocity and pressure along a streamline for such a process. Neglecting elevation differences, Eq. 6.9 becomes

$$\frac{p}{\rho} + \frac{V^2}{2} = \text{constant}$$

If the static pressure is p at a point in the flow where the speed is V, then the stagnation pressure, p_0, where the stagnation speed, V_0, is zero, may be computed from

$$\frac{p_0}{\rho} + \underbrace{\frac{V_0^2}{2}}_{=0} = \frac{p}{\rho} + \frac{V^2}{2}$$

or

$$p_0 = p + \frac{1}{2}\rho V^2 \tag{6.14}$$

Equation 6.14 is a mathematical statement of the definition of stagnation pressure, valid for incompressible flow. The term $\frac{1}{2}\rho V^2$ generally is called the dynamic pressure. Solving for the dynamic pressure gives

$$\frac{1}{2}\rho V^2 = p_0 - p$$

and for the speed

$$V = \sqrt{\frac{2(p_0 - p)}{\rho}} \tag{6.15}$$

Thus, if the stagnation pressure and the static pressure could be measured at a point, Eq. 6.15 would give the local flow speed.

Stagnation pressure is measured in the laboratory using a probe with a hole that faces directly upstream as shown in Fig. 6.3. Such a probe is called a stagnation pressure probe, or pitot tube. Again, the measuring section should be aligned with the local flow direction.

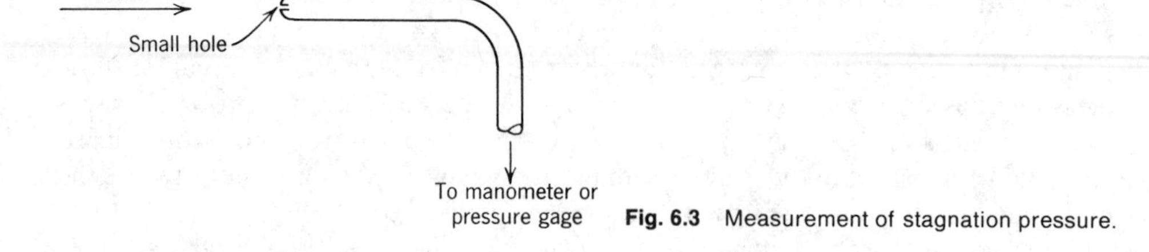

Fig. 6.3 Measurement of stagnation pressure.

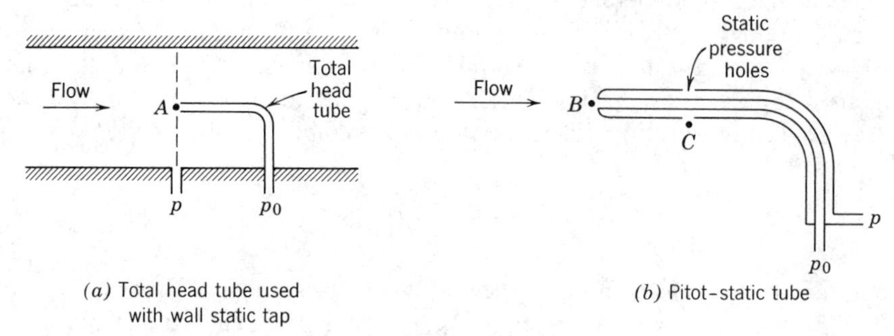

Fig. 6.4 Simultaneous measurement of stagnation and static pressure.

We have seen that static pressure at a point can be measured with a static pressure tap or probe (Fig. 6.2). If we knew the stagnation pressure at the same point, then the flow speed could be computed from Eq. 6.15. Two possible experimental setups are shown in Fig. 6.4.

In Fig. 6.4*a*, the static pressure corresponding to point A is read from the wall static pressure tap. The stagnation pressure is measured directly at A by the total head tube, as shown. (The stem of the total head tube is placed downstream from the measurement location to minimize disturbance of the local flow.)

Two probes often are combined, as in the pitot-static tube shown in Fig. 6.4*b*. The inner tube is used to measure the stagnation pressure at point B, while the static pressure at C is sensed by the small holes in the outer tube. In flow fields where the static pressure variation in the streamwise direction is small, the pitot-static tube may be used to infer the speed at point B in the flow, by assuming $p_B = p_C$, and using Eq. 6.15. (Note that when $p_B \neq p_C$, this procedure will give erroneous results.)

Remember that the Bernoulli equation applies only for incompressible flow ($M \leq 0.3$). The definition and calculation of the stagnation pressure for compressible flow will be discussed in Section 11-3.1.

Example 6.6

A pitot probe is inserted in an air flow (at STP) to measure the flow speed. The tube is inserted so that it points upstream into the flow and the pressure sensed by the probe is the stagnation pressure. The static pressure is measured at the same location in the flow, using a wall pressure tap. If the pressure difference is 30 mm of mercury, determine the flow speed.

EXAMPLE PROBLEM 6.6

GIVEN: A pitot tube inserted in a flow as shown. The flowing fluid is air and the manometer fluid is mercury.

FIND: The flow speed.

SOLUTION:

Basic equation: $$\frac{p}{\rho} + \frac{V^2}{2} + gz = \text{constant}$$

Assumptions: (1) Steady flow
(2) Incompressible flow
(3) Flow along a streamline
(4) Frictionless deceleration along stagnation streamline

Writing Bernoulli's equation along the stagnation streamline, taking $\Delta z = 0$, yields

$$\frac{p_0}{\rho} = \frac{p}{\rho} + \frac{V^2}{2}$$

p_0 is the stagnation pressure at the tube opening where the velocity has been reduced, without friction, to zero. Solving for V gives

$$V = \sqrt{\frac{2(p_0 - p)}{\rho_{\text{air}}}}$$

From the diagram,

$$p_0 - p = \rho_{\text{Hg}} gh = \rho_{\text{H}_2\text{O}} gh(\text{SG}_{\text{Hg}})$$

and

$$V = \sqrt{\frac{2\rho_{\text{H}_2\text{O}} gh(\text{SG}_{\text{Hg}})}{\rho_{\text{air}}}}$$

$$= \sqrt{2 \times \frac{1000 \text{ kg}}{\text{m}^3} \times \frac{9.81 \text{ m}}{\text{sec}^2} \times 30 \text{ mm} \times \frac{\text{m}}{1000 \text{ mm}} \times 13.6 \times \frac{\text{m}^3}{1.23 \text{ kg}}}$$

$V = 80.8$ m/sec ← V

At $T = 20$ C, the speed of sound in air is 343 m/sec. Hence $M = 0.236$ and the assumption of incompressible flow is valid.

{This problem illustrates the use of a pitot tube to determine the speed at a point.}

6-5 RELATION BETWEEN THE FIRST LAW OF THERMODYNAMICS AND THE BERNOULLI EQUATION

The Bernoulli equation, Eq. 6.9, was obtained by integrating Euler's equation along a streamline for steady, incompressible, frictionless flow. Thus Eq. 6.9 was derived from the momentum equation for a fluid particle.

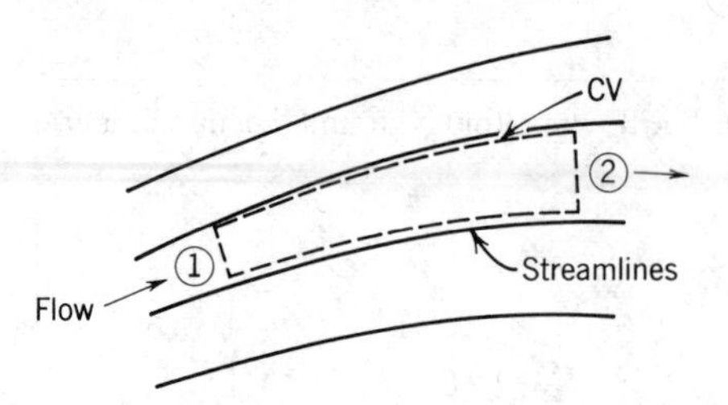

Fig. 6.5 Flow through a stream tube.

An equation identical in form to Eq. 6.9 (although requiring very different restrictions) may be obtained from the first law of thermodynamics. Our objective in this section is to reduce the energy equation to the form of the Bernoulli equation given by Eq. 6.9. Having arrived at this form, we shall compare the restrictions on the two equations. This procedure will help us to understand more clearly the restrictions on the use of Eq. 6.9.

Consider steady flow in the absence of shear forces. We choose a control volume bounded by streamlines along its periphery. Such a control volume, shown in Fig. 6.5, often is called a *stream tube*.

Basic equation:

$$\dot{Q} + \underbrace{\dot{W}_s}_{=0(1)} + \underbrace{\dot{W}_{\text{shear}}}_{=0(2)} + \underbrace{\dot{W}_{\text{other}}}_{=0(3)} = \underbrace{\frac{\partial}{\partial t}\int_{\text{CV}} e\rho\, d\forall}_{=0(4)} + \int_{\text{CS}} (e + pv)\rho \vec{V} \cdot d\vec{A} \tag{4.59}$$

$$e = u + \frac{V^2}{2} + gz$$

Restrictions: (1) $\dot{W}_s = 0$
(2) $\dot{W}_{\text{shear}} = 0$
(3) $\dot{W}_{\text{other}} = 0$
(4) Steady flow
(5) Uniform flow and properties at each section

Under these restrictions, Eq. 4.59 becomes

$$0 = \left(u_1 + p_1 v_1 + \frac{V_1^2}{2} + gz_1\right)\{-|\rho_1 V_1 A_1|\}$$

$$+ \left(u_2 + p_2 v_2 + \frac{V_2^2}{2} + gz_2\right)\{|\rho_2 V_2 A_2|\} - \dot{Q}$$

But from continuity under these restrictions,

$$0 = \underbrace{\frac{\partial}{\partial t}\int_{\text{CV}} \rho\, d\forall}_{=0(4)} + \int_{\text{CS}} \rho \vec{V} \cdot d\vec{A}$$

or

$$0 = \{-|\rho_1 V_1 A_1|\} + \{|\rho_2 V_2 A_2|\}$$

That is,

$$\dot{m} = \rho_1 V_1 A_1 = \rho_2 V_2 A_2$$

Also

$$\dot{Q} = \frac{\delta Q}{dt} = \frac{\delta Q}{dm}\frac{dm}{dt} = \frac{\delta Q}{dm}\dot{m}$$

Thus, from the energy equation,

$$0 = \left[\left(p_2 v_2 + \frac{V_2^2}{2} + g z_2\right) - \left(p_1 v_1 + \frac{V_1^2}{2} + g z_1\right)\right]\dot{m} + \left(u_2 - u_1 - \frac{\delta Q}{dm}\right)\dot{m}$$

or

$$p_1 v_1 + \frac{V_1^2}{2} + g z_1 = p_2 v_2 + \frac{V_2^2}{2} + g z_2 + \left(u_2 - u_1 - \frac{\delta Q}{dm}\right)$$

If the last term in this equation were neglected, we would have the desired form. If we apply the additional restrictions,

(6) Incompressible flow, $v_1 = v_2 = 1/\rho = \text{constant}$
(7) $(u_2 - u_1 - \delta Q/dm) = 0$

then the energy equation reduces to

$$\frac{p_1}{\rho} + \frac{V_1^2}{2} + g z_1 = \frac{p_2}{\rho} + \frac{V_2^2}{2} + g z_2$$

or

$$\frac{p}{\rho} + \frac{V^2}{2} + g z = \text{constant} \tag{6.16}$$

Equation 6.16 is identical in form to the Bernoulli equation, Eq. 6.9. The Bernoulli equation was derived from momentum considerations (Newton's second law), and is valid for steady, incompressible, frictionless flow along a streamline. Equation 6.16 was obtained by applying the first law of thermodynamics to a stream tube control volume, subject to restrictions 1 through 7 above. Thus the Bernoulli equation (Eq. 6.9) and the identical form of the energy equation (Eq. 6.16) were developed from entirely different models, coming from entirely different basic concepts, and involving different restrictions.

Note that restriction 7,

$$(u_2 - u_1) - \frac{\delta Q}{dm} = 0$$

was necessary to obtain the Bernoulli equation from the first law of thermodynamics. This restriction can be satisfied if $\delta Q/dm$ is zero (there is no heat transfer to the fluid) and $u_2 = u_1$ (there is no change in the internal thermal energy of the fluid). The restriction also is satisfied if $(u_2 - u_1)$ and $\delta Q/dm$ are nonzero provided that the values of the two terms are equal. That this is true for incompressible frictionless flow is shown in Example Problem 6.7.

For the special case considered in this section it is true that the first law of thermodynamics reduces to the Bernoulli equation. Since the Bernoulli equation is obtained by integrating Euler's equation (differential form of Newton's second law) for steady, incompressible, frictionless flow along a streamline, for this special case the first law of thermodynamics and Newton's second law do not yield separate information. However, in general, the first law of thermodynamics and Newton's second law are independent equations that must be satisfied separately.

Example 6.7

Consider frictionless, incompressible flow with heat transfer. Show that

$$u_2 - u_1 = \frac{\delta Q}{dm}$$

EXAMPLE PROBLEM 6.7

GIVEN: Frictionless, incompressible flow with heat transfer.

SHOW: $u_2 - u_1 = \dfrac{\delta Q}{dm}$.

SOLUTION:

In general, the internal energy, u, can be expressed as $u = u(T, v)$. For incompressible flow, $v =$ constant, and $u = u(T)$. Thus the thermodynamic state of the fluid is determined by the single thermodynamic property, T. The internal energy change for any process, $u_2 - u_1$, depends only on the temperatures at the end states.

From the Gibbs equation, $T\,ds = du + p\,dv$, valid for a pure substance undergoing any process, we obtain

$$T\,ds = du$$

for incompressible flow, since $dv = 0$. Since the internal energy change, du, between specified end states, is independent of the process, we take a reversible process for which $T\,ds = d(\delta Q/dm) = du$. Therefore,

$$u_2 - u_1 = \frac{\delta Q}{dm}$$

Example 6.8

Water flows steadily from a large open reservoir through a short length of pipe with cross-sectional area, $A = 0.864$ in.^{2} A well-insulated 10 kW heater surrounds the pipe. The flow is assumed to be steady, frictionless, and incompressible. Find the temperature rise of the fluid.

EXAMPLE PROBLEM 6.8

GIVEN: Water flows from a large reservoir through the system shown and discharges to atmospheric pressure.
The heater is 10 kW.
$A_4 = 0.864 \text{ in.}^2$

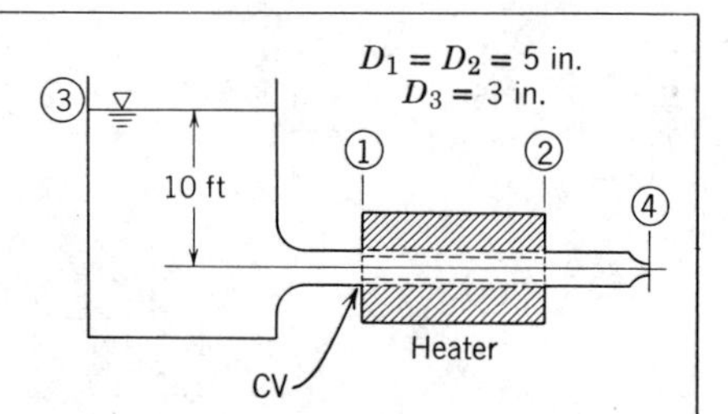

FIND: The temperature rise of the fluid between points ① and ②.

SOLUTION:

Basic equations:

$$\frac{p}{\rho} + \frac{V^2}{2} + gz = \text{constant}$$

$$0 = \cancelto{0(1)}{\frac{\partial}{\partial t}\int_{\text{CV}} \rho \, d\forall} + \int_{\text{CS}} \rho \vec{V} \cdot d\vec{A}$$

$$\dot{Q} + \cancelto{0(4)}{\dot{W}_s} + \cancelto{0(4)}{\dot{W}_{\text{shear}}} = \cancelto{0(1)}{\frac{\partial}{\partial t}\int_{\text{CV}} e\rho \, d\forall} + \int_{\text{CS}} \left(u + pv + \frac{V^2}{2} + gz\right)\rho \vec{V} \cdot d\vec{A}$$

Assumptions:
(1) Steady flow
(2) Frictionless flow
(3) Incompressible flow
(4) No shaft work, no shear work
(5) Flow along a streamline

Under the assumptions listed, the first law of thermodynamics for the CV shown becomes

$$\dot{Q} = \int_{\text{CS}} \left(u + pv + \frac{V^2}{2} + gz\right)\rho \vec{V} \cdot d\vec{A}$$

$$= \int_{A_1} \left(u + pv + \frac{V^2}{2} + gz\right)\rho \vec{V} \cdot d\vec{A} + \int_{A_2} \left(u + pv + \frac{V^2}{2} + gz\right)\rho \vec{V} \cdot d\vec{A}$$

For uniform properties at ① and ②

$$\dot{Q} = -|\rho A_1 V_1|\left(u_1 + p_1 v + \frac{V_1^2}{2} + gz_1\right) + |\rho V_2 A_2|\left(u_2 + p_2 v + \frac{V_2^2}{2} + gz_2\right)$$

From conservation of mass $|\rho A_1 V_1| = |\rho A_2 V_2| = \dot{m}$

$$\dot{Q} = \dot{m}\left[u_2 - u_1 + \left(\frac{p_2}{\rho} + \frac{V_2^2}{2} + gz_2\right) - \left(\frac{p_1}{\rho} + \frac{V_1^2}{2} + gz_1\right)\right]$$

For frictionless, incompressible, steady flow, along a streamline,

$$\frac{p}{\rho} + \frac{V^2}{2} + gz = \text{constant}$$

Therefore,

$$\dot{Q} = \dot{m}(u_2 - u_1)$$

Since for an incompressible fluid, $u_2 - u_1 = c(T_2 - T_1)$, then

$$T_2 - T_1 = \frac{\dot{Q}}{\dot{m}c}$$

From continuity,

$$\dot{m} = \rho A_4 V_4$$

To find V_4, write the Bernoulli equation between the free surface, ③, and ④.

$$\frac{p_3}{\rho} + \frac{V_3^2}{2} + gz_3 = \frac{p_4}{\rho} + \frac{V_4^2}{2} + gz_4$$

Since $p_3 = p_4$ and $V_3 \approx 0$, then

$$V_4 = \sqrt{2g(z_3 - z_4)} = \sqrt{2 \times \frac{32.2\ \text{ft}}{\text{sec}^2} \times 10\ \text{ft}} = 25.4\ \text{ft/sec}$$

and

$$\dot{m} = \rho A_4 V_4 = \frac{1.94\ \text{slug}}{\text{ft}^3} \times 0.864\ \text{in.}^2 \times \frac{\text{ft}^2}{144\ \text{in.}^2} \times \frac{25.4\ \text{ft}}{\text{sec}}$$

$$\dot{m} = 0.296\ \text{slug/sec}$$

Assuming no heat loss to the surroundings, we obtain

$$T_2 - T_1 = \frac{\dot{Q}}{\dot{m}c} = 10\ \text{kW} \times \frac{3413\ \text{Btu}}{\text{kW} \cdot \text{hr}} \times \frac{\text{hr}}{3600\ \text{sec}} \times \frac{\text{sec}}{0.296\ \text{slug}} \times \frac{\text{slug}}{32.2\ \text{lbm}} \times \frac{\text{lbm} \cdot \text{R}}{1\ \text{Btu}}$$

$$T_2 - T_1 = 0.995\ \text{R} \qquad\qquad T_2 - T_1$$

{This problem illustrates that in general the first law of thermodynamics and the Bernoulli equation are independent equations.}

Often it is desirable to represent the mechanical energy level of a flow graphically. The energy equation in the form of Eq. 6.16 suggests such a representation. Dividing Eq. 6.16 by g, we obtain

$$\frac{p}{\rho g} + \frac{V^2}{2g} + z = C = \text{constant} \tag{6.17}$$

Each term in Eq. 6.17 has dimensions of length, or "head" of flowing fluid. The individual terms are

$\frac{p}{\rho g}$ is the head due to local static pressure

$\frac{V^2}{2g}$ is the head due to local dynamic pressure (kinetic energy per unit mg of flowing fluid).

z is the elevation head

C is the constant total head for the flow

The *energy grade line* (EGL) represents the total head height. As shown by Eq. 6.17, the EGL height remains constant for frictionless flow when no work is done on or by the flowing fluid. Fluid would rise to the EGL height in a total head tube placed in the flow.

The *hydraulic grade line* (HGL) height represents the sum of the elevation and static pressure heads, $z + p/\rho g$. In a static pressure tap attached to the flow conduit, fluid would rise to the HGL height. For open-channel flow, the HGL is at the liquid free surface.

The difference in heights between the EGL and the HGL represents the dynamic (velocity) head, $V^2/2g$. The relationship among the EGL, HGL, and velocity head is illustrated schematically in Fig. 6.6 for frictionless flow from a tank through a pipe with a reducer.

The total head of the flow shown in Fig. 6.6 is given by applying Eq. 6.17 at point ①, the free surface in the large reservoir. There the velocity is negligible and the pressure is atmospheric (zero gage). Thus total head is equal to z_1. This defines the height of the

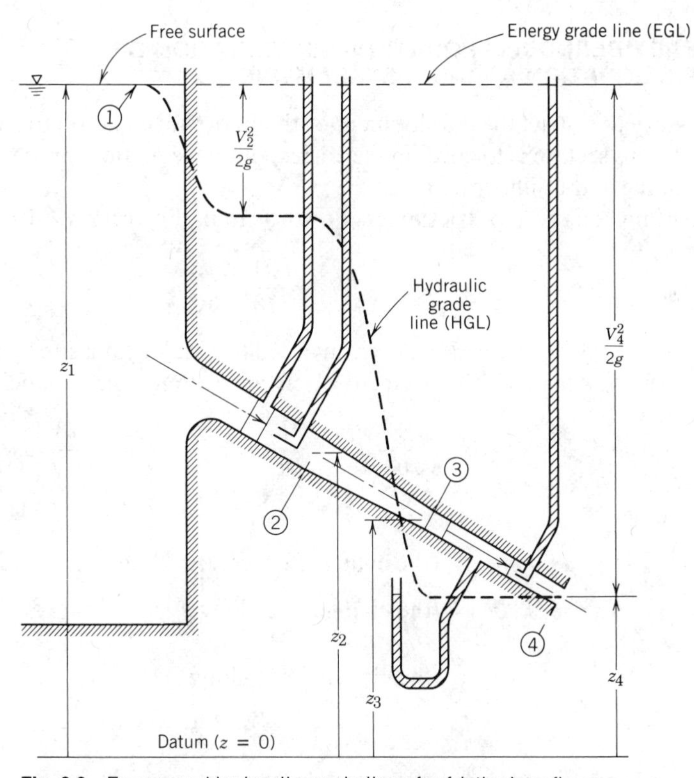

Fig. 6.6 Energy and hydraulic grade lines for frictionless flow.

energy grade line, which remains constant for this flow, since there is no friction or work.

The velocity head increases from zero to $V_2^2/2g$ as the fluid accelerates into the first section of constant-diameter tube. Since the EGL height is constant, the HGL must decrease in height. When the velocity reaches a constant value, then the HGL height stays constant.

The velocity increases again in the reducer between sections ② and ③. As the velocity head increases, the HGL height drops. When the velocity reaches the constant value between sections ③ and ④, the HGL stays constant at a lower height.

At the free discharge at section ④, the static head is zero (gage). There the HGL height is equal to z_4. As shown, the velocity head is $V_4^2/2g$. The sum of the HGL height and velocity head equals the EGL height. (The static head is negative between sections ③ and ④ because the centerline is above the HGL.)

Static taps and total head tubes connected to manometers are shown schematically in Fig. 6.6. The static taps give head readings equal to the HGL height. The total head tubes give readings equal to the EGL height.

The effects of friction and work interactions with a flow will be discussed in detail in Chapter 8. Friction reduces the total head of the flowing fluid, causing a gradual reduction in the EGL height.

**6-6 UNSTEADY BERNOULLI EQUATION—INTEGRATION OF EULER'S EQUATION ALONG A STREAMLINE

It is not necessary to restrict the development of the Bernoulli equation to steady flow. The purpose of this section is to develop the corresponding equation for unsteady flow along a streamline and to illustrate its use.

The momentum equation for frictionless flow was found in Section 6-1 to be

$$-\frac{1}{\rho}\nabla p - g\nabla z = \frac{D\vec{V}}{Dt} \tag{6.3}$$

Equation 6.3 is a vector equation. It can be converted to a scalar equation by taking the dot product with $d\vec{s}$, where $d\vec{s}$ is an element of distance along a streamline. Thus

$$-\frac{1}{\rho}\nabla p \cdot d\vec{s} - g\nabla z \cdot d\vec{s} = \frac{D\vec{V}}{Dt}\cdot d\vec{s} = \frac{DV_s}{Dt}\,ds = V_s\frac{\partial V_s}{\partial s}\,ds + \frac{\partial V_s}{\partial t}\,ds \tag{6.18}$$

The terms become

$$\nabla p \cdot d\vec{s} = dp \qquad \text{(the change in pressure along } s\text{)}$$

$$\nabla z \cdot d\vec{s} = dz \qquad \text{(the change in } z \text{ along } s\text{)}$$

$$\frac{\partial V_s}{\partial s}\,ds = dV_s \qquad \text{(the change in } V_s \text{ along } s\text{)}$$

** This section may be omitted without loss of continuity in the text material.

Substituting into Eq. 6.18, we obtain

$$-\frac{dp}{\rho} - g\,dz = V_s\,dV_s + \frac{\partial V_s}{\partial t}\,ds \tag{6.19}$$

Integrating along a streamline from point 1 to point 2 yields

$$\int_1^2 \frac{dp}{\rho} + \frac{V_2^2 - V_1^2}{2} + g(z_2 - z_1) + \int_1^2 \frac{\partial V_s}{\partial t}\,ds = 0 \tag{6.20}$$

For incompressible flow, the density is constant. For this special case, Eq. 6.20 becomes

---

$$\frac{p_1}{\rho} + \frac{V_1^2}{2} + gz_1 = \frac{p_2}{\rho} + \frac{V_2^2}{2} + gz_2 + \int_1^2 \frac{\partial V_s}{\partial t}\,ds \qquad \text{(along a streamline)} \tag{6.21}$$

Restrictions: (1) Incompressible flow
(2) Frictionless flow
(3) Flow along a streamline

To evaluate the integral term in Eq. 6.21, the variation in $\partial V_s/\partial t$ must be known as a function of s, the distance along the streamline measured from point 1. (For steady flow, $\partial V_s/\partial t = 0$, and Eq. 6.21 reduces to Eq. 6.9.) Equation 6.21 may be applied to any flow in which the restrictions are compatible with the physical situation.

Application of Eq. 6.21 is illustrated in Example Problem 6.9.

Example 6.9

A long pipe is connected to a large reservoir that initially is filled with water to a depth of 3 m. The pipe is 150 mm in diameter and 6 m long. As a first approximation, friction may be neglected. Determine the flow velocity leaving the pipe as a function of time after a cap is removed from its free end. The reservoir is large enough so that the change in its level may be neglected.

EXAMPLE PROBLEM 6.9

GIVEN: Pipe and large reservoir as shown.

FIND: $V_2(t)$.

SOLUTION:

Apply the Bernoulli equation to the unsteady flow along a streamline from point ① to point ②.

Basic equation:

$$\frac{p_1}{\rho} + \underbrace{\frac{V_1^2}{2}}_{\simeq 0(5)} + gz_1 = \frac{p_2}{\rho} + \frac{V_2^2}{2} + \underbrace{gz_2}_{=0(6)} + \int_1^2 \frac{\partial V_s}{\partial t}\,ds$$

Assumptions: (1) Incompressible flow
(2) Frictionless flow
(3) Flow along a streamline from ① to ②
(4) $p_1 = p_2 = p_{atm}$
(5) $V_1^2 \simeq 0$
(6) $z_2 = 0$
(7) $z_1 = h = \text{constant}$
(8) Neglect velocity in reservoir, except for a small region near the inlet to the tube

Then

$$gz_1 = gh = \frac{V_2^2}{2} + \int_1^2 \frac{\partial V_s}{\partial t}\, ds$$

In view of assumption (8), the integral becomes

$$\int_1^2 \frac{\partial V_s}{\partial t}\, ds \approx \int_0^L \frac{\partial V_s}{\partial t}\, ds$$

In the tube, $V_s = V_2$ everywhere, so that

$$\int_0^L \frac{\partial V_s}{\partial t}\, ds = \int_0^L \frac{dV_2}{dt}\, ds = L\frac{dV_2}{dt}$$

Substituting gives

$$gh = \frac{V_2^2}{2} + L\frac{dV_2}{dt}$$

Separating variables, we obtain

$$\frac{dV_2}{2gh - V_2^2} = \frac{dt}{2L}$$

Integrating between limits $V = 0$ at $t = 0$ and $V = V_2$ at $t = t$,

$$\int_0^{V_2} \frac{dV}{2gh - V^2} = \left[\frac{1}{\sqrt{2gh}} \tanh^{-1}\left(\frac{V}{\sqrt{2gh}}\right)\right]_0^{V_2} = \frac{t}{2L}$$

Since $\tanh^{-1}(0) = 0$, we obtain

$$\frac{1}{\sqrt{2gh}} \tanh^{-1}\left(\frac{V_2}{\sqrt{2gh}}\right) = \frac{t}{2L}$$

or

$$\frac{V_2}{\sqrt{2gh}} = \tanh\left(\frac{t}{2L}\sqrt{2gh}\right)$$

$V_2(t)$

For the given conditions,

$$\sqrt{2gh} = \sqrt{2 \times \frac{9.81\ \text{m}}{\text{sec}^2} \times 3\ \text{m}} = 7.67\ \text{m/sec}$$

and

$$\frac{t}{2L}\sqrt{2gh} = \frac{t}{2} \times \frac{1}{6\text{ m}} \times \frac{7.67\text{ m}}{\text{sec}} = 0.639t$$

The results are then $V_2 = 7.67 \tanh(0.639t)$ m/sec.

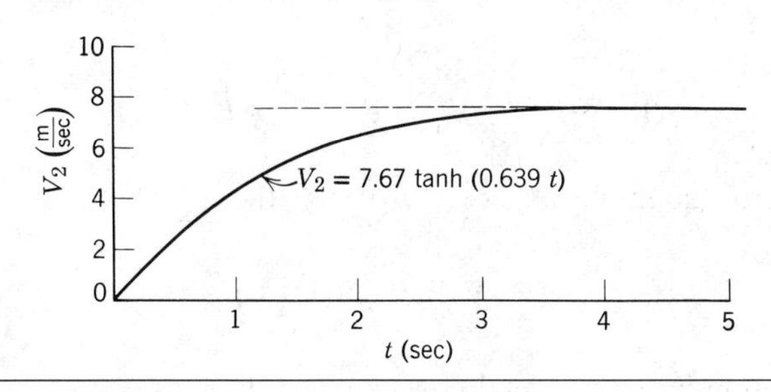

**6-7 IRROTATIONAL FLOW

An *irrotational flow* is one in which fluid elements moving in the flow field do not undergo any rotation. For $\vec{\omega} = 0$, $\nabla \times \vec{V} = 0$ and from Eq. 5.11,

$$\frac{\partial w}{\partial y} - \frac{\partial v}{\partial z} = \frac{\partial u}{\partial z} - \frac{\partial w}{\partial x} = \frac{\partial v}{\partial x} - \frac{\partial u}{\partial y} = 0 \qquad (6.22)$$

In cylindrical coordinates, the irrotationality condition requires that

$$\frac{1}{r}\frac{\partial V_z}{\partial \theta} - \frac{\partial V_\theta}{\partial z} = \frac{\partial V_r}{\partial z} - \frac{\partial V_z}{\partial r} = \frac{1}{r}\frac{\partial r V_\theta}{\partial r} - \frac{1}{r}\frac{\partial V_r}{\partial \theta} = 0 \qquad (6.23)$$

6-7.1 Bernoulli Equation Applied to Irrotational Flow

In Section 6-3.1, we integrated Euler's equation along a streamline for steady, incompressible, inviscid flow to obtain the Bernoulli equation

$$\frac{p}{\rho} + \frac{V^2}{2} + gz = \text{constant} \qquad \text{(along a streamline)} \qquad (6.9)$$

Equation 6.9 can be applied between any two points on the same streamline. The value of the constant will vary, in general, from streamline to streamline.

If, in addition to being inviscid, steady and incompressible, the flow field is also irrotational (the velocity field is such that $2\vec{\omega} = \nabla \times \vec{V} = 0$), we can show that Bernoulli's equation can be applied between any two points in the flow. Then the value

** This section may be omitted without loss of continuity in the text material. (Note that Section 5-2 contains background material needed for study of Sections 6-7.3, 6-7.4, and 6-7.5.)

of the constant in Eq. 6.9 is the same for all streamlines. To illustrate this, we start with Euler's equation in vector form,

$$-\frac{1}{\rho}\nabla p - g\nabla z = (\vec{V}\cdot\nabla)\vec{V} \tag{6.11}$$

Using the vector identity

$$(\vec{V}\cdot\nabla)\vec{V} = \frac{1}{2}\nabla(\vec{V}\cdot\vec{V}) - \vec{V}\times(\nabla\times\vec{V})$$

we see that for irrotational flow, since $\nabla \times \vec{V} = 0$, then

$$(\vec{V}\cdot\nabla)\vec{V} = \frac{1}{2}\nabla(\vec{V}\cdot\vec{V})$$

and Euler's equation for irrotational flow can be written as

$$-\frac{1}{\rho}\nabla p - g\nabla z = \frac{1}{2}\nabla(\vec{V}\cdot\vec{V}) = \frac{1}{2}\nabla(V^2) \tag{6.24}$$

During the interval, dt, a fluid particle moves from the vector position, $\vec{r}$, to the position, $\vec{r} + d\vec{r}$; the displacement, $d\vec{r}$, is an arbitrary infinitesimal displacement in any direction. Taking the dot product of $d\vec{r} = dx\hat{i} + dy\hat{j} + dz\hat{k}$ with each of the terms in Eq. 6.24, we have

$$-\frac{1}{\rho}\nabla p\cdot d\vec{r} - g\nabla z\cdot d\vec{r} = \frac{1}{2}\nabla(V^2)\cdot d\vec{r}$$

and hence

$$-\frac{dp}{\rho} - g\,dz = \frac{1}{2}d(V^2)$$

or

$$\frac{dp}{\rho} + g\,dz + \frac{1}{2}d(V^2) = 0$$

Integrating this equation gives

$$\int\frac{dp}{\rho} + gz + \frac{V^2}{2} = \text{constant} \tag{6.25}$$

For incompressible flow, ρ = constant, and

$$\frac{p}{\rho} + gz + \frac{V^2}{2} = \text{constant} \tag{6.26}$$

Since $d\vec{r}$ was an arbitrary displacement, then for a steady, incompressible, inviscid flow that is also irrotational, Eq. 6.26 is valid between any two points in the flow field.

6-7.2 Velocity Potential

In Section 5-2 we formulated the stream function, ψ, a relation between the streamlines and conservation of mass for two-dimensional, incompressible flow.

We can formulate a relation called the potential function, ϕ, for a velocity field that is irrotational. To do so, we must use the fundamental vector identity[10]

$$\text{curl}(\text{grad}\,\phi) = \nabla \times \nabla\phi = 0 \tag{6.27}$$

which holds if ϕ is any scalar function (of the space coordinates and time) having continuous first and second derivatives.

Then, for an irrotational flow in which $\nabla \times \vec{V} = 0$, a scalar function, ϕ, must exist such that the gradient of ϕ is equal to the velocity vector, $\vec{V}$. In order that the positive direction of flow be in the direction of decreasing ϕ (analogous to the positive direction of heat transfer being defined in the direction of decreasing temperature), we define ϕ such that

$$\vec{V} \equiv -\nabla\phi \tag{6.28}$$

Thus

$$u = -\frac{\partial\phi}{\partial x} \qquad v = -\frac{\partial\phi}{\partial y} \qquad w = -\frac{\partial\phi}{\partial z} \tag{6.29}$$

With the potential function defined in this way, the irrotationality condition, Eq. 6.22, is satisfied identically.

In cylindrical coordinates,

$$\nabla = \hat{i}_r \frac{\partial}{\partial r} + \hat{i}_\theta \frac{1}{r}\frac{\partial}{\partial\theta} + \hat{i}_z \frac{\partial}{\partial z}$$

From Eq. 6.28, then, in cylindrical coordinates

$$V_r = -\frac{\partial\phi}{\partial r} \qquad V_\theta = -\frac{1}{r}\frac{\partial\phi}{\partial\theta} \qquad V_z = -\frac{\partial\phi}{\partial z} \tag{6.30}$$

The velocity potential, ϕ, exists only for an irrotational flow. The stream function, ψ, satisfies the continuity equation for incompressible flow; the stream function is not subject to the restriction of irrotational flow.

The condition of irrotationality may be a valid assumption for those regions of a flow in which viscous forces are negligible.[11] (For example, such a region exists outside the boundary layer in the flow over a solid surface.) The theory for irrotational flow is

[10] The proof of this identity may be found in any book treating vector analysis, or it may be proved by expanding Eq. 6.27 into components.

[11] Examples of rotational and irrotational motion are shown in the film loops: S-FM014A, *Visualization of Vorticity with Vorticity Meter—Part I*; S-FM014B, *Visualization of Vorticity with Vorticity Meter—Part II*.

developed in terms of an imaginary fluid called an ideal fluid whose viscosity is identically zero. Since, in an irrotational flow, the velocity field may be defined by the potential function, ϕ, the theory is often referred to as potential flow theory.

All real fluids possess viscosity, but there are many situations in which the assumption of inviscid flow leads to a considerable simplification in the analysis and, at the same time, gives meaningful results. Because of its utility and mathematical appeal, potential flow has been studied extensively.[12]

6-7.3 Stream Function and Velocity Potential for Two-Dimensional, Irrotational, Incompressible Flow; Laplace's Equation

For a two-dimensional, incompressible, irrotational flow we have expressions for the velocity components, u and v, in terms of both the stream function, ψ, and the velocity potential, ϕ.

$$u = \frac{\partial \psi}{\partial y} \qquad v = -\frac{\partial \psi}{\partial x} \tag{5.4}$$

$$u = -\frac{\partial \phi}{\partial x} \qquad v = -\frac{\partial \phi}{\partial y} \tag{6.29}$$

Substituting for u and v from Eq. 5.4 into the irrotationality condition,

$$\frac{\partial v}{\partial x} - \frac{\partial u}{\partial y} = 0 \tag{6.22}$$

we obtain

$$\frac{\partial^2 \psi}{\partial x^2} + \frac{\partial^2 \psi}{\partial y^2} = 0 \tag{6.31a}$$

Substituting for u and v from Eq. 6.29 into the continuity equation,

$$\frac{\partial u}{\partial x} + \frac{\partial v}{\partial y} = 0 \tag{5.3}$$

we obtain

$$\frac{\partial^2 \phi}{\partial x^2} + \frac{\partial^2 \phi}{\partial y^2} = 0 \tag{6.31b}$$

Equations 6.31a and 6.31b are forms of Laplace's equation—an equation that arises in many areas of the physical sciences and engineering. Any function ψ or ϕ that satisfies Laplace's equation represents a possible two-dimensional, incompressible, irrotational flow field.

[12] Anyone interested in a detailed study of potential flow theory may find the following books of interest: V. L. Streeter, *Fluid Dynamics* (New York: McGraw-Hill, 1948); H. R. Vallentine, *Applied Hydrodynamics* (London: Butterworths, 1959); J. M. Robertson, *Hydrodynamics in Theory and Application* (Englewood Cliffs, N.J.: Prentice-Hall, 1965).

In Section 5-2 we showed that the stream function, ψ, is constant along a streamline. For ψ = constant,

$$d\psi = \frac{\partial \psi}{\partial x}\, dx + \frac{\partial \psi}{\partial y}\, dy$$

The slope of a streamline—a line of constant ψ—is given by

$$\left.\frac{dy}{dx}\right)_{\psi} = -\frac{\partial \psi/\partial x}{\partial \psi/\partial y} = -\frac{-v}{u} = \frac{v}{u} \tag{6.32}$$

Along a line of constant ϕ, $d\phi = 0$, and hence

$$d\phi = \frac{\partial \phi}{\partial x}\, dx + \frac{\partial \phi}{\partial y}\, dy = 0$$

Consequently, the slope of a potential line—a line of constant ϕ—is given by

$$\left.\frac{dy}{dx}\right)_{\phi} = -\frac{\partial \phi/\partial x}{\partial \phi/\partial y} = -\frac{u}{v} \tag{6.33}$$

The last equality of Eq. 6.33 follows from use of Eq. 6.29. Comparing Eqs. 6.32 and 6.33, we see that the slope of a constant ψ line at any point is the negative reciprocal of the slope of the constant ϕ line at that point. Consequently, lines of constant ψ and constant ϕ are orthogonal. This property of potential lines and streamlines is useful in graphical analyses of flow fields.

Example 6.10
Consider the flow field given by $\psi = ax^2 - ay^2$, where $a = 1\ \text{sec}^{-1}$.
(a) Show that the flow is irrotational.
(b) Determine the velocity potential for this flow.

EXAMPLE PROBLEM 6.10

GIVEN: Incompressible flow field with $\psi = ax^2 - ay^2$, where $a = 1\ \text{sec}^{-1}$.

REQUIRED: (a) Show that the flow is irrotational.
(b) Determine the velocity potential for this flow.

SOLUTION:
If the flow is irrotational, then $\omega_z = 0$. Since

$$2\omega_z = \frac{\partial v}{\partial x} - \frac{\partial u}{\partial y} \quad \text{and} \quad u = \frac{\partial \psi}{\partial y}, \quad v = -\frac{\partial \psi}{\partial x},$$

then

$$u = \frac{\partial}{\partial y}(ax^2 - ay^2) = -2ay \quad \text{and} \quad v = -\frac{\partial}{\partial x}(ax^2 - ay^2) = -2ax$$

Thus

$$2\omega_z = \frac{\partial v}{\partial x} - \frac{\partial u}{\partial y} = \frac{\partial}{\partial x}(-2ax) - \frac{\partial}{\partial y}(-2ay) = -2a + 2a = 0$$

Therefore, the flow is irrotational.

The velocity components can be written in terms of the velocity potential as

$$u = -\frac{\partial \phi}{\partial x} \quad \text{and} \quad v = -\frac{\partial \phi}{\partial y}$$

Consequently, $u = -\dfrac{\partial \phi}{\partial x} = -2ay$ and $\dfrac{\partial \phi}{\partial x} = 2ay$

Integrating with respect to x gives $\phi = 2axy + f(y)$, where $f(y)$ is an arbitrary function of y. Then

$$v = -2ax = -\frac{\partial \phi}{\partial y} = -\frac{\partial}{\partial y}[2axy + f(y)]$$

Therefore, $-2ax = -2ax - \dfrac{\partial f(y)}{\partial y} = -2ax - \dfrac{df}{dy}$, so $\dfrac{df}{dy} = 0$ and $f = \text{constant}$.

Therefore, $\phi = 2axy + \text{constant}$ ϕ

We also can show that lines of constant ψ and ϕ are orthogonal.

$$\psi = ax^2 - ay^2$$

For $\psi = \text{constant}$, $d\psi = 0 = 2ax\,dx - 2ay\,dy$, hence $\left.\dfrac{dy}{dx}\right)_{\psi=c} = \dfrac{x}{y}$

For $\phi = \text{constant}$, $d\phi = 0 = 2ay\,dx + 2ax\,dy$, hence $\left.\dfrac{dy}{dx}\right)_{\phi=c} = -\dfrac{y}{x}$

{This problem illustrates the relation among the stream function, the velocity field, and the velocity potential.}

6-7.4 Elementary Plane Flows

A variety of potential flows can be constructed by superposing elementary flow patterns. The ψ and ϕ functions for five elementary two-dimensional flows—a uniform flow, a source, a sink, a vortex, and a doublet—are summarized in Table 6.1. The ψ and ϕ functions can be obtained from the velocity field for each elementary flow.

A *uniform flow* of constant velocity parallel to the x axis satisfies the continuity equation and the irrotationality condition identically. In Table 6.1 we have shown the ψ and ϕ functions for a uniform flow in the positive x direction.

For a uniform flow of constant magnitude, V, inclined at angle, α, to the x axis, then

$$\psi = (V \cos \alpha)\, y - (V \sin \alpha) x$$

$$\phi = -(V \sin \alpha)\, y - (V \cos \alpha) x$$

A simple *source* is a flow pattern in the xy plane from which flow is radially outward from the z axis and symmetrical in all directions. The strength, q, of the source is the volume flow rate per unit depth. At any radius, r, from a source, the tangential velocity, V_θ, is zero; the radial velocity, V_r, is the volume flow rate per unit depth, q, divided by the flow area per unit depth, $2\pi r$. Thus $V_r = q/2\pi r$ for a source. The ψ and ϕ functions for a source are shown in Table 6.1.

In a simple *sink*, flow is radially inward; a sink is a negative source. The ψ and ϕ functions for a sink shown in Table 6.1 are the negatives of the corresponding functions for a source flow.

The origin of either a sink or a source is a singular point, since the radial velocity approaches infinity as the radius approaches zero. Thus, while an actual flow may resemble a source or a sink for some values of r, sources and sinks have no exact physical counterparts. The primary value of the concept of sources and sinks is that, when combined with other elementary flows, they produce flow patterns that adequately represent realistic flows.

A flow pattern in which the streamlines are concentric circles is a vortex; in a *free (irrotational) vortex*, fluid particles do not rotate as they move around the vortex center. The velocity distribution in an irrotational vortex can be determined from Euler's equation and the Bernoulli equation. For irrotational flow, the Bernoulli equation is valid between any two points in the flow field. For flow in a horizontal plane,

$$\frac{1}{\rho}\,dp = -V_\theta\,dV_\theta$$

Euler's equation normal to the streamline is

$$\frac{1}{\rho}\frac{dp}{dr} = \frac{V_\theta^2}{r}$$

Combining these equations yields

$$\frac{dp}{\rho} = -V_\theta\,dV_\theta = \frac{V_\theta^2}{r}\,dr$$

From the last equality

$$V_\theta\,dr + r\,dV_\theta = 0$$

Integrating this equation gives

$$V_\theta r = \text{constant}$$

The strength, K, of the vortex is defined as $K = 2\pi r V_\theta$; the dimensions of K are L^2/t (volume flow rate per unit depth). The irrotational vortex is a reasonable approximation to the flow field in a tornado (except in the region of the origin; the origin is a singular point).

The final "elementary" flow listed in Table 6.1 is the *doublet*. This flow is produced by allowing a source and a sink of numerically equal strengths to approach each other under the condition that as the distance, δs, between them approaches zero, their strengths increase so the product $q\,\delta s/2\pi$ tends to a finite value, Λ, which is termed the strength of the doublet.

Table 6.1 Elementary Plane Flows

Uniform Flow (positive x direction)

$u = U \qquad \psi = Uy$

$v = 0 \qquad \phi = -Ux$

$\Gamma = 0$ around any closed curve

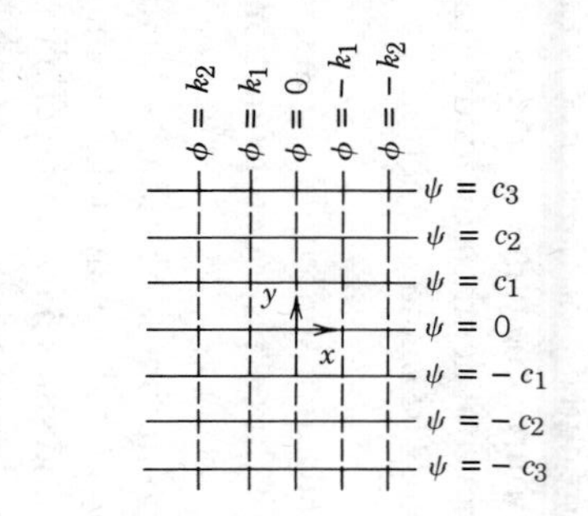

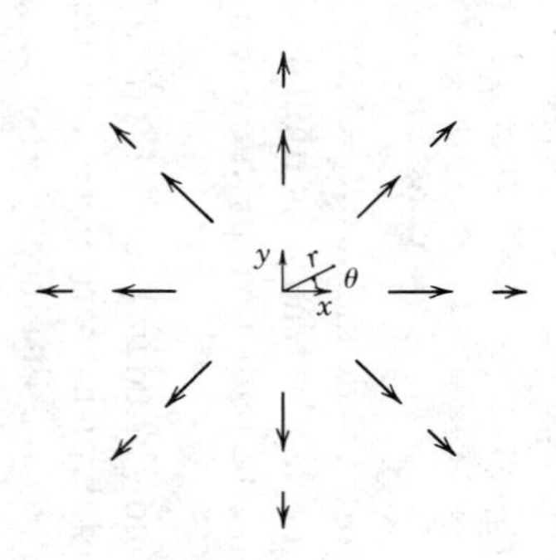

Source Flow (from origin)

$$V_r = \frac{q}{2\pi r} \qquad \psi = \frac{q}{2\pi}\theta$$

$$V_\theta = 0 \qquad \phi = -\frac{q}{2\pi}\ln r$$

Origin is singular point
q is volume flow rate per unit depth
$\Gamma = 0$ around any closed curve

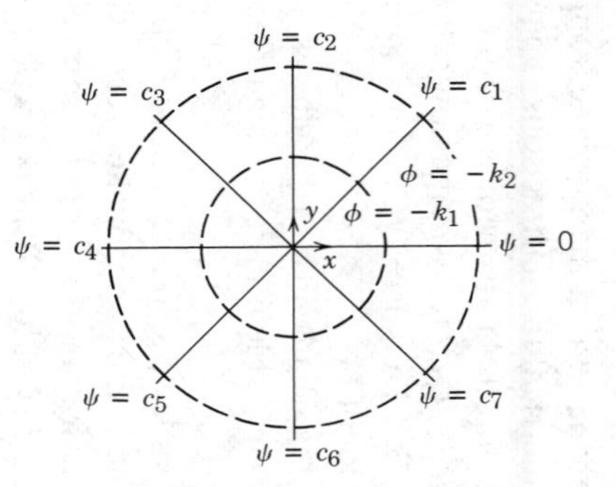

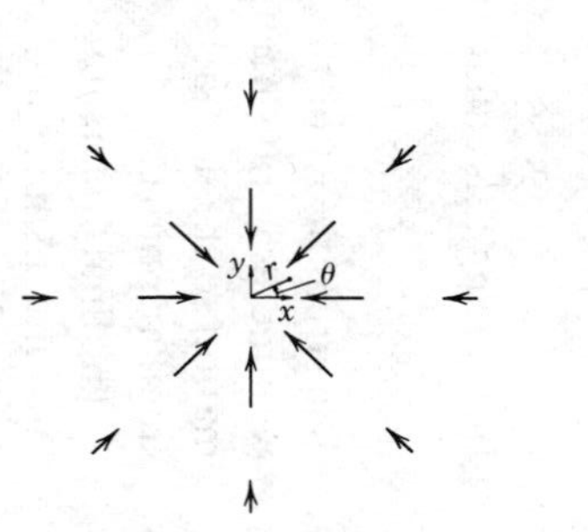

Sink Flow (toward origin)

$$V_r = -\frac{q}{2\pi r} \qquad \psi = -\frac{q}{2\pi}\theta$$

$$V_\theta = 0 \qquad \phi = \frac{q}{2\pi}\ln r$$

Origin is singular point
q is volume flow rate per unit depth
$\Gamma = 0$ around any closed curve

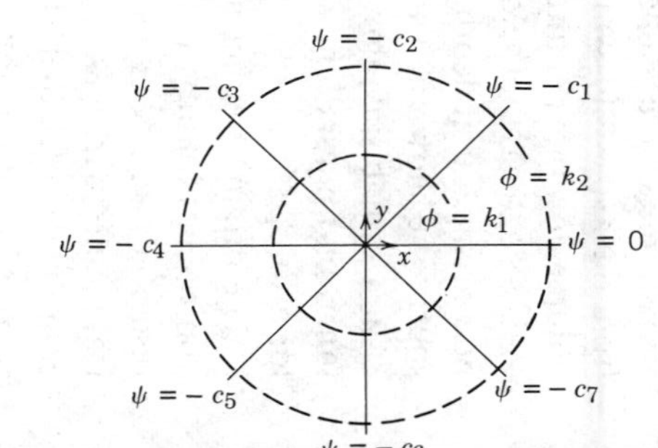

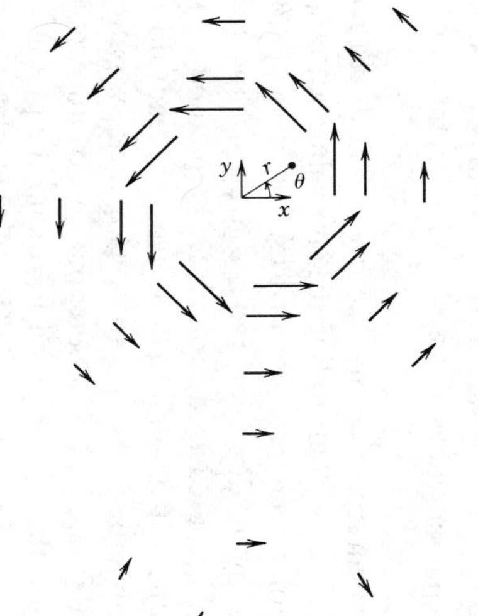

Irrotational Vortex (counterclockwise, center at origin)

$V_r = 0 \qquad \psi = -\frac{K}{2\pi}\ln r$

$V_\theta = \frac{K}{2\pi r} \qquad \phi = -\frac{K}{2\pi}\theta$

Origin is singular point
K is strength of the vortex
$\Gamma = K$ around any closed curve enclosing origin
$\Gamma = 0$ around any closed curve not enclosing origin

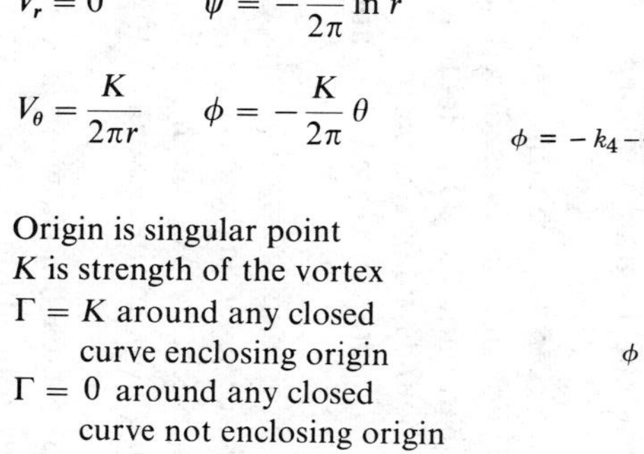

Doublet (center at origin)

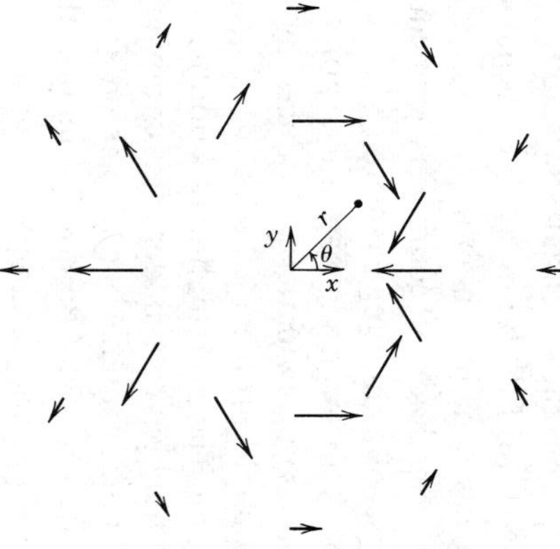

$V_r = -\frac{\Lambda}{r^2}\cos\theta \qquad \psi = -\frac{\Lambda\cos\theta}{r}$

$V_\theta = -\frac{\Lambda}{r^2}\sin\theta \qquad \phi = -\frac{\Lambda\cos\theta}{r}$

Origin is singular point
Λ is strength of the doublet
$\Gamma = 0$ around any closed curve

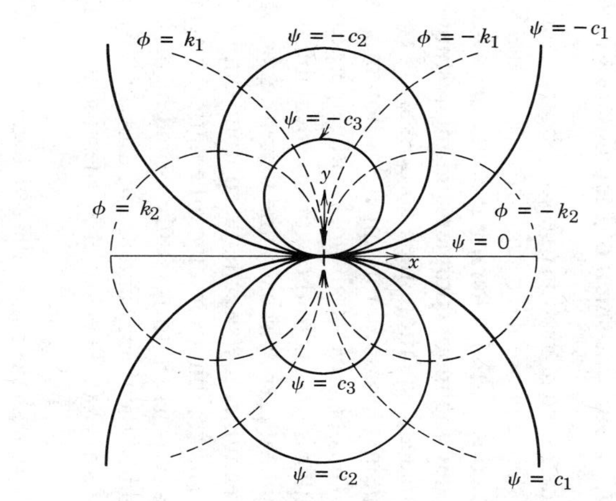

6-7.5 Superposition of Elementary Plane Flows

We showed in Section 6-7.3 that both ϕ and ψ satisfy Laplace's equation for flow that is both incompressible and irrotational. Since Laplace's equation is a linear, homogeneous partial differential equation, solutions may be superposed (added together) to develop more complex and interesting patterns of flow. Thus if ψ_1 and ψ_2 satisfy Laplace's equation, then so does $\psi_3 = \psi_1 + \psi_2$. The elementary plane flows are the building blocks in this superposition process.

The objective of all work with superposition of elementary flows is to produce flow patterns similar to those of practical interest. The ideal flow model postulates an ideal fluid with zero viscosity and therefore zero shear stresses. Since there is no flow across a streamline, any streamline contour can be imagined to represent a solid surface.

The combination of mathematical elegance and utility of potential flow attracted many workers to its study. Some of history's most famous applied mathematicians studied the theory and application of "hydrodynamics," as potential flow was termed before 1900. The list of names includes Bernoulli, Lagrange, d'Alembert, Cauchy, Rankine, and Euler.

Through the end of the nineteenth century, workers in pure hydrodynamics failed to produce results that agreed with experiment. Potential flows produced body shapes with lift but predicted zero drag (the "d'Alembert paradox"). Two influences changed this situation: first Prandtl introduced the boundary-layer concept and began to develop the theory, and second, interest in aeronautics increased dramatically in the early 1900s.

Prandtl showed by mathematical analysis and through elegantly simple experiments that viscous effects are confined to a thin boundary layer on the surface of a body. Even for real fluids, flow outside the boundary layer behaves as though the fluid had zero viscosity. Pressure gradients from the external flow are impressed on the boundary layer.

The most important input to a calculation of the real fluid flow in the boundary layer (and thus the drag on a body) is the pressure distribution. Once the velocity field is known from the potential flow solution, the pressure distribution may be calculated. Knowledge of the boundary layer behavior makes it possible to predict the drag, and in some cases the lift, on an object.

Two methods of combining elementary flows may be used. The direct method consists of combining elementary flows, then calculating the streamline pattern, body shape, velocity field, stagnation point, and pressure distribution directly. Several examples of flow patterns produced by the direct method are given in Table 6.2.

Example Problem 6.11 has been prepared to demonstrate the method of superposition. The combination of a doublet with a uniform flow gives the steady flow of an ideal fluid past a cylinder. This combination was used to develop the flow pattern for ideal flow around a cylinder shown in Fig. 2.12. The potential flow solution gives the velocity field. The Bernoulli equation then can be used to obtain the pressure field. Several additional examples of superposition are included in the problems at the end of the chapter.

Complex flow patterns can be developed by combining elementary flows. Distributed line sources, sinks, and "images" may be used to create bodies of arbitrary

Table 6.2 Superposition of Elementary Plane Flows

Source and Uniform Flow (flow past a half-body)

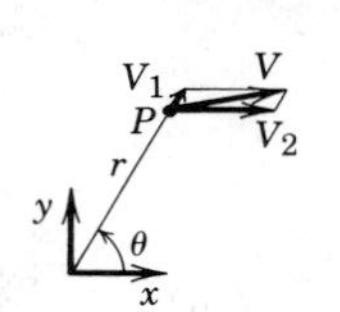

$$\psi = \psi_{so} + \psi_{uf} = \psi_1 + \psi_2 = \frac{q}{2\pi}\,\theta + Uy = \frac{q}{2\pi}\,\theta + Ur\sin\theta$$

$$\phi = \phi_{so} + \phi_{uf} = \phi_1 + \phi_2 = -\frac{q}{2\pi}\ln r - Ux = -\frac{q}{2\pi}\ln r - Ur\cos\theta$$

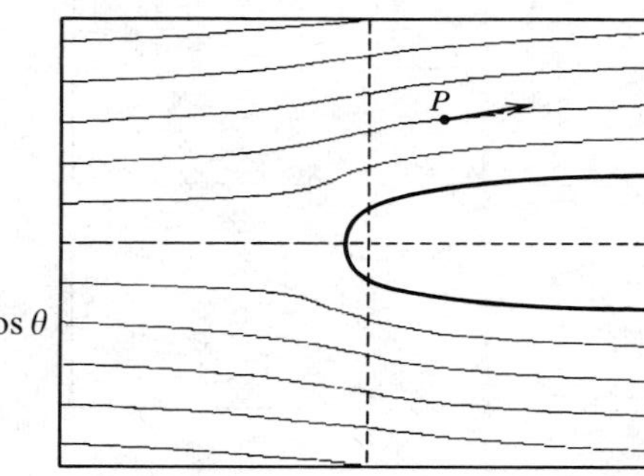

Source and Sink (equal strength, separation distance on x axis = $2a$)

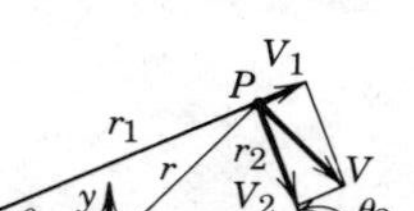

$$\psi = \psi_{so} + \psi_{si} = \psi_1 + \psi_2 = \frac{q}{2\pi}\,\theta_1 - \frac{q}{2\pi}\,\theta_2 = \frac{q}{2\pi}(\theta_1 - \theta_2)$$

$$\phi = \phi_{so} + \phi_{si} = \phi_1 + \phi_2 = -\frac{q}{2\pi}\ln r_1 + \frac{q}{2\pi}\ln r_2 = \frac{q}{2\pi}\ln\frac{r_2}{r_1}$$

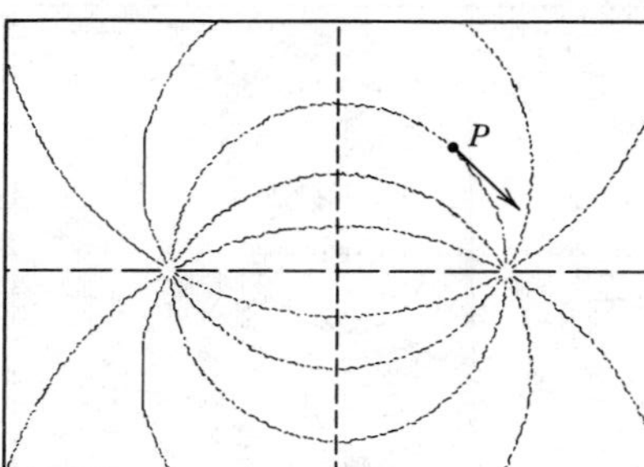

Source, Sink, and Uniform Flow (flow past a Rankine body)

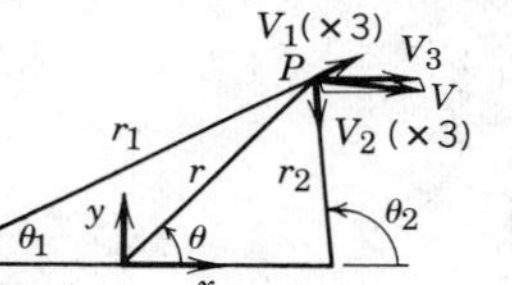

$$\psi = \psi_{so} + \psi_{si} + \psi_{uf} = \psi_1 + \psi_2 + \psi_3 = \frac{q}{2\pi}\,\theta_1 - \frac{q}{2\pi}\,\theta_2 + Uy$$

$$\psi = \frac{q}{2\pi}(\theta_1 - \theta_2) + Ur\sin\theta$$

$$\phi = \phi_{so} + \phi_{si} + \phi_{uf} = \phi_1 + \phi_2 + \phi_3 = -\frac{q}{2\pi}\ln r_1 + \frac{q}{2\pi}\ln r_2 - Ux$$

$$\phi = \frac{q}{2\pi}\ln\frac{r_2}{r_1} - Ur\cos\theta$$

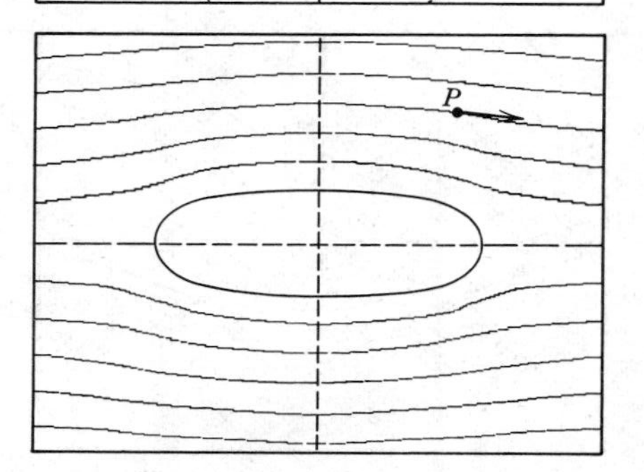

Table 6.2 Superposition of Elementary Plane Flows (*cont'd.*)

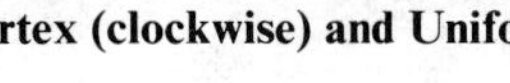

Vortex (clockwise) and Uniform Flow

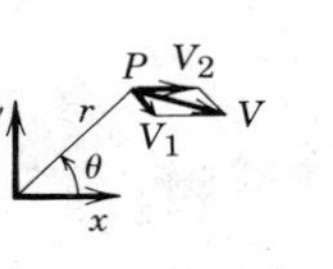

$$\psi = \psi_v + \psi_{uf} = \psi_1 + \psi_2 = \frac{K}{2\pi}\ln r + Uy = \frac{K}{2\pi}\ln r + Ur\sin\theta$$

$$\phi = \phi_v + \phi_{uf} = \phi_1 + \phi_2 = \frac{K}{2\pi}\theta - Ux = \frac{K}{2\pi}\theta - Ur\cos\theta$$

Doublet and Uniform Flow (flow past a cylinder)

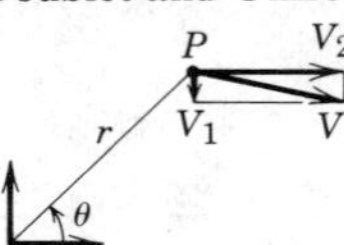

$$\psi = \psi_d + \psi_{uf} = \psi_1 + \psi_2 = -\frac{\Lambda\sin\theta}{r} + Uy = -\frac{\Lambda\sin\theta}{r} + Ur\sin\theta$$

$$\psi = U\left(r - \frac{\Lambda}{Ur}\right)\sin\theta = Ur\left(1 - \frac{a^2}{r^2}\right)\sin\theta \qquad a = \sqrt{\frac{\Lambda}{U}}$$

$$\phi = \phi_d + \phi_{uf} = \phi_1 + \phi_2 = -\frac{\Lambda\cos\theta}{r} - Ux = -\frac{\Lambda\cos\theta}{r} - Ur\cos\theta$$

$$\phi = -U\left(r + \frac{\Lambda}{Ur}\right)\cos\theta = -Ur\left(1 + \frac{a^2}{r^2}\right)\cos\theta$$

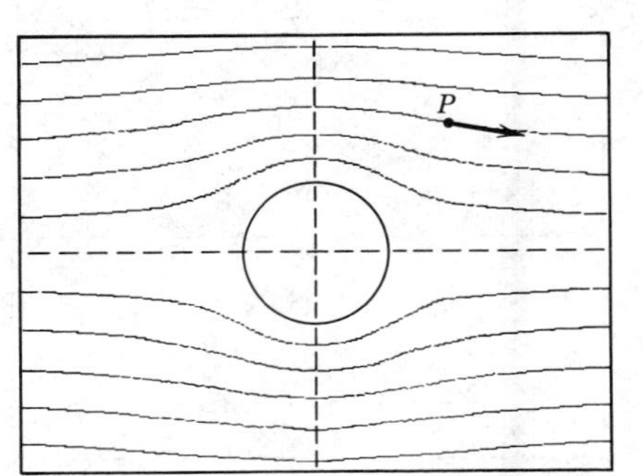

Doublet, Vortex (clockwise), and Uniform Flow (flow past a cylinder with circulation)

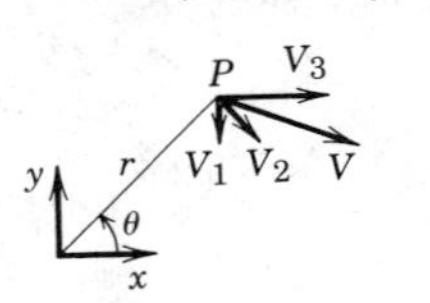

$$\psi = \psi_d + \psi_v + \psi_{uf} = \psi_1 + \psi_2 + \psi_3 = -\frac{\Lambda\sin\theta}{r} + \frac{K}{2\pi}\ln r + Uy$$

$$\psi = -\frac{\Lambda\sin\theta}{r} + \frac{K}{2\pi}\ln r + Ur\sin\theta = Ur\left(1 - \frac{a^2}{r^2}\right)\sin\theta + \frac{K}{2\pi}\ln r$$

$$\phi = \phi_d + \phi_v + \phi_{uf} = \phi_1 + \phi_2 + \phi_3 = -\frac{\Lambda\cos\theta}{r} + \frac{K}{2\pi}\theta - Ux$$

$a = \sqrt{\frac{\Lambda}{U}}$; $K < 4\pi a U$

$$\phi = -\frac{\Lambda\cos\theta}{r} + \frac{K}{2\pi}\theta - Ur\cos\theta = -Ur\left(1 + \frac{a^2}{r^2}\right)\cos\theta + \frac{K}{2\pi}\theta$$

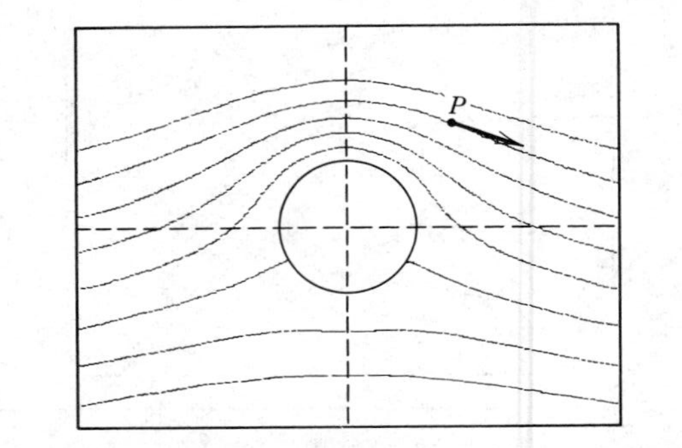

Source and Vortex (spiral vortex)

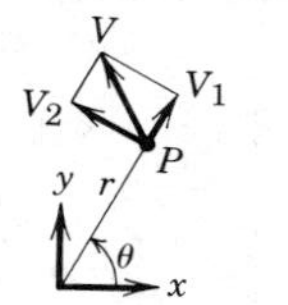

$$\psi = \psi_{so} + \psi_v = \psi_1 + \psi_2 = \frac{q}{2\pi}\,\theta - \frac{K}{2\pi}\ln r$$

$$\phi = \phi_{so} + \phi_v = \phi_1 + \phi_2 = -\frac{q}{2\pi}\ln r - \frac{K}{2\pi}\,\theta$$

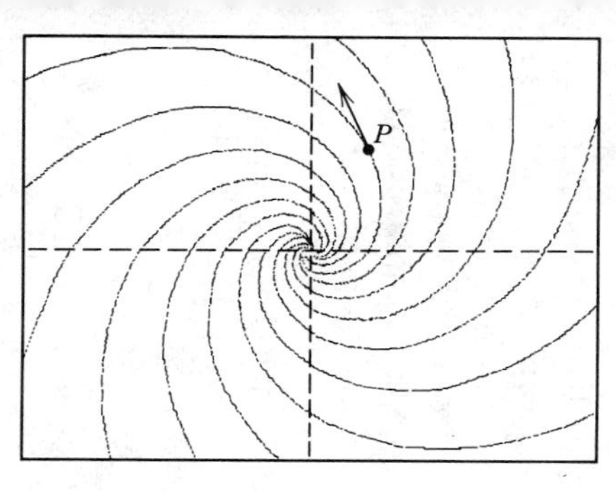

Sink and Vortex

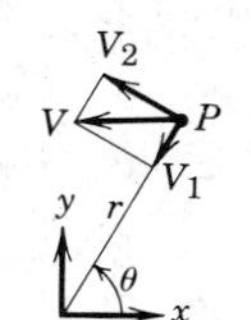

$$\psi = \psi_{si} + \psi_v = \psi_1 + \psi_2 = -\frac{q}{2\pi}\,\theta - \frac{K}{2\pi}\ln r$$

$$\phi = \phi_{si} + \phi_v = \phi_1 + \phi_2 = \frac{q}{2\pi}\ln r - \frac{K}{2\pi}\,\theta$$

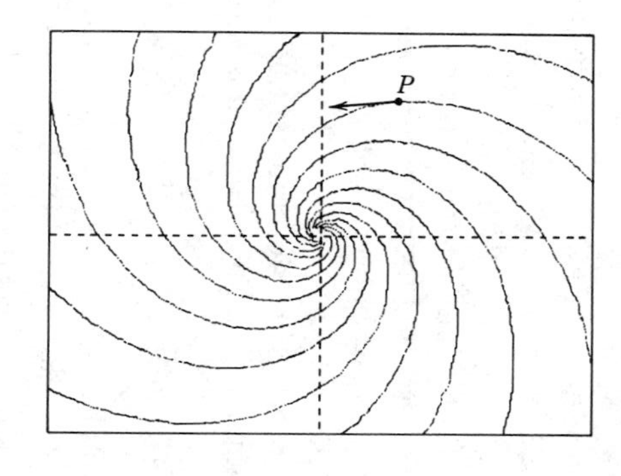

Vortex Pair (equal strength, opposite rotation, separation distance on *x* axis = 2*a*)

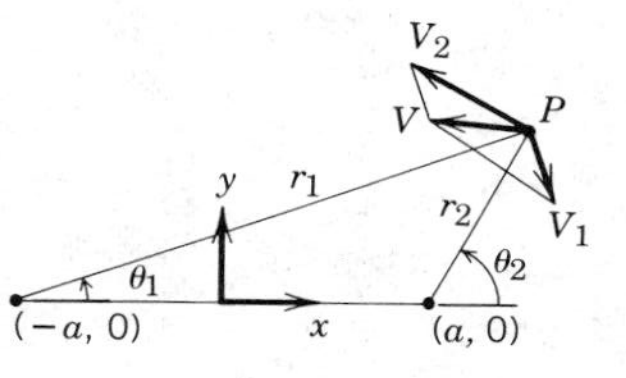

$$\psi = \psi_{v_1} + \psi_{v_2} = \psi_1 + \psi_2 = -\frac{K}{2\pi}\ln r_1 + \frac{K}{2\pi}\ln r_2 = \frac{K}{2\pi}\ln\frac{r_2}{r_1}$$

$$\phi = \phi_{v_1} + \phi_{v_2} = \phi_1 + \phi_2 = -\frac{K}{2\pi}\,\theta_1 + \frac{K}{2\pi}\,\theta_2 = \frac{K}{2\pi}(\theta_2 - \theta_1)$$

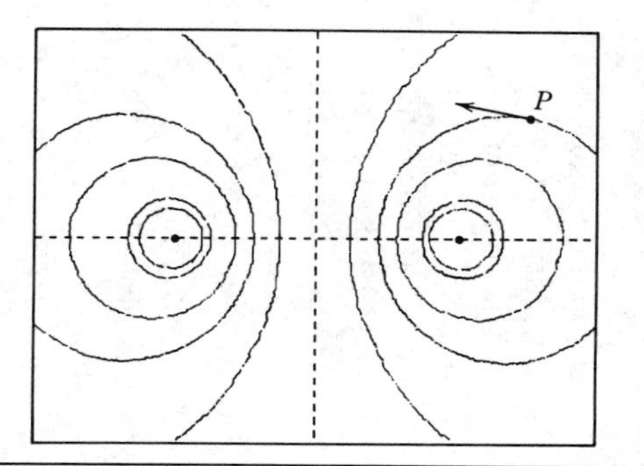

shape. Powerful mathematical techniques, such as complex variables and conformal transformations, can be used to obtain flow fields in interesting geometries.

The inverse method of superposition calculates the body shape to produce a desired pressure distribution. Distributed singularities—vortices, sources, and sinks located on the axis or body surface—are used to model the body. A complex computer code must be used to evaluate the distribution of singularities required to produce the desired pressure distribution.

The use of ever larger computers and development of faster codes today is allowing solution of complex three-dimensional problems using direct or inverse methods.

Example 6.11

For two-dimensional, incompressible, irrotational flow, the superposition of a doublet and a uniform flow represents flow around a circular cylinder. Obtain the stream function and velocity potential for this flow pattern. Find the velocity field, locate the stagnation points and the cylinder surface, and obtain the surface pressure distribution.

EXAMPLE PROBLEM 6.11

GIVEN: Two-dimensional, incompressible, irrotational flow formed from superposition of a doublet and a uniform flow.

FIND:
(a) Stream function and velocity potential.
(b) Velocity field.
(c) Stagnation points.
(d) Cylinder surface.
(e) Surface pressure distribution.

SOLUTION:

Stream functions may be added because the flow field is incompressible and irrotational. Thus from Table 6.1, the stream function for the combination is

$$\psi = \psi_d + \psi_{uf} = -\frac{\Lambda \sin\theta}{r} + Ur\sin\theta \qquad \psi$$

The velocity potential is

$$\phi = \phi_d + \phi_{uf} = -\frac{\Lambda \cos\theta}{r} - Ur\cos\theta \qquad \phi$$

The corresponding velocity components are obtained from Eq. 6.29 as

$$V_r = -\frac{\partial \phi}{\partial r} = -\frac{\Lambda\cos\theta}{r^2} + U\cos\theta$$

$$V_\theta = -\frac{1}{r}\frac{\partial \phi}{\partial \theta} = -\frac{\Lambda\sin\theta}{r^2} - U\sin\theta$$

The velocity field is

$$\vec{V} = V_r \hat{i}_r + V_\theta \hat{i}_\theta = \left(-\frac{\Lambda \cos\theta}{r^2} + U\cos\theta\right)\hat{i}_r + \left(-\frac{\Lambda \sin\theta}{r^2} - U\sin\theta\right)\hat{i}_\theta$$

$\vec{V}$

Stagnation points are where $\vec{V} = V_r \hat{i}_r + V_\theta \hat{i}_\theta = 0$, so

$$V_r = 0 = -\frac{\Lambda\cos\theta}{r^2} + U\cos\theta = \cos\theta\left(U - \frac{\Lambda}{r^2}\right)$$

Thus $V_r = 0$ when $r = \sqrt{\frac{\Lambda}{U}} = a$

$$V_\theta = 0 = -\frac{\Lambda\sin\theta}{r^2} - U\sin\theta = -\sin\theta\left(U + \frac{\Lambda}{r^2}\right)$$

Thus $V_\theta = 0$ when $\theta = 0, \pi$

Stagnation points are $(r, \theta) = (a, 0), (a, \pi)$

Stagnation Points

Note that $V_r = 0$ along $r = a$, so this represents flow around a circular cylinder, as shown in Table 6.2.

Flow is irrotational, so the Bernoulli equation may be applied between any two points. Applying the equation between a point far upstream and a point on the surface of the cylinder (neglecting elevation differences), we obtain

$$\frac{p_\infty}{\rho} + \frac{U^2}{2} + g\not{z} = \frac{p}{\rho} + \frac{V^2}{2} + g\not{z}$$

Thus

$$p - p_\infty = \tfrac{1}{2}\rho(U^2 - V^2)$$

Along the surface, $r = a$,

$$V^2 = V_\theta^2 = \left(-\frac{\Lambda}{r^2} - U\right)^2 \sin^2\theta = 4U^2\sin^2\theta$$

since $\Lambda = Ua^2$. Substituting yields

$$p - p_\infty = \tfrac{1}{2}\rho(U^2 - 4U^2\sin^2\theta) = \tfrac{1}{2}\rho U^2(1 - 4\sin^2\theta)$$

or

$$\frac{p - p_\infty}{\frac{1}{2}\rho U^2} = 1 - 4\sin^2\theta$$

This problem illustrates the techniques used to combine elementary plane flows to form a flow pattern of practical interest.

6-8 SUMMARY OBJECTIVES

After completing study of Chapter 6, you should be able to do the following:

1. Write Euler's equations in (a) vector form, (b) rectangular coordinates, (c) cylindrical coordinates, and (d) streamline coordinates.
2. Integrate Euler's equation along a streamline in steady flow to obtain the Bernoulli equation. State the restrictions on the use of the Bernoulli equation.
3. Define static pressure, stagnation pressure, and dynamic pressure.
4. For steady, incompressible, inviscid flow through a stream tube, state the conditions under which the first law of thermodynamics reduces to the Bernoulli equation.

** 5. Write the unsteady Bernoulli equation for flow along a streamline. State the restrictions on the use of the equation.

** 6. For a steady, inviscid, incompressible flow that is irrotational, show that the Bernoulli equation may be applied between any two points in the flow field.

** 7. For a two-dimensional, irrotational, incompressible flow field:
   (a) given the velocity field, determine the velocity potential, ϕ
   (b) given the velocity potential, determine the velocity field
   (c) show that lines of constant ψ and constant ϕ are orthogonal
   (d) given ψ, determine ϕ (and vice versa).

** 8. Given the stream function, ψ, for a two-dimensional, irrotational flow about a body, determine:
   (a) the equation of the body
   (b) the pressure distribution along the body surface.

9. Solve the problems at the end of the chapter that relate to the material you have studied.

PROBLEMS

6.1 A velocity field is given as $\vec{V} = Ay\hat{i} + Ax\hat{j}$, where $A = 3$ m/sec/m and coordinates x and y are given in meters. The fluid density is 750 kg/m^3. Calculate the acceleration of a fluid particle at the point $(x, y) = (1, 0)$, and determine the pressure gradient at the same point if $\vec{g} = -g\hat{j}$.

6.2 Consider the flow field with velocity given by $\vec{V} = (Axy - Bx^2)\hat{i} + (Axy - By^2)\hat{j}$, where $A = 2/\text{ft}\cdot\text{sec}$ and $B = 1/\text{ft}\cdot\text{sec}$. The density is 2 slug/ft^3, and gravity acts in the negative y direction. Determine the acceleration of a fluid particle and also the pressure gradient at the point $(x, y) = (1, 1)$.

6.3 The x component of velocity in an incompressible flow is given in m/sec by $u = Ax$, where $A = 6\ \text{sec}^{-1}$. At the point $(x, y) = (2, 0)$, the y component of velocity is $v = 0$; $w = 0$ everywhere. Obtain an equation for the y component of velocity, and find the acceleration of a fluid particle at the point $(2, 0)$. Find the pressure gradient at the same point if $\vec{g} = -g\hat{j}$ and the fluid is water.

6.4 An incompressible flow field is given by $\vec{V} = (Ax + By)\hat{i} - Ay\hat{j}$, where $A = 1\ \text{sec}^{-1}$, $B = 2\ \text{sec}^{-1}$, and the coordinates are measured in meters. Find the magnitude and direction of the acceleration of a fluid particle at the point $(x, y) = (1, 2)$. Find the pressure gradient at the same point if $\vec{g} = -g\hat{j}$ and the fluid is water.

** These objectives apply to sections that may be omitted without loss of continuity in the text material.

6.5 A horizontal flow of water is described by the velocity field, $\vec{V} = (Ax + Bt)\hat{i} + (-Ay + Bt)\hat{j}$, where $A = 10\ \text{sec}^{-1}$, $B = 5\ \text{ft}\cdot\text{sec}^{-2}$, x and y are in feet, and t is in seconds. Compute the acceleration of a fluid particle at the point $(x, y) = (1, 5)$ at $t = 10$ sec. Evaluate $\partial p/\partial x$ under the same conditions.

6.6 A velocity field in a fluid with density of $1500\ \text{kg/m}^3$ is given by $\vec{V} = (Ax - By)t\hat{i} - (Ay + Bx)t\hat{j}$, where $A = 1\ \text{sec}^{-2}$, $B = 2\ \text{sec}^{-2}$, x and y are in meters, and t is in seconds. Body forces are negligible. Evaluate ∇p at the point $(x, y) = (1, 2)$ at $t = 1$ sec.

6.7 In a frictionless, incompressible flow, the velocity field in m/sec and the body force are given by $\vec{V} = Ax\hat{i} - Ay\hat{j}$ and $\vec{g} = -g\hat{k}$. The pressure is p_0 at the point $(x, y, z) = (0, 0, 0)$. Obtain an expression for the pressure field, $p(x, y, z)$.

6.8 Example Problem 3.8 read: "As a result of a promotion, you are transferred from your present location. You must transport a fish tank in the back of your station wagon. The tank is 12 × 24 × 12 in. How much water should you leave in the tank to be reasonably sure that it will not spill over during the trip?" Solve this problem using the Euler equations.

6.9 Consider a steady fluid motion with rigid-body swirl about the z axis. Assume $\vec{g} = -g\hat{k}$. Show that Eqs. 6.4 reduce to

$$\frac{\partial p}{\partial r} = \frac{\rho V_\theta^2}{r}, \qquad \frac{\partial p}{\partial \theta} = 0, \qquad \frac{\partial p}{\partial z} = -\rho g$$

6.10 Example Problem 3.9 read: "A cylindrical container, partially filled with liquid, is rotated at constant angular velocity, ω, about its axis. After a short period of time there is no relative motion; the liquid rotates with the cylinder as if the system were a rigid body. Determine the shape of the free surface." Solve this problem using the Euler equations. (The cylinder radius is R.)

6.11 The cylindrical container of Problem 6.10 is initially one-half full of liquid. Determine the maximum angular speed at which the container can be spun without spilling liquid over the top. Use the Euler equations.

6.12 Solve Problem 4.103. Show that the variation of uniform radial velocity is $V_r = Q/2\pi rh$. Obtain expressions for the r component of the acceleration of a fluid particle in the gap and for the pressure variation as a function of radial distance from the holes.

6.13 Consider steady, incompressible flow without friction in the narrow gap between solid parallel disks. Assume that flow is uniform across each plane, r = constant. Show that the Euler equations reduce to $dp = -\rho V\,dV$ for this situation. Show that the same result is obtained by applying the continuity and momentum equations to a segment, $d\theta$, of an annular differential control volume of thickness, dr. (Review Example Problem 3.9 if you need help in formulating the pressure force on your CV.)

6.14 Two circular disks of radius, R, are separated by a distance, b. The upper disk moves toward the lower one at speed, V. The space between the disks is filled with a frictionless, incompressible fluid, which is squeezed out as the disks come together. Assume that, at any radial section, the velocity is uniform across the gap width, b. However, note that b is a function of time. The pressure surrounding the disks is atmospheric. Determine the gage pressure at $r = 0$.

6.15 Steady, frictionless, and incompressible flow from right to left over a stationary circular cylinder of radius, a, is given by the velocity field

$$\vec{V} = U\left[\left(\frac{a}{r}\right)^2 - 1\right]\cos\theta\,\hat{i}_r + U\left[\left(\frac{a}{r}\right)^2 + 1\right]\sin\theta\,\hat{i}_\theta$$

Consider flow along the streamline forming the cylinder surface, $r = a$. Express the components of the pressure gradient in terms of angle, θ.

6.16 The flow field in a forced vortex (rigid-body motion) is given by $\vec{V} = \omega r \hat{i}_\theta$, where $\omega = 10\ \text{sec}^{-1}$, and r is in meters. Assume a frictionless fluid with $\rho = 1000\ \text{kg/m}^3$. Express the radial pressure gradient, $\partial p/\partial r$, as a function of r. Evaluate the pressure change between $r_1 = 1$ m and $r_2 = 2$ m.

6.17 Air at 20 psia and 100 F flows around a smooth corner at the inlet to a diffuser. The air speed is 150 ft/sec, and the radius of curvature of the streamlines is 3 in. Determine the magnitude of the centripetal acceleration in "G's" experienced by a fluid particle rounding the corner. Evaluate the pressure gradient, $\partial p/\partial r$.

6.18 The flow field in a free (irrotational) vortex is given by

$$\vec{V} = \frac{K}{2\pi r}\hat{i}_\theta; \qquad r > 0$$

where r is in meters. Assume a frictionless fluid with $\rho = 1000\ \text{kg/m}^3$, and $K = 20\pi\ \text{m}^2/\text{sec}$. Express the radial pressure gradient, $\partial p/\partial r$, as a function of r, and evaluate the pressure change between $r_1 = 1$ m and $r_2 = 2$ m.

6.19 The radial velocity variation at the midsection of the 180° bend shown is given by $rV_\theta =$ constant. The cross section of the bend is square. Assume that the velocity is not a function of z. Derive an equation for the pressure difference between the outside and the inside of the bend. Express your answer in terms of the mass flow rate, the fluid density, the geometric parameters, R_1 and R_2, and the depth of the bend, h.

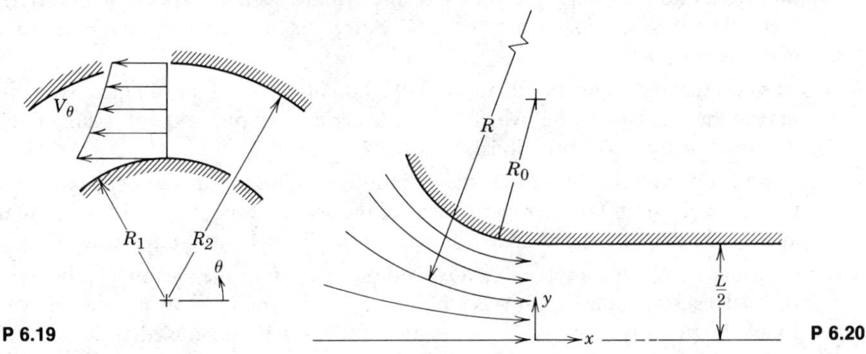

P 6.19 P 6.20

6.20 As a model of the velocity distribution in the curved inlet section of a wind tunnel, the radius of curvature of the streamlines is expressed as

$$R = \frac{(L/2)}{y} R_0$$

As a first approximation, assume the air speed along each streamline is $V = 20$ m/sec. Evaluate the pressure change from $y = 0$ to the tunnel wall at $y = L/2$, if $L = 150$ mm and $R_0 = 0.6$ m.

6.21 The flow area of a horizontal air duct is reduced smoothly from 0.75 to 0.25 ft^2. The flow rate is steady at 1.5 lbm/sec of air at 40 psia, 70 F. Frictionless flow may be assumed. Determine the pressure change over the length of duct in which the area is reduced.

6.22 Water flows in a circular pipe. At one section the diameter is 0.3 m, the static pressure is 260 kPa (gage), the velocity is 3 m/sec, and the elevation is 10 m above ground level. The elevation at a section downstream is 0 m, and the pipe diameter is 0.15 m. Find the gage pressure at the downstream section if frictional effects may be neglected.

6.23 Water flows steadily up the vertical, 0.1 m diameter pipe and out the nozzle, which is 0.05 m in diameter, discharging to atmospheric pressure. The stream velocity at the nozzle exit must be 20 m/sec. Calculate the gage pressure required at section ①, assuming frictionless flow.

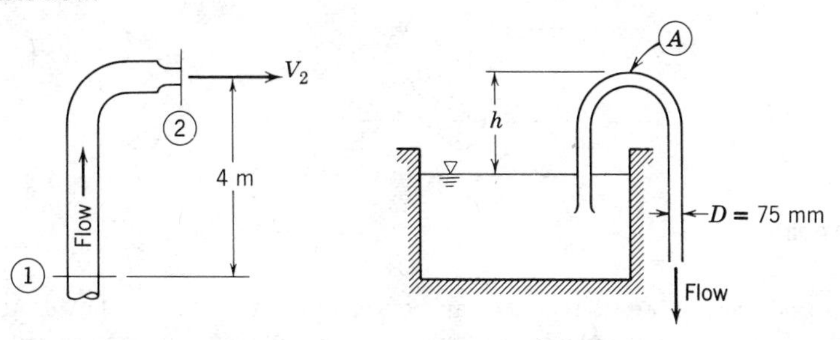

P 6.23 P 6.24

6.24 Water may be considered to flow without friction through the siphon. The water flow rate is 0.03 m^3/sec, its temperature is 20 C, and the pipe diameter is 75 mm. Compute the maximum allowable height, h, so that the pressure at point A is above the vapor pressure of the water.

6.25 Water flows from a very large tank through a 2 in. diameter tube. The dark fluid in the manometer is mercury. Determine the velocity in the pipe and the rate of discharge from the tank.

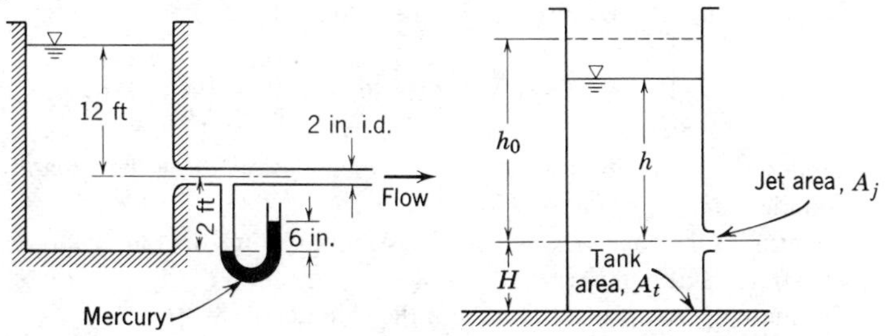

P 6.25 P 6.26

6.26 The tank shown has a well-rounded orifice with area A_j. At time $t = 0$, the water level is at height h_0. Develop an expression for the water height, h, at any later time, t.

6.27 The water level in a large tank is maintained at height, H, above the surrounding level terrain. A rounded orifice placed in the side of the tank discharges a horizontal jet. Neglecting friction, determine the height at which the orifice should be placed so that the water jet strikes the ground at the maximum horizontal distance from the tank.

6.28 A smoothly contoured nozzle is connected to the end of a garden hose. At the nozzle inlet, where the velocity is negligible, the water pressure is 160 kPa (gage). Pressure at the nozzle exit is atmospheric. Assuming that the water remains in a single stream that has negligible aerodynamic drag, estimate the maximum height above the nozzle outlet that the stream could reach.

6.29 A stream of liquid moving at low speed leaves a nozzle pointed directly downward. The velocity may be considered uniform across the nozzle section, and the effects of friction may be ignored. At the nozzle exit, located at elevation z_0, the jet velocity and area are V_0 and A_0, respectively. Determine the variation of jet area with elevation for $z < z_0$.

6.30 The flow system of parallel disks shown contains water. As a first approximation, friction may be neglected. Determine the volume flow rate, and the pressure at point Ⓒ. ($R =$ 300 mm and $r_C = 150$ mm.)

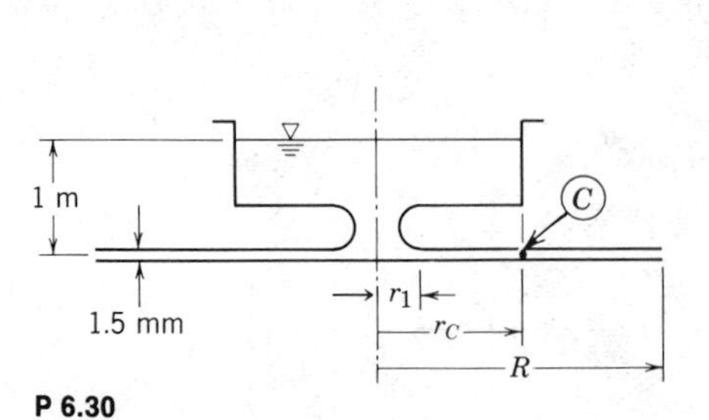

P 6.30

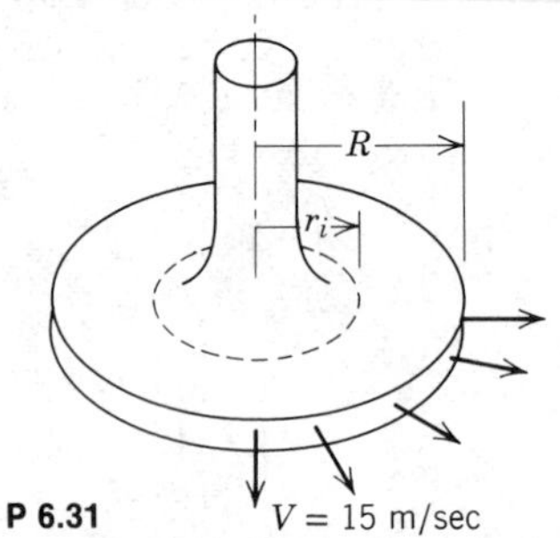

P 6.31

6.31 Consider incompressible flow of air between the parallel disks shown. Assume that the flow is purely radial and uniform at any section, and that viscous effects may be neglected. Obtain an expression for the velocity distribution between the disks if $V = 15$ m/sec at $r = R = 75$ mm. Evaluate the magnitude and direction of the net pressure force that acts on the upper plate between r_i and R, if $r_i = R/2$.

6.32 Obtain an expression for the pressure distribution as a function of radius for the "air hockey" puck of Problem 5.45. Evaluate the net pressure force that acts on the puck.

6.33 Steady, frictionless, and incompressible flow from left to right over a stationary circular cylinder of radius, a, is represented by the velocity field

$$\vec{V} = U\left[1 - \left(\frac{a}{r}\right)^2\right] \cos\theta \hat{i}_r - U\left[1 + \left(\frac{a}{r}\right)^2\right] \sin\theta \hat{i}_\theta$$

(a) Obtain an expression for the pressure distribution along the streamline forming the cylinder surface, $r = a$.

(b) Determine the locations where the static pressure on the cylinder is equal to the freestream static pressure.

(c) Evaluate the net pressure force on the cylinder.

6.34 The flow over a Quonset hut may be approximated by the velocity distribution of Problem 6.33 with $0 \leq \theta \leq \pi$. During a storm the wind speed reaches 100 km/hr; the outside temperature is 5 C. A barometer inside the hut reads 720 mm of mercury; pressure, p_∞, is also 720 mm Hg. The hut has a diameter of 6 m and a length of 18 m. Determine the net force tending to lift the hut off its foundation.

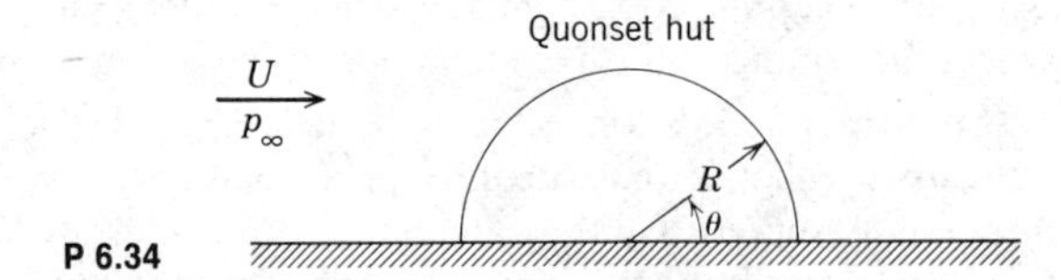

P 6.34

6.35 A horizontal axisymmetric jet of air with 10 mm diameter strikes a stationary vertical disk of 200 mm diameter. The jet speed is 50 m/sec at the nozzle exit. A manometer is connected to the center of the disk. Calculate (a) the manometer deflection if the fluid has SG = 1.75, and (b) the force exerted by the jet on the disk. Sketch the streamline pattern and the distribution of pressure on the face of the disk.

6.36 Water flows at low speed through a circular tube with inside diameter of 50 mm. A smoothly contoured plug of 40 mm diameter is held in the end of the tube where the water discharges to atmosphere. Neglect frictional effects and assume uniform velocity profiles at each section. Determine the pressure measured by the gage and the force required to hold the plug.

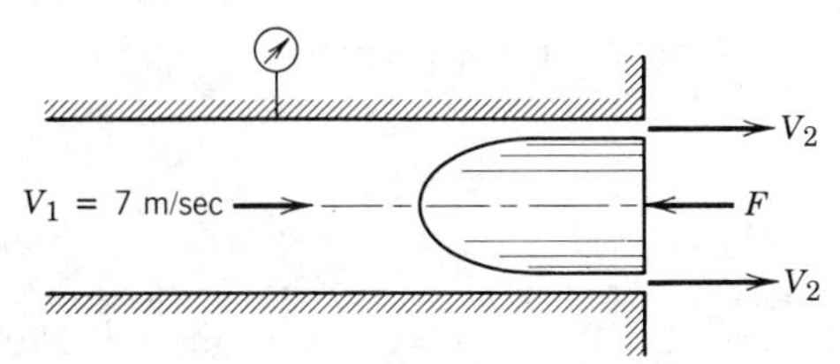

P 6.36

6.37 A tank with a *reentrant* orifice called a *Borda mouthpiece* is shown. The fluid is inviscid and incompressible. The reentrant orifice essentially eliminates flow along the tank walls, so the pressure there is nearly hydrostatic. Calculate the *contraction coefficient*, $C_c = A_j/A_0$. *Hint:* Equate the unbalanced hydrostatic pressure force and momentum flux from the jet.

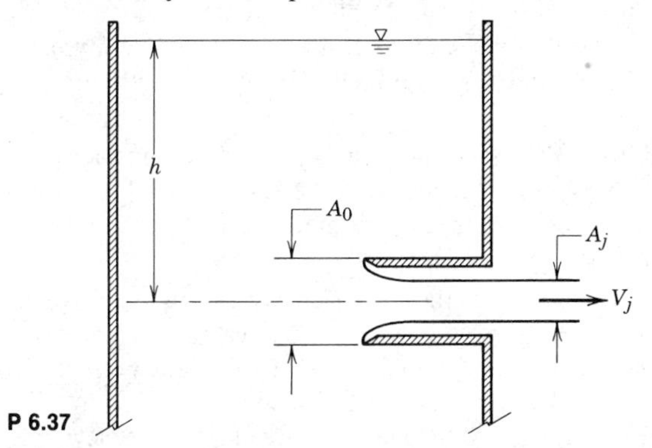

P 6.37

6.38 Consider the steady, frictionless, incompressible flow of air over the wing of an airplane. The air approaching the wing is at 10 psia, 40 F, and has a velocity of 200 ft/sec relative to the wing. At a certain point in the flow, the pressure is −0.40 psi (gage). Calculate the velocity of the air relative to the wing at this point.

6.39 A seawater inlet for reactor cooling is to be located on the outer hull of a nuclear submarine. The maximum submerged speed of the sub is 35 knots. At the location of the inlet, the water speed parallel to the hull is 20 knots when the sub moves at maximum speed. Determine the maximum static pressure that might be expected at the inlet.

6.40 An Indianapolis race car travels at a maximum speed of 350 km/hr. A pressure gage attached to the airfoil reads −50 mm of water (gage). Evaluate the air speed relative to the car at that location.

6.41 The pressure delivered by the turbocharger on an engine used for the Indianapolis 500 mile race is controlled by a pressure relief valve. The valve is set to release when the pressure in the intake manifold reaches 75 in. of mercury above the pressure outside the valve. The pressure relief valve is located at a point on the car where the local air speed is 1.3 times the air speed in the freestream far from the car. For a standard day and a car speed of 220 mph, estimate the static pressure near the pressure relief valve. Specify clearly whether this is a gage or absolute pressure. What is the effect on the boost pressure delivered to the engine?

6.42 A steady force, F, is exerted on the piston shown. The space to the right of the piston is filled with inviscid, incompressible liquid. Obtain an expression for the jet velocity in terms of the geometric and fluid properties given.

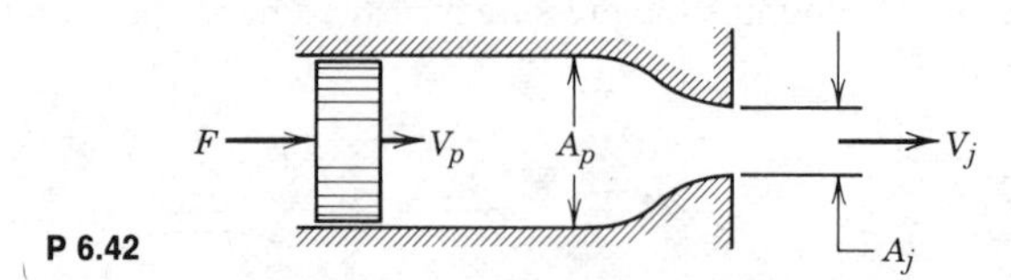

P 6.42

6.43 A pitot-static tube is used to measure the speed of air at standard conditions at a point in a flow system. To ensure that flow may be assumed incompressible for calculations of engineering accuracy, the speed is to be maintained at 100 m/sec or less. Determine the manometer deflection, in millimeters of mercury, that corresponds to the maximum desirable speed.

6.44 An open-circuit wind tunnel draws in air from the atmosphere through a well-contoured nozzle. In the test section, where the flow is straight and nearly uniform, a static pressure tap is drilled into the tunnel wall. A manometer connected to the tap shows that the static pressure within the tunnel is 45 mm of water below atmospheric. Assume that the air is incompressible, and at 25 C, 100 kPa(abs). Calculate the air speed in the wind-tunnel test section.

6.45 A jet of air from a nozzle is blown at right angles against a wall in which a pressure tap is located. A manometer connected to the tap shows a head of 0.14 in. of mercury above atmospheric. Determine the approximate speed of the air leaving the nozzle if it is at 40 F and 14.7 psia.

6.46 Water is discharged from the system shown through pipes ① and ②. The area of ① is 1.0 in.^{2} and the area of ② is 2.0 in.^{2} The heater adds 1.8 Btu per slug of fluid flowing through the heater. Neglect frictional effects and any pressure drop across the heater.
(a) Calculate the volume flow rate at section ①.
(b) If area ③ equals area ④ and both are at the same elevation, compute the change in specific internal energy across the heater.
(c) Calculate the volume flow rate at section ②.

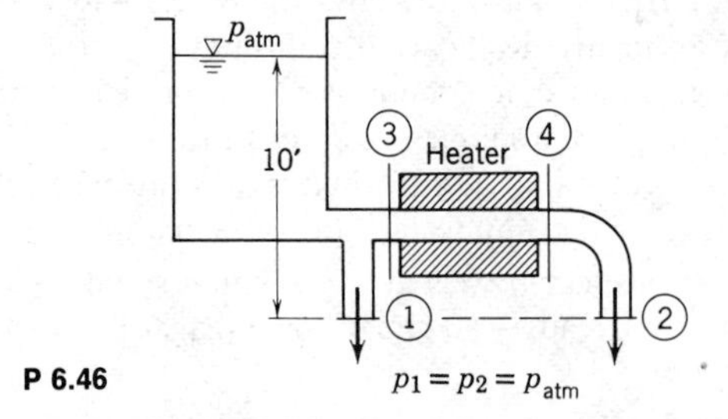

P 6.46

***6.47** The air speed in the inlet manifold of an internal-combustion engine is approximated by the expression $V = V_0(1 + \sin \omega t)$. Assume that $V_0 = 30$ m/sec, $\omega = 50$ Hz, and that the density is half that of standard air. If the length of the inlet passage is $L = 0.3$ m, calculate the pressure variation required to cause the motion. Neglect frictional effects.

* These problems require material from sections that may be omitted without loss of continuity in the text material.

***6.48** Compressed air is used to accelerate water from a tube. Neglect the velocity in the reservoir and assume the flow in the tube is uniform at any section. At a particular instant, it is known that $V = 2\,\text{m/sec}$ and $dV/dt = 0.50\,\text{m/sec}^2$. The cross-sectional area of the tube is $A = 0.02\,\text{m}^2$. Determine the pressure in the tank at this instant.

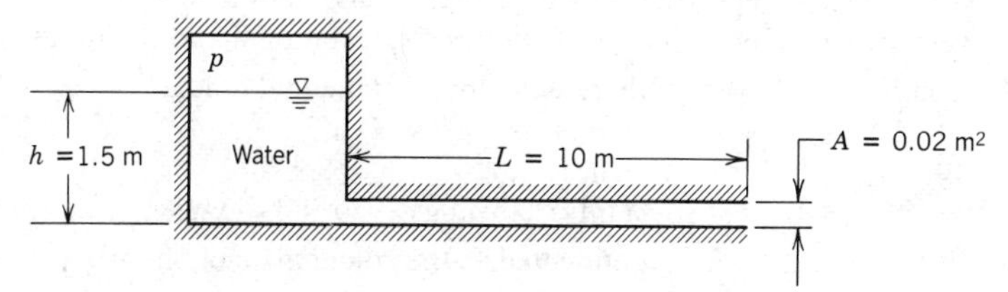

P 6.48

***6.49** Apply the unsteady Bernoulli equation to the U-tube manometer of constant area shown. Assume that the manometer is initially deflected, and then released. Obtain a differential equation for l as a function of time.

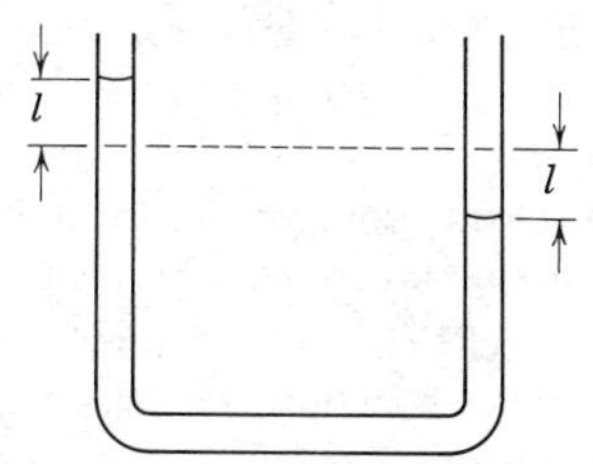

P 6.49

***6.50** Consider the reservoir and disk flow system of Problem 6.30, with the reservoir level maintained constant. Flow between the disks is started from rest at $t = 0$. Evaluate the rate of change of volume flow rate at the instant, $t = 0$, if $r_1 = 50\,\text{mm}$.

***6.51** Consider the tank of Problem 4.43. Using the Bernoulli equation for unsteady flow along a streamline, evaluate the minimum diameter ratio, D/d, required to justify the assumption that flow from the tank is quasi-steady.

***6.52** Consider an incompressible flow of standard air, with velocity field given by $\vec{V} = Ax\hat{i} - Ay\hat{j} + B\hat{k}$, where $A = 10\,\text{sec}^{-1}$, $B = 30\,\text{m/sec}$, and the coordinates are in meters. Neglect gravity. Determine the pressure change between the points $(0, 0, 0)$ and $(3, 1, 0)$.

***6.53** Consider the flow of water represented by the velocity field, $\vec{V} = Ay\hat{i} + Ax\hat{j}$, where $A = 3\,\text{sec}^{-1}$ and the coordinates are in meters. Is it possible to calculate the pressure change between the points $(0, 0, 0)$ and $(1, 1, 1)$? If possible, do so.

***6.54** An incompressible flow field is given as $\vec{V} = Ax^2y\hat{i} - Axy^2\hat{j}$, where $A = 6/\text{m}^2 \cdot \text{sec}$. Evaluate the rotation of the flow. What can be said about the use of the Bernoulli equation for this flow?

***6.55** Consider the flow field of water given by $\vec{V} = Ax^2y^2\hat{i} - Bxy^3\hat{j}$, where $A = 3/\text{m}^3 \cdot \text{sec}$ and $B = 2/\text{m}^3 \cdot \text{sec}$.

(a) Determine the stream function for this flow.

(b) Determine the fluid rotation.

(c) Neglecting gravity, is it possible to calculate the pressure difference between points $(0, 0, 0)$ and $(1, 1, 1)$. If so, do it; if not, why not?

* These problems require material from sections that may be omitted without loss of continuity in the text material.

***6.56** Determine whether the Bernoulli equation can be applied between different radii for the following vortex flows:

(a) $\vec{V} = \omega r \hat{i}_\theta$ (b) $\vec{V} = \dfrac{K}{2\pi r} \hat{i}_\theta$.

***6.57** A flow field is represented by the stream function, $\psi = x^2 - y^2$. Find the corresponding velocity field. Show that this flow field is irrotational and obtain the potential function.

***6.58** Consider the flow field represented by the potential function, $\phi = -Ax + By$.
(a) Is this a possible incompressible flow field?
(b) Is the flow irrotational?
(c) Sketch lines of constant ϕ, labeling $\phi = 0$ and showing the direction of increasing ϕ.

***6.59** Consider the flow field represented by the potential function, $\phi = x^2 - y^2$. Verify that this is an incompressible flow and obtain the corresponding stream function.

***6.60** Consider the flow field represented by the potential function, $\phi = Ax^2 + Bxy - Ay^2$. Verify that this is an incompressible flow and determine the corresponding stream function.

***6.61** An incompressible flow field is characterized by the stream function, $\psi = 3Ax^2y - Ay^3$, where $A = 1\ \text{m}^{-1} \cdot \text{sec}^{-1}$. Sketch a few streamlines in the first quadrant. Show that this flow field is irrotational. Derive the velocity potential for the flow.

***6.62** Beginning with the velocity components for uniform flow shown in Table 6.1, verify the expressions for the stream function and velocity potential.

***6.63** Beginning with the velocity components for a simple source shown in Table 6.1, verify the expressions for the stream function and velocity potential.

***6.64** Beginning with the velocity components for a simple sink shown in Table 6.1, verify the expressions for the stream function and velocity potential.

***6.65** Beginning with the velocity components for a free vortex shown in Table 6.1, verify the expressions for the stream function and velocity potential.

***6.66** Consider the flow field formed by combining a uniform flow in the positive x direction and a source located at the origin. Obtain expressions for the stream function, velocity potential, and velocity field for the combined flow. If $U = 25$ m/sec, determine the source strength if the stagnation point is located at $x = -0.5$ m. Sketch the stagnation streamline. Evaluate the locations of the branches of the stagnation streamline far downstream.

***6.67** Consider the flow field formed by combining a uniform flow in the positive x direction and a source located at the origin. Let $U = 30$ m/sec and $q = 150\ \text{m}^2/\text{sec}$. Use a calculator or simple computer program to locate the points on the stagnation streamline where the velocity reaches its maximum value. Find the gage pressure there if the fluid density is $1.2\ \text{kg/m}^3$.

***6.68** Consider the flow field formed by combining a uniform flow in the positive x direction with a sink located at the origin. Let $U = 50$ m/sec and $q = 90\ \text{m}^2/\text{sec}$. Use a suitably chosen control volume to evaluate the net force per unit depth needed to hold in place (in standard air) the surface shape formed by the stagnation streamline.

***6.69** A crude model of a tornado is formed by combining a sink of strength, $q = 2800\ \text{m}^2/\text{sec}$, and a free vortex of strength, $K = 5600\ \text{m}^2/\text{sec}$. Obtain the stream function and velocity

* These problems require material from sections that may be omitted without loss of continuity in the text material.

potential for this flow field. Estimate the radius beyond which the flow may be treated as incompressible. Find the gage pressure at that radius.

***6.70** A source and a sink of equal strength are placed on the x axis at $x = -a$ and $x = a$, respectively. Obtain the stream function and velocity potential for the resulting flow field. Find points on the y axis between which half the total volume flow rate passes.

***6.71** A source and a sink with strengths of equal magnitude, $q = 3\pi \text{ m}^2/\text{sec}$, are placed on the x axis at $x = -a$ and $x = a$, respectively. A uniform flow with speed, $U = 20 \text{ m/sec}$, in the positive x direction is added to obtain the flow past a Rankine body. Obtain the stream function, velocity potential, and velocity field for the combined flow. Find the value of $\psi = \text{constant}$ on the stagnation streamline. Locate the stagnation points if $a = 0.3 \text{ m}$.

***6.72** Consider again the flow past a Rankine body of Problem 6.71. The half-width, h, of the body in the y direction is given by the transcendental equation

$$\frac{h}{a} = \cot\left(\frac{\pi U h}{q}\right)$$

Evaluate the half-width, h. Find the local velocity and the pressure at the points $(x, y) = (0, \pm h)$. Assume fluid density is that of standard air.

***6.73** A flow field is made from the combination of a uniform flow in the positive x direction with $U = 10 \text{ m/sec}$, and a counterclockwise vortex with strength, $K = 4\pi \text{ m}^2/\text{sec}$, located at the origin. Obtain the stream function, velocity potential, and velocity field for the combined flow. Locate the stagnation point(s) for the flow. Sketch the streamline that passes through the point $(r, \theta) = (5K/\pi U, \pi)$.

***6.74** Consider the flow past a circular cylinder of Example Problem 6.11. Plot the pressure coefficient

$$C_p = \frac{p - p_\infty}{\frac{1}{2}\rho U^2}$$

along the surface of the cylinder versus θ for $0 \le \theta \le \pi$. Find the maximum and minimum values of C_p. Locate the points on the cylinder surface where the static pressure is equal to the freestream value.

***6.75** Consider the flow past a circular cylinder of radius, a, used in Example Problem 6.11. Show that $V_r = 0$ along the lines $(r, \theta) = (r, \pm\pi/2)$. Plot V_θ/U versus radius for $r \ge a$, along the line $(r, \theta) = (r, \pi/2)$. Find the distance beyond which the influence of the cylinder is less than 1 percent of U.

***6.76** Consider the flow past a circular cylinder of radius, a, used in Example Problem 6.11. Integrate the pressure distribution around the cylinder surface to show that the drag force predicted by this ideal flow model is zero.

***6.77** To model flow around a circular cylinder with circulation, add a clockwise vortex to the flow considered in Example Problem 6.11. Find the stream function, velocity potential, and velocity field for this combined flow. Locate the stagnation points on the cylinder surface. Evaluate the range of vortex strengths for which two stagnation points exist.

***6.78** Consider again the ideal fluid flow around a circular cylinder with circulation given in Problem 6.77. Show that the circulation may be expressed as $\Gamma = -K$ by integrating

* These problems require material from sections that may be omitted without loss of continuity in the text material.

around the cylinder surface. Integrate the pressure distribution to evaluate the aerodynamic lift (force in the y direction) on the cylinder. Show that the lift force is $L = -\rho U \Gamma$.

*6.79 A spiral vortex is formed by combining a source and a vortex. Show that the streamlines of this combined flow are logarithmic spirals such that $r = ce^{\pm q\theta/K}$.

*6.80 The flow pattern of a vortex pair is to be investigated. A clockwise vortex is located at $(x, y) = (-a, 0)$, and a counterclockwise one of equal strength is at $(x, y) = (a, 0)$. Find the stream function, velocity potential, and velocity field for this combined flow. Evaluate the pressure distribution along the y axis.

*6.81 A flow field is to be constructed by combining a vortex pair and a uniform flow. A counterclockwise vortex is placed at $(x, y) = (0, a)$, and a clockwise one of equal strength is at $(0, -a)$. The uniform flow is in the positive y direction. Obtain the stream function, velocity potential, and velocity field for this combined flow. Evaluate the speed of the uniform flow needed to form a single closed streamline in the flow with stagnation point at $(0, \pm a/2)$. Sketch this streamline.

* These problems requires material from sections that may be omitted without loss of continuity in the text material.

Chapter 7

DIMENSIONAL ANALYSIS AND SIMILITUDE

The development of fluid mechanics has depended heavily on experimental results because so few real flows can be solved exactly by analytical methods alone. Solutions of real problems involve a combination of analysis and experimental information. First, the real physical flow situation is approximated with a mathematical model that is simple enough to yield a solution. Then experimental measurements are made to check the analytical results. Based on the measurements, refinements in the analysis can be made, and so on. The experimental results are an essential link in this iterative design process. Empirical designs developed without analysis or careful review of available experimental data are often high in cost and poor or inadequate in performance.

However, experimental work in the laboratory is both time-consuming and expensive. One obvious goal is to obtain the most information from the fewest experiments. Dimensional analysis is an important tool that often helps us to achieve this goal. The dimensionless parameters that we obtain also can be used to correlate data for succinct presentation using the minimum possible number of plots.

When experimental testing of a full-size prototype is either impossible or prohibitively expensive (as happens so often), the only feasible way of attacking the problem is through model testing in the laboratory. If we are to predict the prototype behavior from measurements on the model, it is obvious that we cannot run just any test on any model. The model flow and the prototype flow must be related by known scaling laws. We shall investigate the conditions necessary to obtain this similarity of model and prototype flows following the discussion of dimensional analysis.

7-1 NATURE OF DIMENSIONAL ANALYSIS

Most phenomena in fluid mechanics depend in a complex way on geometric and flow parameters. For example, consider the drag force on a stationary smooth sphere immersed in a uniform stream. What experiments must be conducted to determine the drag force on the sphere? In order to answer this question, we must specify the parameters that are important in determining the drag force. Clearly, we would expect the drag force to depend on the size of the sphere (characterized by the diameter, D), the fluid velocity, V, and the fluid viscosity, μ. In addition, the density of the fluid, ρ, also might be important. Representing the drag force by F, we can write the symbolic equation

$$F = f(D, V, \rho, \mu)$$

Although we may have neglected parameters on which the drag force depends, such as surface roughness (or have included parameters on which it does not depend), we have formulated the problem of determining the drag force for a stationary sphere in terms of quantities that are both controllable and measurable in the laboratory.

Let us imagine a series of experiments to determine the dependence of F on the variables D, V, ρ, and μ. After building a suitable experimental facility, the work could begin. To obtain a curve of F versus V for fixed values of ρ, μ, and D, we might need tests at 10 values of V. To explore the diameter effect, each test would be repeated for spheres of 10 different diameters. Then the procedure would be repeated 10 times for ρ and μ in turn. Simple arithmetic shows that 10^4 separate tests would be needed. If each test takes $\frac{1}{2}$ hour and we work 8 hours per day, the testing will require $2\frac{1}{2}$ years to complete. Needless to say, there also would be some difficulty in presenting the data. By plotting F versus V with D as a parameter for each combination of density and viscosity values, all of the data could be presented on a total of 100 sheets of graph paper. The utility of such results would be limited at best.

Fortunately, we can obtain meaningful results with significantly less effort through the use of dimensional analysis. As shown in Example Problem 7.1, all the data for the drag force on a smooth sphere can be plotted as a functional relation between two nondimensional parameters in the form

$$\frac{F}{\rho V^2 D^2} = f_1\left(\frac{\rho V D}{\mu}\right)$$

The form of the function still must be determined experimentally. However, rather than conduct 10^4 experiments, we could establish the nature of the function as accurately with only 10 tests. The time saved in performing only 10 rather than 10^4 tests is obvious. Even more important is the greater experimental convenience. No longer must we find fluids with 10 different values of density and viscosity. Nor must we make 10 spheres of different diameters. Instead, only the *ratio*, $\rho V D/\mu$, must be varied. This can be accomplished by changing the velocity, for example.

The Buckingham Pi theorem is a statement of the relation between a function expressed in terms of dimensional parameters and a related function expressed in terms of nondimensional parameters. Use of the Buckingham Pi theorem allows us to develop the important nondimensional parameters quickly and easily.

7-2 BUCKINGHAM PI THEOREM

Given a physical problem in which the dependent parameter is a function of $n - 1$ independent parameters, we may express the relationship among the variables in functional form as

$$q_1 = f(q_2, q_3, \ldots, q_n)$$

where q_1 is the dependent parameter, and $q_2, q_3, \ldots, q_n$ are the $n - 1$ independent parameters. Mathematically, we can express the functional relationship in the equivalent form

$$g(q_1, q_2, \ldots, q_n) = 0$$

where g is an unspecified function, different from f. For the drag on a sphere we wrote the symbolic equation

$$F = f(D, V, \rho, \mu)$$

We could just as well write

$$g(F, D, V, \rho, \mu) = 0$$

The Buckingham Pi theorem states that: Given a relation among n parameters of the form

$$g(q_1, q_2, \ldots, q_n) = 0$$

then the n parameters may be grouped into $n - m$ independent dimensionless ratios, or Π parameters, expressible in functional form by

$$G(\Pi_1, \Pi_2, \ldots, \Pi_{n-m}) = 0$$

or

$$\Pi_1 = G_1(\Pi_2, \Pi_3, \ldots, \Pi_{n-m})$$

The number m is usually,[1] but not always, equal to the minimum number of independent dimensions required to specify the dimensions of all the parameters $q_1, q_2, \ldots, q_n$.

The theorem does not predict the functional form of G or G_1. The functional relation among the independent, dimensionless Π parameters must be determined experimentally.

A Π parameter is not independent if it can be formed from a product or quotient of the other parameters of the problem. For example, if

$$\Pi_5 = \frac{2\Pi_1}{\Pi_2 \Pi_3} \quad \text{or} \quad \Pi_6 = \frac{\Pi_1^{3/4}}{\Pi_3^2}$$

then neither Π_5 nor Π_6 is independent of the other dimensionless parameters.

Several methods for determining the dimensionless parameters are available. A detailed procedure is developed in the next section.

7-3 DETERMINING THE Π GROUPS

Regardless of the method to be used in determining the dimensionless parameters, one begins by listing all parameters that are known (or believed) to affect the given flow phenomenon. Some experience admittedly is helpful in compiling the list. Students,

[1] See Example Problem 7.3.

who do not have this experience, often are troubled by the need to apply engineering judgment in an apparent massive dose. However, it is difficult to go wrong if a generous selection of parameters is made.

If you suspect that a phenomenon depends on a given parameter, include it. If your suspicion is correct, experiments will show that the parameter must be included to get consistent results. If the parameter is extraneous, an extra Π parameter may result, but experiments will show that it may be eliminated from consideration. Therefore, do not be afraid to include *all* the parameters that you feel are important.

The six steps listed below are a recommended procedure for determining the Π parameters.

Step 1. *List all the parameters involved.* (Let n be the number of parameters.) If all the pertinent parameters are not included, a relation may be obtained, but it will not give the complete story. If parameters that actually have no effect on the physical phenomenon are included, either the process of dimensional analysis will show that these do not enter the relation sought or one or more dimensionless groups will be obtained that experiments will show to be extraneous.

Step 2. *Select a set of fundamental (primary) dimensions, e.g. MLt or FLt.* (Note that for heat transfer problems you may also need T for temperature, and in electrical systems, q for charge.)

Step 3. *List the dimensions of all parameters in terms of primary dimensions.* (Let r be the number of primary dimensions.) Either force or mass may be selected as a primary dimension.

Step 4. *Select from the list of parameters a number of repeating parameters equal to the number of primary dimensions, r, and including all the primary dimensions.* No two repeating parameters should have the same net dimensions differing by only a single exponent; for example, do not include both a length (L) and a moment of inertia of an area (L^4) as repeating parameters. The repeating parameters chosen may appear in all the dimensionless groups obtained; consequently, do *not* include the dependent parameter among those selected in this step.

Step 5. *Set up dimensional equations combining the parameters selected in Step 4 with each of the other parameters in turn to form dimensionless groups.* (There will be $(n - m)$ equations.) Solve the dimensional equations to obtain the $(n - m)$ dimensionless groups.

Step 6. *Check to see that each group obtained is dimensionless.* If mass was initially selected as a primary dimension, it is wise to check the groups using force as a primary dimension, and vice versa.

The functional relationship among the Π parameters must be determined experimentally. The detailed procedure for determining the dimensionless Π parameters is illustrated in Example Problems 7.1 and 7.2.

Example 7.1

As noted in Section 7-1, the drag force, F, on a smooth sphere depends on the relative velocity, V, the sphere diameter, D, the fluid density, ρ, and the fluid viscosity, μ. Obtain a set of dimensionless groups that can be used to correlate experimental data.

EXAMPLE PROBLEM 7.1

GIVEN: $F = f(\rho, V, D, \mu)$ for a smooth sphere.

FIND: An appropriate set of dimensionless groups.

SOLUTION:

(Circled numbers refer to steps in the procedure for determining dimensionless Π parameters.)

① $F \quad V \quad D \quad \rho \quad \mu \qquad n = 5$ parameters

② Select primary dimensions M, L, and t

③ $F \quad V \quad D \quad \rho \quad \mu$

$$\frac{ML}{t^2} \quad \frac{L}{t} \quad L \quad \frac{M}{L^3} \quad \frac{M}{Lt} \qquad r = 3 \text{ primary dimensions}$$

④ $\rho, V, D \qquad m = r = 3$ repeating parameters

⑤ Then $n - m = 2$ dimensionless groups will result. Setting up dimensional equations, we obtain

$$\Pi_1 = \rho^a V^b D^c F = \left(\frac{M}{L^3}\right)^a \left(\frac{L}{t}\right)^b (L)^c \left(\frac{ML}{t^2}\right) = M^0 L^0 t^0$$

Equating the exponents of M, L, and t results in

$$\left.\begin{array}{lll} M: & a + 1 = 0 & a = -1 \\ L: & -3a + b + c + 1 = 0 & c = -2 \\ t: & -b - 2 = 0 & b = -2 \end{array}\right\} \quad \text{Therefore, } \Pi_1 = \frac{F}{\rho V^2 D^2}$$

Similarly,

$$\Pi_2 = \rho^d V^e D^f \mu = \left(\frac{M}{L^3}\right)^d \left(\frac{L}{t}\right)^e (L)^f \left(\frac{M}{Lt}\right) = M^0 L^0 t^0$$

$$\left.\begin{array}{lll} M: & d + 1 = 0 & d = -1 \\ L: & -3d + e + f - 1 = 0 & f = -1 \\ t: & -e - 1 = 0 & e = -1 \end{array}\right\} \quad \text{Therefore, } \Pi_2 = \frac{\mu}{\rho V D}$$

⑥ Check using F, L, t dimensions

$$\Pi_1 = \frac{F}{\rho V^2 D^2} : F \frac{L^4}{Ft^2} \left(\frac{t}{L}\right)^2 \frac{1}{L^2} = [1]$$

where [ ] means "has dimensions of" and

$$\Pi_2 = \frac{\mu}{\rho V D} : \frac{Ft}{L^2} \frac{L^4}{Ft^2} \frac{t}{L} \frac{1}{L} = [1]$$

The functional relationship is $\Pi_1 = f(\Pi_2)$, or

$$\frac{F}{\rho V^2 D^2} = f\left(\frac{\mu}{\rho V D}\right)$$

as noted before. The form of the function, f, must be determined experimentally.

Example 7.2

The pressure drop, Δp, for steady, incompressible viscous flow through a straight horizontal pipe depends on the pipe length, l, the average velocity, $\bar{V}$, the viscosity, μ, the pipe diameter, D, the density, ρ, and the average variation, e, of the inside radius (the average "roughness" height). Determine a set of dimensionless groups that can be used to correlate data.

EXAMPLE PROBLEM 7.2

GIVEN: $\Delta p = f(\rho, \bar{V}, D, l, \mu, e)$ for flow in a circular pipe.

FIND: A suitable set of dimensionless groups.

SOLUTION:

(Circled numbers refer to steps in the procedure for determining dimensionless Π parameters.)

① $\Delta p \quad \rho \quad \mu \quad \bar{V} \quad l \quad D \quad e \qquad n = 7$ parameters

② Choose primary dimensions M, L, and t

③ $\Delta p \quad \rho \quad \mu \quad \bar{V} \quad l \quad D \quad e$

$$\frac{M}{Lt^2} \quad \frac{M}{L^3} \quad \frac{M}{Lt} \quad \frac{L}{t} \quad L \quad L \quad L \qquad r = 3 \text{ primary dimensions}$$

④ $\rho, \bar{V}, D \qquad m = r = 3$ repeating parameters

⑤ Then $n - m = 4$ dimensionless groups will result. Setting up dimensional equations we have:

$$\Pi_1 = \rho^a \bar{V}^b D^c \Delta p = \left(\frac{M}{L^3}\right)^a \left(\frac{L}{t}\right)^b (L)^c \left(\frac{M}{Lt^2}\right) = M^0 L^0 t^0$$

$$\begin{aligned} M&: \quad 0 = a + 1 \\ L&: \quad 0 = -3a + b + c - 1 \\ t&: \quad 0 = -b - 2 \end{aligned} \Bigg\} \quad \begin{aligned} a &= -1 \\ b &= -2 \\ c &= 0 \end{aligned}$$

Therefore, $\Pi_1 = \rho^{-1}\bar{V}^{-2}D^0\,\Delta p = \dfrac{\Delta p}{\rho \bar{V}^2}$

$$\Pi_2 = \rho^d \bar{V}^e D^f \mu = \left(\frac{M}{L^3}\right)^d \left(\frac{L}{t}\right)^e (L)^f \frac{M}{Lt} = M^0 L^0 t^0$$

$$\begin{aligned} M&: \quad 0 = d + 1 \\ L&: \quad 0 = -3d + e + f - 1 \\ t&: \quad 0 = -e - 1 \end{aligned} \Bigg\} \quad \begin{aligned} d &= -1 \\ e &= -1 \\ f &= -1 \end{aligned}$$

Therefore, $\Pi_2 = \dfrac{\mu}{\rho \bar{V} D}$

$$\Pi_3 = \rho^g \bar{V}^h D^i l = \left(\frac{M}{L^3}\right)^g \left(\frac{L}{t}\right)^h (L)^i L = M^0 L^0 t^0$$

$$\begin{aligned} M&: \quad 0 = g \\ L&: \quad 0 = -3g + h + i + 1 \\ t&: \quad 0 = -h \end{aligned} \Bigg\} \quad \begin{aligned} g &= 0 \\ h &= 0 \\ i &= -1 \end{aligned}$$

Therefore, $\Pi_3 = \dfrac{l}{D}$

$$\Pi_4 = \rho^j \bar{V}^k D^l e = \left(\frac{M}{L^3}\right)^j \left(\frac{L}{t}\right)^k (L)^l L = M^0 L^0 t^0$$

$$\begin{aligned} M&: \quad 0 = j \\ L&: \quad 0 = -3j + k + l + 1 \\ t&: \quad 0 = -k \end{aligned} \Bigg\} \quad \begin{aligned} j &= 0 \\ k &= 0 \\ l &= -1 \end{aligned}$$

Therefore, $\Pi_4 = \dfrac{e}{D}$

⑥ Check, using F, L, t dimensions

$$\Pi_1 = \frac{\Delta p}{\rho \bar{V}^2} : \frac{F}{L^2}\frac{L^4}{Ft^2}\frac{t^2}{L^2} = [1] \qquad \Pi_3 = \frac{l}{D} : \frac{L}{L} = [1]$$

$$\Pi_2 = \frac{\mu}{\rho \bar{V} D} : \frac{Ft}{L^2}\frac{L^4}{Ft^2}\frac{t}{L}\frac{1}{L} = [1] \qquad \Pi_4 = \frac{e}{D} : \frac{L}{L} = [1]$$

Finally, the functional relationship is

$$\Pi_1 = f(\Pi_2, \Pi_3, \Pi_4)$$

or

$$\frac{\Delta p}{\rho \bar{V}^2} = f\left(\frac{\mu}{\rho \bar{V} D}, \frac{l}{D}, \frac{e}{D}\right)$$

{Experiments in many laboratories have shown that this relationship correlates the data well. We shall discuss this result in greater detail in Section 8-7.1.}

The procedure outlined above, where m is taken equal to r (the fewest independent dimensions required to specify the dimensions of all parameters involved) almost always produces the correct number of dimensionless Π parameters. In a few cases, trouble arises because the number of primary dimensions differs when variables are expressed in terms of different systems of dimensions. The value of m can be established with certainty by determining the rank of the dimensional matrix; m is equal to the rank of the dimensional matrix. The procedure is illustrated in Example Problem 7.3.

The $n - m$ dimensionless groups obtained from the procedure are independent but not unique. If a different set of repeating parameters is chosen, different groups result. The choice of ρ $[M/L^3]$, $V[L/t]$, and a characteristic length $[L]$ as repeating variables generally leads to a set of dimensionless parameters that have been found to be most suitable for correlating a wide range of experimental data. This is not surprising if one recognizes that inertia forces are important in most fluid mechanics problems. From Newton's second law, $F = ma$; the mass can be written as $m = \rho \forall$ and, since volume has dimensions of L^3, then qualitatively $m \propto \rho L^3$. Since the acceleration can be written as $a = dv/dt = v\, dv/ds$, then qualitatively $a \propto V^2/L$ and the inertia force, F, is proportional to $\rho V^2 L^2$.

If $n - m = 1$, then a single dimensionless Π parameter is obtained. In this case, the Buckingham Pi theorem indicates that the single Π parameter must be a constant.

Example 7.3

When a small tube is dipped into a pool of liquid, surface tension causes a *meniscus* to form at the free surface, which is elevated or depressed depending on the contact angle at the liquid-solid-gas interface. Experiments indicate that the magnitude of this *capillary effect*, Δh, is a function of the tube diameter, D, liquid specific weight, γ, and surface tension, σ. Determine the number of independent Π parameters that can be formed and obtain a set.

EXAMPLE PROBLEM 7.3

GIVEN: $\Delta h = f(D, \gamma, \sigma)$

FIND: (a) Number of independent Π parameters.
(b) Evaluate one set.

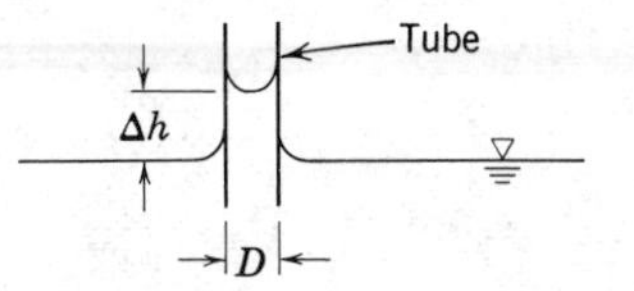

Liquid
(Specific weight = γ
Surface tension = σ)

SOLUTION:
(Circled numbers refer to steps in the procedure for determining dimensionless Π parameters.)

① $\Delta h \quad D \quad \gamma \quad \sigma \qquad n = 4$ parameters

② Choose primary dimensions (use both M, L, t and F, L, t dimensions to illustrate the problem in determining m)

③ (a) M, L, t

$$\begin{array}{cccc} \Delta h & D & \gamma & \sigma \\ L & L & \dfrac{M}{L^2t^2} & \dfrac{M}{t^2} \end{array}$$

$r = 3$ primary dimensions

(b) F, L, t

$$\begin{array}{cccc} \Delta h & D & \gamma & \sigma \\ L & L & \dfrac{F}{L^3} & \dfrac{F}{L} \end{array}$$

$r = 2$ primary dimensions

Thus we ask, "Is m equal to r?" Let us check the dimensional matrix to find out.

	Δh	D	γ	σ
M	0	0	1	1
L	1	1	−2	0
t	0	0	−2	−2

	Δh	D	γ	σ
F	0	0	1	1
L	1	1	−3	−1

The rank of a matrix is equal to the order of its largest nonzero determinant.

$$\begin{vmatrix} 0 & 1 & 1 \\ 1 & -2 & 0 \\ 0 & -2 & -2 \end{vmatrix} = 0 - (1)(-2) + (1)(-2) = 0$$

$$\begin{vmatrix} -2 & 0 \\ -2 & -2 \end{vmatrix} = 4 \neq 0 \qquad \therefore m = 2 \qquad m \neq r$$

$$\begin{vmatrix} 1 & 1 \\ -3 & -1 \end{vmatrix} = -1 + 3 = 2 \neq 0$$

$$\therefore m = 2 \qquad m = r$$

④ (a) $m = 2$. Choose D, γ as repeating parameters.

⑤ (a) $n - m = 2$ dimensionless groups will result.

$$\Pi_1 = D^a \gamma^b \, \Delta h = (L)^a \left(\frac{M}{L^2t^2}\right)^b (L) = M^0L^0t^0$$

$$\left.\begin{aligned} M&: \quad b + 0 = 0 \\ L&: \quad a - 2b + 1 = 0 \\ t&: \quad -2b + 0 = 0 \end{aligned}\right\} \quad b = 0, \quad a = -1$$

④ (b) $m = 2$. Choose D, γ as repeating parameters.

⑤ (b) $n - m = 2$ dimensionless groups will result.

$$\Pi_1 = D^e \gamma^f \, \Delta h = (L)^e \left(\frac{F}{L^3}\right)^f L = F^0L^0t^0$$

$$\left.\begin{aligned} F&: \quad f = 0 \\ L&: \quad e - 3f + 1 = 0 \end{aligned}\right\} \quad e = -1$$

Therefore, $\Pi_1 = \dfrac{\Delta h}{D}$

$$\Pi_2 = D^c \gamma^d \sigma$$

$$= (L)^c \left(\frac{M}{L^2 t^2}\right)^d \frac{M}{t^2} = M^0 L^0 t^0$$

$$\left.\begin{aligned} M:&\quad d + 1 = 0 \\ L:&\quad c - 2d = 0 \\ t:&\quad -2d - 2 = 0 \end{aligned}\right\} \quad \begin{aligned} d &= -1 \\ c &= -2 \end{aligned}$$

Therefore, $\Pi_2 = \dfrac{\sigma}{D^2 \gamma}$

⑥ Check, using F, L, t dimensions

$$\Pi_1 = \frac{\Delta h}{D} : \frac{L}{L} = [1]$$

$$\Pi_2 = \frac{\sigma}{D^2 \gamma} : \frac{F}{L} \frac{1}{L^2} \frac{L^3}{F} = [1]$$

Therefore, $\Pi_1 = \dfrac{\Delta h}{D}$

$$\Pi_2 = D^g \gamma^h \sigma$$

$$= (L)^g \left(\frac{F}{L^3}\right)^h \frac{F}{L} = F^0 L^0 t^0$$

$$\left.\begin{aligned} F:&\quad h + 1 = 0 \\ L:&\quad g - 3h - 1 = 0 \end{aligned}\right\} \quad \begin{aligned} h &= -1 \\ g &= -2 \end{aligned}$$

Therefore, $\Pi_2 = \dfrac{\sigma}{D^2 \gamma}$

Check, using M, L, t dimensions

$$\Pi_1 = \frac{\Delta h}{D} : \frac{L}{L} = [1]$$

$$\Pi_2 = \frac{\sigma}{D^2 \gamma} : \frac{M}{t^2} \frac{1}{L^2} \frac{L^2 t^2}{M} = [1]$$

Therefore, both systems of dimensions yield the same dimensionless Π parameters. The predicted functional relationship is

$$\Pi_1 = f(\Pi_2) \qquad \text{or} \qquad \frac{\Delta h}{D} = f\left(\frac{\sigma}{D^2 \gamma}\right)$$

7-4 DIMENSIONLESS GROUPS OF SIGNIFICANCE IN FLUID MECHANICS

Over the years, several hundred different dimensionless groups that are important in engineering have been identified. Following tradition, each such group has been given the name of a prominent scientist or engineer, usually the one who pioneered its use. Several are so fundamental and occur so frequently in fluid mechanics that we should take time to learn their definitions. Understanding their physical significance also gives insight into the phenomena we study.

Forces encountered in flowing fluids include those due to inertia, viscosity, pressure, gravity, surface tension, and compressibility. The ratio of any two forces will be dimensionless. We have previously shown that the inertia force is proportional to $\rho V^2 L^2$. In order to facilitate taking the ratio of forces we can express each of the remaining forces as follows:

$$\text{Viscous force} = \tau A \propto \mu \frac{du}{dy} A \propto \mu \frac{V}{L} L^2 = \mu V L$$

$$\text{Pressure force} = (\Delta p)A \propto (\Delta p)L^2$$

$$\text{Gravity force} = Mg \propto g\rho L^3$$

$$\text{Surface tension force} = \sigma L$$

$$\text{Compressibility force} = E_v A \propto E_v L^2$$

Inertia forces are important in most fluid mechanics problems. The ratio of the inertia force to each of the other forces listed above leads to five fundamental dimensionless groups encountered in fluid mechanics.

In the 1880s, Osborne Reynolds, the British engineer, studied the transition between laminar and turbulent flow in a tube. He discovered that the parameter (later named after him)

$$Re = \frac{\rho \bar{V} D}{\mu} = \frac{\bar{V} D}{\nu}$$

is a criterion by which the state of a flow may be determined. Later experiments have shown that the *Reynolds number* is a key parameter for other flow cases as well. Thus, in general,

$$Re = \frac{\rho V L}{\mu} = \frac{V L}{\nu}$$

where L is a characteristic length descriptive of the flow field. The Reynolds number is the ratio of inertia forces to viscous forces. "Large" Reynolds number flows generally are turbulent. Flows in which the inertia forces are "small" compared to viscous forces are characteristically laminar flows.

In aerodynamic and other model testing, it is convenient to present pressure data in dimensionless form. The ratio

$$Eu(=C_p) = \frac{\Delta p}{\frac{1}{2}\rho V^2}$$

is formed, where Δp is the local pressure minus the freestream pressure, and ρ and V are properties of the freestream flow. This ratio has been named after Leonhard Euler, the Swiss mathematician who did much of the early analytical work in fluid mechanics. The *Euler number* is the ratio of pressure forces to inertia forces. (The factor $\frac{1}{2}$ is introduced into the denominator to give the dynamic pressure.)

In the study of cavitation phenomena, the pressure difference, Δp, is taken as $\Delta p = p - p_v$, where p, ρ, and V are conditions in the liquid stream, and p_v is the liquid vapor pressure at the test temperature. The resulting dimensionless parameter is referred to as the *cavitation number*.

William Froude was a British naval architect. Together with his son, Robert Edmund Froude, he discovered that the parameter

$$Fr = \frac{V}{\sqrt{gL}}$$

was significant for flows with free surface effects. Squaring the *Froude number* gives

$$Fr^2 = \frac{V^2}{gL} = \frac{\rho V^2 L^2}{\rho g L^3}$$

which may be interpreted as the ratio of inertia forces to gravity forces. The length, L, is a characteristic length descriptive of the flow field. In the case of open-channel flow, the characteristic length is the water depth; Froude numbers less than one indicate subcritical flow and values greater than one indicate supercritical flow.

The *Weber number* is the ratio of inertia forces to surface tension forces. It may be written

$$We = \frac{\rho V^2 L}{\sigma}$$

In the 1870s, the Austrian physicist Ernst Mach introduced the parameter

$$M = \frac{V}{c}$$

where V is the flow speed and c is the local sonic speed. Analysis and experiments have shown that the *Mach number* is a key parameter that characterizes compressibility effects in a flow. The Mach number may be written

$$M = \frac{V}{c} = \frac{V}{\sqrt{\dfrac{dp}{d\rho}}} = \frac{V}{\sqrt{\dfrac{E_v}{\rho}}} \qquad \text{or} \qquad M^2 = \frac{\rho V^2 L^2}{E_v L^2}$$

which may be interpreted as a ratio of inertia forces to forces due to compressibility. For truly incompressible flow (under some conditions even liquids are quite compressible), $c = \infty$ so that $M = 0$.

7-5 FLOW SIMILARITY AND MODEL STUDIES

To be useful, a model test must yield data that can be scaled to obtain the forces, moments, and dynamic loads that would exist on the full-scale prototype. What conditions must be met to ensure the similarity of model and prototype flows?

Perhaps the most obvious requirement is that the model and prototype must be geometrically similar. *Geometric similarity* requires that the model and prototype be the same shape, and that all linear dimensions of the model be related to corresponding dimensions of the prototype by a constant scale factor.

A second requirement is that the model and prototype flows must be *kinematically similar*. Two flows are kinematically similar when the velocities at corresponding points are in the same direction and are related in magnitude by a constant scale factor. Thus two flows that are kinematically similar also have streamline patterns related by a constant scale factor. Since the boundaries form the bounding streamlines, flows that are kinematically similar must be geometrically similar.

In principle, kinematic similarity would require that a wind tunnel of infinite cross section be used to obtain data for drag on an object, in order to model correctly the performance in an infinite flow field. In practice, this restriction may be relaxed considerably, permitting use of reasonable size equipment.

Kinematic similarity requires that the regimes of flow be the same for model and prototype. If compressibility or cavitation effects, which may change even the qualitative patterns of flow, are not present in the prototype flow, they must be avoided in the model flow.

When two flows have force distributions such that identical types of forces are parallel and are related in magnitude by a constant scale factor at all corresponding points, the flows are *dynamically similar.*

The requirements for dynamic similarity are most restrictive: Two flows must possess both geometric and kinematic similarity to be dynamically similar.

To establish the conditions required for complete dynamic similarity, all forces that are important in the flow situation must be considered. Thus the effects of viscous forces, of pressure forces, of surface tension forces, and so on, must be considered. Then test conditions must be established so that all important forces are related by the same scale factor between model and prototype flows. When dynamic similarity exists, data measured in a model flow may be related quantitatively to conditions in the prototype flow. What, then, are the conditions for ensuring dynamic similarity between model and prototype flows?

The Buckingham Pi theorem may be used to obtain the governing dimensionless groups for a flow phenomenon; to achieve dynamic similarity between geometrically similar flows, we must duplicate all but one of these dimensionless groups.

For example, in considering the drag force on a sphere in Example Problem 7.1, we began with

$$F = f(D, V, \rho, \mu)$$

The Buckingham Pi theorem predicted the functional relation

$$\frac{F}{\rho V^2 D^2} = f_1\left(\frac{\rho V D}{\mu}\right)$$

In Section 7-4 we have shown that the dimensionless parameters can be viewed as ratios of forces. Thus, in considering a model flow and a prototype flow about a sphere (the flows are geometrically similar), the flows also will be dynamically similar if

$$\left(\frac{\rho V D}{\mu}\right)_{\text{model}} = \left(\frac{\rho V D}{\mu}\right)_{\text{prototype}}$$

Furthermore, if

$$Re_{\text{model}} = Re_{\text{prototype}}$$

then

$$\left(\frac{F}{\rho V^2 D^2}\right)_{\text{model}} = \left(\frac{F}{\rho V^2 D^2}\right)_{\text{prototype}}$$

and the results determined from the model study can be used to predict the drag on the full-scale prototype.

The actual force due to the fluid on the object is not the same in both cases, but its dimensionless value is. The two tests can be run using different fluids, if desired, as long as the Reynolds numbers are matched. For experimental convenience, test data can be measured in a wind tunnel in air and the results used to predict drag in water, as illustrated in Example Problem 7.4.

Example 7.4

The drag of a sonar transducer is to be predicted, based on wind tunnel test data. The prototype, a 1 ft diameter sphere, is to be towed at 5 knots (nautical miles per hour) in seawater at 5 C. The model is 6 in. in diameter. Determine the required test speed in air. If the drag of the model at test conditions is 5.58 lbf, estimate the drag of the prototype.

EXAMPLE PROBLEM 7.4

GIVEN: Sonar transducer to be tested in a wind tunnel.

FIND:
(a) V_m.
(b) F_p.

$D_p = 1$ ft
$V_p = 5$ knots
F_p
Water at 5 C
$D_m = 6$ in.
V_m
$F_m = 5.58$ lbf
Air

SOLUTION:
Since the prototype operates in water and the model test is to be performed in air, useful results can be expected only if cavitation effects are absent in the prototype flow and compressibility effects are absent from the model test. Under these conditions,

$$\frac{F}{\rho V^2 D^2} = f\left(\frac{\rho V D}{\mu}\right)$$

and the test should be run at

$$Re_{\text{model}} = Re_{\text{prototype}}$$

to ensure dynamic similarity. For seawater at 5 C, $\rho = 1.99$ slug/ft^3 and $\nu = 1.68 \times 10^{-5}$ ft^2/sec. At prototype conditions,

$$V_p = \frac{5 \text{ nmi}}{\text{hr}} \times \frac{6080 \text{ ft}}{\text{nmi}} \times \frac{\text{hr}}{3600 \text{ sec}} = 8.44 \text{ ft/sec}$$

$$Re_p = \frac{V_p D_p}{\nu_p} = \frac{8.44 \text{ ft}}{\text{sec}} \times 1 \text{ ft} \times \frac{\text{sec}}{1.68 \times 10^{-5} \text{ ft}^2} = 5.02 \times 10^5$$

The model test conditions must duplicate this Reynolds number. Thus

$$Re_m = \frac{V_m D_m}{\nu_m} = 5.02 \times 10^5$$

For air at STP, $\rho = 0.00238$ slug/ft^3 and $\nu = 1.56 \times 10^{-4}$ ft^2/sec. The wind tunnel must be operated at

$$V_m = Re_m \frac{\nu_m}{D_m} = 5.02 \times 10^5 \times \frac{1.56 \times 10^{-4} \text{ ft}^2}{\text{sec}} \times \frac{1}{0.5 \text{ ft}}$$

$$V_m = 156 \text{ ft/sec}$$ V_m

This value is low enough for compressibility effects to be negligible.

At these test conditions, the model and prototype flows are dynamically similar. Hence

$$\left.\frac{F}{\rho V^2 D^2}\right)_m = \left.\frac{F}{\rho V^2 D^2}\right)_p$$

and

$$F_p = F_m \frac{\rho_p}{\rho_m} \frac{V_p^2}{V_m^2} \frac{D_p^2}{D_m^2} = \frac{5.58 \text{ lbf}}{} \times \frac{1.99}{0.00238} \times \frac{(8.44)^2}{(156)^2} \times \frac{1}{(0.5)^2}$$

$$F_p = 54.6 \text{ lbf}$$ F_p

If cavitation were expected—if the sonar probe were operated at high speed near the free surface of the seawater—then useful results from a model test in air could not be expected.

{This problem demonstrates the calculation of prototype values from model test data.}

7-5.1 Incomplete Similarity

We have shown that it is necessary to duplicate all but one of the significant dimensionless groups to achieve complete dynamic similarity between geometrically similar flows.

In the simplified situation of Example Problem 7.4, duplicating the Reynolds number between model and prototype ensured dynamically similar flows. Testing in air allowed the Reynolds number to be duplicated exactly (this also could have been accomplished in a water tunnel for this situation). The drag force on a sphere actually depends on the nature of the boundary-layer flow. Therefore, geometric similarity requires that the relative surface roughness of the model and prototype be the same. This means that relative roughness also is a parameter that must be duplicated between model and prototype situations. If we assume that the model was constructed carefully, measured values of drag from model tests could be scaled to predict drag for the operating conditions of the prototype.

In many model studies, to achieve dynamic similarity requires duplication of several dimensionless groups. In some cases complete dynamic similarity between model and prototype may not be attainable. Determining the drag force (resistance) of a surface ship is an example of such a situation. Resistance on a surface ship arises from skin friction on the hull (viscous forces) and surface wave resistance (gravity forces). Complete dynamic similarity requires that both Reynolds and Froude numbers be duplicated between model and prototype.

In general it is not possible to predict wave resistance analytically, so it must be modeled. This requires that

$$Fr_m = \frac{V_m}{(gL_m)^{1/2}} = Fr_p = \frac{V_p}{(gL_p)^{1/2}}$$

To match Froude numbers between model and prototype requires a velocity ratio of

$$\frac{V_m}{V_p} = \left(\frac{L_m}{L_p}\right)^{1/2}$$

to ensure dynamically similar surface wave patterns.

For any model length scale, matching the Froude numbers determines the velocity ratio. Only the kinematic viscosity can be varied to match Reynolds numbers. Thus

$$Re_m = \frac{V_m L_m}{\nu_m} = Re_p = \frac{V_p L_p}{\nu_p}$$

leads to the condition that

$$\frac{\nu_m}{\nu_p} = \frac{V_m}{V_p}\frac{L_m}{L_p}$$

If we use the velocity ratio obtained from matching Froude numbers, equality of Reynolds numbers leads to a kinematic viscosity ratio of

$$\frac{\nu_m}{\nu_p} = \left(\frac{L_m}{L_p}\right)^{1/2}\frac{L_m}{L_p} = \left(\frac{L_m}{L_p}\right)^{3/2}$$

If L_m/L_p equals $\frac{1}{100}$ (a typical length scale for ship model tests), then ν_m/ν_p must be $\frac{1}{1000}$. Figure A.3 shows that mercury is the only liquid with kinematic viscosity less than water. However, it is only about an order of magnitude less, so the kinematic viscosity ratio required to duplicate Reynolds numbers cannot be attained.

Water is the only practical fluid for most model tests of free surface flows. To obtain complete dynamic similarity then would require a full-scale test. However, model studies do provide useful information even though complete similarity cannot be

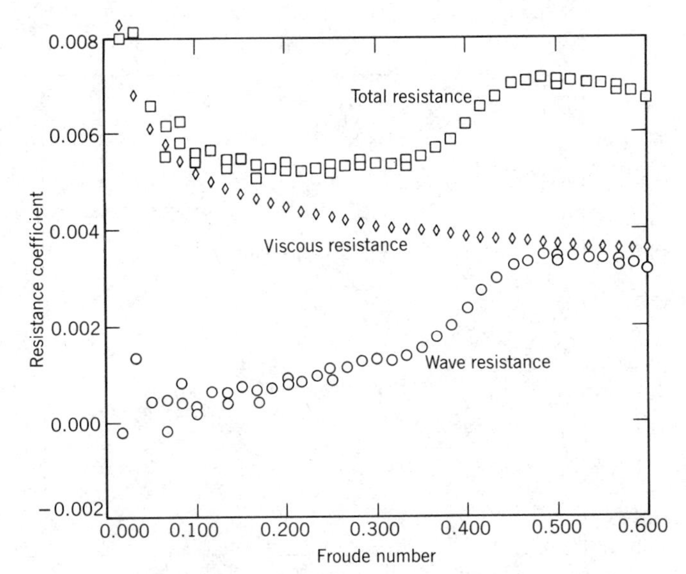

Fig. 7.1 Data from test of 1:80-scale model of U.S. Navy guided missile frigate Oliver Hazard Perry (FFG-7). (Data from U.S. Naval Academy Hydromechanics Laboratory, courtesy of Professor Bruce Johnson.)

obtained. Figure 7.1 shows data from a test of a 1 : 80 scale model of a ship conducted at the U.S. Naval Academy Hydromechanics Laboratory. The plot displays resistance coefficient data versus Froude number. The square points are calculated from values of total resistance measured in the test.

Resistance of the full-scale ship can be calculated from model test results using the following procedure. The pattern of surface waves and thus the wave resistance are matched between model and prototype at corresponding Froude numbers. The viscous drag on the model is estimated using the analytical methods of Chapter 9 (the estimated frictional resistance coefficients are plotted in Fig. 7.1 as inclined squares). The model wave resistance is calculated as the difference between total drag and estimated friction drag (values of estimated wave resistance for the model are plotted as circles).

The prototype wave resistance is calculated using Froude number scaling by equating wave resistance coefficients for model and prototype. The points plotted as circles in Fig. 7.2 for the prototype are identical to model values at corresponding Froude numbers. The skin friction drag calculated analytically for the prototype is shown in Fig. 7.2 by the inclined squares. This estimated viscous drag is added to the scaled wave drag to predict the total prototype drag.

Because the Reynolds number cannot be matched for model tests of surface ships, the boundary-layer behavior is not the same for model and prototype. The model Reynolds number is only $(L_m/L_p)^{3/2}$ as large as the prototype value, so the extent of laminar flow in the boundary layer on the model is too large by a corresponding factor. The method just described assumes that boundary-layer behavior can be scaled. To make this possible, the model boundary layer is "tripped" or "stimulated" to become

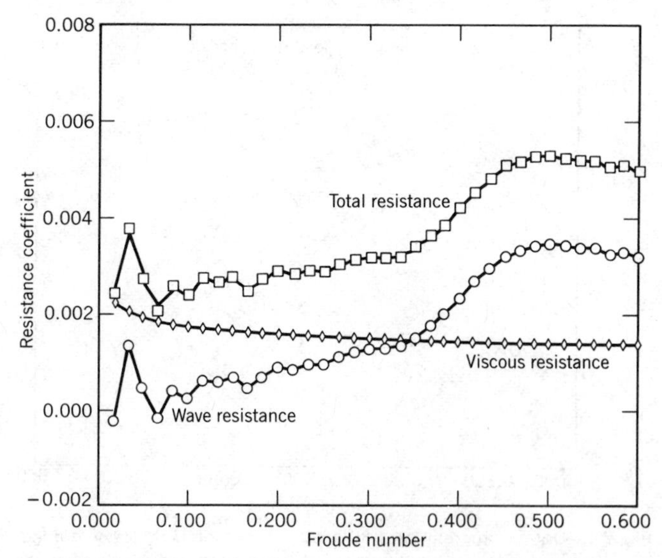

Fig. 7.2 Resistance of full-scale ship predicted from model test results. (Data from U.S. Naval Academy Hydromechanics Laboratory, courtesy of Professor Bruce Johnson.)

turbulent at a location that corresponds to the behavior on the full-scale vessel. "Studs" were used to stimulate the boundary layer for the model test results shown in Fig. 7.1.

A correction factor sometimes is added to the full-scale results calculated from model test data. This factor accounts for roughness, waviness, and unevenness that inevitably are more pronounced on the full-scale ship than on the model. Comparisons between predictions from model tests and measurements made in full-scale trials suggest an overall accuracy within ± 5 percent.

The Froude number is an important parameter in the modeling of rivers and harbors. In these situations it is not practical to obtain complete similarity. Use of a reasonable model scale would lead to extremely small water depths. Viscous forces and surface tension forces would have much larger relative effects on the model flow than in the prototype. Consequently, different length scales are used for the vertical and horizontal directions. Viscous forces in the deeper model flow are increased using artificial roughness elements.

Emphasis on fuel economy has made reduction of aerodynamic drag important for automobiles, trucks, and buses. Most work on development of low-drag configurations is done using model tests. Traditionally, automobile models have been built to $\frac{3}{8}$ scale, at which a model of a full-size automobile has a frontal area of about 0.3 m^2. Thus testing can be done in a wind tunnel with test section area of 6 m^2 or larger. At $\frac{3}{8}$ scale, a wind speed of about 150 mph is needed to model a prototype automobile traveling at the legal speed limit. Thus there is no problem with compressibility effects, but the scale models are expensive and time-consuming to build.

Figure 7.3 shows a large wind tunnel (test section dimensions are 5.4 m high, 10.4 m wide, and 21.3 m long) used by General Motors to test full-scale automobiles at prototype speeds. The large test section allows use of full-scale clay mockups of proposed auto body styles. The relatively low speed permits flow visualization using tufts or "smoke" streams.[2] Using full-size "models," stylists and engineers can work together to achieve optimum results. The car shown under test in Fig. 7.3 is a 1984 Corvette.

It is harder to achieve dynamic similarity in tests of trucks and buses; models must be made to smaller scale than those of automobiles.[3] A large scale for truck and bus testing is 1 : 8. To achieve complete dynamic similarity by matching Reynolds numbers at this scale would require a test speed of 440 mph. This would introduce unwanted compressibility effects, and model and prototype flows would not be kinematically similar. Fortunately, trucks and buses are "bluff" objects. Experiments show that above a certain Reynolds number, their nondimensional drag becomes independent of Reynolds number. Although similarity is not complete, measured test data can be scaled to predict prototype drag force values. The procedure is illustrated in Example Problem 7.5.

[2] A mixture of liquid nitrogen and steam produces "smoke" streaklines that evaporate and thus do not clog the fine mesh screens used to reduce the turbulence level in the tunnel.

[3] The vehicle length is particularly important when testing at large yaw angles to simulate crosswind behavior. Tunnel blockage considerations limit the acceptable model size. See [1] for recommended practices.

Fig. 7.3 Full-scale automobile under test in 5.4 × 10.4 m wind tunnel at General Motors Technical Center, Warren, Michigan. For a detailed description, see the full-color illustration facing page 500. (Photograph courtesy of General Motors.)

Example 7.5

The following wind tunnel test data from a 1:16 scale model of a bus are available:

Air Speed (m/sec)	18.0	21.8	26.0	30.1	35.0	38.5	40.9	44.1	46.7
Drag Force (N)	3.10	4.41	6.09	7.97	10.7	12.9	14.7	16.9	18.9

Using the properties of standard air, calculate and plot the dimensionless aerodynamic drag coefficient,

$$C_D = \frac{F_D}{\frac{1}{2}\rho V^2 A}$$

versus Reynolds number, $Re = \rho V w/\mu$, where w is model width. Find the minimum test speed above which C_D remains constant. Estimate the aerodynamic drag force and power requirement for the prototype vehicle at 100 km/hr. (The width and frontal area of the prototype are 8 ft and 84 ft^2, respectively.)

EXAMPLE PROBLEM 7.5

GIVEN: Data from a wind tunnel test of a model bus. Prototype dimensions are 8 ft width and frontal area of 84 ft^2. Model scale is 1:16. Standard air is the test fluid.

FIND: (a) Calculate and plot aerodynamic drag coefficient,

$$C_D = \frac{F_D}{\frac{1}{2}\rho V^2 A}$$

versus Reynolds number, $Re = \rho V w/\mu$.

(b) Determine the speed above which C_D is constant.

(c) Estimate the aerodynamic drag force and power required for the full-scale vehicle at 100 km/hr.

SOLUTION:

The model width is

$$w_m = \frac{1}{16} w_p = \frac{1}{16} \times 8\text{ ft} \times 0.3048\,\frac{\text{m}}{\text{ft}} = 0.152\text{ m}$$

The model area is

$$A_m = \left(\frac{1}{16}\right)^2 A_p = \left(\frac{1}{16}\right)^2 \times 84\text{ ft}^2 \times (0.3048)^2\,\frac{\text{m}^2}{\text{ft}^2} = 0.0305\text{ m}^2$$

The aerodynamic drag coefficient may be calculated as

$$C_D = \frac{F_D}{\frac{1}{2}\rho V^2 A}$$

$$= 2 \times F_D(\text{N}) \times \frac{\text{m}^3}{1.23\text{ kg}} \times \frac{\text{sec}^2}{(V)^2\,\text{m}^2} \times \frac{1}{0.0305\text{ m}^2} \times \frac{\text{kg}\cdot\text{m}}{\text{N}\cdot\text{sec}^2}$$

$$C_D = \frac{53.3\,F_D(\text{N})}{(V\,(\text{m/sec}))^2}$$

The Reynolds number may be calculated as

$$Re = \frac{\rho V w}{\mu} = \frac{V w}{\nu}$$

$$= \frac{V\ \text{m}}{\text{sec}} \times 0.152\ \text{m} \times \frac{\text{sec}}{1.45 \times 10^{-5}\ \text{m}^2}$$

$$Re = 1.05 \times 10^4\ V\ (\text{m/sec})$$

The calculated values are plotted in the following figure:

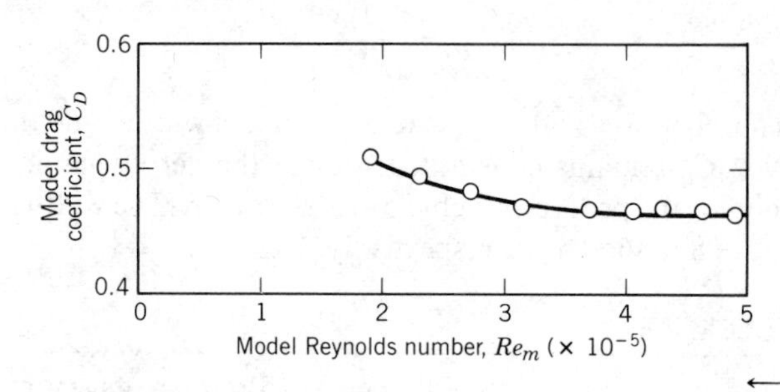

C_{Dm} versus Re_m

The plot shows that the model drag coefficient becomes constant at $C_{Dm} = 0.463$ above $Re_m = 4 \times 10^5$, which corresponds to an air speed of approximately 40 m/sec. Since the drag coefficient is independent of Reynolds number above $Re \approx 4 \times 10^5$, then for the prototype vehicle ($Re \approx 4.5 \times 10^7$), $C_D = 0.463$. The drag force on the full-scale vehicle is

$$F_{D_p} = C_D \frac{1}{2} \rho V_p^2 A_p$$

$$= \frac{0.463}{2} \times \frac{1.23\ \text{kg}}{\text{m}^3} \left(\frac{100\ \text{km}}{\text{hr}} \times \frac{1000\ \text{m}}{\text{km}} \times \frac{\text{hr}}{3600\ \text{sec}} \right)^2$$

$$\times\ 84\ \text{ft}^2 \times (0.3048)^2 \frac{\text{m}^2}{\text{ft}^2} \times \frac{\text{N} \cdot \text{sec}^2}{\text{kg} \cdot \text{m}}$$

$$F_{D_p} = 1.71\ \text{kN}$$

F_{D_p}

The corresponding power required to overcome aerodynamic drag is

$$\mathscr{P}_p = F_{D_p} V_p$$

$$= 1.71 \times 10^3\ \text{N} \times \frac{100\ \text{km}}{\text{hr}} \times \frac{1000\ \text{m}}{\text{km}} \times \frac{\text{hr}}{3600\ \text{sec}} \times \frac{\text{W} \cdot \text{sec}}{\text{N} \cdot \text{m}}$$

$$\mathscr{P}_p = 47.6\ \text{kW}$$

$\mathscr{P}_p$

This example illustrates the application of model test data in a situation where nondimensional drag is independent of Reynolds number when that parameter is above a minimum value. In this situation it is not necessary to duplicate the Reynolds number between model and prototype to obtain useful model test data.

7-5.2 Scaling Laws for Turbomachinery

In some situations of practical importance there may be more than one dependent parameter. Then dimensionless groups must be formed separately for each dependent parameter in turn.

As an example, consider a typical turbomachine such as a centrifugal pump, a blower, or a fan. Performance parameters of interest include the pressure rise (or head) developed, the power input required, and the machine efficiency measured under specific operating conditions. Performance curves are generated by varying an independent parameter such as the volume flow rate. Thus the independent variables are volume flow rate, angular speed, impeller diameter, and fluid properties. Dependent variables are the several performance quantities of interest.

Finding dimensionless parameters for a turbomachine begins from the symbolic equations for the dependence of head, H (energy per unit mass, L^2/t^2, or per unit weight, L), and power, $\mathscr{P}$, on the independent parameters, given by

$$H = g_1(Q, \rho, \omega, D)$$

and

$$\mathscr{P} = g_2(Q, \rho, \omega, D)$$

Straightforward use of the Pi theorem gives the dimensionless *head coefficient* and *power coefficient* as

$$\frac{H}{\omega^2 D^2} = f_1\left(\frac{Q}{\omega D^3}, \frac{\rho \omega D^2}{\mu}\right) \tag{7.1}$$

and

$$\frac{\mathscr{P}}{\rho \omega^3 D^5} = f_2\left(\frac{Q}{\omega D^3}, \frac{\rho \omega D^2}{\mu}\right) \tag{7.2}$$

The dimensionless parameter, $Q/\omega D^3$, in these equations is termed the *flow coefficient*. The dimensionless parameter, $\rho \omega D^2/\mu$ ($\approx \rho V D/\mu$), is a form of Reynolds number.

Head and power in turbomachines are developed by inertia forces. Thus the flow pattern within a machine and its performance change with volume flow rate and speed of rotation. Performance is difficult to predict analytically except at the design point of the machine, so it is measured experimentally. Typical characteristic curves plotted from experimental data for a centrifugal machine tested at constant speed are shown in Fig. 7.4 as functions of volume flow rate. The head and power curves in Fig. 7.4 are faired through points plotted from measured data. The efficiency curve is faired through points calculated from the data.[4] Maximum efficiency usually occurs at the design point.

Complete similarity in pump performance tests would require identical flow coefficients and Reynolds numbers. In practice, it has been found that viscous effects

[4] Efficiency is defined as the ratio of power delivered to the fluid divided by input power, $\eta = \mathscr{P}/\mathscr{P}_{\text{in}}$. For incompressible flow, the energy equation reduces to $\mathscr{P} = \rho Q H$ (when head is expressed as energy per unit mass) or to $\mathscr{P} = \rho g Q H$ (when head is expressed as energy per unit weight).

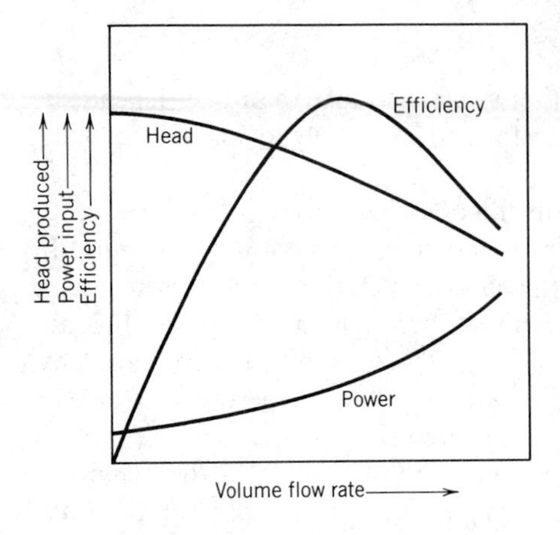

Fig. 7.4 Typical characteristic curves for centrifugal machine tested at constant speed.

are relatively unimportant when two geometrically similar machines operate under "similar" flow conditions. Thus, from Eqs. 7.1 and 7.2, when

$$\frac{Q_1}{\omega_1 D_1^3} = \frac{Q_2}{\omega_2 D_2^3} \tag{7.3}$$

it follows that

$$\frac{H_1}{\omega_1^2 D_1^2} = \frac{H_2}{\omega_2^2 D_2^2} \tag{7.4}$$

and

$$\frac{\mathscr{P}_1}{\rho_1 \omega_1^3 D_1^5} = \frac{\mathscr{P}_2}{\rho_2 \omega_2^3 D_2^5} \tag{7.5}$$

The empirical observation that viscous effects are unimportant under similar flow conditions allows use of Eqs. 7.3 through 7.5 to scale the performance characteristics of machines to different operating conditions, as either the speed or diameter are changed. These useful scaling relationships are known as pump or fan "laws." If operating conditions for one machine are known, operating conditions for any geometrically similar machine can be found by changing D and ω according to Eqs. 7.3 through 7.5.

Another useful pump parameter can be obtained by eliminating the machine diameter from Eqs. 7.3 and 7.4. Designating $\Pi_1 = Q/\omega D^3$ and $\Pi_2 = H/\omega^2 D^2$, then the ratio $\Pi_1^{1/2}/\Pi_2^{3/4}$ is a dimensionless parameter; this parameter is the *specific speed*, N_s,

$$N_s = \frac{\omega Q^{1/2}}{H^{3/4}} \tag{7.6}$$

The specific speed, as defined in Eq. 7.6, is a dimensionless parameter (provided the head, H, is expressed as energy per unit mass). Specific speed may be thought of as the speed required for a machine to produce unit head at unit volume flow rate. Any constant value of specific speed describes all operating conditions of geometrically similar machines with similar flow conditions.

It is customary to characterize a machine by its specific speed at the design point. This value of specific speed has been found to characterize the hydraulic design of a machine. Low specific speeds correspond to relatively low flow rates and high heads, which are produced efficiently by radial-flow machines. High specific speeds (high flow rates and/or low heads) are produced efficiently by axial-flow machines.

Specific speed is a classic example of a parameter obtained from sound reasoning using dimensionless ratios which often is used in dimensional form. The magnitude of the specific speed depends on the units used to calculate it. Typical units used in U.S. engineering practice for pumps are rpm for ω, gpm for Q, and feet (energy per unit weight) for H. In these units, "low" specific speed means $500 < N_s < 4000$ and "high" means $10{,}000 < N_s < 15{,}000$. Example Problem 7.6 illustrates use of the pump scaling laws and specific speed parameter.

Example 7.6

A centrifugal pump has an efficiency of 80 percent at its design point specific speed of 2000 (rpm, gpm, and feet units). The impeller diameter is 8 in. At design point flow conditions, the volume flow rate is 300 gpm of water at 1170 rpm. To obtain a larger flow rate, the pump is to be fitted with a 1750 rpm motor. Use the pump "laws" to find the performance characteristics of the pump at the higher speed. Show that the specific speed remains constant for the higher operating speed. Determine the motor size required.

EXAMPLE PROBLEM 7.6

GIVEN: Centrifugal pump with design specific speed of 2000 (in rpm, gpm, and feet units). Impeller diameter is $D = 8$ in. At the pump's design point flow conditions, $\omega = 1170$ rpm, and $Q = 300$ gpm, with water.

FIND: Use pump "laws" to find for similar flow conditions at 1750 rpm:
(a) Performance characteristics.
(b) Specific speed.
(c) Motor size required.

SOLUTION:
From pump "laws," $Q/\omega D^3 =$ constant, so

$$Q_2 = Q_1 \frac{\omega_2}{\omega_1}\left(\frac{D_2}{D_1}\right)^3 = 300\text{ gpm}\left(\frac{1750}{1170}\right)(1)^3 = 449\text{ gpm} \qquad Q_2$$

The pump head is not specified at $\omega_1 = 1170$ rpm, but it can be calculated from the specific speed, $N_s = 2000$. Using the given units,

$$N_s = \frac{\omega Q^{1/2}}{H^{3/4}}, \qquad \text{so} \qquad H_1 = \left(\frac{\omega_1 Q_1^{1/2}}{N_s}\right)^{4/3} = 21.9\text{ ft}$$

Then $H/\omega^2 D^2 =$ constant, so

$$H_2 = H_1\left(\frac{\omega_2}{\omega_1}\right)^2\left(\frac{D_2}{D_1}\right)^2 = 21.9\text{ ft}\left(\frac{1750}{1170}\right)^2(1)^2 = 49.0\text{ ft} \qquad H_2$$

The pump power output at $\omega_1 = 1170$ rpm is $\mathscr{P}_1 = \rho g Q_1 H_1$, so

$$\mathscr{P}_1 = \frac{1.94\text{ slug}}{\text{ft}^3} \times \frac{32.2\text{ ft}}{\text{sec}^2} \times \frac{300\text{ gal}}{\text{min}} \times 21.9\text{ ft} \times \frac{\text{ft}^3}{7.48\text{ gal}} \times \frac{\text{min}}{60\text{ sec}} \times \frac{\text{lbf}\cdot\text{sec}^2}{\text{slug}\cdot\text{ft}} \times \frac{\text{hp}\cdot\text{sec}}{550\text{ ft}\cdot\text{lbf}}$$

$$\mathscr{P}_1 = 1.66\text{ hp}$$

But $\mathscr{P}/\rho\omega^3 D^5 = \text{constant}$, so

$$\mathscr{P}_2 = \mathscr{P}_1\left(\frac{\rho_2}{\rho_1}\right)\left(\frac{\omega_2}{\omega_1}\right)^3\left(\frac{D_2}{D_1}\right)^5 = 1.66\text{ hp}\,(1)\left(\frac{1750}{1170}\right)^3(1)^5 = 5.55\text{ hp} \qquad \mathscr{P}_2$$

The required input power may be calculated as

$$\mathscr{P}_{\text{in}} = \frac{\mathscr{P}_2}{\eta} = \frac{5.55\text{ hp}}{0.80} = 6.94\text{ hp} \qquad \mathscr{P}_{\text{in}}$$

Thus a 7.5 hp motor (the next largest standard size) must be specified.

The specific speed at $\omega_2 = 1750$ rpm is

$$N_s = \frac{\omega Q^{1/2}}{H^{3/4}} = \frac{1750\,(449)^{1/2}}{(49.0)^{3/4}} = 2000 \qquad N_s$$

This example illustrates application of the pump "laws" and specific speed to scaling of performance data. Pump and fan "laws" are used widely in industry to scale performance data for families of machines from a single performance curve, and to specify drive speed and power in machine applications.

7-5.3 Comments on Model Testing

While outlining the procedures involved in model testing, we have tried not to imply that testing is a simple task that automatically gives results which are easy to interpret, accurate, and complete. As in all experimental work, careful planning and execution are needed to obtain valid results. Models must be constructed carefully and accurately, and they must include sufficient detail in areas critical to the phenomenon being measured. Aerodynamic balances or other force measuring systems must be aligned carefully and calibrated correctly. Mounting methods must be devised that offer adequate rigidity and model motion, yet do not interfere with the phenomenon being measured.

Experimental facilities must be designed and constructed carefully. The quality of the flow in a wind tunnel must be documented. It should be as nearly uniform as possible (unless the desire is to simulate the atmospheric boundary layer or other special profile), free from angularity, and with little swirl content. If they interfere with measurements, boundary layers on tunnel walls must be removed by suction. Pressure gradients in a wind tunnel test section may cause erroneous readings due to horizontal buoyancy in the flow direction.

Special facilities are needed for unusual conditions or for special test requirements, especially to achieve large Reynolds numbers. Many facilities are so large or specialized that they cannot be supported by university laboratories or private industry. A few examples include:

- National Aeronautics and Space Administration (NASA), Ames Research Laboratory, Moffett Field, California
  - 40 × 80 Wind Tunnel with test section 40 ft high and 80 ft wide (12 × 24 m), maximum wind speed of 230 mph, and 36,000 hp electric drive system.
- Naval Ship Research and Development Center (NSRDC), Carderock, Maryland
  - High-Speed Towing Basin with 2968 ft length, 21 ft width, and 16 ft depth. Towing carriage can travel at up to 100 knots while measuring drag loads up to 8000 pounds and side loads to 2000 pounds.
  - 36-Inch Variable Pressure Water Tunnel with 50 knot maximum test speed at pressures between 2 and 60 psia.
  - Anechoic Flow Facility with quiet, low turbulence air flow in 8 ft square by 21 ft long open-jet test section. Flow noise at maximum speed of 200 ft/sec is less than that of conversational speech.
- U.S. Army Corps of Engineers, Sausalito, California
  - San Francisco Bay and Delta Model with slightly more than 1 acre of area, 1 : 1000 horizontal scale and 1 : 100 vertical scale, 13,500 gallons per minute of pumping capacity, use of fresh and salt water, and tide simulation.
- NASA, Langley Research Center, Hampton, Virginia
  - National Transonic Facility (NTF) with cryogenic technology (temperatures as low as −300 F) to reduce gas viscosity, raising Reynolds number by a factor of 6, while halving drive power.

7-6 NONDIMENSIONALIZING THE BASIC DIFFERENTIAL EQUATIONS

Ultimate success in use of the Buckingham Pi theorem is determined by the insight used to select the parameters affecting the problem. If a complete set is chosen, the results will be complete. If one or more important variables are omitted, the results will be without meaning. Additional variables can be included if there is any uncertainty. As more experience is gained with fluid flow phenomena, the selection process becomes easier. Experience also gives more insight into the physical significance of each dimensionless group.

A more rigorous and broader approach to determine the conditions under which two flows are similar is to use the governing differential equations and boundary conditions. Similitude (dynamic similarity) may be present when two physical phenomena are governed by identical differential equations and boundary conditions. Similitude is obtained when the governing equations and boundary conditions have the same dimensionless form. Dynamic similarity is obtained by duplicating the dimensionless coefficients of the equations and boundary conditions between prototype and model.

As an example of nondimensionalizing the basic differential equations, consider steady incompressible two-dimensional flow in the xy plane. The gravity force is assumed to act in the negative y direction.

The equation for conservation of mass is

$$\frac{\partial u}{\partial x} + \frac{\partial v}{\partial y} = 0 \tag{7.7}$$

and the Navier–Stokes equations (Eqs. 5.25) reduce to

$$\rho\left(u\frac{\partial u}{\partial x} + v\frac{\partial u}{\partial y}\right) = -\frac{\partial p}{\partial x} + \mu\left(\frac{\partial^2 u}{\partial x^2} + \frac{\partial^2 u}{\partial y^2}\right) \tag{7.8}$$

$$\rho\left(u\frac{\partial v}{\partial x} + v\frac{\partial v}{\partial y}\right) = -\rho g - \frac{\partial p}{\partial y} + \mu\left(\frac{\partial^2 v}{\partial x^2} + \frac{\partial^2 v}{\partial y^2}\right) \tag{7.9}$$

To nondimensionalize these equations, all lengths are divided by a reference length, L, and all velocities are divided by a reference velocity, V_∞, which is usually taken as the freestream velocity. The pressure is made nondimensional by dividing by ρV_∞^2 (twice the freestream dynamic pressure). Denoting nondimensional quantities by an asterisk, we obtain

$$x^* = \frac{x}{L}, \quad y^* = \frac{y}{L}, \quad u^* = \frac{u}{V_\infty}, \quad v^* = \frac{v}{V_\infty}, \quad \text{and} \quad p^* = \frac{p}{\rho V_\infty^2} \tag{7.10}$$

To illustrate the procedure for nondimensionalizing the equations, consider two typical terms in the equations

$$u\frac{\partial u}{\partial x} = V_\infty\left(\frac{u}{V_\infty}\right)\frac{\partial(u/V_\infty)V_\infty}{\partial(x/L)L} = \frac{V_\infty^2}{L}u^*\frac{\partial u^*}{\partial x^*}$$

and

$$\frac{\partial^2 u}{\partial y^2} = \frac{\partial}{\partial y}\left(\frac{\partial u}{\partial y}\right) = \frac{\partial}{\partial(y/L)L}\left[\frac{\partial(u/V_\infty)V_\infty}{\partial(y/L)L}\right] = \frac{V_\infty}{L^2}\frac{\partial^2 u^*}{\partial y^{*2}}$$

By following this procedure, Eqs. 7.7, 7.8, and 7.9 can be written

$$\frac{V_\infty}{L}\frac{\partial u^*}{\partial x^*} + \frac{V_\infty}{L}\frac{\partial v^*}{\partial y^*} = 0 \tag{7.11}$$

$$\frac{\rho V_\infty^2}{L}\left(u^*\frac{\partial u^*}{\partial x^*} + v^*\frac{\partial u^*}{\partial y^*}\right) = -\frac{\rho V_\infty^2}{L}\frac{\partial p^*}{\partial x^*} + \frac{\mu V_\infty}{L^2}\left(\frac{\partial^2 u^*}{\partial x^{*2}} + \frac{\partial^2 u^*}{\partial y^{*2}}\right) \tag{7.12}$$

$$\frac{\rho V_\infty^2}{L}\left(u^*\frac{\partial v^*}{\partial x^*} + v^*\frac{\partial v^*}{\partial y^*}\right) = -\rho g - \frac{\rho V_\infty^2}{L}\frac{\partial p^*}{\partial y^*} + \frac{\mu V_\infty}{L^2}\left(\frac{\partial^2 v^*}{\partial x^{*2}} + \frac{\partial^2 v^*}{\partial y^{*2}}\right) \tag{7.13}$$

Dividing Eq. 7.11 by V_∞/L and Eqs. 7.12 and 7.13 by $\rho V_\infty^2/L$ gives

$$\frac{\partial u^*}{\partial x^*} + \frac{\partial v^*}{\partial y^*} = 0 \tag{7.14}$$

$$u^* \frac{\partial u^*}{\partial x^*} + v^* \frac{\partial u^*}{\partial y^*} = -\frac{\partial p^*}{\partial x^*} + \frac{\mu}{\rho V_\infty L}\left(\frac{\partial^2 u^*}{\partial x^{*2}} + \frac{\partial^2 u^*}{\partial y^{*2}}\right) \tag{7.15}$$

$$u^* \frac{\partial v^*}{\partial x^*} + v^* \frac{\partial v^*}{\partial y^*} = -\frac{gL}{V_\infty^2} - \frac{\partial p^*}{\partial y^*} + \frac{\mu}{\rho V_\infty L}\left(\frac{\partial^2 v^*}{\partial x^{*2}} + \frac{\partial^2 v^*}{\partial y^{*2}}\right) \tag{7.16}$$

From the nondimensional equations (Eqs. 7.14, 7.15, 7.16), we conclude that the differential equations for two flow systems will be identical if and only if the quantities $\mu/\rho V_\infty L$ and gL/V_∞^2 are the same for both flows. Thus, model studies to determine the drag force on a surface ship require duplication of both the Froude number and the Reynolds number to ensure dynamically similar flows.

For flows around submerged bodies far below the free surface, as for the sphere of Example Problem 7.4, body forces are not important. The governing equations do not include the body force term, ρg. Nondimensionalizing the governing equations in this case shows that the nondimensional equations governing two flows will be identical if the Reynolds number is the same for both flows.

So far we have concentrated on the differential equations governing the flow. It is important to emphasize that in addition to identical nondimensional equations, the nondimensional boundary conditions must be identical if the two flows are to be kinematically similar. This leads to the requirement of geometric similarity between the flows. Nondimensionalizing the boundary conditions may lead to additional requirements that must be satisfied between the two flows. For example, consider the case where the velocity at a specified location is periodic. The boundary condition then specifies

$$u_{bc} = V_\infty \sin \omega t$$

If we nondimensionalize time using the ratio of the reference velocity to the reference length, then

$$t^* = \frac{tV_\infty}{L}$$

The nondimensional boundary condition becomes

$$u_{bc}^* = \frac{u_{bc}}{V_\infty} = \sin\left(\frac{\omega L}{V_\infty} t^*\right)$$

Duplication of the boundary condition requires that the parameter $\omega L/V_\infty$ be the same between the two flows. This parameter is the *Strouhal number*

$$St = \frac{\omega L}{V_\infty}$$

which is named after the German physicist who first discovered its importance while investigating the self-excited "singing" of wires in the wind.

Establishing similitude from the differential equations and boundary conditions that describe the flow is a rigorous procedure. If one begins with the correct equations and performs each step correctly, one can be sure that all appropriate variables have been included.

The governing differential equations often are written in nondimensional form for numerical solution. Scaling is simplified and unit conversion problems are reduced when dimensionless forms of the equations are used. Dimensionless forms frequently permit solutions to be presented in generalized form.

7-7 SUMMARY OBJECTIVES

After completing study of Chapter 7, you should be able to do the following:

1. Define:

Reynolds number	Mach number
Euler number (pressure coefficient)	Strouhal number
Cavitation number	geometric similarity
Froude number	kinematic similarity
Weber number	dynamic similarity

2. State the Buckingham Pi Theorem.

3. Given a physical problem in which the dependent parameter is a function of specified independent parameters, determine a set of independent dimensionless ratios that characterize the problem.

4. State the conditions under which prototype behavior can be predicted from model tests.

5. Predict results for a prototype from model test data.

6. Obtain dimensionless coefficients by nondimensionalizing the governing differential equations.

7. Solve the problems at the end of the chapter that relate to the material you have studied.

REFERENCES

1. "SAE Wind Tunnel Test Procedure for Trucks and Buses," *Recommended Practice*—SAE J1252. Warrendale, Pa.: Society of Automotive Engineers, July 1981.
2. Kowalski, T., "Hydrodynamics of Waterborne Bodies," Chapter 4 in *Introduction to Ocean Engineering*, H. Schenck, ed. New York: McGraw-Hill, 1975.
3. Sedov, L. I., *Similarity and Dimensional Methods in Mechanics*. New York: Academic Press, 1959.
4. Todd, L. H., "Resistance and Propulsion," Chapter VII in *Principles of Naval Architecture*, J. P. Comstock, ed. New York: Society of Naval Architects and Marine Engineers, 1967, pp. 288–462.
5. Baals, D. W., and W. R. Corliss, *Wind Tunnels of NASA*. Washington, D.C.: National Aeronautics and Space Administration, SP-440, 1981.
6. Vincent, M., "The Naval Ship Research and Development Center," Naval Ship Research and Development Center, Carderock, Md., Report 3039 (Revised), November 1971.
7. Birkhoff, G., *Hydrodynamics–A Study in Logic, Fact, and Similitude*, 2nd ed. Princeton, N.J.: Princeton University Press, 1960.
8. Pope, A., and K. L. Goin, *High-Speed Wind Tunnel Testing*. New York: Krieger, 1978.
9. Pankhurst, R. C., and D. W. Holder, *Wind-Tunnel Technique*. London: Pitman, 1965.
10. Waugh, J. G., and G. W. Stubstad, *Hydroballistics Modeling*. San Diego, Calif.: U.S. Naval Undersea Center, undated.

11. Pope, A., and J. J. Harper, *Low-Speed Wind Tunnel Testing.* New York: Wiley-Interscience, 1966.
12. Merzkirch, W., *Flow Visualization.* New York: Academic Press, 1974.
13. Ipsen, D. C., *Units, Dimensions, and Dimensionless Numbers.* New York: McGraw-Hill, 1960.
14. Kline, S. J., *Similitude and Approximation Theory.* New York: McGraw-Hill, 1965.
15. Hansen, A. G., *Similarity Analysis of Boundary Value Problems in Engineering.* Englewood Cliffs, N.J.: Prentice-Hall, 1964.
16. Yalin, M. S., *Theory of Hydraulic Models.* New York: Macmillan, 1971.

PROBLEMS

7.1 The Reynolds number for flow in a pipe is defined as $Re = \rho \bar{V} D/\mu$, where $\bar{V}$ is the average velocity and D is the pipe diameter. Using both M, L, t and F, L, t systems of dimensions, show that the Reynolds number is dimensionless.

7.2 On a standard day, sonic speed at sea level is approximately 1120 ft/sec; at 28,000 ft, it is approximately 1000 ft/sec. A jet aircraft is capable of level flight at a Mach number of 1.8 at sea level, and 2.3 at 28,000 ft. Determine the corresponding air speeds in mph.

7.3 For free-surface flow in a wide channel, the Froude number is defined as $Fr = V/\sqrt{gD}$. Flow in a laboratory flume occurs at a depth of 100 mm with a speed of 0.5 m/sec or at a depth of 12 mm with a speed of 3 m/sec. Calculate the Froude numbers corresponding to these flow conditions.

7.4 The Weber number is defined as $We = \rho V^2 L/\sigma$. For wave motion, the characteristic length used is the wavelength, λ. On a water table where the depth is 0.25 in. and the velocity is 0.75 ft/sec, waves are observed with $\lambda = 6$ in. The water temperature is 68 F. Show that the Weber number is dimensionless. Determine its magnitude for the water table flow.

7.5 Cavitation phenomena are known to depend on the cavitation number, defined as

$$Ca = \frac{p - p_v}{\frac{1}{2}\rho V^2}$$

where p, ρ, and V are conditions in the liquid stream far ahead of a model, and p_v is the liquid vapor pressure at the test temperature. Consider a model test performed in a water tunnel at atmospheric pressure and 20 C. Determine the flow speed at which the cavitation number equals unity.

7.6 The power loss, $\mathscr{P}$, in a journal bearing depends on the length, l, diameter, D, and clearance, c, of the bearing, in addition to its angular speed, ω. The lubricant viscosity and mean pressure are also important. Obtain the dimensionless parameters that characterize this problem. Determine the functional form of the dependence of $\mathscr{P}$ on these parameters.

7.7 The load-carrying capacity, W, of a journal bearing is known to depend on its diameter, D, length, l, and clearance, c, in addition to its angular speed, ω, and the lubricant viscosity, μ. Evaluate the dimensionless parameters that characterize this problem.

7.8 A disk rotates near a fixed surface. The radius of the disk is R, and the space between the disk and the surface is filled with a liquid of viscosity, μ. The spacing between the disk and the surface is h, and the disk rotates at angular speed, ω. Find the dependence between torque on the disk, T, and the other variables.

7.9 A continuous belt moving vertically through a bath of viscous liquid drags a layer of liquid of thickness, h, along with it. The volume flow rate of liquid, Q, is assumed to depend on μ, ρ, g, h, and V, where V is the belt speed. Apply dimensional analysis to predict the form of dependence of Q on other variables. Use ρ, V, and h as repeating variables.

7.10 At very low speeds, the drag on an object is independent of fluid density. Thus the drag, F, on a small sphere is a function only of speed, V, fluid viscosity, and sphere diameter, D. Use dimensional analysis to express the drag as a function of these variables. Use μ, V, and D as repeating variables.

7.11 Experiments show that the pressure drop due to flow through a sudden contraction in a circular duct may be expressed as

$$\Delta p = p_1 - p_2 = f(\rho, \mu, \bar{V}, d, D)$$

You are asked to organize some experimental data. Obtain the resulting dimensionless parameters, using ρ, $\bar{V}$, and D as repeating variables.

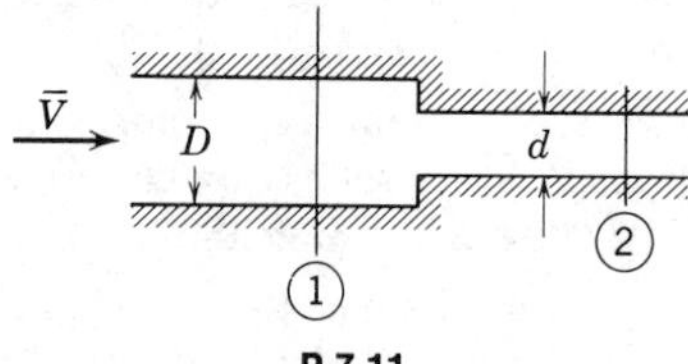

P 7.11

7.12 The boundary-layer thickness, δ, on a smooth flat plate in an incompressible flow without pressure gradients depends on the freestream velocity, U, the fluid density, ρ, the fluid viscosity, μ, and the distance from the leading edge of the plate, x. Express these variables in dimensionless form.

7.13 Assume that the resistance, R, of a flat plate immersed in a fluid depends on the fluid density and viscosity, the velocity, and the width, b, and height, h, of the plate. Find a convenient set of coordinates for organizing data.

7.14 The mean velocity, $\bar{u}$, for turbulent flow in a pipe or a boundary layer may be correlated using the wall shear stress, τ_w, distance from the wall, y, and the fluid properties, ρ and μ. Use dimensional analysis to find one dimensionless parameter containing $\bar{u}$ and one containing y that are suitable for organizing experimental data. Show that the result may be written

$$\frac{\bar{u}}{u_*} = f\left(\frac{yu_*}{\nu}\right)$$

where $u_* = (\tau_w/\rho)^{1/2}$ is the friction velocity.

7.15 Small droplets of liquid are formed when a liquid jet breaks up in spray and fuel injection processes. The resulting droplet diameter, d, is thought to depend on the liquid density, viscosity, and surface tension, as well as the jet speed, V, and diameter, D. How many dimensionless ratios are required to characterize this process? Determine these ratios, using ρ, D, and V as repeating variables.

7.16 The sketch shows an air jet discharging vertically. Experiments show that a ball placed in the jet is suspended in a stable position. The equilibrium height of the ball in the jet is found to depend on D, d, V, ρ, μ, and W, where W is the weight of the ball. Dimensional analysis is suggested to correlate experimental data. Find the Pi parameters that characterize this situation.

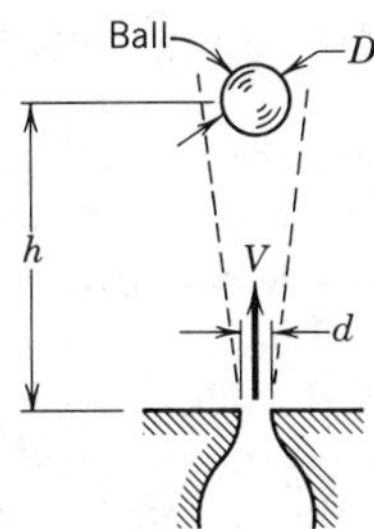

P 7.16

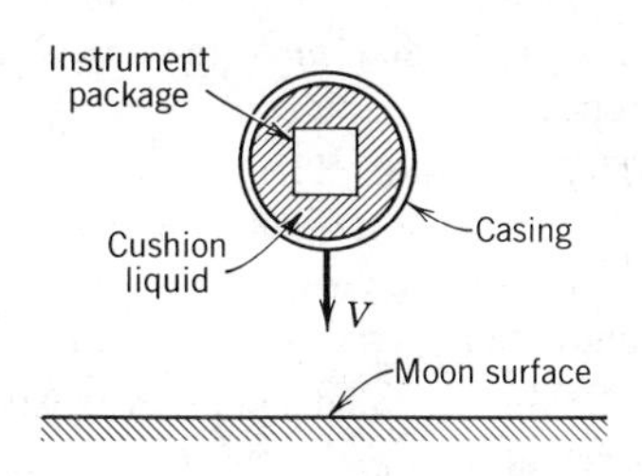

P 7.17

7.17 The instrument package for a moon landing is encased in a viscoelastic liquid as shown. The acceleration, a, of the package is expected to depend on l, a dimension of the package, m, the mass of the package, E, the modulus of elasticity of the liquid, μ, the liquid viscosity, and V, the impact speed. Dimensional analysis is suggested to help design suitable experiments. Determine the dimensionless parameters that result.

7.18 Spin plays an important role in the flight trajectory of golf, Ping-Pong, and tennis balls. Therefore, it is important to know the rate at which spin decreases for a ball in flight. The aerodynamic torque, T, acting on a ball in flight is thought to depend on the flight speed, V, the air density, ρ, the air viscosity, μ, the ball diameter, D, the spin rate (angular speed), ω, and the diameter of the dimples on the ball, d. Determine the dimensionless parameters that result. (Use ρ, V, and D as repeating parameters.)

7.19 Measurements of the liquid height upstream from an obstruction placed in an open-channel flow of a liquid can be used to determine volume flow rate. (Such obstructions, designed and calibrated to measure rate of open-channel flow, are called *weirs*.) Assume the volume flow rate over a weir, Q, is a function of upstream height, h, gravity, g, and channel width, b. Use dimensional analysis to develop an expression for Q.

7.20 The velocity, V, of a free surface gravity wave in deep water is a function of the wavelength, λ, depth, D, water density, ρ, and the acceleration of gravity, g. Use dimensional analysis to find the functional dependence of V on the other variables. Choose ρ, D, and g as repeating variables. Express V in the simplest form possible.

7.21 Capillary waves are formed on a liquid free surface as a result of surface tension. They have short wavelengths. The speed of a capillary wave depends on the surface tension, σ, wavelength, λ, and liquid density, ρ. Use dimensional analysis to express the wave speed as a function of these variables.

7.22 The power, $\mathscr{P}$, required to drive a propeller is known to depend on the following variables: freestream speed, V, propeller diameter, D, angular speed of propeller, ω, fluid viscosity, μ, fluid density, ρ, and speed of sound in the fluid, c.

(a) How many dimensionless groups are required to characterize this situation?

(b) Obtain these dimensionless groups (do *not* use μ as a repeating variable).

7.23 The thrust of a marine propeller is to be measured during "open-water" tests at a variety of angular speeds and forward speeds ("speeds of advance"). The thrust, F_T, is thought to depend on the water density, ρ, the propeller diameter, D, the speed of advance, V, the acceleration of gravity, g, the angular speed of the propeller, ω, the pressure in the liquid, p, and the liquid viscosity, μ. Using ρ, V, and D as repeating parameters, develop a set of dimensionless parameters to characterize the performance of the propeller. (One of the resulting parameters, gD/V^2, is known as the *Froude speed of advance*.)

7.24 The power, $\mathscr{P}$, required to drive a fan is believed to depend on the fluid density, ρ, the volume flow rate, Q, the impeller diameter, D, and angular velocity, ω. Use dimensional analysis to determine the dependence of $\mathscr{P}$ on other variables. Choose ρ, D, and ω as repeating variables.

7.25 The radiator fan in an automobile is a source of considerable noise. The sound power, $\mathscr{P}$ (energy per unit time), radiated by a fan depends on its diameter and angular velocity, and on the fluid density, ρ, and sonic speed, c. Determine the dependence of $\mathscr{P}$ on ω, using dimensional analysis. Choose ρ, ω, and D as repeating variables.

7.26 The vorticity, ζ, at a point in an axisymmetric flow field is thought to depend on the initial circulation, Γ_0, the radius, r, the time, τ, and the fluid kinematic viscosity, ν. Find a set of dimensionless parameters suitable for organizing experimental data.

7.27 Uniform flow of an ideal fluid past a sphere gives the surface velocity distribution, $V_\phi = \frac{3}{2}U \sin\phi$, where ϕ is measured from the centerline. Plot the Euler number (pressure coefficient) versus angle, ϕ, for this flow. Evaluate the magnitude and location of the point(s) where the pressure coefficient is maximum, minimum, and zero.

7.28 The Weber number is an important parameter in the breakup of a liquid jet into droplets. A garden hose discharges 4 liters per minute from a 6 mm diameter nozzle opening. Use the nozzle diameter as the characteristic length and evaluate the Weber number for this flow.

7.29 Kerosine for a burner is atomized into droplets from jets with 100 μm diameter. Evaluate the Weber number based on jet diameter if the flow speed is 20 m/sec.

7.30 The speed of sound in an ideal gas is given by $c = (kRT)^{1/2}$, where k is the ratio of specific heats, R is the ideal gas constant, and T is the absolute temperature (see Table A.6). Compressibility effects become important when the Mach number exceeds about 0.3. Compare the corresponding speeds for flow in air and helium at 300 K.

7.31 Superheated steam behaves as an ideal gas, for which the speed of sound is given by $c = (kRT)^{1/2}$, where k is the ratio of specific heats, R is the ideal gas constant, and T is the absolute temperature (see Table A.6). Evaluate the flow speed at which $M = 0.8$ in superheated steam at 500 K.

7.32 The partial pressure of oxygen dissolved in the blood stream is 22 kPa (abs). Estimate the cavitation number for blood flow (SG = 1.06) at a gage pressure of 100 mm Hg at a location where the mean flow speed is 5 m/sec.

7.33 A model hydrofoil boat is to be tested at 1:20 scale. To duplicate Froude numbers, the test speed will be 13.4 knots, instead of the 60 knot prototype speed. To model cavitation correctly, the cavitation number also must be duplicated. At what ambient pressure must the test be run? Water in the model test basin can be heated to 130 F, compared to 45 F for the prototype.

7.34 An automobile is to travel through standard air at a speed of 100 km/hr. To determine the pressure distribution, a scale model $\frac{1}{5}$ the length of the full-size car is to be tested in water. Determine the water speed that should be used. What factors must be considered to assure kinematic similarity in the tests?

7.35 The drag of an airfoil at zero angle of attack is a function of the flow density, viscosity, and velocity, in addition to a length parameter. A $\frac{1}{10}$-scale model of an airfoil was tested in a wind tunnel at a Reynolds number of 5.5×10^6, based on chord length. Test conditions in the wind tunnel air stream were 15 C and 10 atm absolute pressure. The prototype airfoil has a chord length of 2 m, and it is to be flown in air at standard conditions. Determine the velocity at which the wind tunnel model was tested, and the corresponding prototype velocity.

7.36 The capillary rise, Δh, of a liquid in a circular tube of diameter, D, depends on the surface tension, σ, and the specific weight, γ, of the fluid. The significant variables found from dimensional analysis are

$$\Pi_1 = \Delta h \sqrt{\frac{\gamma}{\sigma}}, \qquad \Pi_2 = \frac{\sigma}{\gamma D^2}$$

The capillary rise for liquid A is 1.0 in. in a tube of 0.010 in. diameter. What will be the rise for liquid B (having the same surface tension but four times the density of A) in a tube of diameter 0.005 in.?

7.37 An airship is to operate at 20 m/sec in air at standard conditions. A model is constructed to $\frac{1}{20}$ scale and tested in a wind tunnel at the same air temperature to determine drag.
(a) What criterion should be considered to obtain dynamic similarity?
(b) If the model is tested at 20 m/sec, what pressure should be used in the wind tunnel?
(c) If the model drag force is 250 N, what will be the drag of the prototype?

7.38 Consider a smooth sphere of diameter, D, immersed in a fluid moving with speed, V. Assume that flow is compressible (the sonic speed is c). Use the Buckingham Pi theorem to show that the drag force, F_D, may be expressed in dimensionless form as

$$\frac{F_D}{\rho V^2 D^2} = f\left(\frac{\rho V D}{\mu}, \frac{V}{c}\right)$$

The drag force on a 3 m diameter weather balloon in air moving at 1.5 m/sec is to be calculated from test data. The test is to be performed in water using a 50 mm diameter model. Under dynamically similar conditions, the model drag force is measured as 3.78 N. Evaluate the model test speed and the drag force to be expected on the full-scale balloon.

7.39 Consider water flow around a circular cylinder of diameter, D, and length, l. In addition to geometry, the drag force is known to depend on liquid speed, V, density, ρ, and viscosity, μ. Express drag force, F_D, in dimensionless form as a function of all relevant variables. The static pressure distribution on a circular cylinder measured in the laboratory can be expressed in terms of the dimensionless pressure coefficient,

$$C_p = \frac{p - p_\infty}{\frac{1}{2}\rho V^2}$$

The lowest pressure coefficient value is $C_p = -2.4$ at the location of the minimum static pressure on the cylinder surface. Estimate the maximum speed at which a cylinder could be towed in water at atmospheric pressure without causing cavitation if the onset of cavitation occurs at a cavitation number of 0.5.

7.40 A $\frac{1}{5}$-scale model of a torpedo is to be tested in a wind tunnel to determine the drag force. The prototype operates in water, has a diameter of 533 mm, and is 6.7 m long. The desired operating speed of the prototype is 28 m/sec. To avoid compressibility effects in the wind tunnel, the maximum speed is limited to 110 m/sec. However, the pressure in the wind tunnel can be varied while holding the temperature constant at 20 C. The drag force, F_D, is a function of speed, V, diameter, D, fluid density, ρ, and fluid viscosity, μ. At what minimum pressure should the wind tunnel be operated to achieve a dynamically similar test? At dynamically similar test conditions, the drag force on the model is measured as 618 N. Evaluate the drag force to be expected on the full-scale torpedo.

7.41 An airplane wing with chord length of 5 ft and span of 30 ft is designed to move through standard air at a speed of 230 ft/sec. A $\frac{1}{10}$-scale model of this wing is to be tested in a water tunnel. What speed is necessary in the water tunnel to achieve dynamic similarity? What will be the ratio of forces measured in the model flow to those on the prototype airfoil?

7.42 In some speed ranges, vortices are shed from the rear of bluff cylinders placed across a flow. The vortices alternately leave the top and bottom of the cylinder, as shown, causing an alternating force normal to the freestream velocity. The vortex shedding frequency, f, is thought to depend on ρ, V, d, and μ.

(a) Use dimensional analysis to develop a functional relationship for f.

(b) Vortex shedding occurs in standard air on two cylinders with a diameter ratio of 2. Determine the velocity ratio for dynamic similarity, and the ratio of vortex shedding frequencies.

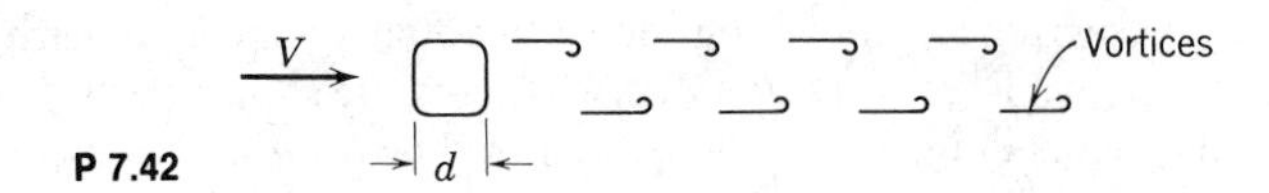

P 7.42

7.43 The fluid dynamic characteristics of a golf ball are to be tested using a model in a wind tunnel. Dependent parameters are the drag force, F_D, and lift force, F_L, that act on the ball. The independent parameters include the ball's diameter, D, speed, V, angular speed, ω, and dimple depth, d, in addition to the air density, ρ, and viscosity, μ. Determine suitable dimensionless parameters and express the functional dependence among them. A golf pro can hit a ball at $V = 240$ ft/sec and $\omega = 9000$ rpm. To model these conditions in a wind tunnel with maximum speed of 80 ft/sec, what diameter model should be used? How fast must this model rotate? (The diameter of a golf ball is 1.68 in.)

7.44 A model test is performed to determine the flight characteristics of a Frisbee. Dependent parameters are the drag force, F_D, and lift force, F_L, that act on the Frisbee. The independent parameters include the Frisbee's diameter, D, speed, V, angular speed, ω, and roughness height, h, in addition to the air density, ρ, and viscosity, μ. Determine suitable dimensionless parameters and express the functional dependence among them. The test is performed (using air) on a $\frac{1}{4}$-scale model Frisbee, which is to be geometrically, kinematically, and dynamically similar to the prototype. The prototype values are $V_p = 20$ ft/sec and $\omega_p = 100$ rpm. What values of V_m and ω_m should be used?

7.45 A $\frac{1}{10}$-scale model of a tractor-trailer rig is to be tested in a wind tunnel. The model has frontal area, $A_m = 1.08$ ft^2. When tested at an air speed of $V_m = 250$ ft/sec, the measured drag force is $F_D = 76.3$ lbf. Evaluate the drag coefficient for the model conditions given. Assuming that the drag coefficient is the same for model and prototype, calculate the drag force on a prototype rig at a highway speed of 55 mph. Determine the air speed at which a model *should* be tested to assure dynamically similar results if the prototype speed is 55 mph. Is this air speed practical? Why or why not? Assume air at STP.

7.46 It is recommended in [1] that the frontal area of a model be less than 5 percent of the wind tunnel test section area located above the ground board. Further, the model height must be less than 30 percent of the net test section height, and the maximum projected width of the model at maximum yaw (20°) must be less than 30 percent of the test section width. The maximum air speed should be less than 300 ft/sec to avoid compressibility effects. A model of a tractor-trailer rig is to be tested in a wind tunnel that has a test section area 1.5 ft high and 2 ft wide above the ground board. The height, width, and length of the full-scale rig are 13 ft 6 in., 8 ft, and 65 ft, respectively. Evaluate the scale ratio of the largest model that meets the recommended criteria. Use the results of Example Problem 7.5 to assess whether an adequate Reynolds number can be achieved in this test facility.

7.47 A circular container partially filled with water is rotated about its axis at constant angular speed, ω. At any time, τ, from the start of the rotation, the velocity, V_θ, at distance, r, from the axis of rotation was found to be a function of τ, ω, and the properties of the liquid. Write the dimensionless parameters that characterize this problem. If, in another experiment, honey is rotated in the same cylinder at the same angular speed, determine from your dimensionless parameters whether honey will attain steady motion as quickly as water. Explain why the Reynolds number would not be an important dimensionless parameter in scaling the steady state motion of liquid in the container.

7.48 An axial-flow pump is required to deliver 25 ft^3/sec of water at a head of 150 ft · lbf/slug. The diameter of the rotor is 1 ft, and it is to be driven at 500 rpm. The prototype is to be modeled on a small test apparatus having a 3 hp, 1000 rpm power supply. For similar performance between the prototype and the model, calculate the head, the volume flow rate, and the diameter of the model.

7.49 A model propeller 2 ft in diameter is tested in a wind tunnel. Air approaches the propeller at 150 ft/sec when it rotates at 2000 rpm. The thrust and torque measured under these conditions are 25 lbf and 15 ft · lbf, respectively. A prototype 10 times as large as the model is to be built. At a dynamically similar operating point, the approach air speed is to be 400 ft/sec. Calculate the speed, thrust, and torque of the prototype propeller under these conditions, neglecting the effect of viscosity but including density.

7.50 Consider again Problem 7.23. Experience shows that for ship-size propellers, viscous effects on scaling are small. Also, when cavitation is not present, the nondimensional parameter containing the pressure can be ignored. Assume that torque, T, and power, $\mathscr{P}$, depend on the same parameters as thrust. For conditions under which effects of μ and p can be neglected, use ρ, ω, and D as repeating parameters to derive scaling "laws" for propellers similar to the pump "laws" of Section 7-5.2, that relate thrust, torque, and power to the angular speed and diameter of the propeller.

7.51 The following data for a model test of a marine propeller are given in Problem 4-6 of [2]:

Parameter	Model	Prototype
Diameter	18 in.	24 ft
Angular speed	960 rpm	240 rpm
Speed of advance	20 ft/sec	
Thrust	35.1 lbf	
Torque	120 in. · lbf	
Fluid	Fresh water	Seawater
Temperature	65 F	59 F

For dynamically similar test conditions, calculate the speed of advance, thrust, and torque of the prototype propeller. (Use the results of Problem 7.50.) Evaluate the thrust power produced, the required power input, and the efficiency of the prototype propeller. Neglect cavitation and Reynolds number considerations.

7.52 Closed-circuit wind tunnels can produce higher speeds than open-circuit tunnels with the same power input because energy is recovered in the diffuser downstream from the test section. The *kinetic energy ratio* is a figure of merit defined as the ratio of the kinetic energy flux in the test section to the drive power. Estimate the kinetic energy ratio for the 40 × 80 wind tunnel at NASA-Ames.

7.53 In tests of models in wind tunnels, forces and moments are measured relative to an axis system aligned with the tunnel centerline. Forces acting on a highway vehicle are expressed most usefully in an axis system referenced to the vehicle (body-axis coordinates). Test data measured at yaw angles must be converted to body-axis coordinates for use. Develop equations to calculate body-axis drag and side force from data measured in wind tunnel coordinates. Assume that the model yaw angle is ψ.

7.54 A 1:16 model of a 20 m long truck is tested in a wind tunnel, where the axial static pressure gradient is -1.2 mm of water per meter at test speed. The frontal area of the prototype is 10 m^2. Estimate the horizontal buoyancy correction for this situation.

7.55 The antenna on a pickup truck is a cylinder of 3 mm diameter. It begins to vibrate in resonance when the truck is driven at 90 km/hr. Experimental data show that the corresponding Strouhal number is about 0.21. Estimate the frequency forcing the vibration.

7.56 The propagation speed of small-amplitude surface waves in a region of uniform depth is given by

$$c^2 = \left(\frac{\sigma}{\rho}\frac{2\pi}{\lambda} + \frac{g\lambda}{2\pi}\right)\tanh\frac{2\pi h}{\lambda}$$

where h is the depth of the undisturbed liquid and λ is the wavelength. Using L as a characteristic length and V_0 as a characteristic velocity, obtain the dimensionless groups that characterize the equation and determine the conditions for similarity.

7.57 By using order of magnitude analysis, the continuity and Navier–Stokes equations can be simplified to the Prandtl boundary-layer equations. For steady, incompressible, and two-dimensional flow, neglecting gravity, the result is

$$\frac{\partial u}{\partial x} + \frac{\partial v}{\partial y} = 0$$

$$u\frac{\partial u}{\partial x} + v\frac{\partial u}{\partial y} = -\frac{1}{\rho}\frac{\partial p}{\partial x} + \nu\frac{\partial^2 u}{\partial y^2}$$

Use L and V_0 as characteristic length and velocity, respectively. Nondimensionalize these equations and identify the similarity parameters that result.

7.58 The slope of the free surface of a steady wave in one-dimensional flow in a shallow liquid layer is described by the equation

$$\frac{\partial h}{\partial x} = -\frac{u}{g}\frac{\partial u}{\partial x}$$

Use a length scale, L, and a velocity scale, V_0, to nondimensionalize this equation. Obtain the dimensionless groups that characterize this flow. Determine the conditions for similarity.

7.59 One-dimensional unsteady flow in a thin liquid layer is described by the equation

$$\frac{\partial u}{\partial t} + u\frac{\partial u}{\partial x} = -g\frac{\partial h}{\partial x}$$

Use a length scale, L, and a velocity scale, V_0, to nondimensionalize this equation. Obtain the dimensionless groups that characterize this flow. Determine the conditions for similarity.

Chapter 8

INTERNAL INCOMPRESSIBLE VISCOUS FLOW

Flows completely bounded by solid surfaces are called internal flows. Thus internal flows include flows through pipes, ducts, nozzles, diffusers, sudden contractions and expansions, valves, and fittings.

Internal flows may be laminar or turbulent. Some laminar flow cases may be solved analytically. In the case of turbulent flow, analytical solutions are not possible, and hence we must rely heavily on semi-empirical theories and on experimental data. The nature of laminar and turbulent flows was discussed in Section 2-5.2. For internal flows the flow regime (laminar or turbulent) is primarily a function of the Reynolds number.

Following a brief introductory section, we consider two cases of fully developed laminar flow. Although most internal flows of engineering interest are turbulent, laminar flow can be important in applications such as lubrication or chemical process flows. Our analysis begins with a differential control volume rather than with the differential equations of motion derived in Chapter 5. This analysis also provides insight into the basic nature of turbulent flows in pipes and ducts, which are considered next. The chapter concludes with a discussion of flow measurements.

8-1 INTRODUCTION

As discussed previously in Section 2-5.2, the pipe flow regime (laminar or turbulent) is determined by the value of the Reynolds number, $Re = \rho \bar{V} D/\mu$. One can demonstrate by the classic Reynolds experiment[1] the qualitative difference between laminar and turbulent flows. In this experiment water flows from a large reservoir through a clear tube. A thin filament of dye injected at the entrance to the tube allows visual observation of the flow. At low flow rates (low Reynolds numbers) the dye injected into the flow remains in a single filament; there is little dispersion of dye because the flow is laminar. A laminar flow is one in which the fluid flows in laminae or layers; there is no macroscopic mixing of adjacent fluid layers.

[1] This experiment is demonstrated in the film *Turbulence*, R. W. Stewart, principal.

As the flow rate through the tube is increased, the dye filament becomes unstable and breaks up into a random motion; the line of dye is stretched and twisted into myriad entangled threads, and it quickly disperses throughout the entire flow field. This behavior of turbulent flow is due to small, high frequency, velocity fluctuations superimposed on the mean motion of a turbulent flow; mixing of fluid particles from adjacent layers of fluid results in rapid dispersion of the dye.

With great care to maintain the flow free from disturbances and with smooth surfaces, it is possible to maintain laminar flow in a pipe to a Reynolds number of about 100,000! However, most engineering flow situations are not so carefully controlled. Under normal conditions, transition occurs at $Re \approx 2300$ for flow in pipes. (Transition Reynolds numbers for other flow situations are given in the Example Problems.)

Figure 8.1 illustrates laminar flow in the entrance region of a circular pipe. The flow is uniform at the pipe entrance with velocity, U_0. Because of the no-slip condition at the wall, we know that the velocity at the wall must be zero along the entire length of the pipe. A boundary layer (Section 2-5.1) develops along the walls of the channel. The solid surface exerts a retarding shear force on the flow; thus the speed of the fluid in the neighborhood of the surface is reduced. At successive sections along the pipe, the effect of the solid surface is felt farther out into the flow.

For incompressible flow, the velocity at the pipe centerline must increase with distance from the inlet in order to satisfy the continuity equation. However, the average velocity at any cross section

$$\bar{V} = \frac{1}{A}\int_{\text{Area}} u\, dA$$

must equal U_0, so

$$\bar{V} = U_0 = \text{constant} \tag{8.1}$$

Sufficiently far from the pipe entrance, the boundary layer developing on the pipe wall reaches the pipe centerline. The distance downstream from the entrance to the location at which the boundary layer reaches the centerline is called the *entrance length*. Beyond the entrance length the velocity profile no longer changes with increasing distance, x, and the flow is fully developed. The actual shape of the fully developed velocity profile depends on whether the flow is laminar or turbulent. In Fig. 8.1 the profile is shown qualitatively for a laminar flow.

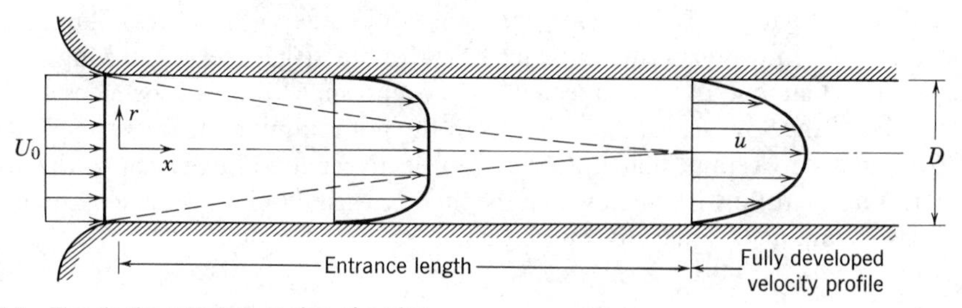

Fig. 8.1 Flow in the entrance region of a pipe.

For laminar flow, the entrance length, L, is a function of the Reynolds number,

$$\frac{L}{D} \simeq 0.06 \frac{\rho \bar{V} D}{\mu}$$

where D is the pipe diameter, $\bar{V}$ is the average velocity, ρ is the fluid density, and μ is the fluid viscosity. Laminar flow in a pipe may be expected only for Reynolds numbers less than 2300. Thus the entrance length for laminar pipe flow may be as long as

$$L \simeq 0.06 \, Re \, D \leq (0.06)(2300)D = 138D$$

or more than 100 pipe diameters. If the flow is turbulent, enhanced mixing among fluid layers[2] causes more rapid growth of the boundary layer. Experiments show that the mean velocity profile becomes fully developed within 25 to 40 pipe diameters from the entrance. However, the details of the turbulent motion may not be fully developed for 80 or more pipe diameters. Fully developed internal flows will be treated in Parts A and B of this chapter.

PART A FULLY DEVELOPED LAMINAR FLOW

There are relatively few viscous flow problems for which we can obtain analytical solutions in closed form, but the method of solution is important. In this section we consider a few classic examples of fully developed laminar flows. Our interest is in obtaining detailed information about the velocity field. Knowledge of the velocity field permits calculation of shear stress, pressure drop, and flow rate.

Rather than use the complete differential equations of motion (Eqs. 5.25) for the flow of a viscous fluid, we shall derive the governing equations from first principles for each flow field of interest. Since we are interested in the details of the flow field, our aim will be to derive differential equations that describe the flow. In every case we shall begin by applying the familiar control volume formulation of Newton's second law to a suitably chosen differential control volume.

8-2 FULLY DEVELOPED LAMINAR FLOW BETWEEN INFINITE PARALLEL PLATES

8-2.1 Both Plates Stationary

Fluid in high-pressure hydraulic systems often leaks through the annular gap between a piston and cylinder. For very small gaps (typically about 0.005 mm), this flow field may be imagined as flow between infinite parallel plates. To calculate the leakage flow rate, we must first determine the velocity field.

Let us consider the fully developed laminar flow between infinite parallel plates. The plates are separated by a distance, a, as shown in Fig. 8.2. The plates are considered

[2] This mixing is illustrated extremely well in the introductory portion of the film *Turbulence*, R. W. Stewart, principal.

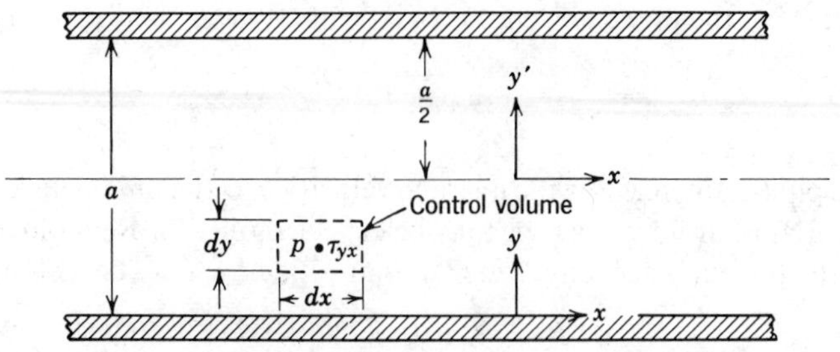

Fig. 8.2 Control volume for analysis of laminar flow between stationary infinite parallel plates.

infinite in the z direction, with no variation of any fluid property in this direction. The flow is further assumed to be steady and incompressible. Before starting our analysis, what do we know about the flow field? For one thing we know that the x component of velocity must be zero at both the upper and lower plates as a result of the no-slip condition at the wall. The boundary conditions are:

$$\text{at} \quad y = 0 \qquad u = 0$$

$$\text{at} \quad y = a \qquad u = 0$$

Since the flow is fully developed, the velocity cannot vary with x and, hence, depends on y only, so that $u = u(y)$. Furthermore, there is no component of velocity in either the y or z directions ($v = w = 0$).

For our analysis we select a differential control volume of size $d\forall = dx\,dy\,dz$ and apply the x component of the momentum equation.

Basic equation:

$$F_{S_x} + \underbrace{F_{B_x}}_{=0(3)} = \underbrace{\frac{\partial}{\partial t}\int_{\text{CV}} u\rho\, d\forall}_{=0(1)} + \int_{\text{CS}} u\rho \vec{V} \cdot d\vec{A} \tag{4.19a}$$

Assumptions: (1) Steady flow
(2) Fully developed flow
(3) $F_{B_x} = 0$

For fully developed flow, the net momentum flux through the control surface is zero. (The momentum flux through the right face of the control surface is equal in magnitude but opposite in sign to the momentum flux through the left face; there is no momentum flux through any of the remaining faces of the control volume.) Since there are no body forces in the x direction, the momentum equation reduces to

$$F_{S_x} = 0 \tag{8.2}$$

The next step is to sum the forces acting on the control volume in the x direction. We recognize that there are normal forces (pressure forces) acting on the left and right faces; there are tangential forces (shear forces) acting on the top and bottom faces.

If the pressure at the center of the element is p, then the force on the left face is

$$\left(p - \frac{\partial p}{\partial x}\frac{dx}{2}\right) dy\, dz$$

and the force on the right face is

$$-\left(p + \frac{\partial p}{\partial x}\frac{dx}{2}\right) dy\, dz$$

If the shear stress at the center of the element is τ_{yx}, then the shear force on the bottom face is

$$-\left(\tau_{yx} - \frac{d\tau_{yx}}{dy}\frac{dy}{2}\right) dx\, dz$$

and the shear force on the top face is

$$\left(\tau_{yx} + \frac{d\tau_{yx}}{dy}\frac{dy}{2}\right) dx\, dz$$

Note that in expanding the shear stress, τ_{yx}, in a Taylor series about the center of the element, we have used the total derivative rather than a partial derivative. We did this because we recognized that τ_{yx} is only a function of y, since $u = u(y)$.

Having evaluated the forces acting on each face of the control volume, we can substitute them into Eq. 8.2; this equation simplifies to

$$-\frac{\partial p}{\partial x} + \frac{d\tau_{yx}}{dy} = 0$$

or

$$\frac{d\tau_{yx}}{dy} = \frac{\partial p}{\partial x} \tag{8.3}$$

Equation 8.3 must be valid for all values of x and y. This requires that

$$\frac{d\tau_{yx}}{dy} = \frac{\partial p}{\partial x} = \text{constant}$$

Integrating this equation, we obtain

$$\tau_{yx} = \left(\frac{\partial p}{\partial x}\right) y + c_1$$

which states that the shear stress varies linearly with y. Since

$$\tau_{yx} = \mu \frac{du}{dy} \tag{2.10}$$

then

$$\mu \frac{du}{dy} = \left(\frac{\partial p}{\partial x}\right) y + c_1$$

and

$$u = \frac{1}{2\mu}\left(\frac{\partial p}{\partial x}\right)y^2 + \frac{c_1}{\mu}y + c_2 \tag{8.4}$$

To evaluate the constants, c_1 and c_2, we must use the boundary conditions. At $y = 0$, $u = 0$. Consequently, $c_2 = 0$. At $y = a$, $u = 0$. Hence

$$0 = \frac{1}{2\mu}\left(\frac{\partial p}{\partial x}\right)a^2 + \frac{c_1}{\mu}a$$

This gives

$$c_1 = -\frac{1}{2}\left(\frac{\partial p}{\partial x}\right)a$$

and hence

$$u = \frac{1}{2\mu}\left(\frac{\partial p}{\partial x}\right)y^2 - \frac{1}{2\mu}\left(\frac{\partial p}{\partial x}\right)ay$$

or

$$u = \frac{a^2}{2\mu}\left(\frac{\partial p}{\partial x}\right)\left[\left(\frac{y}{a}\right)^2 - \left(\frac{y}{a}\right)\right] \tag{8.5}$$

At this point we have the velocity profile. What else can we learn about the flow?

Shear Stress Distribution

The shear stress distribution is given by

$$\tau_{yx} = \left(\frac{\partial p}{\partial x}\right)y + c_1 = \left(\frac{\partial p}{\partial x}\right)y - \frac{1}{2}\left(\frac{\partial p}{\partial x}\right)a = a\left(\frac{\partial p}{\partial x}\right)\left[\frac{y}{a} - \frac{1}{2}\right] \tag{8.6a}$$

Volume Flow Rate

The volume flow rate is given by

$$Q = \int_A \vec{V} \cdot d\vec{A}$$

For a depth l in the z direction

$$Q = \int_0^a ul\,dy$$

or

$$\frac{Q}{l} = \int_0^a \frac{1}{2\mu}\left(\frac{\partial p}{\partial x}\right)(y^2 - ay)\,dy$$

Thus the volume flow rate per depth l is given by

$$\frac{Q}{l} = -\frac{1}{12\mu}\left(\frac{\partial p}{\partial x}\right)a^3 \tag{8.6b}$$

Flow Rate as a Function of Pressure Drop

Since $\partial p/\partial x$ is a constant, then

$$\frac{p_2 - p_1}{L} = \frac{-\Delta p}{L} = \frac{\partial p}{\partial x}$$

Substituting into the expression for volume flow rate gives

$$\frac{Q}{l} = -\frac{1}{12\mu}\left[\frac{-\Delta p}{L}\right]a^3 = \frac{a^3\,\Delta p}{12\mu L} \tag{8.6c}$$

Average Velocity

The average velocity, $\bar{V}$, is given by

$$\bar{V} = \frac{Q}{A} = -\frac{1}{12\mu}\left(\frac{\partial p}{\partial x}\right)\frac{a^3 l}{la} = -\frac{1}{12\mu}\left(\frac{\partial p}{\partial x}\right)a^2 \tag{8.6d}$$

Point of Maximum Velocity

To find the point of maximum velocity, we set du/dy equal to zero and solve for the corresponding value of y. From Eq. 8.5

$$\frac{du}{dy} = \frac{a^2}{2\mu}\left(\frac{\partial p}{\partial x}\right)\left[\frac{2y}{a^2} - \frac{1}{a}\right]$$

Thus

$$\frac{du}{dy} = 0 \quad \text{at} \quad y = \frac{a}{2}$$

At

$$y = \frac{a}{2}, \qquad u = u_{\text{max}} = -\frac{1}{8\mu}\left(\frac{\partial p}{\partial x}\right)a^2 = \frac{3}{2}\,\bar{V} \tag{8.6e}$$

Transformation of Coordinates

In deriving the above relations, the origin of coordinates, $y = 0$, was taken at the bottom plate. We could just as easily have taken the origin at the centerline of the channel. If we denote the coordinates with origin at the channel centerline as x, y', the boundary conditions are $u = 0$ at $y' = \pm a/2$.

To obtain the velocity profile in terms of x, y', we substitute $y = y' + a/2$ into Eq. 8.5. The result is

$$u = \frac{a^2}{2\mu}\left(\frac{\partial p}{\partial x}\right)\left[\left(\frac{y'}{a}\right)^2 - \frac{1}{4}\right] \tag{8.7}$$

This equation shows that the velocity profile we have determined is parabolic, as shown in Fig. 8.3.

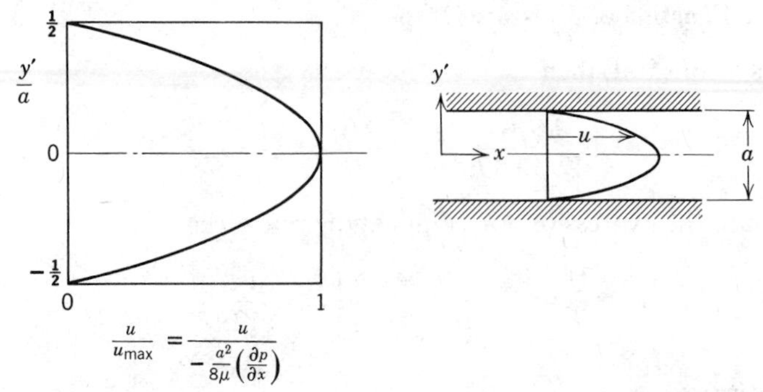

Fig. 8.3 Dimensionless velocity profile for fully developed laminar flow between infinite parallel plates.

All of the results in this section are valid for laminar flow only, since Newton's law of viscosity was used to relate shear stress to velocity gradient. Experiments show that transition to turbulent flow occurs in this flow case for Reynolds number values (defined as $Re = \rho \bar{V} a/\mu$) greater than approximately 1400. Consequently, the Reynolds number should be checked after using Eqs. 8.6 to assure a valid solution.

Example 8.1

A hydraulic system operates at a gage pressure of 20 MPa and 55 C. The hydraulic fluid is SAE 10W oil (SG = 0.92). A control valve consists of a piston 25 mm in diameter, fitted to a cylinder with a mean radial clearance of 0.005 mm. Determine the leakage flow rate if the gage pressure on the low-pressure side of the piston is 1.0 MPa. (The piston is 15 mm long.)

EXAMPLE PROBLEM 8.1

GIVEN: Flow of hydraulic oil between piston and cylinder, as shown.
Fluid is SAE 10W oil at 55 C (SG = 0.92).

FIND: Leakage flow rate, Q.

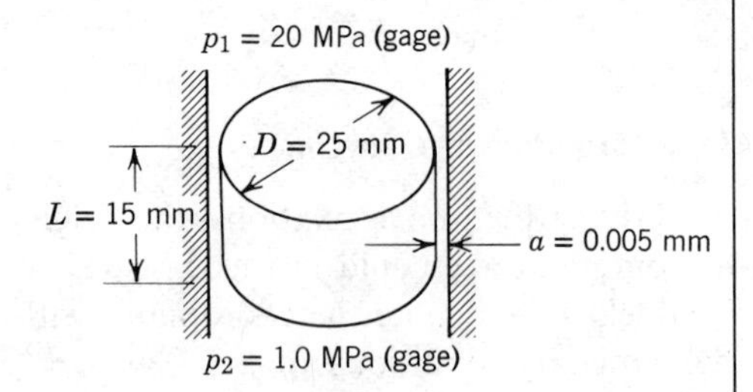

SOLUTION:

The gap width is very small, so the flow may be modeled as flow between parallel plates. Equation 8.6c is applied.

Computing equation:

$$\frac{Q}{l} = \frac{a^3\,\Delta p}{12\mu L} \tag{8.6c}$$

Assumptions: (1) Laminar flow
(2) Steady flow

(3) Incompressible flow
(4) Fully developed flow (note $L/a = 15/0.005 = 3000!$)

The plate width, l, is approximately $l = \pi D$. Thus

$$Q = \frac{\pi D a^3 \Delta p}{12 \mu L}$$

For SAE 10W oil at 55 C, $\mu = 0.018$ kg/m · sec, from Fig. A.2, Appendix A. Thus

$$Q = \frac{\pi}{12} \times 25\text{ mm} \times (0.005)^3\text{ mm}^3 \times (20-1)\,10^6\,\frac{\text{N}}{\text{m}^2} \times \frac{\text{m}\cdot\text{sec}}{0.018\text{ kg}} \times \frac{1}{15\text{ mm}} \times \frac{\text{kg}\cdot\text{m}}{\text{N}\cdot\text{sec}^2}$$

$$Q = 57.6\text{ mm}^3/\text{sec} \qquad Q$$

To assure that flow is laminar, we also should check the flow Reynolds number.

$$\bar{V} = \frac{Q}{A} = \frac{Q}{\pi D a} = \frac{57.6\text{ mm}^3}{\text{sec}} \times \frac{1}{\pi} \times \frac{1}{25\text{ mm}} \times \frac{1}{0.005\text{ mm}} \times \frac{\text{m}}{10^3\text{ mm}} = 0.147\text{ m/sec}$$

and

$$Re = \frac{\rho \bar{V} a}{\mu} = \frac{SG\rho_{H_2O}\bar{V}a}{\mu}$$

$$Re = 0.92 \times 999\,\frac{\text{kg}}{\text{m}^3} \times 0.147\,\frac{\text{m}}{\text{sec}} \times 0.005\text{ mm} \times \frac{\text{m}\cdot\text{sec}}{0.018\text{ kg}} \times \frac{\text{m}}{10^3\text{ mm}} = 0.0375$$

Thus flow is surely laminar, since $Re \ll 1400$.

8-2.2 Upper Plate Moving with Constant Speed, *U*

A second laminar flow case of practical importance is flow in a journal bearing. In such a bearing, an inner cylinder, or journal, rotates inside a stationary member. At light loads, the centers of the two members essentially coincide, and the small clearance gap is symmetric. Since the gap is small, it is reasonable to "unfold" the bearing and consider the flow field between infinite parallel plates.

Let us now consider a case where the upper plate is moving to the right with constant speed, U, as shown in Fig. 8.4. All we have done in going from a stationary upper plate

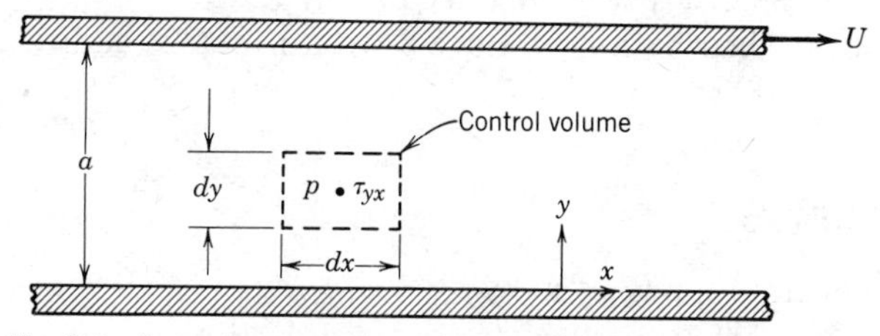

Fig. 8.4 Control volume for analysis of laminar flow between infinite parallel plates: upper plate moving with constant speed, *U*.

to a moving upper plate is to change one of the boundary conditions. The boundary conditions for the moving plate case are:

$$u = 0 \quad \text{at} \quad y = 0$$

$$u = U \quad \text{at} \quad y = a$$

Since only the boundary conditions have changed, there is no need to repeat the entire analysis of Section 8-2.1. The analysis leading to Eq. 8.4 is equally valid for the moving plate case. Thus the velocity distribution is given by

$$u = \frac{1}{2\mu}\left(\frac{\partial p}{\partial x}\right)y^2 + \frac{c_1}{\mu}y + c_2 \tag{8.4}$$

and our only task is to evaluate the constants c_1 and c_2 by using the appropriate boundary conditions.

At $y = 0$, $u = 0$. Consequently, $c_2 = 0$.

At $y = a$, $u = U$. Consequently,

$$U = \frac{1}{2\mu}\left(\frac{\partial p}{\partial x}\right)a^2 + \frac{c_1}{\mu}a$$

Thus

$$c_1 = \frac{U\mu}{a} - \frac{1}{2}\left(\frac{\partial p}{\partial x}\right)a$$

and

$$u = \frac{1}{2\mu}\left(\frac{\partial p}{\partial x}\right)y^2 + \frac{Uy}{a} - \frac{1}{2\mu}\left(\frac{\partial p}{\partial x}\right)ay$$

$$= \frac{Uy}{a} + \frac{1}{2\mu}\left(\frac{\partial p}{\partial x}\right)(y^2 - ay)$$

$$u = \frac{Uy}{a} + \frac{a^2}{2\mu}\left(\frac{\partial p}{\partial x}\right)\left[\left(\frac{y}{a}\right)^2 - \left(\frac{y}{a}\right)\right] \tag{8.8}$$

It is reassuring to note that Eq. 8.8 reduces to Eq. 8.5 for a stationary upper plate. From Eq. 8.8, for zero pressure gradient (for $\partial p/\partial x = 0$) the velocity varies linearly with y. This was the case treated earlier in Chapter 2.

From the velocity distribution of Eq. 8.8 we can obtain additional information about the flow.

Shear Stress Distribution

The shear stress distribution is given by $\tau_{yx} = \mu(du/dy)$,

$$\tau_{yx} = \mu\frac{U}{a} + \frac{a^2}{2}\left(\frac{\partial p}{\partial x}\right)\left[\frac{2y}{a^2} - \frac{1}{a}\right] = \mu\frac{U}{a} + a\left(\frac{\partial p}{\partial x}\right)\left[\frac{y}{a} - \frac{1}{2}\right] \tag{8.9a}$$

Volume Flow Rate

The volume flow rate is given by $Q = \int_A \vec{V} \cdot d\vec{A}$. For a depth l in the z direction

$$Q = \int_0^a ul\, dy$$

or

$$\frac{Q}{l} = \int_0^a \left[\frac{Uy}{a} + \frac{1}{2\mu}\left(\frac{\partial p}{\partial x}\right)(y^2 - ay)\right] dy$$

Thus the volume flow rate per depth l is given by

$$\frac{Q}{l} = \frac{Ua}{2} - \frac{1}{12\mu}\left(\frac{\partial p}{\partial x}\right)a^3 \tag{8.9b}$$

Average Velocity

The average velocity, $\bar{V}$, is given by

$$\bar{V} = \frac{Q}{A} = l\left[\frac{Ua}{2} - \frac{1}{12\mu}\left(\frac{\partial p}{\partial x}\right)a^3\right]\bigg/ la = \frac{U}{2} - \frac{1}{12\mu}\left(\frac{\partial p}{\partial x}\right)a^2 \tag{8.9c}$$

Point of Maximum Velocity

To find the point of maximum velocity, we set du/dy equal to zero and solve for the corresponding value of y. From Eq. 8.8

$$\frac{du}{dy} = \frac{U}{a} + \frac{a^2}{2\mu}\left(\frac{\partial p}{\partial x}\right)\left[\frac{2y}{a^2} - \frac{1}{a}\right]$$

Thus

$$\frac{du}{dy} = 0 \qquad \text{at} \qquad y = \frac{a}{2} - \frac{U/a}{(1/\mu)(\partial p/\partial x)}$$

There is no simple relation between the maximum velocity, u_{max}, and the mean velocity, $\bar{V}$, for this flow case.

Equation 8.8 suggests that the velocity profile may be treated as a combination of a linear and a parabolic velocity profile; the last term in Eq. 8.8 is identical with Eq. 8.5. The result is a family of velocity profiles, depending on the values of U and $(1/\mu)(\partial p/\partial x)$; a few profiles are sketched in Fig. 8.5. (As shown in Fig. 8.5, some reverse flow in the negative x direction can occur when $\partial p/\partial x > 0$.)

Again, all of the results developed in this section are valid for laminar flow only. Experiments show that transition to turbulent flow occurs (for $\partial p/\partial x = 0$) at a Reynolds number of approximately 1500, where $Re = \rho Ua/\mu$ for this flow case. Not much information is available for the case where the pressure gradient is not zero.

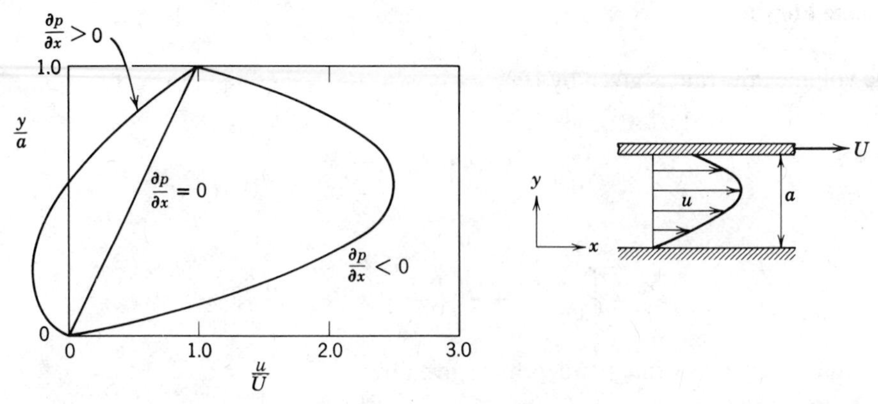

Fig. 8.5 Dimensionless velocity profile for fully developed laminar flow between infinite parallel plates: upper plate moving with constant speed, U.

Example 8.2

A crankshaft journal bearing in an automobile engine is lubricated by SAE 30 engine oil at 210 F. The bearing is 3 in. in diameter, has a diametral clearance of 0.0025 in., and rotates at 3600 rpm. It is 1.25 in. long. The bearing is under no load, so the clearance is symmetric. Determine the torque required to turn the journal, and the power dissipated.

EXAMPLE PROBLEM 8.2

GIVEN: Journal bearing, as shown. Note that the gap width, a, is *half* the diametral clearance. Lubricant is SAE 30 oil at 210 F. Speed: 3600 rpm.

FIND: (a) Torque, T.
(b) Power dissipated.

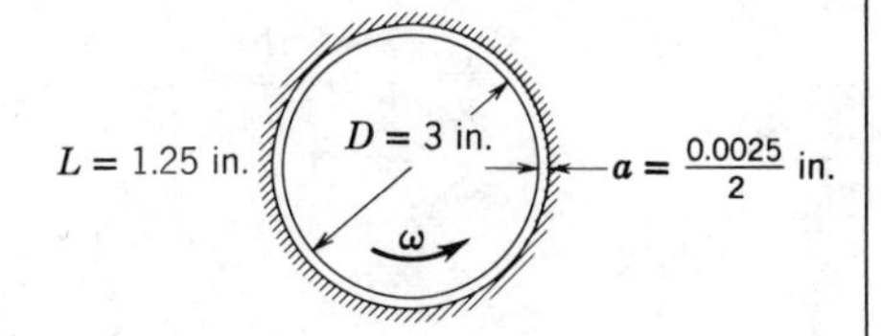

SOLUTION:

Torque on the journal is due to viscous shear in the oil film. The gap width is small, so the flow may be modeled as flow between infinite parallel plates:

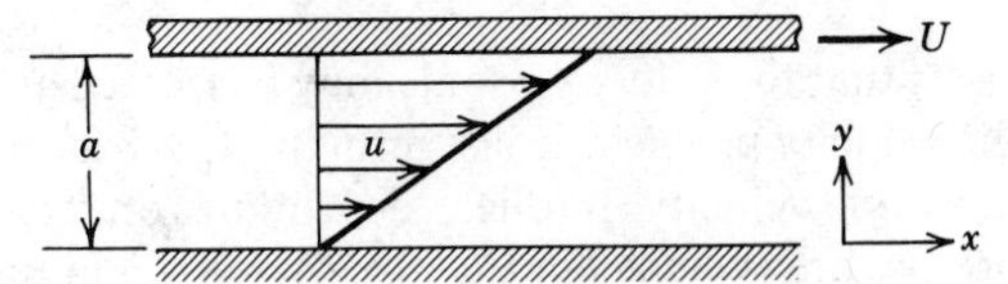

Computing equation:

$$\tau_{yx} = \mu \frac{U}{a} + a\underbrace{\left[\frac{\partial p}{\partial x}\right]}_{=0(6)}\left[\frac{y}{a} - \frac{1}{2}\right] \qquad (8.9a)$$

Assumptions: (1) Laminar flow
(2) Steady flow
(3) Incompressible flow
(4) Fully developed flow
(5) Infinite width ($L/a = 1.25/0.00125 = 1000$, so this is a reasonable assumption)
(6) $\partial p/\partial x = 0$ (flow is symmetric in the actual bearing at no load)

Then

$$\tau_{yx} = \mu \frac{U}{a} = \mu \frac{\omega R}{a} = \mu \frac{\omega D}{2a}$$

For SAE 30 oil at 210 F (99 C), $\mu = 9.6 \times 10^{-3}$ N · sec/m² (2.01×10^{-4} lbf · sec/ft²), from Fig. A.2, Appendix A. Thus

$$\tau_{yx} = \frac{2.01 \times 10^{-4}\ \text{lbf}\cdot\text{sec}}{\text{ft}^2} \times \frac{3600\ \text{rev}}{\text{min}} \times \frac{2\pi\ \text{rad}}{\text{rev}} \times \frac{\text{min}}{60\ \text{sec}} \times 3\ \text{in.} \times \frac{1}{2} \times \frac{1}{0.00125\ \text{in.}}$$

$$\tau_{yx} = 90.9\ \text{lbf/ft}^2$$

Since $\tau_{yx} > 0$, it acts to the *left* on the upper plate, which is a minus y surface. The total shear force is given by the shear stress times the area. It is applied to the journal surface. Therefore,

$$T = FR = \tau_{yx}\pi DLR = \frac{\pi}{2}\tau_{yx}D^2L$$

$$= \frac{\pi}{2} \times \frac{90.9\ \text{lbf}}{\text{ft}^2} \times (3)^2\ \text{in.}^2 \times \frac{\text{ft}^2}{144\ \text{in.}^2} \times 1.25\ \text{in.}$$

$$T = 11.2\ \text{in.}\cdot\text{lbf}$$

T

The power dissipated in the bearing is

$$\dot{W} = FU = FR\omega = T\omega$$

$$= 11.2\ \text{in.}\cdot\text{lbf} \times \frac{3600\ \text{rev}}{\text{min}} \times \frac{\text{min}}{60\ \text{sec}} \times \frac{2\pi\ \text{rad}}{\text{rev}} \times \frac{\text{ft}}{12\ \text{in.}} \times \frac{\text{hp}\cdot\text{sec}}{550\ \text{ft}\cdot\text{lbf}}$$

$$\dot{W} = 0.640\ \text{hp}$$

$\dot{W}$

To assure laminar flow, check the Reynolds number value.

$$Re = \frac{\rho Ua}{\mu} = \frac{\text{SG}\ \rho_{H_2O}Ua}{\mu} = \frac{\text{SG}\ \rho_{H_2O}\omega Ra}{\mu}$$

$$= 0.92 \times \frac{1.94\ \text{slug}}{\text{ft}^3} \times \frac{(3600)2\pi}{60}\frac{\text{rad}}{\text{sec}} \times 1.5\ \text{in.} \times 0.00125\ \text{in.} \times \frac{\text{ft}^2}{2.01 \times 10^{-4}\ \text{lbf}\cdot\text{sec}}$$

$$\times \frac{\text{ft}^2}{144\ \text{in.}^2} \times \frac{\text{lbf}\cdot\text{sec}^2}{\text{slug}\cdot\text{ft}}$$

$$Re = 43.6$$

Therefore, the flow is laminar, since $Re \ll 1500$.

8-3 FULLY DEVELOPED LAMINAR FLOW IN A PIPE

As a final example of fully developed laminar flow cases, let us consider fully developed laminar flow in a pipe. Here the flow is axisymmetric and therefore it is most convenient to work in cylindrical coordinates. We shall again use a differential control volume, but this time, since the flow is axisymmetric, the control volume will be a differential annular ring as shown in Fig. 8.6. The annular differential control volume is of length, dx, and has thickness, dr.

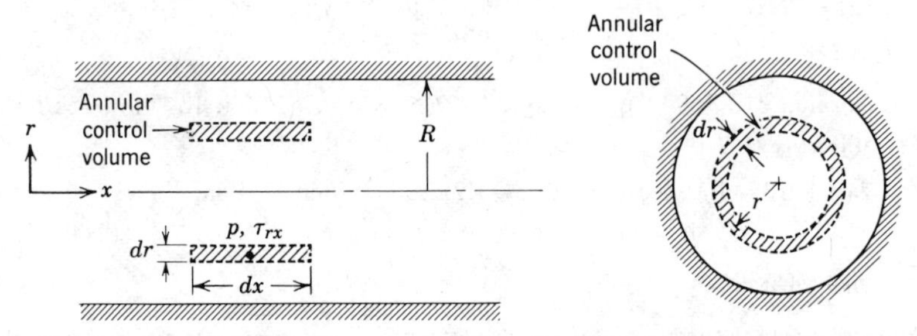

Fig. 8.6 Control volume for analysis of fully developed laminar flow in a pipe.

For a fully developed steady flow, the x component of the momentum equation (Eq. 4.19a) reduces to

$$F_{s_x} = 0$$

The next step is to sum the forces acting on the control volume in the x direction. We know that there are normal forces (pressure forces) acting on the left and right ends of the control volume, and that there are tangential forces (shear forces) acting on the inner and outer cylindrical surfaces.

If the pressure at the center of the annular control volume is p, then the force on the left end is

$$\left(p - \frac{\partial p}{\partial x}\frac{dx}{2}\right)2\pi r\, dr$$

The force on the right end is

$$-\left(p + \frac{\partial p}{\partial x}\frac{dx}{2}\right)2\pi r\, dr$$

If the shear stress at the center of the annular control volume is τ_{rx}, then the shear force on the inner cylindrical surface is

$$-\left(\tau_{rx} - \frac{d\tau_{rx}}{dr}\frac{dr}{2}\right)2\pi\left(r - \frac{dr}{2}\right)dx$$

The shear force on the outer cylindrical surface is

$$\left(\tau_{rx} + \frac{d\tau_{rx}}{dr}\frac{dr}{2}\right)2\pi\left(r + \frac{dr}{2}\right)dx$$

The summation of the x components of the forces acting on the control volume must be zero. This leads to the condition

$$-\frac{\partial p}{\partial x} 2\pi r\, dr\, dx + \tau_{rx}\, dr\, 2\pi\, dx + \frac{d\tau_{rx}}{dr} r\, dr\, 2\pi\, dx = 0$$

Dividing this equation by $2\pi r\, dr\, dx$ gives

$$\frac{\partial p}{\partial x} = \frac{\tau_{rx}}{r} + \frac{d\tau_{rx}}{dr} = \frac{1}{r}\frac{d(r\tau_{rx})}{dr} \tag{8.10}$$

Since τ_{rx} is only a function of r (this is the reason for using the total rather than the partial derivative of τ_{rx} in the force components above), we recognize that Eq. 8.10 holds true for all values of r and x only if each side of the equation is equal to a constant. Equation 8.10 can be written as

$$\frac{1}{r}\frac{d(r\tau_{rx})}{dr} = \frac{\partial p}{\partial x} = \text{constant}$$

or

$$\frac{d(r\tau_{rx})}{dr} = r\frac{\partial p}{\partial x}$$

Integrating this equation, we obtain

$$r\tau_{rx} = \frac{r^2}{2}\left(\frac{\partial p}{\partial x}\right) + c_1$$

or

$$\tau_{rx} = \frac{r}{2}\left(\frac{\partial p}{\partial x}\right) + \frac{c_1}{r}$$

Since

$$\tau_{rx} = \mu\frac{du}{dr}$$

then

$$\mu\frac{du}{dr} = \frac{r}{2}\left(\frac{\partial p}{\partial x}\right) + \frac{c_1}{r}$$

and

$$u = \frac{r^2}{4\mu}\left(\frac{\partial p}{\partial x}\right) + \frac{c_1}{\mu}\ln r + c_2 \tag{8.11}$$

We need to evaluate the constants c_1 and c_2. However, we have only the one boundary condition that $u = 0$ at $r = R$. What do we do? Before throwing in the towel, let us look at the solution for the velocity profile given by Eq. 8.11. Although we do not know the velocity at the pipe centerline, we do know from physical considerations that the velocity must be finite at $r = 0$. The only way that this can be true is for c_1 to be zero. Thus, from physical considerations, we conclude that $c_1 = 0$, and hence

$$u = \frac{r^2}{4\mu}\left(\frac{\partial p}{\partial x}\right) + c_2$$

The constant, c_2, is evaluated by using the available boundary condition at the pipe wall: at $r = R$, $u = 0$. Consequently,

$$0 = \frac{R^2}{4\mu}\left(\frac{\partial p}{\partial x}\right) + c_2$$

This gives

$$c_2 = -\frac{R^2}{4\mu}\left(\frac{\partial p}{\partial x}\right)$$

and hence

$$u = \frac{r^2}{4\mu}\left(\frac{\partial p}{\partial x}\right) - \frac{R^2}{4\mu}\left(\frac{\partial p}{\partial x}\right) = \frac{1}{4\mu}\left(\frac{\partial p}{\partial x}\right)(r^2 - R^2)$$

or

$$u = -\frac{R^2}{4\mu}\left(\frac{\partial p}{\partial x}\right)\left[1 - \left(\frac{r}{R}\right)^2\right] \tag{8.12}$$

Because we have the velocity profile, we can obtain a number of additional features of the flow.

Shear Stress Distribution

The shear stress is given by

$$\tau_{rx} = \mu \frac{du}{dr} = \frac{r}{2}\left(\frac{\partial p}{\partial x}\right) \tag{8.13a}$$

Volume Flow Rate

$$Q = \int_A \vec{V} \cdot d\vec{A} = \int_0^R u 2\pi r \, dr$$

$$= \int_0^R \frac{1}{4\mu}\left(\frac{\partial p}{\partial x}\right)(r^2 - R^2) 2\pi r \, dr$$

$$Q = -\frac{\pi R^4}{8\mu}\left(\frac{\partial p}{\partial x}\right) \tag{8.13b}$$

Flow Rate as a Function of Pressure Drop

In fully developed flow, the pressure gradient, $\partial p/\partial x$, is constant. Therefore, $\partial p/\partial x = (p_2 - p_1)/L = -\Delta p/L$. Substituting into Eq. 8.13b for the volume flow rate gives

$$Q = -\frac{\pi R^4}{8\mu}\left[\frac{-\Delta p}{L}\right] = \frac{\pi \, \Delta p R^4}{8\mu L} = \frac{\pi \, \Delta p D^4}{128\mu L} \tag{8.13c}$$

for laminar flow in a horizontal pipe.

Average Velocity

The average velocity, $\bar{V}$, is given by

$$\bar{V} = \frac{Q}{A} = \frac{Q}{\pi R^2} = -\frac{R^2}{8\mu}\left(\frac{\partial p}{\partial x}\right) \tag{8.13d}$$

Point of Maximum Velocity

To find the point of maximum velocity, we set du/dr equal to zero and solve for the corresponding value of r. From Eq. 8.12

$$\frac{du}{dr} = \frac{1}{2\mu}\left(\frac{\partial p}{\partial x}\right) r$$

Thus

$$\frac{du}{dr} = 0 \qquad \text{at} \qquad r = 0$$

At $r = 0$,

$$u = u_{\text{max}} = U = -\frac{R^2}{4\mu}\left(\frac{\partial p}{\partial x}\right) = 2\bar{V} \tag{8.13e}$$

The velocity profile (Eq. 8.12) may be written in terms of the maximum (centerline) velocity

$$\frac{u}{U} = 1 - \left(\frac{r}{R}\right)^2 \tag{8.14}$$

The parabolic velocity profile, given by Eq. 8.14 for fully developed laminar pipe flow, was sketched in Fig. 8.1.

Example 8.3

A simple and accurate viscometer can be made from a length of capillary tubing. If the flow rate and pressure drop are measured, and the tube geometry is known, the viscosity can be computed from Eq. 8.13c. A test of a certain liquid in a capillary viscometer gave the following data:

Flow rate:	880 mm^3/sec
Tube diameter:	0.50 mm
Tube length:	1 m
Pressure drop:	1.0 MPa

Determine the viscosity of the liquid.

EXAMPLE PROBLEM 8.3

GIVEN: Flow in a capillary viscometer. The flow rate is $Q = 880\ \text{mm}^3/\text{sec}$.

FIND: The fluid viscosity.

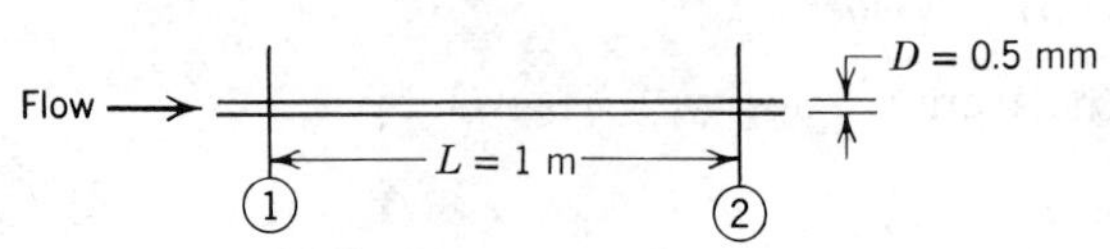

$\Delta p = p_1 - p_2 = 1.0$ MPa

SOLUTION:
Equation 8.13c may be applied.

Computing equation:
$$Q = \frac{\pi\,\Delta p D^4}{128\mu L} \tag{8.13c}$$

Assumptions: (1) Laminar flow
(2) Steady flow
(3) Incompressible flow
(4) Fully developed flow

Then

$$\mu = \frac{\pi\,\Delta p D^4}{128 LQ} = \frac{\pi}{128} \times \frac{1.0 \times 10^6\ \text{N}}{\text{m}^2} \times \frac{(0.50)^4\ \text{mm}^4}{} \times \frac{\text{sec}}{880\ \text{mm}^3} \times \frac{1}{1\ \text{m}} \times \frac{\text{m}}{10^3\ \text{mm}}$$

$$\mu = 1.74 \times 10^{-3}\ \text{N}\cdot\text{sec/m}^2 \qquad \mu$$

Check the Reynolds number. Assume the fluid density to be similar to that of water, 999 kg/m^3.

$$\bar{V} = \frac{Q}{A} = \frac{4Q}{\pi D^2} = \frac{4}{\pi} \times \frac{880\ \text{mm}^3}{\text{sec}} \times \frac{1}{(0.50)^2\ \text{mm}^2} \times \frac{\text{m}}{10^3\ \text{mm}} = 4.48\ \text{m/sec}$$

Then

$$Re = \frac{\rho \bar{V} D}{\mu} = \frac{999\ \text{kg}}{\text{m}^3} \times \frac{4.48\ \text{m}}{\text{sec}} \times \frac{0.50\ \text{mm}}{} \times \frac{\text{m}^2}{1.74 \times 10^{-3}\ \text{N}\cdot\text{sec}} \times \frac{\text{m}}{10^3\ \text{mm}} \times \frac{\text{N}\cdot\text{sec}^2}{\text{kg}\cdot\text{m}}$$

$$Re = 1290$$

Consequently, the flow is laminar, since $Re < 2300$.

Example 8.4

A viscous, incompressible, Newtonian liquid flows in steady, laminar flow down a vertical wall. The thickness, δ, of the liquid film is constant. Since the liquid free surface is exposed to atmospheric pressure, there is no pressure gradient. Apply the momentum equation to a differential control volume, $dx\ dy\ dz$, to derive the velocity distribution in the liquid film.

EXAMPLE PROBLEM 8.4

GIVEN: Fully developed laminar flow of an incompressible, Newtonian liquid down a vertical wall; thickness, δ, of the liquid film is constant and $\partial p/\partial x = 0$.

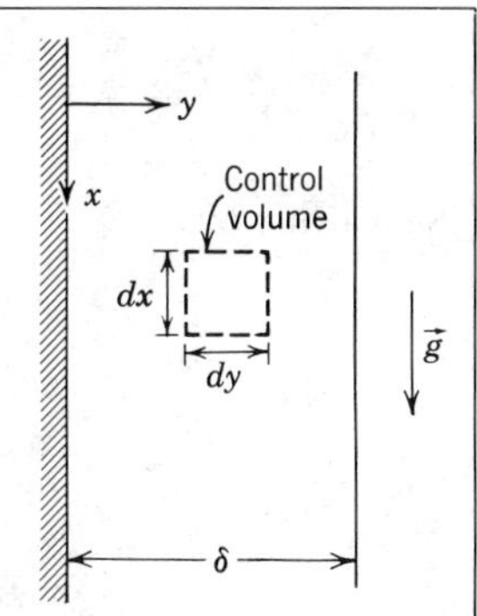

FIND: Expression for the velocity distribution in the film.

SOLUTION:

The x component of the momentum equation for a control volume is

$$F_{S_x} + F_{B_x} = \frac{\partial}{\partial t}\int_{\text{CV}} u\rho\, d\forall + \int_{\text{CS}} u\rho\, \vec{V}\cdot d\vec{A} \tag{4.19a}$$

For steady flow, $\frac{\partial}{\partial t}\int_{\text{CV}} u\rho\, d\forall = 0$

For fully developed flow, $\int_{\text{CS}} u\rho\vec{V}\cdot d\vec{A} = 0$

Thus the momentum equation for the present case reduces to

$$F_{S_x} + F_{B_x} = 0$$

The body force, F_{B_x}, is given by $F_{B_x} = \rho g\, d\forall = \rho g\, dx\, dy\, dz$. The surface forces acting on the differential control volume are shear forces on the vertical surfaces. Since $\partial p/\partial x = 0$, there is no net pressure force acting on the element.

If the shear stress at the center of the differential control volume is τ_{yx}, then,

$$\text{shear stress on left face is } \tau_{yx_L} = \left(\tau_{yx} - \frac{d\tau_{yx}}{dy}\frac{dy}{2}\right),$$

and

$$\text{shear stress on right face is } \tau_{yx_R} = \left(\tau_{yx} + \frac{d\tau_{yx}}{dy}\frac{dy}{2}\right)$$

The direction of the shear stress vectors is taken consistent with the sign convention of Section 2-3.2. Thus on the left face, τ_{yx_L} acts upward, and on the right face, τ_{yx_R} acts downward.

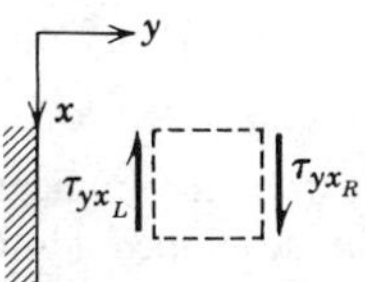

The surface forces are obtained by multiplying each shear stress by the area over which it acts. Substituting into $F_{S_x} + F_{B_x} = 0$, we obtain

$$-\tau_{yx_L}\,dx\,dz + \tau_{yx_R}\,dx\,dz + \rho g\,dx\,dy\,dz = 0$$

or

$$-\left(\tau_{yx} - \frac{d\tau_{yx}}{dy}\frac{dy}{2}\right)dx\,dz + \left(\tau_{yx} + \frac{d\tau_{yx}}{dy}\frac{dy}{2}\right)dx\,dz + \rho g\,dx\,dy\,dz = 0$$

Simplifying gives

$$\frac{d\tau_{yx}}{dy} + \rho g = 0 \qquad \text{or} \qquad \frac{d\tau_{yx}}{dy} = -\rho g$$

Since $\tau_{yx} = \mu \dfrac{du}{dy}$, then $\mu \dfrac{d^2u}{dy^2} = -\rho g$

and

$$\frac{d^2u}{dy^2} = -\frac{\rho g}{\mu}$$

Integrating with respect to y gives

$$\frac{du}{dy} = -\frac{\rho g}{\mu}y + c_1$$

Integrating again, we obtain

$$u = -\frac{\rho g}{\mu}\frac{y^2}{2} + c_1 y + c_2$$

To evaluate constants c_1 and c_2, we must apply appropriate boundary conditions:

(i) $y = 0, \quad u = 0$ (no-slip)

(ii) $y = \delta, \quad \dfrac{du}{dy} = 0$ (neglect air resistance)

From boundary condition (i), $c_2 = 0$

From boundary condition (ii), $0 = -\dfrac{\rho g}{\mu}\delta + c_1$ or $c_1 = \dfrac{\rho g}{\mu}\delta$

Hence

$$u = -\frac{\rho g}{\mu}\frac{y^2}{2} + \frac{\rho g}{\mu}\delta y$$

or

$$u = \frac{\rho g}{\mu}\delta^2\left[\left(\frac{y}{\delta}\right) - \frac{1}{2}\left(\frac{y}{\delta}\right)^2\right] \qquad u(y)$$

From the velocity profile one could determine the volume flow rate, the maximum velocity, and the average velocity.

{The purpose of this problem is to illustrate the application of the momentum equation to a differential control volume for a fully developed laminar flow.}

PART B FLOW IN PIPES AND DUCTS

Our main purpose in this section is to evaluate the pressure changes that result from incompressible flow in pipes, ducts, and flow systems. The pressure changes in a flow system result from changes in elevation or flow velocity (due to area changes) and from friction. In a frictionless flow, the Bernoulli equation could be used to account for the effects of changes in elevation and flow velocity. Thus the prime concern in the analysis of real flows is to account for the effect of friction. The effect of friction is to decrease the pressure, causing a pressure "loss" compared to the ideal, frictionless flow case. To simplify analysis, the "loss" will be divided into *major losses* (due to friction in fully developed flow in constant-area portions of the system) and *minor losses* (due to flow through valves, tees, elbows, and frictional effects in other nonconstant-area portions of the system).

In developing the relations for the major losses due to friction in constant-area ducts, we shall deal with fully developed flows in which the velocity profile is unvarying in the direction of flow. Our attention will focus on turbulent flows, since the pressure drop for fully developed laminar flow in a pipe can be calculated from the results of Section 8-3. The pressure drop that occurs at the entrance of a pipe will be treated as a minor loss.

Since ducts of circular cross section are most common in engineering applications, the basic analysis will be performed for circular geometries. The results can be extended to other geometries by introducing the hydraulic diameter, which is treated in Section 8-8.3. (Compressible flow in ducts will be treated in Chapter 12.)

8-4 SHEAR STRESS DISTRIBUTION IN FULLY DEVELOPED PIPE FLOW

In fully developed steady flow in a horizontal pipe, be it laminar or turbulent, the pressure drop is balanced only by shear forces at the pipe wall. This can be seen by applying the momentum equation to a cylindrical control volume in the flow, Fig. 8.7. The x component of the momentum equation is

Basic equation: $= 0(1) = 0(2)$ $= 0(3, 4)$

$$F_{S_x} + F_{B_x} = \frac{\partial}{\partial t}\int_{\text{CV}} u\rho \, d\forall + \int_{\text{CS}} u\rho \vec{V} \cdot d\vec{A} \tag{4.19a}$$

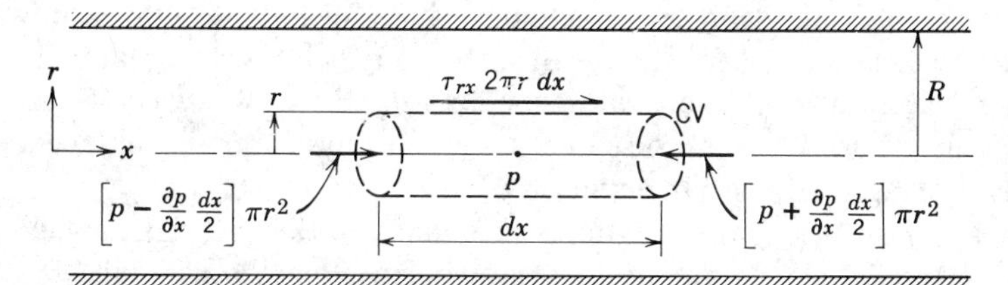

Fig. 8.7 Control volume for analysis of shear stress distribution in fully developed flow in a circular pipe.

Assumptions: (1) Horizontal pipe, $F_{B_x} = 0$
(2) Steady flow
(3) Incompressible flow
(4) Fully developed flow

Then

$$F_{S_x} = 0$$

The surface forces acting on the control volume are shown in Fig. 8.7. The pressure at the center of the element is p; the pressure at each end of the element is obtained from a Taylor series expansion of p about the center of the element. The shear force acts on the circumferential surface of the element. The direction of the stress has been assumed such that the stress is positive. Thus

$$F_{S_x} = \left(p - \frac{\partial p}{\partial x}\frac{dx}{2}\right)\pi r^2 - \left(p + \frac{\partial p}{\partial x}\frac{dx}{2}\right)\pi r^2 + \tau_{rx} 2\pi r\, dx = 0$$

or

$$-\frac{\partial p}{\partial x} dx\, \pi r^2 + \tau_{rx} 2\pi r\, dx = 0$$

Therefore,

$$\tau_{rx} = \frac{r}{2}\frac{\partial p}{\partial x} \tag{8.15}$$

Thus we see that the shear stress on the fluid varies linearly across the pipe, from zero at the centerline to a maximum at the pipe wall. If we denote the wall value of shear stress as τ_w, Eq. 8.15 shows that at the surface of the pipe

$$\tau_w = -[\tau_{rx}]_{r=R} = -\frac{R}{2}\frac{\partial p}{\partial x} \tag{8.16}$$

Equation 8.16 relates the wall shear stress to the axial pressure gradient. The momentum equation was used to derive it, but no assumption was made about a relation between the shear stress and the velocity field. Consequently, Eq. 8.16 is applicable to both laminar and turbulent pipe flow.

If we could relate the shear stress field to the mean velocity field, we could determine analytically the pressure drop over a length of pipe for fully developed flow. Such a relation between the stress field and the mean velocity field exists for laminar flow and was used in Section 8-3. The resulting equation, Eq. 8.13c, was first discovered experimentally by Jean Louis Poiseuille, a French physician, and independently by Gotthilf H. L. Hagen, a German engineer, in the 1850s.

In turbulent flow, no simple relation exists between the shear stress field and the mean velocity field. Velocity fluctuations in turbulent flow (discussed in Section 2-5.2) exchange momentum between adjacent layers of fluid, thereby causing apparent shear stresses that must be added to the stress caused by the mean velocity gradients. For fully developed turbulent channel flow, the total shear stress is given by

$$\tau = \mu \frac{d\bar{u}}{dy} - \rho \overline{u'v'} \tag{8.17}$$

In Eq. 8.17, y is the distance from the pipe wall, $\bar{u}$ is the mean velocity, u' and v' are the fluctuating components of velocity in the x and y directions, respectively, and $\overline{u'v'}$ is the time average of the product of u' and v'. The notion of an apparent stress was first introduced by Osborne Reynolds; the term $-\rho\overline{u'v'}$ is referred to as the *Reynolds stress.*

Dividing Eq. 8.17 by ρ gives

$$\frac{\tau}{\rho} = \nu \frac{d\bar{u}}{dy} - \overline{u'v'} \tag{8.18}$$

The term τ/ρ arises frequently in the consideration of turbulent flows; it has dimensions of velocity squared. The quantity $(\tau_w/\rho)^{1/2}$ is called the *friction velocity* and is denoted by the symbol u_*. In Fig. 8.8, experimental measurements of $\overline{u'v'}$ for fully developed turbulent pipe flow at two Reynolds numbers are presented. The turbulent shear stress has been nondimensionalized with the square of the friction velocity. In the region very close to the wall (the *wall layer*), viscous shear is dominant. The turbulent stress goes to zero at the wall because the no-slip condition requires that the velocity at the wall be zero. Since the Reynolds stress is zero at the wall, then from Eq. 8.17, the wall shear stress is given by $\tau_w = \mu(d\bar{u}/dy)_{y=0}$. The total shear stress varies linearly across the pipe radius, so the turbulent shear is dominant over the center region of the pipe. In the region between the wall layer and the central portion of the pipe both viscous and turbulent shear are important.

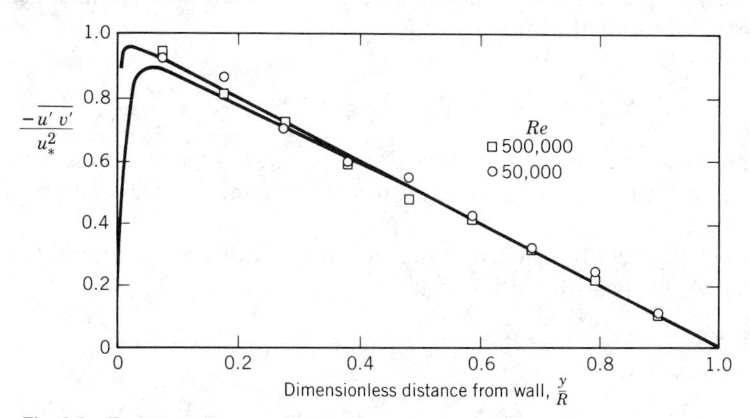

Fig. 8.8 Turbulent shear stress (Reynolds stress) for fully developed turbulent flow in a pipe. (Data from [1].)

8-5 TURBULENT VELOCITY PROFILES IN FULLY DEVELOPED PIPE FLOW

Except for flows of very viscous fluids in small diameter ducts, internal flows in general are turbulent. As noted in the discussion of shear stress distribution in fully developed pipe flow (Section 8-4), in turbulent flow there is no universal relationship between the stress field and the mean velocity field. Thus, for turbulent flows we are forced to rely on experimental data.

The velocity profile for fully developed turbulent flow through a smooth pipe is shown in Fig. 8.9.

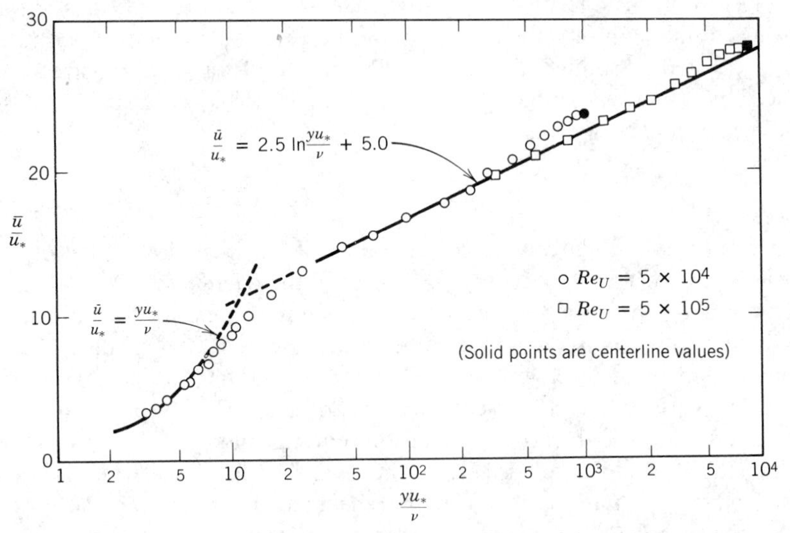

Fig. 8.9 Turbulent velocity profile for fully developed flow in a smooth pipe. (Data from [1].)

In the region very close to the wall where viscous shear is dominant, the mean velocity profile follows the linear viscous relation

$$u^+ = \frac{\bar{u}}{u_*} = \frac{yu_*}{\nu} = y^+ \tag{8.19}$$

where y is distance measured from the wall ($y = R - r$; R is the pipe radius), and $\bar{u}$ is the mean velocity. Equation 8.19 is valid for $0 \leq y^+ \leq 5$; this region is referred to as the *viscous sublayer.*

In the region where both viscous and turbulent shear are important, the velocity profile follows the logarithmic relation

$$\frac{\bar{u}}{u_*} = 2.5 \ln \frac{yu_*}{\nu} + 5.0 \tag{8.20}$$

There is considerable scatter in the numerical constants of Eq. 8.20; the values given represent averages over many experiments [2]. From Fig. 8.9 we see that the logarithmic profile gives a reasonably good approximation of the velocity profile all the way to the pipe centerline.

In the central region where turbulent shear is dominant, the velocity profile data are well correlated by the equation

$$\frac{U - \bar{u}}{u_*} = 2.5 \ln \frac{R}{y} \tag{8.21}$$

where U is the centerline velocity. Equation 8.21 is referred to as the velocity defect law.

The velocity profile for turbulent flow through a smooth pipe can be represented by the empirical "power-law" equation

$$\frac{\bar{u}}{U} = \left(\frac{y}{R}\right)^{1/n} = \left(1 - \frac{r}{R}\right)^{1/n} \tag{8.22}$$

where the exponent, n, varies with the Reynolds number. In Fig. 8.10 the data of Laufer [1] are shown on a plot of $\ln y/R$ versus $\ln \bar{u}/U$; the slope of the straight line through the data gives the value of n.

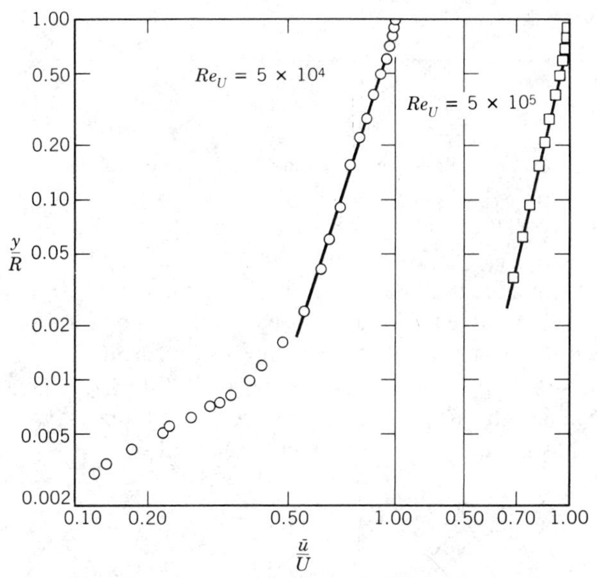

Fig. 8.10 Power-law velocity profiles for fully developed turbulent flow in a smooth pipe. (Data from [1].)

The power-law profile is not applicable close to the wall ($y/R < 0.04$); the profile gives an infinite velocity gradient at the wall. Although the profile fits the data close to the centerline, it fails to give a zero slope at the centerline. The variation of the exponent, n, in the power-law profile with Reynolds number (based on pipe diameter, D, and centerline velocity, U) is shown in Fig. 8.11.

Since the average velocity is $\bar{V} = Q/A$, and

$$Q = \int_A \vec{V} \cdot d\vec{A}$$

the ratio of the average velocity to the centerline velocity may be calculated for the power-law profiles of Eq. 8.22. The result is

$$\frac{\bar{V}}{U} = \frac{2n^2}{(n+1)(2n+1)} \tag{8.23}$$

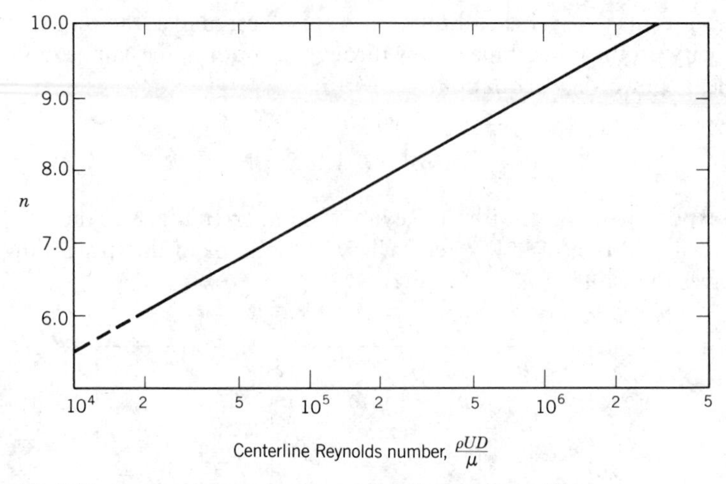

Fig. 8.11 Exponent for power-law profile. (Adapted from [3].)

From Eq. 8.23 we see that as n increases (due to increasing Reynolds number) the ratio of the average velocity to the centerline velocity increases; with increasing Reynolds number the velocity profile becomes more blunt or "fuller" (for $n = 6$, $\bar{V}/U = 0.79$; for $n = 10$, $\bar{V}/U = 0.87$). A value of 7 often is used for the exponent; this gives rise to the term "a one-seventh power profile" for fully developed turbulent flow.

Velocity profiles for $n = 6$ and $n = 10$ are shown in Fig. 8.12. The parabolic profile for fully developed laminar flow has been included for comparison. It is clear that the turbulent profile has a much steeper slope near the wall.

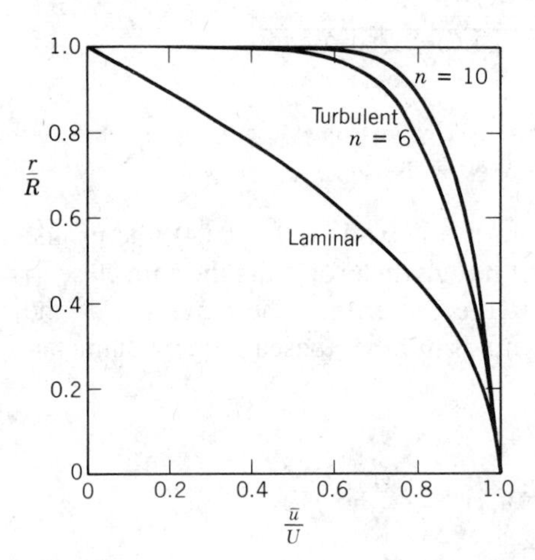

Fig. 8.12 Velocity profiles for fully developed pipe flow.

8-6 ENERGY CONSIDERATIONS IN PIPE FLOW

Thus far in our discussion of viscous flow, we have derived all results by applying the momentum equation for a control volume. We have, of course, also used the control volume formulation of the conservation of mass. Nothing has been said about the conservation of energy—the first law of thermodynamics. Additional insight into the nature of the pressure losses in internal viscous flows can be obtained from the energy equation. Consider, for example, the steady flow through the section of a piping system including a reducing elbow shown in Fig. 8.13. The control volume boundaries are shown as dashed lines. They are normal to the flow at sections ① and ② and coincide with the inside pipe wall elsewhere.

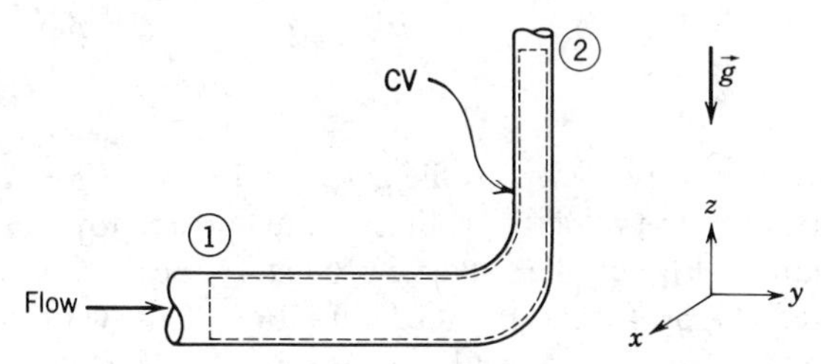

Fig. 8.13 Control volume and coordinates for energy analysis of flow through a 90° reducing albow.

Basic equation:

$$\dot{Q} + \overset{=0(1)}{\dot{W}_s} + \overset{=0(2)}{\dot{W}_{\text{shear}}} + \overset{=0(1)}{\dot{W}_{\text{other}}} = \overset{=0(3)}{\frac{\partial}{\partial t}\int_{\text{CV}} e\rho\, d\forall} + \int_{\text{CS}} (e + pv)\rho \vec{V} \cdot d\vec{A} \tag{4.59}$$

$$e = u + \frac{V^2}{2} + gz$$

Assumptions:
(1) $\dot{W}_s = 0$, $\dot{W}_{\text{other}} = 0$
(2) $\dot{W}_{\text{shear}} = 0$ (although shear stresses are present at the walls of the elbow, the velocities are zero at the walls)
(3) Steady flow
(4) Incompressible flow
(5) Internal energy and pressure uniform across sections ① and ②

Then the energy equation reduces to

$$\dot{Q} = \dot{m}(u_2 - u_1) + \dot{m}\left(\frac{p_2}{\rho} - \frac{p_1}{\rho}\right) + \dot{m}g(z_2 - z_1)$$

$$+ \int_{A_2} \frac{V_2^2}{2}\rho V_2\, dA_2 - \int_{A_1} \frac{V_1^2}{2}\rho V_1\, dA_1 \tag{8.24}$$

Note that we have not assumed the velocity to be uniform at sections ① and ②, since we know for viscous flows the velocity at a section cannot be uniform. However, it is convenient to introduce the average velocity into Eq. 8.24 so that we can eliminate the integrals. To do this, we define a kinetic energy coefficient.

8-6.1 Kinetic Energy Coefficient

The kinetic energy coefficient, α, is defined such that

$$\int_A \frac{V^2}{2} \rho V \, dA = \alpha \int_A \frac{\bar{V}^2}{2} \rho V \, dA = \alpha \dot{m} \frac{\bar{V}^2}{2} \tag{8.25a}$$

or

$$\alpha = \frac{\int_A \rho V^3 \, dA}{\dot{m} \bar{V}^2} \tag{8.25b}$$

For laminar flow in a pipe (velocity profile given by Eq. 8.12), $\alpha = 2.0$.

In turbulent pipe flow, the velocity profile is quite flat, as shown in Fig. 8.12. We can use Eq. 8.25b together with Eqs. 8.22 and 8.23 to determine the value of α. Substituting the power-law velocity profile of Eq. 8.22 into Eq. 8.25b, we obtain (after a little algebra) the expression

$$\alpha = \left[\frac{U}{\bar{V}}\right]^3 \frac{2n^2}{(3 + n)(3 + 2n)} \tag{8.26}$$

The value of $\bar{V}/U$ is determined from Eq. 8.23. For $n = 6$, $\alpha = 1.08$; for $n = 10$, $\alpha = 1.03$. Since the exponent, n, in the power-law profile is a function of the Reynolds number, α also varies with the Reynolds number. Because α is reasonably close to unity for large Reynolds number, a value of unity often is assumed for pipe flow calculations. However, for developing flows at moderate Reynolds numbers the change of kinetic energy may be significant.

8-6.2 Head Loss

Using the definition of α, the energy equation (Eq. 8.24) can be written

$$\dot{Q} = \dot{m}(u_2 - u_1) + \dot{m}\left(\frac{p_2}{\rho} - \frac{p_1}{\rho}\right) + \dot{m}g(z_2 - z_1) + \dot{m}\left(\frac{\alpha_2 \bar{V}_2^2}{2} - \frac{\alpha_1 \bar{V}_1^2}{2}\right)$$

Dividing by the mass rate of flow gives

$$\frac{\delta Q}{dm} = u_2 - u_1 + \frac{p_2}{\rho} - \frac{p_1}{\rho} + gz_2 - gz_1 + \frac{\alpha_2 \bar{V}_2^2}{2} - \frac{\alpha_1 \bar{V}_1^2}{2}$$

Rearranging this equation, we have

$$\left(\frac{p_1}{\rho} + \alpha_1 \frac{\bar{V}_1^2}{2} + gz_1\right) = \left(\frac{p_2}{\rho} + \alpha_2 \frac{\bar{V}_2^2}{2} + gz_2\right) + (u_2 - u_1) - \frac{\delta Q}{dm} \tag{8.27}$$

In Eq. 8.27, the term

$$\left(\frac{p}{\rho} + \alpha \frac{\bar{V}^2}{2} + gz\right)$$

represents the mechanical energy per unit mass at a flow cross section. The term $u_2 - u_1 - \delta Q/dm$ is equal to the difference in mechanical energy per unit mass between sections ① and ②. It represents the (irreversible) conversion of mechanical energy at section ① to unwanted thermal energy $(u_2 - u_1)$ and loss of energy via heat transfer $(-\delta Q/dm)$. We identify this group of terms as the total head loss, h_{l_T}. Then

$$\left(\frac{p_1}{\rho} + \alpha_1 \frac{\bar{V}_1^2}{2} + gz_1\right) - \left(\frac{p_2}{\rho} + \alpha_2 \frac{\bar{V}_2^2}{2} + gz_2\right) = h_{l_T} \tag{8.28}$$

Head loss has dimensions of energy per unit mass $[FL/M]$. In the $FLtT$ system of dimensions, this is equivalent to dimensions of $[L^2/t^2]$.

If the flow were frictionless, under the assumptions used to derive Eq. 8.28, the velocity at a section would be uniform $(\alpha_1 = \alpha_2 = 1)$ and Bernoulli's equation would predict zero head loss.

In incompressible frictionless flow a change in internal energy only can occur through the effects of heat transfer; there is no conversion of mechanical energy $(p/\rho + V^2/2 + gz)$ to internal energy. For viscous flow in a pipe, one effect of friction may be to increase the internal energy of the flow, as shown by Eq. 8.27.

You might wonder why the energy loss, h_{l_T}, is called a "head" loss. As the empirical science of hydraulics developed during the nineteenth century, it was common practice to express the energy balance in terms of energy per unit *weight* of flowing liquid (e.g. water) rather than energy per unit *mass* as in Eq. 8.28. To obtain dimensions of energy per unit weight, we divide each term in Eq. 8.28 by the acceleration of gravity, g. Then the net dimensions of h_{l_T} are $[(L^2/t^2)(t^2/L)] = [L]$, or feet of flowing liquid. Since the term head loss is in common use, we shall also use it here. Remember that its physical interpretation is a loss in mechanical energy, expressed per unit mass of flowing fluid.

Equation 8.28 can be used to calculate the pressure difference between any two points in a piping system, provided the head loss, h_{l_T}, is known. We shall consider calculation of h_{l_T} in the next section.

8-7 CALCULATION OF HEAD LOSS

Total head loss, h_{l_T}, is regarded as the sum of major losses, h_l, due to frictional effects in fully developed flow in constant-area tubes, and minor losses, h_{l_m}, due to entrances, fittings, area changes, and so on. Consequently, we consider the major and minor losses separately.

8-7.1 Major Losses: Friction Factor

The energy balance, expressed by Eq. 8.28, can be used to evaluate the major head loss. For fully developed flow through a constant-area pipe, $h_{l_m} = 0$, and $\alpha_1(\bar{V}_1^2/2) = \alpha_2(\bar{V}_2^2/2)$. Then Eq. 8.28 becomes

$$\frac{p_1 - p_2}{\rho} = g(z_2 - z_1) + h_l \tag{8.29}$$

If the pipe is horizontal, then $z_2 = z_1$ and

$$\frac{p_1 - p_2}{\rho} = \frac{\Delta p}{\rho} = h_l \tag{8.30}$$

Thus the major head loss can be expressed as the pressure loss for fully developed flow through a horizontal pipe of constant area.

Since head loss represents the energy converted by frictional effects from mechanical to thermal energy, head loss for fully developed flow in a constant-area duct depends only on the details of the flow through the duct. Head loss is independent of pipe orientation.

a. Laminar Flow

In laminar flow the pressure drop may be computed analytically for fully developed flow in a horizontal pipe. Thus, from Eq. 8.13c,

$$\Delta p = \frac{128\mu L Q}{\pi D^4} = \frac{128\mu L \bar{V}(\pi D^2/4)}{\pi D^4} = 32\frac{L}{D}\frac{\mu \bar{V}}{D}$$

Substituting in Eq. 8.30 gives

$$h_l = 32\frac{L}{D}\frac{\mu \bar{V}}{\rho D} = \frac{L}{D}\frac{\bar{V}^2}{2}\left(64\frac{\mu}{\rho \bar{V} D}\right) = \left(\frac{64}{Re}\right)\frac{L}{D}\frac{\bar{V}^2}{2} \tag{8.31}$$

(We shall see the reason for writing h_l in this form shortly.)

b. Turbulent Flow

In turbulent flow we cannot evaluate the pressure drop analytically, so we must resort to experimental results and use dimensional analysis to correlate the experimental data. In fully developed turbulent flow, the pressure drop, Δp, due to friction in a horizontal constant-area pipe is known to depend on the pipe diameter, D, the pipe length, L, the pipe roughness, e, the average flow velocity, $\bar{V}$, the fluid density, ρ, and the fluid viscosity, μ. In functional form

$$\Delta p = \Delta p(D, L, e, \bar{V}, \rho, \mu)$$

We applied dimensional analysis to this problem in Example Problem 7.2. The results were a correlation of the form

$$\frac{\Delta p}{\rho \bar{V}^2} = f\left(\frac{\mu}{\rho \bar{V} D}, \frac{L}{D}, \frac{e}{D}\right)$$

We recognize that $\mu/\rho \bar{V} D = 1/Re$, so we could just as well write

$$\frac{\Delta p}{\rho \bar{V}^2} = \phi\left(Re, \frac{L}{D}, \frac{e}{D}\right)$$

Substituting from Eq. 8.30, we see that

$$\frac{h_l}{\bar{V}^2} = \phi\left(Re, \frac{L}{D}, \frac{e}{D}\right)$$

Although dimensional analysis predicts the functional relationship, we must resort to experiment to obtain actual values.

Experiments show that the nondimensional head loss is directly proportional to L/D. Hence we can write

$$\frac{h_l}{\bar{V}^2} = \frac{L}{D}\,\phi_1\left(Re, \frac{e}{D}\right)$$

Since the function ϕ_1 is still undetermined, it is permissible to introduce a constant into the left side of the above equation. The number $\frac{1}{2}$ is introduced into the denominator such that the head loss is nondimensionalized on the kinetic energy per unit mass of flow. Then

$$\frac{h_l}{\frac{1}{2}\bar{V}^2} = \frac{L}{D}\,\phi_2\left(Re, \frac{e}{D}\right)$$

The unknown function $\phi_2(Re, e/D)$ is defined as the friction factor, f,

$$f \equiv \phi_2\left(Re, \frac{e}{D}\right)$$

and

$$h_l = f\,\frac{L}{D}\,\frac{\bar{V}^2}{2} \tag{8.32}$$

The friction factor, f, is determined experimentally. The results, published by L. F. Moody, are shown in Fig. 8.14.

To determine head loss for fully developed flow with known conditions, the flow Reynolds number is evaluated first. The value of relative roughness, e/D, for the flow is obtained from Fig. 8.15. Then the friction factor, f, is read from the appropriate curve in Fig. 8.14, at the known values of Re and e/D. Finally, the head loss is found using Eq. 8.32.

Several features of Fig. 8.14 require some discussion. The friction factor for laminar flow may be obtained by comparing Eqs. 8.31 and 8.32

$$h_l = \left(\frac{64}{Re}\right)\frac{L}{D}\,\frac{\bar{V}^2}{2} = f\,\frac{L}{D}\,\frac{\bar{V}^2}{2}$$

Consequently, for laminar flow

$$f_{\text{laminar}} = \frac{64}{Re} \tag{8.33}$$

Thus, in laminar flow, the friction factor is a function of Reynolds number only; it is independent of roughness. Although we took no notice of roughness in deriving

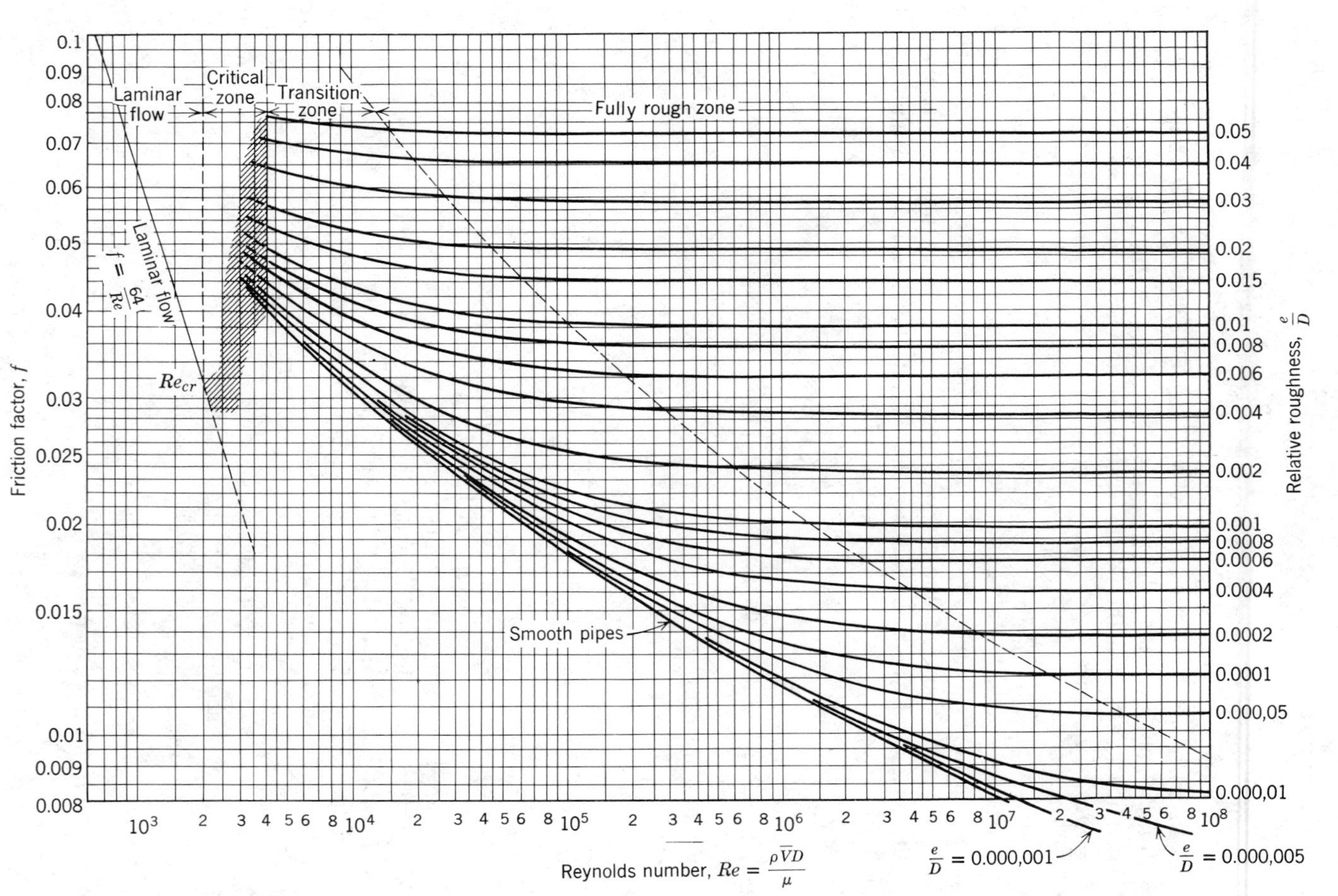

Fig. 8.14 Friction factor for fully developed flow in circular pipes. (Data from [4], used by permission.)

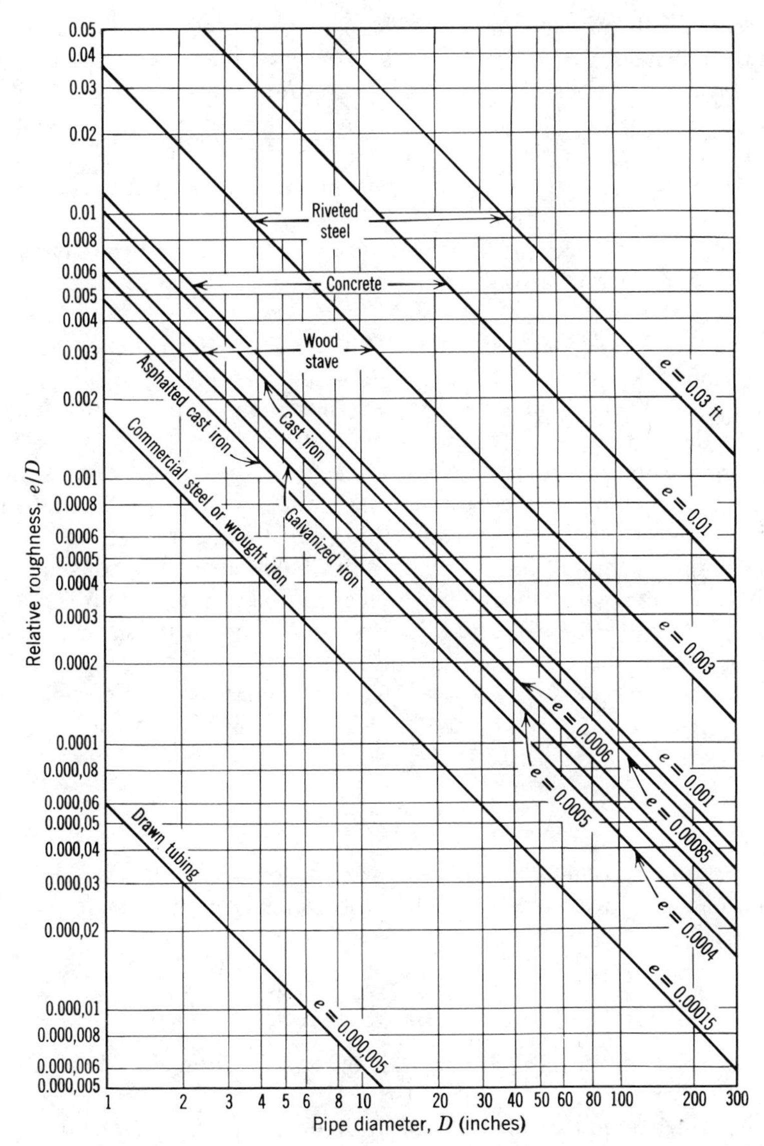

Fig. 8.15 Relative roughness values for pipes of common engineering materials. (Data from [4], used by permission.)

Eq. 8.31, experimental results verify that the friction factor is a function only of the Reynolds number in laminar flow.

The Reynolds number in a pipe may be changed most easily by varying the average flow velocity. If the flow in a pipe is originally laminar, increasing the velocity until the critical Reynolds number is reached causes transition to occur; the laminar flow gives way to turbulent flow. The effect of transition on the velocity profile was discussed in Section 8-5. Figure 8.12 shows that the velocity gradient at the tube wall is much larger for turbulent flow than for laminar flow. This change in velocity profile causes the wall shear stress to increase sharply, with the same effect on the friction factor.

As the Reynolds number is increased above the transition value, the velocity profile continues to become fuller, as noted in Section 8-5; the friction factor at first tends to follow the smooth pipe curve, along which friction factor is a function of Reynolds number only. However, as the Reynolds number increases, the velocity profile becomes still fuller. The size of the thin viscous sublayer near the tube wall decreases. As roughness elements begin to poke through this layer, the effect of roughness becomes important, and the friction factor becomes a function of both the Reynolds number *and* the relative roughness.

At very large Reynolds number, most of the roughness elements on the tube wall protrude through the viscous sublayer; the drag and, hence, the pressure loss, depend only on the size of the roughness elements. This flow regime is termed the "fully rough" flow regime; the friction factor depends only on e/D in this regime.

To summarize the preceding discussion, we see that as the Reynolds number is increased, the friction factor decreases as long as the flow remains laminar. At transition, f increases sharply. In the turbulent flow regime, the friction factor decreases gradually along the smooth pipe curve, and finally levels out at a constant value for extremely large Reynolds number. (The only exception to these trends is that friction factors for pipes with $e/D \gtrsim 0.001$ in turbulent flow fall above the smooth pipe curve.)

In order to use the computer to solve problems, it is necessary to have a mathematical formulation for the friction factor, f, in terms of the Reynolds number, Re, and the relative roughness, e/D. The Blasius correlation for turbulent flow in smooth pipes, valid for $Re \leq 10^5$ is

$$f = \frac{0.3164}{Re^{0.25}} \tag{8.34}$$

When this relation is combined with the expression for wall shear stress (Eq. 8.16), the expression for head loss (Eq. 8.30), and the definition of friction factor (Eq. 8.32), a useful expression for the wall shear stress is obtained

$$\tau_w = 0.03325\,\rho \bar{V}^2 \left(\frac{\nu}{R\bar{V}} \right)^{0.25} \tag{8.35}$$

This equation will be used later in our study of turbulent boundary-layer flow over a flat plate (Chapter 9).

The most widely used formula for friction factor is that due to Colebrook

$$\frac{1}{f^{0.5}} = -2.0 \log \left(\frac{e/D}{3.7} + \frac{2.51}{Re\, f^{0.5}} \right) \tag{8.36}$$

Equation 8.36 is transcendental, so iteration is needed to evaluate f. Miller [5] suggests that a single iteration will produce a result within 1 percent if the initial estimate is calculated from

$$f_0 = 0.25 \left[\log \left(\frac{e/D}{3.7} + \frac{5.74}{Re^{0.9}} \right) \right]^{-2} \tag{8.37}$$

Equation 8.37 is from [6].

Fig. 8.16 Pipe section removed after 40 years of service as a water line, showing formation of scale.

Figure 8.15 also needs some explanation. All of the e/D values given are those for new pipes, in relatively good condition. Over long periods of service, corrosion takes place and, particularly in hard water areas, lime deposits and rust scale form on pipe walls. Corrosion can weaken pipes, eventually leading to failure. Deposit formation increases wall roughness appreciably, and also decreases the effective diameter. These factors combine to cause e/D to increase by a factor of 2 or 3 for old pipes. An example is shown in Fig. 8.16.

Curves presented in Figs. 8.14 and 8.15 represent average values for data obtained from numerous experiments. Values taken from the curves should be considered accurate within approximately ± 10 percent, which is sufficient for many engineering analyses. If more accuracy is needed, actual test data should be used.

8-7.2 Minor Losses

The flow in a piping system may be required to pass through a variety of fittings, bends, or abrupt changes in area. Additional head losses are encountered primarily as a result of flow separation. (Energy eventually is dissipated by violent mixing in the separated zones.) These losses will be minor (hence the term *minor losses*) if the piping system includes long lengths of constant-area pipe. The minor head loss may be expressed as

$$h_{l_m} = K \frac{\bar{V}^2}{2} \tag{8.38a}$$

where the *loss coefficient*, K, must be determined experimentally for each situation.

Minor head loss also may be expressed as

$$h_{l_m} = f \frac{L_e}{D} \frac{\bar{V}^2}{2} \tag{8.38b}$$

where L_e is an *equivalent length* of straight pipe.

For flow through pipe bends and fittings, the loss coefficient, K, is found to vary with pipe size (diameter) in much the same manner as does the friction factor, f, for flow through a straight pipe. Consequently, the equivalent length, L_e/D, tends toward a constant for different sizes of a given type of fitting.

Experimental data for minor losses are plentiful, but they are scattered among a variety of sources. Different sources may give different values for the same flow configuration. The data presented here should be considered as representative for some commonly encountered situations; in each case the source of the data is identified.

a. Inlets and Exits

A poorly designed inlet to a pipe can cause an appreciable head loss. If the inlet has sharp corners, flow separation occurs at the corners, and a *vena contracta* is formed.[3] The fluid must accelerate locally to pass through the reduced flow area at the vena contracta. Losses in mechanical energy result from the unconfined mixing as the flow stream decelerates again to fill the pipe. Three basic inlet geometries are shown in Table 8.1. From the table it is clear that the loss coefficient is reduced significantly when the inlet is rounded even slightly. For a well-rounded inlet ($r/D \geq 0.15$) the entrance loss coefficient is almost negligible. Example Problem 8.9 illustrates a procedure for experimentally determining the loss coefficient for a pipe inlet.

Table 8.1 Minor Loss Coefficients for Pipe Entrances (Data from [7])

Entrance Type	Minor Loss Coefficient, K^a
Reentrant	0.78
Square-edged	0.5
Rounded	r/D: 0.02, 0.06, ≥0.15; K: 0.28, 0.15, 0.04

r/D	0.02	0.06	≥0.15
K	0.28	0.15	0.04

[a] Based on $h_{l_m} = K(\bar{V}^2/2)$, where $\bar{V}$ is the mean velocity in the pipe.

[3] This behavior is well illustrated in film loop S-FM016, *Flow from a Reservoir to a Duct.*

The kinetic energy per unit mass of flow, $\alpha\bar{V}^2/2$, is completely dissipated by mixing when flow discharges from a duct into a large reservoir or plenum chamber. The situation corresponds to flow through an abrupt expansion with $AR = 0$ (Fig. 8.17). The minor loss coefficient thus equals α. No improvement in minor loss coefficient for an exit is possible; however, addition of a diffuser can reduce the value of $\bar{V}^2/2$ considerably (see Example Problem 8.10).

b. Enlargements and Contractions

Minor loss coefficients for sudden expansions and contractions in circular ducts are given in Fig. 8.17. Note that both loss coefficients are based on the larger value of $\bar{V}^2/2$. Thus losses for a sudden expansion are based on $\bar{V}_1^2/2$, and those for a contraction are based on $\bar{V}_2^2/2$.

Losses due to area change can be reduced somewhat by installing a nozzle or diffuser between the two sections of straight pipe. Data for nozzles are given in Table 8.2.

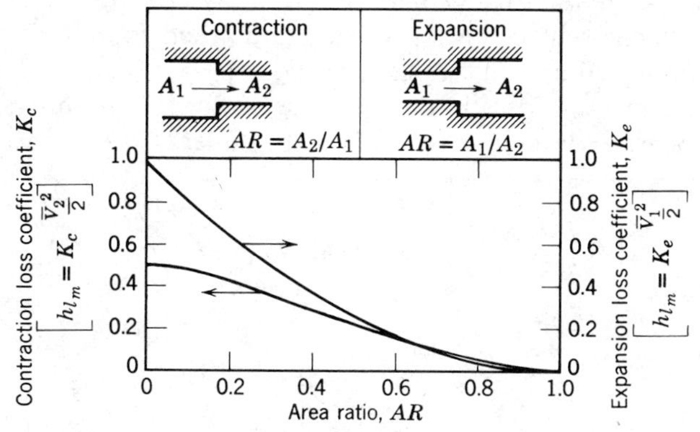

Fig. 8.17 Loss coefficients for flow through sudden area changes. (Data from [8].)

Table 8.2 Loss Coefficients for Gradual Contractions: Round and Rectangular Ducts (Data from [9])

	Loss Coefficient, K^a							
		Included Angle, θ, Degrees						
	A_2/A_1	10	15–40	50–60	90	120	150	180
Flow A_1 θ A_2	0.50	0.05	0.05	0.06	0.12	0.18	0.24	0.26
	0.25	0.05	0.04	0.07	0.17	0.27	0.35	0.41
	0.10	0.05	0.05	0.08	0.19	0.29	0.37	0.43

[a] Based on $h_{l_m} = K(\bar{V}_2^2/2)$.

Losses in diffusers depend on a number of geometric and flow variables. Diffuser data most commonly are presented in terms of a pressure recovery coefficient, C_p, defined as the ratio of static pressure rise to inlet dynamic pressure,

$$C_p \equiv \frac{p_2 - p_1}{\frac{1}{2}\rho \bar{V}_1^2} \tag{8.39}$$

Data for conical diffusers with fully developed turbulent pipe flow at the inlet are presented in Fig. 8.18 as a function of geometry. From the performance map of Fig. 8.18 we see that optimum diffuser geometries may be defined. For a given area ratio, AR, there is a value of N/R_1 above which no increase in pressure recovery would be expected; this is particularly clear for $AR < 1.40$ in Fig. 8.18. Similarly, for a given value of dimensionless length, N/R_1, there is an optimum value of the area ratio for maximum pressure recovery.

Performance maps for plane wall and annular diffusers [11] and for radial diffusers [12] are available in the literature.

Diffuser pressure recovery is essentially independent of Reynolds number for inlet Reynolds numbers greater than 7.5×10^4 [13]. Diffuser pressure recovery with uniform inlet flow is somewhat better than that for fully developed inlet flow. Performance maps for plane wall, conical, and annular diffusers for a variety of inlet flow conditions are presented in a diffuser data book [14].

Since the static pressure rises in the direction of flow in a diffuser, the flow may separate from the walls. For some geometries, the outlet flow is distorted; sometimes

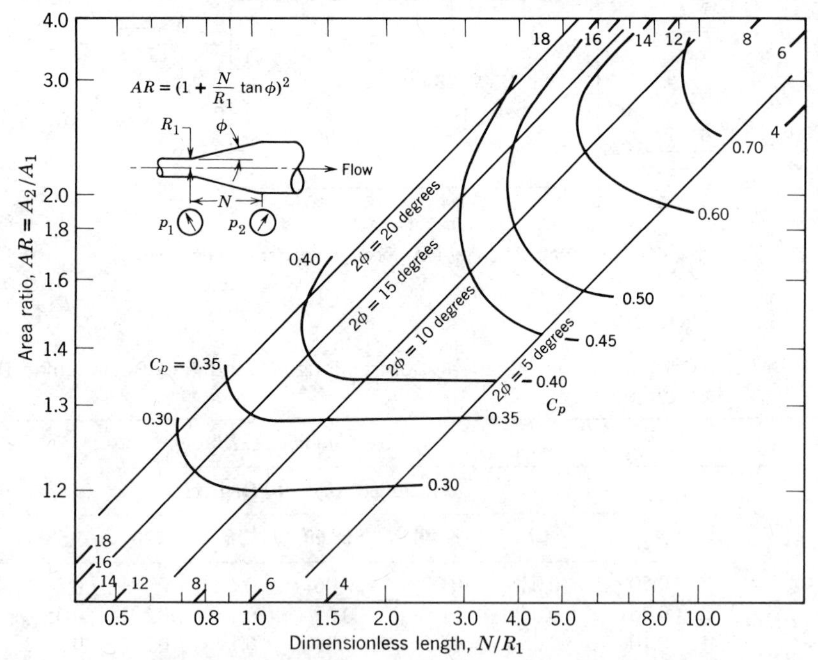

Fig. 8.18 Pressure recovery data for conical diffusers with fully developed turbulent pipe flow at inlet. (Data from [10].)

pulsations occur. The flow regime behavior of plane wall diffusers is illustrated well in the film *Flow Visualization* (S. J. Kline, principal).[4] For wide angle diffusers, vanes or splitters can be used to suppress stall and improve pressure recovery [15].

The definition of C_p may be related to head loss. If gravity is neglected, and $\alpha_1 = \alpha_2 = 1.0$, Eq. 8.28 reduces to

$$\left[\frac{p_1}{\rho} + \frac{\bar{V}_1^2}{2}\right] - \left[\frac{p_2}{\rho} + \frac{\bar{V}_2^2}{2}\right] = h_{l_T} = h_{l_m}$$

Thus

$$h_{l_m} = \frac{\bar{V}_1^2}{2} - \frac{\bar{V}_2^2}{2} - \frac{p_2 - p_1}{\rho}$$

$$h_{l_m} = \frac{\bar{V}_1^2}{2}\left[\left(1 - \frac{\bar{V}_2^2}{\bar{V}_1^2}\right) - \frac{p_2 - p_1}{\frac{1}{2}\rho\bar{V}_1^2}\right] = \frac{\bar{V}_1^2}{2}\left[\left(1 - \frac{\bar{V}_2^2}{\bar{V}_1^2}\right) - C_p\right]$$

From continuity, $A_1\bar{V}_1 = A_2\bar{V}_2$, so

$$h_{l_m} = \frac{\bar{V}_1^2}{2}\left[1 - \left(\frac{A_1}{A_2}\right)^2 - C_p\right]$$

or

$$h_{l_m} = \frac{\bar{V}_1^2}{2}\left[\left(1 - \frac{1}{(AR)^2}\right) - C_p\right] \tag{8.40}$$

For a frictionless flow, $h_{l_m} = 0$; for this case Eq. 8.40 gives the ideal pressure recovery coefficient, denoted C_{p_i}, as

$$C_{p_i} = 1 - \frac{1}{(AR)^2} \tag{8.41}$$

This same result can be obtained by applying the Bernoulli equation, together with the continuity equation, to frictionless flow through the diffuser. Thus the head loss for flow through an actual diffuser may be written

$$h_{l_m} = (C_{p_i} - C_p)\frac{\bar{V}_1^2}{2} \tag{8.42}$$

c. Pipe Bends

The head loss of a bend is larger than that for fully developed flow through a straight section of equal length. The additional loss is primarily the result of secondary flow,[5] and is expressed most conveniently by an equivalent length of straight pipe. The equivalent length depends on the relative radius of curvature of the bend as shown in

[4] Film loop S-FM049, *Flow Regimes in Subsonic Diffusers*, was edited from this film. See also film loop S-FM065, *Wide Angle Diffuser with Suction.*

[5] Secondary flows are shown in film loop S-FM019, *Secondary Flow in a Bend.*

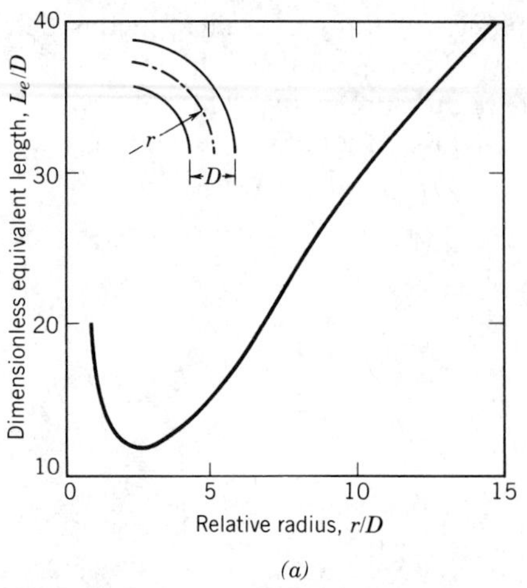

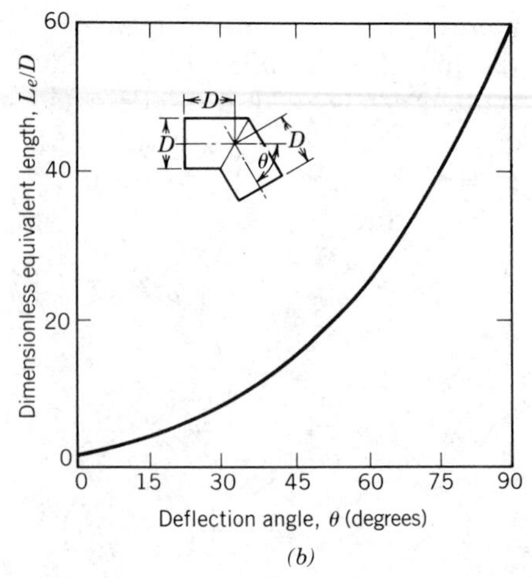

Fig. 8.19 Representative resistance values (L_e/D) for (a) 90° pipe bends and flanged elbows, and (b) miter bends. (Data from [7].)

Fig. 8.19*a*, for 90° bends. An approximate procedure for computing the resistance of bends with other turning angles is given in [7].

Because they are simple and inexpensive to erect in the field, miter bends often are used, especially in large pipe systems. Design data for miter bends are given in Fig. 8.19*b*.

d. Valves and Fittings

Losses for flow through valves and fittings also may be expressed in terms of an equivalent length of straight pipe. Some representative data are given in Table 8.3.

The resistance values given for valves are all for the fully open position. Significant increases in flow losses occur for partially open conditions. Valve design may vary significantly among manufacturers. Whenever possible, resistance values furnished by the supplier should be used if accurate calculations are needed.

Fittings are inserted in a piping system with threaded or flanged connections: for small pipe diameters, threaded connections are most common; large pipe systems frequently are fabricated with welded or flanged connections.

In practice, insertion losses for fittings and valves vary considerably, depending on the care used in fabricating the pipe system. If burrs from cutting pipe sections are allowed to remain, they cause local flow obstructions, which increase losses appreciably.

Although the losses discussed in this section were termed "minor losses," they can be a large fraction of the overall system loss. Thus a system for which calculations are to be made should be checked carefully to make sure all losses have been identified and their magnitudes estimated. If calculations are made carefully, the results will be of satisfactory engineering accuracy. Computed values may be expected to predict actual losses within ±10 percent.

Table 8.3 Representative Dimensionless Equivalent Lengths (L_e/D) for Valves and Fittings (Data from [7])

Fitting Type	Equivalent Length,[a] L_e/D
Valves (fully open)	
Gate valve	8
Globe valve	340
Angle valve	150
Ball valve	3
Lift check valve: globe lift	600
: angle lift	55
Foot valve with strainer: poppet disk	420
: hinged disk	75
Standard elbow: 90°	30
: 45°	16
Return bend, close pattern	50
Standard tee: flow through run	20
: flow through branch	60

[a] Based on $h_{l_m} = f \dfrac{L_e}{D} \dfrac{\bar{V}^2}{2}$

8-8 SOLUTION OF PIPE FLOW PROBLEMS

Once the total head loss has been calculated using the methods of Section 8-7, pipe flow problems can be solved using the energy equation, Eq. 8.28. The same basic techniques are used, even for complex piping systems, but let us first consider single-path pipe flow problems.

8-8.1 Single-Path Systems

Equation 8.28 is the computing equation for pipe systems. The pressure drop across a pipe system is a function of flow rate, elevation change, and total head loss. The total head loss consists of major losses due to friction in constant-area sections (Eq. 8.32) and minor losses due to fittings, area changes, and so forth (Eqs. 8.38). The pressure drop could be written in the functional form

$$\Delta p = \phi_3(L, Q, D, e, \Delta z, \text{system configuration}, \rho, \mu)$$

The fluid properties may be assumed constant for incompressible pipe flow. The roughness, elevation change, and system configuration depend on the pipe system layout. Once these have been fixed (for a given system and fluid), the dependence reduces to

$$\Delta p = \phi_4(L, Q, D) \tag{8.43}$$

Equation 8.43 relates four variables. Any one of these may be the unknown quantity in a practical flow situation. Thus four general cases are possible:

(a) L, Q, and D known, Δp unknown.
(b) Δp, Q, and D known, L unknown.
(c) Δp, L, and D known, Q unknown.
(d) Δp, L, and Q known, D unknown.

Cases (a) and (b) may be solved directly by applying the continuity and energy equations, and using loss data from Section 8-7. Solutions for cases (c) and (d) make use of the same equations and data, but require iteration. Each case is discussed below and illustrated by an example.

a. L, Q, and D Known, Δp Unknown

A friction factor is obtained from the Moody chart or empirical equations using Re and e/D values computed from the given data. The total head loss is computed from Eqs. 8.32 and 8.38. Equation 8.28 is then used to evaluate the pressure drop, Δp. The procedure is illustrated in Example Problem 8.5.

b. Δp, Q, and D Known, L Unknown

The total head loss is calculated from Eq. 8.28. A friction factor is obtained from the Moody diagram or empirical equations using Re and e/D values computed from the given data. The unknown length is determined by solving Eq. 8.32. The procedure is illustrated in Example Problem 8.6.

c. Δp, L, and D Known, Q Unknown

Equation 8.28 is combined with the defining equations for head loss; the result is an expression for $\bar{V}$ (or Q) in terms of the friction factor, f. Most pipe flows of engineering interest have relatively large values of Reynolds number. Thus, even though the Reynolds number (and hence f) cannot be calculated because Q is not known, a good first guess for the friction factor is the value in the fully rough region of Fig. 8.14. Using the assumed value for f, a first approximation for $\bar{V}$ is calculated. The Reynolds number is computed for this value of $\bar{V}$, and a new value for f and a second approximation for $\bar{V}$ are obtained. Since f is a rather weak function of Reynolds number, more than two iterations seldom are required for convergence. A flow of this type is evaluated in Example Problem 8.7.

d. Δp, L, and Q Known, D Unknown

When a fluid-handling device is available and the geometry of the piping system is known, the problem is to determine the smallest (and hence least costly) pipe size that can deliver the desired flow rate. Since the pipe diameter is unknown, neither the Reynolds number nor relative roughness can be computed directly, and an iterative solution is required.

Calculations begin by assuming a trial value of pipe diameter. The Reynolds number and relative roughness then are calculated using the assumed value of D. A value for the friction factor is obtained from Fig. 8.14. Then the head loss is computed from Eqs. 8.32 and 8.38, and Eq. 8.28 is solved for pressure drop. The resulting trial value is compared to the system requirement.

If the trial value of Δp is too large, calculations are repeated for a larger assumed value of D. If the trial value of Δp is less than the criterion, a smaller assumed value of D should be checked.

In choosing a pipe diameter, it is logical to work with values that are available commercially. Pipe is manufactured in a limited number of standard sizes. Some data for standard pipe sizes are given in Table 8.4. For data on extra strong or double extra strong pipes, consult a handbook, e.g. [7]. Pipe larger than 12 in. nominal diameter is produced in multiples of 2 in. up to a nominal diameter of 36 in. and in multiples of 6 in. for still larger sizes.

Table 8.4 Standard Sizes for Carbon Steel, Alloy Steel, and Stainless Steel Pipe (Data from [7])

Nominal Pipe Size (in.)	Inside Diameter (in.)	Nominal Pipe Size (in.)	Inside Diameter (in.)
$\frac{1}{8}$	0.269	$2\frac{1}{2}$	2.469
$\frac{1}{4}$	0.364	3	3.068
$\frac{3}{8}$	0.493	4	4.026
$\frac{1}{2}$	0.622	5	5.047
$\frac{3}{4}$	0.824	6	6.065
1	1.049	8	7.981
$1\frac{1}{2}$	1.610	10	10.020
2	2.067	12	12.000

Example 8.5

A 100 m length of smooth horizontal pipe is attached to a large reservoir. What depth, d, must be maintained in the reservoir to produce a volume flow rate of 0.03 m^3/sec of water? The inside diameter of the smooth pipe is 75 mm. The loss coefficient, K, for the square-edged inlet is 0.5. The water discharges to the atmosphere.

EXAMPLE PROBLEM 8.5

GIVEN: Water flow at 0.03 m^3/sec through a 75 mm diameter pipe, with $L = 100$ m, attached to a constant-level reservoir. Inlet loss coefficient, $K = 0.5$.

FIND: Reservoir depth, d, to maintain the flow.

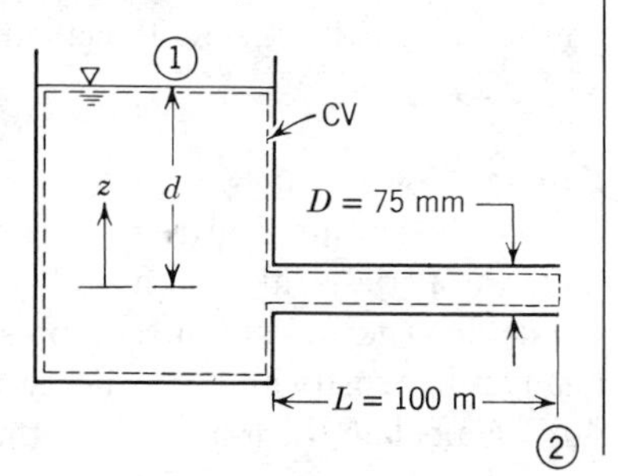

SOLUTION:
Computing equation:

$$\left(\frac{p_1}{\rho} + \alpha_1 \frac{\bar{V}_1^2}{2} + gz_1\right) - \left(\frac{p_2}{\rho} + \alpha_2 \frac{\bar{V}_2^2}{2} + gz_2\right) = h_{l_T} = h_l + h_{l_m} \tag{8.28}$$

where

$$h_l = f\frac{L}{D}\frac{\bar{V}^2}{2} \qquad \text{and} \qquad h_{l_m} = K\frac{\bar{V}^2}{2}$$

For the given problem, $p_1 = p_2 = p_{\text{atm}}$, $\bar{V}_1 \simeq 0$, $\bar{V}_2 = \bar{V}$, and $\alpha_2 \simeq 1.0$. If it is assumed that $z_2 = 0$, then $z_1 = d$. Simplifying Eq. 8.28 gives

$$gd - \frac{\bar{V}^2}{2} = f\frac{L}{D}\frac{\bar{V}^2}{2} + K\frac{\bar{V}^2}{2}$$

Then

$$d = \frac{1}{g}\left[f\frac{L}{D}\frac{\bar{V}^2}{2} + K\frac{\bar{V}^2}{2} + \frac{\bar{V}^2}{2}\right] = \frac{\bar{V}^2}{2g}\left[f\frac{L}{D} + K + 1\right]$$

Since $\bar{V} = \dfrac{Q}{A} = \dfrac{4Q}{\pi D^2}$, then

$$d = \frac{8Q^2}{\pi^2 D^4 g}\left[f\frac{L}{D} + K + 1\right]$$

Assuming water at 20 C, $\rho = 999\ \text{kg/m}^3$, and $\mu = 1.0 \times 10^{-3}\ \text{kg/m}\cdot\text{sec}$. Thus

$$Re = \frac{\rho \bar{V} D}{\mu} = \frac{4\rho Q}{\pi \mu D}$$

$$Re = \frac{4}{\pi} \times \frac{999\ \text{kg}}{\text{m}^3} \times \frac{0.03\ \text{m}^3}{\text{sec}} \times \frac{\text{m}\cdot\text{sec}}{1.0 \times 10^{-3}\ \text{kg}} \times \frac{1}{0.075\ \text{m}} = 5.09 \times 10^5$$

For smooth pipe, from Fig. 8.14, $f = 0.0131$. Then

$$d = \frac{8Q^2}{\pi^2 D^4 g}\left[f\frac{L}{D} + K + 1\right]$$

$$= \frac{8}{\pi^2} \times \frac{(0.03)^2\ \text{m}^6}{\text{sec}^2} \times \frac{1}{(0.075)^4\ \text{m}^4} \times \frac{\text{sec}^2}{9.81\ \text{m}}\left[(0.0131)\frac{100\ \text{m}}{0.075\ \text{m}} + 0.5 + 1\right]$$

$d = 44.6\ \text{m}$ ← d

{This problem illustrates the method for calculating the total head loss.}

Example 8.6

A compressed air drill requires an air supply of 0.25 kg/sec at a gage pressure of 650 kPa at the drill. The hose from the air compressor to the drill is 40 mm inside diameter. The maximum compressor discharge gage pressure is 690 kPa. Neglect changes in density and any effects due to hose curvature. Air leaves the compressor at 40 C. Calculate the longest hose that may be used.

EXAMPLE PROBLEM 8.6

GIVEN: Air flow through a line of length, L, and diameter, $D = 40$ mm.

$p_1 = 690$ kPa $\quad p_2 = 650$ kPa

$T_1 = 40$ C $\quad \dot{m} = 0.25$ kg/sec $\quad \rho \approx$ constant

FIND: Allowable length of hose.

SOLUTION:
Computing equation:

$$\left(\frac{p_1}{\rho} + \alpha_1 \frac{\bar{V}_1^2}{2} + gz_1\right) - \left(\frac{p_2}{\rho} + \alpha_2 \frac{\bar{V}_2^2}{2} + gz_2\right) = h_{l_T} = h_l + h_{l_m} \tag{8.28}$$

where

$$h_l = f\frac{L}{D}\frac{\bar{V}^2}{2} \qquad h_{l_m} = K\frac{\bar{V}^2}{2}$$

For $\rho = c$, then $\bar{V}_1 = \bar{V}_2$, since $A_1 = A_2$. Since p_1 and p_2 are given, neglect minor losses. Assume that $\alpha_1 = \alpha_2$. Neglect changes in elevation; assume $z_1 = z_2$. Then Eq. 8.28 can be written

$$\frac{p_1 - p_2}{\rho} = f\frac{L}{D}\frac{\bar{V}^2}{2} \qquad \text{or} \qquad L = \frac{(p_1 - p_2)}{\rho}\frac{2D}{f\bar{V}^2}$$

The density is

$$\rho = \rho_1 = \frac{p_1}{RT_1} = \frac{7.91 \times 10^5\ \text{N}}{\text{m}^2} \times \frac{\text{kg} \cdot \text{K}}{287\ \text{N} \cdot \text{m}} \times \frac{1}{313\ \text{K}} = 8.81\ \text{kg/m}^3$$

From continuity

$$\bar{V} = \frac{\dot{m}}{\rho A} = \frac{4\dot{m}}{\pi \rho D^2} = \frac{4}{\pi} \times \frac{0.25\ \text{kg}}{\text{sec}} \times \frac{\text{m}^3}{8.81\ \text{kg}} \times \frac{1}{(0.04)^2\ \text{m}^2} = 22.6\ \text{m/sec}$$

For air at 40 C, $\mu = 1.8 \times 10^{-5}$ kg/m · sec, so

$$Re = \frac{\rho \bar{V} D}{\mu} = \frac{8.81\ \text{kg}}{\text{m}^3} \times \frac{22.6\ \text{m}}{\text{sec}} \times 0.04\ \text{m} \times \frac{\text{m} \cdot \text{sec}}{1.8 \times 10^{-5}\ \text{kg}} = 4.42 \times 10^5$$

Assume smooth pipe; then from Fig. 8.14, $f = 0.0134$. Substituting gives

$$L = \frac{(p_1 - p_2)}{\rho}\frac{2D}{f\bar{V}^2}$$

$$= \frac{0.40 \times 10^5\ \text{N}}{\text{m}^2} \times 2 \times 0.04\ \text{m} \times \frac{\text{m}^3}{8.81\ \text{kg}} \times \frac{1}{0.0134} \times \frac{\text{sec}^2}{(22.6)^2\ \text{m}^2} \times \frac{\text{kg} \cdot \text{m}}{\text{N} \cdot \text{sec}^2}$$

$L = 53.1$ m $\qquad L$

This problem illustrates the method of solving for an unknown pipe length. Note that the relative change in density for this problem, $\Delta\rho/\rho \simeq \Delta p/p$, is only about 5 percent. Thus the assumption of incompressible flow is reasonable.

Example 8.7

A fire protection system is supplied from a water tower and standpipe 80 ft tall. The longest pipe in the system is 600 ft long, and is made of cast iron about 20 years old. The pipe contains one gate valve; other minor losses may be neglected. The pipe diameter is 4 in. Determine the maximum rate of flow through this pipe, in gallons per minute.

EXAMPLE PROBLEM 8.7

GIVEN: Fire protection system, as shown.

FIND: Q, gpm.

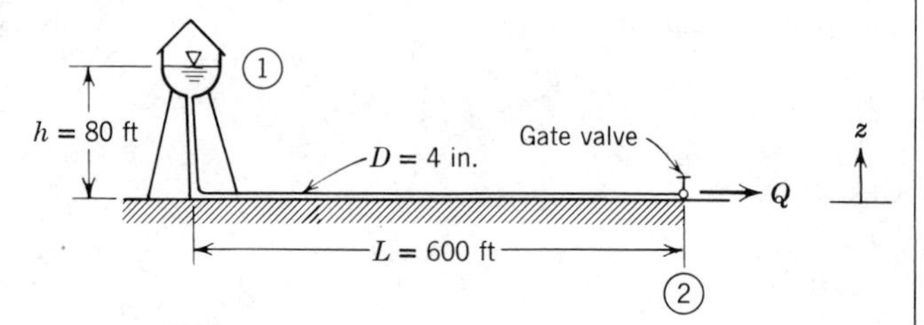

SOLUTION:

Computing equations:

$$\left(\frac{p_1}{\rho} + \alpha_1 \frac{\bar{V}_1^2}{2} + gz_1\right) - \left(\frac{p_2}{\rho} + \alpha_2 \frac{\bar{V}_2^2}{2} + gz_2\right) = h_{l_T} \quad (8.28)$$

($\bar{V}_1^2/2 \approx 0$ (2))

$$h_{l_T} = f \frac{L}{D} \frac{\bar{V}_2^2}{2} + h_{l_m} = f \frac{(L + L_e)}{D} \frac{\bar{V}_2^2}{2}$$

Assumptions: (1) $p_1 = p_2 = p_{\text{atm}}$
(2) $\bar{V}_1 \simeq 0$, and $\alpha_2 \simeq 1.0$

For a fully open gate valve, from Table 8.3, $L_e/D = 8$. Then

$$h_{l_T} = f \frac{L}{D} \frac{\bar{V}_2^2}{2} + 8f \frac{\bar{V}_2^2}{2} = g(z_1 - z_2) - \frac{\bar{V}_2^2}{2}$$

or

$$\frac{\bar{V}_2^2}{2}\left[f\left(\frac{L}{D} + 8\right) + 1\right] = g(z_1 - z_2)$$

Solving for $\bar{V}_2$, we obtain

$$\bar{V}_2 = \left[\frac{2g(z_1 - z_2)}{f(L/D + 8) + 1}\right]^{1/2}$$

To be conservative, assume the standpipe is the same diameter as the horizontal pipe. Then

$$\frac{L}{D} = \frac{600 \text{ ft} + 80 \text{ ft}}{4 \text{ in.}} \times \frac{12 \text{ in.}}{\text{ft}} = 2040$$

Also

$$z_1 - z_2 = h = 80 \text{ ft}$$

Since $\bar{V}_2$ is not known, we cannot compute Re. But we can assume a value of friction factor in the fully rough flow region. From Fig. 8.15, $e/D \simeq 0.0025$ for cast iron pipe. Since the pipe is quite old, choose $e/D = 0.005$. Then, from Fig. 8.14, guess $f \simeq 0.03$. Then a first approximation to $\bar{V}_2$ is

$$\bar{V}_2 = \left[2 \times \frac{32.2 \text{ ft}}{\text{sec}^2} \times 80 \text{ ft} \times \frac{1}{0.03(2040 + 8) + 1}\right]^{1/2} = 9.08 \text{ ft/sec}$$

Now check the value assumed for f.

$$Re = \frac{\rho \bar{V} D}{\mu} = \frac{\bar{V} D}{\nu} = \frac{9.08 \text{ ft}}{\text{sec}} \times \frac{\text{ft}}{3} \times \frac{\text{sec}}{1.2 \times 10^{-5} \text{ ft}^2} = 2.52 \times 10^5$$

For $e/D = 0.005$, $f = 0.031$ from Fig. 8.14. Using this value, we obtain

$$\bar{V}_2 = \left[2 \times \frac{32.2 \text{ ft}}{\text{sec}^2} \times 80 \text{ ft} \times \frac{1}{0.031(2040 + 8) + 1}\right]^{1/2} = 8.94 \text{ ft/sec}$$

Thus convergence is satisfactory. The volume flow rate is

$$Q = \bar{V}_2 A = \bar{V}_2 \frac{\pi D^2}{4} = \frac{8.94 \text{ ft}}{\text{sec}} \times \frac{\pi}{4}\left(\frac{1}{3}\right)^2 \text{ft}^2 \times \frac{7.48 \text{ gal}}{\text{ft}^3} \times \frac{60 \text{ sec}}{\text{min}}$$

$$Q = 350 \text{ gpm}$$ Q

This problem illustrates the procedure for solving pipe flow problems in which the flow rate is unknown. Note that the velocity and, hence, the flow rate, is essentially proportional to $1/\sqrt{f}$. Doubling the value of e/D to account for aging reduced the flow rate by about 10 percent.

Example 8.8

Spray heads in an agricultural spraying system are to be supplied with water through 500 ft of drawn aluminum tubing from an engine-driven pump. In its most efficient operating range, the pump output is 1500 gpm at a discharge pressure not exceeding 65 psig. For satisfactory operation, the sprinklers must operate at 30 psig or higher pressure. Minor losses and elevation changes may be neglected. Determine the smallest standard pipe size that can be used.

EXAMPLE PROBLEM 8.8

GIVEN: Water supply system, as shown.

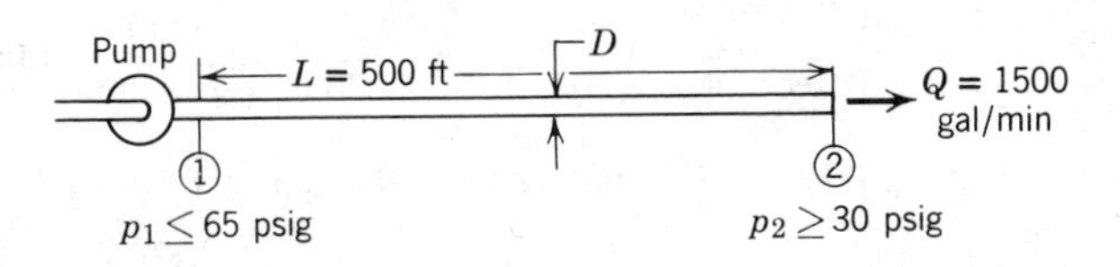

FIND: Smallest standard D.

SOLUTION:

Δp, L, and Q are known. D is unknown, so iteration will be required to determine the minimum standard diameter that satisfies the pressure drop constraint at the given flow rate.

The maximum allowable pressure drop is

$$\Delta p_{max} = p_{1\,max} - p_{2\,min} = (65 - 30)\text{ psi} = 35\text{ psi}$$

Computing equations:

$$\left(\frac{p_1}{\rho} + \alpha_1 \frac{\bar{V}_1^2}{2} + gz_1\right) - \left(\frac{p_2}{\rho} + \alpha_2 \frac{\bar{V}_2^2}{2} + gz_2\right) = h_{l_T} \tag{8.28}$$

$$h_{l_T} = h_l + h_{l_m} = f\frac{L}{D}\frac{\bar{V}_2^2}{2} \qquad (h_{l_m} = 0(3))$$

Assumptions:
(1) Steady flow
(2) Incompressible flow
(3) $h_{l_T} = h_l$, i.e. $h_{l_m} = 0$
(4) $z_1 = z_2$
(5) $\bar{V}_1 = \bar{V}_2 = \bar{V}$; $\alpha_1 \simeq \alpha_2$

Then

$$\Delta p = p_1 - p_2 = f\frac{L}{D}\frac{\rho\bar{V}^2}{2}$$

Since trial values of D are to be assumed, it is convenient to substitute $\bar{V} = Q/A = 4Q/\pi D^2$ so that

$$\Delta p = f\frac{L}{D}\frac{\rho}{2}\left(\frac{4Q}{\pi D^2}\right)^2 = \frac{8fL\rho Q^2}{\pi^2 D^5} \tag{1}$$

The Reynolds number is needed to find f. In terms of Q,

$$Re = \frac{\rho\bar{V}D}{\mu} = \frac{\bar{V}D}{\nu} = \frac{4Q}{\pi D^2}\frac{D}{\nu} = \frac{4Q}{\pi\nu D}$$

Finally, Q must be converted to cubic feet per second.

$$Q = \frac{1500\text{ gal}}{\text{min}} \times \frac{\text{min}}{60\text{ sec}} \times \frac{\text{ft}^3}{7.48\text{ gal}} = 3.34\text{ ft}^3/\text{sec}$$

For an initial guess, take nominal 4 in. (4.026 in. i.d.) pipe:

$$Re = \frac{4Q}{\pi\nu D} = \frac{4}{\pi} \times \frac{3.34\text{ ft}^3}{\text{sec}} \times \frac{\text{sec}}{1.2 \times 10^{-5}\text{ ft}^2} \times \frac{1}{4.026\text{ in.}} \times \frac{12\text{ in.}}{\text{ft}} = 1.06 \times 10^6$$

For drawn tubing, $e/D = 0.000{,}016$ (Fig. 8.15), so $f \simeq 0.012$ (Fig. 8.14), and

$$\Delta p = \frac{8fL\rho Q^2}{\pi^2 D^5} = \frac{8}{\pi^2} \times 0.012 \times 500\text{ ft} \times \frac{1.94\text{ slug}}{\text{ft}^3} \times \frac{(3.34)^2\text{ ft}^6}{\text{sec}^2}$$

$$\times \frac{1}{(4.026)^5\text{ in.}^5} \times \frac{1728\text{ in.}^3}{\text{ft}^3} \times \frac{\text{lbf}\cdot\text{sec}^2}{\text{slug}\cdot\text{ft}}$$

$$\Delta p = 172\text{ lbf/in.}^2 > \Delta p_{max}$$

Since this value is too large, try $D = 6$ in. (actually 6.065 in. i.d.):

$$Re = \frac{4}{\pi} \times \frac{3.34\ \text{ft}^3}{\text{sec}} \times \frac{\text{sec}}{1.2 \times 10^{-5}\ \text{ft}^2} \times \frac{1}{6.065\ \text{in.}} \times \frac{12\ \text{in.}}{\text{ft}} = 7.01 \times 10^5$$

For drawn tubing, $e/D = 0.000{,}010$ (Fig. 8.15), so $f \simeq 0.013$ (Fig. 8.14), and

$$\Delta p = \frac{8}{\pi^2} \times 0.013 \times 500\ \text{ft} \times \frac{1.94\ \text{slug}}{\text{ft}^3} \times \frac{(3.34)^2\ \text{ft}^6}{\text{sec}^2} \times \frac{1}{(6.065)^5\ \text{in.}^5} \times \frac{(12)^3\ \text{in.}^3}{\text{ft}^3} \times \frac{\text{lbf} \cdot \text{sec}^2}{\text{slug} \cdot \text{ft}}$$

$$\Delta p = 24.0\ \text{lbf/in.}^2 < \Delta p_{max}$$

Since this value is less than the allowable pressure drop, we should check a 5 in. (nominal) pipe. With an actual i.d. of 5.047 in.,

$$Re = \frac{4}{\pi} \times \frac{3.34\ \text{ft}^3}{\text{sec}} \times \frac{\text{sec}}{1.2 \times 10^{-5}\ \text{ft}^2} \times \frac{1}{5.047\ \text{in.}} \times \frac{12\ \text{in.}}{\text{ft}} = 8.43 \times 10^5$$

For drawn tubing, $e/D = 0.000{,}012$ (Fig. 8.15), so $f \simeq 0.012$ (Fig. 8.14), and

$$\Delta p = \frac{8}{\pi^2} \times 0.012 \times 500\ \text{ft} \times \frac{1.94\ \text{slug}}{\text{ft}^3} \times \frac{(3.34)^2\ \text{ft}^6}{\text{sec}^2} \times \frac{1}{(5.047)^5\ \text{in.}^5} \times \frac{(12)^3\ \text{in.}^3}{\text{ft}^3} \times \frac{\text{lbf} \cdot \text{sec}^2}{\text{slug} \cdot \text{ft}}$$

$$\Delta p = 55.5\ \text{lbf/in.}^2 > \Delta p_{max}$$

Thus the criterion for pressure drop is satisfied for a minimum nominal diameter of 6 in. pipe. *D*

This problem illustrates the procedure for solving pipe flow problems when the diameter is unknown. Note from Eq. 1 that the pressure drop in turbulent pipe flow is proportional to f/D^5. The variation of f is small, so Δp at constant flow rate is approximately proportional to $1/D^5$.

We have solved each of Example Problems 8.7 and 8.8 by direct iteration. Several specialized forms of friction factor versus Reynolds number diagrams have been introduced to solve problems of this type without the need for iteration. For examples of these specialized diagrams, see [16] and [17].

Example Problems 8.9 and 8.10 illustrate the evaluation of minor loss coefficients and the application of a diffuser to reduce exit kinetic energy from a flow system.

Example 8.9

Reference 18 reports results of measurements made to determine entrance losses for flow from a reservoir to a pipe with various degrees of entrance rounding. A copper pipe 10 ft long with 1.5 in. i.d. was used for the tests. The pipe discharged to atmosphere.

For a square-edged entrance a discharge of 0.566 ft^3/sec was measured when the reservoir level was 85.1 ft above the pipe centerline. From these data evaluate the loss coefficient for a square-edged entrance.

EXAMPLE PROBLEM 8.9

GIVEN: Pipe with square-edged entrance discharging from reservoir as shown.

FIND: K_{entrance}.

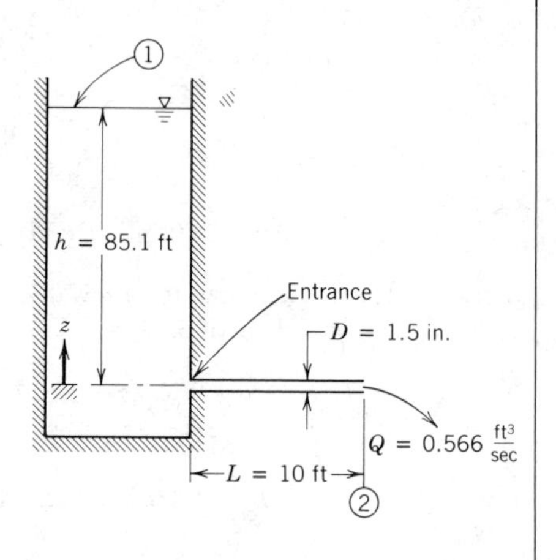

SOLUTION:
Apply the energy equation for steady, incompressible pipe flow.

Computing equations:
$$\frac{p_1}{\rho} + \alpha_1 \frac{\bar{V}_1^2}{2} + gz_1 = \frac{p_2}{\rho} + \alpha_2 \frac{\bar{V}_2^2}{2} + gz_2 + h_{l_T}$$
(with $\bar{V}_1^2/2 \approx 0(2)$ and $gz_2 = 0$)

$$h_{l_T} = f\frac{L}{D}\frac{\bar{V}_2^2}{2} + K_{\text{entrance}}\frac{\bar{V}_2^2}{2}$$

Assumptions: (1) $p_1 = p_2 = p_{\text{atm}}$
(2) $\bar{V}_1 \approx 0$

Dividing by g gives
$$z_1 = h = \alpha_2 \frac{\bar{V}_2^2}{2g} + f\frac{L}{D}\frac{\bar{V}_2^2}{2g} + K_{\text{entrance}}\frac{\bar{V}_2^2}{2g}$$

or

$$K_{\text{entrance}} = \frac{2gh}{\bar{V}_2^2} - f\frac{L}{D} - \alpha_2 \tag{1}$$

The average velocity is

$$\bar{V}_2 = \frac{Q}{A} = \frac{4Q}{\pi D^2} = \frac{4}{\pi} \times \frac{0.566 \text{ ft}^3}{\text{sec}} \times \frac{1}{(1.5)^2 \text{ in.}^2} \times \frac{144 \text{ in.}^2}{\text{ft}^2} = 46.1 \text{ ft/sec}$$

Assume $T = 75$ F (24 C), so $\nu = 8.8 \times 10^{-7}$ m^2/sec (Fig. A.3). Then

$$Re = \frac{\bar{V}D}{\nu} = \frac{46.1 \text{ ft}}{\text{sec}} \times 1.5 \text{ in.} \times \frac{\text{sec}}{8.8 \times 10^{-7} \text{ m}^2} \times \frac{\text{ft}}{12 \text{ in.}} \times \frac{(0.3048)^2 \text{ m}^2}{\text{ft}^2} = 6.08 \times 10^5$$

For drawn tubing, $e/D = 0.000{,}04$ (Fig. 8.15), so $f = 0.013$ (Fig. 8.14). From Fig. 8.11, $n = 8.7$

$$\frac{\bar{V}}{U} = \frac{2n^2}{(n+1)(2n+1)} = 0.848 \tag{8.23}$$

$$\alpha = \left(\frac{U}{\bar{V}}\right)^3 \frac{2n^2}{(3+n)(3+2n)} = 1.04 \tag{8.26}$$

Substituting into Eq. 1, we obtain

$$K_{\text{entrance}} = 2 \times 32.2\ \frac{\text{ft}}{\text{sec}^2} \times 85.1\ \text{ft} \times \frac{\text{sec}^2}{(46.1)^2\ \text{ft}^2} - (0.013)\ \frac{10\ \text{ft}}{1.5\ \text{in.}} \times \frac{12\ \text{in.}}{\text{ft}} - 1.04$$

$$K_{\text{entrance}} = 0.499$$

K_{entrance}

This value compares favorably with that shown in Table 8.1. The hydraulic and energy grade lines are shown below. The large head loss in a square-edged entrance is due primarily to separation at the sharp inlet corner and formation of a vena contracta immediately downstream from the corner. The effective flow area reaches a minimum at the vena contracta, so the flow velocity is a maximum there. The flow expands again following the vena contracta to fill the pipe. The uncontrolled expansion following the vena contracta is responsible for most of the head loss. (See Example Problem 8.12.)

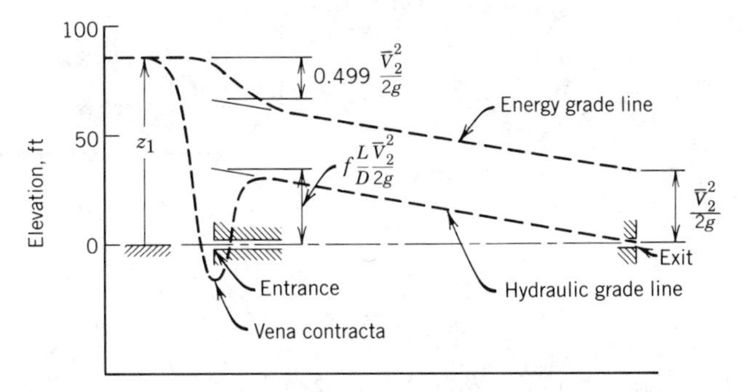

Rounding the inlet corner reduces the extent of separation significantly. This reduces the velocity increase through the vena contracta and consequently reduces the head loss due to the entrance. A "well-rounded" inlet almost eliminates flow separation; the flow pattern approaches that shown in Fig. 8.1. The added head loss in a well-rounded inlet compared to fully developed flow is the result of higher wall shear stresses in the entrance length.

Example 8.10

Water rights granted to each citizen by the Emperor of Rome gave permission to install into the public water main a calibrated, circular, tubular bronze nozzle [19]. Some citizens were clever enough to take unfair advantage of a law that regulated flow rate by such an indirect method. They installed diffusers on the outlet of the nozzles to increase their discharge. Assume the static head available from the main is $z_0 = 1.5$ m and the nozzle exit diameter is $D = 25$ mm. (The discharge is to atmospheric pressure.) Determine the increase in flow rate when a diffuser with $N/R_1 = 3.0$ and $AR = 2.0$ is attached to the end of the nozzle.

EXAMPLE PROBLEM 8.10

GIVEN: Nozzle attached to water main as shown.

FIND: Increase in discharge when diffuser with $N/R_1 = 3.0$ and $AR = 2.0$ is installed.

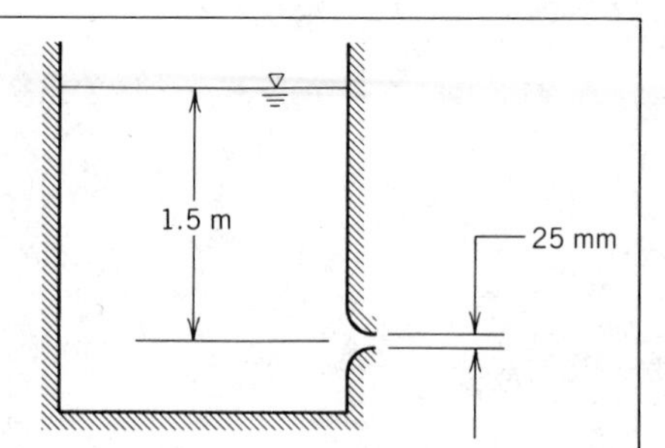

SOLUTION:
Apply the energy equation for steady, incompressible pipe flow.

Computing equation: $$\frac{p_0}{\rho} + \alpha_0 \frac{\bar{V}_0^2}{2} + gz_0 = \frac{p_1}{\rho} + \alpha_1 \frac{\bar{V}_1^2}{2} + gz_1 + h_{l_T}$$

Assumptions: (1) $\bar{V}_0 \approx 0$
(2) $\alpha \approx 1$

For the nozzle alone,

$$\cancel{\frac{p_0}{\rho}} + \alpha_0 \underbrace{\cancel{\frac{\bar{V}_0^2}{2}}}_{\approx 0(1)} + gz_0 = \cancel{\frac{p_1}{\rho}} + \underbrace{\cancel{\alpha_1}}_{\approx 1(2)} \frac{\bar{V}_1^2}{2} + \underbrace{\cancel{gz_1}}_{=0} + h_{l_T}$$

$$h_{l_T} = K_{\text{entrance}} \frac{\bar{V}_1^2}{2} \approx 0.04 \frac{\bar{V}_1^2}{2} \quad \text{(from Table 8.1)}$$

Thus

$$gz_0 = \frac{\bar{V}_1^2}{2} + 0.04 \frac{\bar{V}_1^2}{2} = 1.04 \frac{\bar{V}_1^2}{2}$$

$$\bar{V}_1 = \sqrt{\frac{2gz_0}{1.04}} = \sqrt{\frac{2}{1.04} \times \frac{9.81 \text{ m}}{\text{sec}^2} \times 1.5 \text{ m}} = 5.32 \text{ m/sec}$$

$$Q = \bar{V}_1 A_1 = \bar{V}_1 \frac{\pi D^2}{4} = \frac{5.32 \text{ m}}{\text{sec}} \times \frac{\pi}{4} \times (0.025)^2 \text{ m}^2 = 0.00261 \text{ m}^3/\text{sec}$$ $\longleftarrow Q$

For the nozzle with diffuser attached,

$$\cancel{\frac{p_0}{\rho}} + \alpha_0 \underbrace{\cancel{\frac{\bar{V}_0^2}{2}}}_{\approx 0(1)} + gz_0 = \cancel{\frac{p_2}{\rho}} + \underbrace{\cancel{\alpha_2}}_{\approx 1(2)} \frac{\bar{V}_2^2}{2} + \underbrace{\cancel{gz_2}}_{=0} + h_{l_T}$$

$$h_{l_T} = K_{\text{entrance}} \frac{\bar{V}_1^2}{2} + K_{\text{diffuser}} \frac{\bar{V}_1^2}{2}$$

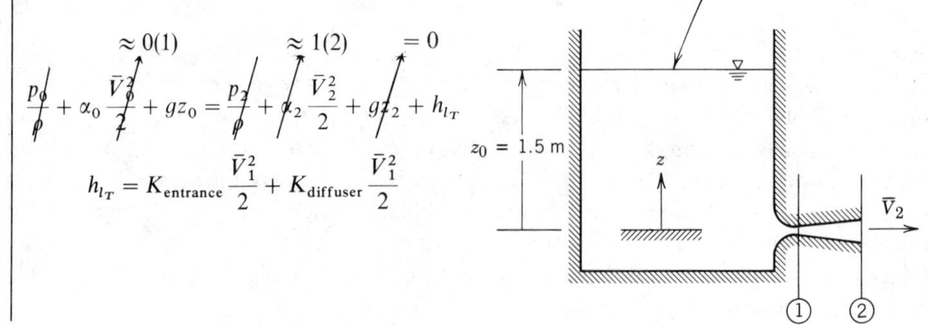

or

$$gz_0 = \frac{\bar{V}_2^2}{2} + (K_{\text{entrance}} + K_{\text{diffuser}})\frac{\bar{V}_1^2}{2} \qquad (1)$$

Figure 8.18 gives data for $C_p = \dfrac{p_2 - p_1}{\frac{1}{2}\rho \bar{V}_1^2}$ for diffusers.

To obtain K_{diffuser}, apply energy equation from ① to ②.

$$\frac{p_1}{\rho} + \alpha_1 \frac{\bar{V}_1^2}{2} + g\cancel{z}_1 = \frac{p_2}{\rho} + \alpha_2 \frac{\bar{V}_2^2}{2} + g\cancel{z}_2 + K_{\text{diffuser}} \frac{\bar{V}_1^2}{2}$$

Solving, we obtain

$$K_{\text{diffuser}} = 1 - \frac{\bar{V}_2^2}{\bar{V}_1^2} - \frac{p_2 - p_1}{\frac{1}{2}\rho \bar{V}_1^2} = 1 - \left(\frac{A_1}{A_2}\right)^2 - C_p = 1 - \frac{1}{(AR)^2} - C_p$$

where from continuity, $\bar{V}_1 A_1 = \bar{V}_2 A_2$. From Fig. 8.18, $C_p = 0.45$, so

$$K_{\text{diffuser}} = 1 - \frac{1}{(2.0)^2} - 0.45 = 0.75 - 0.45 = 0.3$$

Thus, substituting in Eq. 1,

$$gz_0 = \frac{\bar{V}_2^2}{2} + (0.04 + 0.30)\frac{\bar{V}_1^2}{2} = \frac{\bar{V}_1^2}{2}\left[\frac{1}{(AR)^2} + 0.34\right] = 0.59\frac{\bar{V}_1^2}{2}$$

For this system

$$\bar{V}_1 = \sqrt{\frac{2gz_0}{0.59}} = \sqrt{\frac{2}{0.59} \times \frac{9.81 \text{ m}}{\text{sec}^2} \times 1.5 \text{ m}} = 7.06 \text{ m/sec}$$

and

$$Q_d = \bar{V}_1 A_1 = \bar{V}_1 \frac{\pi D^2}{4} = \frac{7.06 \text{ m}}{\text{sec}} \times \frac{\pi}{4} \times (0.025)^2 \text{ m}^2 = 0.00347 \text{ m}^3\text{/sec} \qquad Q_d$$

The flow rate increase that results from adding the diffuser is

$$\frac{\Delta Q}{Q} = \frac{Q_d - Q}{Q} = \frac{Q_d}{Q} - 1 = \frac{0.00347}{0.00261} - 1 = 0.330 \quad \text{or} \quad 33 \text{ percent} \qquad \frac{\Delta Q}{Q}$$

Addition of the diffuser increases the flow rate through the nozzle significantly. The diffuser exhausts to atmospheric pressure. The pressure rise through the diffuser thus reduces the pressure in the nozzle exit plane, causing the flow rate to increase.

Energy and hydraulic grade lines are sketched in the following figures to approximate scale for the bare nozzle and for the nozzle-diffuser combination. The bare nozzle exhausts to atmospheric pressure so that the hydraulic grade line drops to zero elevation at the nozzle exit. The only loss in the nozzle is 0.04 times the velocity head, so the energy grade line elevation decreases slightly as shown.

Hydrostatic pressure at the nozzle exit plane is reduced below atmospheric by the diffuser. As shown in the sketch, the pressure at the diffuser discharge plane increases again to atmospheric. The kinetic energy leaving the diffuser exit is one-fourth of the inlet value, since the velocity is reduced by a factor of 2. The energy grade line drops by 0.04 times the inlet

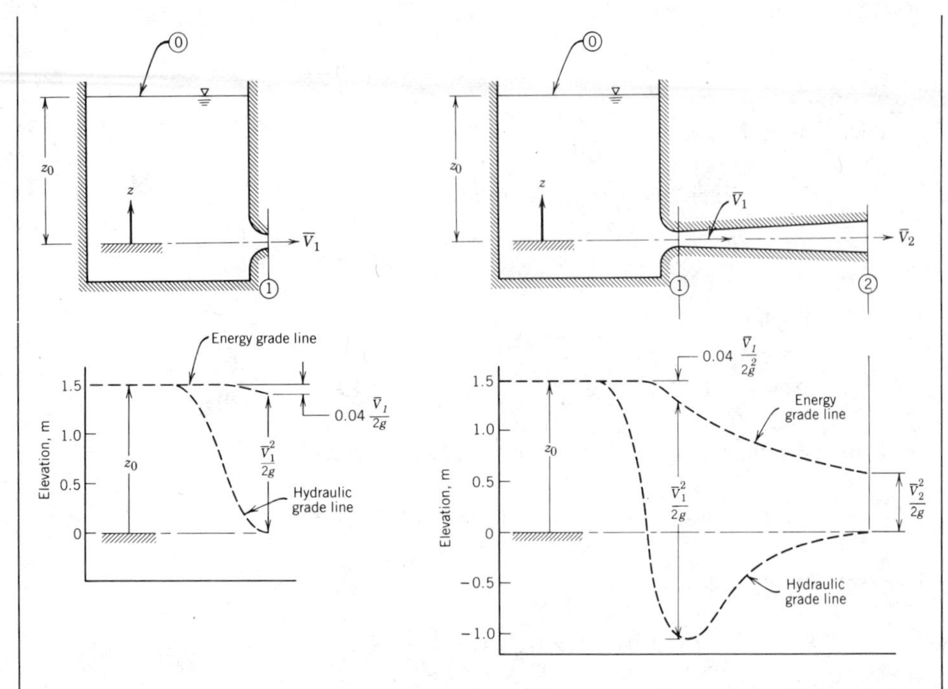

velocity head at the nozzle exit plane. The diffuser introduces the additional loss, $K_{\text{diffuser}} = 0.3$, which further drops the level of the energy grade line.

{Water Commissioner Frontinus standardized conditions for all in 97 A.D. He required that the tube attached to the nozzle of each customer's pipe be the same diameter for at least 50 lineal feet from the public water main (see Problem 8.95).}

**8-8.2 Multiple-Path Systems

In many practical situations, such as water supply or fire protection systems, complex pipe networks must be analyzed. The basic techniques developed in Section 8-8.1 also may be used to analyze flow in multiple-path pipe systems. The procedure is analogous to that used in solving direct current electric circuits, but with nonlinear elements.

The representative pipe system shown in Fig. 8.20 has two nodes, labeled A and B, and three branches. The total flow rate into the system must be distributed among the branches. Consequently, the flow rate through each branch is unknown. However, the pressure drop for each branch is the same, $p_A - p_B$. This information is sufficient to permit an iterative solution for the flow rate in each branch, as shown in Example Problem 8.11.

** This section may be omitted without loss of continuity in the text material.

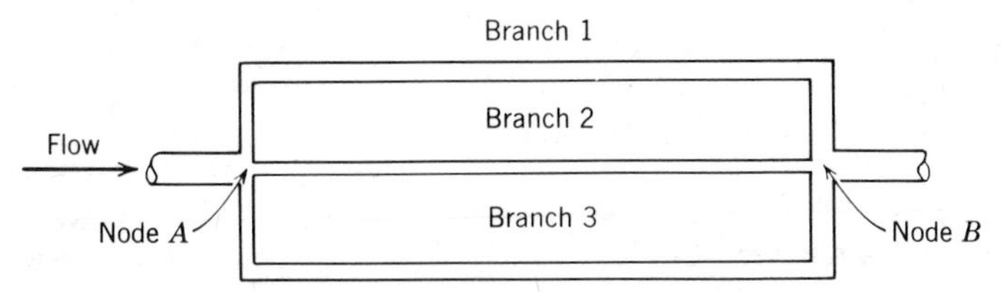

Fig. 8.20 Simple multiple-path pipe flow system.

The fluid flow rate and pressure drop are, respectively, analogous to the current and voltage in an electric circuit. However, the simple linear relation between voltage and current given by Ohm's law does not apply to the fluid flow system. Instead, the flow pressure drop is approximately proportional to the square of the flow rate. This nonlinearity makes iterative solutions necessary, and the resulting calculations can be quite lengthy and tedious. A number of schemes have been developed for use with digital computers. For an example, see [20].

Example 8.11

The irrigation system of Example 8.8 is to be extended, holding the total flow rate constant at 1500 gpm, by adding three branches, each made from 3 in. pipe, as shown in the sketch. The sprinkler at the end of each branch has a nozzle of 1.5 in. minimum diameter. Minor losses at elbows should be included, but elevation terms may be neglected. Determine the flow rate in each branch, the pressure required at Ⓐ, and the pressure applied to each sprinkler nozzle.

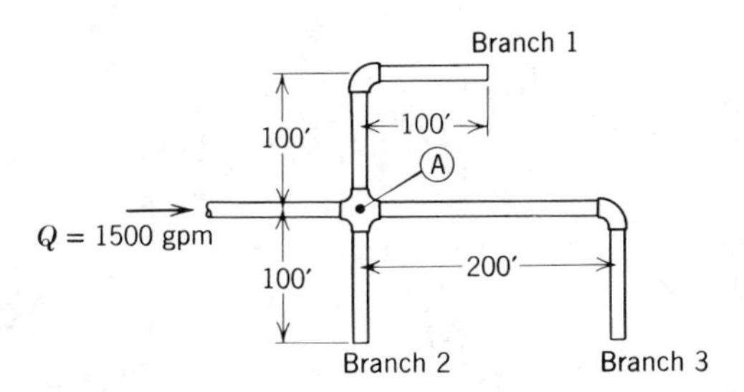

EXAMPLE PROBLEM 8.11

GIVEN: Pipe network shown schematically in the following figure.

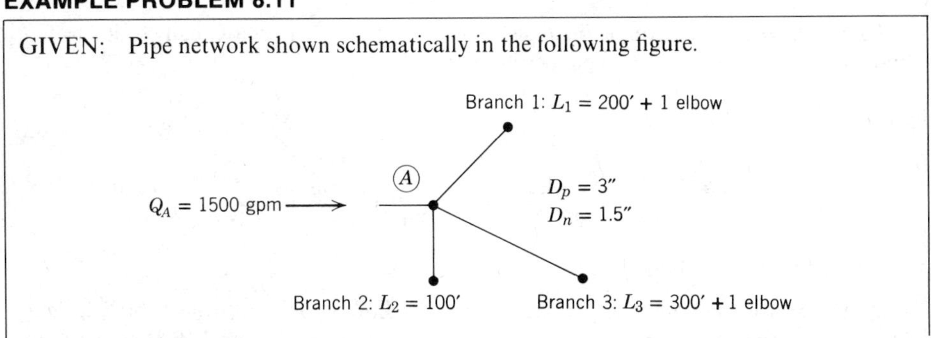

FIND: (a) Q_1, Q_2, Q_3.
(b) p_A.
(c) p_1, p_2, p_3 at nozzle inlets.

SOLUTION:
Apply Eq. 8.28 to each branch. Choose subscript n for nozzle exit, subscript p for pipe.

Computing equations:

$$\frac{p_A}{\rho} + \alpha_A \frac{\bar{V}_A^2}{2} + g z_A = \frac{p_n}{\rho} + \alpha_n \frac{\bar{V}_n^2}{2} + g z_n + h_{l_T} \qquad h_{l_T} = f\frac{L}{D}\frac{\bar{V}_p^2}{2} + f\frac{L_e}{D}\frac{\bar{V}_p^2}{2}$$

($\bar{V}_A^2/2 \approx 0(1)$; terms $\bar{V}_A^2/2$, gz_A, p_n/ρ, gz_n struck out by assumptions 1–3.)

Assumptions: (1) $\bar{V}_A^2 \ll \bar{V}_n^2$; $\alpha_n \simeq 1.0$
(2) $z_A \simeq z_n$
(3) $p_n = p_{\text{atmosphere}}$
(4) Neglect loss due to flow split at Ⓐ.

Then, for each branch of the flow system, using gage pressure at Ⓐ, we obtain

$$\frac{p_A}{\rho} = \frac{\bar{V}_n^2}{2} + f\frac{L}{D}\frac{\bar{V}_p^2}{2} + f\frac{L_e}{D}\frac{\bar{V}_p^2}{2}$$

But for incompressible flow, $\bar{V}_p A_p = \bar{V}_n A_n$, so $\bar{V}_n^2 = \bar{V}_p^2(A_p/A_n)^2 = \bar{V}_p^2(D_p/D_n)^4$, and

$$\frac{p_A}{\rho} = \frac{\bar{V}_p^2}{2}\left[\left(\frac{D_p}{D_n}\right)^4 + f\left(\frac{L}{D} + \frac{L_e}{D}\right)\right]$$

or

$$\bar{V}_p = \sqrt{\frac{2p_A}{\rho}}\left[\frac{1}{\left(\frac{D_p}{D_n}\right)^4 + f\left(\frac{L}{D} + \frac{L_e}{D}\right)}\right]^{1/2}$$

The corresponding flow rate for each branch is

$$Q_p = \bar{V}_p A_p = A_p\sqrt{\frac{2p_A}{\rho}}\left[\frac{1}{\left(\frac{D_p}{D_n}\right)^4 + f\left(\frac{L}{D} + \frac{L_e}{D}\right)}\right]^{1/2} \tag{1}$$

To find f, we must determine the pipe Reynolds number. If flow is equally split as a first approximation, then $Q_p \simeq 500$ gpm per branch, and

$$Re = \frac{\bar{V}_p D_p}{\nu} = \frac{4Q_p}{\pi \nu D_p}$$

$$= \frac{4}{\pi} \times \frac{500\ \text{gal}}{\text{min}} \times \frac{\text{sec}}{1.2 \times 10^{-5}\ \text{ft}^2} \times \frac{1}{3\ \text{in.}} \times \frac{\text{min}}{60\ \text{sec}} \times \frac{\text{ft}^3}{7.48\ \text{gal}} \times \frac{12\ \text{in.}}{\text{ft}}$$

$$Re = 4.73 \times 10^5$$

From Fig. 8.14, for smooth pipe, $f \simeq 0.0133$. For an elbow, $L_e/D = 30$, from Table 8.3. Using these values, we can obtain a first approximation for Q_p for each branch. Substituting into

Eq. 1 yields

$$Q_{p_1} \simeq A_p\sqrt{\frac{2p_A}{\rho}}\left[\frac{1}{\left(\frac{3.0}{1.5}\right)^4 + 0.0133\left(\frac{200\text{ ft}}{} \times \frac{1}{3\text{ in.}} \times \frac{12\text{ in.}}{\text{ft}} + 30\right)}\right]^{1/2}$$

$$Q_{p_1} \simeq 0.192 A_p\sqrt{\frac{2p_A}{\rho}}$$

$$Q_{p_2} \simeq A_p\sqrt{\frac{2p_A}{\rho}}\left[\frac{1}{\left(\frac{3.0}{1.5}\right)^4 + 0.0133\left(\frac{100\text{ ft}}{} \times \frac{1}{3\text{ in.}} \times \frac{12\text{ in.}}{\text{ft}}\right)}\right]^{1/2}$$

$$Q_{p_2} \simeq 0.217 A_p\sqrt{\frac{2p_A}{\rho}}$$

and

$$Q_{p_3} \simeq A_p\sqrt{\frac{2p_A}{\rho}}\left[\frac{1}{\left(\frac{3.0}{1.5}\right)^4 + 0.0133\left(\frac{300\text{ ft}}{} \times \frac{1}{3\text{ in.}} \times \frac{12\text{ in.}}{\text{ft}} + 30\right)}\right]^{1/2}$$

$$Q_{p_3} \simeq 0.176 A_p\sqrt{\frac{2p_A}{\rho}}$$

From continuity,

$$Q_A = Q_{p_1} + Q_{p_2} + Q_{p_3} \simeq (0.192 + 0.217 + 0.176) A_p\sqrt{\frac{2p_A}{\rho}} \simeq 0.585 A_p\sqrt{\frac{2p_A}{\rho}}$$

Thus

$$\frac{Q_{p1}}{Q_A} \simeq \frac{0.192 A_p\sqrt{\frac{2p_A}{\rho}}}{0.585 A_p\sqrt{\frac{2p_A}{\rho}}} = 0.328$$

or

$$Q_{p_1} \simeq (0.328) 1500\text{ gpm} = 492\text{ gpm}$$

Similarly

$$\frac{Q_{p2}}{Q_A} \simeq \frac{0.217}{0.585} = 0.371; \quad Q_{p_2} \simeq 557\text{ gpm}$$

$$\frac{Q_{p3}}{Q_A} \simeq \frac{0.176}{0.585} = 0.301; \quad Q_{p_3} \simeq 452\text{ gpm}$$

Better approximations for the Reynolds number and friction factor for each branch now may be calculated.

$$Re_1 = \frac{Q_{p_1}}{500\text{ gpm}} \times 4.73 \times 10^5 \simeq \frac{492}{500} \times 4.73 \times 10^5 = 4.65 \times 10^5$$

From Fig. 8.14, $f_1 \simeq 0.0133$. Similarly,

$$Re_2 \simeq \frac{557}{500} \times 4.73 \times 10^5 = 5.27 \times 10^5; \quad f_2 \simeq 0.0130$$

and

$$Re_3 \simeq \frac{452}{500} \times 4.73 \times 10^5 = 4.28 \times 10^5; \quad f_3 \simeq 0.0136$$

Substituting these values into Eq. 1, we obtain

$$Q_{p_1} \simeq 0.192 A_p \sqrt{\frac{2p_A}{\rho}}$$

$$Q_{p_2} \simeq 0.217 A_p \sqrt{\frac{2p_A}{\rho}}$$

and

$$Q_{p_3} \simeq 0.175 A_p \sqrt{\frac{2p_A}{\rho}}$$

From continuity,

$$Q_A = Q_{p_1} + Q_{p_2} + Q_{p_3} \simeq (0.192 + 0.217 + 0.175) A_p \sqrt{\frac{2p_A}{\rho}} \simeq 0.584 A_p \sqrt{\frac{2p_A}{\rho}}$$

Solving for the individual flow rates,

$$Q_{p_1} \simeq \frac{0.192}{0.584} \times 1500 \text{ gpm} = 493 \text{ gpm}$$

$$Q_{p_2} \simeq \frac{0.217}{0.584} \times 1500 \text{ gpm} = 557 \text{ gpm}$$

and

$$Q_{p_3} \simeq \frac{0.175}{0.584} \times 1500 \text{ gpm} = 449 \text{ gpm}$$

Q_1, Q_2, Q_3

Solving Eq. 1 for p_A gives

$$p_A = \frac{\rho}{2}\left(\frac{Q_p}{A_p}\right)^2 \left[\left(\frac{D_p}{D_n}\right)^4 + f\left(\frac{L}{D} + \frac{L_e}{D}\right)\right]$$

Substituting values for Branch ①,

$$p_A = \frac{1}{2} \times 1.94 \frac{\text{slug}}{\text{ft}^3} \left(493 \frac{\text{gal}}{\text{min}} \times \frac{4}{\pi} \frac{1}{(3)^2 \text{ in.}^2} \times \frac{\text{ft}^3}{7.48 \text{ gal}} \times \frac{\text{min}}{60 \text{ sec}} \times \frac{144 \text{ in.}^2}{\text{ft}^2}\right)^2$$

$$\times \left[\left(\frac{3.0}{1.5}\right)^4 + 0.0133\left(200 \text{ ft} \times \frac{1}{3 \text{ in.}} \times \frac{12 \text{ in.}}{\text{ft}} + 30\right)\right] \frac{\text{lbf} \cdot \text{sec}^2}{\text{slug} \cdot \text{ft}} \times \frac{\text{ft}^2}{144 \text{ in.}^2}$$

$$p_A = 91.2 \text{ lbf/in.}^2$$

The values for Branches ② and ③ by similar calculations are 91.4 and 90.5 psig, respectively. (The slight differences are due to rounding the flow rate values.) Thus

$$p_A \approx 91.0 \text{ psig} \qquad p_A$$

The pressure at the inlet to each sprinkler nozzle may be calculated from the energy equation. The equation between Ⓐ and the nozzle inlet, section ⓘ, is

$$\frac{p_A}{\rho} + \frac{\bar{V}_A^2}{2} + gz_A = \frac{p_i}{\rho} + \frac{\bar{V}_p^2}{2} + gz_i + h_{l_T} \qquad h_{l_T} = f\frac{L}{D}\frac{\bar{V}_p^2}{2} + f\frac{L_e}{D}\frac{\bar{V}_p^2}{2}$$

From continuity, $\bar{V}_A = Q_A/A_A$ and $\bar{V}_p = Q_p/A_p$, so

$$p_i = p_A + \frac{\rho}{2}\left[\left(\frac{Q_A}{A_A}\right)^2 - \left(f\frac{L}{D} + f\frac{L_e}{D} + 1\right)\left(\frac{Q_p}{A_p}\right)^2\right]$$

or

$$p_i = p_A + \frac{\rho}{2}\left(\frac{Q_A}{A_A}\right)^2\left\{1 - \left[f\left(\frac{L}{D} + \frac{L_e}{D}\right) + 1\right]\left(\frac{Q_p}{Q_A}\right)^2\left(\frac{A_A}{A_p}\right)^2\right\}$$

Substituting values for Branch ① gives

$$p_{i_1} = \frac{91.0 \text{ lbf}}{\text{in.}^2} + \frac{1}{2} \times \frac{1.94 \text{ slug}}{\text{ft}^3}\left(\frac{1500 \text{ gal}}{\text{min}} \times \frac{4}{\pi} \frac{1}{(6)^2 \text{ in.}^2} \times \frac{\text{ft}^3}{7.48 \text{ gal}} \times \frac{\text{min}}{60 \text{ sec}} \times \frac{144 \text{ in.}^2}{\text{ft}^2}\right)^2$$

$$\times \left\{1 - \left[0.0133\left(\frac{200 \text{ ft}}{} \times \frac{1}{3 \text{ in.}} \times \frac{12 \text{ in.}}{\text{ft}} + 30\right) + 1\right]\left(\frac{493}{1500}\right)^2\left(\frac{6}{3}\right)^4\right\} \frac{\text{lbf} \cdot \text{sec}^2}{\text{slug} \cdot \text{ft}}$$

$$\times \frac{\text{ft}^2}{144 \text{ in.}^2}$$

$$p_{i_1} = 52.3 \text{ lbf/in.}^2$$

Similar calculations for Branches ② and ③ give

$$p_{i_2} = 66.3 \text{ lbf/in.}^2 \quad \text{and} \quad p_{i_3} = 43.3 \text{ lbf/in.}^2 \qquad p_i$$

{This problem illustrates the general method used to solve multiple-path pipe flow problems.}

**8-8.3 Noncircular Ducts

The empirical correlations for pipe flow also may be used for computations involving noncircular ducts, provided their cross sections are not too exaggerated. Thus ducts of square or rectangular cross section may be treated if the ratio of height to width is less than about 3 or 4.

** This section may be omitted without loss of continuity in the text material.

The correlations for turbulent pipe flow developed in Section 8-7 are extended for use with noncircular geometries by introducing the *hydraulic diameter*, defined as,

$$D_h \equiv \frac{4A}{P} \tag{8.44}$$

in place of the diameter, D. In Eq. 8.44, A is the cross-sectional area, and P is the *wetted perimeter*, the length of wall in contact with the flowing fluid at any cross section. The factor 4 is introduced so that the hydraulic diameter will equal the duct diameter for a circular geometry. For a circular duct, $A = \pi D^2/4$ and $P = \pi D$ so that

$$D_h = \frac{4A}{P} = \frac{4\left(\frac{\pi}{4}\right)D^2}{\pi D} = D$$

For a rectangular duct of width, b, and height, h, then $A = bh$ and $P = 2(b + h)$, so

$$D_h = \frac{4bh}{2(b + h)}$$

If the *aspect ratio*, *ar*, is defined as $ar = h/b$, then

$$D_h = \frac{2h}{1 + ar}$$

for rectangular ducts. For a square duct, $ar = 1$, and $D_h = h$.

As noted, the hydraulic diameter concept can be applied in the approximate range $\frac{1}{3} < ar < 3$. Under these conditions, the correlations of Section 8-7 give acceptably accurate results for rectangular ducts. Since such ducts are easy and cheap to fabricate from sheet metal, they are commonly used in air conditioning, heating, and ventilating applications. Extensive data on losses for air flow are available (e.g. see [9, 21]).

Losses due to secondary flows increase rapidly for more extreme geometries, so the correlations are not applicable to wide, flat ducts, or to ducts of triangular or other irregular shapes. Experimental data must be used when precise design information is required for specific situations.

PART C FLOW MEASUREMENT

The choice of a flow meter is influenced by the required accuracy and range, cost, complication, ease of reading or data reduction, and service life. The simplest and cheapest device that gives the desired accuracy should be chosen.

8-9 DIRECT METHODS

Tanks can be used to determine flow rate for steady liquid flows by measuring the volume or mass of liquid collected during a known time interval. If the time interval is long enough to be measured accurately, extremely precise values of flow rate may be determined in this way.

Compressibility must be considered in volume measurements for gas flows. The densities of gases generally are too small to permit accurate direct measurement of mass flow rate. However, a volume sample often can be collected by displacing a "bell," or inverted jar over water (if the pressure is held constant by counterweights). If volume or mass measurements are set up carefully, no calibration is required; this is a great advantage of direct methods.

In specialized applications, particularly for remote or recording uses, positive displacement flow meters may be specified. Common examples include household water and natural gas meters, which are calibrated to read directly in units of product, or gasoline metering pumps, which measure total flow and automatically compute the cost. Many positive-displacement meters are available commercially. Manufacturers' literature or references (e.g. [5]) should be consulted for design and installation details.

8-10 RESTRICTION FLOW METERS FOR INTERNAL FLOWS

Most restriction flow meters for internal flow (except the laminar flow element, Section 8-10.4) are based on acceleration of a fluid stream through some form of nozzle, as shown schematically in Fig. 8.21. Flow separation at the sharp edge of the nozzle throat causes a recirculation zone to form, as shown by the dashed lines downstream from the nozzle. The mainstream flow continues to accelerate from the nozzle throat to form a *vena contracta* at section ② and then decelerates again to fill the duct. At the vena contracta the flow area passes through a minimum, the flow streamlines are essentially straight, and the pressure is uniform across the channel section.

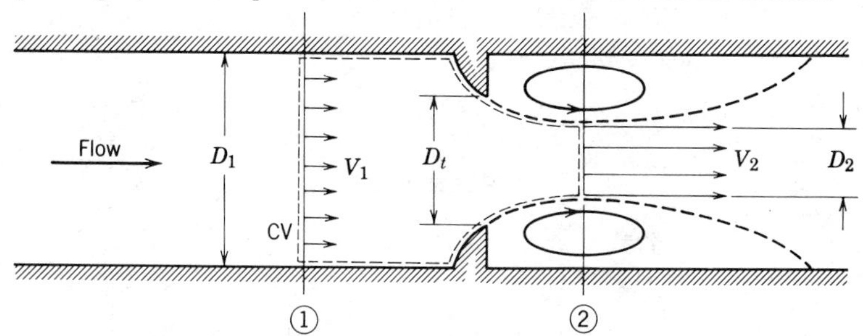

Fig. 8.21 Internal flow through a generalized nozzle, showing control volume used for analysis.

The theoretical flow rate may be related to the pressure differential between sections ① and ② by applying the continuity and Bernoulli equations. Then empirical correction factors may be applied to obtain the actual flow rate.

Basic equations:

$$0 = \cancelto{0(1)}{\frac{\partial}{\partial t}} \int_{\text{CV}} \rho \, d\forall + \int_{\text{CS}} \rho \vec{V} \cdot d\vec{A} \tag{4.13}$$

$$\frac{p_1}{\rho} + \frac{V_1^2}{2} + g\not{z}_1 = \frac{p_2}{\rho} + \frac{V_2^2}{2} + g\not{z}_2 \tag{6.9}$$

Assumptions:
(1) Steady flow
(2) Incompressible flow
(3) Flow along a streamline
(4) No friction
(5) Uniform velocity at sections ① and ②
(6) No streamline curvature at section ① or ②, so pressure is uniform across those sections
(7) $z_1 = z_2$

Then, from the Bernoulli equation,

$$p_1 - p_2 = \frac{\rho}{2}(V_2^2 - V_1^2) = \frac{\rho V_2^2}{2}\left[1 - \left(\frac{V_1}{V_2}\right)^2\right]$$

and from continuity

$$0 = \{-|\rho V_1 A_1|\} + \{|\rho V_2 A_2|\}$$

or

$$V_1 A_1 = V_2 A_2 \qquad \text{so} \qquad \left(\frac{V_1}{V_2}\right)^2 = \left(\frac{A_2}{A_1}\right)^2$$

Substituting gives

$$p_1 - p_2 = \frac{\rho V_2^2}{2}\left[1 - \left(\frac{A_2}{A_1}\right)^2\right]$$

Solving for the theoretical velocity, V_2,

$$V_2 = \sqrt{\frac{2(p_1 - p_2)}{\rho[1 - (A_2/A_1)^2]}} \tag{8.45}$$

The theoretical flow rate is then given by

$$\dot{m}_{\text{theoretical}} = \rho V_2 A_2 = \rho \sqrt{\frac{2(p_1 - p_2)}{\rho[1 - (A_2/A_1)^2]}}\, A_2$$

or

$$\dot{m}_{\text{theoretical}} = \frac{A_2}{\sqrt{1 - (A_2/A_1)^2}}\sqrt{2\rho(p_1 - p_2)} \tag{8.46}$$

Equation 8.46 shows the general relationship between mass flow rate and pressure drop for a restriction flow meter: Mass flow rate is proportional to the square root of the pressure differential across the meter taps. This relationship limits the flow rates that can be measured accurately to approximately a 4:1 range.

Several effects limit the utility of Eq. 8.46 for calculating the actual mass flow rate through a meter. The actual flow area at section ② is unknown when the vena contracta is pronounced (e.g. for orifice plates when D_t is a small fraction of D_1). The velocity profiles approach the uniform flow assumed only at very large Reynolds numbers. Frictional effects can become important (especially downstream from the meter) when the meter contours are abrupt. Finally, the location of pressure taps influences the differential pressure reading, $p_1 - p_2$.

The theoretical equation is adjusted for Reynolds number and diameter ratio by defining an empirical discharge coefficient

$$C \equiv \frac{\text{actual mass flow rate}}{\text{theoretical mass flow rate}} \tag{8.47}$$

Using the discharge coefficient, the actual mass flow rate is expressed as

$$\dot{m}_{\text{actual}} = \frac{CA_t}{\sqrt{1 - (A_t/A_1)^2}} \sqrt{2\rho(p_1 - p_2)} \tag{8.48}$$

or with $\beta = D_t/D_1$, then $(A_t/A_1)^2 = (D_t/D_1)^4 = \beta^4$, so

$$\dot{m}_{\text{actual}} = \frac{CA_t}{\sqrt{1 - \beta^4}} \sqrt{2\rho(p_1 - p_2)} \tag{8.49}$$

In Eq. 8.49, the factor $1/\sqrt{1 - \beta^4}$ is termed the velocity of approach factor. The discharge coefficient and velocity of approach factor frequently are combined into a single flow coefficient,

$$K \equiv \frac{C}{\sqrt{1 - \beta^4}} \tag{8.50}$$

In terms of the flow coefficient, the actual mass flow rate is expressed as

$$\dot{m}_{\text{actual}} = KA_t\sqrt{2\rho(p_1 - p_2)} \tag{8.51}$$

For standardized metering elements, test data [5, 22] have been used to develop empirical equations that predict discharge and flow coefficients from meter bore, pipe diameter, and flow Reynolds number. The accuracy of the equations (within specified ranges) usually is adequate so that the meter can be used without calibration. If the Reynolds number, pipe size, or bore diameter are outside the specified range of the equation, the coefficients should be measured experimentally.

For the turbulent flow regime (pipe Reynolds number greater than 4000) the discharge coefficient may be expressed by an equation of the form [5]

$$C = C_\infty + \frac{b}{Re_{D_1}^n} \tag{8.52}$$

The corresponding form for the flow coefficient is

$$K = K_\infty + \frac{1}{\sqrt{1 - \beta^4}} \frac{b}{Re_{D_1}^n} \tag{8.53}$$

In Eqs. 8.52 and 8.53, subscript ∞ denotes coefficient values at infinite Reynolds number; constants b and n allow for scaling to finite Reynolds numbers. Correlating equations and curves of coefficients versus Reynolds number are given for specific metering elements in the next three subsections, following a general comparison of their characteristics.

As we have noted, selection of a flow meter type depends on factors such as cost, accuracy, need for calibration, and ease of installation and maintenance. Some of these factors are compared for orifice plate, flow nozzle, and venturi meters in Table 8.5.

Table 8.5 Characteristics of Orifice, Flow Nozzle, and Venturi Flow Meters

Flow Meter Type	Diagram	Head Loss	Cost
Orifice	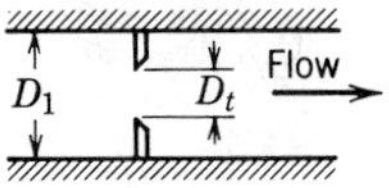 D_1, D_t, Flow	High	Low
Flow nozzle	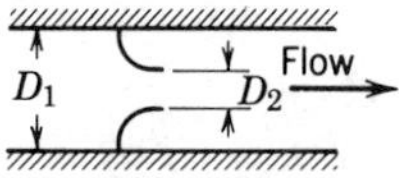 D_1, D_2, Flow	Intermediate	Intermediate
Venturi	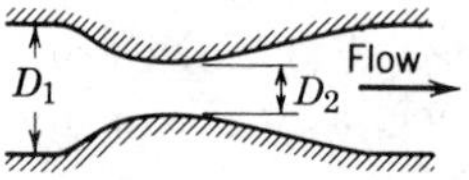 D_1, D_2, Flow	Low	High

Flow meter coefficients reported in the literature have been measured with a fully developed turbulent velocity distribution at the meter inlet (Section ①). If a flow meter is to be installed downstream from a valve, elbow, or other disturbance, a straight section of pipe must be placed in front of the meter. Approximately 10 diameters of straight pipe are required for venturi meters, and up to 40 diameters for orifice plate or flow nozzle meters. When a meter has been properly installed, the flow rate may be computed from Eqs. 8.49 or 8.51, after choosing an appropriate value for the empirical discharge coefficient, C, or flow coefficient, K, defined in Eqs. 8.47 and 8.50. Some design data for incompressible flow are given in the next few sections. The same basic methods can be extended to compressible flows, but these will not be treated here. For complete details, see [5] or [22].

8-10.1 The Orifice Plate

The orifice plate (Fig. 8.22) is a thin plate that may be clamped between pipe flanges. Since its geometry is simple, it is low in cost and easy to install or replace. The sharp edge of the orifice will not foul with scale or suspended matter. However, suspended matter can build up at the inlet side of a concentric orifice in a horizontal pipe; an eccentric orifice may be placed flush with the bottom of the pipe to avoid this difficulty. The primary disadvantages of the orifice are its limited capacity and the high permanent head loss due to the uncontrolled expansion downstream from the metering element.

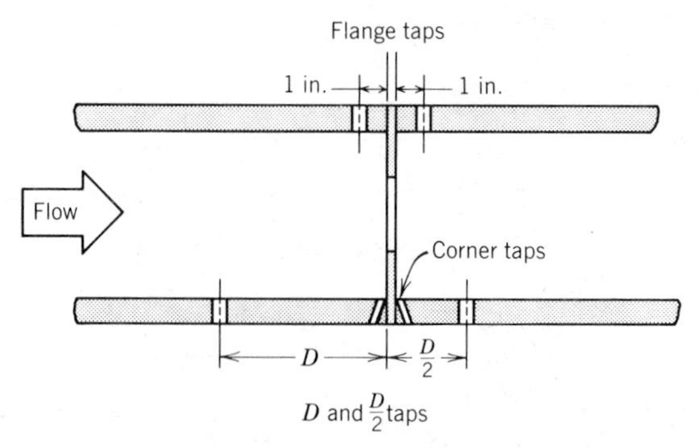

Fig. 8.22 Orifice geometry and pressure tap locations [5].

Pressure taps for orifices may be placed in several locations as shown in Fig. 8.22 (see [5] or [22] for additional details). Since the location of the pressure taps influences the empirically determined flow coefficient, one must select handbook values of C or K consistent with the location of pressure taps.

The correlating equation recommended for a concentric orifice with corner taps [5] is

$$C = 0.5959 + 0.0312\,\beta^{2.1} - 0.184\,\beta^{8} + \frac{91.71\,\beta^{2.5}}{Re_{D_1}^{0.75}} \tag{8.54}$$

Equation 8.54 predicts orifice discharge coefficients within ± 0.6 percent for $0.2 < \beta < 0.75$ and for $10^4 < Re_{D_1} < 10^7$. Some values of flow coefficients calculated from Eqs. 8.54 and 8.50 are presented in Fig. 8.23. Flow coefficient values are relatively insensitive to Reynolds number for values of Re_{D_1} above 10^5 when $\beta > 0.5$.

A similar correlating equation is available for orifice plates with D and $D/2$ taps. Flange taps require a different correlation for every line size. Pipe taps, located at $2\frac{1}{2}$ and 8 D, no longer are recommended for accurate work.

Example Problem 8.12, which appears later in this section, illustrates the application of flow coefficient data to orifice sizing.

8-10.2 The Flow Nozzle

Flow nozzles may be used as metering elements in either plenums or ducts, as shown in Fig. 8.24; the nozzle section is approximately a quarter ellipse. Design details and recommended locations for pressure taps may be found in [22].

The correlating equation recommended for an ASME long-radius flow nozzle [5] is

$$C = 0.9975 - \frac{6.53\,\beta^{0.5}}{Re_{D_1}^{0.5}} \tag{8.55}$$

Equation 8.55 predicts discharge coefficients for flow nozzles within ± 2.0 percent for $0.25 < \beta < 0.75$ for $10^4 < Re_{D_1} < 10^7$. Some values of flow coefficients calculated from Eq. 8.55 and Eq. 8.50 are presented in Fig. 8.25. (Values for which K is greater than one are the result of velocity of approach factors exceeding one.)

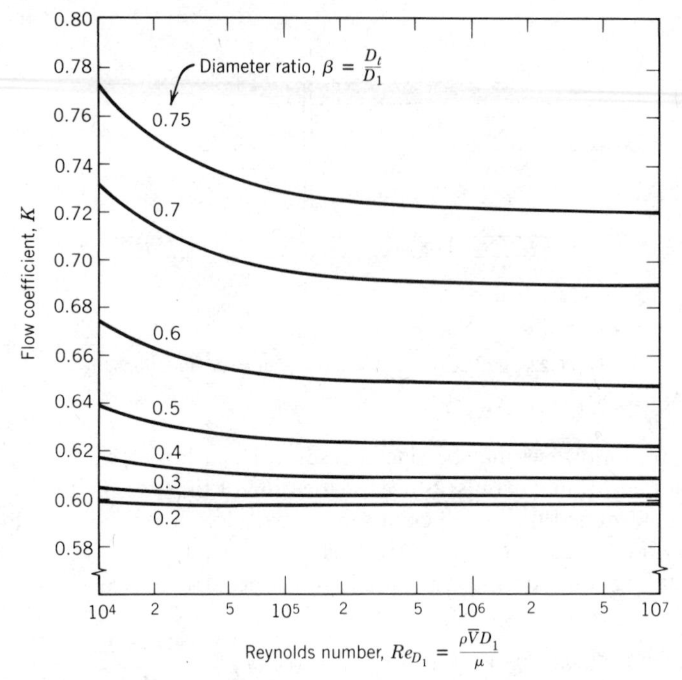

Fig. 8.23 Flow coefficients for concentric orifices with corner taps.

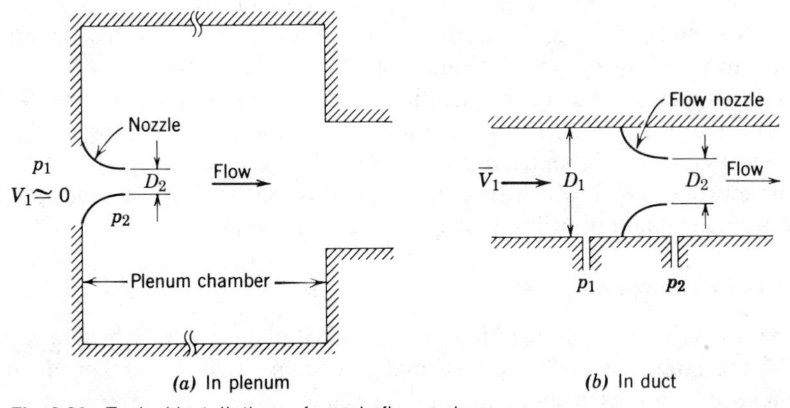

(a) In plenum *(b)* In duct

Fig. 8.24 Typical installations of nozzle flow meters.

a. Pipe Installation

For pipe installation, a value of K read from Fig. 8.25 must be used. Figure 8.25 shows that K is essentially constant for large Reynolds number ($Re_{D_1} > 10^5$). Thus at high flow rates, the flow rate may be computed directly using Eq. 8.51. At lower flow rates, where K is a weak function of Reynolds number, iteration may be required.

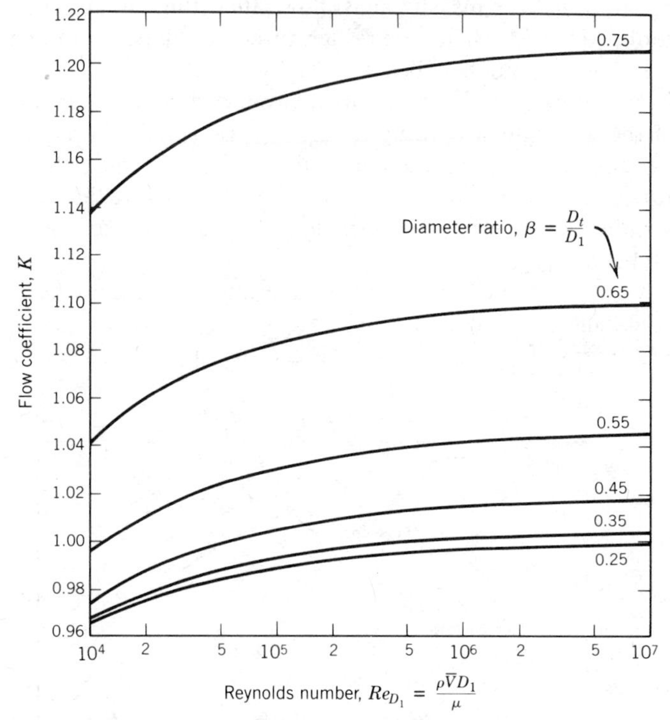

Fig. 8.25 Flow coefficients for ASME long-radius flow nozzles.

b. Plenum Installation

For plenum installation, nozzles may be fabricated from spun aluminum, molded fiberglass, or other inexpensive materials. Thus they are simple and cheap to make and install. Since the plenum pressure is equal to p_2, the location of the downstream pressure tap is not critical. Meters suitable for a wide range of flow rates may be made by installing several nozzles in a plenum. At low flow rates, most of them may be plugged. For larger flow rates, more nozzles may be used.

For plenum nozzles, typical flow coefficient values are in the range, $0.95 < K < 0.99$; the larger values apply at high Reynolds numbers. Thus the mass rate of flow can be computed within approximately ± 2 percent using Eq. 8.51 with $K = 0.97$.

8-10.3 The Venturi

The venturi meter was sketched in Table 8.5. It is generally made from a casting that is machined to close tolerances to assure the performance of the standard design. As a result, venturi meters are bulky, heavy, and expensive. The conical section downstream from the throat gives excellent pressure recovery; the overall head loss therefore is low. The venturi meter also is self-cleaning due to its smooth internal contour.

Experimental data show that discharge coefficients for venturi meters are in the range of $0.980 - 0.995$ at high Reynolds numbers ($Re_{D_1} > 2 \times 10^5$). The value of

$C = 0.99$ thus can be used to measure mass flow rate within about ± 1 percent at high Reynolds number [5]. Manufacturers' literature should be consulted for specific information at Reynolds numbers below 10^5.

The orifice plate, flow nozzle, and venturi all produce pressure drops proportional to the square of the flow rate, according to Eq. 8.51. In practice, a meter size must be chosen to accommodate the largest flow rate expected. Because the pressure drop versus flow rate relationship is nonlinear, the range of flow rate that can be measured accurately is limited. Flow meters with single throats usually are considered for flow rates over a 4:1 range [5].

The unrecoverable loss in head across a metering element may be expressed as a fraction of the differential pressure, Δp, across the element. Pressure losses are displayed as functions of diameter ratio in Fig. 8.26 [5].

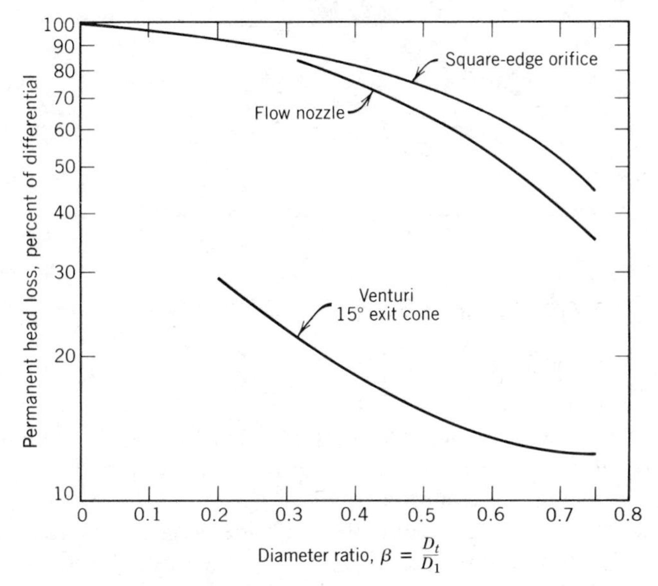

Fig. 8.26 Permanent head loss produced by various flow metering elements [5].

8-10.4 The Laminar Flow Element

The laminar flow element[6] is designed to produce a pressure differential directly proportional to flow rate. The laminar flow element (LFE) contains a metering section subdivided into many passages, each small enough in diameter to assure fully developed laminar flow. As shown in Section 8-3, the pressure drop in laminar duct flow is directly proportional to flow rate. Because the pressure drop versus flow rate

[6] Patented and manufactured by Meriam Instrument Co., 10920 Madison Ave., Cleveland, Ohio.

relationship is linear, the LFE may be used with reasonable accuracy over a 10 : 1 range in flow rate. The relationship between the pressure drop and flow rate for laminar flow also depends on the fluid viscosity, which is a strong function of temperature. Therefore, the fluid temperature must be known to obtain accurate metering with a LFE.

A laminar flow element costs approximately as much as a venturi, but it is much lighter and smaller. Thus the LFE is becoming widely used in applications where compactness and extended range are important.

Example 8.12

An air flow rate of 1 m^3/sec at standard conditions is expected in a 0.25 m diameter duct. An orifice meter is to be installed to measure the rate of flow. The manometer available to make the measurement has a maximum range of 300 mm of water. What diameter orifice plate should be used with corner taps? Analyze the head loss if the flow area at the vena contracta is $A_2 = 0.65\ A_t$. Compare with data from Fig. 8.26.

EXAMPLE PROBLEM 8.12

GIVEN: Flow through duct and orifice as shown.

$Q = 1$ m^3/sec

$D_1 = 0.25$ m

D_t

Air

① ② ③

$(p_1 - p_2)_{max} = 300$ mm H_2O

FIND:
(a) D_t.
(b) Head loss between sections ① and ③.
(c) Compare with data from Fig. 8.26.

SOLUTION:

The orifice plate may be designed using Eq. 8.51 and data from Fig. 8.23

Computing equation: $$\dot{m}_{\text{actual}} = KA_t\sqrt{2\rho(p_1 - p_2)} \qquad (8.51)$$

Since $A_t/A_1 = (D_t/D_1)^2 = \beta^2$,

$$\dot{m}_{\text{actual}} = K\beta^2 A_1\sqrt{2\rho(p_1 - p_2)}$$

or

$$K\beta^2 = \frac{\dot{m}_{\text{actual}}}{A_1\sqrt{2\rho(p_1 - p_2)}} = \frac{\rho Q}{A_1\sqrt{2\rho(p_1 - p_2)}} = \frac{Q}{A_1}\sqrt{\frac{\rho}{2(p_1 - p_2)}}$$

$$= \frac{Q}{A_1}\sqrt{\frac{\rho}{2g\rho_{H_2O}\Delta h}}$$

$$= \frac{1\ \text{m}^3}{\text{sec}} \times \frac{4}{\pi}\frac{1}{(0.25)^2\ \text{m}^2}\left[\frac{1}{2} \times \frac{1.23\ \text{kg}}{\text{m}^3} \times \frac{\text{sec}^2}{9.81\ \text{m}} \times \frac{\text{m}^3}{999\ \text{kg}} \times \frac{1}{0.30\ \text{m}}\right]^{1/2}$$

$$K\beta^2 = 0.295 \qquad \text{or} \qquad K = \frac{0.295}{\beta^2} \qquad (1)$$

Since K is a function of both β and Re_{D_1}, we must iterate to find β. The duct Reynolds number is

$$Re_{D_1} = \frac{\rho \bar{V}_1 D_1}{\mu} = \frac{\rho (Q/A_1) D_1}{\mu} = \frac{4Q}{\pi \nu D_1}$$

$$Re_{D_1} = \frac{4}{\pi} \times \frac{1\ \text{m}^3}{\text{sec}} \times \frac{\text{sec}}{1.45 \times 10^{-5}\ \text{m}^2} \times \frac{1}{0.25\ \text{m}} = 3.51 \times 10^5$$

Guess $\beta = 0.75$. From Fig. 8.23, K should be 0.72. From Eq. 1,

$$K = \frac{0.295}{(0.75)^2} = 0.524$$

Thus our guess for β is too large. Guess $\beta = 0.70$. From Fig. 8.23, K should be 0.69. From Eq. 1,

$$K = \frac{0.295}{(0.70)^2} = 0.602$$

Thus our guess for β is still too large. Guess $\beta = 0.65$. From Fig. 8.23, K should be 0.67. From Eq. 1,

$$K = \frac{0.295}{(0.65)^2} = 0.698$$

There is satisfactory agreement with $\beta \simeq 0.66$ and

$$D_t = \beta D_1 = 0.66(0.25\ \text{m}) = 0.165\ \text{m}$$

D_t

To evaluate the head loss, apply Eq. 8.28 between sections ① and ③.

Computing equation:

$$\left[\frac{p_1}{\rho} + \alpha_1 \frac{\bar{V}_1^2}{2} + g z_1\right] - \left[\frac{p_3}{\rho} + \alpha_3 \frac{\bar{V}_3^2}{2} + g z_3\right] = h_{l_T} \qquad (8.28)$$

Assumptions: (1) $\alpha_1 \bar{V}_1^2 = \alpha_3 \bar{V}_3^2$
(2) Neglect Δz

Then

$$h_{l_T} = \frac{p_1 - p_3}{\rho} = \frac{p_1 - p_2 - (p_3 - p_2)}{\rho} \qquad (2)$$

The pressure at ③ may be found by applying the x component of the momentum equation to a control volume between sections ② and ③.

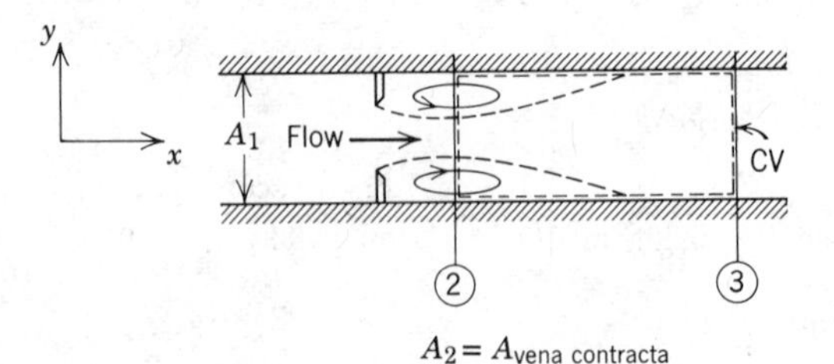

$A_2 = A_{\text{vena contracta}}$

$$= 0(3) = 0(4)$$

Basic equation: $$F_{S_x} + F_{B_x} = \frac{\partial}{\partial t}\int_{CV} u\rho\, d\forall + \int_{CS} u\rho \vec{V} \cdot d\vec{A} \tag{4.19a}$$

Assumptions: (3) $F_{B_x} = 0$
(4) Steady flow
(5) Uniform flow at sections ② and ③
(6) Pressure uniform across duct at sections ② and ③
(7) No friction

Then

$$(p_2 - p_3)A_1 = u_2\{-|\rho \bar{V}_2 A_2|\} + u_3\{|\rho \bar{V}_3 A_3|\} = (u_3 - u_2)\rho Q = (\bar{V}_3 - \bar{V}_2)\rho Q$$

or

$$p_3 - p_2 = (\bar{V}_2 - \bar{V}_3)\frac{\rho Q}{A_1}$$

Now $\bar{V}_3 = Q/A_1$, and

$$\bar{V}_2 = \frac{Q}{A_2} = \frac{Q}{0.65 A_t} = \frac{Q}{0.65\beta^2 A_1}$$

Thus

$$p_3 - p_2 = \frac{\rho Q^2}{A_1^2}\left[\frac{1}{0.65\beta^2} - 1\right]$$

$$= \frac{1.23 \text{ kg}}{\text{m}^3} \times \frac{(1)^2 \text{ m}^6}{\text{sec}^2} \times \frac{4^2}{\pi^2}\frac{1}{(0.25)^4 \text{ m}^4}\left[\frac{1}{0.65(0.66)^2} - 1\right]\frac{\text{N}\cdot\text{sec}^2}{\text{kg}\cdot\text{m}}$$

$$p_3 - p_2 = 1290 \text{ N/m}^2$$

The diameter ratio, β, was selected to give the maximum manometer deflection at maximum flow rate. Thus

$$p_1 - p_2 = \rho_{H_2O} g\, \Delta h = \frac{999 \text{ kg}}{\text{m}^3} \times \frac{9.81 \text{ m}}{\text{sec}^2} \times 0.30 \text{ m} \times \frac{\text{N}\cdot\text{sec}^2}{\text{kg}\cdot\text{m}} = 2940 \text{ N/m}^2$$

Substituting into Eq. 2 gives

$$h_{l_T} = \frac{p_1 - p_3}{\rho} = \frac{p_1 - p_2 - (p_3 - p_2)}{\rho}$$

$$h_{l_T} = \frac{(2940 - 1290) \text{ N}}{\text{m}^2} \times \frac{\text{m}^3}{1.23 \text{ kg}} = 1340 \text{ N}\cdot\text{m/kg} \qquad h_{l_T}$$

To compare with Fig. 8.26, express the permanent pressure loss as a fraction of the meter differential

$$\frac{p_1 - p_3}{p_1 - p_2} = \frac{(2940 - 1290) \text{ N/m}^2}{2940 \text{ N/m}^2} = 0.561$$

This is satisfactory agreement!

{This problem illustrates flow meter calculations and shows use of the momentum equation to compute the pressure rise in a sudden expansion.}

8-11 LINEAR FLOW METERS

Several flow meter types produce outputs that are directly proportional to flow rate. These meters produce signals without the need to measure a differential pressure. The most common linear flow meters are discussed briefly in the following paragraphs.

Float meters may be used to indicate the flow rate directly for liquids or gases. An example is shown in Fig. 8.27. In operation, the ball or float is carried upward in the tapered clear tube by the flowing fluid until the drag force and float weight are in equilibrium. Such meters (often called rotameters) are available with factory calibration for a number of common fluids and flow rate ranges.

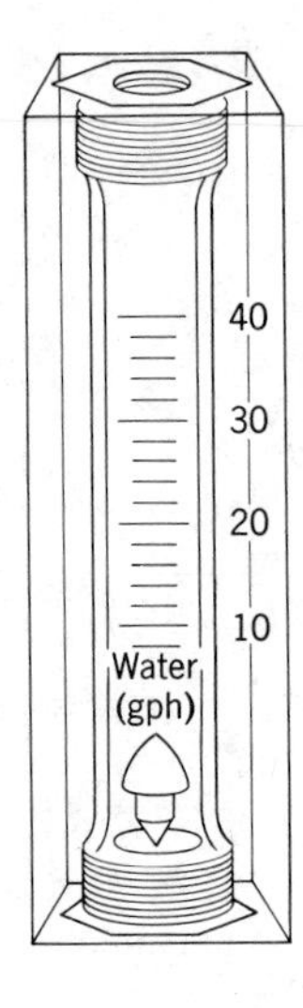

Fig. 8.27 Float-type variable-area flow meter. (Courtesy of Dwyer Instrument Co., Michigan City, Indiana.)

A free-running vaned impeller may be mounted in a cylindrical section of tube (Fig. 8.28) to make a *turbine flow meter*. With proper design, the rate of rotation of the impeller may be made closely proportional to the volume flow rate over a wide range.

Rotational speed of the turbine element can be sensed using a magnetic or modulated carrier pickup external to the meter. This sensing method therefore requires no penetrations or seals in the duct. Thus turbine flow meters can be used safely to measure flow rates in corrosive or toxic fluids. The electrical signal can be displayed, recorded, or integrated to provide total flow information.

Vortex shedding from a bluff obstruction may be used to meter flow. As noted in Chapter 7, the Strouhal number, $St = fL/V$, is approximately constant ($St \approx 0.21$). Thus frequency of vortex shedding is proportional to flow velocity. The vortex shedding from the obstruction results in velocity and pressure changes around and downstream from the obstruction. Pressure, thermal, or ultrasonic sensors may be used to detect the vortex shedding frequency, and thus to infer the fluid velocity. (The velocity profile does affect the constancy of the shedding frequency.) Vortex flow meters can be used over a 20 : 1 range of flow rates [5].

The electromagnetic flow meter uses the principle of magnetic induction. A magnetic

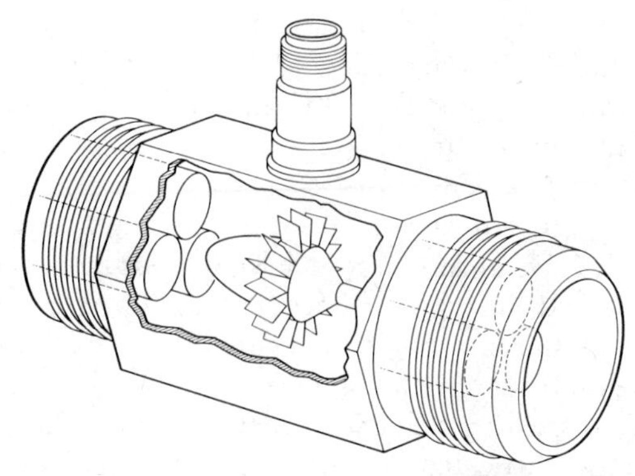

Fig. 8.28 Turbine flow meter. (Courtesy of Potter Aeronautical Corp., Union, New Jersey.)

field is created across a pipe. When a conductive fluid passes through the field, a voltage is generated at right angles to the field and the velocity vectors. Electrodes placed on a pipe diameter are used to detect the resulting signal voltage. The signal voltage is proportional to the average axial velocity when the profile is axisymmetric.

Magnetic flow meters may be used with liquids that have electrical conductivities above 100 microsiemens per meter (1 siemen = 1 ampere per volt). The minimum flow speed should be above about 0.3 m/sec, but there are no restrictions on flow Reynolds number. The flow range normally quoted is 10:1 [5].

Ultrasonic flow meters also respond to average velocity at a pipe cross section. Two principal types of ultrasonic meters are common: Propagation time is measured for clean liquids and reflection frequency shift (Doppler effect) for flows carrying particulates. The speed of an acoustic wave increases in the flow direction and decreases when transmitted against the flow. An acoustic path inclined to the pipe axis is used to infer flow velocity. Multiple paths are used in practice to estimate the volume flow rate accurately.

Doppler effect ultrasonic flow meters depend on reflection of sonic waves (in the MHz range) from scattering particles in the fluid. When the particles move at flow speed, the frequency shift is proportional to flow speed; for a suitably chosen path, output is proportional to volume flow rate. One or two transducers may be used; the meter may be clamped to the outside of the pipe. Ultrasonic meters may require calibration in place. Flow rate range is 10:1 [5].

8-12 TRAVERSING METHODS

In some situations, e.g. in air handling or refrigeration equipment, it is impractical or impossible to install fixed flow meters. In such cases it may be possible to obtain flow rate data using traversing techniques.

To make a flow rate measurement by traverse, the duct cross section is subdivided into segments of equal area. The velocity is measured at the center of each area segment using a pitot tube, a total head tube, or a suitable anemometer. The volume flow rate for each segment is approximated by the product of the measured velocity and the segment area. The flow rate through the entire duct is the sum of these segmental flow rates. Details of recommended procedures for flow rate measurements by the traverse method may be found in [23].

Use of pitot or pitot-static tubes for traverse measurements requires direct access to the flow field. Pitot tubes give uncertain results when pressure gradients or streamline curvature are present, and their response times are slow. Two types of anemometers—thermal anemometers and laser Doppler anemometers—overcome these difficulties partially, although they introduce new complications.

Thermal anemometers use tiny elements (either hot-wire or hot-film elements) that are heated electrically. Sophisticated electronic feedback circuits are used to maintain the temperature of the element constant and to sense the input heating rate. The heating rate may be related to the local flow velocity through heat transfer correlations or by calibration. The primary advantage of thermal anemometers is the small size of the sensing element. Sensors as small as 0.002 mm in diameter and 0.1 mm long are available commercially. Because the thermal mass of such tiny elements is extremely small, their response to fluctuations in flow velocity is rapid. Frequency responses to the 50 kHz range have been quoted [24]. Thus thermal anemometers are ideal for measurement of turbulence quantities. Insulating coatings may be applied to permit their use in conductive or corrosive gases or liquids.

Because of their fast response and small size, thermal anemometers are used extensively for research. Numerous schemes have been published for treating the resulting data. Digital processing techniques, including fast Fourier transforms, can be applied to the signals to obtain mean values and moments, and to analyze frequency content and correlations.

Laser Doppler anemometers (LDAs) are becoming widely used for specialized applications where direct physical access to the flow field is difficult or impossible. One or more laser beams are focused to a small volume in the flow at the location of interest. Laser light is scattered from particles that are present in the flow (dust or particulates) or introduced for this purpose. A frequency shift is caused by the local flow speed (Doppler effect). Scattered light and a reference beam are collected by receiving optics. The frequency shift is proportional to the flow speed; this relationship may be calculated so that there is no need for calibration. Since velocity is measured directly, the signal is unaffected by changes in temperature, density, or composition in the flow field. The primary disadvantages of LDAs are that the optical equipment is expensive and fragile, and that extremely careful alignment is needed.

8-13 SUMMARY OBJECTIVES

After completing study of Chapter 8, you should be able to do the following:

1. Define:

internal flow	kinetic energy flux coefficient
entrance length	momentum flux coefficient
fully developed flow	head loss
friction velocity	major, minor head loss
Reynolds stress	**hydraulic diameter

2. For fully developed laminar flows, apply the control volume formulation of Newton's second law to a suitably chosen differential control volume to determine: the velocity distribution, the shear stress distribution, the volume flow rate, the average velocity, and the location of the maximum velocity.

3. For fully developed flow in a pipe, determine the wall shear stress and the shear stress variation in the flow in terms of the pressure gradient.

4. For fully developed turbulent flow in a pipe, with velocity distribution represented by a power-law profile, determine $\bar{V}/U$.

5. Write the first law of thermodynamics in a form suitable for the solution of pipe flow problems.

6. Use the Moody diagram to determine the friction factor for fully developed pipe flow.

7. Solve single-path pipe flow system problems for each of the four cases discussed in Section 8-8.1; sketch pressure distribution, hydraulic, and energy grade lines.

****8.** Use the basic techniques of Section 8-8.1 to analyze multiple-path pipe flow systems.

9. Determine the mass flow rate from the pressure differential measured using an orifice plate, flow nozzle, or venturi meter.

10. Solve the problems at the end of the chapter that relate to the material you have studied.

REFERENCES

1. Laufer, J., "The Structure of Turbulence in Fully Developed Pipe Flow," U.S. National Advisory Committee for Aeronautics (NACA), Technical Report 1174, 1954.
2. Tennekes, H., and J. L. Lumley, *A First Course in Turbulence.* Cambridge, Mass.: The MIT Press, 1972.
3. Hinze, J. O., *Turbulence,* 2nd ed. New York: McGraw-Hill, 1975.
4. Moody, L. F., "Friction Factors for Pipe Flow," *Transactions of the ASME, 66,* 8, November 1944, pp. 671–684.
5. Miller, R. W., *Flow Measurement Engineering Handbook.* New York: McGraw-Hill, 1983.
6. Swamee, P. K., and A. K. Jain, "Explicit Equations for Pipe-Flow Problems," *Proceedings of the ASCE, Journal of the Hydraulics Division, 102,* HY5, May 1976, pp. 657–664.
7. "Flow of Fluids through Valves, Fittings, and Pipe," Crane Company, New York, N.Y., Technical Paper No. 410, 1982.
8. Streeter, V. L., ed., *Handbook of Fluid Dynamics.* New York: McGraw-Hill, 1961.
9. *ASHRAE Handbook—Fundamentals.* Atlanta, Ga.: American Society of Heating, Refrigerating, and Air-Conditioning Engineers, Inc., 1981.
10. Cockrell, D. J., and C. I. Bradley, "The Response of Diffusers to Flow Conditions at Their Inlet," Paper No. 5, *Symposium on Internal Flows,* University of Salford, Salford, England, April 1971, pp. A32–A41.
11. Sovran, G., and E. D. Klomp, "Experimentally Determined Optimum Geometries for Rectilinear Diffusers with Rectangular, Conical, or Annular Cross Sections," in *Fluid Mechanics of Internal Flow,* G. Sovran, ed. Amsterdam: Elsevier, 1967, pp. 270–319.

** These objectives apply to sections that may be omitted without loss of continuity in the text material.

12. Feiereisen, W. J., R. W. Fox, and A. T. McDonald, "An Experimental Investigation of Incompressible Flow without Swirl in *R*-Radial Diffusers," Proceedings, Second International Japan Society of Mechanical Engineers Symposium on Fluid Machinery and Fluidics, Tokyo, Japan, September 4–9, 1972, pp. 81–90.
13. McDonald, A. T., and R. W. Fox, "An Experimental Investigation of Incompressible Flow in Conical Diffusers," *International Journal of Mechanical Sciences*, *8*, 2, February 1966, pp. 125–139.
14. Runstadler, P. W., Jr., "Diffuser Data Book," Creare, Inc., Hanover, N.H., Technical Note 186, 1975.
15. Reneau, L. R., J. P. Johnston, and S. J. Kline, "Performance and Design of Straight, Two-Dimensional Diffusers," *Transactions of the ASME, Journal of Basic Engineering*, *89D*, 1, March 1967, pp. 141–150.
16. Daily, J. W., and D. R. F. Harleman, *Fluid Dynamics*. Reading, Mass.: Addison-Wesley, 1966.
17. White, F. M., *Fluid Mechanics*. New York: McGraw-Hill, 1979.
18. Hamilton, J. B., "The Suppression of Intake Losses by Various Degrees of Rounding," University of Washington, Seattle, Wash., Experiment Station Bulletin 51, 1929.
19. Herschel, C., *The Two Books on the Water Supply of the City of Rome, from Sextus Julius Frontinus* (ca. 40–103 A. D.), Boston, 1899, p. 205.
20. Lam, C. F., and M. L. Wolla, "Computer Analysis of Water Distribution Systems: Part 1, Formulation of Equations," *Proceedings of the ASCE, Journal of the Hydraulics Division*, *98*, HY2, February 1972, pp. 335–344.
21. *Aerospace Applied Thermodynamics Manual*. New York (now Warrendale, Pa.): Society of Automotive Engineers, 1969.
22. Bean, H. S., ed., *Fluid Meters, Their Theory and Application*. New York: American Society of Mechanical Engineers, 1971.
23. ISO 7145, *Determination of Flowrate of Fluids in Closed Conduits or Circular Cross Sections—Method of Velocity Measurement at One Point in the Cross Section*, ISO UDC 532.57.082.25:532.542, 1st ed. Geneva: International Standards Organization, 1982.
24. Goldstein, R. J., ed., *Fluid Mechanics Measurements*. Washington, D.C.: Hemisphere, 1983.

PROBLEMS

8.1 Standard air enters a 0.3 m diameter duct. The volume flow rate is 2 m^3/min. Determine whether the flow is laminar or turbulent. Estimate the entrance length required for fully developed flow to be established.

8.2 The Reynolds number for fully developed flow in a circular tube usually is defined as $Re = \rho \bar{V} D/\mu$. Transition to turbulent flow for engineering systems seldom is delayed past $Re = 2300$. On a log-log plot of average velocity versus tube diameter, plot lines that correspond to $Re = 2300$ for (a) standard air, and (b) water at 15 C.

8.3 Approximately how far from the entrance of a 10 mm diameter pipe would flow be fully developed if the Reynolds number is 1500, based on average velocity?

8.4 The velocity profile for flow between stationary parallel plates is given by

$$u = ay(h - y)$$

where a is a constant, h is the total gap width between plates, and y is the distance measured upward from the lower plate. Determine the ratio $\bar{V}/u_{max}$.

8.5 An incompressible fluid flows between two infinite stationary parallel plates. The velocity profile is given by

$$u = u_{max}(Ay^2 + By + C)$$

where A, B, and C are constants, and y is measured from the center of the gap. The total gap width is h units. Use appropriate boundary conditions to express the constants in terms of h. Develop an expression for volume flow rate per unit depth.

8.6 A viscous oil flows steadily between parallel plates. The flow is laminar and fully developed. The velocity profile is given by

$$u = -\frac{h^2}{8\mu}\frac{\partial p}{\partial x}\left[1 - \left(\frac{2y}{h}\right)^2\right]$$

where the total gap width between the plates, $h = 3$ mm, and y is measured from the centerline of the gap. The oil viscosity is 0.5 N · sec/m^2, and the pressure gradient is -1200 N/m^2/m. Find the magnitude and direction of the shear stress on the upper plate, and the volume flow rate through the channel, per meter of width.

8.7 A fluid flows steadily between two parallel plates. The distance between the plates is h. The velocity profile for fully developed laminar flow is given by

$$u = -\frac{h^2}{8\mu}\frac{\partial p}{\partial x}\left[1 - \left(\frac{2y}{h}\right)^2\right]$$

(a) Derive an equation for the shear stress as a function of y. Plot this function.
(b) For $\mu = 2.4 \times 10^{-5}$ lbf · sec/ft^2, $\partial p/\partial x = -4.0$ lbf/ft^2/ft, and $h = 0.05$ in., calculate the maximum shear stress in lbf/ft^2.

8.8 Viscous oil flows steadily between parallel plates. The flow is fully developed and laminar. The pressure gradient is -8 lbf/ft^2/ft and the channel half-width is $h = 0.06$ in. Calculate the magnitude and direction of the wall shear stress at the upper plate surface. Find the volume flow rate through the channel.

8.9 The basic component of a pressure gage tester consists of a piston-cylinder apparatus as shown. The piston, 6 mm in diameter, is loaded to develop a pressure of known magnitude. (The piston length is 25 mm.) Calculate the mass, M, required to produce 1.5 MPa (gage) in the cylinder. Determine the leakage flow rate as a function of radial clearance, a, for this load if the liquid is SAE 30 oil at 20 C. Specify the maximum allowable radial clearance so the vertical movement of the piston due to leakage will be less than 1 mm/min.

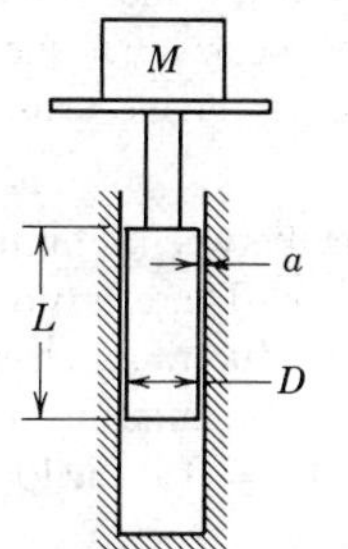

P 8.9

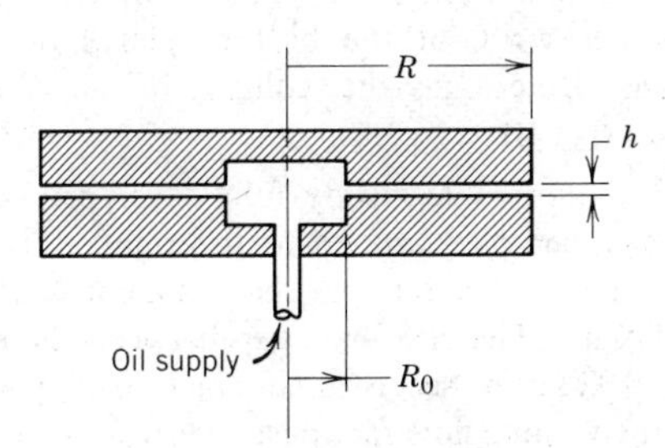

P 8.10

8.10 Viscous liquid at volume flow rate, Q, is pumped through the central opening and the narrow gap between the parallel disks shown. The flow rate is low, so the flow is laminar and the pressure gradient due to convective acceleration in the gap is negligible compared to the gradient due to viscous forces (this is termed *creeping flow*). Obtain a general expression for the variation of average velocity in the gap between the disks. For creeping flow the velocity profile at any cross section in the gap is the same as for fully developed flow between stationary parallel plates. Evaluate the pressure gradient, dp/dr, as a function of radius. Determine the net force required to hold the upper plate in the position shown.

8.11 Consider the simple power-law model for a non-Newtonian fluid given by Eq. 2.11

$$\tau_{yx} = k\left(\frac{du}{dy}\right)^n$$

Extend the analysis of Section 8-2.1 to show that the velocity profile for fully developed laminar flow of a power-law fluid between parallel plates separated by distance, $2h$, may be written

$$u = \left(\frac{h}{k}\frac{\Delta p}{L}\right)^{1/n}\frac{nh}{n+1}\left[1 - \left(\frac{y}{h}\right)^{\frac{n+1}{n}}\right]$$

where y is the coordinate measured from the channel centerline.

8.12 Evaluate the volume flow rate for fully developed laminar flow of a power-law fluid between stationary parallel plates (Problem 8.11). Show that the velocity profile may be written

$$\frac{u}{\bar{V}} = \frac{2n+1}{n+1}\left[1 - \left(\frac{y}{h}\right)^{\frac{n+1}{n}}\right]$$

where y is the coordinate measured from the channel centerline.

8.13 A sealed journal bearing is formed from concentric cylinders. The inner and outer radii are 25 and 26 mm; the journal length is 100 mm, and it turns at 2800 rpm. The gap is filled with oil in laminar motion. The velocity profile is linear across the gap. The torque needed to turn the journal is 0.2 N · m. Calculate the viscosity of the oil. Will the torque increase or decrease with time? Why?

8.14 Water at 60 C flows between two large flat plates. The lower plate moves to the left at a speed of 0.3 m/sec. The plate spacing is 3 mm, and the flow is laminar. Determine the pressure gradient required to produce zero net flow at a cross section.

8.15 Water at 60 F flows between parallel plates with gap width, $b = 0.01$ ft. The upper plate moves with speed $U = 1$ ft/sec in the positive x direction. The pressure gradient is $\partial p/\partial x = -0.06$ lbf/ft^2/ft. Locate the point of maximum velocity and determine its magnitude (let $y = 0$ at the bottom plate). Sketch the velocity and shear stress distributions. Determine the volume of flow that passes a given cross section ($x =$ constant) in 10 sec.

8.16 Consider incompressible, steady, fully developed laminar flow between infinite parallel plates. The upper plate moves to the right at $U = 3$ mm/sec. There is no pressure variation in the x direction, but there is a constant body force due to an electric field, $\rho B_x = 800$ N/m^3. The clearance gap between the plates is $h = 0.1$ mm and the liquid viscosity is 0.02 kg/m · sec. Evaluate the velocity profile, $u(y)$, if $y = 0$ at the lower plate. Compute the volume flow rate past a vertical section.

8.17 The record-write head for a computer disk-storage system floats above the spinning disk on a very thin film of air (the film thickness is 0.5 μm). The head location is 150 mm from the disk centerline; the disk spins at 3600 rpm. The record-write head is 10 mm square. Determine for standard air in the gap between the head and disk (a) the Reynolds number of the flow, (b) the viscous shear stress, and (c) the power required to overcome viscous shear.

8.18 The clamping force to hold a part in a metal-turning operation is provided by high pressure oil supplied by a pump. Oil leaks axially through an annular gap seal with diameter, D, length, L, and radial clearance, a. The inner member of the annulus rotates at angular speed, ω. Power is required both to pump the oil and to overcome viscous dissipation in the annular gap. Develop expressions in terms of the specified geometry for the pump power, $\mathscr{P}_p$, and the viscous dissipation power, $\mathscr{P}_v$. Show that the total power requirement is minimized when the radial clearance, a, is chosen such that $\mathscr{P}_v = 3\mathscr{P}_p$.

8.19 A viscous-shear pump is made from a stationary housing with a close-fitting rotating drum inside. The clearance is small compared to the diameter of the drum, so flow in the annular space may be treated as flow between parallel plates. Fluid is dragged around the annulus by viscous forces. Evaluate the performance characteristics of the shear pump (pressure differential, input power, and efficiency) as functions of volume flow rate. Assume that the depth normal to the diagram is b.

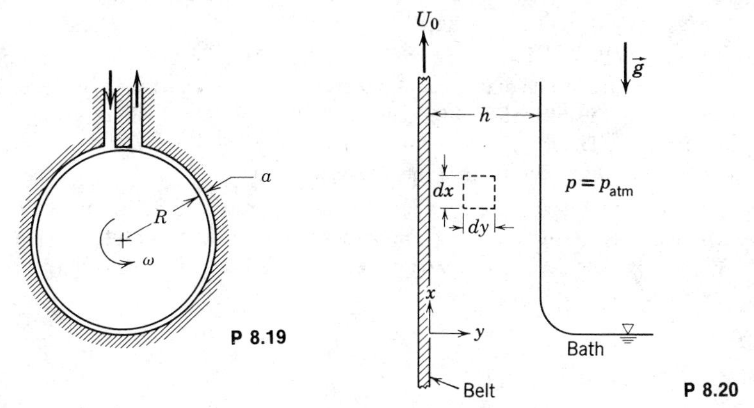

8.20 A continuous belt passing upward through a chemical bath at speed, U_0, picks up a liquid film of thickness, h, density, ρ, and viscosity, μ. Gravity tends to make the liquid drain down, but the movement of the belt keeps the liquid from running off completely. Assume that the flow is fully developed laminar flow with zero pressure gradient, and that the atmosphere produces no shear stress at the outer surface of the film. State clearly the boundary conditions to be satisfied by the velocity at $y = 0$ and $y = h$. Obtain an expression for the velocity profile.

8.21 A film of water (at 15 C) in steady, laminar motion runs down a long slope, inclined 30° below the horizontal. The thickness of the film is 0.8 mm. Assume that flow is fully developed, and at zero pressure gradient. Determine the surface shear stress, and the volume flow rate per unit width.

8.22 The velocity distribution for fully developed laminar flow in a pipe is given by

$$u = -\frac{R^2}{4\mu}\frac{\partial p}{\partial x}\left[1 - \left(\frac{r}{R}\right)^2\right]$$

Determine the radial distance from the pipe axis at which the velocity equals the average velocity.

8.23 Consider first water and then SAE 10 W lubricating oil flowing at 40 C in a 6 mm diameter tube. Determine the maximum flow rate (and the corresponding pressure gradient, $\partial p/\partial x$, for each fluid at which laminar flow would be expected.

8.24 A water injection line is made from smooth capillary tubing with inside diameter, $D = 0.25$ mm. Determine the maximum volume flow rate at which flow is laminar. Evaluate the pressure drop required to produce this flow rate through a section of tubing with length, $L = 0.75$ m.

8.25 A liquid drug with the viscosity and density of water is to be administered through a hypodermic needle. The inside diameter of the needle is 0.25 mm and its length is 50 mm. Determine (a) the maximum volume flow rate for which the flow will be laminar, (b) the pressure drop required to deliver the maximum flow rate, and (c) the corresponding wall shear stress.

8.26 A tube 17.6 in. long, with an inside diameter of 0.030 in., is used as a capillary viscometer. Calibration tests are made using water at 60 F, and a flow rate of 1 cm^3/sec is measured for an applied pressure drop of 10 psi. (Assume that the pressure drop in the entrance length is twice that for the same length of fully developed flow.) Determine the percentage error in viscosity that would result if Eq. 8.13c were used directly to compute it, without considering the entrance length.

8.27 Cutting oil flows vertically downward under gravity in a long circular tube. The tube is long compared to its diameter so that end effects are negligible. The specific gravity and viscosity of the oil are SG $= 0.88$ and $\mu = 0.002$ lbf · sec/ft^2. The pressure is atmospheric everywhere. Determine the volume flow rate of oil that will be delivered if the tube diameter is 0.5 in. Verify that the flow is laminar.

8.28 Consider fully developed laminar flow due to gravity of a viscous liquid in a vertical circular tube. Assume that the pressure is atmospheric at both the tube inlet and outlet. Show that the relationship between tube diameter and Reynolds number may be expressed as

$$D = \left(\frac{32\,Re\,\nu^2}{g}\right)^{1/3}$$

Evaluate the maximum tube diameter for laminar flow of (a) water, and (b) SAE 30 oil at 20 C.

8.29 Consider fully developed laminar flow in a circular pipe. Use a cylindrical control volume as shown. Indicate the forces acting on the control volume. Using the momentum equation, develop an expression for the velocity distribution.

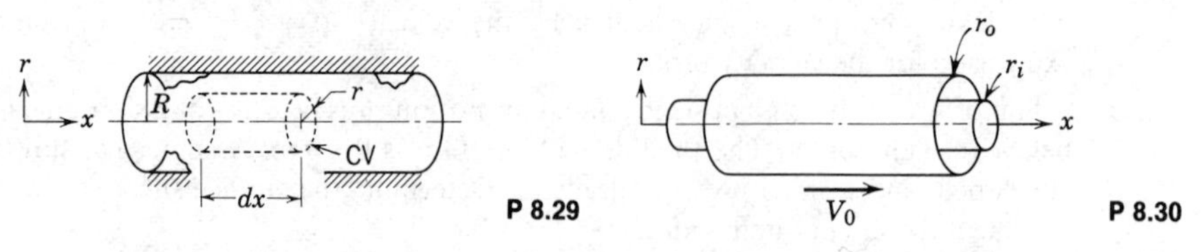

P 8.29 P 8.30

8.30 Consider fully developed laminar flow in the annulus between two concentric pipes. The inner pipe is stationary, and the outer pipe moves in the x direction with speed, V_0. Assume the axial pressure gradient to be zero ($\partial p/\partial x = 0$). Obtain a general expression for the shear stress, τ, as a function of the radius, r, in terms of a constant, C_1. Obtain a general expression for the velocity profile, $V(r)$, in terms of two constants, C_1 and C_2. Evaluate the constants, C_1 and C_2.

8.31 Resistance to fluid flow can be defined by analogy to Ohm's law for electric current. Thus resistance to flow is given by the ratio of pressure drop (driving potential) to volume flow rate (current). Show that resistance to laminar flow is given by

$$\text{resistance} = \frac{128\mu L}{\pi D^4}$$

which is independent of flow rate. Find the maximum pressure drop for which this relation is valid for a tube 50 mm long and 0.25 mm inside diameter for both kerosine and castor oil at 40 C.

8.32 Consider the simple power-law model for a non-Newtonian fluid given by Eq. 2.11

$$\tau_{yx} = k\left(\frac{du}{dy}\right)^n$$

Extend the analysis of Problem 8.29 to show that the velocity profile for fully developed laminar flow of a power-law fluid in a circular tube may be written

$$u = \left(\frac{R}{2k}\frac{\Delta p}{L}\right)^{1/n} \frac{nR}{n+1}\left[1 - \left(\frac{r}{R}\right)^{\frac{n+1}{n}}\right]$$

Hint: Use the cylindrical control volume of Problem 8.29 for your analysis.

8.33 Evaluate the volume flow rate for fully developed laminar flow of a power-law fluid in a circular tube (Problem 8.32). Show that the velocity profile may be written

$$\frac{u}{\bar{V}} = \left(\frac{3n+1}{n+1}\right)\left[1 - \left(\frac{r}{R}\right)^{\frac{n+1}{n}}\right]$$

as suggested in Problem 2.30.

8.34 A horizontal pipe carries fluid in fully developed turbulent flow. The static pressure difference measured between two sections is 3 psi. The distance between the sections is 25 ft and the pipe diameter is 6 in. Calculate the shear stress, τ_w, that acts on the walls.

8.35 The pressure drop between two taps separated in the streamwise direction by 3 m in a horizontal, fully developed channel flow of water is 1.78 kPa. The cross section of the channel is a 30 × 240 mm rectangle. Calculate the average wall shear stress.

8.36 Kerosine at 70 F flows in a smooth tube with an inside diameter of 1 in. The flow Reynolds number is 4000. For laminar flow, the pressure gradient is found to be $\partial p/\partial x = -0.2\ \text{lbf/ft}^2\text{/ft}$, while for turbulent flow, $\partial p/\partial x = -0.5\ \text{lbf/ft}^2\text{/ft}$. Plot the variation of shear stress as a function of radius for both flow conditions.

8.37 Consider the velocity profile for fully developed laminar pipe flow, Eq. 8.14, and the empirical "power-law" profile for turbulent pipe flow, Eq. 8.22. Assume that $n = 7$ for the turbulent profile. Determine the value of r/R for each profile at which u is equal to the average velocity, $\bar{V}$.

8.38 Consider the empirical power-law velocity profile for fully developed turbulent pipe flow,

$$\frac{\bar{u}}{U} = \left(1 - \frac{r}{R}\right)^{1/n} = \left(\frac{y}{R}\right)^{1/n}$$

Using the method of least squares, show that a best-fit value of n may be calculated from measured data as

$$\frac{1}{n} = \frac{\sum \ln(\bar{u}/U)\ln(y/R)}{\sum[\ln(y/R)]^2}$$

Hint: Express the profile as $\ln(\bar{u}/U) = (1/n)\ln(y/R)$.

8.39 Laufer [1] measured the following data for mean velocity in fully developed turbulent pipe flow at $Re_U = 50{,}000$:

$\bar{u}/U$	0.996	0.981	0.963	0.937	0.907	0.866	0.831
y/R	0.898	0.794	0.691	0.588	0.486	0.383	0.280

$\bar{u}/U$	0.792	0.742	0.700	0.650	0.619	0.551
y/R	0.216	0.154	0.093	0.062	0.041	0.024

Plot the data on log-log graph paper. Evaluate a power-law velocity profile exponent graphically. Compare with a value calculated by the method of least squares using the result of Problem 8.38.

8.40 Laufer [1] measured the following data for mean velocity in fully developed turbulent pipe flow at $Re_U = 500{,}000$:

$\bar{u}/U$	0.997	0.988	0.975	0.959	0.934	0.908
y/R	0.898	0.794	0.691	0.588	0.486	0.383

$\bar{u}/U$	0.874	0.847	0.818	0.771	0.736	0.690
y/R	0.280	0.216	0.154	0.093	0.062	0.037

Plot the data on log-log graph paper. Evaluate a power-law velocity profile exponent graphically. Compare with a value calculated by the method of least squares using the result of Problem 8.38.

8.41 The velocity profile for turbulent flow through smooth pipes is often represented by the empirical equation

$$\frac{u}{U} = \left(1 - \frac{r}{R}\right)^{1/n}$$

Show that the ratio of average to centerline velocities is given by

$$\frac{\bar{V}}{U} = \frac{2n^2}{(n+1)(2n+1)}$$

Using the values for n given in Fig. 8.11, plot $\bar{V}/U$ as a function of Reynolds number.

8.42 Figure 8.11 is a plot of power-law velocity profile exponent, n, versus centerline Reynolds number, Re_U, for fully developed turbulent pipe flow. Equation 8.23 relates mean velocity, $\bar{V}$, to centerline velocity, U, for values of n. Prepare a plot of $\bar{V}/U$ versus log Re_U.

8.43 A momentum flux coefficient, β, is defined as

$$\int_A u\rho u\, dA = \beta \int_A \bar{V}\rho u\, dA = \beta \dot{m}\bar{V}$$

Evaluate β for a laminar velocity profile, Eq. 8.14, and for a "power-law" turbulent velocity profile, Eq. 8.22 (choose $n = 7$).

8.44 Consider fully developed laminar flow of water between infinite parallel plates. The maximum flow speed, plate spacing, and width are 6 m/sec, 2 mm, and 30 mm, respectively. Find the flux of kinetic energy at a cross section.

8.45 The kinetic energy flux coefficient, α, is defined by Eq. 8.25b. Using the "power-law" turbulent velocity profile, show that

$$\alpha = \left[\frac{U}{\bar{V}}\right]^3 \frac{2n^2}{(3+n)(3+2n)}$$

Evaluate α for $n = 7$.

8.46 Evaluate the kinetic energy flux coefficient, α, for the flow of Problem 8.44.

8.47 Consider fully developed laminar flow in a circular tube. Evaluate the kinetic energy flux coefficient for this flow.

8.48 Water flows in a constant-area pipeline; the pipe diameter is 50 mm and the average flow speed is 1.5 m/sec. At the pipe inlet the gage pressure is 590 kPa. The outlet of the pipe is at an elevation 25 m higher than the inlet; the outlet pressure is atmospheric. Determine the head loss between the inlet and outlet of the pipe.

8.49 The pipe of Problem 8.48 is placed on a horizontal surface. The flow rate and outlet pressure are to remain the same. Compute the inlet pressure for this new condition.

8.50 Water at 70 F flows steadily in a long horizontal pipe with inside diameter of 36 in. The average speed is $\bar{V} = 16.3$ ft/sec and $e/D = 0.0004$. Evaluate f for this flow.

8.51 Laufer [1] measured the following data for mean velocity near the wall in fully developed turbulent pipe flow at $Re_U = 50{,}000$ in air:

$\frac{\bar{u}}{U}$	0.343	0.318	0.300	0.264	0.228	0.221	0.179	0.152	0.140
$\frac{y}{R}$	0.0082	0.0075	0.0071	0.0061	0.0055	0.0051	0.0041	0.0034	0.0030

Plot the data on linear graph paper. Evaluate a best-fit value for $d\bar{u}/dy$. ($U = 9.8$ ft/sec, and $R = 5.9$ in.) Compare the wall shear stress evaluated from $\tau_w = \mu\, d\bar{u}/dy$ with that calculated from a friction factor from the Moody diagram.

8.52 A 15 m length of new wrought iron pipe of 25 mm i.d. is to be used in a horizontal position to convey water at 15 C. The average speed of the water in the pipe is 5 m/sec. Determine the volume flow rate and friction factor for the flow.

8.53 Water flows through a galvanized iron pipe at a flow rate of 0.2 m^3/sec. The inside diameter of the pipe is 150 mm, and the water temperature is 20 C. Compute the friction factor for the flow.

8.54 A smooth, 3 in. diameter pipe carries water horizontally at 150 F at a mass flow rate of 0.006 slug/sec. The pressure drop is observed to be 0.065 lbf/ft^2 per 100 ft of pipe. From the Moody chart, the friction factor could be chosen as 0.021 or 0.042. Which is correct?

8.55 The curves plotted on the Moody chart are derived from the empirical correlation

$$\frac{1}{f^{1/2}} = -2.0 \log\left(\frac{e/D}{3.7} + \frac{2.51}{Re\, f^{1/2}}\right) \tag{8.36}$$

As noted in Section 8-7.1, an initial guess calculated from

$$f_0 = 0.25\left[\log\left(\frac{e/D}{3.7} + \frac{5.74}{Re^{0.9}}\right)\right]^{-2} \tag{8.37}$$

produces results accurate to 1 percent with a single iteration [5]. Write a subroutine for calculator or computer. Use it to validate the accuracy of this claim for $Re = 10^4$ and 10^7 for $e/D = 0$ and 0.010.

8.56 An empirical correlation for friction factor in turbulent flow in smooth pipes was developed by H. Blasius in 1911. He found that

$$f = \frac{0.316}{Re^{1/4}} \qquad Re \leq 100{,}000$$

correlated data well. Show that, in turbulent flow, the predicted pressure drop is proportional to $(\bar{V})^{7/4}$ when the Blasius correlation is used. How does pressure drop depend on tube diameter at a given flow rate?

8.57 A correlating equation for flow in the fully rough flow regime is

$$f = \frac{1}{4[0.57 - \log(e/D)]^2}$$

as originally obtained by von Kármán. Compare values obtained from this equation with those from the fully rough region of the Moody chart for values of e/D equal to 0.01, 0.001, and 0.0001.

8.58 The Moody diagram gives the Darcy friction factor, f, in terms of Reynolds number and relative roughness. The *Fanning friction factor* for pipe flow is defined as

$$f_F = \frac{\tau_w}{\frac{1}{2}\rho \bar{V}^2}$$

where τ_w is the wall shear stress in the pipe. Obtain a relation between the Darcy and Fanning friction factors for fully developed pipe flow. Show that $f = 4f_F$.

8.59 Water flows through a 1 in. diameter tube that suddenly enlarges to a diameter of 2 in. The flow rate through the enlargement is 20 gallons per minute. Calculate the pressure rise across the enlargement.

8.60 Air at standard conditions flows through a sudden expansion in a circular duct. The upstream and downstream duct diameters are 3 and 9 in., respectively. The pressure downstream is 0.25 in. of water *higher* than that upstream. Determine the average speed of the air approaching the expansion, and the volume flow rate.

8.61 Water flows through a 50 mm diameter tube that suddenly contracts to a 25 mm diameter. The pressure drop across the contraction is 3.4 kPa. Determine the volume flow rate.

8.62 Flow through a sudden contraction is shown. The minimum flow area at the vena contracta is given in terms of the area ratio by the contraction coefficient,

$$C_c = \frac{A_c}{A_2} = 0.62 + 0.38\left(\frac{A_2}{A_1}\right)^3$$

Because flow accelerates from A_1 to A_c, losses are quite small. However, losses are not negligible for the "sudden" expansion from A_c to A_2. Use these assumptions to evaluate (a) the contraction coefficient, and (b) the minor loss coefficient, for a sudden contraction with $AR = A_2/A_1 = 0.5$. Compare with data from Fig. 8.17.

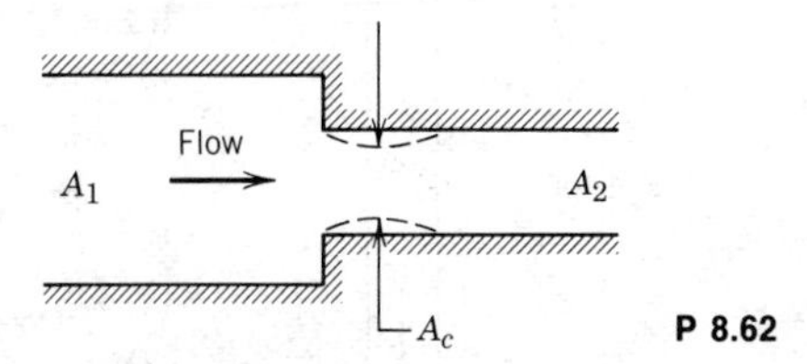

P 8.62

8.63 A clean room must be supplied with $800\ m^3/hr$ of air at standard conditions. The goemetry of the supply duct is shown. Evaluate the gage pressure in the clean room. Sketch the pressure distribution along the supply duct. What improvements to the duct system could you recommend? How would these improvements reduce losses?

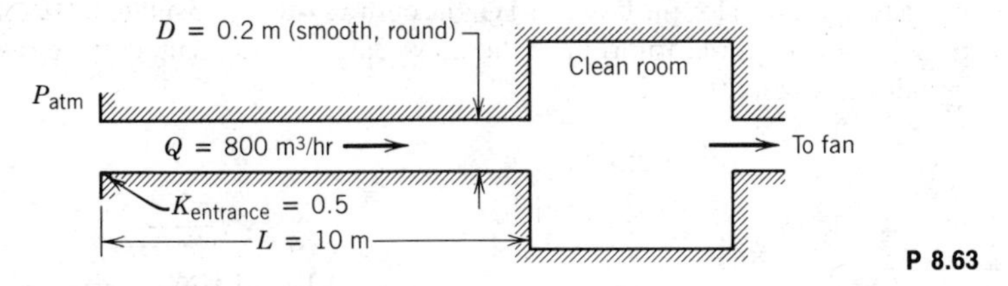

P 8.63

8.64 Air flows out of a clean room test chamber through a 150 mm diameter duct. The original duct had a square-edged entrance, but this has been replaced with a well-rounded one. The pressure in the chamber is 2.5 mm of water above ambient. Losses due to friction are negligible compared to the entrance and exit losses. Determine the increase in volume flow rate that results from the change in entrance contour.

8.65 Space has been found for a conical diffuser 0.45 m long in the clean room ventilation system described in Problem 8.64. The best diffuser of this size is to be used. Assume that data from Fig. 8.18 may be used. Determine the appropriate diffuser angle and area ratio for this installation and predict the volume flow rate that will be delivered after it is installed.

8.66 Water at 20 C flows through a 0.1 m (internal diameter) concrete drainage pipe at a rate of 15 kg/sec. Determine the pressure drop per 100 m of horizontal pipe.

8.67 Air at 15 C flows through a straight, 0.3 m diameter, smooth duct 50 m long. The flow rate is $0.6\ m^3/sec$, and the pressure is the same at both ends of the duct. Determine the change in elevation between the inlet and outlet.

8.68 Water at 78 F flows in a pipe whose inside diameter is 1.2 in. The flow rate is $0.04\ ft^3/sec$. Determine the slope that the pipe must have to maintain constant pressure along its length. If the temperature remains constant, determine the heat transfer per 100 ft of pipe.

8.69 An Ocean Thermal Energy Conversion (OTEC) plant draws cold seawater (at $T = 4$ C) into a cold water pipe far below the surface, as shown. The pipe inlet is located 1000 m below sea level. The hydrostatic pressure at that depth is $p_1 = 9.9$ MPa (gage). The cold water temperature stays nearly constant. The mean velocity in the cold water pipe is

$\bar{V} = 1.83$ m/sec, and the pipe diameter is $D = 28.2$ m. Its effective roughness height is $e = 0.01$ m. Estimate the static pressure at sea level in the cold water pipe.

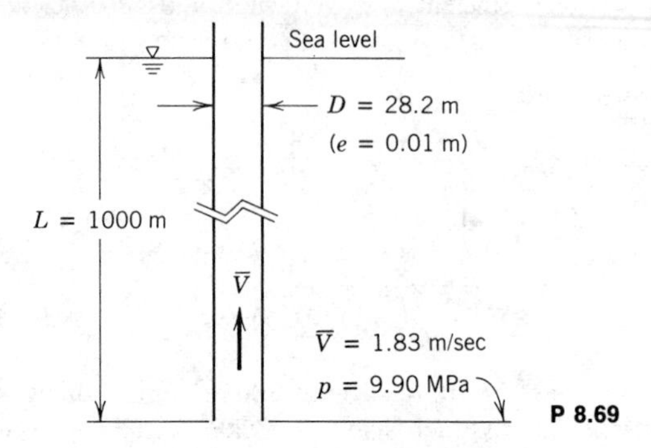

P 8.69

8.70 Water flows from a large reservoir as shown. The pipe is cast iron, with inside diameter of 0.2 m. The flow rate is 0.14 m^3/sec, and the discharge is to atmospheric pressure. The mean temperature for the flow is 10 C; the entire system is insulated. Determine the gage pressure, p_1, required to produce the flow. Calculate the temperature rise between the liquid surface and the exit.

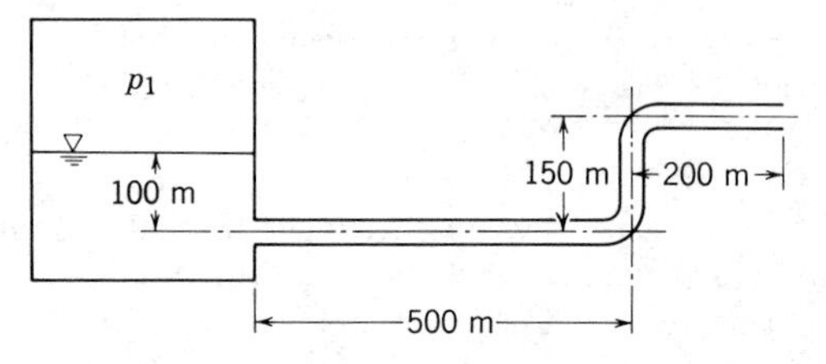

P 8.70

8.71 Water from a pump flows through a 0.25 m diameter pipe for a distance of 5 km from the pump discharge to a reservoir open to the atmosphere. The level of the water in the reservoir is 7 m above the pump discharge, and the average speed of the fluid in the pipe is 3 m/sec. Calculate the pressure at the pump discharge.

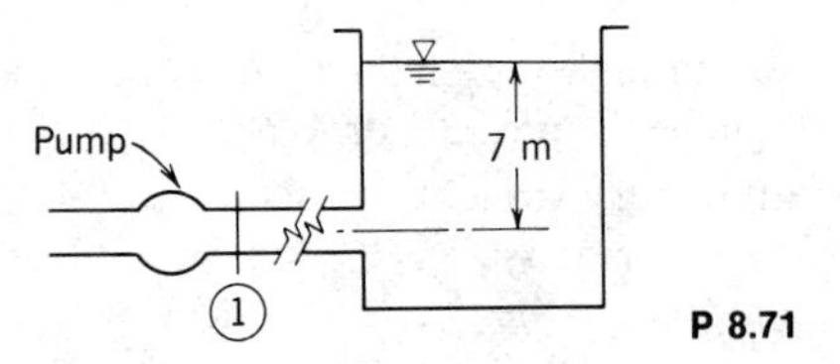

P 8.71

8.72 Water is to flow by gravity from one reservoir to a lower one through a straight, inclined pipe. The flow rate required is 0.007 m^3/sec, the pipe diameter is 50 mm, and the total length is 250 m. Each reservoir is open to the atmosphere. Neglecting minor losses, calculate the difference in level required to maintain this flow rate.

8.73 In the water flow system shown, reservoir B is a variable elevation reservoir. Determine the water level in reservoir B so that no water flows into or out of the reservoir. The speed in the 12 in. diameter pipe is 10 ft/sec. Neglect all minor losses.

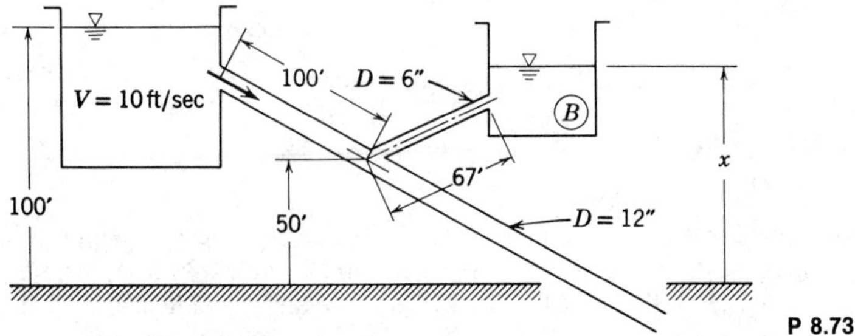

P 8.73

8.74 Two reservoirs are connected by three clean cast-iron pipes in series, $L_1 = 600$ m, $D_1 = 0.3$ m, $L_2 = 900$ m, $D_2 = 0.4$ m, $L_3 = 1500$ m, and $D_3 = 0.45$ m. When the discharge is 0.11 m^3/sec of water at 15 C, determine the difference in elevation between the reservoirs.

8.75 Lightweight crude oil (SG = 0.855 with viscosity similar to SAE 30 oil) is pumped horizontally through a 1 mile length of 12 in. diameter pipe. The average roughness size is 0.01 in. The flow rate is 4500 gpm. Calculate the horsepower required to drive the pump if it is 75 percent efficient.

8.76 Kerosine at 60 C flows through a pipe system in a refinery at the rate of 2.3 m^3/min. The pipe is commercial steel, with inside diameter of 0.15 m. The gage pressure in the reactor vessel is 90 kPa. Determine the total length, L, of the straight pipe in the system.

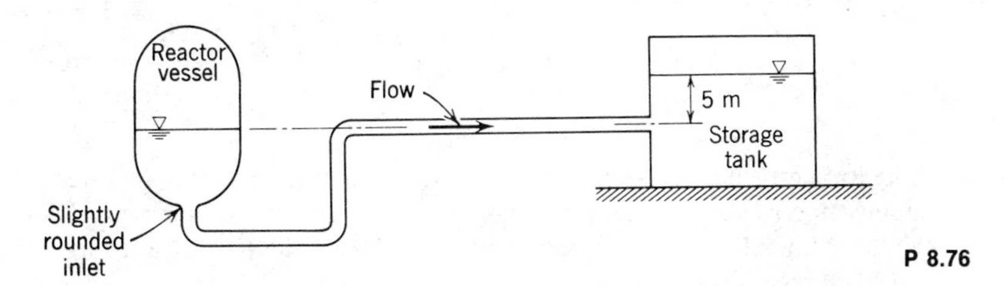

P 8.76

8.77 Two cubic feet of water per second flow through a pipe system made from 6 in. i.d. cast-iron pipe. Minor losses for the system are such that $K = 2.0$. Find the length, L, of straight pipe in the system.

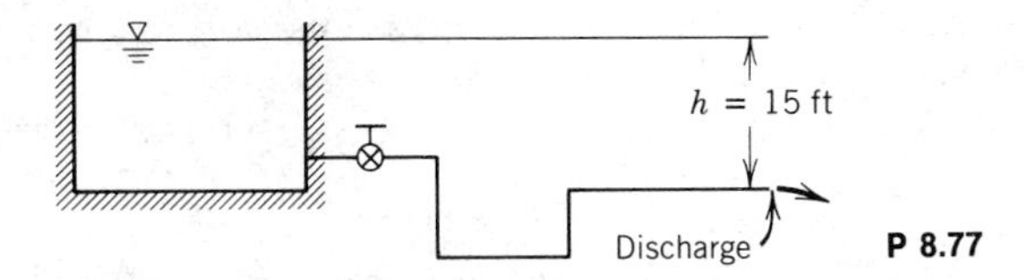

P 8.77

8.78 The spray nozzle is designed to produce a flat, radial sheet of water that sprays through 180° of arc. The nozzle discharge radius is $R = 50$ mm; the water sheet thickness is $t = 0.5$ mm. The water inlet pipe has 20 mm i.d., and is 7 m long. The nozzle is supplied through a section of commercial steel pipe with an average roughness height of 0.05 mm. The loss coefficient for the nozzle is $K = 2.0$, based on $\bar{V}_2$. Determine the volume flow rate at which flow in the supply pipe is in the fully rough zone on the Moody diagram. Calculate the supply pressure, p_1, required to supply this volume flow rate.

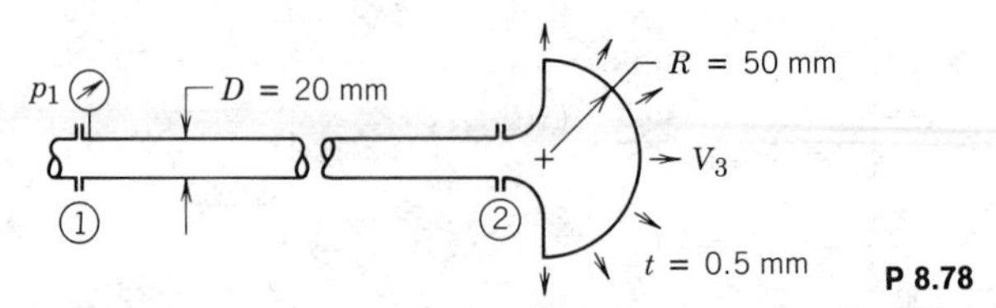

P 8.78

8.79 Gasoline flows in a long, underground pipeline at a constant temperature of 15 C. Two pumping stations at the same elevation are located 13 km apart. The pressure drop between the stations is 1.4 MPa. The pipeline is made from 0.6 m diameter pipe. Although made from commercial steel, age and corrosion have raised the pipe roughness to approximately that for galvanized iron. The specific gravity of the gasoline is 0.68. Compute the volume rate of flow of gasoline through the pipe.

8.80 Water flows steadily in a horizontal 5 in. diameter cast-iron pipe. The pipe is 500 ft long and the pressure drop between sections ① and ② is 23 psi. Find the volume flow rate of water through the pipe.

8.81 Water flows from the tank shown. Assume the flow is quasi-steady. Determine the flow rate at the instant shown. Estimate the temperature rise for the water. How could you improve the flow system if a larger flow rate were desired?

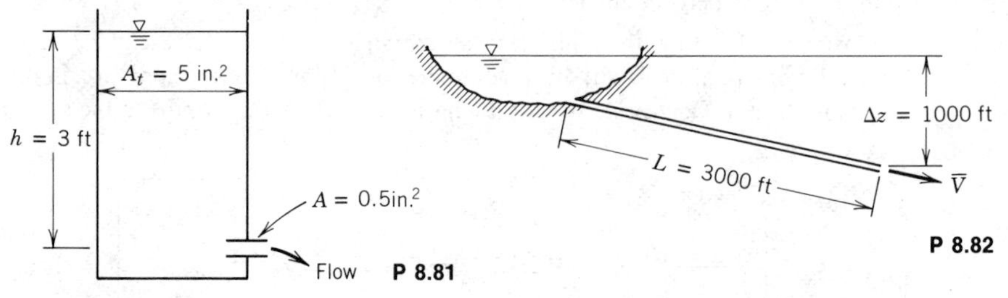

P 8.81

P 8.82

8.82 A hydraulic turbine is to be supplied with water from a mountain stream through a supply pipe as shown. The pipe diameter is 1 ft and the average roughness height is 0.05 in. Minor losses can be neglected. Flow leaves the pipe at atmospheric pressure. Calculate the discharge velocity.

8.83 A fire nozzle is supplied through 300 ft of 1.5 in. diameter, smooth, rubber-lined hose. Water from a hydrant is supplied to a booster pump on board the pumper truck at 50 psig. At design conditions, the pressure at the nozzle inlet is 100 psig, and the pressure drop along the hose is 33 psi per 100 ft of length. Determine (a) the design flow rate, (b) the nozzle exit velocity, assuming no losses in the nozzle, and (c) the power required to drive the booster pump, if its efficiency is 70 percent.

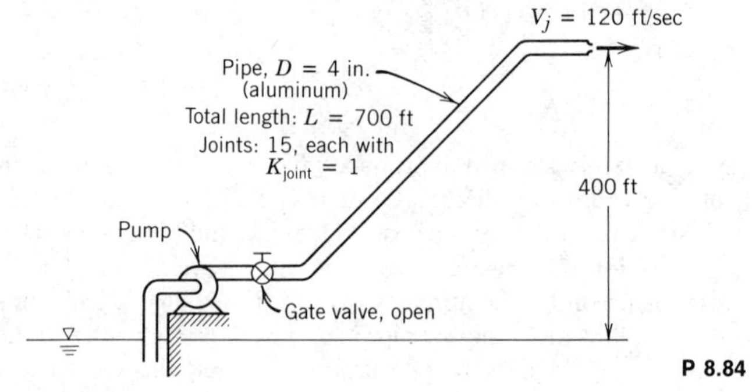

P 8.84

8.84 Cooling water is pumped from a reservoir to rock drills on a construction job using the pipe system shown. The flow rate must be 600 gallons per minute and water must leave the spray nozzle at 120 ft/sec. Calculate the minimum supply pressure needed at the inlet section. Estimate the required power input if the pump efficiency is 70 percent.

8.85 Water is pumped at the rate of 2 ft^3/sec from a reservoir 20 ft above a pump to a free discharge 90 ft above the pump. The pressure on the intake side of the pump is 5 psig and the pressure on the discharge side is 50 psig. All pipes are commercial steel and 6 in. in diameter. Determine (a) the head supplied by the pump, and (b) the total head loss between the pump and point of free discharge.

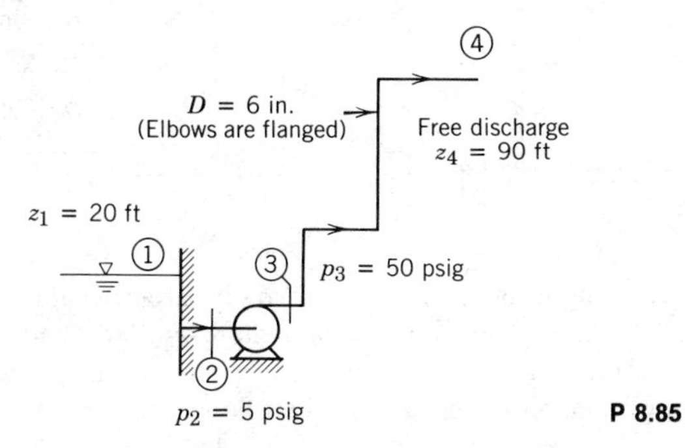

P 8.85

8.86 Air conditioning for the Purdue University campus is provided by chilled water pumped through a main supply pipe. The pipe makes a loop 3 miles in length. The pipe diameter is 2 ft and the material is steel. The maximum design volume flow rate is 11,200 gallons per minute. The circulating pump is driven by an electric motor. The efficiencies of pump and motor are $\eta_p = 0.80$ and $\eta_m = 0.90$, respectively. Electricity cost is \$0.067/kW · hr. Determine (a) the pressure drop, (b) the minimum required pumping power, and (c) the annual cost of electrical energy for pumping.

8.87 Heavy crude oil (SG = 0.925 and $\nu = 1.1 \times 10^{-3}$ ft^2/sec) is pumped through a pipeline laid on flat ground. The line is made from steel pipe with 24 in. i.d. and has a wall thickness of $\frac{1}{2}$ in. The allowable tensile stress in the pipe wall is limited to 40,000 psi by corrosion considerations. At the same time it is important to keep the oil under pressure to ensure that gases remain in solution. The minimum recommended pressure is 75 psia. The pipeline carries a flow of 400,000 barrels (in the petroleum industry, a "barrel" is 42 gal) per day. Determine the maximum spacing between pumping stations. Compute the power added to the oil at each pumping station.

8.88 Crude oil flows through a level section of the Alaskan pipeline at a rate of 1.6 million barrels per day (1 barrel = 42 gal). The pipe inside diameter is 48 in.; its roughness is equivalent to that of galvanized iron. The maximum allowable pressure in the pipe is 1200 psi, and the minimum pressure required to keep dissolved gases in solution in the crude oil is 50 psi. The crude oil has SG = 0.93. Its viscosity at the pumping temperature of 140 F is $\mu = 3.5 \times 10^{-4}$ lbf · sec/ft^2. For these conditions, determine the maximum possible spacing between pumping stations. If the pump efficiency is 85 percent, determine the power that must be supplied at each pumping station.

8.89 The Alaskan pipeline runs from Prudhoe Bay to Valdez, a total distance of 798 mi. Both terminal points are at sea level. The pipe is 48 in. i.d. commercial steel. The capacity of the line is 2 million barrels of crude oil per day (1 barrel of oil = 42 gal). The specific gravity of the oil is 0.93, and its viscosity at the pumping temperature of 140 F is

$\mu = 3.5 \times 10^{-4}$ lbf · sec/ft^2. Determine the total pumping power required if the pump efficiency is 85 percent. Express this result as a fraction of the chemical energy conveyed by the oil stream. (Assume that crude oil has a heating value of 18,000 Btu/lbm.)

8.90 For flow through the siphon shown, assume that the only head loss occurs at the tube inlet, which may be considered a reentrant entrance. The tube i.d. is 75 mm. The fluid is water at 15 C. Determine the volume flow rate through the siphon. Compute the pressure at point *A*.

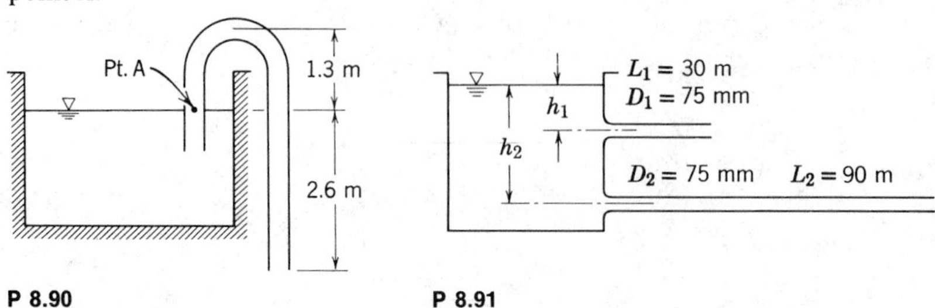

P 8.90 **P 8.91**

8.91 A tank has two lengths of 75 mm galvanized pipe attached at the levels shown. Flow may be assumed to be in the fully rough region, and the entrances are well-rounded. Determine the ratio of h_2 to h_1 that will cause the same flow rate in each pipe. Compute the minimum value of h_1 that will produce flow in the fully rough region.

8.92 Two reservoirs containing water are connected by a constant-area, galvanized iron pipe that has one right-angle bend. The surface pressure at the upper reservoir is atmospheric whereas the gage pressure at the lower reservoir surface is 70 kPa. The pipe diameter is 75 mm. Assume that the only significant losses occur in the pipe and bend. Determine the magnitude and direction of the volume flow rate.

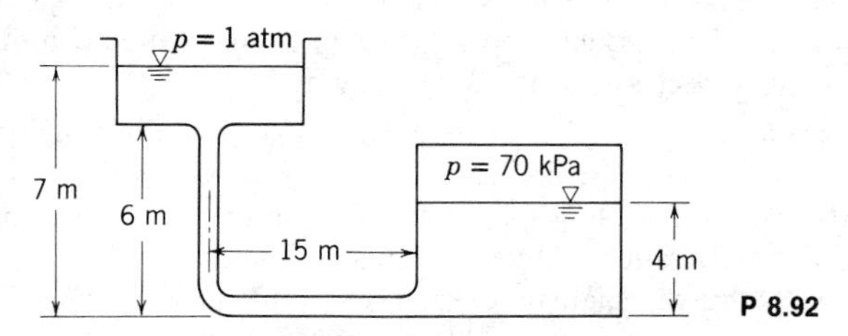

P 8.92

8.93 You are watering your lawn with an *old* hose. Because of the buildup of lime deposits over the years, the 0.75 in. i.d. hose now has an average roughness height of 0.022 in. One 50 ft length of the hose, attached to your spigot, delivers 20 gal of water (60 F) per min. Compute the pressure at the spigot, in psi. Assuming the pressure at the spigot remains constant, estimate the delivery if two 50 ft lengths of the hose are connected together.

8.94 A circular tank 1.5 m in diameter is filled to a depth of 5 m with water at 15 C. A smooth tube 7 m long is attached to a well-rounded entrance at the tank bottom. The 50 mm diameter tube discharges to atmosphere 5 m below the tank bottom. Estimate the time required for the tank level to drop 1.5 m. (You may wish to use the computer to solve this problem. If so, the Blasius correlation for friction factor for turbulent flow,

$$f = \frac{0.316}{Re^{1/4}}$$

might be useful.)

8.95 Consider again the Roman water supply discussed in Example Problem 8.10. Assume that the 50 ft length of horizontal constant-diameter pipe required by law has been installed. The relative roughness of the pipe is 0.01. Estimate the flow rate of water delivered by the pipe under the inlet conditions of the example. What would be the effect of adding a diffuser to the end of the 50 ft pipe?

8.96 At a section 4 ft downstream from a duct inlet, the fully developed turbulent flow of air discharges into a large plenum chamber. The minor loss coefficient for the duct entrance is $K_e = 0.20$. The duct wall is galvanized iron. At exit the kinetic energy flux coefficient is 1.05. Determine the gage pressure that must be maintained in the plenum to produce a mean flow speed of 70 ft/sec in the duct. Sketch the pressure distribution along the duct, from p_{atm} far upstream, to the plenum.

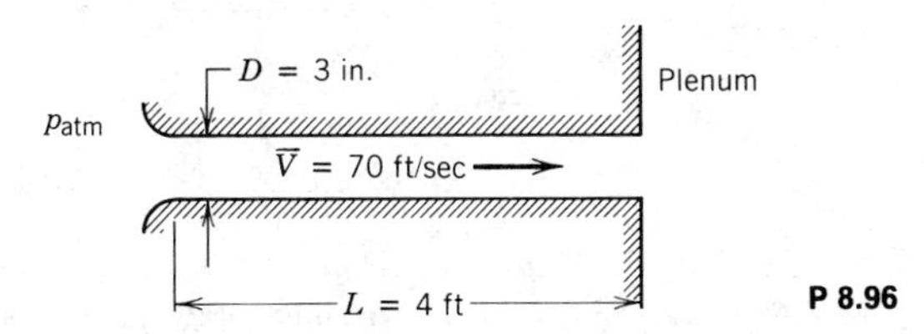

P 8.96

8.97 A hydraulic press is powered by a remote high-pressure pump. The gage pressure at the pump outlet is 20 MPa, whereas the pressure required for the press is 19 MPa (gage), at a flow rate of 0.032 m^3/min. The press and pump are to be connected by 50 m of smooth, drawn steel tubing. The fluid is SAE 10 W oil at 40 C. Determine the minimum tubing diameter that may be used.

8.98 A pump is located 15 ft to one side and 12 ft above a reservoir. The pump is designed for a flow rate of 100 gpm. For satisfactory operation, the suction head at the pump inlet must not be lower than −20 ft of water. Determine the smallest standard commercial steel pipe that will give the required performance.

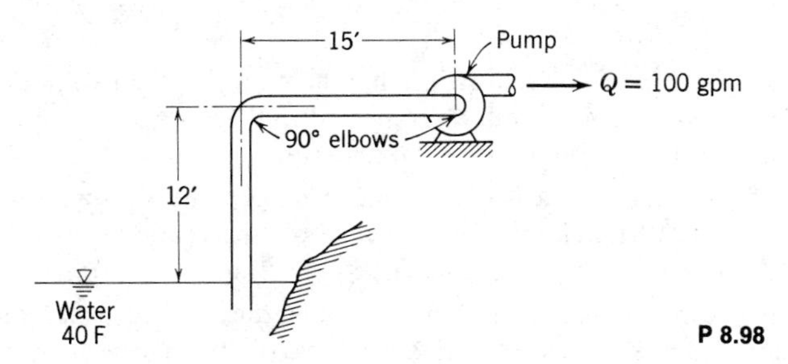

P 8.98

8.99 The jet pump of Problem 4.84 is connected to a 50 m length of horizontal galvanized pipe ($D = 75$ mm). The secondary-stream flow rate depends on the total flow delivered. Evaluate the pressures at sections ① and ② and the total flow rate delivered by the system.

8.100 A new industrial plant requires a water flow rate of 5.7 m^3/min. The gage pressure in the water main, located in the street 50 m from the plant, is 800 kPa. The supply line will require installation of 4 elbows in a total length of 65 m. The gage pressure required in the plant is 500 kPa. What size galvanized iron line should be installed?

8.101 The total cost of a pipeline for petroleum products consists of right-of-way acquisition, construction, and operating costs. Construction cost is roughly proportional to the amount of material excavated—to D^2—for a system of given length. Operating cost may be expressed per unit mass of product delivered. Show that operating cost per unit mass, if flow is in the fully rough zone, is proportional to $1/D^5$. Explain how an optimum pipeline size may be calculated.

8.102 A swimming pool has a partial-flow filtration system. Water at 75 F is pumped from the pool through the system shown. The pump delivers 30 gpm at 60 psig. The pipe is galvanized iron ($e/D \simeq 0.009$). The pressure loss through the filter is approximately $\Delta p = 0.6Q^2$, where Δp is in psi and Q is in gallons per minute. Determine the flow rate through each branch of the system.

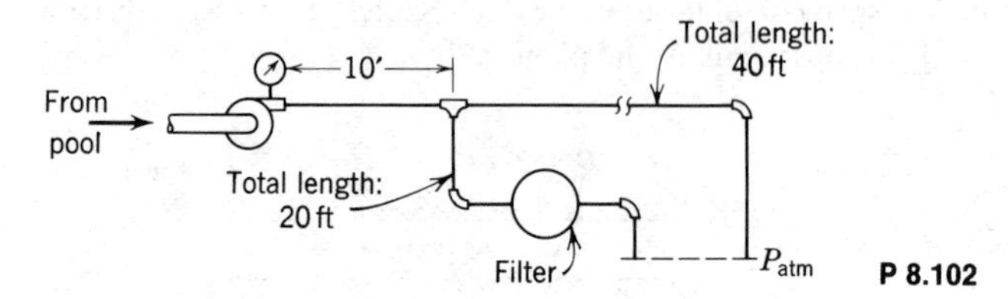

P 8.102

8.103 The results of Example Problem 8.11 indicated that the pipes used were too small to produce the same pressure at each nozzle inlet. Rework the problem using the same geometry but with $3\frac{1}{2}$ in. smooth pipes.

8.104 A spray system for a sewage treatment plant is shown. Water at 60 F is pumped through a spray arm. The effective flow area of each nozzle is 0.25 in.^{2} The pipe inside diameter is 1 in., and the material is galvanized iron. Determine the flow rate of water through the spray arm.

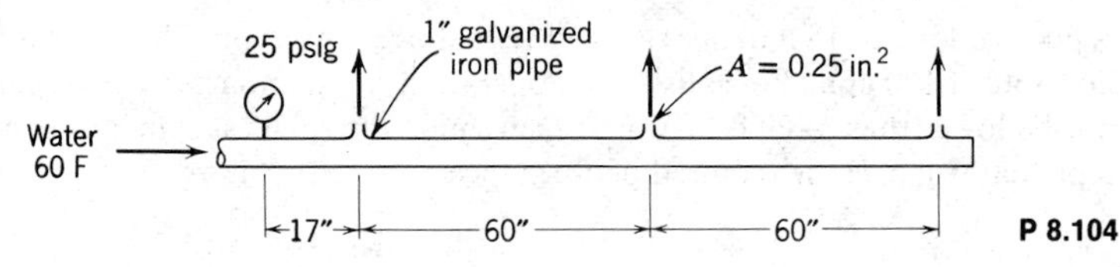

P 8.104

8.105 In a certain air-conditioning installation, a flow rate of 35 m^3/min of air at standard conditions is required. A smooth sheet metal duct 0.3 m square is to be used. Determine the pressure drop for a 30 m horizontal duct run.

8.106 Consider flow of standard air at 35 m^3/min. Compare the pressure drop per unit length of a round duct with that for rectangular ducts of aspect ratio 1, 2, and 3. Assume that all ducts are smooth, with cross-sectional areas of 0.1 m^2.

8.107 Determine the minimum size smooth rectangular duct with an aspect ratio of 2 that will pass 80 m^3/min of standard air with a head loss of 30 mm of water per 30 m of duct.

8.108 The head versus capacity curve for a certain fan may be approximated by the equation

$$h = 30 - 10^{-7}(Q^2)$$

where h is the output static head in inches of water and Q is the air flow rate in ft^3/min. The fan outlet dimensions are 8 × 16 in. Determine the air flow rate delivered by the fan into a 200 ft straight length of 8 × 16 in. rectangular duct.

8.109 The mass flow rate in a water flow system determined by collecting the discharge over a timed interval is 0.3 kg/sec. The scales used can be read to the nearest 0.05 kg, and the stopwatch is accurate to 0.2 sec. Estimate the precision with which the flow rate can be calculated for time intervals of (a) 10 sec, and (b) 1 min.

8.110 An "ergometer" is used to measure the rate at which air is consumed by a research subject running on a treadmill. The atmospheric temperature and pressure are 23 C and 752 mm of mercury, respectively. An inverted bell, which is used to measure air flow rate, is 0.45 m in diameter. During a 30 sec test run, the bell rises 43 mm. Determine the rate at which the subject consumes oxygen.

8.111 Water at 150 F flows through a 3 in. diameter orifice installed in a 6 in. i.d. pipe. The flow rate is 300 gpm. Determine the pressure difference between the corner taps.

8.112 Kerosine at 40 C flows through a 0.3 m diameter line in a refinery. The flow rate is not expected to exceed 120 kg/sec. A manometer with a range of 1 m of water is available for use with an orifice meter. Specify a recommended orifice diameter for use with this system. What minimum rate of flow could be measured within 10 percent accuracy if the manometer least count is 1 mm of water?

8.113 Air flow rate in a test of an internal combustion engine is to be measured using a flow nozzle installed in a plenum. The engine displacement is 1.6 liters and its maximum operating speed is 6000 rpm. The maximum pressure drop across the nozzle should not exceed 0.25 m of water, to avoid loading the engine. The manometer can be read to ± 0.5 mm of water. Determine the flow nozzle diameter that should be specified. Find the minimum rate of air flow that can be metered to ± 2 percent using this setup.

8.114 Consider a flow nozzle installation in a pipe. Apply the basic equations to the control volume indicated, to show that the head loss across the meter can be expressed in dimensionless form as the head loss coefficient,

$$C_l = \frac{p_1 - p_3}{p_1 - p_2} = \frac{1 - A_2/A_1}{1 + A_2/A_1}$$

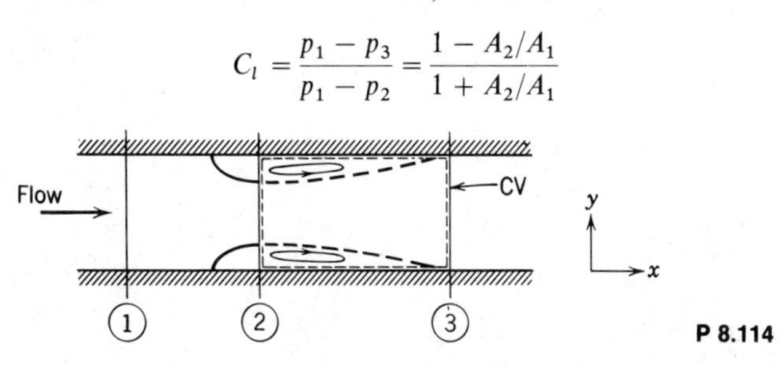

P 8.114

8.115 A venturi meter with a 75 mm diameter throat is placed in a 150 mm diameter line carrying water at 25 C. The pressure drop between the upstream tap and the venturi throat is 300 mm of mercury. Compute the rate of flow.

8.116 Air flows through the venturi meter described in Problem 8.115. Assume that the upstream pressure is 400 kPa, and that the temperature is everywhere constant at 20 C. Determine the maximum possible mass flow rate of air for which the assumption of incompressible flow is a valid engineering approximation. Compute the corresponding pressure reading on a mercury manometer.

8.117 Oil (SG = 0.88) flows through a 70 × 30 mm venturi meter. The differential pressure is 410 mm of mercury. Find the volume flow rate of oil.

8.118 Gasoline (SG = 0.80) flows through a 2 × 1 in. venturi meter. The differential pressure is 380 mm of mercury. Find the volume flow rate of gasoline.

8.119 Consider a horizontal 2 × 1 in. venturi with water flow. For a differential pressure reading of 5 psi, calculate the volume flow rate of water.

8.120 Drinking straws are to be used to improve the air flow in a pipe-flow experiment. Packing a section of the air pipe with drinking straws to form a "laminar flow element" might allow the air flow rate to be measured directly, and simultaneously would act as a flow straightener. To evaluate this idea, determine (a) the Reynolds number for flow in each drinking straw, (b) the friction factor for flow in each straw, and (c) the gage pressure at the exit from the drinking straws. (For laminar flow in a tube, the entrance loss coefficient is $K_{ent} \approx 1.4$ and $\alpha = 2.0$.) Comment on the utility of this idea.

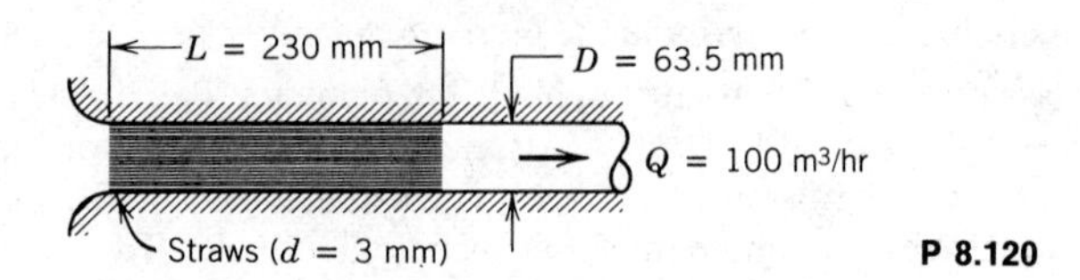

P 8.120

Chapter 9

EXTERNAL INCOMPRESSIBLE VISCOUS FLOW

External flows are flows over bodies immersed in an unbounded fluid. The flow over a semi-infinite flat plate (Fig. 2.11) and the flow over a cylinder (Fig. 2.12*a*) are examples of external flows. These were discussed qualitatively in Chapter 2. Our objective in this chapter is to quantify the behavior of viscous, incompressible fluids in external flow.

A number of phenomena that occur in external flow over a body are illustrated in the sketch of viscous flow over an airfoil (Fig. 9.1). The freestream flow divides at the stagnation point and flows around the body. Fluid at the surface takes on the velocity of the body as a result of the no-slip condition. Boundary layers form on both the upper and lower surfaces of the body. (The boundary-layer thickness on both surfaces in Fig. 9.1 is exaggerated greatly for clarity.) Flow in the boundary layer initially is laminar. Transition to turbulent flow occurs at some distance from the stagnation point, depending on freestream conditions, surface roughness, and pressure gradient. The transition points are indicated by "T" in the figure. The turbulent boundary layer following transition grows more rapidly than the laminar layer. A slight displacement of the streamlines of the external flow is caused by the thickening boundary layers on the surface. In a region of increasing pressure (an "adverse pressure gradient") flow

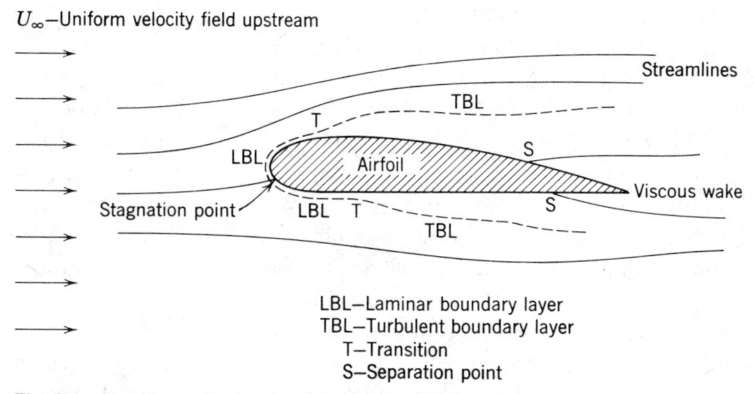

Fig. 9.1 Details of viscous flow around an airfoil.

separation may occur. Separation points are indicated by "S" in the figure. Fluid that was in the boundary layers on the body surface forms the viscous *wake* behind the separation points.

Part A of this chapter is devoted to boundary-layer flows. Following a discussion of the boundary-layer concept, the exact solution for laminar boundary-layer flow over a flat plate (zero pressure gradient) is presented. Since exact solutions for turbulent boundary layers do not exist, approximate solutions must be used. The momentum integral equation for nonzero pressure gradient is derived from first principles as the basis for the approximate solutions; approximate solutions for both laminar and turbulent flow over flat plates are considered. Although approximate solutions for boundary-layer flows with pressure gradients are beyond the scope of this book, the effect of pressure gradients on boundary-layer flows is discussed.

The airfoil of Fig. 9.1 experiences a net force as a result of the shear and pressure forces acting on its surfaces. The component of the net force parallel to the uniform upstream flow, U_∞, is called the drag force; the component of the net force perpendicular to U_∞ is called the lift. The presence of flow separation precludes the analytical determination of lift and drag. In Part B of this chapter approximate analyses and correlations of experimental data are presented for the determination of drag and lift on a number of bodies of interest.

PART A BOUNDARY LAYERS

9-1 THE BOUNDARY-LAYER CONCEPT

The concept of a boundary layer was first introduced by Ludwig Prandtl [1], a German aerodynamicist, in 1904.

Prior to Prandtl's historic breakthrough, the science of fluid mechanics had been developing in two rather different directions. Theoretical hydrodynamics evolved from Euler's equation (Eq. 6.2, published by Leonhard Euler in 1755) of motion for a nonviscous fluid. Since the results of hydrodynamics contradicted many experimental observations, practicing engineers developed their own empirical art of hydraulics. This was based on experimental data and differed significantly from the purely mathematical approach of theoretical hydrodynamics.

Although the complete equations describing the motion of a viscous fluid (the Navier–Stokes equations, Eqs. 5.24, developed by Navier, 1827, and independently by Stokes, 1845) were known prior to Prandtl, the mathematical difficulties in solving these equations (except for a few simple cases) prohibited a theoretical treatment of viscous flows. Prandtl showed that many viscous flows can be analyzed by dividing the flow into two regions, one close to solid boundaries, the other covering the rest of the flow. Only in the thin region adjacent to a solid boundary (the boundary layer) is the effect of viscosity important. In the region outside of the boundary layer, the effect of viscosity is negligible and the fluid may be treated as inviscid.

The boundary-layer concept provided the link that had been missing between theory and practice. Furthermore, the boundary-layer concept permitted the solution of viscous flow problems that would have been impossible through application of the Navier–Stokes equations to the complete flow field.[1] Thus the introduction of the boundary-layer concept marked the beginning of the modern era of fluid mechanics.

The development of a boundary layer on a solid surface was discussed in Section 2-5.1. The development of a laminar boundary layer on a flat plate was illustrated in Fig. 2.11. In the boundary layer both viscous and inertia forces are important. Consequently, it is not surprising that the Reynolds number (which represents the ratio of inertia to viscous forces) is significant in characterizing boundary-layer flows. The characteristic length used in the Reynolds number is either the length in the flow direction over which the boundary layer has developed or some measure of the boundary-layer thickness.

As for flow in a duct, flow in a boundary layer may be laminar or turbulent. There is no unique value of the Reynolds number at which transition from laminar to turbulent flow occurs in a boundary layer. Among the factors that affect boundary-layer transition are pressure gradient, surface roughness, heat transfer, body forces, and freestream disturbances. Detailed consideration of these effects is beyond the scope of this book.

In many real flow situations, a boundary layer develops over a long, essentially flat surface. Examples include flow over ship and submarine hulls, aircraft wings, and atmospheric motions over flat terrain. Since the basic features of all these flows are illustrated in the simpler case of flow over a flat plate, let us consider this first.

For incompressible flow over a smooth flat plate (zero pressure gradient) in the absence of heat transfer, transition from laminar to turbulent flow in the boundary layer can be delayed to a Reynolds number, $Re_x = \rho U x/\mu$, between 3 and 4 million if external disturbances are minimized. For calculation purposes, under typical flow conditions, transition usually is considered to occur at a length Reynolds number of 5×10^5. For air at standard conditions, with a freestream velocity, $U = 30$ m/sec, this corresponds to a length, x, along the plate of $x \approx 0.24$ m.

A qualitative picture of the boundary-layer growth over a flat plate is shown in Fig. 9.2. The boundary layer is laminar for a short distance downstream from the

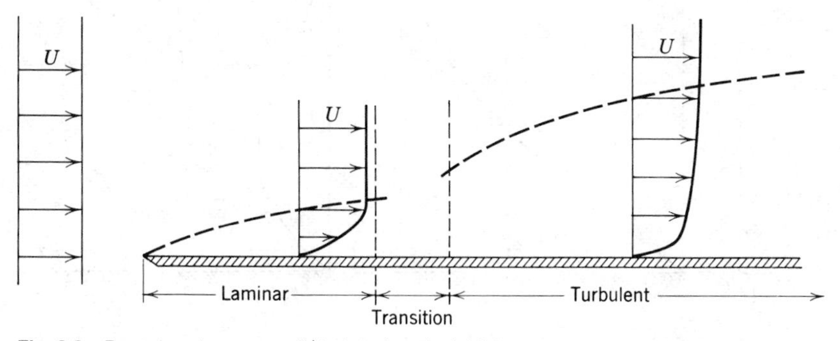

Fig. 9.2 Boundary layer on a flat plate (vertical thickness exaggerated greatly).

[1] Today, computer solutions of the Navier–Stokes equations are common.

leading edge; transition occurs over a region of the plate rather than at a single line across the plate. The transition region extends downstream to the location where the boundary-layer flow becomes completely turbulent. In the qualitative picture of Fig. 9.2, we have shown the turbulent boundary layer growing at a faster rate than the laminar layer. In later sections of this chapter we shall show that this is indeed true.

9-2 BOUNDARY-LAYER THICKNESSES

The boundary layer is the region adjacent to a solid surface in which viscous forces are important. The boundary-layer *disturbance thickness*, δ, usually is defined as the distance from the surface to the point where the velocity is within 1 percent of the freestream velocity. Since the velocity profile merges smoothly and asymptotically into the freestream, the boundary-layer thickness, δ, is difficult to measure.

The effect of viscous forces in the boundary layer is to retard the flow. The mass flow rate adjacent to a solid surface is less than the mass flow rate that would pass through the same region in the absence of a boundary layer. The decrease in flow rate due to the influence of viscous forces is $\int_0^\infty \rho(U - u)\,dy$. If viscous forces were absent, the velocity at a section would be U. Displacing the boundary by a distance, δ^*, would result in a mass flow deficiency of $\rho U \delta^*$. The *displacement thickness*, δ^*, is the distance by which the solid boundary would have to be displaced in a frictionless flow to give the same mass deficit as exists in the boundary layer. Thus, as illustrated in Fig. 9.3*a*,

$$\rho U \delta^* = \int_0^\infty \rho(U - u)\,dy$$

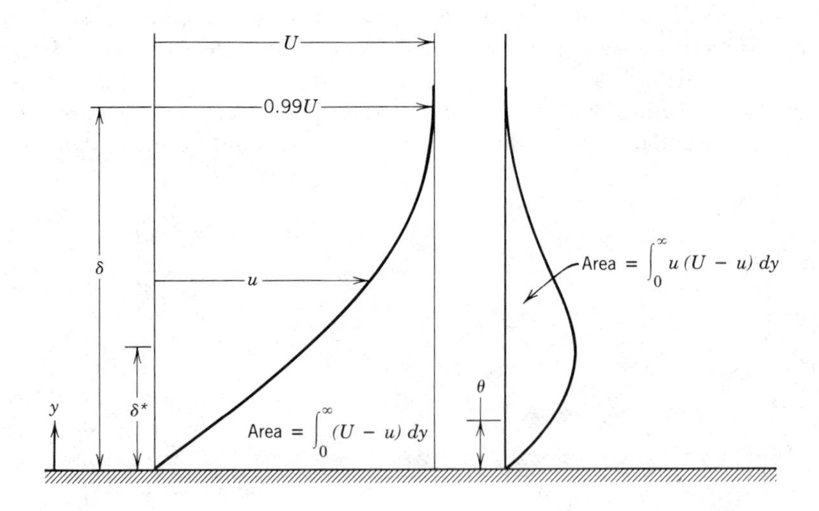

(a) $U\delta^* = \int_0^\infty (U - u)\,dy$ (b) $U^2\theta = \int_0^\infty u(U - u)\,dy$

Fig. 9.3 Boundary-layer thickness definitions.

ρ = constant

For incompressible flow, ρ = constant and

$$\delta^* = \int_0^\infty \left(1 - \frac{u}{U}\right) dy \approx \int_0^\delta \left(1 - \frac{u}{U}\right) dy \tag{9.1}$$

since $u \approx U$ and the integrand is essentially zero for $y \geq \delta$. Application of the displacement-thickness concept is illustrated in Example Problem 9.1.

Flow retardation within the boundary layer also results in a reduction in momentum flux at a section compared to inviscid flow. The momentum deficiency of the actual mass flow, $\int_0^\infty \rho u \, dy$, through the boundary layer is $\int_0^\infty \rho u(U - u) \, dy$. If viscous forces were absent, it would be necessary to move the solid boundary outward to obtain a momentum deficiency; denoting this distance (the momentum thickness) as θ, the momentum deficiency would be $\rho U^2 \theta$. The *momentum thickness*, θ, is defined as the thickness of a layer of fluid of velocity, U, for which the momentum flux is equal to the deficit of momentum flux through the boundary layer. Thus, as illustrated in Fig. 9.3*b*,

$$\rho U^2 \theta = \int_0^\infty \rho u(U - u) \, dy$$

For incompressible flow, ρ = constant and

$$\theta = \int_0^\infty \frac{u}{U}\left(1 - \frac{u}{U}\right) dy \approx \int_0^\delta \frac{u}{U}\left(1 - \frac{u}{U}\right) dy \tag{9.2}$$

Again, the integrand is essentially zero for $y \geq \delta$.

The displacement and momentum thicknesses, δ^* and θ, are termed *integral* thicknesses. Their definitions, Eqs. 9.1 and 9.2, are in terms of integrals across the boundary layer. Because they are defined in terms of integrals for which the integrand vanishes in the freestream, they are appreciably easier to evaluate accurately from experimental data than the boundary-layer disturbance thickness, δ. This fact, coupled with their physical significance, accounts for their common use in specifying boundary-layer thickness.

Example 9.1

A laboratory wind tunnel has a test section that is 305 mm square. Boundary-layer velocity profiles are measured at two cross sections and displacement thicknesses are evaluated from the measured profiles. At section ①, where the freestream speed is $U_1 = 26$ m/sec, the displacement thickness is $\delta_1^* = 1.5$ mm. At section ②, located downstream from section ①, $\delta_2^* = 2.1$ mm. Calculate the change in static pressure between sections ① and ②. Express the result as a fraction of the freestream dynamic pressure at section ①. Assume standard atmosphere conditions.

EXAMPLE PROBLEM 9.1

GIVEN: Flow of standard air in laboratory wind tunnel. Test section is $L = 305$ mm square. Displacement thicknesses are $\delta_1^* = 1.5$ mm and $\delta_2^* = 2.1$ mm. Freestream speed is $U_1 = 26$ m/sec.

FIND: Change in static pressure between sections ① and ②. (Express as a fraction of freestream dynamic pressure at section ①.)

SOLUTION:
Use the displacement thickness concept to find the effective flow area for the freestream flow outside the thin wall boundary layers. Replace the actual boundary-layer velocity profiles with uniform velocity profiles as sketched in the following figures.

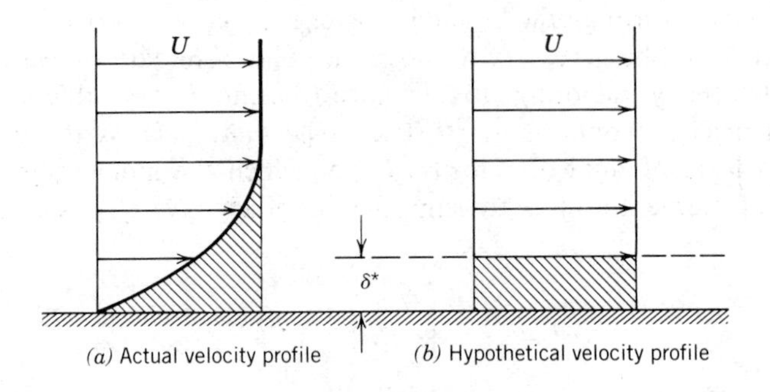

(a) Actual velocity profile (b) Hypothetical velocity profile

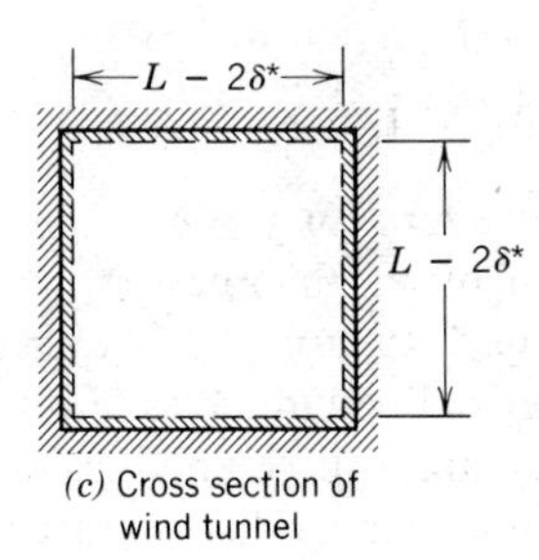

(c) Cross section of wind tunnel

Apply the continuity and Bernoulli equations to freestream flow outside the boundary-layer displacement thickness, where viscous effects are negligible.

Basic equations:

$$0 = \cancel{\frac{\partial}{\partial t}\int_{CV} \rho \, d\forall}^{\,=\,0(1)} + \int_{CS} \rho \vec{V} \cdot d\vec{A}$$

$$\frac{p_1}{\rho} + \frac{V_1^2}{2} + g\cancel{z_1} = \frac{p_2}{\rho} + \frac{V_2^2}{2} + g\cancel{z_2}$$

Assumptions: (1) Steady flow
(2) Incompressible flow
(3) Flow uniform at each section outside δ^*
(4) Flow along a streamline from sections ① to ②
(5) No frictional effects in freestream
(6) Neglect elevation changes

From the Bernoulli equation we obtain

$$p_1 - p_2 = \frac{1}{2}\rho(V_2^2 - V_1^2) = \frac{1}{2}\rho(U_2^2 - U_1^2) = \frac{1}{2}\rho U_1^2\left[\left(\frac{U_2}{U_1}\right)^2 - 1\right]$$

or

$$\frac{p_1 - p_2}{\frac{1}{2}\rho U_1^2} = \left(\frac{U_2}{U_1}\right)^2 - 1$$

From continuity, $V_1 A_1 = U_1 A_1 = V_2 A_2 = U_2 A_2$, so $\dfrac{U_2}{U_1} = \dfrac{A_1}{A_2}$, where $A = (L - 2\delta^*)^2$ is the effective flow area. Substituting gives

$$\frac{p_1 - p_2}{\frac{1}{2}\rho U_1^2} = \left(\frac{A_1}{A_2}\right)^2 - 1 = \left[\frac{(L - 2\delta_1^*)^2}{(L - 2\delta_2^*)^2}\right]^2 - 1$$

$$\frac{p_1 - p_2}{\frac{1}{2}\rho U_1^2} = \left[\frac{305 - 2(1.5)}{305 - 2(2.1)}\right]^4 - 1 = 0.0161 \quad \text{or} \quad 1.61 \text{ percent}$$

$$\frac{p_1 - p_2}{\frac{1}{2}\rho U_1^2}$$

This problem illustrates the application of the displacement-thickness concept. The real, viscous boundary-layer flow is modeled as a uniform, inviscid flow displaced from the boundary through distance, δ^*.

**9-3 LAMINAR FLAT-PLATE BOUNDARY LAYER: EXACT SOLUTION

The solution for the laminar boundary layer on a flat plate was obtained by H. Blasius [2], one of Prandtl's students, in 1908. For two-dimensional, steady flow with zero pressure gradient, the governing equations of motion reduce to [3]

$$\frac{\partial u}{\partial x} + \frac{\partial v}{\partial y} = 0 \tag{9.3}$$

$$u\frac{\partial u}{\partial x} + v\frac{\partial u}{\partial y} = \nu\frac{\partial^2 u}{\partial y^2} \tag{9.4}$$

with boundary conditions

$$\text{at} \quad y = 0, \qquad u = 0$$

$$\text{at} \quad y = \infty, \qquad u = U \tag{9.5}$$

Blasius reasoned that the velocity profile, u/U, should be similar, for all values of x, when plotted versus a nondimensional distance from the wall; the boundary-layer thickness, δ, was a natural choice for nondimensionalizing the distance from the wall. Thus the solution is of the form

$$\frac{u}{U} = g(\eta) \qquad \text{where} \quad \eta = \frac{y}{\delta} \tag{9.6}$$

** This section may be omitted without loss of continuity in the text material.

Based on the solution of Stokes [4], Blasius reasoned that $\delta \sim \sqrt{\nu x/U}$ and set

$$\eta = y\sqrt{\frac{U}{\nu x}} \tag{9.7}$$

Introducing the stream function, ψ, where

$$u = \frac{\partial \psi}{\partial y} \quad \text{and} \quad v = -\frac{\partial \psi}{\partial x} \tag{5.4}$$

satisfies the continuity equation (Eq. 9.3) identically; substituting for u and v into Eq. 9.4 reduces the equation to one in which ψ is the single dependent variable. Defining a dimensionless stream function as

$$f(\eta) = \frac{\psi}{\sqrt{\nu x U}} \tag{9.8}$$

makes $f(\eta)$ the dependent variable and η the independent variable in Eq. 9.4. With ψ defined by Eq. 9.8 and η defined by Eq. 9.7 we can evaluate each of the terms in Eq. 9.4.

The velocity components are given by

$$u = \frac{\partial \psi}{\partial y} = \frac{\partial \psi}{\partial \eta}\frac{\partial \eta}{\partial y} = \sqrt{\nu x U}\,\frac{df}{d\eta}\sqrt{\frac{U}{\nu x}} = U\frac{df}{d\eta} \tag{9.9}$$

and

$$v = -\frac{\partial \psi}{\partial x} = -\left[\sqrt{\nu x U}\,\frac{\partial f}{\partial x} + \frac{1}{2}\sqrt{\frac{\nu U}{x}}\,f\right]$$

$$= -\left[\sqrt{\nu x U}\,\frac{df}{d\eta}\left(-\frac{1}{2}\eta\frac{1}{x}\right) + \frac{1}{2}\sqrt{\frac{\nu U}{x}}\,f\right]$$

$$v = \frac{1}{2}\sqrt{\frac{\nu U}{x}}\left[\eta\frac{df}{d\eta} - f\right] \tag{9.10}$$

By differentiating the velocity components, it also may be shown that

$$\frac{\partial u}{\partial x} = -\frac{U}{2x}\eta\frac{d^2 f}{d\eta^2}$$

$$\frac{\partial u}{\partial y} = U\sqrt{U/\nu x}\,\frac{d^2 f}{d\eta^2}$$

and

$$\frac{\partial^2 u}{\partial y^2} = \frac{U^2}{\nu x}\frac{d^3 f}{d\eta^3}$$

Substituting these expressions into Eq. 9.4, we obtain

$$2\frac{d^3 f}{d\eta^3} + f\frac{d^2 f}{d\eta^2} = 0 \tag{9.11}$$

with boundary conditions:

$$\text{at} \quad \eta = 0, \qquad f = \frac{df}{d\eta} = 0$$

$$\text{at} \quad \eta = \infty, \qquad \frac{df}{d\eta} = 1 \tag{9.12}$$

The second-order, partial differential equations governing the growth of the laminar boundary layer on a flat plate (Eqs. 9.3 and 9.4) have been transformed to a nonlinear, third-order ordinary differential equation (Eq. 9.11) with boundary conditions given by Eq. 9.12. It is not possible to solve Eq. 9.11 in closed form. Blasius solved this equation using a series expansion; the same equation later was solved more precisely, using numerical methods, by Howarth [5]. The numerical values of f, $df/d\eta$ and $d^2f/d\eta^2$ given in Table 9.1 are from [5].

The velocity profile is obtained in dimensionless form by plotting u/U versus η, using values obtained from Table 9.1. The resulting profile is plotted in Fig. 9.3*a*. Velocity profiles measured experimentally are in excellent agreement with the analytical solution. Profiles from all locations on a flat plate are *similar*; they collapse to a single profile when plotted in nondimensional coordinates.

Table 9.1 The function $f(\eta)$ for the laminar boundary layer along a flat plate at zero incidence. (After L. Howarth [5].)

$\eta = y\sqrt{\frac{U_\infty}{\nu x}}$	f	$f' = \frac{u}{U_\infty}$	f''
0	0	0	0.33206
0.4	0.02656	0.13277	0.33147
1.0	0.16557	0.32979	0.32301
1.4	0.32298	0.45627	0.30787
2.0	0.65003	0.62977	0.26675
2.4	0.92230	0.72899	0.22809
3.0	1.39682	0.84605	0.16136
3.4	1.74696	0.90177	0.11788
4.0	2.30576	0.95552	0.06424
4.4	2.69238	0.97587	0.03897
5.0	3.28329	0.99155	0.01591
5.4	3.68094	0.99616	0.00793
6.0	4.27964	0.99898	0.00240
6.4	4.67938	0.99961	0.00098
7.0	5.27926	0.99992	0.00022
7.4	5.67924	0.99998	0.00007
8.0	6.27923	1.00000	0.00001
8.4	6.67923	1.00000	0.00000

From Table 9.1 we see that at $\eta = 5.0$, $u/U = 0.992$. Defining the boundary layer thickness, δ, as the value of y for which $u/U = 0.99$, then from Eq. 9.7,

$$\delta \approx \frac{5.0}{\sqrt{U/\nu x}} = \frac{5.0x}{\sqrt{Re_x}} \tag{9.13}$$

The boundary-layer thickness, δ, is indicated on the velocity profile plot of Fig. 9.3*a*.

The wall shear stress may be expressed as

$$\tau_w = \mu \frac{\partial u}{\partial y}\bigg]_{y=0} = \mu U \sqrt{U/\nu x}\, \frac{d^2 f}{d\eta^2}\bigg]_{\eta=0}$$

Then

$$\tau_w = 0.332U\sqrt{\rho\mu U/x} \tag{9.14}$$

and the wall shear stress coefficient, C_f, is given by

$$C_f = \frac{\tau_w}{\frac{1}{2}\rho U^2} = \frac{0.664}{\sqrt{Re_x}} \tag{9.15}$$

Each of the results for boundary-layer thicknesses, δ, wall shear stress, τ_w, and skin friction coefficient, C_f, Eqs. 9.13 through 9.15, depends on the length Reynolds number, Re_x, to the one-half power. The boundary-layer thickness increases as $x^{1/2}$, and the wall shear stress and skin friction coefficient vary as $1/x^{1/2}$. These results characterize the behavior of the laminar boundary layer on a flat plate.

Example 9.2

Use the numerical results of Howarth (presented in Table 9.1) to evaluate the following quantities for laminar boundary-layer flow on a flat plate:

(a) δ^*/δ (evaluate for $\eta = 5$ and as $\eta \to \infty$).

(b) v/U at the boundary-layer edge.

(c) Compare the slope of a streamline at the boundary-layer edge with the slope of δ versus x.

EXAMPLE PROBLEM 9.2

GIVEN: Numerical solution for laminar flat-plate boundary layer, Table 9.1.

FIND: (a) δ^*/δ (evaluate for $\eta = 5$ and as $\eta \to \infty$).
(b) v/U at boundary-layer edge.
(c) Compare the slope of a streamline at the boundary-layer edge with the slope of δ versus x.

SOLUTION:

The displacement thickness is defined by Eq. 9.1 as

$$\delta^* = \int_0^\delta \left(1 - \frac{u}{U}\right) dy$$

From Eq. 9.7, $\eta = y\sqrt{\dfrac{U}{\nu x}}$, so $y = \eta\sqrt{\dfrac{\nu x}{U}}$ and $dy = d\eta\sqrt{\dfrac{\nu x}{U}}$

Thus

$$\delta^* = \int_0^{\eta_{\max}} \left(1 - \frac{u}{U}\right)\sqrt{\frac{\nu x}{U}}\,d\eta = \sqrt{\frac{\nu x}{U}}\int_0^{\eta_{\max}}\left(1 - \frac{u}{U}\right)d\eta$$

But from Eq. 9.13

$$\delta \approx \frac{5}{\sqrt{U/\nu x}}, \qquad \text{so} \qquad \sqrt{\frac{\nu x}{U}} = \frac{\delta}{5}$$

Thus

$$\frac{\delta^*}{\delta} = \frac{1}{5}\int_0^{\eta_{\max}}\left(1 - \frac{u}{U}\right)d\eta$$

Substituting from Eq. 9.9, we obtain

$$\frac{\delta^*}{\delta} = \frac{1}{5}\int_0^{\eta_{\max}}\left(1 - \frac{df}{d\eta}\right)d\eta$$

Integrating gives

$$\frac{\delta^*}{\delta} = \frac{1}{5}\Big[\eta - f(\eta)\Big]_0^{\eta_{\max}}$$

Evaluating at $\eta = 5$, we obtain

$$\frac{\delta^*}{\delta} = \frac{1}{5}(5.0 - 3.28329) = 0.34334$$

$\dfrac{\delta^*}{\delta}(\eta = 5)$

The quantity, $\eta - f(\eta)$, becomes constant for $\eta > 8$. Evaluating at $\eta = 8$ gives

$$\frac{\delta^*}{\delta} = \frac{1}{5}(8.0 - 6.27923) = 0.34415$$

$\dfrac{\delta^*}{\delta}(\eta \to \infty)$

Thus $\delta^*_{\eta\to\infty}$ is 0.236 percent larger than $\delta^*_{\eta=5}$.

From Eq. 9.10,

$$v = \frac{1}{2}\sqrt{\frac{\nu U}{x}}\left(\eta\frac{df}{d\eta} - f\right), \qquad \text{so} \qquad \frac{v}{U} = \frac{1}{2}\sqrt{\frac{\nu}{Ux}}\left(\eta\frac{df}{d\eta} - f\right) = \frac{1}{2\sqrt{Re_x}}\left(\eta\frac{df}{d\eta} - f\right)$$

Evaluating at the boundary-layer edge ($\eta = 5$), we obtain

$$\frac{v}{U} = \frac{1}{2\sqrt{Re_x}}[5(0.99155) - 3.28329] = \frac{0.83723}{\sqrt{Re_x}} \approx \frac{0.84}{\sqrt{Re_x}}$$

$\dfrac{v}{U}(\eta = 5)$

Thus v is only 0.84 percent of U at $Re_x = 10^4$, and only about 0.12 percent of U at $Re_x = 5 \times 10^5$.

The slope of a streamline at the boundary-layer edge is

$$\left.\frac{dy}{dx}\right)_{\text{streamline}} = \frac{v}{u} = \frac{v}{U} \approx \frac{0.84}{\sqrt{Re_x}}$$

The slope of the boundary-layer edge may be obtained from Eq. 9.13.

$$\delta \approx \frac{5}{\sqrt{U/\nu x}} = 5\sqrt{\frac{\nu x}{U}}, \quad \text{so} \quad \frac{d\delta}{dx} = 5\sqrt{\frac{\nu}{U}}\frac{1}{2}x^{-1/2} = 2.5\sqrt{\frac{\nu}{Ux}} = \frac{2.5}{\sqrt{Re_x}}$$

$$\text{Thus} \left.\frac{dy}{dx}\right)_{\text{streamline}} = \frac{0.84}{2.5}\frac{d\delta}{dx} = 0.336\frac{d\delta}{dx}$$

This result shows that streamlines penetrate the boundary-layer edge, as sketched:

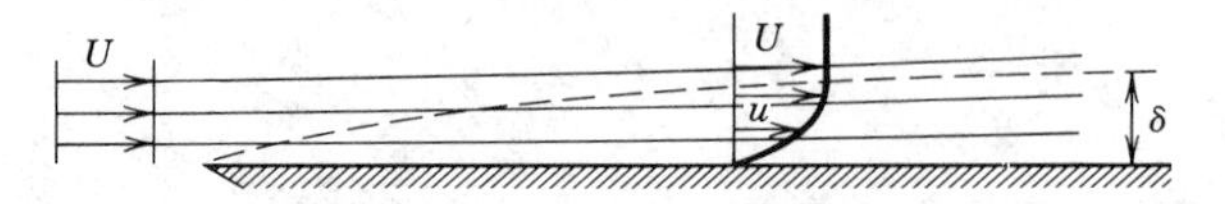

{This problem illustrates application of results from the numerical solution of the laminar boundary layer on a flat plate.}

9-4 MOMENTUM INTEGRAL EQUATION

The exact solution of Blasius provided an expression for the boundary-layer thickness, $\delta(x)$, and the wall shear stress, $\tau_w(x)$. The velocity profiles were found to be similar when plotted nondimensionally as u/U versus y/δ. A closed-form solution for the velocity profile was not possible; a numerical solution was necessary.

Approximate methods may be used to obtain solutions for laminar boundary-layer flow on a flat plate in closed form. The same approximate methods may be used to solve for the characteristics of turbulent boundary-layer development. Since exact solutions for turbulent boundary layers do not exist, approximate solution techniques are necessary in this case. In this section we shall develop an analysis that will enable us to determine a good approximation for the thickness of a laminar or turbulent boundary layer as a function of distance along a body. We shall again apply the integral equations to a differential control volume. Our aim is to develop an equation that will enable us to predict (at least approximately) the manner in which the boundary layer grows as a function of distance along a body. We shall derive a relation that may be applied to both laminar and turbulent flow; the relation will not be restricted to zero pressure gradient flows.

Consider the incompressible, steady flow over a solid surface. The boundary-layer thickness, δ, grows in some manner with increasing distance, x. For our analysis we choose a differential control volume of length, dx, width, dz, and height, $\delta(x)$, as shown in Fig. 9.4.

We wish to determine the boundary-layer thickness, δ, as a function of x. There will be mass flow across surfaces *ab* and *cd* of differential control volume, *abcd*. What about surface *bc*? Will there be a mass flow across this surface? In our earlier discussion of boundary layers (Chapter 2), and in Example Problem 9.2, we found that the edge of the boundary layer is not a streamline. Thus there will be mass flow across surface *bc*. Since control surface *ad* is adjacent to a solid boundary, there will not be flow across

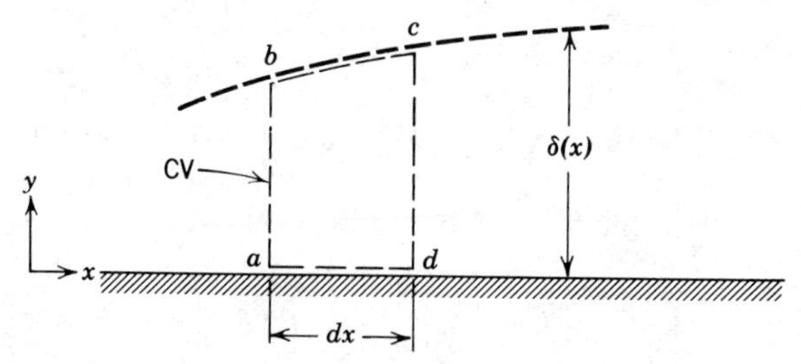

Fig. 9.4 Differential control volume in a boundary layer.

ad. Before considering the forces acting on the control volume and the momentum fluxes through the control surface, let us apply the continuity equation to determine the mass flux through each portion of the control surface.

a. Continuity Equation

Basic equation:

$$0 = \overset{=0(1)}{\cancel{\frac{\partial}{\partial t}}} \int_{\text{CV}} \rho \, d\forall + \int_{\text{CS}} \rho \vec{V} \cdot d\vec{A} \tag{4.13}$$

Assumptions: (1) Steady flow
(2) Two-dimensional flow

Then

$$0 = \int_{\text{CS}} \rho \vec{V} \cdot d\vec{A} = \dot{m}_{ab} + \dot{m}_{bc} + \dot{m}_{cd}$$

or

$$\dot{m}_{bc} = -\dot{m}_{ab} - \dot{m}_{cd}$$

Now let us evaluate these terms:

Surface	*Mass Flux*
ab	Surface *ab* is located at x. Since the flow is two-dimensional (no variation with z), the mass flux through *ab* is $$\dot{m}_{ab} = -\left\{\int_0^\delta \rho u \, dy\right\} dz$$
cd	Surface *cd* is located at $x + dx$. Expanding $\dot{m}$ in a Taylor series about the location, x, we obtain $$\dot{m}_{x+dx} = \dot{m}_x + \left.\frac{\partial \dot{m}}{\partial x}\right]_x dx$$

and hence

$$\dot{m}_{cd} = \left\{ \int_0^\delta \rho u \, dy + \frac{\partial}{\partial x}\left[\int_0^\delta \rho u \, dy\right] dx \right\} dz$$

bc Thus for surface *bc* we obtain

$$\dot{m}_{bc} = -\left\{ \frac{\partial}{\partial x}\left[\int_0^\delta \rho u \, dy\right] dx \right\} dz$$

Now let us consider the momentum fluxes and forces associated with control volume *abcd*. These are related by the momentum equation.

b. Momentum Equation

Apply the x component of the momentum equation to control volume *abcd*:

Basic equation:

$$F_{S_x} + \cancel{F_{B_x}}^{=0(3)} = \cancel{\frac{\partial}{\partial t}}^{=0(1)} \int_{CV} u\rho \, d\forall + \int_{CS} u\rho \vec{V} \cdot d\vec{A} \qquad (4.19a)$$

Assumption: (3) $F_{B_x} = 0$

Then

$$F_{S_x} = \text{mf}_{ab} + \text{mf}_{bc} + \text{mf}_{cd}$$

where mf represents momentum flux.

To apply this equation to differential control volume *abcd*, we must obtain expressions for the momentum flux through the control surface and also the surface forces acting on the control volume. Let us consider the momentum flux first and again consider each segment of the control surface.

Surface *Momentum Flux* (mf)

ab Surface *ab* is located at x. Since the flow is two-dimensional, the x momentum flux through *ab* is

$$\text{mf}_{ab} = -\left\{ \int_0^\delta u\rho u \, dy \right\} dz$$

cd Surface *cd* is located at $x + dx$. Expanding the x momentum flux (mf) in a Taylor series about the location, x, we obtain

$$\text{mf}_{x+dx} = \text{mf}_x + \left.\frac{\partial \text{mf}}{\partial x}\right]_x dx$$

or

$$\text{mf}_{cd} = \left\{ \int_0^\delta u\rho u \, dy + \frac{\partial}{\partial x}\left[\int_0^\delta u\rho u \, dy\right] dx \right\} dz$$

bc Since the mass crossing surface *bc* has velocity U, the momentum flux across *bc* is given by

$$\text{mf}_{bc} = U\dot{m}_{bc}$$

$$\text{mf}_{bc} = -U\left\{\frac{\partial}{\partial x}\left[\int_0^\delta \rho u\,dy\right]dx\right\}dz$$

From the above we can evaluate the net momentum flux through the control surface,

$$\int_{\text{CS}} u\rho\vec{V}\cdot d\vec{A} = -\left\{\int_0^\delta u\rho u\,dy\right\}dz + \left\{\int_0^\delta u\rho u\,dy\right\}dz + \left\{\frac{\partial}{\partial x}\left[\int_0^\delta u\rho u\,dy\right]dx\right\}dz - U\left\{\frac{\partial}{\partial x}\left[\int_0^\delta \rho u\,dy\right]dx\right\}dz$$

Collecting terms, we find that

$$\int_{\text{CS}} u\rho\vec{V}\cdot d\vec{A} = \left\{\frac{\partial}{\partial x}\left[\int_0^\delta u\rho u\,dy\right]dx - U\frac{\partial}{\partial x}\left[\int_0^\delta \rho u\,dy\right]dx\right\}dz$$

Now that we have a suitable expression for the x momentum flux through the control surface, let us consider the surface forces acting on the control volume in the x direction. For convenience the differential control volume has been redrawn in Fig. 9.5. To analyze the x component forces acting on the control volume, we recognize that normal forces act on three surfaces of the control surface. In addition, a shear force acts on surface *ad*. Since the velocity gradient goes to zero at the edge of the boundary layer, no shear force acts along surface *bc*.

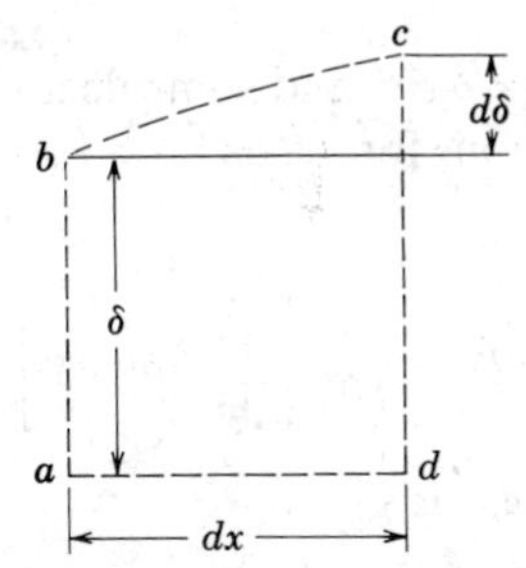

Fig. 9.5 Differential control volume.

Surface	*Force*
ab	If the pressure at x is p, then the force acting on surface *ab* is given by

$$F_{ab} = p\delta\,dz$$

(The boundary layer is very thin; its thickness has been greatly exaggerated in all the sketches we have made. Because it is thin, pressure variations in the y direction may be neglected and within the boundary layer, $p = p(x)$.)

cd Expanding in a Taylor series, the pressure at $x + dx$ is given by

$$p_{x+dx} = p + \left.\frac{dp}{dx}\right]_x dx$$

The force on surface *cd* is then given by

$$F_{cd} = -\left(p + \left.\frac{dp}{dx}\right]_x dx\right)(\delta + d\delta)\, dz$$

bc The average pressure acting over surface *bc* is

$$p + \frac{1}{2}\left.\frac{dp}{dx}\right]_x dx$$

Then the x component of the normal force acting over *bc* is given by

$$F_{bc} = \left(p + \frac{1}{2}\left.\frac{dp}{dx}\right]_x dx\right) d\delta\, dz$$

ad The shear force acting on *ad* is given by

$$F_{ad} = -\tau_w\, dx\, dz$$

Summing the x component of each force acting on the control volume, we obtain

$$F_{S_x} = \left\{-\frac{dp}{dx}\,\delta\, dx - \frac{1}{2}\frac{dp}{dx}\, \overbrace{dx\, d\delta}^{\simeq 0} - \tau_w\, dx\right\} dz$$

where we note that $dx\, d\delta \ll \delta\, dx$, and so neglect the second term.

Substituting the expressions for $\int_{CS} u\rho\vec{V}\cdot d\vec{A}$ and F_{S_x} into the momentum equation, we obtain

$$\left\{-\frac{dp}{dx}\,\delta\, dx - \tau_w\, dx\right\} dz = \left\{\frac{\partial}{\partial x}\left[\int_0^\delta u\rho u\, dy\right] dx - U\frac{\partial}{\partial x}\left[\int_0^\delta \rho u\, dy\right] dx\right\} dz$$

Dividing this equation by $dx\, dz$ gives

$$-\delta\frac{dp}{dx} - \tau_w = \frac{\partial}{\partial x}\int_0^\delta u\rho u\, dy - U\frac{\partial}{\partial x}\int_0^\delta \rho u\, dy \tag{9.16}$$

Equation 9.16 is a "momentum integral" equation that gives a relation between the x components of the forces acting in a boundary layer and the momentum flux. Since the velocity within the boundary layer approaches the freestream velocity asymptotically, it is desirable to recast the equation in a more convenient form for computations. The pressure gradient, dp/dx, can be determined by applying the Bernoulli equation to the inviscid flow outside the boundary layer; $dp/dx = -\rho U\, dU/dx$. If we recognize that

$\delta = \int_0^\delta dy$, then Eq. 9.16 can be written as

$$\tau_w = -\frac{\partial}{\partial x}\int_0^\delta u\rho u\, dy + U\frac{\partial}{\partial x}\int_0^\delta \rho u\, dy + \frac{dU}{dx}\int_0^\delta \rho U\, dy$$

Since

$$U\frac{\partial}{\partial x}\int_0^\delta \rho u\, dy = \frac{\partial}{\partial x}\int_0^\delta \rho u U\, dy - \frac{dU}{dx}\int_0^\delta \rho u\, dy$$

then

$$\tau_w = \frac{\partial}{\partial x}\int_0^\delta \rho u(U - u)\, dy + \frac{dU}{dx}\int_0^\delta \rho(U - u)\, dy$$

and

$$\tau_w = \frac{\partial}{\partial x} U^2 \int_0^\delta \rho\frac{u}{U}\left(1 - \frac{u}{U}\right) dy + U\frac{dU}{dx}\int_0^\delta \rho\left(1 - \frac{u}{U}\right) dy$$

Using the definitions of displacement thickness, δ^*, (Eq. 9.1) and momentum thickness, θ, (Eq. 9.2), then

$$\frac{\tau_w}{\rho} = \frac{d}{dx}(U^2\theta) + \delta^* U\frac{dU}{dx} \tag{9.17}$$

Equation 9.17 is the momentum integral equation; this equation gives an ordinary differential equation for boundary-layer thickness, provided that a suitable form is assumed for the velocity profile and that the wall shear stress can be related to other variables. Once the boundary-layer thickness is determined, the momentum thickness, the displacement thickness, and the wall shear stress can be calculated.

Equation 9.17 was obtained by applying the basic equations (continuity and momentum) to a differential control volume. Reviewing the assumptions we made in the derivation, we see that the equation is subject to the restrictions:

1. Steady flow.
2. Incompressible flow.
3. Two-dimensional flow.
4. No body forces.

We have not made any specific assumption relating the wall shear stress, τ_w, to the velocity field. Thus Eq. 9.17 is valid for either a laminar or turbulent boundary-layer flow. In order to use this equation to estimate the boundary-layer thickness as a function of x, we must:

1. Determine a first approximation to the pressure gradient, dp/dx. This is determined from inviscid flow theory (the pressure gradient that would exist in the absence of a boundary layer). The pressure in the boundary layer is related to the freestream velocity, U, using the Bernoulli equation.
2. Assume a reasonable velocity profile shape inside the boundary layer.
3. Relate the wall shear stress to the velocity field.

9-5 USE OF THE MOMENTUM INTEGRAL EQUATION FOR ZERO PRESSURE GRADIENT FLOW

To illustrate the method of using Eq. 9.17, we consider the special case of flow over a flat plate, for which U = constant. From Bernoulli's equation we see that for this case, p = constant, and thus $dp/dx = 0$.

The momentum integral equation then reduces to

$$\tau_w = \rho U^2 \frac{d\theta}{dx} = \rho U^2 \frac{d}{dx} \int_0^\delta \frac{u}{U}\left(1 - \frac{u}{U}\right) dy \tag{9.18}$$

The velocity distribution, u/U, in the boundary layer normally is specified as a function of y/δ. (Note that u/U is dimensionless and δ is a function of x only.) Consequently, it is convenient to change the variable of integration from y to y/δ. Defining

$$\eta = \frac{y}{\delta}$$

then

$$dy = \delta \, d\eta$$

and the momentum integral equation for zero pressure gradient is written

$$\tau_w = \rho U^2 \frac{d\theta}{dx} = \rho U^2 \frac{d\delta}{dx} \int_0^1 \frac{u}{U}\left(1 - \frac{u}{U}\right) d\eta \tag{9.19}$$

We wish to solve this equation for the boundary-layer thickness as a function of x. To do this, we must:

1. Assume a velocity distribution in the boundary layer—a functional relationship of the form

$$\frac{u}{U} = f\left(\frac{y}{\delta}\right)$$

(a) The assumed velocity distribution should satisfy certain physical boundary conditions:

$$\text{at} \quad y = 0, \qquad u = 0$$

$$\text{at} \quad y = \delta, \qquad u = U$$

$$\text{at} \quad y = \delta, \qquad \frac{\partial u}{\partial y} = 0$$

(b) Note that once the velocity distribution has been assumed, then the numerical value of the integral in Eq. 9.19 is simply

$$\int_0^1 \frac{u}{U}\left(1 - \frac{u}{U}\right) d\eta = \frac{\theta}{\delta} = \text{constant} = \beta$$

and the momentum integral equation becomes

$$\tau_w = \rho U^2 \frac{d\delta}{dx} \beta$$

2. Obtain an expression for τ_w in terms of δ. This will then permit us to solve for $\delta(x)$ as illustrated below.

9-5.1 Laminar Flow

For laminar flow over a flat plate, a reasonable assumption for the velocity profile is a polynomial in y:

$$u = a + by + cy^2$$

The physical boundary conditions are:

$$\text{at} \quad y = 0, \qquad u = 0$$

$$\text{at} \quad y = \delta, \qquad u = U$$

$$\text{at} \quad y = \delta, \qquad \frac{\partial u}{\partial y} = 0$$

Evaluating the constants, a, b, and c, gives

$$\frac{u}{U} = 2\left(\frac{y}{\delta}\right) - \left(\frac{y}{\delta}\right)^2 = 2\eta - \eta^2 \tag{9.20}$$

The wall shear stress is given by

$$\tau_w = \mu \left.\frac{\partial u}{\partial y}\right)_{y=0}$$

Substituting the assumed velocity profile, Eq. 9.20, into this expression for τ_w gives

$$\tau_w = \mu \left.\frac{\partial u}{\partial y}\right]_{y=0} = \mu \left.\frac{U\partial(u/U)}{\delta\partial(y/\delta)}\right]_{y/\delta=0} = \left.\frac{\mu U}{\delta}\frac{d(u/U)}{d\eta}\right]_{\eta=0}$$

or

$$\tau_w = \left.\frac{\mu U}{\delta}\frac{d}{d\eta}(2\eta - \eta^2)\right]_{\eta=0} = \frac{2\mu U}{\delta}$$

We are now in a position to apply the momentum integral equation

$$\tau_w = \rho U^2 \frac{d\delta}{dx}\int_0^1 \frac{u}{U}\left(1 - \frac{u}{U}\right) d\eta \tag{9.19}$$

Substituting for τ_w and u/U, we obtain

$$\frac{2\mu U}{\delta} = \rho U^2 \frac{d\delta}{dx}\int_0^1 (2\eta - \eta^2)(1 - 2\eta + \eta^2)\, d\eta$$

or

$$\frac{2\mu U}{\delta\rho U^2} = \frac{d\delta}{dx}\int_0^1 (2\eta - 5\eta^2 + 4\eta^3 - \eta^4)\, d\eta$$

Integrating and substituting limits yields

$$\frac{2\mu}{\delta \rho U} = \frac{2}{15} \frac{d\delta}{dx}$$

or

$$\delta \, d\delta = \frac{15\mu}{\rho U} \, dx$$

which is a differential equation for δ. Integrating again gives

$$\frac{\delta^2}{2} = \frac{15\mu}{\rho U} x + c$$

If it is assumed that $\delta = 0$ at $x = 0$, then $c = 0$ and thus

$$\delta = \sqrt{\frac{30\mu x}{\rho U}}$$

or

$$\frac{\delta}{x} = \sqrt{\frac{30\mu}{\rho U x}} = \frac{5.48}{\sqrt{Re_x}} \tag{9.21}$$

Equation 9.21 shows that the ratio of laminar boundary-layer thickness to distance along a flat plate varies inversely with the square root of length Reynolds number. It has the same form as the exact solution derived from the complete differential equations of motion by H. Blasius in 1908. Remarkably, Eq. 9.21 is only in error (the constant is too large) by about 10 percent compared to the exact solution (Section 9-3).

Once we know the boundary-layer thickness, all details of the flow may be determined. The wall shear stress, or "skin friction," coefficient is defined as

$$C_f \equiv \frac{\tau_w}{\frac{1}{2}\rho U^2} \tag{9.22}$$

Substituting from the velocity profile and Eq. 9.21 gives

$$C_f = \frac{\tau_w}{\frac{1}{2}\rho U^2} = \frac{2\mu(U/\delta)}{\frac{1}{2}\rho U^2} = \frac{4\mu}{\rho U \delta} = 4\frac{\mu}{\rho U x}\frac{x}{\delta} = 4\frac{1}{Re_x}\frac{\sqrt{Re_x}}{5.48}$$

Finally,

$$C_f = \frac{0.730}{\sqrt{Re_x}} \tag{9.23}$$

Once the variation of τ_w is known, the viscous drag on the surface can be evaluated by integration over the area of the flat plate, as illustrated in Example Problem 9.3.

Equation 9.21 can be used to calculate the thickness of the laminar boundary layer at transition. At $Re = 5 \times 10^5$, with $U = 30\,\text{m/sec}$, $x = 0.24\,\text{m}$ for air at standard conditions. Thus

$$\frac{\delta}{x} = \frac{5.48}{\sqrt{Re_x}} = \frac{5.48}{\sqrt{5 \times 10^5}} = 0.00775$$

and the boundary-layer thickness is

$$\delta = 0.00775x = 0.00775(0.24\text{ m}) = 1.86\text{ mm}$$

The boundary-layer thickness at transition is less than 1 percent of the development length, x. These calculations confirm that viscous effects are confined to a very thin layer near the surface of a body.

Example 9.3

Consider two-dimensional laminar boundary-layer flow along a flat plate. The velocity profile in the boundary layer is assumed to be sinusoidal

$$\frac{u}{U} = \sin\left(\frac{\pi}{2}\frac{y}{\delta}\right)$$

Find expressions for:
(a) The rate of growth of δ as a function of x
(b) The displacement thickness, δ^*, as a function of x
(c) The total friction force on a plate of length, L, and width, b

EXAMPLE PROBLEM 9.3

GIVEN: Two-dimensional, laminar boundary-layer flow along a flat plate. The boundary-layer velocity profile is

$$\frac{u}{U} = \sin\left(\frac{\pi}{2}\frac{y}{\delta}\right) \quad \text{for } 0 \leq y \leq \delta$$

and

$$\frac{u}{U} = 1 \quad \text{for } y > \delta$$

y
$\delta(x)$
x

FIND: (a) $\delta(x)$.
(b) $\delta^*(x)$.
(c) Total friction force on a plate of length, L, and width, b.

SOLUTION:
For flat plate flow, U = constant, $dp/dx = 0$ and

$$\tau_w = \rho U^2 \frac{d\theta}{dx} = \rho U^2 \frac{d\delta}{dx} \int_0^1 \frac{u}{U}\left(1 - \frac{u}{U}\right) d\eta \tag{9.19}$$

Substituting $\frac{u}{U} = \sin \frac{\pi}{2} \eta$, we obtain

$$\tau_w = \rho U^2 \frac{d\delta}{dx} \int_0^1 \sin \frac{\pi}{2} \eta \left(1 - \sin \frac{\pi}{2} \eta \right) d\eta$$

$$= \rho U^2 \frac{d\delta}{dx} \int_0^1 \left(\sin \frac{\pi}{2} \eta - \sin^2 \frac{\pi}{2} \eta \right) d\eta$$

$$= \rho U^2 \frac{d\delta}{dx} \frac{2}{\pi} \left[-\cos \frac{\pi}{2} \eta - \frac{1}{2} \frac{\pi}{2} \eta + \frac{1}{4} \sin \pi\eta \right]_0^1$$

$$= \rho U^2 \frac{d\delta}{dx} \frac{2}{\pi} \left[0 + 1 - \frac{\pi}{4} + 0 + 0 - 0 \right]$$

$$\tau_w = 0.137 \rho U^2 \frac{d\delta}{dx} = \beta \rho U^2 \frac{d\delta}{dx}; \qquad \beta = 0.137$$

Now

$$\tau_w = \mu \frac{\partial u}{\partial y} \bigg]_{y=0} = \mu \frac{U}{\delta} \frac{\partial (u/U)}{\partial (y/\delta)} \bigg]_{y=0} = \mu \frac{U}{\delta} \frac{\pi}{2} \cos \frac{\pi}{2} \eta \bigg]_{\eta=0} = \frac{\pi \mu U}{2\delta}$$

Therefore,

$$\tau_w = \frac{\pi \mu U}{2\delta} = 0.137 \rho U^2 \frac{d\delta}{dx}$$

Separating variables gives

$$\delta \, d\delta = 11.5 \frac{\mu}{\rho U} dx$$

Integrating, we obtain

$$\frac{\delta^2}{2} = 11.5 \frac{\mu}{\rho U} x + c$$

But $c = 0$, since $\delta = 0$ at $x = 0$, so

$$\delta = \sqrt{23.0 \frac{x\mu}{\rho U}}$$

or

$$\frac{\delta}{x} = 4.80 \sqrt{\frac{\mu}{\rho U x}} = \frac{4.80}{\sqrt{Re_x}} \qquad \delta(x)$$

The displacement thickness, δ^*, is given by

$$\delta^* = \delta \int_0^1 \left(1 - \frac{u}{U} \right) d\eta$$

$$= \delta \int_0^1 \left(1 - \sin \frac{\pi}{2} \eta \right) d\eta = \delta \left[\eta + \frac{2}{\pi} \cos \frac{\pi}{2} \eta \right]_0^1$$

$$\delta^* = \delta \left[1 - 0 + 0 - \frac{2}{\pi} \right] = \delta \left[1 - \frac{2}{\pi} \right]$$

Since, from part (a),

$$\frac{\delta}{x} = \frac{4.80}{\sqrt{Re_x}}$$

then

$$\frac{\delta^*}{x} = \left(1 - \frac{2}{\pi}\right)\frac{4.80}{\sqrt{Re_x}} = \frac{1.74}{\sqrt{Re_x}}$$

$\delta^*(x)$

The total friction force on one side of the plate is given by

$$F = \int_{A_p} \tau_w \, dA$$

Since $dA = b\,dx$ and $0 \leq x \leq L$, then

$$F = \int_0^L \tau_w \, b\,dx = \int_0^L \rho U^2 \frac{d\theta}{dx} b\,dx = \rho U^2 b \int_0^{\theta_L} d\theta = \rho U^2 b \theta_L$$

$$\theta_L = \int_0^{\delta_L} \frac{u}{U}\left(1 - \frac{u}{U}\right) dy = \delta_L \int_0^1 \frac{u}{U}\left(1 - \frac{u}{U}\right) d\eta = \beta \delta_L$$

From part (a), $\beta = 0.137$ and $\delta_L = \dfrac{4.80L}{\sqrt{Re_L}}$, so

$$F = \frac{0.658\,\rho U^2 bL}{\sqrt{Re_L}}$$

F

{This problem illustrates the application of the momentum integral equation to a flat plate, laminar, boundary-layer flow.}

9-5.2 Turbulent Flow

Details of the turbulent velocity profile for boundary layers at zero pressure gradient are very similar to those for turbulent flow in pipes and channels. Data for turbulent boundary layers plot on the universal velocity profile using coordinates of $\bar{u}/u_*$ versus y/δ. However, the universal velocity profile is too complex mathematically for easy use with the momentum integral equation. The momentum integral equation is approximate; a suitable velocity profile for turbulent boundary layers is the empirical power-law profile. An exponent of $\frac{1}{7}$ is used to model the velocity profile

$$\frac{u}{U} = \left(\frac{y}{\delta}\right)^{1/7} = \eta^{1/7}$$

However, this profile does not hold in the immediate vicinity of the wall, since at the wall it predicts $du/dy = \infty$. Consequently, we cannot use this profile in the definition of τ_w to obtain an expression for τ_w in terms of δ as we did for laminar boundary-layer flow. For turbulent boundary-layer flow we use the expression developed for pipe flow

$$\tau_w = 0.03325\rho\bar{V}^2\left[\frac{\nu}{R\bar{V}}\right]^{0.25} \tag{8.35}$$

For a $\frac{1}{7}$-power profile in a pipe, $\bar{V}/U = 0.8$. Substituting $\bar{V} = 0.8U$ and $R = \delta$ into Eq. 8.35, we obtain

$$\tau_w = 0.0225\,\rho U^2\left(\frac{\nu}{U\delta}\right)^{1/4} \tag{9.24}$$

We are now in a position to apply the momentum integral equation

$$\tau_w = \rho U^2 \frac{d\delta}{dx}\int_0^1 \frac{u}{U}\left(1 - \frac{u}{U}\right) d\eta \tag{9.19}$$

Substituting for τ_w and u/U and integrating, we obtain

$$0.0225\left(\frac{\nu}{U\delta}\right)^{1/4} = \frac{d\delta}{dx}\int_0^1 \eta^{1/7}(1 - \eta^{1/7})\,d\eta = \frac{7}{72}\frac{d\delta}{dx}$$

Thus we obtain a differential equation for δ:

$$\delta^{1/4}\,d\delta = 0.231\left(\frac{\nu}{U}\right)^{1/4} dx$$

Integrating gives

$$\frac{4}{5}\delta^{5/4} = 0.231\left(\frac{\nu}{U}\right)^{1/4} x + c$$

If it is assumed that $\delta \simeq 0$ at $x = 0$ (this is equivalent to assuming turbulent flow from the leading edge), then $c = 0$ and

$$\delta = 0.370\left(\frac{\nu}{U}\right)^{1/5} x^{4/5}$$

or

$$\frac{\delta}{x} = 0.370\left(\frac{\nu}{Ux}\right)^{1/5} = \frac{0.370}{Re_x^{1/5}} \tag{9.25}$$

Using Eq. 9.24, we obtain the skin friction coefficient in terms of δ:

$$C_f = \frac{\tau_w}{\frac{1}{2}\rho U^2} = 0.0450\left(\frac{\nu}{U\delta}\right)^{1/4}$$

Substituting for δ, we obtain

$$C_f = \frac{\tau_w}{\frac{1}{2}\rho U^2} = \frac{0.0577}{Re_x^{1/5}} \tag{9.26}$$

Experiments show that Eq. 9.26 predicts turbulent skin friction on a flat plate within about 3 percent for $5 \times 10^5 < Re_x < 10^7$. This agreement is remarkable in view of the approximate nature of our analysis.

Application of the momentum integral equation for turbulent boundary-layer flow is illustrated in Example Problem 9.4.

Use of the momentum integral equation is an approximate technique to predict boundary-layer development; the equation predicts trends correctly. Parameters of the laminar boundary layer depend on $Re_x^{1/2}$; those for the turbulent boundary layer depend on $Re_x^{1/5}$. The turbulent boundary layer develops more rapidly than the laminar boundary layer. The agreement we have obtained with experimental results shows that use of the momentum integral equation is an effective method, which gives us considerable insight into the general behavior of boundary layers.

Example 9.4

Water flows at $U = 1$ m/sec past a flat plate with $L = 1$ m in the flow direction. The boundary layer is tripped so it becomes turbulent at the leading edge. Evaluate the disturbance thickness, δ, displacement thickness, δ^*, and wall shear stress at location, $x = L$. Compare with values that would be found if laminar flow could be maintained to the same position. Assume a $\frac{1}{7}$-power turbulent velocity profile.

EXAMPLE PROBLEM 9.4

GIVEN: Flat-plate boundary-layer flow.

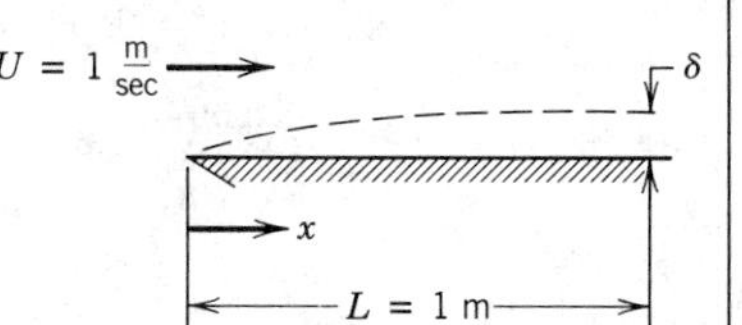

FIND: For turbulent flow from the leading edge:
(a) Disturbance thickness, δ.
(b) Displacement thickness, δ^*.
(c) Wall shear stress, τ_w.
Compare with results for laminar flow from leading edge.

SOLUTION:

Apply results from the momentum integral equation.

Computing equations:

$$\frac{\delta}{x} = \frac{0.370}{(Re_x)^{1/5}} \tag{9.25}$$

$$\delta^* = \int_0^\infty \left(1 - \frac{u}{U}\right) dy \tag{9.1}$$

$$C_f = \frac{\tau_w}{\frac{1}{2}\rho U^2} = \frac{0.0577}{(Re_x)^{1/5}} \tag{9.26}$$

At $x = L$, with $\nu = 1.00 \times 10^{-6}$ m²/sec for water,

$$Re_L = \frac{UL}{\nu} = \frac{1\text{ m}}{\text{sec}} \times 1\text{ m} \times \frac{\text{sec}}{10^{-6}\text{ m}^2} = 10^6$$

From Eq. 9.25,

$$\delta_L = \frac{0.370}{(Re_L)^{1/5}} L = \frac{0.370}{(10^6)^{1/5}} 1\text{ m} = 0.0233\text{ m} \quad \text{or} \quad \delta_L = 23.3\text{ mm}$$

δ_L

Using Eq. 9.1 with $u/U = (y/\delta)^{1/7} = \eta^{1/7}$, we obtain

$$\delta_L^* = \int_0^\infty \left(1 - \frac{u}{U}\right) dy = \delta_L \int_0^1 \left(1 - \frac{u}{U}\right) d\left(\frac{y}{\delta}\right) = \delta_L \int_0^1 (1 - \eta^{1/7})\, d\eta = \delta_L \left[\eta - \frac{7}{8}\eta^{8/7}\right]_0^1 = \frac{\delta_L}{8}$$

$$\delta_L^* = \frac{23.3 \text{ mm}}{8} = 2.91 \text{ mm} \qquad \delta_L^*$$

From Eq. 9.26,

$$C_f = \frac{0.0577}{(10^6)^{1/5}} = 0.00364$$

$$\tau_w = C_f \frac{1}{2} \rho U^2 = \frac{0.00364}{} \times \frac{1}{2} \times \frac{999 \text{ kg}}{\text{m}^3} \times \frac{(1)^2 \text{ m}^2}{\text{sec}^2} \times \frac{\text{N} \cdot \text{sec}^2}{\text{kg} \cdot \text{m}} = 1.82 \text{ N/m}^2 \qquad \tau_w(L)$$

For laminar flow, use Blasius solution values. From Eq. 9.13,

$$\delta_L = \frac{5.0}{\sqrt{Re_L}} L = \frac{5.0}{(10^6)^{1/2}} \times 1 \text{ m} = 0.005 \text{ m} \quad \text{or} \quad 5.00 \text{ mm}$$

From Example 9.2, $\delta^*/\delta = 0.344$, so

$$\delta^* = 0.344\, \delta = 0.344 \times 5.00 \text{ mm} = 1.72 \text{ mm}$$

From Eq. 9.15, $C_f = \dfrac{0.664}{\sqrt{Re_x}}$, so

$$\tau_w = C_f \frac{1}{2} \rho U^2 = \frac{0.664}{\sqrt{10^6}} \times \frac{1}{2} \times \frac{999 \text{ kg}}{\text{m}^3} \times \frac{(1)^2 \text{ m}^2}{} \times \frac{\text{N} \cdot \text{sec}^2}{\text{kg} \cdot \text{m}} = 0.332 \text{ N/m}^2$$

Comparing values, we obtain

$$\text{Disturbance thickness, } \frac{\delta_{\text{turbulent}}}{\delta_{\text{laminar}}} = \frac{23.3 \text{ mm}}{5.00 \text{ mm}} = 4.66$$

$$\text{Displacement thickness, } \frac{\delta^*_{\text{turbulent}}}{\delta^*_{\text{laminar}}} = \frac{2.91 \text{ mm}}{1.72 \text{ mm}} = 1.69$$

$$\text{Wall shear stress, } \frac{\tau_{w,\,\text{turbulent}}}{\tau_{w,\,\text{laminar}}} = \frac{1.82 \text{ N/m}^2}{0.332 \text{ N/m}^2} = 5.48$$

{This problem illustrates use of the momentum integral equation for turbulent boundary layers. The results, when compared to those for laminar flow, indicate much more rapid growth due to the higher wall shear stress for the turbulent boundary layer.}

9-6 PRESSURE GRADIENTS IN BOUNDARY-LAYER FLOW

We have restricted our analyses of boundary-layer flows to flow over a flat plate for which the pressure gradient is zero. The momentum integral equation for this case was given as

$$\tau_w = \rho U^2 \frac{d\theta}{dx} = \rho U^2 \frac{d}{dx} \int_0^\delta \frac{u}{U}\left(1 - \frac{u}{U}\right) dy \tag{9.18}$$

Recall that in deriving this equation, no assumption was made regarding the flow in the boundary layer; the equation is valid for both laminar and turbulent boundary layers. Equation 9.18 indicates that the wall shear stress is balanced by a decrease in fluid momentum. Thus the velocity profiles change as we move along the plate. The boundary-layer thickness continues to increase and the fluid close to the wall is continually being slowed down (losing momentum).

One question of interest is, "Will the fluid close to the wall ever be brought to rest?" For the case of $dp/dx = 0$, is it possible that $\partial u/\partial y)_{y=0} = 0$?[2]

If we consider the wall shear stress distributions obtained for flat plates, we found that for laminar flow

$$\frac{\tau_w(x)}{\rho U^2} = \frac{\text{constant}}{\sqrt{Re_x}}$$

and for turbulent flow

$$\frac{\tau_w(x)}{\rho U^2} = \frac{\text{constant}}{(Re_x)^{1/5}}$$

Recalling that $\tau_w = \mu(\partial u/\partial y)_{y=0}$, we can then say that for any finite length plate, $\partial u/\partial y)_{y=0}$ will never be zero. The point on a solid boundary at which $\partial u/\partial y = 0$ is defined as the point of *separation*. Consequently, we can conclude that for $dp/dx = 0$, the flow will not separate; the fluid layer in the neighborhood of a solid surface cannot be brought to zero velocity.

The pressure gradient is said to be adverse if the pressure increases in the direction of flow (if $\partial p/\partial x > 0$). When $\partial p/\partial x < 0$ (when the pressure decreases in the direction of flow), the pressure gradient is said to be favorable.

Consider the flow through a channel of variable cross section shown in Fig. 9.6. To simplify our discussion, we consider the flow along the straight wall.

If we consider the forces acting on a fluid particle close to the solid boundary, we see that there is a net retarding shear force on the particle no matter what the sign of the pressure gradient. For $\partial p/\partial x = 0$, the result is a decrease in momentum, but as we have already shown it is not sufficient to bring the particle to rest. Since $\partial p/\partial x < 0$ in region 1, the pressure behind the particle (aiding its motion) is greater than that opposing its motion; the particle is "sliding down a pressure hill," without danger of being slowed to zero velocity. However, in attempting to flow through region 3, the particle encounters an adverse pressure gradient, $\partial p/\partial x > 0$, and the particle must "climb a pressure hill." The fluid particle could be brought to rest, thus causing the neighboring fluid to be deflected away from the boundary; when this occurs, the flow is said to separate from the surface. Just downstream from the point of separation, the flow direction in the separated region is opposite to the main flow direction. The low energy fluid in the separated region is forced back upstream by the increased pressure downstream.

[2] Note that if $\partial u/\partial y)_{y=0} = 0$, then the fluid layer near the wall will have zero velocity, since

$$u_{0+dy} = u_0 + \left.\frac{\partial u}{\partial y}\right)_{y=0} dy$$

and $u_0 = 0$ from the no-slip condition.

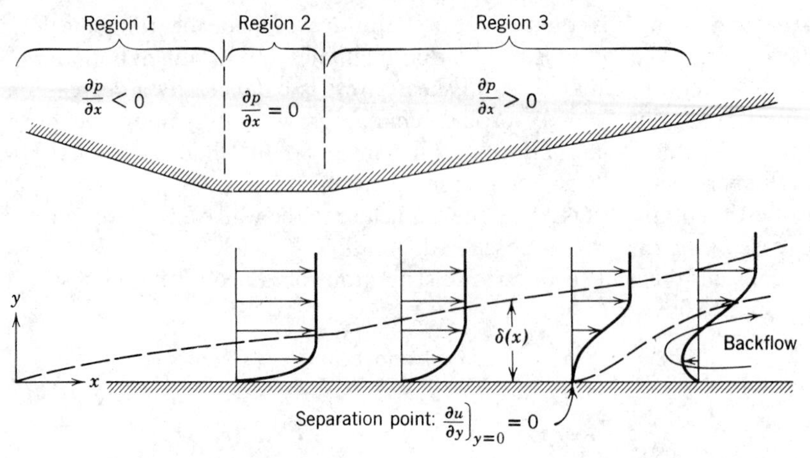

Fig. 9.6 Boundary-layer flow with pressure gradient (boundary-layer thickness exaggerated for clarity).

Thus we see that an adverse pressure gradient, $\partial p/\partial x > 0$, is a necessary condition for separation. Does this mean that if $\partial p/\partial x > 0$, we will have separation? No, it does not. We have not shown that $\partial p/\partial x > 0$ will always lead to separation, but rather we have reasoned that separation cannot occur unless $\partial p/\partial x > 0$. This conclusion can be shown rigorously using the complete differential equations of motion for boundary-layer flow ([3], p. 132).

The nondimensional velocity profiles for laminar and turbulent boundary-layer flow over a flat plate are shown in Fig. 9.7*a*. The turbulent profile is much fuller (more blunt) than the laminar profile. At the same freestream speed the momentum flux within the turbulent boundary layer is greater than that through the laminar layer (Fig. 9.7*b*).

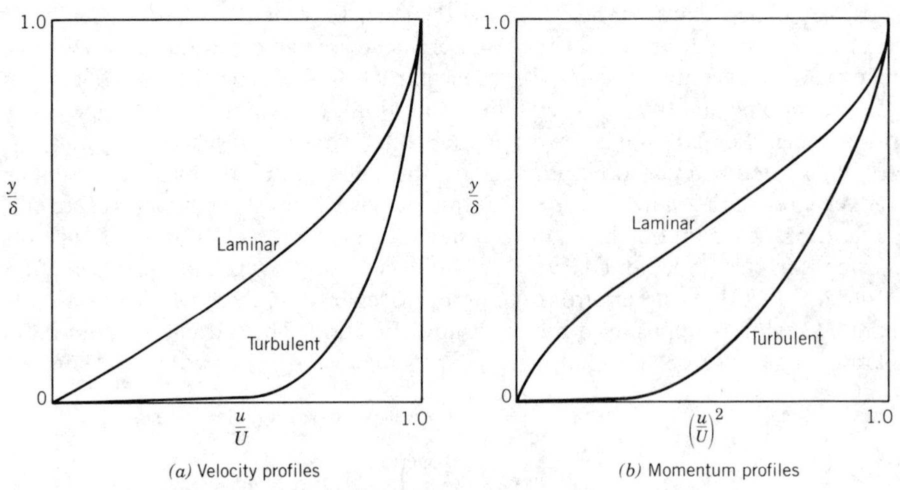

Fig. 9.7 Nondimensional profiles for flat plate boundary-layer flow.

Separation occurs when the momentum of fluid layers near the surface is reduced to zero by the combined action of pressure and viscous forces. As shown in Fig. 9.7*b*, the momentum of the fluid near the surface is significantly greater for the turbulent profile. Consequently, the turbulent layer is better able to resist separation in an adverse pressure gradient. We shall discuss some consequences of this behavior in Section 9-7.3.

Adverse pressure gradients cause significant changes in velocity profiles for both laminar and turbulent boundary-layer flows. Approximate solutions for nonzero pressure gradient flow may be obtained from the momentum integral equation

$$\frac{\tau_w}{\rho} = \frac{d}{dx}(U^2\theta) + \delta^* U \frac{dU}{dx} \tag{9.17}$$

Expanding the first term, we can write

$$\frac{\tau_w}{\rho} = U^2 \frac{d\theta}{dx} + (\delta^* + 2\theta) U \frac{dU}{dx}$$

or

$$\frac{\tau_w}{\rho U^2} = \frac{C_f}{2} = \frac{d\theta}{dx} + (H + 2)\frac{\theta}{U}\frac{dU}{dx} \tag{9.27}$$

where $H = \delta^*/\theta$ is a velocity-profile "shape factor." The shape factor increases in an adverse pressure gradient. For turbulent boundary-layer flow, H increases from 1.3 for a zero pressure gradient to approximately 2.5 at separation. For laminar flow with zero pressure gradient, $H = 2.6$; at separation $H = 3.5$.

The freestream velocity distribution, $U(x)$, must be known before Eq. 9.27 can be applied. Since $dp/dx = -\rho U\, dU/dx$, specifying $U(x)$ is equivalent to specifying the pressure gradient. We can obtain a first approximation for $U(x)$ from ideal flow theory for an inviscid flow under the same conditions. As pointed out in Chapter 6, for frictionless irrotational flow (potential flow), the stream function, ψ, and the velocity potential, ϕ, satisfy Laplace's equation. These can be used to determine $U(x)$ over the body surface.

Much effort has been devoted to calculation of velocity distributions over bodies of known shape (the "direct" problem) and to the determination of body shapes to produce a desired pressure distribution (the "inverse" problem). Smith and co-workers [6] have developed calculation methods that use singularities distributed over the body surface to solve the direct problem for two-dimensional or axisymmetric body shapes. A type of finite-element method that uses singularities defined on discrete surface panels (the "panel" method [7]) recently has gained increased popularity for application to three-dimensional flows.

Once the velocity distribution, $U(x)$, is known, Eq. 9.27 can be integrated to determine $\theta(x)$ if H and C_f can be correlated with θ. A detailed discussion of various calculation methods for nonzero pressure gradient flows is beyond the scope of this book. Numerous solutions for laminar flows are given in [8]. Calculation methods for turbulent boundary-layer flow based on the momentum integral equation are reviewed in [9].

Because of the importance of turbulent boundary layers in engineering flow situations, the state-of-the-art of calculation schemes is advancing rapidly. Numerous calculation schemes have been proposed [10, 11]; most such schemes for turbulent flow use models to predict turbulent shear stress and then solve the boundary-layer equations numerically [12, 13]. Continuing development in the size and speed of digital computers is beginning to make possible the solution of the full Navier–Stokes equations using numerical methods.

PART B FLUID FLOW ABOUT IMMERSED BODIES

Whenever there is relative motion between a solid body and the fluid in which it is immersed, the body experiences a net force, $\vec{F}$, due to the action of the fluid. In general, the infinitesimal force, $d\vec{F}$, acting on an element of surface area, will be neither normal nor parallel to the element. This can be seen clearly when one considers the nature of the surface forces that contribute to the net force, $\vec{F}$. If the body is moving through a viscous fluid, then both shear and pressure forces act on the body,

$$\vec{F} = \int_{\text{body surface}} d\vec{F} = \int_{\text{body surface}} d\vec{F}_{\text{shear}} + \int_{\text{body surface}} d\vec{F}_{\text{pressure}}$$

The resultant force, $\vec{F}$, can be resolved into components parallel and perpendicular to the direction of motion. The component of force parallel to the direction of motion is called the drag force, F_D, and the force component perpendicular to the direction of motion is called the lift force, F_L.

Recognizing that

$$d\vec{F}_{\text{shear}} = \vec{\tau}_w \, dA$$

and

$$d\vec{F}_{\text{pressure}} = -p \, d\vec{A}$$

one might be inclined to think that drag and lift could be evaluated analytically. This proves not to be so; there are very few cases in which the lift and drag can be determined without recourse to experimental results. As we have seen, the presence of an adverse pressure gradient often leads to separation; flow separation prohibits the analytical determination of the force acting on a body. Therefore, for most shapes of interest, we must resort to the use of experimentally measured coefficients for lift and drag computations.

9-7 DRAG

The drag force is the component of force on a body acting parallel to the direction of motion. In discussing the need for experimental results in fluid mechanics (Chapter 7), we considered the problem of determining the drag force, F_D, on a smooth sphere of diameter, d, moving through a viscous, incompressible fluid with speed, V; the fluid density and viscosity were ρ and μ, respectively. The drag force, F_D, was written in the functional form

$$F_D = f_1(d, V, \mu, \rho)$$

Application of the Buckingham Pi theorem resulted in two dimensionless Π parameters that can be written in functional form as

$$\frac{F_D}{\rho V^2 d^2} = f_2\left(\frac{\rho V d}{\mu}\right)$$

Note that d^2 is proportional to the cross-sectional area ($A = \pi d^2/4$) and, hence, we could write

$$\frac{F_D}{\rho V^2 A} = f_3\left(\frac{\rho V d}{\mu}\right) = f_3(Re) \tag{9.28}$$

Although Eq. 9.28 was obtained for a sphere, the form of the equation is valid for incompressible flow over any body; the characteristic length used in the Reynolds number depends on the body shape.

The drag coefficient, C_D, is defined as

$$C_D \equiv \frac{F_D}{\frac{1}{2}\rho V^2 A} \tag{9.29}$$

The number $\frac{1}{2}$ has been inserted (as was done in the defining equation for the friction factor) to form the familiar dynamic pressure. Then Eq. 9.28 can be written as

$$C_D = f(Re) \tag{9.30}$$

We have not considered compressibility or free surface effects in this discussion of the drag force. Had these been included, we would have obtained the functional form

$$C_D = f(Re, Fr, M)$$

At this point we shall consider the drag force on several bodies for which Eq. 9.30 is valid. The total drag force is the sum of the friction drag and the pressure drag. However, the drag coefficient is a function only of the Reynolds number.

9-7.1 Flow over a Flat Plate Parallel to the Flow: Friction Drag

This flow situation has been considered in detail in Section 9-5. Since the pressure gradient is zero, the total drag is equal to the friction drag. Thus

$$F_D = \int_{\text{plate surface}} \tau_w \, dA$$

and

$$C_D = \frac{F_D}{\frac{1}{2}\rho V^2 A} = \frac{\int_{PS} \tau_w \, dA}{\frac{1}{2}\rho V^2 A} \tag{9.31}$$

The drag coefficient for a flat plate parallel to the flow depends on the shear stress distribution along the plate.

For laminar flow over a flat plate, the shear stress coefficient was given by

$$C_f = \frac{\tau_w}{\frac{1}{2}\rho U^2} = \frac{0.664}{\sqrt{Re_x}} \tag{9.15}$$

The drag coefficient for flow with freestream velocity, V, over a flat plate of length, L, and width, b, is obtained by substituting for τ_w from Eq. 9.15 into Eq. 9.31. Thus

$$C_D = \frac{1}{A}\int_A 0.664\, Re_x^{-0.5}\, dA = \frac{1}{bL}\int_0^L 0.664\left(\frac{V}{\nu}\right)^{-0.5} x^{-0.5} b\, dx$$

$$= \frac{0.664}{L}\left(\frac{\nu}{V}\right)^{0.5}\left[\frac{x^{0.5}}{0.5}\right]_0^L = 1.328\left(\frac{\nu}{VL}\right)^{0.5}$$

$$C_D = \frac{1.328}{\sqrt{Re_L}} \tag{9.32}$$

If the boundary layer is turbulent from the leading edge, the shear stress coefficient, based on the approximate analysis of Section 9-5.2, is given by

$$C_f = \frac{\tau_w}{\frac{1}{2}\rho U^2} = \frac{0.0577}{Re_x^{1/5}} \tag{9.26}$$

Substituting for τ_w from Eq. 9.26 into Eq. 9.31 we obtain

$$C_D = \frac{1}{A}\int_A 0.0577\, Re_x^{-0.2}\, dA = \frac{1}{bL}\int_0^L 0.577\left(\frac{V}{\nu}\right)^{-0.2} x^{-0.2} b\, dx$$

$$= \frac{0.0577}{L}\left(\frac{\nu}{V}\right)^{0.2}\left[\frac{x^{0.8}}{0.8}\right]_0^L = 0.072\left(\frac{\nu}{VL}\right)^{0.2}$$

$$C_D = \frac{0.072}{Re_L^{1/5}} \tag{9.33}$$

A better fit with experimental data is obtained if the constant in Eq. 9.33 is changed to 0.074.

$$C_D = \frac{0.074}{Re_L^{1/5}} \tag{9.34}$$

Equation 9.34 is valid for $Re_L < 10^7$. For $Re_L < 10^9$ the empirical equation, due to Schlichting,

$$C_D = \frac{0.455}{(\log Re_L)^{2.58}} \tag{9.35}$$

fits experimental data very well.

For a boundary layer that is initially laminar and undergoes transition at some location on the plate, the turbulent drag coefficient must be adjusted to account for the laminar flow over the initial length. The adjustment is made by subtracting the quantity A/Re_L from the value of C_D determined for completely turbulent flow. The value of A depends on the Reynolds number at transition; A is given by

$$A = Re_{tr}(C_{D_{\text{turbulent}}} - C_{D_{\text{laminar}}}) \tag{9.36}$$

For a transition Reynolds number of 5×10^5, the drag coefficient may be calculated by making the adjustment to Eq. 9.34, in which case

$$C_D = \frac{0.074}{Re_L^{1/5}} - \frac{1740}{Re_L} \tag{9.37a}$$

or to Eq. 9.35, in which case

$$C_D = \frac{0.455}{(\log Re_L)^{2.58}} - \frac{1610}{Re_L} \tag{9.37b}$$

The variation in drag coefficient for a flat plate parallel to the flow is shown in Fig. 9.8.

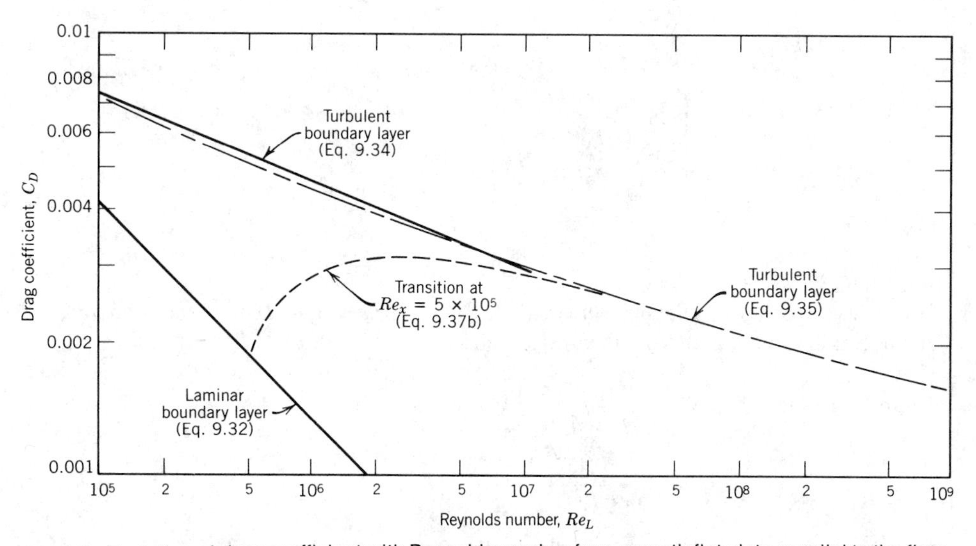

Fig. 9.8 Variation of drag coefficient with Reynolds number for a smooth flat plate parallel to the flow.

In the plot of Fig. 9.8, transition was assumed to occur at $Re_x = 5 \times 10^5$ for flows in which the boundary layer was initially laminar. The Reynolds number at which transition occurs depends on a combination of factors, such as surface roughness and freestream disturbances. Transition tends to occur earlier (at lower Reynolds number) as surface roughness or freestream turbulence is increased. For transition at other values of Re, the constant in the second term of Eqs. 9.37 is modified using Eq. 9.36. Figure 9.8 shows that the drag coefficient is less, for a given length of plate, when laminar flow is maintained over the longest possible distance. However, at large $Re_L (> 10^7)$ the contribution of the laminar drag is negligible.

Example 9.5

A supertanker is 360 m long, has a beam width of 70 m, and a draft of 25 m. Estimate the force and power required to overcome skin friction drag at a cruising speed of 13 kt in seawater at 10 C.

EXAMPLE PROBLEM 9.5

GIVEN: Supertanker cruising at $U = 13$ kt.

FIND: (a) Force
(b) Power
required to overcome skin friction drag.

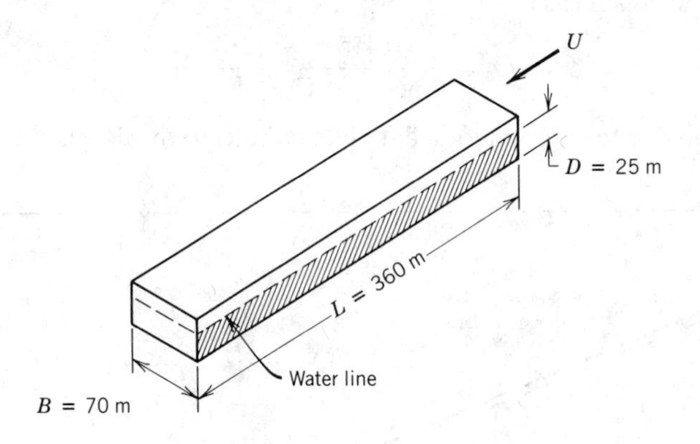

SOLUTION:

Model the tanker hull as a flat plate of length, L, and width, $b = B + 2D$ in contact with water. Estimate skin friction drag from the drag coefficient.

Computing equations: $$C_D = \frac{F_D}{\frac{1}{2}\rho U^2 A} \qquad (9.31)$$

$$C_D = \frac{0.455}{(\log Re_L)^{2.58}} - \frac{1610}{Re_L} \qquad (9.37b)$$

The ship speed is 13 kt (nautical miles per hour), so

$$U = \frac{13 \text{ nm}}{\text{hr}} \times \frac{6076 \text{ ft}}{\text{nm}} \times \frac{0.3048 \text{ m}}{\text{ft}} \times \frac{\text{hr}}{3600 \text{ sec}} = 6.69 \text{ m/sec}$$

From Appendix A, at 10 C, $\nu \approx 1.4 \times 10^{-6}$ m^2/sec for seawater. Then

$$Re_L = \frac{UL}{\nu} = \frac{6.69 \text{ m}}{\text{sec}} \times 360 \text{ m} \times \frac{\text{sec}}{1.4 \times 10^{-6} \text{ m}^2} = 1.72 \times 10^9$$

Assuming Eq. 9.37b is valid,

$$C_D = \frac{0.455}{(\log 1.72 \times 10^9)^{2.58}} - \frac{1610}{1.72 \times 10^9} = 0.00147$$

and from Eq. 9.31

$$F_D = C_D A \frac{1}{2} \rho U^2$$

$$= \frac{0.00147}{} \times \frac{(360\text{ m})(70 + 50)\text{m}}{} \times \frac{1}{2} \times \frac{1020\text{ kg}}{\text{m}^3} \times \frac{(6.69)^2\text{ m}^2}{\text{sec}^2} \times \frac{\text{N} \cdot \text{sec}^2}{\text{kg} \cdot \text{m}}$$

$F_D = 1.45$ MN ← F_D

The corresponding power is

$$\mathscr{P} = F_D U = \frac{1.45 \times 10^6\text{ N}}{} \times \frac{6.69\text{ m}}{\text{sec}} \times \frac{\text{W} \cdot \text{sec}}{\text{N} \cdot \text{m}}$$

$\mathscr{P} = 9.70$ MW ← $\mathscr{P}$

{ This power requirement (~13,000 hp) is substantial. Although the drag coefficient is very low, the wetted surface area is large (~10.7 acres). Because the Reynolds number is large, the effect of laminar flow is negligible; transition occurs at $x \approx 0.1$ m. }

9-7.2 Flow over a Flat Plate Normal to the Flow: Pressure Drag

In flow over a flat plate normal to the flow (Fig. 9.9), we see that the wall shear stress does not contribute to the drag force. The drag is given by

$$F_D = \int_{\text{surface}} p \, dA$$

For this geometry the flow separates from the edges of the plate; there is backflow in the low energy wake of the plate. Although the pressure over the rear surface of the plate is essentially constant, its magnitude cannot be determined analytically. Consequently, we must resort to experiments to determine the drag coefficient.

The drag coefficient for a finite plate normal to the flow depends on the ratio of plate width to height and on the Reynolds number. For values of *Re* (based on height) greater than about 1000, the drag coefficient is essentially independent of the Reynolds number. The variation of C_D with the ratio of plate width to height (b/h) is shown in Fig. 9.10. (The ratio b/h is defined as the *aspect ratio* of the plate.) For $b/h = 1.0$, the

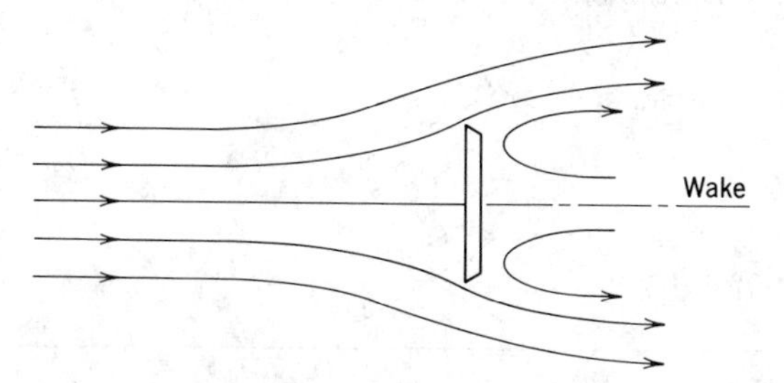

Fig. 9.9 Flow over a flat plate normal to the flow.

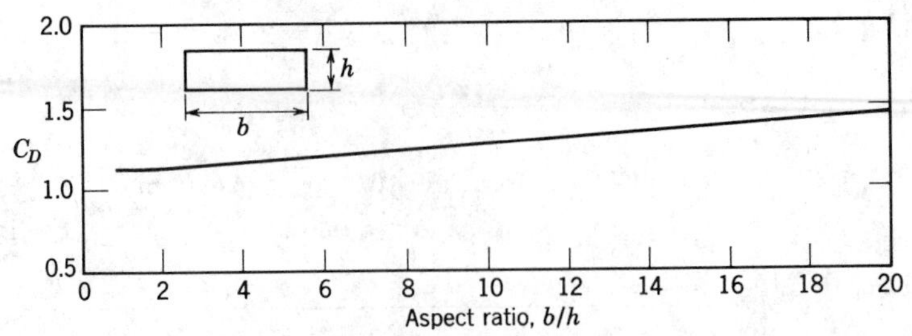

Fig. 9.10 Variation of drag coefficient with aspect ratio for a flat plate of finite width normal to the flow with $Re_h > 1000$ [14].

drag coefficient is a minimum at $C_D = 1.18$; this value is just slightly higher than that for a circular disk ($C_D = 1.17$) at large Reynolds number.

The drag coefficient for all objects with sharp edges is essentially independent of Reynolds number (for $Re > 1000$) because the separation points are fixed by the geometry of the object. Drag coefficients for a few selected objects are given in Table 9.2.

Table 9.2 Drag Coefficient Data for Selected Objects ($Re \gtrsim 1000$)[a]

Object	Diagram	$C_D(Re \gtrsim 10^3)$
Square cylinder	$b/h = \infty$ $b/h = 1$	2.05 1.05
Disk		1.17
Ring		1.20[b]
Hemisphere (open end facing flow)		1.42
Hemisphere (open end facing downstream)		0.38
C-section (open side facing flow)		2.30
C-section (open side facing downstream)		1.20

[a] Data from [14].
[b] Based on ring area.

9-7.3 Flow over a Sphere and Cylinder: Friction and Pressure Drag

We have looked at two special flow cases in which either friction or pressure drag was the sole form of drag present. In the former case the drag coefficient was a strong function of the Reynolds number, while in the latter case, C_D was essentially independent of Reynolds number for $Re \gtrsim 1000$.

In the case of flow over a sphere, both friction drag and pressure drag contribute to the total drag. The drag coefficient for flow over a sphere is shown in Fig. 9.11 as a function of Reynolds number.

At very low Reynolds number,[3] $Re \leq 1$, there is no flow separation from a sphere; the wake is laminar and the drag is predominantly friction drag. Stokes has shown analytically, for very low Reynolds number flows where inertia forces may be neglected, that drag force on a sphere of diameter, d, moving at speed, V, through a fluid of viscosity, μ, is given by

$$F_D = 3\pi\mu V\, d$$

The drag coefficient, C_D, defined by Eq. 9.29, is then

$$C_D = \frac{24}{Re}$$

As shown in Fig. 9.11 this expression agrees with experimental values at low Reynolds number but begins to deviate significantly from the experimental data for $Re > 1.0$.

As the Reynolds number is increased up to about 1000, the drag coefficient drops continuously. As a result of flow separation the drag is a combination of friction and pressure drag. The relative contribution of friction drag decreases with increasing Reynolds number; at $Re \simeq 1000$, the friction drag is approximately 5 percent of the total drag.

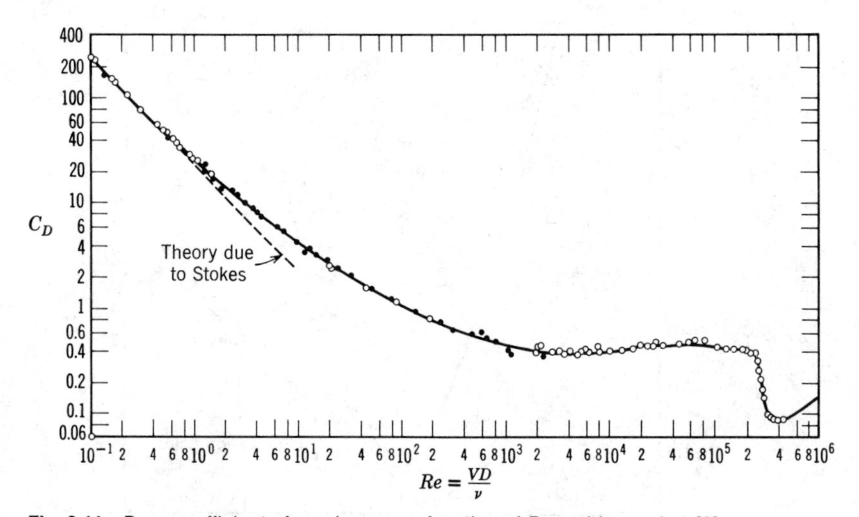

Fig. 9.11 Drag coefficient of a sphere as a function of Reynolds number [3].

[3] See the film, *The Fluid Dynamics of Drag*, A. H. Shapiro, principal, or [15] for a good discussion of drag on spheres and other shapes. Another excellent film is *Low Reynolds Number Flows*, Sir G. I. Taylor, principal. See also [16].

In the range of Reynolds number, $10^3 < Re < 2 \times 10^5$, the drag coefficient curve is relatively flat. The drag coefficient undergoes a rather sharp drop at a Reynolds number of approximately 2×10^5. Experiments show that for $Re < 2 \times 10^5$ the boundary layer on the forward portion of the sphere is laminar. Separation of the boundary layer occurs just upstream of the sphere midsection; a relatively wide turbulent wake is present downstream from the sphere. In the separated region behind the sphere, the pressure is essentially constant and lower than the pressure over the forward portion of the sphere (Fig. 9.12). It is this pressure difference that is the main contributor to the drag.

For Reynolds numbers larger than about 2×10^5, transition occurs in the boundary layer on the forward portion of the sphere. The point of separation then moves downstream from the center of the sphere and the size of the wake is decreased. The net pressure force on the sphere is reduced (Fig. 9.12), and the drag coefficient decreases abruptly.

A turbulent boundary layer, since it has more momentum than a laminar boundary layer, can better resist an adverse pressure gradient, as discussed in Section 9-6.

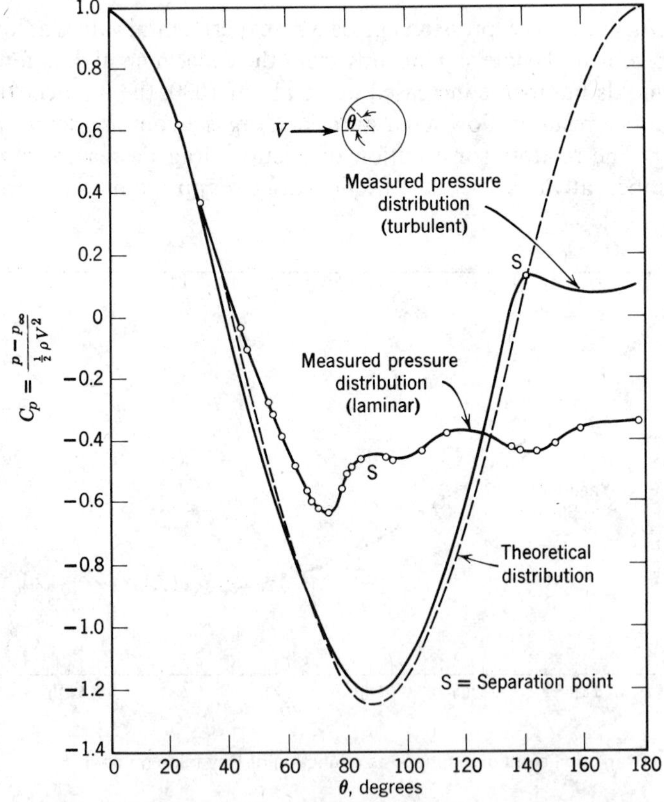

Fig. 9.12 Pressure distribution around a sphere for laminar and turbulent boundary-layer flow, compared to inviscid flow [16].

Consequently, turbulent boundary-layer flow is desirable on a blunt body because it delays separation and thus reduces the pressure drag.

Transition in the boundary layer is affected by roughness of the sphere surface and turbulence in the flow stream. Therefore, the reduction in drag associated with a turbulent boundary layer does not occur at a unique value of Reynolds number. Experiments with smooth spheres in a flow with low turbulence level show that transition may be delayed to a critical Reynolds number, Re_D, of about 4×10^5. For rough surfaces and/or highly turbulent freestream flow, transition can occur at a critical Reynolds number as low as 1×10^5.

The drag coefficient with turbulent boundary-layer flow is about 5 times less than that for laminar flow near the critical Reynolds number. The corresponding reduction in drag force can affect the range of a sphere (e.g. a golf ball) appreciably. The "dimples" on a golf ball are designed to "trip" the boundary layer and, thus, to guarantee turbulent boundary-layer flow and minimum drag. To illustrate this effect more graphically, we obtained samples of golf balls without dimples a few years ago. One of our students volunteered to hit some drives with the smooth balls. In 50 tries with each type of ball, the average distance with the standard balls was 215 yards; the average with the smooth balls was only 125 yards!

Adding roughness elements to a sphere also can suppress local oscillations in location of the transition between laminar and turbulent flow in the boundary layer. These oscillations can lead to variations in drag and to random fluctuations in lift (see Section 9-8). In baseball the "knuckle ball" pitch is intended to behave erratically, to confuse the batter. By throwing the ball with almost no spin, the pitcher relies on the seams to cause transition in an unpredictable fashion, as the ball moves on its way to the batter. This causes the desired variation in the flight path of the ball.

The drag coefficient for flow over a circular cylinder is shown in Fig. 9.13. The variation of C_D with Reynolds number shows the same characteristics as observed in the flow over a sphere, but the values of C_D are about twice as large.

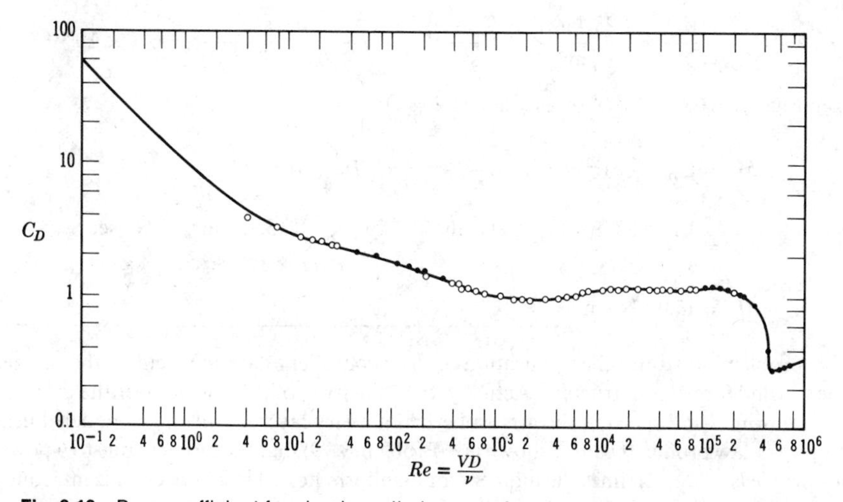

Fig. 9.13 Drag coefficient for circular cylinders as a function of Reynolds number [3].

Example 9.6

A cylindrical chimney 1 m in diameter and 25 m tall is exposed to a uniform 50 km/hr wind at standard atmospheric conditions. End effects and gusts may be neglected. Estimate the bending moment at the base of the chimney due to wind forces.

EXAMPLE PROBLEM 9.6

GIVEN: Cylindrical chimney, $D = 1$ m, $L = 25$ m in uniform flow with

$$V = 50 \text{ km/hr} \qquad p = 101 \text{ kPa} \qquad T = 15 \text{ C}$$

Neglect end effects.

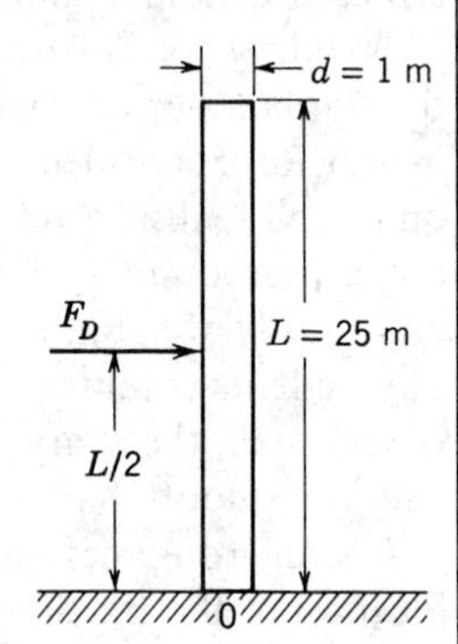

FIND: Bending moment at bottom of chimney.

SOLUTION:

The drag coefficient is given by $C_D = \dfrac{F_D}{\frac{1}{2}\rho V^2 A}$,

and thus $F_D = C_D A \dfrac{1}{2}\rho V^2$.

Since the force per unit length is uniform over the entire length, the resultant force, F_D, will act at the midpoint of the chimney. Hence the moment about the chimney base is

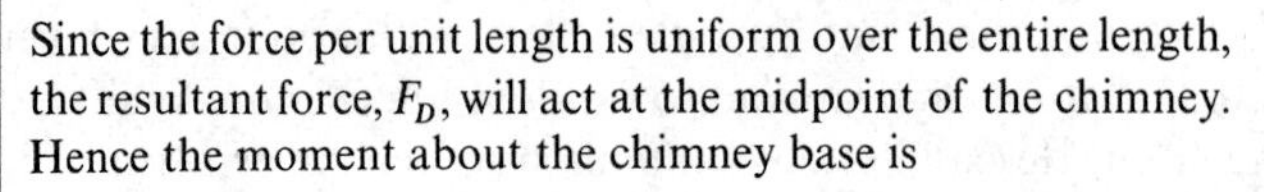

$$M_0 = F_D \frac{L}{2} = C_D A \frac{1}{2} \rho V^2 \frac{L}{2} = C_D A \frac{L}{4} \rho V^2$$

$$V = \frac{50 \text{ km}}{\text{hr}} \times \frac{10^3 \text{ m}}{\text{km}} \times \frac{\text{hr}}{3600 \text{ sec}} = 13.9 \text{ m/sec}$$

For air at standard conditions, $\rho = 1.23$ kg/m^3, and $\mu = 1.78 \times 10^{-5}$ kg/m · sec. Thus

$$Re = \frac{\rho V D}{\mu} = \frac{1.23 \text{ kg}}{\text{m}^3} \times \frac{13.9 \text{ m}}{\text{sec}} \times 1 \text{ m} \times \frac{\text{m} \cdot \text{sec}}{1.78 \times 10^{-5} \text{ kg}} = 9.61 \times 10^5$$

From Fig. 9.13, $C_D \approx 0.35$. For a cylinder, $A = DL$, so

$$M_0 = C_D A \frac{L}{4} \rho V^2 = C_D D L \frac{L}{4} \rho V^2 = C_D D \frac{L^2}{4} \rho V^2$$

$$= \frac{1}{4} \times 0.35 \times 1 \text{ m} \times (25)^2 \text{ m}^2 \times \frac{1.23 \text{ kg}}{\text{m}^3} \times \frac{(13.9)^2 \text{ m}^2}{\text{sec}^2} \times \frac{\text{N} \cdot \text{sec}^2}{\text{kg} \cdot \text{m}}$$

$M_0 = 13.0$ kN · m $\qquad M_0$

This problem illustrates the application of drag coefficient data to calculate the moment due to wind force on a structure. Actually, the velocity profile for wind over the ground is not uniform. Wind speed in the atmospheric boundary layer frequently is modeled using the power-law profile. It can be shown (see Problem 9.79) that the moment due to a power-law profile is $n/(n + 1)$ times the moment for a uniform flow. Thus little error is introduced by assuming uniform flow.

Example 9.7

A class AA fuel dragster weighing 1600 lbf attains a speed of 240 mph in the quarter mile. Immediately after passing through the timing lights, the driver opens the drag chute. The chute area is 25 ft^2, and it has a constant drag coefficient of 1.2. Air and rolling resistance of the car may be neglected. The local air density is 0.0024 slug/ft^3. Find the time required for the machine to decelerate to 100 mph.

EXAMPLE PROBLEM 9.7

GIVEN: Dragster weighing 1600 lbf, moving with speed, $V = 240$ mph, is slowed by the drag force on a chute.

Chute area, $A = 25\ \text{ft}^2$ Drag coefficient, $C_D = 1.2$ $\rho_{air} = 0.0024\ \text{slug/ft}^3$

Air and rolling resistance of the car may be neglected.

FIND: Time required for the machine to decelerate to 100 mph.

SOLUTION:

Taking the car as a system and writing Newton's second law in the direction of motion gives

$$-F_D = ma = m\frac{dV}{dt}$$

F_D V x

$V_0 = 240$ mph
$V_f = 100$ mph

Since $C_D = \dfrac{F_D}{\frac{1}{2}\rho V^2 A}$, then $F_D = \dfrac{1}{2}C_D\rho V^2 A$

Substituting into Newton's second law gives

$$-\frac{1}{2}C_D\rho V^2 A = m\frac{dV}{dt}$$

Separating variables and integrating, we obtain

$$-\frac{1}{2}C_D\rho\frac{A}{m}\int_0^t dt = \int_{V_0}^{V_f}\frac{dV}{V^2}$$

$$-\frac{1}{2}C_D\rho\frac{A}{m}t = -\frac{1}{V}\bigg]_{V_0}^{V_f} = -\frac{1}{V_f} + \frac{1}{V_0} = -\frac{(V_0 - V_f)}{V_f V_0}$$

Finally,

$$t = \frac{(V_0 - V_f)}{V_f V_0}\frac{2m}{C_D\rho A} = \frac{(V_0 - V_f)}{V_f V_0}\frac{2W}{C_D\rho A g}$$

$$= \frac{(240 - 100)\ \text{mph}}{} \times 2 \times 1600\ \text{lbf} \times \frac{1}{100\ \text{mph}} \times \frac{\text{hr}}{240\ \text{mi}} \times \frac{1}{1.2} \times \frac{\text{ft}^3}{0.0024\ \text{slug}}$$

$$\times \frac{1}{25\ \text{ft}^2} \times \frac{\text{sec}^2}{32.2\ \text{ft}} \times \frac{\text{slug}\cdot\text{ft}}{\text{lbf}\cdot\text{sec}^2} \times \frac{\text{mi}}{5280\ \text{ft}} \times \frac{3600\ \text{sec}}{\text{hr}}$$

$t = 5.49$ sec t

All experimental data presented in this section are for single objects immersed in an unbounded fluid stream. We must add a few comments about more realistic flow situations where interactions occur with nearby surfaces or objects.

The objective of wind tunnel tests is to simulate the conditions of an unbounded flow. Limitations on equipment size make this goal unreachable in practice. Frequently it is necessary to apply corrections to derive results applicable to unbounded flow conditions.

Drag can be reduced significantly when two or more objects moving in tandem interact. This phenomenon is well known to bicycle riders and those interested in automobile racing, where "drafting" is a common practice. Drag reductions of 80 percent may be achieved with optimum spacing [17]. Drag also can be increased significantly when the spacing is not optimum.

Drag can be affected by adjacent neighbors as well. Small particles falling under gravity travel more slowly when they have neighbors than when they are isolated. This phenomenon, which is illustrated in the film, *Low Reynolds Number Flows*,[4] has important applications to mixing and sedimentation processes.

Experimental data for drag coefficients on objects must be selected and applied carefully. Due regard must be given to the differences between the actual conditions and the more controlled conditions under which measurements were made.

9-7.4 Streamlining

The extent of the separated flow region behind many of the objects discussed in the previous section can be reduced or eliminated by streamlining, or fairing, the body shape. The objective of streamlining is to reduce the adverse pressure gradient that occurs behind the point of maximum thickness on the body, as pointed out in Section 2-5.1. This delays boundary-layer separation and thus reduces the pressure drag. However, addition of a faired tail section increases the surface area of the body; this causes skin friction drag to increase. The optimum streamlined shape is thus the shape that gives minimum total drag. These effects are discussed at length in the film series, *The Fluid Dynamics of Drag.*[5]

The pressure gradient around a "teardrop" shape (a "streamlined" cylinder) is less severe than that around a cylinder of circular section. The trade-off between pressure and friction drag for this case is illustrated by the results presented in Fig. 9.14, for tests at $Re_c = 4 \times 10^5$. (This Reynolds number is typical of that for a strut on an early aircraft.) From the figure, the minimum drag coefficient is $C_D \simeq 0.06$, which occurs when $t/c \simeq 0.25$. This value is approximately 20 percent of the minimum drag coefficient for a circular cylinder of the same thickness! Consequently, a streamlined strut about 5 times the thickness of a cylindrical strut could be used with no penalty in aerodynamic drag.

The maximum thickness (and therefore minimum pressure) for the shapes shown in Fig. 9.14 is located approximately 25 percent of the chord distance from the leading

[4] Sir G.I. Taylor, principal.

[5] A. H. Shapiro, principal.

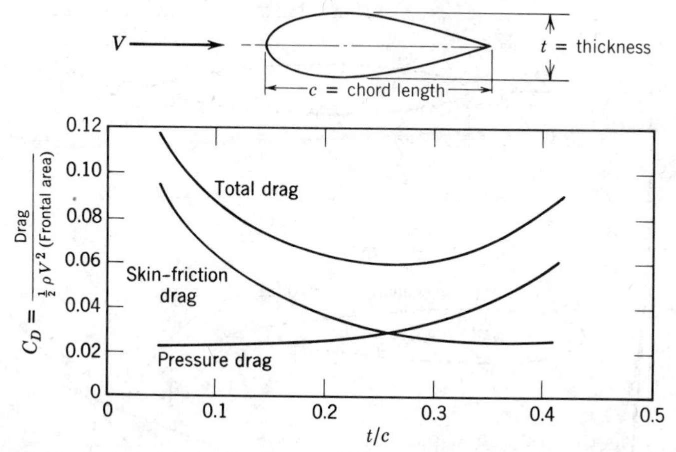

Fig. 9.14 Drag coefficient on a streamlined strut as a function of thickness ratio, showing contribution of skin friction and pressure to total drag [18].

edge. Most of the drag on these shapes is due to skin friction in the turbulent boundary layers on the tapered rear section. Interest in low-drag airfoils increased during the 1930s. The National Advisory Committee for Aeronautics (NACA) developed several series of "laminar-flow" airfoils for which transition was postponed to 60 or 65 percent chord length aft from the airfoil nose.

Pressure distribution and drag data[6] for two symmetric airfoils of infinite span and 15 percent thickness at zero angle of attack are presented in Fig. 9.15. Transition on the conventional (NACA 0015) airfoil takes place where the pressure gradient becomes adverse, at $x/c = 0.13$, near the point of maximum thickness. Thus most of the airfoil surface is covered with a turbulent boundary layer; the drag coefficient is $C_D \simeq 0.0061$. The point of maximum thickness has been moved aft on the airfoil (NACA 66_2–015) designed for laminar flow. The boundary layer is maintained in the laminar state by the favorable pressure gradient to $x/c = 0.63$. Thus the bulk of the flow is laminar; $C_D \simeq 0.0035$ for this section, based on planform area. The drag coefficient based on frontal area is $C_{D_f} = C_D/0.15 = 0.0233$, or about one-third of that for the shapes shown in Fig. 9.14.

Tests in special wind tunnels have shown that laminar flow can be maintained up to length Reynolds numbers as high as 30 million by appropriate profile shaping. Because they have favorable drag characteristics, laminar-flow airfoils are used in the design of most modern subsonic aircraft.

Recent advances have made possible development of low-drag shapes even better than the NACA 60-series shapes. Experiments [20] led to the development of a pressure distribution that prevented separation while maintaining the turbulent boundary layer in a condition that produces negligible skin friction. Improved methods

[6] Note that drag coefficients for airfoils are based on the planform area, i.e. $C_D = F_D/\frac{1}{2}\rho V^2 A_p$, where A_p is the maximum projected wing area.

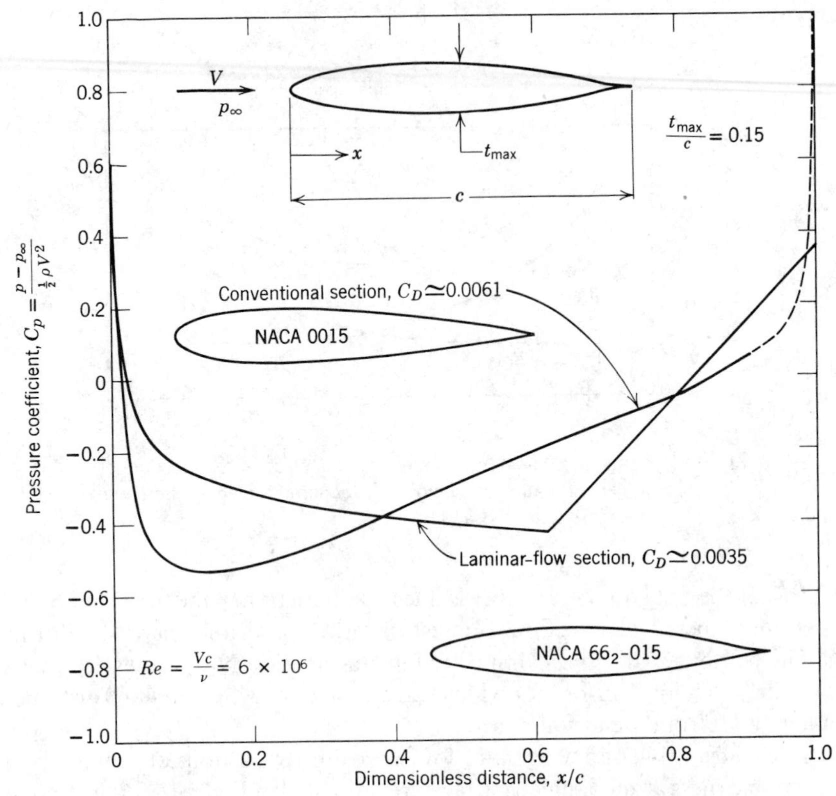

Fig. 9.15 Theoretical pressure distributions at zero angle of attack for two symmetric airfoil sections of 15 percent thickness ratio. (Data from [19].)

for calculating body shapes that produced a desired pressure distribution [21] led to development of nearly optimum shapes for thick struts with low drag. Figure 9.16 shows an example of the results.

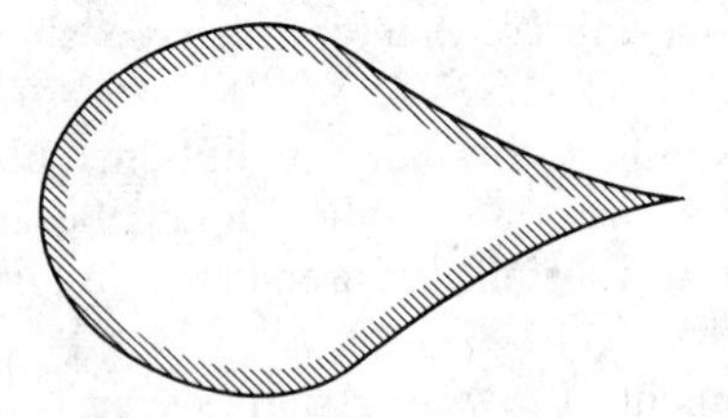

Fig. 9.16 Nearly optimum shape for low-drag strut.

Reduction of aerodynamic drag also is important for road vehicle applications. Recent interest in fuel economy has provided significant economic incentive to balance efficient aerodynamic performance with attractive design for automobiles. Drag reduction also has become important for buses and trucks.

Practical considerations limit the overall length of road vehicles. Fully streamlined tails are impractical for all but land-speed record cars. Consequently, it is not possible to achieve results comparable to those for optimum airfoil shapes. However, it is possible to optimize both front and rear contours within given constraints on overall length [22–24].

Much attention has been focused on front contours. Studies on buses have shown that drag reductions up to 30 percent are possible with careful attention to front contour [24]. Thus it is possible to reduce the drag coefficient of a bus from about 0.65 to less than 0.5 with practical designs. Highway tractor-trailer rigs have higher drag coefficients—values from 0.90 to 1.1 have been reported. Commercially available add-on devices offer improvements in drag of up to 15 percent, particularly for windy conditions where yaw angles are nonzero. The typical fuel saving is half the percentage by which aerodynamic drag is reduced.

Front contours and details are important for automobiles. A low nose and smoothly rounded contours are the primary features that promote low drag. Radii of "A-pillar" and windshield header, and blending of accessories to reduce parasite and interference drag have received increased attention. As a result, drag coefficients have been reduced from about 0.55 to about 0.35 for recent production vehicles.

Recent advances in computational methods have led to development of computer-generated optimum shapes. A number of designs have been proposed, with claims of C_D values below 0.2 for vehicles complete with running gear.

9-8 LIFT

Lift is the component of resultant aerodynamic force perpendicular to the fluid motion. A common example of dynamic lift is flow over an airfoil.[7] The lift coefficient, C_L, is defined as

$$C_L \equiv \frac{F_L}{\frac{1}{2}\rho V^2 A_p} \tag{9.38}$$

The lift and drag coefficients for an airfoil are functions of both Reynolds number and angle of attack; the angle of attack, α, is the angle between the airfoil chord and the freestream velocity vector. The *chord* of an airfoil is the straight line joining the mean thickness line between the airfoil leading edge and the trailing edge. When the airfoil has a symmetric section, the *mean line* and the chord line both are straight lines and they coincide. An airfoil with a curved mean line is said to be *cambered.*

The area at right angles to the flow changes with angle of attack. Consequently, the planform area, A_p (the maximum projected area of the wing), is used to define lift and drag coefficients for an airfoil.

Lift and drag coefficient data for typical conventional and laminar-flow profiles are plotted in Fig. 9.17, for a Reynolds number of 9×10^6, based on chord length.

[7] Flow over an airfoil is shown in the film loop S-FM045, *Velocities near an Airfoil.*

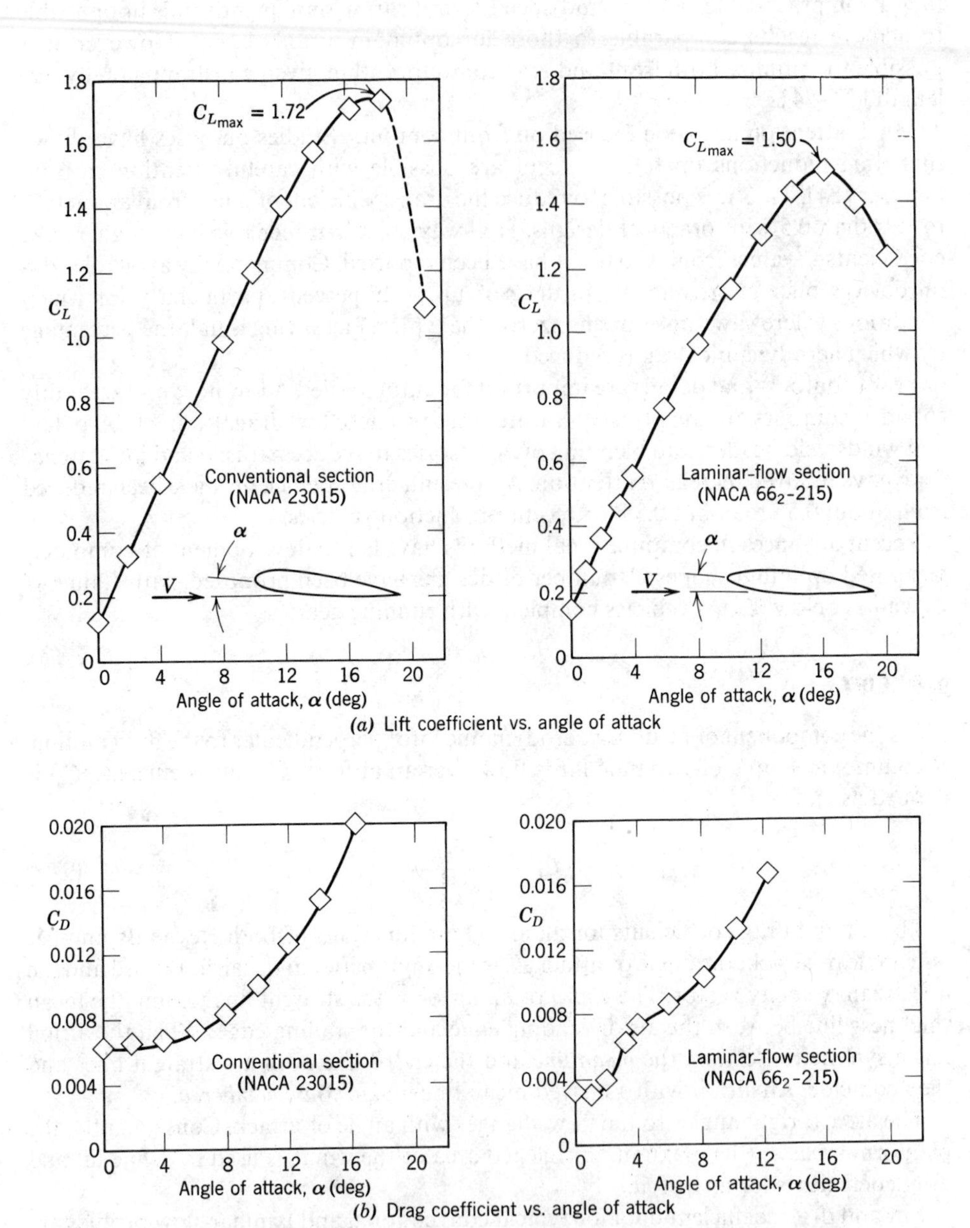

Fig. 9.17 Lift and drag coefficients versus angle of attack for two airfoil sections of 15 percent thickness ratio. (Data from [19].)

The section shapes in Fig. 9.17 are designated as follows:

Conventional—23015

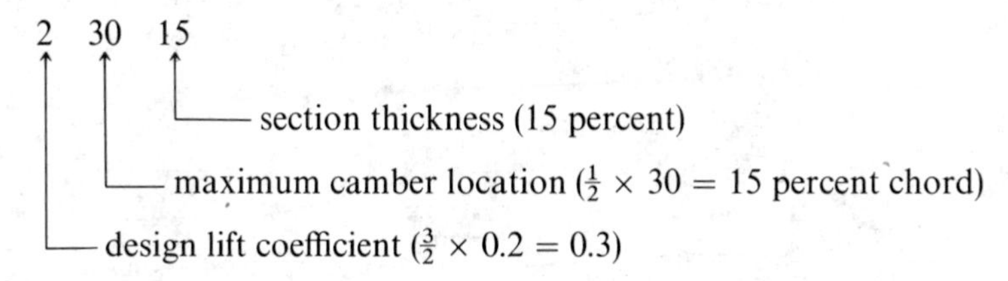

Laminar Flow—66_2-215

6 6 2 - 2 15

- 15 — section thickness (15 percent)
- 2 — design lift coefficient (0.2)
- 2 (subscript) — maximum lift coefficient for favorable pressure gradient (0.2)
- 6 (second) — location of minimum pressure ($x/c \approx 0.6$)
- 6 (first) — series designation (laminar flow)

Both sections are cambered to give lift at zero angle of attack. As the angle of attack is increased, the lift coefficients increase smoothly until a maximum value is reached. Further increases in angle of attack produce a sudden decrease in C_L. The airfoil is said to have *stalled* when C_L drops in this fashion.

Airfoil stall results when flow separation occurs over a major portion of the upper surface of the airfoil. As the angle of attack is increased, the stagnation point moves back along the lower surface of the airfoil, as shown schematically in Fig. 9.18. The flow on the upper surface then must accelerate sharply to round the nose of the airfoil.[8] The minimum pressure becomes lower, and it moves forward on the upper surface. A severe adverse pressure gradient appears following the point of minimum pressure; finally, it causes the flow to separate completely from the upper surface; the airfoil stalls.

Movement of the minimum pressure point and accentuation of the adverse pressure gradient are responsible for the sudden increase in C_D at an angle of attack of about 1.5° for the laminar-flow section, which is apparent in Fig. 9.17. The sudden rise in C_D is due to early transition from laminar to turbulent boundary-layer flow on the upper surface. Aircraft with laminar-flow sections are designed to cruise in the low drag region.

Because laminar-flow sections have very sharp leading edges, all of the effects we have described are exaggerated, and they stall at lower angles of attack than conventional sections, as shown in Fig. 9.17. The maximum possible lift coefficient, $C_{L_{max}}$, also is less for laminar-flow sections.

[8] Flow patterns and pressure distributions for airfoil sections are shown in the film loops, S-FM117, *Subsonic Flow Patterns and Pressure Distributions for an Airfoil* and S-FM118, *Laminar-Flow versus Conventional Airfoils.*

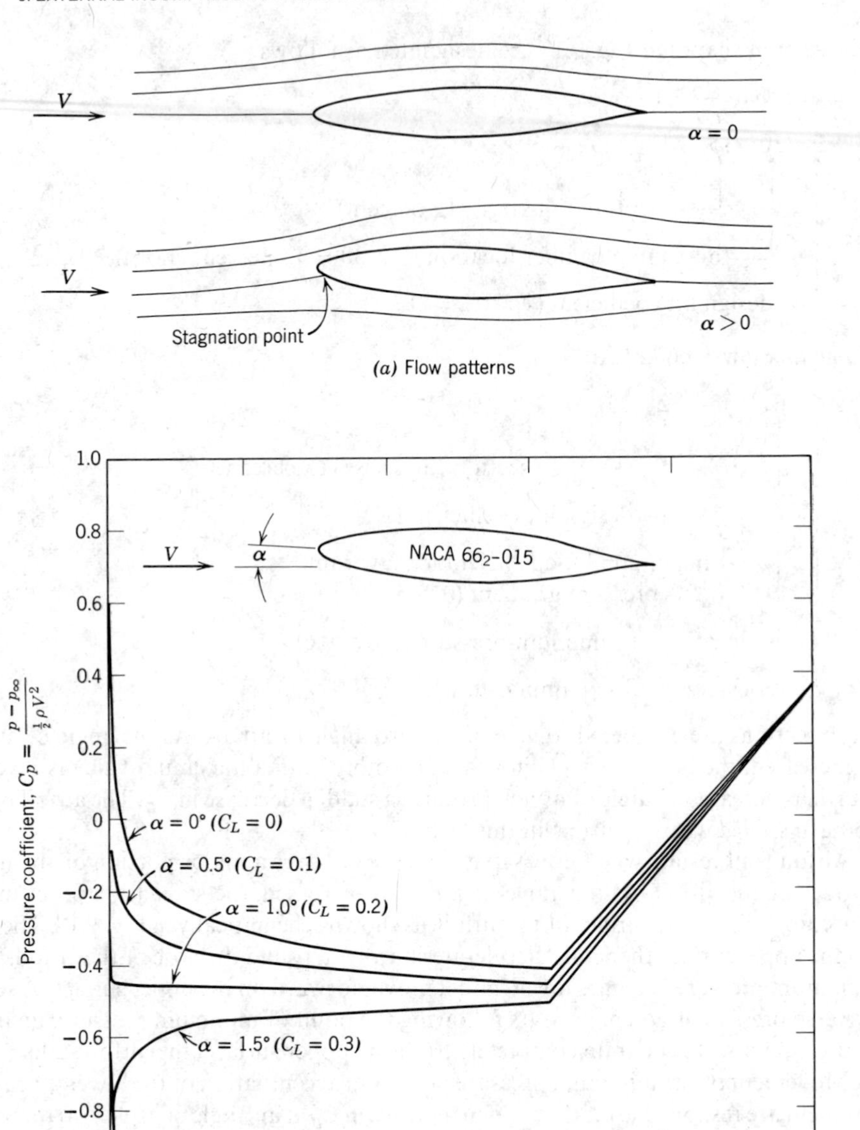

Fig. 9.18 Effect of angle of attack on flow pattern and theoretical pressure distribution for a symmetric laminar-flow airfoil of 15 percent thickness ratio. (Data from [19].)

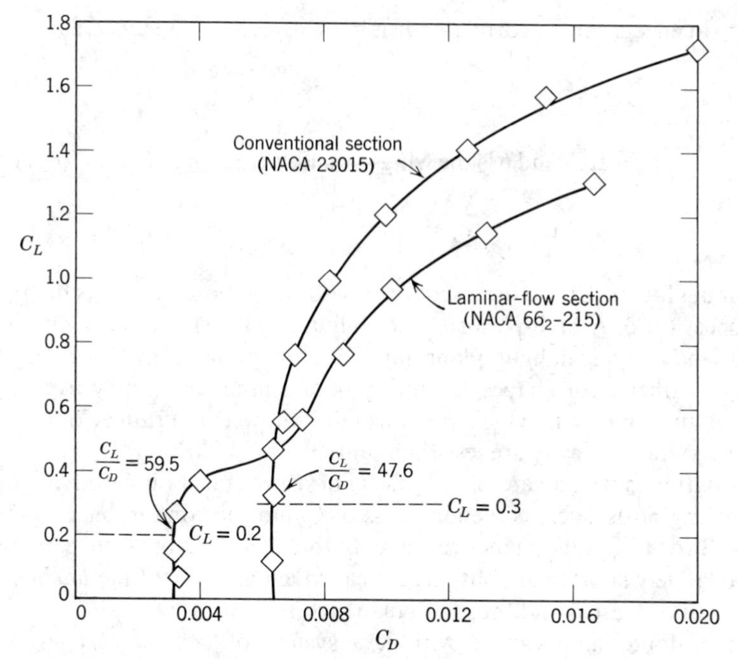

Fig. 9.19 Lift-drag polars for two airfoil sections of 15 percent thickness ratio. (Data from [19].)

Polar plots (plots of C_L versus C_D) often are used to present airfoil data. A polar plot is given in Fig. 9.19 for the two sections we have discussed. The lift/drag ratio, C_L/C_D, is shown at the design lift coefficient for both sections. For an aircraft of given mass at fixed speed, the power required for level flight is inversely proportional to the lift/drag ratio. The advantage of the laminar-flow section is clear.

Recent improvements in modeling and computational capabilities have made it possible to design airfoil sections that develop high lift while maintaining very low drag [21]. Boundary-layer calculation codes are used with inverse methods for calculating potential flow to develop pressure distributions and the resulting body shapes that postpone transition to the most rearward location possible. The turbulent boundary layer following transition is maintained in a state of incipient separation with nearly zero skin friction by appropriate shaping of the pressure distribution.

Such computer-designed airfoils have been used on racing cars to develop very high negative lift (downforce) to improve high-speed stability and cornering performance [21]. Airfoil sections especially designed for operation at low Reynolds number were used for the wings and propeller on the Kremer prize-winning man-powered "Gossamer Condor" [25], which now hangs in the National Air and Space Museum in Washington, D.C.

All of the data presented so far have been for *sections*, slices from airfoils of infinite span. End effects on wings of finite span reduce lift and increase drag. Thus the lift/drag ratios that can be achieved in practice are less than those obtained from tests of airfoil sections.

Finite-span effects can be correlated using the *aspect ratio*, defined as

$$ar \equiv \frac{b^2}{A_p} \tag{9.39}$$

where A_p is planform area and b is the wingspan. For a rectangular planform of span, b,

$$ar = \frac{b^2}{A_p} = \frac{b^2}{bc} = \frac{b}{c}$$

The maximum lift/drag ratio for a modern low-drag section may be as high as 400 for infinite aspect ratio. A high-performance sailplane (glider) with $ar = 40$ might have $L/D = 40$, and a typical light plane ($ar \sim 12$) might have $L/D \sim 20$ or so. Two examples of rather poor shapes are lifting bodies used for reentry from the upper atmosphere, and water skis, which are *hydrofoils* of low aspect ratio. For both of these shapes, L/D values typically are less than unity.

Mother nature is well aware of the effects of aspect ratio on aerodynamic performance. Soaring birds, such as the albatross or California condor, have thin wings of long span. Birds that must maneuver quickly to catch their prey, such as owls, have wings of relatively short span, but large area, which gives low *wing loading* (ratio of weight to planform area) and thus high maneuverability.

A wing of finite span carries with it a system of *trailing vortices*, as shown schematically in Fig. 9.20, whenever it generates lift. Trailing vortices result from leakage flows around the wing tips[9] from the high pressure below to the low pressure above the wing. Trailing vortices may be very strong and persistent, and they may present a hazard to light planes for 5 to 10 *miles* behind a large plane. Velocities exceeding 200 mph have been measured in trailing vortices from large, heavy aircraft.[10]

Induced downwash velocities on a lifting wing reduce the effective angle of attack, reducing lift. (At fixed geometric angle of attack, the wing "sees" a flow at

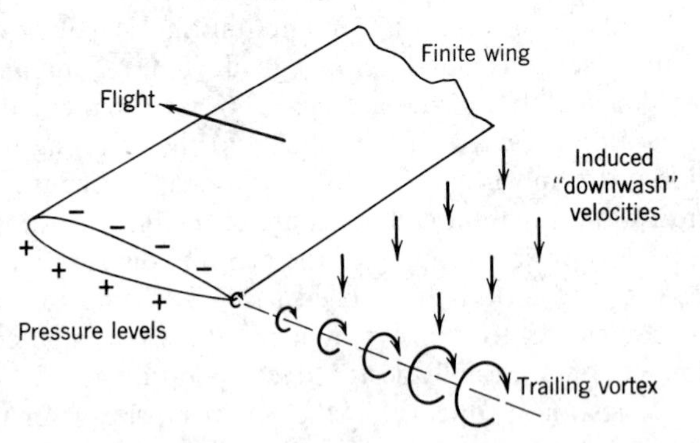

Fig. 9.20 Schematic representation of the trailing vortex system of a finite wing.

[9] Formation of trailing vortices is shown clearly in the film loops, S-FM024, *Wing Tip Vortex*, and S-FM071, *Flow near Tip of Lifting Wing*.

[10] Sforza, P. M., "Aircraft Vortices: Benign or Baleful?" *Space/Aeronautics*, *53*, 4, April 1970, pp. 42–49. See also the University of Iowa film, *Form Drag, Lift, and Propulsion*.

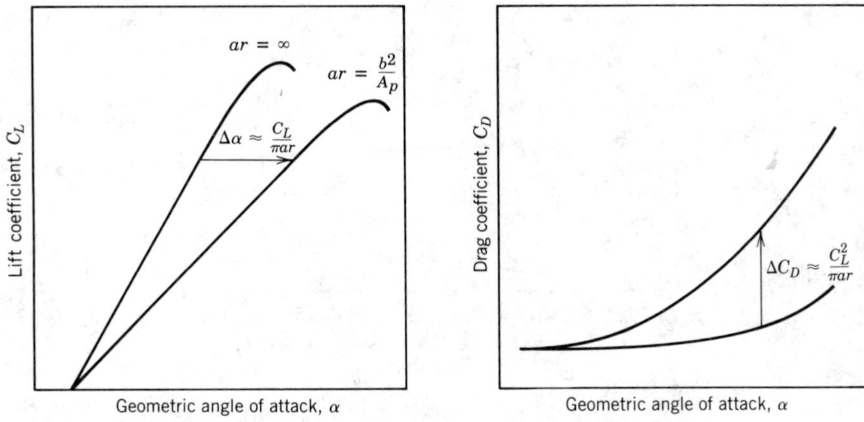

Fig. 9.21 Effect of finite aspect ratio on lift and drag coefficients for a wing.

approximately the mean of the upstream and downstream flow directions.) To maintain lift, the geometric angle of attack must be increased. This causes drag to increase compared to the infinite aspect ratio case. These effects are illustrated schematically in Fig. 9.21.

Theory and experiment show that downwash velocities reduce the effective angle of attack in proportion to the lift coefficient. Compared to an airfoil section with $ar = \infty$, the geometric angle of attack must be increased by

$$\Delta\alpha \approx \frac{C_L}{\pi ar} \tag{9.40}$$

to achieve the lift coefficient of the section. This causes an increase in drag coefficient given by

$$\Delta C_D \approx C_L\,\Delta\alpha \approx \frac{C_L^2}{\pi ar} \tag{9.41}$$

This increase in drag due to lift is termed *induced drag.*

The effective aspect ratio includes the effect of planform shape. For most planform shapes the effective aspect ratio is within 15 percent of the geometric aspect ratio. When written in terms of effective aspect ratio, the drag of a wing of finite span becomes [19]

$$C_D = C_{D,\infty} + C_{D,i} = C_{D,\infty} + \frac{C_L^2}{\pi ar} \tag{9.42}$$

where $C_{D,\infty}$ is the section drag coefficient at C_L, $C_{D,i}$ is the induced drag coefficient at C_L, and ar is the effective aspect ratio.

Drag on airfoils arises from viscous and pressure forces. Viscous drag changes with Reynolds number but only slightly with angle of attack. These relationships and some commonly used terminology are illustrated in Fig. 9.22.

A useful approximation to the drag polar for a wing (Fig. 9.20) or a complete aircraft may be obtained by adding the induced drag to the drag at zero lift. The drag at any lift

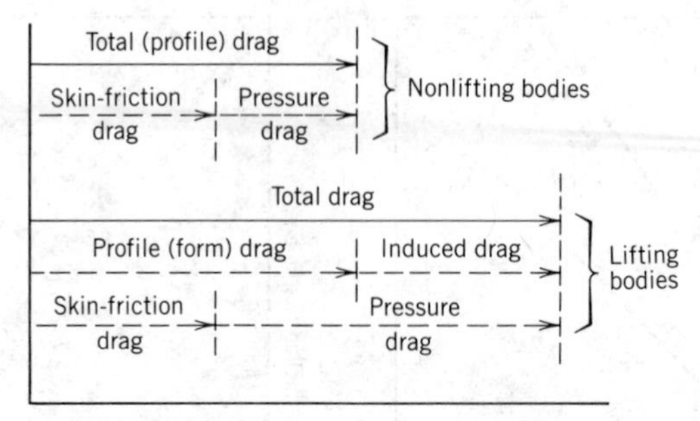

Fig. 9.22 Drag breakdown on nonlifting and lifting bodies.

coefficient is obtained from

$$C_D = C_{D,0} + C_{D,i} = C_{D,0} + \frac{C_L^2}{\pi ar} \tag{9.43}$$

where $C_{D,0}$ is the drag coefficient at zero lift and ar is the effective aspect ratio.

As we have seen, aircraft can be fitted with low-drag airfoils to give excellent performance at cruise conditions. However, since the maximum lift coefficient is low for thin airfoils, additional effort must be expended to obtain acceptably low landing speeds. In steady state flight conditions, the lift must equal the aircraft weight. Thus, from Eq. 9.38,

$$W = F_L = C_L \tfrac{1}{2}\rho V^2 A$$

The minimum flight speed is obtained when $C_L = C_{L_{\max}}$. Solving for $V_{\min}$,

$$V_{\min} = \sqrt{\frac{2W}{\rho C_{L_{\max}} A}} \tag{9.44}$$

According to Eq. 9.44, the minimum landing speed can be reduced by increasing either $C_{L_{\max}}$ or the wing area. Two basic techniques are available for controlling these variables: variable geometry wing sections (e.g. through the use of flaps) or boundary-layer control techniques.

Flaps are movable portions of a wing surface that may be extended during landing and takeoff to increase the effective wing area. The effects on lift and drag of two typical flap configurations are shown in Fig. 9.23, as applied to an NACA 23012 airfoil section. The maximum lift coefficient for this section is increased from 1.52 in the "clean" condition to 3.48 with double-slotted flaps. From Eq. 9.44 the corresponding reduction in landing speed would be 34 percent.

Figure 9.23 shows that section drag is increased substantially by high-lift devices. From Fig. 9.23*b*, the section drag at $C_{L_{\max}}$ ($C_D \simeq 0.28$) with double-slotted flaps is about 5 times larger than the section drag at $C_{L_{\max}}$ ($C_D \simeq 0.055$) for the clean airfoil. The induced drag due to lift must be added to the section drag to obtain the total drag. Because the induced-drag increment is proportional to C_L^2, Eq. 9.43, total drag rises sharply at low aircraft speeds. At speeds near the stall speed, drag may increase sharply enough to exceed the thrust available from the engines. To avoid this dangerous region of unstable operation, the Federal Aviation Administration (FAA) limits operation of commercial aircraft to speeds in excess of 1.2 times the stall speed.

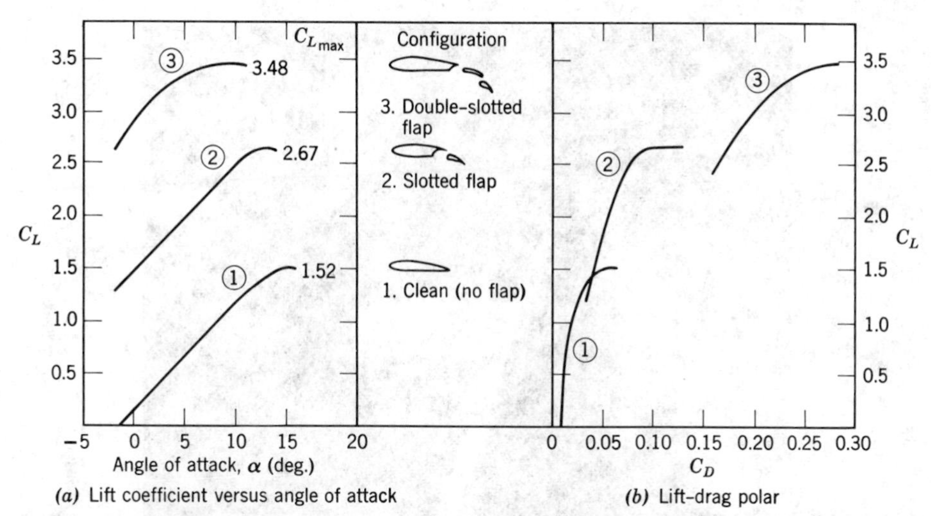

Fig. 9.23 Effect of flaps on aerodynamic characteristics of NACA 23012 airfoil section. (Data from [19].)

Fig. 9.24(a) Application of high-lift boundary-layer control devices to jet transport aircraft for reduction of landing speed. The wing of the Boeing 727 is one of the most mechanized of those in commercial service today. During the approach to landing, huge triple-slotted, trailing edge flaps roll out from under the wing and deflect downward to increase lift. A section of the leading edge near the outboard end of the wing slides forward to open a slot that keeps the air flow close to the wing's upper surface. At the leading edge of the wing near the root, a Kruger flap drops down from under the wing, increasing the effective radius of the leading edge to prevent flow separation. After touchdown, spoilers (not shown) pop up in front of each flap to kill the lift and ensure that the plane remains on the ground despite the lift-augmenting devices. (Photograph courtesy of Boeing Airplane Company.)

Fig. 9.24(b) On the Boeing 707 both the inboard and outboard trailing edge flaps (shown here fully deflected) are equipped with double slots near the pivot. The slots duct air from the under surface to the upper surface of the flap, creating a streamwise jet that causes the air flow to cling more closely to the upper surface. Two rows of short blades in front of the flaps are vortex generators whose purpose is to encourage the slow-moving layers of air next to the surface to follow the wing surface more closely. The raised panels just above the flaps are lift-destroying devices called spoilers. Normally, they are used to control lift for landing and maneuvering. They are raised here to help the plane descend from high altitude. (Photograph courtesy of Boeing Airplane Company.)

Although the details of boundary-layer control techniques[11] are beyond the scope of this book, the basic purpose of all of them is to delay separation or reduce drag, by adding momentum to the boundary layer through blowing, or by removing low-momentum boundary-layer fluid by suction. Many examples of practical boundary-layer control systems may be seen on commercial transport aircraft at your local airport. One of the more sophisticated systems, on the Boeing 727 transport, is shown in Fig. 9.24. On this aircraft, leading edge devices are used in conjunction with triple-slotted trailing edge flaps, to achieve a value of $C_{L_{max}}$ in excess of 3.6.

[11] See the excellent film, *Boundary-Layer Control* (D. C. Hazen, principal), for a review of these techniques. The film loops, S-FM119, *Reduction of Airfoil Friction Drag by Suction* and S-FM120, *Some Methods for Increasing Lift Coefficient*, are edited from this film.

Example 9.8

Jet engines burn fuel at a rate proportional to thrust delivered. Therefore, the optimum cruise condition for a jet aircraft is at maximum speed for a given thrust. In steady level flight, thrust and drag are equal. Optimum cruise occurs at the speed when the ratio of drag force to air speed is minimized.

A Boeing 727-200 jet transport has wing planform area, $A_p = 1600$ square feet, and effective aspect ratio, $ar = 6.5$. Stall speed at sea level for this aircraft with flaps up and an all-up weight of 150,000 pounds is 175 mph. Below $M = 0.6$, drag due to compressibility effects is negligible, so Eq. 9.43 may be used to estimate the total drag on the aircraft. The value of $C_{D,0}$ for the aircraft is constant at 0.0182.

Evaluate the performance envelope for this aircraft at sea level by plotting the drag force versus speed between stall and $M = 0.6$. Use this graph to estimate the optimum cruise speed for the aircraft at sea level conditions. Comment on the stall and optimum cruise speeds for the aircraft at an altitude of 30,000 ft on a standard day.

EXAMPLE PROBLEM 9.8

GIVEN: Boeing 727-200 jet transport at sea-level conditions.

$$W = 150{,}000 \text{ lbf}, \; A = 1600 \text{ ft}^2, \; ar = 6.5, \text{ and } C_{D,0} = 0.0182$$

Stall speed is $V_{\text{stall}} = 175$ mph, and compressibility effects on drag are negligible for $M \leq 0.6$ (sonic speed at sea level is $c = 759$ mph).

FIND:
(a) Evaluate and plot drag force versus speed from V_{stall} to $M = 0.6$.
(b) Estimate optimum cruise speed at sea level.
(c) Comment on performance at altitude of 30,000 ft.

SOLUTION:

For steady, level flight, weight equals lift and thrust equals drag.

Computing equations:
$$F_L = C_L A \frac{1}{2} \rho V^2 = W \qquad C_D = C_{D,0} + \frac{C_L^2}{\pi ar}$$

$$F_D = C_D A \frac{1}{2} \rho V^2 = T \qquad M = \frac{V}{c}$$

At sea level, $\rho = 0.00238$ slug/ft^3, and $c = 759$ mph.

At stall speed, $C_L = \dfrac{W}{\frac{1}{2}\rho V^2 A} = \dfrac{2W}{\rho V^2 A}$, so

$$C_L = 2 \times 150{,}000 \text{ lbf} \times \frac{\text{ft}^3}{0.00238 \text{ slug}} \left[\frac{\text{hr}}{175 \text{ mi}} \times \frac{\text{mi}}{5280 \text{ ft}} \times \frac{3600 \text{ sec}}{\text{hr}}\right]^2 \frac{1}{1600 \text{ ft}^2} \times \frac{\text{slug} \cdot \text{ft}}{\text{lbf} \cdot \text{sec}^2}$$

$$C_L = \frac{3.65 \times 10^4}{[V(\text{mph})]^2} = \frac{3.65 \times 10^4}{(175)^2} = 1.19$$

$$C_D = C_{D,0} + \frac{C_L^2}{\pi ar} = 0.0182 + \frac{(1.19)^2}{\pi(6.5)} = 0.0875$$

and

$$F_D = W\frac{C_D}{C_L} = 150{,}000 \text{ lbf}\left(\frac{0.0875}{1.19}\right) = 11{,}000 \text{ lbf}$$

At $M = 0.6$, $V = Mc = (0.6)759$ mph $= 455$ mph, so $C_L = 0.176$ and

$$C_D = 0.0182 + \frac{(0.176)^2}{\pi(6.5)} = 0.0197 \quad \text{and} \quad F_D = 150{,}000 \text{ lbf}\left(\frac{0.0197}{0.176}\right) = 16{,}800 \text{ lbf}$$

Similar calculations lead to the following table:

V(mph)	175	200	300	400	455
C_L	1.19	0.913	0.406	0.228	0.176
C_D	0.0875	0.0590	0.0263	0.0207	0.0197
F_D(lbf)	11,000	9690	9720	13,600	16,800

These data may be plotted as:

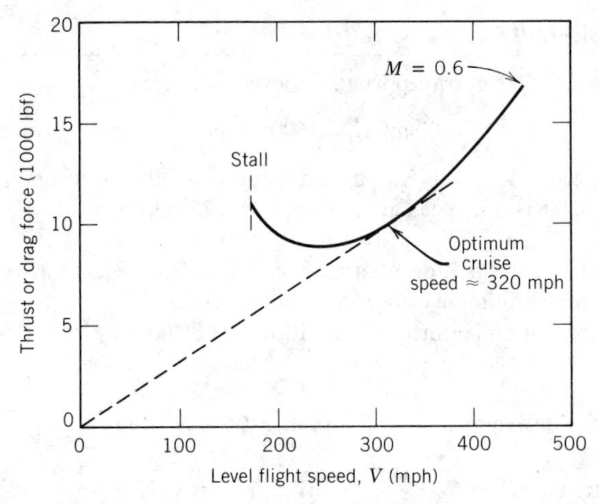

From the plot, the optimum cruise speed at sea level is estimated to be about 320 mph.

At 30,000 ft (9140 m), the density is only about 0.375 times sea level density. The speeds for corresponding forces are calculated from

$$F_L = C_L A\frac{1}{2}\rho V^2 \quad \text{or} \quad V = \sqrt{\frac{2F_L}{C_L\rho A}} \quad \text{or} \quad \frac{V_{30}}{V_{SL}} = \sqrt{\frac{\rho_{SL}}{\rho_{30}}} = \sqrt{\frac{1}{0.375}} = 1.63$$

Thus speeds increase 63 percent at altitude: $V_{\text{stall}} \approx 285$ mph
$V_{\text{cruise}} \approx 522$ mph

Aerodynamic lift is an important consideration in the design of high-speed land vehicles such as racing cars and land-speed record machines. A road vehicle generates lift by virtue of its shape [26]. A representative centerline pressure distribution measured in a wind tunnel for an automobile is shown in Fig. 9.25 [27].

The pressure is low around the nose due to streamline curvature as the flow rounds the nose. The pressure reaches a maximum at the base of the windshield, again due to

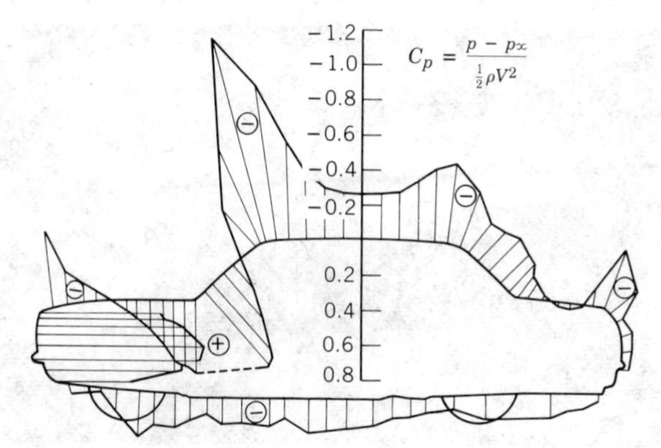

Fig. 9.25 Pressure distribution along the centerline for an automobile [27].

streamline curvature. Low-pressure regions also occur at the windshield header and over the top of the automobile. The air speed across the top is approximately 30 percent higher than the freestream air speed. The same effect occurs around the "A-pillars" at the windshield edges. The drag increase due to an added object such as an antenna, spotlight, or mirror thus would be $(1.3)^2 \approx 1.7$ times the drag the object would experience in an undisturbed flow field. Thus the *parasite drag* of an added component can be much higher than would be predicted from its drag coefficient measured in free flow.

At high speeds, aerodynamic lift forces can unload tires, causing serious reductions in steering control and reducing stability to a dangerous extent. Lift forces on early racing cars were counteracted somewhat by "spoilers," at considerable penalty in drag. In 1965 Jim Hall introduced the use of movable inverted airfoils on his Chaparral sports cars to develop aerodynamic downforce and provide aerodynamic braking [28]. Since then the developments in application of aerodynamic devices have been rapid. Aerodynamic design is used to reduce lift on all modern racing cars, as exemplified in Fig. 9.26. Liebeck airfoils are used frequently for high-speed automobiles. Their high lift coefficients and relatively low drag allow downforce equal to or greater than the car weight to be developed at racing speeds. "Ground effect" cars use venturi-shaped ducts under the car and movable side skirts to seal leakage flows. The result of aerodynamic loading is significantly higher cornering speeds and lower lap times.

Another method of boundary-layer control is to use moving surfaces to reduce skin friction effects on the boundary layer. This method is hard to apply to practical devices, because of geometric and weight complications, but it is very important in recreation. Most golfers, tennis players, Ping-Pong enthusiasts, and baseball pitchers can attest to this!

It has long been known that a spinning projectile in flight is affected by a force perpendicular to the direction of motion and to the spin axis. This effect, known as the Magnus effect, is responsible for the systematic drift of artillery shells. Tennis and Ping-Pong players use spin to control the trajectory and bounce of a shot. In golf, a drive can leave the tee at 275 ft/sec or more, with a backspin of 9000 rpm! This spin provides

Fig. 9.26 Contemporary sports-racing car, showing aerodynamic design features. For a detailed description, see the full-color illustration facing page 000. (Photograph courtesy of B. F. Goodrich.)

significant aerodynamic lift that substantially increases the carry of a drive. It is also largely responsible for hooking and slicing, when shots are not hit squarely. The baseball pitcher use spin to throw a curve ball.

Flow about a spinning sphere is shown in Fig. 9.27*a*. The spin alters the pressure distribution and also affects the location of boundary-layer separation. Separation is delayed on the upper surface of the sphere in Fig. 9.27*a*, and it occurs earlier on the lower surface. The pressure is reduced on the upper surface and increased on the lower surface; the wake is deflected downward as shown. The pressure forces cause a lift in the direction shown; spin in the opposite direction would produce negative lift—a downward force. The force is directed perpendicular to both V and the spin axis.

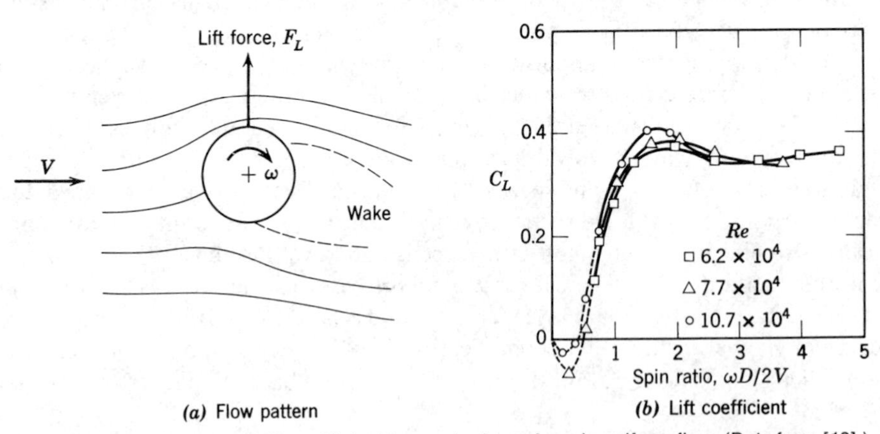

Fig. 9.27 Flow pattern and lift coefficient for a spinning sphere in uniform flow. (Data from [18].)

Example 9.9

A tennis ball, with mass of 57 g and diameter of 64 mm, is hit at 25 m/sec with topspin of 7500 rpm. Calculate the aerodynamic lift acting on the ball. Evaluate the radius of curvature of its path in a vertical plane. Compare with the value for no spin.

EXAMPLE PROBLEM 9.9

GIVEN: Tennis ball in flight, with $m = 57\,\text{g}$ and $D = 64\,\text{mm}$, hit with $V = 25\,\text{m/sec}$ and topspin of 7500 rpm.

FIND: (a) Aerodynamic lift acting on ball.
(b) Radius of curvature of path in vertical plane.
(c) Compare with value for no spin.

SOLUTION: Use data from Fig. 9.27 to find lift: $C_L = f\left(\frac{\omega D}{2V}, Re_D\right)$.

From given data (for standard air, $\nu = 1.45 \times 10^{-5}\,\text{m}^2/\text{sec}$),

$$\frac{\omega D}{2V} = \frac{1}{2} \times \frac{7500\,\text{rev}}{\text{min}} \times 0.064\,\text{m} \times \frac{\text{sec}}{25\,\text{m}} \times \frac{2\pi\,\text{rad}}{\text{rev}} \times \frac{\text{min}}{60\,\text{sec}} = 1.01$$

$$Re_D = \frac{VD}{\nu} = \frac{25\,\text{m}}{\text{sec}} \times 0.064\,\text{m} \times \frac{\text{sec}}{1.45 \times 10^{-5}\,\text{m}^2} = 1.1 \times 10^5$$

From Fig. 9.27, $C_L \approx 0.3$, so

$$F_L = C_L A \frac{1}{2} \rho V^2 = C_L \frac{\pi D^2}{4} \frac{1}{2} \rho V^2 = \frac{\pi}{8} C_L D^2 \rho V^2$$

$$F_L = \frac{\pi}{8} \times 0.3 \times (0.064)^2\,\text{m}^2 \times \frac{1.23\,\text{kg}}{\text{m}^3} \times \frac{(25)^2\,\text{m}^2}{\text{sec}^2} \times \frac{\text{N} \cdot \text{sec}^2}{\text{kg} \cdot \text{m}} = 0.371\,\text{N}$$

F_L

Because the ball is hit with topspin, this force acts down.

Use Newton's second law to evaluate curvature of path. In the vertical plane,

$$\sum F_z = -F_L - mg = ma_z = -m\frac{V^2}{R} \quad \text{or} \quad R = \frac{V^2}{g + F_L/m}$$

$$R = \frac{(25)^2\,\text{m}^2}{\text{sec}^2}\left[\frac{1}{\frac{9.81\,\text{m}}{\text{sec}^2} + 0.371\,\text{N} \times \frac{1}{0.057\,\text{kg}} \times \frac{\text{kg} \cdot \text{m}}{\text{N} \cdot \text{sec}^2}}\right] = 38.3\,\text{m (with spin)}$$

R

$$R = \frac{(25)^2\,\text{m}^2}{\text{sec}^2} \times \frac{\text{sec}^2}{9.81\,\text{m}} = 63.7\,\text{m (without spin)}$$

R

Thus topspin has a significant effect on trajectory of the shot!

9-9 SUMMARY OBJECTIVES

After completing study of Chapter 9, you should be able to do the following:

1. Define:
   - external flow
   - boundary-layer thickness
   - displacement thickness
   - momentum thickness
   - pressure gradient (favorable, adverse)
   - separation
   - skin friction coefficient
   - drag
   - lift
   - angle of attack
   - wing span
   - chord

2. Beginning with the momentum integral equation for zero pressure gradient flow, develop expressions for $\delta(x)$, $\delta^*(x)$, $\tau_w(x)$, $C_f(x)$; determine the total friction force on a flat plate placed parallel to the flow.
3. Use the displacement-thickness concept to approximate the pressure drop in the entrance region of a channel.
4. Determine the drag and lift forces for bodies in external flow.
5. Calculate the moment on an object due to aerodynamic drag forces.
6. Solve the problems at the end of the chapter that relate to the material you have studied.

REFERENCES

1. Prandtl, L., "Fluid Motion with Very Small Friction (in German)," Proceedings of the Third International Congress on Mathematics, Heidelberg, 1904; English translation available as NACA TM 452, March 1928.
2. Blasius, H., "The Boundary Layers in Fluids with Little Friction (in German)," *Zeitschrift fur Mathematik und Physik*, *56*, 1, 1908, pp. 1–37; English translation available as NACA TM 1256, February 1950.
3. Schlichting, H., *Boundary-Layer Theory*, 7th ed. New York: McGraw-Hill, 1979.
4. Stokes, G. G., "On the Effect of the Internal Friction of Fluids on the Motion of Pendulums," *Cambridge Philosophical Transactions*, *IX*, 8, 1851.
5. Howarth, L., "On the Solution of the Laminar Boundary-Layer Equations," *Proceedings of the Royal Society of London*, *A164*, 1938, pp. 547–579.
6. Hess, J. L., and A. M. O. Smith, "Calculation of Potential Flow about Arbitrary Bodies," in *Progress in Aeronautical Sciences*, Vol. 8, D. Kuchemann, et al., eds. Elmsford, N.Y.: Pergamon Press, 1966.
7. Kraus, W., "Panel Methods in Aerodynamics," in *Numerical Methods in Fluid Dynamics*, H. J. Wirz and J. J. Smolderen, eds., Washington, D.C.: Hemisphere, 1978.
8. Rosenhead, L., ed., *Laminar Boundary Layers*. London: Oxford University Press, 1963.
9. Rotta, J. C., "Turbulent Boundary Layers in Incompressible Flow," in *Progress in Aeronautical Sciences*, A. Ferri, et al., eds., New York: Pergamon Press, 1960, pp. 1–220.
10. Kline, S. J., et al., eds., *Proceedings, Computation of Turbulent Boundary Layers—1968 AFOSR-IFP-Stanford Conference*, Vol. I: Methods, Predictions, Evaluation, and Flow Structure, and Vol. II: Compiled Data. Stanford, Calif.: Thermosciences Division, Department of Mechanical Engineering, Stanford University, 1969.
11. Kline, S. J., et al., eds., *Proceedings*, *1980–81 AFOSR-HTTM-Stanford Conference on Complex Turbulent Flows*: *Comparison of Computation and Experiment*, three volumes. Stanford, Calif.: Thermosciences Division, Department of Mechanical Engineering, Stanford University, 1982.
12. Cebeci, T., and P. Bradshaw, *Momentum Transfer in Boundary Layers*. Washington, D.C.: Hemisphere, 1977.
13. Bradshaw, P., T. Cebeci, and J. H. Whitelaw, *Engineering Calculation Methods for Turbulent Flow*. New York: Academic Press, 1981.
14. Hoerner, S. F., *Fluid Dynamic Drag*, 2nd ed. Midland Park, N.J.: Published by the author, 1965.

15. Shapiro, A. H., *Shape and Flow, The Fluid Dynamics of Drag*. New York: Anchor, 1961 (paperback).
16. Fage, A., "Experiments on a Sphere at Critical Reynolds Numbers," Great Britain, *Aeronautical Research Council, Reports and Memoranda*, No. 1766, 1937.
17. Morel, T., and M. Bohn, "Flow over Two Circular Disks in Tandem," *Transactions of the ASME, Journal of Fluids Engineering*, *102*, 1, March 1980, pp. 104–111.
18. Goldstein, S., ed., *Modern Developments in Fluid Dynamics*, Vols. I and II. Oxford: Clarendon Press, 1938. (Reprinted in paperback by Dover, New York, 1967.)
19. Abbott, I. H., and A. E. von Doenhoff, *Theory of Wing Sections, Including a Summary of Airfoil Data*. New York: Dover, 1959 (paperback).
20. Stratford, B. S., "An Experimental Flow with Zero Skin Friction," *Journal of FluidMechanics*, *5*, Pt. 1, January 1959, pp. 17–35.
21. Liebeck, R. H., "Design of Subsonic Airfoils for High Lift," *AIAA Journal of Aircraft*, *15*, 9, September 1978, pp. 547–561.
22. Morel, T., "Effect of Base Slant on Flow in the Near Wake of an Axisymmetric Cylinder," *Aeronautical Quarterly*, *XXXI*, Pt. 2, May 1980, pp. 132–147.
23. Hucho, W. H., "The Aerodynamic Drag of Cars—Current Understanding, Unresolved Problems, and Future Prospects," in *Aerodynamic Drag Mechanisms of Bluff Bodies and Road Vehicles*, G. Sovran, T. Morel, and W. T. Mason, eds. New York: Plenum, 1978.
24. McDonald, A. T., and G. M. Palmer, "Aerodynamic Drag Reduction of Intercity Buses," *Transactions, Society of Automotive Engineers*, *89*, Section 4, 1980, pp. 4469–4484 (SAE Paper No. 801404).
25. Grosser, M., *Gossamer Odyssey*. Boston: Houghton Mifflin, 1981.
26. Carr, G. W., "The Aerodynamics of Basic Shapes for Road Vehicles. Part 3: Streamlined Bodies," Motor Industries Research Association, Warwickshire, England, Report No. 1970/4, 1969.
27. Goetz, H., "The Influence of Wind Tunnel Tests on Body Design, Ventilation, and Surface Deposits of Sedans and Sports Cars," SAE Paper No. 710212, 1971.
28. Hall, J., "What's Jim Hall Really Like?" *Automobile Quarterly*, *VIII*, 3, Spring 1970, pp. 282–293.
29. Chow, C.-Y., *An Introduction to Computational Fluid Mechanics*. New York: Wiley, 1980.

PROBLEMS

9.1 A submarine moves at 20 knots through seawater at 7 C. Assume that near the bow, the boundary layer behaves as though on a flat plate. Determine the distance from the bow where transition from laminar to turbulent flow in the boundary layer might be expected.

9.2 A model of a river towboat is to be tested at 1:13.5 scale. The boat is designed to travel at 8 mph in fresh water at 10 C. Estimate the distance from the bow where transition occurs. Where should transition be stimulated on the model towboat?

9.3 An airplane cruises at 300 knots at an altitude of 10 km on a standard day. Assume that the boundary layers on the wing surfaces behave as on a flat plate. Determine the expected extent of laminar flow in the wing boundary layers.

9.4 Aircraft and missiles flying at high altitude may have regions of laminar flow that become turbulent at lower altitudes at the same speed. Explain a possible mechanism for this effect. Support your answer with calculations based on standard atmosphere data.

9.5 The Reynolds number is given by $Re = \rho VL/\mu$, where L is a characteristic dimension for the flow field. Consider boundary-layer development in parallel flow past a flat plate, for which the characteristic dimension is x, the distance measured from the leading edge. Prepare a log-log plot of velocity versus distance for $0.5 < V < 50$ m/sec and $0.01 < x < 10$ m. Show lines for which $Re_x = 5 \times 10^5$ and 1×10^6 if the fluid is water.

9.6 The Reynolds number is given by $Re = \rho VL/\mu$, where L is a characteristic dimension for the flow field. Consider boundary-layer development in parallel flow past a flat plate, for which the characteristic dimension is x, the distance measured from the leading edge. Prepare a log-log plot of velocity versus distance for $3 < V < 300$ m/sec and $0.01 < x < 10$ m. Show lines of $Re_x = 5 \times 10^5$ and 1×10^6 for standard air at (a) sea level conditions, and (b) 10 km altitude.

9.7 The most general sinusoidal velocity profile for laminar boundary-layer flow on a flat plate is

$$u = A \sin(By) + C$$

State three boundary conditions applicable to the laminar boundary-layer velocity profile. Evaluate the constants, A, B, and C.

***9.8** The velocity profile for a laminar boundary layer is to be approximated by the expression

$$\frac{u}{U} = a + b\eta + c\eta^m, \qquad \text{where } \eta = \frac{y}{\delta}$$

Evaluate a, b, and c for the situation where $m = 1.5$. Compare the resulting profile with the Blasius profile, Table 9.1.

9.9 Velocity profiles in laminar boundary layers often are approximated by the equations

Linear: $\dfrac{u}{U} = \dfrac{y}{\delta}$ Cubic: $\dfrac{u}{U} = \dfrac{3}{2}\left(\dfrac{y}{\delta}\right) - \dfrac{1}{2}\left(\dfrac{y}{\delta}\right)^3$

Parabolic: $\dfrac{u}{U} = 2\left(\dfrac{y}{\delta}\right) - \left(\dfrac{y}{\delta}\right)^2$ Sinusoidal: $\dfrac{u}{U} = \sin\left(\dfrac{\pi}{2}\dfrac{y}{\delta}\right)$

Compare the shapes of these velocity profiles by plotting y/δ (on the ordinate) versus u/U (on the abscissa).

9.10 The velocity profile in a turbulent boundary layer often is approximated by the "power-law" equation

$$\frac{u}{U} = \left(\frac{y}{\delta}\right)^{1/7}$$

Compare the shape of this profile with the parabolic laminar boundary-layer profile (Problem 9.9) by plotting y/δ (on the ordinate) versus u/U (on the abscissa) for both profiles.

9.11 Transition from laminar to turbulent boundary-layer flow actually occurs over a finite length of surface, during which the velocity profile and wall shear stress adjust from laminar to turbulent forms. A useful approximation during transition is that the momentum thickness of the boundary layer remains constant. Assuming constant momen-

* These problems require material from sections that may be omitted without loss of continuity in the text material.

tum thickness, show that

$$\delta_{\text{turbulent}} = \tfrac{144}{105}\delta_{\text{laminar}}$$

for transition from a parabolic laminar velocity profile to a "$\frac{1}{7}$-power" turbulent velocity profile.

9.12 Evaluate the ratio, δ^*/δ, for each of the laminar boundary-layer velocity profiles given in Problem 9.9.

9.13 Evaluate the ratio, δ^*/δ, for the "power-law" form used to represent the turbulent velocity profile,

$$\frac{u}{U} = \left(\frac{y}{\delta}\right)^{1/7}$$

Compare with the value for the cubic, laminar, boundary-layer velocity profile given in Problem 9.9.

9.14 Consider a laminar boundary layer on a flat plate with a velocity profile given by

$$\frac{u}{U} = \frac{3}{2}\eta - \frac{\eta^3}{2}; \qquad \eta = \frac{y}{\delta}$$

For this profile

$$\frac{\delta}{x} = \frac{4.64}{\sqrt{Re_x}}$$

Determine an expression for δ^*/x.

9.15 Air at standard conditions flows over a flat plate. The flow is uniform at the leading edge of the plate. The velocity profile in the boundary layer is of the form

$$\frac{u}{U} = 2\eta - \eta^2 + C_1 \qquad \text{where } \eta \equiv y/\delta$$

At a section, $U = 20$ m/sec, $L = 0.20$ m, and $\delta = 5.7$ mm. Is the flow at this section laminar or turbulent? Why? What boundary conditions must the equation for u/U satisfy, and what is the value of C_1? Determine δ^* and τ_w at the given section.

9.16 Evaluate the ratio, θ/δ, for each of the laminar boundary-layer velocity profiles given in Problem 9.9.

9.17 Evaluate the ratio, θ/δ, for the $\frac{1}{7}$-power-law profile used to represent the turbulent velocity profile. Compare with the value for the cubic, laminar boundary-layer velocity profile given in Problem 9.9.

9.18 Obtain an expression for θ/x for the conditions of Problem 9.14.

9.19 Evaluate $H \equiv \delta^*/\theta$ for each of the laminar boundary-layer velocity profiles given in Problem 9.9.

9.20 Evaluate $H \equiv \delta^*/\theta$ for the power-law profile used to represent the turbulent velocity profile, using $1/n = \frac{1}{7}$. Compare with the value for the cubic, laminar boundary-layer velocity profile given in Problem 9.9.

9.21 Air flows in the entrance region of a square duct, as shown. The velocity at the inlet is uniform, $U_0 = 30$ m/sec, and the duct is 80 mm square. At a section 0.3 m downstream from the entrance, the displacement thickness, δ^*, on each wall measures 1.0 mm. Determine the pressure change between sections ① and ②.

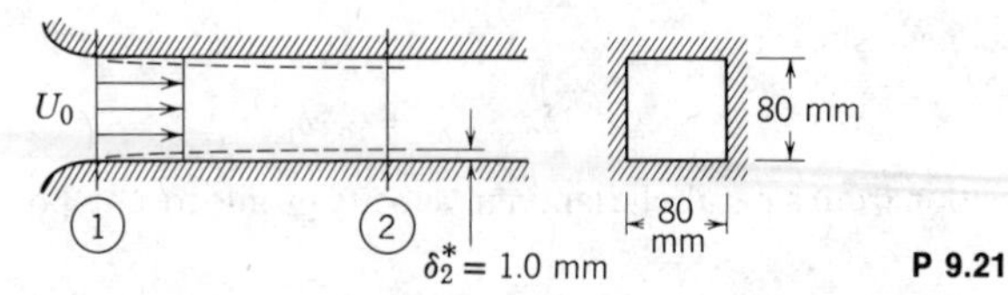

P 9.21

9.22 Air enters a 3 in. diameter circular duct through a smoothly contoured inlet. The flow is steady and the duct area is constant. The velocity is uniform at section ①, where the static pressure is -0.567 in. of water (gage). At section ② the velocity varies linearly, from 0 at the wall to V_2 at a distance of 0.15 in. from the wall. Determine (a) the volume flow rate of air through the duct, (b) the core velocity at section ②, and (c) the displacement thickness at section ②.

9.23 Laboratory wind tunnels have test sections that are 1 ft square and 2 ft long. With a nominal air speed of $U_1 = 80$ ft/sec at the test section inlet, turbulent boundary layers form on the top, bottom, and side walls of the tunnel. The boundary-layer thickness is $\delta_1 = 0.8$ in. at the inlet, and $\delta_2 = 1.2$ in. at the outlet from the test section. The boundary-layer velocity profiles are of power-law form, with $u/U = (y/\delta)^{1/7}$. Evaluate the freestream velocity, U_2, at the exit from the wind-tunnel test section. Determine the change in static pressure along the test section.

9.24 Flow of air develops in a flat horizontal duct following a well-rounded entrance section. The duct height is $h = 300$ mm. Turbulent boundary layers grow on the duct walls, but the flow is not yet fully developed. Assume that the velocity profile in each boundary layer is given by $u/U = (y/\delta)^{1/7}$. The inlet flow is uniform at $\bar{V} = 10$ m/sec at section ①. At section ②, the boundary-layer thickness on each wall of the channel is $\delta_2 = 100$ mm. Show that for this flow, $\delta^* = \delta/8$. Evaluate the static gage pressure at section ②. Calculate the average wall shear stress between the channel entrance and section ②.

9.25 Air at standard conditions is flowing over a thin flat plate which is 1 m long and 0.3 m wide. The flow is uniform at the leading edge of the plate. The velocity profile in the boundary layer is assumed to be linear, and the freestream velocity is $U = 30$ m/sec. Treat the flow as two-dimensional: assume that flow conditions are independent of z. Using control volume *abcd*, shown by the dashed lines, compute the mass flow rate across surface *ab*. Determine the magnitude and direction of the x component of the force required to hold the plate stationary.

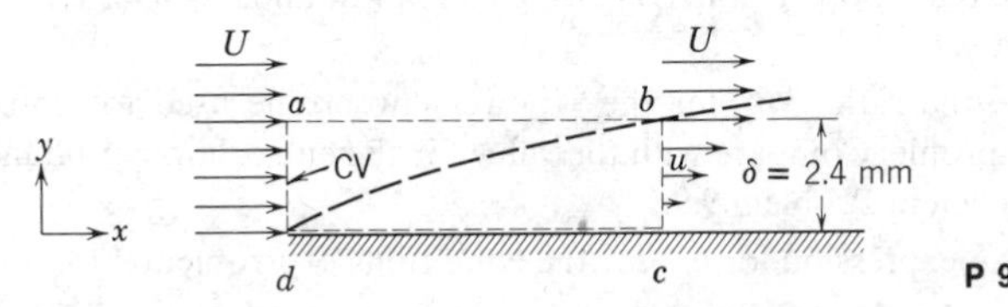

P 9.25

9.26 Rework Problem 9.25 with the same information, except assume the parabolic velocity profile,

$$\frac{u}{U} = 2\left(\frac{y}{\delta}\right) - \left(\frac{y}{\delta}\right)^2$$

at section *bc*, and $\delta = 3.8$ mm.

9.27 Water flows over a flat plate at a freestream speed of 0.5 ft/sec. There is no pressure gradient and the laminar boundary layer is 0.25 in. thick. Assume a sinusoidal velocity profile

$$\frac{u}{U} = \sin\left(\frac{\pi}{2}\frac{y}{\delta}\right)$$

Derive an equation for the shear stress at any location within the boundary layer. For the flow conditions given, compute the local wall shear stress and skin friction coefficient.

***9.28** Using numerical results for the Blasius exact solution for laminar boundary-layer flow on a flat plate, plot the dimensionless velocity profile, u/U, versus the dimensionless distance from the surface, y/δ. Compare with the approximate parabolic velocity profile of Problem 9.9.

***9.29** Using graphical or numerical integration, evaluate the ratio, δ^*/δ, for the Blasius exact solution for the laminar boundary layer on a flat plate (Table 9.1). Show that the result is

$$\frac{\delta^*}{x} = \frac{1.73}{\sqrt{Re_x}}$$

***9.30** Using graphical or numerical integration, evaluate the ratio, θ/δ, for the Blasius exact solution for the laminar boundary layer on a flat plate (Table 9.1). Show that the result is

$$\frac{\theta}{x} = \frac{0.664}{\sqrt{Re_x}}$$

***9.31** Evaluate the distribution of shear stress in a laminar boundary layer on a flat plate from numerical results obtained by Blasius (Table 9.1). Plot τ/τ_w versus y/δ. Compare results from the exact solution of Blasius with results derived from the approximate parabolic velocity profile,

$$\frac{u}{U} = 2\left(\frac{y}{\delta}\right) - \left(\frac{y}{\delta}\right)^2$$

***9.32** Evaluate the distribution of shear stress in a laminar boundary layer on a flat plate from numerical results obtained by Blasius (Table 9.1). Plot τ/τ_w versus y/δ. Compare results from the exact solution of Blasius with results derived from the approximate sinusoidal velocity profile,

$$\frac{u}{U} = \sin\left(\frac{\pi}{2}\frac{y}{\delta}\right)$$

***9.33** Evaluate the distribution of shear stress in a laminar boundary layer on a flat plate from numerical results obtained by Blasius (Table 9.1). Plot τ/τ_w versus y/δ. Compare results from the exact solution of Blasius with results derived from the approximate cubic velocity profile,

$$\frac{u}{U} = \frac{3}{2}\left(\frac{y}{\delta}\right) - \frac{1}{2}\left(\frac{y}{\delta}\right)^3$$

***9.34** Evaluate the vertical component of velocity in a laminar boundary layer on a flat plate from numerical results obtained by Blasius (Table 9.1). Plot v/U versus y/δ for $Re_x = 10^5$.

* These problems require material from sections that may be omitted without loss of continuity in the text material.

9.35 Consider horizontal, steady, incompressible flow in a boundary layer with distributed wall suction. The wall suction velocity is constant with $v = -v_0$ at $y = 0$. There is no pressure gradient. Use a differential control volume to show that

$$\frac{d\theta}{dx} = \frac{\tau_w}{\rho U^2} - \frac{v_0}{U}$$

where U is the freestream speed and τ_w is the wall shear stress.

9.36 When a cubic velocity profile for the laminar boundary layer,

$$\frac{u}{U} = \frac{3}{2}\eta - \frac{1}{2}\eta^3; \qquad \eta = \frac{y}{\delta}$$

is used in the momentum integral equation, the variation of δ with x is found to be

$$\frac{\delta}{x} = \frac{4.64}{\sqrt{Re_x}}$$

For a laminar boundary layer with cubic profile, obtain an expression for the local skin friction coefficient,

$$C_f = \frac{\tau_w}{\frac{1}{2}\rho U^2}$$

in terms of distance and flow properties.

9.37 The velocity profile in a laminar boundary-layer flow at zero pressure gradient is to be approximated by the linear expression,

$$\frac{u}{U} = \eta; \qquad \eta = \frac{y}{\delta}$$

Use the momentum integral equation with this profile to obtain an expression for the ratio, δ/x, and the skin friction coefficient, C_f.

9.38 The velocity profile in a laminar boundary-layer flow at zero pressure gradient is to be approximated by the cubic expression

$$\frac{u}{U} = \frac{3}{2}\eta - \frac{1}{2}\eta^3; \qquad \eta = \frac{y}{\delta}$$

Use the momentum integral equation with this profile to obtain an expression for the ratio, δ/x, and the skin friction coefficient, C_f.

9.39 Calculate the displacement thickness for a laminar boundary layer where $Re_x = 10^5$. Use a velocity profile that satisfies at least the conditions, $u = 0$ at $y = 0$, $u = U$ at $y = \delta$, and $\partial u/\partial y = 0$ at $y = \delta$. Recall that $\delta/x \approx 5/\sqrt{Re_x}$ according to the Blasius solution.

9.40 Water flows over the top surface of a flat plate, forming a laminar boundary layer. The boundary-layer velocity profile is approximated by the cubic polynomial expression

$$\frac{u}{U} = \frac{3}{2}\left(\frac{y}{\delta}\right) - \frac{1}{2}\left(\frac{y}{\delta}\right)^3$$

For this profile it is known that

$$\frac{\delta}{x} = \frac{4.64}{\sqrt{Re_x}}$$

The plate is 0.5 ft long in the flow direction and 3 ft wide. The freestream flow speed is 4 ft/sec. Determine the maximum value of δ for the plate. Where does the minimum wall shear stress occur? Illustrate with a sketch of τ_w versus x. Calculate the minimum wall shear stress. Set up an algebraic expression for the drag force on the plate.

9.41 A thin flat plate is installed in a water tunnel as a splitter. The plate is 0.3 m long and 1 m wide. The freestream speed is 2 m/sec. Laminar boundary layers form on both sides of the plate. The boundary-layer velocity profile is approximated by

$$\frac{u}{U} = 2\left(\frac{y}{\delta}\right) - \left(\frac{y}{\delta}\right)^2$$

for which

$$\frac{\delta}{x} = \frac{5.48}{\sqrt{Re_x}}$$

Determine the total viscous drag force on the plate assuming that pressure drag is negligible.

9.42 Air at standard conditions flows over a flat plate. The freestream speed is 15 m/sec. Find δ and τ_w, at $x = 1$ m from the leading edge for (a) completely laminar flow (assume a parabolic velocity profile), and (b) completely turbulent flow (assume a "$\frac{1}{7}$-power" velocity profile).

9.43 The velocity profile in a turbulent boundary-layer flow at zero pressure gradient is to be approximated by the "$\frac{1}{6}$-power" profile expression,

$$\frac{u}{U} = \eta^{1/6}; \qquad \eta = \frac{y}{\delta}$$

Use the momentum integral equation with this profile to obtain an expression for the ratio, δ/x, and the skin friction coefficient, C_f. Compare with the results obtained in Section 9-5.2 for the "$\frac{1}{7}$-power" profile.

9.44 Repeat Problem 9.43, using the "$\frac{1}{8}$-power" profile expression.

9.45 A uniform flow of standard air at 60 m/sec enters a plane-wall diffuser with negligible boundary-layer thickness. The inlet width is 75 mm. The diffuser walls diverge slightly to accommodate the boundary-layer growth so that the pressure gradient is negligible. Flat-plate boundary-layer behavior may be assumed. Explain why the Bernoulli equation is applicable to this flow. Estimate the diffuser width 1.2 m downstream from the entrance.

9.46 Calculate the drag force on a flat plate with dimensions of 0.75 m × 0.75 m when it is aligned in a flow of air where the freestream speed is 1.8 m/sec.

9.47 Two hypothetical boundary-layer velocity profiles are shown. Calculate the momentum flux of each profile. If the two profiles were subjected to the same pressure gradient conditions, which would be most likely to separate first? Why?

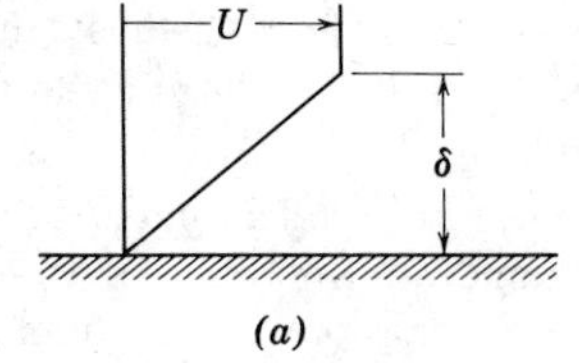

(a)

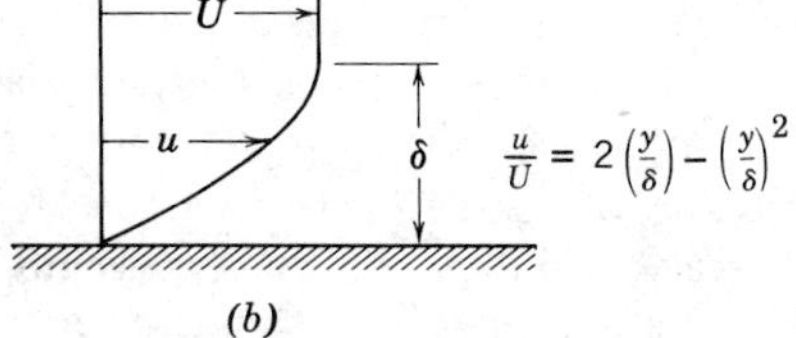

(b)

P 9.47

9.48 We wish to compare the flow of an ideal fluid ($\mu = 0$) and a real fluid in a plane-wall diffuser. Consider first the straight-channel case where $\phi = 0$. What can be said of the pressure gradient for the real and ideal fluids? Which fluid gives the higher value of p_2? Now consider a case where ϕ is not equal to zero but is small enough to avoid separation. Again, what can be said of the pressure gradient for real and ideal fluids? Which case results in the highest exit pressure?

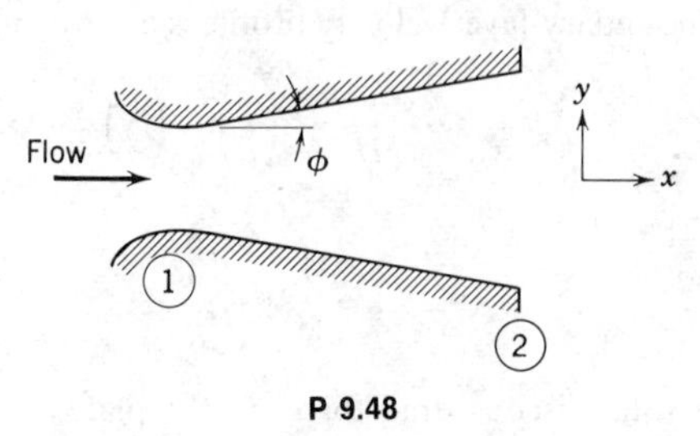

P 9.48

9.49 Water at 15 C flows over a flat plate at a speed of 1 m/sec. The plate is 0.4 m long and 1 m wide. The boundary layer on each surface of the plate is laminar. Assume that the velocity profile is approximated by a linear expression, for which

$$\frac{\delta}{x} = \frac{3.46}{\sqrt{Re_x}}$$

Determine the drag force on the plate.

9.50 Derive an equation for the drag force of the turbulent portion of a boundary layer on a flat plate when the fluid is air at standard conditions. The freestream velocity is U_∞, the plate has length, L, and width, b, and transition occurs at $Re_x = 5 \times 10^5$. Your solution should be such that given numerical values for U_∞, L, and b, you could solve for a numerical value of the drag force.

9.51 Equation 9.26 gives the local shear stress, τ_w, in terms of Re_x for turbulent boundary-layer flow at zero pressure gradient. The total skin-friction coefficient is defined as

$$\bar{C}_f = \frac{\text{drag}}{\frac{1}{2}\rho U^2 bL}$$

where b is the width and L is the length of a flat plate. Show, using Eq. 9.26, that

$$\bar{C}_f = \frac{0.072}{(UL/\nu)^{1/5}}$$

for turbulent boundary-layer flow with the "$\frac{1}{7}$-power" velocity profile. (Experiments show that the constant should be "adjusted" to a value of 0.074.)

9.52 A towboat for river barges is to be tested in a towing tank. The towboat model is built at a scale ratio of 1:13.5. Dimensions of the model are overall length, 11.1 ft, beam, 3.11 ft, and draft, 0.62 ft. (The model displacement in fresh water is 858 lb.) Estimate the average length of wetted surface on the hull. Calculate the skin friction drag force on the prototype at a speed relative to the water of 8 mph.

9.53 Resistance of a barge is to be determined from model test data. The model is constructed to a scale ratio of 1:13.5, and has length, beam, and draft of 22.0, 4.00, and 0.667 ft, respectively. The test is to simulate performance of the prototype at 8 mph. At what speed should the model be tested? Are boundary layers on the prototype laminar or turbulent? Where should boundary-layer trips be placed on the model? Estimate the skin friction drag forces for the model and prototype barges.

9.54 A flat-bottomed barge, which is 25 m long, 10 m wide, and submerged to a depth of 1.5 m, is to be pushed up a river at the rate of 8 km/hr. Estimate the power required to overcome skin friction if the water temperature is 15 C.

9.55 Consider the ship model test data presented in Figs. 7.1 and 7.2. Calculate the skin friction drag coefficients and the corresponding forces for model and prototype at $Fr = 0.5$. Verify the coefficients presented in Fig. 7.2 at $Fr = 0.5$.

9.56 A sheet of plastic material $\frac{3}{8}$ in. thick, with specific gravity, SG = 1.5, is dropped into a large tank containing water. The sheet is 2 ft high and 3 ft wide. It falls vertically. Estimate the terminal speed of the sheet, assuming that the only drag is due to skin friction, and that the boundary layers are turbulent from the leading edge.

9.57 A supertanker has a displacement of approximately 600,000 metric tons. The ship has length, $L = 300$ m, beam (width), $b = 80$ m, and draft (depth), $D = 25$ m. The ship steams at 14 knots through seawater at a mean temperature of 4 C. For these conditions, estimate (a) the thickness of the boundary layer at the stern of the ship, (b) the total skin-friction drag acting on the ship, (c) the power required to overcome the drag force, (d) the kinetic energy contained in the boundary layers on the ship, and (e) the minimum distance required to bring the ship to a stop.

9.58 As a part of the 1976 bicentennial celebration an enterprising group hung a giant American flag (59 m high and 112 m wide) from the suspension cables of the Verrazano Narrows Bridge. They apparently were reluctant to make holes in the flag to alleviate the wind force, and hence they effectively had a flat plate normal to the flow. The flag tore loose from its mountings when the wind speed reached 16 km/hr. Estimate the wind force acting on the flag at this wind speed. Should they have been surprised that the flag blew down?

9.59 A rotary mixer is constructed from two circular disks as shown. The mixer is rotated at 60 rpm in a large vessel containing a brine solution (SG = 1.1). The drag on the rods, and motion induced in the liquid, may be neglected. Estimate the minimum torque and power required to drive the mixer.

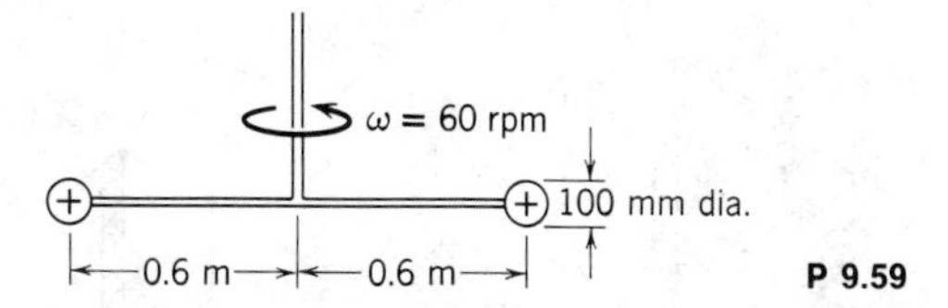

P 9.59

9.60 The vertical component of the landing speed of a parachute is to be less than 6 m/sec. The parachute may be treated as an open hemisphere. The total mass of chute and jumper is 120 kg. Determine the minimum diameter of the open parachute.

9.61 Ballistic data obtained on a firing range show that the speed of a 44 magnum revolver bullet is reduced from 250 m/sec to 210 m/sec by aerodynamic drag as it travels over a horizontal distance of 150 m. The diameter and mass of the bullet are 11.2 mm and 15.6 g, respectively. Evaluate the average drag coefficient for the bullet. (Assume standard atmosphere conditions.)

9.62 The resistance to motion of a good bicycle on smooth pavement is nearly all due to aerodynamic drag. Assume that the total mass of rider and bike is $M = 100$ kg. The frontal area measured from a photograph is $A = 0.46\ \text{m}^2$. Experiments on a hill where the road grade is 8 percent show that terminal speed is $V_t = 15$ m/sec. From these data, the drag coefficient is estimated as $C_D = 1.2$. Verify this calculation of drag coefficient. Estimate the distance needed for the bike and rider to decelerate from 15 to 10 m/sec after reaching level road.

9.63 An anemometer to measure wind speed is made from four hemispherical cups of 50 mm diameter as shown. The center of each cup is placed at $R = 75$ mm from the pivot. The anemometer is to start rotating when wind speed is above 1 km/hr. Calculate the maximum frictional torque that can be present in the pivot. Estimate the error in speed measurement caused by the pivot frictional torque at a wind speed of 10 km/hr.

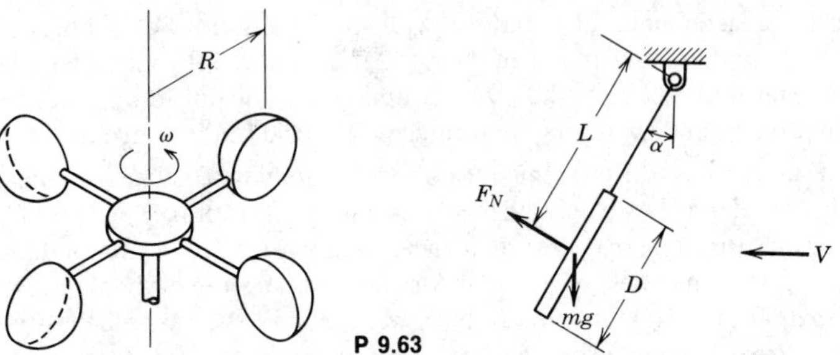

P 9.63 P 9.64

9.64 A circular disk is hung in an air stream from a pivoted strut as shown. Experiments show that the normal force coefficient,

$$C_N = \frac{F_N}{\frac{1}{2}\rho V^2 A}$$

is approximately constant at $C_N = 1.2$ over the range $0 < \alpha < 40$ degrees. In a wind-tunnel experiment performed in air at 50 ft/sec with a 1 in. diameter disk, α was measured at 40°. Determine for these conditions the weight of the disk. Assume drag on the strut and friction in the pivot are negligible.

9.65 It is proposed that a windmill be constructed by cutting a 55-gal drum into 2 longitudinal "C" sections and mounting the 2 halves on arms as shown. The diameter of each drum is 0.6 m and the length is 0.75 m. Estimate the starting torque of this apparatus.

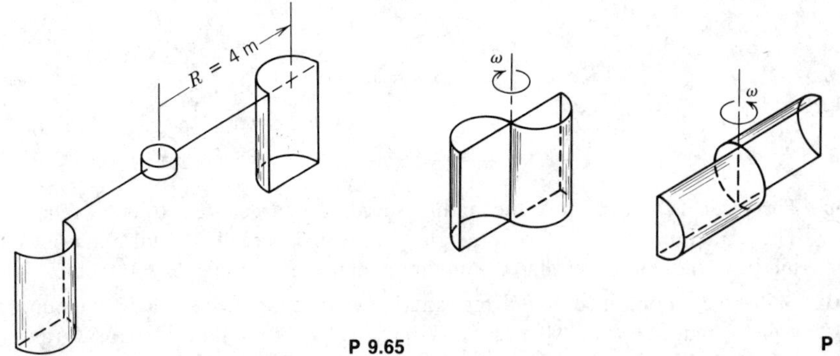

P 9.65 P 9.66

9.66 It has been proposed that surplus 55-gal oil drums be used to make simple windmills for underdeveloped countries. Two possible configurations are shown. Estimate which would be better, why, and by how much. The diameter and length of a 55-gal drum are $D = 24$ in. and $H = 29$ in.

9.67 A simple but effective anemometer to measure wind speed can be made from a thin plate hinged to deflect in the wind. Consider a thin plate made from brass (SG = 8.5) that is

20 mm high and 10 mm wide. Derive a relationship for wind speed as a function of deflection angle, θ. (Use the assumption that the aerodynamic force on the plate depends only on the velocity component normal to the surface for $\theta < 40°$.) What thickness of brass should be used to give $\theta = 30°$ at a wind speed of 10 m/sec?

9.68 An F-4 aircraft is slowed after landing by dual parachutes deployed from the rear. Each parachute is 20 ft in diameter and may be assumed to have the same drag coefficient as an open hemisphere facing upstream. The F-4 weighs 32,000 lbf, and lands at 160 knots. Estimate the time and distance required to decelerate the aircraft to 100 knots, assuming that the brakes are not used and the drag of the aircraft is negligible.

9.69 A top-notch athlete can ride a bicycle at a sustained speed of 37 km/hr on a calm day at maximum exertion. (The total mass of rider and bike is $M = 80$ kg. The rolling resistance force from the tires is $F_R = 4$ N. The drag coefficient and frontal area of the bike and rider are $C_D = 1.2$ and $A = 0.25\ m^2$.) The athlete has made a bet that he can ride at a ground speed of 30 km/hr into a headwind that blows at 10 km/hr. Determine the maximum power output that the athlete can sustain. Evaluate the athlete's prospects to win this bet.

9.70 Experimental data [13] suggest that the maximum and minimum drag area (C_DA) for a skydiver varies from about $0.85\ m^2$ for a prone, spread-eagle position to about $0.11\ m^2$ for vertical fall. Estimate the terminal speeds for a 75 kg skydiver in each position. Calculate the time and distance needed for the skydiver to reach 95 percent of terminal speed at an altitude of 3000 m on a standard day.

9.71 A field hockey ball has diameter, $D = 73$ mm, and mass, $m = 160$ g. When struck well, it leaves the stick with initial speed, $U_0 = 50$ m/sec. The ball is an essentially smooth sphere. Estimate the distance traveled in horizontal flight before the speed of the ball is reduced 10 percent by aerodynamic drag.

9.72 A VW van drives down a highway at 60 mph in standard air. The frontal area of the van is $36\ ft^2$ and the drag coefficient is 0.42. How much power is required to overcome aerodynamic drag forces?

9.73 A vehicle is built to try for the land speed record at the Bonneville Salt Flats, elevation 4400 ft. The engine delivers 500 hp to the rear wheels, and careful streamlining has resulted in a drag coefficient of 0.3, based on a $15\ ft^2$ frontal area. Compute the theoretical maximum ground speed of the car (a) in still air, and (b) with a 20 mph head wind.

9.74 A typical large American sedan has a frontal area of $23.4\ ft^2$, and a drag coefficient of 0.5. Plot a curve of horsepower required to overcome drag versus road speed in standard air. If rolling resistance is 1.5 percent of curb weight (4500 lbf), determine the speed at which the aerodynamic force exceeds frictional resistance. How much power is required to cruise at 55 mph, and at 70 mph?

9.75 A tractor-trailer rig has frontal area, $A = 102\ ft^2$, and drag coefficient, $C_D = 0.9$. Rolling resistance is 8 lbf per 1000 lbf of vehicle weight. The specific fuel consumption of the diesel engine is 0.38 lbm of fuel per horsepower hour, and drivetrain efficiency is 82 percent. The density of diesel fuel is 6.9 lbm/gal. Estimate the fuel economy of the rig at 55 mph if its gross weight is 72,000 lbf. An air deflector reduces aerodynamic drag 6 percent. The truck travels 120,000 miles per year. Calculate the fuel saved by the air deflector per year.

9.76 According to an advertisement, the Porsche 944 has the following characteristics: $C_D = 0.35$, $A = 1.83\ m^2$, and maximum power, $\mathscr{P} = 143$ bhp. The ad further states that the vehicle requires 13.9 hp to cruise at 55 mph. Use these data to estimate (a) the maximum acceleration capability at 55 mph, and (b) the top speed of the car. (Assume that rolling resistance is 1 percent of car weight.)

9.77 An object of mass, m, with cross-sectional area equal to half the size of the chute falls down a mail chute. The motion is steady. The wake area is $\frac{3}{4}$ the size of the chute at its maximum area. Use the assumption of constant pressure in the wake. Apply the continuity, Bernoulli, and momentum equations to develop an expression for terminal speed of the object in terms of its mass and other quantities.

9.78 An object falls in air down a long vertical chute. The speed of the object is constant at 3 m/sec. The flow pattern around the object is shown. The static pressure is uniform across sections ① and ②; pressure is atmospheric at section ①. The effective flow area at section ② is 20 percent of the chute area. Frictional effects between sections ① and ② are negligible. Evaluate the flow speed relative to the object at section ②. Calculate the static pressure at section ②. Determine the mass of the object.

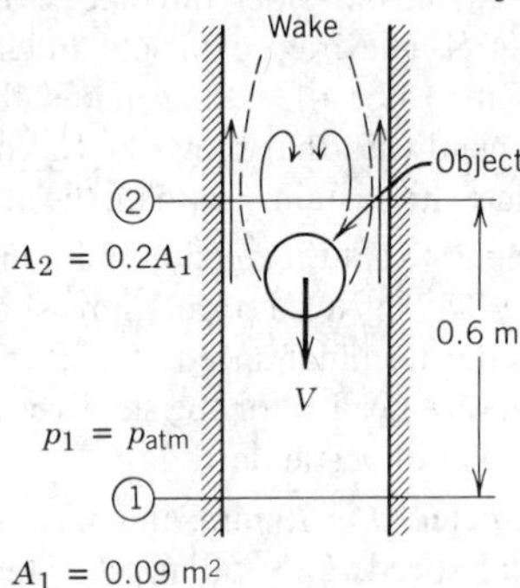

P9.78

9.79 Consider a cylindrical flag pole of height, H. For constant drag coefficient, evaluate the drag force and bending moment on the pole if the wind speed varies as $u/U = (y/H)^{1/7}$, where y is distance measured from the ground. Compare with corresponding values obtained for a uniform wind profile with constant speed, U.

9.80 A pitot-static probe with stem diameter, $d = 6$ mm, is inserted a distance, $L = 300$ mm, into a wind-tunnel air flow, where the uniform speed is 25 m/sec. Calculate the drag force and bending moment acting on the probe.

9.81 At low Reynolds number ($Re < 1$), drag of a sphere can be predicted analytically from the complete equations of motion. The result is called Stokes flow, and the drag is given by $F_D = 3\pi\mu DV$, where μ is the fluid viscosity, and D and V are the sphere diameter and relative velocity, respectively. Show that for Stokes flow, $C_D = 24/Re$. A small sphere ($D = 6$ mm) is observed to fall through castor oil at a terminal speed of 60 mm/sec. The temperature is 20 C. Compute the drag coefficient for the sphere. Determine the density of the sphere. If dropped in water, would the sphere fall slower or faster? Why?

9.82 The Stokes drag law for smooth spheres, $F_D = 3\pi\mu VD$, is to be verified experimentally by dropping steel ball bearings in glycerin. Evaluate the largest diameter steel ball (SG = 7.8) for which $Re < 10$. Calculate the height of glycerin column needed to reach 95 percent of terminal speed.

9.83 Consider small oil droplets (SG = 0.85) rising in water. Develop a relation for calculating terminal speed of a droplet, in m/sec, as a function of droplet diameter, in mm, assuming Stokes flow. For what range of droplet diameter is Stokes flow a reasonable assumption?

9.84 Hydrogen bubbles frequently are used as markers for flow visualization. Consider bubbles with diameters from 0.01 to 0.1 mm. Estimate their terminal speeds in water. (Be sure to consider the effect of surface tension on the pressure of hydrogen within the bubble.)

9.85 A dust particle falling in air is observed to settle at a speed of 2 mm/sec. The specific gravity of the particle is 4.5. Estimate its size.

9.86 In the ink-jet printing process, small spherical droplets of ink (SG = 1.2) are sprayed from a nozzle. Each droplet is electrically charged, and its trajectory is controlled by passing it through an electric field. The diameter of a typical droplet is $D = 60\ \mu\text{m}$, and the initial velocity is $V_0 = 17.5$ m/sec. The drag coefficient for a single spherical droplet in this speed range is given approximately by

$$C_D \approx \frac{10.4}{\sqrt{Re_D}}$$

Consider a single spherical droplet of ink moving horizontally in otherwise undisturbed air. Evaluate the speed of the droplet when it reaches the paper $L = 30$ mm from the nozzle.

9.87 The plot shows values of pressure difference versus angle measured for air flow around a cylinder at $Re = 80{,}000$. Use these data to estimate C_D for this flow. Compare with data from Fig. 9.13. How can you explain the difference?

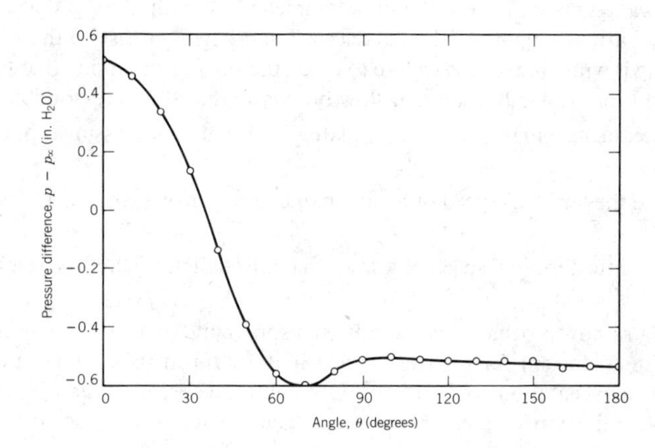

P 9.87

9.88 The following curve fit for the drag coefficient of a sphere as a function of Reynolds number has been proposed by Chow [29]:

$C_D = 24/Re$	$Re \leq 1$
$C_D = 24/Re^{0.646}$	$1 < Re \leq 400$
$C_D = 0.5$	$400 < Re \leq 3 \times 10^5$
$C_D = 0.000366\, Re^{0.4275}$	$3 \times 10^5 < Re \leq 2 \times 10^6$
$C_D = 0.18$	$Re > 2 \times 10^6$

Use data from Fig. 9.11 to evaluate the magnitude and location of the maximum error between the curve fit and data.

9.89 A spherical hydrogen-filled balloon 0.6 m in diameter exerts an upward force of 1.3 N on a restraining string when held stationary in standard air with no wind. With a wind speed of 3 m/sec, the string holding the balloon makes an angle of 60° with the horizontal. Calculate the drag coefficient of the balloon under these conditions, neglecting the weight of the string.

9.90 An antique airplane carries 15 m of external guy wires stretched normal to the direction of motion. The diameter of the wires is 6 mm. Basing calculations on two-dimensional flow around the wires, what power saving may be effected by removing the wires if the speed is 150 km/hr in standard air at sea level?

9.91 A water tower consists of a 12 m diameter sphere on top of a vertical tower 30 m tall. Estimate the bending moment exerted on the base of the tower due to the aerodynamic force imposed by a 100 km/hr wind on a standard day. Interference at the joint between the sphere and tower may be neglected.

9.92 The CB antenna on a car is 8 mm in diameter and 2 m long. Estimate the torque that tends to snap it off if the car is driven at 125 km/hr on a standard day.

9.93 A spherical balloon contains helium and ascends through standard air. The mass of the balloon and its payload is 150 kg. Determine the required diameter if it is to ascend at 3 m/sec.

9.94 A thin-walled plastic sphere 10 mm in diameter and with mass of 0.05 g, immersed in a glycerin bath at a depth of 1 m, is released and begins to rise to the surface. Calculate how long it willl take for the sphere to reach the surface, assuming that terminal speed is attained rapidly. Assume laminar flow but verify that this is reasonable.

9.95 Compute the terminal speed of 10 mm diameter hailstones (assume spheres) in standard air.

9.96 Compute the terminal speed of a $\frac{1}{8}$ in. diameter raindrop (assume spherical) in standard air.

9.97 Determine the terminal speed of a smooth tennis ball in still air. Its weight is 2 oz and its diameter is 2.5 in.

9.98 A falling raindrop behaves essentially as a sphere moving through an infinite medium. The drag coefficient for a sphere is nearly constant at 0.5 for the range of Reynolds number between about 10^3 and 10^5. Obtain an algebraic relationship for the terminal speed of a raindrop for constant drag coefficient. Evaluate numerically to obtain the form

$$V\,(\text{m/sec}) = K\sqrt{D\,(\text{mm})}$$

where K is a dimensional constant and D is the drop diameter in millimeters. Would you expect this equation to over- or under-predict the raindrop speed outside the given range of Reynolds number? Why?

9.99 A cast iron "12-pounder" cannon ball rolls off the deck of a ship and falls into the ocean at a location where the depth is 1000 m. Estimate the time that elapses before the cannon ball hits the sea bottom. (For cast iron, SG = 7.8.)

9.100 Air bubbles rise from the regulator of a scuba diver who swims at a depth of 10 m in seawater. Consider a bubble of 10 mm diameter at this depth, which begins to rise. Evaluate the terminal speed of this bubble as a function of depth. Estimate the time needed for the bubble to rise to the surface. Use numerical or graphical integration if necessary.

9.101 Coastdown tests, performed on a level road on a calm day, can be used to measure the aerodynamic drag and rolling resistance coefficients for a full-scale vehicle. Rolling resistance is estimated from dV/dt measured at low speed, where aerodynamic drag is small. The rolling resistance then is deducted from dV/dt measured at high speed to determine the aerodynamic drag. The following data were obtained during a test with vehicle weight, $W = 25{,}000$ lbf, and frontal area, $A = 79\ \text{ft}^2$:

V (mph)	5	55
$\dfrac{dV}{dt}\left(\dfrac{\text{mph}}{\text{sec}}\right)$	-0.150	-0.475

Evaluate the aerodynamic drag coefficient for this vehicle.

9.102 Motion of a small rocket was analyzed in Example Problem 4.12 under the assumption that aerodynamic drag was negligible. This was not realistic at the final calculated speed of 369 m/sec. Write a simple computer or calculator program to evaluate the rocket speed as a function of time, assuming $C_D = 0.3$ and a rocket diameter of 700 mm. Compare with results from Example Problem 4.12.

9.103 A light plane tows an advertising banner over a football stadium on a Saturday afternoon. The banner is 1 m tall and 30 m long. According to Hoerner [14], the drag coefficient based on area (Lh) for such a banner is approximated by $C_D = 0.05\ L/h$, where L is the banner length, and h is the banner height. Estimate the power required to tow the banner at $V = 90$ km/hr. Compare with the drag of a rigid flat plate. Why is the drag larger for the banner?

9.104 The guy wires on the antique plane in Problem 9.90 are to be streamlined rather than removed. Data from Fig. 9.14 may be used to select an optimum faired shape for the guys. Evaluate the maximum power saving that results from fairing the wires.

9.105 The bicycle of Problem 9.69 is equipped with a fairing to reduce aerodynamic drag. The fairing reduces C_D to 0.90 but increases the frontal area to $0.30\ \text{m}^2$. Calculate the rider's top speed with the fairing installed.

9.106 Air moving over an automobile is accelerated to speeds higher than the travel speed as shown in Fig. 9.25. This causes changes in interior pressure when windows are opened or closed. Use the data of Fig. 9.25 to estimate the pressure reduction when a window is opened slightly at a speed of 100 km/hr. What is the air speed in the freestream near the window opening?

9.107 A light plane has a 10 m effective wingspan and a 1.8 m chord. It was originally designed to use a conventional (NACA 23015) airfoil section. With this airfoil its cruising speed on a standard day near sea level is 225 km/hr. A conversion to a laminar-flow (NACA 66_2–215) section airfoil is proposed. Determine the cruising speed that could be achieved with the new airfoil section for the same power.

9.108 An aircraft is flying in level flight at a speed of 250 km/hr through air at standard conditions. The lift coefficient at this speed is 0.4 and the drag coefficient is 0.065. The mass of the aircraft is 850 kg. Calculate the effective lift area for the craft.

9.109 Consider a kite of mass 0.2 kg as a flat plate with an area of $1\ \text{m}^2$. It is flown in standard air moving horizontally at 10 m/sec. The kite makes an angle of 5° with the horizontal. Assume the lift coefficient is given by the equation, $C_L = 2\pi \sin\alpha$, where α is the angle of attack. If the string makes an angle of 60° with the horizontal, determine the tension in the string.

9.110 The foils of a hydrofoil type watercraft have an effective area of $0.7\ \text{m}^2$. Their coefficients of lift and drag are 1.6 and 0.5, respectively. The total mass of the craft in running trim is

1800 kg. Determine the minimum speed at which the craft is supported by the hydrofoils. At this speed, find the power required to overcome water resistance. If the craft is fitted with a 110 kW engine, estimate its top speed.

9.111 An airplane with an effective lift area of 25 m^2 is fitted with airfoils of NACA 23012 section (Fig. 9.23). The maximum flap setting that can be used at takeoff corresponds to configuration ② in Fig. 9.23. Determine the maximum gross mass possible for the airplane if its takeoff speed is 150 km/hr (neglect added lift due to ground effect).

9.112 An airplane with mass of 4500 kg is flown at constant elevation and speed on a circular path at 250 km/hr. The flight circle has a radius of 1000 m. The plane has an effective lifting area of 22 m^2, and is fitted with NACA 23015 section airfoils. Determine the drag on the aircraft, and the power required.

9.113 Jim Hall's Chaparral 2F sports-racing cars used airfoils mounted above the rear suspension to enhance stability and improve braking performance. The airfoil is 6 ft wide (span) and has a chord of 1 ft. Its angle of attack is variable between 0 and minus 12 degrees. Assume lift and drag coefficient data are given by curves (for conventional section) in Fig. 9.17. Consider a car speed of 120 mph on a calm day. For an airfoil deflection of 12° down, calculate (a) the maximum downward force, and (b) the maximum increase in braking thrust produced by the airfoil.

9.114 A tennis ball with mass of 57 g and diameter of 64 mm is dropped in standard sea-level air. Calculate the terminal speed of the ball. Estimate the time and distance required for the ball to reach 95 percent of its terminal speed.

9.115 A golf ball (diameter, $D = 43$ mm) is hit from a sand trap at 20 m/sec with backspin of 12,000 rpm. The mass of the ball is 48 g. Evaluate the lift and drag forces acting on the ball. Express your results as fractions of the body force due to gravity acting on the ball.

9.116 Performance of jet aircraft is considered in Example Problem 9.8. Show analytically that at the optimum cruise speed, $C_{D,i} = \frac{1}{3}C_{D,0}$.

9.117 Performance of jet aircraft is considered in Example Problem 9.8. Show analytically that at the speed for maximum endurance, $C_{D,i} = C_{D,0}$.

9.118 The wing loading of the Gossamer Condor is 0.4 lbf/ft^2 of wing area. Crude measurements showed drag was approximately 6 lbf at a speed of 12 mph. The total weight of the Condor was 200 lbf. The effective aspect ratio of the Condor is 17. Estimate the minimum power required to fly the aircraft. Compare to the 0.39 hp that pilot Brian Allen could sustain for 2 hr.

9.119 The glide angle for unpowered flight is such that lift, drag, and weight are in equilibrium. Show that the glide slope angle, θ, is such that

$$\tan\theta = \frac{C_D}{C_L}$$

The minimum glide slope occurs at the speed where C_L/C_D is a maximum. For the conditions of Example Problem 9.8, evaluate the minimum glide slope angle for a Boeing 727-200. How far could this aircraft glide from an initial altitude of 10 km on a standard day?

9.120 A baseball pitcher throws a ball at 90 km/hr. Home plate is 18 m away from the pitcher's mound. What spin should be placed on the ball for maximum horizontal deviation from a straight path? (A baseball has $m = 145$ g and $D = 74$ mm.) How far will the ball deviate from a straight line?

Chapter 10

FLOW IN OPEN CHANNELS

Many flows in nature occur with a *free surface*. Because free surface flows differ in several important respects from flows in closed conduits, they are treated separately in this chapter. Rainwater runoff, and flows in rivers, aqueducts, irrigation canals, and drainage ditches are familiar examples where the free surface is at atmospheric pressure. Geometric properties of common open-channel shapes are presented in Section 10-1.

Surface waves can form in flows with a free surface. The propagation speed of a single, or solitary, wave is analogous in many respects to the propagation of a sound wave in a compressible fluid medium. The propagation of disturbances in open-channel flow depends on the value of the Froude number of the flow (Section 10-2). Changes in channel cross section or depth, and their effects on the mean flow velocity, also are distinguishing features of free surface flows.

In contrast to flow in a closed conduit where the flow is sustained by a pressure difference, the principal driving force for open-channel flow is gravity. The gravity force is opposed by a friction force on the solid boundaries of the channel.

Most flows of interest are large in physical scale, so the Reynolds numbers generally are large. Consequently, open-channel flow seldom is laminar. As in the case of turbulent flow in pipes, we must rely on empirical correlations to relate frictional effects to the average velocity of flow. The empirical correlation is included through a head loss term in the energy equation (Section 10-3). Additional complications in many practical cases include the presence of sediment or other particulate matter in the flow and the erosion of earthen channels or structures by water action.

In this chapter we shall analyze several aspects of steady open-channel flow using the basic control volume equations of Chapter 4. In Section 10-4 we consider flows in which the effects of area change predominate and frictional forces may be neglected. When the flow cross section does not vary in the direction of the flow, the flow is said to be at normal depth, or uniform (Section 10-5). For flow at normal depth, the liquid surface is parallel to the channel bed. This is analogous to fully developed flow in a pipe. When the liquid depth is not constant, we have varied flow. In Section 10-6 we consider gradually varied flow. The major objective in the analysis of gradually varied flow is to predict the shape of the free surface. When conditions require the flow to change in depth abruptly, this is accomplished through a hydraulic jump (Section 10-7). The chapter concludes with a brief discussion of flow measurement techniques for use in open channels (Section 10-8).

10-1 CHARACTERISTICS OF OPEN CHANNELS

Any conduit with a liquid free surface is classified as an open channel; examples of natural channels are abundant. Man-made channels are given many different names, including canal, flume, or culvert, which are defined and applied rather loosely. A *canal* usually is excavated below ground level, and may be unlined or lined. Canals generally are long and of very mild slope; they are used to carry irrigation or storm water or for navigation. A *flume* usually is built above ground level to carry water across a depression. A *culvert*—which usually is designed to flow only part full—is a short covered channel used to drain water under highway or railroad embankments.

Channels may be constructed in a variety of cross-sectional shapes; usually regular geometric shapes are used. A channel with constant slope and cross section is termed *prismatic*. Lined canals often are built with rectangular or trapezoidal sections; smaller troughs or ditches sometimes are triangular. Culverts and tunnels generally are circular or elliptical in section. Natural channels are highly irregular and nonprismatic, but often they are approximated by trapezoid or paraboloid sections. Geometric properties of common open-channel shapes are summarized in Table 10.1. The depth of flow, y, is the perpendicular distance measured from the channel bed to the free surface. The flow area, A, is the cross section of the flow perpendicular to the flow direction. The wetted perimeter, P, is the length of the solid channel surface in contact with the liquid. The hydraulic radius, R_h, is defined as

$$R_h = \frac{A}{P} \tag{10.1}$$

Note that for flow in noncircular closed conduits (Section 8-8.3) the hydraulic diameter was defined as

$$D_h = 4\frac{A}{P} \tag{8.44}$$

Thus, for a circular pipe, the hydraulic diameter, from Eq. 8.44, is equal to the pipe diameter. From Eq. 10.1 the hydraulic radius is half the pipe radius. The hydraulic radius as defined by Eq. 10.1 is commonly used to analyze open-channel flows, so it will be used throughout this chapter.

For nonrectangular channels, the *hydraulic depth* is defined as $y_h = A/b_s$, where b_s is the surface width. The hydraulic depth represents the average depth of the channel at any cross section. It gives the depth of an equivalent rectangular channel.

Typical contours of streamwise velocity for a number of open-channel sections are shown in Fig. 10.1. Although the profiles are not uniform, the approach followed is to assume uniform flow at a section. The kinetic energy coefficient, α, is taken as unity.

Note from Fig. 10.1 that the measured maximum velocity occurs below the free surface. Since there is negligible shear stress due to air drag, one would expect the

Full-scale automobile under test in 5.4 $\times$ 10.4 m wind tunnel at General Motors Technical Center, Warren, Michigan. The photograph shows streaklines for flow over a full-size 1984 Chevrolet Corvette. The streaklines are formed from a mixture of liquid nitrogen and steam. Colors are introduced by multiple exposures made

Contemporary sports-racing car, showing aerodynamic design features. This photograph shows the B.F. Goodrich Mazda-powered Lola T616 in action. This automobile, built for endurance racing in the IMSA GTP class, has a turbo-charged engine that develops 300 hp. The car is capable of speeds approaching 200 mph. To achieve this performance requires careful attention to streamlining for low drag and to aerodynamic downforce for stability and high cornering speeds. The photo shows the carefully streamlined mirrors, flush inlet ducts, and other details needed to achieve low drag. The low front contour, underbody details, and the rear wing create downforce for stability and high-speed cornering performance. (Photograph courtesy of B.F. Goodrich)

Table 10.1 Geometric Properties of Common Open-Channel Shapes

Shape	Section	Flow Area, A	Wetted Perimeter, P	Hydraulic Radius, R_h
Trapezoidal	b_s, y, α, b	$y(b + y\cot\alpha)$	$b + \dfrac{2y}{\sin\alpha}$	$\dfrac{y(b + y\cot\alpha)}{b + \dfrac{2y}{\sin\alpha}}$
Triangular	b_s, y, α	$y^2\cot\alpha$	$\dfrac{2y}{\sin\alpha}$	$\dfrac{y\cos\alpha}{2}$
Rectangular	y, b	by	$b + 2y$	$\dfrac{by}{b + 2y}$
Wide Flat	y, $b>>y$	by	b	y
Circular	b_s, α, y, D	$(\alpha - \sin\alpha)\dfrac{D^2}{8}$	$\dfrac{\alpha D}{2}$	$\dfrac{D}{4}\left(1 - \dfrac{\sin\alpha}{\alpha}\right)$

maximum velocity to occur at the free surface. Secondary flows are responsible for distorting the axial velocity profile.[1]

The nonuniform profile of axial velocity causes strong secondary flows at the base of an obstruction, such as a bridge pier. The location of the maximum velocity is well

[1] The film, *Secondary Flow*, E. S. Taylor, principal, illustrates several examples of secondary flow phenomena.

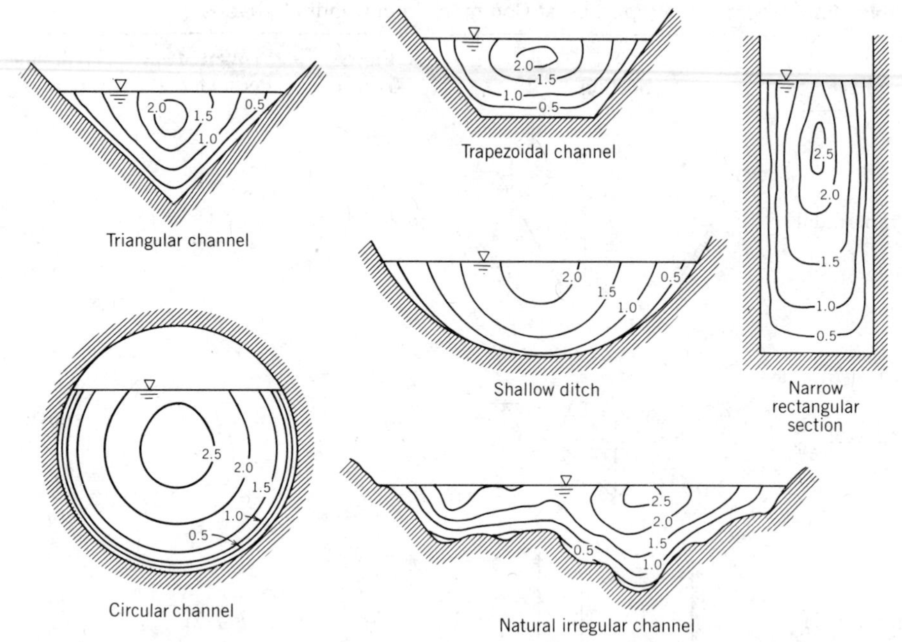

Fig. 10.1 Typical contours of equal velocity in open-channel sections. (From [1], used by permission.)

above the channel floor; the high stagnation pressure at this location causes a recirculation zone to form at the base of the front of the obstruction. As this swirling flow moves around the sides of the obstruction, it forms a horseshoe vortex. The vortex core is stretched and the swirl velocities are increased along the sides of the obstruction.[2] The high velocities present in the vortex can seriously erode the bottom of a natural channel along the sides of a pier.

10-2 PROPAGATION OF SURFACE WAVES

Consider an open channel with a movable end wall, containing a liquid initially at rest. If the end wall is given a sudden displacement as in Fig. 10.2*a*, a small wave forms and travels down the channel. The speed of wave propagation, the wave *celerity*, is denoted by c.

10-2.1 Wave Speed

The wave speed may be computed by applying the basic equations. As seen by a fixed observer, the wave propagation is an unsteady flow phenomenon. However, the flow appears steady to an observer *on* a differential control volume that moves with the wave, Fig. 10.2*b*.

[2] Formation and stretching of vortices are shown in the film, *Vorticity*, A. H. Shapiro, principal.

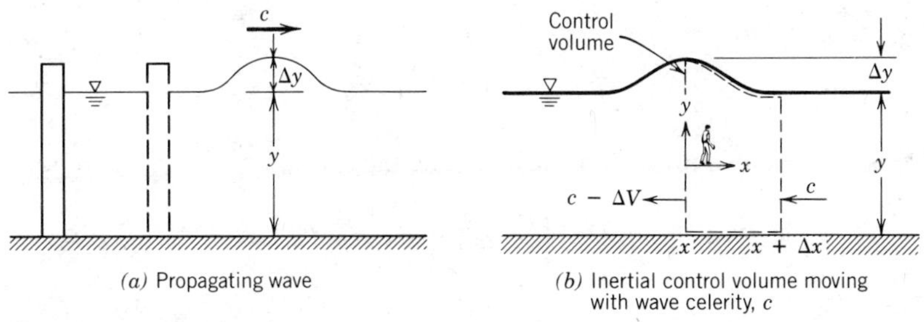

Fig. 10.2 Generation of a small solitary wave in a quiescent liquid.

a. Continuity Equation

Basic equation:
$$0 = \overset{=0(1)}{\cancel{\frac{\partial}{\partial t}}} \int_{\text{CV}} \rho \, d\forall + \int_{\text{CS}} \rho \vec{V} \cdot d\vec{A} \tag{4.14}$$

Assumptions: (1) Steady flow
(2) Uniform flow at a section

Then
$$0 = \{+|\rho b(y + \Delta y)(c - \Delta V)|\} + \{-|\rho byc|\} \tag{10.2a}$$
$$0 = yc - y\Delta V + c\Delta y - \Delta y \Delta V - yc$$

Solving for ΔV,

$$\Delta V = c \frac{\Delta y}{y + \Delta y} \tag{10.2b}$$

b. Momentum Equation

Basic equation:

$$F_{S_x} + \overset{=0(6)}{\cancel{F_{B_x}}} = \overset{=0(1)}{\cancel{\frac{\partial}{\partial t}}} \int_{\text{CV}} u\rho \, d\forall + \int_{\text{CS}} u\rho \vec{V} \cdot d\vec{A} \tag{4.19a}$$

Assumptions: (3) Hydrostatic pressure variation (this will be exactly true if streamline curvature effects are negligible, i.e. for small Δy), $\frac{dp}{dy} = -\rho g$
(4) Incompressible, ρ = constant
(5) No viscous or surface tension effects
(6) $F_{B_x} = 0$

In the absence of viscous and surface tension effects, F_{S_x} will be due to pressure forces only. Since the pressure variation is hydrostatic at both vertical faces of the control volume,

$$\frac{dp}{dy} = -\rho g \qquad \text{and} \qquad p = \rho g(y_s - y)$$

where y_s is the distance to the free surface. The magnitude of the pressure force is

$$F_{S_x} = \int p\,dA = \int_0^{y_s} pb\,dy = \int_0^{y_s} \rho g(y_s - y)b\,dy$$

$$F_{S_x} = \rho g b\left[yy_s - \frac{y^2}{2}\right]_0^{y_s} = \frac{\rho g b y_s^2}{2}$$

Then

$$F_{S_x}]_{x+\Delta x} = -\frac{\rho g b}{2}\,y^2$$

$$F_{S_x}]_x = \frac{\rho g b}{2}\,(y + \Delta y)^2$$

Substituting for the pressure forces into the momentum equation, we obtain

$$\frac{\rho g b}{2}(y + \Delta y)^2 - \frac{\rho g b}{2}\,y^2 = -|c - \Delta V|\{+|\rho b(y + \Delta y)(c - \Delta V)|\} - |c|\{-|\rho b y c|\}$$

The two terms { } in this equation are equal by continuity, Eq. 10.2a, so the momentum equation reduces to

$$gy\,\Delta y + g\frac{(\Delta y)^2}{2} = \Delta V y c$$

$$g\left(1 + \frac{\Delta y}{2y}\right)\Delta y = \Delta V c \tag{10.3}$$

Combining Eqs. 10.2b and 10.3, we obtain

$$c^2 = g\left(1 + \frac{\Delta y}{2y}\right)(y + \Delta y)$$

or

$$c^2 = gy\left(1 + \frac{\Delta y}{2y}\right)\left(1 + \frac{\Delta y}{y}\right) \tag{10.4}$$

In our development of Eq. 10.4, we assumed a hydrostatic pressure variation in the liquid. A careful study of the details of wave motions [2] indicates that this is a good assumption, provided that the wavelength, λ, is long compared to the liquid depth. Such waves are called *shallow water waves.*

Thus, for the case $\Delta y \ll y$, there will be negligible variation in propagation speed across the wave and our assumption of hydrostatic pressure variation is reasonable. We may use

$$c = \pm\sqrt{gy} \tag{10.5}$$

to represent the speed of such waves.

The celerity, c, depends on the local depth, y. Consequently, c will be larger at the peak of the wave than at the leading or trailing edge. Thus real waves of finite amplitude get steeper as they travel. This fact is responsible for the "breaking" of waves on a beach.[3]

Example 10.1

Calculate and plot the speeds of propagation of isolated waves for water depths in the range from 10 mm to 10 km. Comment on the significance of the speed of propagation for the average ocean depth of 4 km.

EXAMPLE PROBLEM 10.1

GIVEN: Isolated wave propagation.

FIND: Evaluate and plot speeds of propagation for depths from 10 mm to 10 km. Comment on speed for average ocean depth of 4 km.

SOLUTION:

From Eq. 10.5, $c = \pm\sqrt{gy}$, where y is water depth. The equation will plot as a straight line on log-log coordinates. At $y = 4$ km,

$$c = \sqrt{gy} = \left[\frac{9.81\ \text{m}}{\text{sec}^2} \times 4\ \text{km} \times \frac{1000\ \text{m}}{\text{km}}\right]^{1/2} = 198\ \text{m/sec} \qquad c$$

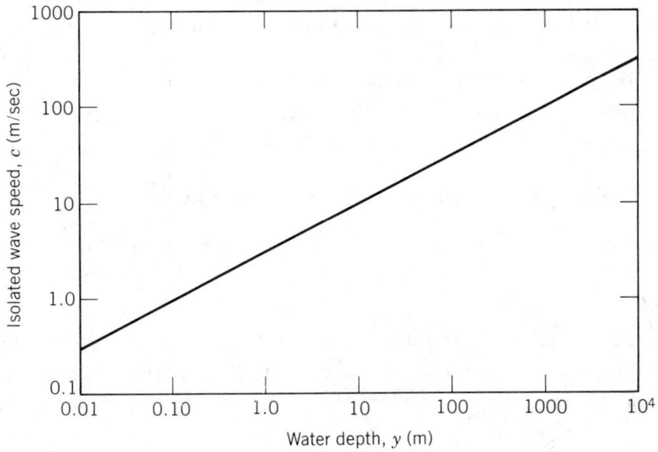

At the average ocean depth of 4 km, the wave speed is 198 m/sec (713 km/hr). Tidal waves (tsunamis) generated by underwater earthquakes propagate rapidly; they can do tremendous damage near a coastline.

[3] These and other surface wave phenomena are illustrated in the film, *Waves in Fluids*, A. E. Bryson, principal.

10-2.2 The Froude Number

Most phenomena of interest take place in moving streams. The analysis of Section 10-2.1 would apply equally well to a small wave propagating on the surface of a stream moving with speed, V. The wave speed seen by a fixed observer would be

$$V_w = V \pm c = V \pm \sqrt{gy}$$

Consequently V_w can take on any value depending on the magnitudes of V and $\sqrt{gy}$. The wave speed, V_w, can be negative—a wave can move upstream—only when $V < \sqrt{gy}$. Thus we see that the character of the flow changes at the condition when $V = \sqrt{gy}$, or when

$$\frac{V}{\sqrt{gy}} = Fr = 1 \tag{10.6}$$

where Fr is the Froude number introduced in Chapter 7. When $Fr = 1$, the character of the wave motion changes.

Open-channel flows[4] may be classified on the basis of Froude number:

$Fr < 1$	Flow is *subcritical, tranquil*, or *streaming*. Disturbances can travel upstream; downstream conditions can affect the flow upstream.
$Fr = 1$	Flow is *critical*.
$Fr > 1$	Flow is *supercritical, rapid*, or *shooting*. No disturbance can travel upstream; downstream conditions cannot be felt upstream.

These regimes of flow behavior are qualitatively analogous to the subsonic, sonic, and supersonic regimes of gas flow discussed in Chapter 11.

10-3 ENERGY EQUATION FOR OPEN-CHANNEL FLOW

As in the case of pipe flow, friction in open-channel flows results in a loss of mechanical energy; this can be characterized by a head loss. The effect of friction is particularly important in long channels.

Consider flow through a long rectangular channel of width, b, and bed slope, $S_b = \tan\theta$, where S_b is assumed to be small. The flow depth may vary. To derive a suitable form of the energy equation for open-channel flow, we assume uniform flow at each section. The control volume used for the analysis is shown in Fig. 10.3. Note that the coordinate, z, indicates distances measured in the vertical direction; distances measured normal to the bed are denoted by y.

[4] The Froude number for nonrectangular channels must be based on the hydraulic depth so that

$$Fr = \frac{V}{\sqrt{gy_h}}$$

For a rectangular channel, $y_h = y$, so this equation reduces to the expression given above.

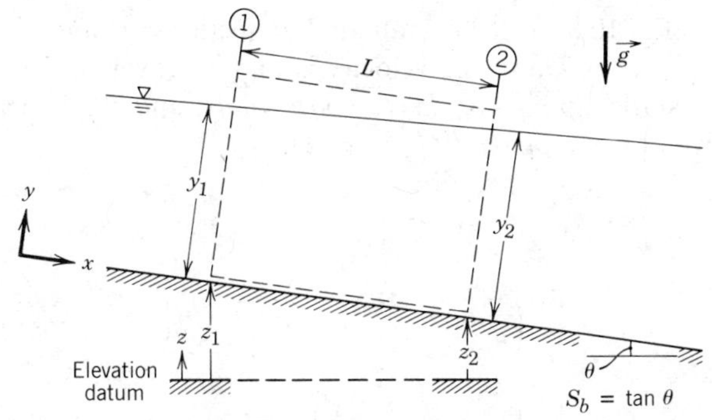

Fig. 10.3 Control volume and coordinate definitions for energy analysis of open-channel flow.

The energy equation for a control volume is

$$\dot{Q} + \underbrace{\dot{W}_s}_{=0(6)} + \underbrace{\dot{W}_{\text{shear}}}_{=0(6)} + \underbrace{\dot{W}_{\text{other}}}_{=0(6)} = \underbrace{\frac{\partial}{\partial t}}_{=0(1)}\int_{\text{CV}} e\rho\, d\forall + \int_{\text{CS}}\left(u + pv + \frac{V^2}{2} + gz\right)\rho \vec{V}\cdot d\vec{A} \tag{4.59}$$

The equation is to be applied to the control volume of Fig. 10.3 under the assumptions:

(1) Steady flow
(2) Incompressible flow
(3) Uniform flow at a section
(4) Depth varies gradually so that pressure distribution is hydrostatic
(5) Bed slope is small, $\theta \sim \sin\theta \sim \tan\theta = S_b$
(6) $\dot{W}_s = \dot{W}_{\text{shear}} = \dot{W}_{\text{other}} = 0$

Under these assumptions, the energy equation reduces to

$$\dot{Q} = \int_{\text{CS}}\left(u + pv + \frac{V^2}{2} + gz\right)\rho \vec{V}\cdot d\vec{A} \tag{10.7}$$

As a result of friction, mechanical energy will be dissipated between sections ① and ②. As in the case of pipe flow, the dissipation is characterized by a head loss. To accomplish this, Eq. 10.7 is rewritten as

$$\int_{\text{CS}}\left(pv + \frac{V^2}{2} + gz\right)\rho \vec{V}\cdot d\vec{A} + \int_{\text{CS}} u\,\rho \vec{V}\cdot d\vec{A} - \dot{Q} = 0$$

$$\int_{\text{CS}}\left(pv + \frac{V^2}{2} + gz\right)\rho \vec{V}\cdot d\vec{A} + (u_2 - u_1)\dot{m} - \frac{\delta Q}{dm}\dot{m} = 0$$

$$\int_{\text{CS}}\left(pv + \frac{V^2}{2} + gz\right)\rho \vec{V}\cdot d\vec{A} + \dot{m}h_\ell = 0 \tag{10.8}$$

The integral of Eq. 10.8 must be evaluated at sections ① and ②. At section ①, $dA = b\,dy$, $z = z_1 + y\cos\theta$, and the velocity is uniform over the area. The pressure variation is hydrostatic; $dp = -\rho g\,dz = -\rho g\cos\theta\,dy$, and the hydrostatic pressure distribution is given by $p = \rho g\cos\theta(y_1 - y)$. Thus

$$\int_{CS_1}\left(pv + \frac{V^2}{2} + gz\right)\rho\vec{V}\cdot d\vec{A}$$

$$= \int_0^{y_1}\left(g\cos\theta(y_1 - y) + \frac{V_1^2}{2} + g(z_1 + y\cos\theta)\right)\{-|\rho V_1 b\,dy|\}$$

$$= \int_0^{y_1}\left(g\cos\theta\,y_1 + \frac{V_1^2}{2} + gz_1\right)\{-|\rho V_1 b\,dy|\}$$

Since all of the terms under the integral sign are constant,

$$\int_{CS_1}\left(pv + \frac{V^2}{2} + gz\right)\rho\vec{V}\cdot d\vec{A} = -\dot{m}_1\left(\frac{V_1^2}{2} + gz_1 + g\cos\theta\,y_1\right)$$

The form of the integral at section ② is identical to that at section ① (except the sign will be positive). Substituting into Eq. 10.8 gives

$$\left(\frac{V_2^2}{2} + gz_2 + g\cos\theta\,y_2\right) - \left(\frac{V_1^2}{2} + gz_1 + g\cos\theta\,y_1\right) + h_\ell = 0 \tag{10.9}$$

For small slopes, $\cos\theta \approx 1$. To obtain head loss in dimensions of length (head loss per unit weight rather than per unit mass), Eq. 10.9 is divided by g. Thus

$$\frac{V_1^2}{2g} + y_1 + z_1 = \frac{V_2^2}{2g} + y_2 + z_2 + h_\ell \tag{10.10}$$

For flow without friction there is no head loss and the energy equation (Eq. 10.10) becomes

$$\frac{V_1^2}{2g} + y_1 + z_1 = \frac{V_2^2}{2g} + y_2 + z_2$$

10-3.1 Specific Energy

The energy equation for open-channel flow (Eq. 10.10) contained the sum of the terms, $V^2/2g + y$, on both sides. This sum is defined as the *specific energy* (or *specific head*) and denoted by the symbol, E

$$E = \frac{V^2}{2g} + y \tag{10.11}$$

where the flow depth, y, is measured normal to the channel bed.

For uniform flow at a section, the velocity can be written in terms of the volume flow rate. From continuity, $Q = AV$, and

$$E = \frac{Q^2}{2gA^2} + y \tag{10.12}$$

For a given flow rate, the specific energy, E, is a function of the depth, y. The depth also appears in the expression for the area (Table 10.1). To illustrate the relation between flow depth and specific energy at constant volume flow rate, consider the flow in a rectangular channel of width, b. Then $A = by$ and

$$E = \frac{Q^2}{2gb^2y^2} + y \tag{10.13}$$

The variation of depth as a function of specific energy for a given flow rate is plotted in Fig. 10.4; curves for several values of Q (actually $Q^2/2gb^2$) are shown.

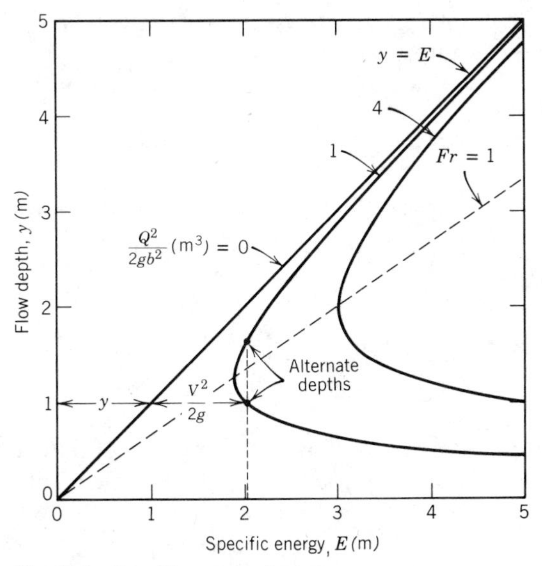

Fig. 10.4 Specific energy diagram.

When $Q = 0$, then $y = E$; this limiting case is a 45° line on the plot. For a given flow rate ($Q > 0$) and specific energy, there are two possible values of depth, y. These two depths are called *alternate depths*. The curve for constant Q gives the locus of all possible values of depth and the corresponding specific energy that satisfy Eq. 10.13. As Q increases, the curves are displaced to the right.

For any value of E, the horizontal distance from the vertical axis ($E = 0$) to the line, $y = E$, gives the depth, y. The distance from the line, $y = E$, to the Q curve is then equal to the kinetic energy, $V^2/2g$; this is shown in Fig. 10.4.

For each curve representing a given flow rate in Fig. 10.4, there is a value of depth that gives a minimum E. We may determine this value of y by differentiating Eq. 10.13; E will be a minimum when

$$\frac{dE}{dy} = -\frac{Q^2}{gb^2y^3} + 1 = 0 \tag{10.14}$$

Solving for y, we obtain

$$y = \left[\frac{Q^2}{gb^2}\right]^{1/3}$$

Substituting this result ($Q^2/gb^2 = y^3$) into Eq. 10.13 gives

$$E_{\text{min}} = \tfrac{1}{2}y + y = \tfrac{3}{2}y$$

The locus of minimum values of E is thus a straight line with $y = \frac{2}{3}E_{\text{min}}$.

We can solve for the velocity at E_{min} from Eq. 10.14,

$$V^2 = \frac{Q^2}{b^2 y^2} = gy$$

Thus, at the minimum value of E, $Fr = V/\sqrt{gy} = 1.0$, and the condition corresponds to critical flow. The depth at E_{min} is termed the critical depth, y_c. Thus, for flow in a rectangular channel of width, b,

$$y_c = \left[\frac{Q^2}{gb^2}\right]^{1/3} \qquad \text{and} \qquad E_{\text{min}} = \frac{3}{2}\,y_c$$

From Eq. 10.11

$$E_{\text{min}} = \frac{3}{2}\,y_c = \frac{V_c^2}{2g} + y_c$$

so

$$\frac{V_c^2}{2g} = \frac{y_c}{2} \tag{10.15}$$

We can investigate the nature of the flow on the branches of the curve above and below the critical depth by writing an expression for the Froude number. From continuity

$$Q = V_c b y_c = V b y \tag{10.16}$$

Then, using Eqs. 10.15 and 10.16,

$$Fr = \frac{V}{\sqrt{gy}} = \frac{V_c y_c}{y\sqrt{gy}} = \frac{\sqrt{gy_c}\,y_c}{y\sqrt{gy}} = \left[\frac{y_c}{y}\right]^{3/2} \tag{10.17}$$

On the upper branch of the curve,

$y > y_c$, so $Fr < 1$; the flow is subcritical

On the lower branch of the curve,

$y < y_c$, so $Fr > 1$; the flow is supercritical

For nonrectangular channels, the channel depth varies across the width. For these cases, evaluation of the critical depth (and the corresponding critical speed and area) usually requires a trial and error solution. At the minimum specific energy, differentiating Eq. 10.12 with respect to y at constant Q gives

$$\frac{dE}{dy} = 0 = -\frac{Q^2}{gA^3}\frac{dA}{dy} + 1 \tag{10.18}$$

Since $dA = b_s\,dy$, where b_s is the channel width at the free surface, then

$$\frac{gA_c^3}{b_{sc}Q^2} = 1 \qquad \text{at} \qquad E = E_{\min} \tag{10.19}$$

Both A_c and b_{sc} are functions of the critical depth, y_c. For a specific cross-sectional shape, Eq. 10.19 can be solved numerically for the critical depth, y_c.

The critical area, A_c, corresponding to $Fr = 1$ is

$$A_c = \left[\frac{b_s Q^2}{g}\right]^{1/3} \tag{10.20}$$

From continuity, $Q = V_c A_c$. Solving Eq. 10.19 for V_c, we obtain

$$V_c = \left[\frac{Q^2}{A_c^2}\right]^{1/2} = \left[\frac{gA_c}{b_{sc}}\right]^{1/2} = [g y_{hc}]^{1/2} \tag{10.21}$$

where y_{hc} is the hydraulic depth at critical conditions. Thus the Froude number is unity for flow at critical conditions, which correspond to minimum specific energy.

For a rectangular channel, $A_c = by_c$, and $b_{sc} = b$; Eq. 10.21 gives $V_c = (gy_c)^{1/2}$ (which is the same result as obtained from Eq. 10.15).

Near the minimum value of E, the rate of change of y with E is nearly infinite. Even small changes in E, due to channel irregularities or disturbances, can cause pronounced changes in fluid depth. Thus surface waves usually form when a flow is near critical conditions. Long runs of near-critical flow consequently are avoided in practice.

10-4 FRICTIONLESS FLOW: EFFECT OF AREA CHANGE

We begin by considering two simple flow cases in which the channel bed is horizontal, the effects of channel cross section (area change) predominate, and the effect of friction may be neglected. Since the flow is assumed to be frictionless, the energy equation (Eq. 10.8) reduces to a form of the Bernoulli equation. These flow cases are analyzed using the Bernoulli and continuity equations.

10-4.1 Flow over a Bump

Consider frictionless flow in a horizontal rectangular channel of constant width, b, with a bump in the channel bed, as illustrated in Fig. 10.5. The bump has height, $h(x)$, above the horizontal bed of the channel; the water depth, $y(x)$, is measured from the local channel bottom surface. The flow is assumed to be uniform at each section. We are interested in investigating the shape of the free surface as the flow passes over the bump.

Since the flow is steady, incompressible, and frictionless, we may apply the Bernoulli equation along a streamline,

$$\frac{p}{\rho} + \frac{V^2}{2} + gz = \text{constant} \tag{6.9}$$

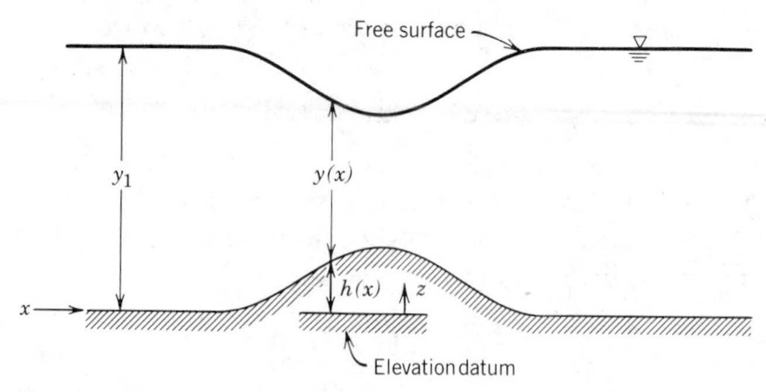

Fig. 10.5 Open-channel flow over a bump in a horizontal channel bed.

Applying the equation along the free surface streamline between upstream location ① and a point above the bump, we obtain

$$\frac{p_1}{\rho g} + \frac{V_1^2}{2g} + y_1 = \frac{p}{\rho g} + \frac{V^2}{2g} + y + h$$

Since the pressure is atmospheric along the free surface, $p_1 = p = p_{\text{atm}}$, and

$$\frac{V_1^2}{2g} + y_1 = \frac{V^2}{2g} + y + h = \text{constant} \tag{10.22}$$

The volume flow rate is a constant. Thus, from continuity for steady, incompressible, uniform flow at a section,

$$\frac{Q}{b} = V_1 y_1 = Vy$$

Substituting for V_1 and V into Eq. 10.22 yields

$$\frac{Q^2}{2gb^2y_1^2} + y_1 = \frac{Q^2}{2gb^2y^2} + y + h = \text{constant} \tag{10.23}$$

We can obtain an expression for the variation of the free surface depth by differentiating Eq. 10.23

$$\frac{-Q^2}{gb^2y^3}\frac{dy}{dx} + \frac{dy}{dx} + \frac{dh}{dx} = 0$$

Solving for the slope of the free surface, we obtain

$$\frac{dy}{dx} = \frac{dh/dx}{\left[\dfrac{Q^2}{gb^2y^3} - 1\right]} = \frac{dh/dx}{\left[\dfrac{V^2}{gy} - 1\right]} = \frac{dh/dx}{Fr^2 - 1} \tag{10.24}$$

From Eq. 10.24 we see that the slope of the free surface depends on the value of the local Froude number. For $Fr < 1$ an increase in bed elevation causes a decrease in water depth; a decrease in bed elevation causes water depth to increase. For $Fr > 1$ an increase in bed elevation causes an increase in water depth; a decrease in bed elevation causes water depth to decrease. These results are summarized in Fig. 10.6.

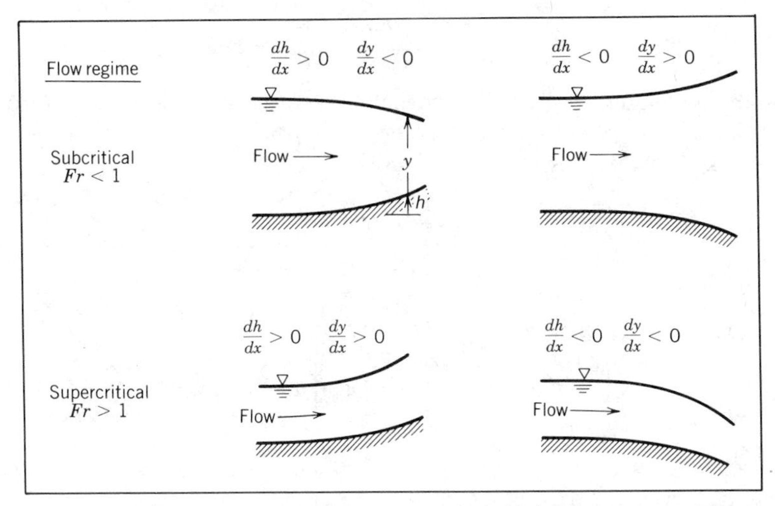

Fig. 10.6 Effect of bed elevation changes on water depth in open-channel flow.

When the Froude number is equal to unity, Eq. 10.24 predicts an infinite water surface slope, unless dh/dx equals zero. Since the free surface slope cannot be infinite, then dh/dx must be zero when $Fr = 1$; a Froude number of unity can only exist at the location where $dh/dx = 0$. If critical flow is attained, then downstream of the critical flow location the flow may be subcritical or supercritical depending on downstream conditions. If critical flow does not occur where $dh/dx = 0$, then the flow downstream from this location will be of the same type as the flow upstream from the location.

Example 10.2

Water flows in a horizontal rectangular channel. The flow speed and depth at section ① are 0.5 m/sec and 0.3 m, respectively. The flow passes over a smooth bump on the channel floor. Evaluate the flow speed and depth directly over the peak of the bump. Assume that the peak height is 0.03 m and neglect friction.

EXAMPLE PROBLEM 10.2

GIVEN: Water flow in a horizontal rectangular channel. Neglect frictional effects.

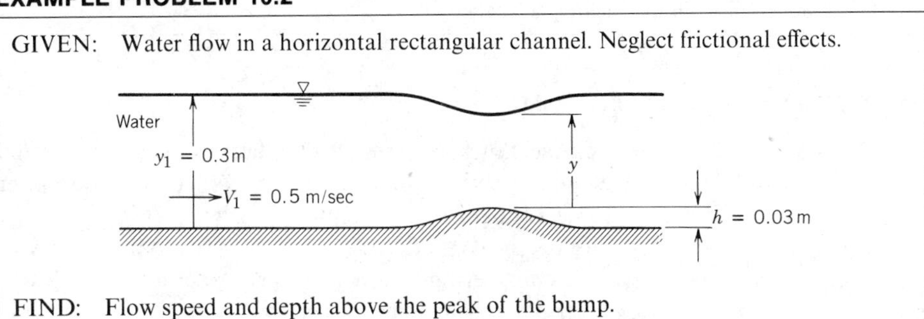

FIND: Flow speed and depth above the peak of the bump.

SOLUTION:
For steady, incompressible flow along a streamline without friction, the Bernoulli equation may be applied. Along the free surface the pressure is constant, so

$$\frac{p_{atm}}{\rho g} + \frac{V_1^2}{2g} + y_1 = \frac{p_{atm}}{\rho g} + \frac{V^2}{2g} + y + h$$

For flow that is uniform at each section, the continuity equation reduces to $V_1 y_1 = Vy$. These two equations permit us to solve for V and y; the solution process may be visualized using the specific energy diagram as follows. From Eq. 10.22,

$$E_1 = \frac{V_1^2}{2g} + y_1 = E + h = \frac{V^2}{2g} + y + h \qquad \text{or} \qquad E = E_1 - h$$

Evaluating, we obtain

$$E_1 = \frac{1}{2} \times \frac{(0.5)^2\ \text{m}^2}{\text{sec}^2} \times \frac{\text{sec}^2}{9.807\ \text{m}} + 0.3\ \text{m}$$

$$E_1 = 0.3127\ \text{m}$$

and

$$E = E_1 - h = (0.3127 - 0.03)\ \text{m} = 0.2827\ \text{m}$$

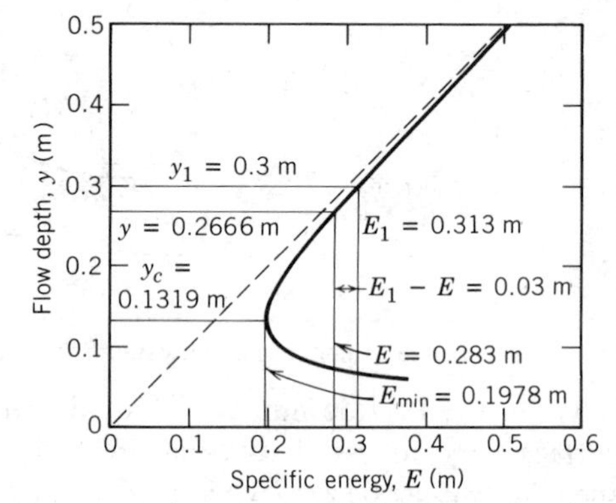

These points are plotted on the specific energy diagram. By iteration it may be shown that the flow depth that gives this value of E is

$$y = 0.2666\ \text{m} \qquad y$$

The change in surface level is

$$\Delta = y + h - y_1 = (0.2666 + 0.0300 - 0.3000)\ \text{m} = -3.4\ \text{mm}$$

The flow speed is

$$V = V_1 \frac{y_1}{y} = \frac{0.5\ \text{m}}{\text{sec}} \times \frac{0.3000\ \text{m}}{0.2666\ \text{m}} = 0.563\ \text{m/sec} \qquad V$$

{The specific energy diagram suggests that the flow could traverse a bump approximately 0.115 m in height without reaching critical speed.}

10-4.2 Flow through a Sluice Gate

As a second example of open-channel flow where the effect of friction may be neglected, consider flow through a sluice gate. A sluice gate is a form of control structure often used to regulate discharge. Flow beneath a sluice gate is shown in Fig. 10.7. At a location well upstream from the gate, the water depth, y_0, is constant and the flow speed is negligible. This is equivalent to considering flow from a large reservoir. The flow at section ① is assumed to be uniform across the depth, y_1.

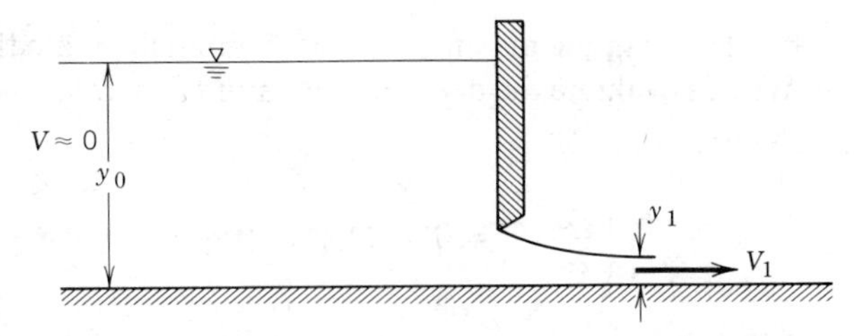

Fig. 10.7 Flow through a sluice gate.

Since the flow is steady, incompressible, and frictionless, we may apply the Bernoulli equation along a streamline. Applying the equation along the surface streamline between sections ⓪ and ①, then

$$y_0 = \frac{V_1^2}{2g} + y_1 \tag{10.25}$$

We can express V in terms of the volume flow rate. Since $Q = AV$ and $A = by$, where b is the width of gate, then

$$y_0 = \frac{Q^2}{2gb^2y_1^2} + y_1 \tag{10.26}$$

Equation 10.26 gives a relation between the depth, y_1, and the volume flow rate

$$y_1^2(y_0 - y_1) = \frac{Q^2}{2gb^2} \tag{10.27}$$

This relation is shown in Fig. 10.8; depth, y_1, has been nondimensionalized on reservoir depth, y_0. Figure 10.8 shows that for a given flow rate there are two possible values of

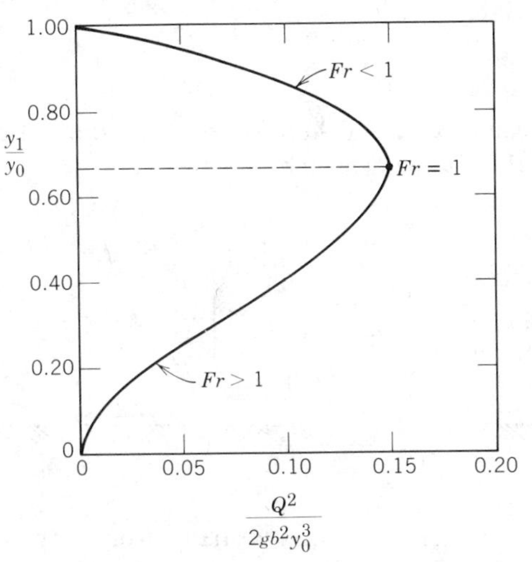

Fig. 10.8 Relation between downstream depth and flow rate for flow through a sluice gate.

depth for any flow rate less than the maximum. At maximum flow rate there is a single value of the depth. We can evaluate the depth at maximum flow rate by differentiating Eq. 10.27 with respect to y_1,

$$\frac{d}{dy_1}\left[\frac{Q^2}{2gb^2}\right] = 0 = 2y_1 y_0 - 3y_1^2$$

At maximum flow rate

$$y_1 = \frac{2}{3} y_0$$

and from Eq. 10.27

$$\frac{Q_{max}^2}{gb^2} = 2\left[\frac{2}{3} y_0\right]^2 \left[y_0 - \frac{2}{3} y_0\right] = \frac{8}{27} y_0^3$$

The Froude number at the maximum flow rate is given by

$$Fr^2 = \frac{V_{max}^2}{gy_1} = \frac{Q_{max}^2}{gb^2 y_1^3} = \frac{8}{27} y_0^3 \left[\frac{3}{2y_0}\right]^3 = 1$$

Thus, in Fig. 10.8, maximum flow rate occurs at $y_1/y_0 = \frac{2}{3}$. The upper portion of the curve corresponds to subcritical flow ($Fr < 1$) and the lower portion of the curve corresponds to supercritical flow ($Fr > 1$).

Flow leaving a sluice gate passes through a vena contracta as shown in Fig. 10.7. It is not possible analytically to relate the depth, y_1, to the gate opening. Experimental data must be used to determine the gate setting needed to pass a desired flow rate, as discussed in Section 10-8.

Example 10.3

Water flow under a sluice gate was considered in Example Problem 4.6, under the conditions shown in the following sketch:

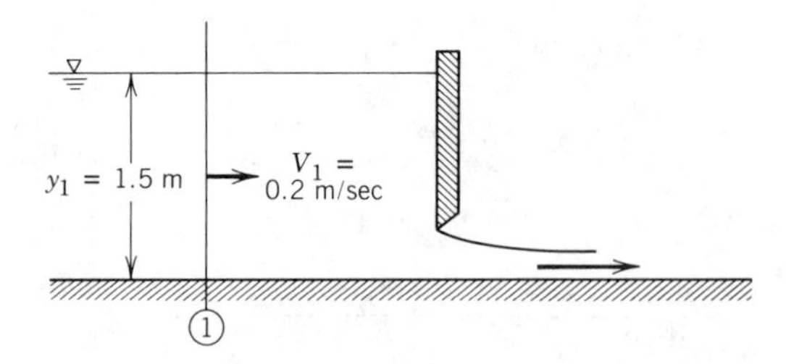

Evaluate the water depth, y_0, at a location upstream from the gate where the flow speed is negligible. Find the exact values of downstream depth and speed at this flow rate required by the condition that the specific energy remain constant. Find the depth and speed that would provide maximum flow rate. Calculate this flow rate.

EXAMPLE PROBLEM 10.3

GIVEN: Water flow under a sluice gate at the conditions shown.

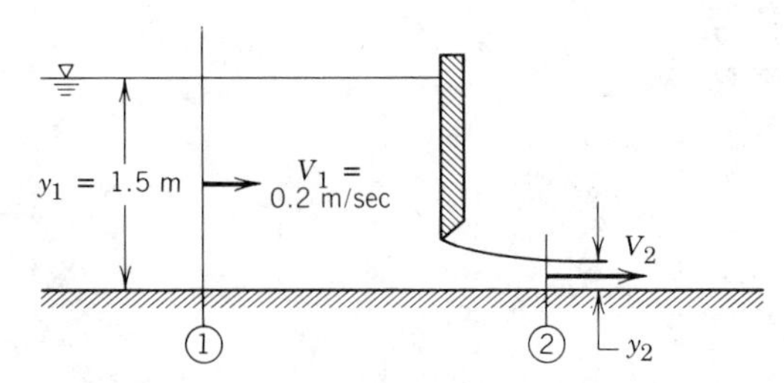

FIND: (a) Water depth, y_0, where flow speed is negligible.
(b) Water depth, y_2, and speed, V_2, downstream at this flow rate.
(c) Water depth, y, and speed, V, at downstream conditions that produce the maximum flow rate possible.
(d) Maximum flow rate.

SOLUTION:
The specific energy upstream is

$$E_1 = \frac{V_1^2}{2g} + y_1 = \frac{1}{2} \times \frac{(0.2)^2}{} \frac{\text{m}^2}{\text{sec}^2} \times \frac{\text{sec}^2}{9.807 \text{ m}} + 1.5 \text{ m} = 1.502 \text{ m}$$

For constant specific energy, $E_0 = \frac{V_0^2}{2g} + y_0 = E_1$, so $y_0 = 1.502$ m (with $\frac{V_0^2}{2g} \approx 0$) ← y_0

For constant specific energy, the downstream conditions must be at the alternate depth that corresponds to the rate of flow specified, as shown on the specific energy diagram,

$$E_2 = y_0 = 1.502 \text{ m} = \frac{Q^2}{2gb^2y_2^2} + y_2$$

By iteration, for $Q/b = 0.3 \text{ m}^2/\text{sec}$,

$y_2 = 0.05632$ m ← y_2

The corresponding speed is

$$V_2 = V_1 \frac{y_1}{y_2}$$

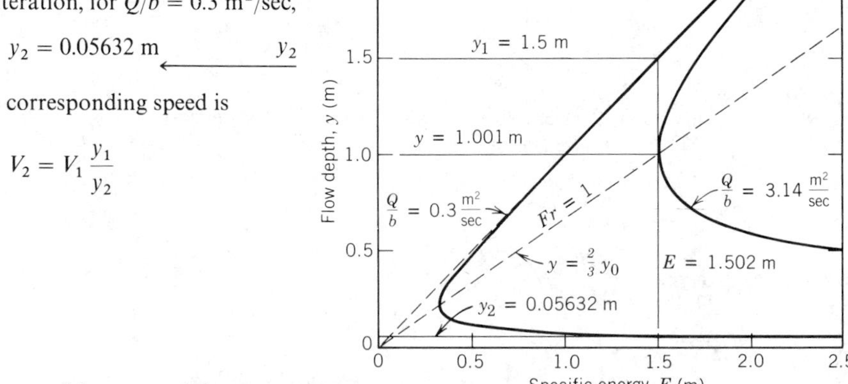

$$V_2 = \frac{0.2 \text{ m}}{\text{sec}} \times \frac{1.5 \text{ m}}{0.05632 \text{ m}} = 5.327 \text{ m/sec}$$ ← V_2

These conditions define the alternate depth, as shown above. (To three significant figures, these results agree with the data given in Example Problem 4.6.)

From the solution to Eq. 10.27, at maximum flow rate,

$$y = \tfrac{2}{3} y_0 = \tfrac{2}{3} \times 1.502 \text{ m} = 1.001 \text{ m} \qquad y$$

and

$$V = \sqrt{gy} = \left[9.807 \frac{\text{m}}{\text{sec}^2} \times 1.001 \text{ m} \right]^{1/2} = 3.133 \text{ m/sec} \qquad V$$

The maximum flow rate is

$$\frac{Q}{b} = Vy = 3.133 \frac{\text{m}}{\text{sec}} \times 1.001 \text{ m} = 3.14 \text{ m}^3\text{/sec/m (m}^2\text{/sec)} \qquad Q$$

The corresponding curve of y versus E also is shown above.

10-5 FLOW AT NORMAL DEPTH: UNIFORM FLOW

Fully developed flow through a prismatic channel (a channel with constant slope and cross section) at constant depth, y_n, is termed flow at *normal depth* or *uniform flow*. The slope of the channel bottom, or bed, is denoted by S_b and is assumed to be small. To analyze the flow, we assume uniform flow at each cross section and apply the basic equations to the control volume of Fig. 10.9.

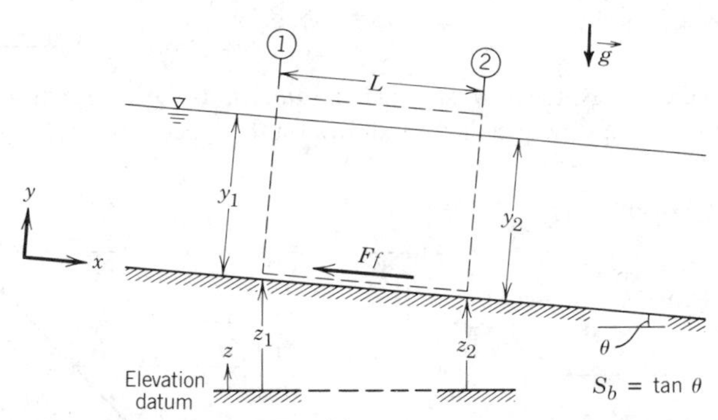

Fig. 10.9 Control volume and coordinates used for analysis of flow at normal depth.

10-5.1 Basic Equations

a. Continuity Equation

Basic equation:

$$0 = \overbrace{\frac{\partial}{\partial t} \int_{\text{CV}} \rho \, d\forall}^{=0(1)} + \int_{\text{CS}} \rho \vec{V} \cdot d\vec{A} \tag{4.13}$$

Assumptions: (1) Steady flow
(2) Incompressible flow
(3) Uniform flow at a section
(4) Normal depth, $y_1 = y_2 = y_n$

The continuity equation then gives

$$0 = -|\rho V_1 A_1| + |\rho V_2 A_2|$$

Since $A_1 = A_2$, then

$$V_1 = V_2$$

b. Momentum Equation

Basic equation:

$$F_{S_x} + F_{B_x} = \overset{=0(1)}{\frac{\partial}{\partial t}\int_{\mathrm{CV}} u\rho\, d\mathcal{V}} + \int_{\mathrm{CS}} u\rho \vec{V} \cdot d\vec{A} \tag{4.19a}$$

Assumptions: (5) Hydrostatic pressure distribution
(6) Small slope, $\theta \sim \sin\theta \sim \tan\theta = S_b$

Since the flow has constant depth and uniform velocity, the momentum flux through the control surface is zero. The net pressure force on the control volume in the x direction is zero because the pressure distribution is hydrostatic. The x component of the body force is the component of the weight of liquid in the control volume in the x direction. Thus the momentum equation reduces to

$$-F_f + W \sin\theta = 0$$

or

$$F_f = W \sin\theta \tag{10.28}$$

The friction force may be expressed as the product of the wall shear stress, τ_w, and the channel surface area on which the stress acts. Equation 10.28 shows that for flow at normal depth, the component of the gravity force driving the flow is balanced by the friction force acting on the channel walls.

c. Energy Equation

The energy equation for open-channel flow was derived in Section 10-3

$$\frac{V_1^2}{2g} + y_1 + z_1 = \frac{V_2^2}{2g} + y_2 + z_2 + h_\ell \tag{10.10}$$

For the case of flow at normal depth, $y_1 = y_2 = y_n$, and $V_1 = V_2$. Thus, from Eq. 10.10,

$$h_\ell = z_1 - z_2 = LS_b \tag{10.29}$$

For flow at normal depth, the head loss due to friction is equal to the change in elevation of the bed. The specific energy, E, is constant at each section normal to the bed. The energy grade line, the hydraulic grade line, and the channel bed are all parallel.

10-5.2 The Manning Correlation for Velocity

Open-channel flows in practice invariably are turbulent. Since no simple constitutive relation is available to relate shear stress and velocity gradient, we must rely on empirical correlations. As in the case of fully developed turbulent pipe flow, the head loss may be written in terms of a friction factor (Eq. 8.32). For application to open-channel flow the pipe diameter, D, is written in terms of the hydraulic radius ($D = 2R = 4R_h$) and the head loss is expressed as head loss per unit weight (h_ℓ has units of length). Thus, for open-channel flow,

$$h_\ell = f \frac{L}{4R_h} \frac{V^2}{2g} \tag{10.30}$$

From Eq. 10.29, $h_\ell / L = S_b$, and the velocity for flow at normal depth is

$$V = \left[\frac{8g}{f}\right]^{1/2} \sqrt{R_h S_b} \tag{10.31}$$

For most channel flows the friction factor is only a function of surface roughness; it is independent of Reynolds number. This is analogous to flow in the fully rough regime for turbulent pipe flow. For a given roughness, Eq. 10.31 is then

$$V = C\sqrt{R_h S_b} \tag{10.32}$$

Equation 10.32 is known as the Chezy equation. Empirical values of C were determined by Manning. He suggested that

$$C = \left[\frac{8g}{f}\right]^{1/2} = \frac{R_h^{1/6}}{n} \tag{10.33}$$

where n is a roughness coefficient having different values for different types of boundary roughness. With this expression for C, the velocity for flow at normal depth becomes

$$V = \frac{R_h^{2/3} S_b^{1/2}}{n} \tag{10.34}$$

and the volume flow rate is written

$$Q = AV = \frac{R_h^{2/3} S_b^{1/2}}{n} A \tag{10.35}$$

Values of n for some typical surfaces are given in Table 10.2. These values of n, taken to be dimensionless (which they are not; from Eq. 10.34 we see that n has dimensions of $L^{-1/3}t$), were based on measurements in SI units.[5]

The relationship among variables in Eq. 10.35 can be viewed in a number of ways.

[5] Equations 10.34 and 10.35 are valid for SI units. In British units the numerator of these equations must be multiplied by $(0.3048)^{-1/3} \approx 1.49$. With this modification the value of n can be considered to have the same numerical value in both systems of units.

Table 10.2 Values of Manning Roughness Coefficient, n, for Representative Surfaces

Type of Channel Surface	Representative[a] n-value
Lucite, glass, or plastic film	0.010
Wood or finished concrete	0.013
Unfinished concrete, well-laid brickwork, concrete or cast-iron pipe	0.015
Riveted or spiral steel pipe	0.017
Smooth, uniform earth channel	0.022
Corrugated metal flumes, typical canals, rivers free from large stones and heavy weeds	0.025
Canals and rivers with many stones and weeds	0.035

[a] See [1] for a more complete table.

The volume flow rate through a prismatic channel of given slope and roughness is a function of both the channel size and the channel shape. This is illustrated in Example Problem 10.4. For a specified flow rate through a prismatic channel of given slope and roughness, the depth of the uniform flow is a function of both the channel size and the channel shape. There is only one depth for uniform flow at a given flow rate; the depth may be greater than, less than, or equal to the critical depth. This is illustrated in Example Problem 10.5.

Example 10.4

Open channels of square and semicircular shapes are being considered to carry flow on a slope of 0.001; the channel walls are to be poured concrete with $n = 0.015$. Evaluate the flow rate delivered by the channels for maximum dimensions between 0.5 and 2.0 m. Compare the channels on the basis of volume flow rate for a given cross-sectional area.

EXAMPLE PROBLEM 10.4

GIVEN: Square and semicircular channels.
$S_b = 0.001, \quad n = 0.015$
Sizes between 0.5 and 2.0 m across.

FIND: (a) Evaluate flow rate as a function of size.
(b) Compare channels on the basis of volume flow rate, Q, versus cross-sectional area, A.

SOLUTION:
Apply Eq. 10.35 for flow at normal depth in a long channel.

Computing equation: $Q = \dfrac{R_h^{2/3} S_b^{1/2}}{n} A$ (10.35)

For the square channel

$P = 3L$, $A = L^2$, so $R_h = \dfrac{L}{3}$

Substituting into Eq. 10.35, we obtain

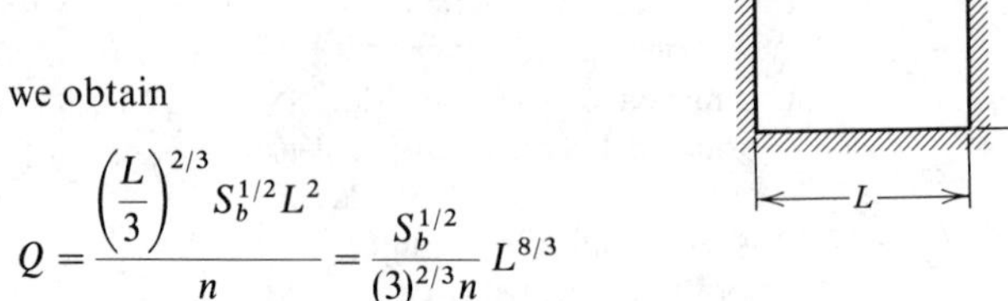

$$Q = \frac{\left(\dfrac{L}{3}\right)^{2/3} S_b^{1/2} L^2}{n} = \frac{S_b^{1/2}}{(3)^{2/3} n} L^{8/3}$$

For $L = 1$ m,

$$Q = \frac{(0.001)^{1/2}}{(3)^{2/3}(0.015)} (1)^{8/3} = 1.01 \text{ m}^3/\text{sec}$$

$Q_\square$

Tabulating for a range of sizes yields

L(m)	0.5	1.0	1.5	2.0
A(m^2)	0.25	1.00	2.25	4.00
Q(m^3/sec)	0.160	1.01	2.99	6.44

For the semicircular channel

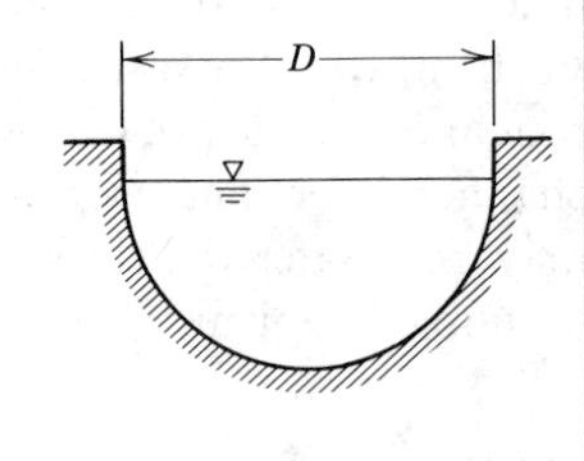

$P = \dfrac{\pi D}{2}$, $A = \dfrac{\pi D^2}{8}$, so $R_h = \dfrac{\pi D^2}{8} \dfrac{2}{\pi D} = \dfrac{D}{4}$

Substituting into Eq. 10.35, we obtain

$$Q = \frac{\left(\dfrac{D}{4}\right)^{2/3} S_b^{1/2}}{n} \frac{\pi D^2}{8} = \frac{S_b^{1/2} \pi}{4^{5/3}(2) n} D^{8/3}$$

For $D = 1$ m,

$$Q = \frac{(0.001)^{1/2} \pi}{(2)(4)^{5/3}(0.015)} (1)^{8/3} = 0.329 \text{ m}^3/\text{sec}$$

$Q_\smile$

Tabulating for a range of sizes yields

D(m)	0.5	1.0	1.5	2.0
A(m^2)	0.0982	0.393	0.884	1.57
Q(m^3/sec)	0.0517	0.329	0.969	2.09

For both channels, volume flow rate varies as

$$Q \sim L^{8/3} \quad \text{or} \quad Q \sim A^{4/3}$$

since $A \sim L^2$. The plot of flow rate versus cross-sectional area shows that the semicircular channel is more "efficient."

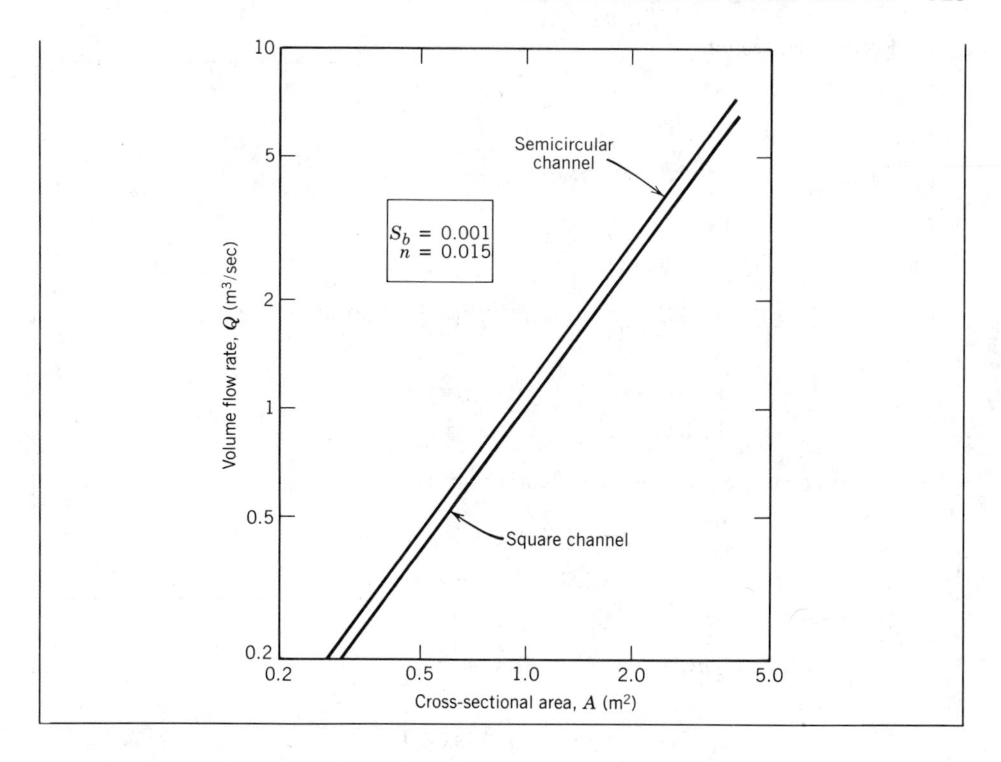

Example 10.5

An above-ground flume is to be built from timber to convey water from a mountain lake to a small hydroelectric plant. The flume is to deliver water at flow rate, $Q = 2\ \text{m}^3/\text{sec}$; the slope is $S_b = 0.002$, and $n = 0.013$. Evaluate the required flume size for (a) a rectangular section with $y/b = 0.5$, and (b) an equilateral triangular section.

EXAMPLE PROBLEM 10.5

GIVEN: Flume to be built from timber.

$S_b = 0.002$, $n = 0.013$, $Q = 2.00\ \text{m}^3/\text{sec}$.

FIND: Required flume size for
(a) Rectangular section with $y/b = 0.5$.
(b) Equilateral triangular section.

SOLUTION:
Assume flume is long so that flow is uniform (at normal depth). Then Eq. 10.35 applies.

Computing equation:
$$Q = \frac{R_h^{2/3} S_b^{1/2}}{n} A \qquad (10.35)$$

The choice of channel shape fixes the relationship between R_h and A, so Eq. 10.35 may be solved for the normal depth, y_n, thus determining the channel size required.

(a) Rectangular section

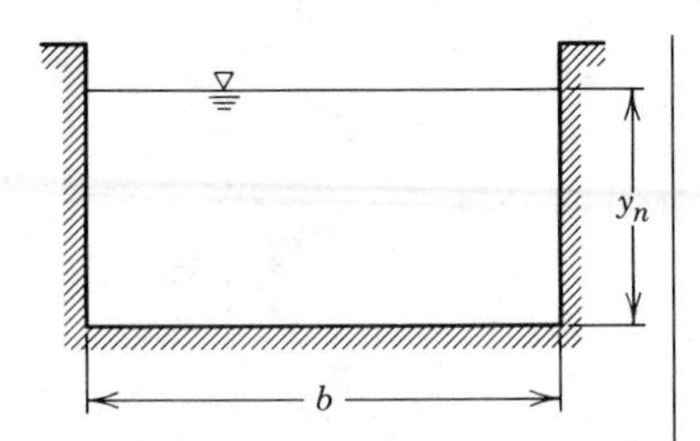

$P = 2y_n + b$; $y_n/b = 0.5$

$$\left.\begin{aligned} P &= 2y_n + 2y_n = 4y_n \\ A &= y_n b = y_n(2y_n) = 2y_n^2 \end{aligned}\right\} R_h = \frac{A}{P} = \frac{2y_n^2}{4y_n} = 0.5y_n$$

Substituting into Eq. 10.35, we obtain

$$Q = \frac{R_h^{2/3} S_b^{1/2}}{n} A = \frac{(0.5y_n)^{2/3} S_b^{1/2} (2y_n^2)}{n} = \frac{2(0.5)^{2/3} S_b^{1/2} y_n^{8/3}}{n}$$

$$y_n = \left[\frac{nQ}{2(0.5)^{2/3} S_b^{1/2}}\right]^{3/8} = \left[\frac{0.013(2.00)}{2(0.5)^{2/3}(0.002)^{1/2}}\right]^{3/8} = 0.748 \text{ m}$$

The required dimensions for the rectangular channel are

$$y_n = 0.748 \text{ m} \qquad A = 1.12 \text{ m}^2$$

$$b = 1.50 \text{ m} \qquad P = 3.00 \text{ m}$$

Also $Fr = \dfrac{V}{\sqrt{gy_n}} = \dfrac{Q}{A\sqrt{gy_n}}$

$$Fr = \frac{2.00 \text{ m}^3}{\text{sec}} \times \frac{1}{1.12 \text{ m}^2} \times \frac{1}{\left[9.81 \dfrac{\text{m}}{\text{sec}^2} \times 0.748 \text{ m}\right]^{1/2}} = 0.659$$

Fr

(b) Equilateral triangular section

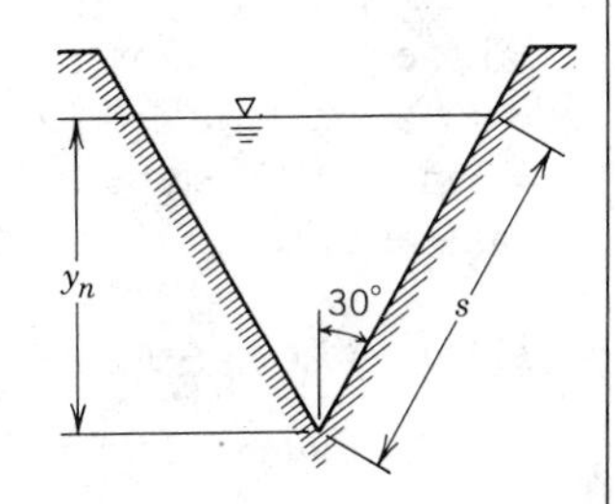

$$\left.\begin{aligned} P &= 2s = \frac{2y_n}{\cos 30^\circ} \\ A &= \frac{y_n s}{2} = \frac{y_n^2}{2\cos 30^\circ} \end{aligned}\right\} R_h = \frac{A}{P} = \frac{y_n}{4}$$

Substituting into Eq. 10.35 yields

$$Q = \frac{R_h^{2/3} S_b^{1/2}}{n} A = \frac{\left(\dfrac{y_n}{4}\right)^{2/3} S_b^{1/2} \left(\dfrac{y_n^2}{2\cos 30^\circ}\right)}{n} = \frac{S_b^{1/2} y_n^{8/3}}{2(4)^{2/3} \cos 30^\circ \, n}$$

$$y_n = \left[\frac{2(4)^{2/3} \cos 30^\circ \, nQ}{S_b^{1/2}}\right]^{3/8} = \left[\frac{2(4)^{2/3} \cos 30^\circ (0.013)(2.00)}{(0.002)^{1/2}}\right]^{3/8} = 1.42 \text{ m}$$

The required dimensions for the triangular channel are

$$y_n = 1.42 \text{ m} \qquad P = 3.28 \text{ m}$$

$$A = 1.16 \text{ m}^2 \qquad b_s = 1.64 \text{ m}$$

Also $V = \frac{Q}{A} = \frac{2.0\ \text{m}^3}{\text{sec}} \times \frac{1}{1.16\ \text{m}^2} = 1.72\ \text{m/sec}$

and $Fr = \frac{V}{\sqrt{g y_h}} = \frac{V}{\sqrt{g A/b_s}}$

$$Fr = \frac{1.72\ \text{m}}{\text{sec}} \frac{1}{\left[\frac{9.81\ \text{m}}{\text{sec}^2} \times 1.16\ \text{m}^2 \times \frac{1}{1.64\ \text{m}}\right]^{1/2}} = 0.653 \qquad Fr$$

Comparing results, we see that the rectangular flume would be cheaper to build; its perimeter is about 8.5 percent less than that of the triangular flume.

{This example shows the effect of channel shape on the size required to deliver a given flow rate at a specified bed slope and roughness coefficient. At specified S_b and n, flow may be subcritical, critical, or supercritical, depending on Q.}

10-5.3 Optimum Channel Cross Section

For a given slope and roughness, the optimum channel cross section requires a minimum flow area for a given flow rate. From Eq. 10.35

$$\frac{Q}{A} = \frac{R_h^{2/3} S_b^{1/2}}{n} \tag{10.36}$$

Thus the optimum cross section is the one for which the hydraulic radius, R_h, is a maximum. Since $R_h = A/P$, R_h is a maximum when the wetted perimeter is a minimum. Solving Eq. 10.36 for A (with $R_h = A/P$) then yields

$$A = \left[\frac{nQ}{S_b^{1/2}}\right]^{3/5} P^{2/5} \tag{10.37}$$

From Eq. 10.37, the flow area will be a minimum when the wetted perimeter is a minimum.

The wetted perimeter, P, is a function of channel shape. For any given prismatic channel shape (rectangular, trapezoidal, triangular, circular, etc.), the channel cross section can be optimized. Optimum cross sections for common channel shapes are given in Table 10.3. The determination of the optimum cross section for a trapezoidal channel is illustrated in Example Problem 10.6.

Once the optimum cross section for a given channel shape has been determined, expressions for the normal depth, y_n, and area, A, as functions of flow rate can be obtained from Eq. 10.35. These expressions are included in Table 10.3.

Table 10.3 Properties of Optimum Open-Channel Sections

Shape	Section	Optimum Geometry	Normal Depth, y_n	Cross-Sectional Area, A
Trapezoidal		$\alpha = 60°$ $b = \dfrac{2}{\sqrt{3}} y_n$	$0.968\left[\dfrac{Qn}{S_b^{1/2}}\right]^{3/8}$	$1.622\left[\dfrac{Qn}{S_b^{1/2}}\right]^{3/4}$
Rectangular		$b = 2y_n$	$0.917\left[\dfrac{Qn}{S_b^{1/2}}\right]^{3/8}$	$1.682\left[\dfrac{Qn}{S_b^{1/2}}\right]^{3/4}$
Triangular		$\alpha = 45°$	$1.297\left[\dfrac{Qn}{S_b^{1/2}}\right]^{3/8}$	$1.682\left[\dfrac{Qn}{S_b^{1/2}}\right]^{3/4}$
Wide flat		None	$1.00\left[\dfrac{(Q/b)n}{S_b^{1/2}}\right]^{3/8}$	—
Circular		$D = 2y_n$	$1.00\left[\dfrac{Qn}{S_b^{1/2}}\right]^{3/8}$	$1.583\left[\dfrac{Qn}{S_b^{1/2}}\right]^{3/4}$

Example 10.6

A trapezoidal channel is to be optimized so that the excavation is minimized for a given discharge at normal depth. Find the optimum ratio of channel side length to bottom width and the optimum side slope angle.

EXAMPLE PROBLEM 10.6

GIVEN: Trapezoidal channel section.

FIND: (a) Optimum ratio of channel side length to bottom width.

(b) Optimum side slope angle.

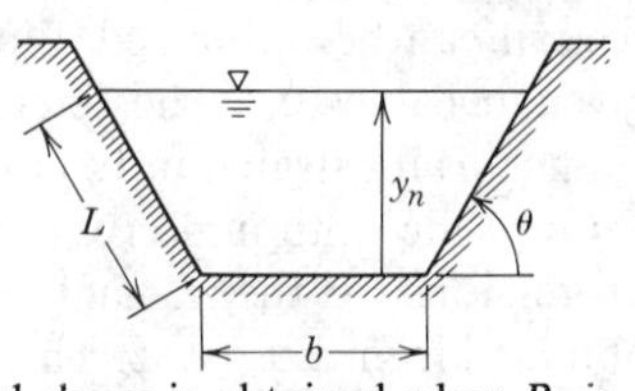

SOLUTION:

From the discussion of Section 10-5.3, the optimum channel shape is obtained when R_h is maximized, or when P is minimized for a given flow area. The area for the section is

$$A = by_n + L\cos\theta\, L\sin\theta$$

But $L = \dfrac{y_n}{\sin\theta}$, so

$$A = by_n + y_n^2\frac{\cos\theta}{\sin\theta} = by_n + y_n^2\cot\theta \qquad (1)$$

The wetted perimeter is

$$P = b + 2L = b + 2\frac{y_n}{\sin\theta}$$

The bottom width may be eliminated using Eq. 1 to obtain

$$P = \frac{A}{y_n} - y_n\cot\theta + \frac{2y_n}{\sin\theta} \qquad (2)$$

For any side slope angle, the perimeter can be minimized by differentiating Eq. 2 with respect to y_n. Thus

$$\frac{dP}{dy_n} = -\frac{A}{y_n^2} - \cot\theta + \frac{2}{\sin\theta} = 0$$

and

$$y_n^2 = \frac{A\sin\theta}{2 - \cos\theta} \qquad (3)$$

(For a rectangular channel, $\theta = \pi/2$ and $A = by_n$. Equation 3 becomes

$$y_n^2 = \frac{by_n}{2} \qquad \text{or} \qquad y_n = \frac{b}{2}$$

as shown in Table 10.2.)

To find the optimum side slope angle, differentiate Eq. 3 with respect to θ. Thus

$$2y_n\frac{dy_n}{d\theta} = \frac{A\cos\theta}{2 - \cos\theta} + \frac{A\sin^2\theta}{(2-\cos\theta)^2} = A\left[\frac{\cos\theta(2-\cos\theta) - \sin^2\theta}{(2-\cos\theta)^2}\right] = 0$$

which reduces to $2\cos\theta - 1 = 0$, $\cos\theta = \frac{1}{2}$, or $\theta = 60^\circ$ ← θ

Substituting into Eq. 3 gives

$$y_n^2 = \frac{A(\sqrt{3}/2)}{2 - \frac{1}{2}} = A\frac{\sqrt{3}}{3} = \frac{A}{\sqrt{3}} \qquad \text{or} \qquad A = \sqrt{3}\,y_n^2$$

Finally, substituting into Eq. 1 gives

$$\sqrt{3}\,y_n^2 = by_n + y_n^2\frac{1}{\sqrt{3}} = by_n + \frac{\sqrt{3}}{3}y_n^2 \text{ or } b = \frac{2}{3}\sqrt{3}\,y_n = \frac{2y_n}{\sqrt{3}}$$

so that

$$L = \frac{y_n}{\sin\theta} = \frac{2y_n}{\sqrt{3}} = b$$ ← L

Thus the optimum trapezoidal channel has sides and bottom of equal length and a side slope angle of 60°; it is half of a hexagon.

10-5.4 Critical Normal Flow

Equation 10.35 indicates that for a given volume flow rate through a prismatic channel of fixed roughness there is only one bed slope for flow at normal depth. Solving Eq. 10.35 for S_b,

$$S_b = \left[\frac{nQ}{R_h^{2/3}A}\right]^2 \tag{10.38}$$

If the channel bed slope is such that the normal depth for a given flow rate is exactly equal to the critical depth, that slope is called the critical slope. Denoting the critical bed slope as S_c, then

$$S_c = \left[\frac{nQ}{R_{hc}^{2/3}A_c}\right]^2 = \frac{n^2Q^2}{R_{hc}^{4/3}A_c^2} \tag{10.39}$$

From Eq. 10.21, for critical flow,

$$\frac{Q^2}{A_c^2} = \frac{gA_c}{b_s}$$

Thus we can write

$$S_c = \frac{n^2 g A_c}{b_s R_{hc}^{4/3}} \tag{10.40}$$

For a wide rectangular channel of width, b, with $b \gg y_c$, then $b_s = b$, $A_c = by_c$, and $R_{hc} = y_c$. Substituting into Eq. 10.40, we obtain

$$S_c = \frac{n^2 g}{y_c^{1/3}} \tag{10.41}$$

In terms of critical depth and critical slope,

if $y_n > y_c$, then $S_b < S_c$: slope is mild (subcritical)

if $y_n = y_c$, then $S_b = S_c$: slope is critical

if $y_n < y_c$, then $S_b > S_c$: slope is steep (supercritical)

Example 10.7

For critical normal flow in a wide flat channel made from unfinished concrete, plot the bed slope as a function of flow depth for the range, $0.01 \text{ m} < y < 10 \text{ m}$. Compare with the corresponding curve for critical normal flow in an optimum rectangular channel.

EXAMPLE PROBLEM 10.7

GIVEN: Critical normal flow in (a) a wide flat channel, and (b) an optimum rectangular channel, made from unfinished concrete.

FIND: Plot S_c versus y_c for both cases.

SOLUTION:

For unfinished concrete, Table 10.2 gives $n = 0.015$. For the wide channel, apply Eq. 10.41 directly.

Computing equation:
$$S_c = \frac{n^2 g}{y_c^{1/3}} \tag{10.41}$$

Thus

$$S_c = (0.015)^2(9.81)y_c^{-1/3} = 0.00221 y_c^{-1/3} \tag{1}$$

For the optimum rectangular channel, start with Eq. 10.34, and use $y_n/b = 0.5$ for optimum conditions.

Computing equation:
$$V = \frac{R_h^{2/3} S_b^{1/2}}{n} \tag{10.34}$$

For optimum conditions, $P = b + 2y_n = 4y_n$, $A = by_n = 2y_n^2$, and

$$R_h = \frac{A}{P} = \frac{2y_n^2}{4y_n} = 0.5y_n$$

For critical normal flow, $y_n = y_c$, and $V = V_c = \sqrt{gy_c}$. Substituting into Eq. 10.34,

$$V_c = \sqrt{gy_c} = g^{1/2} y_c^{1/2} = \frac{(0.5y_c)^{2/3} S_c^{1/2}}{n}$$

Thus

$$S_c^{1/2} = \frac{ng^{1/2}}{(0.5)^{2/3} y_c^{1/6}} \quad \text{or} \quad S_c = \frac{n^2 g}{(0.5)^{4/3} y_c^{1/3}}$$

and

$$S_c = \frac{(0.015)^2(9.81)}{(0.5)^{4/3}} y_c^{-1/3} = 0.00556 y_c^{-1/3} \tag{2}$$

Values calculated from Eqs. 1 and 2 are shown in the following figure:

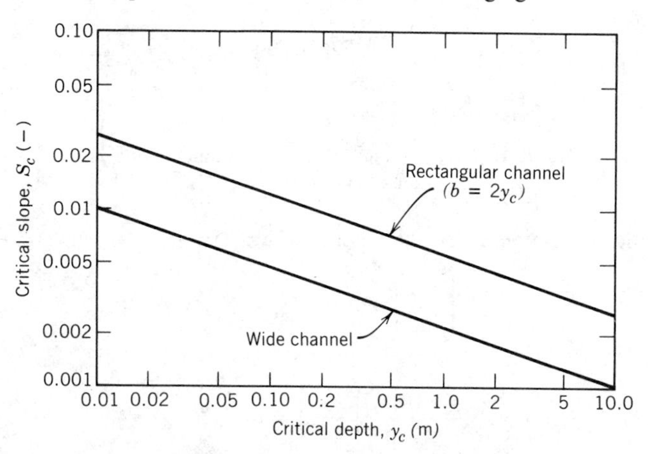

There is only one bed slope for flow at normal depth for given flow rate, channel geometry, and roughness coefficient. If the normal depth is the critical depth, then the slope is as given above. Comparing Eqs. 1 and 2 shows that for given depth (and hence the same flow rate per unit width, since $V_c = \sqrt{gy_c}$) the critical slope for the rectangular channel is 2.52 times larger than for the flat channel, due to the extra friction from the channel sides.

10-6 FLOW WITH GRADUALLY VARYING DEPTH

When open-channel flow encounters a change in bed slope or is approaching normal depth, the flow depth changes gradually. Flow with gradually varying depth must be analyzed by applying the energy equation to a differential control volume; the result is a differential equation that relates changes in depth to distance along the flow. The resulting equation may be solved numerically if one assumes that the head loss at each section is the same as that for a flow at normal depth with the same velocity and hydraulic radius of the section. The water depth and the channel bed height are assumed to change slowly. As in the case of flow at normal depth, the velocity is assumed uniform and the pressure distribution is assumed hydrostatic at each section.

The energy equation (Eq. 10.10) for open-channel flow was applied to a finite control volume in Section 10-3. We shall adapt this equation to the differential control volume of length, dx, shown in Fig. 10.10. Thus

$$\frac{V^2}{2g} + y + z = \frac{V^2}{2g} + d\left[\frac{V^2}{2g}\right] + y + dy + z + dz + dh_\ell$$

The change in bed elevation, dz, can be written in terms of the bed slope as $dz = -S_b\,dx$. The head loss, dh_ℓ, can be written in terms of the slope of the energy grade line as $dh_\ell = S\,dx$, where S is positive because the EGL drops in the direction of flow.

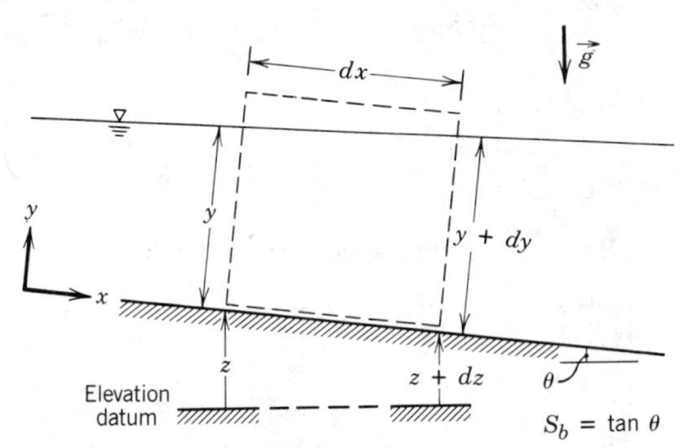

Fig. 10.10 Differential control volume used for analysis of gradually varied flow.

Thus the differential energy equation can be written as

$$d\left[\frac{V^2}{2g}\right] + dy = (S_b - S)\,dx$$

or

$$\frac{dy}{dx} + \frac{d}{dx}\left[\frac{V^2}{2g}\right] = S_b - S \qquad (10.42)$$

For flow in a rectangular channel of width, b, $Q = Vby$, and $V = Q/by$, so

$$\frac{d}{dx}\left[\frac{V^2}{2g}\right] = \frac{d}{dx}\left[\frac{Q^2}{2gb^2y^2}\right] = -2\frac{Q^2}{2gb^2y^3}\frac{dy}{dx} = -\frac{V^2}{gy}\frac{dy}{dx}$$

Since $Fr = V/\sqrt{gy}$, then

$$\frac{d}{dx}\left[\frac{V^2}{2g}\right] = -Fr^2 \frac{dy}{dx}$$

Substituting into Eq. 10.42, we obtain

$$\frac{dy}{dx} = \frac{S_b - S}{1 - Fr^2} \tag{10.43}$$

The slope, S, of the energy grade line is determined by assuming that the rate of head loss at a section is the same as that for a flow at normal depth with the same velocity and hydraulic radius of the section.

From Eq. 10.34 for flow at normal depth,

$$V = \frac{R_h^{2/3} S^{1/2}}{n}$$

Solving for the slope of the energy grade line yields

$$S = \frac{n^2 V^2}{R_h^{4/3}} \tag{10.44}$$

Equation 10.43, with S given by Eq. 10.44, is a first-order, nonlinear, ordinary differential equation that describes the variation of the water surface profile for gradually varying flows. The sign of the slope of the water surface profile depends on whether the flow is subcritical or supercritical ($Fr < 1$ or $Fr > 1$) and on the relative magnitudes of S and S_b.

10-6.1 Classification of Surface Profiles

The general behavior of surface profiles can be studied most easily for wide rectangular channels; for such channels $b \gg y$ and $R_h \approx y$. For this case the surface profile is governed by

$$\frac{dy}{dx} = \frac{S_b - S}{1 - Fr^2} = \frac{S_b\left[1 - \left(\frac{S}{S_b}\right)\right]}{1 - Fr^2}$$

where

$$S = \frac{n^2 V^2}{y^{4/3}} = \frac{n^2 Q^2}{b^2 y^{10/3}} \tag{10.45}$$

For a given flow rate through a wide rectangular channel of width, b, and specified roughness, the channel bed slope is uniquely related to the normal depth. From Eq. 10.35

$$Q = \frac{R_h^{2/3} S_b^{1/2}}{n} A = \frac{y_n^{2/3} S_b^{1/2}}{n} b y_n$$

Solving for S_b, we obtain

$$S_b = \frac{n^2 Q^2}{b^2 y_n^{10/3}} \tag{10.46}$$

Thus

$$\frac{S}{S_b} = \left[\frac{y_n}{y}\right]^{10/3} \tag{10.47}$$

The Froude number can be written in terms of the critical depth. From Eq. 10.17,

$$Fr = \left[\frac{y_c}{y}\right]^{3/2} \tag{10.17}$$

The slope of the water surface for flow in a wide rectangular channel then can be written as

$$\frac{dy}{dx} = S_b \frac{[1 - (y_n/y)^{10/3}]}{[1 - (y_c/y)^3]} \tag{10.48}$$

Equation 10.48 shows that the sign of dy/dx depends only on the relation among the depths, y, y_n, and y_c. In discussing critical normal flow (Section 10-5.4), we noted that

if $y_n > y_c$, then $S_b < S_c$: slope is mild (M)

if $y_n = y_c$, then $S_b = S_c$: slope is critical (C)

if $y_n < y_c$, then $S_b > S_c$: slope is steep (S)

In addition, the channel bed may be horizontal (H) in which case $S_b = 0$. Finally, the bed slope may be adverse (A) with $S_b < 0$.

For a given bed slope (designated by the letter, M, C, S, H, or A), the shape of the surface profile depends on the actual depth, y, relative to y_n and y_c. Numbers used to designate the three possibilities are

Curve 1: $y > y_n$ and $y > y_c$

Curve 2: $y_n > y > y_c$ or $y_c > y > y_n$

Curve 3: $y < y_n$ and $y < y_c$

The possible surface profiles and their properties for each of the five possible channel bed slopes are summarized in Table 10.4.

a. Surface Profiles in Channels with Mild Slope ($S_b < S_c$)

For mild channel slopes, $y_n > y_c$. For the $M1$ profile, $y > y_n$, and hence both the numerator and denominator of Eq. 10.48 are positive; thus $dy/dx > 0$. As y increases, dy/dx approaches S_b and the free surface approaches the horizontal. The $M1$ backwater curve is typical of that found upstream from a dam or control structure. Since the actual water surface lies above the normal depth, flooding may occur near the dam if this is not accounted for in the design.

Table 10.4 Surface Profiles for Gradually Varied Flow

Surface Profiles	Curve	Depth	Flow	Surface Slope
Mild slope, $S_b < S_c$	$M1$	$y > y_n > y_c$	Subcritical	Positive
	$M2$	$y_n > y > y_c$	Subcritical	Negative
	$M3$	$y_n > y_c > y$	Supercritical	Positive
Steep slope, $S_b > S_c$	$S1$	$y > y_c > y_n$	Subcritical	Positive
	$S2$	$y_c > y > y_n$	Supercritical	Negative
	$S3$	$y_c > y_n > y$	Supercritical	Positive
Critical slope, $S_b = S_c$	$C1$	$y > y_c = y_n$	Subcritical	Positive
	$C3$	$y < y_c = y_n$	Supercritical	Positive
Horizontal slope, $S_b = 0$	$H2$	$y > y_c$	Subcritical	Negative
	$H3$	$y < y_c$	Supercritical	Positive
Adverse slope, $S_b < 0$	$A2$	$y > y_c$	Subcritical	Negative
	$A3$	$y < y_c$	Supercritical	Positive

For the $M2$ profile, $y_n > y > y_c$, and the flow is subcritical. The numerator of Eq. 10.48 is negative and the denominator is positive; thus $dy/dx < 0$ and the depth decreases in the direction of flow. As y approaches y_c, Eq. 10.48 indicates that $dy/dx \to \infty$, which is not possible. With strong surface curvature the assumptions of straight streamlines and hydrostatic pressure variation (inherent in Eq. 10.48) no longer are valid. Consequently, the profile is shown dashed as the flow depth approaches the critical depth. The $M2$ drawdown curve can occur upstream from a section where the channel slope changes from mild to critical or supercritical. An example is the flow over a spillway crest.

For the $M3$ profile, $y_n > y_c > y$, and the flow is supercritical. Both the numerator and denominator of Eq. 10.48 are negative, so dy/dx is positive and the depth increases in the direction of flow. The $M3$ profile, also a backwater curve, occurs in supercritical flow, e.g. downstream from a spillway or sluice gate. As the critical flow depth is approached, a sudden transition from supercritical to subcritical flow occurs. This sudden transition is a *hydraulic jump*, which will be analyzed and discussed in detail in Section 10-7.

b. Surface Profiles in Channels with Steep Slope ($S_b > S_c$)

For steep channel slopes, $y_n < y_c$. For the $S1$ profile, $y > y_c$, and hence both the numerator and denominator of Eq. 10.48 are positive; thus $dy/dx > 0$. Furthermore, as y becomes much larger than y_c, then $dy/dx \to S_b$. Consequently, the $S1$ curve approaches the horizontal.

The $S2$ profile with $y_c > y > y_n$ is a drawdown curve with $dy/dx < 0$. The flow is supercritical. As y approaches y_n, the slope of the curve approaches zero; the depth approaches the normal depth. The $S2$ profile may occur downstream from a transition in channel slope from mild to steep.

For the $S3$ profile, $y < y_n < y_c$. The flow is supercritical and the slope of the surface profile is positive. The depth increases in the direction of flow. As y approaches y_n, the surface slope approaches zero. This backwater curve may result downstream of a sluice gate if the depth of flow is less than the normal depth on a steep slope.

c. Surface Profiles in Channels with Critical Slope ($S_b = S_c$)

For critical channel slopes, $y_n = y_c$. The slope of the surface profile is positive for both the $C1$ and $C3$ curves. For the $C1$ curve, $y > y_c$; the flow is subcritical and the slope approaches S_b for large values of y. Consequently, the $C1$ curve approaches the horizontal. When $y < y_c$, the flow is supercritical. As y approaches y_c, by application of L'Hospital's rule to Eq. 10.48, it can be shown that the slope of the curve approaches a value of $\frac{10}{9}$.

There is no $C2$ curve for gradually varying flow. When $y = y_n = y_c$, we have critical flow at normal depth.

d. Surface Profiles in Horizontal Channels ($S_b = 0$)

For flow in a horizontal channel, $S_b = 0$, and the normal depth is infinite. Consequently, an $H1$ profile cannot exist. The $H2$ and $H3$ profiles correspond to the $M2$ and

$M3$ curves for $S_b = 0$. Equation 10.48 becomes indeterminate for $S_b = 0$ and $y_n \to \infty$. The profile slope can be determined from Eq. 10.43. For the $H2$ profile, $y > y_c$, the flow is subcritical and the slope is negative. For the $H3$ profile, $y < y_c$, the flow is supercritical and the slope is positive. For both profiles, Eq. 10.43 predicts infinite slope as $y \to y_c$ ($Fr \to 1$).

Since flow cannot continue indefinitely on a horizontal bed, these profiles can be found only in horizontal sections of more complex channels. The drawdown profile, $H2$, might be found across the top of a broad-crested weir (Section 10-8.2), and the profile $H3$ might be found on the flat floor of a stilling basin downstream from a spillway or below a sluice gate.

e. Surface Profiles in Channels with Adverse Slope ($S_b < 0$)

When the channel bed has an adverse slope, S_b is negative; there is no real value of y_n. From Eq. 10.43 we conclude that the $A2$ and $A3$ profiles are similar to the $H2$ and $H3$ profiles.

As in the case of a horizontal channel, sustained flow is not possible over a long reach of adverse slope. A hydraulic jump may occur as profile $A3$ develops.

10-6.2 Calculation of Surface Profiles

Table 10.4 illustrated the general nature of surface profiles. To determine the actual profile for a given flow rate and initial conditions, we must integrate the governing differential equation numerically

$$\frac{dy}{dx} + \frac{d}{dx}\left[\frac{V^2}{2g}\right] = S_b - S \tag{10.42}$$

This equation can be written in terms of the specific energy, E,

$$\frac{dE}{dx} = S_b - S$$

Writing the equation in finite difference form, we obtain

$$\Delta x = \frac{\Delta E}{(S_b - S)_m} = \frac{E(y + \Delta y) - E(y)}{(S_b - S)_m} \tag{10.49}$$

where the subscript, m, denotes the mean properties over a channel reach of length, Δx. Equation 10.49 can be used to solve for the channel section length, Δx, which corresponds to a given or assumed Δy, provided that S can be determined. If we assume that the local energy loss is equal to that for flow at normal depth, we can obtain S from the Manning correlation, since S_b for flow at normal depth equals S. From Eq. 10.44,

$$S_m = \frac{n^2 V_m^2}{R_{h_m}^{4/3}} \tag{10.50}$$

where $V_m = \frac{1}{2}[V(y + \Delta y) + V(y)]$ and $R_{h_m} = \frac{1}{2}[R_h(y + \Delta y) + R_h(y)]$.

Computation of the surface profile must begin at a point where the coordinates are known. Calculations may proceed upstream or downstream. To use Eq. 10.49, a depth, y_2, is assumed at an adjacent point, Δx away from the initial point. For the depth, y_2, the velocity, V_2, is calculated from continuity. The average velocity, V_m, and hydraulic radius are calculated and the mean slope of the energy grade line is determined from Eq. 10.50. The corresponding distance, Δx, is then computed directly from Eq. 10.49.

Obviously, hand calculations using Eq. 10.49 would be laborious, since small increments of y must be used to permit the use of mean properties. The calculations are easy to program for computer solution; this permits evaluation for small steps. The denominator of Eq. 10.49 vanishes when flow at normal depth is approached (at normal depth, $S = S_b$). Consequently, the program should include a test of the denominator.

Calculation of surface profiles is illustrated in Example Problem 10.8.

Example 10.8

Water flows in a 5 m wide rectangular channel made from unfinished concrete with $n = 0.015$. The channel contains a long reach on which S_b is constant at $S_b = 0.020$. At one section, flow is at depth, $y_1 = 1.5$ m, with speed, $V_1 = 4.0$ m/sec. Estimate the channel location where the flow reaches a depth of 0.9 m.

EXAMPLE PROBLEM 10.8

GIVEN: Water flow in a long rectangular channel.

$S_b = 0.020$, $n = 0.015$, and $b = 5.0$ m.

At section ①, $y_1 = 1.5$ m, and $V_1 = 4.0$ m/sec.

FIND: Location in channel where flow depth is $y_2 = 0.9$ m.

SOLUTION:

The location where $y_2 = 0.9$ m may be upstream or downstream from section ①, depending on y_1, y_c, and y_n.

The flow rate is

$$Q = V_1 b y_1 = \frac{4\ \text{m}}{\text{sec}} \times 5\ \text{m} \times 1.5\ \text{m} = 30\ \text{m}^3/\text{sec}$$

At critical conditions $V = V_c = \sqrt{g y_c}$, so

$$Q = V_1 b y_1 = V_c b y_c = \sqrt{g y_c}\, b y_c = b\sqrt{g}\, y_c^{3/2}$$

Thus

$$y_c = \left(\frac{Q}{b\sqrt{g}}\right)^{2/3} = \left[\frac{30\ \text{m}^3}{\text{sec}} \times \frac{1}{5\ \text{m}} \times \frac{\text{sec}}{\sqrt{9.81}\ \text{m}^{1/2}}\right]^{2/3} = 1.54\ \text{m}$$

Since $y < y_c$, flow is supercritical.

$$Fr_1 = \frac{V_1}{\sqrt{g y_1}} = \frac{4\ \text{m}}{\text{sec}} \times \frac{1}{\left[\frac{9.81\ \text{m}}{\text{sec}^2} \times 1.5\ \text{m}\right]^{1/2}} = 1.04 > 1$$

The normal depth at this slope, roughness, and discharge may be found from

$$Q = \frac{R_h^{2/3} S_b^{1/2}}{n} A = \frac{\left[\dfrac{y_n}{1 + 2y_n/b}\right]^{2/3} S_b^{1/2}}{n} b y_n = 47.14\, y_n \left(\frac{y_n}{1 + 2y_n/b}\right)^{2/3}$$

By iteration it may be found that $y_n = 0.858$ m.

Thus $y_c > y > y_n$, so flow is on an $S2$ curve. We expect the depth to *decrease* in the direction of flow; y should approach $y_n = 0.858$ m asymptotically.

Equation 10.49 may be used to solve numerically for the channel reach, Δx, required to change the surface level by an assumed amount, Δy. Thus, since $S_b = \text{constant}$, $(S_b - S)_m = S_b - S_m$, and Eq. 10.49 becomes

$$\text{Computing equation:} \quad \Delta x = \frac{E(y + \Delta y) - E(y)}{S_b - S_m}$$

To apply Eq. 10.49, we must express E as a function of y and we must evaluate S_m. Since $V = Q/by$, then

$$E = \frac{V^2}{2g} + y = \frac{Q^2}{2gb^2y^2} + y$$

and

$$S_m = \frac{n^2 V_m^2}{R_{h_m}^{4/3}}$$

where $V_m = \frac{1}{2}[V(y + \Delta y) - V(y)]$, and $R_{h_m} = \frac{1}{2}[R_h(y + \Delta y) - R_h(y)]$.

Calculating equations for quantities in terms of y are

$$V = \frac{Q}{by} = \frac{V_1 b y_1}{by} = V_1 \frac{y_1}{y} = 4\,\frac{\text{m}}{\text{sec}} \times 1.5\text{ m} \times \frac{1}{y} = \frac{6.000}{y}$$

$$E = \frac{Q^2}{2gb^2y^2} + y = \frac{1}{2} \times \frac{(30)^2\ \text{m}^6}{\text{sec}^2} \times \frac{\text{sec}^2}{9.807\text{ m}} \times \frac{1}{(5)^2\text{ m}} \times \frac{1}{y^2} + y = \frac{1.835}{y^2} + y$$

$$R_h = \frac{A}{P} = \frac{by}{b + 2y} = \frac{y}{1 + 2y/b} = \frac{y}{1 + 2y \times \dfrac{1}{5\text{ m}}} = \frac{y}{1 + 0.4\,y}$$

Calculations may be tabulated as shown in the following table. (Four significant figures are shown to reduce roundoff errors; all intermediate results were retained in the calculator during the computations.)

y (m)	V (m/sec)	E (m)	R_h (m)	R_{h_m} (m)	V_m (m/sec)	S_m (−)	Δx (m)	x (m)
1.5	4.000	2.316	0.9375					0.0
				0.9177	4.143	0.004330	1.321	
1.4	4.286	2.336	0.8974					1.32
				0.8766	4.451	0.005132	3.378	
1.3	4.615	2.386	0.8553					4.70
				0.8333	4.808	0.006632	6.624	
1.2	5.000	2.475	0.8108					11.3
				0.7877	5.227	0.008452	12.32	
1.1	5.455	2.617	0.7639					23.6
				0.7394	5.727	0.01104	24.39	
1.0	6.000	2.835	0.7143					48.0
				0.6884	6.333	0.01485	64.15	
0.9	6.667	3.166	0.6618					112.2

A graph of the results follows. The plot shows the water depth dropping rapidly at first, then approaching $y_n = 0.858$ m asymptotically farther downstream.

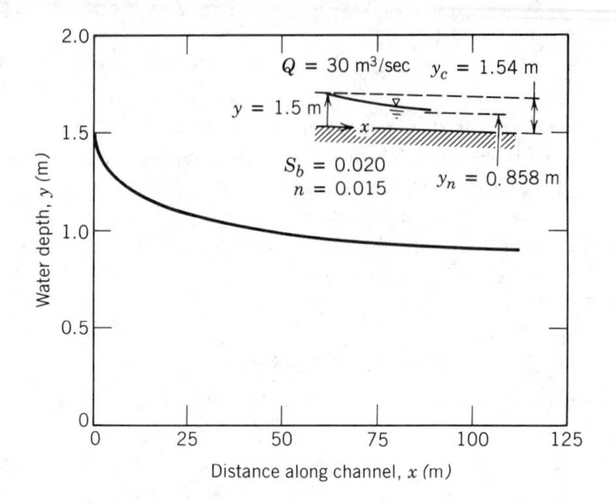

Accuracy of the results depends on (a) the accuracy of the mathematical model, and (b) the technique used for numerical integration. The assumptions of negligible streamline curvature (and hydrostatic pressure distribution) are questionable from $y = 1.5$ to $y = 1.3$ m, where the depth changes rapidly. However, the channel distance during this interval is only 4.7 m, which is a small part of the total reach. Repeating the calculations with half the depth interval ($\Delta y = 0.05$ m) increased the total reach to 117.7 m, a 5 percent increase. The effect of step size is larger where the rate of change of depth is smaller.

10-7 THE HYDRAULIC JUMP

We have shown that open-channel flow may be subcritical ($Fr < 1$) or supercritical ($Fr > 1$). For subcritical flow, disturbances caused by a change in bed slope or flow cross section may move upstream and downstream; the result is a smooth adjustment of the flow. When the flow at a section is supercritical and downstream conditions require a change to subcritical flow, the need for the change cannot be telegraphed upstream. Thus a gradual change with a smooth transition through the critical point is not possible. The transition from supercritical to subcritical flow occurs abruptly through a hydraulic jump. The abrupt change in depth involves a significant loss of mechanical energy through turbulent mixing.

The general features of a hydraulic jump are sketched in Fig. 10.11. We shall analyze the jump phenomenon by applying the basic equations to the control volume shown in the sketch. Experiments show that the jump occurs over a relatively short distance; the maximum value of L/y_2 is approximately 6.1 [3]. In view of this short length it is reasonable to assume that friction forces acting on the control volume are negligible compared to the pressure forces. Although hydraulic jumps can occur on inclined surfaces, let us assume a horizontal bed and select a control volume of width, b, for simplicity.

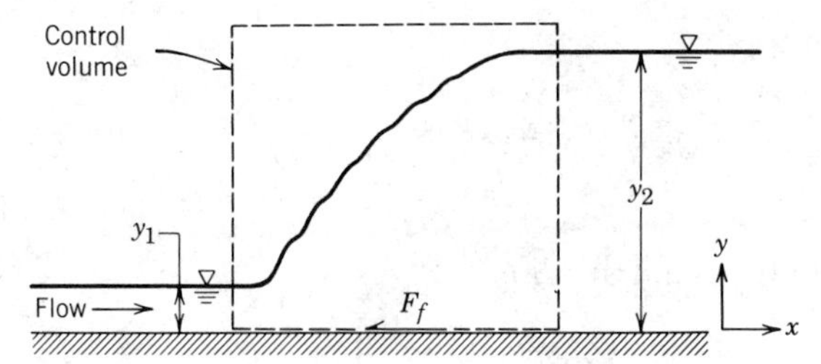

Fig. 10.11 Schematic of hydraulic jump, showing control volume used for analysis.

10-7.1 Basic Equations

a. Continuity Equation

Basic equation:

$$0 = \overset{=0(1)}{\frac{\partial}{\partial t}\int_{CV}} \rho \, d\forall + \int_{CS} \rho \vec{V} \cdot d\vec{A} \tag{4.13}$$

Assumptions: (1) Steady flow
(2) Uniform flow at a section
(3) Incompressible flow

Then

$$0 = \{-|\rho V_1 b y_1|\} + \{|\rho V_2 b y_2|\} \quad \text{or} \quad V_1 y_1 = V_2 y_2 \tag{10.51}$$

b. Momentum Equation

Basic equation:

$$F_{S_x} + \overset{=0(4)}{F_{B_x}} = \overset{=0(1)}{\frac{\partial}{\partial t}\int_{CV}} u\rho \, d\forall + \int_{CS} u\rho \vec{V} \cdot d\vec{A} \tag{4.19a}$$

Assumptions: (4) $F_{B_x} = 0$
(5) Hydrostatic pressure distribution, so that

$$\frac{dp}{dy} = -\rho g, \text{ and } p = p_{\text{atm}} + \rho g(y_s - y)$$

(6) Neglect the friction force, F_f

The momentum equation becomes

$$\int_0^{y_1} [p_{\text{atm}} + \rho g(y_1 - y)]b \, dy - \int_0^{y_2} [p_{\text{atm}} + \rho g(y_2 - y)]b \, dy$$

$$+ p_{\text{atm}} b(y_2 - y_1) - \overset{\approx 0(6)}{F_f} = V_1\{-|\rho V_1 b y_1|\} + V_2\{|\rho V_2 b y_2|\}$$

or

$$p_{\text{atm}} b y_1 + \rho g\left(y_1^2 - \frac{y_1^2}{2}\right)b - p_{\text{atm}} b y_2 - \rho g\left(y_2^2 - \frac{y_2^2}{2}\right)b + p_{\text{atm}} b(y_2 - y_1) = -\rho V_1^2 b y_1 + \rho V_2^2 b y_2$$

Finally, after simplifying and dividing by $\rho g b$

$$\frac{V_1^2}{g} y_1 + \frac{y_1^2}{2} = \frac{V_2^2}{g} y_2 + \frac{y_2^2}{2} \tag{10.52}$$

c. Energy Equation

The energy equation for open-channel flow was derived in Section 10-3

$$\frac{V_1^2}{2g} + y_1 + z_1 = \frac{V_2^2}{2g} + y_2 + z_2 + h_\ell \tag{10.10}$$

For the hydraulic jump of Fig. 10.11 the channel bed is horizontal and the energy equation becomes

$$\frac{V_1^2}{2g} + y_1 = \frac{V_2^2}{2g} + y_2 + h_\ell \tag{10.53}$$

In terms of the specific energy, Eq. 10.53 becomes

$$E_1 = E_2 + h_\ell \tag{10.54}$$

Since h_ℓ is positive, Eq. 10.54 shows that the flow leaving a hydraulic jump has a lower specific energy than the entering flow.

The continuity and momentum equations, Eqs. 10.51 and 10.52, can be solved for y_2 in terms of upstream conditions (Section 10-7.2). Then the head loss across the jump also can be computed in terms of upstream conditions (Section 10-7.3).

10-7.2 Depth Increase across a Hydraulic Jump

To determine the downstream or *sequent* depth in terms of conditions upstream from the hydraulic jump, we begin by eliminating V_2 from the momentum equation. From continuity, $V_2 = V_1 y_1/y_2$ and hence Eq. 10.52 can be written as

$$\frac{V_1^2 y_1}{g} + \frac{y_1^2}{2} = \frac{V_1^2 y_1}{g}\left(\frac{y_1}{y_2}\right) + \frac{y_2^2}{2}$$

Rearranging this equation gives

$$y_2^2 - y_1^2 = \frac{2V_1^2 y_1}{g}\left(1 - \frac{y_1}{y_2}\right) = \frac{2V_1^2 y_1}{g}\left(\frac{y_2 - y_1}{y_2}\right)$$

Dividing both sides by the common factor, $(y_2 - y_1)$, we obtain

$$y_2 + y_1 = \frac{2V_1^2 y_1}{g y_2}$$

Multiplying through by y_2/y_1^2 gives

$$\left(\frac{y_2}{y_1}\right)^2 + \left(\frac{y_2}{y_1}\right) = \frac{2V_1^2}{gy_1} = 2\,Fr_1^2$$

Solving for y_2/y_1 using the quadratic formula (and choosing the positive root, since y_2/y_1 must be positive), we find that

$$\frac{y_2}{y_1} = \frac{1}{2}\left(\sqrt{1 + 8\,Fr_1^2} - 1\right) \tag{10.55}$$

Thus the ratio of downstream to upstream depths across a hydraulic jump is only a function of the upstream Froude number. The depths, y_1 and y_2 are referred to as *conjugate* depths. From Eq. 10.55, we see that an increase in depth requires an upstream Froude number greater than one. Experimental confirmation of Eq. 10.55 is shown in Fig. 10.12.

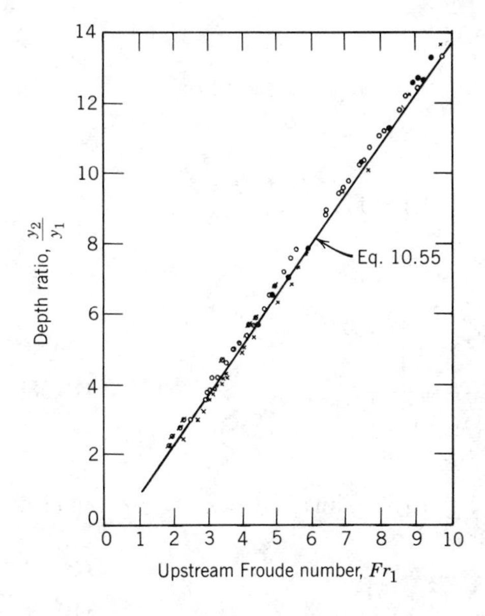

Fig. 10.12 Comparison of theory and experimental results for depth ratio across a hydraulic jump. (Data from [3], used by permission.)

Hydraulic jumps often are desired to dissipate energy below spillways to prevent erosion of artificial or natural channel bottom or sides. The head loss due to a hydraulic jump is treated in the next section.

10-7.3 Head Loss across a Hydraulic Jump

The head loss across a hydraulic jump may be calculated from the energy equation

$$h_\ell = E_1 - E_2 = \frac{V_1^2}{2g} + y_1 - \left[\frac{V_2^2}{2g} + y_2\right] \tag{10.54}$$

From continuity, $V_2 = V_1 y_1/y_2$, and hence

$$h_\ell = \frac{V_1^2}{2g}\left[1 - \left(\frac{y_1}{y_2}\right)^2\right] + (y_1 - y_2)$$

or

$$\frac{h_\ell}{y_1} = \frac{Fr_1^2}{2}\left[1 - \left(\frac{y_1}{y_2}\right)^2\right] + \left[1 - \frac{y_2}{y_1}\right] \tag{10.56}$$

Solving Eq. 10.55 for Fr_1 in terms of y_2/y_1 and substituting into Eq. 10.56, we obtain (after considerable algebraic manipulation),

$$\frac{h_\ell}{y_1} = \frac{1}{4}\frac{\left[\frac{y_2}{y_1} - 1\right]^3}{\frac{y_2}{y_1}} \tag{10.57}$$

Since h_ℓ is positive, Eq. 10.57 shows that y_2/y_1 must be greater than one.

The specific energy, E_1, can be written as

$$E_1 = \frac{V_1^2}{2g} + y_1 = y_1\left[\frac{V_1^2}{2gy_1} + 1\right] = y_1\frac{(Fr_1^2 + 2)}{2}$$

Nondimensionalizing h_ℓ on E_1, then

$$\frac{h_\ell}{E_1} = \frac{1}{2}\frac{\left[\frac{y_2}{y_1} - 1\right]^3}{\frac{y_2}{y_1}\left[Fr_1^2 + 2\right]} \tag{10.58}$$

The depth ratio is given in terms of Fr_1 by Eq. 10.55. Thus h_ℓ/E_1 can be written as a function of entering Froude number. The result is

$$\frac{h_\ell}{E_1} = \frac{[\sqrt{1 + 8Fr_1^2} - 3]^3}{8[\sqrt{1 + 8Fr_1^2} - 1][Fr_1^2 + 2]} \tag{10.59}$$

The effect of the entering Froude number on head loss is shown in dimensionless form in Fig. 10.13. Experimental data correlate well with the prediction of Eq. 10.59; Fig. 10.13 shows that more than 70 percent of the energy of the entering stream is dissipated in jumps with $Fr_1 > 9$.

Inspection of Eq. 10.59 shows that if $Fr_1 = 1$, then $h_\ell = 0$. Imaginary values are predicted for $Fr_1 < 1$. Since h_ℓ must be positive in any real flow, a hydraulic jump can occur only in supercritical flow with $Fr > 1$. Flow downstream from a jump always is subcritical.

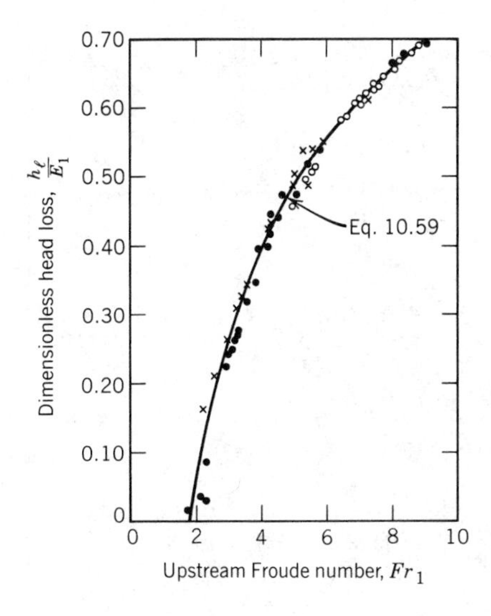

Fig. 10.13 Head loss across a hydraulic jump as a function of approach Froude number. (Data from [3], used by permission.)

Example 10.9

A hydraulic jump occurs on a horizontal bed immediately downstream from the sluice gate of Example Problem 10.3. Calculate the sequent depth following the hydraulic jump. Evaluate the head loss across the jump.

EXAMPLE PROBLEM 10.9

GIVEN: Flow under a sluice gate at the conditions shown.

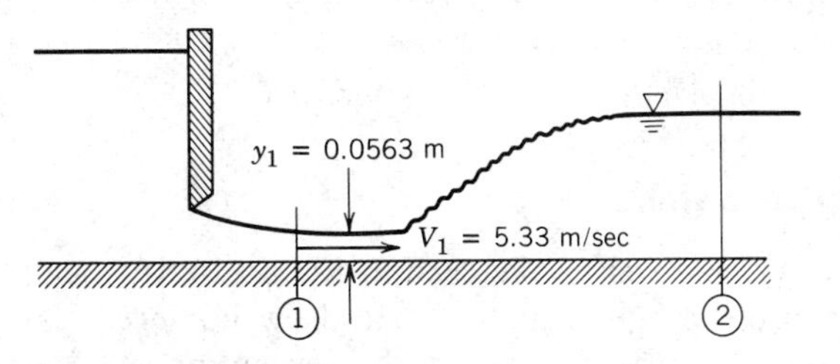

FIND: (a) Calculate the sequent depth downstream.
(b) Evaluate the head loss across the jump.

SOLUTION:

Apply Eqs. 10.55 and 10.59 directly.

Computing equations:
$$\frac{y_2}{y_1} = \frac{1}{2}[\sqrt{1 + 8Fr_1^2} - 1] \tag{10.55}$$

$$\frac{h_\ell}{E_1} = \frac{[\sqrt{1 + 8Fr_1^2} - 3]^3}{8[\sqrt{1 + 8Fr_1^2} - 1][Fr_1^2 + 2]} \tag{10.59}$$

The Froude number of the flow entering the jump is

$$Fr_1 = \frac{V_1}{\sqrt{gy_1}} = \frac{5.33 \text{ m}}{\text{sec}} \times \frac{1}{\left[\frac{9.81 \text{ m}}{\text{sec}^2} \times 0.0563 \text{ m}\right]^{1/2}} = 7.17$$

Then

$$\frac{y_2}{y_1} = \frac{1}{2}[\sqrt{1 + 8(7.17)^2} - 1] = 9.65$$

and

$$y_2 = 9.65\, y_1 = 9.65 \times 0.0563 \text{ m} = 0.543 \text{ m} \qquad y_2$$

Also

$$\frac{h_\ell}{E_1} = \frac{[\sqrt{1 + 8(7.17)^2} - 3]^3}{8[\sqrt{1 + 8(7.17)^2} - 1][(7.17)^2 + 2]} = 0.629$$

Since

$$E_1 = \frac{V_1^2}{2g} + y_1 = \frac{1}{2} \times \frac{(5.33)^2 \text{ m}^2}{\text{sec}^2} \times \frac{\text{sec}^2}{9.81 \text{ m}} + 0.0563 \text{ m} = 1.50 \text{ m}$$

then

$$h_\ell = 0.629\, E_1 = (0.629)\, 1.50 \text{ m} = 0.944 \text{ m} \qquad h_\ell$$

This hydraulic jump dissipates nearly 63 percent of the specific energy of the incoming flow.

10-8 MEASUREMENTS IN OPEN-CHANNEL FLOW

Two basic methods are available to measure flow rates in open channels. At a critical section, the flow speed is equal to the critical speed, so the flow rate may be calculated from a depth measurement. At an obstruction in a channel (a *weir*), flow depth correlates with the rate of flow.

10-8.1 Sharp-Crested Weirs

Flow over a sharp-crested weir is sketched in Fig. 10.14. It is tempting to apply the Bernoulli equation from far upstream to flow at the weir *nappe*. Near the weir crest, the streamlines are highly curved, making the assumptions of uniform flow and hydrostatic pressure variation poor. Consequently, accurate calculation of flow over a sharp-crested weir requires use of an empirically determined discharge coefficient.

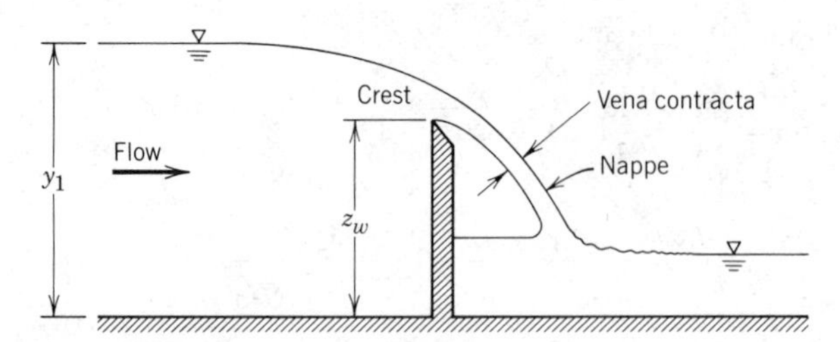

Fig. 10.14 Section view of flow over a sharp-crested weir.

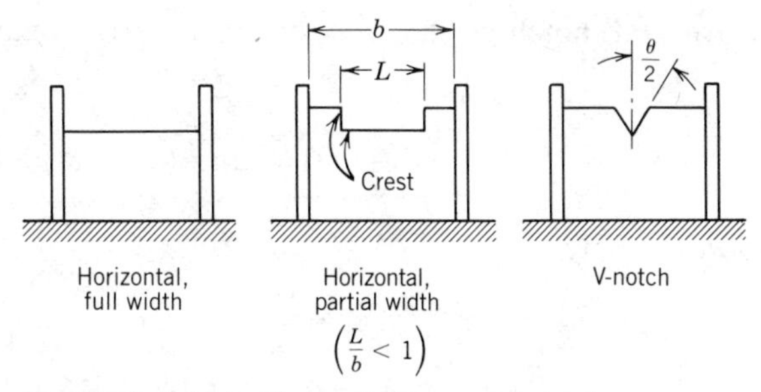

Fig. 10.15 Principal types of sharp-crested weirs.

Many experiments have been run on a variety of weir configurations. Figure 10.15 illustrates three principal types of sharp-crested weirs.

The flow area across a horizontal weir is proportional to the height difference, $y_1 - z_w$. Thus

$$A \sim b\left(\frac{L}{b}\right)(y_1 - z_w)$$

If we neglect the upstream velocity, the velocity across the weir may be determined approximately from the Bernoulli equation as

$$V \approx \sqrt{g(y_1 - z_w)}$$

Combining these equations yields

$$Q \sim b\left(\frac{L}{b}\right)\sqrt{g}(y_1 - z_w)^{3/2}$$

Introducing the empirically determined discharge coefficient, C_d, to account for velocity nonuniformity, stream contraction, and frictional effects, then

$$Q = C_d b\left(\frac{L}{b}\right)\sqrt{g}(y_1 - z_w)^{3/2} \tag{10.60}$$

Many correlations have been suggested for the discharge coefficient. A simple equation of the form [4]

$$C_d = 0.59 + 0.08\left[\frac{y_1}{z_w} - 1\right] \tag{10.61}$$

gives satisfactory results for many cases.

The nappe (Fig. 10.14) of a full-width horizontal weir must be aerated (ventilated to the atmosphere). If the nappe is not aerated, entrainment forms a low-pressure region, causing the discharge to increase for a given head.

Aeration is automatic when the weir spans only part of the channel. However, additional contraction of the stream is caused by the sharp vertical edges of the crest. For approximate calculations [4, 5], the length, L, used in Eq. 10.60 may be reduced by $0.2\,(y_1 - z_w)$.

The flow area over a V-notch crest is proportional to the square of the height difference

$$A \sim (y_1 - z_w) \tan \frac{\theta}{2} (y_1 - z_w) = \tan \frac{\theta}{2} (y_1 - z_w)^2$$

The velocity is approximately

$$V \approx \sqrt{g(y_1 - z_w)}$$

which suggests

$$Q = C_d \tan \frac{\theta}{2} \sqrt{g} (y_1 - z_w)^{5/2} \tag{10.62}$$

Experiments show that $C_d \approx 0.44$ for a V-notch weir [4].

The selection of a weir for a given situation depends upon the range of flow rate to be measured, the accuracy desired, and whether the weir can be calibrated after installation. Equations 10.60 and 10.62 are applicable only for $y_1 - z_w$ greater than about 0.06 m ($\sim$0.2 ft). This suggests that a V-notch weir should be chosen for low flow rates. For repeatability, the weir crest must be kept sharp. Periodic removal of scale or rust is simple in the laboratory but may be more difficult in the field.

When calibrated in place using volume measurements or weigh tanks, the accuracy of a weir in use is limited only by the accuracy of measuring head. Errors in head measurement can be reduced using a stilling well and hook gage, as shown in Fig. 10.16. The surface can be located with extreme precision by sighting across it nearly horizontally and noting the reflection of the hook point in the surface.

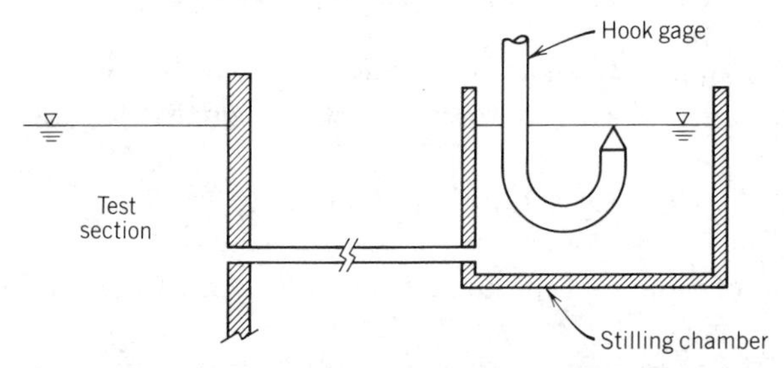

Fig. 10.16 Use of a stilling chamber and hook gage for precise liquid level measurements.

Measurement uncertainties for weirs depend on many factors. Flow rates can be measured to ± 5 percent. Detailed information in [4] and [5] should be consulted if more precise results are needed.

Example 10.10

A horizontal, sharp-crested weir is to be used in a laboratory setup to control and measure flow rate, which is expected to vary between 200 and 300 gallons of water per minute. The width, b, is to be such that the level change between these flow rates is held to 0.10 ft.

EXAMPLE PROBLEM 10.10

GIVEN: Horizontal weir, water flow rate between 200 and 300 gpm.

FIND: Width, b, such that $\Delta z_{hi} - \Delta z_{lo} < 0.10$ ft

SOLUTION:

From Section 10-8.1,

$$Q = C_d b\left(\frac{L}{b}\right)\sqrt{g}(y_1 - z_w)^{3/2} \tag{10.60}$$

and

$$C_d = 0.59 + 0.08\left[\frac{y_1}{z_w} - 1\right] \tag{10.61}$$

Assumptions: (1) $L/b = 1$
(2) Neglect the second term as a first approximation, by assuming $\Delta z = y_1 - z_w \ll z_w$

Then $C_d = 0.59$, and

$$\Delta z_{hi} - \Delta z_{lo} = \left[\frac{1}{0.59\, b\sqrt{g}}\right]^{2/3}\left[Q_{hi}^{2/3} - Q_{lo}^{2/3}\right]$$

Finally,

$$b = \frac{1}{0.59\sqrt{g}}\,\frac{\left[Q_{hi}^{2/3} - Q_{lo}^{2/3}\right]^{3/2}}{\left[\Delta z_{hi} - \Delta z_{lo}\right]^{3/2}}$$

The flow rates are

$$Q_{hi} = \frac{300\ \text{gal}}{\text{min}} \times \frac{\text{ft}^3}{7.48\ \text{gal}} \times \frac{\text{min}}{60\ \text{sec}} = 0.668\ \text{ft}^3/\text{sec}$$

and

$$Q_{lo} = \frac{200\ \text{gal}}{\text{min}} \times \frac{\text{ft}^3}{7.48\ \text{gal}} \times \frac{\text{min}}{60\ \text{sec}} = 0.446\ \text{ft}^3/\text{sec}$$

Substituting

$$b = \frac{\text{sec}}{0.59(32.2)^{1/2}\text{ft}^{1/2}}\,\frac{\left[(0.668)^{2/3} - (0.446)^{2/3}\right]^{3/2}}{(0.10)^{3/2}\ \text{ft}^{3/2}}\,\frac{\text{ft}^3}{\text{sec}}$$

$$b = 0.724\ \text{ft}$$

b

Equation 10.60 is valid if the elevation change, Δz, is greater than 0.2 ft. Solving for Δz at the lowest flow rate, we obtain

$$\Delta z = \left[\frac{Q}{C_d b\sqrt{g}}\right]^{2/3}$$

$$= \left[\frac{0.446\ \text{ft}^3}{\text{sec}} \times \frac{1}{0.59} \times \frac{1}{0.724\ \text{ft}} \times \frac{\text{sec}}{(32.2)^{1/2}\ \text{ft}^{1/2}}\right]^{2/3}$$

$$\Delta z = 0.324\ \text{ft}$$

which satisfies the criterion.

For the second term in Eq. 10.61 to be negligible,

$$0.08\left[\frac{y_1}{z_w} - 1\right] = (0.08)\frac{y_1 - z_w}{z_w} = (0.08)\frac{\Delta z}{z_w} < 0.01$$

so z_w must be

$$z_w > 8\ \Delta z = 8 \times 0.324\ \text{ft} = 2.59\ \text{ft}$$

10-8.2 Broad-Crested Weirs

Flow over a broad-crested weir is sketched in Fig. 10.17. When the tailwater level downstream is sufficiently low, critical depth, y_c, is reached at some point on the top of the weir. There $Fr = 1$, so

$$V_c = \sqrt{g y_c}$$

For a wide channel, the discharge is given by

$$Q \approx AV = b y_c \sqrt{g y_c} = b\sqrt{g}\, y_c^{3/2}$$

or

$$Q = C_d b\sqrt{g}\, y_c^{3/2} \tag{10.63}$$

where C_d is an empirical discharge coefficient. Experimental data [4] suggest that when $(y_1 - z_w)/L > 0.2$ the discharge coefficient is between 0.95 and 0.98.

When a broad-crested weir is long and the tailwater is low enough, a *free overfall* occurs [5]. As shown in Fig. 10.18, the water depth at the brink, y_b, is less than the critical depth, y_c. Also, y_c is located some distance, L_c, upstream from the brink.

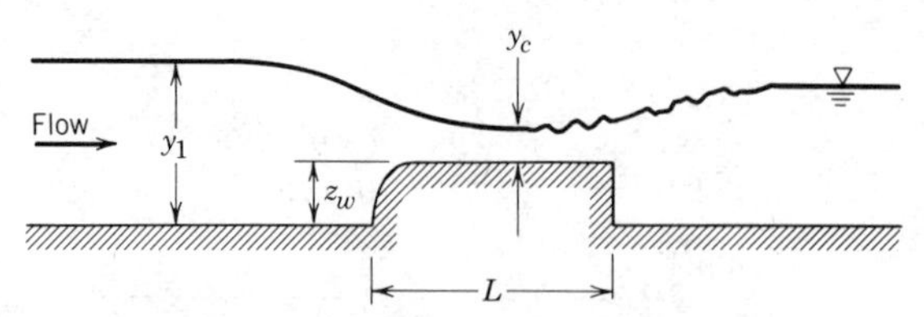

Fig. 10.17 Flow over a broad-crested weir.

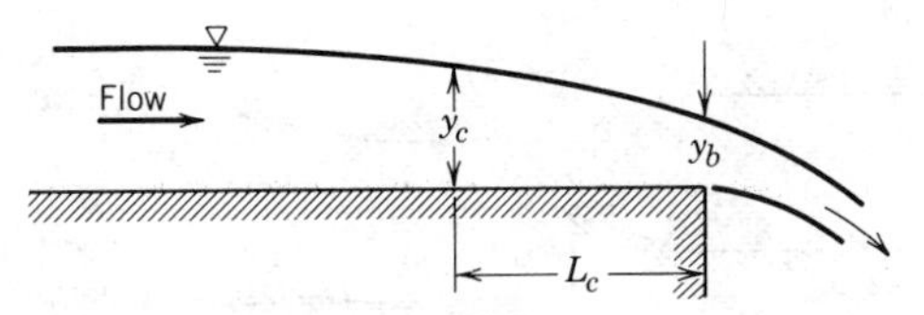

Fig. 10.18 Flow over a broad-crested weir with a free overfall.

Experiments [5] show that

$$y_b \approx 0.72 y_c \qquad \text{and} \qquad L_c \approx 3.5 y_c$$

Thus, if the brink depth, y_b, is known, then y_c may be computed.

10-8.3 Sluice Gates

A sluice gate often is used to regulate discharge. Flow beneath a typical sluice gate is shown in Fig. 10.19 for both low and high tailwater conditions.

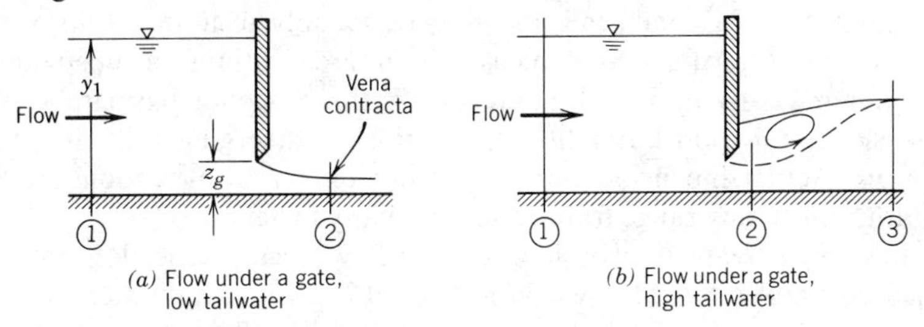

Fig. 10.19 Flow under a sluice gate.

For low tailwater, as shown in Fig. 10.19*a*, the vena contracta is the section of minimum area in the discharge stream. There streamlines are straight and parallel, so the pressure distribution is hydrostatic. If we assume frictionless flow, the Bernoulli equation may be applied between sections ① and ②. By introducing an empirical discharge coefficient, C_d, the flow rate may be written

$$Q = C_d b z_g \sqrt{2 g y_1} \tag{10.64}$$

Experiments [5] show that $0.6 < C_d < 0.9$.

For high tailwater, as shown in Fig. 10.19*b*, analysis of flow under a sluice gate is impossible. The general form of Eq. 10.64 may be used with an appropriate value of C_d, e.g. from [5].

10-8.4 Critical Flumes

Accurate flow measurements can be made using weirs. However, practical problems that limit their field use include:

- Fouling by silt or debris
- Deterioration of sharp edges
- Large head loss

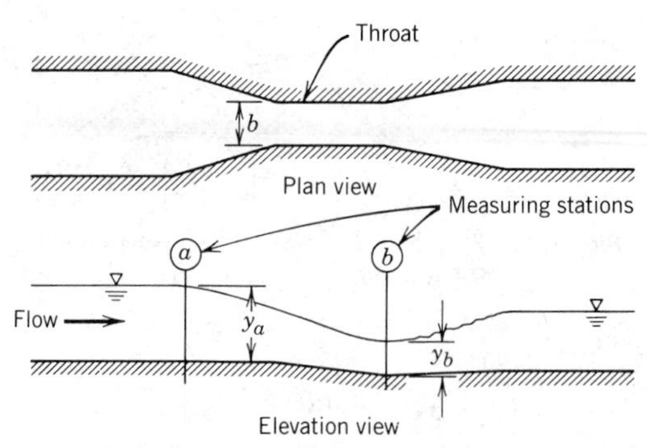

Fig. 10.20 Parshall flume for measuring open-channel flow rate.

These problem can be overcome by using a critical flow meter, such as the Parshall flume, shown in Fig. 10.20.

The Parshall flume is widely used to measure irrigation water flows because it is self-cleaning, requires only a small head, and gives reasonably accurate results over a wide range of flow rate [6]. A Parshall flume is built in three sections: the upstream section has a flat floor and converging walls, the center or throat section has parallel walls and a floor that slopes downward, and the outlet section has diverging walls and a floor that slopes upward. Actual dimensions and the location of measuring sections are based on throat width, which may range from 75 mm to many meters.

When the downstream level is low, critical flow occurs. The flow rate may be determined from tables if y_a is known. When $y_b > 0.7y_a$, both depths must be measured. Tables and charts are available for many sizes of Parshall flumes [6].

10-9 SUMMARY OBJECTIVES

After completing study of Chapter 10, you should be able to do the following:

1. Define the terms:

Froude number	energy grade line
prismatic channel	hydraulic grade line
wetted perimeter	Manning roughness coefficient
hydraulic radius	optimum channel cross section
secondary flow	critical depth
wave celerity	mild slope
subcritical flow	critical slope
critical flow	steep slope
supercritical flow	hydraulic jump
specific energy	sequent depth
alternate depths	conjugate depths
uniform flow	free overfall
normal depth	

2. Calculate the cross-sectional area, wetted perimeter, and hydraulic radius for common open-channel shapes.

3. Derive an equation for the propagation speed of an isolated free-surface wave and show that for small amplitudes

$$c = \sqrt{gy}$$

4. Apply the control volume form of the energy equation to open-channel flow. Discuss the variation in specific energy for frictionless flow and for flow at normal depth.
5. Evaluate the effect of changes in channel bottom contour on frictionless flow. Sketch changes in surface profile for subcritical and supercritical flows over a bump and a depression in the channel bottom.
6. Derive the conditions required to maximize the flow rate through a sluice gate assuming frictionless flow.
7. Use the Manning correlation for velocity to evaluate normal depth for uniform flow on a bed of constant slope. Calculate the critical slope for a given flow rate.
8. Define and calculate the optimum channel cross section or flow depth for flow at normal depth.
9. Apply the basic equations to a differential control volume to evaluate the rate of change of surface profile for gradually varied flow. Classify and calculate surface profiles for various bed slopes in subcritical and supercritical flows.
10. Analyze flow through a hydraulic jump to calculate the downstream surface level and head loss across the jump.
11. Determine the volume flow rate from elevation measurements for sharp-edged and broad-crested weirs and sluice gates.
12. Solve the problems at the end of the chapter that relate to the material you have studied.

REFERENCES

1. Chow, V. T., *Open Channel Hydraulics.* New York: McGraw-Hill, 1959.
2. White, F. M., "Hydrodynamics of Coastal Areas," in *Introduction to Ocean Engineering*, H. Schenck, ed. New York: McGraw-Hill, 1975, pp. 8–42.
3. Peterka, A. J., "Hydraulic Design of Stilling Basins and Energy Dissipators," U.S. Department of the Interior, Bureau of Reclamation, Engineering Monograph No. 25, (Revised) July 1963.
4. Ackers, P., W. R. White, J. A. Perkins, and A. J. M. Harrison, *Weirs and Flumes for Flow Measurement.* New York: Wiley, 1978.
5. King, H. W., and E. F. Brater, *Handbook of Hydraulics*, 5th ed. New York: McGraw-Hill, 1963.
6. Scott, V. L., and C. E. Houston, "Measuring Irrigation Water," Agricultural Extension Service, University of California, Davis, Circular 473, January 1959.
7. Todd, D. K., ed., *The Water Encyclopedia.* Port Washington, N.Y.: Water Information Center, 1970.
8. Leopold, L. B., and W. B. Langbein, "River Meanders," *Scientific American, 214*, 6, June 1966, pp. 60–70.
9. Lynch, D. K., "Tidal Bores," *Scientific American, 247*, 4, October 1982, pp. 146–156.
10. Fowler, R. G., "Current Problems of Fluid Mechanics," *Physics Today, 19*, 6, June 1966, pp. 37–42.

11. Varley, E., R. Venkataraman, and E. Cumberbatch, "The Propagation of Large Amplitude Tsunamis across a Basin of Varying Depth. Part I, Off-Shore Behavior," *J. Fluid Mechanics*, *49*, Pt. 4, 29 October 1971, pp. 771–802.

PROBLEMS

10.1 Experiments show that transition to turbulent flow in a broad open channel occurs at $Re = yV_{\max}/\nu \approx 1200$. Derive an expression for the velocity profile assuming fully developed laminar flow. Show that to maintain laminar flow down an inclined surface the flow rate per unit width, Q/b, must be less than 800ν.

10.2 Derive an expression for hydraulic radius for a trapezoidal channel of bottom width, b, liquid depth, y, and side slope angle, θ. Verify the equation given in Table 10.1. Plot the ratio, R_h/y, for $b = 2$ m with side slope angles of 30° and 60° for the range, $0.5 < y < 3$ m.

10.3 Use the equation given in Table 10.1 for the hydraulic radius of a circular channel to evaluate and plot the ratio, R_h/D, for liquid depths in the range between 0 and D.

10.4 A wave from a passing boat in a lake is observed to travel at 10 mph. Determine the approximate water depth at this location.

10.5 Solution of the complete differential equations for wave motion without surface tension shows that wave speed is given by

$$V_w = \sqrt{\frac{g\lambda}{2\pi}\tanh\left(\frac{2\pi y}{\lambda}\right)}$$

where λ is the wavelength. Show that when $\lambda/y \ll 1$, the wave speed becomes proportional to $\sqrt{\lambda}$. In the limit as $\lambda/y \to \infty$, $V_w = \sqrt{gy}$. Determine the value of λ/y above which $V_w > 0.99\sqrt{gy}$.

10.6 Solution of the complete differential equations for wave motion in quiescent liquid including the effect of surface tension shows that the wave speed is given by

$$V_w = \sqrt{\left(\frac{g\lambda}{2\pi} + \frac{2\pi\sigma}{\rho\lambda}\right)\tanh\left(\frac{2\pi y}{\lambda}\right)}$$

where λ is wavelength, y is liquid depth, and σ is surface tension. Prepare a plot of wave speed versus wavelength for the range $1 < \lambda < 100$ mm for (a) water, and (b) mercury. Assume that $y = 7$ mm for both liquids.

10.7 Water flows in a rectangular channel at a depth of 2 ft. If the flow speed is (a) 4 ft/sec, and (b) 12 ft/sec, compute the corresponding Froude numbers.

10.8 A pebble is dropped into a stream of water that flows in a rectangular channel at a depth of 8 ft. In one second, a ripple caused by the stone is carried 20 ft downstream. What is the speed of the flowing water?

10.9 A pebble is dropped into a stream of water of uniform depth. A wave is observed to travel upstream 5 ft in 1 sec, and 13 ft downstream in the same time. Determine the flow speed and depth.

10.10 Surface waves are caused by a sharp object that just touches the free surface of a stream of flowing water, forming the wave pattern shown. The stream depth is 6 in. Determine the flow speed and Froude number.

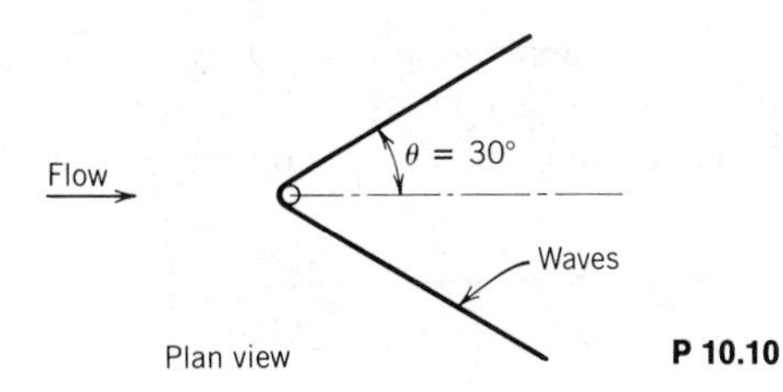

P 10.10

10.11 The Froude number characterizes flow with a free surface. On a log-log plot of speed versus depth for the ranges $0.1 < V < 3$ m/sec and $0.001 < y < 1$ m, plot the line, $Fr = 1$, and indicate regions that correspond to tranquil and rapid flow.

10.12 A submerged body traveling horizontally beneath a liquid surface at a Froude number (based on body length) near 0.5 produces a strong surface wave pattern if it is submerged less than half its length. (The wave pattern of a surface ship also is pronounced at this Froude number.) On a log-log plot of speed versus body (or ship) length for the ranges $1 < V < 30$ m/sec and $1 < x < 300$ m, plot the line $Fr = 0.5$.

10.13 A rectangular channel carries a discharge of 10 ft^3/sec per foot of width. Determine the minimum specific energy possible for this flow. Compute the corresponding flow depth and speed.

10.14 Flow in the channel of Problem 10.13 ($E_{min} = 2.19$ ft) is to be at twice the minimum specific energy. Compute the alternate depths for this value of E.

10.15 Water flows in a wide channel of uniform slope as shown. Determine the elevations of the hydraulic and energy grade lines above the datum for sections ① and ②.

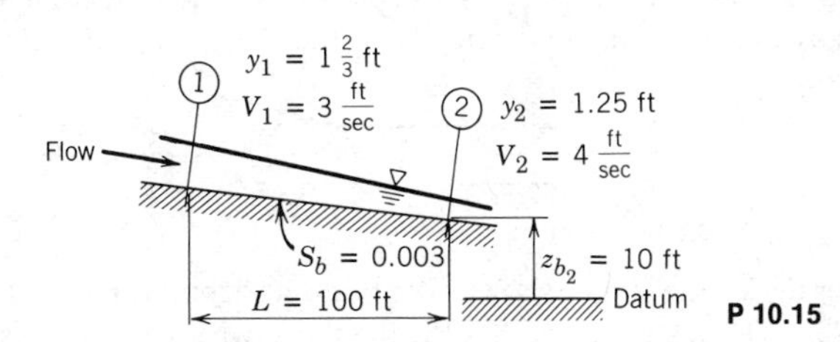

P 10.15

10.16 Consider flow in a channel with arbitrary cross section. For uniform velocity show that the flux of energy across each section is given by

$$ef = \left(\frac{p_{atm}}{\rho} + \frac{V^2}{2} + gy + gz_b\right)\dot{m}$$

where y is the flow depth.

10.17 A long rectangular channel 10 ft wide is observed to have a wavy surface at a depth of about 6 ft. Estimate the rate of discharge.

10.18 Water flows at 300 ft^3/sec in a trapezoidal channel with bottom width of 8 ft. The sides are sloped at 2:1. Find the critical depth for this channel.

10.19 For a channel of nonrectangular cross section, the specific energy can be written in terms of area as

$$E = \frac{Q^2}{2gA^2} + y$$

Critical depth occurs at minimum specific energy. Obtain a general equation for critical depth in a triangular channel in terms of Q, g, and α.

10.20 For a channel of nonrectangular cross section, the specific energy can be written in terms of area as

$$E = \frac{Q^2}{2gA^2} + y$$

Critical depth occurs at minimum specific energy. Obtain a general equation relating the critical depth, y, in a channel of trapezoidal section to Q, g, b, and α.

10.21 A rectangular channel 10 ft wide carries 100 cfs on a horizontal bed at a depth of 1.0 ft. A smooth bump across the channel rises 4 in. above the channel bottom as shown. Find the elevation of the liquid free surface above the bump.

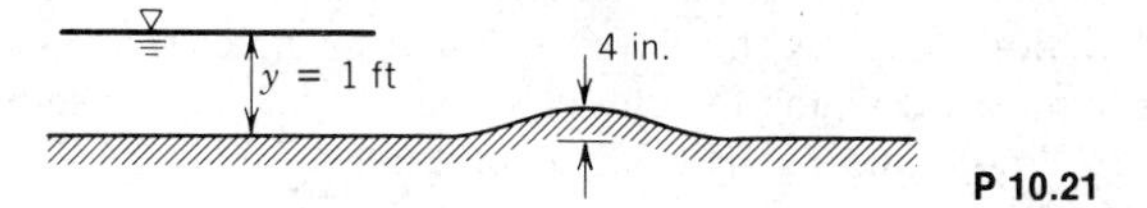

P 10.21

10.22 A rectangular channel 10 ft wide carries a discharge of 20 ft^3/sec at a depth of 0.9 ft. A smooth bump 0.2 ft high is placed on the floor of the channel. Estimate the local change in flow depth caused by the bump.

10.23 At a section of a 10 ft wide rectangular channel the depth is 0.3 ft for a discharge of 20 ft^3/sec. A smooth bump 0.1 ft high is placed on the floor of the channel. Determine the local change in flow depth caused by the bump.

10.24 Water, at a speed of 3 ft/sec and a depth of 2 ft, approaches a smooth rise in a wide channel as shown. Estimate the depth of the stream after the 0.5 ft rise.

$V = 3\ \frac{\text{ft}}{\text{sec}}$ $y = 2$ ft 0.5 ft

P 10.24

10.25 Consider the Venturi flume shown. The bed is horizontal and the flow may be considered frictionless. The upstream depth is 1 ft and the downstream depth is 0.75 ft. The upstream breadth is 2 ft and the breadth of the throat is 1 ft. Estimate the flow rate through the flume.

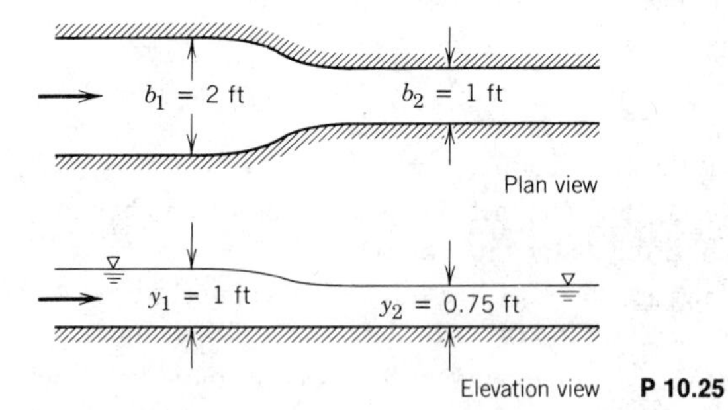

P 10.25

10.26 A sluice gate is partially opened in a rectangular channel that is 3 m wide and carries water at a flow rate of 8.5 m^3/sec. The upstream depth is 2 m. Find the downstream depth and Froude number.

10.27 Water issues from a sluice gate at a depth of 0.6 m. The discharge per unit width is 6.0 m^3/sec/m. Estimate the water level at a location far upstream, where the flow speed is negligible. Calculate the maximum rate of flow per unit width that could be delivered through the sluice gate.

10.28 A horizontal rectangular channel 3 ft wide contains a sluice gate. Upstream of the gate the flow depth is 6 ft; the depth downstream is 0.9 ft. Estimate the volume flow rate in the channel.

10.29 A rectangular flume built of timber, with a slope of 1 ft per 1000 ft, is 6 ft wide. Water flows at a normal depth of 3 ft. Compute the discharge.

10.30 A rectangular flume built of timber is 3 ft wide. The flume is to handle a flow of 90 ft^3/sec at a normal depth of 6 ft. Determine the slope required.

10.31 The flume of Problem 10.29 is fitted with a new plastic film liner. Find the corresponding depth of flow if the discharge remains constant at 85.5 ft^3/sec.

10.32 A channel with square cross section is to carry 20 m^3/sec of water at normal depth on a slope of 0.003. Compare the dimensions of the channel required for (a) unfinished, and (b) finished concrete.

10.33 Water flows in a trapezoidal channel at a normal depth of 1.2 m. The bottom width is 2.4 m and the sides slope at 1:1 (45°). The flow rate is 7.1 m^3/sec. The channel is excavated from firm, smooth earth. Find the bed slope.

10.34 Discharge through the channel of Problem 10.33 is increased to 15 m^3/sec. Find the corresponding normal depth, if the bed slope is 0.00193.

10.35 The channel of Problem 10.33 has a bed slope of 0.00193. Find the normal depth for the given discharge after a new plastic liner is installed.

10.36 A triangular channel with side angles of 45° is to carry a discharge of 10 m^3/sec at a slope of 0.001. The channel is unfinished concrete. Find the required dimensions of the channel.

10.37 A semicircular trough of corrugated steel with diameter, $D = 1$ m, carries water at depth, $y_n = 0.25$ m. The slope is 0.01. Find the discharge.

10.38 Find the discharge at which the channel of Problem 10.37 flows full.

10.39 Consider again the semicircular channel of Problem 10.37. Find the normal depth that corresponds to a discharge of 0.3 m^3/sec.

10.40 Consider a symmetric open channel of triangular cross section. Show that for a given flow area, the wetted perimeter is minimized when the sides meet at a right angle.

10.41 An above-ground rectangular flume is to be constructed of timber. For a drop of 10 ft/mile, what will be the depth and width for the most economical solution if it is to discharge 40 cfs?

10.42 An irrigation canal is to be designed to carry water at 250 m^3/sec on a bed slope of 0.0004. Calculate the required dimensions of the best trapezoidal channel with side slopes of 45° that is to be excavated in smooth earth. Estimate the reduction in excavation that would be possible if the canal were lined with plastic film.

10.43 Consider flow in a triangular channel at depth, y_n. Obtain expressions for A, P, and R_h, in terms of y_n and side angle, α. By plotting P vs. α for a given A, confirm that the optimum channel configuration has $\alpha = 45°$. How far off optimum are channels with side angles of 60°?

10.44 A wide flat unfinished concrete channel discharges water at 20 ft^3/sec per foot of width. Find the critical slope.

10.45 An optimum rectangular unfinished concrete storm sewer channel is to be designed to carry a maximum flow rate of 100 $\mathrm{ft^3/sec}$ at critical normal flow conditions. Determine the channel width and slope.

10.46 A trapezoidal canal lined with brick has side slopes of 2:1 and a bottom width of 10 ft. It carries 600 $\mathrm{ft^3/sec}$ at critical speed. Determine the critical slope.

10.47 Consider flow in a rectangular channel. Show that for flow at critical depth and optimum aspect ratio ($b = 2y$), the volume flow rate and bed slope are given by the expressions:

$$Q = 6.26 y_c^{5/2} \qquad \text{and} \qquad S_c = 24.7 \frac{n^2}{y_c^{1/3}}$$

10.48 Water flows at 1400 $\mathrm{ft^3/sec}$ in a 15 ft wide rectangular channel. The bottom slope is constant, and the channel roughness is such that normal depth is 6.0 ft. At a certain section, the depth is 2.8 ft. How will the depth change with distance downstream?

10.49 Classify the water surface profile of Problem 10.52. Show all necessary calculations.

10.50 A rectangular flume of timber is 5 ft wide and carries 60 cfs of water. The bed slope is 0.0006, and at a certain section the depth is 3 ft. Find the distance (in one reach) to the section where the depth is 2.5 ft. Is this section upstream or downstream?

10.51 Suppose that the slope of the flume in Problem 10.50 is changed so that with the same flow rate the depth varies from 3 ft at one section to 4 ft at a section 1000 ft downstream. Using a single reach, estimate the new bed slope of the flume. Sketch the flume and the water surface to assure that the answer is reasonable.

10.52 Water discharges from under a gate onto a wide channel of adverse slope as shown. The depth at the vena contracta is 0.6 m, and the speed is 12 m/sec. For a channel roughness of 0.015, estimate the depth at the downstream end of the adverse slope. Use a single reach.

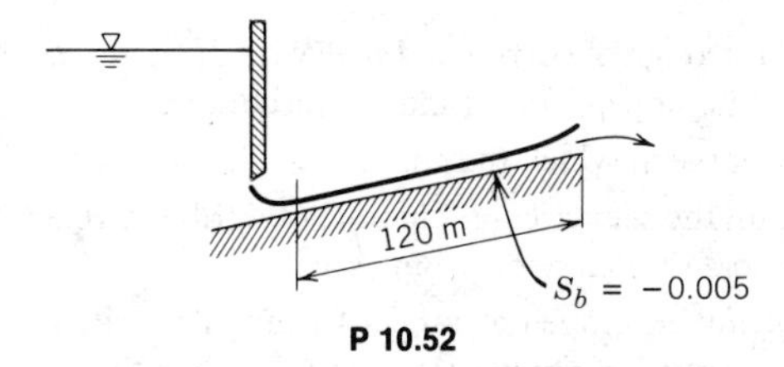

P 10.52

10.53 Consider a rectangular channel 15 ft wide, made from unfinished concrete. The normal depth of water flow in the channel is 3 ft at a Froude number of 2. At one section in the channel, the depth is 4.5 ft. Estimate the channel length required for the flow depth to change to 3.1 ft. Does the change occur upstream or downstream from the section with 4.5 ft depth?

10.54 A horizontal wooden flume is 5 m wide and 300 m in length. The flow depth at the flume exit is 2.0 m. If the discharge is 40 $\mathrm{m^3/sec}$, estimate the flow depth at the flume inlet.

10.55 A dam partially restricts the flow of a small river. Assume that the river can be modeled as a wide channel with $S_b = 0.001$ and $n = 0.025$. The flow rate per unit width is 3 $\mathrm{m^3/sec/m}$, and the water depth immediately upstream from the dam is 5 m. Find the upstream distance to the location where $y = 2$ m.

10.56 Water from a sluice gate issues onto a wide horizontal floor of unfinished concrete. The depth is 0.6 m and the discharge per unit width is 6 $\mathrm{m^2/sec}$. Estimate the distance to the downstream location where the flow depth reaches 1.2 m. (Assume that no hydraulic jump occurs.)

10.57 Water flows at 40 m^3/sec in a 5 m wide rectangular channel. The bottom slope is constant and the roughness is $n = 0.017$; normal depth is $y_n = 2.0$ m. At a certain section, the depth is 0.9 m. Evaluate S_b. Estimate the channel length needed to raise the water depth to 1.6 m.

10.58 A wide channel carries a discharge of 20 ft^3/sec per foot of width at a depth of 1 ft at the toe of a hydraulic jump. Determine the sequent depth of the jump and the head loss across it.

10.59 A hydraulic jump occurs in a wide horizontal channel. The discharge is 30 ft^3/sec per foot of width. The upstream depth is 1.3 ft. Determine the sequent depth for the jump.

10.60 A tidal bore—an abrupt translating wave or moving hydraulic jump—often forms when the tide flows into the wide estuary of a river. In one case, a bore is observed to have a height of 12 ft above the undisturbed level of the river that is 8 ft deep. The bore travels upstream at 18 mph. Determine the approximate speed of the current of the undisturbed river.

10.61 A positive surge wave, or moving hydraulic jump, can be produced in the laboratory by suddenly opening a sluice gate. Consider a surge of depth y_2 advancing into a quiescent channel of depth y_1, as shown. Obtain an expression for the surge speed in terms of y_1 and y_2.

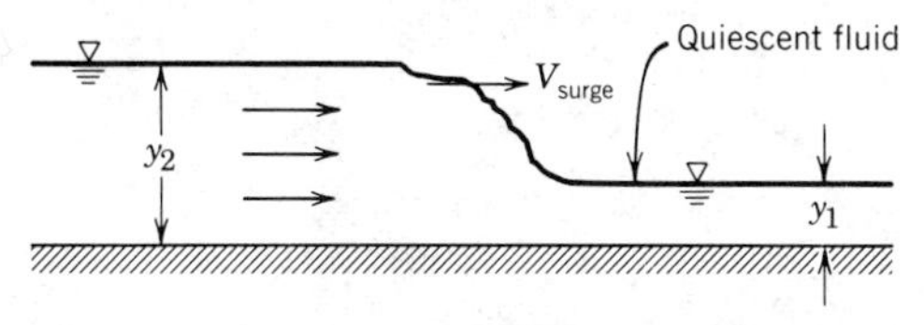

P 10.61

10.62 A hydraulic jump occurs in a rectangular channel. The flow rate is 200 ft^3/sec, and the depth before the jump is 1.2 ft. Determine the depth behind the jump and the head loss, if the channel is 10 ft wide.

10.63 The hydraulic jump may be used as a crude flow meter. Suppose that in a horizontal rectangular channel 5 ft wide the observed depths before and after a hydraulic jump are 0.66 and 3.0 ft. Find the rate of flow and the head loss.

10.64 A hydraulic jump occurs on a horizontal apron downstream from a wide spillway at a location where the depth is 0.9 m and the speed is 25 m/sec. Estimate the depth and speed downstream from the jump. Evaluate the specific energy downstream from the jump compared to that upstream.

10.65 A hydraulic jump occurs in a rectangular channel. The flow rate is 6.5 m^3/sec and the depth before the jump is 0.4 m. Determine the depth behind the jump and the head loss, if the channel is 1 m wide.

10.66 Consider the flow conditions of the hydraulic jump of Example Problem 10.9. Calculate the increase in water temperature that would result if all the head loss due to the jump were converted to thermal energy. Would it be practical to measure the jump head loss with a thermometer?

10.67 The crest of a broad-crested weir is 1 ft below the level of an upstream reservoir, where the water depth is 8 ft. What is the maximum flow rate per unit width that could pass over the weir?

10.68 Water flows over a horizontal sharp-crested weir 5 ft wide in a channel of 10 ft width. The height of the weir is 3 ft at the crest. Upstream from the weir, the flow depth is 4 ft. Determine the flow rate.

10.69 Water flows over a 60° V-notch weir. The height differential is $y_1 - z_w = 0.6$ ft. Determine the discharge.

10.70 Water flows under a sluice gate from a reservoir, where the depth is 8 ft. The gate is raised 1.5 ft above the channel floor. Estimate the flow rate per unit width if $C_d = 0.8$.

Chapter 11

INTRODUCTION TO COMPRESSIBLE FLOW

In Chapter 4 we developed control volume formulations of the basic equations. For incompressible flow the two variables of principal interest were pressure and velocity. The continuity and momentum equations provided the two independent relations needed to solve for these variables. In Chapter 8 the energy equation was used to identify the losses in mechanical energy due to friction in duct flows.

"Compressible" flow implies appreciable variations in density throughout a flow field. Compressibility becomes important at high flow speeds. Large changes in velocity involve large pressure changes; for gas flows, these pressure changes are accompanied by significant variations in both density and temperature. Since two additional variables are encountered in treating compressible flow, two additional equations are needed. Both the energy equation and an equation of state must be applied to solve compressible flow problems.

In our study of compressible fluid flow, we shall deal primarily with the steady, one-dimensional flow of an ideal gas. Although many real flows of interest are more complex, these restrictions will allow us to concentrate on the effects of basic flow processes.

A review of the thermodynamics necessary for the study of compressible flows, including the equation of state and Tds equations, is presented in the next section.

11-1 REVIEW OF THERMODYNAMICS

The pressure, density, and temperature of a substance may be related by an equation of state. Although many substances are quite complex in behavior, experience shows that most gases of engineering interest, at moderate pressure and temperature, are well represented by the ideal gas equation of state,

$$p = \rho R T \tag{11.1}$$

where R is a constant for each gas;[1] R is given by

$$R = \frac{R_u}{M_m}$$

where R_u is the universal gas constant, $R_u = 8314\ \text{N} \cdot \text{m/kgmole} \cdot \text{K}$ ($1544\ \text{ft} \cdot \text{lbf/lbmole} \cdot \text{R}$) and M_m is the molecular mass of the gas. Although no real substance behaves exactly as an ideal gas,[2] Eq. 9.1 is in error by less than 1 percent for air at room temperature for pressures as high as 30 atm. For air at 1 atm, the equation is less than 1 percent in error for temperatures as low as 140 K.

The ideal gas has other features that are simple and useful. In general, the internal energy of a substance may be expressed as $u = u(v, T)$. Then

$$du = \left(\frac{\partial u}{\partial T}\right)_v dT + \left(\frac{\partial u}{\partial v}\right)_T dv$$

where $v = 1/\rho$, the specific volume. The specific heat at constant volume is defined as $c_v \equiv (\partial u/\partial T)_v$ so that

$$du = c_v\, dT + \left(\frac{\partial u}{\partial v}\right)_T dv$$

For any substance that follows the ideal gas equation of state, $p = \rho RT$, then $(\partial u/\partial v)_T = 0$, and hence $u = u(T)$. Consequently,

$$du = c_v\, dT \tag{11.2}$$

for an ideal gas; this means that internal energy and temperature changes may be related if c_v is known. Furthermore, since $u = u(T)$, then $c_v = c_v(T)$.

The enthalpy of a substance is defined as $h \equiv u + p/\rho$. For an ideal gas, $p = \rho RT$, and hence $h = u + RT$. Since, for an ideal gas, $u = u(T)$, then h also must be a function of temperature alone.

To obtain a relation between h and T, we express h in its most general form as

$$h = h(p, T)$$

Then

$$dh = \left(\frac{\partial h}{\partial T}\right)_p dT + \left(\frac{\partial h}{\partial p}\right)_T dp$$

and

$$dh = c_p\, dT + \left(\frac{\partial h}{\partial p}\right)_T dp$$

from the definition, $c_p \equiv (\partial h/\partial T)_p$. We have shown that for an ideal gas h is a function

[1] For air, $R = 287\ \text{N} \cdot \text{m/kg} \cdot \text{K}$ ($53.3\ \text{ft} \cdot \text{lbf/lbm} \cdot \text{R}$).

[2] See, e.g. M. J. Zucrow and J. D. Hoffman, *Gas Dynamics, Vol. 1*. New York: Wiley, 1976, Chapter 1.

of T only. Consequently, $(\partial h/\partial p)_T = 0$ and

$$dh = c_p\, dT \tag{11.3}$$

Again, since h is a function of T alone, Eq. 11.3 requires that c_p be a function of T only for an ideal gas.

The specific heats for an ideal gas have been shown to be functions of temperature only. Their difference is a constant for each gas. From

$$h = u + RT$$

we can write

$$dh = du + R\, dT$$

Combining this with Eq. 11.3, and using Eq. 11.2, we can write

$$dh = c_p\, dT = du + R\, dT = c_v\, dT + R\, dT$$

Then

$$c_p - c_v = R \tag{11.4}$$

The ratio of specific heats is defined as

$$k \equiv \frac{c_p}{c_v} \tag{11.5}$$

By using the definition of k, Eq. 11.4 can be solved for either c_p or c_v in terms of k and R. Thus

$$c_p = \frac{kR}{k-1} \tag{11.6a}$$

and

$$c_v = \frac{R}{k-1} \tag{11.6b}$$

For an ideal gas, the specific heats are functions of temperature only. Within reasonable temperature ranges, the specific heats of an ideal gas may be treated as constants for calculations of engineering accuracy. Under these conditions

$$u_2 - u_1 = \int_{u_1}^{u_2} du = \int_{T_1}^{T_2} c_v\, dT = c_v(T_2 - T_1) \tag{11.7a}$$

$$h_2 - h_1 = \int_{h_1}^{h_2} dh = \int_{T_1}^{T_2} c_p\, dT = c_p(T_2 - T_1) \tag{11.7b}$$

These equations obviously may be used to advantage in simplifying analyses.

Values of M_m, c_p, c_v, R, and k for common gases are given in Appendix A, Table A.6.

The property entropy is extremely useful in the analysis of compressible flows. State diagrams, particularly the temperature-entropy (Ts) diagram, are valuable aids in the physical interpretation of analytical results. Since we shall make extensive use of Ts diagrams in solving compressible flow problems, let us review briefly some useful relationships involving the property entropy.[3]

Entropy is defined by the equation

$$\Delta S \equiv \int_{\text{rev}} \frac{\delta Q}{T} \qquad \text{or} \qquad dS = \left(\frac{\delta Q}{T}\right)_{\text{rev}} \tag{11.8}$$

The inequality of Clausius, deduced from the second law, states that

$$\oint \frac{\delta Q}{T} \leq 0$$

As a consequence of the second law these results can be extended to

$$dS \geq \frac{\delta Q}{T} \qquad \text{or} \qquad T\,dS \geq \delta Q \tag{11.9a}$$

For *reversible* processes, the equality holds, and

$$T\,ds = \frac{\delta Q}{dm} \qquad \text{(reversible process)} \tag{11.9b}$$

The inequality holds for *irreversible* processes, and

$$T\,ds > \frac{\delta Q}{dm} \qquad \text{(irreversible process)} \tag{11.9c}$$

For an *adiabatic* process, $\delta Q/dm \equiv 0$. Thus

$$ds = 0 \qquad \text{(reversible adiabatic process)}$$

and

$$ds > 0 \qquad \text{(irreversible adiabatic process)}$$

A useful relationship among properties (p, v, T, s, u) can be obtained from considering the first and second laws together. The result is the Gibbs, or $T\,ds$ equation,

$$T\,ds = du + p\,dv \tag{11.10a}$$

This is a relationship among properties, valid for all processes between equilibrium states. Although it is derived from the first and second laws, in itself it is a statement of neither.

An alternate form of Eq. 11.10a can be obtained by substituting

$$du = d(h - pv) = dh - p\,dv - v\,dp$$

to obtain

$$T\,ds = dh - v\,dp \tag{11.10b}$$

[3] For a detailed discussion see, e.g. G. J. Van Wylen and R. E. Sonntag, *Fundamentals of Classical Thermodynamics (SI Version)*, 2nd ed. New York: Wiley, 1978, Chapters 6 and 7.

For an ideal gas, the entropy change can be evaluated from the $T\,ds$ equations as

$$ds = \frac{du}{T} + \frac{p}{T}\,dv = c_v\,\frac{dT}{T} + R\,\frac{dv}{v}$$

$$ds = \frac{dh}{T} - \frac{v}{T}\,dp = c_p\,\frac{dT}{T} - R\,\frac{dp}{p}$$

For constant specific heats, these equations may be integrated to yield

$$s_2 - s_1 = c_v \ln \frac{T_2}{T_1} + R \ln \frac{v_2}{v_1}$$

$$s_2 - s_1 = c_p \ln \frac{T_2}{T_1} - R \ln \frac{p_2}{p_1}$$

For the special case of an isentropic process, $ds = 0$, and the $T\,ds$ equations reduce to

$$0 = du + p\,dv$$

$$0 = dh - v\,dp$$

For an ideal gas, we have

$$0 = c_v\,dT + p\,dv$$

$$0 = c_p\,dT - v\,dp$$

Solving for dT gives

$$dT = \frac{v\,dp}{c_p} = -\frac{p\,dv}{c_v}$$

or

$$\frac{dp}{p} + \frac{c_p}{c_v}\frac{dv}{v} = \frac{dp}{p} + k\,\frac{dv}{v} = 0$$

Integrating (for k = constant) gives

$$\ln p + k \ln v = \ln c$$

or

$$\ln p + \ln v^k = \ln c$$

Taking antilogarithms, this equation reduces to

$$pv^k = \text{constant} \tag{11.11a}$$

or

$$\frac{p}{\rho^k} = \text{constant} \tag{11.11b}$$

Equations 11.11 are property relations for an ideal gas undergoing an isentropic process.

Qualitative information that is useful in drawing state diagrams also can be obtained from the $T\,ds$ equations. To complete our review of thermodynamic fundamentals, the slopes of lines of constant pressure and of constant volume on the Ts diagram are evaluated in Example Problem 11.2.

Example 11.1

Air flows through a long duct of constant area at a rate of 0.15 kg/sec. A short section of the duct is cooled by liquid nitrogen that surrounds the duct. The rate of heat loss in this section is 15.0 kJ/sec from the air. The absolute pressure, temperature, and velocity at the entrance to the cooled section are 188 kPa, 440 K, and 210 m/sec, respectively. At the outlet, the absolute pressure and temperature are 213 kPa and 351 K. Compute the duct cross-sectional area, and the changes in enthalpy, internal energy, and entropy for this flow.

EXAMPLE PROBLEM 11.1

GIVEN: Air flows steadily through a short section of constant-area duct that is cooled by liquid nitrogen.

$T_1 = 440$ K

$p_1 = 188$ kPa (abs)

$V_1 = 210$ m/sec

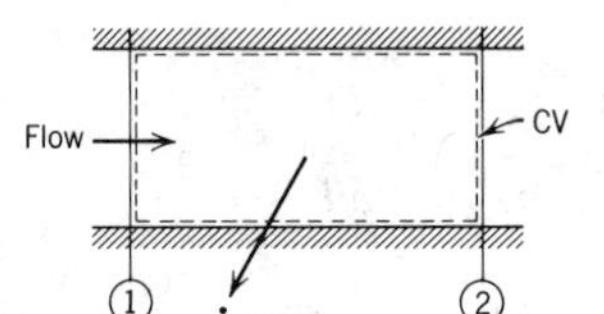

$T_2 = 351$ K

$p_2 = 213$ kPa (abs)

FIND: (a) The duct area. (b) Δh. (c) Δu. (d) Δs.

SOLUTION:

The duct area may be found from the continuity equation.

Basic equation:

$$0 = \overset{=0(1)}{\cancel{\frac{\partial}{\partial t}}\int_{\text{CV}} \rho\, d\forall} + \int_{\text{CS}} \rho \vec{V} \cdot d\vec{A} \tag{4.13}$$

Assumptions: (1) Steady flow
(2) Uniform flow at each section
(3) Ideal gas

Then

$$0 = \{-|\rho_1 V_1 A_1|\} + \{|\rho_2 V_2 A_2|\}$$

or

$$\dot{m} = \rho_1 V_1 A = \rho_2 V_2 A$$

since $A = A_1 = A_2 =$ constant. Using the ideal gas relation, $p = \rho RT$, we find

$$\rho_1 = \frac{p_1}{RT_1} = \frac{1.88 \times 10^5\ \text{N}}{\text{m}^2} \times \frac{\text{kg} \cdot \text{K}}{287\ \text{N} \cdot \text{m}} \times \frac{1}{440\ \text{K}} = 1.49\ \text{kg/m}^3$$

From continuity

$$A = \frac{\dot{m}}{\rho_1 V_1} = \frac{0.15 \text{ kg}}{\text{sec}} \times \frac{\text{m}^3}{1.49 \text{ kg}} \times \frac{\text{sec}}{210 \text{ m}} = 4.79 \times 10^{-4} \text{ m}^2 \longleftarrow \quad A$$

For an ideal gas, $dh = c_p\, dT$, so

$$\Delta h = h_2 - h_1 = \int_{T_1}^{T_2} c_p\, dT = c_p(T_2 - T_1)$$

$$\Delta h = \frac{1.00 \text{ kJ}}{\text{kg} \cdot \text{K}} \times (351 - 440) \text{ K} = -89.0 \text{ kJ/kg} \longleftarrow \quad \Delta h$$

Also $du = c_v\, dT$, so

$$\Delta u = u_2 - u_1 = \int_{T_1}^{T_2} c_v\, dT = c_v(T_2 - T_1)$$

$$\Delta u = \frac{0.717 \text{ kJ}}{\text{kg} \cdot \text{K}} \times (351 - 440) \text{ K} = -63.8 \text{ kJ/kg} \longleftarrow \quad \Delta u$$

The entropy change may be obtained from the $T\,ds$ equation

$$T\,ds = dh - v\,dp$$

$$ds = \frac{dh}{T} - \frac{v\,dp}{T} = c_p \frac{dT}{T} - R\frac{dp}{p}$$

or

$$\Delta s = s_2 - s_1 = \int_{T_1}^{T_2} c_p \frac{dT}{T} - \int_{p_1}^{p_2} R\frac{dp}{p} = c_p \ln \frac{T_2}{T_1} - R \ln \frac{p_2}{p_1}$$

$$= \frac{1.00 \text{ kJ}}{\text{kg} \cdot \text{K}} \ln\left(\frac{351}{440}\right) - \frac{0.287 \text{ kJ}}{\text{kg} \cdot \text{K}} \ln\left(\frac{2.13 \times 10^5}{1.88 \times 10^5}\right)$$

$$\Delta s = -0.262 \text{ kJ/kg} \cdot \text{K} \longleftarrow \quad \Delta s$$

{The purpose of this problem was to review the calculation of thermodynamic properties for an ideal gas.}

Example 11.2

For an ideal gas, determine the slope of (a) a constant volume line, and (b) a constant pressure line, in the Ts plane.

EXAMPLE PROBLEM 11.2

GIVEN: An ideal gas.

FIND: (a) Slope of constant volume line in Ts plane.
(b) Slope of constant pressure line in Ts plane.

SOLUTION:

The $T\,ds$ equations may be applied.

$$T\,ds = du + p\,dv \tag{11.10a}$$

$$T\,ds = dh - v\,dp \tag{11.10b}$$

Substituting for an ideal gas, $du = c_v\,dT$ and $dh = c_p\,dT$, we obtain

$$T\,ds = c_v\,dT + p\,dv$$

$$T\,ds = c_p\,dT - v\,dp$$

For a constant volume process, $dv = 0$. From the first equation

$$\left.\frac{dT}{ds}\right)_{\substack{\text{constant}\\ \text{volume}}} = \left.\frac{\partial T}{\partial s}\right)_v = \frac{T}{c_v} \qquad \text{constant volume slope}$$

For a constant pressure process, $dp = 0$. From the second equation

$$\left.\frac{dT}{ds}\right)_{\substack{\text{constant}\\ \text{pressure}}} = \left.\frac{\partial T}{\partial s}\right)_p = \frac{T}{c_p} \qquad \text{constant pressure slope}$$

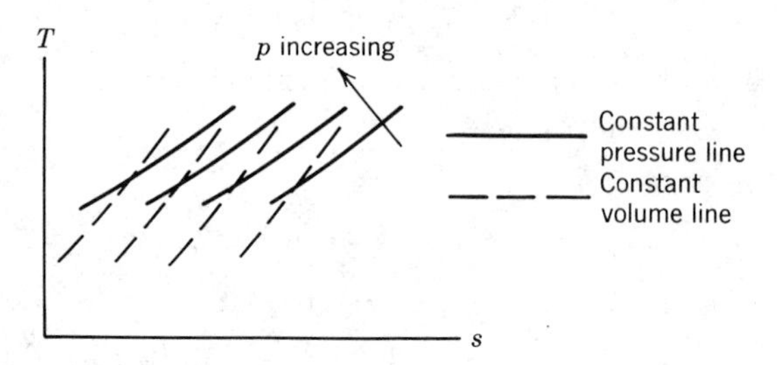

Note that the slope of each line is proportional at any point to the absolute temperature. Furthermore, at any point a constant volume line has a slope $c_p/c_v = k$ times larger than a line of constant pressure.

11-2 PROPAGATION OF SOUND WAVES

11-2.1 Speed of Sound

The terms *supersonic* and *subsonic* are familiar terms; they refer to velocities that are, respectively, greater than and less than the speed of sound. The speed of sound (a pressure wave of infinitesimal strength) is an important characteristic parameter for compressible flow. We have previously (Chapters 2 and 7) introduced the Mach number, $M = V/c$, the ratio of the local flow speed to the local speed of sound, as an important nondimensional parameter characterizing compressible flows. Before studying compressible flows, it is logical to obtain a general relation for calculating the sonic speed.

Consider the propagation of a sound wave of infinitesimal strength into an undisturbed medium as shown in Fig. 11.1*a*. We are interested in relating the speed of wave propagation, c, to fluid property changes across the wave. If the pressure and density in the undisturbed medium ahead of the wave are denoted by p and ρ, the

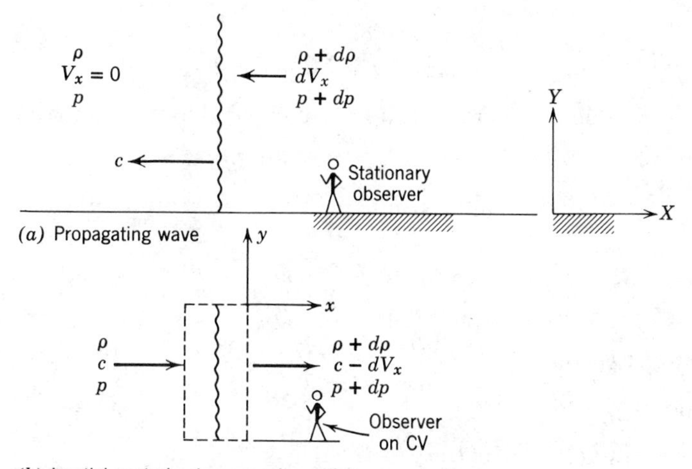

(b) Inertial control volume moving with wave, velocity c

Fig. 11.1 Propagating sound wave showing control volume chosen for analysis.

passage of the wave will cause them to undergo infinitesimal changes and to become $p + dp$ and $\rho + d\rho$. Since the wave propagates into a stationary fluid, the velocity ahead of the wave, V_x, is zero. The magnitude of the velocity behind the wave, $V_x + dV_x$, will then simply be dV_x; in Fig. 11.1*a*, the direction of the motion behind the wave has been assumed to the left.

Unfortunately, the flow of Fig. 11.1*a* appears unsteady to a stationary observer, observing the motion of the wave from a fixed point on the ground. However, the flow appears steady to an observer located *on* an inertial control volume moving with a segment of the wave as shown in Fig. 11.1*b*. The velocity approaching the control volume is c, and the velocity leaving, $c - dV_x$.

The basic equations may be applied to the differential control volume shown in Fig. 11.1*b* (we use V_x for the x component of velocity to avoid confusion with internal energy, u).

a. Continuity Equation

Basic equation:

$$0 = \overbrace{\frac{\partial}{\partial t}\int_{\text{CV}} \rho \, d\forall}^{=0(1)} + \int_{\text{CS}} \rho \vec{V} \cdot d\vec{A} \tag{4.13}$$

Assumptions: (1) Steady flow
(2) Uniform flow at each section

Then

$$0 = \{-|\rho c A|\} + \{|(\rho + d\rho)(c - dV_x)A|\} \tag{11.12a}$$

or

$$0 = -\rho c A + \rho c A - \rho\, dV_x A + d\rho c A - \underbrace{d\rho\, dV_x A}_{\simeq 0}$$

Neglecting the product of differentials, $d\rho\, dV_x$, compared to either $d\rho$ or dV_x, we obtain

$$0 = -\rho A\, dV_x + cA\, d\rho$$

or

$$dV_x = \frac{c}{\rho}\, d\rho \qquad (11.12b)$$

b. Momentum Equation

Basic equation:

$$F_{S_x} + \overset{=0(3)}{F_{B_x}} - \int_{CV} \overset{=0(4)}{a_{rf_x}}\rho\, d\Psi = \overset{=0(1)}{\frac{\partial}{\partial t}\int_{CV}} V_x \rho\, d\Psi + \int_{CS} V_x \rho \vec{V}_{xyz} \cdot d\vec{A} \qquad (4.35a)$$

Assumptions: (3) $F_{B_x} = 0$
(4) $a_{rf_x} = 0$

The surface forces acting on the infinitesimal control volume are

$$F_{S_x} = dR_x + pA - (p + dp)A$$

where dR_x represents all forces applied to the horizontal portions of the control surface shown in Fig. 11.1*b*. We consider only a portion of a moving sound wave, so $dR_x = 0$ because there is no relative motion along the wave. Thus the surface force simplifies to

$$F_{S_x} = -A\, dp$$

Substituting into the basic equation gives

$$-A\, dp = c\{-|\rho c A|\} + (c - dV_x)\{|(\rho + d\rho)(c - dV_x)A|\}$$

Using the continuity equation in the form of Eq. 11.12a reduces this to

$$\begin{aligned} -A\, dp &= c\{-|\rho c A|\} + (c - dV_x)\{|\rho c A|\} \\ &= (-c + c - dV_x)\{|\rho c A|\} \\ -A\, dp &= -\rho c A\, dV_x \end{aligned}$$

or

$$dV_x = \frac{1}{\rho c}\, dp \qquad (11.12c)$$

Combining Eqs. 11.12b and 11.12c, we obtain

$$dV_x = \frac{c}{\rho}\, d\rho = \frac{1}{\rho c}\, dp$$

from which

$$dp = c^2\, d\rho$$

or

$$c^2 = \frac{dp}{d\rho}$$

To evaluate the derivative of a thermodynamic property, we must specify the property to be held constant during differentiation. For the present case, the limit at vanishing strength of a sound wave will be

$$\lim_{\text{strength}\to 0} \frac{dp}{d\rho} = \left.\frac{\partial p}{\partial \rho}\right)_{s=\text{constant}}$$

Perhaps a more physical justification for the assumption of isentropic propagation is that an infinitesimal pressure change is reversible. Hence, since there is too little time for heat transfer, the process is reversible and adiabatic. Thus the speed of propagation of a sound wave is given by

$$c = \sqrt{\left.\frac{\partial p}{\partial \rho}\right)_s}$$

Data for solid and liquid media usually are reported as the bulk modulus,

$$E_v = \frac{dp}{d\rho/\rho} = \rho\,\frac{dp}{d\rho} \tag{3.8}$$

For these media

$$c = \sqrt{E_v/\rho} \tag{11.13}$$

For an ideal gas, the pressure and density in isentropic flow are related by

$$\frac{p}{\rho^k} = \text{constant} \tag{11.11b}$$

as shown previously. Taking logarithms, we obtain

$$\ln p - k \ln \rho = \ln c$$

and differentiating

$$\frac{dp}{p} - k\,\frac{d\rho}{\rho} = 0$$

Therefore,

$$\left.\frac{\partial p}{\partial \rho}\right)_s = k\,\frac{p}{\rho}$$

But $p/\rho = RT$, so finally

$$c = \sqrt{kRT} \tag{11.14}$$

for an ideal gas.

The important feature of sound propagation in an ideal gas, as shown by Eq. 11.14, is that the speed of sound is a function of temperature only. The variation in atmospheric temperature with altitude on a standard day was discussed in Chapter 3. The properties are summarized in Table A.3. The corresponding variation in c is computed in Example Problem 11.3 and plotted as a function of altitude.

Example 11.3

Compute the speed of sound at sea level in standard air. Evaluate the speed of sound and plot for altitudes to 15 km.

EXAMPLE PROBLEM 11.3

GIVEN: Air under standard atmospheric conditions.

FIND: (a) The speed of sound at sea level.
(b) Plot the speed of sound for altitudes to 15 km.

SOLUTION:
Assume an ideal gas.

Computing equation: $c = \sqrt{kRT}$

From Table 3.1, the temperature at sea level on a standard day is 288 K. Thus

$$c = \left(1.4 \times 287 \frac{\text{N} \cdot \text{m}}{\text{kg} \cdot \text{K}} \times 288\ \text{K} \times \frac{\text{kg} \cdot \text{m}}{\text{N} \cdot \text{sec}^2}\right)^{1/2} = 340\ \text{m/sec}$$

Temperatures at various altitudes may be found from Table A.3. The resulting sound speeds are plotted in the following figure.

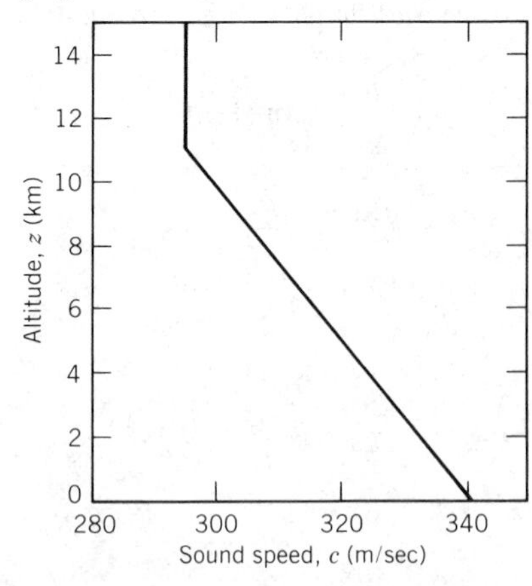

From the plot we see that the speed of sound in air on a standard day varies from 340 m/sec at sea level to 295 m/sec at 11 km altitude.

11-2.2 Types of Flow—The Mach Cone

Flows for which $M < 1$ are termed *subsonic*, while those for which $M > 1$ are *supersonic*. Flow fields that have both subsonic and supersonic regions are termed *transonic*. (The transonic regime occurs for Mach numbers between about 0.9 and 1.2.) Although most flows within our experience are subsonic, there are important practical cases where $M \geq 1$ occurs in a flow field. Perhaps the most obvious are supersonic aircraft, transonic flows in aircraft compressors and fans, and steam turbines. Yet another flow regime, *hypersonic* flow ($M \gtrsim 5$), is of interest in missile and reentry vehicle design. Some important qualitative differences between subsonic and supersonic flows can be deduced from the properties of a simple moving sound source.

Consider a point source that emits instantaneous infinitesimal disturbances. These disturbances propagate in all directions with speed, c. At any time, t, the location of the wave front from the disturbance emitted at time, t_0, will be represented by a sphere with radius, $c(t - t_0)$, whose center coincides with the location of the disturbance at time, t_0.

We are interested in determining the nature of the disturbance propagation for different speeds of the moving source. Four cases are shown in Fig. 11.2:

1. $V = 0$. The sound pattern propagates uniformly in all directions. At the instant Δt after emission, any given sound pulse is located at radius $c\,\Delta t$ from the source. At the instant $2\,\Delta t$, the radius is $c(2\,\Delta t)$. Each wave front is spherical; all wave fronts are concentric spheres.
2. $0 < V < c$. The concentricity of the wave pattern is lost. Individual wave fronts are spherical, but each successive sound is emitted from a different position, $V\,\Delta t$ distant from the previous position.

If you imagine the circles shown to be the amplitude peaks of a sinusoidal tone, the same qualitative picture holds for a moving source of continuous sound. If this source moves with constant speed, V, the whole pattern shown in Fig. 11.2*b* is carried along with the emitter. Thus a stationary observer would hear more peaks per unit time as the source approaches than after it passes. This is known as the "Doppler effect." (Have you ever heard a fast-moving train whistle through a crossing?)

3. $V = c$. The locus of the leading surfaces of all waves will be a plane at the source, perpendicular to the path of motion. No sound wave can travel in front of the source. Consequently, an observer in front of the source will not hear it approaching.
4. $V > c$. In this case, the locus of the leading surfaces of the sound waves will be a cone. Again, no sound will be heard in front of this cone.

The cone angle can be related to the Mach number at which the source moves. From the geometry of Fig. 11.2*d*,

$$\sin\alpha = \frac{c}{V} = \frac{1}{M}$$

or

$$\alpha = \sin^{-1}\left(\frac{1}{M}\right) \tag{11.15}$$

The cone depicted in Fig. 11.2*e* is termed the Mach cone; angle, α, is the Mach angle. The regions inside and outside the cone are sometimes called the *zone of action* and the *zone of silence*, respectively.

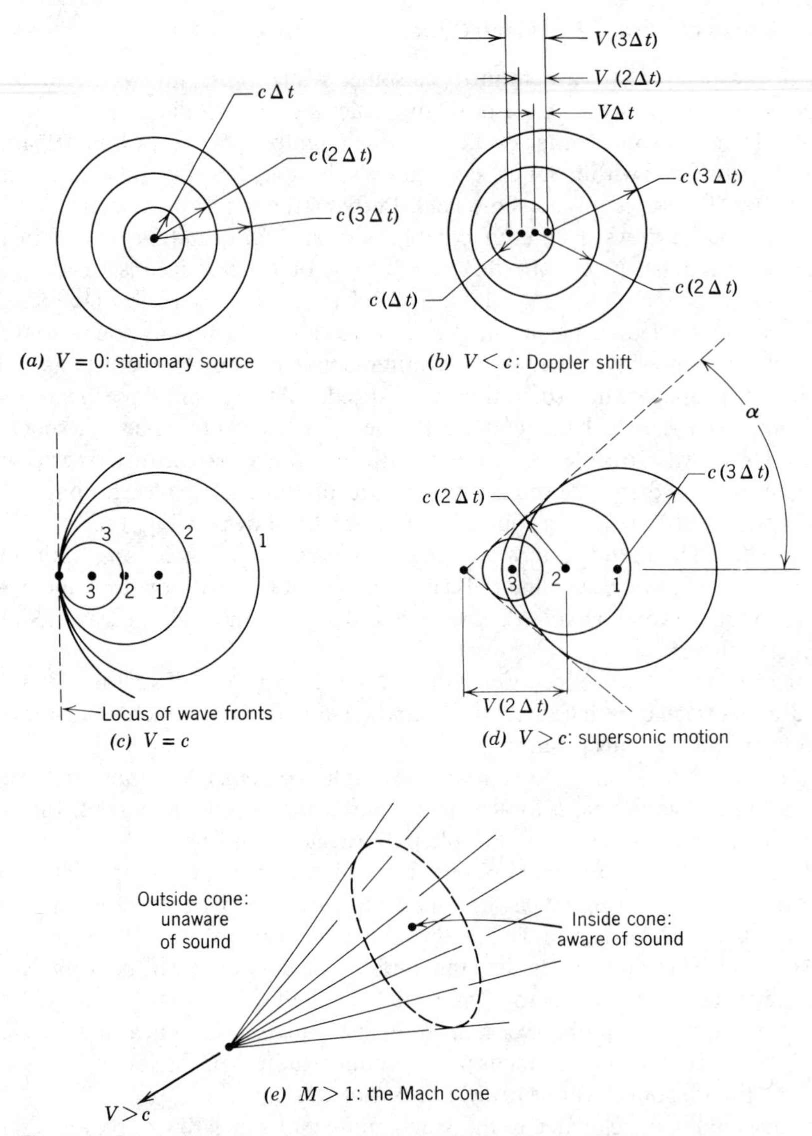

Fig. 11.2 Propagation of sound waves from a moving source: The Mach cone.

11-3 REFERENCE STATE: LOCAL ISENTROPIC STAGNATION PROPERTIES

If we wish to describe the state of a fluid at any point in a flow field, we must specify two independent intensive thermodynamic properties (usually pressure and temperature), plus the fluid velocity at the point.[4]

[4] The state of a pure substance, in the absence of motion, gravity, and surface, magnetic, or electrical effects, is defined by two independent intensive thermodynamic properties.

In our discussion of compressible flow, we shall find it convenient to use the stagnation state as a reference state. The stagnation state is characterized by a condition of zero velocity; the stagnation properties at any point in a flow field are those properties that would exist at that point if the velocity were reduced to zero. Consider a point in a flow field at which the temperature is T, the pressure, p, and the velocity, V. The stagnation state at that point in the flow field would be characterized by the stagnation pressure, p_0, stagnation temperature, T_0, and zero velocity.

Before we can calculate the stagnation properties, we must specify the process by which the fluid is imagined to decelerate to zero velocity.

For incompressible flow (Chapter 6), integration of Euler's equation for frictionless, incompressible flow along a streamline led to the Bernoulli equation

$$\frac{p}{\rho} + \frac{V^2}{2} + gz = \text{constant} \tag{6.9}$$

For incompressible flow, a frictionless deceleration process to zero velocity led to the stagnation pressure, p_0, given by

$$p_0 = p + \tfrac{1}{2}\rho V^2 \tag{6.14}$$

For compressible flow we again use a frictionless deceleration process; in addition we specify that the deceleration process be adiabatic. In short, we specify an isentropic deceleration process to define local isentropic stagnation properties:

> *Local isentropic stagnation properties* are those properties that would be obtained at any point in a flow field if the fluid at that point were decelerated from local conditions to zero velocity following a frictionless, adiabatic (an *isentropic*) process.

Isentropic stagnation properties are reference properties that may be evaluated at any point in a flow field. The variations in these properties from point to point in a flow field give information about the flow process between points. This will become apparent in our discussion of one-dimensional flow cases. However, before proceeding let us first develop equations for computing the local isentropic stagnation properties for the flow of an ideal gas.

11-3.1 Local Isentropic Stagnation Properties for the Flow of an Ideal Gas

The stagnation properties at a point are the properties that would be obtained if the fluid at that point were decelerated isentropically to zero velocity. In order to calculate the stagnation properties, we must look at the deceleration process. At the beginning of the process the conditions are those corresponding to the actual flow at the point (velocity, V, pressure, p, and temperature, T, etc.); at the end of the process the velocity is zero and the conditions are the corresponding stagnation properties (stagnation pressure, p_0, and stagnation temperature, T_0, etc.).

We need to develop an expression describing the relationship among the fluid properties during the process. Since both the initial and final properties are specified, we develop the relationship among properties in differential form. Then we can integrate to obtain expressions for the stagnation conditions in terms of the initial conditions (the conditions corresponding to the actual flow at the point).

The deceleration process is shown schematically in Fig. 11.3. We are interested in determining the stagnation properties for the flow at point ①. To determine a relationship among the fluid properties during the deceleration process, we apply the continuity and momentum equations to the stationary differential stream tube control volume shown.

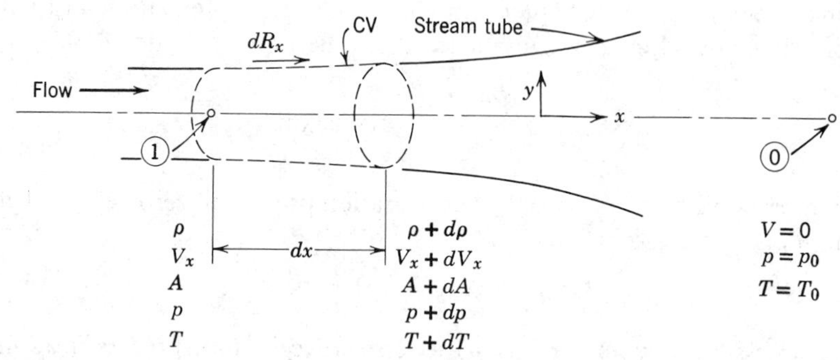

Fig. 11.3 Compressible flow in an infinitesimal stream tube.

a. Continuity Equation

Basic equation:

$$0 = \frac{\partial}{\partial t}\int_{CV} \rho \, d\forall + \int_{CS} \rho \vec{V} \cdot d\vec{A} \qquad (4.13)$$

(first term $= 0(1)$)

Assumptions: (1) Steady flow
(2) Uniform flow at each section

Then

$$0 = \{-|\rho V_x A|\} + \{|(\rho + d\rho)(V_x + dV_x)(A + dA)|\}$$

or

$$\rho V_x A = (\rho + d\rho)(V_x + dV_x)(A + dA) \qquad (11.16a)$$

b. Momentum Equation

Basic equation:

$$F_{S_x} + F_{B_x} - \int_{CV} a_{rf_x} \rho \, d\forall = \frac{\partial}{\partial t}\int_{CV} V_x \rho \, d\forall + \int_{CS} V_x \rho \vec{V}_{xyz} \cdot d\vec{A} \qquad (4.35a)$$

($F_{B_x} = 0(3)$, $\int_{CV} a_{rf_x}\rho \, d\forall = 0(4)$, $\frac{\partial}{\partial t}\int_{CV} V_x \rho \, d\forall = 0(1)$)

Assumptions: (3) $F_{B_x} = 0$
(4) $a_{rf_x} = 0$
(5) Frictionless flow

The surface forces acting on the infinitesimal control volume are

$$F_{S_x} = dR_x + pA - (p + dp)(A + dA)$$

The force, dR_x, is applied along the streamtube boundary as shown in Fig. 11.3. There the average pressure is $p + dp/2$, and the area component in the x direction is dA. There is no friction. Thus

$$F_{S_x} = \left(p + \frac{dp}{2}\right) dA + pA - (p + dp)(A + dA)$$

or

$$F_{S_x} = p\,dA + \overset{\simeq 0}{\frac{dp\,dA}{2}} + pA - pA - dp\,A - p\,dA - \overset{\simeq 0}{dp\,dA}$$

Substituting this result into the momentum equation,

$$-dp\,A = V_x\{-|\rho V_x A|\} + (V_x + dV_x)\{|(\rho + d\rho)(V_x + dV_x)(A + dA)|\}$$

which may be simplified using Eq. 11.16a to obtain

$$-dp\,A = (-V_x + V_x + dV_x)(\rho V_x A)$$

Finally,

$$dp = -\rho V_x\,dV_x = -\rho d\left(\frac{V_x^2}{2}\right)$$

or

$$\frac{dp}{\rho} + d\left(\frac{V_x^2}{2}\right) = 0 \qquad (11.16b)$$

Equation 11.16b is a relation among properties during the deceleration process. In developing this relation, we have specified a frictionless deceleration process. Before we can integrate this relation between the initial and final (stagnation) states, we must specify the relation that exists between the pressure, p, and the density, ρ, along the process path.

Since the deceleration process is isentropic, then p and ρ for an ideal gas are related by the expression

$$\frac{p}{\rho^k} = \text{constant} \qquad (11.11b)$$

Our task now is to integrate Eq. 11.16b subject to this relation. Along the stagnation streamline there is only a single component of velocity; V_x is the magnitude of the velocity. Hence we can drop the subscript and write

$$\frac{dp}{\rho} + d\left(\frac{V^2}{2}\right) = 0 \tag{11.16c}$$

From $p/\rho^k = \text{constant} = C$, we can write

$$p = C\rho^k \qquad \text{and} \qquad \rho = p^{1/k}C^{-1/k}$$

Then, from Eq. 11.16c

$$-d\left(\frac{V^2}{2}\right) = \frac{dp}{\rho} = p^{-1/k}C^{1/k}\,dp$$

We can integrate this equation between the initial state and the corresponding stagnation state

$$-\int_V^0 d\left(\frac{V^2}{2}\right) = C^{1/k}\int_p^{p_0} p^{-1/k}\,dp$$

to obtain

$$\frac{V^2}{2} = C^{1/k}\frac{k}{k-1}\left[p^{(k-1)/k}\right]_p^{p_0} = C^{1/k}\frac{k}{k-1}\left[p_0^{(k-1)/k} - p^{(k-1)/k}\right]$$

$$\frac{V^2}{2} = C^{1/k}\frac{k}{k-1}p^{(k-1)/k}\left[\left(\frac{p_0}{p}\right)^{(k-1)/k} - 1\right]$$

Since $C^{1/k} = p^{1/k}/\rho$, then

$$\frac{V^2}{2} = \frac{k}{k-1}\frac{p^{1/k}}{\rho}p^{(k-1)/k}\left[\left(\frac{p_0}{p}\right)^{(k-1)/k} - 1\right] = \frac{k}{k-1}\frac{p}{\rho}\left[\left(\frac{p_0}{p}\right)^{(k-1)/k} - 1\right]$$

Since we are seeking an expression for the stagnation pressure, we can rewrite this equation as

$$\left(\frac{p_0}{p}\right)^{(k-1)/k} = 1 + \frac{k-1}{k}\frac{\rho}{p}\frac{V^2}{2}$$

and

$$\frac{p_0}{p} = \left[1 + \frac{k-1}{k}\frac{\rho V^2}{2p}\right]^{k/(k-1)}$$

For an ideal gas $p = \rho RT$, and hence

$$\frac{p_0}{p} = \left[1 + \frac{k-1}{2}\frac{V^2}{kRT}\right]^{k/(k-1)}$$

Also, for an ideal gas the sonic speed is $c = \sqrt{kRT}$, and thus

$$\frac{p_0}{p} = \left[1 + \frac{k-1}{2}\frac{V^2}{c^2}\right]^{k/(k-1)}$$

$$\frac{p_0}{p} = \left[1 + \frac{k-1}{2}M^2\right]^{k/(k-1)} \tag{11.17a}$$

Equation 11.17a enables us to calculate the isentropic stagnation pressure at any point in a flow field of an ideal gas, provided that we know the static pressure and the Mach number at that point.

Since the process of decelerating the flow at a point to determine the corresponding isentropic stagnation properties was an isentropic process, we can readily obtain expressions for other isentropic stagnation properties by applying the relation

$$\frac{p}{\rho^k} = \text{constant}$$

between end points of the process. Thus

$$\frac{p_0}{p} = \left(\frac{\rho_0}{\rho}\right)^k$$

and

$$\frac{\rho_0}{\rho} = \left(\frac{p_0}{p}\right)^{1/k}$$

From the ideal gas equation of state, $p = \rho RT$. Then

$$\frac{T_0}{T} = \frac{p_0}{p}\frac{\rho}{\rho_0} = \frac{p_0}{p}\left(\frac{p_0}{p}\right)^{-1/k} = \left(\frac{p_0}{p}\right)^{(k-1)/k}$$

Using Eq. 11.17a, we can summarize the equations for determining the isentropic stagnation properties of an ideal gas as

$$\frac{p_0}{p} = \left[1 + \frac{k-1}{2}M^2\right]^{k/(k-1)} \tag{11.17a}$$

$$\frac{T_0}{T} = 1 + \frac{k-1}{2}M^2 \tag{11.17b}$$

$$\frac{\rho_0}{\rho} = \left[1 + \frac{k-1}{2}M^2\right]^{1/(k-1)} \tag{11.17c}$$

From Eqs. 11.17 the ratio of each isentropic stagnation property to the corresponding static property at any point in a flow field for an ideal gas can be found if the local Mach number is known. In fact, these ratios could be calculated once and for all and tabulated. The calculation procedure is illustrated in Example Problem 11.4; ratios of local isentropic stagnation properties to the corresponding static properties for an ideal gas are tabulated in Table D.1 of Appendix D.

The Mach number range for validity of the assumption of incompressible flow is investigated in Example Problem 11.5.

Example 11.4

Air flows steadily through the duct shown from 350 kPa (abs), 60 C, and 183 m/sec at the initial state to a Mach number of 1.3 at the outlet, where the local isentropic stagnation conditions are known to be 385 kPa (abs) and 350 K. Compute the isentropic stagnation pressure and temperature at the inlet and the static pressure and temperature at the duct outlet. Locate the inlet and outlet static state points on a Ts diagram, and indicate the stagnation processes.

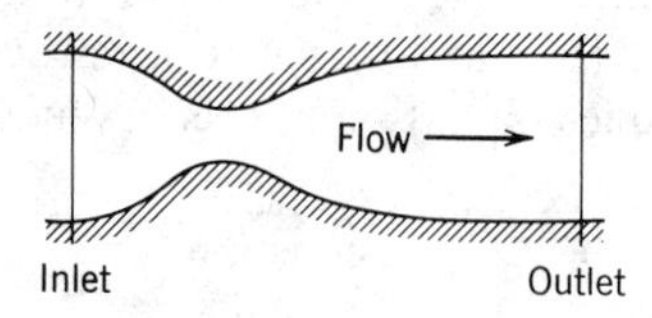

EXAMPLE PROBLEM 11.4

GIVEN: Steady flow of air through a duct as shown in the sketch.

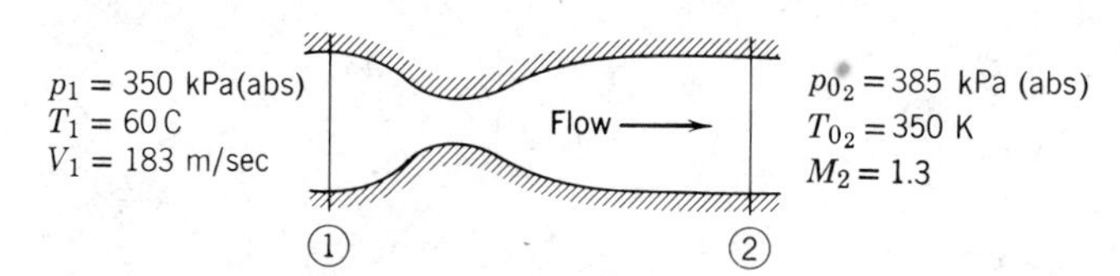

FIND: (a) p_{0_1}. (b) T_{0_1}. (c) p_2. (d) T_2.
(e) Locate state points ① and ② on a Ts diagram and indicate the stagnation processes.

SOLUTION:
To evaluate local isentropic stagnation conditions at section ①, we must calculate the Mach number, $M_1 = V_1/c_1$. For an ideal gas, $c = \sqrt{kRT}$. Then

$$c_1 = \sqrt{kRT_1} = \left[1.4 \times \frac{287\ \text{N}\cdot\text{m}}{\text{kg}\cdot\text{K}} \times (273 + 60)\ \text{K} \times \frac{\text{kg}\cdot\text{m}}{\text{N}\cdot\text{sec}^2}\right]^{1/2} = 366\ \text{m/sec}$$

and

$$M_1 = \frac{V_1}{c_1} = \frac{183}{366} = 0.5$$

Local isentropic stagnation properties may be evaluated from Eqs. 11.17. Thus

$$\frac{p_{0_1}}{p_1} = \left[1 + \frac{k-1}{2} M_1^2\right]^{k/(k-1)} = [1 + 0.2(0.5)^2]^{3.5} = 1.186$$

and

$$p_{0_1} = 1.186 p_1 = (1.186)(350\ \text{kPa}) = 415\ \text{kPa (abs)}$$

p_{0_1}

$$\frac{T_{0_1}}{T_1} = 1 + \frac{k-1}{2} M_1^2 = 1.05$$

and

$$T_{0_1} = 1.05 T_1 = (1.05)(333\ \text{K}) = 350\ \text{K}$$

T_{0_1}

At section ②, Eqs. 11.17 may be applied again. Thus

$$\frac{p_{0_2}}{p_2} = \left[1 + \frac{k-1}{2} M_2^2\right]^{k/(k-1)} = [1 + 0.2(1.3)^2]^{3.5} = 2.77$$

and

$$p_2 = \frac{p_{0_2}}{2.77} = \frac{385 \text{ kPa}}{2.77} = 139 \text{ kPa (abs)} \qquad p_2$$

$$\frac{T_{0_2}}{T_2} = 1 + \frac{k-1}{2} M_2^2 = 1.338$$

and

$$T_2 = \frac{T_{0_2}}{1.338} = \frac{350 \text{ K}}{1.338} = 262 \text{ K} \qquad T_2$$

The entropy change must be evaluated to locate state ② with respect to state ①. Using the $T\,ds$ equation

$$T\,ds = dh - v\,dp$$

or

$$ds = \frac{dh}{T} - \frac{v\,dp}{T} = c_p \frac{dT}{T} - R\frac{dp}{p}$$

Integrating gives

$$s_2 - s_1 = \int_{T_1}^{T_2} c_p \frac{dT}{T} - \int_{p_1}^{p_2} R\frac{dp}{p} = c_p \ln\frac{T_2}{T_1} - R\ln\frac{p_2}{p_1}$$

$$s_2 - s_1 = 1.00 \frac{\text{kJ}}{\text{kg}\cdot\text{K}} \times \ln\left(\frac{262}{333}\right) - 0.287 \frac{\text{kJ}}{\text{kg}\cdot\text{K}} \times \ln\left(\frac{1.39}{3.50}\right) = 0.0252 \text{ kJ/kg}\cdot\text{K}$$

Therefore, state ② lies to the right of state ① on the Ts plane as shown in the following sketch:

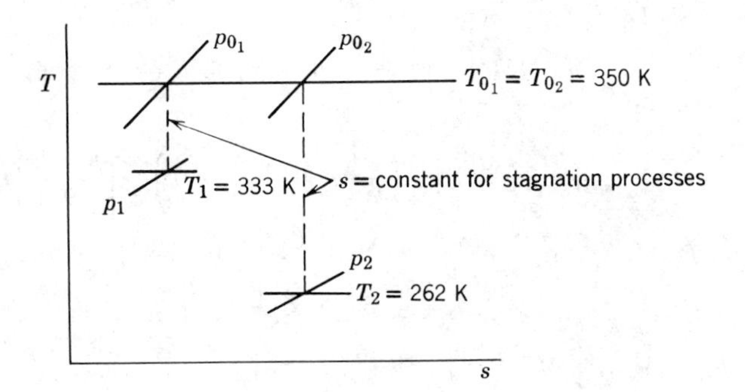

{ The process the fluid follows between states ① and ② is not specified. However, it need not be known. A unique isentropic stagnation process is defined at each state point. Note that $s_{0_2} - s_{0_1} = s_2 - s_1$. }

Example 11.5

We have derived equations for the ratio, p_0/p, for both compressible and "incompressible" flow. By writing both equations in terms of Mach number, compare their behavior. Find the Mach number below which the two equations agree within engineering accuracy.

EXAMPLE PROBLEM 11.5

GIVEN: The incompressible and compressible forms of the equations for stagnation pressure, p_0.

Incompressible
$$p_0 = p + \tfrac{1}{2}\rho V^2 \tag{6.14}$$

Compressible
$$\frac{p_0}{p} = \left[1 + \frac{k-1}{2}M^2\right]^{k/(k-1)} \tag{11.17a}$$

FIND: (a) Behavior of both equations as a function of Mach number.
(b) Mach number below which calculated values of $(p_0 - p)/p_0$ agree within engineering accuracy.

SOLUTION:
First, let us write Eq. 6.14 in terms of Mach number. Using the ideal gas equation of state, and $c^2 = kRT$,

$$\frac{p_0}{p} = 1 + \frac{\rho V^2}{2p} = 1 + \frac{V^2}{2RT} = 1 + \frac{kV^2}{2kRT} = 1 + \frac{kV^2}{2c^2}$$

Thus

$$\frac{p_0}{p} = 1 + \frac{k}{2}M^2 \tag{1}$$

for "incompressible" flow.

Equation 11.17a may be expanded using the binomial theorem,

$$(1+x)^n = 1 + nx + \frac{n(n-1)}{2!}x^2 + \cdots, |x| < 1$$

For Eq. 11.17a, $x = [(k-1)/2]M^2$, and $n = k/(k-1)$. Thus the series converges for $[(k-1)/2]M^2 < 1$, and

$$\frac{p_0}{p} = 1 + \left(\frac{k}{k-1}\right)\left[\frac{k-1}{2}M^2\right] + \left(\frac{k}{k-1}\right)\left(\frac{k}{k-1} - 1\right)\frac{1}{2!}\left[\frac{k-1}{2}M^2\right]^2$$
$$+ \left(\frac{k}{k-1}\right)\left(\frac{k}{k-1} - 1\right)\left(\frac{k}{k-1} - 2\right)\frac{1}{3!}\left[\frac{k-1}{2}M^2\right]^3 + \cdots$$

$$\frac{p_0}{p} = 1 + \frac{k}{2}M^2 + \frac{k}{8}M^4 + \frac{k(2-k)}{48}M^6 + \cdots$$

$$\frac{p_0}{p} = 1 + \frac{k}{2}M^2\left[1 + \frac{1}{4}M^2 + \frac{(2-k)}{24}M^4 + \cdots\right] \tag{2}$$

for compressible flow.

In the limit, as $M \to 0$, the term in brackets in Eq. 2 approaches 1.0. Thus, for flow at low Mach number, the incompressible and compressible equations give the same result. The variation of p_0/p with Mach number is shown in the sketch below. As Mach number is increased, the compressible equation gives larger values for the ratio, p_0/p.

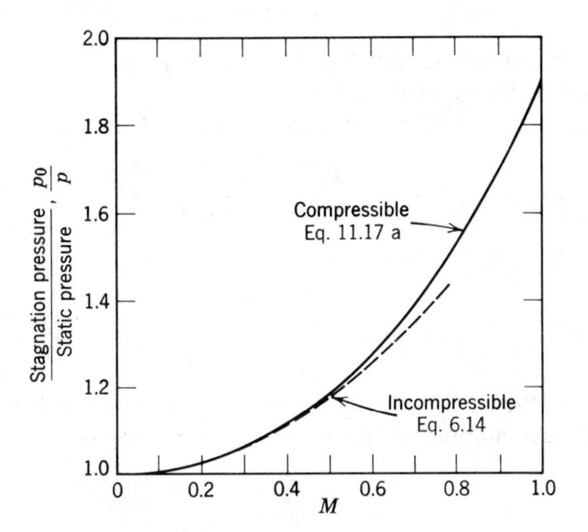

Equations 1 and 2 may be compared quantitatively most simply by writing

$$\frac{p_0}{p} - 1 = \frac{k}{2}M^2 \qquad \text{("incompressible")}$$

$$\frac{p_0}{p} - 1 = \frac{k}{2}M^2\left[1 + \frac{1}{4}M^2 + \frac{(2-k)}{24}M^4 + \cdots\right] \qquad \text{(compressible)}$$

The term in brackets is approximately equal to 1.02 at $M = 0.3$, and to 1.04 at $M = 0.4$. Thus, for calculations of engineering accuracy, flow may be considered incompressible if $M < 0.3$. The two agree within 5 percent for $M \gtrsim 0.45$.

11-3.2 Critical Conditions

Stagnation conditions are extremely useful as reference conditions for thermodynamic properties; this is not true for velocity, since $V = 0$ by definition at stagnation. A useful reference value for velocity is the critical speed—the speed at a Mach number of unity. Even if there is no point in a given flow field where the Mach number is equal to unity, such a hypothetical condition still is useful as a reference condition.

If we use an asterisk to denote conditions at $M = 1$, then by definition

$$V^* \equiv c^* \tag{11.18}$$

At critical conditions, Eqs. 11.17 for stagnation properties become (for $k = 1.4$)

$$\frac{p_0^*}{p^*} = \left[1 + \frac{k-1}{2}\right]^{k/(k-1)} = 1.893$$

$$\frac{T_0^*}{T^*} = 1 + \frac{k-1}{2} = 1.200$$

$$\frac{\rho_0^*}{\rho^*} = \left[1 + \frac{k-1}{2}\right]^{1/(k-1)} = 1.577$$

The critical speed may be written in terms of either the critical temperature, T^*, or the critical stagnation temperature, T_0^*.

For an ideal gas, $c^* = \sqrt{kRT^*}$, and thus $V^* = \sqrt{kRT^*}$. Since

$$T^* = \frac{T_0^*}{1 + (k-1)/2} = \frac{2}{k+1} T_0^*$$

then

$$V^* = c^* = \sqrt{\frac{2k}{k+1} R T_0^*} \tag{11.19}$$

We shall use both the stagnation conditions and the critical conditions as reference conditions in the next chapter when we consider a variety of one-dimensional compressible flows.

11-4 SUMMARY OBJECTIVES

After completing study of Chapter 11, you should be able to do the following:

1. Define:

Mach number	Mach angle
subsonic flow	zone of action
supersonic flow	zone of silence
transonic flow	local isentropic stagnation properties
hypersonic flow	critical conditions
Mach cone	

2. For an ideal gas, write expressions for
   (a) The change in internal energy and enthalpy
   (b) The $T\,ds$ equations
   (c) The relation between pressure and density for an isentropic process
3. Derive an equation for the speed of sound in a medium and show that for an ideal gas, $c = \sqrt{kRT}$.
4. Write (and derive) expressions for the local isentropic stagnation properties (temperature, pressure, density) for the flow of an ideal gas.
5. Solve the problems at the end of the chapter that relate to the material you have studied.

PROBLEMS

11.1 Five kg of air in a closed system expands reversibly with constant entropy from 300 kPa (abs), 60 C, to 150 kPa (abs). Calculate the air temperature after the expansion. Show the process state points on a Ts diagram.

11.2 Ten lbm of air is cooled in a closed tank from 500 to 100 F. The initial pressure is 400 psia. Compute the changes in entropy, internal energy, and enthalpy. Show the process state points on a *Ts* diagram.

11.3 A turbine operating on air has inlet conditions of 200 kPa (abs), 95 C, and an exhaust pressure of 100 kPa (abs). If the expansion is adiabatic and $\Delta KE = 0$, is an exhaust temperature of 5 C possible? Is it possible if $\Delta KE \neq 0$? Show the process state points on a *Ts* diagram.

11.4 Air is compressed irreversibly from 100 kPa (abs), 5 C, to 200 kPa (abs), 115 C. Calculate the change in entropy of the air. Show the process state points on a *Ts* diagram.

11.5 Is an adiabatic expansion of air from 300 kPa (abs), 60 C, to 150 kPa (abs), 27 C, possible? Justify your answer. Show the process state points on a *Ts* diagram.

11.6 In a closed system, a gas undergoes a cycle made up of the following processes: 1–2 reversible isothermal compression, 2–3 reversible constant volume heating, 3–4 reversible constant pressure expansion, and 4–1 reversible adiabatic expansion.
(a) Sketch *pv* and *Ts* diagrams.
(b) State whether each of the following quantities is positive, zero, negative, or indeterminate in sign:

$$\oint \delta W, \quad \oint \delta Q, \quad \oint dS, \quad \oint dU, \quad \oint dH$$

11.7 An ideal gas is heated at constant volume from state ① to state ②, expanded isothermally to state ③, expanded adiabatically to state ④, which is at the same pressure as state ①, and then restored to state ① by a constant pressure process. All four processes are reversible.
(a) Sketch *pv* and *Ts* diagrams of the cycle.
(b) State whether each of the following quantities is greater than zero, less than zero, equal to zero, or of indeterminate sign:

$$\oint \delta Q, \quad \oint \delta W, \quad \oint du, \quad \oint ds$$

11.8 A fluid passing through a steady flow system can be heated reversibly from state ① to state ② in either of two ways. In one case, $T = a + bs$; in the other case, $T = c + es^2$, where a, b, c, and e are constants. In which case is the heat transfer greater? Show the state points on a *Ts* diagram for each process.

11.9 A tank contains 10 m^3 of compressed air at 15 C. The gage pressure in the tank is 4.50 MPa. Evaluate the energy required to fill the tank by compressing air from standard atmosphere conditions for (a) isothermal compression, and (b) isentropic compression followed by cooling at constant pressure. What is the peak temperature of the isentropic compression process? Calculate the energy removed during cooling for process (b). Assume ideal gas behavior and reversible processes. Label state points on a *Ts* diagram for each process.

11.10 Air enters a turbine in steady flow at 0.5 kg/sec with negligible velocity. Inlet conditions are 1300 C and 2.0 MPa (abs). The air is expanded through the turbine to atmospheric pressure. If the actual temperature and velocity at the turbine exit are 500 C and 200 m/sec, determine the power produced by the turbine. Label state points on a *Ts* diagram for this process.

11.11 Natural gas with the thermodynamic properties of methane flows in an underground pipeline of 0.6 m diameter. The gage pressure at the inlet to a compressor station is 0.5 MPa; the outlet pressure is 8.0 MPa (gage). The gas temperature and speed at the inlet are 13 C and 320 m/sec, respectively. The compressor efficiency is $\eta_c = 0.85$. Calculate the mass flow rate of natural gas through the pipeline. Label state points on a Ts diagram for compressor inlet and outlet. Evaluate the gas temperature and speed at the compressor outlet and the power required to drive the compressor.

11.12 Air is cooled at constant pressure from 858 K and 4.5 MPa (gage) to 15 C. Show the process on a Ts diagram. Calculate the change in specific entropy for the air if it behaves as an ideal gas. Evaluate the heat transferred per unit mass if the process is reversible.

11.13 Air is compressed isothermally from standard conditions to 4.5 MPa (gage). Show the process on a Ts diagram. Calculate the change in specific entropy for the process if the air behaves as an ideal gas. Evaluate the heat transferred per unit mass if the process is reversible.

11.14 Air is expanded in a steady flow process through a turbine. Initial conditions are 1300 C and 2.0 MPa (abs). Final conditions are 500 C and atmospheric pressure. Show this process on a Ts diagram. Evaluate the changes in internal energy, enthalpy, and specific entropy for this process.

11.15 An aircraft flies at a speed of 960 km/hr through air at a pressure of 82 kPa and temperature of 0 C. Calculate the Mach number of the craft.

11.16 Air at a temperature of 20 C is flowing with a speed of 150 m/sec. A bullet is fired into the air stream with a speed of 800 m/sec. (The direction of the bullet is opposite to that of the air.) Calculate:
(a) The Mach number of the air flow
(b) The Mach number of the bullet if it were fired in still air
(c) The Mach number of the bullet with respect to the moving air

11.17 An airplane flies at a speed of 180 m/sec at an altitude of 500 m, where the temperature is 20 C. The plane climbs to 15 km, where the temperature is -56 C and levels off at a speed of 320 m/sec. Calculate the Mach number of flight in both cases.

11.18 The Boeing 727 aircraft of Example Problem 9.8 cruises at 520 mph at an altitude of 33,000 ft on a standard day. Calculate the cruise Mach number of the aircraft. If the maximum allowable operating Mach number for the aircraft is 0.9, what is the corresponding flight speed?

11.19 What is the speed of sound in carbon dioxide at 150 C? What is the acoustic speed in water at 20 C?

11.20 The speed of sound in steel (SG = 7.8) is observed to be about 5.3 km/sec. Estimate the bulk modulus for steel. Compare your estimate with the bulk modulus for mercury. Compute the speed of sound in mercury.

11.21 Heptane and octane are used to define the 0 and 100 octane rating points for gasoline. Use data from Appendix A to estimate the bulk modulus of gasoline. Compare the speeds of sound in gasoline and in water.

11.22 The transonic flow regime begins at about $M = 0.9$. Compare the flow speeds for which the Mach number equals 0.9 in (a) air, and (b) low-pressure steam at 75 C, assuming that the steam behaves as an ideal gas.

11.23 Mixtures of helium and oxygen are used instead of air for long-term undersea work to minimize the risk of causing the "bends" from release of nitrogen bubbles in the bloodstream upon decompression. The speech of "aquanauts" who breathe this mixture is strange, partly because the speed of sound is changed in the mixture. Estimate the gas constant, specific heat ratio, and speed of sound at standard conditions in a mixture that contains 20 percent oxygen and 80 percent helium by volume.

11.24 Published data indicate that the F-5G aircraft can accelerate from a Mach number of 0.9 to 1.2 in 30 sec, and from a Mach number of 0.9 to 1.6 in 80 sec at 10 km altitude on a standard day. Express the average rate of acceleration of the aircraft in g's.

11.25 Published data indicate that the F-5G aircraft is capable of making sustained horizontal turns at a rate of 6°/sec at $M = 0.7$ at 30,000 ft altitude. At $M = 1.6$ the aircraft can sustain a turn at 3.5°/sec. Calculate the radius of curvature and normal acceleration produced by these turns.

11.26 Actual performance characteristics of the Lockheed SR-71 "Blackbird" reconnaissance aircraft are classified. However, it is thought to cruise at $M = 3.3$ at 85,000 ft altitude. Evaluate the speed of sound and flight speed for these conditions. Compare to the muzzle speed of a 30–06 rifle bullet (700 m/sec).

11.27 The temperature varies linearly from sea level to approximately 10 km altitude in the standard atmosphere. Evaluate the *lapse rate*—the rate of decrease of temperature with altitude—in the standard atmosphere. Derive an expression for the rate of change of sonic speed with altitude in an ideal gas under standard atmosphere conditions. Evaluate at sea level and at 10 km altitude.

11.28 A photograph of a bullet shows a Mach angle of 28°. Determine the speed of the bullet for standard air.

11.29 Air at 25 C is flowing at a Mach number of 1.9. Determine the air speed and the Mach angle.

11.30 Measurements in very high-speed flow often are made by firing a projectile upstream into a high-speed gas flow. A projectile is fired at 4,500 m/sec into a stream of helium that has $M = 3.5$ at a temperature of -20 C. Calculate the Mach number of the projectile with respect to the gas stream. What Mach angle would be expected on a photograph?

11.31 A projectile is fired into a gas in which the pressure is 50 psia and the density is 0.27 lbm/ft^3. It is observed experimentally that a Mach cone emanates from the projectile with a total angle of 20°. What is the speed of the projectile with respect to the gas?

11.32 The National Transonic Facility (NTF) is a high-speed wind tunnel designed to operate with air at cryogenic temperatures to reduce viscosity, thus raising the unit Reynolds number (Re/x) and reducing pumping power requirements. Operation is envisioned at temperatures of -270 F and below. A schlieren photograph taken in the NTF shows a Mach angle of 57° at a location where $T = -270$ F and $p = 1.3$ psia. Evaluate the local Mach number and flow speed. Calculate the unit Reynolds number for the flow.

11.33 An F-4 aircraft makes a high-speed pass over an airfield on a day when the temperature is 35 C. The aircraft flies at $M = 1.4$ and 200 m altitude. Calculate the air speed of the aircraft. How long after it passes directly overhead does its Mach cone pass a point on the ground?

11.34 An F-5G aircraft passes overhead at an altitude of 3 km. The aircraft flies at $M = 1.35$; assume that the air temperature is constant at 30 C. Find the air speed of the aircraft. A headwind blows at 10 m/sec. How long after the aircraft passes directly overhead does its sound reach a point on the ground?

11.35 A supersonic aircraft flies at an altitude of 10,000 ft at a speed of 3000 ft/sec on a standard day. How long after passing directly above a ground observer will it be before the sound of the aircraft is heard?

11.36 For the conditions of Problem 11.34, find the location at which the sound wave that first strikes the point on the ground was emitted.

11.37 The Concorde supersonic transport cruises at $M = 2.2$ at an altitude of 17 km on a standard day. How long after the aircraft passes directly above a ground observer will it be before the sound of the aircraft is heard?

11.38 The pressure at the nose of an aircraft in flight was found to be 44.3 kPa (abs). (The velocity of air relative to the craft was zero at this point.) Estimate the Mach number, speed, and altitude of the craft if the pressure and temperature of the undisturbed air were 27.6 kPa (abs) and -50 C, respectively.

11.39 Air flows in a duct where the static temperature is 50 F and the static pressure is 10 psia. Calculate the local isentropic stagnation pressure if the air speed is (a) 200 ft/sec, and (b) 2000 ft/sec.

11.40 A plane is flying at an altitude where the air temperature is 10 C. At a point on the plane where the relative air velocity is zero, the temperature is found to be 49 C. Determine the Mach number and speed of the plane.

11.41 Consider flow of standard air with a speed of 600 m/sec. What is the local isentropic stagnation pressure? The stagnation enthalpy? The stagnation temperature?

11.42 Air flows steadily through a section (① denotes inlet and ② denotes exit) of an insulated constant-area duct. Properties change along the duct as a result of friction.

(a) Beginning with the control volume form of the first law of thermodynamics, show that the equation can be reduced to

$$h_1 + \frac{V_1^2}{2} = h_2 + \frac{V_2^2}{2} = \text{constant}$$

(b) Denoting the constant by h_0 (the stagnation enthalpy), show that for adiabatic flow of an ideal gas with friction

$$\frac{T_0}{T} = 1 + \frac{k-1}{2} M^2$$

(c) For this flow does $T_{0_1} = T_{0_2}$? $p_{0_1} = p_{0_2}$?

11.43 A body moves through standard air at 200 m/sec. What is the pressure at a point on the body where the velocity of the air relative to the body is zero? Assume (a) compressible flow, and (b) incompressible flow.

11.44 The maximum density in a compressible flow field occurs at stagnation conditions. Evaluate the Mach number values for flow of air at which ρ_0 and ρ differ by 2 percent and 5 percent. Comment on the significance of these results.

11.45 According to a recent article in *Popular Science* magazine, the maximum air speed in Ford's Variable Venturi carburetor is approximately 430 ft/sec. For standard day atmospheric conditions, evaluate the minimum static pressure in the carburetor venturi. What percentage error in calculated static pressure would result if this calculation were made assuming incompressible flow?

11.46 For aircraft flying at supersonic speeds, the lift and drag coefficients are functions of Mach number only. A supersonic transport with wingspan of 75 m is to fly at an air speed of 780 m/sec at an altitude of 20 km on a standard day. Performance of the aircraft is to be measured from model test data using a scale model with 0.9 m wingspan in a supersonic wind tunnel. The wind tunnel is to be supplied from a large reservoir of compressed air, which can be heated if desired. The static temperature of air in the test section is to be 10 C to avoid freezing of moisture. At what air speed should the wind tunnel tests be run to duplicate the Mach number of the prototype? What must be the stagnation temperature in the reservoir? What pressure is required in the reservoir if the test section pressure is to be 10 kPa (abs)?

11.47 A DC-10 aircraft cruises at an altitude of 12 km on a standard day. A pitot-static tube located on the nose of the aircraft measures the stagnation and static pressures as 29.6 kPa and 19.4 kPa, respectively. Calculate (a) the flight Mach number of the aircraft, (b) the speed of the aircraft, and (c) the stagnation temperature that would be sensed by a probe on the aircraft.

11.48 The Anglo-French "Concorde" supersonic transport cruises at $M = 2.2$ at an altitude of 20 km. Evaluate the speed of sound, aircraft flight speed, and Mach angle for these conditions. Compare the aircraft speed to the muzzle speed of a .22-caliber rifle bullet (460 m/sec). What is the maximum air temperature at stagnation points on the aircraft structure?

11.49 Published data indicate that the F-5G aircraft has a maximum speed of $M = 2.1$ at 36,000 ft altitude on a standard day. Calculate the aircraft flight speed for these conditions. Determine the local isentropic stagnation pressure that corresponds to these flight conditions. Because the aircraft speed is supersonic, a normal shock occurs in front of a total-head tube. The stagnation pressure decreases by 32.6 percent across the shock. Evaluate the stagnation pressure that would be sensed by a probe on the aircraft. What is the maximum air temperature at stagnation points on the aircraft structure?

11.50 Actual performance characteristics of the Lockheed SR-71 "Blackbird" reconnaissance aircraft are classified. However, it is thought to cruise at $M = 3.3$ at 26 km altitude. Calculate the aircraft flight speed for these conditions. Determine the local isentropic stagnation pressure that corresponds to these flight conditions. Because the aircraft speed is supersonic, a normal shock occurs in front of a total-head tube. The stagnation pressure decreases by 74.7 percent across the shock. Evaluate the stagnation pressure that would be sensed by a probe on the aircraft. What is the maximum air temperature at stagnation points on the aircraft structure?

11.51 A jet transport aircraft cruises at $M = 0.85$ at 12.5 km altitude on a standard day. Evaluate the stagnation pressure sensed by a probe on the aircraft. What velocity would be calculated from the incompressible Bernoulli equation? By what percentage is this value in error compared to the true speed of the aircraft?

11.52 A transonic wind tunnel operates at test section Mach number, $M = 1.2$. Operating conditions in the test section are $T = -34$ C and $p = 41.8$ kPa (abs). Calculate the local isentropic stagnation conditions for this flow. A normal shock forms ahead of a pitot-static tube placed in the flow stream. The total-head tube measures a stagnation pressure of 100.6 kPa (abs). Determine the loss in stagnation pressure across the normal shock. By what percentage is the local isentropic stagnation pressure reduced compared to the upstream value?

11.53 All modern high-speed aircraft use "air data computers" to calculate air speed from measured values of dynamic pressure. Evaluate the subsonic Mach number above which the incompressible form of the Bernoulli equation predicts a speed error of 2 percent compared to the true air speed calculated including compressibility effects. Assume flight at 10 km altitude on a standard day. Is the result you obtain independent of freestream conditions?

11.54 A supersonic wind tunnel operates at test section Mach number, $M = 1.8$. Operating conditions in the tunnel test section are $T = -80$ C and $p = 258$ kPa (gage). A normal shock that stands in front of a total-head tube reduces the stagnation pressure by 18.7 percent. Calculate (a) the local isentropic stagnation conditions for the test section conditions, and (b) the pressure sensed by the total-head tube.

11.55 Consider the steady adiabatic flow of air through a long straight pipe with cross-sectional area 0.05 m^2. At the inlet (section ①) the air is at an absolute pressure of 200 kPa, 60 C, and velocity of 146 m/sec. At a section downstream the air is at 95.6 kPa (abs) and has a velocity of 280 m/sec. Determine p_{0_1}, p_{0_2}, T_{0_1}, T_{0_2}, and the entropy change for the flow. Show the static and stagnation state points on a Ts diagram.

11.56 Air flows steadily through a constant-area duct. At section ①, the air is at 60 psia, 600 R, with velocity 500 ft/sec. As a result of heat transfer and friction, the air at section ② downstream is at 40 psia, 800 R. Determine p_{0_1}, p_{0_2}, T_{0_1}, T_{0_2}, and the entropy change for the flow. Show the static and stagnation state points on a Ts diagram.

11.57 Air enters a long, insulated duct at $M_1 = 0.2$, $T_1 = 286$ K, and $p_1 = 98.5$ kPa (abs). At a location downstream, the properties are $M_2 = 0.6$, $T_2 = 268.9$ K, and $p_2 = 31.3$ kPa (abs). (Four significant figures are given to minimize roundoff errors.) Evaluate the local isentropic stagnation conditions (a) at the inlet section, and (b) at the outlet section of the duct. Calculate the change in specific entropy along the duct. Plot static and stagnation state points on a Ts diagram.

11.58 Air enters a combustion chamber at $M_1 = 0.2$, $T_1 = 580$ K, and $p_1 = 1.0$ MPa (abs). Heat addition in the combustion chamber causes the fluid properties at the exit to be $M_2 = 0.4$, $T_2 = 1727$ K, and $p_2 = 862.7$ kPa (abs). (Four significant figures are given to minimize roundoff errors.) Evaluate the local isentropic stagnation conditions (a) at the inlet to, and (b) at the outlet from the combustion chamber. Calculate the change in specific entropy across the combustor. Plot static and stagnation state points on a Ts diagram.

11.59 Air passes through a normal shock in a supersonic wind tunnel. Upstream conditions are $M_1 = 1.8$, $T_1 = 270$ K, and $p_1 = 10.0$ kPa (abs). Downstream conditions are $M_2 = 0.6165$, $T_2 = 413.6$ K, and $p_2 = 36.13$ kPa (abs). (Four significant figures are given to minimize roundoff errors.) Evaluate the local isentropic stagnation conditions (a) upstream from, and (b) downstream from the normal shock. Calculate the change in specific entropy across the shock. Plot static and stagnation state points on a Ts diagram.

11.60 Consider again the gas pipeline compressor of Problem 11.11. Calculate the local isentropic stagnation conditions for the flow of gas (a) at the compressor inlet, and (b) at the compressor outlet. Calculate the change in specific entropy across the compressor. Show the static and stagnation state points on a Ts diagram. (The compressor outlet temperature and Mach number are 572 K and 0.0762, respectively.)

11.61 Air enters a turbine at $M_1 = 0.4$, $T_1 = 2350$ F, and $p_1 = 90.0$ psia. Conditions leaving the turbine are $M_2 = 0.8$, $T_2 = 1200$ F, and $p_2 = 3.00$ psia. (Four significant figures are given to minimize roundoff errors.) Evaluate the local isentropic stagnation conditions (a) at the turbine inlet, and (b) at the turbine outlet. Calculate the change in specific entropy across the turbine. Plot the static and stagnation state points on a Ts diagram.

11.62 The *critical pressure ratio*, p^*/p_0^*, is the ratio of static pressure to local isentropic stagnation pressure at the condition where $V = V^* = c^*$. For an ideal gas the critical pressure ratio is a function only of the ratio of specific heats, k. Show that for air,

$p^*/p_0^* = 0.5283$. Evaluate for comparison the critical pressure ratios for gases with the maximum and minimum values of k given in Table A.6.

11.63 The tires on a modern lightweight bicycle are inflated to 800 kPa (gage). Consider a day when a tire is in equilibrium with the ambient at 37 C. Calculate the critical conditions (temperature, pressure, and flow speed) that correspond to these stagnation conditions.

11.64 An air tank at an automobile service station holds air at 800 kPa (gage) and 30 C. Calculate the critical conditions (temperature, pressure, and flow speed) that correspond to these stagnation conditions.

11.65 A CO_2 cartridge is used to propel a toy rocket. Gas in the cartridge is pressurized to 45 MPa (gage) and is at 25 C. Calculate the critical conditions (temperature, pressure, and flow speed) that correspond to these stagnation conditions.

11.66 A fire extinguisher filled with carbon dioxide gas is pressurized to 35 MPa (gage) and stored in a location where the ambient temperature is 30 C. Calculate the critical conditions (temperature, pressure, and flow speed) that correspond to these stagnation conditions.

11.67 The gas storage reservoir for a high-speed wind tunnel contains helium at 2500 K and 6.0 MPa (gage). Calculate the critical conditions (temperature, pressure, and flow speed) that correspond to these stagnation conditions.

11.68 Consider the natural gas pipeline of Problem 11.11. Calculate the critical conditions (temperature, pressure, and flow speed) that correspond to the compressor inlet conditions.

11.69 The hot gas stream at the turbine inlet of a JT9-D jet engine is at 2350 F and 140 kPa (abs). The Mach number at that location is 0.32. Calculate the critical conditions (temperature, pressure, and flow speed) that correspond to these conditions. Assume the fluid properties are those of pure air.

11.70 Stagnation conditions in a solid propellant rocket motor are $T_0 = 3500$ K and $p_0 = 40$ MPa (gage). Critical conditions occur in the throat of the rocket nozzle where the Mach number is equal to one. Evaluate the temperature, pressure, and flow speed at the throat of the rocket nozzle. Assume ideal gas behavior with $R = 60$ J/kg · K and $k = 1.2$ for the exhaust gases.

Chapter 12

STEADY ONE-DIMENSIONAL COMPRESSIBLE FLOW

Fluid properties in compressible flow are affected by area change, friction, heat transfer, and normal shocks. In this chapter each of these effects is considered separately for steady, one-dimensional, compressible flow.

Isentropic flow, in which area is the independent variable (friction and heat transfer are neglected), is considered first for a general fluid. Then the isentropic flow of an ideal gas and applications to nozzles are considered in greater detail.

Following isentropic flow, adiabatic flow in a constant-area duct with friction and frictionless flow in a constant-area duct with heat transfer are considered. A discussion of normal shocks concludes the chapter.

12-1 BASIC EQUATIONS FOR ISENTROPIC FLOW

Consider the steady, one-dimensional, isentropic flow of any compressible fluid through a channel of arbitrary cross section; a portion of such a duct is shown in Fig. 12.1. To develop the governing equations for this flow, we apply the basic equations, derived in Chapter 4, to the finite, fixed control volume of Fig. 12.1. The

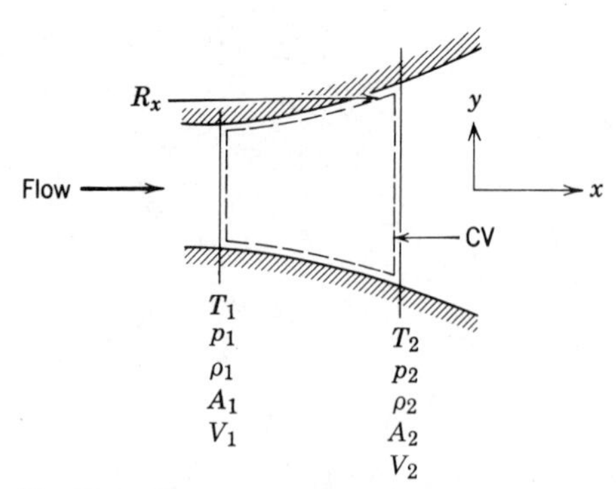

Fig. 12.1 Control volume for analysis of a general isentropic flow.

properties at sections ① and ② are labeled with the appropriate subscripts; R_x is the x component of the surface force acting on the control volume.

a. Continuity Equation

Basic equation:

$$0 = \overset{=0(1)}{\cancel{\frac{\partial}{\partial t}\int_{\text{CV}} \rho \, d\forall}} + \int_{\text{CS}} \rho \vec{V} \cdot d\vec{A} \tag{4.13}$$

Assumptions: (1) Steady flow
(2) One-dimensional flow

Then

$$0 = \{-|\rho_1 V_1 A_1|\} + \{|\rho_2 V_2 A_2|\}$$

Using scalar magnitudes and dropping absolute value signs gives the familiar form

$$\rho_1 V_1 A_1 = \rho_2 V_2 A_2 = \rho VA = \dot{m} = \text{constant} \tag{12.1a}$$

b. Momentum Equation

Basic equation:

$$F_{S_x} + \overset{=0(3)}{\cancel{F_{B_x}}} = \overset{=0(1)}{\cancel{\frac{\partial}{\partial t}\int_{\text{CV}} V_x \rho \, d\forall}} + \int_{\text{CS}} V_x \rho \vec{V} \cdot d\vec{A} \tag{4.19a}$$

Assumptions: (3) $F_{B_x} = 0$

The surface force will be due to pressure forces at the surfaces ① and ②, and to the distributed pressure force, R_x, along the channel walls. Substituting gives

$$R_x + p_1 A_1 - p_2 A_2 = V_1\{-|\rho_1 V_1 A_1|\} + V_2\{|\rho_2 V_2 A_2|\}$$

Using scalar magnitudes and dropping absolute value signs, we obtain

$$R_x + p_1 A_1 - p_2 A_2 = \dot{m} V_2 - \dot{m} V_1 \tag{12.1b}$$

c. First Law of Thermodynamics

Basic equation:

$$\overset{=0(4)}{\cancel{\dot{Q}}} + \overset{=0(5)}{\cancel{\dot{W}_s}} + \overset{=0(6)}{\cancel{\dot{W}_{\text{shear}}}} + \overset{=0(6)}{\cancel{\dot{W}_{\text{other}}}} = \overset{=0(1)}{\cancel{\frac{\partial}{\partial t}\int_{\text{CV}} e\rho \, d\forall}} + \int_{\text{CS}} (e + pv)\rho \vec{V} \cdot d\vec{A} \tag{4.59}$$

where

$$e = u + \frac{V^2}{2} + \cancel{gz} \quad (\simeq 0(7))$$

Assumptions: (4) $\dot{Q} = 0$ (isentropic, i.e. frictionless adiabatic flow)
(5) $\dot{W}_s = 0$
(6) $\dot{W}_{\text{shear}} = \dot{W}_{\text{other}} = 0$
(7) Effects of gravity are negligible

Under these assumptions, the first law reduces to

$$0 = \left(u_1 + p_1 v_1 + \frac{V_1^2}{2}\right)\{-|\rho_1 V_1 A_1|\} + \left(u_2 + p_2 v_2 + \frac{V_2^2}{2}\right)\{|\rho_2 V_2 A_2|\}$$

But we know from continuity that the mass flow rate terms in brackets are equal, so they may be canceled. We may also substitute $h \equiv u + pv$, to obtain

$$h_1 + \frac{V_1^2}{2} = h_2 + \frac{V_2^2}{2} = h + \frac{V^2}{2} = \text{constant} \tag{12.1c}$$

The combination $h + V^2/2$ occurs often in compressible flow problems. It is convenient to define the stagnation enthalpy, h_0, as

$$h_0 \equiv h + \frac{V^2}{2}$$

Physically, the stagnation enthalpy is the enthalpy that would be reached if the fluid were decelerated adiabatically to zero velocity. We note that the stagnation enthalpy is constant throughout an adiabatic flow field.

d. Second Law of Thermodynamics

Basic equation:

$$\int_{\text{CS}} \frac{1}{T} \cancel{\left(\frac{\dot{Q}}{A}\right)} dA \leq \cancel{\frac{\partial}{\partial t}} \int_{\text{CV}} s\rho \, d\forall + \int_{\text{CS}} s\rho \vec{V} \cdot d\vec{A} \quad (=0(4)\ \ =0(1)) \tag{4.61}$$

Then, for a reversible adiabatic process,

$$0 = s_1\{-|\rho_1 V_1 A_1|\} + s_2\{|\rho_2 V_2 A_2|\}$$

Since the mass flow rate terms { } are equal by continuity,

$$s_1 = s_2 = s = \text{constant} \tag{12.1d}$$

e. Equation of State

Equations of state are relations among intensive thermodynamic properties. These relations may be in the form of tables, charts, or algebraic equations. Since, for a pure substance, it is possible to specify any intensive thermodynamic property in terms of any other two intensive thermodynamic properties, we can write

$$h = h(s, p) \tag{12.1e}$$

and

$$\rho = \rho(s, p) \tag{12.1f}$$

as equations of state.

Before summarizing the simplified forms of the basic equations for steady, one-dimensional, isentropic flow of any compressible fluid, let us turn to a representation of an isentropic flow on an *hs* diagram. At some point in the isentropic flow field (call it state ①), the flow properties are h_1, s_1, p_1, V_1, and so on. Clearly, in isentropic flow, the flow may proceed to state ② (with properties $h_2, s_2 = s_1, p_2, V_2$, etc.) or to state ③ (with properties $h_3, s_3 = s_1, p_3, V_3$, etc.) as shown in Fig. 12.2.

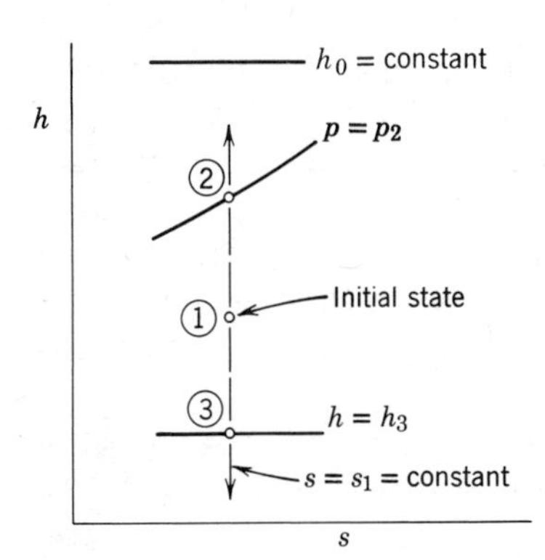

Fig. 12.2 Representation of isentropic flow in the *hs* plane.

How are the isentropic stagnation properties at states ①, ②, and ③ related? To answer this question, consider the results obtained from the first law analysis for isentropic flow. From Eq. 12.1c and the definition of stagnation enthalpy, we have

$$h_1 + \frac{V_1^2}{2} = h_2 + \frac{V_2^2}{2} = h + \frac{V^2}{2} = h_0 = \text{constant}$$

Thus all states in an isentropic flow have the same stagnation enthalpy. In addition, by definition, all states in an isentropic flow (including the stagnation state) have the same entropy. Thus all stagnation states in an isentropic flow have the same stagnation enthalpy and stagnation entropy. Since at the stagnation state the velocity is zero, it follows that the stagnation properties are constant for all points in an isentropic flow.

For isentropic flow, the first law, in the form

$$h + \frac{V^2}{2} = h_0 = \text{constant}$$

suggests another interpretation for the stagnation enthalpy. The stagnation enthalpy, h_0, represents the total energy per unit mass of flowing fluid. The kinetic energy per unit

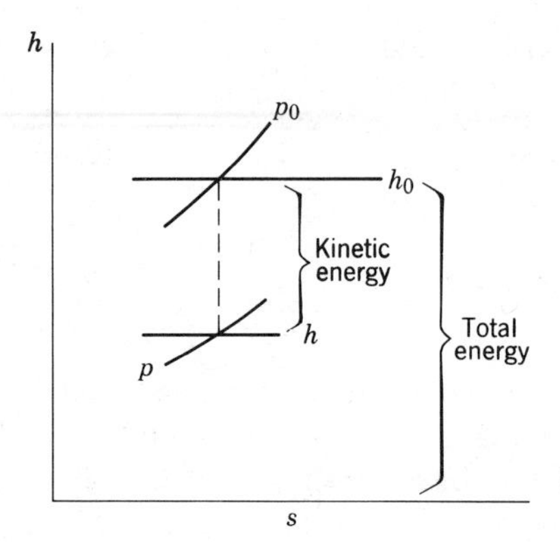

Fig. 12.3 Schematic *hs* diagram illustrating interpretation of energy per unit mass in a flow.

mass of the flow is represented by the enthalpy difference, $h_0 - h$. This is illustrated graphically in Fig. 12.3.

Equations 12.1a-f are the simplified forms of the basic equations that describe steady, one-dimensional, isentropic flow of any compressible fluid. There are six independent equations. If all the properties of the flow are known at state ①, then we have a total of seven unknowns (p_2, A_2, V_2, ρ_2, h_2, s_2, and R_x) in these six equations. Consequently, an isentropic flow can proceed to a variety of states from state ①. Each such state must have $s_2 = s_1$, satisfying Eq. 12.1d. This leaves, in effect, six unknowns and five equations and the problem is indeterminate. In order to determine conditions at state ②, one of the six unknowns must be specified.

What causes fluid property changes in isentropic flow? You undoubtedly recognize that it is the effect of area variation. In the next section we analyze the effect of area variation on flow properties in isentropic flow.

12-2 EFFECT OF AREA VARIATION ON PROPERTIES IN ISENTROPIC FLOW

In considering the effect of area variation on flow properties in isentropic flow, we shall concern ourselves primarily with velocity and pressure. We wish to determine the effect of a change in the area, A, on the velocity, V, and the pressure, p; i.e. for a change, dA, in the area, A, are dV and dp positive or negative?

To answer these questions, it is convenient to work with the differential forms of the governing equations. These were derived for the differential control volume of Fig. 11.3 (Section 11-3.1). The differential momentum equation for isentropic flow reduces to

$$\frac{dp}{\rho} + d\left(\frac{V^2}{2}\right) = 0 \tag{11.16c}$$

or

$$dp = -\rho V\,dV$$

Dividing by ρV^2, we obtain

$$\frac{dp}{\rho V^2} = -\frac{dV}{V} \tag{12.2}$$

A convenient differential form of the continuity equation can be obtained from Eq. 12.1a

$$\rho A V = \text{constant}$$

Taking the natural logarithm of both sides, yields

$$\ln \rho + \ln A + \ln V = \ln C$$

Differentiating,

$$\frac{d\rho}{\rho} + \frac{dA}{A} + \frac{dV}{V} = 0 \tag{12.3}$$

Solving Eq. 12.3 for dA/A gives

$$\frac{dA}{A} = -\frac{dV}{V} - \frac{d\rho}{\rho}$$

Substituting from Eq. 12.2,

$$\frac{dA}{A} = \frac{dp}{\rho V^2} - \frac{d\rho}{\rho}$$

or

$$\frac{dA}{A} = \frac{dp}{\rho V^2}\left[1 - \frac{V^2}{dp/d\rho}\right]$$

Now recall that for an isentropic process $dp/d\rho = \partial p/\partial \rho)_s = c^2$, so

$$\frac{dA}{A} = \frac{dp}{\rho V^2}\left[1 - \frac{V^2}{c^2}\right] = \frac{dp}{\rho V^2}[1 - M^2] \tag{12.4}$$

From Eq. 12.4 we see that for $M < 1$ an area change causes a pressure change of the same sign (positive dA means positive dp for $M < 1$); for $M > 1$ an area change causes a pressure change of opposite sign.

Substituting from Eq. 12.2 into Eq. 12.4, we obtain

$$\frac{dA}{A} = \frac{-dV}{V}[1 - M^2] \tag{12.5}$$

From Eq. 12.5 we see that for $M < 1$ an area change causes a velocity change of opposite sign (positive dA means negative dV for $M < 1$); for $M > 1$ an area change causes a velocity change of the same sign.

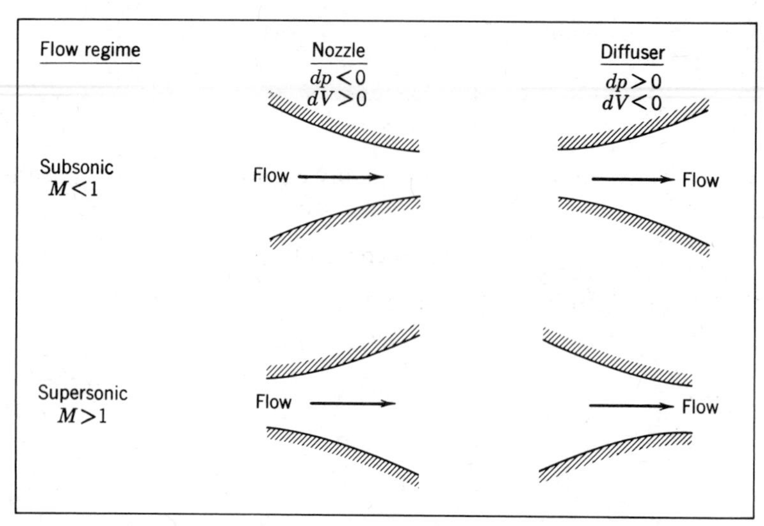

Fig. 12.4 Nozzle and diffuser shapes as a function of initial Mach number.

These results are summarized in Fig. 12.4. For subsonic flows ($M < 1$) flow acceleration in a *nozzle* requires a passage of diminishing cross section; the area must decrease to cause a velocity increase. This produces a passage shaped like that shown in the upper left of Fig. 12.4, and this result is in accord with our experience. A subsonic *diffuser* requires that the passage area increase to cause a velocity decrease. Again this result agrees with our experience.

In supersonic flows ($M > 1$), the effects of area change are different. According to Eq. 12.5, a *supersonic nozzle* must be built with an area increase in the flow direction. A *supersonic diffuser* must be a converging channel. Although these predictions may be contrary to our experience, laboratory experiments show that they are valid. We also can recall seeing divergent nozzles designed to produce supersonic flow on missiles and launch vehicles.

What of the remaining case, $M = 1$? Further inspection of Eq. 12.5 shows that at $M = 1$, $dA/dV = 0$. The fact that $dA/dV = 0$ means that the passage area must pass through a minimum or maximum at $M = 1$. Inspection of Fig. 12.4 shows that $M = 1$ can be reached only in a *throat* or section of minimum area.

To accelerate flow from rest to supersonic speed ($M > 1$) requires first a subsonic *converging nozzle*. Under the proper conditions, the flow will be at $M = 1$ at the throat, where the area is a minimum. Further acceleration is possible if a supersonic *diverging nozzle* segment is added downstream from a throat. Isentropic flow in converging nozzles will be treated in Section 12-3.4, and isentropic flow in converging-diverging nozzles will be covered in Section 12-3.5.

To decelerate flow from supersonic ($M > 1$) to subsonic speed requires first a supersonic (converging) diffuser. In theory, the flow speed could be reduced isentropically to $M = 1$ at a throat where the area is a minimum, and further isentropic deceleration could take place in a diverging subsonic diffuser section. In practice, supersonic flow cannot be decelerated to exactly $M = 1$ at a throat because the sonic flow near the throat is unstable in a rising (adverse) pressure gradient. (Disturbances that always are present in a real subsonic flow propagate upstream, disturbing the sonic flow at the throat, causing shock waves to form and travel upstream, where they are disgorged from the inlet of the supersonic diffuser.)

The throat area of a real supersonic diffuser must be slightly larger than that required to reduce the flow to $M = 1$. Under the proper downstream conditions, a weak normal shock forms in the diverging channel just downstream from the throat. Flow leaving the shock is subsonic and decelerates in the diverging channel. Thus deceleration from supersonic to subsonic flow cannot occur isentropically in practice, since the weak normal shock causes an entropy increase. Normal shocks will be analyzed in Section 12.6.

For accelerating flows (favorable pressure gradients) the idealization of isentropic flow is generally a realistic model of the actual flow behavior. For decelerating flows, the idealization of isentropic flow may not be a realistic flow model because of the adverse pressure gradients and the attendant possibility of flow separation, as discussed for incompressible flow in Chapter 9.

12-3 ISENTROPIC FLOW OF AN IDEAL GAS

12-3.1 Basic Equations

In Section 12-1 we applied the basic equations to a finite control volume for the steady, one-dimensional, isentropic flow of any compressible fluid. The only modification required to restrict our discussion to an ideal gas is in the equation of state. For an ideal gas the equation of state is $p = \rho RT$. In addition, for the isentropic flow of an ideal gas we have the process equation, $p/\rho^k = \text{constant}$. Then, for the isentropic flow of an ideal gas, we can summarize the basic equations as follows:

Continuity:
$$\rho_1 V_1 A_1 = \rho_2 V_2 A_2 = \rho VA = \dot{m} \tag{12.1a}$$

Momentum:
$$R_x + p_1 A_1 - p_2 A_2 = \dot{m} V_2 - \dot{m} V_1 \tag{12.1b}$$

First law:
$$h_1 + \frac{V_1^2}{2} = h_2 + \frac{V_2^2}{2} = h + \frac{V^2}{2} \tag{12.1c}$$

Second law:
$$s_1 = s_2 = s \tag{12.1d}$$

Equation of state:
$$p = \rho RT \tag{11.1}$$

Process equation:
$$p/\rho^k = \text{constant} \tag{11.11b}$$

These are the governing equations for steady, one-dimensional, isentropic flow of an ideal gas. If all the properties at state ① are known, then we have eight unknowns (ρ_2, A_2, V_2, p_2, h_2, s_2, T_2, and R_x) in these six equations. However, we have the known relationship between h and T for an ideal gas, $dh = c_p dT$. For an ideal gas with constant specific heats,

$$\Delta h = h_2 - h_1 = c_p \Delta T = c_p(T_2 - T_1) \tag{11.7b}$$

Thus, as in the general case (Section 12-1), the problem is indeterminate. One condition (other than s_2) must be specified at state ② before conditions at state ② can be completely determined.

12-3.2 Reference Conditions for Isentropic Flow of an Ideal Gas

Expressions for local isentropic stagnation properties for an ideal gas were developed in Chapter 11 (Section 11-3.1). For completeness these expressions are repeated here.

Stagnation pressure: $$\frac{p_0}{p} = \left[1 + \frac{k-1}{2} M^2\right]^{k/(k-1)} \tag{11.17a}$$

Stagnation temperature: $$\frac{T_0}{T} = 1 + \frac{k-1}{2} M^2 \tag{11.17b}$$

Stagnation density: $$\frac{\rho_0}{\rho} = \left[1 + \frac{k-1}{2} M^2\right]^{1/(k-1)} \tag{11.17c}$$

As shown in Section 12-1, the stagnation properties are constant throughout a steady, isentropic flow field.

The critical conditions—the values of the flow properties at which the Mach number is unity—were introduced in Section 11-3.2. Since the stagnation properties are constant in an isentropic flow, then from Eqs. 11.17 we can write, for $k = 1.4$,

$$\frac{p_0}{p^*} = \left[1 + \frac{k-1}{2}\right]^{k/(k-1)} = 1.893; \qquad \frac{p^*}{p_0} = 0.5283$$

$$\frac{T_0}{T^*} = 1 + \frac{k-1}{2} = 1.200; \qquad \frac{T^*}{T_0} = 0.8333$$

$$\frac{\rho_0}{\rho^*} = \left[1 + \frac{k-1}{2}\right]^{1/(k-1)} = 1.577$$

In addition, from Eq. 11.19, we have

$$V^* = c^* = \sqrt{\frac{2k}{k+1} RT_0}$$

In Section 12-2 we saw that it was necessary for a passage to have a section of minimum area (a throat) to accelerate a flow isentropically from rest to a Mach number greater than unity. Furthermore, in such a flow the Mach number is unity at the throat. If the area at which the Mach number is unity is designated by A^*, then it is possible to express the contour of a passage in terms of the area ratio, A/A^*.

Since for steady, one-dimensional flow, the continuity equation can be written

$$\rho AV = \text{constant} = \rho^* A^* V^*$$

then

$$\frac{A}{A^*} = \frac{\rho^*}{\rho}\frac{V^*}{V} = \frac{\rho^*}{\rho}\frac{c^*}{Mc} = \frac{1}{M}\frac{\rho^*}{\rho}\sqrt{\frac{T^*}{T}}$$

$$\frac{A}{A^*} = \frac{1}{M}\frac{\rho^*}{\rho_0}\frac{\rho_0}{\rho}\sqrt{\frac{T^*/T_0}{T/T_0}}$$

$$\frac{A}{A^*} = \frac{1}{M}\frac{\left[1+\frac{k-1}{2}M^2\right]^{1/(k-1)}}{\left[1+\frac{k-1}{2}\right]^{1/(k-1)}}\left[\frac{1+\frac{k-1}{2}M^2}{1+\frac{k-1}{2}}\right]^{1/2}$$

$$\frac{A}{A^*} = \frac{1}{M}\left[\frac{1+\frac{k-1}{2}M^2}{1+\frac{k-1}{2}}\right]^{(k+1)/2(k-1)} \tag{12.6}$$

From Eq. 12.6 we see that a choice of M gives a unique value of A/A^*. The variation of A/A^* with M is shown in Fig. 12.5. Note that the curve is double-valued; for a given value of A/A^* other than unity, there are two possible values of Mach number. This is consistent with the results of Section 12-2 (see Fig. 12.4), where it was found that a converging-diverging passage with a section of minimum area is required to accelerate a flow from subsonic to supersonic speed.

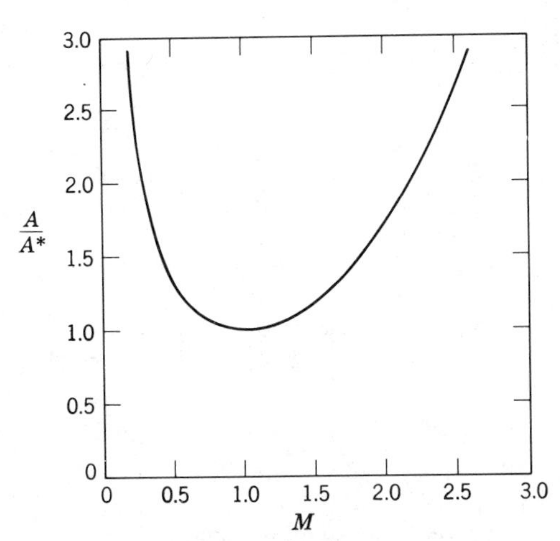

Fig. 12.5 Variation of A/A^* with Mach number in isentropic flow for $k = 1.4$.

Example 12.1

Air flows isentropically in a channel. At section ①, the Mach number is 0.3, the area is 0.001 m², and the absolute pressure and the temperature are 650 kPa and 62 C, respectively. At section ②, the Mach number is 0.8. Sketch the channel shape, plot a Ts diagram for the process, and evaluate the properties at section ②.

EXAMPLE PROBLEM 12.1

GIVEN: Isentropic flow of air in a channel. At sections ① and ②, the following data are given: $M_1 = 0.3$, $T_1 = 62$ C, $p_1 = 650$ kPa (abs), $A_1 = 0.001$ m^2, and $M_2 = 0.8$.

FIND:
(a) Sketch the channel shape.
(b) Plot a Ts diagram for the process.
(c) Properties at section ②.

SOLUTION:
To accelerate a subsonic flow requires a converging nozzle. The channel shape must be as shown.

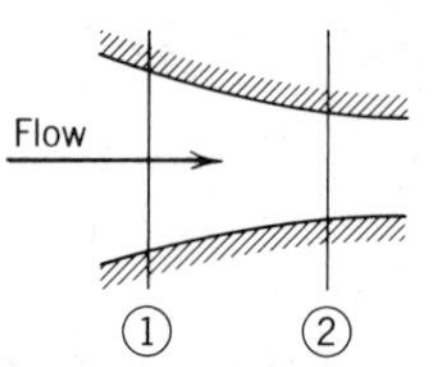

On the Ts plane, the process follows a line, s = constant. Stagnation conditions remain fixed for isentropic flow. Consequently, the stagnation temperature at section ② can be calculated (for air, $k = 1.4$) from

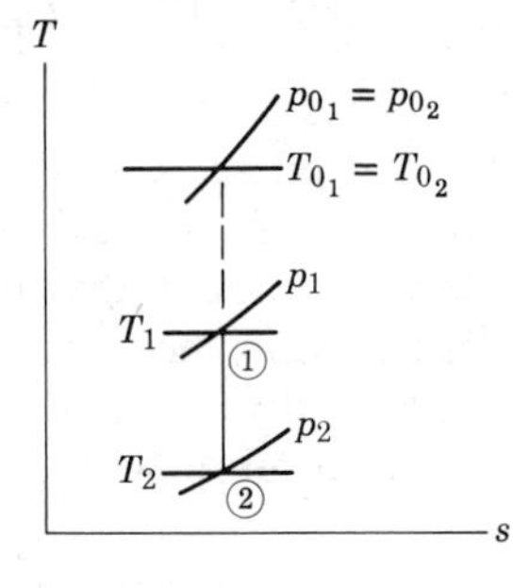

$$T_{0_2} = T_{0_1} = T_1\left[1 + \frac{k-1}{2} M_1^2\right]$$

$$= (62 + 273)\text{ K}\left[1 + \frac{1.4 - 1}{2}(0.3)^2\right]$$

$$T_{0_2} = T_{0_1} = 341 \text{ K} \qquad T_{0_1}, T_{0_2}$$

and

$$T_2 = \frac{T_{0_2}}{\left[1 + \frac{k-1}{2} M_2^2\right]} = \frac{341 \text{ K}}{[1 + 0.2(0.8)^2]} = 302 \text{ K or } 29 \text{ C} \qquad T_2$$

Also, for an ideal gas,

$$c_2 = \sqrt{kRT_2} = \left[1.4 \times 287 \frac{\text{N}\cdot\text{m}}{\text{kg}\cdot\text{K}} \times 302 \text{ K} \times \frac{\text{kg}\cdot\text{m}}{\text{N}\cdot\text{sec}^2}\right]^{1/2} = 348 \text{ m/sec} \qquad c_2$$

From the definition of Mach number,

$$V_2 = M_2 c_2 = (0.8)348 \text{ m/sec} = 278 \text{ m/sec} \qquad V_2$$

Using the isentropic relation, p/ρ^k = constant, we obtain

$$\frac{p_2}{p_1} = \left(\frac{T_2}{T_1}\right)^{k/(k-1)} = \left(\frac{302 \text{ K}}{335 \text{ K}}\right)^{3.5} = 0.696$$

and

$$p_2 = 0.696 p_1 = (0.696)650 \text{ kPa (abs)} = 452 \text{ kPa (abs)} \qquad p_2$$

Also,

$$\rho_2 = \frac{p_2}{RT_2} = \frac{4.52 \times 10^5\ \text{N}}{\text{m}^2} \times \frac{\text{kg} \cdot \text{K}}{287\ \text{N} \cdot \text{m}} \times \frac{1}{302\ \text{K}} = 5.21\ \text{kg/m}^3$$

ρ_2

From the continuity equation

$$\dot{m} = \rho_1 V_1 A_1 = \rho_2 V_2 A_2 = \text{constant}$$

so that

$$A_2 = A_1 \frac{\rho_1}{\rho_2} \frac{V_1}{V_2} = A_1 \left(\frac{T_1}{T_2}\right)^{1/(k-1)} \frac{M_1 c_1}{M_2 c_2} = A_1 \left(\frac{T_1}{T_2}\right)^{1/(k-1)} \frac{M_1}{M_2} \sqrt{\frac{kRT_1}{kRT_2}}$$

or

$$A_2 = A_1 \frac{M_1}{M_2} \left(\frac{T_1}{T_2}\right)^{(k+1)/2(k-1)} = 0.001\ \text{m}^2 \times \frac{0.3}{0.8} \left(\frac{335}{302}\right)^3 = 5.12 \times 10^{-4}\ \text{m}^2$$

A_2

Thus, $A_2 < A_1$, as expected. Finally,

$$p_{0_2} = p_2 \left[1 + \frac{k-1}{2} M_2^2\right]^{k/(k-1)} = 452\ \text{kPa}[1 + 0.2(0.8)^2]^{3.5}$$

$$p_{0_2} = 689\ \text{kPa (abs)}$$

p_{0_2}

The stagnation pressure should be constant for isentropic flow. Checking gives

$$p_{0_1} = p_1 \left[1 + \frac{k-1}{2} M_1^2\right]^{k/(k-1)} = 650\ \text{kPa}[1 + 0.2(0.3)^2]^{3.5}$$

$$p_{0_1} = 692\ \text{kPa (abs)}$$

{The discrepancy between the calculated values of p_{0_1} and p_{0_2} is due to use of temperature values rounded to three significant figures in the calculation of p_{0_2}. Thus we note again for isentropic flow that $T_{0_1} = T_{0_2}$ and $p_{0_1} = p_{0_2}$.}

**12-3.3 Tables for Computation of Isentropic Flow of an Ideal Gas

In the previous section we saw (Eqs. 11.17a, 11.17b, 11.17c, and 12.6) that the properties at a point in a compressible flow of an ideal gas may be related to appropriate reference conditions by functions of the local Mach number. This makes it possible to tabulate or plot them as functions of Mach number, M, for a given value of k.

Table D.1 of Appendix D lists values of T/T_0, p/p_0, ρ/ρ_0, and A/A^* as functions of M for isentropic flow of an ideal gas with $k = 1.4$.[1] The use of tables can reduce the labor of calculations significantly.

Since the reference conditions remain constant in isentropic flow, the ratio of properties at two points in a flow may readily be found from the tables. Use of the tables is illustrated in Example Problem 12.2.

** This section may be omitted without loss of continuity in the text material.

[1] Tables for other common values of k have been published. See, e.g. J. H. Keenan and J. Kaye, *Gas Tables*. New York: Wiley, 1948.

Example 12.2

Air flows isentropically in a channel. At section ①, the Mach number is 0.3, the area is $0.001\ \text{m}^2$, and the absolute pressure and the temperature are 650 kPa and 62 C, respectively. Evaluate the properties at section ②, where the Mach number is 0.8, using the isentropic flow tables. (Note that these are the data of Example 12.1.)

EXAMPLE PROBLEM 12.2

GIVEN: Isentropic flow of air in a channel. At sections ① and ②, the following data are given: $M_1 = 0.3$, $T_1 = 62$ C, $p_1 = 650$ kPa (abs), $A_1 = 0.001\ \text{m}^2$, and $M_2 = 0.8$.

FIND: Properties at section ②, using isentropic flow tables.

SOLUTION:

The *Ts* diagram was sketched in Example Problem 12.1. From Table D.1, Appendix D, we find

M	T/T_0	p/p_0	ρ/ρ_0	A/A^*
0.3	0.9823	0.9395	0.9564	2.035
0.8	0.8865	0.6560	0.7400	1.038

For isentropic flow, $T_{0_1} = T_{0_2} = T_0$. Thus

$$\frac{T_2}{T_1} = \frac{T_2}{T_0}\frac{T_0}{T_1} = \frac{(T/T_0)_2}{(T/T_0)_1} = \frac{0.8865}{0.9823} = 0.9025$$

$$T_2 = 0.9025 T_1 = 0.9025(273 + 62)\ \text{K} = 302\ \text{K}$$ ← T_2

Also $p_{0_2} = p_{0_1} = p_0$, so

$$\frac{p_2}{p_1} = \frac{p_2}{p_0}\frac{p_0}{p_1} = \frac{(p/p_0)_2}{(p/p_0)_1} = \frac{0.6560}{0.9395} = 0.6982$$

$$p_2 = 0.6982 p_1 = 0.6982(650\ \text{kPa}) = 454\ \text{kPa (abs)}$$ ← p_2

and

$$\rho_2 = \frac{p_2}{RT_2} = \frac{4.54 \times 10^5\ \text{N}}{\text{m}^2} \times \frac{\text{kg}\cdot\text{K}}{287\ \text{N}\cdot\text{m}} \times \frac{1}{302\ \text{K}} = 5.24\ \text{kg/m}^3$$ ← ρ_2

The stagnation properties are

$$T_{0_2} = T_{0_1} = \frac{T_1}{(T/T_0)_1} = \frac{(273 + 62)\ \text{K}}{0.9823} = 341\ \text{K}$$ ← T_{0_2}

and

$$p_{0_2} = p_{0_1} = \frac{p_1}{(p/p_0)_1} = \frac{650\ \text{kPa}}{0.9395} = 692\ \text{kPa (abs)}$$ ← p_{0_2}

The area may be computed using the ratio, A/A^*. Thus, since $A^* =$ constant,

$$\frac{A_2}{A_1} = \frac{A_2}{A^*}\frac{A^*}{A_1} = \frac{(A/A^*)_2}{(A/A^*)_1} = \frac{1.038}{2.035} = 0.5101$$

$$A_2 = 0.5101A_1 = 0.5101\,(0.001\ \text{m}^2) = 5.10 \times 10^{-4}\ \text{m}^2$$

A_2

The velocity at section ② may be calculated from $V_2 = M_2c_2$.

12-3.4 Isentropic Flow in a Converging Nozzle

In this section we investigate the operation of a converging nozzle under various back pressures. Flow through the converging nozzle shown in Fig. 12.6 is supplied from a large plenum chamber, where conditions are assumed to be stagnation conditions; the flow is induced by a vacuum pump downstream and is controlled by the valve shown.

The back pressure, p_b, to which the nozzle discharges, is controlled by the valve. The upstream stagnation conditions (T_0, p_0, etc.) are maintained constant. The pressure in the exit plane of the nozzle is denoted as p_e. We wish to investigate the effect of variations in back pressure on the pressure distribution through the nozzle, on the mass flow rate, and on the exit plane pressure. The results are illustrated graphically in Fig. 12.6. Let us look at each of the cases shown.

When the valve is closed, there is no flow through the nozzle. The pressure is p_0 throughout, as shown by condition (*i*) in Fig. 12.6*a*.

If the back pressure, p_b, is now reduced to a value slightly less than p_0, there will be flow through the nozzle with a decrease in pressure in the direction of flow as shown by condition (*ii*). Flow at the exit plane will be subsonic with the exit plane pressure equal to the back pressure.

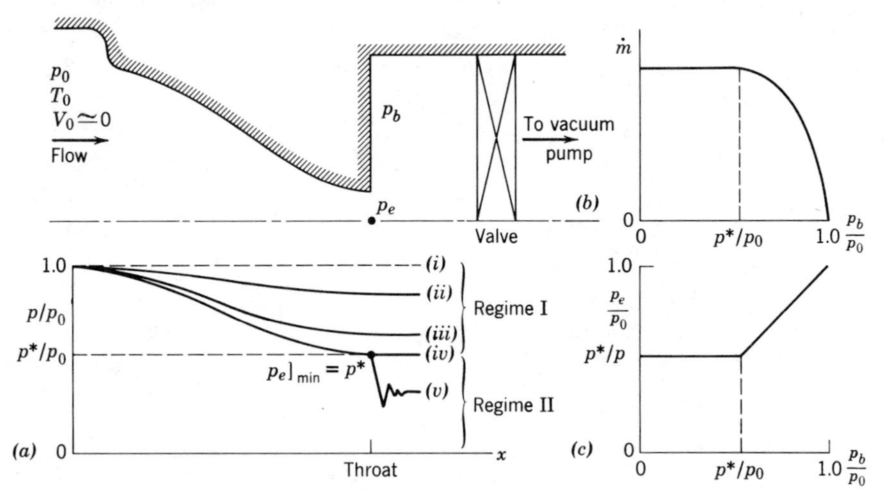

Fig. 12.6 Converging nozzle operating at various back pressures.

What happens as we continue to decrease the back pressure? The flow rate will continue to increase and the exit plane pressure will continue to decrease as shown by condition (*iii*) in Fig. 12.6. Will these trends continue indefinitely as the back pressure is lowered?

Recall from our previous discussion that in a converging section the Mach number cannot increase beyond unity in isentropic flow. Thus, with continued decrease of the back pressure, the flow at the exit plane of the nozzle eventually will reach a Mach number of unity. The corresponding pressure is the critical pressure, p^*. Condition (*iv*) illustrates the condition where M_e equals unity and p_b/p_0 equals p^*/p_0.

From Eq. 11.17a, with $M = 1$, the critical pressure ratio for an ideal gas is given by

$$\frac{p^*}{p_0} = \left(\frac{2}{k+1}\right)^{k/(k-1)}$$

For $k = 1.4$, $p^*/p_0 = 0.528$.

What happens when the back pressure is reduced further to a value below p^*, such as condition (*v*)? Since the Mach number at the throat is unity ($V_e = c_e$), information about conditions in the exhaust duct cannot be transmitted upstream. Consequently, reductions in p_b below p^* have no effect on flow conditions in the nozzle; thus neither the pressure distribution through the nozzle, the nozzle exit pressure, nor the mass flow rate are affected by lowering p_b below p^*. When p_b is less than or equal to p^*, the nozzle is said to be *choked*.

For p_b less than p^*, the flow leaving the nozzle will expand to match the lower back pressure as shown for condition (*v*) in Fig. 12.6*a*. This unconfined expansion process is three-dimensional; the pressure distribution cannot be predicted by one-dimensional theory. Experiments show that a series of shocks form in the exit stream, resulting in an increase in entropy.

Flow through a converging nozzle may be divided into two regimes:

1. In Regime I, $1 \geq p_b/p_0 \geq p^*/p_0$. The flow to the throat is isentropic; $p_e = p_b$.
2. In Regime II, $p_b/p_0 < p^*/p_0$. The flow to the throat is isentropic, but a nonisentropic expansion occurs in the flow leaving the nozzle; $p_e = p^* > p_b$.

The flow processes corresponding to Regime II are shown on a *Ts* diagram in Fig. 12.7.

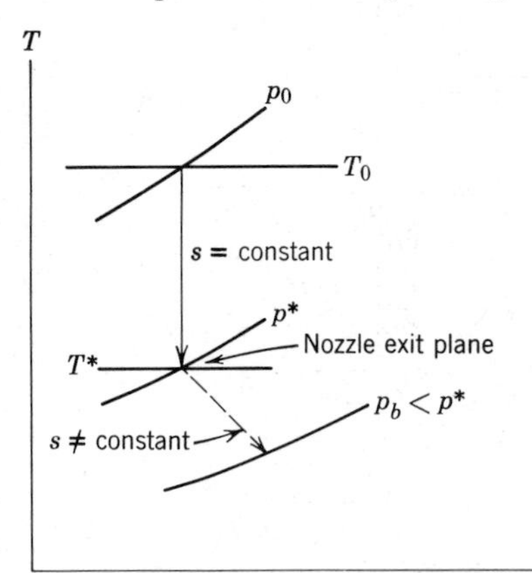

Fig. 12.7 Schematic *Ts* diagram for choked flow through a converging nozzle.

Two problems involving converging nozzles are solved in Example Problems 12.3 and 12.4.

Although isentropic flow is an idealization, it often is a very good approximation for the actual behavior of nozzles. Since a nozzle is a device that accelerates a flow, the internal pressure gradient is favorable. This tends to keep the wall boundary layers thin and to minimize the effects of friction.

Example 12.3

A converging nozzle with a throat area of 0.001 m^2 is operated with air at a back pressure of 591 kPa (abs). The nozzle is fed from a large plenum chamber where the absolute stagnation pressure and the temperature are 1.0 MPa and 60 C, respectively. The exit Mach number and mass flow rate are to be determined using (i) isentropic flow relations, and (ii) tables for isentropic flow.

EXAMPLE PROBLEM 12.3

GIVEN: Air flow through a converging nozzle at the conditions shown: Flow is isentropic.

FIND:
(a) M_e. (b) $\dot{m}$.

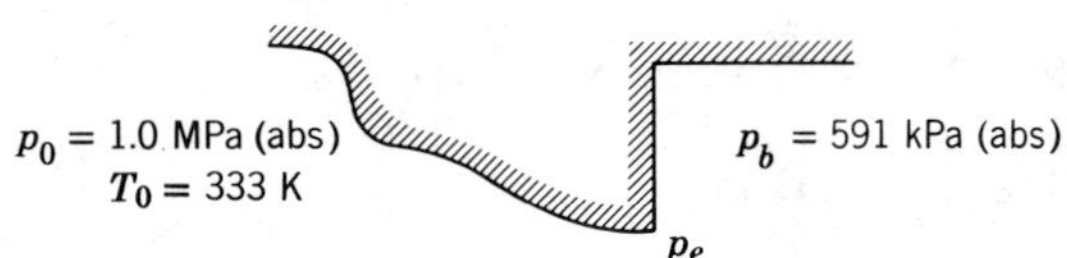

SOLUTION:
The first step is to check for choking. The pressure ratio is

$$\frac{p_b}{p_0} = \frac{5.91 \times 10^5}{1.0 \times 10^6} = 0.591 > 0.528$$

so the flow is not choked. Thus $p_b = p_e$, and the flow is isentropic, as sketched on the Ts diagram.

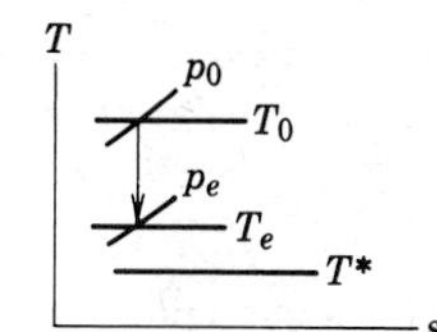

(*i*) *Isentropic Flow Relations*

Since p_0 = constant, the Mach number, M_e, may be found from the pressure ratio,

$$\frac{p_0}{p_e} = \left[1 + \frac{k-1}{2} M_e^2\right]^{k/(k-1)}$$

Solving for M_e, since $p_e = p_b$, we obtain

$$1 + \frac{k-1}{2} M_e^2 = \left(\frac{p_0}{p_b}\right)^{(k-1)/k}$$

and

$$M_e = \left\{\left[\left(\frac{p_0}{p_b}\right)^{(k-1)/k} - 1\right]\frac{2}{k-1}\right\}^{1/2} = \left\{\left[\left(\frac{1.0 \times 10^6}{5.91 \times 10^5}\right)^{0.286} - 1\right]\frac{2}{1.4-1}\right\}^{1/2} = 0.9 \qquad M_e$$

The mass flow rate will be given by

$$\dot{m} = \rho_e V_e A_e = \rho_e M_e c_e A_e$$

Thus we need the temperature, T_e, to find ρ_e and c_e. Since $T_0 = \text{constant}$,

$$\frac{T_0}{T_e} = 1 + \frac{k-1}{2} M_e^2$$

or

$$T_e = \frac{T_0}{1 + \frac{k-1}{2} M_e^2} = \frac{(273 + 60)\ \text{K}}{1 + 0.2(0.9)^2} = 287\ \text{K}$$

$$c_e = \sqrt{kRT_e} = \left[1.4 \times \frac{287\ \text{N}\cdot\text{m}}{\text{kg}\cdot\text{K}} \times 287\ \text{K} \times \frac{\text{kg}\cdot\text{m}}{\text{N}\cdot\text{sec}^2}\right]^{1/2} = 340\ \text{m/sec}$$

and

$$\rho_e = \frac{p_e}{RT_e} = \frac{5.91 \times 10^5\ \text{N}}{\text{m}^2} \times \frac{\text{kg}\cdot\text{K}}{287\ \text{N}\cdot\text{m}} \times \frac{1}{287\ \text{K}} = 7.18\ \text{kg/m}^3$$

Finally,

$$\dot{m} = \rho_e M_e c_e A_e = \frac{7.18\ \text{kg}}{\text{m}^3} \times 0.9 \times \frac{340\ \text{m}}{\text{sec}} \times 0.001\ \text{m}^2 = 2.20\ \text{kg/sec}$$ $\dot{m}$

(*ii*) ***Tables for Isentropic Flow*

The pressure ratio is

$$\frac{p_b}{p_0} = \frac{5.91 \times 10^5}{1.0 \times 10^6} = 0.591 > 0.528$$

so the flow is not choked. Thus $p_e = p_b$, and the flow is isentropic. From Table D.1, Appendix D, $p/p_0 = 0.591$ at $M = 0.90$. Thus

$$M_e = 0.90$$ M_e

The simplest procedure in using the tables is to express the desired result in terms of property ratios, which then may be found from the tables. Thus

$$\dot{m} = \rho_e V_e A_e = \rho_e M_e c_e A_e = \rho_e M_e \sqrt{kRT_e} A_e$$

$$\dot{m} = \rho_0 \frac{\rho_e}{\rho_0} M_e \sqrt{kRT_0} \sqrt{\frac{T_e}{T_0}} A_e = \frac{p_0}{RT_0} \frac{\rho_e}{\rho_0} M_e \sqrt{kRT_0} \sqrt{\frac{T_e}{T_0}} A_e$$

or finally,

$$\dot{m} = \frac{\rho_e}{\rho_0} \sqrt{\frac{T_e}{T_0}}\, p_0 M_e \sqrt{\frac{k}{RT_0}}\, A_e$$

Using values for $M_e = 0.9$ gives

$$\dot{m} = (0.6870)(0.8606)^{1/2} \left(1.0 \times 10^6\ \frac{\text{N}}{\text{m}^2}\right)(0.9)$$

$$\times \left[1.4 \times \frac{\text{kg}\cdot\text{K}}{287\ \text{N}\cdot\text{m}} \times \frac{1}{333\ \text{K}} \times \frac{\text{kg}\cdot\text{m}}{\text{N}\cdot\text{sec}^2}\right]^{1/2} (0.001\ \text{m}^2)$$

$$\dot{m} = 2.20\ \text{kg/sec}$$ $\dot{m}$

Example 12.4

Air flows isentropically through a converging nozzle. At a section where the nozzle area is 0.013 ft^2, the local pressure, temperature, and Mach number are 60 psia, 40 F, and 0.52, respectively. The back pressure is 30 psia. The Mach number at the throat, the mass flow rate, and the throat area are to be determined, using (*i*) isentropic flow relations and (*ii*) tables for isentropic flow.

EXAMPLE PROBLEM 12.4

GIVEN: Air flow through a converging nozzle at the conditions shown:

$M_1 = 0.52$

$T_1 = 40$ F

$p_1 = 60$ psia

$A_1 = 0.013$ ft^2

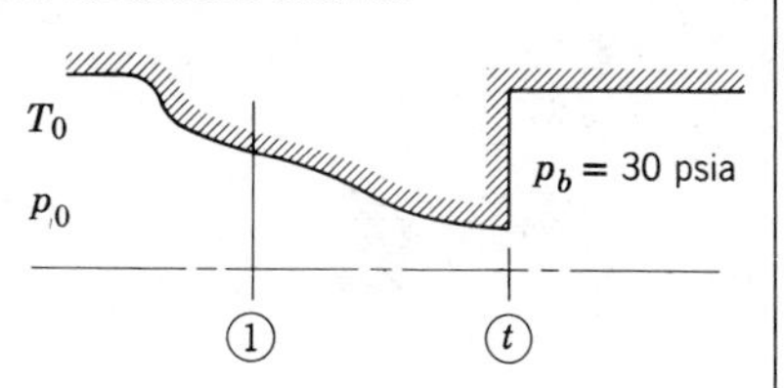

FIND: (a) M_t. (b) $\dot{m}$. (c) A_t.

SOLUTION:

(*i*) *Isentropic Flow Relations*

We must first check for choking, to determine if flow is isentropic down to p_b. To check, we evaluate the stagnation conditons.

$$\frac{p_0}{p_1} = \left[1 + \frac{k-1}{2} M_1^2\right]^{k/(k-1)} = [1 + 0.2(0.52)^2]^{3.5} = 1.20$$

and

$$p_0 = 1.20 p_1 = (1.2)60 \text{ psia} = 72.0 \text{ psia}$$

The back pressure ratio is

$$\frac{p_b}{p_0} = \frac{30.0}{72.0} = 0.417 < 0.528$$

so the flow is choked! For choked flow,

$$M_t = 1.0$$

M_t

The Ts diagram is

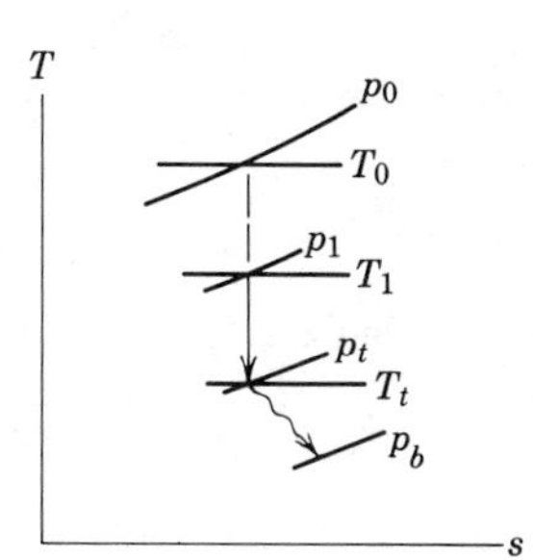

The mass flow rate may be found from conditions at section ①, using $\dot{m} = \rho_1 V_1 A_1$.

$$V_1 = M_1 c_1 = M_1 \sqrt{kRT_1}$$

$$= 0.52 \left[1.4 \times \frac{53.3 \text{ ft} \cdot \text{lbf}}{\text{lbm} \cdot \text{R}} \times (460 + 40) \text{ R} \times \frac{32.2 \text{ lbm}}{\text{slug}} \times \frac{\text{slug} \cdot \text{ft}}{\text{lbf} \cdot \text{sec}^2} \right]^{1/2}$$

$$V_1 = 570 \text{ ft/sec}$$

$$\rho_1 = \frac{p_1}{RT_1} = \frac{60 \text{ lbf}}{\text{in.}^2} \times \frac{\text{lbm} \cdot \text{R}}{53.3 \text{ ft} \cdot \text{lbf}} \times \frac{1}{500 \text{ R}} \times \frac{144 \text{ in.}^2}{\text{ft}^2} = 0.324 \text{ lbm/ft}^3$$

$$\dot{m} = \rho_1 V_1 A_1 = \frac{0.324 \text{ lbm}}{\text{ft}^3} \times \frac{570 \text{ ft}}{\text{sec}} \times 0.013 \text{ ft}^2 = 2.40 \text{ lbm/sec}$$

$\dot{m}$

The throat area may be computed by applying continuity between section ① and the throat; i.e. $\dot{m} = \rho_1 V_1 A_1 = \rho_t V_t A_t$, so

$$A_t = A_1 \frac{\rho_1}{\rho_t} \frac{V_1}{V_t} = A_1 \frac{p_1}{RT_1} \frac{RT_t}{p_t} \frac{M_1 \sqrt{kRT_1}}{M_t \sqrt{kRT_t}} = A_1 \frac{p_1}{p_t} \frac{M_1}{M_t} \sqrt{\frac{T_t}{T_1}}$$

For isentropic flow, $T_0 =$ constant, so

$$\frac{T_t}{T_1} = \frac{T_t}{T_0} \frac{T_0}{T_1} = \frac{1 + \frac{k-1}{2} M_1^2}{1 + \frac{k-1}{2} M_t^2} = \frac{1 + (0.2)(0.52)^2}{1.2} = 0.878$$

Also,

$$\frac{p_0}{p_t} = \left[1 + \frac{k-1}{2} M_t^2 \right]^{k/(k-1)} = (1.2)^{3.5} = 1.89$$

so that

$$p_t = \frac{p_0}{1.89} = \frac{72.0 \text{ psia}}{1.89} = 38.1 \text{ psia}$$

Substituting gives

$$A_t = A_1 \frac{p_1}{p_t} \frac{M_1}{M_t} \sqrt{\frac{T_t}{T_1}} = 0.013 \text{ ft}^2 \times \frac{60.0 \text{ psia}}{38.1 \text{ psia}} \times \frac{0.52}{1.0} \sqrt{0.878} = 9.98 \times 10^{-3} \text{ ft}^2$$

A_t

(*ii*) ***Tables for Isentropic Flow*

We must first check for choking. From Table D.1, Appendix D, at $M_1 = 0.52$

$$\frac{p_1}{p_0} = 0.8317; \qquad p_0 = \frac{p_1}{0.8317} = \frac{60 \text{ psia}}{0.8317} = 72.1 \text{ psia}$$

From the table at $M = 1.0$, the minimum isentropic pressure ratio in a converging nozzle is

$$\frac{p}{p_0} = 0.5283$$

From the conditions given

$$\frac{p_b}{p_0} = \frac{30}{72.1} = 0.416 < 0.5283$$

so the flow is choked! For choked flow,

$$M_t = 1.0 \qquad\qquad M_t$$

The Ts diagram was given above. The mass flow rate calculation is the same as in the previous solution. From Table D.1, at $M_1 = 0.52$, $A_1/A^* = 1.303$. For choked flow, $A_t = A^*$. Thus

$$A_t = A^* = \frac{A_1}{1.303} = \frac{0.013 \text{ ft}^2}{1.303} = 9.98 \times 10^{-3} \text{ ft}^2 \qquad\qquad A_t$$

12-3.5 Isentropic Flow in a Converging-Diverging Nozzle

Having considered isentropic flow in a converging nozzle, we turn now to isentropic flow in a converging-diverging (C-D) nozzle. As in the previous case, the flow through the converging-diverging passage of Fig. 12.8 is induced by a vacuum pump downstream, and is controlled by the valve shown. The upstream stagnation conditions are assumed constant. The pressure in the exit plane of the nozzle is denoted as p_e; the nozzle discharges to back pressure, p_b. We wish to investigate the effect of variations in back pressure on the pressure distribution through the nozzle. The results are illustrated graphically in Fig. 12.8. Let us consider each of the cases shown.

With the valve initially closed, there is no flow through the nozzle; the pressure is constant at p_0. Opening the valve slightly (p_b slightly less than p_0) produces the pressure distribution curve (i). If the flow rate is low enough, at all points on this curve

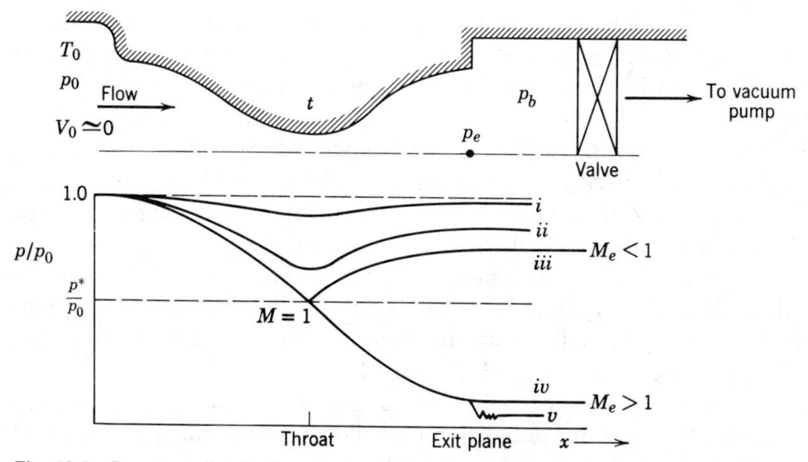

Fig. 12.8 Pressure distributions for isentropic flow in a converging-diverging nozzle.

the flow will be subsonic and essentially incompressible. Under these conditions, the C-D nozzle will behave like a venturi, with flow accelerating in the converging portion, until a point of maximum velocity and minimum pressure is reached at the throat, then decelerating in the diverging portion to the nozzle exit.

As the valve is opened farther and the flow rate is increased, a more sharply defined pressure minimum occurs, as shown by curve (*ii*). Although compressibility effects become important, the flow is still subsonic everywhere, and deceleration takes place in the diverging section. Finally, as the valve is opened farther, curve (*iii*) results. At the section of minimum area the flow finally reaches $M = 1$ and the nozzle is choked—the flow rate is the maximum possible for the given nozzle and stagnation conditions.

All of the flows giving rise to pressure distributions (*i*), (*ii*), and (*iii*) are isentropic; each curve is associated with a unique rate of mass flow. Finally, when curve (*iii*) is reached, critical conditions are present at the throat. For this flow rate, the flow is choked, and

$$\dot{m} = \rho^* V^* A^*$$

where $A^* = A_t$.

In our discussion of the effect of area variation on isentropic flow we noted that a diverging section was required to accelerate a flow to supersonic speed from $M = 1$ at a throat. At this point, then, we ask the question, "What back pressure, p_b, is necessary to accelerate the flow isentropically in the diverging portion of the nozzle?"

To accelerate flow in the diverging section requires a pressure decrease. This condition is illustrated by curve (*iv*) in Fig. 12.8. The flow will accelerate isentropically in the nozzle provided the exit pressure is set at the value p_{iv}. Thus we see that with a throat Mach number of unity, there are two possible isentropic flow conditions in the converging-diverging nozzle. This is consistent with the results of Fig. 12.5, where we found two Mach numbers for a given value of A/A^* in isentropic flow.

Lowering the back pressure below that of condition (*iv*), say to condition (*v*), has no effect on the flow in the nozzle. The flow is isentropic from the plenum chamber to the nozzle exit [same as condition (*iv*)] and then undergoes a three-dimensional irreversible expansion to the lower back pressure. A nozzle operating under these conditions is said to be *underexpanded*, since additional expansion takes place outside the nozzle.

A converging-diverging nozzle generally is intended to produce supersonic flow at the exit plane. If the back pressure is set at p_{iv}, flow will be isentropic through the nozzle, and supersonic at the nozzle exit. Nozzles operating at $p_b = p_{iv}$ [corresponding to curve (*iv*) in Fig. 12.8] are said to be at *design conditions.*

Flow leaving a C-D nozzle is supersonic when the back pressure is at or below the design value. The exit Mach number is fixed once the area ratio, A_e/A^*, is specified. All other exit plane properties (for isentropic flow) are uniquely related to the stagnation properties by the fixed exit plane Mach number.

The assumption of isentropic flow for a real nozzle at design conditions is a reasonable one. However, the one-dimensional flow model is inadequate for the design of relatively short nozzles to produce uniform supersonic exit flow.

Rocket-propelled vehicles use C-D nozzles to accelerate the exhaust gases to the maximum possible velocity to produce high thrust. A propulsion nozzle is subject to varying ambient discharge conditions during flight through the atmosphere, so it is impossible to attain the maximum theoretical thrust over the complete operating range. Because only a single supersonic Mach number can be obtained for a given area ratio, nozzles for supersonic wind tunnels often are built with interchangeable sections or variable geometry.

You undoubtedly have noticed that nothing has been said about the operation of converging-diverging nozzles with back pressure in the range $p_{iii} > p_b > p_{iv}$. For such cases the flow cannot expand isentropically to p_b. Under these conditions a shock (which may be treated as an irreversible discontinuity involving entropy increase) occurs somewhere within the flow. Following a discussion of normal shocks in Section 12-6, we shall return to complete the discussion of converging-diverging nozzle flows. Nozzles operating with $p_{iii} > p_b > p_{iv}$ are said to be *overexpanded* because the pressure at some point in the nozzle is less than the back pressure. Obviously, an overexpanded nozzle could be made to operate at a new design condition by cutting off a portion of the diverging section.

One other comment should be made at this point. Real compressible fluid flows are affected by friction, heating or cooling, and the possible presence (in supersonic flow) of shock waves. We have treated isentropic flow first because it is a useful idealized model for many real flow processes and because it gives us valuable insight into the behavior of fluids in compressible flow. In the next two sections we take up the effects of friction and heat transfer separately to gain insight into the effect of each factor on flow behavior. Following this we return to a discussion of the normal shock and complete our study of converging-diverging nozzle flows. In later courses it will be possible to explore real flows and the results of combining several of these effects.

Example 12.5

Air flows isentropically in a converging-diverging nozzle, with exit area of 0.001 m^2. The nozzle is fed from a large plenum where the stagnation temperature and the stagnation pressure are 350 K and 1.0 MPa (abs), respectively. The exit pressure is 954 kPa (abs), and the Mach number at the throat is 0.68. Flow conditions at the throat and the exit Mach number are to be determined.

EXAMPLE PROBLEM 12.5

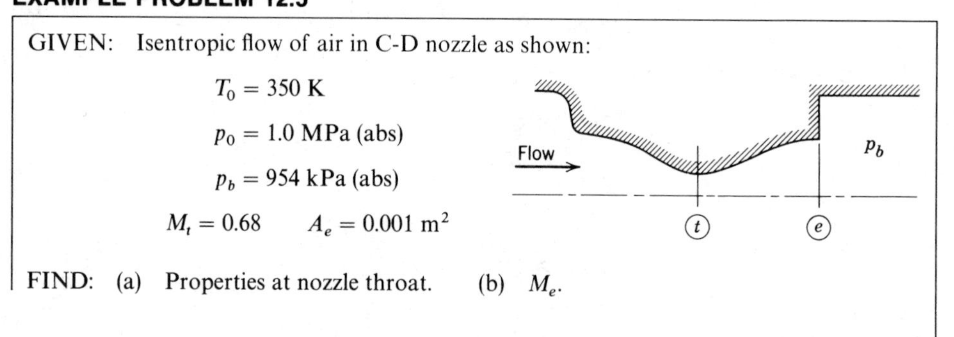

GIVEN: Isentropic flow of air in C-D nozzle as shown:

$T_0 = 350$ K

$p_0 = 1.0$ MPa (abs)

$p_b = 954$ kPa (abs)

$M_t = 0.68 \qquad A_e = 0.001\ \text{m}^2$

FIND: (a) Properties at nozzle throat. (b) M_e.

SOLUTION:

(i) Isentropic Flow Relations

The stagnation temperature is constant for isentropic flow. Thus, since

$$\frac{T_0}{T} = 1 + \frac{k-1}{2} M^2$$

$$T_t = \frac{T_0}{1 + \dfrac{k-1}{2} M_t^2} = \frac{350 \text{ K}}{1 + 0.2(0.68)^2} = 320 \text{ K} \qquad T_t$$

Also, since p_0 is constant for isentropic flow, then

$$p_t = p_0 \left(\frac{T_t}{T_0}\right)^{k/(k-1)} = p_0 \left[\frac{1}{1 + \dfrac{k-1}{2} M_t^2}\right]^{k/(k-1)}$$

$$p_t = 1.0 \times 10^6 \text{ Pa} \left[\frac{1}{1 + 0.2(0.68)^2}\right]^{3.5} = 734 \text{ kPa (abs)} \qquad p_t$$

$$\rho_t = \frac{p_t}{RT_t} = \frac{7.34 \times 10^5 \text{ N}}{\text{m}^2} \times \frac{\text{kg} \cdot \text{K}}{287 \text{ N} \cdot \text{m}} \times \frac{1}{320 \text{ K}} = 7.99 \text{ kg/m}^3 \qquad \rho_t$$

and

$$V_t = M_t c_t = M_t \sqrt{kRT_t}$$

$$V_t = 0.68 \left[\frac{1.4 \times 287 \text{ N} \cdot \text{m}}{\text{kg} \cdot \text{K}} \times 320 \text{ K} \times \frac{\text{kg} \cdot \text{m}}{\text{N} \cdot \text{sec}^2}\right]^{1/2} = 244 \text{ m/sec} \qquad V_t$$

Since $M_t < 1$, flow at the exit must be subsonic. Therefore, $p_e = p_b$. The stagnation properties are constant, so

$$\frac{p_0}{p_e} = \left[1 + \frac{k-1}{2} M_e^2\right]^{k/(k-1)}$$

Solving for M_e gives

$$M_e = \left\{\left[\left(\frac{p_0}{p_e}\right)^{(k-1)/k} - 1\right] \frac{2}{k-1}\right\}^{1/2}$$

$$M_e = \left\{\left[\left(\frac{1.0 \times 10^6}{9.54 \times 10^5}\right)^{0.286} - 1\right](5)\right\}^{1/2} = 0.26 \qquad M_e$$

The Ts diagram for this flow is

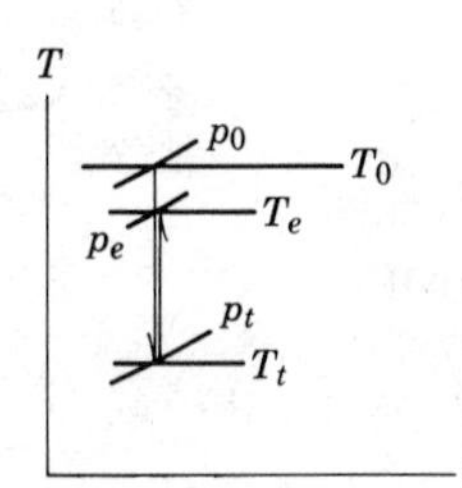

(*ii*) ***Tables for Isentropic Flow*

The stagnation properties are constant for isentropic flow. From Table D.1, Appendix D, at $M = 0.68$,

$$\frac{T}{T_0} = 0.9154; \qquad T_t = 0.9154T_0 = (0.9154)(350 \text{ K}) = 320 \text{ K} \qquad T_t$$

$$\frac{p}{p_0} = 0.7338; \qquad p_t = 0.7338p_0 = (0.7338)(1.0 \times 10^6 \text{ Pa}) = 734 \text{ kPa (abs)} \qquad p_t$$

$$\frac{\rho}{\rho_0} = 0.8016; \qquad \rho_t = 0.8016\rho_0 = 0.8016\frac{p_0}{RT_0}$$

$$\rho_t = 0.8016 \times \frac{1.0 \times 10^6 \text{ N}}{\text{m}^2} \times \frac{\text{kg} \cdot \text{K}}{287 \text{ N} \cdot \text{m}} \times \frac{1}{350 \text{ K}} = 7.98 \text{ kg/m}^3 \qquad \rho_t$$

and

$$\frac{A}{A^*} = 1.110; \qquad A_t = 1.110A^*$$

but at this point A^* is not known.

At the exit, $p_e = 954$ kPa (abs). Thus $p_e/p_0 = 0.954$, and from Table D.1,

$$M_e = 0.26 \qquad M_e$$

Since A_e is known, we can compute A^*. From Table D.1, at $M = 0.26$, $A/A^* = 2.317$. Thus

$$A^* = \frac{A_e}{2.317} = \frac{0.001 \text{ m}^2}{2.317} = 4.32 \times 10^{-4} \text{ m}^2$$

and

$$A_t = 1.110A^* = (1.110)(4.32 \times 10^{-4} \text{ m}^2) = 4.80 \times 10^{-4} \text{ m}^2 \qquad A_t$$

Note that the solution for A_t using the tables was relatively effortless. To find A_t using the isentropic relations, we could have applied continuity between the throat and exit planes. Since this would have required calculation of all properties at the exit, it would have been a lengthy process.

Example 12.6

The nozzle of Example 12.5 has a design back pressure of 72.8 kPa (abs) but is operated at a back pressure of 50.0 kPa (abs). Flow within the nozzle may be assumed isentropic. Determine the exit Mach number and mass flow rate. Use (*i*) isentropic flow relations, and (*ii*) tables for isentropic flow.

EXAMPLE PROBLEM 12.6

GIVEN: Air flow through C-D nozzle as shown:

$T_0 = 350$ K

$p_0 = 1.0$ MPa (abs)

$p_e(\text{design}) = 72.8$ kPa (abs)

$p_b = 50.0$ kPa (abs)

$A_e = 0.001\ \text{m}^2$

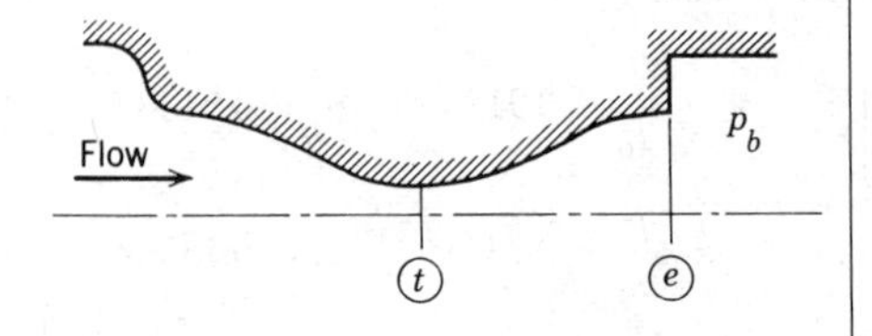

FIND: (a) M_e. (b) $\dot{m}$.

SOLUTION:
The operating back pressure is *below* the design value. Consequently, the nozzle is underexpanded, and the Ts diagram and pressure distribution will be as shown:

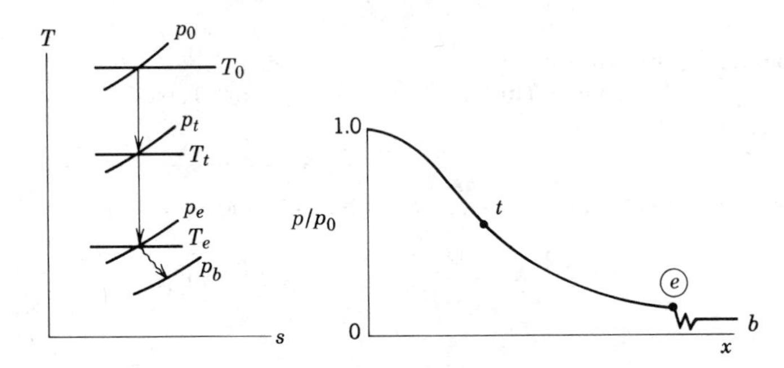

Flow *within* the nozzle will be isentropic, but the irreversible expansion from p_e to p_b will cause an entropy increase; $p_e = p_e(\text{design}) = 72.8$ kPa (abs).

(*i*) *Isentropic Flow Relations*

Since stagnation properties are constant for isentropic flow, the exit Mach number can be computed from the pressure ratio. Thus

$$\frac{p_0}{p_e} = \left[1 + \frac{k-1}{2} M_e^2\right]^{k/(k-1)}$$

or

$$M_e = \left\{\left[\left(\frac{p_0}{p_e}\right)^{(k-1)/k} - 1\right]\frac{2}{k-1}\right\}^{1/2} = \left\{\left[\left(\frac{1.0 \times 10^6}{7.28 \times 10^4}\right)^{0.286} - 1\right]\frac{2}{0.4}\right\}^{1/2} = 2.36$$

$\longleftarrow M_e$

The mass flow rate is given by

$$\dot{m} = \rho_e V_e A_e = \rho_e M_e c_e A_e = \rho_e M_e \sqrt{kRT_e} A_e = \frac{p_e}{RT_e} M_e \sqrt{kRT_e} A_e$$

or

$$\dot{m} = p_e M_e \sqrt{\frac{k}{RT_e}} A_e$$

Since T_0 is constant,

$$\frac{T_0}{T_e} = 1 + \frac{k-1}{2} M_e^2; \qquad T_e = \frac{T_0}{1 + \frac{k-1}{2} M_e^2} = \frac{350 \text{ K}}{1 + 0.2(2.36)^2} = 166 \text{ K}$$

Then

$$\dot{m} = p_e M_e \sqrt{\frac{k}{RT_e}} A_e$$

$$= \frac{7.28 \times 10^4 \text{ N}}{\text{m}^2} \times 2.36 \left[1.4 \times \frac{\text{kg} \cdot \text{K}}{287 \text{ N} \cdot \text{m}} \times \frac{1}{166 \text{ K}} \times \frac{\text{kg} \cdot \text{m}}{\text{N} \cdot \text{sec}^2} \right]^{1/2} 0.001 \text{ m}^2$$

$$\dot{m} = 0.931 \text{ kg/sec}$$

$\dot{m}$

(*ii*) ****Tables for Isentropic Flow*

From Table D.1, Appendix D, at $p/p_0 = p_e/p_0 = 0.0728$,

$$M_e = 2.36$$

M_e

Also,

$$\frac{T_e}{T_0} = 0.4731; \qquad T_e = 0.4731\, T_0 = 0.4731(350 \text{ K}) = 166 \text{ K}$$

{This temperature checks the value obtained above. The mass flow rate calculation also would be as above.}

12-4 ADIABATIC FLOW IN A CONSTANT-AREA DUCT WITH FRICTION

To obtain an overall view of the problem of frictional adiabatic flow, apply the basic equations to the steady uniform flow of an ideal gas with constant specific heats through the finite control volume shown in Fig. 12.9.

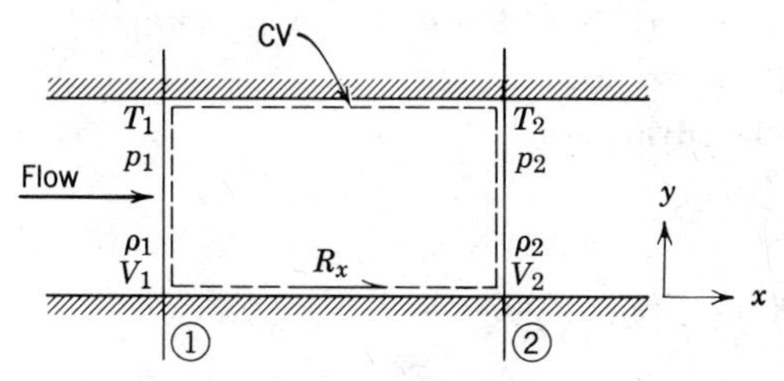

Fig. 12.9 Control volume used for integral analysis of frictional adiabatic flow.

12-4.1 Basic Equations

a. Continuity Equation

Basic equation:

$$0 = \overset{=0(1)}{\cancel{\frac{\partial}{\partial t}}} \int_{\mathrm{CV}} \rho \, d\forall + \int_{\mathrm{CS}} \rho \vec{V} \cdot d\vec{A} \tag{4.13}$$

Assumptions: (1) Steady flow
(2) Uniform flow at each section

Then

$$0 = \{-|\rho_1 V_1 A_1|\} + \{|\rho_2 V_2 A_2|\}$$

The area is constant, so

$$\rho_1 V_1 = \rho_2 V_2 \equiv G = \frac{\dot{m}}{A} \tag{12.7a}$$

b. Momentum Equation

Basic equation:

$$F_{S_x} + \overset{=0(3)}{\cancel{F_{B_x}}} = \overset{=0(1)}{\cancel{\frac{\partial}{\partial t}}} \int_{\mathrm{CV}} V_x \rho \, d\forall + \int_{\mathrm{CS}} V_x \rho \vec{V} \cdot d\vec{A} \tag{4.19a}$$

Assumption: (3) $F_{B_x} = 0$

The surface force is due to pressure forces at sections ① and ②, and to the friction force, R_x, of the duct wall *on* the flow. Substituting, recognizing that $A_2 = A_1 = A$,

$$R_x + p_1 A - p_2 A = V_1\{-|\rho_1 V_1 A|\} + V_2\{|\rho_2 V_2 A|\}$$

Using scalar magnitudes and dropping absolute value signs

$$R_x + p_1 A - p_2 A = \dot{m} V_2 - \dot{m} V_1 \tag{12.7b}$$

c. First Law of Thermodynamics

Basic equation:

$$\overset{=0(4)}{\cancel{\dot{Q}}} + \overset{=0(5)}{\cancel{\dot{W}_s}} + \overset{=0(6)}{\cancel{\dot{W}_{\text{shear}}}} + \overset{=0(6)}{\cancel{\dot{W}_{\text{other}}}} = \overset{=0(1)}{\cancel{\frac{\partial}{\partial t}}} \int_{\mathrm{CV}} e \rho \, d\forall + \int_{\mathrm{CS}} (e + pv) \rho \vec{V} \cdot d\vec{A} \tag{4.59}$$

where

$$e = u + \frac{V^2}{2} + \cancelto{\simeq 0(7)}{gz}$$

Assumptions: (4) $\dot{Q} = 0$ (adiabatic flow)
(5) $\dot{W}_s = 0$
(6) $\dot{W}_{\text{shear}} = \dot{W}_{\text{other}} = 0$
(7) Effects of gravity are negligible

With these restrictions, the equation becomes

$$0 = \left(u_1 + \frac{V_1^2}{2} + p_1 v_1\right)\{-|\rho_1 V_1 A|\} + \left(u_2 + \frac{V_2^2}{2} + p_2 v_2\right)\{|\rho_2 V_2 A|\}$$

Since the mass flow rate terms in { } are identical by continuity,

$$u_1 + \frac{V_1^2}{2} + p_1 v_1 = u_2 + \frac{V_2^2}{2} + p_2 v_2$$

or

$$h_1 + \frac{V_1^2}{2} = h_2 + \frac{V_2^2}{2} \tag{12.7c}$$

We also could write

$$h_{0_1} = h_{0_2}$$

which is a physical consequence of our assumption of adiabatic flow.

d. Second Law of Thermodynamics

Basic equation:

$$\int_{CS} \frac{1}{T} \cancelto{= 0(4)}{\left(\frac{\dot{Q}}{A}\right)} dA \leq \cancelto{= 0(1)}{\frac{\partial}{\partial t}} \int_{CV} s\rho \, d\mathbf{V} + \int_{CS} s\rho \vec{V} \cdot d\vec{A} \tag{4.61}$$

Then because the flow is frictional and, hence, irreversible,

$$0 < s_1\{-|\rho_1 V_1 A|\} + s_2\{|\rho_2 V_2 A|\} = \dot{m}(s_2 - s_1)$$

The control volume form of the second law then tells us that $s_2 - s_1 > 0$.

This fact is of little help in calculating the actual entropy change between any two points in an adiabatic frictional flow. To calculate the entropy change, we rely on the $T\,ds$ equations. Since

$$T\,ds = dh - v\,dp$$

for an ideal gas we can write

$$ds = c_p \frac{dT}{T} - R\frac{dp}{p}$$

For constant specific heats the equation can be integrated to give

$$s_2 - s_1 = c_p \ln \frac{T_2}{T_1} - R \ln \frac{p_2}{p_1} \tag{12.7d}$$

e. Equations of State

For an ideal gas, the equation of state is given by

$$p = \rho RT \tag{12.7e}$$

Equations 12.7a to 12.7e are the governing equations for the steady, one-dimensional, adiabatic, frictional flow of an ideal gas in a constant-area duct. If all the properties at state ① are known, then we have seven unknowns ($T_2, p_2, \rho_2, V_2, h_2, s_2$, and R_x) in these five equations. However, we have the known relationship between h and T for an ideal gas, $dh = c_p\,dT$. For an ideal gas with constant specific heats,

$$\Delta h = h_2 - h_1 = c_p\,\Delta T = c_p(T_2 - T_1) \tag{12.7f}$$

We thus have the situation of six equations and seven unknowns.

If all conditions at state ① are known, how many possible states ② are there? The mathematics of the situation (six equations and seven unknowns) indicate that there are an infinite number of possible states ②.

With an infinite number of possible states ② for a given state ①, what is to be expected if all possible states ② are plotted on a *Ts* diagram? It follows that the locus of all possible states ② reachable from state ① is a continuous curve passing through state ①.

How might we determine this curve? Perhaps the simplest way is to assume different values of T_2. For an assumed value of T_2 we could calculate the corresponding values of all other properties at state ② and also R_x.

12-4.2 The Fanno Line

The results of these calculations are shown qualitatively on the *Ts* plane of Fig. 12.10. The locus of all possible downstream states is referred to as the *Fanno line.*[2] Detailed calculations show some interesting features of Fanno line flow. At the point of maximum entropy, the Mach number is unity. On the upper branch of the curve the Mach number is always less than unity, and it increases monotonically as we proceed to the right along the curve. At every point on the lower portion of the curve the Mach number is greater than unity; the Mach number decreases monotonically as we move to the right along the curve.

[2] A convenient way to remember this name is to think of Frictional Adiabatic flow.

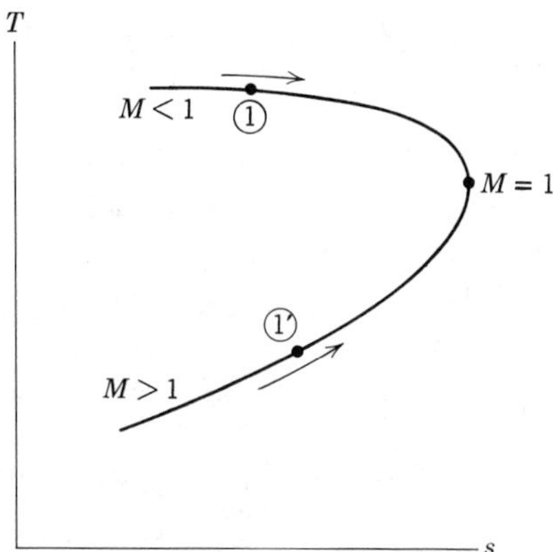

Fig. 12.10 Schematic *Ts* diagram for frictional adiabatic (Fanno line) flow in a constant-area duct.

For any initial state in a Fanno line flow, any point on the Fanno line represents a mathematically possible downstream state. Indeed, we determined the locus of all possible downstream states by assuming values of T_2 and calculating the corresponding properties. Although the Fanno line represents all mathematically possible downstream states, are they all physically attainable downstream states? A moment's reflection will indicate that they are not. Why not? The second law requires that the entropy must increase. Consequently, we must always move to the right along the Fanno line of Fig. 12.10. Indeed, it is the effect of friction that causes the flow properties to change from the initial state. Referring again to Fig. 12.10, we see that for an initially subsonic flow (state ①), the effect of friction is to increase the Mach number toward unity. For a flow that is initially supersonic (state ①′), the effect of friction is to decrease the Mach number toward unity.

In developing the simplified form of the first law for Fanno line flow, we found that the stagnation enthalpy remains constant for the flow. Consequently, when the fluid is an ideal gas with constant specific heats, the stagnation temperature also must remain constant. What happens to the stagnation pressure? Friction causes the local isentropic stagnation pressure to decrease for all Fanno line flows as shown in Fig. 12.11. Recall that the second law requires $s_2 - s_1 > 0$. Since the entropy must increase in the direction of flow, the flow process must proceed to the right on the *Ts* diagram. In Fig. 12.11, a path from state ① to state ② is shown on the subsonic portion of the curve. The corresponding values of local isentropic stagnation pressure, p_{0_1} and p_{0_2}, clearly show that $p_{0_2} < p_{0_1}$. An identical result is obtained for flow on the supersonic branch of the curve from state ①′ to state ②′. Again $p_{0_{2'}} < p_{0_{1'}}$. Thus p_0 decreases for any Fanno line flow.

At this point it may be beneficial to summarize the effects of friction on flow properties in Fanno line flow. The summary is presented in Table 12.1.

In deducing the effect of friction on flow properties for Fanno line flow, we have used the shape of the Fanno line on the *Ts* diagram and the basic governing equations (Eqs. 12.7a to 12.7f). You are encouraged to follow through the logic indicated in the right-hand column of the table.

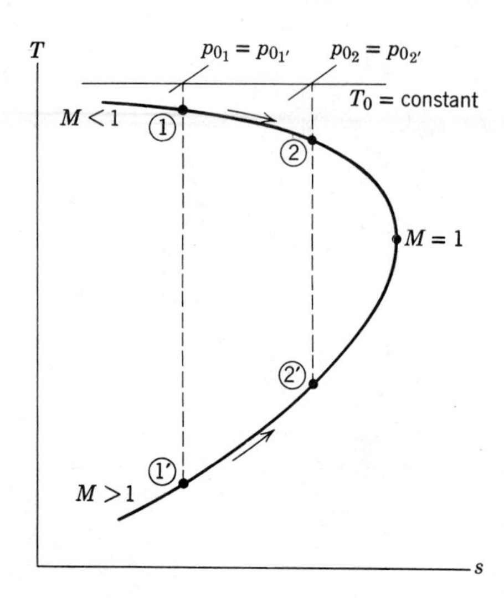

Fig. 12.11 Schematic of Fanno line flow on Ts plane, showing reduction in local isentropic stagnation pressure caused by friction.

Table 12.1 Summary of Effects of Friction on Properties in Fanno Line Flow

Property	Subsonic $M < 1$	Supersonic $M > 1$	Obtained from:
Stagnation temperature, T_0	Constant	Constant	Energy equation
Entropy, s	Increases	Increases	Second law
Stagnation pressure, p_0	Decreases	Decreases	T_0 = constant; s increases
Temperature, T	Decreases	Increases	Shape of Fanno line
Velocity, V	Increases	Decreases	Energy equation, and trend of T
Mach number, M	Increases	Decreases	Trends of V, T, and definition of M
Density, ρ	Decreases	Increases	Continuity equation, and effect on V
Pressure, p	Decreases	Increases	Equation of state, and effects on ρ, T

We have noted that the entropy must increase in the direction of flow; it is the effect of friction that causes the change in flow properties along the Fanno line curve. From Fig. 12.11 we see that there is a maximum entropy point corresponding to $M = 1$ for each Fanno line. The maximum entropy point is reached by increasing the amount of friction (through addition of duct length), just enough to produce a Mach number of unity (choked flow) at the exit. How do we compute this critical length of duct?

To compute the critical duct length, we must analyze the flow in detail, accounting for the effect of friction. This analysis requires that we begin with a differential control volume, develop expressions in terms of Mach number, and integrate along the duct to the section where $M = 1$. This analysis is completed in Section 12-4.3, where tables for Fanno line flow are developed. The algebra required for the detailed analysis tends to obscure the physics of the flow; the general trends in properties caused by friction can be demonstrated using finite control volumes and the basic governing equations. This approach is illustrated in Example Problem 12.7.

Example 12.7

Air flow is induced in an insulated tube of 7.16 mm diameter by a vacuum pump. The air is drawn from a room where the absolute pressure is 101 kPa and the temperature is 23 C, through a smoothly contoured, converging nozzle. At section ①, where the nozzle joins the constant-area tube, the static pressure is 98.5 kPa (abs). At section ②, located some distance downstream in the constant-area tube, the air temperature is 14 C. Determine the mass flow rate, the local isentropic stagnation pressure at section ②, and the friction force on the duct wall between sections ① and ②.

EXAMPLE PROBLEM 12.7

GIVEN: Air flow in insulated tube.

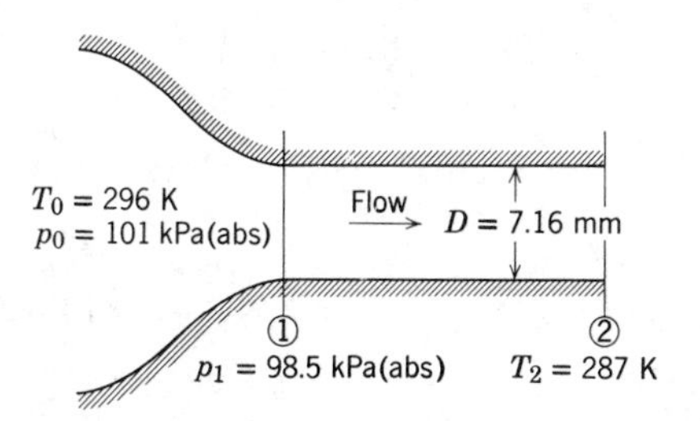

FIND: (a) $\dot{m}$.
(b) Stagnation pressure at section ②.
(c) Force on duct wall.

SOLUTION:

The mass flow rate can be obtained from properties at section ①. For isentropic flow through the converging nozzle, local isentropic stagnation properties remain constant. Thus

$$\frac{p_{0_1}}{p_1} = \left(1 + \frac{k-1}{2} M_1^2\right)^{k/(k-1)}$$

and

$$M_1 = \left\{\frac{2}{k-1}\left[\left(\frac{p_{0_1}}{p_1}\right)^{(k-1)/k} - 1\right]\right\}^{1/2} = \left\{\frac{2}{0.4}\left[\left(\frac{1.01 \times 10^5}{9.85 \times 10^4}\right)^{0.286} - 1\right]\right\}^{1/2} = 0.190$$

$$T_1 = \frac{T_{0_1}}{1 + \dfrac{k-1}{2} M_1^2} = \frac{296 \text{ K}}{1 + 0.2(0.190)^2} = 294 \text{ K}$$

For an ideal gas,

$$\rho_1 = \frac{p_1}{RT_1} = \frac{9.85 \times 10^4}{} \frac{\text{N}}{\text{m}^2} \times \frac{\text{kg} \cdot \text{K}}{287\ \text{N} \cdot \text{m}} \times \frac{1}{294\ \text{K}} = 1.17\ \text{kg/m}^3$$

$$V_1 = M_1 c_1 = M_1 \sqrt{kRT_1} = (0.19)\left[1.4 \times \frac{287\ \text{N} \cdot \text{m}}{\text{kg} \cdot \text{K}} \times 294\ \text{K} \times \frac{\text{kg} \cdot \text{m}}{\text{N} \cdot \text{sec}^2}\right]^{1/2} = 65.3\ \text{m/sec}$$

$$A_1 = A = \frac{\pi D^2}{4} = \frac{\pi}{4}(7.16 \times 10^{-3})^2\ \text{m}^2 = 4.03 \times 10^{-5}\ \text{m}^2$$

From continuity

$$\dot{m} = \rho_1 V_1 A_1 = \frac{1.17\ \text{kg}}{\text{m}^3} \times \frac{65.3\ \text{m}}{\text{sec}} \times 4.03 \times 10^{-5}\ \text{m}^2$$

$$\dot{m} = 3.08 \times 10^{-3}\ \text{kg/sec} \qquad \longleftarrow \dot{m}$$

Flow is adiabatic and frictional so T_0 is constant, and

$$T_{0_2} = T_{0_1} = 296\ \text{K} \qquad \longleftarrow T_{0_2}$$

Then

$$\frac{T_{0_2}}{T_2} = 1 + \frac{k-1}{2} M_2^2$$

Solving for M_2 gives

$$M_2 = \left[\frac{2}{k-1}\left(\frac{T_{0_2}}{T_2} - 1\right)\right]^{1/2} = \left[\frac{2}{0.4}\left(\frac{296}{287} - 1\right)\right]^{1/2} = 0.396 \qquad \longleftarrow M_2$$

$$V_2 = M_2 c_2 = M_2 \sqrt{kRT_2} = (0.396)\left[1.4 \times \frac{287\ \text{N} \cdot \text{m}}{\text{kg} \cdot \text{K}} \times 287\ \text{K} \times \frac{\text{kg} \cdot \text{m}}{\text{N} \cdot \text{sec}^2}\right]^{1/2}$$

$$V_2 = 134\ \text{m/sec} \qquad \longleftarrow V_2$$

From continuity, $\rho_1 V_1 = \rho_2 V_2$, so

$$\rho_2 = \rho_1 \frac{V_1}{V_2} = \frac{1.17\ \text{kg}}{\text{m}^3} \times \frac{65.3}{134} = 0.570\ \text{kg/m}^3 \qquad \longleftarrow \rho_2$$

and

$$p_2 = \rho_2 R T_2 = \frac{0.570\ \text{kg}}{\text{m}^3} \times \frac{287\ \text{N} \cdot \text{m}}{\text{kg} \cdot \text{K}} \times 287\ \text{K} = 47.0\ \text{kPa (abs)} \qquad \longleftarrow p_2$$

The local isentropic stagnation pressure is

$$p_{0_2} = p_2\left(1 + \frac{k-1}{2} M_2^2\right)^{k/(k-1)} = 4.70 \times 10^4\ \text{Pa}[1 + 0.2(0.396)^2]^{3.5}$$

$$p_{0_2} = 52.4\ \text{kPa (abs)} \qquad \longleftarrow p_{0_2}$$

The friction force may be found by applying the momentum equation to the control volume shown:

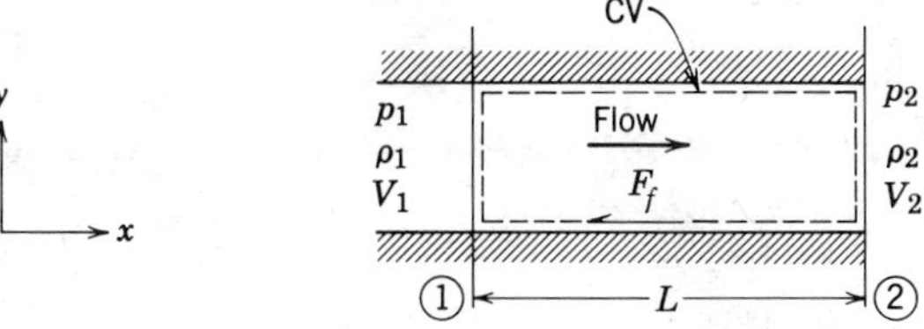

Basic equation:

$$F_{S_x} + \cancelto{=0(1)}{F_{B_x}} = \cancelto{=0(2)}{\frac{\partial}{\partial t}} \int_{\text{CV}} V_x \rho \, dV + \int_{\text{CS}} V_x \rho \vec{V} \cdot d\vec{A} \tag{4.19a}$$

Assumptions: (1) $F_{B_x} = 0$
(2) Steady flow
(3) Uniform flow at each section

Then

$$-F_f + p_1 A - p_2 A = V_1\{-|\rho_1 V_1 A|\} + V_2\{|\rho_2 V_2 A|\} = \dot{m}(V_2 - V_1)$$

and

$$-F_f = (p_2 - p_1)A + \dot{m}(V_2 - V_1)$$

$$-F_f = (4.70 - 9.85)10^4 \, \frac{\text{N}}{\text{m}^2} \times 4.03 \times 10^{-5} \, \text{m}^2$$

$$+ \, 3.08 \times 10^{-3} \, \frac{\text{kg}}{\text{sec}} \, (134 - 65.3) \, \frac{\text{m}}{\text{sec}} \times \frac{\text{N} \cdot \text{sec}^2}{\text{kg} \cdot \text{m}}$$

or

$$F_f = 1.86 \text{ N} \qquad \text{(to the left, as shown)}$$

This is the force exerted on the control volume by the duct wall. The force of the *fluid* on the *duct* is

$$K_x = F_f = 1.86 \text{ N} \qquad \text{(to the right)}$$

K_x

**12-4.3 Tables for Computation of Fanno Line Flow of an Ideal Gas

The primary independent variable in Fanno line flow is the friction force, F_f. Knowledge of the total friction force between any two points in a Fanno line flow would enable us to predict downstream conditions when conditions upstream were known. The total friction force is the integral of the wall shear stress over the duct surface area. Since the wall shear stress varies along the duct, we must develop a

** This section may be omitted without loss of continuity in the text material.

differential equation and then integrate to find property variations. To set up the differential equation, we use the differential control volume shown in Fig. 12.12 for our analysis.

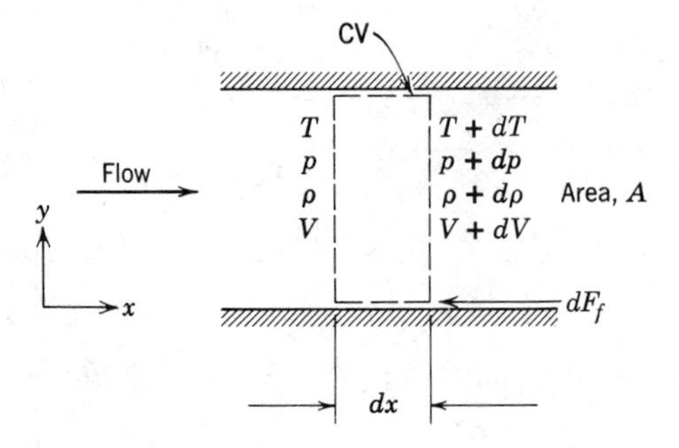

Fig. 12.12 Differential control volume used for analysis of Fanno line flow.

a. Continuity Equation

Basic equation:

$$0 = \overbrace{\frac{\partial}{\partial t} \int_{\mathrm{CV}} \rho \, d\forall}^{=0(1)} + \int_{\mathrm{CS}} \rho \vec{V} \cdot d\vec{A} \tag{4.13}$$

Assumptions: (1) Steady flow
(2) Uniform flow at each section

Then

$$0 = \{-|\rho V A|\} + \{|(\rho + d\rho)(V + dV)A|\}$$

or

$$\rho V A = (\rho + d\rho)(V + dV)A$$

Simplifying and canceling A gives

$$0 = -\cancel{\rho V} + \cancel{\rho V} + \rho \, dV + V \, d\rho + \overbrace{\cancel{d\rho} \, dV}^{\simeq 0}$$

which reduces to

$$\rho \, dV + V \, d\rho = 0 \tag{12.8a}$$

since products of differentials are negligible.

b. Momentum Equation

Basic equation:

$$F_{S_x} + \overbrace{\cancel{F_{B_x}}}^{=0(3)} = \overbrace{\cancel{\frac{\partial}{\partial t}} \int_{\mathrm{CV}} V_x \rho \, d\forall}^{=0(1)} + \int_{\mathrm{CS}} V_x \rho \vec{V} \cdot d\vec{A} \tag{4.19a}$$

Assumption: (3) $F_{B_x} = 0$

The momentum equation becomes

$$-dF_f + pA - (p + dp)A = V\{-|\rho VA|\} + (V + dV)\{|(\rho + d\rho)(V + dV)A|\}$$

which simplifies, using continuity, to give

$$-\frac{dF_f}{A} - dp = \rho V\,dV \tag{12.8b}$$

c. First Law of Thermodynamics

Basic equation:

$$\overset{=0(4)}{\cancel{\dot{Q}}} + \overset{=0(5)}{\cancel{\dot{W}_s}} + \overset{=0(6)}{\cancel{\dot{W}_{shear}}} + \overset{=0(6)}{\cancel{\dot{W}_{other}}} = \overset{=0(1)}{\frac{\partial}{\partial t}\int_{CV} e\rho\, d\mathcal{V}} + \int_{CS} (e + pv)\rho \vec{V} \cdot d\vec{A} \tag{4.59}$$

where

$$e = u + \frac{V^2}{2} + \overset{\simeq 0(7)}{\cancel{gz}}$$

Assumptions: (4) Adiabatic flow, $\dot{Q} = 0$
(5) $\dot{W}_s = 0$
(6) $\dot{W}_{shear} = \dot{W}_{other} = 0$
(7) Effects of gravity are negligible

Under these restrictions, we obtain

$$0 = \left(u + \frac{V^2}{2} + pv\right)\{-|\rho VA|\} + \left[u + du + \frac{V^2}{2} + d\left(\frac{V^2}{2}\right) + pv + d(pv)\right]$$
$$\times \{|(\rho + d\rho)(V + dV)A|\}$$

Noting from continuity that the flow rate terms { } are equal and substituting $h = u + pv$, we obtain

$$dh + d\left(\frac{V^2}{2}\right) = 0 \tag{12.8c}$$

To complete our formulation, we must relate the friction force, dF_f, to the flow variables at each cross section. We note that

$$dF_f = \tau_w dA_w = \tau_w P dx \tag{12.9}$$

where P is the wetted perimeter of the duct. To obtain an expression for τ_w in terms of flow variables at each cross section, we assume that changes in flow variables with x are gradual and use the correlations developed in Chapter 8 for fully developed, incompressible duct flow. For incompressible flow, the local wall shear stress can be written in terms of flow properties and friction factor. From Eqs. 8.16, 8.30, and 8.32 we have for incompressible flow

$$\tau_w = -\frac{R}{2}\frac{dp}{dx} = \frac{\rho R}{2}\frac{dh_l}{dx} = \frac{f\rho V^2}{8} \tag{12.10}$$

where f is the friction factor for pipe flow, Fig. 8.14. We assume that this correlation of experimental data also applies to compressible flow. This assumption, when checked against experimental data, shows surprisingly good agreement for subsonic flows; data for supersonic flow are sparse.

Ducts of other than circular shape can be included in our analysis by introducing the hydraulic diameter

$$D_h = \frac{4A}{P} \tag{8.44}$$

(Recall that the factor of 4 was included in Eq. 8.44 so that D_h would reduce to D, the diameter, for circular ducts.)

If we combine Eqs. 8.44, 12.9, and 12.10, we obtain

$$dF_f = \tau_w P\, dx = f\,\frac{\rho V^2}{8}\frac{4A}{D_h}\, dx$$

or

$$dF_f = \frac{fA}{D_h}\frac{\rho V^2}{2}\, dx \tag{12.11}$$

Substituting this result into the momentum equation (Eq. 12.8b), we obtain

$$-\frac{f}{D_h}\frac{\rho V^2}{2}\, dx - dp = \rho V\, dV$$

or, after dividing by p,

$$\frac{dp}{p} = -\frac{f}{D_h}\frac{\rho V^2}{2p}\, dx - \frac{\rho V\, dV}{p}$$

Noting that $p/\rho = RT = c^2/k$, and $V\, dV = d(V^2/2)$, we obtain

$$\frac{dp}{p} = -\frac{f}{D_h}\frac{kM^2}{2}\, dx - \frac{k}{c^2}\, d\left(\frac{V^2}{2}\right)$$

and finally,

$$\frac{dp}{p} = -\frac{f}{D_h}\frac{kM^2}{2}\, dx - \frac{kM^2}{2}\frac{d(V^2)}{V^2} \tag{12.12}$$

In order to obtain an equation relating M and x, we need to eliminate dp/p and $d(V^2)/V^2$ from Eq. 12.12. From the definition of Mach number, $M = V/c$, then $V^2 = M^2c^2 = M^2kRT$ and

$$\frac{d(V^2)}{V^2} = \frac{dT}{T} + \frac{d(M^2)}{M^2} \tag{12.13a}$$

From the continuity equation, $d\rho/\rho = -dV/V$ and

$$\frac{d\rho}{\rho} = -\frac{1}{2}\frac{d(V^2)}{V^2}$$

From the ideal gas equation of state, $p = \rho RT$, we can write

$$\frac{dp}{p} = \frac{d\rho}{\rho} + \frac{dT}{T}$$

Combining these three equations, we obtain

$$\frac{dp}{p} = \frac{1}{2}\frac{dT}{T} - \frac{1}{2}\frac{d(M^2)}{M^2} \tag{12.13b}$$

Substituting Eqs. 12.13 into Eq. 12.12 gives

$$\frac{1}{2}\frac{dT}{T} - \frac{1}{2}\frac{d(M^2)}{M^2} = -\frac{f}{D_h}\frac{kM^2}{2}dx - k\frac{M^2}{2}\frac{dT}{T} - \frac{kM^2}{2}\frac{d(M^2)}{M^2}$$

This equation can be simplified to

$$\left(\frac{1 + kM^2}{2}\right)\frac{dT}{T} = -\frac{f}{D_h}\frac{kM^2}{2}dx + \left(\frac{1 - kM^2}{2}\right)\frac{d(M^2)}{M^2} \tag{12.14}$$

We have been successful in reducing the number of variables somewhat. However, to relate M and x, we must obtain an expression for dT/T in terms of M. Such an expression can be obtained most readily from the stagnation temperature equation

$$\frac{T_0}{T} = 1 + \frac{k-1}{2}M^2 \tag{11.17}$$

Since the stagnation temperature is constant for Fanno line flow,

$$T\left(1 + \frac{k-1}{2}M^2\right) = \text{constant}$$

and

$$\frac{dT}{T} + \frac{M^2\dfrac{(k-1)}{2}}{\left(1 + \dfrac{k-1}{2}M^2\right)}\frac{d(M^2)}{M^2} = 0$$

Substituting for dT/T into Eq. 12.14 yields

$$\frac{M^2 \frac{(k-1)}{2}\left(\frac{1+kM^2}{2}\right)}{\left(1+\frac{k-1}{2}M^2\right)} \frac{d(M^2)}{M^2} = \frac{f}{D_h}\frac{kM^2}{2}\,dx - \left(\frac{1-kM^2}{2}\right)\frac{d(M^2)}{M^2}$$

Combining terms, we obtain

$$\frac{(1-M^2)}{\left(1+\frac{k-1}{2}M^2\right)} \frac{d(M^2)}{kM^4} = \frac{f}{D_h}\,dx \tag{12.15}$$

We have obtained a differential equation that relates changes in M with x. All we have to do now is integrate the equation to find M as a function of x.

Integration of Eq. 12.15 between states ① and ② would produce a complicated function of both M_1 and M_2. The function would have to be evaluated numerically for each new combination of M_1 and M_2 encountered in a problem. Calculations can be simplified considerably by use of the critical conditions (conditions where, by definition, $M = 1$). All Fanno line flows tend toward $M = 1$, so the integration is carried out between a section where the Mach number is M and the section where sonic conditions occur (the critical conditions). The Mach number will reach unity when the maximum possible length of duct is used, as shown schematically in Fig. 12.13.

The task is to perform the integration

$$\int_M^1 \frac{(1-M^2)}{kM^4\left(1+\frac{k-1}{2}M^2\right)}\,d(M^2) = \int_0^{L_{\max}} \frac{f}{D_h}\,dx \tag{12.16}$$

The left side may be evaluated using integration by parts. On the right side the friction factor, f, may vary with x, since the Reynolds number will vary along the duct. Note, however, that since the ρV product is constant along the duct (from continuity) the variation in Reynolds number is caused solely by variations in fluid viscosity.

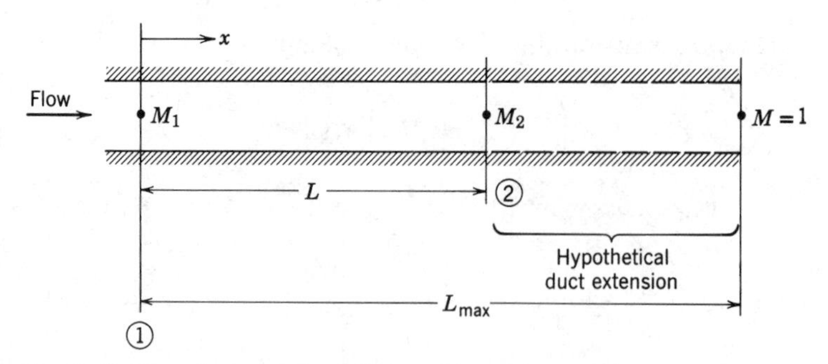

Fig. 12.13 Coordinates and notation used for analysis of Fanno line flow.

If we define a mean friction factor, $\bar{f}$, over the duct length as

$$\bar{f} = \frac{1}{L_{\max}} \int_0^{L_{\max}} f\,dx$$

then integration of Eq. 12.16 leads to

$$\frac{1 - M^2}{kM^2} + \frac{k+1}{2k} \ln\left[\frac{(k+1)M^2}{2\left(1 + \frac{k-1}{2}M^2\right)}\right] = \frac{\bar{f}L_{\max}}{D_h} \tag{12.17}$$

Equation 12.17 gives the maximum value of $\bar{f}L/D_h$ corresponding to any given initial Mach number. For any given Mach number we can compute the corresponding value of $\bar{f}L_{\max}/D_h$. These values are tabulated in Table D.2 of Appendix D.

Since the parameter, $\bar{f}L_{\max}/D_h$, is a function only of M, the duct length, L, required for the flow Mach number to change from an initial value, M_1, to a final value, M_2 (as illustrated in Fig. 12.13), may be found from

$$\frac{\bar{f}L}{D_h} = \left(\frac{\bar{f}L_{\max}}{D_h}\right)_{M_1} - \left(\frac{\bar{f}L_{\max}}{D_h}\right)_{M_2}$$

The critical conditions are appropriate reference conditions to use in tabulating fluid properties as a function of local Mach number. Thus, for example, since T_0 is constant, we can write

$$\frac{T}{T^*} = \frac{T/T_0}{T^*/T_0} = \frac{1}{1 + \frac{k-1}{2}M^2} \bigg/ \frac{1}{1 + \frac{k-1}{2}} = \frac{\left(\frac{k+1}{2}\right)}{\left(1 + \frac{k-1}{2}M^2\right)} \tag{12.18a}$$

Similarly,

$$\frac{V}{V^*} = \frac{M\sqrt{kRT}}{\sqrt{kRT^*}} = M\sqrt{\frac{T}{T^*}} = \left[\frac{\left(\frac{k+1}{2}\right)M^2}{1 + \frac{k-1}{2}M^2}\right]^{1/2} \tag{12.18b}$$

From continuity

$$\frac{\rho}{\rho^*} = \frac{V^*}{V} = \left[\frac{1 + \frac{k-1}{2}M^2}{\left(\frac{k+1}{2}\right)M^2}\right]^{1/2} \tag{12.18c}$$

From the ideal gas equation of state

$$\frac{p}{p^*} = \frac{\rho}{\rho^*}\frac{T}{T^*} = \frac{1}{M}\left[\frac{\left(\frac{k+1}{2}\right)}{1 + \frac{k-1}{2}M^2}\right]^{1/2} \tag{12.18d}$$

The ratio of local stagnation pressure to the reference stagnation pressure is given by

$$\frac{p_0}{p_0^*} = \frac{p_0}{p}\frac{p}{p^*}\frac{p^*}{p_0^*}$$

$$= \left(1 + \frac{k-1}{2}M^2\right)^{k/(k-1)} \frac{1}{M}\left[\frac{\left(\frac{k+1}{2}\right)}{1 + \frac{k-1}{2}M^2}\right]^{1/2} \frac{1}{\left(1 + \frac{k-1}{2}\right)^{k/(k-1)}}$$

or

$$\frac{p_0}{p_0^*} = \frac{1}{M}\left[\left(\frac{2}{k+1}\right)\left(1 + \frac{k-1}{2}M^2\right)\right]^{(k+1)/2(k-1)} \tag{12.18e}$$

The ratios in Eqs. 12.18 are tabulated as functions of Mach number in Table D.2 of Appendix D.

Example 12.8

Air flow is induced in a smooth insulated tube of 7.16 mm diameter by a vacuum pump. The air is drawn from a room where the absolute pressure is 760 mm Hg and the temperature is 23 C through a smoothly contoured, converging nozzle. At section ①, where the nozzle joins the constant-area tube, the static gage pressure is − 18.9 mm Hg. At section ②, located some distance downstream in the constant-area tube, the static pressure is −412 mm Hg (gage). The duct walls are smooth; the average friction factor, $\bar{f}$, may be taken as the value at section ①. Determine the length of duct required for choking from section ①, the Mach number at section ②, and the duct length, L_{12}, between sections ① and ②.

EXAMPLE PROBLEM 12.8

GIVEN: Air flow (with friction) in an insulated constant-area tube.

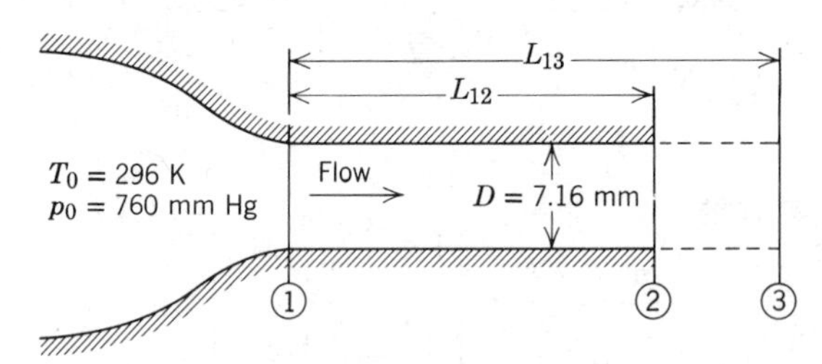

Gage pressures: $p_1 = -18.9$ mm Hg, $p_2 = -412$ mm Hg. $M_3 = 1.0$

FIND: (a) L_{13}. (b) M_2. (c) L_{12}.

SOLUTION:

Flow in the constant-area tube is frictional and adiabatic, a Fanno line flow. To find the friction factor, we need to know the flow conditions at section ①. If it is assumed that flow in the nozzle is isentropic, local properties at the nozzle exit may be computed using isentropic relations. Thus

$$\frac{p_{0_1}}{p_1} = \left(1 + \frac{k-1}{2} M_1^2\right)^{k/(k-1)}$$

Solving for M_1 we obtain

$$M_1 = \left\{\frac{2}{k-1}\left[\left(\frac{p_{0_1}}{p_1}\right)^{(k-1)/k} - 1\right]\right\}^{1/2} = \left\{\frac{2}{0.4}\left[\left(\frac{760}{760-18.9}\right)^{0.286} - 1\right]\right\}^{1/2} = 0.190$$

$$T_1 = \frac{T_{0_1}}{1 + \frac{k-1}{2} M_1^2} = \frac{296 \text{ K}}{1 + 0.2(0.190)^2} = 294 \text{ K}$$

$$V_1 = M_1 c_1 = M_1 \sqrt{kRT_1} = 0.190\left[1.4 \times \frac{287 \text{ N}\cdot\text{m}}{\text{kg}\cdot\text{K}} \times 294 \text{ K} \times \frac{\text{kg}\cdot\text{m}}{\text{N}\cdot\text{sec}^2}\right]^{1/2}$$

$$V_1 = 65.3 \text{ m/sec}$$

$$p_1 = g\rho_{Hg}h_1 = gSG\rho_{H_2O}h_1$$

$$= \frac{9.81 \text{ m}}{\text{sec}^2} \times 13.6 \times \frac{999 \text{ kg}}{\text{m}^3} \times (760 - 18.9)10^{-3} \text{ m} \times \frac{\text{N}\cdot\text{sec}^2}{\text{kg}\cdot\text{m}}$$

$$p_1 = 98.8 \text{ kPa (abs)}$$

$$\rho_1 = \frac{p_1}{RT_1} = \frac{9.88 \times 10^4 \text{ N}}{\text{m}^2} \times \frac{\text{kg}\cdot\text{K}}{287 \text{ N}\cdot\text{m}} \times \frac{1}{294 \text{ K}} = 1.17 \text{ kg/m}^3$$

At $T = 294$ K (21 C), $\mu = 1.9 \times 10^{-5}$ kg/m · sec from Fig. A.2, Appendix A. Thus

$$Re_1 = \frac{\rho_1 V_1 D_1}{\mu_1} = \frac{1.17 \text{ kg}}{\text{m}^3} \times \frac{65.3 \text{ m}}{\text{sec}} \times 0.00716 \text{ m} \times \frac{\text{m}\cdot\text{sec}}{1.9 \times 10^{-5} \text{ kg}} = 2.88 \times 10^4$$

From Fig. 8.14, for smooth pipe, $f = 0.0235$. From Table D.2 at $M_1 = 0.19$, $p/p^* = 5.745$, and $\bar{f}L_{max}/D_h = 16.38$. Thus, assuming $\bar{f} = f_1$,

$$L_{13} = (L_{max})_1 = \left(\frac{\bar{f}L_{max}}{D_h}\right)_1 \frac{D_h}{f_1} = 16.38 \times 0.00716 \text{ m} \times \frac{1}{0.0235} = 4.99 \text{ m}$$

L_{13}

Since p^* is constant for Fanno line flow, conditions at section ② can be determined from the pressure ratio, $(p/p^*)_2$. Thus

$$\left(\frac{p}{p^*}\right)_2 = \frac{p_2}{p^*} = \frac{p_2}{p_1}\frac{p_1}{p^*} = \frac{p_2}{p_1}\left(\frac{p}{p^*}\right)_1 = \left(\frac{760-412}{760-18.9}\right) 5.745 = 2.698$$

From Table D.2, at $(p/p^*)_2 = 2.698$, $M_2 \simeq 0.40$

M_2

At $M_2 = 0.40$, $\bar{f}L_{max}/D_h = 2.309$ (Table D.2). Thus

$$L_{23} = (L_{max})_2 = \left(\frac{\bar{f}L_{max}}{D_h}\right)_2 \frac{D_h}{f_1} = 2.309 \times 0.00716 \text{ m} \times \frac{1}{0.0235} = 0.704 \text{ m}$$

Finally,

$$L_{12} = L_{13} - L_{23} = (4.99 - 0.704) \text{ m} = 4.29 \text{ m}$$

L_{12}

{This is the same physical system as that analyzed in Example Problem 12.7. Use of the tables simplifies the calculations and makes it possible to determine the duct length.}

12-5 FRICTIONLESS FLOW IN A CONSTANT-AREA DUCT WITH HEAT TRANSFER

To explore the effects of heat transfer on a compressible flow, let us apply the basic equations to the steady, one-dimensional, frictionless flow of an ideal gas with constant specific heats through the finite control volume shown in Fig. 12.14.

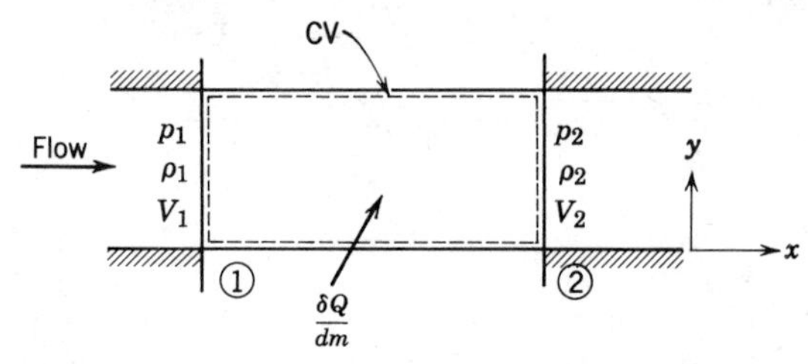

Fig. 12.14 Control volume used for integral analysis of frictionless flow with heat transfer.

12-5.1 Basic Equations

a. Continuity Equation

Basic equation:

$$0 = \overbrace{\frac{\partial}{\partial t} \int_{\mathrm{CV}} \rho \, d\mathcal{V}}^{=0(1)} + \int_{\mathrm{CS}} \rho \vec{V} \cdot d\vec{A} \tag{4.13}$$

Assumptions: (1) Steady flow
(2) Uniform flow at each section

Then

$$0 = \{-|\rho_1 V_1 A_1|\} + \{|\rho_2 V_2 A_2|\}$$

The area is constant, so

$$\rho_1 V_1 = \rho_2 V_2 = G = \frac{\dot{m}}{A} \tag{12.19a}$$

b. Momentum Equation

Basic equation:

$$F_{S_x} + \overbrace{F_{B_x}}^{=0(3)} = \overbrace{\frac{\partial}{\partial t} \int_{\mathrm{CV}} V_x \rho \, d\mathcal{V}}^{=0(1)} + \int_{\mathrm{CS}} V_x \rho \vec{V} \cdot d\vec{A} \tag{4.19a}$$

Assumption: (3) $F_{B_x} = 0$

Since there is no friction between the duct walls and the flow, and $A_2 = A_1 = A$, then

$$p_1 A - p_2 A = V_1\{-|\rho_1 V_1 A|\} + V_2\{|\rho_2 V_2 A|\}$$

This can be written as

$$p_1 A - p_2 A = \dot{m} V_2 - \dot{m} V_1 \tag{12.19b}$$

or

$$p_1 + \rho_1 V_1^2 = p_2 + \rho_2 V_2^2$$

c. First Law of Thermodynamics

Basic equation:

$$\dot{Q} + \underbrace{\dot{W}_s}_{=0(4)} + \underbrace{\dot{W}_{\text{shear}}}_{=0(5)} + \underbrace{\dot{W}_{\text{other}}}_{=0(5)} = \underbrace{\frac{\partial}{\partial t}}_{=0(1)} \int_{\text{CV}} e\rho \, d\forall + \int_{\text{CS}} (e + pv)\rho \vec{V} \cdot d\vec{A} \tag{4.59}$$

where

$$e = u + \frac{V^2}{2} + \underbrace{gz}_{\simeq 0(6)}$$

Assumptions: (4) $\dot{W}_s = 0$
(5) $\dot{W}_{\text{shear}} = \dot{W}_{\text{other}} = 0$
(6) Effects of gravity are negligible

With these restrictions

$$\dot{Q} = \left(u_1 + \frac{V_1^2}{2} + p_1 v_1\right)\{-|\rho_1 V_1 A|\} + \left(u_2 + \frac{V_2^2}{2} + p_2 v_2\right)\{|\rho_2 V_2 A|\}$$

or

$$\dot{Q} = \dot{m}\left(h_2 + \frac{V_2^2}{2} - h_1 - \frac{V_1^2}{2}\right)$$

But

$$\frac{\delta Q}{dm} = \frac{1}{\dot{m}} \dot{Q}$$

so

$$\frac{\delta Q}{dm} + h_1 + \frac{V_1^2}{2} = h_2 + \frac{V_2^2}{2} \tag{12.19c}$$

or

$$\frac{\delta Q}{dm} = h_{0_2} - h_{0_1}$$

We see that heat transfer causes the stagnation enthalpy, and hence the stagnation temperature, to change.

d. Second Law of Thermodynamics

Basic equation:

$$\int_{CS} \frac{1}{T}\frac{\dot{Q}}{A}\,dA \leq \overset{=0(1)}{\frac{\partial}{\partial t}\int_{CV} s\rho\, d\forall} + \int_{CS} s\rho\vec{V}\cdot d\vec{A} \tag{4.61}$$

$$\int_{CS} \frac{1}{T}\frac{\dot{Q}}{A}\,dA \leq \dot{m}(s_2 - s_1)$$

The flow is frictionless, and the heat transfer may be carried out over an arbitrarily small difference in temperature. Since under these conditions the process could be considered reversible, the equality in Eq. 4.61 would hold. However, even with heat transfer uniform over the control surface, we would need to know the temperature distribution over the area to evaluate the integral on the left side of Eq. 4.61. Consequently, the control volume form of the second law does not enable us to calculate the actual entropy change between any two points in the flow. The rate of heat transfer, $\dot{Q}$, may be positive (heat addition to the flow) or negative (heat rejection from the flow). Therefore, the entropy change in a frictionless flow with heat transfer may be either positive or negative.

To calculate the entropy change, we rely on the $T\,ds$ equations. Since

$$T\,ds = dh - v\,dp$$

for an ideal gas we can write

$$ds = c_p\frac{dT}{T} - R\frac{dp}{p}$$

For constant specific heats the equation can be integrated to give

$$s_2 - s_1 = c_p \ln\frac{T_2}{T_1} - R\ln\frac{p_2}{p_1} \tag{12.19d}$$

e. Equations of State

For an ideal gas, the equation of state is given by

$$p = \rho R T \tag{12.19e}$$

Equations 12.19a to 12.19e are the governing equations for the steady, one-dimensional, frictionless flow of an ideal gas in a constant-area duct with heat transfer. If all properties at state ① are known, then we have seven unknowns ($\rho_2, V_2, p_2, h_2, s_2, T_2$, and $\delta Q/dm$) in these five equations. However, we have the known relationship between h and T for an ideal gas, $dh = c_p\,dT$. For an ideal gas with constant specific heats,

$$\Delta h = h_2 - h_1 = c_p\Delta T = c_p(T_2 - T_1) \tag{12.19f}$$

We thus have the situation of six equations and seven unknowns.

12-5.2 The Rayleigh Line

If all conditions at state ① are known, how many possible states ② are there? The mathematics of the situation (six equations and seven unknowns) indicate that there are an infinite number of possible states ②.

With an infinite number of possible states ② for a given state ①, what is to be expected if all possible states ② are plotted on a *Ts* diagram? It follows that the locus of all possible states ②, reachable from state ①, is a continuous curve passing through state ①.

How can we determine this curve? Perhaps the simplest way is to assume different values of T_2. For an assumed value of T_2 we could then calculate the corresponding values of all other properties at state ② and also $\delta Q/dm$.

The results of these calculations are shown qualitatively on the *Ts* plane in Fig. 12.15. The locus of all possible downstream states is referred to as the *Rayleigh line*. The calculations show some interesting features of Rayleigh line flow. At the point of maximum temperature (point *a* of Fig. 12.15), the Mach number for an ideal gas is $1/\sqrt{k}$. At the point of maximum entropy, the Mach number is unity. On the upper branch of the curve the Mach number is always less than unity, and it increases monotonically as we proceed to the right along the curve. At every point on the lower portion of the curve the Mach number is greater than unity, and it decreases monotonically as we move to the right along the curve. Regardless of the initial Mach number, with heat addition the flow state proceeds to the right and with heat rejection the flow state proceeds to the left along the Rayleigh line.

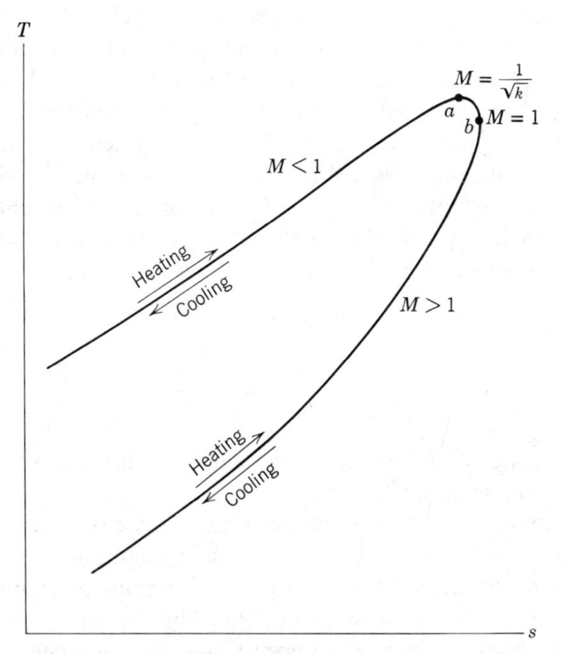

Fig. 12.15 Schematic *Ts* diagram for frictionless flow in a constant-area duct with heat transfer (Rayleigh line flow).

For any initial state in a Rayleigh line flow, any point on the Rayleigh line represents a mathematically possible downstream state. Indeed, we determined the locus of all possible downstream states by assuming values of T_2 and calculating the corresponding properties. Although the Rayleigh line represents all mathematically possible states, are they all physically attainable downstream states? A moment's reflection will indicate that they are. Since we are considering a flow with heat transfer, the second law does not impose any restrictions on the sign of the entropy change.

The effects of heat transfer on properties in the steady, frictionless, compressible flow of an ideal gas may be found from the basic equations, Eqs. 12.19a to 12.19f, and the Rayleigh line, Fig. 12.15. These effects are summarized in Table 12.2; the basis of the indicated trends is discussed in the next few paragraphs.

The direction of entropy change is always determined by the heat transfer; entropy increases with heating and decreases with cooling. Similarly, the first law, Eq. 12.19c, shows that heating increases the stagnation enthalpy and cooling decreases it; since $\Delta h_0 = c_p \, \Delta T_0$, the same effect is found on the stagnation temperature.

The effect on temperature of heating and cooling may be deduced from the shape of the Rayleigh line in Fig. 12.15. We see that for $M < 1/\sqrt{k}$ (for air, $1/\sqrt{k} = 0.85$) or for $M > 1$, heating causes T to increase, and in the same regions cooling causes T to decrease. However, we also see the unexpected result that for $1/\sqrt{k} < M < 1$, *heat addition* causes the stream temperature to *decrease*, and *heat rejection* causes the stream temperature to *increase*!

For subsonic flow, the Mach number increases monotonically with heating, until the value $M = 1$ is reached. For a given set of inlet conditions, all possible downstream states lie on a single Rayleigh line. Therefore, the point $M = 1$ determines the maximum possible heat addition without choking. If the flow is initially supersonic, heating will reduce the Mach number. Again the maximum possible heat addition without choking is that which reduces the Mach number to $M = 1.0$.

The effect of heat transfer on static pressure is obtained from the shapes of the Rayleigh line and of constant pressure lines on the Ts plane (see Fig. 12.16). For $M < 1$, pressure falls with heating and for $M > 1$, pressure increases with heating, as shown by the shapes of the constant-pressure lines. Once the pressure variation has been found, the effect on velocity may be found from the momentum equation

$$p_1 A - p_2 A = \dot{m} V_2 - \dot{m} V_1 \tag{12.19b}$$

or

$$p + \left(\frac{\dot{m}}{A}\right) V = \text{constant}$$

Thus, since $\dot{m}/A = \text{constant}$, trends in p and V must be opposite. From the continuity equation the trend in ρ is opposite to that in V.

The local isentropic stagnation pressure always decreases with heating. This is illustrated schematically in Fig. 12.16. A reduction in stagnation pressure has obvious practical implications for heating processes, such as combustion chambers. Adding the same amount of energy per unit mass (same change in T_0) causes a larger change in p_0 for supersonic flow; because the heating occurs at a lower temperature in supersonic flow, the entropy increase is larger.

Table 12.2 Summary of Heat Transfer Effects on Fluid Properties

Property	Heating $M < 1$	Heating $M > 1$	Cooling $M < 1$	Cooling $M > 1$	Obtained from:
Entropy, s	Increase	Increase	Decrease	Decrease	Second law
Stagnation temperature, T_0	Increase	Increase	Decrease	Decrease	First law, and $\Delta h_0 = c_p \Delta T_0$
Temperature, T	$\left(M < \frac{1}{\sqrt{k}}\right)$ Increase $\left(\frac{1}{\sqrt{k}} < M < 1\right)$ Decrease	Increase	$\left(M < \frac{1}{\sqrt{k}}\right)$ Decrease $\left(\frac{1}{\sqrt{k}} < M < 1\right)$ Increase	Decrease	Shape of Rayleigh line
Mach number, M	Increase	Decrease	Decrease	Increase	Trend on Rayleigh line
Pressure, p	Decrease	Increase	Increase	Decrease	Trend on Rayleigh line
Velocity, V	Increase	Decrease	Decrease	Increase	Momentum equation, and effect on p
Density, ρ	Decrease	Increase	Increase	Decrease	Continuity, and effect on V
Stagnation pressure, p_0	Decrease	Decrease	Increase	Increase	Fig. 12.16

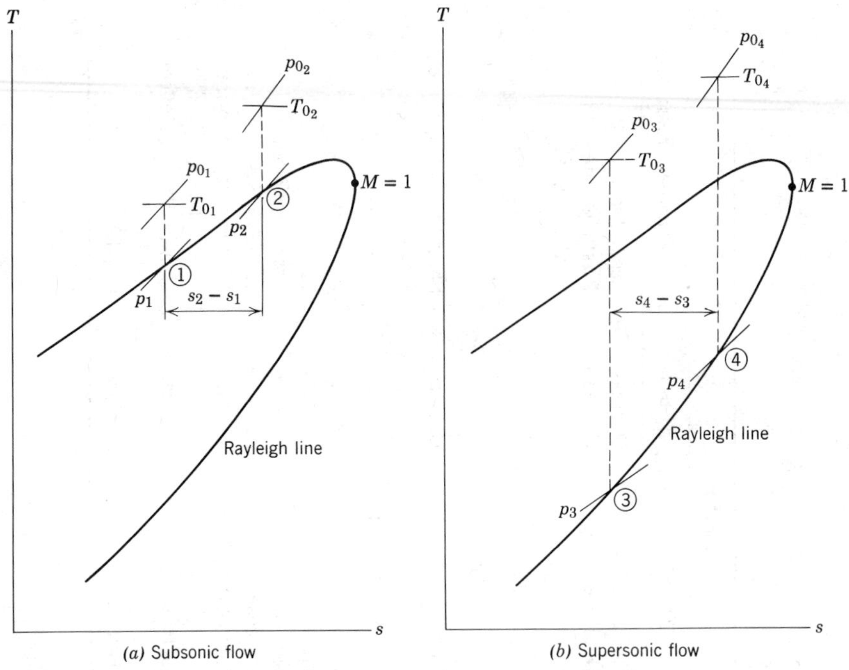

Fig. 12.16 Reduction in stagnation pressure due to heating for two flow cases.

Example 12.9

Air flows with negligible friction through a duct with cross-sectional area 0.25 ft^2. At section ①, the flow properties are $T_1 = 600$ R, $p_1 = 20$ psia, and $V_1 = 360$ ft/sec. At section ②, the pressure is 10 psia. The flow is heated between sections ① and ②. Determine the properties at section ②, the energy added, and the entropy change. Finally, plot the process on a Ts diagram.

EXAMPLE PROBLEM 12.9

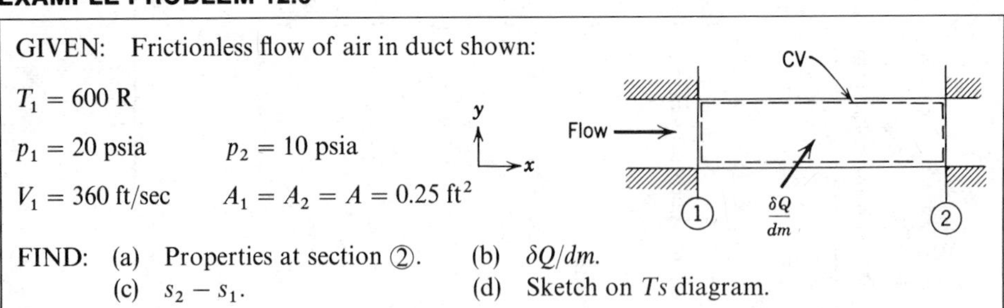

GIVEN: Frictionless flow of air in duct shown:

$T_1 = 600$ R

$p_1 = 20$ psia $\qquad p_2 = 10$ psia

$V_1 = 360$ ft/sec $\qquad A_1 = A_2 = A = 0.25$ ft^2

FIND: (a) Properties at section ②. (b) $\delta Q/dm$.
(c) $s_2 - s_1$. (d) Sketch on Ts diagram.

SOLUTION:

Apply the x component of the momentum equation, using the coordinates and control volume shown.

Basic equation:

$$F_{S_x} + \cancel{F_{B_x}} = \cancel{\frac{\partial}{\partial t}} \int_{CV} V_x \rho \, d\mathcal{V} + \int_{CS} V_x \rho \vec{V} \cdot d\vec{A} \qquad (4.19a)$$

(with $F_{B_x} = 0(1)$ and $\frac{\partial}{\partial t} = 0(2)$)

Assumptions: (1) $F_{B_x} = 0$
(2) Steady flow
(3) Uniform flow at each section

Then

$$p_1 A - p_2 A = V_1\{-|\rho_1 V_1 A|\} + V_2\{|\rho_2 V_2 A|\} = \dot{m}(V_2 - V_1)$$

or

$$p_1 - p_2 = \frac{\dot{m}}{A}(V_2 - V_1) = \rho_1 V_1 (V_2 - V_1)$$

Solving for V_2 gives

$$V_2 = \frac{p_1 - p_2}{\rho_1 V_1} + V_1$$

For an ideal gas,

$$\rho_1 = \frac{p_1}{RT_1} = \frac{20 \text{ lbf}}{\text{in.}^2} \times \frac{144 \text{ in.}^2}{\text{ft}^2} \times \frac{\text{lbm} \cdot \text{R}}{53.3 \text{ ft} \cdot \text{lbf}} \times \frac{1}{600 \text{ R}} = 0.0901 \text{ lbm/ft}^3$$

$$V_2 = \frac{(20 - 10) \text{ lbf}}{\text{in.}^2} \times \frac{144 \text{ in.}^2}{\text{ft}^2} \times \frac{\text{ft}^3}{0.0901 \text{ lbm}} \times \frac{\text{sec}}{360 \text{ ft}} \times \frac{32.2 \text{ lbm}}{\text{slug}} \times \frac{\text{slug} \cdot \text{ft}}{\text{lbf} \cdot \text{sec}^2} + \frac{360 \text{ ft}}{\text{sec}}$$

$$V_2 = 1790 \text{ ft/sec} \qquad V_2$$

From continuity, $G = \rho_1 V_1 = \rho_2 V_2$, so

$$\rho_2 = \rho_1 \frac{V_1}{V_2} = 0.0901 \frac{\text{lbm}}{\text{ft}^3}\left(\frac{360}{1790}\right) = 0.0181 \text{ lbm/ft}^3 \qquad \rho_2$$

Solving for T_2, we obtain

$$T_2 = \frac{p_2}{\rho_2 R} = \frac{10 \text{ lbf}}{\text{in.}^2} \times \frac{144 \text{ in.}^2}{\text{ft}^2} \times \frac{\text{ft}^3}{0.0181 \text{ lbm}} \times \frac{\text{lbm} \cdot \text{R}}{53.3 \text{ ft} \cdot \text{lbf}} = 1490 \text{ R} \qquad T_2$$

The local isentropic stagnation temperature is given by

$$T_{0_2} = T_2\left(1 + \frac{k-1}{2} M_2^2\right)$$

$$c_2 = \sqrt{kRT_2} = 1890 \text{ ft/sec}; \qquad M_2 = \frac{V_2}{c_2} = \frac{1790}{1890} = 0.947$$

$$T_{0_2} = 1490 \text{ R}[1 + 0.2(0.947)^2] = 1760 \text{ R} \qquad T_{0_2}$$

and

$$p_{0_2} = p_2 \left(\frac{T_{0_2}}{T_2} \right)^{k/(k-1)} = 10 \text{ psia} \left(\frac{1760}{1490} \right)^{3.5} = 17.9 \text{ psia} \qquad p_{0_2}$$

The heat transfer is determined from the energy equation.

Basic equation:

$$\dot{Q} + \dot{W}_s^{=0(4)} + \dot{W}_{\text{shear}}^{=0(5)} + \dot{W}_{\text{other}}^{=0(5)} = \frac{\partial}{\partial t}\int_{\text{CV}} e\rho \, d\forall^{=0(2)} + \int_{\text{CS}} (e + pv)\rho \vec{V} \cdot d\vec{A} \qquad (4.59)$$

where

$$e = u + \frac{V^2}{2} + gz^{\simeq 0(6)}$$

Assumptions: (4) $\dot{W}_s = 0$
(5) $\dot{W}_{\text{shear}} = \dot{W}_{\text{other}} = 0$
(6) Neglect changes in z

Then

$$\dot{Q} = \left(u_1 + p_1 v_1 + \frac{V_1^2}{2} \right)\{-|\rho_1 V_1 A|\} + \left(u_2 + p_2 v_2 + \frac{V_2^2}{2} \right)\{|\rho_2 V_2 A|\}$$

$$\dot{Q} = \dot{m}\left(h_2 + \frac{V_2^2}{2} - h_1 - \frac{V_1^2}{2} \right) = \dot{m}(h_{0_2} - h_{0_1}) = \dot{m}c_p(T_{0_2} - T_{0_1})$$

and

$$\frac{\delta Q}{dm} = \frac{1}{\dot{m}}\dot{Q} = c_p(T_{0_2} - T_{0_1})$$

$$T_{0_1} = T_1\left(1 + \frac{k-1}{2} M_1^2 \right)$$

$$c_1 = \sqrt{kRT_1} = 1200 \text{ ft/sec}; \qquad M_1 = \frac{V_1}{c_1} = \frac{360}{1200} = 0.3$$

$$T_{0_1} = (600 \text{ R})[1 + 0.2(0.3)^2] = 611 \text{ R}$$

so

$$\frac{\delta Q}{dm} = 0.240 \frac{\text{Btu}}{\text{lbm} \cdot \text{R}} (1760 - 611) \text{ R} = 276 \text{ Btu/lbm} \qquad \delta Q/dm$$

Using the $T\,ds$ equation, $T\,ds = dh - v\,dp$, we obtain for an ideal gas with constant specific heats,

$$s_2 - s_1 = c_p \ln \frac{T_2}{T_1} - R \ln \frac{p_2}{p_1} = c_p \ln \frac{T_2}{T_1} - (c_p - c_v) \ln \frac{p_2}{p_1}$$

Then

$$s_2 - s_1 = \frac{0.240 \quad \text{Btu}}{\text{lbm} \cdot \text{R}} \times \ln\left(\frac{1490}{600}\right) - \frac{(0.240 - 0.171) \quad \text{Btu}}{\text{lbm} \cdot \text{R}} \times \ln\left(\frac{10}{20}\right)$$

$$s_2 - s_1 = 0.266 \text{ Btu/lbm} \cdot \text{R}$$ ← $s_2 - s_1$

The process follows a Rayleigh line:

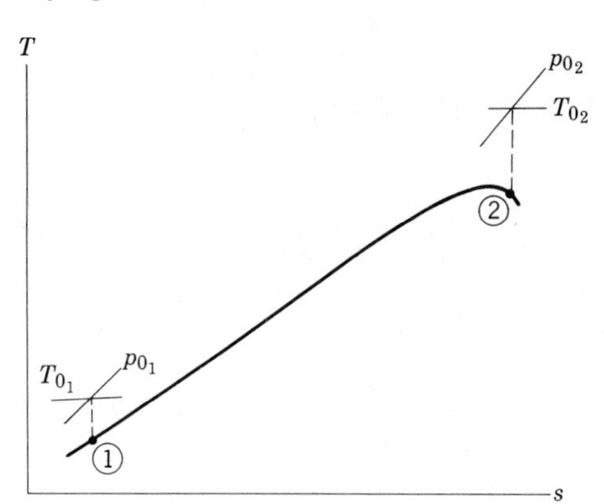

To complete our analysis, let us examine the change in p_0 by comparing p_{0_2} with p_{0_1}.

$$p_{0_1} = p_1\left(\frac{T_{0_1}}{T_1}\right)^{k/(k-1)} = 20.0 \text{ psia}\left(\frac{611}{600}\right)^{3.5} = 21.3 \text{ psia}$$ ← p_{0_1}

Comparing the two values, we see that p_{0_2} is less than p_{0_1}.

{In general, the stagnation pressure is decreased by heating and increased by cooling in Rayleigh line flow.}

**12-5.3 Tables for Computation of Rayleigh Line Flow of an Ideal Gas

In Section 12-5.1 we wrote the basic equations for Rayleigh line flow between two arbitrary states ① and ② in the flow. To facilitate the solution of problems, it is convenient to tabulate dimensionless properties in terms of local Mach number as we did for Fanno line flow. The reference state is again taken as the critical conditions, the state at which the Mach number is unity; properties at the critical condition are denoted by (*).

The dimensionless properties (such as T/T^* and p/p^*) may be obtained by writing the basic equations between a point in the flow where the properties are M, T, p, etc. (unsubscripted), and the critical state ($M = 1$ with properties denoted as T^*, p^*, etc.).

The pressure ratio, p/p^*, may be obtained from the momentum equation

$$pA - p^*A = \dot{m}V^* - \dot{m}V \tag{12.19b}$$

** This section may be omitted without loss of continuity in the text material.

or

$$p + \rho V^2 = p^* + \rho^* V^{*2}$$

Substituting $\rho = p/RT$ and factoring out pressures yields

$$p\left[1 + \frac{V^2}{RT}\right] = p^*\left[1 + \frac{V^{*2}}{RT^*}\right]$$

Noting that $V^2/RT = k(V^2/kRT) = kM^2$, then we find

$$p[1 + kM^2] = p^*[1 + k]$$

and finally,

$$\frac{p}{p^*} = \frac{1 + k}{1 + kM^2} \tag{12.20a}$$

From the ideal gas equation of state

$$\frac{T}{T^*} = \frac{p}{p^*}\frac{\rho^*}{\rho}$$

From the continuity equation

$$\frac{\rho^*}{\rho} = \frac{V}{V^*} = M\frac{c}{c^*} = M\sqrt{\frac{T}{T^*}}$$

Then substituting for ρ^*/ρ, we obtain

$$\frac{T}{T^*} = \frac{p}{p^*} M\sqrt{\frac{T}{T^*}}$$

Squaring, and substituting from Eq. 12.20a gives

$$\frac{T}{T^*} = \left[\frac{p}{p^*} M\right]^2 = \left[M\left(\frac{1 + k}{1 + kM^2}\right)\right]^2 \tag{12.20b}$$

From continuity, using Eq. 12.20b,

$$\frac{\rho^*}{\rho} = \frac{V}{V^*} = \frac{M^2(1 + k)}{1 + kM^2} \tag{12.20c}$$

The dimensionless stagnation temperature, T_0/T_0^*, can be determined from

$$\frac{T_0}{T_0^*} = \frac{T_0}{T}\frac{T}{T^*}\frac{T^*}{T_0^*} = \left(1 + \frac{k - 1}{2}M^2\right)\left[M\left(\frac{1 + k}{1 + kM^2}\right)\right]^2 \frac{1}{\left(1 + \frac{k - 1}{2}\right)}$$

$$\frac{T_0}{T_0^*} = \frac{2(k + 1)M^2\left(1 + \frac{k - 1}{2}M^2\right)}{(1 + kM^2)^2} \tag{12.20d}$$

Similarly,

$$\frac{p_0}{p_0^*} = \frac{p_0}{p}\frac{p}{p^*}\frac{p^*}{p_0^*} = \left(1 + \frac{k-1}{2}M^2\right)^{k/(k-1)}\left(\frac{1+k}{1+kM^2}\right)\frac{1}{\left(1+\frac{k-1}{2}\right)^{k/(k-1)}}$$

$$\frac{p_0}{p_0^*} = \frac{1+k}{1+kM^2}\left[\left(\frac{2}{k+1}\right)\left(1+\frac{k-1}{2}M^2\right)\right]^{k/(k-1)} \tag{12.20e}$$

Since the ratios in Eqs. 12.20a to 12.20e are functions of Mach number only, they may be computed once and presented in tabular form. These ratios are tabulated as functions of Mach number in Table D.3 of Appendix D.

Example 12.10

Air flows with negligible friction in a constant-area duct. At section ①, the flow properties are $T_1 = 60$ C, $p_1 = 135$ kPa (abs), and $V_1 = 732$ m/sec. Heat is added to the flow between section ① and section ②, where the Mach number is 1.2. Determine the flow properties at section ②, the heat transfer per unit mass, the entropy change, and sketch the process on a *Ts* diagram. Use tables.

EXAMPLE PROBLEM 12.10

GIVEN: Frictionless flow of air as shown:

$T_1 = 333$ K $\quad M_2 = 1.2$

$p_1 = 135$ kPa (abs)

$V_1 = 732$ m/sec

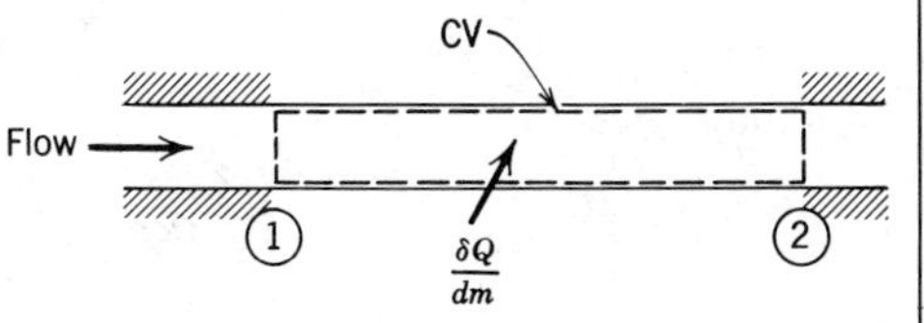

FIND: (a) Properties at section ②. (b) $\delta Q/dm$.
(c) $s_2 - s_1$. (d) Sketch on a *Ts* diagram.

SOLUTION:

To obtain property ratios from tables, we need both Mach numbers.

$$c_1 = \sqrt{kRT_1} = \left[1.4 \times \frac{287\ \text{N}\cdot\text{m}}{\text{kg}\cdot\text{K}} \times 333\ \text{K} \times \frac{\text{kg}\cdot\text{m}}{\text{N}\cdot\text{sec}^2}\right]^{1/2} = 366\ \text{m/sec}$$

$$M_1 = \frac{V_1}{c_1} = \frac{732\ \text{m}}{\text{sec}} \times \frac{\text{sec}}{366\ \text{m}} = 2.00$$

From Table D.3, Appendix D,

M	T_0/T_0^*	p_0/p_0^*	T/T^*	p/p^*	V/V^*
2.00	0.7934	1.503	0.5289	0.3636	1.455
1.20	0.9787	1.019	0.9119	0.7958	1.146

Using these data and recognizing that critical properties are constant, we obtain

$$\frac{T_2}{T_1} = \frac{T_2/T^*}{T_1/T^*} = \frac{0.9119}{0.5289} = 1.72; \qquad T_2 = 1.72\, T_1 = (1.72)\, 333\ \text{K} = 573\ \text{K} \qquad T_2$$

$$\frac{p_2}{p_1} = \frac{p_2/p^*}{p_1/p^*} = \frac{0.7958}{0.3636} = 2.19; \qquad p_2 = 2.19 p_1 = (2.19)\, 135\ \text{kPa} = 296\ \text{kPa (abs)} \qquad p_2$$

$$\frac{V_2}{V_1} = \frac{V_2/V^*}{V_1/V^*} = \frac{1.146}{1.455} = 0.788; \qquad V_2 = 0.788 V_1 = \frac{(0.788)\, 732\ \text{m}}{\text{sec}} = 577\ \text{m/sec} \qquad V_2$$

$$\rho_2 = \frac{p_2}{RT_2} = \frac{2.96 \times 10^5\ \text{N}}{\text{m}^2} \times \frac{\text{kg} \cdot \text{K}}{287\ \text{N} \cdot \text{m}} \times \frac{1}{573\ \text{K}} = 1.80\ \text{kg/m}^3 \qquad \rho_2$$

The heat transfer may be determined from the energy equation, which reduces to (see Example Problem 12.9)

$$\frac{\delta Q}{dm} = h_{0_2} - h_{0_1} = c_p(T_{0_2} - T_{0_1})$$

From the isentropic flow tables (Table D.1), at $M = 2.0$,

$$\frac{T}{T_0} = \frac{T_1}{T_{0_1}} = 0.5556; \qquad T_{0_1} = \frac{T_1}{0.5556} = \frac{333\ \text{K}}{0.5556} = 599\ \text{K}$$

and at $M = 1.2$

$$\frac{T}{T_0} = \frac{T_2}{T_{0_2}} = 0.7764; \qquad T_{0_2} = \frac{T_2}{0.7764} = \frac{573\ \text{K}}{0.7764} = 738\ \text{K} \qquad T_{0_2}$$

Substituting gives

$$\frac{\delta Q}{dm} = c_p(T_{0_2} - T_{0_1}) = 1.00 \frac{\text{kJ}}{\text{kg} \cdot \text{K}} (738 - 599)\ \text{K} = 139\ \text{kJ/kg} \qquad \delta Q/dm$$

The entropy change may be found from the $T\,ds$ equation, $T\,ds = dh - v\,dp$. For an ideal gas with constant specific heats,

$$s_2 - s_1 = c_p \ln \frac{T_2}{T_1} - R \ln \frac{p_2}{p_1}$$

$$= \frac{1.00\ \text{kJ}}{\text{kg} \cdot \text{K}} \ln\left(\frac{573}{333}\right) - \frac{287\ \text{N} \cdot \text{m}}{\text{kg} \cdot \text{K}} \ln\left(\frac{2.96 \times 10^5}{1.35 \times 10^5}\right) \frac{\text{kJ}}{1000\ \text{N} \cdot \text{m}}$$

$$s_2 - s_1 = 0.317\ \text{kJ/kg} \cdot \text{K} \qquad s_2 - s_1$$

Finally, let us check the effect on p_0. From Table D.1, at $M = 2.0$,

$$\frac{p}{p_0} = \frac{p_1}{p_{0_1}} = 0.1278; \qquad p_{0_1} = \frac{p_1}{0.1278} = \frac{135\ \text{kPa}}{0.1278} = 1.06\ \text{MPa (abs)}$$

and at $M = 1.2$,

$$\frac{p}{p_0} = \frac{p_2}{p_{0_2}} = 0.4124; \qquad p_{0_2} = \frac{p_2}{0.4124} = \frac{196\ \text{kPa}}{0.4124} = 718\ \text{kPa (abs)} \qquad p_{0_2}$$

Thus $p_{0_2} < p_{0_1}$, as expected for a heating process.

The process follows the supersonic branch of a Rayleigh line:

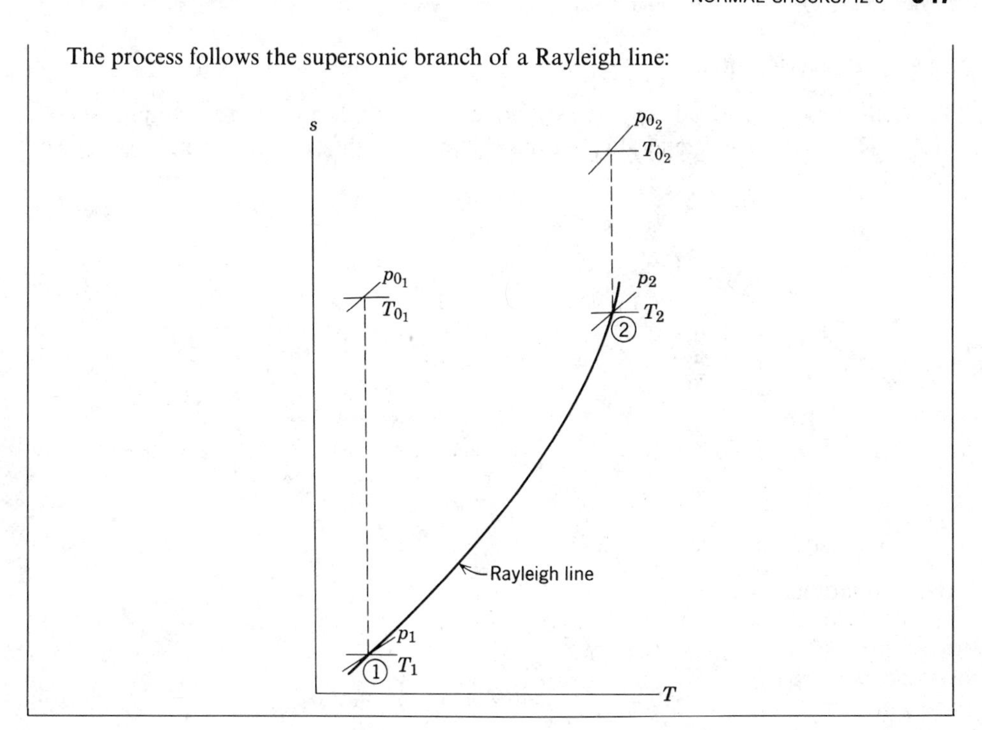

12-6 NORMAL SHOCKS

We have previously mentioned normal shocks in the section on nozzle flow. In practice, these irreversible discontinuities can occur in any supersonic flow field, in either internal flow or external flow.[3] Knowledge of property changes across shocks and of shock behavior is important in understanding the design of supersonic diffusers, e.g. for inlets on high performance aircraft, and supersonic wind tunnels. Accordingly, the purpose of this section is to analyze the normal shock process.

Before applying the basic equations to normal shocks, it is important that we have in mind a clear physical picture of the shock itself. Although it is physically impossible to have discontinuities in fluid properties, the normal shock is nearly discontinuous. The thickness of a shock is in the order of 0.2 microns (10^{-5} in.), or roughly 4 times the mean free path of the gas molecules. Across this small distance, large changes in pressure, temperature, and other properties occur. Local fluid accelerations can reach tens of millions of g's! These considerations justify treating the normal shock as an abrupt discontinuity; we are interested in changes occurring across the shock rather than in the details of its structure.

[3] The film, *Channel Flow of a Compressible Fluid*, D. Coles, principal, shows several examples of shock formation in internal flow.

12-6.1 Basic Equations

To begin our analysis, let us apply the basic equations to the thin control volume shown in Fig. 12.17, where for generality we have depicted a shock standing in a passage of arbitrary shape.

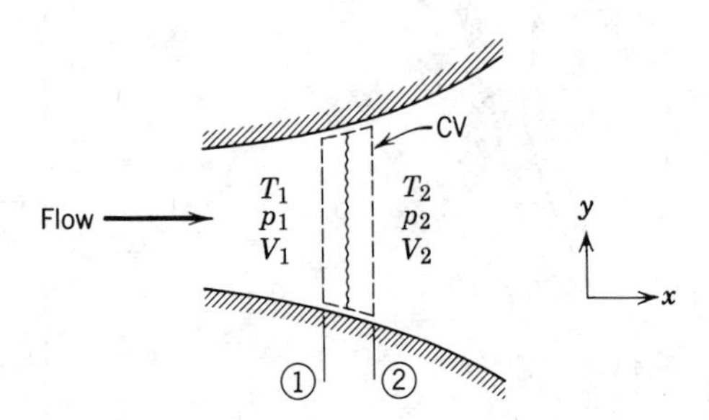

Fig. 12.17 Control volume used for analysis of normal shock.

a. Continuity Equation

Basic equation:

$$0 = \overset{=0(1)}{\cancel{\frac{\partial}{\partial t}}} \int_{CV} \rho \, d\forall + \int_{CS} \rho \vec{V} \cdot d\vec{A} \qquad (4.13)$$

Assumptions: (1) Steady flow
(2) Uniform flow at each section
(3) $A_1 = A_2 = A$, because the shock is so extraordinarily thin

Then

$$0 = \{-|\rho_1 V_1 A|\} + \{|\rho_2 V_2 A|\}$$

Writing the result in terms of scalar magnitudes, we obtain

$$\rho_1 V_1 = \rho_2 V_2 = G = \frac{\dot{m}}{A} \qquad (12.21a)$$

b. Momentum Equation

Basic equation:

$$F_{S_x} + \overset{=0(5)}{\cancel{F_{B_x}}} = \overset{=0(1)}{\cancel{\frac{\partial}{\partial t}}} \int_{CV} V_x \rho \, d\forall + \int_{CS} V_x \rho \vec{V} \cdot d\vec{A} \qquad (4.19a)$$

Assumptions: (4) Negligible friction force at duct walls because the shock is so thin
(5) $F_{B_x} = 0$

Under these conditions,

$$F_{S_x} = p_1 A - p_2 A = V_1\{-|\rho_1 V_1 A|\} + V_2\{|\rho_2 V_2 A|\}$$

Using scalar magnitudes and dropping absolute value signs, we obtain

$$p_1 A - p_2 A = \dot{m}V_2 - \dot{m}V_1 \tag{12.21b}$$

or

$$p_1 + \rho_1 V_1^2 = p_2 + \rho_2 V_2^2$$

c. First Law of Thermodynamics

Basic equation:

$$\overset{=0(6)}{\cancel{\dot{Q}}} + \overset{=0(7)}{\cancel{\dot{W}_s}} + \overset{=0(8)}{\cancel{\dot{W}_{\text{shear}}}} + \overset{=0(8)}{\cancel{\dot{W}_{\text{other}}}} = \overset{=0(1)}{\cancel{\frac{\partial}{\partial t}}}\int_{\text{CV}} e\rho\, d\forall + \int_{\text{CS}} (e + pv)\rho \vec{V} \cdot d\vec{A} \tag{4.59}$$

where

$$e = u + \frac{V^2}{2} + \overset{\simeq 0(9)}{\cancel{gz}}$$

Assumptions: (6) $\dot{Q} = 0$ (adiabatic flow)
(7) $\dot{W}_s = 0$
(8) $\dot{W}_{\text{shear}} = \dot{W}_{\text{other}} = 0$
(9) Effects of gravity are negligible

Then

$$0 = \left(u_1 + p_1 v_1 + \frac{V_1^2}{2}\right)\{-|\rho_1 V_1 A|\} + \left(u_2 + p_2 v_2 + \frac{V_2^2}{2}\right)\{|\rho_2 V_2 A|\}$$

However, from continuity, the mass flow rate terms in brackets are equal. We may also substitute $h = u + pv$, to obtain

$$h_1 + \frac{V_1^2}{2} = h_2 + \frac{V_2^2}{2} \tag{12.21c}$$

or, in terms of stagnation enthalpy,

$$h_{0_1} = h_{0_2}$$

Physically, we should expect the total energy of the flow to remain constant, since there is no energy addition.

d. Second Law of Thermodynamics

Basic equation:

$$\int_{\text{CS}} \frac{1}{T} \overset{=0(6)}{\cancel{\left(\frac{\dot{Q}}{A}\right)}} dA \leq \overset{=0(1)}{\cancel{\frac{\partial}{\partial t}}} \int_{\text{CV}} s\rho\, d\forall + \int_{\text{CS}} s\rho \vec{V} \cdot d\vec{A} \tag{4.61}$$

Then

$$0 \leq s_1\{-|\rho_1 V_1 A|\} + s_2\{|\rho_2 V_2 A|\}$$

Flow through the normal shock is irreversible because of the almost discontinuous property changes across the shock. Consequently, the inequality in the above equation holds. The control volume form of the second law then tells us that $s_2 - s_1 > 0$.

This fact is of little help in calculating the actual entropy change across the shock. To calculate the entropy change, we rely on the $T\,ds$ equations. Since

$$T\,ds = dh - v\,dp$$

then for an ideal gas we can write

$$ds = c_p \frac{dT}{T} - R\frac{dp}{p}$$

For constant specific heats this equation can be integrated to give

$$s_2 - s_1 = c_p \ln \frac{T_2}{T_1} - R \ln \frac{p_2}{p_1} \tag{12.21d}$$

e. Equation of State

For an ideal gas, the equation of state is given by

$$p = \rho R T \tag{12.21e}$$

Equations 12.21a to 12.21e are the governing equations for the flow of an ideal gas through a normal shock. If all the properties at state ① (immediately upstream of the shock) are known, then we have six unknowns $(T_2, p_2, \rho_2, V_2, h_2, s_2)$ in these five equations. However, we have the known relationship between h and T for an ideal gas, $dh = c_p\,dT$. For an ideal gas with constant specific heats,

$$\Delta h = h_2 - h_1 = c_p\,\Delta T = c_p(T_2 - T_1) \tag{12.21f}$$

We thus have the situation of six equations and six unknowns.

Then, if all conditions at state ① (immediately ahead of the shock) are known, how many possible states ② (immediately behind the shock) are there? The mathematics of the situation (six equations and six unknowns) indicates that there is a unique state ② for a given state ①.

We can obtain a physical picture of the flow through a normal shock by using some of the notions developed in the consideration of Fanno line and Rayleigh line flows. For convenience let us first rewrite the governing equations for a normal shock.

$$\rho_1 V_1 = \rho_2 V_2 = G = \frac{\dot{m}}{A} \tag{12.21a}$$

$$p_1 A - p_2 A = \dot{m} V_2 - \dot{m} V_1 \tag{12.21b}$$

$$h_1 + \frac{V_1^2}{2} = h_2 + \frac{V_2^2}{2} \tag{12.21c}$$

$$s_2 - s_1 = c_p \ln \frac{T_2}{T_1} - R \ln \frac{p_2}{p_1} \tag{12.21d}$$

$$p = \rho RT \tag{12.21e}$$

$$h_2 - h_1 = c_p(T_2 - T_1) \tag{12.21f}$$

Flow through a normal shock must satisfy Eqs. 12.21a to 12.21f. Since all conditions at state ① are known, we can locate state ① on a Ts diagram. If we were to draw a Fanno line curve through state ①, we would have a locus of mathematical states that satisfy Eqs. 12.21a and 12.21c to 12.21f. (The Fanno line curve does not satisfy Eq. 12.21b.) Drawing a Rayleigh line curve through state ① gives a locus of mathematical states that satisfy Eqs. 12.21a, 12.21b, and 12.21d to 12.21f. (The Rayleigh line curve does not satisfy Eq. 12.21c.) These curves are shown in Fig. 12.18.

The normal shock must satisfy all six of Eqs. 12.21a to 12.21f. Consequently, for a given state ①, the end state (state ②) of the normal shock must lie on both the Fanno line and the Rayleigh line passing through state ①. Hence, the intersection of the two lines at state ② represents the conditions downstream from the shock, corresponding to the given upstream conditions at state ①. In Fig. 12.18 the flow through the shock has been indicated as occurring from state ① to state ②. This is the only possible direction of the shock process as dictated by the second law ($s_2 > s_1$).

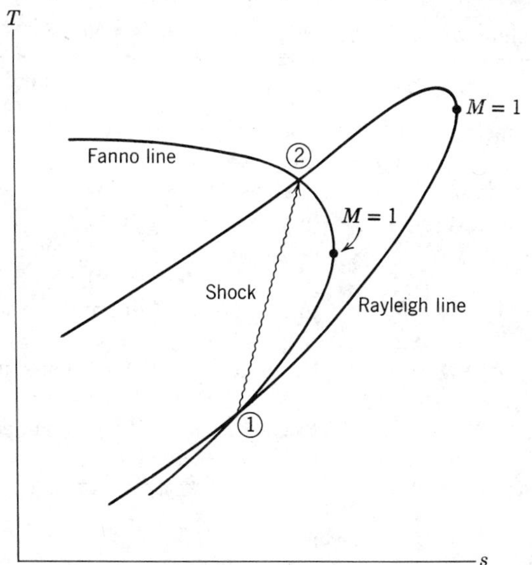

Fig. 12.18 Intersection of Fanno line and Rayleigh line as a solution of the normal shock equations.

From Fig. 12.18 we note also that the flow through a normal shock involves a change from supersonic to subsonic speeds. Normal shocks can occur only in a flow that is initially supersonic.

As an aid in summarizing the effects of a normal shock on the flow properties, a schematic of the normal shock process is illustrated on the Ts plane of Fig. 12.19. This

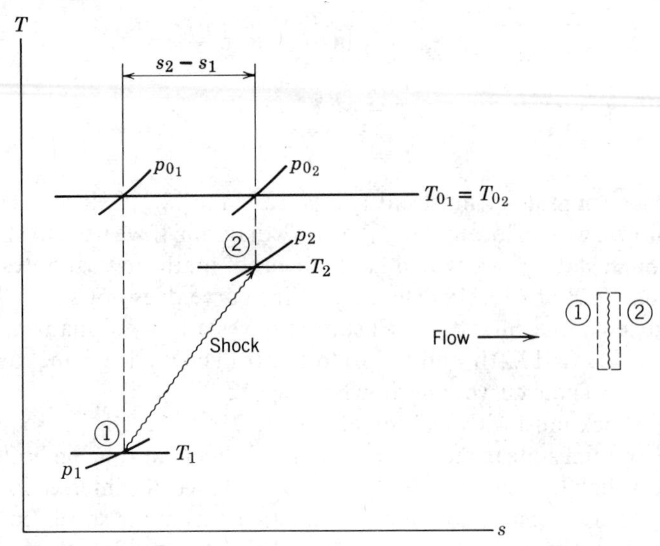

Fig. 12.19 Schematic of normal shock process on the *Ts* plane.

Table 12.3 Summary of Property Changes across a Normal Shock

Property	Effect	Obtained from
Stagnation temperature, T_0	Constant	Energy equation
Entropy, s	Increase	Second law
Stagnation pressure, p_0	Decrease	Ts diagram
Temperature, T	Increase	Ts diagram
Velocity, V	Decrease	Energy equation, and effect on T
Density, ρ	Increase	Continuity equation, and effect on V
Pressure, p	Increase	Momentum equation, and effect on V
Mach number, M	Decrease	$M = V/c$, and effect on V and T

figure, together with the governing basic equations, is the basis for Table 12.3. You are encouraged to follow through the logic indicated in the table.

In theory, the solution of six equations in six unknowns poses no difficulty, but in practice the algebra becomes involved. It makes sense to recast the equations in a form suitable for tabulating property ratios across the shock. This we shall do in the next section. To illustrate the direct application of the control volume equations to a normal shock, consider Example Problem 12.11, in which the problem is overspecified (one property downstream from the shock is given).

Example 12.11

A normal shock stands in a duct. The fluid is air, which may be considered an ideal gas. Properties upstream from the shock are $T_1 = 5$ C, $p_1 = 65.0$ kPa (abs), and $V_1 = 668$ m/sec. The temperature at section ②, downstream from the shock, is $T_2 = 469$ K. Determine the property values at section ② and compare them with the upstream values. Sketch the process on a Ts diagram.

EXAMPLE PROBLEM 12.11

GIVEN: Normal shock in a duct as shown:

$T_1 = 5\text{ C} \qquad T_2 = 469\text{ K}$

$p_1 = 65.0\text{ kPa (abs)}$

$V_1 = 668\text{ m/sec}$

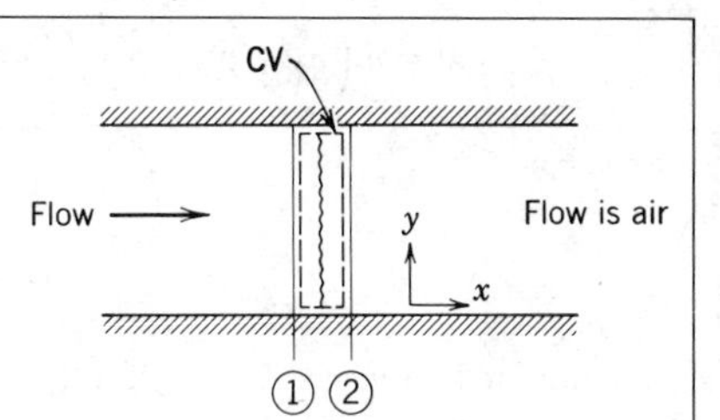

FIND: (a) Properties at section ②.
(b) Ts diagram.

SOLUTION:
Let us first compute the remaining properties at section ①. For an ideal gas

$$\rho_1 = \frac{p_1}{RT_1} = \frac{6.5 \times 10^4\text{ N}}{\text{m}^2} \times \frac{\text{kg}\cdot\text{K}}{287\text{ N}\cdot\text{m}} \times \frac{1}{278\text{ K}} = 0.815\text{ kg/m}^3$$

$$c_1 = \sqrt{kRT_1} = \left[1.4 \times \frac{287\text{ N}\cdot\text{m}}{\text{kg}\cdot\text{K}} \times 278\text{ K} \times \frac{\text{kg}\cdot\text{m}}{\text{N}\cdot\text{sec}^2}\right]^{1/2} = 334\text{ m/sec}$$

$$M_1 = \frac{V_1}{c_1} = \frac{668}{334} = 2.00$$

$$T_{0_1} = T_1\left(1 + \frac{k-1}{2}M_1^2\right) = 278\text{ K}[1 + 0.2(2.0)^2] = 500\text{ K}$$

$$p_{0_1} = p_1\left(1 + \frac{k-1}{2}M_1^2\right)^{k/(k-1)} = 65.0\text{ kPa}[1 + 0.2(2.0)^2]^{3.5} = 509\text{ kPa (abs)}$$

V_2 may be evaluated by applying the energy equation to the control volume shown.

Basic equation:

$$\overset{=0(1)}{\dot{Q}} + \overset{=0(2)}{\dot{W}_s} + \overset{=0(3)}{\dot{W}_{\text{shear}}} + \overset{=0(3)}{\dot{W}_{\text{other}}} = \overset{=0(4)}{\frac{\partial}{\partial t}\int_{\text{CV}} e\rho\, d\forall} + \int_{\text{CS}} (e + pv)\rho \vec{V} \cdot d\vec{A} \tag{4.59}$$

where

$$e = u + \frac{V^2}{2} + \overset{\simeq 0(6)}{gz}$$

Assumptions: (1) $\dot{Q} = 0$
(2) $\dot{W}_s = 0$
(3) $\dot{W}_{\text{shear}} = \dot{W}_{\text{other}} = 0$
(4) Steady flow
(5) Uniform flow at each section
(6) Neglect gravity term
(7) $A_1 = A_2 = A$

Then

$$0 = \left(u_1 + p_1 v_1 + \frac{V_1^2}{2}\right)\{-|\rho_1 V_1 A|\} + \left(u_2 + p_2 v_2 + \frac{V_2^2}{2}\right)\{|\rho_2 V_2 A|\}$$

$$0 = \dot{m}\left(u_2 + p_2 v_2 + \frac{V_2^2}{2} - u_1 - p_1 v_1 - \frac{V_1^2}{2}\right)$$

or

$$h_1 + \frac{V_1^2}{2} = h_2 + \frac{V_2^2}{2}$$

Solving for V_2 gives

$$V_2 = [V_1^2 + 2(h_1 - h_2)]^{1/2} = [V_1^2 + 2c_p(T_1 - T_2)]^{1/2}$$

$$= \left[\frac{(668)^2\ \text{m}^2}{\text{sec}^2} + 2 \times \frac{1000\ \text{N}\cdot\text{m}}{\text{kg}\cdot\text{K}}\ (278 - 469)\ \text{K} \times \frac{\text{kg}\cdot\text{m}}{\text{N}\cdot\text{sec}^2}\right]^{1/2}$$

$$V_2 = 253\ \text{m/sec}$$

V_2

Continuity reduces to $G = \rho_1 V_1 = \rho_2 V_2$, so

$$\rho_2 = \rho_1 \frac{V_1}{V_2} = 0.815\ \frac{\text{kg}}{\text{m}^3}\left(\frac{668}{253}\right) = 2.15\ \text{kg/m}^3$$

ρ_2

The pressure can be obtained two ways:
(1) From the ideal gas equation of state

$$p_2 = \rho_2 R T_2 = \frac{2.15\ \text{kg}}{\text{m}^3} \times \frac{287\ \text{N}\cdot\text{m}}{\text{kg}\cdot\text{K}} \times 469\ \text{K} = 289\ \text{kPa (abs)}$$

p_2

(2) From the momentum equation

Basic equation:

$$F_{S_x} + \cancelto{0(8)}{F_{B_x}} = \cancelto{0(4)}{\frac{\partial}{\partial t}\int_{\text{CV}} V_x \rho\, d\forall} + \int_{\text{CS}} V_x \rho \vec{V}\cdot d\vec{A} \qquad (4.19\text{a})$$

Assumption: (8) $F_{B_x} = 0$

Then

$$p_1 A - p_2 A = V_1\{-|\rho_1 V_1 A|\} + V_2\{|\rho_2 V_2 A|\} = \dot{m}(V_2 - V_1)$$

or

$$p_1 - p_2 = \rho_1 V_1 (V_2 - V_1)$$

Solving gives

$$p_2 = p_1 - \rho_1 V_1 (V_2 - V_1)$$

$$p_2 = \frac{6.5 \times 10^4 \text{ N}}{\text{m}^2} - \frac{0.815 \text{ kg}}{\text{m}^3} \times \frac{668 \text{ m}}{\text{sec}} \frac{(253 - 668) \text{ m}}{\text{sec}} \times \frac{\text{N} \cdot \text{sec}^2}{\text{kg} \cdot \text{m}} = 291 \text{ kPa (abs)}$$

(The two calculated pressure values are in close, but not exact, agreement due to roundoff errors.)

For adiabatic flow, $T_0 =$ constant. Thus

$$T_{0_2} = T_{0_1} = 500 \text{ K}$$

T_{0_2}

The local isentropic stagnation pressure at section ② is

$$p_{0_2} = p_2 \left(\frac{T_{0_2}}{T_2} \right)^{k/(k-1)} = 289 \text{ kPa} \left(\frac{500}{469} \right)^{3.5} = 362 \text{ kPa (abs)}$$

p_{0_2}

Comparing, we see that

$$T_{0_2} = T_{0_1}$$

$$p_{0_2} < p_{0_1}$$

$$T_2 > T_1$$

$$p_2 > p_1$$

and

$$V_2 < V_1$$

in accord with Table 12.3.

The entropy change may be computed from the $T\,ds$ equation

$$T\,ds = dh - v\,dp$$

For an ideal gas

$$ds = c_p \frac{dT}{T} - R \frac{dp}{p}$$

Integrating, for constant specific heats, yields

$$s_2 - s_1 = c_p \ln \frac{T_2}{T_1} - R \ln \frac{p_2}{p_1}$$

Since $s_{0_2} - s_{0_1} = s_2 - s_1$, the entropy change is easiest to evaluate at stagnation conditions because $T_{0_2} = T_{0_1}$. At stagnation conditions we have

$$s_2 - s_1 = s_{0_2} - s_{0_1} = \overbrace{c_p \ln \frac{T_{0_2}}{T_{0_1}}}^{=0} - R \ln \frac{p_{0_2}}{p_{0_1}} = -\frac{287 \text{ N} \cdot \text{m}}{\text{kg} \cdot \text{K}} \ln \left(\frac{3.62 \times 10^5}{5.09 \times 10^5} \right)$$

$$s_2 - s_1 = 0.0978 \text{ kJ/kg} \cdot \text{K}$$

$s_2 - s_1$

Finally, the Ts diagram may be sketched:

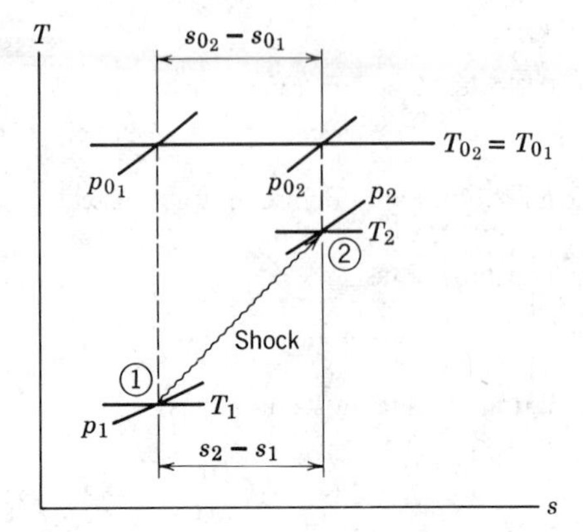

{Note once again that $s_{0_2} - s_{0_1} = s_2 - s_1$, since $s_{0_1} = s_1$ and $s_{0_2} = s_2$.}

**12-6.2 Tables for Computation of Normal Shocks in an Ideal Gas

The basic equations for the flow through a normal shock indicate that for given conditions ahead of the shock there is a unique downstream state. Therefore, it is possible to develop expressions for property ratios across the shock in terms of the Mach number, M_1, ahead of the shock; these results can be tabulated as functions of M_1.

To obtain the results, we proceed in three steps. First, we obtain property ratios (e.g. T_2/T_1 and p_2/p_1) in terms of M_1 and M_2. Then we develop a relation between M_1 and M_2. Finally, we use this relation to obtain expressions for the property ratios in terms of the upstream Mach number, M_1.

The temperature ratio can be expressed as

$$\frac{T_2}{T_1} = \frac{T_2}{T_{0_2}} \frac{T_{0_2}}{T_{0_1}} \frac{T_{0_1}}{T_1}$$

Since the stagnation temperature is constant across the shock, we have

$$\frac{T_2}{T_1} = \frac{1 + \dfrac{k-1}{2} M_1^2}{1 + \dfrac{k-1}{2} M_2^2} \tag{12.22a}$$

A velocity ratio may be obtained by using

$$\frac{V_2}{V_1} = \frac{M_2 c_2}{M_1 c_1} = \frac{M_2}{M_1} \frac{\sqrt{kRT_2}}{\sqrt{kRT_1}} = \frac{M_2}{M_1} \sqrt{\frac{T_2}{T_1}}$$

** This section may be omitted without loss of continuity in the text material.

or

$$\frac{V_2}{V_1} = \frac{M_2}{M_1}\left[\frac{1 + \frac{k-1}{2}M_1^2}{1 + \frac{k-1}{2}M_2^2}\right]^{1/2} \tag{12.22b}$$

A ratio of densities may be obtained from the continuity equation

$$\rho_1 V_1 = \rho_2 V_2 \tag{12.21a}$$

Substituting from Eq. 12.22b gives

$$\frac{\rho_2}{\rho_1} = \frac{V_1}{V_2} = \frac{M_1}{M_2}\left[\frac{1 + \frac{k-1}{2}M_2^2}{1 + \frac{k-1}{2}M_1^2}\right]^{1/2} \tag{12.22c}$$

Finally, we can obtain a pressure ratio from the momentum equation

$$p_1 A - p_2 A = \dot{m}V_2 - \dot{m}V_1 \tag{12.21b}$$

or

$$p_1 + \rho_1 V_1^2 = p_2 + \rho_2 V_2^2$$

Substituting $\rho = p/RT$ and factoring out pressures gives

$$p_1\left[1 + \frac{V_1^2}{RT_1}\right] = p_2\left[1 + \frac{V_2^2}{RT_2}\right]$$

Since

$$\frac{V^2}{RT} = k\frac{V^2}{kRT} = kM^2$$

then

$$p_1[1 + kM_1^2] = p_2[1 + kM_2^2]$$

and finally,

$$\frac{p_2}{p_1} = \frac{1 + kM_1^2}{1 + kM_2^2} \tag{12.22d}$$

In order to solve for M_2 in terms of M_1, we need to obtain another expression for one of the property ratios given by Eqs. 12.22a to 12.22d.

From the ideal gas equation of state the temperature ratio may be written

$$\frac{T_2}{T_1} = \frac{p_2/\rho_2 R}{p_1/\rho_1 R} = \frac{p_2}{p_1}\frac{\rho_1}{\rho_2}$$

Substituting from Eqs. 12.22c and 12.22d yields

$$\frac{T_2}{T_1} = \left[\frac{1 + kM_1^2}{1 + kM_2^2}\right]\frac{M_2}{M_1}\left[\frac{1 + \dfrac{k-1}{2}M_1^2}{1 + \dfrac{k-1}{2}M_2^2}\right]^{1/2} \tag{12.23}$$

Equations 12.22a and 12.23 are two equations for T_2/T_1. We can combine them and solve for M_2 in terms of M_1. Combining and canceling gives

$$\left[\frac{1 + \dfrac{k-1}{2}M_1^2}{1 + \dfrac{k-1}{2}M_2^2}\right]^{1/2} = \frac{M_2}{M_1}\left[\frac{1 + kM_1^2}{1 + kM_2^2}\right]$$

Squaring, we obtain

$$\frac{1 + \dfrac{k-1}{2}M_1^2}{1 + \dfrac{k-1}{2}M_2^2} = \frac{M_2^2}{M_1^2}\left[\frac{1 + 2kM_1^2 + k^2M_1^4}{1 + 2kM_2^2 + k^2M_2^4}\right]$$

which may be solved explicitly for M_2^2. Two solutions are obtained:

$$M_2^2 = M_1^2 \tag{12.24a}$$

and

$$M_2^2 = \frac{M_1^2 + \dfrac{2}{k-1}}{\dfrac{2k}{k-1}M_1^2 - 1} \tag{12.24b}$$

Obviously, the first of these is trivial. The second expresses the unique dependence of M_2 on M_1. This relation is tabulated in Appendix D, Table D.4. Now, having a relationship between M_2 and M_1, we can solve for the property ratios across a shock. Knowing M_1, M_2 is obtained from Eq. 12.24b. The property ratios subsequently can be determined from Eqs. 12.22a to 12.22d. These property ratios are tabulated as functions of M_1 in Table D.4 of Appendix D.

Since the stagnation temperature remains constant, the stagnation temperature ratio across the shock is unity. The ratio of stagnation pressures is evaluated as

$$\frac{p_{0_2}}{p_{0_1}} = \frac{p_{0_2}}{p_2}\frac{p_2}{p_1}\frac{p_1}{p_{0_1}} = \frac{p_2}{p_1}\left[\frac{1 + \dfrac{k-1}{2}M_2^2}{1 + \dfrac{k-1}{2}M_1^2}\right]^{k/(k-1)} \tag{12.25}$$

Combining Eqs. 12.22d and 12.24b, we obtain (after considerable algebra)

$$\frac{p_2}{p_1} = \frac{1 + kM_1^2}{1 + kM_2^2} = \frac{2k}{k+1} M_1^2 - \frac{k-1}{k+1} \tag{12.26}$$

Using Eqs. 12.24b and 12.26, we find that Eq. 12.25 becomes

$$\frac{p_{0_2}}{p_{0_1}} = \frac{\left[\dfrac{\dfrac{k+1}{2} M_1^2}{1 + \dfrac{k-1}{2} M_1^2}\right]^{k/(k-1)}}{\left[\dfrac{2k}{k+1} M_1^2 - \dfrac{k-1}{k+1}\right]^{1/(k-1)}} \tag{12.27}$$

The stagnation pressure ratio as a function of upstream Mach number is included in Table D.4 of Appendix D.

Use of the tables in the solution of a problem involving a normal shock is illustrated in Example Problem 12.12.

Example 12.12

A normal shock stands in a duct. The fluid is air, which can be considered an ideal gas. Properties upstream from the shock are $T_1 = 5$ C, $p_1 = 65.0$ kPa (abs), and $V_1 = 668$ m/sec. Determine the properties downstream and $s_2 - s_1$. Include a Ts plot.

(Note that these upstream conditions are the same as those given in Example Problem 12.11. However, with the tables available, we do not need to know any properties downstream to complete a solution.)

EXAMPLE PROBLEM 12.12

GIVEN: Normal shock in a duct as shown:

$T_1 = 278$ K

$p_1 = 65.0$ kPa (abs)

$V_1 = 668$ m/sec

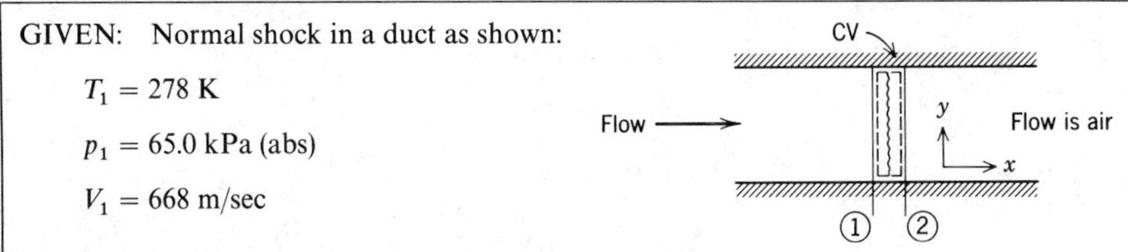

FIND: (a) Properties at section ②.
(b) $s_2 - s_1$.
(c) Ts plot.

SOLUTION:

To use the shock tables, we need to know M_1. For an ideal gas,

$$c_1 = \sqrt{kRT_1} = \left[1.4 \times 287 \frac{\text{N} \cdot \text{m}}{\text{kg} \cdot \text{K}} \times 278 \text{ K} \times \frac{\text{kg} \cdot \text{m}}{\text{N} \cdot \text{sec}^2}\right]^{1/2} = 334 \text{ m/sec}$$

$$M_1 = \frac{V_1}{c_1} = \frac{668}{334} = 2.0$$

Normal shock property ratios are given in Table D.4, Appendix D. At $M_1 = 2.0$,

M_1	M_2	p_{0_2}/p_{0_1}	T_2/T_1	p_2/p_1	ρ_2/ρ_1
2.00	0.5774	0.7209	1.688	4.500	2.667

From these data

$$T_2 = 1.688T_1 = (1.688)278\text{ K} = 469\text{ K} \qquad T_2$$

$$p_2 = 4.500p_1 = (4.500)65.0\text{ kPa} = 293\text{ kPa (abs)} \qquad p_2$$

For an ideal gas,

$$\rho_2 = \frac{p_2}{RT_2} = \frac{2.93 \times 10^5\text{ N}}{\text{m}^2} \times \frac{\text{kg}\cdot\text{K}}{287\text{ N}\cdot\text{m}} \times \frac{1}{469\text{ K}} = 2.18\text{ kg/m}^3 \qquad \rho_2$$

and

$$V_2 = M_2c_2 = M_2\sqrt{kRT_2} = (0.5774)\left[1.4 \times \frac{287\text{ N}\cdot\text{m}}{\text{kg}\cdot\text{K}} \times 469\text{ K} \times \frac{\text{kg}\cdot\text{m}}{\text{N}\cdot\text{sec}^2}\right]^{1/2}$$

$$V_2 = 251\text{ m/sec} \qquad V_2$$

Local isentropic stagnation properties at section ① may be evaluated using tables for isentropic flow. From Table D.1, Appendix D, at $M = 2.0$,

$$\frac{T}{T_0} = \frac{T_1}{T_{0_1}} = 0.5556; \qquad T_{0_1} = \frac{T_1}{0.5556} = \frac{278\text{ K}}{0.5556} = 500\text{ K}$$

$$\frac{p}{p_0} = \frac{p_1}{p_{0_1}} = 0.1278; \qquad p_{0_1} = \frac{p_1}{0.1278} = \frac{65.0\text{ kPa}}{0.1278} = 509\text{ kPa (abs)}$$

The stagnation temperature is constant in adiabatic flow. Thus

$$T_{0_2} = T_{0_1} = 500\text{ K} \qquad T_{0_2}$$

Using the property ratios for a normal shock, we obtain

$$p_{0_2} = p_{0_1}\frac{p_{0_2}}{p_{0_1}} = 509\text{ kPa }(0.7209) = 367\text{ kPa (abs)} \qquad p_{0_2}$$

The entropy change across the shock may be found from the $T\,ds$ equation

$$T\,ds = dh - v\,dp$$

For an ideal gas,

$$ds = c_p\frac{dT}{T} - R\frac{dp}{p}$$

Integrating for constant specific heats gives

$$s_2 - s_1 = c_p \ln\frac{T_2}{T_1} - R\ln\frac{p_2}{p_1}$$

But $s_{0_2} - s_{0_1} = s_2 - s_1$, so

$$s_{0_2} - s_{0_1} = s_2 - s_1 = c_p \ln \frac{T_{0_2}}{T_{0_1}} - R \ln \frac{p_{0_2}}{p_{0_1}} = -\frac{0.287}{} \frac{\text{kJ}}{\text{kg} \cdot \text{K}} \ln(0.7209)$$

(where $c_p \ln \frac{T_{0_2}}{T_{0_1}} = 0$)

$$s_2 - s_1 = 0.0939 \text{ kJ/kg} \cdot \text{K}$$

$s_2 - s_1$

The Ts diagram is

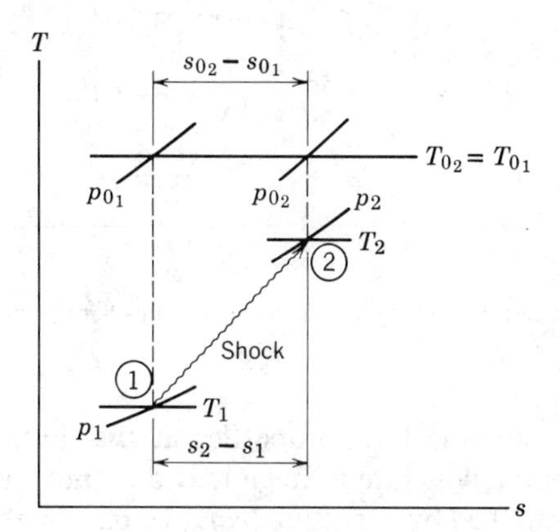

Comparing the solutions presented in Example Problems 12.11 and 12.12, we see that using the tables simplifies the computations appreciably. In addition, the consistency of the calculated values is improved as a result of using tabulated values accurate to four significant figures. You must be careful when using the tables to check your work as you proceed. Each calculated result should be examined to be sure both its trend and magnitude are reasonable. This process is simplified and made almost automatic when a Ts diagram is drawn for each problem.

12-6.3 Flow in a Converging-Diverging Nozzle

Since we have considered normal shocks, we now can complete our discussion of flow in a converging-diverging nozzle operating under varying back pressures, begun in Section 12-3.5. The pressure distribution through the nozzle for different back pressures is shown in Fig. 12.20.

Four different regimes of flow are possible. In Regime I the flow is subsonic throughout. The flow rate increases with decreasing back pressure. At condition (*iii*), which forms the dividing line between Regimes I and II, the flow at the throat is sonic, i.e. $M_t = 1$.

As the back pressure is lowered below that of condition (*iii*), a normal shock appears downstream from the throat. There is a pressure rise across the shock. Since the flow is subsonic ($M < 1$) behind the shock, the flow decelerates, with an accompanying increase in pressure, through the diverging portion of the channel. As the back pressure is lowered further, the shock moves downstream until it appears at the exit plane of the nozzle (condition *vii*). In Regime II, as in Regime I, the exit flow is subsonic and

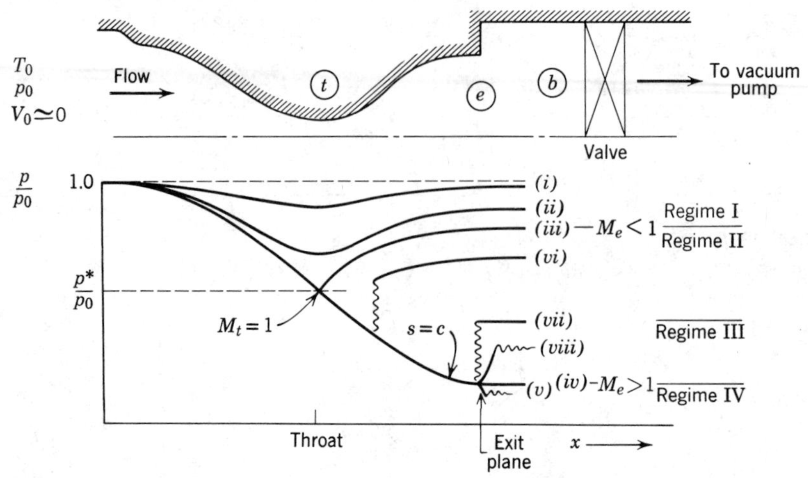

Fig. 12.20 Pressure distributions for flow in a converging-diverging nozzle as a function of back pressure.

consequently $p_e = p_b$. Since the flow properties at the throat are constant for all conditions in Regime II, the flow rate in Regime II does not vary with back pressure.

In Regime III, as exemplified by condition (*viii*), the back pressure is higher than the exit pressure but not sufficiently high to sustain a normal shock in the exit plane. The flow adjusts to the back pressure through a series of oblique compression shocks that cannot be treated by one-dimensional theory.

As was previously noted in Section 12-3.5, condition (*iv*) represents the design condition.[4] In Regime IV the flow adjusts to the lower back pressure through a series of oblique expansion waves that cannot be treated by one-dimensional theory.

12-7 SUMMARY OBJECTIVES

After completing study of Chapter 12, you should be able to do the following:

1. Write the basic equations for the steady, one-dimensional isentropic flow of (a) any compressible fluid, and (b) an ideal gas, through a channel of arbitrary cross section. Use these equations, together with the appropriate Ts plot, to solve isentropic flow problems.
2. Determine the effect of area change on fluid properties for isentropic flow. Sketch flow passages for the following: subsonic nozzle, subsonic diffuser, supersonic nozzle, and supersonic diffuser.
3. For flow in (a) a converging nozzle, and (b) a converging-diverging nozzle, plot the pressure distribution through the nozzle as a function of the nozzle back pressure. State the conditions under which the nozzle is choked.
4. Write the basic equations for the steady, one-dimensional, adiabatic flow of an ideal gas with

[4] Flow behavior in converging-diverging nozzles and applications to supersonic wind tunnels are shown in the film, *Channel Flow of a Compressible Fluid*, D. Coles, principal.

constant specific heats through a constant-area duct. Use these equations, together with the appropriate Ts plot, to solve Fanno line flow problems.

5. Write the basic equations for the steady, one-dimensional, frictionless flow of an ideal gas with heat transfer through a constant-area duct. Use these equations, together with the appropriate Ts plot, to solve Rayleigh line flow problems.
6. Write the basic equations for the steady, one-dimensional flow of an ideal gas through a normal shock. Use these equations, together with the appropriate Ts plot, to solve normal shock problems.

**7. Use the tables for computation of (a) isentropic, (b) Fanno line, (c) Rayleigh line, and (d) normal shock flow of an ideal gas.

8. Solve the problems at the end of the chapter that relate to the material you have studied.

PROBLEMS

12.1 Steam flows steadily and isentropically through a nozzle. At a section the steam is at 700 F and 240 psia, and the flow speed is 600 ft/sec. The steam is to be accelerated to the maximum possible speed such that the exit stream quality is 1.0. Determine the corresponding exit speed.

12.2 Steam flows steadily and isentropically through a nozzle. At a section the steam is at 700 F and 240 psia, and the flow speed is 600 ft/sec. The steam is to be accelerated to the maximum possible speed such that the exit stream quality is 0.9. Determine the corresponding exit speed.

12.3 Steam flows steadily and isentropically through a nozzle. At a section the steam is at 350 C and an absolute pressure of 1.0 MPa, and the flow speed is 200 m/sec. The steam is to be accelerated to the maximum possible speed such that the exit stream quality is 1.0. Determine the corresponding exit speed.

12.4 Steam flows steadily and isentropically through a nozzle. At a section the steam is at 350 C and an absolute pressure of 1.0 MPa, and the flow speed is 200 m/sec. The steam is to be accelerated to the maximum possible speed such that the exit stream quality is 0.9. Determine the corresponding exit speed.

12.5 Steam flows steadily and isentropically through a nozzle. At an upstream section where the flow speed is negligible, the temperature and pressure are 900 F and 900 psia, respectively. At a section where the nozzle diameter is 0.188 in., the steam pressure is 600 psia. Determine the speed and Mach number at this section and the mass flow rate of steam. Sketch the passage shape.

12.6 Steam flows steadily and isentropically through a nozzle. At an upstream section where the flow speed is negligible, the temperature and pressure are 880 F and 875 psia. At a section where the nozzle diameter is 0.50 in., the steam pressure is 290 psia. Determine the speed and Mach number at this section and the mass flow rate of steam. Sketch the passage shape.

12.7 Steam flows steadily and isentropically through a nozzle. At an upstream section where the flow speed is negligible, the temperature and absolute pressure are 475 C and 6.0 MPa. At a section where the nozzle diameter is 6 mm, the absolute pressure of the steam is 4.0 MPa. Determine the speed and Mach number at this section and the mass flow rate of steam. Sketch the passage shape.

** This objective applies to sections that may be omitted without loss of continuity in the text material.

12.8 Steam flows steadily and isentropically through a nozzle. At an upstream section where the flow speed is negligible, the temperature and absolute pressure are 475 C and 6.0 MPa. At a section where the nozzle diameter is 12 mm, the absolute pressure of the steam is 2.0 MPa. Determine the speed and Mach number at this section and the mass flow rate of steam. Sketch the passage shape.

12.9 An F-4 aircraft makes a low-level pass over an airfield at sea level on a standard day. A pitot tube on the aircraft senses a stagnation pressure of 23.6 psia. Determine the Mach number at which the aircraft flies. Evaluate the speed of the aircraft.

12.10 The aircraft of Problem 12.9 flies at $M = 0.851$. Air is slowed in the engine inlet system to 475 ft/sec relative to the aircraft. Determine the temperature of the air at this location. If the deceleration process were modeled as isentropic, what would be the static pressure at this section?

12.11 A nozzle is designed to expand air isentropically to atmospheric pressure from a large tank in which properties are held constant at 5 C and 304 kPa (abs). The desired flow rate is 1 kg/sec. Determine the exit area of the nozzle. Sketch a plot of Mach number and pressure as a function of distance along the nozzle.

12.12 At a section upstream of the throat in a converging-diverging nozzle the flow is at a pressure of 30 psia, and a temperature of 90 F; its speed is 575 ft/sec. For the isentropic flow of air determine the Mach number at the point where the pressure is 12 psia.

12.13 Air flows steadily and isentropically into an aircraft inlet at a rate of 100 kg/sec. At the section where the area is 0.464 m^2, $M = 3$, $T = -60$ C, and the absolute pressure is 15.0 kPa. Determine the speed and cross-sectional area downstream where $T = 138$ C. Sketch the flow passage.

12.14 Air flows steadily and isentropically through a passage. At section ① where the cross-sectional area is 0.02 m^2, the air is at 40.0 kPa (abs), 60 C, and the Mach number is 2.0. At a section ② downstream, the speed is 519 m/sec. Calculate the Mach number at section ②. Sketch the shape of the passage between sections ① and ②.

12.15 Air at an absolute pressure of 60.0 kPa and 27 C enters a passage at 486 m/sec. The cross-sectional area at the entrance is 0.02 m^2. At section ② downstream, the pressure is 78.8 kPa (abs). Assuming isentropic flow, calculate the Mach number at section ②.

12.16 A supersonic diffuser decelerates air isentropically from a Mach number of 3 to a Mach number of 1.4. If the static pressure at the diffuser inlet is 30.0 kPa (abs), calculate the static pressure rise in the diffuser, and the ratio of inlet to outlet area of the diffuser.

12.17 Air, with a stagnation temperature of 600 R, flows isentropically through a converging nozzle. At the point in the flow where the temperature is 571 R, the pressure is 100 psia. Determine the speed and the stagnation pressure at the downstream location where $T = 532$ R.

12.18 Air flows isentropically through a converging nozzle into a receiver where the pressure is 33 psia. If the pressure is 50 psia, and the speed is 500 ft/sec at the nozzle location where the Mach number is 0.4, determine the pressure, speed, and Mach number at the nozzle throat.

12.19 Air flows isentropically through a converging nozzle into a receiver in which the absolute pressure is 240 kPa. The air enters the nozzle with negligible speed at a pressure of 406 kPa (abs) and a temperature of 95 C. Determine the flow rate through the nozzle for a nozzle throat area of 0.01 m^2.

12.20 Air flows isentropically through a converging nozzle attached to a large tank where the absolute pressure is 171 kPa and the temperature is 27 C. At the inlet section the Mach number is 0.2. The nozzle discharges to the atmosphere; the discharge area is 0.015 m^2. Determine the magnitude and direction of the force that must be applied to hold the nozzle in place.

12.21 A stream of air flowing in a duct (area = 1 $in.^2$) is at a pressure of 20 psia, has a Mach number of 0.6, and flows at a rate of 0.5 lbm/sec.
(a) Determine the local isentropic stagnation temperature.
(b) If the cross-sectional area of the passage were reduced downstream, determine the maximum percentage reduction of area allowable without reducing the flow rate (assume isentropic flow).
(c) Determine the speed and pressure at the minimum area location.

12.22 Air flowing isentropically through a converging nozzle discharges to the atmosphere. At the section where the absolute pressure is 179 kPa, the temperature is 39 C, and the air speed is 177 m/sec. Determine the nozzle throat pressure.

12.23 Air flows from a large tank ($p = 650$ kPa (abs), $T = 550$ C) through a converging nozzle, with a throat area of 600 mm^2, and discharges to the atmosphere. Determine the mass rate of flow for isentropic flow through the nozzle.

12.24 A large tank supplies air to a convergent nozzle that discharges to atmospheric pressure. Assume the flow to be reversible and adiabatic.
(a) For what range of tank pressures will the flow at the nozzle exit be sonic ($M = 1$)?
(b) If the tank pressure is 600 kPa (abs) and the temperature is 600 K, what is the mass flow rate through the nozzle if the exit area is 1.29×10^{-3} m^2?

12.25 Helium in a very large tank is maintained at 800 kPa (abs), 250 C. Helium leaves the tank steadily and isentropically through a converging nozzle that discharges to the atmosphere. The nozzle throat area is 0.002 m^2. Determine the helium density at the nozzle throat.

12.26 Carbon dioxide at 50 MPa (abs) and 150 C discharges isentropically from a large tank to the atmosphere through a converging nozzle whose throat area is 1.0 mm^2. Find the temperature at the exit, the pressure difference between nozzle exit and atmosphere, and the mass flow rate.

12.27 Air, with a stagnation pressure of 650 kPa (abs) and a stagnation temperature of 350 K, is flowing isentropically through a converging nozzle. At the section in the nozzle where the area is 2.6×10^{-3} m^2, the Mach number is 0.5. The nozzle discharges to a back pressure of 270 kPa (abs). Determine the exit area of the nozzle.

12.28 A converging nozzle discharges helium to atmosphere. Conditions at section ① in the nozzle result in a pressure of 350 kPa (abs), a temperature of 20 C, and a speed of 201 m/sec. The mass flow rate is to be 0.15 kg/sec. Assume that the flow is frictionless and adiabatic. Determine the Mach number at the nozzle exit, and the exit area of the nozzle.

12.29 Critical flow through a choked orifice may be used to meter a constant mass flow rate of gas. Calculate the flow area and diameter for a critical orifice to pass 0.03 kg/sec if stagnation conditions are 15 C and 101 kPa (abs).

12.30 A converging nozzle is bolted to the side of a large tank. Air inside the tank is maintained at a constant pressure of 50 psia and a temperature of 100 F. The inlet area of the nozzle is 10 $in.^2$ and the exit area is 1 $in.^2$ The nozzle discharges to the atmosphere. For isentropic flow in the nozzle, determine the total force on the bolts, and indicate whether the bolts are in tension or compression.

12.31 A large tank initially is evacuated to 27 in. Hg (vacuum). (Ambient conditions are 29.4 in. Hg at 70 F.) At $t = 0$, an orifice of 0.25 in. diameter is opened in the tank wall; the vena contracta area is 65 percent of the geometric area. Calculate the mass flow rate at which air initially enters the tank. Show the process on a Ts diagram. Make a schematic plot of mass flow rate as a function of time. Explain why the plot is nonlinear.

12.32 A cylinder of gas used for welding contains helium at 3000 psig and room temperature. The cylinder is knocked over, its valve is broken off, and gas escapes through a converging passage. The minimum flow area is 0.10 in.^{2} at the outlet section where the gas flow is uniform. Find (a) the mass flow rate at which gas leaves the cylinder, and (b) the instantaneous acceleration of the cylinder (assume the cylinder axis is horizontal and that its mass is 125 lb). Show static and stagnation states and the process path on a Ts diagram.

12.33 A converging nozzle is connected to a large tank that contains compressed air at 75 F. The nozzle exit area is 1.5 in.^{2} The exhaust is discharged to the atmosphere. To obtain a satisfactory shadow photograph of the flow pattern leaving the nozzle exit, the pressure in the exit plane must be greater than 45 psig. What pressure is required in the tank? How much air must be supplied if the system is to run continuously? Show static and stagnation state points on a Ts diagram.

12.34 A jet transport aircraft with pressurized cabin cruises at 11 km altitude. The cabin temperature and pressure initially are at 25 C and 2.5 km altitude. The interior volume of the cabin is 25 m^3. Air escapes through a small hole with effective flow area of 0.002 m^2. Calculate the time required for the cabin pressure to decrease by 40 percent.

12.35 Consider the isentropic flow of helium through a converging-diverging wind-tunnel nozzle. The stagnation pressure at the nozzle entrance is 700 kPa (abs) and the stagnation temperature is 60 C. At a section downstream from the throat the pressure is 528 kPa (abs) and the area is 1.2×10^{-3} m^2. Determine the Mach number, the temperature, the stagnation pressure at this section, and the mass flow rate.

12.36 A converging-diverging nozzle is attached to a very large tank of air in which the pressure is 20 psia and the temperature is 40 F. The nozzle exhausts to the atmosphere where the pressure is 14.7 psia. The exit area of the nozzle is 2 in.^{2} What is the flow rate through the nozzle? Assume the flow to be isentropic.

12.37 Air enters a converging-diverging nozzle with negligible speed at an absolute pressure of 1.0 MPa and a temperature of 60 C. If the flow is isentropic and the exit temperature is -11 C, what is the Mach number at exit?

12.38 A converging-diverging nozzle with a throat area of 2 in.^{2} is connected to a large tank in which air is kept at a pressure of 80 psia and a temperature of 60 F. If the nozzle is to operate at design conditions (flow is isentropic) and the ambient pressure outside the nozzle is 12.9 psia, calculate the exit area of the nozzle and the mass flow rate.

12.39 Air is to be expanded through a converging-diverging nozzle by a frictionless adiabatic process from a pressure of 1.10 MPa (abs) and a temperature of 115 C to a pressure of 141 kPa (abs). Determine the throat and exit areas for a well-designed shockless nozzle if the mass flow rate is 2 kg/sec.

12.40 Air, at a stagnation pressure of 7.20 MPa (abs) and a stagnation temperature of 1100 K, flows isentropically through a converging-diverging nozzle having a throat area of 0.01 m^2. Determine (a) the speed at the downstream section where the Mach number is 4.0, and (b) the mass flow rate.

12.41 A large tank supplies helium to a converging-diverging nozzle. The pressure in the tank remains constant at 8.00 MPa (abs). The tank temperature remains constant at 1000 K. The flow throughout the nozzle is isentropic. The nozzle is designed to discharge to atmospheric pressure with an exit Mach number, $M = 3.5$. The exit area of the nozzle is 100 mm^2. What is the mass flow rate through the nozzle?

12.42 Air flows isentropically through a converging-diverging nozzle attached to a large tank in which the pressure is 100 psia and the temperature is 500 R. The nozzle is operating at design conditions for which the nozzle exit pressure, p_e, is equal to the surrounding atmospheric pressure, p_a. The exit area of the nozzle, $A_e = 4.0$ in.^{2}

(a) Calculate the flow rate through the nozzle.

(b) If the temperature of the air in the tank is increased to 2000 R (all pressures remaining the same), how will the flow rate be affected?

12.43 Nitrogen at a pressure and temperature of 371 kPa (abs) and 400 K enters a nozzle with negligible speed. The exhaust jet is directed against a large flat plate that is perpendicular to the jet axis. The flow leaves the nozzle at atmospheric pressure. The exit area is 0.003 m^2. Find the force required to hold the plate.

12.44 At a point upstream of the throat in a converging-diverging nozzle, the air speed is 172 m/sec; $p = 200$ kPa (abs), and $T = 22$ C. The flow is isentropic and is supersonic at the nozzle exit. If the nozzle throat area is 0.01 m^2, determine the flow rate.

12.45 Air is flowing isentropically in a converging-diverging nozzle. At the section in the converging portion where the area is 1250 mm^2, the pressure is 600 kPa (abs), the temperature is 22 C, and the Mach number is 0.50. Determine the area in the diverging section where the Mach number is 2.

12.46 Air flows steadily and isentropically through a converging-diverging nozzle. At the throat the air is at 140 kPa (abs), 60 C. The throat cross-sectional area is 0.05 m^2. At a certain section in the diverging part of the nozzle, the pressure is 70.0 kPa (abs). Calculate the speed and the area at this section.

12.47 A small, solid fuel rocket motor is tested on a thrust stand. The chamber pressure and temperature are 600 psia and 6000 R, respectively. The propulsion nozzle is designed to expand the exhaust gases isentropically to a back pressure of 10.0 psia. The nozzle exit area is 0.60 ft^2. The gas may be treated as ideal with $k = 1.2$ and $R = 60.0$ ft · lbf/lbm · R. Determine the mass flow rate of propellant gas and the thrust force exerted against the test stand.

12.48 A liquid rocket motor is fueled with hydrogen and oxygen. The chamber temperature and absolute pressure are 3300 K and 6.90 MPa, respectively. The nozzle is to be designed to expand the exhaust gases isentropically to a design back pressure corresponding to an altitude of 10 km on a standard day. The thrust produced by the motor is to be 100 kN at the design conditions. Treat the exhaust gases as water vapor and assume ideal gas behavior. Determine the propellant mass flow rate needed to produce the desired thrust, the nozzle exit area, and the area ratio, A_e/A_t.

12.49 A small rocket motor fueled with hydrogen and oxygen is tested on a thrust stand at a simulated altitude of 10 km. The motor is operated at chamber stagnation conditions of 1500 K and 8.0 MPa (gage). The combustion product is water vapor, which may be treated as an ideal gas. Expansion occurs through a converging-diverging nozzle with design Mach number of 3.5 and exit area of 700 mm^2. Evaluate the pressure at the nozzle exit plane. Calculate the mass flow rate of exhaust gas. Determine the force exerted by the rocket motor on the thrust stand.

12.50 A CO_2 cartridge is used to propel a small rocket cart. Compressed gas stored at 6000 psig and 70 F is expanded through a smoothly contoured converging nozzle with 0.020 in. throat diameter. The back pressure is atmospheric. Calculate the pressure at the nozzle throat. Evaluate the mass flow rate of carbon dioxide through the nozzle. Determine the thrust available to propel the cart. How much would the thrust increase if a diverging section were added to the nozzle to expand the gas to atmospheric pressure? Show stagnation states, static states, and the processes on a Ts diagram.

12.51 A converging-diverging nozzle has a throat diameter of 10 mm. Measurements show that a turbulent boundary layer forms on the nozzle walls. The velocity profile in the boundary layer is modeled closely by the 1/7th power-law expression. The boundary-layer thickness is 1.0 mm at the throat where $M = 1.0$, and 2.5 mm at the exit plane where $M = 2.0$. The nozzle is supplied with helium from a tank at 25 C and 1.0 MPa (abs). Calculate (a) the pressure at the nozzle throat, (b) the mass flow rate through the nozzle, and (c) the diameter at the nozzle exit plane. Show static and stagnation state points for the nozzle throat and exit plane on a Ts diagram.

***12.52** A converging-diverging nozzle has a throat diameter of 10 mm and an exit plane diameter of 20 mm. Measurements show that a turbulent boundary layer forms on the nozzle walls. The velocity profile in the boundary layer is modeled closely by the 1/7th power-law expression. The boundary-layer thickness is 1.0 mm at the throat where $M = 1.0$, and 2.5 mm at the exit plane. Evaluate the effective flow areas at the nozzle throat and exit plane. For flow of an ideal gas with $k = 1.4$, estimate (a) the subsonic, and (b) the supersonic Mach numbers in the nozzle exit plane. How much do the boundary layers alter the mass flow rate and the Mach number in the exit plane? Show static and stagnation state points for the nozzle throat and exit plane on a Ts diagram.

12.53 Air flows through a constant-area duct. At section ① conditions are 550 K, 973 kPa (abs), and $M_1 = 0.20$. At section ② flow is at 1100 K, 910 kPa (abs), and $M_2 = 0.30$.

(a) Is the flow isentropic? Justify your answer.

(b) Calculate the stagnation enthalpy, stagnation temperature, and stagnation pressure at sections ① and ②.

12.54 Air flows adiabatically with friction through a duct of 0.3 m square cross section. At one section, the pressure, temperature, and speed are 60.0 kPa (abs), 50 C, and 180 m/sec. At the exit from the duct, the pressure and temperature are 31.7 kPa (abs) and 19 C. Calculate the exit Mach number and the stagnation temperature midway between the first section and the exit.

12.55 Air flows steadily and adiabatically in a horizontal, 50 mm diameter pipe. At section ① the pressure is 340 kPa (abs) and the temperature is 53 C. At section ② (downstream) the temperature is 114 C and the speed is 591 m/sec. Determine the speed and Mach number at section ①.

12.56 Air is flowing through a well-insulated, constant-area channel. At a section where the speed is 144 m/sec, $T = 50$ C and $p = 600$ kPa (abs). Determine (a) the temperature and stagnation pressure at the downstream section where the density is 3.05 kg/m^3, and (b) the entropy increase.

12.57 Air flows steadily and adiabatically from a large tank through a converging nozzle connected to a constant-area duct. The nozzle may be considered frictionless. Air in the tank is at $p = 1.00$ MPa (abs), $T = 125$ C. The absolute pressure at the nozzle exit (duct

* Problems marked with an asterisk are designed to be solved using tables.

inlet) is 784 kPa. Determine the pressure at the end of the duct length, L, if the temperature there is 65 C. Find the entropy increase.

12.58 Consider compressible flow in a long straight duct. The inlet to the duct is from the atmosphere where $T = 25$ C. The flow is considered to be adiabatic and the pipe is long enough that the flow is choked. Find the speed and temperature at the pipe exit.

12.59 Air flows steadily and adiabatically through a constant-area duct. The density at the inlet is 3.51 kg/m^3. If the exit Mach number is unity and the exit temperature and pressure are 227 C and 110 kPa (abs), respectively, determine the inlet Mach number and the entropy change from inlet to exit.

12.60 Consider adiabatic flow of air in a constant-area pipe with friction. At one section of the pipe, $p_0 = 100$ psia, $T_0 = 500$ R, and $M = 0.70$. If the cross-sectional area of the pipe is 1 ft^2 and the Mach number at the exit is $M_2 = 1$, find the friction force exerted on the fluid by the pipe.

12.61 Air flows through a smooth well-insulated 4 in. diameter pipe at a rate of 600 lbm/min. At one section the air is at 100 psia, 80 F. Determine the minimum pressure and the maximum speed that can occur in the pipe.

12.62 Air flows through an insulated duct with constant area of 0.03 m^2. At section ①, the static temperature, static pressure, and Mach number are 277 K, 690 kPa (abs), and 0.60. At section ② downstream, the temperature is 260 K. Draw a Ts diagram for the flow between states ① and ②, showing both the static and stagnation states. Calculate M_2 and p_2. Evaluate the frictional force exerted by the air on the duct wall.

12.63 A dental drill is powered by a miniature air turbine. Air is supplied by a compressor at a stagnation pressure of 50 psia. Air flows to the turbine through a long 0.060 in. i.d. tube. Air enters the tube through a smooth nozzle; flow within the tube is adiabatic. The Mach number at the turbine inlet is 0.608. Model flow through the turbine as isentropic. The pressure drop across the turbine is 10 psi. Air leaves the turbine at atmospheric pressure. Because the exhaust stream is used to cool the cutting area, its temperature must be held at 30 F. Determine the air temperature at the turbine inlet. Calculate the stagnation temperature at the compressor outlet (nozzle inlet). Evaluate the loss in stagnation pressure and the change in entropy through the supply tube. Show static and stagnation state points and the process path on a Ts diagram.

12.64 Measurements are made of compressible flow in a long smooth 7.16 mm i.d. tube. Air is drawn from the surroundings (20 C and 101 kPa) by a vacuum pump downstream. Pressure readings along the tube become steady when the downstream pressure is reduced to 626 mm Hg (vacuum) or below. For these conditions, determine (a) the maximum mass flow rate possible through the tube, (b) the stagnation pressure of the air leaving the tube, and (c) the entropy change of the air in the tube. Show static and stagnation state points and the process on a Ts diagram.

12.65 Room air (75 F and 14.7 psia) is to be drawn through a converging nozzle into a Fanno line demonstration apparatus. Flow in the nozzle may be modeled as isentropic. Air from the nozzle enters a 0.50 in. diameter constant-area duct at $M_1 = 0.3$. Flow in the duct is adiabatic but frictional. Evaluate properties T_{0_1}, p_{0_1}, T_1, p_1, ρ_1, V_1, and $\dot{m}$. Assuming that flow is choked at section ②, evaluate T_2, V_2, ρ_2, p_2, T_{0_2}, and p_{0_2}. Compute the change in specific entropy. Show all static and stagnation state points on a Ts diagram.

12.66 Air is drawn from the atmosphere (20 C and 101 kPa) through a converging nozzle into a long insulated 20 mm diameter tube of constant area. Flow in the nozzle is isentropic.

The Mach number at the inlet to the constant-area tube is 0.15. Evaluate the mass flow rate through the tube. Calculate T^* and p^* for the isentropic process. Calculate T^* and p^* for flow through the constant-area tube. Show the corresponding static and stagnation state points on a Ts diagram. Show schematically how these results would be changed if the Mach number at the tube inlet were raised to 0.30.

12.67 A converging-diverging nozzle supplies air to a well-insulated, constant-area duct. At the inlet to the duct, $M = 2.0$, $p = 19.4$ psia, and $T = 278$ R. At the duct exit the Mach number is unity and the stagnation pressure is 90 psia. Determine the pressure and temperature at the duct exit and the entropy change.

12.68 A converging-diverging nozzle discharges air into an insulated pipe with area, $A = 650\ \text{mm}^2$. At the pipe inlet, $p = 128$ kPa (abs), $T = 39$ C, and $M = 2.00$. For shockless flow to a Mach number of unity at the pipe exit, calculate the exit temperature, the net force of the fluid on the pipe, and the entropy change.

12.69 Air flows steadily and adiabatically in a horizontal tube of 50 mm diameter. Measurements at section ① show that the temperature and pressure are 53 C and 340 kPa (abs). At section ② downstream, the temperature is 114 C and the flow speed is 591 m/sec. Find (a) the flow speed and Mach number at section ①, and (b) the magnitude and direction of the friction force exerted by the air on the duct wall. Draw a Ts diagram for the process, showing all static and stagnation state points.

***12.70** For the conditions of Problem 12.57, find the length, L, of commercial steel pipe of 50 mm diameter between sections ① and ②.

***12.71** Solve Problem 12.58 for the duct length, L, if the duct diameter is 12 mm and the surface is smooth. At the duct inlet, $p = 94.1$ kPa (abs).

***12.72** For the conditions of Problem 12.60, determine the duct length. Assume that the duct is circular and made from commercial steel. Plot the variations of pressure and Mach number versus distance along the duct.

***12.73** Air flows in an insulated duct with a speed of 140 m/sec. The temperature and absolute pressure are 200 C and 2.00 MPa, respectively.

(a) Find the temperature in this duct where the pressure has dropped to 1.26 MPa (abs) as a result of friction.

(b) If the duct ($e/D = 0.0003$) has a diameter of 150 mm, find the distance between the two points.

***12.74** Consider the flow described in Example Problem 12.8. Using the tables for Fanno line flow of an ideal gas, plot the static pressure, temperature, and Mach number versus L/D measured from the tube inlet; continue until the choked state is reached.

***12.75** Using coordinates T/T_0 and $(s - s^*)/c_p$, where s^* is the entropy at the condition where $M = 1$, plot the Fanno line starting from the inlet conditions specified in Example Problem 12.8. Use tables for Fanno line flow, and proceed to the section where $M = 1$.

***12.76** Using coordinates T/T^* and $(s - s^*)/c_p$, where s^* is the entropy at the condition where $M = 1$, plot the Fanno line for air flow ($k = 1.4$) for Mach numbers in the range $0.1 < M < 3.0$.

***12.77** The duct of Problem 12.67 has relative roughness, $e/D = 0.0002$, and has a hydraulic diameter of 0.5 ft. Determine the duct length, L.

***12.78** For the conditions of Problem 12.68, determine the duct length. Assume that the duct is

* Problems marked with an asterisk are designed to be solved using tables.

circular and made from commercial steel. Plot the variations of pressure and Mach number versus distance along the duct.

***12.79** Beginning with the inlet conditions of Problem 12.68, and using coordinates T/T_0 and $(s - s^*)/c_p$, plot the supersonic and subsonic branches of the Fanno line for the flow.

***12.80** A smooth constant-area duct assembly ($D = 150$ mm) is to be fed by a converging-diverging nozzle from a tank containing air at 295 K and 1.0 MPa (abs). Shock-free operation is desired. The Mach number at the duct inlet is to be 2.1 and the Mach number at the duct outlet is to be 1.4. The entire assembly will be insulated. Find (a) the pressure required at the duct outlet, (b) the duct length required, and (c) the change in specific entropy along the duct. Show the static and stagnation state points and the process line on a *Ts* diagram.

12.81 Air flows steadily through a constant-area duct. At section ①, the air is at 60 psia, 600 R, with a speed of 500 ft/sec. As a result of heat transfer and friction, the air at section ② downstream is at 40 psia, 800 R. Calculate the heat transfer per pound of air between sections ① and ②, and the stagnation pressure at section ②.

12.82 Consider the frictionless flow of air in a constant-area duct. At section ①, $M_1 = 0.50$, $p_1 = 1.10$ MPa (abs), and $T_{0_1} = 333$ K. Through the effect of heat transfer, the Mach number at section ② is $M_2 = 0.90$ and the stagnation temperature, $T_{0_2} = 478$ K.

(a) Determine the amount of heat transfer per unit mass to or from the fluid between sections ① and ②.

(b) Determine the pressure difference, $p_1 - p_2$.

12.83 Air flows in a constant-area duct without friction. At section ① in the duct, $p_1 = 97.3$ psia and $T_1 = 992$ R, and the speed is $V_1 = 309$ ft/sec. Through the effect of heat transfer the speed at section ② downstream is $V_2 = 652$ ft/sec. Determine the pressure, temperature, stagnation pressure, and stagnation temperature at section ②, and the heat transfer per unit mass between sections ① and ②.

12.84 At a position 3 m from the exit of a constant-area duct (area $= 0.02$ m^2), air is at an absolute pressure of 126 kPa and a temperature of 260 C. The air flows steadily through the duct (without friction) at the rate of 1.83 kg/sec. The air leaves the duct subsonically at atmospheric pressure. Determine the Mach number, temperature, and stagnation temperature at the exit of the duct and the heat transfer over the 3 m of duct length.

12.85 Air flows without friction through a short duct of constant area. At the duct entrance, $M_1 = 0.30$, $T_1 = 50$ C, and $\rho_1 = 2.16$ kg/m^3. As a result of heating, the Mach number and density at the tube outlet are $M_2 = 0.60$ and $\rho_2 = 0.721$ kg/m^3. Determine the heat transfer per unit mass and the entropy change for the process.

12.86 Air flows at the rate of 1.42 kg/sec through a duct of 100 mm diameter. At the inlet section the temperature and absolute pressure are 52 C and 60.0 kPa, respectively. At the section downstream where the flow is choked, $T_2 = 45$ C. Determine the heat transfer per unit mass, the entropy change, and the change in stagnation pressure for the process, assuming frictionless flow.

12.87 Air flows without friction through a duct of 75 mm diameter. At the duct inlet, the Mach number, stagnation temperature, and absolute pressure are $M_1 = 0.2$, $T_{0_1} = 278$ K, and $p_1 = 275$ kPa. Heat is added at the rate of 219 kJ/kg of flowing fluid, with the result that

* Problems marked with an asterisk are designed to be solved using tables.

the temperature at the duct outlet is 489 K. Determine the stagnation temperature and Mach number at the duct outlet, and the entropy change for the process.

12.88 Air flows steadily and without friction in a duct of 0.5 ft^2 cross-sectional area. The inlet conditions are $M_1 = 0.30$, $T_{0_1} = 400$ R, and $p_{0_1} = 12.0$ psia. The exit stagnation temperature and static pressure are 797 R and 9.43 psia, respectively. Determine the rate of heat transfer and the exit Mach number.

12.89 A constant-area duct is fed with air from a converging-diverging nozzle. At the entrance to the duct the following properties are known: $p_{0_1} = 800$ kPa (abs), $T_{0_1} = 700$ K, and $M_1 = 3.0$. A short distance down the duct (at section ②) the density is 0.334 kg/m^3. Assuming frictionless flow, determine the speed, pressure, and Mach number at section ②, and the heat transfer between the inlet and section ②.

12.90 Consider frictionless flow of air in a duct of constant area, $A = 0.087$ ft^2. At one section the static properties are 500 R and 15.0 psia and the Mach number is 0.2. At a section downstream, the static pressure is 10.0 psia. Draw a Ts diagram showing the static and stagnation states. Calculate the flow speed and temperature at the downstream location. Evaluate the rate of heat transfer for the process.

***12.91** Air enters the inlet manifold of a spark ignition engine at 0.308 kg/sec with $T_{0_1} =$ 300 K. At a certain location, $M_1 = 0.30$, and $p_1 = 94.0$ kPa (abs). The air stream is *cooled* by the evaporation of gasoline. After the evaporation section, $M_2 = 0.28$. The effects of friction and area change are negligible. Treat the fluid as pure air. Find the flow area. Calculate the heat transferred per unit mass of air, the change in stagnation pressure for the process, and the entropy change. On a Ts diagram show the static and stagnation states and the process direction.

***12.92** Air is drawn from the atmosphere (where it is at 288 K and 101 kPa) through a smooth converging nozzle into the inlet manifold of a racing engine. Flow is isentropic through the nozzle. Air leaves the nozzle and enters the manifold at $M = 0.30$. In the manifold, alcohol fuel is evaporated in a short section of constant-area duct where friction is negligible. The effect is to remove 60.3 kJ/kg of thermal energy from the air stream. The fluid properties are those of pure air. Find (a) the Mach number, and (b) the pressure, at the outlet of the evaporation section. Show static and stagnation state points and the process on a Ts diagram.

12.93 A combustor from a JT8D jet engine (as used on the Douglas DC-9 aircraft) has an air flow rate of 15 lbm/sec. The area is constant and frictional effects are negligible. Properties at the combustor inlet are 1260 R, 235 psia, and 609 ft/sec. At the combustor outlet, $T = 1850$ R and $M = 0.476$. The heating value of the fuel is 18,000 Btu/lbm; the air-fuel ratio is large enough so properties are those of air. Calculate the pressure at the combustor outlet. Determine the rate of energy addition to the air stream. Find the mass flow rate of fuel required; compare it to the air flow rate. Show the process on a Ts diagram, indicating static and stagnation states and the process direction.

12.94 A fuel-air mixture with the thermodynamic properties of pure air enters a combustor of constant area. The stagnation temperature of the inlet stream is constant at 400 K. Friction is negligible. When heat is added to the flow at the rate of 1.4 MJ/kg, flow at the duct exit is choked. At this condition the static pressure at the inlet is 144.6 kPa (abs) and the pressure at the duct exit is 64.7 kPa (abs). Calculate (a) the combustor outlet temperature, (b) the combustor inlet Mach number, and (c) the loss in stagnation

* Problems marked with an asterisk are designed to be solved using tables.

pressure through the combustor. Show static and stagnation state points on a Ts diagram; indicate the process direction.

***12.95** Air flows without friction in a constant-area duct. At section ①, $M_1 = 0.50$, $p_1 = 1.10$ MPa (abs), and $T_{0_1} = 335$ K. Through the effect of heat transfer, the Mach number is raised to $M_2 = 0.9$ at section ②. Determine the heat transfer per unit mass between sections ① and ②, and the pressure, p_2.

***12.96** Air flows without friction in a constant-area duct. The properties at section ① are $T_1 = 992$ R and $p_1 = 97.3$ psia, and the speed is $V_1 = 309$ ft/sec. Heat transfer causes the speed to increase to 652 ft/sec at section ②. Determine the Mach number, temperature, pressure, stagnation temperature, and stagnation pressure at section ②.

***12.97** Air flows steadily and without friction at 1.83 kg/sec through a duct with cross-sectional area of 0.02 m^2. At the duct inlet, the temperature and absolute pressure are 260 C and 126 kPa, respectively. The exit flow discharges subsonically to atmospheric pressure. Determine the Mach number, temperature, and stagnation temperature at the duct outlet, and the heat transfer rate.

***12.98** A converging-diverging nozzle feeds a short duct of constant area. At the duct inlet, the local isentropic stagnation conditions are $T_{0_1} = 700$ K and $p_{0_1} = 800$ kPa (abs), and the Mach number is $M_1 = 3.0$. A short distance down the duct at section ②, the density is $\rho_2 = 0.334$ kg/m^3. Assuming frictionless flow in the duct, determine the speed, pressure, and Mach number at section ②, and the heat transfer between the inlet and section ②.

***12.99** Air flows without friction in a short section of constant-area duct. At the duct inlet, $M_1 = 0.30$, $T_1 = 50$ C, and $\rho_1 = 2.16$ kg/m^3. At the duct outlet, $M_2 = 0.60$. Determine the heat transfer per unit mass, the entropy change, and the change in stagnation pressure for the process.

***12.100** In the frictionless flow of air through a 100 mm diameter duct, 1.42 kg/sec enters at a temperature of 52 C and an absolute pressure of 60.0 kPa. Determine the amount of heat that must be added to choke the flow, and the fluid properties at the choked state.

***12.101** Air flows without friction through a duct of 75 mm diameter. The inlet Mach number, stagnation temperature, and absolute pressure are $M_1 = 0.20$, $T_{0_1} = 278$ K, and $p_1 = 275$ kPa, respectively. Heat is added at a rate equivalent to 219 kJ/kg of flowing air. Determine the stagnation temperature and Mach number at the duct outlet, and the changes in entropy and stagnation pressure for the process.

***12.102** Air flows steadily in a constant-area, frictionless duct of 0.5 ft^2 cross-sectional area. The inlet conditions are $M_1 = 0.3$, $T_{0_1} = 400$ R, and $p_{0_1} = 10$ psia; the exit stagnation temperature is $T_{0_2} = 797$ R. Determine the amount of heat transfer, and the exit Mach number and pressure.

***12.103** Consider steady, one-dimensional flow of air in a combustor with constant area of 0.5 ft^2, where hydrocarbon fuel added to the air stream burns. The process is equivalent to simple heating because the amount of fuel is small compared to the amount of air; it occurs over a short distance so that friction is negligible. Properties at the combustor inlet are 818 R, 200 psia, and $M = 0.3$. The flow speed at the combustor outlet must not exceed 2000 ft/sec. Find the properties at the combustor outlet and the heat addition rate. Show the process on a Ts diagram, indicating static and stagnation state points before and after the heat addition.

* Problems marked with an asterisk are designed to be solved using tables.

***12.104** Flow in a gas turbine combustor is modeled as steady, one-dimensional, frictionless heating of air in a channel of constant area. For a certain process, the inlet conditions are 960 F, 225 psia, and $M = 0.4$. Calculate the maximum possible heat addition. Find all fluid properties at the outlet section and the reduction in stagnation pressure. Show the process on a Ts diagram, indicating all static and stagnation state points.

***12.105** A jet aircraft cruises at $M = 0.85$ at 10 km altitude on a standard day. Air is slowed to $M_1 = 0.3$ relative to the aircraft by the inlet system. Flow in the inlet system is adiabatic, but friction causes the absolute stagnation pressure to drop by 5 percent. Air from the inlet is compressed to $p_2 = 30p_1$ in the engine's compressor section. Flow in the compressor is adiabatic, but the stagnation temperature of the stream increases by 750 K due to work addition. The Mach number of the flow leaving the compressor and entering the combustor is $M_2 = 0.4$. Heat is added in the combustor, where the maximum temperature must be held to $T_3 = 1630$ K or less. Evaluate for these conditions (a) the maximum possible heat addition in the combustor, (b) the combustor exit Mach number for part (a), and (c) the change in stagnation pressure across the combustor. On a Ts diagram for the process show static and stagnation states and indicate the process direction.

***12.106** Air from an aircraft inlet system enters the engine combustion chamber where heat is added during a frictionless process in a tube with constant area of 0.01 m^2. The local isentropic stagnation temperature and Mach number entering the combustor are 427 K and 0.3. The mass flow rate is 0.5 kg/sec. When the rate of heat transfer is set at 404 kW, the flow leaves the combustor at 1026 K and 22.9 kPa (abs). Determine for this process (a) the Mach number at the combustor outlet, (b) the static pressure at the combustor inlet, and (c) the change in local isentropic stagnation pressure during the heat addition process. Show static and stagnation state points and indicate the process direction on a Ts diagram.

***12.107** Using coordinates T/T^* and $(s - s^*)/c_p$, where s^* is the entropy at the condition where $M = 1$, plot the Rayleigh line for air flow ($k = 1.4$) for Mach numbers in the range $0.4 < M < 3.0$.

***12.108** Beginning with the inlet conditions of Problem 12.68, and using coordinates T/T_{0_1} and $(s - s^*)/c_p$, plot the supersonic and subsonic branches of the Rayleigh line for the flow.

12.109 In long, constant-area pipelines, such as those used for natural gas, the temperature may be considered constant. Assume that gas leaves a pumping station at 50 psia and 70 F with a Mach number of 0.10 and density, $\rho = 0.255$ lbm/ft^3. At the section along the pipe where the pressure has dropped to 20 psia:

(a) Calculate the Mach number of the flow.
(b) Is heat added to or removed from the gas over the length between the pressure taps? Justify your answer.
(c) Sketch the process on a Ts diagram. Indicate (qualitatively) T_{0_1}, T_{0_2}, and p_{0_2}.

12.110 Natural gas (molecular mass, $M_m = 18$, $k = 1.3$) is to be pumped through a 36 in. i.d. pipe connecting two compressor stations 40 miles apart. At the upstream station the pressure is not to exceed 90 psig, and at the downstream station it is to be at least 10 psig. Calculate the maximum allowable rate of flow (ft^3/day at 70 F and 1 atm) assuming that there is sufficient heat transfer through the pipe to maintain the gas at 70 F.

* Problems marked with an asterisk are designed to be solved using tables.

12.111 An air stream with temperature, $T_1 = 0$ C, absolute pressure, $p_1 = 60.0$ kPa, and speed, $V_1 = 497$ m/sec, undergoes a normal shock. The temperature downstream from the shock is 87 C. Determine the Mach number, speed, and stagnation pressure downstream from the shock.

12.112 Air approaches a normal shock at 0 C, 60 kPa (abs), and 497 m/sec. The air speed immediately downstream from the shock is 267 m/sec. Evaluate all flow properties downstream from the shock and the entropy change across the shock. Show static and stagnation states and the process direction on a *Ts* diagram.

12.113 Air with stagnation temperature, $T_{0_1} = 333$ K, and stagnation pressure, $p_{0_1} = 600$ kPa (abs), approaches a normal shock at Mach number, $M_1 = 2.0$. The speed downstream from the shock is $V_2 = 204$ m/sec. Determine the static pressure downstream from the shock.

12.114 Air approaches a normal shock with speed, $V_1 = 951$ m/sec. The stagnation temperature and absolute pressure of the air stream are $T_{0_1} = 700$ K and $p_1 = 125$ kPa, respectively; the Mach number is $M_1 = 3.0$. The absolute pressure downstream from the shock is 1.29 MPa. Determine the downstream speed and temperature.

12.115 A normal shock stands in a constant-area duct. Air approaches the shock with $T_{0_1} = 1000$ R and $p_{0_1} = 100$ psia, at a Mach number of 3.0. The Mach number downstream from the shock is known to be $M_2 = 0.475$. Determine the static pressure downstream from the shock.

12.116 Air undergoes a normal compression shock. The upstream temperature, absolute pressure, and speed are $T_1 = 35$ C, $p_1 = 229$ kPa, and $V_1 = 704$ m/sec, respectively. The density downstream from the shock is $\rho_2 = 6.91$ kg/m^3. Determine the temperature and stagnation pressure of the air stream leaving the shock.

12.117 An air stream approaches a normal shock at Mach number, $M_1 = 2.64$. The upstream stagnation pressure and density are $p_{0_1} = 3.00$ MPa (abs) and $\rho_1 = 1.65$ kg/m^3, respectively. The ratio of static pressure to stagnation pressure immediately behind the shock is 0.843. Determine the downstream Mach number and temperature.

12.118 A normal shock occurs in air at a section where the flow speed is 924 m/sec. The temperature and absolute pressure at this point are $T_1 = 10$ C and $p_1 = 35.0$ kPa. The absolute pressure downstream from the shock is $p_2 = 301$ kPa. Determine the speed and Mach number downstream from the shock.

12.119 Air approaches a normal compression shock with temperature, absolute pressure, and speed of $T_1 = 18$ C, $p_1 = 101$ kPa, and $V_1 = 766$ m/sec. The temperature immediately downstream from the shock is $T_2 = 551$ K. Determine the velocity immediately downstream from the shock and the pressure change across the shock. Calculate the corresponding pressure change for a frictionless, shockless deceleration between the same speeds.

12.120 A supersonic wind tunnel is to be operated at a test section Mach number of 2.2. Upstream from the test section, the nozzle throat area is 0.07 m^2. Air is supplied at stagnation conditions of 500 K and 1.0 MPa (abs). At one flow condition while the tunnel is being brought up to speed a normal shock stands at the nozzle exit plane. The flow is steady and the temperature downstream from the shock is 472 K. For this *starting* condition, immediately downstream from the shock find (a) the Mach number, (b) the static pressure, (c) the stagnation pressure, and (d) the minimum area theoretically possible for the second throat downstream from the test section. On a *Ts* diagram show static and stagnation state points and the process direction.

12.121 A supersonic aircraft flies at $M = 2.7$ at 20 km altitude on a standard day. Air enters the engine inlet system where it is slowed isentropically to $M = 1.3$. A normal shock occurs at that location. (The Mach number immediately downstream from the shock is 0.786.) The resulting subsonic flow is decelerated further to a Mach number of 0.40. The subsonic diffusion is adiabatic but not isentropic; the final pressure is 104 kPa (abs). Evaluate (a) the stagnation temperature for the flow, (b) the pressure change across the shock, (c) the entropy change, and (d) the final stagnation pressure. Show the process on a Ts diagram, indicating all static and stagnation states.

12.122 A supersonic aircraft cruises at $M = 2.2$ at 12 km altitude. A pitot tube is used to sense stagnation pressure for calculating air speed. A normal shock stands in front of the pitot tube; the Mach number following the shock is 0.5471. Evaluate local isentropic stagnation conditions in front of the shock. Estimate the pressure sensed by the pitot tube. Show all static and stagnation points and the process on a Ts diagram.

12.123 A total-pressure probe is placed in a supersonic wind-tunnel where the freestream air temperature is 530 K and the Mach number is 2.0. A normal shock stands in front of the probe. The Mach number and static pressure behind the shock are 0.5774 and 5.76 psia. Find (a) the downstream stagnation pressure and stagnation temperature, and (b) all fluid properties upstream from the shock. Show static and stagnation state points and the process on a Ts diagram.

12.124 Air flows steadily through a long, insulated constant-area pipe. At section ①, the Mach number is 2.0, and the temperature and pressure are 140 F and 35.9 psia, respectively. At section ② downstream from a normal shock (the shock stands in the duct between sections ① and ②), the speed is 1080 ft/sec. Determine the density and Mach number at section ②. Make a qualitative sketch of the pressure distribution along the pipe.

12.125 A blast wave propagates outward from an explosion. At large radii, curvature is small and the wave may be treated as a strong normal shock. (The pressure and temperature rise associated with the blast wave decrease slowly as the wave travels outward.) At one instant a blast wave front travels at $M = 1.60$ with respect to undisturbed air at standard conditions. The static pressure immediately behind the wave is 286 kPa (abs). Find (a) the speed of the air behind the blast wave with respect to the wave, and (b) the speed of the air behind the blast wave as seen by an observer on the ground. Draw a Ts diagram for the process as seen by an observer on the wave, indicating static and stagnation state points and property values.

***12.126** An air stream with temperature, $T_1 = 0$ C, absolute pressure, $p_1 = 60.0$ kPa, and speed, $V_1 = 497$ m/sec, undergoes a normal shock. Determine the Mach number, speed, and stagnation pressure downstream from the shock.

***12.127** Air with stagnation temperature, $T_{0_1} = 333$ K, and stagnation pressure, $p_{0_1} = 600$ kPa (abs), approaches a normal shock at Mach number, $M_1 = 2.0$. Determine the static pressure downstream from the shock, and the decrease in stagnation pressure across the shock.

***12.128** Air approaches a normal shock with speed, $V_1 = 951$ m/sec. The stagnation temperature and absolute pressure of the air stream are $T_{0_1} = 700$ K and $p_1 = 125$ kPa, respectively; the Mach number is $M_1 = 3.0$. Determine the speed and temperature of the air leaving the shock, and the entropy change across the shock.

***12.129** A normal shock stands in a constant-area duct. Air approaches the shock with

* Problems marked with an asterisk are designed to be solved using tables.

$T_{0_1} = 1000$ R, $p_{0_1} = 100$ psia, at a Mach number of 3.0. Determine the static pressure downstream from the shock. Compare the downstream pressure with the value that would be reached by decelerating isentropically to the same subsonic Mach number.

***12.130** Air undergoes a normal compression shock. The upstream temperature, absolute pressure, and speed are $T_1 = 35$ C, $p_1 = 229$ kPa, and $V_1 = 704$ m/sec. Determine the temperature and stagnation pressure of the air stream leaving the shock.

***12.131** An air stream approaches a normal shock at Mach number, $M_1 = 2.64$. The upstream stagnation pressure and density are $p_{0_1} = 3.00$ MPa (abs) and $\rho_1 = 1.65$ kg/m^3. Determine the downstream Mach number and temperature, and the entropy change across the shock.

***12.132** A normal shock occurs in air at a section where the flow speed is 924 m/sec. The temperature and absolute pressure at this point are $T_1 = 10$ C and $p_1 = 35.0$ kPa. Determine the speed and Mach number downstream from the shock, and the change in stagnation pressure across the shock.

***12.133** Air approaches a normal compression shock with temperature, absolute pressure, and speed of $T_1 = 18$ C, $p_1 = 101$ kPa, and $V_1 = 766$ m/sec. Determine the speed immediately downstream from the shock, and the pressure change across the shock. Calculate the corresponding pressure change for a frictionless, shockless deceleration between the same speeds.

***12.134** Air flows steadily through a long insulated pipe of 2 in. diameter. At section ①, the Mach number is 2.0, and the temperature and pressure are 140 F and 35.9 psia, respectively. A normal shock occurs at section ② in the duct where $M_2 = 1.88$. At section ④, some distance downstream from the shock, the speed is 1080 ft/sec. Determine the flow properties at sections ② and ③, immediately upstream and downstream from the shock, and the Mach number and pressure at section ④. Determine the length of the pipe and plot the pressure distribution along the pipe.

***12.135** A supersonic aircraft flies at $M = 2.7$ at an altitude of 20 km on a standard day. Air approaching a forward-facing total-head (pitot) tube passes through a normal shock and then is brought isentropically to rest relative to the aircraft. Determine the final temperature of the air and the pressure sensed by the total-head tube.

***12.136** A supersonic aircraft flies at an altitude of 22 km on a standard day. A forward-facing stagnation tube senses a total pressure of 55.3 kPa (abs). Determine the Mach number and speed of the aircraft.

***12.137** Stagnation pressure and temperature probes are located on the nose of a supersonic aircraft that flies at 35,000 ft. A normal shock stands in front of the probes. The temperature probe indicates a stagnation temperature of 420 F behind the shock. Calculate the Mach number and air speed of the plane. Find the static and stagnation pressures behind the shock. Show the process and the static and stagnation state points on a Ts diagram.

***12.138** A supersonic aircraft cruises at $M = 2.7$ at 60,000 ft altitude. A normal shock stands in front of a pitot tube on the aircraft, which senses a stagnation pressure of 10.4 psia. Calculate the static pressure and temperature behind the shock. Evaluate the loss in stagnation pressure through the shock. Determine the change in specific entropy across the shock. Show static and stagnation states and the process on a Ts diagram.

* Problems marked with an asterisk are designed to be solved using tables.

*12.139 A supersonic aircraft cruises at $M = 2.2$ at 12 km altitude. A pitot tube is used to sense pressure for calculating air speed. A normal shock stands in front of the tube. Evaluate the local isentropic stagnation conditions in front of the shock. Estimate the stagnation pressure sensed by the pitot tube. Show static and stagnation state points and the process on a *Ts* diagram.

*12.140 The Concorde supersonic transport flies at $M = 2.2$ at an altitude of 20 km. Air is decelerated isentropically by the engine inlet system to a local Mach number of 1.3. The air passes through a normal shock and is decelerated further to $M = 0.4$ at the engine compressor section. Assume as a first approximation that this subsonic diffusion process is isentropic, and use standard atmosphere data for freestream conditions. Determine the temperature, pressure, and stagnation pressure of the air entering the engine compressor.

*12.141 The subsonic portion of the inlet diffuser described in Problem 12.140 has an isentropic efficiency of 0.97; the actual static pressure rise is 97 percent of the value that could be obtained from an isentropic deceleration between the same initial and final velocities. The supersonic diffusion may be assumed to be isentropic. The entire flow is adiabatic. Determine the overall reduction in stagnation pressure for the inlet flow and the static pressure of the air stream entering the engine compressor at $M = 0.4$.

12.142 A supersonic diffuser for an aircraft is designed to operate with a flight speed of $M = 3.0$ and inlet area of 3.0 m^2 at 30 km altitude. A normal shock occurs downstream from the diffuser throat at a section where the Mach number is 1.4. (The Mach number immediately behind the shock is 0.7397.) The final Mach number leaving the diffuser is to be 0.4. Sketch the passage shape required. Compute T_0, T, p_0, and p at the inlet, the throat, before and after the shock, and the diffuser outlet, assuming that all processes are isentropic except across the shock. Show the processes and the static and stagnation states on a *Ts* diagram.

12.143 A supersonic diffuser for an aircraft is designed to operate with a flight speed of $M = 2.2$ and inlet area of 1.5 m^2 at 20 km altitude. A normal shock occurs downstream from the diffuser throat at a section where the Mach number is 1.3. (The Mach number immediately behind the shock is 0.7860.) The final Mach number leaving the diffuser is to be 0.3. Sketch the passage shape required. Compute T_0, T, p_0, and p at the inlet, the throat, before and after the shock, and the diffuser outlet section, assuming that all processes are isentropic except across the shock. Show the processes and the static and stagnation states on a *Ts* diagram.

12.144 A missile powered by a ramjet engine is to fly at $M = 3.5$ at 30 km altitude. Air can be decelerated isentropically to $M = 1.4$ in the inlet system before a normal shock occurs. Calculate the pressure rise that theoretically could be obtained in this process. If the efficiency of the diffusion process actually is 85 percent, calculate the loss in stagnation pressure for the process. (Diffuser efficiency is defined as $\eta = \Delta h(\text{isentropic})/\Delta h(\text{actual})$ between the same pressure levels.) Show static and stagnation states for the theoretical and actual processes on a *Ts* diagram.

12.145 Air flows adiabatically from a reservoir, where the temperature and absolute pressure are 60 C and 600 kPa, through a converging-diverging nozzle. The design Mach number of the nozzle is 2.94. A normal shock occurs at the location in the nozzle where $M = 2.42$. (The Mach number immediately behind the shock is $M = 0.521$.) Assuming

* Problems marked with an asterisk are designed to be solved using tables.

isentropic flow before and after the shock, determine the back pressure downstream from the nozzle, if the temperature there is 54.6 C. Sketch the pressure distribution.

12.146 A normal shock occurs in the diverging section of a converging-diverging nozzle at the location where the area is 4.0 in.^{2} and the local Mach number is 2.50. (The Mach number immediately downstream from the shock is 0.513.) The stagnation conditions for the upstream flow are $T_0 = 1000$ R and $p_0 = 100$ psia. The nozzle exit area is 6.0 in.^{2} and the exit temperature is 981 R. Assume that the flow is isentropic except across the shock. Determine the nozzle exit pressure, throat area, and mass flow rate.

12.147 A converging-diverging nozzle is designed to expand air isentropically to atmospheric pressure from a large tank where the temperature and absolute pressure are 150 C and 790 kPa. A normal shock stands in the diverging section at a location where the absolute pressure is 160 kPa and the cross-sectional area is 600 mm^2. The Mach number immediately behind the shock is 0.641. Determine the nozzle back pressure, exit area, and throat area.

12.148 Air flows through a converging-diverging nozzle designed to give an exit Mach number, $M = 2.80$. The upstream stagnation conditions are atmospheric; the back pressure is maintained by a vacuum pump. Determine the back pressure required to cause a normal shock to stand in the exit plane and the flow speed after the shock. (The pressure ratio across a normal shock occurring at $M = 2.80$ is 8.98.)

12.149 A normal shock occurs in the diverging section of a converging-diverging nozzle at the location where the area is 4.0 in.^{2} and the local Mach number is 2.00. (The Mach number immediately downstream from the shock is 0.5774.) Stagnation conditions for the upstream flow are 1000 R and 100 psia. The nozzle exit area is 6.0 in.^{2} and the exit temperature is 981 R. Assume that flow is isentropic except across the shock. Find the nozzle exit pressure. Show the processes on a Ts diagram and indicate the static and stagnation state points.

***12.150** A converging-diverging nozzle is attached to a large tank of air in which the temperature and pressure are 300 K and 250 kPa (abs). At the nozzle throat (section of minimum area) the pressure is 132 kPa (abs). In the diverging section the pressure falls to 68.1 kPa before rising suddenly across a normal shock. At the nozzle exit the pressure is 180 kPa. Find the Mach number immediately behind the shock. Determine the pressure immediately downstream from the shock. Calculate the entropy change across the shock. Sketch the Ts diagram for this flow, indicating static and stagnation state points for conditions at the nozzle throat, both sides of the shock, and the exit plane.

***12.151** A converging-diverging nozzle with throat area, $A_t = 1.0$ in.^{2}, is attached to a large tank in which the pressure and temperature are maintained at 100 psia and 600 R. The nozzle exit area is 1.58 in.^{2} Determine the exit Mach number at design conditions. Referring to Fig. 12.20, determine the back pressures corresponding to the boundaries of Regimes I, II, III, and IV. Sketch the corresponding plot for this nozzle.

***12.152** A converging-diverging nozzle with area ratio, $A_e/A_t = 4.0$, is designed to expand air isentropically to atmospheric pressure. Determine the exit Mach number at design conditions, and the required value of inlet stagnation pressure. Referring to Fig. 12.20, determine the back pressures that correspond to the boundaries of Regimes I, II, III, and IV. Sketch the plot of pressure ratio versus axial distance for this nozzle.

***12.153** Air flows adiabatically from a reservoir, where the temperature and absolute pressure are 60 C and 600 kPa, through a converging-diverging nozzle of area ratio, $A_e/A_t = 4.0$.

* Problems marked with an asterisk are designed to be solved using tables.

A normal shock occurs at the location in the nozzle where $M = 2.42$. Assuming isentropic flow before and after the shock, determine the back pressure downstream from the nozzle. Sketch the pressure distribution.

***12.154** A normal shock occurs in the diverging section of a converging-diverging nozzle at the location where the area is 4.0 in.^{2} and the local Mach number is 2.50. The stagnation conditions for the upstream flow are $T_0 = 1000$ R and $p_0 = 100$ psia. The nozzle exit area is 6.0 in.^{2} Assume that the flow is isentropic except across the shock. Determine the nozzle exit pressure, throat area, and mass flow rate.

***12.155** A converging-diverging nozzle is designed to expand air isentropically to atmospheric pressure from a large tank where the temperature and absolute pressure are 150 C and 790 kPa. A normal shock stands in the diverging section at a location where the pressure is 160 kPa (abs) and the cross-sectional area is 600 mm^2. Determine the nozzle back pressure, exit area, and throat area.

***12.156** A converging-diverging nozzle with a design pressure ratio of $p_e/p_0 = 0.1278$ is operated with a back pressure condition such that $p_b/p_0 = 0.830$, causing a normal shock to stand in the diverging section. Determine the Mach number at which the shock occurs.

***12.157** Air flows through a converging-diverging nozzle with area ratio, $A_e/A_t = 3.5$. The upstream stagnation conditions are atmospheric; the back pressure is maintained by a vacuum pump. Determine the back pressure required to cause a normal shock to stand in the nozzle exit plane, and the flow speed leaving the shock.

***12.158** Air flows through a converging-diverging nozzle with area ratio, $A_e/A_t = 3.5$. The upstream stagnation conditions are atmospheric; the back pressure is maintained by a vacuum system. Determine the range of back pressures for which a normal shock will occur within the nozzle, and the corresponding mass flow rate if $A_t = 500$ mm^2.

***12.159** Air flows through a converging-diverging nozzle with area ratio, $A_e/A_t = 1.87$. The upstream stagnation conditions are $T_{0_1} = 240$ F and $p_{0_1} = 100$ psia. The back pressure is maintained at 40 psia. Determine the Mach number and flow speed in the nozzle exit plane.

***12.160** A converging-diverging nozzle with area ratio, $A_e/A_t = 1.633$, is designed to operate with atmospheric pressure at the exit plane. Determine the range(s) of stagnation pressures for which the nozzle will be free from normal shocks.

* Problems marked with an asterisk are designed to be solved using tables.

Appendix A

FLUID PROPERTY DATA

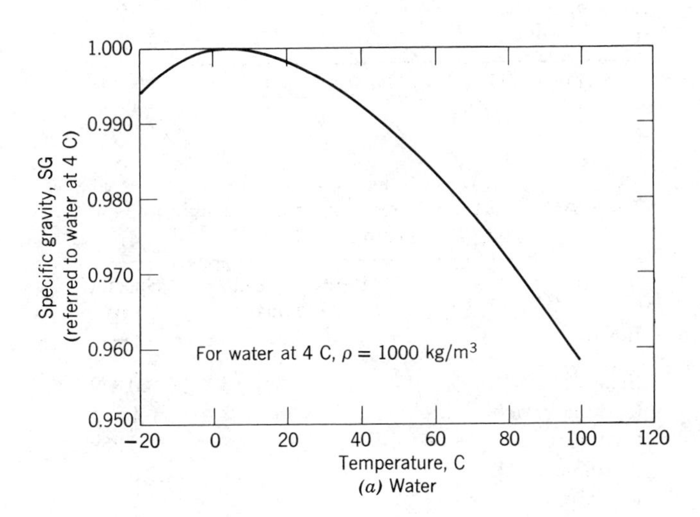

(a) Water

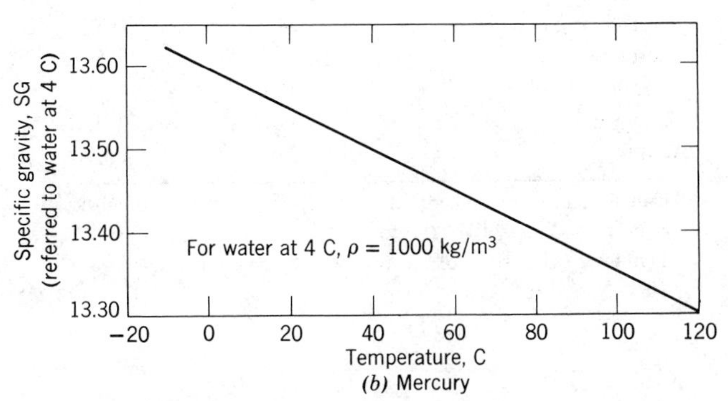

(b) Mercury

Fig. A.1 Specific gravity of water and mercury as functions of temperature. (Data from [1].)

Table A.1 Specific Gravities of Several Common Manometer Liquids at 20 C (Data from [1, 2, 3].)

Fluid	Specific Gravity[a]
E. V. Hill blue oil	0.797
Meriam red oil	0.827
Benzene	0.879
Dibutyl phthalate	1.04
Monochloronaphthalene	1.20
Carbon tetrachloride	1.595
Bromoethylbenzene (Meriam blue)	1.75
Tetrabromoethane	2.95
Mercury	13.55

[a] Specific gravity, $SG \equiv \rho/\rho_{H_2O}$ (at 4 C); ρ_{H_2O} (at 4 C) = 1000 kg/m^3 (1.94 slug/ft^3).

Table A.2 Physical Properties of Common Liquids at 20 C (Data from [1, 4, 5].)

Liquid	Isentropic Bulk Modulus[a] (GN/m^2)	Specific Gravity (–)
Benzene	1.48	0.879
Carbon tetrachloride	1.36	1.595
Castor oil	2.11	0.969
Gasoline	—	0.72
Glycerin	4.59	1.26
Heptane	0.886	0.684
Kerosine	1.43	0.82
Lubricating oil	1.44	0.88
Mercury	28.5	13.55
Octane	0.963	0.702
Seawater[b]	2.42	1.025
Water	2.24	0.998

[a] Calculated from speed of sound; 1 GN/m^2 = 10^9 N/m^2 (1 N/m^2 = 1.45×10^{-4} lbf/in.^{2}).

[b] Dynamic viscosity of seawater at 20 C is $\mu = 1.08 \times 10^{-3}$ N · sec/m^2.

Table A.3 Properties of the U.S. Standard Atmosphere (Data from [8].)

Geometric Altitude (meters)	Temperature (K)	p/p_0 (—)	ρ/ρ_0 (—)
−500	291.4	1.061	1.049
0	288.2	1.000[a]	1.000[b]
500	284.9	0.9421	0.9529
1,000	281.7	0.8870	0.9075
1,500	278.4	0.8345	0.8638
2,000	275.2	0.7846	0.8217
2,500	271.9	0.7372	0.7812
3,000	268.7	0.6920	0.7423
3,500	265.4	0.6492	0.7048
4,000	262.2	0.6085	0.6689
4,500	258.9	0.5700	0.6343
5,000	255.7	0.5334	0.6012
6,000	249.2	0.4660	0.5389
7,000	242.7	0.4057	0.4817
8,000	236.2	0.3519	0.4292
9,000	229.7	0.3040	0.3813
10,000	223.3	0.2615	0.3376
11,000	216.8	0.2240	0.2978
12,000	216.7	0.1915	0.2546
13,000	216.7	0.1636	0.2176
14,000	216.7	0.1399	0.1860
15,000	216.7	0.1195	0.1590
16,000	2'6.7	0.1022	0.1359
17,000	216.7	0.08734	0.1162
18,000	216.7	0.07466	0.09930
19,000	216.7	0.06383	0.08489
20,000	216.7	0.05457	0.07258
22,000	218.6	0.03995	0.05266
24,000	220.6	0.02933	0.03832
26,000	222.5	0.02160	0.02797
28,000	224.5	0.01595	0.02047
30,000	226.5	0.01181	0.01503
40,000	250.4	0.002834	0.003262
50,000	270.7	0.0007874	0.0008383
60,000	255.8	0.0002217	0.0002497
70,000	219.7	0.00005448	0.00007146
80,000	180.7	0.00001023	0.00001632
90,000	180.7	0.000001622	0.000002588

[a] $p_0 = 1.01325 \times 10^5$ N/m^2 abs (= 14.696 psia).
[b] $\rho_0 = 1.2250$ kg/m^3 (=0.002377 slug/ft^3).

A-1 SURFACE TENSION

The values of surface tension, σ, for most organic compounds are remarkably similar at room temperature; the typical range is 25 to 40 mN/m. Water is higher, at about 73 mN/m at 20 C. Liquid metals have values in the range between 300 and 600 mN/m; liquid mercury has a value of about 480 mN/m at 20 C. Surface tension decreases with temperature; the decrease is nearly linear with absolute temperature. Surface tension at the critical temperature is zero.

Values of σ are usually reported for surfaces in contact with the pure vapor of the liquid being studied or with air. At low pressures both values are about the same.

Table A.4 Surface Tension of Common Liquids at 20 C (Data from [1, 4–7].)

Liquid	Surface Tension, σ (mN/m)[a]	Contact Angle, θ (degrees)
(a) In contact with air		Air / Liquid / θ
Benzene	28.9	
Carbon tetrachloride	27.0	
Glycerin	63.0	
Hexane	18.4	
Kerosine	26.8	
Lube oil	25–35	
Mercury	484	140
Methanol	22.6	
Octane	21.8	
Water	72.8	~0
(b) In contact with water		Water / Liquid / θ
Benzene	35.0	
Carbon tetrachloride	45.0	
Hexane	51.1	
Mercury	375	140
Methanol	22.7	
Octane	50.8	

[a] 1 mN/m = 10^{-3} N/m.

A-2 THE PHYSICAL NATURE OF VISCOSITY

Viscosity is a measure of internal fluid friction, i.e. resistance to deformation. The mechanism of gas viscosity is reasonably well understood, but the theory is poorly developed for liquids. We can gain some insight into the physical nature of viscous flow by discussing these mechanisms briefly.

The viscosity of a Newtonian fluid is fixed by the state of the material. Thus $\mu = \mu(T, p)$. Temperature is the more important variable, so let us consider it first. Excellent empirical equations for prediction of viscosity as a function of temperature are available.

A-2.1 Effect of Temperature on Viscosity

a. Gases

All gas molecules are in continuous random motion. When there is bulk motion due to flow, the bulk motion is superimposed on the random motions. It is then distributed throughout the fluid by molecular collisions. Analyses based on kinetic theory predict

$$\mu \propto \sqrt{T}$$

The kinetic theory prediction is in fair agreement with experimental trends, but the constant of proportionality and one or more correction factors must be determined. This limits practical application of this simple equation.

If two or more experimental datum points are available, the data may be correlated using the empirical Sutherland correlation

$$\mu = \frac{bT^{1/2}}{1 + S/T} \tag{A.1}$$

The constants, b and S, may be determined most simply by writing

$$\mu = \frac{bT^{3/2}}{S + T}$$

or

$$\frac{T^{3/2}}{\mu} = \left(\frac{1}{b}\right)T + \frac{S}{b}$$

(Compare this with $y = mx + c$.) From a plot of $T^{3/2}/\mu$ versus T, one obtains the slope, $1/b$, and the intercept, S/b. For air,

$$b = 1.458 \times 10^{-6} \frac{\text{kg}}{\text{m} \cdot \text{sec} \cdot \text{K}^{1/2}}$$

$$S = 110.4 \text{ K}$$

These values were used with Eq. A.1 to compute viscosities for the standard atmosphere in [8].

b. Liquids

Viscosities for liquids cannot be estimated well theoretically. The phenomenon of momentum transfer by molecular collisions seems overshadowed in liquids by the effects of the interacting force fields among the closely packed liquid molecules.

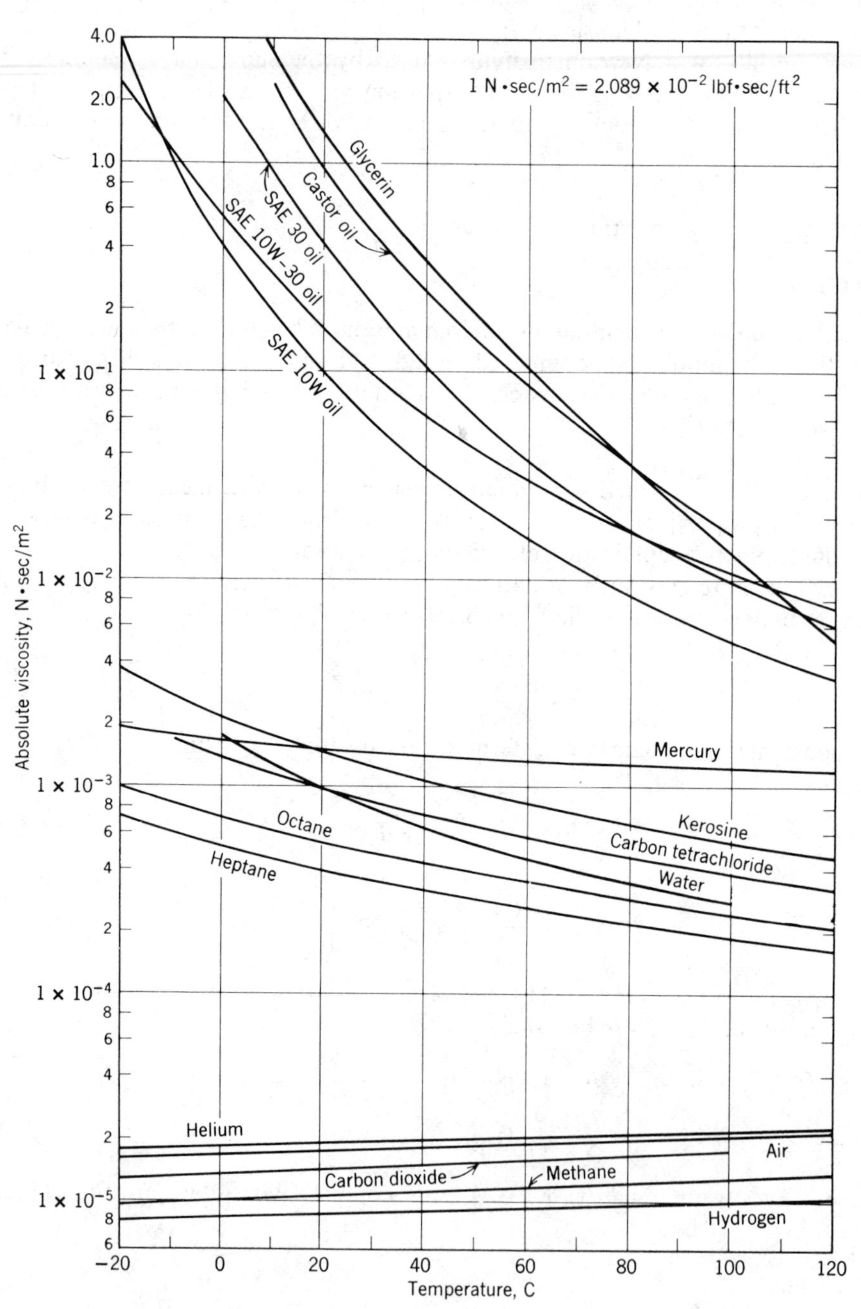

Fig. A.2 Dynamic (absolute) viscosity of common fluids as a function of temperature. (Data from [1], [5], and [9].)

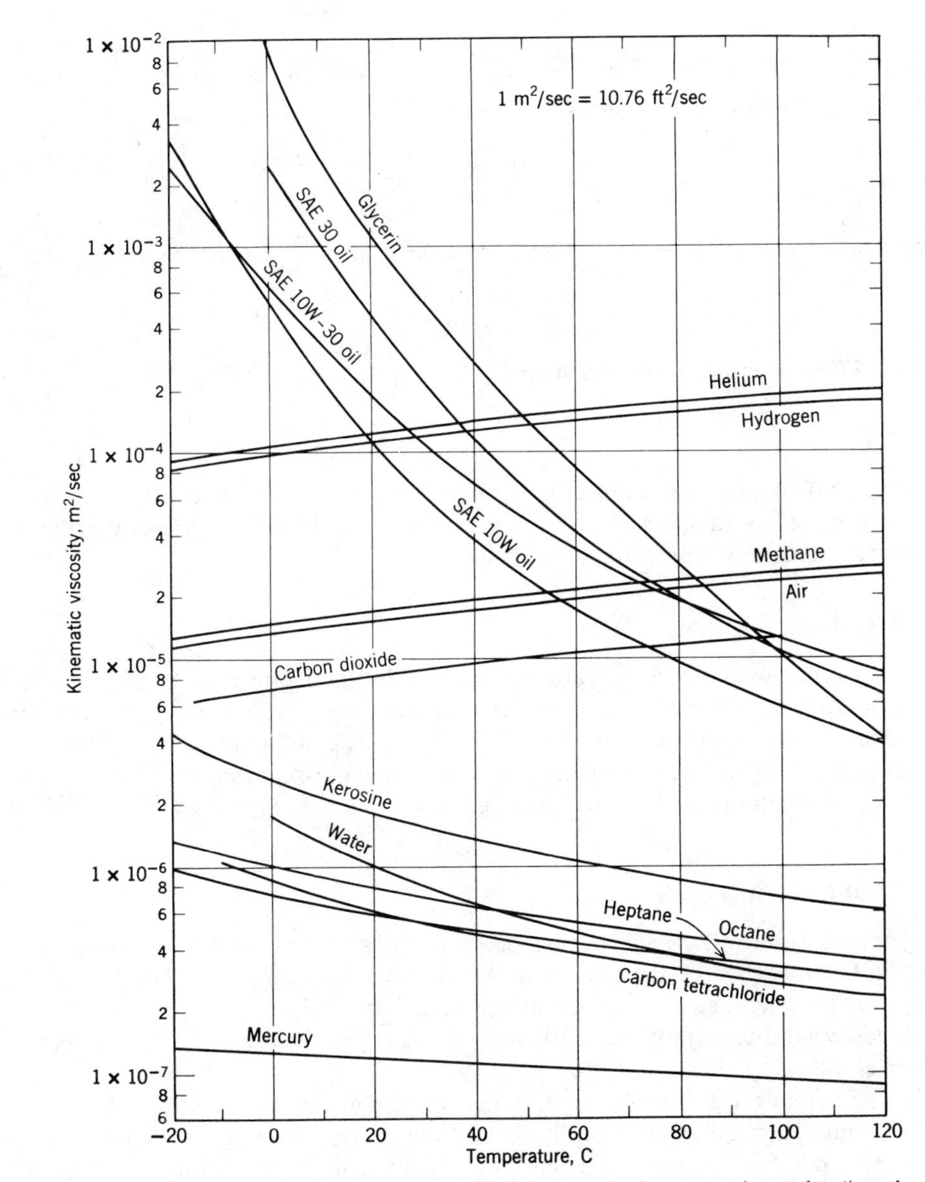

Fig. A.3 Kinematic viscosity of common fluids (at atmospheric pressure) as a function of temperature. (Data from [1], [5], and [9].)

Liquid viscosities are affected drastically by temperature. This dependence on absolute temperature is well represented by the empirical equation

$$\mu = Ae^{B/T} \tag{A.2}$$

Equation A.2 requires two datum points to fit A and B. This may be done easily by plotting $\log \mu$ versus $1/T$ in the form

$$\log \mu = \log A + \frac{B}{T} \log e = \log A + \frac{0.434B}{T}$$

Log A may be identified as the intercept and $0.434B$ as the slope of the resulting plot.

A-2.2 Effect of Pressure on Viscosity

a. Gases

The viscosity of gases is essentially independent of pressure for values between a few hundredths of an atmosphere and a few atmospheres. However, viscosity at high pressures increases with pressure (or density).

b. Liquids

The viscosities of most liquids are not affected by moderate pressures, but large increases have been found at very high pressures. For example, the viscosity of water at 10,000 atm is twice the value at 1 atm. More complex compounds show a viscosity increase of several orders of magnitude over the same pressure range.

More information may be found in [10].

A-3 LUBRICATING OILS

Engine and transmission lubricating oils are classified by viscosity according to standards established by the Society of Automotive Engineers [11]. The allowable ranges of viscosity for several grades are given in Table A.5.

Viscosity numbers with W (e.g. 20W) are classified by viscosity at 0 F. Those without W are classified by viscosity at 210 F.

Multigrade oils (e.g. 10W-40) are formulated to minimize viscosity variation with temperature. High polymer "viscosity index" improvers are used in blending these multigrade oils. Such additives are highly non-Newtonian; they may suffer permanent viscosity loss due to shearing.

Special charts used to estimate the viscosity of petroleum products as a function of temperature are available. The charts were used to develop the data for typical lubricating oils plotted in Figs. A.2 and A.3. For details, see [12].

Table A.5 Allowable Viscosity Ranges for SAE Lubricant Classifications (Data from [11].)

Lubricant Type	SAE Viscosity Number	Viscosity Range (centistokes)[a]			
		At 0 F		At 210 F	
		Minimum	Maximum	Minimum	Maximum
Crankcase	5 *W*		1,300	3.9	
	10 *W*	1,300	2,600	3.9	
	20 *W*	2,600	10,500	3.9	
	20			5.7	9.6
	30			9.6	12.9
	40			12.9	16.8
	50			16.8	22.7
Transmission and axle	75		15,000		
	80	15,000	100,000		
	90			75	120
	140			120	200
	250			200	
Automatic transmisssion fluid	Type *A*	39[b]	43[b]	7	8.5

[a] 1 centistoke = 1 cSt = 10^{-6} m^2/sec (= 1.08×10^{-5} ft^2/sec).
[b] At 100 F.

Table A.6 Thermodynamic Properties of Common Gases at STP[a] (Data from [8, 13, 14].)

Gas	Chemical Symbol	Molecular Mass, M_m	R^b $\left(\frac{J}{kg \cdot K}\right)$	c_p $\left(\frac{J}{kg \cdot K}\right)$	c_v $\left(\frac{J}{kg \cdot K}\right)$	$k = \frac{c_p}{c_v}$ (—)	R^b $\left(\frac{ft \cdot lbf}{lbm \cdot R}\right)$	c_p $\left(\frac{Btu}{lbm \cdot R}\right)$	c_v $\left(\frac{Btu}{lbm \cdot R}\right)$
Air	—	28.98	286.9	1,004	717.4	1.40	53.33	0.2399	0.1713
Carbon dioxide	CO_2	44.01	188.9	840.4	651.4	1.29	35.11	0.2007	0.1556
Carbon monoxide	CO	28.01	296.8	1,039	742.1	1.40	55.17	0.2481	0.1772
Helium	He	4.003	2,077	5,225	3,147	1.66	386.1	1.248	0.7517
Hydrogen	H_2	2.016	4,124	14,180	10,060	1.41	766.5	3.388	2.402
Methane	CH_4	16.04	518.3	2,190	1,672	1.31	96.32	0.5231	0.3993
Nitrogen	N_2	28.01	296.8	1,039	742.0	1.40	55.16	0.2481	0.1772
Oxygen	O_2	32.00	259.8	909.4	649.6	1.40	48.29	0.2172	0.1551
Steam[c]	H_2O	18.02	461.4	~2,000	~1,540	~1.30	85.78	~0.478	~0.368

[a] STP = standard temperature and pressure, $T = 15$ C (59 F) and $p = 101.325$ kPa abs (14.696 psia).
[b] $R \equiv R_u/M_m$; $R_u = 8314.3$ J/kgmol · K (1545.3 ft · lbf/lbmol · R); 1 Btu = 778.2 ft · lbf.
[c] Water vapor behaves as an ideal gas when superheated by 55 C (100 F) or more.

REFERENCES

1. *Handbook of Chemistry and Physics*, 62nd ed. Cleveland, Ohio: Chemical Rubber Publishing Co., 1981–1982.
2. "Meriam Standard Indicating Fluids," Pamphlet No. 920GEN: 430-1, The Meriam Instrument Co., 10920 Madison Avenue, Cleveland, Ohio 44102.
3. E. Vernon Hill, Inc., P.O. Box 7053, Corte Madera, Calif. 94925.
4. *Handbook of Tables for Applied Engineering Science*. Cleveland, Ohio: Chemical Rubber Publishing Co., 1970.
5. Vargaftik, N. B., *Tables on the Thermophysical Properties of Liquids and Gases*, 2nd ed. Washington, D.C.: Hemisphere Publishing Corp., 1975.
6. Trefethen, L., "Surface Tension in Fluid Mechanics," in *Illustrated Experiments in Fluid Mechanics*. Cambridge, Mass.: The M.I.T. Press, 1972.
7. Streeter, V. L., ed., *Handbook of Fluid Dynamics*. New York: McGraw-Hill, 1961.
8. *The U.S. Standard Atmosphere (1962)*. Washington, D.C.: U.S. Government Printing Office, 1962. Pages 10, 14 and Table I, pp. 39–53.
9. Touloukian, Y. S., Saxena, S. C., and Hestermans, P., *Thermophysical Properties of Matter, the TPRC Data Series, Vol. 11—Viscosity*. New York: Plenum Publishing Corp., 1975.
10. Reid, R. C. and Sherwood, T. K., *The Properties of Gases and Liquids*, 2nd ed. New York: McGraw-Hill, 1966.
11. "Crankcase Oil Viscosity Classification, Recommended Practice SAE J300b," *SAE Handbook*, 1976 ed. Warrendale, Pa.: Society of Automotive Engineers, 1976, p. 3.5.
12. ASTM Standard D 341-77, "Viscosity-Temperature Charts for Liquid Petroleum Products," American Society for Testing and Materials, 1916 Race Street, Philadelphia, Pa. 19103.
13. NASA, *Compressed Gas Handbook* (Revised). Washington, D.C.: National Aeronautics and Space Administration, SP-3045, 1970.
14. ASME, *Thermodynamic and Transport Properties of Steam*. New York: American Society of Mechanical Engineers, 1967.

Appendix B

EQUATIONS OF MOTION IN CYLINDRICAL COORDINATES

The continuity equation in cylindrical coordinates for constant density is

$$\frac{1}{r}\frac{\partial}{\partial r}(rv_r) + \frac{1}{r}\frac{\partial}{\partial \theta}(v_\theta) + \frac{\partial}{\partial z}(v_z) = 0 \tag{B.1}$$

Normal and shear stresses in cylindrical coordinates for constant density and viscosity are

$$\begin{aligned}
\sigma_{rr} &= -p + 2\mu\frac{\partial v_r}{\partial r} & \tau_{r\theta} &= \mu\left[r\frac{\partial}{\partial r}\left(\frac{v_\theta}{r}\right) + \frac{1}{r}\frac{\partial v_r}{\partial \theta}\right] \\
\sigma_{\theta\theta} &= -p + 2\mu\left(\frac{1}{r}\frac{\partial v_\theta}{\partial \theta} + \frac{v_r}{r}\right) & \tau_{\theta z} &= \mu\left(\frac{\partial v_\theta}{\partial z} + \frac{1}{r}\frac{\partial v_z}{\partial \theta}\right) \\
\sigma_{zz} &= -p + 2\mu\frac{\partial v_z}{\partial z} & \tau_{zr} &= \mu\left(\frac{\partial v_r}{\partial z} + \frac{\partial v_z}{\partial r}\right)
\end{aligned} \tag{B.2}$$

The Navier–Stokes equations in cylindrical coordinates for constant density and viscosity are

r component:

$$\rho\left(\frac{\partial v_r}{\partial t} + v_r\frac{\partial v_r}{\partial r} + \frac{v_\theta}{r}\frac{\partial v_r}{\partial \theta} - \frac{v_\theta^2}{r} + v_z\frac{\partial v_r}{\partial z}\right)$$

$$= \rho g_r - \frac{\partial p}{\partial r} + \mu\left\{\frac{\partial}{\partial r}\left(\frac{1}{r}\frac{\partial}{\partial r}[rv_r]\right) + \frac{1}{r^2}\frac{\partial^2 v_r}{\partial \theta^2} - \frac{2}{r^2}\frac{\partial v_\theta}{\partial \theta} + \frac{\partial^2 v_r}{\partial z^2}\right\} \tag{B.3a}$$

θ component:

$$\rho\left(\frac{\partial v_\theta}{\partial t} + v_r\frac{\partial v_\theta}{\partial r} + \frac{v_\theta}{r}\frac{\partial v_\theta}{\partial \theta} + \frac{v_r v_\theta}{r} + v_z\frac{\partial v_\theta}{\partial z}\right)$$

$$= \rho g_\theta - \frac{1}{r}\frac{\partial p}{\partial \theta} + \mu\left\{\frac{\partial}{\partial r}\left(\frac{1}{r}\frac{\partial}{\partial r}[rv_\theta]\right) + \frac{1}{r^2}\frac{\partial^2 v_\theta}{\partial \theta^2} + \frac{2}{r^2}\frac{\partial v_r}{\partial \theta} + \frac{\partial^2 v_\theta}{\partial z^2}\right\} \tag{B.3b}$$

z component:

$$\rho\left(\frac{\partial v_z}{\partial t} + v_r \frac{\partial v_z}{\partial r} + \frac{v_\theta}{r}\frac{\partial v_z}{\partial \theta} + v_z \frac{\partial v_z}{\partial z}\right)$$

$$= \rho g_z - \frac{\partial p}{\partial z} + \mu\left\{\frac{1}{r}\frac{\partial}{\partial r}\left(r\frac{\partial v_z}{\partial r}\right) + \frac{1}{r^2}\frac{\partial^2 v_z}{\partial \theta^2} + \frac{\partial^2 v_z}{\partial z^2}\right\} \quad \text{(B.3c)}$$

Appendix C

FILMS AND FILM LOOPS FOR FLUID MECHANICS

Listed below by supplier are titles of sound films and silent film "loops" enclosed in plastic cartridges.

1. Encyclopaedia Britannica Educational Corporation
780 South Lapeer Road
Lake Orion, Michigan 48035

a. Sound films[1] (length as noted):

Aerodynamic Generation of Sound (44 min, principals: M. J. Lighthill, J. E. Ffowcs-Williams)
Boundary Layer Control (25 min, principal: D. C. Hazen)
Cavitation (31 min, principal: P. Eisenberg)
Channel Flow of a Compressible Fluid (29 min, principal: D. E. Coles)
Deformation of Continuous Media (38 min, principal: J. L. Lumley)
Eulerian and Lagrangian Descriptions in Fluid Mechanics (27 min, principal: J. L. Lumley)
Flow Instabilities (27 min, principal: E. L. Mollo-Christensen)
Flow Visualization (31 min, principal: S. J. Kline)
The Fluid Dynamics of Drag[2] (4 parts, 120 min, principal: A. H. Shapiro)
Fundamentals of Boundary Layers (24 min, principal: F. H. Abernathy)
Low-Reynolds-Number Flows (33 min, principal: Sir G. I. Taylor)
Magnetohydrodynamics (27 min, principal: J. A. Shercliff)
Pressure Fields and Fluid Acceleration (30 min, principal: A. H. Shapiro)
Rarefied Gas Dynamics (33 min, principals: F. C. Hurlbut, F. S. Sherman)
Rheological Behavior of Fluids (22 min, principal: H. Markovitz)
Rotating Flows (29 min, principal: D. Fultz)
Secondary Flow (30 min, principal: E. S. Taylor)
Stratified Flow (26 min, principal: R. R. Long)
Surface Tension in Fluid Mechanics (29 min, principal: L. M. Trefethen)

[1] Detailed summaries of these films have been prepared by the National Committee for Fluid Mechanics Films. See *Illustrated Experiments in Fluid Mechanics* (Cambridge, Mass.: The M.I.T. Press, 1972).
[2] The contents of this film are summarized and illustrated in *Shape and Flow, The Fluid Dynamics of Drag* (New York: Anchor Books, 1961).

Turbulence (29 min, principal: R. W. Stewart)
Vorticity (2 parts, 44 min, principal: A. H. Shapiro)
Waves in Fluids (33 min, principal: A. E. Bryson)

b. Silent film loops[3] (all are 3–5 min in length):

S-FM001 *Some Regimes of Boundary-Layer Transition*
S-FM002 *Structure of the Turbulent Boundary Layer*
S-FM003 *Shear Deformation of Viscous Fluids*
S-FM004 *Separated Flows—Part I*
S-FM005 *Separated Flows—Part II*
S-FM006 *Boundary-Layer Formation*
S-FM007 *Propagating Stall in Airfoil Cascade*
S-FM008 *The Occurrence of Turbulence*
S-FM009 *Aerodynamic Heating as Shown by Temperature-Sensitive Paints*
S-FM010 *Generation of Circulation and Lift for an Airfoil*
S-FM011 *The Magnus Effect*
S-FM012 *Flow Separation and Vortex Shedding*
S-FM013 *The Bathtub Vortex*
S-FM014A *Visualization of Vorticity with Vorticity Meter—Part I*
S-FM014B *Visualization of Vorticity with Vorticity Meter—Part II*
S-FM015 *Incompressible Flow through Area Contractions and Expansions*
S-FM016 *Flow from a Reservoir to a Duct*
S-FM017 *Flow Patterns in Venturis, Nozzles, and Orifices*
S-FM018 *Secondary Flow in a Teacup*
S-FM019 *Secondary Flow in a Bend*
S-FM020 *The Horseshoe Vortex*
S-FM021 *Techniques of Visualization for Low Speed Flows—Part I*
S-FM002 *Techniques of Visualization for Low Speed Flows—Part II*
S-FM023 *Tollmien–Schlichting Waves*
S-FM024 *Wing Tip Vortex*
S-FM025 *Low Speed Jets: Stability and Mixing*
S-FM026 *Tornadoes in Nature and the Laboratory*
S-FM027 *Interaction of Oblique Shock with Flat-Plate Boundary Layer*
S-FM028 *Transonic Flow Past a Symmetric Airfoil*
S-FM029 *Supersonic Zones on Airfoils in Subsonic Flow*
S-FM030 *Shock and Boundary-Layer Interaction on Transonic Airfoil*
S-FM031 *Instabilities in Circular Couette Flow*
S-FM032 *Examples of Turbulent Flow between Concentric Rotating Cylinders*
S-FM033 *Stagnation Pressure*
S-FM034 *The Coanda Effect*
S-FM035 *Radial Flow between Parallel Disks*
S-FM036 *Venturi Passage*
S-FM037 *Streamline Curvature and Normal Pressure Gradient*
S-FM038 *Streamwise Pressure Gradient in Inviscid Flow*
S-FM039 *Interpretation of Flow Using Wall Tufts*
S-FM043 *Buoyancy-Induced Waves in Rotating Fluid*

[3] For summaries, see *Illustrated Experiments in Fluid Mechanics* (footnote 1).

S-FM044 *Elastoid-Inertia Oscillations in a Rotating Fluid*
S-FM045 *Velocities near an Airfoil*
S-FM046 *Current-Induced Instabilities of a Mercury Jet*
S-FM047 *Pathlines, Streaklines, Streamlines, and Timelines in Steady Flow*
S-FM048 *Pathlines, Streaklines, Streamlines, and Timelines in Unsteady Flow*
S-FM049 *Flow Regimes in Subsonic Diffusers*
S-FM050 *Flow over an Upstream-Facing Step*
S-FM051 *Simple Supersonic Inlet (Axially Symmetric Geometry)*
S-FM052 *Supersonic Conical-Spike Inlet*
S-FM054 *Leading-Edge Separation Bubble in Two-Dimensional Flow*
S-FM055 *Bow Waves in Hypersonic Flow*
S-FM056 *Three-Dimensional Boundary-Layer Separation*
S-FM057 *Effect of Axial Jet on Afterbody Separation*
S-FM058 *Effect of Jet Blowing over Airfoil Flap*
S-FM059 *Leading-Edge Vortices on Delta Wing in Subsonic Flow*
S-FM060 *Breakdown of Leading-Edge Vortices on Delta Wing in Subsonic Flow*
S-FM061 *Ablation of Ice Models in a Water Tunnel*
S-FM062 *Interactions between Oblique Shocks and Expansion Waves*
S-FM063 *Slot Blowing to Suppress Shock-Induced Separation*
S-FM065 *Wide-Angle Diffuser with Suction*
S-FM066 *Thin Bodies of Revolution at Incidence*
S-FM067 *Flow through Right-Angle Bends*
S-FM068 *Flow through Ported Chambers*
S-FM069 *Flow through Tee-Elbow*
S-FM070 *The Sink Vortex*
S-FM071 *Flow near Tip of Lifting Wing*
S-FM072 *Examples of Surface Tension*
S-FM073 *Surface Tension and Contact Angles*
S-FM074 *Formation of Bubbles*
S-FM075 *Surface Tension and Curved Surfaces*
S-FM076 *Breakup of Liquid into Drops*
S-FM077 *Motions Caused by Composition Gradients along Liquid Surfaces*
S-FM078 *Motions Caused by Electrical and Chemical Effects on Liquid Surfaces*
S-FM079 *Motions Caused by Temperature Gradients along Liquid Surfaces*
S-FM080 *Hele-Shaw Analog to Potential Flows, Part I: Sources and Sinks in Uniform Flow*
S-FM081 *Hele-Shaw Analog to Potential Flows, Part II: Sources and Sinks*
S-FM082 *Water Jet Instability in Electric Field*
S-FM083 *Induced* $\vec{J} \times \vec{B}$ *Forces in Solids and Liquids*
S-FM084 *An MHD Pump*
S-FM086 *Suppression of Vorticity by MHD Forces*
S-FM087 *The Hartmann Layer*
S-FM088 *Laminar Boundary Layers*
S-FM089 *Turbulent Boundary Layers*
S-FM090 *Supersonic Flow past Diamond Airfoil*
S-FM091 *Modes of Sloshing in Tanks*
S-FM092 *Stages of Boundary-Layer Instability and Transition*
S-FM093 *Supersonic Spike Inlet with Variable Geometry*
S-FM097 *Flow through Fans and Propellers*
S-FM098 *Aerodynamic Heating and Ablation of Missile Shapes*

S-FM099 *Passage of Shock Waves over Bodies*
S-FM100 *Passage of Shock Waves through Constrictions*
S-FM101 *Reflections of Shock Waves*
S-FM102 *Passage of a Shock Wave through a Circular Orifice*
S-FM103 *Effect of Knudsen Number on Flow Past a Blunt Body*
S-FM104 *Effect of Knudsen Number on a Jet*
S-FM105 *Ripple-Tank Radiation Patterns of Source, Dipole, and Quadrupole*
S-FM107 *Deformation in Fluids Illustrated by a Rectilinear Shear Flow*
S-FM108 *Small-Amplitude Waves*
S-FM109 *Source Moving at Speeds below and above Wave Speeds*
S-FM112 *Examples of Low-Reynolds-Number Flows*
S-FM113 *Hydrodynamic Lubrication*
S-FM114 *Sedimentation at Low Reynolds Number*
S-FM115 *Kinematic Reversibility of Low-Reynolds-Number Flows*
S-FM116 *Swimming Propulsion at Low Reynolds Number*
S-FM117 *Subsonic Flow Patterns and Pressure Distributions for an Airfoil*
S-FM118 *Laminar-Flow versus Conventional Airfoils*
S-FM119 *Reduction of Airfoil Friction Drag by Suction*
S-FM120 *Some Methods for Increasing Lift Coefficient*
S-FM122 *Nonlinear Shear Stress Behavior in Steady Flows*
S-FM123 *Normal-Stress Effects in Viscoelastic Fluids*
S-FM124 *Memory Effects in Viscoelastic Fluids*
S-FM125 *Examples of Cavitation*
S-FM126 *Cavitation on Hydrofoils*
S-FM127 *Cavitation Bubble Dynamics*
S-FM128 *Cavity Flows*
S-FM129 *Compressible Flow through Convergent-Divergent Nozzle*
S-FM130 *Starting of Supersonic Wind Tunnel with Variable-Throat Diffuser*
S-FM134 *Laminar and Turbulent Pipe Flow*
S-FM135 *Averages and Transport in Turbulence*
S-FM136 *Structure of Turbulence*
S-FM137 *Effects of Density Stratification on Turbulence*
S-FM138 *Taylor Columns in Rotating Flows (at Low Rossby Number)*
S-FM139 *Small-Amplitude Gravity Waves in an Open Channel*
S-FM140 *Flattening and Steepening of Large-Amplitude Gravity Waves*
S-FM141 *The Hydraulic Surge Wave*
S-FM142 *The Hydraulic Jump*
S-FM143 *Free-Surface Flow over a Towed Obstacle*
S-FM144 *Flow of a Two-Liquid Stratified Fluid past an Obstacle*
S-FM145 *Flow of a Continuously Stratified Fluid past an Obstacle*
S-FM146 *Examples of Flow Instability (Part I)*
S-FM147 *Examples of Flow Instability (Part II)*
S-FM148 *Experimental Study of a Flow Instability*

2. University of Iowa
The Audiovisual Center
Iowa City, Iowa 52240

The following six films were prepared as a series, in the order listed. They can be viewed individually without serious loss of continuity.

Introduction to the Study of Fluid Motion (24 min, principal: H. Rouse). This orientation film shows a variety of familiar flow phenomena. Use of scale models for empirical study of complex phenomena is illustrated and the significance of the Euler, Froude, Mach, and Reynolds numbers as similarity parameters is shown using several sequences of model and prototype flows.

Fundamental Principles of Flow (23 min, principal: H. Rouse). The basic concepts and physical relationships needed to analyze fluid motions are developed in this film. The continuity, momentum, and energy equations are derived and used to analyze a jet propulsion device.

Fluid Motion in a Gravitational Field (23 min, principal: H. Rouse). Buoyancy effects and free-surface flows are illustrated in this film. The Froude number is shown to be a fundamental parameter for flows with a free surface. Wave motions are shown for open-channel and density-stratified flows.

Characteristics of Laminar and Turbulent Flow (26 min, principal: H. Rouse). Dye, smoke, suspended particles, and hydrogen bubbles are used to visualize laminar and turbulent flows. Instabilities that lead to turbulence are shown; production and decay of turbulence and mixing are described.

Form Drag, Lift, and Propulsion (24 min, principal: H. Rouse). The effects of boundary-layer separation on flow patterns and pressure distributions are shown for several body shapes. The basic characteristics of lifting shapes, including effects of aspect ratio, are discussed, and the results are applied to analysis of the performance of propellers and torque converters.

Effects of Fluid Compressibility (17 min, principal: H. Rouse). The hydraulic analogy between open-channel liquid flow and compressible gas flow is used to show representative wave patterns. Schlieren optical flow visualization is used in a supersonic wind tunnel to show patterns of flow past several bodies at subsonic and supersonic speeds.

3. Ohio State University
   Film Distribution Supervisor
   Motion Picture Division
   1885 Neil Avenue
   Columbus, Ohio 44223

   Tacoma Narrows Bridge Collapse. The film contains spectacular original footage from this disaster, which occurred in 1940.

4. University of Minnesota
   Saint Anthony Falls Hydraulics Laboratory
   Mississippi River and 3rd Avenue, SE
   Minneapolis, Minn. 55414

   Some Phenomena of Open-Channel Flow (33 min, silent). Many features of open-channel flow are demonstrated. Applications of the specific energy and pressure-momentum curves are illustrated.

   Fluid Mechanics—The Boundary Layer (30 min, sound). The film contains a series of demonstrations of physical principles. It is planned for use as a summary after regular classroom discussion of boundary-layer phenomena.

5. U.S. Army Corps of Engineers
   c/o Modern Talking Picture Service
   86 Weldon Parkway
   Maryland Heights, Mo. 63043

   Speaking of Models (25 min, sound). This color film was produced by the Waterways Experiment Station of the U.S. Army Corps of Engineers. It illustrates the use of hydraulic models to study problems of silt deposit, channel design, breakwater placement, and towboat maneuvering. Several methods of flow visualization are demonstrated.

6. American Institute of Aeronautics and Astronautics
   Director of Public Information
   1290 Avenue of the Americas
   New York, N.Y. 10019

   America's Wings (29 min). Individuals who made significant contributions to development of aircraft for high-speed flight are interviewed; they discuss and explain their contributions. This is an effective film for a relatively sophisticated audience.

7. Engineering Societies Library
   United Engineering Center
   345 East 47th Street
   New York, N.Y. 10017

   Flow Separation and Formation of Vortices (16 min)
   Hydraulic Analogy for Flow Studies (24 min)
   Turbulent Flow (An Introductory Lecture) (35 min)
   Visual Cavitation Studies of Mixed-Flow Pump Impellers (22 min)
   Titles are added periodically.

Appendix D

TABLES FOR COMPUTATION OF COMPRESSIBLE FLOW

Table D.1 Isentropic Flow Functions (one-dimensional flow, ideal gas, $k = 1.4$)

M	T/T_0	p/p_0	ρ/ρ_0	A/A^*
0.00	1.0000	1.0000	1.0000	∞
0.02	0.9999	0.9997	0.9998	28.94
0.04	0.9997	0.9989	0.9992	14.48
0.06	0.9993	0.9975	0.9982	9.666
0.08	0.9987	0.9955	0.9968	7.262
0.10	0.9980	0.9930	0.9950	5.822
0.12	0.9971	0.9900	0.9928	4.864
0.14	0.9961	0.9864	0.9903	4.182
0.16	0.9949	0.9823	0.9873	3.673
0.18	0.9936	0.9777	0.9840	3.278
0.20	0.9921	0.9725	0.9803	2.964
0.22	0.9904	0.9669	0.9762	2.708
0.24	0.9886	0.9607	0.9718	2.496
0.26	0.9867	0.9541	0.9670	2.317
0.28	0.9846	0.9470	0.9619	2.166
0.30	0.9823	0.9395	0.9564	2.035
0.32	0.9799	0.9315	0.9506	1.922
0.34	0.9774	0.9231	0.9445	1.823
0.36	0.9747	0.9143	0.9380	1.736
0.38	0.9719	0.9052	0.9313	1.659
0.40	0.9690	0.8956	0.9243	1.590
0.42	0.9659	0.8857	0.9170	1.529
0.44	0.9627	0.8755	0.9094	1.474
0.46	0.9594	0.8650	0.9016	1.425
0.48	0.9560	0.8541	0.8935	1.380
0.50	0.9524	0.8430	0.8852	1.340

Table D.1 (*Continued*)

M	T/T_0	p/p_0	ρ/ρ_0	A/A^*
0.50	0.9524	0.8430	0.8852	1.340
0.52	0.9487	0.8317	0.8766	1.303
0.54	0.9449	0.8201	0.8679	1.270
0.56	0.9410	0.8082	0.8589	1.240
0.58	0.9370	0.7962	0.8498	1.213
0.60	0.9328	0.7840	0.8405	1.188
0.62	0.9286	0.7716	0.8310	1.166
0.64	0.9243	0.7591	0.8213	1.145
0.66	0.9199	0.7465	0.8115	1.127
0.68	0.9154	0.7338	0.8016	1.110
0.70	0.9108	0.7209	0.7916	1.094
0.72	0.9061	0.7080	0.7814	1.081
0.74	0.9013	0.6951	0.7712	1.068
0.76	0.8964	0.6821	0.7609	1.057
0.78	0.8915	0.6691	0.7505	1.047
0.80	0.8865	0.6560	0.7400	1.038
0.82	0.8815	0.6430	0.7295	1.030
0.84	0.8763	0.6300	0.7189	1.024
0.86	0.8711	0.6170	0.7083	1.018
0.88	0.8659	0.6041	0.6977	1.013
0.90	0.8606	0.5913	0.6870	1.009
0.92	0.8552	0.5785	0.6764	1.006
0.94	0.8498	0.5658	0.6658	1.003
0.96	0.8444	0.5532	0.6551	1.001
0.98	0.8389	0.5407	0.6445	1.000
1.00	0.8333	0.5283	0.6339	1.000
1.02	0.8278	0.5160	0.6234	1.000
1.04	0.8222	0.5039	0.6129	1.001
1.06	0.8165	0.4919	0.6024	1.003
1.08	0.8108	0.4801	0.5920	1.005
1.10	0.8052	0.4684	0.5817	1.008
1.12	0.7994	0.4568	0.5714	1.011
1.14	0.7937	0.4455	0.5612	1.015
1.16	0.7880	0.4343	0.5511	1.020
1.18	0.7822	0.4232	0.5411	1.025
1.20	0.7764	0.4124	0.5311	1.030
1.22	0.7706	0.4017	0.5213	1.037
1.24	0.7648	0.3912	0.5115	1.043
1.26	0.7590	0.3809	0.5019	1.050
1.28	0.7532	0.3708	0.4923	1.058
1.30	0.7474	0.3609	0.4829	1.066

Table D.1 Isentropic Flow Functions (*Continued*)

M	T/T_0	p/p_0	ρ/ρ_0	A/A^*
1.30	0.7474	0.3609	0.4829	1.066
1.32	0.7416	0.3512	0.4736	1.075
1.34	0.7358	0.3417	0.4644	1.084
1.36	0.7300	0.3323	0.4553	1.094
1.38	0.7242	0.3232	0.4463	1.104
1.40	0.7184	0.3142	0.4374	1.115
1.42	0.7126	0.3055	0.4287	1.126
1.44	0.7069	0.2969	0.4201	1.138
1.46	0.7011	0.2886	0.4116	1.150
1.48	0.6954	0.2804	0.4032	1.163
1.50	0.6897	0.2724	0.3950	1.176
1.52	0.6840	0.2646	0.3869	1.190
1.54	0.6783	0.2570	0.3789	1.204
1.56	0.6726	0.2496	0.3711	1.219
1.58	0.6670	0.2423	0.3633	1.234
1.60	0.6614	0.2353	0.3557	1.250
1.62	0.6558	0.2284	0.3483	1.267
1.64	0.6502	0.2217	0.3409	1.284
1.66	0.6447	0.2152	0.3337	1.301
1.68	0.6392	0.2088	0.3266	1.319
1.70	0.6337	0.2026	0.3197	1.338
1.72	0.6283	0.1966	0.3129	1.357
1.74	0.6229	0.1907	0.3062	1.376
1.76	0.6175	0.1850	0.2996	1.397
1.78	0.6121	0.1794	0.2931	1.418
1.80	0.6068	0.1740	0.2868	1.439
1.82	0.6015	0.1688	0.2806	1.461
1.84	0.5963	0.1637	0.2745	1.484
1.86	0.5911	0.1587	0.2686	1.507
1.88	0.5859	0.1539	0.2627	1.531
1.90	0.5807	0.1492	0.2570	1.555
1.92	0.5756	0.1447	0.2514	1.580
1.94	0.5705	0.1403	0.2459	1.606
1.96	0.5655	0.1360	0.2405	1.633
1.98	0.5605	0.1318	0.2352	1.660
2.00	0.5556	0.1278	0.2301	1.688
2.02	0.5506	0.1239	0.2250	1.716
2.04	0.5458	0.1201	0.2200	1.745
2.06	0.5409	0.1164	0.2152	1.775
2.08	0.5361	0.1128	0.2105	1.806
2.10	0.5314	0.1094	0.2058	1.837

Table D.1 (*Continued*)

M	T/T_0	p/p_0	ρ/ρ_0	A/A^*
2.10	0.5314	0.1094	0.2058	1.837
2.12	0.5266	0.1060	0.2013	1.869
2.14	0.5219	0.1027	0.1968	1.902
2.16	0.5173	0.09956	0.1925	1.935
2.18	0.5127	0.09650	0.1882	1.970
2.20	0.5081	0.09352	0.1841	2.005
2.22	0.5036	0.09064	0.1800	2.041
2.24	0.4991	0.08784	0.1760	2.078
2.26	0.4947	0.08514	0.1721	2.115
2.28	0.4903	0.08252	0.1683	2.154
2.30	0.4859	0.07997	0.1646	2.193
2.32	0.4816	0.07751	0.1610	2.233
2.34	0.4773	0.07513	0.1574	2.274
2.36	0.4731	0.07281	0.1539	2.316
2.38	0.4689	0.07057	0.1505	2.359
2.40	0.4647	0.06840	0.1472	2.403
2.42	0.4606	0.06630	0.1440	2.448
2.44	0.4565	0.06426	0.1408	2.494
2.46	0.4524	0.06229	0.1377	2.540
2.48	0.4484	0.06038	0.1347	2.588
2.50	0.4444	0.05853	0.1317	2.637
2.52	0.4405	0.05674	0.1288	2.687
2.54	0.4366	0.05500	0.1260	2.737
2.56	0.4328	0.05332	0.1232	2.789
2.58	0.4289	0.05169	0.1205	2.842
2.60	0.4252	0.05012	0.1179	2.896
2.62	0.4214	0.04859	0.1153	2.951
2.64	0.4177	0.04711	0.1128	3.007
2.66	0.4141	0.04568	0.1103	3.065
2.68	0.4104	0.04429	0.1079	3.123
2.70	0.4068	0.04295	0.1056	3.183
2.72	0.4033	0.04166	0.1033	3.244
2.74	0.3998	0.04039	0.1010	3.306
2.76	0.3963	0.03917	0.09885	3.370
2.78	0.3928	0.03800	0.09671	3.434
2.80	0.3894	0.03685	0.09462	3.500
2.82	0.3860	0.03574	0.09259	3.567
2.84	0.3827	0.03467	0.09059	3.636
2.86	0.3794	0.03363	0.08865	3.706
2.88	0.3761	0.03262	0.08674	3.777
2.90	0.3729	0.03165	0.08489	3.850

Table D.1 Isentropic Flow Functions (*Continued*)

M	T/T_0	p/p_0	ρ/ρ_0	A/A^*
2.90	0.3729	0.03165	0.08489	3.850
2.92	0.3697	0.03071	0.08308	3.924
2.94	0.3665	0.02980	0.08130	3.999
2.96	0.3633	0.02891	0.07957	4.076
2.98	0.3602	0.02805	0.07788	4.155
3.00	0.3571	0.02722	0.07623	4.235
3.10	0.3422	0.02345	0.06852	4.657
3.20	0.3281	0.02023	0.06165	5.121
3.30	0.3147	0.01748	0.05554	5.629
3.40	0.3019	0.01512	0.05009	6.184
3.50	0.2899	0.01311	0.04523	6.790
3.60	0.2784	0.01138	0.04089	7.450
3.70	0.2675	0.009903	0.03702	8.169
3.80	0.2572	0.008629	0.03355	8.951
3.90	0.2474	0.007532	0.03044	9.799
4.00	0.2381	0.006586	0.02766	10.72
4.10	0.2293	0.005769	0.02516	11.71
4.20	0.2208	0.005062	0.02292	12.79
4.30	0.2129	0.004449	0.02090	13.95
4.40	0.2053	0.003918	0.01909	15.21
4.50	0.1980	0.003455	0.01745	16.56
4.60	0.1911	0.003053	0.01597	18.02
4.70	0.1846	0.002701	0.01463	19.58
4.80	0.1783	0.002394	0.01343	21.26
4.90	0.1724	0.002126	0.01233	23.07
5.00	0.1667	0.001890	0.01134	25.00

Table D.2 Fanno Line Flow Functions (one-dimensional flow, ideal gas, $k = 1.4$)

M	p_0/p_0^*	T/T^*	p/p^*	V/V^*	$\bar{f}L_{max}/D_h$
0.00	∞	1.200	∞	0.0000	∞
0.02	28.94	1.200	54.77	0.02191	1778
0.04	14.48	1.200	27.38	0.04381	440.5
0.06	9.666	1.199	18.25	0.06570	193.0
0.08	7.262	1.199	13.68	0.08758	106.7
0.10	5.822	1.198	10.94	0.1094	66.92
0.12	4.864	1.197	9.116	0.1313	45.41
0.14	4.182	1.195	7.809	0.1531	32.51
0.16	3.673	1.194	6.829	0.1748	24.20
0.18	3.278	1.192	6.066	0.1965	18.54
0.20	2.964	1.191	5.456	0.2182	14.53
0.22	2.708	1.189	4.955	0.2398	11.60
0.24	2.496	1.186	4.538	0.2614	9.387
0.26	2.317	1.184	4.185	0.2829	7.688
0.28	2.166	1.182	3.882	0.3044	6.357
0.30	2.035	1.179	3.619	0.3257	5.299
0.32	1.922	1.176	3.389	0.3470	4.447
0.34	1.823	1.173	3.185	0.3682	3.752
0.36	1.736	1.170	3.004	0.3894	3.180
0.38	1.659	1.166	2.842	0.4104	2.706
0.40	1.590	1.163	2.696	0.4313	2.309
0.42	1.529	1.159	2.563	0.4522	1.974
0.44	1.474	1.155	2.443	0.4729	1.692
0.46	1.425	1.151	2.333	0.4936	1.451
0.48	1.380	1.147	2.231	0.5141	1.245
0.50	1.340	1.143	2.138	0.5345	1.069
0.52	1.303	1.138	2.052	0.5548	0.9174
0.54	1.270	1.134	1.972	0.5750	0.7866
0.56	1.240	1.129	1.898	0.5951	0.6736
0.58	1.213	1.124	1.828	0.6150	0.5757
0.60	1.188	1.119	1.763	0.6348	0.4908
0.62	1.166	1.114	1.703	0.6545	0.4172
0.64	1.145	1.109	1.646	0.6740	0.3533
0.66	1.127	1.104	1.592	0.6934	0.2979
0.68	1.110	1.098	1.541	0.7127	0.2498
0.70	1.094	1.093	1.493	0.7318	0.2081
0.72	1.081	1.087	1.448	0.7508	0.1722
0.74	1.068	1.082	1.405	0.7696	0.1411
0.76	1.057	1.076	1.365	0.7883	0.1145
0.78	1.047	1.070	1.326	0.8068	0.09167
0.80	1.038	1.064	1.289	0.8251	0.07229

Table D.2 Fanno Line Flow Functions (*Continued*)

M	p_0/p_0^*	T/T^*	p/p^*	V/V^*	$\bar{f}L_{max}/D_h$
0.80	1.038	1.064	1.289	0.8251	0.07229
0.82	1.030	1.058	1.254	0.8433	0.05593
0.84	1.024	1.052	1.221	0.8614	0.04226
0.86	1.018	1.045	1.189	0.8793	0.03097
0.88	1.013	1.039	1.158	0.8970	0.02180
0.90	1.0089	1.033	1.129	0.9146	0.01451
0.92	1.0056	1.026	1.101	0.9320	0.008913
0.94	1.0031	1.020	1.074	0.9493	0.004815
0.96	1.0014	1.013	1.049	0.9663	0.002057
0.98	1.0003	1.007	1.024	0.9832	0.0004947
1.00	1.0000	1.000	1.000	1.000	0.0000
1.02	1.0003	0.9933	0.9771	1.017	0.0004587
1.04	1.0013	0.9866	0.9551	1.033	0.001769
1.06	1.0029	0.9798	0.9338	1.049	0.003838
1.08	1.0051	0.9730	0.9134	1.065	0.006585
1.10	1.0079	0.9662	0.8936	1.081	0.009935
1.12	1.011	0.9593	0.8745	1.097	0.01382
1.14	1.015	0.9524	0.8561	1.113	0.01819
1.16	1.020	0.9455	0.8383	1.128	0.02298
1.18	1.025	0.9386	0.8210	1.143	0.02814
1.20	1.030	0.9317	0.8044	1.158	0.03364
1.22	1.037	0.9247	0.7882	1.173	0.03942
1.24	1.043	0.9178	0.7726	1.188	0.04547
1.26	1.050	0.9108	0.7574	1.203	0.05174
1.28	1.058	0.9038	0.7427	1.217	0.05820
1.30	1.066	0.8969	0.7285	1.231	0.06483
1.32	1.075	0.8899	0.7147	1.245	0.07161
1.34	1.084	0.8829	0.7012	1.259	0.07850
1.36	1.094	0.8760	0.6882	1.273	0.08550
1.38	1.104	0.8690	0.6755	1.286	0.09259
1.40	1.115	0.8621	0.6632	1.300	0.09974
1.42	1.126	0.8551	0.6512	1.313	0.1069
1.44	1.138	0.8482	0.6396	1.326	0.1142
1.46	1.150	0.8413	0.6282	1.339	0.1215
1.48	1.163	0.8345	0.6172	1.352	0.1288
1.50	1.176	0.8276	0.6065	1.365	0.1361
1.52	1.190	0.8208	0.5960	1.377	0.1434
1.54	1.204	0.8139	0.5858	1.389	0.1506
1.56	1.219	0.8072	0.5759	1.402	0.1579
1.58	1.234	0.8004	0.5662	1.414	0.1651
1.60	1.250	0.7937	0.5568	1.425	0.1724

Table D.2 *(Continued)*

M	p_0/p_0^*	T/T^*	p/p^*	V/V^*	$\bar{f}L_{max}/D_h$
1.60	1.250	0.7937	0.5568	1.425	0.1724
1.62	1.267	0.7870	0.5476	1.437	0.1795
1.64	1.284	0.7803	0.5386	1.449	0.1867
1.66	1.301	0.7736	0.5299	1.460	0.1938
1.68	1.319	0.7670	0.5213	1.471	0.2008
1.70	1.338	0.7605	0.5130	1.482	0.2078
1.72	1.357	0.7539	0.5048	1.494	0.2147
1.74	1.376	0.7474	0.4969	1.504	0.2216
1.76	1.397	0.7410	0.4891	1.515	0.2284
1.78	1.418	0.7345	0.4815	1.526	0.2352
1.80	1.439	0.7282	0.4741	1.536	0.2419
1.82	1.461	0.7218	0.4668	1.546	0.2485
1.84	1.484	0.7155	0.4597	1.556	0.2551
1.86	1.507	0.7093	0.4528	1.566	0.2616
1.88	1.531	0.7030	0.4460	1.576	0.2680
1.90	1.555	0.6969	0.4394	1.586	0.2743
1.92	1.580	0.6907	0.4329	1.596	0.2806
1.94	1.606	0.6847	0.4265	1.605	0.2868
1.96	1.633	0.6786	0.4203	1.615	0.2930
1.98	1.660	0.6726	0.4142	1.624	0.2990
2.00	1.688	0.6667	0.4083	1.633	0.3050
2.02	1.716	0.6608	0.4024	1.642	0.3109
2.04	1.745	0.6549	0.3967	1.651	0.3168
2.06	1.775	0.6491	0.3911	1.660	0.3225
2.08	1.806	0.6433	0.3856	1.668	0.3282
2.10	1.837	0.6376	0.3802	1.677	0.3339
2.12	1.869	0.6320	0.3750	1.685	0.3394
2.14	1.902	0.6263	0.3698	1.694	0.3449
2.16	1.935	0.6208	0.3648	1.702	0.3503
2.18	1.970	0.6152	0.3598	1.710	0.3556
2.20	2.005	0.6098	0.3549	1.718	0.3609
2.22	2.041	0.6043	0.3502	1.726	0.3661
2.24	2.078	0.5990	0.3455	1.734	0.3712
2.26	2.115	0.5936	0.3409	1.741	0.3763
2.28	2.154	0.5883	0.3364	1.749	0.3813
2.30	2.193	0.5831	0.3320	1.756	0.3862
2.32	2.233	0.5779	0.3277	1.764	0.3911
2.34	2.274	0.5728	0.3234	1.771	0.3959
2.36	2.316	0.5677	0.3193	1.778	0.4006
2.38	2.359	0.5626	0.3152	1.785	0.4053
2.40	2.403	0.5576	0.3111	1.792	0.4099

Table D.2 Fanno Line Flow Functions (*Continued*)

M	p_0/p_0^*	T/T^*	p/p^*	V/V^*	$\bar{f}L_{max}/D_h$
2.40	2.403	0.5576	0.3111	1.792	0.4099
2.42	2.448	0.5527	0.3072	1.799	0.4144
2.44	2.494	0.5478	0.3033	1.806	0.4189
2.46	2.540	0.5429	0.2995	1.813	0.4233
2.48	2.588	0.5381	0.2958	1.819	0.4277
2.50	2.637	0.5333	0.2921	1.826	0.4320
2.52	2.687	0.5286	0.2885	1.832	0.4362
2.54	2.737	0.5239	0.2850	1.839	0.4404
2.56	2.789	0.5193	0.2815	1.845	0.4445
2.58	2.842	0.5147	0.2781	1.851	0.4486
2.60	2.896	0.5102	0.2747	1.857	0.4526
2.62	2.951	0.5057	0.2714	1.863	0.4565
2.64	3.007	0.5013	0.2682	1.869	0.4604
2.66	3.065	0.4969	0.2650	1.875	0.4643
2.68	3.123	0.4925	0.2619	1.881	0.4681
2.70	3.183	0.4882	0.2588	1.887	0.4718
2.72	3.244	0.4839	0.2558	1.892	0.4755
2.74	3.306	0.4797	0.2528	1.898	0.4792
2.76	3.370	0.4755	0.2499	1.903	0.4827
2.78	3.434	0.4714	0.2470	1.909	0.4863
2.80	3.500	0.4673	0.2441	1.914	0.4898
2.82	3.567	0.4632	0.2414	1.919	0.4932
2.84	3.636	0.4592	0.2386	1.925	0.4966
2.86	3.706	0.4553	0.2359	1.930	0.5000
2.88	3.777	0.4513	0.2333	1.935	0.5033
2.90	3.850	0.4474	0.2307	1.940	0.5065
2.92	3.924	0.4436	0.2281	1.945	0.5097
2.94	3.999	0.4398	0.2256	1.950	0.5129
2.96	4.076	0.4360	0.2231	1.954	0.5160
2.98	4.155	0.4323	0.2206	1.959	0.5191
3.00	4.235	0.4286	0.2182	1.964	0.5222
3.50	6.79	0.3478	0.1685	2.064	0.5864
4.00	10.72	0.2857	0.1336	2.138	0.6331
4.50	16.56	0.2376	0.1083	2.194	0.6676
5.00	25.00	0.2000	0.08944	2.236	0.6938

Table D.3 Rayleigh Line Flow Functions (one-dimensional flow, ideal gas, $k = 1.4$)

M	T_0/T_0^*	p_0/p_0^*	T/T^*	p/p^*	V/V^*
0.00	0.0000	1.268	0.0000	2.400	0.0000
0.02	0.001918	1.268	0.002301	2.399	0.0009595
0.04	0.007648	1.267	0.009175	2.395	0.003831
0.06	0.01712	1.265	0.02053	2.388	0.008597
0.08	0.03021	1.262	0.03621	2.379	0.01522
0.10	0.04678	1.259	0.05602	2.367	0.02367
0.12	0.06661	1.255	0.07970	2.353	0.03388
0.14	0.08947	1.251	0.1070	2.336	0.04578
0.16	0.1151	1.246	0.1374	2.317	0.05931
0.18	0.1432	1.241	0.1708	2.296	0.07438
0.20	0.1736	1.235	0.2066	2.273	0.09091
0.22	0.2057	1.228	0.2445	2.248	0.1088
0.24	0.2395	1.221	0.2841	2.221	0.1279
0.26	0.2745	1.214	0.3250	2.193	0.1482
0.28	0.3104	1.206	0.3667	2.163	0.1696
0.30	0.3469	1.199	0.4089	2.131	0.1918
0.32	0.3837	1.190	0.4512	2.099	0.2149
0.34	0.4206	1.182	0.4933	2.066	0.2388
0.36	0.4572	1.174	0.5348	2.031	0.2633
0.38	0.4935	1.165	0.5755	1.996	0.2883
0.40	0.5290	1.157	0.6152	1.961	0.3137
0.42	0.5638	1.148	0.6535	1.925	0.3395
0.44	0.5975	1.139	0.6903	1.888	0.3656
0.46	0.6301	1.131	0.7254	1.852	0.3918
0.48	0.6614	1.122	0.7587	1.815	0.4181
0.50	0.6914	1.114	0.7901	1.778	0.4445
0.52	0.7199	1.106	0.8196	1.741	0.4708
0.54	0.7470	1.098	0.8470	1.704	0.4970
0.56	0.7725	1.090	0.8723	1.668	0.5230
0.58	0.7965	1.083	0.8955	1.632	0.5489
0.60	0.8189	1.075	0.9167	1.596	0.5745
0.62	0.8398	1.068	0.9359	1.560	0.5998
0.64	0.8592	1.061	0.9530	1.525	0.6248
0.66	0.8771	1.055	0.9682	1.491	0.6494
0.68	0.8935	1.049	0.9814	1.457	0.6737
0.70	0.9085	1.043	0.9929	1.424	0.6975
0.72	0.9221	1.038	1.003	1.391	0.7209
0.74	0.9344	1.033	1.011	1.359	0.7439
0.76	0.9455	1.028	1.017	1.327	0.7665
0.78	0.9553	1.023	1.022	1.296	0.7885
0.80	0.9639	1.019	1.025	1.266	0.8101

Table D.3 Rayleigh Line Flow Functions (*Continued*)

M	T_0/T_0^*	p_0/p_0^*	T/T^*	p/p^*	V/V^*
0.80	0.9639	1.019	1.025	1.266	0.8101
0.82	0.9715	1.016	1.028	1.236	0.8313
0.84	0.9781	1.012	1.029	1.207	0.8519
0.86	0.9836	1.010	1.028	1.179	0.8721
0.88	0.9883	1.007	1.027	1.152	0.8918
0.90	0.9921	1.005	1.025	1.125	0.9110
0.92	0.9951	1.003	1.021	1.098	0.9297
0.94	0.9973	1.002	1.017	1.073	0.9480
0.96	0.9988	1.001	1.012	1.048	0.9658
0.98	0.9997	1.000	1.006	1.024	0.9831
1.00	1.000	1.000	1.000	1.000	1.000
1.02	0.9997	1.000	0.9930	0.9770	1.016
1.04	0.9990	1.001	0.9855	0.9546	1.032
1.06	0.9977	1.002	0.9776	0.9328	1.048
1.08	0.9960	1.003	0.9691	0.9115	1.063
1.10	0.9939	1.005	0.9603	0.8909	1.078
1.12	0.9915	1.007	0.9512	0.8708	1.092
1.14	0.9887	1.010	0.9417	0.8512	1.106
1.16	0.9856	1.012	0.9320	0.8322	1.120
1.18	0.9823	1.016	0.9220	0.8137	1.133
1.20	0.9787	1.019	0.9119	0.7958	1.146
1.22	0.9749	1.023	0.9015	0.7783	1.158
1.24	0.9709	1.028	0.8911	0.7613	1.171
1.26	0.9668	1.033	0.8805	0.7447	1.182
1.28	0.9624	1.038	0.8699	0.7287	1.194
1.30	0.9580	1.044	0.8592	0.7130	1.205
1.32	0.9534	1.050	0.8484	0.6978	1.216
1.34	0.9487	1.056	0.8377	0.6830	1.226
1.36	0.9440	1.063	0.8270	0.6686	1.237
1.38	0.9392	1.070	0.8161	0.6546	1.247
1.40	0.9343	1.078	0.8054	0.6410	1.256
1.42	0.9293	1.086	0.7947	0.6278	1.266
1.44	0.9243	1.094	0.7841	0.6149	1.275
1.46	0.9193	1.103	0.7735	0.6024	1.284
1.48	0.9143	1.112	0.7629	0.5902	1.293
1.50	0.9093	1.122	0.7525	0.5783	1.301
1.52	0.9042	1.132	0.7422	0.5668	1.310
1.54	0.8992	1.142	0.7319	0.5555	1.318
1.56	0.8942	1.153	0.7217	0.5446	1.325
1.58	0.8892	1.164	0.7117	0.5339	1.333
1.60	0.8842	1.176	0.7017	0.5236	1.340

Table D.3 *(Continued)*

M	T_0/T_0^*	p_0/p_0^*	T/T^*	p/p^*	V/V^*
1.60	0.8842	1.176	0.7017	0.5236	1.340
1.62	0.8792	1.188	0.6919	0.5135	1.348
1.64	0.8743	1.200	0.6822	0.5036	1.355
1.66	0.8694	1.213	0.6726	0.4941	1.361
1.68	0.8645	1.226	0.6631	0.4847	1.368
1.70	0.8597	1.240	0.6538	0.4756	1.375
1.72	0.8549	1.255	0.6446	0.4668	1.381
1.74	0.8502	1.269	0.6355	0.4581	1.387
1.76	0.8455	1.284	0.6265	0.4497	1.393
1.78	0.8409	1.300	0.6177	0.4415	1.399
1.80	0.8363	1.316	0.6089	0.4335	1.405
1.82	0.8317	1.332	0.6004	0.4257	1.410
1.84	0.8273	1.349	0.5919	0.4181	1.416
1.86	0.8228	1.367	0.5836	0.4107	1.421
1.88	0.8185	1.385	0.5754	0.4035	1.426
1.90	0.8141	1.403	0.5673	0.3964	1.431
1.92	0.8099	1.422	0.5594	0.3896	1.436
1.94	0.8057	1.442	0.5516	0.3828	1.441
1.96	0.8015	1.462	0.5439	0.3763	1.446
1.98	0.7974	1.482	0.5364	0.3699	1.450
2.00	0.7934	1.503	0.5289	0.3636	1.455
2.02	0.7894	1.525	0.5216	0.3575	1.459
2.04	0.7855	1.547	0.5144	0.3516	1.463
2.06	0.7816	1.569	0.5074	0.3458	1.467
2.08	0.7778	1.592	0.5004	0.3401	1.471
2.10	0.7741	1.616	0.4936	0.3345	1.475
2.12	0.7704	1.640	0.4868	0.3291	1.479
2.14	0.7667	1.665	0.4802	0.3238	1.483
2.16	0.7631	1.691	0.4737	0.3186	1.487
2.18	0.7596	1.717	0.4673	0.3136	1.490
2.20	0.7561	1.743	0.4611	0.3086	1.494
2.22	0.7527	1.771	0.4549	0.3038	1.497
2.24	0.7493	1.799	0.4488	0.2991	1.501
2.26	0.7460	1.827	0.4429	0.2945	1.504
2.28	0.7428	1.856	0.4370	0.2899	1.507
2.30	0.7395	1.886	0.4312	0.2855	1.510
2.32	0.7364	1.917	0.4256	0.2812	1.513
2.34	0.7333	1.948	0.4200	0.2770	1.517
2.36	0.7302	1.979	0.4145	0.2728	1.520
2.38	0.7272	2.012	0.4091	0.2688	1.522
2.40	0.7242	2.045	0.4038	0.2648	1.525

Table D.3 Rayleigh Line Flow Functions (*Continued*)

M	T_0/T_0^*	p_0/p_0^*	T/T^*	p/p^*	V/V^*
2.40	0.7242	2.045	0.4038	0.2648	1.525
2.42	0.7213	2.079	0.3986	0.2609	1.528
2.44	0.7184	2.114	0.3935	0.2571	1.531
2.46	0.7156	2.149	0.3885	0.2534	1.533
2.48	0.7128	2.185	0.3836	0.2497	1.536
2.50	0.7101	2.222	0.3787	0.2462	1.539
2.52	0.7074	2.259	0.3739	0.2427	1.541
2.54	0.7047	2.298	0.3692	0.2392	1.543
2.56	0.7021	2.337	0.3646	0.2359	1.546
2.58	0.6995	2.377	0.3601	0.2326	1.548
2.60	0.6970	2.418	0.3556	0.2294	1.551
2.62	0.6945	2.459	0.3512	0.2262	1.553
2.64	0.6921	2.502	0.3469	0.2231	1.555
2.66	0.6896	2.545	0.3427	0.2201	1.557
2.68	0.6873	2.589	0.3385	0.2171	1.559
2.70	0.6849	2.634	0.3344	0.2142	1.561
2.72	0.6826	2.680	0.3304	0.2113	1.563
2.74	0.6804	2.727	0.3264	0.2085	1.565
2.76	0.6782	2.775	0.3225	0.2058	1.567
2.78	0.6760	2.824	0.3186	0.2031	1.569
2.80	0.6738	2.873	0.3149	0.2004	1.571
2.82	0.6717	2.924	0.3111	0.1978	1.573
2.84	0.6696	2.975	0.3075	0.1953	1.575
2.86	0.6675	3.028	0.3039	0.1927	1.577
2.88	0.6655	3.081	0.3004	0.1903	1.578
2.90	0.6635	3.136	0.2969	0.1879	1.580
2.92	0.6615	3.191	0.2934	0.1855	1.582
2.94	0.6596	3.248	0.2901	0.1832	1.583
2.96	0.6577	3.306	0.2868	0.1809	1.585
2.98	0.6558	3.365	0.2835	0.1787	1.587
3.00	0.6540	3.424	0.2803	0.1765	1.588
3.50	0.6158	5.328	0.2142	0.1322	1.620
4.00	0.5891	8.227	0.1683	0.1026	1.641
4.50	0.5698	12.50	0.1354	0.08177	1.656
5.00	0.5556	18.63	0.1111	0.06667	1.667

Table D.4 Normal Shock Flow Functions (one-dimensional flow, ideal gas, $k = 1.4$)

M_1	M_2	p_{0_2}/p_{0_1}	T_2/T_1	p_2/p_1	ρ_2/ρ_1
1.00	1.000	1.000	1.000	1.000	1.000
1.02	0.9805	1.000	1.013	1.047	1.033
1.04	0.9620	0.9999	1.026	1.095	1.067
1.06	0.9444	0.9998	1.039	1.144	1.101
1.08	0.9277	0.9994	1.052	1.194	1.135
1.10	0.9118	0.9989	1.065	1.245	1.169
1.12	0.8966	0.9982	1.078	1.297	1.203
1.14	0.8820	0.9973	1.090	1.350	1.238
1.16	0.8682	0.9961	1.103	1.403	1.272
1.18	0.8549	0.9946	1.115	1.458	1.307
1.20	0.8422	0.9928	1.128	1.513	1.342
1.22	0.8300	0.9907	1.141	1.570	1.376
1.24	0.8183	0.9884	1.153	1.627	1.411
1.26	0.8071	0.9857	1.166	1.686	1.446
1.28	0.7963	0.9827	1.178	1.745	1.481
1.30	0.7860	0.9794	1.191	1.805	1.516
1.32	0.7760	0.9757	1.204	1.866	1.551
1.34	0.7664	0.9718	1.216	1.928	1.585
1.36	0.7572	0.9676	1.229	1.991	1.620
1.38	0.7483	0.9630	1.242	2.055	1.655
1.40	0.7397	0.9582	1.255	2.120	1.690
1.42	0.7314	0.9531	1.268	2.186	1.724
1.44	0.7235	0.9477	1.281	2.253	1.759
1.46	0.7157	0.9420	1.294	2.320	1.793
1.48	0.7083	0.9360	1.307	2.389	1.828
1.50	0.7011	0.9298	1.320	2.458	1.862
1.52	0.6941	0.9233	1.334	2.529	1.896
1.54	0.6874	0.9166	1.347	2.600	1.930
1.56	0.6809	0.9097	1.361	2.673	1.964
1.58	0.6746	0.9026	1.374	2.746	1.998
1.60	0.6684	0.8952	1.388	2.820	2.032
1.62	0.6625	0.8876	1.402	2.895	2.065
1.64	0.6568	0.8799	1.416	2.971	2.099
1.66	0.6512	0.8720	1.430	3.048	2.132
1.68	0.6458	0.8640	1.444	3.126	2.165
1.70	0.6406	0.8557	1.458	3.205	2.198
1.72	0.6355	0.8474	1.473	3.285	2.230
1.74	0.6305	0.8389	1.487	3.366	2.263
1.76	0.6257	0.8302	1.502	3.447	2.295
1.78	0.6210	0.8215	1.517	3.530	2.327
1.80	0.6165	0.8127	1.532	3.613	2.359

Table D.4 Normal Shock Flow Functions (*Continued*)

M_1	M_2	p_{0_2}/p_{0_1}	T_2/T_1	p_2/p_1	ρ_2/ρ_1
1.80	0.6165	0.8127	1.532	3.613	2.359
1.82	0.6121	0.8038	1.547	3.698	2.391
1.84	0.6078	0.7947	1.562	3.783	2.422
1.86	0.6036	0.7857	1.577	3.870	2.454
1.88	0.5996	0.7766	1.592	3.957	2.485
1.90	0.5956	0.7674	1.608	4.045	2.516
1.92	0.5918	0.7581	1.624	4.134	2.546
1.94	0.5880	0.7488	1.639	4.224	2.577
1.96	0.5844	0.7395	1.655	4.315	2.607
1.98	0.5808	0.7302	1.671	4.407	2.637
2.00	0.5774	0.7209	1.687	4.500	2.667
2.02	0.5740	0.7115	1.704	4.594	2.696
2.04	0.5707	0.7022	1.720	4.689	2.725
2.06	0.5675	0.6928	1.737	4.784	2.755
2.08	0.5643	0.6835	1.754	4.881	2.783
2.10	0.5613	0.6742	1.770	4.978	2.812
2.12	0.5583	0.6649	1.787	5.077	2.840
2.14	0.5554	0.6557	1.805	5.176	2.868
2.16	0.5525	0.6464	1.822	5.277	2.896
2.18	0.5498	0.6373	1.839	5.378	2.924
2.20	0.5471	0.6281	1.857	5.480	2.951
2.22	0.5444	0.6191	1.875	5.583	2.978
2.24	0.5418	0.6100	1.892	5.687	3.005
2.26	0.5393	0.6011	1.910	5.792	3.032
2.28	0.5368	0.5921	1.929	5.898	3.058
2.30	0.5344	0.5833	1.947	6.005	3.085
2.32	0.5321	0.5745	1.965	6.113	3.110
2.34	0.5297	0.5658	1.984	6.222	3.136
2.36	0.5275	0.5572	2.002	6.331	3.162
2.38	0.5253	0.5486	2.021	6.442	3.187
2.40	0.5231	0.5402	2.040	6.553	3.212
2.42	0.5210	0.5318	2.059	6.666	3.237
2.44	0.5189	0.5234	2.079	6.779	3.261
2.46	0.5169	0.5152	2.098	6.894	3.285
2.48	0.5149	0.5071	2.118	7.009	3.310
2.50	0.5130	0.4990	2.137	7.125	3.333
2.52	0.5111	0.4910	2.157	7.242	3.357
2.54	0.5092	0.4832	2.177	7.360	3.380
2.56	0.5074	0.4754	2.198	7.479	3.403
2.58	0.5056	0.4677	2.218	7.599	3.426
2.60	0.5039	0.4601	2.238	7.720	3.449

Table D.4 *(Continued)*

M_1	M_2	p_{0_2}/p_{0_1}	T_2/T_1	p_2/p_1	ρ_2/ρ_1
2.60	0.5039	0.4601	2.238	7.720	3.449
2.62	0.5022	0.4526	2.259	7.842	3.471
2.64	0.5005	0.4452	2.280	7.965	3.494
2.66	0.4988	0.4379	2.301	8.088	3.516
2.68	0.4972	0.4307	2.322	8.213	3.537
2.70	0.4956	0.4236	2.343	8.338	3.559
2.72	0.4941	0.4166	2.364	8.465	3.580
2.74	0.4926	0.4097	2.386	8.592	3.601
2.76	0.4911	0.4028	2.407	8.721	3.622
2.78	0.4897	0.3961	2.429	8.850	3.643
2.80	0.4882	0.3895	2.451	8.980	3.664
2.82	0.4868	0.3829	2.473	9.111	3.684
2.84	0.4854	0.3765	2.496	9.243	3.704
2.86	0.4840	0.3701	2.518	9.376	3.724
2.88	0.4827	0.3639	2.540	9.510	3.743
2.90	0.4814	0.3577	2.563	9.645	3.763
2.92	0.4801	0.3517	2.586	9.781	3.782
2.94	0.4788	0.3457	2.609	9.918	3.801
2.96	0.4776	0.3398	2.632	10.06	3.820
2.98	0.4764	0.3340	2.656	10.19	3.839
3.00	0.4752	0.3283	2.679	10.33	3.857
3.10	0.4695	0.3012	2.799	11.05	3.947
3.20	0.4644	0.2762	2.922	11.78	4.031
3.30	0.4596	0.2533	3.049	12.54	4.112
3.40	0.4552	0.2322	3.180	13.32	4.188
3.50	0.4512	0.2130	3.315	14.13	4.261
3.60	0.4474	0.1953	3.454	14.95	4.330
3.70	0.4440	0.1792	3.596	15.81	4.395
3.80	0.4407	0.1645	3.743	16.68	4.457
3.90	0.4377	0.1510	3.893	17.58	4.516
4.00	0.4350	0.1388	4.047	18.50	4.571
4.10	0.4324	0.1276	4.205	19.45	4.624
4.20	0.4299	0.1173	4.367	20.41	4.675
4.30	0.4277	0.1080	4.532	21.41	4.723
4.40	0.4255	0.09948	4.702	22.42	4.768
4.50	0.4236	0.09170	4.875	23.46	4.812
4.60	0.4217	0.08459	5.052	24.52	4.853
4.70	0.4199	0.07809	5.233	25.61	4.893
4.80	0.4183	0.07214	5.418	26.71	4.930
4.90	0.4167	0.06670	5.607	27.85	4.966
5.00	0.4152	0.06172	5.800	29.00	5.000

Appendix E

ANALYSIS OF EXPERIMENTAL UNCERTAINTY

E-1 INTRODUCTION

Results of experimental tests often are used for engineering analysis and design. Not all data are equally good; the validity of data should be documented before test results are used for design. Uncertainty analysis is the procedure used to quantify data validity and accuracy.

Analysis of uncertainty also is useful during experiment design. Careful study may indicate potential sources of unacceptable error and suggest improved measurement methods.

E-2 TYPES OF ERROR

Errors always are present when experimental measurements are made. Aside from gross blunders by the experimenter, experimental error may be of two types. Fixed (or systematic) error causes repeated measurements to be in error by the same amount for each trial. Fixed error is the same for each reading and can be removed by proper calibration or correction. Random error (nonrepeatability) is different for every reading and hence cannot be removed. The factors that introduce random error are uncertain by their nature. The objective of uncertainty analysis is to estimate the probable random error in experimental results.

We assume that equipment has been constructed correctly and calibrated properly to eliminate fixed errors. We assume that instrumentation has adequate resolution and that fluctuations in readings are not excessive. We assume also that care is used in making and recording observations so that only random errors remain.

E-3 ESTIMATION OF UNCERTAINTY

Our goal is to estimate the uncertainty of experimental measurements and calculated results due to random errors. The procedure has three steps:

1. Estimate the uncertainty interval for each measured quantity.

2. State the confidence limit on each measurement.
3. Analyze the propagation of uncertainty into results calculated from experimental data.

Below we outline the procedure for each step and illustrate applications with examples.

Step 1. *Estimate the measurement uncertainty interval.* Designate the measured variables in an experiment as $x_1, x_2, \ldots, x_n$. One possible way to find the uncertainty interval for each variable would be to repeat each measurement many times. The result would be a distribution of data values for each variable. Random errors in measurement usually produce a *normal* (*Gaussian*) frequency distribution of measured values. The data scatter for a normal distribution is characterized by the standard deviation, σ. The uncertainty interval for each measured variable, x_i, may be stated as $\pm n\sigma_i$, where $n = 1, 2$, or 3.

For normally distributed data, over 99 percent of measured values of x_i lie within $\pm 3\sigma_i$ of the mean value, 95 percent lie within $\pm 2\sigma_i$, and 50 percent lie within $\pm \sigma_i$ of the mean value of the data set [1]. Thus it would be possible to quantify expected errors within any desired *confidence limit* if a statistically significant set of data were available.

The method of repeated measurements usually is impractical. In most applications it is impossible to obtain enough data for a statistically significant sample owing to the excessive time and cost involved. However, the normal distribution suggests several important concepts:

1. Small errors are more likely than large ones.
2. Plus and minus errors are about equally likely.
3. No finite maximum error can be specified.

A more typical situation in engineering work is a "single-sample" experiment, where only one measurement is made for each datum point [2]. A reasonable estimate of the measurement uncertainty due to random error in a single-sample experiment usually is plus or minus half the smallest scale division (the *least count*) of the instrument. However, this approach also must be used with caution, as illustrated in the following example.

Example E. 1

The observed height of the mercury barometer column is $h = 752.6$ mm. The least count on the vernier scale is 0.1 mm, so one might estimate the probable measurement error as ± 0.05 mm.

A measurement probably could not be made this precisely. The barometer sliders and meniscus must be aligned by eye. The slider has a least count of 1 mm. As a conservative estimate, a measurement could be made to the nearest millimeter. The probable value of a single measurement then would be expressed as 752.6 ± 0.5 mm. The relative uncertainty in barometric height would be stated as

$$u_h = \pm \frac{0.5 \text{ mm}}{752.6 \text{ mm}} = \pm 0.000664 \qquad \text{or} \qquad \pm 0.0664 \text{ percent}$$

Comments:

1. An uncertainty interval of ± 0.1 percent corresponds to a result specified to three significant figures; this precision is sufficient for most engineering work.
2. The measurement of barometer height was precise, as shown by the uncertainty estimate. But was it accurate? At typical room temperatures, the observed barometer reading must be reduced by a temperature correction of nearly 3 mm! This is an example of a fixed error that requires a correction factor.

Step 2. *State the confidence limit on each measurement.* The uncertainty interval of a measurement should be stated at specified odds. For example, one may write $h = 752.6 \pm 0.5$ mm (20 to 1). This means that one is willing to bet 20 to 1 that the height of the mercury column actually is within ± 0.5 mm of the stated value. It should be obvious [3] that "... the specification of such odds can only be made by the experimenter based on ... total laboratory experience. There is no substitute for sound engineering judgment in estimating the uncertainty of a measured variable."

The confidence interval statement is based on the concept of standard deviation for a normal distribution. Odds of 100 to 1 correspond to $\pm 3\sigma$; 99 percent of all future readings are expected to fall within the interval. Odds of 20 to 1 correspond to $\pm 2\sigma$ and odds of 2 to 1 correspond to $\pm \sigma$ confidence limits. Odds of 20 to 1 typically are used for engineering work.

Step 3. *Analyze the propagation of uncertainty in calculations.* Suppose that measurements of independent variables, $x_1, x_2, \ldots, x_n$, are made in the laboratory. The relative uncertainty of each independently measured quantity is estimated as u_i. The measurements are used to calculate some result, R, for the experiment. We wish to analyze how errors in the x_i's *propagate* into the calculation of R from measured values.

In general, R may be expressed mathematically as $R = R(x_1, x_2, \ldots, x_n)$. The effect on R of an error in measuring an individual x_i may be estimated by analogy to the derivative of a function [4]. A variation, δx_i, in x_i would cause R to vary according to

$$\delta R_i = \frac{\partial R}{\partial x_i} \delta x_i$$

For applications, it is convenient to normalize this equation by R to obtain

$$\frac{\delta R_i}{R} = \frac{1}{R} \frac{\partial R}{\partial x_i} \delta x_i = \frac{x_i}{R} \frac{\partial R}{\partial x_i} \frac{\delta x_i}{x_i} \tag{E.1}$$

Equation E.1 might be used to estimate the uncertainty interval in the result due to variations in x_i. To do this, substitute the uncertainty interval for x_i

$$u_{R_i} = \frac{x_i}{R} \frac{\partial R}{\partial x_i} u_{x_i} \tag{E.2}$$

How do we estimate the uncertainty in R due to the combined effects of uncertainty intervals in all the x_i's? The uncertainty interval in each variable has a range of values. It is unlikely that all will have adverse values at the same time. It can be shown [2] that the best representation is

$$u_R = \pm\left[\left(\frac{x_1}{R}\frac{\partial R}{\partial x_1}u_1\right)^2 + \left(\frac{x_2}{R}\frac{\partial R}{\partial x_2}u_2\right)^2 + \cdots + \left(\frac{x_n}{R}\frac{\partial R}{\partial x_n}u_n\right)^2\right]^{1/2} \tag{E.3}$$

Example E.2

Obtain an expression for the uncertainty in determining the volume of a cylinder from measurements of its radius and height. The volume of a cylinder in terms of radius and height is

$$\forall = \forall(r, h) = \pi r^2 h$$

Differentiating, we obtain

$$d\forall = \frac{\partial \forall}{\partial r}dr + \frac{\partial \forall}{\partial h}dh = 2\pi rh\,dr + \pi r^2\,dh$$

since

$$\frac{\partial \forall}{\partial r} = 2\pi rh \qquad \text{and} \qquad \frac{\partial \forall}{\partial h} = \pi r^2$$

From Eq. E.2 the fractional uncertainty due to radius is

$$u_{\forall,r} = \frac{\delta\forall_r}{\forall} = \frac{r}{\forall}\frac{\partial \forall}{\partial r}u_r = \frac{r}{\pi r^2 h}(2\pi rh)u_r = 2u_r$$

and the uncertainty due to height is

$$u_{\forall,h} = \frac{\delta\forall_h}{\forall} = \frac{h}{\forall}\frac{\partial \forall}{\partial h}u_h = \frac{h}{\pi r^2 h}(\pi r^2)u_h = u_h$$

The combined uncertainty in volume is then

$$u_\forall = \pm[(2u_r)^2 + (u_h)^2]^{1/2} \tag{E.4}$$

Comment: The coefficient, 2, in Eq. E.4 shows that the uncertainty in measuring cylinder radius has a larger effect than the uncertainty in measuring height. This is true because the radius is squared in the equation for volume.

E-4 APPLICATIONS TO DATA

Applications to data obtained from laboratory measurements are illustrated in the following examples.

Example E.3

The mass flow rate of water through a tube is to be determined by collecting water in a beaker. The mass flow rate is calculated from the net mass of water collected divided by the time interval

$$\dot{m} = \frac{\Delta m}{\Delta t} \tag{E.5}$$

where $\Delta m = m_f - m_e$. Error estimates for the measured quantities are

$$\text{Mass of full beaker, } m_f = 400 \pm 2 \text{ g (20 to 1)}$$

$$\text{Mass of empty beaker, } m_e = 200 \pm 2 \text{ g (20 to 1)}$$

$$\text{Collection time interval, } \Delta t = 10 \pm 0.2 \text{ sec (20 to 1)}$$

The relative uncertainties in measured quantities are

$$u_{m_f} = \pm \frac{2 \text{ g}}{400 \text{ g}} = \pm 0.005$$

$$u_{m_e} = \pm \frac{2 \text{ g}}{200 \text{ g}} = \pm 0.01$$

$$u_{\Delta t} = \pm \frac{0.2 \text{ sec}}{10 \text{ sec}} = \pm 0.02$$

The relative uncertainty in the measured value of net mass is calculated from Eq. E.3 as

$$\begin{aligned} u_{\Delta m} &= \pm \left[\left(\frac{m_f}{\Delta m} \frac{\partial \Delta m}{\partial m_f} u_{m_f} \right)^2 + \left(\frac{m_e}{\Delta m} \frac{\partial \Delta m}{\partial m_e} u_{m_e} \right)^2 \right]^{1/2} \\ &= \pm \{[(2)(1)(\pm 0.005)]^2 + [(1)(-1)(\pm 0.01)]^2\}^{1/2} \\ u_{\Delta m} &= \pm 0.0141 \end{aligned}$$

Because $\dot{m} = \dot{m}\,(\Delta m, \Delta t)$, we may write Eq. E.3 as

$$u_{\dot{m}} = \pm \left[\left(\frac{\Delta m}{\dot{m}} \frac{\partial \dot{m}}{\partial \Delta m} u_{\Delta m} \right)^2 + \left(\frac{\Delta t}{\dot{m}} \frac{\partial \dot{m}}{\partial \Delta t} u_{\Delta t} \right)^2 \right]^{1/2} \tag{E.6}$$

The required partial derivative terms are

$$\frac{\Delta m}{\dot{m}} \frac{\partial \dot{m}}{\partial \Delta m} = 1 \quad \text{and} \quad \frac{\Delta t}{\dot{m}} \frac{\partial \dot{m}}{\partial \Delta t} = -1$$

Substituting into Eq. E.6 gives

$$\begin{aligned} u_{\dot{m}} &= \pm \{[(1)(\pm 0.0141)]^2 + [(-1)(\pm 0.02)]^2\}^{1/2} \\ u_{\dot{m}} &= \pm 0.0245 \quad \text{or} \quad \pm 2.45 \text{ percent (20 to 1)} \end{aligned}$$

Comment: The 2 percent uncertainty interval in time measurement makes the most important contribution to the uncertainty interval in the result.

Example E.4

The Reynolds number is to be calculated for flow of water in a tube. The computing equation for the Reynolds number is

$$Re = \frac{4\dot{m}}{\pi \mu D} = Re(\dot{m}, D, \mu) \tag{E.7}$$

We have considered the uncertainty interval in calculating the mass flow rate. What about uncertainties in μ and D? The tube diameter is given as $D = 6.35$ mm. Do we assume that it is exact? The diameter might be measured to the nearest 0.1 mm. If so, the relative uncertainty in diameter would be estimated as

$$u_D = \pm\frac{0.05 \text{ mm}}{6.35 \text{ mm}} = \pm 0.00787 \qquad \text{or} \qquad \pm 0.787 \text{ percent}$$

The viscosity of water depends on temperature. The temperature is estimated as $T = 24 \pm 0.5$ C. How will the uncertainty in temperature affect the uncertainty in μ? One way to estimate this is to write

$$u_{\mu(T)} = \pm\frac{\delta\mu}{\mu} = \frac{1}{\mu}\frac{d\mu}{dT}(\pm\delta T) \tag{E.8}$$

The derivative can be estimated from tabulated viscosity data near the nominal temperature of 24 C. Thus

$$\frac{d\mu}{dT} \approx \frac{\Delta\mu}{\Delta T} = \frac{\mu(25 \text{ C}) - \mu(23 \text{ C})}{(25 - 23) \text{ C}}$$

$$= (0.000890 - 0.000933)\frac{\text{N} \cdot \text{sec}}{\text{m}^2} \times \frac{1}{2 \text{ C}}$$

$$\frac{d\mu}{dT} = -2.15 \times 10^{-5} \text{ N} \cdot \text{sec/m}^2 \cdot \text{C}$$

It follows from Eq. E.8 that the uncertainty in viscosity due to temperature is

$$u_{\mu(T)} = \frac{1}{0.000911}\frac{\text{m}^2}{\text{N} \cdot \text{sec}} \times -2.15 \times 10^{-5}\frac{\text{N} \cdot \text{sec}}{\text{m}^2 \cdot \text{C}} \times (\pm 0.5 \text{ C})$$

$$u_{\mu(T)} = \pm 0.0118 \qquad \text{or} \qquad \pm 1.18 \text{ percent}$$

Tabulated viscosity data themselves also have some uncertainty. If this is ± 1.0 percent, an estimate for the overall uncertainty in viscosity is

$$u_\mu = \pm[(0.01)^2 + (\pm 0.0118)^2]^{1/2} = \pm 0.0155 \qquad \text{or} \qquad \pm 1.55 \text{ percent}$$

The uncertainties in mass flow rate, tube diameter, and viscosity needed to compute the uncertainty interval for the calculated Reynolds number now are known. The required partial derivative terms determined from Eq. E.7 are

$$\frac{\dot{m}}{Re}\frac{\partial Re}{\partial \dot{m}} = \frac{\dot{m}}{Re}\frac{4}{\pi\mu D} = \frac{Re}{Re} = 1$$

$$\frac{\mu}{Re}\frac{\partial Re}{\partial \mu} = \frac{\mu}{Re}(-1)\frac{4\dot{m}}{\pi\mu^2 D} = -\frac{Re}{Re} = -1$$

$$\frac{D}{Re}\frac{\partial Re}{\partial D} = \frac{D}{Re}(-1)\frac{4\dot{m}}{\pi\mu D^2} = -\frac{Re}{Re} = -1$$

Substituting into Eq. E.3 gives

$$u_{Re} = \pm\{[(1)(\pm 0.0245)]^2 + [(-1)(\pm 0.0155)]^2 + [(-1)(\pm 0.00787)]^2\}^{1/2}$$

$$u_{Re} = \pm 0.0300 \qquad \text{or} \qquad \pm 3.00 \text{ percent}$$

Comment: Examples E.3 and E.4 illustrate two points important for experiment design. First, the mass of water collected, Δm, is calculated from two measured values, m_f and m_e. For any stated uncertainty interval in the measurements of m_f and m_e, the *relative* uncertainty in Δm can be decreased by making Δm larger. This might be accomplished by using larger containers or a longer measuring interval, Δt, which also would reduce the relative uncertainty in the measured value of Δt. Second, the uncertainty in tabulated property data may be significant. The data uncertainty also is increased by the uncertainty in measurement of fluid temperature.

Example E.5

Air speed is calculated from pitot tube measurements in a wind tunnel. From the Bernoulli equation

$$V = \left(\frac{2gh\rho_{\text{water}}}{\rho_{\text{air}}}\right)^{1/2} \tag{E.9}$$

where h is the observed height of the manometer column.

The only new element in this example is the square root. The variation in V due to the uncertainty interval in h is

$$\frac{h}{V}\frac{\partial V}{\partial h} = \frac{h}{V}\frac{1}{2}\left(\frac{2gh\rho_{\text{water}}}{\rho_{\text{air}}}\right)^{-1/2}\frac{2g\rho_{\text{water}}}{\rho_{\text{air}}}$$

$$\frac{h}{V}\frac{\partial V}{\partial h} = \frac{h}{V}\frac{1}{2}\frac{1}{V}\frac{2g\rho_{\text{water}}}{\rho_{\text{air}}} = \frac{1}{2}\frac{V^2}{V^2} = \frac{1}{2}$$

Using Eq. E.3, we calculate the uncertainty in V as

$$u_V = \pm\left[\left(\frac{1}{2}u_h\right)^2 + \left(\frac{1}{2}u_{\rho_{\text{water}}}\right)^2 + \left(-\frac{1}{2}u_{\rho_{\text{air}}}\right)^2\right]^{1/2}$$

If $u_h = \pm 0.01$ and the other uncertainties are negligible,

$$u_V = \pm\left\{\left[\frac{1}{2}(\pm 0.01)\right]^2\right\}^{1/2}$$

$$u_V = \pm 0.00500 \qquad \text{or} \qquad \pm 0.500 \text{ percent}$$

Comment: The square root reduces the uncertainty interval in the calculated velocity to half that of u_h.

E-5 SUMMARY

A statement of the probable uncertainty of data is an important part of reporting experimental results completely and clearly. Estimating uncertainty in experimental results requires care, experience, and judgment, in common with many endeavors in engineering. We have emphasized the need to quantify the uncertainty of measurements, but space allows including only a few examples. Much more information is available in the references that follow. We urge you to consult them when designing experiments or analyzing data.

E-6 REFERENCES

1. Pugh, E. M., and G. H. Winslow, *The Analysis of Physical Measurements.* Reading, Mass. Addison-Wesley, 1966.
2. Kline, S. J., and F. A. McClintock, "Describing Uncertainties in Single-Sample Experiments," Mechanical Engineering, *75*, 1, January 1953, pp. 3–9.
3. Doebelin, E. O., *Measurement Systems: Analysis and Design.* New York: McGraw-Hill, 1969.
4. Young, H. D., *Statistical Treatment of Experimental Data.* New York: McGraw-Hill, 1962.
5. Holman, J. P., *Experimental Methods for Engineers.* New York: McGraw-Hill, 1978.

Appendix F

SI UNITS, PREFIXES, AND CONVERSION FACTORS

Table F.1 SI Units and Prefixes[a]

SI Units	Quantity	Unit	SI Symbol	Formula
SI base units:	Length	meter	m	—
	Mass	kilogram	kg	—
	Time	second	sec	—
	Temperature	kelvin	K	—
SI supplementary unit:	Plane angle	radian	rad	—
SI derived units:	Energy	joule	J	N · m
	Force	newton	N	kg · m/sec^2
	Power	watt	W	J/sec
	Pressure	pascal	Pa	N/m^2
	Work	joule	J	N · m
SI prefixes	**Multiplication Factor**		**Prefix**	**SI Symbol**
	1 000 000 000 000 = 10^{12}		tera	T
	1 000 000 000 = 10^{9}		giga	G
	1 000 000 = 10^{6}		mega	M
	1 000 = 10^{3}		kilo	k
	0.01 = 10^{-2}		centi[b]	c
	0.001 = 10^{-3}		milli	m
	0.000 001 = 10^{-6}		micro	μ
	0.000 000 001 = 10^{-9}		nano	n
	0.000 000 000 001 = 10^{-12}		pico	p

[a] Source: ASTM Standard for Metric Practice E 380–82, 1982.
[b] To be avoided where possible.

UNIT CONVERSIONS

The data needed to solve problems are not always available in consistent units. Thus it is often necessary to convert from one system of units to another.

In principle, all derived units can be expressed in terms of basic units. Then only conversion factors for basic units would be required.

In practice, many engineering quantities are expressed in terms of defined units, for example, the horsepower, British thermal unit (Btu), quart, or nautical mile. Definitions for such quantities are necessary, and additional conversion factors are useful in calculations.

Basic SI units and necessary conversion factors, plus a few definitions and convenient conversion factors are given in Table F.2.

Table F.2 Conversion Factors and Definitions

Fundamental Conversion Factor	English Unit	Exact SI Value	Approximate SI Value
Length	1 in.	0.0254 m	—
Mass	1 lbm	0.453 592 37 kg	0.4536 kg
Temperature	1 F	5/9 K	—

Definitions:

Acceleration of gravity:	$g = 9.8066\ \mathrm{m/sec^2}$ $(= 32.174\ \mathrm{ft/sec^2})$
Energy:	Btu (British thermal unit) ≡ amount of energy required to raise the temperature of 1 lbm of water 1 F (1 Btu = 778.2 ft · lbf)
	kilocalorie ≡ amount of energy required to raise 1 kg of water 1 K (1 kcal = 4187 J)
Length:	1 mile = 5280 ft; 1 nautical mile = 6076.1 ft
Power:	1 horsepower ≡ 550 ft · lbf/sec
Pressure:	1 bar ≡ 10^5 Pa
Temperature:	degree Fahrenheit, $T_F = \frac{9}{5} T_C + 32$ (where T_C is degrees Celsius)
	degree Rankine, $T_R = T_F + 459.67$
	Kelvin, $T_K = T_C + 273.15$ (exact)
Viscosity:	1 Poise ≡ 0.1 kg/m · sec
	1 Stoke ≡ 0.0001 $\mathrm{m^2/sec}$
Volume:	1 gal ≡ 231 $\mathrm{in.^3}$ (1 $\mathrm{ft^3}$ = 7.48 gal)

Useful Conversion Factors:

	1 lbf = 4.448 N
	1 $\mathrm{lbf/in.^2}$ = 6895 Pa
	1 Btu = 1055 J
	1 hp = 746 W = 2545 Btu/hr
	1 kW = 3413 Btu/hr
	1 quart = 0.000946 $\mathrm{m^3}$ = 0.946 liter
	1 kcal = 3.968 Btu

ANSWERS TO SELECTED EVEN-NUMBERED PROBLEMS

Chapter 1

1.4 $\forall = 0.254 \text{ ft}^3$
1.6 $5 \times 10^3 \text{ ft} \cdot \text{lbf/slug}; 31.7 \text{ ft} \cdot \text{lbf/ft}^3$
1.8 $W = 394 \text{ kJ}$
1.10 $T_2 \approx 269\ F;\ p_2 \approx 41$ psia; $W_{out} = 5390$ Btu
1.12 $\vec{V} = -A\omega \sin(\omega t)\hat{i} + B\omega \cos(\omega t)\hat{j}$; $\vec{a} = -A\omega^2 \cos(\omega t)\hat{i} - B\omega^2 \sin(\omega t)\hat{j}$; V_{max} at $\omega t = \pi/2, 3\pi/2$; a_{max} at $\omega t = 0, \pi$
1.14 $t = 3\ W/gk$
1.16 $V_t = 53.9$ m/sec; $s = 345$ m; $s = 134$ m
1.20 $V = \sqrt{mg/k}\tan\left(\tan^{-1}\sqrt{\frac{k}{mg}}V_0 - \sqrt{\frac{k}{mg}}gt\right)$;
$$y_{max} = \frac{m}{2k}ln\left(1 + \frac{kV_0^2}{mg}\right);$$
$$t_{max} = \sqrt{\frac{m}{gk}}\tan^{-1}\sqrt{\frac{k}{mg}}V_0;\ t_{max} = \frac{V_0}{g}$$
1.22 $V_{up} \approx 1.0$ m/sec; $V_{down} \approx 2.1$ m/sec
1.26 $1 \text{ Pa} = 1.45 \times 10^{-4} \text{ lbf/in.}^2$
1.28 $1 \text{ m}^3\text{/sec} = 35.3 \text{ ft}^3\text{/sec}$; $1 \text{ ft}^3\text{/sec} = 449$ gal/min; 1 gal/hr = 0.472 kg/min; $1 \text{ ft}^3\text{/min} = 4.59$ lbm/hr
1.30 1 home run = 117 m; 1 fastball = 2.92 sec; 1 baseball = 1.41 N; mass = baseball · fastball^2/home run = 0.103 kg
1.32 $\rho = 1130 \pm 21.4 \text{ kg/m}^3$ (20 to 1 odds); $SG = 1.13 \pm 0.0214$ (20 to 1 odds)
1.34 $W = 59.9$ lbf; $\forall = 0.964 \text{ ft}^3$
1.36 $\forall = 15{,}600$ gal
1.38 $\Delta\gamma = -1.75 \text{ lbf/ft}^3$
1.40 $\rho = 917 \text{ kg/m}^3$; $SG = 0.918$
1.42 (a) $x^3y - xy^3 + 2z$; (b) $\hat{i}(y^2z + 2xy) + \hat{j}(2x^2 - xyz) - \hat{k}2x^2y^2$; (c) $\hat{i}y$; (d) $\hat{i}x - \hat{j}y$; (e) $3y$; (f) $-\hat{k}y$
1.44 (a) $x^3yz + x^2yz^2$; (b) $-\hat{i}xyz^3 + \hat{j}(xz^3 - x^4y^2) + \hat{k}x^3y^2z$; (c) $\hat{i}x + \hat{j}xy$; (d) $-\hat{k}x^2$; (e) $z(1 + x)$; (f) 0

Chapter 2

2.2 (1) 1-D, steady; (2) 1-D, steady; (3) 1-D, unsteady; (4) 2-D, steady; (5) 2-D, unsteady; (6) 2-D steady; (7) 3-D, steady; (8) 3-D, unsteady
2.4 Steady, 3-*D* with respect to xyz, 2-*D* with respect to $r\theta z$.
2.6 $y = 2x^{-b/a}$
2.8 2-*D*; Unsteady; $dy/dx = 2$
2.10 $y = cx^{-b/a}$
2.14 $x = x_0 e^{[(t-\tau) + 0.1(t^2 - \tau^2)]}$, $y = y_0 e^{(t-\tau)}$
2.16 $\vec{F}_B = 1.15\hat{j}$ lbf
2.18 $\vec{F}_B = 261\hat{i} + 312\hat{k}$ N
2.22 $p - p_a = 291/D$; $m = 4.41 \times 10^{-13}\ pD^3$ (p in kPa, D in mm)

2.24 1 $N \cdot sec/m^2 = 2.089 \times 10^{-2}$ $lbf \cdot sec/ft^2$
2.26 $\mu = 2.27 \times 10^{-8}\, T^{1/2}/(1 + 198.7/T)$
2.28 $\mu = 1.75 \times 10^{-3}\, N \cdot sec/m^2$; $1.00 \times 10^{-3}\, N \cdot sec/m^2$; $6.51 \times 10^{-4}\, N \cdot sec/m^2$
2.32 $n = 0.661$; $k = 0.121\, N \cdot sec^{0.661}/m^2$; $\tau_{yx} = 0.0264\, N/m^2$; $\eta_{cream}/\mu_{H_2O} = 264$
2.34 $\tau_{yx} = 4.6\, N/m^2$; $\tau_{slurry}/\tau_{H_2O} = 9.2$
2.38 $T = 15.3$ ft · lbf
2.40 $\mu = 0.0208\, N \cdot sec/m^2$
2.42 $\dot{\gamma} = \omega/\theta$; $T = \frac{2}{3}\pi R^3 \tau_{yx}$
2.44 $F = 34.8$ lbf
2.46 $F = 1.01$ N
2.48 $U = 9.84$ m/sec
2.50 $\bar{V}_{max} = 5.07 \times 10^{-2}$ m/sec
2.52 $Re = 1444$; Laminar

Chapter 3

3.2 $D = 0.254$ m; $p = 154$ kPa (gage)
3.4 $m = 14.7$ kg; $t = 11.9$ mm
3.6 (a) 0.3 percent; (b) −0.2 percent
3.8 $\delta h = 5.08 \times 10^{-5}$ in.
3.10 $p_a = 1.18$ psig
3.12 $\ell = 1.6$ m
3.14 $\ell = 0.546$ m
3.16 $p_1 = 2.45$ kPa (gage)
3.18 $\theta = 11.1°$
3.20 $\Delta h = -5.584/R$ (Δh and R in mm)
3.22 $p = 13.0$ psi
3.24 $y = \frac{1}{2}\left[h + H + \frac{p_a}{\rho g} - \sqrt{\left(h + H + \frac{p_a}{\rho g}\right)^2 - 4hH}\right]$
3.26 $p - p_{atm} = 9.74 \times 10^3$ psi; $\rho = 65.8$ lbm/ft^3
3.28 No
3.30 $\Delta\rho/\rho = 0.414$ percent; $\Delta\rho/\rho = 0.251$ percent
3.32 $\Delta z = 1390$ m
3.34 $p/p_0 = [(1 + mz)/(1 + mz_0)]^{-g/mRT_0}$
3.36 $\Delta\mathscr{P}/\mathscr{P} = 25.8$ percent
3.38 $D = 5.97$ mm
3.40 $p = 45.1$ kPa (gage); $T_{sat} \approx 110$ C
3.42 $|\vec{F}_R| = 25.7$ kN
3.44 $|\vec{F}_R| = 2616$ lbf; $|\vec{F}_{net}| = 62.4$ lbf
3.46 $|\vec{F}_R| = 44.1$ kN; $|\vec{R}| = 66.7$ kN
3.48 $|\vec{F}_R| = 1940$ lbf
3.50 $|\vec{R}| = 52.6$ kN
3.52 $|\vec{F}| = 1800$ lbf
3.54 $|\vec{R}| = 1.74$ kN, to the right
3.56 $|\vec{F}_A| = 32.7$ kN
3.58 $d = 2.66$ m
3.60 $b = 6.15$ m
3.62 $|\vec{F}_A|/b = 15.6$ kN/m
3.64 $F_{R_x} = 39.2$ kN (to right); $y' = \frac{2}{3}$ m
3.66 $L = 1.46$ m
3.68 $F_{R_x} = 7.35$ kN (to left); $y' = \frac{1}{3}$ m
3.70 $F_{R_y} = 582$ lbf (upward); $x' = 0.858$ ft
3.72 $F_v = 12{,}460$ lbf (downward); $z' = 1.93$ ft
3.74 $F_{R_y} = \rho g \omega R^2 \pi/4$ (upward); $x' = 4R/3\pi$
3.76 $F_R = 370$ kN at $\alpha = 57.6°$ to horizontal
3.78 $F_v = 532$ kN
3.80 $\gamma_s = 51.2$ lbf/ft^3; No
3.82 $SG = SG_{H_2O}\, W_{air}/(W_{air} - W_{net})$
3.84 Tank is not neutrally buoyant
3.86 $D = 27.4$ m; $\forall_{sw}/\forall_{oil} = 0.507$
3.88 $R = 8.79$ mN
3.90 $D = 82.7$ m; $M = 637$ kg
3.92 Claims are valid; Lift is increased 45 percent
3.94 $\omega = 1.81$ rad/sec
3.96 Slope = 0.22
3.98 $\alpha = 13.3°$
3.100 $|\vec{F}|_{rear} = 107$ N; $|\vec{F}|_{front} = 54.1$ N; $|\vec{F}|_{bot} = 156$ N
3.102 $\partial p/\partial r = \rho\omega^2 r$
3.104 $\theta = 14.3°$ (toward center of curvature)
3.106 $a_r = -r\omega^2$; $\partial p/\partial r = \rho r \omega^2$; $p = 7.19$ MPa
3.108 $\Delta p = \rho\omega^2 R^2/2$; $\omega = 7.16$ rad/sec

Chapter 4

4.2 $V = 87.5$ km/hr
4.4 $s = 2730$ ft; $t = 26.6$ sec
4.6 $F = 961$ N; $F = 586$ N
4.8 $a_r = -31.7$ m/sec^2; $V = 89.0$ m/sec (199 mph)
4.10 $\vec{T}_{avg} = -3.25 \times 10^{-4}\, \hat{k}$ N · m
4.12 $\vec{a}_{XYZ} = -(V_0^2 r_0^2/R^3 + \omega^2 R)\hat{i}_r + 2\omega V_0 r_0/R\hat{i}_\theta$
4.14 $t = 9.38$ sec; $s = 156$ m

4.16 $\Delta U = 9.00 \times 10^5$ Btu; $\Delta U = 0$
4.18 $\Delta u = 77.5$ kJ/kg
4.20 $\Delta t = 419$ sec
4.22 $s_2 - s_1 = -0.291$ kJ/kg · K
4.24 $Q = 1.50$ ft^3/sec
4.26 $d\vec{A} = dydz\,\hat{i} - dxdz\,\hat{j}$; $\int \vec{V}\cdot d\vec{A} = -12.0$ m^3/sec; $\int \vec{V}(\vec{V}\cdot d\vec{A}) = 16\hat{i} - 24\hat{j} - 12\hat{k}$ m^4/sec^2
4.28 $\vec{V}_3 = 4.04\hat{i} - 2.34\hat{j}$ m/sec
4.30 $D = 55.6$ mm
4.32 $\bar{V} = V_{max}/3$
4.34 $u_{max} = 7.50$ m/sec
4.36 $C = U/3$
4.38 $\dot{m}/w = \rho^2 g \sin\theta h^3/6\mu$
4.40 $\partial\rho_0/\partial t = 2.50 \times 10^{-3}$ slug/ft^3 · sec
4.42 $\partial h/\partial t = -56.6$ mm/sec
4.44 $t = 28.6$ sec
4.46 $d\rho/dt = -(\rho - \rho_{H_2O})\, VA/\forall$;
$$t = \frac{\forall}{VA} \ln\left[\frac{\rho_i - \rho_{H_2O}}{\rho_f - \rho_{H_2O}}\right]$$
4.48 $t = 1400$ min
4.50 $\partial\vec{P}_{cv}/\partial t = 2\rho\,\hat{i}$ N
4.52 $mf = 349\hat{i} - 13.5\hat{j} N$
4.54 Ratio = 1.2
4.56 $M = 409$ kg
4.58 $T = 59.9$ N
4.60 $T = 0$
4.62 $F = 321$ N
4.64 $h_2/h = 0.5\,(1 + \sin\theta)$
4.66 $F_x = 0.0230$ lbf
4.68 $T = 15.6$ kN
4.70 $F_{max} = 97.0$ lbf
4.72 Scale reading = 213 lbf; $W = 203$ lbf
4.74 $F = 496$ kN, tension
4.76 $\vec{F} = -340\hat{i} + 1.66\hat{j}$ kN
4.78 $F = 0.446$ lbf
4.80 $t = 1.19$ mm; $K_x = 3.63$ kN, tension
4.84 $V = 6.60$ m/sec; $p_2 - p_1 = 84.2$ kPa
4.86 $F = 5.11$ kN
4.88 $Q = 0.141$ m^3/sec; $\vec{F} = -1.65\hat{i} - 1.34\hat{j}$ kN
4.90 $Q = 0.424$ m^3/sec; $F_y = 4.05$ kN
4.92 $F_D = 0.558$ N
4.94 $F/w = 0.0393$ N/m
4.96 $F_D/w = 54.1$ N/m
4.98 $u_{max} = 30$ ft/sec; $p_1 - p_2 = 0.190$ lbf/ft^2
4.102 $V = Qx/whL$
4.104 $p(x) = p(0) - \rho\left(\frac{Qx}{whL}\right)^2$
4.106 $V = V_0 x/h$
4.108 $V = V_0 - \frac{qx}{A}$; $p(x) = p(0) + \frac{2\rho q x V_0}{A}\left(1 - \frac{qx}{2V_0 A}\right)$
4.110 $V = V_0 + \frac{qx}{A}$, $p(x) = p(0) - \frac{2\rho q x V_0}{A}\left(1 + \frac{qx}{2V_0 A}\right)$
4.112 $\frac{\partial h}{\partial t} + \frac{\partial uh}{\partial x} = 0$; $-\frac{\tau}{\rho h} + g\frac{\partial h}{\partial x} = \frac{\partial u}{\partial t} + u\frac{\partial u}{\partial x}$
4.114 $V/V_0 = \sqrt{1 + 2gz/V_0^2}$; $A/A_0 = V_0/V$
4.116 $h_1 = \left[h_2^2 + \frac{2Q^2}{gb^2 h_2}\right]^{1/2}$
4.118 $r = [2gh]^{1/4}[A/\pi s]^{1/2}$; $\forall = \frac{2A}{3}\sqrt{2gs}\, n^{3/2}$
4.120 $\frac{dh}{dt} = \frac{Q}{A_1} - \sqrt{2gh}\,\frac{A_2}{A_1}$; $t \approx 730$ sec
4.122 $\mu_s < 0.20$
4.124 $h = 89.6$ mm; $F = 0.242$ N
4.126 $V = [V_0^2 + 2gh]^{1/2}$; $F = 1.49$ N
4.128 $\vec{F} = -822\hat{i} + 220\hat{j}$ N
4.130 $F_x = \rho(V - U)^2 A(1 - \cos\theta)$; $\dot{W} = \rho(V - U)^2\, U\, A(1 - \cos\theta)$
4.132 $\vec{F} = -570\hat{i} + 329\hat{j}$ lbf
4.134 $F = 1.73$ kN
4.136 $V = 95.4$ ft/sec
4.138 $F = 4240$ N
4.140 $F = 15.5$ kN
4.142 $T = \frac{1}{2}\rho\,(V_4^2 - V_1^2)A$
4.144 $\theta = 19.7°$
4.146 $A = 2\,Ma/3\rho(V - U)^2$; $A = 0.901$ in.^{2}; $t = 3$ sec
4.148 $t = 24.7$ sec
4.150 $a_{rf_x} = \frac{2\rho V^2 A}{M}\left[\frac{1}{1 + \frac{2\rho V A_t}{M}}\right]^2$; $t = M/2\rho VA$
4.152 $dU/dt = 5.99$ m/sec^2; $U/U_t = 0.667$

4.154 $dU/dt = 13.7\ \text{m/sec}^2$; $U_t = 15.8\ \text{m/sec}$
4.156 $dU/dt = 14.2\ \text{m/sec}^2$; $U_t = 15.2\ \text{m/sec}$
4.158 $U = 16.5\ \text{m/sec}$
4.160 $a = -4\dfrac{\rho VA}{M}U_0 e^{-4\rho VAt/M}$; $U = U_0 e^{-4\rho VAt/M}$
4.162 $a_y = -20.1\ \text{ft/sec}^2$
4.164 $U = 83.9\ \text{ft/sec}$
4.166 $M = M_0 - \dot{m}t$; $a = -\mu_k g$; $U = U_0 - \mu_k gt$; $X = X_0 + U_0 t - \mu_k g t^2/2$
4.168 $t = 129\ \text{sec}$
4.170 $M = 186\ \text{lbm}$
4.172 $\dfrac{U}{U_t} = \dfrac{1-\left(1-\dfrac{\dot{m}t}{M_0}\right)^{2V_e/U_t}}{1+\left(1-\dfrac{\dot{m}t}{M_0}\right)^{2V_e/U_t}}$; $U_t = \left(\dfrac{V_e\dot{m}}{k}\right)^{1/2}$
4.174 $\dfrac{U}{V} = ln\left(\dfrac{M_0}{M_0 - \rho VAt}\right)$
4.176 $a = 17.3\ g$
4.178 $V = 1780\ \text{ft/sec}$; $Y = 15{,}000\ \text{ft}$
4.180 $I = 2.21\ \text{lbf}\cdot\text{sec}$; $I_{sp} = 80.2\ \text{sec}$; $V_{max} = 139\ \text{m/sec}$; $Y_{max} = 1100\ \text{m}$
4.182 $U = 13.9\ \text{ft/sec}$
4.184 $F = aM_0 - \rho V_j^2 A + 3\rho V_j A(at) - \dfrac{3}{2}\rho A(at)^2$
4.186 $a_x = -180\ \text{ft/sec}^2$
4.188 $k = 0.004\ \text{N}\cdot\text{sec}^2/\text{m}^2$; $t = 21.9\ \text{sec}$
4.190 $F = 44.4\ \text{kN}$; $T = 920\ \text{kN}\cdot\text{m}$
4.192 $\vec{F} = 969\hat{i} - 2.07\hat{j} + 492\hat{k}\ \text{lbf}$
4.194 $T = 29.4\ \text{N}\cdot\text{m}$; $\vec{M} = 51.0\hat{i} + 1.40\hat{j}\ \text{N}\cdot\text{m}$
4.196 $Q = 4140\ \text{gal/min}$
4.198 $T = 0.0161\ \text{N}\cdot\text{m}$
4.200 $\omega = 57.7\ \text{rpm}$; $A = 1720\ \text{m}^2$
4.202 $\dot{Q} = -146\ \text{Btu/sec}$
4.204 $p_1 - p_2 = 75.4\ \text{kPa}$
4.206 $\eta = 0.348$
4.208 $\eta = 0.50$; $\eta = 0$
4.210 $\eta = 0.571$
4.212 $Q = 0.0166\ \text{m}^3/\text{sec}$; $z_{max} = 61.4\ \text{m}$; $F = 561\ \text{N}$
4.214 $\Delta me = -1.88\ \text{N}\cdot\text{m/kg}$; $\Delta T = 4.49 \times 10^{-4}\text{K}$

Chapter 5

5.2 $x = cy$; $x = e^{[\ln y + A(\ln y)^2/2]}$
5.4 $y = y_0 x/x_0$; $z = z_0(x/x_0)^2$; $z = z_0(y/y_0)^2$
5.6 (*a*)
5.8 (*a*); (*b*)
5.10 $u = -2xy - 2x + f(y)$
5.12 $v/U = 0.00363$
5.14 $V_\theta = -\Lambda\sin\theta/r^2 + f(r)$
5.18 $\vec{V} = Ax\hat{i} - Ay\hat{j}$
5.20 $\psi = Uy^2/2h$; $y = h/\sqrt{2}$
5.22 $\psi = -y^3 z - 2z^2$
5.24 $Q = 4\ \text{m}^3/\text{sec/m}$
5.28 $|\vec{V}| = U$ at $\theta = \pm 30°$; $\pm 150°$
5.30 $\psi = -\dfrac{1}{2}\omega r^2$; $Q/b = 0.060\ \text{m}^3/\text{sec/m}$
5.32 $\vec{a}_p = \dfrac{1}{3}(16\hat{i} + 32\hat{j} + 16\hat{k})\ \text{m/sec}^2$
5.34 $\vec{a}_p = (x\hat{i} + y\hat{j})\ \text{sec}^{-2}$
5.36 $v = v_0(1 - y/h)$; $a_{p_x} = v_0^2 x/h^2$
5.40 $A = 2.72\ \text{ft/sec}$; $a_x = -11.1\ \text{ft/sec}^2$; $a_y = -19.2\ \text{ft/sec}^2$
5.42 $\vec{a}_p = (Q/2\pi h)^2 r^{-3}\hat{i}_r$
5.44 $V_z = v_0(1 - z/h)$; $a_r = v_0^2 r/4h^2$
5.46 $\vec{a}_p = 44.5\hat{i}\ \text{ft/sec}^2$; $\vec{a}_p = 103\hat{i}\ \text{ft/sec}^2$
5.48 $a_x = 4[20 + 2\sin(\omega t)]^2 + 1.2\cos(\omega t)$
5.50 Yes; Yes
5.52 Yes; Yes
5.54 $\vec{V} = -2y\hat{i} - 2x\hat{j}$; $\phi = 2xy$
5.56 Yes; $\psi = -\dfrac{q}{2\pi}\theta - \dfrac{K}{2\pi}\ln r$
5.58 $\Gamma = -0.1\ \text{m}^2/\text{sec}$
5.60 $df/d\mathcal{V} = -3.75\ \text{N/m}^3$
5.62 $df/d\mathcal{V} = -0.00670\ \text{lbf/ft}^3$

Chapter 6

6.2 $\vec{a} = 2\hat{i} + 2\hat{j}\ \text{ft/sec}^2$; $\nabla p = -(4\hat{i} + 68.4\hat{j})\ \text{lbf/ft}^2/\text{ft}$
6.4 $a = 2.24\ \text{m/sec}^2$ at $\theta = 63.4°$ above x axis; $\nabla p = -(1.0\hat{i} + 11.8\hat{j})\ \text{kN/m}^2/\text{m}$
6.6 $\nabla p = -(3.0\hat{i} + 9.0\hat{j})\ \text{kN/m}^2/\text{m}$
6.12 $a_r = -\left[\dfrac{Q}{2\pi h}\right]^2\dfrac{1}{r^3}$; $\dfrac{\partial p}{\partial r} = \rho\left[\dfrac{Q}{2\pi h}\right]^2\dfrac{1}{r^3}$
6.14 $p_{gage} = \dfrac{3}{8}\dfrac{\rho V^2 R^2}{b^2}$

6.16 $\frac{\partial p}{\partial r} = \rho\omega^2 r$; $p_2 - p_1 = 150$ kN/m^2

6.18 $\frac{\partial p}{\partial r} = \frac{\rho K^2}{4\pi^2 r^3}$; $p_2 - p_1 = 37.5$ kN/m^2

6.20 $p_{L/2} - p_0 = -30.6$ N/m^2

6.22 $p_2 = 291$ kPa(gage)

6.24 $h = 7.72$ m

6.26 $h = \left[h_0^{1/2} - \frac{1}{2}\left(\frac{2g}{AR^2 - 1}\right)^{1/2} t \right]^2$

6.28 $h = 16.3$ m

6.30 $Q = 1.25 \times 10^{-2}$ m^3/sec; $p_c = 71.6$ kPa

6.32 $F = 3.06$ mN

6.34 $F = 83.3$ kN

6.36 $p = 164$ kPa(gage); $F = 152$ N

6.38 $V = 330$ ft/sec

6.40 $V = 101$ m/sec

6.42 $V_j = \left[\frac{2F}{\rho A_p} \frac{1}{[1 - (A_j/A_p)^2]} \right]^{1/2}$

6.44 $V = 27.5$ m/sec

6.46 $Q_1 = 0.176$ ft^3/sec; $u_4 - u_3 = 1.8$ Btu/slug; $Q = 0.353$ ft^3/sec

6.48 $p = 2.09$ kPa(gage)

6.50 $dQ/dt = 0.0516$ m^3/sec/sec

6.52 $\Delta p = -615$ N/m^2

6.54 $\vec{\omega} = -\frac{A}{2}(x^2 + y^2)\hat{k}$

6.56 (a) No; (b) Yes

6.58 (a) Yes; (b) No

6.60 $\psi = \frac{1}{2} b(x^2 - y^2) - 2axy$

6.66 $\psi = Ur\sin\theta + \frac{q}{2\pi}\theta$;

$\phi = -Ur\cos\theta - \frac{q}{2\pi}\ln r$;

$\vec{V} = \left(U + \frac{q}{2\pi}\frac{x}{r^2}\right)\hat{\imath} + \frac{q}{2\pi}\frac{y}{r^2}\hat{\jmath}$;

$q = 25\pi$ m^2/sec;

$y = \pm\pi/2$

6.68 $R_x = 5.51$ kN/m

6.70 $y = \pm a$

6.72 $h = 0.1615$ m; $\vec{V} = 44.3\hat{\imath}$ m/sec; $p - p_\infty = -957$ N/m^2

6.74 $C_{p_{\max}} = 1.0$; $C_{p_{\min}} = -3.0$; $p = p_\infty$ at $\theta = 30°, 150°, 210°, 270°$

6.80 $\psi = \frac{K}{2\pi}\ln\frac{r_2}{r_1}$; $\phi = \frac{K}{2\pi}(\theta_2 - \theta_1)$;

$\vec{V} = \frac{K}{2\pi}\left[\left(\frac{\sin\theta_1}{r_1} - \frac{\sin\theta_2}{r_2}\right)\hat{\imath} + \left(\frac{\cos\theta_2}{r_2} - \frac{\cos\theta_1}{r_1}\right)\hat{\jmath}\right]$;

$p_{x=0} - p_\infty = -\frac{\rho}{2}\left[\frac{K}{\pi}\frac{a}{(a^2 + y^2)}\right]^2$

Chapter 7

7.2 $V = 1370$ mph; $V = 1570$ mph

7.4 $We = 109$

7.6 $\mathscr{P}/p\omega D^3 = f(\mu\omega/p, c/D, l/D)$

7.8 $T = R^3\mu\omega\, f(h/R)$

7.10 $F/\mu VD = \text{constant}$

7.12 $\delta/x = f(\rho Ux/\mu)$

7.16 $\mu/\rho Vd$, h/d, D/d

7.18 $T/\rho V^2 D^3$, $\mu/\rho VD$, $\omega D/V$, d/D

7.20 $V = \sqrt{gD}\, f(\lambda/D)$

7.22 $\mathscr{P}/\rho D^2 V^3$, $\omega D/V$, $\mu/\rho VD$, c/V

7.24 $\mathscr{P}/\rho D^5\omega^3 = f(Q/D^3\omega)$

7.26 $\zeta r^2/\Gamma_0$, $\Gamma_0\tau/r^2$, ν/Γ_0

7.28 $We = 459$

7.30 $V_{\text{air}} = 104$ m/sec; $V_{\text{He}} = 305$ m/sec

7.32 $Ca = 7.0$

7.34 $V = 9.58$ m/sec

7.36 $\Delta h = 0.50$ in.

7.38 $V = 6.21$ m/sec; $F_D = 0.978$ N

7.40 $p = 1.88$ MPa(abs); $F_D = 43.7$ kN

7.42 $fd/V = g(\rho Vd/\mu)$; $V_1/V_2 = 1/2$; $f_1/f_2 = 1/4$

7.44 $F_D/\rho V^2 D^2 = f_1(Re, \omega D/V, d/D)$; $F_L/\rho V^2 D^2 = f_2(Re, \omega D/V, d/D)$; $V = 80$ ft/sec; $\omega = 1600$ rpm

7.46 Scale ratio = 1/50; Adequate Re not possible

7.48 $H = 145$ ft · lbf/slug; $Q = 5.92$ ft^3/sec; $D = 0.491$ ft

7.50 $F_t/\rho\omega^2 D^4 = f_1(g/\omega^2 D, \omega D/V)$; $T/\rho\omega^2 D^5 = f_2(g/\omega^2 D, \omega D/V)$; $\mathscr{P}/\rho\omega^3 D^5 = f_3(g/\omega^2 D, \omega D/V)$

7.52 KE ratio = 7.38

7.54 $F_B = 0.574$ N

7.56 $\sigma/\rho L V_0^2$, gL/V_0^2

7.58 V_0^2/gL

Chapter 8

8.4 $\bar{V}/u_{\max} = 2/3$

8.6 $\tau_{yx} = -1.80$ N/m^2 (to right); $Q/b = 5.40 \times 10^{-6}$ m^3/sec/m

8.8 $\tau_{yx} = -0.04$ lbf/ft^2 (to right); $Q/b = 6.66 \times 10^{-5}$ ft^3/sec/ft

8.10 $V = Q/2\pi rh$; $dp/dr = -6\mu Q/\pi rh^3$; $F = 12\mu Q(R^2/h^3)$ $\left\{\frac{R_0}{R}\frac{1}{2}\left[\ln\left(\frac{R_0}{R}\right) - \frac{1}{4}\right] + \frac{1}{4}\right\}$

8.12 $Q/b = \left(\frac{h}{k}\frac{\Delta p}{L}\right)^{1/n} \frac{nh^2}{2n+1}$

8.14 $\partial p/\partial x = -94.0$ N/m^2/m

8.16 $u = U\frac{y}{h} + \frac{\rho B_x}{\mu}\frac{1}{2}(hy - y^2)$; $Q/b = 1.50 \times 10^{-4}$ m^3/sec/m

8.18 $\mathscr{P}_p = \pi D a^3 \Delta p^2/12\mu L$; $\mathscr{P}_v = \pi\mu\omega^2 D^3 L/4a$

8.20 $u = \frac{\rho g}{\mu}\left(\frac{y^2}{2} - hy\right) + U_0$

8.22 $r = 0.707\ R$

8.24 $Q = 27.1$ cm^3/min; $\Delta p = 471$ kPa

8.26 Error = 15.5 percent

8.28 $D = 1.96$ mm (water); 115 mm (SAE 30 oil)

8.30 $\tau = C_1/r$, $u = \frac{C_1}{\mu}\ln r + C_2$, $C_1 = \frac{\mu V_0}{\ln(r_0/r_i)}$, $C_2 = \frac{-V_0 \ln r_i}{\ln(r_0/r_i)}$

8.34 $\tau_w = 2.16$ lbf/ft^2

8.40 $n = 8.3$ (graphical); 9.2 (least squares)

8.44 KE flux = 6000 ft · lb/sec/ft

8.46 $\alpha = 2.5$

8.48 $h_l = 345$ J/kg

8.50 $f = 0.016$

8.52 $Q = 2.45 \times 10^{-3}$ m^3/sec; $f = 0.025$

8.54 $f = 0.042$

8.56 $\Delta p \sim D^{-5}$

8.60 $\bar{V} = 76.2$ ft/sec; $Q = 224$ ft^3/min

8.62 $K = 0.247$

8.64 $\Delta Q = 0.0184$ m^3/sec (20.2 percent increase)

8.66 $p_1 - p_2 = 67.4$ kPa

8.68 $S = -4.85°$; $\dot{Q} = -0.0271$ Btu/sec

8.70 $p_1 = 1.40$ MPa(gage); $\Delta T = 0.215$ K

8.72 $\Delta z = 51.8$ m

8.74 $\Delta z = 8.13$ m

8.76 $L = 212$ m

8.78 $Q = 15.8$ m^3/sec; $p_1 = 2.65$ MPa(gage)

8.80 $Q = 1.49$ ft^3/sec

8.82 $V = 27.5$ ft/sec

8.84 $p \geq 362$ psig; $\dot{W} = 182$ hp

8.86 $\Delta p = 47.0$ psi; $\dot{W} = 307$ hp; Cost = \$117,000/year

8.88 $L = 45.2$ mi; $\dot{W} = 43{,}800$ hp

8.90 $Q = 0.0223$ m^3/sec; $p_a = -25.5$ kPa(gage)

8.92 $Q = 0.0134$ m^3/sec, to the left

8.94 $\Delta t \approx 168$ sec

8.96 $p = -1.84$ in. H_2O (gage)

8.98 $D = 2.5$ in.

8.100 $D = 6$ in.

8.102 $Q = 8$ gpm; $Q = 22$ gpm

8.104 $Q = 106$ gpm

8.108 $Q \simeq 10{,}000$ ft^3/min

8.110 $\dot{m} = 0.0592$ g/sec

8.112 $D = 225$ mm; $\dot{m} = 8.48$ kg/sec

8.116 $\dot{m} = 2.10$ kg/sec; $\Delta h = 185$ mm Hg

8.118 $Q = 88$ gpm

8.120 $Re = 1810$; $f = 0.0354$; $p = -30.0$ mm H_2O (gage)

Chapter 9

9.2 $x_p = 196$ mm; $x_m = 14.5$ mm

9.8 $a = 0$, $b = 3$, $c = -2$

9.12 δ^*/δ: linear, 1/2; parabolic, 1/3; cubic, 3/8; sinusoidal, 0.363

9.14 $\delta^*/x = 1.74/\sqrt{Re_x}$

9.16 $\theta/\delta = 0.167$ (linear); 0.133 (parabolic); 0.139 (cubic); 0.137 (sinusoidal)

9.18 $\theta/x = 0.645/\sqrt{Re_x}$

9.20 $H = 1.29$; $H_{\text{laminar}} = 2.70$

9.22 $Q = 147$ ft^3/min; $V_2 = 55.2$ ft/sec; $\delta_2^* = 0.075$ in.

9.24 $p_2 = -73.1$ Pa (gage); $\bar{\tau}_w = 0.300$ N/m^2

9.26 $\dot{m}_{ab} = 0.0140$ kg/sec; $F_x = 0.168$ N (to left)

9.36 $C_f = 0.647/\sqrt{Re_x}$

9.38 $\delta/x = 4.64/\sqrt{Re_x}$; $C_f = 0.647/\sqrt{Re_x}$

9.40 $\delta_{\text{max}} = 0.00569$ ft; $\tau_{w,\text{min}} = 0.0245$ lbf/ft^2

9.42 $\delta_{\text{lam}} = 5.48$ mm; $\tau_{w,\text{lam}} = 0.101$ N/m^2; $\delta_{\text{turb}} = 23.7$ mm; $\tau_{w,\text{turb}} = 0.502$ N/m^2

9.44 $\delta/x = 0.398/(Re_x)^{1/5}$; $C_f = 0.0567/(Re_x)^{1/5}$

9.46 $F_D = 4.80$ mN (per side)

9.48 For $\phi = 0$, $p_2 < p_{2i}$; for $\phi > 0$ (no separation), $p_2 < p_{2i}$

9.50 $F_D = 0.0361\ \rho U_\infty^2 (U_\infty/\nu)^{-1/5}\ b(L^{4/5} - x_t^{4/5})$

9.52 $\bar{L} \simeq 7.13$ ft; $F_D \simeq 1740$ lbf
9.54 $\mathscr{P} = 3.86$ kW
9.56 $V = 11.0$ ft/sec
9.58 $F_D = 92.3$ kN
9.60 $D = 6.90$ m
9.62 $s = 119$ m
9.64 $W = 0.0303$ lbf
9.66 Horizontal is 20 percent better.
9.68 $t = 2.12$ sec; $x = 450$ ft
9.70 $V_t = 43.5$, 121 m/sec; $t = 8.11$, 22.6 sec; $\Delta z = 224$ m, 1.73 km
9.72 $\mathscr{P} = 17.2$ hp
9.74 $V = 47.5$ mph; $\mathscr{P} = 23.2$ hp; $\mathscr{P} = 40.0$ hp
9.76 $a_{max} = 2.54$ m/sec^2; $V_{max} = 60.3$ m/sec
9.78 $V_{2_{rel}} = 15$ m/sec; $p_{2g} = -133$ N/m^2; $m = 9.04$ kg
9.80 $F_D = 0.830$ N; $M = 0.125$ N · m
9.82 $D < 17.2$ mm; $t = 0.254$ sec
9.84 $V_t = 5.45$ mm/sec, 0.0545 mm/sec
9.86 $V = 10.3$ m/sec
9.88 Error in region where $C_D = 0.5$
9.90 $\mathscr{P} = 4.80$ kW
9.92 $T = 11.9$ N · m
9.94 $\Delta t = 19.9$ sec
9.96 $V_t = 30.1$ ft/sec
9.98 $V_t = \left[\frac{8g\rho_{H_2O}D}{3\rho_{air}}\right]^{1/2}$; $K = 4.61$ m/sec · mm$^{1/2}$
9.100 $t \simeq 18$ sec
9.102 $V = 302$ m/sec at $t = 10$ sec
9.104 $\Delta\mathscr{P} = 4.56$ kW
9.106 $p_{gage} = -584$ N/m^2; $V = 149$ km/hr
9.108 $A_p = 7.03$ m^2
9.110 $V_{min} = 5.62$ m/sec; $\mathscr{P} = 31.0$ kW; $V_{max} = 19.9$ m/sec
9.112 $F_D = 502$ N; $\mathscr{P} = 34.9$ kW
9.114 $V_t = 23.8$ m/sec; $t = 4.44$ sec; $s = 67.1$ m
9.118 $\mathscr{P}_{min} = 0.180$ hp
9.120 $\omega = 9860$ rpm; $\Delta s = 1.19$ m

Chapter 10

10.4 $y = 6.68$ ft
10.8 $V = 3.95$ ft/sec
10.10 $V = 8.02$ ft/sec; $Fr = 2$
10.14 $y = 0.645$ ft; $y = 4.30$ ft
10.18 $y_c = 3.28$ ft
10.20 $(b + 2y \cot\alpha)/(by + y^2 \cot\alpha)^3 = g/Q^2$
10.22 $y = 0.61$ ft (32 percent drop)
10.24 $y = 1.32$ ft
10.26 $y = 0.507$ m; $Fr = 2.51$
10.28 $Q = 49.5$ ft^3/sec
10.30 $S_b = 1.49 \times 10^{-3}$
10.32 $b = 2.36$ m; $b = 2.49$ m
10.34 $y_n = 1.79$ m
10.36 $y_n = 2.32$ m
10.38 $Q = 0.623$ m^3/sec
10.42 $y_n = 7.79$ m; $b = 6.46$ m; 44.6 percent
10.44 $S_c = 2.48 \times 10^{-3}$
10.46 $S_c = 3.82 \times 10^{-3}$
10.48 Depth increases (S3 curve)
10.50 $\Delta x = 892$ ft, downstream
10.52 $y = 0.966$ m
10.54 $y = 2.86$ m (one reach)
10.56 $\Delta x = 221$ m (one reach)
10.58 $y = 4.51$ ft; $h_\ell = 2.40$ ft
10.60 $V = 4.89$ mph
10.62 $y = 3.99$ ft; $h_\ell = 1.13$ ft
10.64 $y = 10.3$ m; $V = 2.18$ m/sec; $E_d/E_u = 0.321$
10.66 $\Delta T = 0.00221$ K
10.68 $Q = 16.8$ ft^3/sec
10.70 $Q/b = 27.2$ ft^3/sec/ft

Chapter 11

11.2 $\Delta S = -0.923$ Btu/R; $\Delta U = -684$ Btu; $\Delta H = -960$ Btu
11.4 $s_2 - s_1 = 134$ J/kg · K
11.6 <0; >0; 0; 0; 0
11.8 Case 2
11.10 $\dot{W} = 392$ kW
11.12 $s_2 - s_1 = -1.09$ kJ/kg · K; $\delta Q/dm = -570$ kJ/kg
11.14 $\Delta u = -574$ kJ/kg; $\Delta h = -803$ kJ/kg; $\Delta s = 144$ J/kg · K
11.16 $M = 0.437$; $M = 2.33$; $M = 2.77$
11.18 $M = 0.776$; $V = 603.$ mph
11.20 $E_v = 219$ GPa; $c = 1450$ m/sec
11.22 $V = 306$ m/sec; $V = 374$ m/sec
11.24 $a = 0.305$ g's; $a = 0.267$ g's
11.26 $c = 299$ m/sec; $V = 987$ m/sec; $V/V_b = 1.41$
11.28 $V = 725$ m/sec

11.30 $M = 8.74$; $\alpha = 6.57°$
11.32 $M = 1.19$; $V = 806$ ft/sec; $Re/x = 9.90 \times 10^6$ m^{-1}
11.34 $V = 471$ m/sec; $t = 5.90$ sec
11.36 $\Delta x = 3.31$ km
11.38 $M = 0.851$; $V = 255$ m/sec; $z \approx 10$ km
11.40 $M = 0.83$; $V = 280$ m/sec
11.42 $T_{0_1} = T_{0_2}$; $p_{0_1} > p_{0_2}$
11.44 $M = 0.199$; $M = 0.314$
11.46 $V = 890$ m/sec; $T_0 = 677$ K; $p_0 = 212$ kPa
11.48 $c = 295$ m/sec; $V = 620$ m/sec; $\alpha = 28.4°$; $V/V_b = 1.35$; $T_0 = 408$ K
11.50 $V = 987$ m/sec; $p_0 = 125$ kPa; $p_0 = 31.7$ kPa; $T_0 = 707$ K
11.52 $T_0 = 308$ K; $p_0 = 101.4$ kPa; $\Delta p_0 = 800$ Pa; 0.79 percent
11.54 $T_0 = 318$ K; $p_0 = 2.06$ MPa(abs); $p_{\text{probe}} = 1.67$ MPa(abs)
11.56 $p_{0_1} = 67.6$ psia; $T_{0_1} = 621$ R; $p_{0_2} = 56.6$ psia; $T_{0_2} = 883$ R; $s_2 - s_1 = 0.0968$ Btu/lbm · R
11.58 $T_{0_1} = 584.6$ K; $p_{0_1} = 1.028$ MPa; $T_{0_2} = 1782$ K; $p_{0_2} = 0.9632$ MPa; $s_2 - s_1 = 1138$ J/kg · K
11.60 $T_{0_1} = 309$ K; $p_{0_1} = 0.838$ MPa(abs); $T_{0_2} = 573$ K; $p_{0_2} = 8.13$ MPa(abs); $s_2 - s_1 = 175$ J/kg · K
11.62 $p^*/p_0^* = 0.4881$ (He); $p^*/p_0^* = 0.5475$ (CO_2)
11.64 $T^* = 253$ K; $p^* = 476$ kPa(abs); $V^* = 319$ m/sec
11.66 $T^* = 265$ K; $p^* = 19.2$ MPa(abs); $V^* = 254$ m/sec
11.68 $T^* = 268$ K; $p^* = 456$ kPa(abs); $V^* = 426$ m/sec
11.70 $T^* = 3180$ K; $p^* = 14.1$ MPa(abs); $V^* = 478$ m/sec

Chapter 12

12.2 $V = 4210$ ft/sec
12.4 $V = 1280$ m/sec
12.6 $V = 2620$ ft/sec; $M = 1.36$, $\dot{m} = 1.76$ lbm/sec
12.8 $V = 797$ m/sec; $M = 1.35$; $\dot{m} = 0.706$ kg/sec
12.10 $T = 575$ R; $p = 21.1$ psia
12.12 $M = 1.35$
12.14 $M_2 = 1.20$
12.16 $\Delta p = 315$ kPa; $A_1/A_2 = 3.79$
12.18 $p_t = 33$ psia; $V_t = 1060$ ft/sec; $M_t = 0.90$
12.20 $R_x = 1560$ N (to the left)
12.22 $p_t = 112$ kPa
12.24 $p_0 \geq 191$ kPa; $\dot{m} = 1.28$ kg/sec
12.26 $T = 369$ K; $\Delta p = 2.64$ MPa; $\dot{m} = 1.14 \times 10^{-2}$ kg/sec
12.28 $M = 1.0$; $A = 450$ mm^2
12.30 $R_x = 304$ lbf, tension
12.32 $\dot{m} = 2.73$ lbm/sec; $a = 99.8$ ft/sec^2
12.34 $t = 23.6$ sec
12.36 $\dot{m} = 0.856$ lbm/sec
12.38 $A = 2.99$ in.^{2}; $\dot{m} = 3.74$ lbm/sec
12.40 $V = 1300$ m/sec; $\dot{m} = 87.4$ kg/sec
12.42 $\dot{m} = 6.05$ lbm/sec; Will decrease by a factor of 2
12.44 $\dot{m} = 5.44$ kg/sec
12.46 $V = 504$ m/sec; $A = 0.0596$ m^2
12.48 $\dot{m} = 33.0$ kg/sec; $A_e = 0.158$ m^2; $A_e/A_t = 18.0$
12.50 $p_t = 3293$ psia; $\dot{m} = 0.0524$ lbm/sec; Thrust = 2.37 lbf; 36.3 percent
12.52 $A_{\text{eff},t} = 74.66$ mm^2; $A_{\text{eff},e} = 294.8$ mm^2; $M_e = 0.149$; $M_e = 2.927$; $\Delta \dot{m} = -4.9$ percent
12.54 $M = 0.897$; $T_0 = 339$ K
12.56 $T = 287$ K; $p_0 = 423$ kPa; $\Delta s = 132$ J/kg · K
12.58 $T = 248$ K; $V = 316$ m/sec
12.60 $F = 822$ lbf
12.62 $M_2 = 0.844$; $p_2 = 475$ kPa; $F = 2650$ N
12.64 $\dot{m} = 0.00321$ kg/sec; $p_0 = 33.8$ kPa(abs); $\Delta s = 314$ J/kg · K
12.66 $\dot{m} = 0.0192$ kg/sec; $T^* = 244$ K; $p^* = 53.4$ kPa(abs); $T^* = 294$ K; $p^* = 13.6$ kPa(abs)
12.68 $T = 468$ K; $F = 60$ N; $\Delta s = 149$ J/kg · K
12.70 $L = 1.27$ m
12.72 $L = 18.8$ ft
12.78 $L = 0.405$ m
12.80 $p = 190.8$ kPa; $L = 5.02$ m; $\Delta s = 0.326$ kJ/kg · K
12.82 $\delta Q/dm = 145$ kJ/kg; $p_1 - p_2 = 405$ kPa

12.84 $M = 0.498$; $T = 1480$ K; $T_0 = 1550$ K; $\dot{Q} = 1.85$ MJ/sec
12.86 $\delta Q/dm = 18$ kJ/kg; $\Delta s = 53.2$ J/kg · K; $\Delta p_0 = 2.0$ kPa
12.88 $\dot{Q} = 1080$ Btu/sec; $M = 0.501$
12.90 $V = 1524$ ft/sec; $T = 2310$ R; $\dot{Q} = 744$ Btu/sec
12.92 $M = 0.26$; $p = 97.9$ kPa
12.94 $T = 1495$ K; $M = 0.23$; $\Delta p_0 = 27.5$ kPa
12.96 $M_2 = 0.30$; $T_2 = 1964$ R; $p_2 = 91.2$ psia; $T_{0_2} = 1998$ R; $p_{0_2} = 97.1$ psia
12.98 $V_2 = 865$ m/sec; $p_2 = 46.7$ kPa; $M_2 = 1.96$; $\delta Q/dm = 162$ kJ/kg
12.100 $\delta Q/dm = 17$ kJ/kg; $T = 318$ K; $p = 46.3$ kPa; $T_0 = 382$ K; $p_0 = 87.7$ kPa
12.102 $\dot{Q} = 898$ Btu/sec; $M = 0.50$; $p = 7.84$ psia
12.104 $\delta Q/dm = 313$ Btu/lbm; $T_0 = 2769$ R; $p_0 = 217$ psia; $T = 2308$ R; $p = 114.7$ psia; $\Delta p_0 = 34.2$ psia
12.106 $M = 1.0$; $p = 48.8$ kPa; $\Delta p_0 = 8.6$ kPa
12.110 $Q = 1.84 \times 10^8$ ft^3/day
12.112 $T = 360.5$ K; $M = 0.702$; $p = 147.5$ kPa; $p_0 = 205$ kPa
12.114 $V = 247$ m/sec; $T = 671$ K
12.116 $T = 521$ K; $p_0 = 1.29$ MPa
12.118 $V = 256$ m/sec; $M = 0.490$
12.120 $M = 0.545$; $p = 513.5$ kPa; $p_0 = 628.4$ kPa; $A = 0.111$ m^2
12.122 $T_0 = 426.5$ K; $p_0 = 207$ kPa; $p_0 = 130$ kPa
12.124 $\rho_2 = 0.359$ lbm/ft^3; $M_2 = 0.701$
12.126 $M = 0.701$; $V = 267$ m/sec; $p_0 = 205$ kPa
12.128 $V = 247$ m/sec; $T = 670$ K; $\Delta s = 0.315$ kJ/kg · K
12.130 $T = 520$ K; $p_0 = 1.29$ MPa
12.132 $V = 257$ m/sec; $M = 0.493$; $\Delta p_0 = 512$ kPa
12.134 $T_2 = 633$ R; $p_2 = 39.2$ psia; $T_{0_2} = T_{0_3} = 1080$ R; $p_{0_2} = 255$ psia; $M_3 = 0.60$; $T_3 = 1010$ R; $p_3 = 155$ psia; $p_{0_3} = 198$ psia; $M_4 = 0.70$; $p_4 = 131$ psia; $L = 5.93$ ft
12.136 $M = 3.20$, $V = 949$ m/sec
12.138 $p = 8.74$ psia; $T = 478$ R; $\Delta p_0 = 14.1$ psi; $\Delta s = 0.591$ Btu/lbm · R
12.140 $T = 414$ K; $p = 51.9$ kPa; $p_0 = 57.9$ kPa
12.142 $T_0 = 634.2$ K, $T = 226.5$ K, $p_0 = 43.97$ kPa, $p = 1.197$ kPa; $T_0 = 634.2$ K, $T = 528.5$ K, $p_0 = 43.97$ kPa, $p = 23.23$ kPa; $T_0 = 634.2$ K, $T = 455.6$ K, $p_0 = 43.97$ kPa, $p = 13.82$ kPa; $T_0 = 634.2$ K, $T = 571.6$ K, $p_0 = 42.14$ kPa, $p = 29.3$ kPa; $T_0 = 634.2$ K, $T = 614.5$ K, $p_0 = 42.14$ kPa, $p = 37.74$ kPa
12.144 $\Delta p = 27.5$ kPa; $\Delta p_0 = 27$ kPa
12.146 $p = 46.7$ psia; $A_t = 1.52$ in.^{2}; $\dot{m} = 2.55$ lbm/sec
12.148 $p = 33.4$ kPa; $V = 162$ m/sec
12.150 $M = 0.701$; $p = 167.4$ kPa; $\Delta s = 20.9$ J/kg · K
12.152 $M = 2.94$; $p_0 = 3.39$ MPa; $p = 3.35$ MPa; $p_b = 1.00$ MPa; $p = 101$ kPa
12.154 $p = 46.7$ psia; $A_t = 1.52$ in.^{2}; $\dot{m} = 2.55$ lbm/sec
12.156 $M = 1.50$
12.158 $33.4 \text{ kPa} < p_b < 99.6$ kPa; $\dot{m} = 0.121$ kg/sec
12.160 $p_{atm} < p_0 < 112$ kPa and $p_0 > 743$ kPa

INDEX

Absolute metric (system of units), 11
Absolute pressure, 58
Absolute viscosity, 29
Acceleration:
  convective, 216
  gravitational, 11
  local, 216
  of particle in velocity field, 214, 216
Accelerometer, 93
Adiabatic flow, *see* Fanno line flow
Adiabatic process, 564
Adverse pressure gradient, 37, 451
Aging of pipes, 365
Alternate depths, 511
Anemometer:
  hot-film, 404
  hot-wire, 404
Angle of attack, 469
Angular deformation, 224
Angular momentum, *see* Moment of momentum
Archimedes' principle, 73
Area, centroid of, 65
  moment of inertia of, 65
  product of inertia of, 65
Area ratio, 600
  isentropic flow, 601
Aspect ratio:
  airfoil, 474
  flat plate, 459
  rectangular duct, 390
Atmosphere:
  isothermal, 85
  standard, 58
Average velocity, 332

Barometer, 58
Barotropic fluid, 57
Barrels, U.S. petroleum industry, 16
Basic equation of fluid statics, 49
Basic equations for control volume, 95
  conservation of mass, 104
  first law of thermodynamics, 158
  moment of momentum, for inertial control volume, 143
    for rotating control volume, 152
  Newton's second law (linear momentum), for control volume moving with constant velocity, 128
    for control volume with arbitrary acceleration, 139
    for control volume with rectilinear acceleration, 130
    for differential control volume, 123
    for nonaccelerating control volume, 111
    second law of thermodynamics, 165
Basic laws for system, conservation of mass, 95
  first law of thermodynamics, 96
  moment of momentum, 96
  Newton's second law (linear momentum), 96
    differential form, 228
    second law of thermodynamics, 97
Basic pressure-height relation, 53
Bearing, journal, 339
Bernoulli equation, 126, 246, 513
  applications, 249
  cautions on use of, 255
  irrotational flow, 269
  relation to first law of thermodynamics, 259
  restrictions on use of, 255
  unsteady flow, 266
Bingham plastic, 31
Blasius' solution, 431
Blood, 48
Blower, 145
Body force, 25
Borda mouthpiece, 289
Boundary layer, 35, 426
  displacement thickness, 428
  effect of pressure gradient on, 450
  flat plate, 427
  integral thicknesses, 429
  laminar:
    approximate solution, 443
    exact solution, 431
  momentum integral equation for, 436
  momentum profiles, 452
  momentum thickness, 429
  separation, 451
  shape factor, 453
  thickness, 428
  transition, 427
  turbulent, 427
  velocity profiles, 452
Boundary-layer:
  control, 477
  thicknesses, 428

British gravitational (system of units), 11
Buckingham Pi theorem, 296
Bulk (compressibility) modulus, 57, 571, 682
Buoyancy force, 73

Camber, 469
Canal, 502
Capillary effect, 83, 301
Capillary viscometer, 347
Capillary wave, 325
Cavitation number, 304
Celerity, 504
Center of pressure, 62, 64
Chezy equation, 522
Choking, 602, 622, 638
Chord, 469
Circulation, 222
Coanda effect, 175
Compressible flow, 39, 561
  tables for computation of, 700
Compressor, 145
Concentric-cylinder viscometer, 46
Confidence limit, 718
Conical diffuser, 368
Conjugate depth, 543
Conservation:
  of energy, *see* First law of thermodynamics
  of mass, 95, 201
    cylindrical coordinates, 207
    rectangular coordinates, 201
Consistency index, 31
Constitutive equations, 5
Contact angle, 684
Continuity, *see* Conservation of mass
Continuity equation, differential form, 201
  cylindrical coordinates, 207
  rectangular coordinates, 201
Continuum, 18
Contraction coefficient, 289, 414
Control surface, 7
Control volume, 7
  rate of work done on, 159
Convective acceleration, 216
Converging-diverging nozzle, *see* Nozzle
Converging nozzle, *see* Nozzle
Conversion factors, 725
Critical conditions, compressible flow, 583
Critical depth, 512
Critical flow in open channel, 508
Critical pressure ratio, 583
Critical Reynolds number, *see* Transition
Critical speed:
  compressible flow, 584
  open-channel flow, 508
Culvert, 502
Curl, 222
Cylinder:
  flow around, 36
  inviscid flow around, 36, 282

Deformation:
  angular, 213, 224
  linear, 213, 227
  rate of 28, 225
Del operator:
  cylindrical coordinates, 209
  rectangular coordinates, 204
Density, 18
Density field, 20
Derivative, substantial, 216
Design conditions, *see* Nozzle
Differential equation, nondimesionalizing, 319
Diffuser, 367, 381, 598
  optimum geometries, 368
  pressure recovery in, 368
Dilatant, 31
Dilation, volume, 226
Dimension, 10
Dimensional homogeneity, 10
Dimensional matrix, 301
Dimensions of flow field, 21
Discharge coefficient, 393
  flow nozzle, 395
  orifice plate, 395
  sluice gate, 551
  venturi meter, 397
  weir, 546, 550
Displacement thickness, 428
Disturbance thickness, *see* Boundary layer
Doppler effect, 404, 573
Doublet, 275
  strength of, 275
Downwash, 474
Drag, 37, 426, 454
  form, 476
  friction, 455, 461
  pressure, 459, 461
  profile, 476
Drag coefficient, 455
  airfoil, 470
  cylinder, 463
  flat plate normal to flow, 459
  flat plate parallel to flow, 456
  induced, 475
  parasite, 481
  selected objects, 460
  sphere, 461
  streamlined strut, 467
  vehicle, 313
Dynamic pressure, 256, 257
Dynamic similarity, 306
Dynamic viscosity, 29
Dyne, 11

Efficiency:
  propeller, 199
  propulsive, 199
  pump, 315
  windmill, 199
Elementary plane flows, *see* Potential flow theory
Energy equation, for pipe flow, 356. *See also* First law of thermodynamics
Energy grade line, 265, 381, 384, 521
English Engineering (system of units), 11
Enthalpy, 562
Entrance length, 332
Entropy, 563
Equation of state, 5
  ideal gas, 5
Equations of motion, *see* Navier-Stokes equations
Equivalent length, 366
  bends, 369, 370
  fittings and valves, 370, 371
  miter bends, 370
Ergometer, 423
Euler equations, 231, 241
  along streamline, 244
  cylindrical coordinates, 242

normal to streamline, 244
rectangular coordinates, 241
streamline coordinates, 242
Eulerian method of description, 9, 217
Euler number, 304
Euler turbine equation, 146
Experimental uncertainty, 2, 716
Extensive property, 97
External flow, 40, 425

Fan, 145
"laws," 316
Fanno line flow, 617
basic equations for, 618
choking length, 622, 630
effects on properties, 622
tables for computation of, 625, 705
Ts diagram, 621
Field representation, 20
Films and film loops, 694
First law of thermodynamics, 96, 158
Fittings, losses in, *see* Head loss, in valves and fittings
Flap, 476
Flat plate, flow over, 427
Float-type flow meter, 402
Flow behavior index, 31
Flow coefficient, 393
flow nozzle, 395, 397
orifice plate, 395, 396
turbomachine, 315
Flow field, dimensions of, 21
Flow measurement, 390
internal flow, 390
direct methods, 390
linear flow meters, 402
electromagnetic, 402
float type, 402
rotameter, 402
turbine, 402
ultrasonic, 403
vortex shedding, 402
restriction flow meters, 390
flow nozzle, 395
laminar flow element, 398
orifice plate, 394
venturi, 397
traversing methods, 403
laser Doppler anemometer, 404
thermal anemometer, 404
open-channel flow, 546
critical flumes, 551
sluice gates, 551
weirs, 546, 550
Flow meter, *see* Flow measurement
Flow nozzle, 395
Flow visualization, 22, 311, 496
Fluid, 3
Fluidic device, 176
Fluid particle, 20
Fluid properties, 681
Fluid statics:
basic equation of, 49
pressure height relation, 53
Flume, 502, 551
Force:
body, 25
buoyancy, 73
compressibility, 304
drag, 454
gravity, 304
hydrostatic, 62
on curved submerged surface, 69
on plane submerged surface, 61
inertia, 303
lift, 454
pressure, 50, 304, 454
shear, 454
surface, 25, 454
surface tension, 304
viscous, 303
Forced vortex, 223
Francis turbine, 197
Free surface, 79, 501
Free vortex, 223, 275
Friction drag, *see* Drag
Friction factor, 359, 361, 362
data correlation for, 364
Fanning, 414
fully rough flow regime correlation, 414
smooth pipe correlation, 364
Frictionless flow:
compressible adiabatic, *see* Isentropic flow
compressible with heat transfer, *see* Rayleigh line flow
incompressible, 33, 241
open channel, *see* Open-channel flow
Friction velocity, 324, 353
Froude number, 305, 508
Froude speed of advance, 325
Fully developed flow, 332
laminar, 333
turbulent, 353
Fully rough flow regime, 362, 364

g_c, 10
Gage pressure, 58
Gas constant:
ideal gas equation of state, 5, 562, 690
universal, 562
Geometric similarity, 305
Gibbs equations, 564
Grade line, 265
energy, 265, 381, 384
hydraulic, 265, 381, 384
Gradient, 51
Gradually varied flow, 532
surface profile classification, 533
Gravitational attraction, law of, 81, 168
Gravity, acceleration of, 11
Guide vanes, 145

Head, 264, 359
Head coefficient, 315
Head loss, 351, 358
in diffusers, 369
in enlargements and contractions, 367
in exits, 367
in gradual contractions, 367
in inlets, 366
major, 351, 359
minor, 351, 359, 365
in miter bends, 370
in nozzles, 367
in open-channel flow, 509, 522
permanent, 398
in pipe bends, 369
in pipe entrances, 366
in pipes, 359
in sudden area changes, 367

Head loss (*Continued*)
  total, 351, 359
  in valves and fittings, 370
Head loss coefficient, 365
Heat transfer, sign convention for, 97
Hook gage, 548
Hot-film anemometer, 404
Hot-wire anemometer, 404
Hydraulic depth, 502
Hydraulic diameter, 390, 628
Hydraulic grade line, 265, 381, 384, 521
Hydraulic jump, 200, 540
  basic equations for, 541
  depth increase across, 542, 543
  head loss across, 543, 545
Hydraulic radius, 502
Hydraulic systems, 61
Hydrometer, 90
Hydrostatic force, 62
  on curved submerged surfaces, 69
  on plane submerged surfaces, 61
Hydrostatic pressure distribution, 118
Hypersonic flow, 573

Ice, 91
Ideal fluid, 241
Ideal gas, 5, 562
Incomplete similarity, 308
Incompressible flow, 39, 204, 209
Incompressible fluid, 53
Induced drag, 475
Inertial control volume, 111, 128
Inertial coordinate system, 111, 131
Intensive property, 97
Internal energy, 562
Internal flow, 40, 331
Inviscid flow, 33, 331
Irreversible process, 564
Irrotational flow, 221, 269
Irrotationality condition, 269
Irrotational vortex, 275
Isentropic flow, 592
  basic equations for, 592
    ideal gas, 599
  in converging-diverging nozzle, 611
  in converging nozzle, 605
  effect of area variation on, 596, 601
  reference conditions for, 600
  tables for computation of, 603, 700
Isentropic process, 565
Isentropic stagnation properties, 574, 575
  for ideal gas, 575, 579

Jet pump, 179
Journal bearing, 339

Kinematic similarity, 305
Kinematics of fluid motion, 213
Kinematic viscosity, 29
Kinetic energy coefficient, 358
Kinetic energy ratio, 329

Lagrangian method of description, 8, 217
Laminar boundary layer, 431
  flat plate, approximate solution, 443
  flat plate, exact solution, 431
Laminar flow, 38, 331
  between parallel plates, 333
    both plates stationary, 333
    one plate moving, 339
  in pipe, 344
Laminar flow element (LFE), 398
Laplace's equation, 272
Lapse rate, 587
Lift, 426, 454, 469
Lift coefficient, 469
  airfoil, 470
  spinning sphere, 482
Lift/drag ratio, 473
Linear deformation, 213, 227
Linear momentum, *see* Newton's second law of motion
Local acceleration, 216
Loss, major and minor, *see* Head loss
Loss coefficient, *see* Head loss
Lubricating oil, 688

Mach angle, 573
Mach cone, 573
Mach number, 40, 305
Magnus effect, 481
Major loss, *see* Head loss
Manning correlation, 522
Manning roughness coefficient, 523
Manometer, 54
  reservoir, 55
Material derivative, 216
Mean line, 469
Measurement, flow, *see* Flow measurement
Mechanical energy, 359
Mechanical flow meter, *see* Flow measurement
Meniscus, 301
Meter, flow, *see* Flow measurement
Methods of description:
  Eulerian, 9, 217
  Lagrangian, 8, 217
Mile, nautical, 725
Minor loss, *see* Head loss
Minor loss coefficient, *see* Head loss coefficient
Model studies, 305
Model test facilities, 319
Modulus of elasticity, 57
Molecular mass, 562, 690
Moment of momentum, 96, 143
  fixed control volume, 143
  rotating control volume, 152
Momentum:
  angular, *see* Moment of momentum
  linear, *see* Newton's second law of motion
Momentum equation:
  for control volume moving with constant velocity, 128
  for control volume with arbitrary acceleration, 139
  for control volume with rectilinear acceleration, 130
  for differential control volume, 123
  for differential form, 228
  for inertial control volume, 111
  for inviscid flow, 241
Momentum flux, 113
Momentum flux coefficient, 413
Momentum integral equation, 436
  for zero pressure gradient flow, 442
Momentum thickness, 429
Moody diagram, 362

Nappe, 546
National Transonic Facility (NTF), 319, 587
Nautical mile, 725

Navier-Stokes equations, 230
  cylindrical coordinates, 692
  rectangular coordinates, 231
Network, pipe, 384
Newton, 11
Newtonian fluid, 29, 230
Newton's second law of motion, 96
Noncircular duct, 389
Noninertial reference frame, 140
Non-Newtonian fluid, 29, 31
  apparent viscosity, 32
  consistency index, 31
  flow behavior index, 31
  power-law model, 31
  rheopectic, 33
  thixotropic, 33
  time-dependent, 31
  viscoelastic, 33
Normal depth, 501, 520
Normal shock, 647
  basic equations for, 648
  effects on properties, 652
  tables for computation of, 656, 713
  Ts diagram, 651
Normal stress, 25
No-slip condition, 3, 34
Nozzle, 250, 598
  choked flow in, 606, 612
  converging, 598, 605
  converging-diverging, 598, 611, 661
  design conditions, 612
  incompressible flow through, 250
  normal shock in, 662
  overexpanded, 613
  underexpanded, 612

One-dimensional flow, 21
Open-channel flow, 40, 501
  critical normal flow, 530
  effect of area change, 513
  energy equation for, 508
  frictionless flow in, 513
    effect of bed elevation, 515
  gradually varied flow, 532
  hydraulic jump, 540
  measurements in, 546
  normal depth, 520
  surface profiles in, 533, 535, 537
  surface wave propagation, 504
  uniform flow, 520
Open channels:
  characteristics of, 502
  geometric properties of, 503
  optimum cross sections, 527, 528
  prismatic, 502
  regimes of flow in, 508
  velocity contours in, 504
Orifice, reentrant, 289
Orifice plate, 394
Overexpanded nozzle, 613

Parshall flume, 552
Particle derivative, 216
Pascal, 15, 724
Pathline, 23
Pelton wheel, 198
Permanent head loss, *see* Head loss
Physical properties, 681
Pipe:
  aging, 365
  compressible flow in, *see* Fanno line flow
  head loss, *see* Head loss
  laminar flow in, 344
  noncircular, 389
  relative roughness, 361, 364
  standard sizes, 373
Pipe systems, 371
  networks, 384
  single-path, 371
Pi theorem, 296
Pitot-static tube, 258
Pitot tube, 257
Planform area, 467, 469
Poise, 29
Polar plot, lift-drag, 473
Potential, velocity, 271
Potential flow theory, 272
  elementary plane flows, 274
    doublet, 276
    sink, 276
    source, 275
    uniform flow, 274
    vortex, 276
  superposition of elementary plane flows, 278
Potential function, 271
Power coefficient, 315
Power-law model, non-Newtonian fluid, 31
Power-law velocity profile, 355
Prandtl boundary layer equations, 330
Pressure, 49, 58
  absolute, 58
  center of, 62, 64
  dynamic, 256, 257
  gage, 58
  isentropic stagnation, *see* Isentropic stagnation properties
  stagnation, 256, 257
  static, 256
  thermodynamic, 230
Pressure coefficient, 304
Pressure distribution:
  airfoil, 468, 472
  automobile, 481
  converging-diverging nozzle, 611, 662
  converging nozzle, 605
  cylinder, inviscid flow, 282
  diffuser, 384
  entrance length of pipe, 381
  sphere, 462
Pressure drag, *see* Drag
Pressure field, 49
Pressure force, 50
Pressure gradient, 51, 451
  effect on boundary layer, 450
Pressure recovery coefficient, 368
  ideal, 369
Pressure tap, 256, 257, 395
Primary dimension, 10, 298
Prismatic channel, 502
Profile, velocity *see* Velocity profile
Propeller, 176, 199
Properties, fluid, 681
Propulsive, efficiency, 199
Psuedoplastic, 31
Pump, 145
  "laws," 316

Rankine propeller theory, 176
Rate of deformation, 28, 225
Rayleigh line flow, 634
  basic equations for, 634
  choking, 638
  effects on properties, 639
  maximum heat addition, 638
  tables for computation of, 643, 709
  Ts diagram, 637
Reentrant entrance, 289
Reference frame, noninertial, 140
Relative roughness, 361, 363
Repeating parameter, 298
Reversible process, 564
Reynolds experiment, 331
Reynolds number, 40, 304
  critical, *see* Transition
Reynolds stress, 353
Rheopectic, 33
Rigid-body, motion of fluid, 74
Rotation, 213, 219
Rotor, turbomachine, 144
Roughness, relative, 361, 363
Roughness coefficient, Manning, 523
Runner, turbomachine, 144

Secondary dimension, 10
Secondary flow, 369, 503
Second law of thermodynamics, 97, 165
Separation, 37, 451
Sequent depth, 542
Shaft work, 159
Shallow water wave, 506
Shape factor, velocity profile, 453
Shear rate, 29, 225
Shear stress, 3, 25
  distribution in pipe, 351
Shear work, 159
Shock, normal, *see* Normal shock
Significant figures, 2
Similarity:
  dynamic, 306
  geometric, 305
  incomplete, 308
  kinematic, 305
Similar velocity profiles, 433
Similitude, 319
Sink, 275
  strength of, 275
Siphon, 251
SI units, 11, 724
  prefixes, 724
Skin friction coefficient, 433, 444, 448
Slope:
  bed, 530
  classification of, 530
Slug, 11
Sluice gate, 118, 252, 516, 551
  flow rate *vs*. downstream depth, 517
Solitary wave, 505
Sound, speed of, 568, 572
Source, 275
  strength of, 275
Span, wing, 474
Specific energy, 510
Specific energy diagram, 511
Specific gravity, 681, 682
Specific head, 510
Specific heat:
  constant pressure, 562, 690
  constant volume, 562, 690
Specific heat ratio, 563, 690
Specific speed, 316
Specific volume, 562
Specific weight, 53
Speed of sound, 568
  ideal gas, 572
Stability, 73
Stage, 145
Stagnation enthalpy, 594
Stagnation point, 36, 282, 425
Stagnation pressure, 256, 257
  isentropic, *see* Isentropic stagnation properties
Stagnation pressure probe, 257
Stagnation properties, *see* Isentropic stagnation properties
Stagnation state, 575
Stagnation temperature, 579
Stall, wing, 471
Standard atmosphere, 58
  properties of, 59, 683
Standard cubic foot (of gas), 13
Standard pipe sizes, 373
State:
  equation of, 5
  thermodynamic, 574
Static fluid, pressure variation in, 53
Static pressure, 256
Static pressure probe, 256
Static pressure tap, 256
Steady flow, 20, 204, 209
Stoke, 29
Stokes' drag law, 461
Stokes' theorem, 223
STP (standard temperature and pressure), 18
Streakline, 23
Stream function, 210, 212
Streamline, 23
  equation of, 23, 210
Streamline coordinates, 243, 246
Streamline curvature, 244, 481
Streamlining, 38, 466
Stream tube, 260
Stress, 25
  components, 26, 230, 692
  compressive, 50
  normal, 25, 230, 692
  notation, 27
  shear, 25, 230, 692
  sign convention, 27
  yield, 32
Stresses, Newtonian fluid, 230
Stress field, 25
Strouhal number, 321
Substantial derivative, 216
Sudden expansion, 367
Superposition, of elementary plane flows, 278
  direct method of, 278
  inverse method of, 282
Surface force, 25
Surface profile, open-channel flow, 533, 535
  calculation of, 537
Surface tension, 684
Surge wave, 559
System, 5
System derivative, 97
  relation to control volume, 103

Systems:
of dimensions, 10
of units, 11

Taylor series expansion, 50, 352, 437
Tds equations, 564
Terminal speed, 9
Thermodynamic pressure, *see* Pressure
Thermodynamics, review of, 561
Thixotropic, 33
Three-dimensional flow, 21
Throat, nozzle, 598
Tidal bore, 559
Timeline, 22
Total head tube, 258
Trailing vortex, 474
Transition, 332, 427, 457
Translation, 213
Transonic flow, 573
Ts diagram, 563
Tsunami, 507
Turbine, 144
impulse, 144, 147
reaction, 144, 147
Turbine flow meter, 403
Turbomachine, 144
axial flow, 145
flow coefficient, 315
head coefficient, 315
mixed flow, 145
power coefficient, 315
radial flow, 145
scaling laws for, 315
specific speed, 316
Turbulent boundary layer, approximate solution for flat plate, 447
Turbulent flow, 38, 332
Turbulent pipe flow, 351
fluctuating velocity, 353
mean velocity, 353
shear stress distribution, 353
velocity profile, 354
logarithmic, 354
power-law, 353
velocity defect, 355
viscous sublayer, 353
wall layer, 354
Two-dimensional flow, 21

Uncertainty, experimental, 716
Underexpanded nozzle, 612
Uniform flow:
in open channel, 520
at a section, 22, 106, 520
Uniform flow field, 22
Units, 10, 11, 724
Universal gas constant, 562
Unsteady Bernoulli equation, 266
Unsteady flow, 21

Vector, differentiation of, 209
Velocity field, 20
Velocity measurement, *see* Flow measurement
Velocity of approach factor, 393
Velocity of polygon, 146
Velocity potential, 271
Velocity profile, 34
in pipe flow, laminar, 346
turbulent, 354
Vena contracta, 366, 381, 391
Venturi flowmeter, 397
Viscoelastic, 33
Viscometer:
capillary, 46, 347
concentric cylinder, 46
cone-and-plate, 47
Saybolt, 46
Viscosity, 28, 29
absolute (or dynamic), 29, 686
kinematic, 29, 687
physical nature of, 684
Viscous flow, 33
Viscous sublayer, 354, 364
Visualization, flow, 22, 311
Volume dilation, 226
Volume flow rate, 105
Vortex:
forced, 223
free, 223, 275
irrotational, 275
strength of, 275
trailing, 474
Vortex generator, 478
Vortex shedding, 328
Vorticity, 222

Wake, 37, 426
Wall shear stress, 352, 353, 434, 444, 448
Wave speed, 504
Weber number, 305
Weight, 12
Weir, 325, 546, 550
Wetted perimeter, 390, 502, 627
Wheel, turbomachine, 144
Windmill, 176, 199
Wind tunnel, 311, 319
Wing loading, 474
Wing span, 474
Work, rate of, 159
shaft, 159
shear, 159
sign convention for, 97

Yield stress, 32

Zone:
of action, 573
of silence, 573